The Tropic and Subtropic Zones of the World

This map of the world shows the tropic belt, 5200 km. (3,200 miles) wide, with the countries and islands which, together with the subtropics, an area of varying breadth with long summers and mild winters, are the native habitat of most of the exotic plants used for indoor decoration. The Tropics of Cancer and Capricorn border the tropic zone at 23.5 degrees north and south of the equator. The region between or near these parallels is marked by its warm climate and luxuriant vegetation, modified of course by altitude, prevailing winds and precipitation. Guided by the lines of latitude on this map, and the tables of elevations, temperature and rainfall under 'Plant Geography', a fair idea can be formed about climatic background and geographical distribution of plants, which in turn gives us a clue to their requirements.

SCALE ALONG EQUATOR

MILES
0 500 1000 1500 2000 2500

KILOMETRES
0 500 1000 1500 2000 2500

Longitude East of Greenwich

Exotic Plant Manual

Fascinating Plants to live with
– their requirements, propagation and use –

by Alfred Byrd Graf

Second Edition

All about Plants Indoors

for Decoration or as a Hobby;

in Warm Regions the Sheltered Outdoors.

4200 Illustrations

with Pictograph Keys to Plant Care

Plant Geography

and Ecology

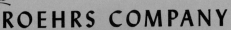

ROEHRS COMPANY
East Rutherford, New Jersey 07073, U. S. A.

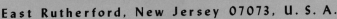

Contents

International Standard
Book Number
ISBN 0-911266-04-6

Library of Congress
Card No. 72-81651

© *Copyright 1970*
by A. B. Graf

Art Director:
Norman E. Carley
Lithography:
DeVries Brothers, New Jersey
Sweeney, Krist and Dimm, Oregon
C. S. Hammond & Co., New York
Bindery:
A. Horowitz & Son, New Jersey

Printed in U.S.A.

Cover Illustration:
Tropical Plants in Terracotta
Bowl: (from bottom left, clockwise) Dracaena goldieana, Dracaena sanderiana, Nepenthes x
atrosanguinea, Anthurium andraeanum album and andraeanum, Xanthosoma lindenii
'Magnificum', Hoffmannia
ghiesbreghtii, Aglaonema costatum; as ground covers Ficus
radicans 'Variegata', Selaginella kraussiana brownii, Peperomia obtusifolia variegata.

Exotic Plants - an Introduction

Climate has subtle effects. In the tropics where plants and flowers form a constant background to life, they are taken for granted, and no great effort is needed to cultivate them. Beyond the tropical zones, however, the seasons become more defined, with marked extremes in climate, and plant life is more highly appreciated. The impulse to be living with nature is strong, and plants of exotic origin or flavor seem to appeal to us most. If freezing outdoor climate has no place for them, we try to bring these Exotics indoors.

In times past only the wealthy could afford the time and means to cultivate collections of outlandish plants at home. But today, everyone in possession of a window, or other place with fair illumination, may be intrigued by the unlimited possibilities to experiment with plants that have been brought into cultivation from foreign climes. These opportunities become even more promising if a little home greenhouse is available. Most indoor plants are exotic and of tropical or near-tropical origin and, depending on their climatic background, more or less adaptable to the living room. If too demanding in their requirements the more humid conditions of the greenhouse may be needed for their best success, or where warm enough, may prefer the sheltered patio.

This EXOTIC PLANT MANUAL is intended to be a handy pictorial reference for visual plant identification and to suggest the best in cultivated tropicalia. It brings 3600 photographs of those exotic indoor plants which are most likely found in commercial cultivation – unlike the coverage of my larger Cyclopedia "Exotica" with its more than 12,000 illustrations. They are grouped for their glamour or other characters, for suitability in warm or cool interiors, or in sections covering succulents, carnivorous plants, aquatics, herbs or vines; holiday and fancy blooming plants, bulbs and bedding plants, and ornamental fruit, growing in containers. Important families such as aroids, begonias, bromeliads, ferns, orchids, gesneriads and palms, are shown apart. A chapter on plant curios may prove interesting, as well as the discussion on the trends in artificial plants.

By means of a novel and easy-to-understand pictograph symbol, each plant photo is keyed to a "Code to Care", and both botanical and common names are used as captions. The descriptive text for each plant shows its native habitat, synonym if any, and family relationship, and brings concise details on growth habit, foliage, inflorescence, and any features of peculiar interest.

In nomenclature, I have conscientiously tried to determine the latest valid, botanical names and have been guided by the reports from the Bailey Hortorium, and qualified taxonomists elsewhere. In botanical literature, I depended on the latest editions of Bailey's 'Hortus' and 'Manual of Cultivated Plants', the Royal Horticultural Society's 'Dictionary of Gardening', Engler's 'Pflanzenreich', Parey's 'Blumengaertnerei', and other recognized international works. I have spent many worried hours in trying to establish the proper name for a plant, and I am aware that even now, unintentionally, some of the captions may be subject to correction.

For home growing as a hobby, or general decoration, the photographs of plants included in these pages cover every normal need or fancy. How one may use these plants indoors is amply illustrated by examples of various decorative schemes, or little growing nooks that would enliven and embellish any home or office.

Because the climatic backgrounds of exotic plants very much determine their needs in cultivation, sections on 'Exotic Origins' and 'Plant Geography' tell about regional floras and growing conditions around the earth, as experienced and recorded during my travels throughout the world.

For a fuller understanding of the factors surrounding plant life, twelve introductory chapters endeavour to explain the needs of exotic plants and the importance of all fundamentals affecting them indoors, such as light, temperature, water and humidity; soils, soil-less culture; insect control and ailments; also demonstrating in pictures all possible methods of propagation.

Ploughing deeper, further topics covered are biology, physiology, physical laws and genetics, as well as a history of the introduction of exotics, and the psychological reasons for their cultivation – aspects which may be of value to those taking pleasure in exploring all the details. Made available for use to the inquiring mind – as a fountain beckons to the traveler – you may choose to drink from it or not depending on your thirst and mood!

The SECOND EDITION of this MANUAL made it possible to embody more than a hundred corrections and updatings. Changes were made in 13 botanical names, such as in Asparagus, to bring them in line with latest taxonomic findings; photographs of several meritorious new plants of recent introduction were added. The chapters on pesticides and propagation were augmented. Toward the projected public conversion to metrics some measurements are now given in both the traditional English as well as the world-wide International system of standards. Finally, technical improvements such as larger type on 175 pages of descriptive text for better readability, made possible by a slight increase in format, should help to make this MANUAL more attractive and useful for authentic everyday reference.

Alfred B. Graf

Second Edition 1972

Plants and Flowers - an integral part of life the world over

Flowers bring joy, plants lasting pleasure. There is a link between flowers and happiness and man that is transposed into our innermost feelings if we attune ourselves to the refreshing influence of nature's beauty and unceasing powers so keenly demonstrated in its wondrous cycles from seed to germination, and through growth to flower and ultimate fruit. Aside from the arts, nothing can give us the recuperative lift, relaxing peace and deep satisfaction in this mechanized, unnerving and disturbed world as to find our way back to nature and to take the time to subject ourselves to its healing therapy.

Slumbering within us is the yearning toward nature which lives deep in our subconscious mind, and going back in history when man was close to everything primeval and had an instinctive knowledge of the universe surrounding him. Then, the waxing and the waning of the moon were of deep significance—the shedding of the leaves and the unfolding of buds and flowers established a measure of time and the rhythm in his life. The first religious feelings were stirred in the belief that spirits lived in certain trees and gods on lordly mountains, and the goddesses of fertility and of the harvest in their bounty blessed the earth.

It took man thousands of years to discover the making of fire to cook his food; more ages to invent ax and arrow, to hunt with and to till the soil, and to finally exchange his cave for a dwelling which he built. Throughout this time of evolution when man lived in the open, he saw around him flowers and trees appealing to his senses—his sight, touch, taste and smell. The behavior of these plants was always peaceful and soothing, and his mind became attached to them as friendly beings, willingly lending themselves to his needs, giving him food and shelter and pleasing impressions for all his senses. Throughout history flowers, plants and trees became interwoven with man's daily life. They developed into symbols for his sentiments and beliefs, his fears and superstitions. In Egyptian, Greek, Roman and Nordic mythology; in Biblical scriptures; in Oriental beliefs and Occidental lore the human mind assigned magic powers to many plants, others were chosen because of their fragrance and beauty for religious ceremonies or as tokens of reverence and love, customs still preserved to this day.

Even people so remote that they have never heard of Greece or China, use flowers and plants for adornment and the beautification of their villages, as I have found wherever I have traveled North or South of the equator.

I met the gay Melanesian youngsters pictured (page 5), on the South Pacific island of Viti Levu giving expression to their exuberent happiness in an uncomplicated life by adorning themselves with showy Hibiscus blossoms found there by the wayside day after sunny day. We are all more or less familiar with the lavish fashion of Polynesian maidens wearing leis of frangipani and skirts of 'Ti' leaves. In the Kunai grass and bamboo villages straddling the razorback mountains of interior New Guinea, where the stone adze is still in daily use, I have marveled at their neat plantings of native Impatiens and flame-colored

Cordylines. In Java and Bali treasures in collected Orchids and Platyceriums grow on thatched roofs and fencing. The Malayan junks at sea have their bows decorated with Tuberoses and Goldenrods. In India women to the lowest caste wear a cluster of flowers in the nape of their shining hair. Gentle Thai and Burmese ladies can be seen praying at pagodas offering woven garlands and clasping flowers between their imploring hands. Even in sophisticated Japan every little home has its tokonoma, a place of reverence set aside for arrangements of flowers or a dish with lovingly trained miniature bonsai.

A small world of durable exotics planted in a glazed ceramic bowl for long lasting pleasure.

Around the Mediterranean the houses or patios are studded with plants or shrubs in terracotta jars, a custom dating from Roman and Moorish times. And which European traveler has not been impressed with the lavish display of bright flowers in window boxes or on balconies in all the Alpine countries. Further north in Europe, especially Germany and the Lowlands, to bring a few flowers as a hostess gift for coffee or an after-supper visit, or to take a little potplant to join the many more found on every broad window sill is a time-honored habit and custom never neglected.

In our hemisphere the fondness for flowers is especially apparent in all Latin-American countries. From Mexico south to Chile every typical patio is lovingly cared for. Even the smallest casitas abound in plants grown in tin cans and pots, likely collected in the immediate vicinity. And on religious occasions and fiestas flowers are used for decorations everywhere.

Happy young Fijians of the South Pacific adorn themselves with tropical Hibiscus blossoms.

Plants and flowers in our own homes have become an integral part of our modern concept of living also. That this is particularly noticeable in windows of crowded apartments in Metropolitan centers is quite significant and would seem to demonstrate that the more removed from nature a person is, the more he tries to recapture it.

We are attracted to flowers because we love things beautiful. But their deeper appeal lies in the mystery of their being alive, the miracle which man has been unable to duplicate. Nor is he capable of breathing into a plant the living tropism which makes it sprout, turning to the sun and unfold leaf after curious leaf and the fragile loveliness of its blooms. And Backster's polygraph tests in New York have indicated that plants can actually feel emotions and read our minds, responding favorably to affection, and even the sonic vibration of music, but also will quickly detect threats and ill intent, reacting accordingly.

House plants cannot exist without us. They appeal to the best within us which wants to make us care for someone or for something. Transplanted from a strange land, a living plant calls upon our sense of wonder, curiosity and responsibility. We feel involved in its very existence and try to give it understanding for its needs. When it does respond we thrill to every newly forming shoot with a feeling of achievement and deep satisfaction. This is why plastic plants will never replace real life.

This link with nature and the wonders of its creation offer an escape from the daily pressures and fretful pace of the machine age. Indoor plants add fresh beauty and living charm to a home all year. They bring us lasting pleasure and carry a message of worldwide understanding as they make us pause and think of their exotic origins in balmy climes amongst people beyond our horizon. And flowers say the sweetest things—in a language that is universal and understood by all.

Exotic Origins of Indoor Plants

Climatic Backgrounds

The Tropics, because of the romance surrounding them, are frequently not truly understood. We imagine them as overflowing with color and exotic treasures, coddled by comfortable warmth, humidity and sunshine. This is only true in part. The equatorial rain forest is not colorful. Typically it is a luxuriating scene of varied foliage in shades of green—an occasional bright spot the exception. Nor is the climate so ideal. Steaming humidity is characteristic of tropical lowland forests, occupied by plants quite temperature-sensitive. Such growing conditions can only be duplicated in a controlled, humid-warm greenhouse, the soil moist yet well drained, the strong sun safely filtered. But tropical plants are not necessarily all heat-loving plants. Many a day or night have I shivered at high elevations under the equator, or even in lowland jungle if the humidity rises sharply during the night, and where 50° F coupled with heavy mist can make one's teeth chatter.

On the other extreme, there are deserts where rock and sand luxuriate without effort, but plants and animals must use their whole force merely to exist. This is how cacti and many succulents have emerged in their peculiar forms, and this makes them ideal windowsill subjects, requiring a minimum of care.

Climatic backgrounds govern the relative wealth of species of tropicals in the various floristic regions of the world. The tropical belt extending 1600 miles on either side of the equator, and the outer subtropical zones bordering it north and south varyingly for 300 to 700 miles, are the home of most of our indoor plants. Depending on elevation, ocean currents, and precipitation, various characteristic climates prevail, but they share a relative uniformity of mean monthly temperatures throughout the year, thus differing from the strong contrasts between summer and winter in temperate regions.

Plants that inhabit a region where temperature fluctuates widely, and accustomed to rigorous climate—except those of extreme altitude, adjust themselves easier to varying conditions in the home. They will have greater tolerance than tropical lowland species that are always coddled by warmth and therefore require the more equable humid-warm surroundings available in a controlled greenhouse or enclosed terrarium. However, many fragile tropicals may be weaned to tolerate less favorable growing conditions, and those plant lovers that accept the challenge often prove to everyone's surprise that a plant will adapt itself, much better than expected, to a mere subsistence level, thereby helping to control and maintain their intended size in decorative schemes.

Floristic Regions of the World

Our cultivated plants have been collected from all parts of the world. Having traveled into its tropical and subtropical regions, one realizes the magnitude of effort and sacrifice which must have gone toward identifying and collecting, often under great difficulty, so many of the different subjects which nature has produced. Consider that there are now recognized an estimated 350,000 species of flowering plants alone, in some 12,500 genera (as against 85,000 species known in Linnaeus time in 1753), plus more than 9,000 kinds of ferns, scattered over remarkably big areas.

Richest in tropical plants appears to be the Western hemisphere with an estimated 112,000 species, of which 60,000 are found in the Amazon basin alone. In the Eastern hemisphere, Tropical Africa has nearly 40,000 kinds; in addition there is the rich near-tropical Cape flora with 16,000 species. The Monsoon region of Tropical Asia is immeasurably wealthy in beautiful, though often delicate plants but their number can only be estimated at upwards of 50,000 species of which 35,000 are recorded for the Malaysian area including New Guinea. Over three quarters of all higher land plants are denizens of equatorial regions.

The Americas

The Western hemisphere is the original home of many of our best indoor plants. Nowhere is there as great a variety nor such wealth in endemic plant types as in Tropical America. The important Aroid family inhabiting its forests and savannahs has given us many handsome and dependable house plants, notably Philodendron, Dieffenbachia, Syngonium and Spathiphyllum. And who would not be impressed by the exotic inflorescence of an Anthurium or the gay foliage of fancy-leaved Caladiums? Orchids are represented by an unusually large number of genera, many of them epiphytic. Typically American are the ever-fascinating Bromeliads or Natural vase plants, usually dwelling on trees or rocks. From the Western hemisphere also is the succulent family of Cactus with 2000 species widely scattered, growing as epiphytes in dripping rain forests, braving the deserts, and climbing the Cordilleras even to the zone of snow. In other succulents Agave, Yuccas, Echeveria and many Sedums are endemic.

Great treeferns (Dicksonia squarrosa) luxuriate in close proximity to the icefields of the Southern Alps in the Fiordland of New Zealand.

Other attractive plant treasures include Marantas and Calatheas, Heliconias and Miconias, Begonias, Hoffmannias, Peperomias, Fittonias, and Alternantheras; the gesneriads Achimenes, Columnea, Episcia, Kohleria, Sinningia and Smithiantha; the popular bulbs Hippeastrum and Hymenocallis; the vines Allamanda, Bougainvillea, Dipladenia, and Passiflora; a multitude of ferns and treeferns; amongst palms the useful small Chamaedoreas. Numerous bedding plants also spring from American parents, including Fuchsias, Lantanas, Petunias, Salvias and Zinnias.

Europe
Because of the nature of its climate, few of our indoor plants stem from Europe. The areas adjacent to·the Mediterranean, being subtropical but dry, are characteristic for their many fragrant shrubs with rigid leaves, typified by the myrtle, Ruscus, baytree, Smilax, Santolina, rosemary and olive. Other old acquaintances originating from this area are Cyclamen and carnations, Acanthus, oleander and figs; in bulbs Scilla, hyacinth, and Narcissus. Chamaerops humilis is the only palm that is native in Europe.

Africa
The ancient and undisturbed continent of Africa is as remarkable for what is absent from its flora as for what it possesses. It has fewer bamboos, orchids, aroids and palm genera than other tropical areas. Most of Northern and Southern Africa, and parts of the East are dry grasslands or desert, with the largest assortment of diverse succulents

other than cacti, in the old world. Most striking of these and a feature of the landscape are the cactus-like Tree-Euphorbias and the tree-like Aloes. Popular succulents of smaller habit are the Sansevierias of the steppe, Crassulas, Gasterias, Haworthias, more Aloe, Kalanchoe, Senecio and hundreds of Mesembryanthemum, or Mid-day flowers. Quite remarkable are the Stone mimicry plants of the South, especially Lithops, imitating the gravel amidst which they exist. Pelargonium and Ericas, as well as Proteas and Strelitzias are indigenous to the extreme South. Many well-known bulbs are African: Agapanthus, Amaryllis, Clivia, Crinum, Gladiolus, Lachenalia, and the "Calla" lilies. From East Africa come Streptocarpus and the well-beloved Saintpaulias. Unusual in palms are Hyphaene with their branched trunks; also many Cycads are characteristic in Southern Africa. Through the center and westward to the Gulf of Guinea extends the belt of rain forest especially along the rivers, and known for its Dracaenas and great timber trees, laden with epiphytes including many peculiar orchids.

Asia

The highly interesting Monsoon region extends from the Himalayas through Southeast Asia, the Philippines, and the beautiful, lush tropical islands of Indonesia. These "East Indies" are immeasurably rich in beautiful, though often delicate plants. Beneficial rains produce a moist-warm climate and with it a profusion of forests rich in striking forms of Ficus and many palms. Epiphytic orchids and ferns abound, as do Nepenthes, beautiful Begonias, Strobilanthes, gingers, jasmines and Hoya. Typical Aroids are Aglaonema, Alocasia, Colocasia, Epipremnum, Homalomena, and Scindapsus. Bright-colored Crotons, Acalypha, Hibiscus and Coleus thrive exuberantly under the tropic sun. The subtropical Northeast, Southern China and Japan have given us many plants suitable for cooler locations, best known of which are many types of Azaleas; also Aspidistra, Aucuba, boxwood, Camellia, Euonymus, Fatsia, Hydrangea, Ligustrum, Pittosporum and Podocarpus; the legendary Citrus, and many bamboos; in palms Trachycarpus and Rhapis. Westward, Asia Minor is the historic home of many good bulbous species, as the tulip, hyacinth, Crocus and many Iris and lilies.

Australia and the South Pacific

Queensland is an extension of the Indian monsoon region and in its forests we find the now well-known Brassaia or "Umbrella tree"; Heimerliodendron, Araucaria, cycads, and some palms including clambering Calamus, a rattan; the undergrowth harbors a wide variety of ferns. The drier areas southeast are the habitat of the rich "New Holland"

flora abounding in woody shrubs and trees, such as the many Eucalyptus, Grevillea, Eugenia, Callistemon, Epacris, Melaleuca, Leptospermum, and more than 600 species of gloriously flowering Acacias.

New Zealand has a distinctly insular climate and is rich in evergreen vegetation. Tree-like Araliads are typical as are Phormium and Cordyline, but epiphytes are few. Ferns abound and have no equal in the world in wealth of form, from treeferns to the delicate "Filmy-ferns".

The native flora of the Pacific islands varies according to rainfall, but is comparatively unremarkable. South Seas shores are always fringed with the ubiquitous coconut palms. Endemic plants include Cordylines, Pandanus, Crotons, Dizygotheca, Polyscias and some ferns. Most of the showiest shrubs and trees so often identified with South Sea islands are actually introduced from other tropical areas of the world, giving a deceptively beautiful but erroneous impression of its endemic vegetation.

Naturalistic setting of tropical and subtropical exotics at the International Flower Show, New York Coliseum.

Cultivated Exotics in History –
Past, Present, and Future

The cultivation of imported and unusual plants has been undertaken since very early times. Through their carvings in stone, we know of plants grown in containers by the ancient empires of the Sumerians and of Egypt 3500 years ago. The Hanging Gardens of Babylon, rebuilt by Nebuchadnezzar II in 605 B.C., one of the Seven Wonders of the Ancient World, represent the best classic example of the growing of plants in stone vessels and on terraces, though the assortment was probably very limited.

In China, ornamental horticulture dates back 2500 years to Confucius and Lao-tse, and in its classical concept, including the growing of little trees in containers, spread to Korea and Japan—though in modified form.

Greece and Rome

The ancient Greeks were not swept by garden fervor, being more interested in the arts, architecture, and philosophy. But for the festivals of Adonis, the beloved of Aphrodite, they decorated his statues by creating tiny exotic plantings using seed, bulbs or small shrubs, cultivated under the humid environment of bell-shaped glass jars. During the 4th century B.C.—the Golden Age of Greece—the "father of botany", Theophrastus, a student of Aristotle, classified and described 450 plants then known in botany's first classic texts, his "History of Plants".

The Roman scrolls of Virgil, Horace and Seneca, written 2000 years ago, sufficiently prove that amongst the nations of antiquity a very strong attachment existed to cultivating strange plants and fruit. The Romans used stone containers for the growing of many kinds of ornamental plants in the inner courts of their homes, the peristyle, open-air-rooms also serving as a garden, and some times heated with hot air flues. During the first century, they also harnessed solar warmth, or piped vapor heat in hothouses with windows made of thin talc or mica, and from 290 A.D. of glass, where foreign plants were nurtured and roses forced to bloom.

Middle Ages

With the disintegration of the Roman Empire and disappearance of its luxurious way of life, following the capture of Rome by the Goths in 410 and again by the Vandals in 455 A.D., the medieval gardens that followed were primarily functional. During these Romanesque and subsequent Gothic periods the monasteries, in cloistered quadrangles, not only preserved the faith, but kept alive the knowledge of plants and the tradition of the growing of fruits, vegetables and healing herbs, such as decreed by Charlemagne in 812.

Plant culture in Europe made considerable gains during the crusades, which ended in 1291, when knights brought back flowers, bulbs, and trees—the rose, jasmine and carnation into their castle gardens from the Near East. The High Middle age from the 12th to the 14th centuries is remembered as the age of chivalry, when romantic love was celebrated and flowers bloomed in every garden, and pomegranates and lemons were favorite exotics cultivated.

Moorish and Mogul Gardens

The Arab influence in Spain between 712 and 1492, for nearly eight centuries and ending with the last Moorish kingdom of Granada in the 15th century, left a tradition of the enclosed garden with reflecting pools, and plants in terra-cotta jars, carried over in patios or outdoor living rooms to this day. These gardens gave the Islamic conquerors a tantalizing mirage of paradise—gardens of delight with cool, splashing water and shade, and sherbet served by lovely houris.

In similar concept, but thousands of miles away from Spain and two hundred years later at Agra, the capital of the Mogul Empire, emperor Shah Jehan created the perfumed gardens of the Taj Mahal with reflecting pools and fountains, above the river Jumna. His palace in the Red Fort was transversed by channels of running water, an early and practical application of the principle of air conditioning.

Renaissance – the Age of Awakening

In Western culture, gardening as an art began with the Renaissance, an epoch of immense curiosity and revival of learning in the 15th and 16th centuries. Following the discovery of America in 1492, of India in 1498 and Java in 1511, tropical plants from the West Indies and the East Indies were introduced into Southern Europe by ship captains and early explorers. The successful expeditions of Columbus first introduced the cactus, the agave and the pineapple from

The noble Orangerie of Louis XIV at Versailles, built in 1683 for the wintering of 1500 citrus trees and palms of the Royal gardens – a forerunner of the Victorian wintergarden.

the New World. Wealthy merchants and ruling princes, particularly the Medici, began to develop collections of exotic plants. In 1545, the first Botanic Garden was established in Padua, possibly inspired by the brilliant gardens of emperor Montezuma of Mexico, whose Aztec subjects worshipped flowers.

From the middle of the 16th century a taste for exotic flowering plants and trees began to prevail in France and England amongst the aristocracy. In Germany, again in the 16th century, even the average burgher had quietly initiated a new custom, the growing of decorative plants in clay pottery, and setting them into his windows to make them look attractive, and to impress his neighbors. This may be considered the beginning of the potplant habit in the home.

Orangeries of the Baroque Period

Beginning in the 17th century and into the 18th, born of the Baroque way of life, glass-windowed orangeries were erected in increasing numbers all over Europe. At first warmed by charcoal braziers, Dutch tile stoves were used later on for heating them. In 1683, at Versailles, Le Nôtre completed for Louis XIV the noblest orangery in Europe, which could hold for winter storage 1200 movable orange trees in tubs, as well as 300 other kinds of tender shrubs, laurel, figs, pomegranates and palms. About the same time (1687), the first warm planthouses for imported plants were built at the Leyden Botanical Gardens in Holland. After 1700, with the development of cheaper methods of producing flat glass, and increasingly from 1750 on, more and more real glasshouses were built, heated by continuous hot-air flues. Such houses proved far more adequate for the cultivation of the many exotics that now began to arrive in Europe.

During the early part of the 18th century upwards of 5000 species of plants were introduced. But the greatest botanical era was from the latter 18th century to the beginning of the 19th, when plants from India, the Americas, Africa and Australia were collected by now famous explorer-botanists and sent into the care of eager gardeners. The earliest Australian plants arrived only after 1770.

Taxonomy and Nomenclature

With the flood of newly arriving plants, the old method of naming them by complexes of descriptive terms proved inadequate. The Swedish botanist Linnaeus, by 1753, worked out a new system of taxonomy, or classification, giving each plant a set of two scientific names: the first one for the genus, and the other a qualifying epithet to designate its species, thus laying the groundwork for present nomenclature. The system Linnaeus conceived was artificial in that he classified plants on the basis of sexual characteristics such as the number of stamens and their relation to each other, but did not take into account the natural relationship of plants as revealed by evolutionary studies, and later had to be modified. But before the 19th century it was the generally accepted belief that all species of plants had come into being as the result of a special act of creation, and that each had continued from the beginning as a fixed and unvarying species.

Victorian period

Parlor gardening indoors became a fad during Victorian times late in the 19th century not only in England but also in North America. Even people of limited income were amassing oversized tropical plants in every parlor, hall, and drawing room, while lobbies and public rooms of every hotel and theater had its Boston fern and potted palm. However, the custom of having plants in offices was initiated only with the employment of women in business when these brought flowers of their own to decorate their desks. From this hesitant beginning has grown the now acknowledged need for decorative plants in every modern building.

Indoor plants today and tomorrow

Architects and interior decorators consider the use of ornamental foliage plants in containers essential for accent and modern decor. About the home, there has developed wide interest on the part of many people to grow plants, mostly of smaller size, as a hobby—often favoring certain families, and the vogue of using and caring for plants has added greatly to the enrichment of everyday living. A promise and vision of a new way of life with nature comes from a different direction. The trend in architectural home design is to break up the separating wall of garden and building, the outside and the inside, by bringing indoors part of the garden and the principle of living with plants. The sides of the house are being filled in with glass panels, often in form of sliding glass doors, and through these transparent walls the outside becomes an integral part of interior architecture. The Japanese have long been masters in creating illusions by uniting their living room with the concentrated image of nature in their gardens by removing whole walls thus leading the eye and imagination over lakes and streams into forest and distant mountains beyond.

By the use of glass we are enabled to maintain such a visual connection with nature throughout the seasons. The imaginative architect Frank Lloyd Wright was the foremost proponent of this tenet of interweaving building and garden, and throughout his life he sought and preached the integration of man and nature, and to have the interior graciously at one with external scenery.

Still further in consequent thinking went the visionary Buckminster Fuller, who endeavored to remove entirely the definition of inside and outside. Disregarding aesthetic considerations, his approach was that we no longer live in houses but in climatically controlled gardens covered by light-transparent domes. Such geodesic designs have already been constructed, for various purposes, and one of its best examples is the ethereal Climatron at St. Louis, Missouri. Here an enormous area has been spanned and equipped with such controls as will permit the selection of any climate desired. Within such an area, transformed into a living-garden in which both plants and people feel comfortably at home in perfect harmony, the distinction between inside and outside ceases to exist. Once more, as at the beginning of history, man may elect to live in a garden, and become one with nature.

The tropical jungle moved to Saint Louis, sheltered by a greenhouse of the modern age – the Climatron – a geodesic dome 175 ft. across, built of aluminum tubing and organic plexiglass, a hard methacrylate resin product with high ability to transmit light. Inside (right) the rain forest luxuriates, dominated by the Cuban Royal palm, Roystonea regia.

Biology, Physiology and Genetics of the Green Plant

Biology, the Science of Life, is the branch of knowledge which deals with living organisms; Physiology is that division of biology which deals with the processes, activities, and phenomena of Life; Genetics is the study of heredity. The biological sciences may be classified according to the nature of the organisms considered: as Botany, dealing with plants in all their aspects; Bacteriology, with bacteria; Zoology, with animals.

As living organisms, higher plants differ from animals by their lack of any mechanisms for locomotion and inability to move about; by their cellulose cell walls, and usually by their possession of the green pigment, chlorophyll. While typical animals take in solid food of plant or animal origin, green plants depend solely on inorganic substances and radiant energy. Since the carbon needed by animals cannot be absorbed directly from CO_2, its chief source, all higher forms of life are dependent on the photochemical assimilation of carbon dioxide from the air by the green cells of plants to produce food in the form of carbohydrates.

The Living Cell

Living things are formed of cells, microscopic bodies distinguishing living from non-living protoplasm, and as long as they are alive are capable of functioning as an independent unit in the body. They assimilate food stuff and transform it into cellular material; they respire; they respond to stimuli, and they reproduce in the sense that they are able to divide and multiply. Live cells contain a semi-liquid material, the protoplasm, enclosed by a semi-permeable membrane, and they make up the tissues of root, stem, and leaf. Many materials—foods, mineral substances, water, atmospheric gases—are continually moving, in the form of molecules, from one living cell to another.

Plant Metabolism and Photosynthesis

In plant metabolism, the sum of the processes incidental to the life of the plant, chemical changes take place in living cells by which energy is provided for the vital activities of living and growing, and new material is assimilated to build and maintain. Plants use radiant energy from the sun to create food out of air and water, thus initiating the cycle that supports all life on earth. This is commonly known as the process of photosynthesis.

Sunlight is absorbed by the green pigment chlorophyll in the leaf or stem and used to convert hydrogen molecules from water and carbon from atmospheric carbon dioxide into simple sugar, the basic food of all plant life. As sugar molecules accumulate they are converted into starches and other complex carbohydrates. These constitute one of the three major groups of organic substances of which living matter is composed, and the food which it needs to grow; the other groups being the proteins and the fats.

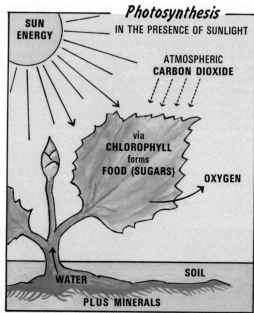

Photosynthesis

SUN ENERGY — IN THE PRESENCE OF SUNLIGHT

ATMOSPHERIC CARBON DIOXIDE

via CHLOROPHYLL forms FOOD (SUGARS)

OXYGEN

WATER

SOIL

PLUS MINERALS

The Leaf is Life

Leaves are the photosynthetic centers on which plants depend for their lives, receiving light from the sun and carbon dioxide from the air in daytime. The carbon dioxide which is normally .03% of the air is absorbed through the breathing pores, or stomata, which are usually wide open during sunlight hours and partially closed in darkness. In this process the all-important oxygen, sustaining all human and animal life, is produced by green plants. The functions of photosynthesis are reversed during the dark period when the plant uses oxygen and gives off carbon dioxide, in quantities not harmful, to balance its metabolism and daily cycle of existence. Water moves to the leaf surface where most of it escapes through the stomata while products of photosynthesis are transferred to the growing shoots and roots. It could be said that one green tree produces enough oxygen each day to sustain one life—although most of our available oxygen is derived from green algae in the world's oceans.

The Functions of the Roots

Roots not only anchor a plant to the soil, they absorb minerals and water from the soil, they transport these materials to the stem, and they serve as food-accumulation organs. The outermost cells of the root of a plant are in direct contact with the soil, and the air and water absorbed within it. Mineral salts are absorbed from the soil by osmosis, or diffusion, when the root is actively growing and has an abundant supply of carbohydrates and an adequate supply of oxygen in porous soil.

Photosynthesis versus Respiration

Since photosynthesis, which is an energy-storing process, takes place only in the chlorophyll-containing cells and only when sufficient light is available, whereas respiration, which is an energy-releasing process, takes place in all living cells day and night, it follows that in order for the plant to increase in size (grow) it must produce more through photosynthesis than is broken down through respiration.

The factors which govern both of these processes are light, temperature, water, aeration, availability of minerals, etc. These will be discussed later.

Respiration Produces Energy

Some activities—such as photosynthesis, assimilation and the manufacture of complex food-stuffs and other materials from basic substances—build up new compounds and hence are constructive. Other reactions, the most important of which are respiration and digestion, are destructive, since complex food materials are broken down into simpler compounds. These processes are interdependent, and are both essential to cell survival and growth.

Like an engine which burns gasoline to release power, the energy required by a plant is obtained through the process of respiration. In respiration, sugar is transformed into simpler substances, a reaction directly opposite from photosynthesis. In photosynthesis, in the presence of light or in daytime, energy is accumulated; respiration is the reverse process where as a result of the breaking down of the sugars the energy released is utilized for the maintenance and growth of the plant.

Genetics — the Study of Heredity

Mendel in 1865, in his Laws of heredity demonstrated that a hybrid is not necessarily intermediate in type between its parents. He proved that height, color or general habit of a plant depend on the presence of determining factors (genes), which reappear in more or less orderly

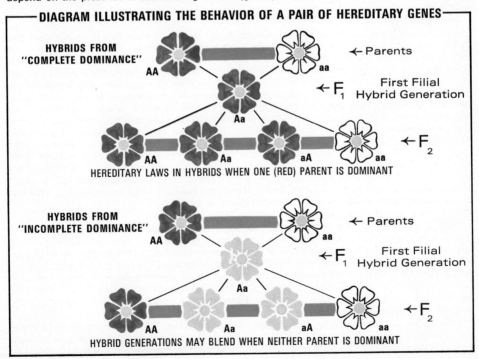

DIAGRAM ILLUSTRATING THE BEHAVIOR OF A PAIR OF HEREDITARY GENES

HYBRIDS FROM "COMPLETE DOMINANCE"
AA   aa   ← Parents
Aa   ←F_1 First Filial Hybrid Generation
AA   Aa   aA   aa   ←F_2
HEREDITARY LAWS IN HYBRIDS WHEN ONE (RED) PARENT IS DOMINANT

HYBRIDS FROM "INCOMPLETE DOMINANCE"
AA   aa   ← Parents
Aa   ←F_1 First Filial Hybrid Generation
AA   Aa   aA   aa   ←F_2
HYBRID GENERATIONS MAY BLEND WHEN NEITHER PARENT IS DOMINANT

proportions in future hybrid generations. Mendel's experiments showed that when a true-breeding tall variety of the garden pea was crossed with a dwarf variety, the offsprings (F_1) were all tall when the tall character of one of the parents proved dominant ("complete" dominance). Self-pollinating any one of these F_1 tall plants, the seed obtained from them, the second or F_2 generation, produced 3 tall to 1 dwarf. Self-pollinating these F_2 plants, the third generation (F_3) produced from the dwarfs only dwarfs, whereas the talls were of two kinds: 1 tall out of 3 produced only talls, while the other 2 talls resulted in both talls and dwarfs in ratio 3 : 1.

It was found later that an F_1 hybrid of two parents of "incomplete" dominance may blend features of both parents thus becoming intermediate between the two. A red-flowered species crossed with a white-flowered form might then result in a hybrid flowering pink. And seedlings from the F_1 plant, being the F_2 generation, would produce on the average 1 red, 2 pink, and 1 white. "Complete" dominance tends to be a feature found in hybrids of two forms within the same species; "incomplete" dominance in crosses between two different species.

These formulas become more complicated when other parental characters enter, but Mendel's Laws are guidelines to allow the plant breeder to foresee the limitations in hybridization and possible results.

Plant Breeding and Reproduction

The horticulturist or geneticist is always looking for plants with the most desirable genes for plant breeding. These genes, or hereditary carriers of plant characteristics, are contained in the chromosomes, which are complex protein bodies of microscopic size. Chromosomes are definite in number for any given species of plant normally existing in pairs and therefore known as diploids. In variations the diploid set is often doubled and becomes a tetraploid which may or may not have desirable improvements of color, size, vigor, or disease resistance. Sometimes plants with three sets of chromosomes are found and hence are known as triploids. The chromosome numbers of the species within each genus are usually alike, but those of different genera may not be. Cross fertilization is usually compatible only with like numbers of chromosomes or their multiples. Hybrids are formed when the pollen containing the chromosomes bearing the genes of one parent is applied to the stigma of the seed parent carrying its own set of chromosomes. The pollen grains grow down the stigma to the ovary, where the cells of each parent fuse and then divide and multiply to form a new individual.

Environmental Response of Plants

Plants respond to external stimuli or changes in environment, light, atmosphere, temperature and chemicals; together with a supply of water and nutrients these are the requirements for growth of any plant.

Light is Vital

Light cannot be measured for its intensity alone, but growth or flowering is usually regulated by length of day or night. The metabolism of a plant—its rate of growth—is in direct relation not only to temperature but also the quality and duration of light received. During fall and winter when the days are short and temperature is low, growth virtually ceases. Most plants will not bloom unless the intensity of light approximates that in their native habitat. Foliage plants are quite tolerant of low light intensity, but if the environment of a flowering plant is too dark, sufficient supplementary light must be provided for successful blooming; a sunny window or an incandescent spotlight is usually best. Many flowering plants are sensitive to daylength, and this is known as photoperiodism. Long day (short night) as well as short day (long night) plants can be guided into blooming practically at will. By giving additional hours of artificial light on short days to long-day plants as Gardenias, Calceolarias and many spring flowering annuals, early flowering is induced. Covering short-day plants such as Chrysanthemums with a black cloth for several hours when the days are long will bring them into bloom as their blooming is dependent on day-length as well as temperature. Intermittent light applied during the night is as effective as continuous light, in obtaining short night response in plants.

The Influence of Temperature

Plant activities and growth generally increase with higher temperatures, and are retarded if the temperature is low. However, excessive heat may result in injury from desiccation and a rate of respiration so high that the consumption of food materials tends to exceed production by photosynthesis. If temperatures are steadily too high, the plant uses up the stored food manufactured during daytime and may exhaust itself, unless followed by an alternate period of lower temperature to restore its energy. Temperature affects growth through its influence upon all metabolic activities: photosynthesis, respiration, digestion, transpiration, absorption of water, and root growth.

Water and Atmosphere

Water is the medium for life to exist, and constitutes more than 90% of the plant's make-up. It carries its food supply, and in fact helps to keep the plant more or less rigid. Water continuously enters the plant through the roots and most of it is exhaled or transpired through the stomata in the green parts while part of it is used in the process of photosynthesis. In greenhouses we often see large plants flourishing in small pots, with soil comparatively dry. This is understandable when we consider that in a humid greenhouse, the normal process of evaporation from the plant into the atmosphere, which hinges on the drying power of the air, or in reverse, upon the moisture carried by it, is assisted by sufficient air moisture, expressed scientifically as the relative humidity.

Atmosphere

Water vapor from the internal tissues of living plants is transpired mainly through the leaves. This will also permit evaporative cooling of the plant. That such moisture loss actually occurs is demonstrated easily by covering a plant with a glass container to let the air inside become saturated and as a result, condensation of water will form on the glass. This is especially noticeable if the temperature is lowered at night and increased in daytime. By keeping the atmosphere around the plant thus confined, the leaves will remain turgid. If on the other hand, a delicate plant is exposed to drying air currents, the loss of water through the pores may be so great that transpiration exceeds the rate of water absorption through the roots and it will wilt; if the wilting is severe or prolonged part of the leaf tissues may be permanently injured (leaf burn).

Nutrient Elements

The elements essential to plant growth are carbohydrates, fats, proteins and minerals which are put together from the air and the substances found in soil. Green plants require at least fourteen elements for normal growth and reproduction. Important minerals obtained from the soil are nitrogen, phosphorus, potassium, calcium, sulfur, magnesium, iron, and boron, each of which plays one or more roles in plant metabolism. Some of these are found in organic matter which is broken down with the aid of soil bacteria in decay. The use of radio-active elements indicate that plants absorb minerals not only through their roots, but also through the foliage, fruit, twigs and even flowers. In soils deficient in humus, the adding of vitamins may improve the size of the plant and flowers.

Growth Regulators

Growth is governed not only by the more usual environmental factors but also by chemical stimulants and other growth regulators produced in plants in the form of hormones. The introduction of the first rooting compounds in the 1930's initiated a new concept of growth controls in horticulture. Various specific hormone extracts and hormone-like growth-regulating chemicals made synthetically, when applied in minute concentrations to growing tips, seed, bulbs or cuttings, can stimulate or retard cell development and division. They cause selective changes such as stem elongation or reduction and multiple branching; bud initiation or formation; promote the setting of fruit and fruit enlargement; control flower drop and prevent premature fruit-drop; stimulate seed germination; break or prolong the dormancy in plants and bulbs; induce cuttings to root. On the other hand, when used in higher concentrations they may be toxic and are used as weed-killers, to defoliate crops, or to inhibit growth. Constant research is being carried on in the use of hormones as growth regulators, and their application may have effects as yet unknown.

Scientific experiments have long been made showing that X-ray treatments will produce genetic changes in plants, as have treatments with Colchicine, a drug derived from the autumn crocus (Colchicum). More recently it has been demonstrated that in many ornamental as well as food crops, exposure to atomic radiation may influence plant behavior, inducing useful mutations characterized by faster growth, heavier foliage, more brilliant flowers, earlier fruiting, higher yields, and disease resistance.

The effect of certain gases, particularly ethylene, hydrazine and acetylene, in upsetting plant metabolism, such as initiating the flowering in bromeliads, inducing pineapple to set fruit, tomatoes to go "to sleep" or hydrangeas to shed leaves in speeding dormancy is well known and applied in practice. A basket of apples or a carbide lamp are means of generating ethylene. It might be mentioned here that natural gas is not harmful to plants, but the ethylene often present as a by-product in manufactured gas is harmful. Carbon dioxide is the gas that is vitally important to plant life, but is rarely lacking outdoors. In a closed room or a greenhouse however where many plants compete for the available CO_2, the variation is much greater; at night, when the plants are using sugars in the process of respiration, much carbon dioxide is released into the air while oxygen is taken in, but as soon as photosynthesis begins at dawn, it is used up very quickly. If the windows or vents remain closed, growth may be limited, and remarkable responses have been recorded when supplemental CO_2 is added to the trapped atmosphere especially during winter. Aside from decaying mulches, carbon dioxide may be derived from dry ice, liquid CO_2 in cylinders, the combustion of bottled propane, natural gas in a Bunsen burner, the open flame of wood alcohol in a lamp, or kerosene burned in stoves, introduced during daylight or sunny hours. Air movement is beneficial for the reason that it brings more CO_2 molecules into contact with the plant, improving its chances for healthy growth.

How to Care for Plants Indoors

The requirements for growth of any plant are: temperature, light, water, and plant nutrients. The latter may be said to include carbon dioxide of the atmosphere, but since this is rarely lacking, we need not usually worry about it. Nutrients lead us to the soil problem, but plants are quite adaptable to most mixes that happen to be available.

The matter of light and temperature requirement has much to do with the class of the plant. Exotics generally come from warm countries, and we are prone to think of them as requiring really torrid temperatures, but surprisingly they generally do best in rooms which we would term cool or downright cold, such as a poorly heated glass-enclosed porch. The benefits of cooler nights are needed for the translocation of manufactured sugars from the leaves to the roots and other growing parts of the plant.

The key to the problem is water, the medium for life, which constitutes about 90% of the plant's make-up, and carries its food supply. Water is absorbed through the roots and is transpired through the green parts. The rate of absorption depends on the root system and the soil, and the size and nature of the container.

We have no difficulty in imagining the sides of a tropical canyon next to a waterfall, clothed with delicate foliage, jewelled with drops of spray. Here the air has no drying power, it is saturated with moisture, its humidity is 100%. The beautiful plants flourishing in this habitat are creatures of just such atmospheric humidity. Take them into a well-heated living room where the air carries only a small fraction of the moisture it can hold, and evaporation far outruns the capacity of the plant to replace the loss of moisture. Wilting, drying out, and death of the plant easily follow. This is the unfortunate history of many plants injudiciously or thoughtlessly brought into the home.

However, from xerophytic regions with little moisture come plants that are invulnerable to drought, and from tropical undergrowth come others tolerant of gloom, some of which show remarkable adjustment to living room conditions. From these classes, with thousands of varieties, by experiment and experience, are selected the popular house plants, varying with locality, climate, fashion and favor.

Environment

According to their respective climatic backgrounds, plants intended for indoors are often divided into (A) House plants and (B) Greenhouse plants, suitable for Warm, Intermediate, or Cool surroundings, as indicated on the "Key to Care" chart on page 19.

"House plants" are such exotics as are satisfactory for home and interior decoration, tolerant of the reduced light and artificially dry atmosphere of the living room or office.

Those designated as a "Greenhouse plant" require the higher relative humidity prevailing in a glass house, but may also be grown within the home in a glass terrarium (converted aquarium) or a shelter of translucent plastic where extra atmospheric moisture is provided. In humid-warm regions a moist, slatted shade house will do just as well for many "Conservatory" plants.

"Warm" greenhouse plants are known as "Stove plants" in England—and usually grown best there in humid-warm greenhouses because living rooms are frequently kept much cooler than in the United States.

"Intermediate" plants usually enjoy fresh air and good light—a place near a window open in the summertime, a sun-porch glassed-in during winter, or a light, well ventilated home greenhouse; in warmer climate the sheltered patio, or under a tree outdoors.

"Cool" plants of course also like the same basic environment as intermediate (temperate) plants but could be left almost entirely outdoors except during periods of freezing weather. If kept indoors they require good light.

Flowering vs. Foliage Plants

Blooming plants such as we receive at holidays require much more water and stronger light than the average foliage plant for the continuous opening of their buds and longer enjoyment of their blooms. Foliage plants, not having to support a demanding inflorescence, get along on less light and are much easier to maintain. It is usually safer to keep them on the dry side than too wet, especially if they are not too well established in their pots or tubs.

The modern trend toward decorating indoors or on the patio with specimen exotic evergreens and palms in fancy bowls and jardinieres of highly glazed ceramics or fiberglass brings back memories of the Victorian era of pleasant and relaxed living. Decorative container plants as shown at International Flower Show, New York: background – Dracaena massangeana, Chamaedorea erumpens, Dracaena marginata; center – Araucaria, Chamaerops, Dieffenbachia, Croton; foreground – Scindapsus, Spathiphyllum, Pittosporum, Chamaedorea elegans, Brassaia, Aglaonema.

Maintenance of Decorator Plants

Clean foliage improves the appearance of a decorative plant and assists it in maintaining its life processes. To wipe the foliage with soft cloth will cleanse the breathing stomates, and a touch of milk adds luster to the leaves. Commercial leaf-shiners or oils should be used with caution as they may clog the pores or otherwise cause damage. It is safer to dilute leaf-polish up to 50%.

If a plant stands a long time, say a year, in the same pot, and the roots have filled the same to capacity, mild feedings once a month will sufficiently sustain it. It may be more desirable to confine a house plant, intended for decoration, to as nearly as possible its original size by merely "maintaining" it, watering it sparingly and withholding growth-stimulating feeding. Such a plant will subsist on a minimum of care although in the "weaning" process from the humid greenhouse to a relatively dry living room a few leaves may initially drop before the subject becomes adjusted to its new surroundings. But after that, a plant so conditioned will be less troublesome, not always crying for attention, and continue for a long time to grace the spot for which it was intended.

How a CODE to CARE can help

A key to plant care such as devised for this Manual can at best be only a general guide. Based on the climatic background of our cultivated exotics, the pictograph symbols assigned to each plant try to spell out the conditions prevailing in their native habitats, to provide whatever would make them happiest in the plant lover's care. However, many plants are quite flexible in their requirements and have shown a remarkable capacity for tolerance and adaptation to unfavorable environment. Anyone interested in plants may attempt with confidence their cultivation at home. Many a beginner has succeeded with growing a fancy plant on a window sill which would worry an expert, proving only that the so-dalled "Green thumb" is no miracle gift but merely the result of careful observation and patient understanding of the plant's needs.

This CODE to CARE should be of special help to the amateur gardener. Once having understood and mastered the basic fundamentals there is nothing to prevent the adventurous from trying his own best methods and with a little feeling for the plant he will probably be successful. To grow plants is an art, and not an exact science—yet for this very reason it is so fascinating.

EXPLANATION OF THE FIVE ELEMENTS COMBINED
IN THE PICTOGRAPH SYMBOLS USED IN THIS MANUAL

Environment (see page 16)

HOUSE PLANT: can be used for home and interior decoration, as it tolerates the reduced light of the living room or office with its artificially dry atmosphere.

GREENHOUSE PLANT: Requires relative high humidity such as can be maintained in home greenhouse, under plastic shelter, or a glass terrarium.

Temperature (see pages 27-30)

W **Warm:** 62–65° F (16–18° C) at night, can rise to 80 or 85° F (27–30° C) in daytime before vents must be opened.

T **Temperate:** 50–55° F (10–13° C) at night, rising to 65 or 70° F (18–21° C) on a sunny day or higher with air.

C **Cool:** 40–45° F (5–7° C) at night, 55–60° F (13–15° C) when sunny, with air; 50° F (10° C) in cloudy weather.

TEMPERATURE CONVERSION

°F 0 10 20 30 40 50 60 70 80 90 +100

°C -10 0 10 20 30

Fahrenheit-Centigrade

Light (see pages 22-26)

Maximum Light (sunny South window). "Maximum light" plants theoretically have a preference of 4000–8000 footcandles (for average daylength) for growth, but for mere maintenance will tolerate between 500–2000 fc. A South window facing direct sun is ideal.

Partial Shade (East or West window). Plants classified "Partial shade" would like 1000–3000 footcandles but will survive at 100–1000 fc. A simple indicator of diffused sunlight is to pass your hand over your plants and barely see its shadow. A clear East or West window is best, but a Southern exposure must be lightly shaded from direct sun by Venetian blinds, a bamboo screen, or curtain. For mere maintenance of most plants in this group in the home, light intensity may go as low as 25 fc.

Shady or away from sun (North window). Delicate exotics designated "Shady" are safest in a North window, with diffused light, and would normally receive 50–500 footcandles. There are very few plants which do not want some sunlight by preference; shade lovers are limited mostly to delicate plants from the forest floor, and ferns. Under artificial illumination, light intensity may be as low as 10 fc. High humidity is important to the well-being of plants in this group.

Soil (see pages 37-42)

Loam, rich garden soil, or decomposed granite with some rotted manure or some humus; also, in heavy soils add some sand or grit to prevent caking.

Humusy soil, rich in organic matter; sphagnum peatmoss is excellent for moisture-holding capacity; also leafmold. Add a small amount of loam; perlite for drainage.

Watering (see pages 31-36)

Moderately Dry. Drench thoroughly then allow to dry between waterings. This admits air into the soil structure which, in turn, promotes development of a healthy white root system; wiry, thick roots being characteristic in this group. Watering means soaking the root-ball penetratingly holding the pot if necessary in a bucket, sink or tub of tepid water until air bubbles cease to rise.

Evenly moist but not constantly wet. Plants so classified generally have delicate, hair-like, fibrous roots, subject to rot if kept too wet, and equally easily burning and shrivelling if too dry, especially in hot weather. Maintain uniform moisture throughout the root-ball, without letting the soil become water-soaked and "sour".

18

**PICTOGRAPH SYMBOLS AT BASE OF EACH PHOTO, OR THE NUMBERS
IN THE DESCRIPTIVE TEXT INDICATE PLANT REQUIREMENTS AND TOLERANCE.**

ADAPTABLE AS HOUSE PLANT				BETTER AS GREENHOUSE PLANT		
suitable for growing or decoration in Home, Office, or sheltered Patio; tolerates dry atmosphere				in need of humid or airy conditions; Glass Terrarium or Plastic frame. Not recommended for the living room.		

WARM 62–80° F 16–26° C	TEMPERATE 50–65° F up 10–18° C	COOL 40–60° F up 5–15° C		WARM 62–80° F 16–26° C	TEMPERATE 50–65° F up 10–18° C	COOL 40–60° F up 5–15° C

PLANTS IN LOAMY GARDEN SOIL, HUMUS ADDED

W	T	C	Maximum Light; Sunny 4,000-8,000 Footcandles KEEP ON DRY SIDE	W	T	C
1	13	25		51	63	75
W	T	C	KEEP EVENLY MOIST	W	T	C
2	14	26		52	64	76
W	T	C	Partial Shade; Diffused Sun 1,000-3,000 Footcandles KEEP ON DRY SIDE	W	T	C
3	15	27		53	65	77
W	T	C	KEEP EVENLY MOIST	W	T	C
4	16	28		54	66	78
W	T	C	Shady; Away From Sun 50-1,000 Footcandles KEEP ON DRY SIDE	W	T	C
5	17	29		55	67	79
W	T	C	KEEP EVENLY MOIST	W	T	C
6	18	30		56	68	80

HUMUSY; SOIL RICH IN HUMUS; OSMUNDA OR FIRBARK ON EPIPHYTES, SUCH AS ORCHIDS

W H	T H	C H	Maximum Light; Sunny 4,000-8,000 Footcandles KEEP ON DRY SIDE	W H	T H	C H
7	19	31		57	69	81
W H	T H	C H	KEEP EVENLY MOIST	W H	T H	C H
8	20	32		58	70	82
W H	T H	C H	Partial Shade; Diffused Sun 1,000-3,000 Footcandles KEEP ON DRY SIDE	W H	T H	C H
9	21	33		59	71	83
W H	T H	C H	KEEP EVENLY MOIST	W H	T H	C H
10	22	34		60	72	84
W H	T H	C H	Shady; Away From Sun 50-1,000 Footcandles KEEP ON DRY SIDE	W H	T H	C H
11	23	35		61	73	85
W H	T H	C H	KEEP EVENLY MOIST	W H	T H	C H
12	24	36		62	74	86

For Gardening indoors the window is best. Not only is natural daylight better than artificial light, but the exposure from different directions offers an ideal place for almost any type of plant. Four simulated windows show what can be used and where.

Seldom, however, are all four sides in a home equally available and suitable for plants. But the suggested types offer the opportunity to make a wise selection for any window. This should not, however, prevent us from trying any plant in whatever spot that may happen to be otherwise ideal. A north window can, if necessary, be artificially lighted, and a south window easily shaded.

For ease of care, attractive looks, and flourishing growth of plants try grouping them in boxes. For convenient access to the window, these could be mounted on legs with rollers. Fill a window box with moist peatmoss, and plunge your pots rim-deep. If the peatmoss is kept moist, plants absorb by capillary action, whatever they may need and practically take care of themselves, though some pots may have to be watered individually on occasion. The roots begin to ramble, and the big plants and the little ones, the climbers and the creepers, flowering plants and foliage together form a mutually beneficial ecology with a microclimate that favors luxuriant, natural growth. In close symbiosis, they enjoy each other's company, and will soon create the image of a natural tropic mini-landscape, cover the other's blemishes, and sickly plants take on new life. Miraculously, plants begin to flower, and even delicate exotics seem to forget their being super-sensitive. The secret lies in the complementary give-and-take and intimate climate set up by such an harmonious community of partners, and a window filled with beautiful exotics brings both restfulness and life into the home.

NORTH WINDOW   *(Sunny day in winter near glass, 150–500 footcandles; average 300 fc).*
Sun-less windows are not for most flowering plants. But foliage plants including tender tropicals do well in a fully light north window. Weather-stripping or a protective sheet of plastic to prevent cold drafts is advisable during the cold season. A recommended list of satisfactory subjects would be:

For WARM conditions—Aroids such as Aglaonema, Dieffenbachia, Philodendron, Scindapsus and Syngoniums; Cissus, Dichorisandra, Dizygotheca, Dracaenas; ferns including Platycerium; Ficus elastica and lyrata; Fittonias, Hoffmannias, Marantas, Peperomias and Pileas; even Sansevierias.

For INTERMEDIATE temperatures—Amomum, Araucaria, Aspidistra, spider plants (Chlorophytum), hollyferns (Cyrtomium), Helxine, Saxifraga, Tolmiea, Tradescantias; also most palms including Kentias (Howeia).

At a COLD window—Ardisia, Fatshedera, Fatsia (Aralia), Laurus, Ophiopogon, Rohdea and Skimmia; also all kinds of ivies (Hedera helix).

SOUTH WINDOW
(Midday sun, near glass 5500 fc winter, to 8000 fc summer; cloudy day 1000–2000 fc).
An exposure to the potent southern sun offers a wonderful chance to grow the many plants that are sun-lovers. This includes practically all the holiday blooming plants; flowering shrubs and vines; miniature subtropical fruit trees; many bulbs; and of course the exciting world of succulents.

Plants responding to both sun and WARMTH are—Acalyphas and Croton; Achyranthes, Coleus and Iresine; Amaryllis and other tropical bulbs; Beloperone, Gardenia, Hibiscus and Strelitzia; Tropical clamberers such as Allamanda, Bougainvillea, Clerodendrum, Gloriosa, Jasmine, and Passiflora; flowering succulents like Euphorbia splendens and pulcherrima (Poinsettia), and Kalanchoe; also sunny Dendrobium and Oncidium orchids.

INTERMEDIATE and COOLER plants in need of sun but with the window open when it is warm outside—are Abutilon, Campanula isophylla, and most bedding stock; Geraniums, Petunias, wax begonias, Lantanas; Azaleas, Chrysanthemums, heather, Hydrangeas, Pot roses; Lachenalia and Veltheimia, also Dutch bulbs; fruited Jerusalem cherry; Kumquats and other Citrus.

Most cacti and other succulents enjoy the sun; want to be warm in daytime but cool at night.

EAST or WEST WINDOWS
(Winter sun, near glass, 2000–4000 fc, off-sun 200–500 fc; average 2000 fc).
The partial sunshine of several hours daily at an East or West window is ideal for many flowering and foliage plants that normally grow in partial shade and dislike the intense sun of windows facing South. East windows are generally cooler and receive the clear morning sun; Western exposures tend to be warmer but often with hazy or moderated light.

WARM PLANTS doing well East or West with the benefit of a limited amount of sun are—Anthurium scherzerianum, Alpinia, Aphelandra, Brassaia, Brunfelsia, Caladium, Calathea, Clivia, Costus, Eucharis; Birdsnest and Boston ferns; Ficus benjamina, Hoyas, Impatiens, Jacobinia, oleander, Pandanus, Pseuderanthemum, Spathiphyllum, Torenia, Zebrinas. Great variety is also offered in the many Begonias, bromeliads, and warm orchids. More or less hairy gesneriads, such as African violets (Saintpaulia), Achimenes, Aeschynanthus, Columneas, gloxinias, Rechsteinerias, Smithianthas, and Streptocarpus enjoy the good light East or West but in sunny hours this light must be diffused to prevent burning.

INTERMEDIATE or COOLER PLANTS for partial sun are—Araucaria, Aucuba, Asparagus, Camellias, Cinerarias, Crinum, Cyclamen, Euonymus, Fuchsia, Grevillea, Kaempferia, Marica, Myrtus, Osmanthus, Oxalis, Plectranthus, Podocarpus, Primula, Vinca, and variegated ivies.

Northward for shade plants *(protect from chills): (top) Guzmania musaica; (center) Maranta leuconeura massangeana, Spathiphyllum floribundum, Ficus radicans 'Variegata', Begonia masoniana, Hoya carnosa variegata, Begonia rex 'Peace', Anthurium crystallinum, Philodendron squamiferum, Dieffenbachia picta 'Rud. Roehrs'; (bottom) Maranta leuconeura kerchoveana, Cryptanthus zonatus zebrinus, Scindapsus aureus 'Marble Queen', Fittonia verschaffeltii argyroneura, Calathea makoyana.*

←◄◄◄

Southern exposure for sun-lovers: *(top) Cotyledon undulata, Gymnocalycium mihanovichii friederickii, Kalanchoe marmorata, Opuntia microdasys, Pachyphytum 'Blue Haze', Haworthia margaritifera; (center) Aeonium haworthii, Euphorbia lactea cristata, Crassula 'Tricolor Jade', Sansevieria 'Hahnii', Cephalocereus senilis, Echeveria glauca pumila, Ananas comosus variegatus, Vriesea carinata; (bottom) Aechmea fasciata, Opuntia ficus-indica 'Burbank's Spineless', Echeveria 'Setoliver', Kalanchoe tomentosa, Mammillaria geminispina, Echinocereus dasyacanthus, Kalanchoe daigremontiana hybrid, Espostoa lanata, Crassula rupestris, Sansevieria trifasciata 'Golden hahnii', Cereus 'Peruvianus hybrid' (specimen), x Gastrolea 'Spotted Beauty'.* ►►►→

East for cool morning sun: *(top) Hedera helix 'Manda's Star', Hedera helix 'Manda's Crested'; (center) Begonia semperflorens; (bottom) Ficus elastica 'Decora', Hedera helix 'Hahn's Variegated', Dracaena marginata, Hedera canariensis 'Variegata', Euonymus japonicus medio-pictus, albo-marginata, microphyllus variegata; Eurya japonica 'Variegata', Dracaena godseffiana, Cissus antarctica.* ↓

Moderate Western light- *warm and diffused: (top) Episcia cupreata 'Acajou', Saintpaulia ionantha hybrid, Howeia forsteriana, Episcia reptans 'Lady Lou', Syngonium wendlandii; (bottom) Dieffenbachia picta, Philodendron elegans, Peperomia caperata 'Emerald Ripple', Alocasia x amazonica, Calathea zebrina, Aphelandra squarrosa 'Louisae', Campelia zanonia 'Mexican Flag'. Philodendron 'Lynette'.* ↓

Light and its Importance to Plants

Light is the most important environmental factor influencing photosynthesis, and its importance to the metabolism of the plant cannot be underestimated; the reasons are explained under Plant Physiology. Without light life cannot exist. Light triggers the processes that manufacture sugar in the leaf, and without its presence all activity stops. In the absence of light, the plant utilizes reserve foods for energy. If conditions are unfavorable for photosynthesis and production of new sugars but favorable for respiration, as during periods of low light intensity combined with high temperature, the plant may use all of its stored food and perish.

Light is that form of radiant energy which is capable of producing the sensation of seeing. All light has three main characteristics: quality or color, brightness or intensity, and duration. The most abundant and efficient source of light is the sun. The solar spectrum sends its rays in all the colors of the rainbow, but not all the colors are utilized by plant life.

The Color of Light Waves — the Angstrom

The color of light is determined by its wavelength. The different wavelengths are measured by the Angstrom unit which is equal to 10 millimicrons, or one ten-millionth part of a millimeter; 10 Angstroms are 1 nanometer. The human eye is tuned to respond only to the energy within the visible spectrum, a narrow band of wavelengths between 3800 Angstroms and 7600 A.

Light at the shortwave end of the visible spectrum, from 3800–4500 A produces the sensation of violet; the longest visible waves, from 6400 to 7600 A appear as red. Between these lie the blue (4500–4900 A), green (4900–5600 A), yellow (5600–5900 A) and orange (5900–6400 A)— the colors of the prism. The mixture of all colors gives us what we call "white" or daylight.

Color Temperature — degrees Kelvin

Color temperature is a term used, for example in color photography, to describe the sum of color we see, such as the yellow morning light or the blue midday sky. This visible effect is measured in degrees Kelvin, a temperature scale which has its zero point at −273° Centigrade. In illustration we may imagine an incandescent body such as tungsten changing colors with increasing temperature or radiation and to compare these colors to a scale by which we can rate the "color temperature". Accordingly, the color of a candle is 1800° Kelvin; a general incandescent lamp, 2500–3050° K; the Warm-white fluorescent lamp 3000° K; the White 3500° K; a Cool-white tube, 4500° K; the bluer Daylight fluorescent lamp, 6500° K. Of the "ideal" light sources—sun and sky—sunlight varies from about 1800° K at sunrise to 5300° K at noon. Sky light ranges from 7000° K (uniform overcast) to 28,000° K for clear blue sky. Daylight (outdoor) color film is usually balanced for 5600° K; artificial light (indoor) color film for 3200°–3400° K (Photoflood).

The Rainbow

Plants are thought to need the whole range of colors in the spectrum but certain wavelengths in the blues and reds are more important to them than the green and yellow, which are predominant in sunlight. Specific wavelengths or combinations govern their behavior; 4350 Angstroms (blue) and 6500 A (red) control photosynthesis; 3700 A (ultraviolet), 4550 A and 4750 A (blue) for phototropism (bending to the light); 6600 A (red) and 7300 A (far red) controls stemlength, leaf size, photoperiodism (natural day-night schedule), and other photomorphogenic responses—changes in development, or inhibition of flowering influenced by light.

Sunlight benefits plants most. A sunny place by the window is ideal for many indoor plants such as this Brassaia (Schefflera) actinophylla – the Queensland umbrella tree – an enduring decorative plant for the living room as it enjoys being warm and dry, even tolerating a darker location.

An exhibition hall at the De Young Museum in San Francisco, where an attractive corner planting of large Dracaena warneckei, Philodendron selloum, Monstera deliciosa, and Rhapis palms frame an unobtrusive marble bench where the visitor may rest and contemplate a French sculpture of the goddess 'Pomona'. Mellow sunshine diffused through a glass ceiling, and supplementary incandescent spotlights both accent and benefit these showy foliage plants with warm light.

Natural Light

Adequate daylight is the best type of light for house plants. They require light, but not all need the same amount. Most foliage plants do perfectly well in limited sunlight and react unfavorably to intense sun. Some highly colored foliage plants such as Coleus and Crotons retain their brilliant colors best in bright sunlight. Sun-lovers, too, are most succulents and many flowering plants including geraniums, Abutilon, Hibiscus and Gardenias. Sun rooms and glass-enclosed porches—and in a limited way the windows facing South—are naturally ideal for light-hungry plants. These include most of those timed to flower for our holidays like Roses, Hydrangeas, Azaleas, lilies, Chrysanthemums, poinsettias, and bulbous stock. Blooming plants may, of course, be placed into any other part of the home for temporary enjoyment, but to prolong their flowering period should be rotated to the lightest and somewhat cooler places.

Light Intensity — the Footcandle

To properly measure the energy of a light source significant to plant life, readings should be made in watts per 100 Angstrom bands. However, the old footcandle method of reading light direct from the source is relatively simple, although the current light meters are color-blind and primarily read the yellow-green bands of the spectrum—the light made for reading. One footcandle is the amount of light cast by a candle on a white surface 1 foot away in a completely dark room. In European practice, the light intensity is expressed in the metric term of Lux: 1 footcandle = 10.76 Lux. Bright moonlight may measure 6–7 Lux or 0.02 fc.   One lumen light output illuminating 1 sq. ft. of surface equals 1 footcandle. A 100 watt light bulb emits about 1700 lumens.

When considering light requirements, bear in mind that many plants that need shade in summer prefer more light in winter. Light indoors is much less intense than outdoors, and rapidly diminishes further away from the window, and is less than 60 fc in the average home. A light meter reading of daylight in winter 1 foot distant from a North window gave me nearly 500 footcandles but 3 feet away only 150 fc. Curtains will reduce the light intensity even more. Windows South, or corner windows normally allow more light than windows facing West or North, unless shaded by a tree or wall. The sunlight (winter) 1 foot from a South window without shade may reach 5600 fc against 250–400 fc of light in the northerly direction. Outdoor reading in mid-winter, in New York City (latitude 40.7° N) are 5–9000 fc with a 9½ hour day, whereas in mid-summer and a daylength of over 15 hrs. we record 10,000 fc or more; in the shade of a tree 100–1000 fc.

Light Requirements of Plants

Light, which gives energy to the plant, is like the force of electricity running a motor. Just like volts indicate the amount of electricity received by the motor, footcandles indicate the amount of light received by an object. Below a certain minimum of about 200 fc plants fail to grow actively, although 10 footcandles in many species will sustain life. But too much light can be harmful too.

When you can barely see the shadow of your hand if held between the sun and plant – this is a simple method to determine diffused light, safe for begonias (left *Begonia gogoensis*, right *Beg. 'Spaulding'*).

Why Poinsettias refuse to bloom in the living room: Poinsettias are "long-night" plants and as little as 1½ foot-candles of light during the night in September can delay or inhibit flowering; keep in a closet for 14 hours nightly until bracts form. Photo'ed at Christmas, a greenhouse grown plant at left had the normal darkness of more than 12 hours in early October; the plant on right was in my office with lights and too warm, and formed only rudimentary bracts.

With this "two-decker" arrangement of fluorescent lamps over trays with peatmoss, gravel, sand, or vermiculite, many seedlings or low-intensity light plants such as gesneriads may be grown successfully even in the warm basement of your home.

A windowsill in a modern German home that really invites house plants: a plate of polished black marble extends the full width of the living room, the heat radiator underneath. Airy windows and short curtains for optimum light combine to provide near greenhouse conditions for anything in pots.

As we can burn up a motor, it is possible to burn up a plant by unneeded light. Depending on the type of plant, a single leaf can assimilate up to 2000 fc of light efficiently, and the total plant can use considerably more because many of the leaves will be shaded. But excessive light intensity and resulting high leaf temperature will cause the photosynthetic system to break down and food manufacture is impaired. Most foliage plants enjoy optimum levels of 1000–2000 footcandles; low energy flowering plants like gloxinias 1000–2000 fc; high energy flowering plants including Azaleas 3000–5000 fc, Chrysanthemums and Roses to 10,000 fc.

People find more than 100 footcandles indoors excessive for reading; a level of 15–20 footcandles, equivalent to fair reading light, is sufficient to many foliage plants for survival if combined with reduced moisture. However, in evaluating quantity of light, intensity alone is not all that matters but footcandles should be multiplied by daily duration. Ten hours of light at 30 fc is as effective as 15 hours at 20 fc.

Footcandle Readings of Artificial Light

Required light levels expressed in the 'Key to Care' as **"Full sun"**, **"Partial shade"**, and **"Shady"** can be determined more accurately with the aid of a **Direct-reading** light meter.

Meter readings I made in office and living room indicate that a 100 watt incandescent lamp measured 60–70 footcandles 2 feet away; 25–35 fc 3 ft. distant; 15–20 fc 4 ft. away. A 150 watt lamp at 3 ft. 60 fc, 4 ft. 30 fc, 6 ft. 10 fc. Incandescent reflector spot 300 watt 3 ft. 180 fc, 6 ft. 50 fc. Two 40 watt fluorescent tubes 1 foot directly above 240 fc, 2 ft. 120 fc, 3 ft. 60 to 80 fc, 4 ft. 40–50 fc, 5 ft. 30 fc. For comparative brightness, fluorescent lamps produce or emit approximately 3 to 4 times the output of light (lumens) per watt over Tungsten filament (incandescent) bulbs.

If no direct-reading light meter is available, a photo-electric meter for reflected light may be utilized to convert photographic settings to footcandles. Set the film speed dial to ASA 200 and take the reading toward a sheet of white paper. A shutter speed of 1/500 sec. with various F-stop readings indicates footcandles about as follows: F 22 = 5000 fc, F 16 = 2500 fc, F 11 = 1200 fc, F 8 = 550 fc, F 6.3 = 300 fc, F 4.5 = 150 fc.

Artificial Illumination

The single illuminant most capable of duplicating sunlight is an incandescent source. Combinations of fluorescent and incandescent lamps are the next best simulated daylight sources.

Foliage reflects green and yellow but absorbs blue, orange and red. Under artificial illumination good growth has been obtained by using a combination of orange and red (6000–7000 Angstroms) and blue (4000–5000 A) fluorescent lamps, in combination with some 10% of visible radiation from incandescent bulbs to supply the additional potence of far red needed particularly by energy-hungry blooming plants and fruit.

This *Philodendron oxycardium* in a planter lamp is quite happy under artificial light. A 75 watt incandescent bulb would provide 80–120 fc of light at 1 ft (or 20 fc at 3 ft).

A small Green world in our home – this cozy nook encloses a little indoor pool. And moisture-loving tropical plants including Episcias and Begonias thrive here under fluorescent light.

Most fluorescent tubes are high in blue (5000–6000 Angstroms), but being cooler may be placed nearer to the growing plants. A 3500 degree Kelvin tube lamp is a good balance light. Where the close proximity of a light fixture near the plants is not objectionable, low energy flowering plants such as Saintpaulias (African violets) and gloxinias, and foliage plants, will grow well under it. The cost per footcandle of light received by the plant is less than that of incandescent light, and these tubes have their place in the growing of smaller plants, seed cultures, and vegetative propagation. But the decorative effect of fluorescent lamps is flat and cold, while incandescent lighting is orange in characteristic and all the flower colors with a large amount of yellow and into red appear rich and vibrant.

Fluorescent tubes have been hung perpendicular between the foliage of taller plants with great effect, their ballast wired to some nearby closet. This makes these lamps much less unattractive and brings the light close to all the leaves where it can do the most good.

Lighting in the Home

It would be so easy to combine the comfort of a reading lamp in the home with the vast amount of good this will accomplish in the maintenance of plants if every day at dusk a 100 watt bulb from an overhead bridge lamp could shine on a planter or the window garden. This, costing about 10 cents a week, provides not only welcome added light but will give accent to any plant or planting. For fluorescent illumination a simple rule to follow when planning a house plant corner is to provide 15–20 watts of fluorescent light per square foot of growing area.

Contrary to popular belief, 24 hours of continuous light will do no harm to growing plants not photosensitive provided that the temperature is changed and lowered for the night. However, such lengthening of the day by additional illumination during hours of natural darkness may play tricks on some flowering plants that are sensitive to daylength. Poinsettias or Chrysanthemums, which are long night (short day) plants, would continue in vegetative growth as long as illumination is over 12 hours and even a fraction of light shining on a poinsettia during its normal bud initiation time end of September would prevent it from coming into flower.

Plants for poor light

House plants recommended as most tolerant to poor light, down to 5 or 10 footcandles, and yet long lasting—include Aglaonema, Araucaria, Aspidistra, bromeliads, Dieffenbachia, Dracaena marginata, sanderiana and warneckei; Ficus decora, Hoya, Philodendron, Chamaedorea and Kentia palms; Pittosporum, Podocarpus, and many succulents including Sansevieria. Somewhat better light is needed by Begonias, Boston ferns and others, Brassaias, Cissus, ivies, and Scindapsus, most Ficus relish light and require at least 50 fc and more. But more light will benefit all of these, and an incandescent spotlight flooding any interior plantings with warm and glowing light will enrich all colors and dramatically accent their decorative effect.

Temperature controls Growth

Temperature is a word that in our daily lives may convey a sensation of comfort and warmth, or of shivering and cold. In the living room we normally feel comfortable at 72° F (22° C). This is characteristically tropical. For this reason tropical plants can be made to feel at home in our dwellings also. However, not all tropicals are heat-lovers. This depends on the altitude as well as the latitude of their native habitats.

Cold Climate in the Tropics

According to elevation, there exist subtropical, temperate, and alpine climates even under the equator. Temperatures at increasing altitude on tropical mountains are identical with temperatures at sea level in latitudes correspondingly more distant from the equator. Generally speaking, for every 1000 feet in elevation, the temperature drops 3 degrees F as compared with that at sea level. The mean temperature at sea level near the equator is 81° F. At 2000–4000 feet elevation at latitude zero (equator) the average may be 75° F, which corresponds with a similar mean temperature at latitude 15 to 23 at sea level, near the beginning of the subtropical belt; the 70° F average in altitudes from 4000–6000 feet at the equator, is 70° F also at latitude 23 to 34 at sea level. The difference is that at higher elevations we find more of a rhythm of warmer days alternating with cooler nights. Plants which inhabit a region where temperature fluctuates widely—learn to have greater tolerance than those which live where conditions are more uniform. Most of our best house plants come from elevations of 2000–5000 ft. in the tropics and the bordering near-tropic zones. It is natural that such exotics also prefer to be cooler at night when in cultivation. Fortunately it is common practice in the home to lower the temperature about 10 degrees for the night, and as a result we sleep more restfully. For plants, the change to a cooler level provides a more favorable balance for their process of respiration—the breaking down of sugars for energy. Most growth takes place at night. Natives of low elevations in the tropics that are accustomed to high temperatures all the time get along well in rooms where there is no marked temperature change. Typical is the Dieffenbachia picta which I found thriving in steamy heat not far from the shores of the mighty Amazon near Manaus, or the many Philodendrons, Marantas, Caladiums and bromeliads of Northern Brazil and the Guyanas.

Covered peristyle of the interior court of a suburban residence in Westchester, New York. Ideally suited for naturalistic planting with attractive and colorful exotic plants, and West Indian treeferns (Cyathea arborea) for tropical accent, the area features a pool and flagstone walks as well as a patio area for dining and restful reading. Roofing of frosted glass admits diffused light, and the plants are thriving in each other's company under optimum conditions of humidity and temperature.

This alcove window is a charming spot for house plants and permits the use of plate glass shelves. An interesting variety of plants are grouped into a pretty window garden, the pots in deep saucers to keep from drying out too quickly. The bottom group closest to the warming radiator stands in a tray with peatmoss or sand which, if kept wet, evaporates coddling humidity.

Plants for cool locations

Some of our decorative plants feel comfortable in surroundings that we would consider too cool for comfort. They have been collected in high tropical mountain regions of the Cordilleras of Colombia, Ecuador and Peru; the Himalayas in Asia; or Mt. Kilimanjaro in Africa. Many cool-type broad-leaved evergreens come from warm-temperate areas with insular climate such as New Zealand or Japan, notably Podocarpus, Pittosporum, Araucaria, Aucuba, bamboos and many araliads including ivies, Fatsias, Panax and others. These species have proven ideal for decoration in cooler parts of the house or office buildings. The various temperature requirements should be comparatively easy to determine if we know the climatic background of a plant, and accordingly, will try to provide them with similar surroundings when growing as a house plant to make them feel at home and happy.

Variation in Room temperatures

Room temperatures are not uniform throughout the house. Checking with a thermometer will prove that some rooms are cooler than others, and locations in the same room may vary tremendously. Positions close to radiators or hot air vents are often hot and very dry and distressing to plants. A shallow tray with pebbles, vermiculite, or peat to set the plants, and with a trace of water at the bottom would help to counteract excessive dryness of the air. On the other hand, exposure to cold close to a window can be just as harmful. There is always a difference of temperature near the glass and the interior of a room, and cool flowering plants such as Cyclamen, Azaleas, Begonias, Cinerarias, Primulas, and Dutch bulbs love such a cool window. But in freezing weather, ice may form inside within a foot of the window glass, and differences 15 to 20 deg. lower than room temperatures are quite common. Heat radiates from the foliage to the cold glass, causing a chilling and possible freezing of the leaves. Pull down the curtains, or place some layers of newspapers between the window and plants that could be damaged.

Plants, mainly woody shrubs and trees from temperate regions that normally go dormant in the wintertime, also perennials and bulbs, respond to higher temperatures after having gone into a period of dormancy. If by special treatment and the use of extra warmth through hot water, steam heat or electricity we bring them into flower in advance of their natural season, we call this "Forcing".

Forcing of Hardy plants

Aside from bulbs and tubers, plants frequently forced are perennials, shrubs and hardy trees, often intended for early spring flower shows. Commonly forced perennials, preferably potgrown, are Dicentra, Astilbe, Iris, Primula, Phlox. They should take about 3–4 weeks to flower at 50° F, depending on whether they are early or late bloomers outdoors.

Many shrubs can be successfully scheduled for early bloom but certain preparations are necessary to ensure success. Woody shrubs and trees to be forced should be selected from the open ground a year ahead. Proper pruning is important to obtain well-budded branches. Deutzias, Spiraeas, Philadelphus bloom on year-old wood, and tips only should be cut after blooming; Prunus triloba should be cut hard. Either grow them in pots for one summer, or root-prune several times, and lift them with good soil-balls in autumn to be placed in temporary storage in a cold shed or cellar. Keep the roots moist and covered with soil and leaves or moss and hold at a temperature near 35° F until needed. Moderate freezing is not detrimental. A sufficient period of dormancy to precede forcing is all-important, and in fact will be decisive in the length of time of forcing response and final success. Before November, nothing except plants held in cold storage from the previous year, a few bulbs and some Prunus should be attempted. December 15 to January 1 is as early as it is safe to begin forcing most hardy plants; it will be found that as the days lengthen the results will be more satisfactory. At first the plants must be kept cool, about 45° F. Syringe woody subjects several times a day preferrably with warm water, until the buds swell. After growth starts, they may be given more warmth if so desired, ideally 65–70° F. For timing, a month or six weeks is a good time to allow in February or March at lower temperatures ranging from 45–55° F. With advancing season, the results are quicker. Popular forcing subjects are dogwood (Cornus florida), the flowering plum (Prunus triloba), Japanese quince, lilacs, Deutzia, Spiraeas, flowering crabapples, Forsythia, Magnolias and, of course, Azaleas and Rhododendron.

Hippeastrum hybrids, commonly known as "Dutch amaryllis", can best be forced for Christmas and January if the bulbs are large enough and well established from the previous season or with their roots preserved, and well rested in late fall. Pot with the upper half of bulb above the soil. When sprouting, set them dark for longer flower stalks, or paper tubes may be used to draw up the stems.

The early stage in the process of forcing of hardy shrubs and trees. After bringing the dormant trees inside, the rootballs must be kept moist, the branches frequently misted with tepid water, and the temperature increased very gradually.

Timing for early bloom

Correct timing of flowering shrubs being forced depends on how well they are prepared and on their bud-set, sufficient rest, on temperature and sunlight. From records I have kept over many years on forcing, mainly in preparation for the International Spring Flower Shows in New York, I have listed the estimated number of weeks in which various forced shrubs can be expected to bloom in a greenhouse, allowing a week or so at 45° F to start, then giving increased warmth:

Started Jan. 15: Evergreen Rhododendrons (hardy Dexter's and other early hybrids); tender hybrids (Pink Pearl etc.); deciduous Rhod. luteum (Azalea pontica); Rhod. mollis, dogwood, birch, 6–9 weeks; lilacs, crab apples, Japanese cherries, Deutzias, 4–5 weeks; Magnolias, Flowering almond, pussy willow (Salix caprea), 2 weeks.

Started Feb. 15: Rhododendron, 5 wk.; crabapples, dogwood, Philadelphus, lilac, Rhod. mollis, 4 w.; hardy Azaleas 3–4 w.; Jap. cherries, Prunus triloba, Jap. quince, Spiraea, Deutzia, Kalmia, birches, Jap. maple, 3 weeks; Flowering almond, Flowering peach, Forsythias, 2 weeks; Magnolia, pussywillow (Salix caprea), 1 week.

Started March 15: Dogwood, deciduous Azaleas, Daphne, $2\frac{1}{2}$ w., Jap. cherries, Prunus triloba, crabapples, Jap. quince, Deutzia, Spiraea, 2 weeks; Flowering peach, Magnolias, 1 week.

Lilacs (Syringa vulgaris) can be forced for November or even earlier, but require much preparation such as winter pruning a year back, timely digging, or pot growing for more flowering wood, and dark storage, for success. Early forcing requires high temperatures of up to 110° F and frequent misting and will benefit from exposure to ether or hanging the tops for 12 hrs. into a warm water bath of 100° F. From December on forcing temperature need not be more than 70–80° F. The bushes may be wrapped into tarred paper to prevent the formation of foliage, or such growth should be broken out 10 days after forcing begins. Flowers will be ready in 3–4 weeks. For Easter, potgrown plants will come along at 55° F for 1 month. A greenhouse is practically a necessity for such "forcing" techniques.

The Forcing of Flower Bulbs (see page 387)

Dutch-type bulbs such as tulips, hyacinths and narcissus should be boxed or potted as they are received—not later than November, and stored for at least 6–8 weeks in a cool (35–40° F), dark place to develop sufficient root-growth. Or they may be buried in the earth and covered with leaves to keep soil temperature fairly even. From January on the early varieties can be set into a room at not over 60° F for the first 8–10 days, keeping them shaded. Thereafter they may be placed in a bright location at regular room temperature and watered to keep moist. Early tulips should bloom in 3 weeks; the later varieties when brought inside in February should bloom in a month. Hyacinths take 2 to 4 weeks depending on temperature; Narcissus 2–3 weeks.

Some attractive lilies also are willing bloomers indoors. If planted from pre-cooled stock (at 34–40° F) they can be planted from early December and such types as the colorful Oregon hybrids will come into bloom in 10–12 weeks at 55 to 60 degree minimum temperature; Easter lilies take about 16–18 weeks at 60° F if timed for Easter.

Many of our house plants grow from tubers or bulbs that are tender and tropical. This includes the popular Hippeastrum, callas, Caladium, and gloxinias as well as the South African Freesias, Haemanthus, Lachenalia and Veltheimia. These and others all need a period of rest after blooming, and should be kept a few degrees cooler during this time until active growth is to begin again.

By bringing budded plant material or even branches that are cut, indoors to force into bloom out of season, a home may always be filled with color and good cheer while snow still covers the wintry outdoors. In the swelling of buds far ahead of nature's return we may doubly anticipate the glory of the awakening of spring.

Hyacinths may be grown in water, in hyacinth glasses, keeping them in the dark until the base of the glass is well filled with roots, then bringing them into light to bloom. Heat-prepared bulbs may be flowered as early as Christmas.

Water and Atmospheric Humidity

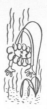

Of all the environmental factors that affect plant life, none is more important than water. Water is formed, sometimes through lightning, by combining 11.2% hydrogen, a gas lighter than any other known substance, with 88.8% of oxygen, the most abundant of all the elements on the earth's surface. This is the life-sustaining liquid that forms 90% of the plant's body and carries its supply of food. Water is the standard of specific gravities, 1 cubic centimeter weighing 1 gram; 1 liter or 1000 cc is the volume of 1 kilogram. It also is the standard of specific heats; water freezes at 32° F (0° C) and becomes ice; it will boil at 212° F (100° C) and become steam.

What could better conjure up visions of Paradise on earth than the swaying coconut palms, and graceful Polynesian maidens here wearing leis of frangipani and skirts of Cordyline leaves, on an island of the South Seas? The lush growth of the true tropics is entirely due to a blend of comfortably warm climate with sufficient atmospheric humidity and which results in harmonious interdependence and well-being of exotic plants and people.

Soft or Hard Water

There are two types of water—"hard" and "soft". We recognize soft water by the ease with which suds form in a soap solution. The degree of softness or hardness of water has a profound effect on plant life. Soft water is generally free of salts, and the soft rain water is ideal for watering, and kind to plants. This would be water that is acid in reaction, with a hydrogen-ion concentration below pH 7, the point of neutrality. Differences in the acidity of the soil-water affect the availability of many plant nutrients. Mildly acid water favors the growth of soil bacteria which break down organic matter in the soil, and is important particularly to ericaceous plants as Azaleas and Erica.

Alkaline or hard water, measuring above pH 7, is "hard" because it contains soluble salts of calcium and magnesium, also chlorides and sulphates. When the salt concentration in the soil water becomes higher than inside the plant cells, osmotic pressure will tend to draw the water out from the plant, injuring its roots and stunting the growth. On the foliage, salts present in water will form grayish scale deposits over a period of time. This not only makes the leaves unsightly, but cuts down the amount of light the plant is able to receive. Hard water may be softened chemically by water conditioners, a simple, effective method is to boil it—or use snow or rain water.

City water usually contains Chlorine, a deadly poison which checks plant growth. Chlorinated water can be aerated by letting it stand in containers with exposed surface for 24–48 hours.

WATER

Plants should be watered as and when they need it and not just at regular intervals. Some thrive well in moist soil, while others do better in a comparatively dry medium. Potted plants require more moisture than those in boxes because of greater evaporation through the porous clay unless the pot is wrapped in plastic or tin foil. Plastic pots or tubs retard drying out longer, for some plants too long. More water is needed when plants are growing actively than they do when semi-dormant. But many apartment dwellers, and well-meaning plant lovers give their plants too much water and kill them because of their very kindness. The texture of the plant stems and leaves and of the root system often indicate how much to water. The harder the foliage and the heavier the roots the less water is likely to be needed to maintain it, as in the rubber plant (Ficus elastica) or the Kentia palm (Howeia forsteriana). The amount of foliage affects the plant's capacity for using water also. Plants which have been cut back, or when they are resting or are unhealthy, must be kept drier until they are again in active growth.

Water a pot when the surface begins to get dry. Generally, plants with coarse roots and growing in heavier, loamy soil should be allowed to get on the dry side—then water well by thoroughly soaking the pot. But root-balls full of very fine fibrous roots, and growing in humusy soil mix, must be kept more evenly moist. To determine the moisture condition simply "feel" the soil; if it feels dry and hard to the touch, and it looks light-colored, then it is high time to water; if the soil feels damp or muddy and looks dark, then better wait a day or longer.

One can probe the dryness of the soil by using a thin white dowel stick and pushing it into the rootball: if the stick comes up clean—water; if smeared with soil then it is wet enough. Or insert a pencil to tell the moisture need by the hardness of the soil. A time-honored test also is to rap a clay pot with the fingernail or wooden stick: if it rings hollow the pot is dry; if dull and without metallic sound it is still wet.

During the cold season, with steam heat in the living room, more water will be needed for most plants than from spring to fall. In winter, watering may be necessary every day; while in summer, when the radiators are off, 2 or 3 times a week will probably be sufficient.

To help in keeping a plant evenly moist, place a saucer underneath it, and supply the water from the bottom. By capillary action the root ball will draw and distribute the water according to its needs. Watering from the bottom has the advantage that oxygen diffusion is not stopped from the top of the pot. However if soil is held constantly wet, the roots use up all the oxygen in the air spaces of the soil and roots will suffocate and rot. During resting periods or when temperatures drop a soil-ball should be kept more on the dry side.

The most practical means to determine whether a plant needs watering, is to test soil moisture by "feeling". If the soil is still moist and looks dark – it is likely wet enough. If it feels dry and hard, and looks gray – it is time to water, making sure to saturate the rootball completely.

Temperature and humidity are closely related. The rate of transpiration in a plant depends largely upon the difference between the concentration of the water vapor molecules within the leaf and that in the air outside it. Most of the water entering the plant escapes, or transpires again as water vapor, increasingly so when temperatures are high. This evaporation is astounding: the average plant evaporates enough water in a growing season to cover all of its leaves $\frac{1}{2}$ inch deep with water. Such moisture transpiration through the pores also permits the evaporative cooling of the plant. The leaf temperature is kept down which means that the plant can operate at peak efficiency. The water actually used is only a small fraction of the amount absorbed from the soil.

A well ventilated conservatory is conceded to provide the best atmospheric environment for the growing of tender exotics; it also offers peaceful relaxation especially during winter. Baskets of hanging begonias luxuriate in this "Wintergarden" at Banff Springs Hotel, in Canadian Rockies.

Humidity of the Air

Humidity is moisture, or water in a gaseous form, which has evaporated into the air. Most indoor plants are tropical or subtropical and in their habitats are accustomed to high humidity.

The function of humidity is primarily to keep the plant from drying out through too rapid transpiration of water from the leaves. Moist air is needed by plants according to their type in a greater or lesser percentage in the atmosphere.

When we speak of "high humidity"—this is air with a high degree of wetness, a condition which we can attain in a greenhouse or glass terrarium. "Low humidity" is air low in moisture usually prevailing in a heated room or often under air-conditioning.

Relative Humidity

However, it isn't humidity as such that makes people (and plants) comfortable or uncomfortable—it is the "relative" humidity. This is the term used to measure the amount of water vapor in the air as compared to the amount the air could possibly contain at a given temperature. Air is like a sponge that absorbs water. Cold air acts like a small sponge that is capable of holding only a little water. But when that air is heated, it swells into a huge sponge with a giant thirst that pulls moisture from everything it contacts.

Dew on leaves in the morning shows that the chilled air of the night will not hold so much moisture as the warmer daytime air and such excess has fallen out as dew, or remains as fog. The warmer the temperature, the more humidity is required to saturate it.

Humidity is measured in relation to its surrounding temperature. In fog, or when it rains, the atmosphere is saturated with water vapor, the relative humidity is 100%. If this thick fog, saturated 100% say at a cool temperature of 45° F, were brought into the house and warmed to 75° F and thereby expanded, the actual moisture content would remain the same but in relation to the volume of air it would sink to 36%.

Delicate moist-tropical plants want to be surrounded by more humidity than the average living room provides. This includes such as Alocasia, Nepenthes, Anthuriums, some Marantacae, gesneriads and orchids, and dainty ferns. A glass-enclosed container or "Terrarium" with waterproof base can become a little conservatory for such subjects, setting them above a tray of water, damp moss, peat, vermiculite, or sand, and lighted for not over 18 hours by concealed lamps. The miniature greenhouse on left uses 5 cool white 20-watt fluorescent tubes plus four 10-watt incandescent bulbs. In so avoiding excessive respiration and subsequent dehydration, it is not astonishing to see softer exotics thrive in these turgid surroundings.

Humidity in the Home

People feel most comfortable at a temperature between 68–72° F, with a moderately high relative humidity of 40 to 60%. This is conceded to be the ideal condition for health, comfort and efficiency. When the air is more dehydrated it is greedy for moisture and sucks it up from everything including our body or a plant. Rapid evaporation from the skin or leaf makes a body shiver or a plant uncomfortable. A humid room feels warmer than a dry room, and a plant "feels" this also.

Sufficient humidity in the air is badly lacking in most steam-heated apartments. The average relative humidity in nine out of ten American homes during heating season is only 13%, whereas even the Sahara has a higher average of 25%; in Death Valley 23%.

Cacti and some succulents from arid regions will tolerate a relative low of 10%. Exotics with their membranous leaves including ferns usually require greater attention to humidity requirements than those with thick, leathery foliage, or with leaves gray-hairy. In experiments with a number of tropical foliage plants however, it has been found that best **growth** was obtained at 76° F and 70% relative humidity in daytime; 65° F temperature and 70% humidity at night. But for maintenance only of a conditioned plant, the humidity may well drop to 25% or even less without harm. Decorative plants tolerant of low humidity include Dieffenbachias, Philodendron, Scindapsus, Araucarias, Brassaias, and Cissus.

To be successful in flowering Saintpaulias, Gardenias or orchids and to keep ferns, or Malayan plants such as Alocasias, the humidity should be somewhere between 50 and 70%. Considering that in greenhouses they are accustomed to perhaps 60–80%, and in their native habitat 70 to 95% plus circulating air, it is easy to see why these types of plants refuse to flourish in the average home without providing sufficient humidity.

In our own best interest as well as that of our house plants, additional humidity in the home in winter should be provided in some manner. Air humidifiers are of course available but not always practical. Evaporating pans or buckets with water on the radiator, will be of help, also setting the plants on trays with moist sand, and misting them. In all, to evaporate a gallon of water per room per day would provide effective winter-time humidification in that it should raise the humidity from 15% to about 32% or 40%.

This home-made, water proof tray wrapped with tinfoil, filled with moist vermiculite or other porous media, on the window sill above the radiator, ideally solves the problem of providing a humid-warm setting for Polystichum fern, Rex begonias, and Saintpaulia.

Self-watering devices for the home: left – an Aphelandra louisae standing on a saucer and placed inside a transparent polythene plastic bag which will condense and conserve moisture for weeks. Right – a Dieffenbachia 'Rud. Roehrs' drawing water by capillarity through a spun-glass wick from a nearby pitcher.

Automatic watering

Semi-automatic wick-watering can be a great boon to plants. One unravelled end of a wick of cotton, asbestos rope or spun glass is inserted through the bottom of the pot, or embedded into the side of the soil ball, while the other end is submerged into water. If the plant is standing elevated on a wire screen supported by lath in a waterproof tray holding $\frac{1}{2}$ inch of water, this is ideal. By capillary action the wick will conduct the water to the soil. Wick irrigation can be modified for use when going away on vacation by using a larger supply of water in a pail or pitcher which can supply a plant for weeks with needed moisture. A group of plants can also be bedded into wet peatmoss in the bathtub for some period of absence. Or individual plants may be set inside a larger pot filled with peat. A simpler and perhaps more efficient way is to enclose the plant, pot and all but not air tight into a plastic bag where it will set up its own system of humidifying and hold all moisture for a long period of time.

Commercial growers have adapted a promising new system of "trickle" watering by leading thin (1/16 in. O.D.) plastic medical tubing of equal lengths to the soil surface of the pot from plastic hoses. This works on the principle that pressure through these "spaghetti" tubes is equally distributed to all pots and the watering is in form of a steady trickle. A modification of this principle might well be adapted for home use also, and such devices, self timing, are now available.

Subirrigation was practiced in Xochimilco, Mexico, since the Aztec Empire.

Large container plants in Southern California are watered dependably and with thorough penetration of the soil ball from octopus-like leaders through individual 1/4 in. O.D. tubing which must all be of even length for equal distribution of water pressure.

Efficient commercial "trickle watering": a poinsettia hooked up to a 3/4 inch polythene water pipe by individual 1/16 inch O.D. plastic tubing of equal lengths ("spaghetti"), the water-flow controlled by automatic moist-scale.

Double-bottom self-watering dishes: the hollow space within the double walls is filled with water which seeps into the soil from below. Cover bottom with coarse gravel to keep perforations from filling with soil. Left, Plant master glazed tray; right, plastic hollow-wall container with variegated Plectranthus coleoides.

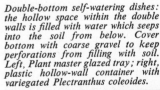

The assembly building of the United Nations in New York is completely air-conditioned; besides, adequate provisions have been made in its architectural design for recessed spotlights in the ceilings. The plantings of Dieffenbachia and Spathiphyllum (left) and Monstera deliciosa (Philodendron pertusum, right) prove by their well-being and happy growth how pleased they are with their controlled climate under artificial light.

Air-conditioning and Plants

Air-conditioning is the process of washing, humidifying and dehumidifying air before it enters a room. It is sometimes roughly effected by introducing an atomized mist spray of water to both cool and moisten the enclosed air; but more precisely by using an air-conditioning apparatus by which air is drawn over water receptacles to extract heat and humidity and remove impurities. In this manner air is purified and its temperature and moisture definitely controlled before it is introduced into the space to be conditioned.

Plants themselves would normally be most effective as air-conditioners. They filter and improve the atmosphere by mixing clean with impure air, giving off pure oxygen and adding to the humidity of the atmosphere.

Controlled conditions in an air-conditioned building hold the temperature comparatively cool when it is hot outside, at the same time lowering excessive humidity for comfort. In winter, the temperature is brought to an agreeable 72° F but in warming up the air, moisture is removed and sinks to a harmful low. Fortunately, provision is usually made whereby in such a case moisture is added automatically to the atmosphere, maintaining its humidity at an ideal 43%. This keeps glass from fogging or dripping, furniture from warping or cracking, and is more healthful for people too, as well as increasing their efficiency. Indirectly but no less important, the plants indoors will also benefit as most of them thrive well at a relative humidity near 50%.

Most modern buildings in New York and other cities are theoretically air-conditioned but in everyday operation there are many fluctuations that may hurt plants. Quite frequently air-conditioning units, even during the hot season, are turned off on weekends or at night and temperatures may rise into the 90's while humidity drops to arid levels. Even when operating, moisture is not always evenly maintained but may vary from a high 70% down to 20%. When it is low and the temperature is high, container plants naturally require more frequent attention to watering. Air-conditioning is generally not set low enough to chill plants normally grown indoors unless they stand close to windows or directly in front of an air duct where the temperature may register 40° F. If plants do not do well in air-conditioned rooms, complaints are probably due to dehydration from cold drafts or air movement.

Do not be afraid to try plants in air-conditioning, as many are not that sensitive. Nearly all the Philodendrons adapt themselves to mechanical climate control well, and we have kept Araucarias and Brassaias (Schefflera) satisfactorily right on top of air outlets or heating ducts.

Ideally, warm decorative plants do best in home or office with the usual automatic daytime control of 72–74° F, but benefit immensely if the temperature for the night can be lowered to 62–65° F. This will slow down their metabolism, stiffen their growth and renew their will to live, and lets plants in flower keep ever so much longer.

Their behavior under air-conditioning has demonstrated that decorative container plants generally adapt themselves with the same pleasure to climate-controlled environment as people do, and in fact—often feel more comfortable than under natural conditions in the tropics.

Soil is a mixture of weathered particles of rock or gravel and decaying organic matter containing moisture and air covering the earth. It is a medium quite complex, and serves as support and anchor for plants and as a reservoir for fertilizers and water.

Characteristics of Inorganic Mineral Media. Half of the soil consists of solids, the other half of openings which contain water and air. Soil texture refers to the size of the individual particles of decomposed rock and minerals which are classified respectively as sand, silt, or clay.

CLAY is an earthy material, sticky and firmly coherent when moist but hard and harsh when dry. Clay may be termed the "active" part of the soil. It is composed largely of very fine grains, of crystalline minerals smaller than 1/500 mm. Some clays are rigid and non-expanding crystal; others are capable of internal expansion and shrinking during wetting and drying and which accounts for the appearance of cracks in soil when it dries out. Clay particles are negatively charged and will attract positively charged fertilizer ions (electrically charged atoms) and hold them tightly for exchange with other elements, to be absorbed by the roots. Clay also acts as a buffer resisting rapid change of acidity because calcium tends to adhere firmly to its particles at the same time aggregating them for better drainage.

SILT consists of particles larger than clay and usually is found in deposits of mud or fine earth. Flour-like, but resembling sand in character, it is less able to cohere to other mineral soil constituents, which gives it a tendency to "run" or ooze when wetted. Because the particles are so fine so as to block passage, water and air do not move well through their small pore spaces giving trouble to the grower. Nor can silt hold fertilizer as clay can. It gives soils the feeling of flour.

SAND is a collection of fine particles of stone, and these may be a thousand times as large as clay. Because of their size, this will also increase the spaces between smaller soil particles and facilitate movement of air and water. Sand is mostly quartz, and does not decompose, nor is it chemically active as is clay. Sand gives soils a gritty feeling.

LOAM is a collective name for natural mixtures of clay, silt, and sand, and according to their proportion may be called sandy, silty, or clay loam. Such loams are usually found in cultivated topsoils or sod, and which include an already incorporated and desirable portion of organic matter making good "garden" soil.

PERLITE is a modern, horticulturally sterile potting (and propagating) medium widely accepted under various trade names. It is a volcanic mineral exploded by heat into lightweight white pebbles filled with air bubbles which can attract and hold moisture uniformly, thereby furthering a healthy root system.

Capillarity and Porosity. In the structure of good soils, with the aid of clay and organic matter, the primary particles tend to be grouped or glued together in "aggregate" granules which in turn makes the soil more porous and friable. The pores are larger and allow more space for air and water to pass in and out. When the particles are loose and "dispersed", the soil structure can become compact and impervious to water and air, resulting in poor drainage.

Within the soil granules and where they touch, the pore spaces are small and capable of holding water by cohesion against the force of gravity, and these are known as "capillary" pores. When a

Plants in close grouping with others support each other by setting up a microclimate that provides humidity and an even level of moisture in the soil, especially if the plants are plunged deep into a box with peat as shown here at this window.

pot of soil is placed in water, the moisture will rise in capillary pores as ink would spread in a blotter. Water moves in the direction from wet soil to dry soil. The smaller, capillary pores hold the water that is absorbed by the roots of plants. Clay soils are wet and heavy because of the many capillary pores that are formed by the very small size of the clay particles.

The larger spaces between granules are known as "non-capillary" pores, and water will not rise in them by capillarity. They provide the channels through which air can circulate in or out of the soil, and for the "free" water that passes through by force of gravity so that it may drain away quickly. Sandy soils have numerous larger, non-capillary pores and this is why they dry rapidly.

Air is present in the soil in all pores from which water has evaporated or been absorbed. Close to 20% of the soil atmosphere is oxygen which is used in respiration by roots and microorganisms. In turn, carbon dioxide in the soil air arises from respiration of roots and the microorganisms, leading to the liberation of fertilizer materials. Roots usually do not grow into poorly aerated soil such as those in which the soil is saturated with water. In soils which are overwatered, roots are injured or killed from lack of oxygen and not from too much moisture.

Soil Acidity.

Soils are considered acid (sour) or alkaline (sweet) in reaction. They are considered acid when positive hydrogen ions in the soil solution are more numerous than the negative ions. When they balance, as in distilled water, the reaction is neutral. The chemical scale to measure the degree of acidity or alkalinity uses the symbol pH, or "the potential acidity due to Hydrogen ions". pH 7 is neutral. Higher readings indicate that the soil is progressively more alkaline; lower values continue to be more acid. Each degree up or down the pH scale represents a ten-fold increase or decrease in soil acidity. A soil solution of pH 6 is ten times more acid than one with pH 7. The growth of plants is affected by soil pH primarily through the changes in the availability of various nutrient elements in the soil. Calcium is tied up in insoluble forms at low pH, while aluminum and iron as well as potassium and nitrate nitrogen become soluble. High "pH" reduces the availability of boron, iron, manganese and phosphorus.

Most garden soils in areas of greater rainfall are acid, usually pH 6 to 7; likely higher in our Western, more arid regions. Slightly acid, around pH 6.5 is best for most trop. foliage plants. Gardenias, Camellias, Ericas and Azaleas because of the availability of more ferrous iron grow better in somewhat more acid soil; cacti and most legumes lean toward the alkaline side because of the more plentiful supply of calcium available.

Microorganisms react markedly to the pH of the soil. Fungi do the work of releasing nitrogen from the organic matter in very acid soils. The beneficial bacteria that perform the vital task of changing ammonia into nitrates are most abundant in soils approaching neutrality and above pH 6. The organisms that fix atmospheric nitrogen do not flourish below pH 5.7.

Soil pH may be tested with a simple Home soil test kit. If a need for correction is indicated, ground limestone is used to raise the pH of acid soil (try 2 teaspoons of lime to 1 quart soil); dusting sulfur will acidify or lower the pH of alkaline soil (1 teaspoon sulphur to 1 qt. soil). Gypsum will add calcium without altering the acidity. Acid liquid fertilizers such as ammonium sulphate also help to increase the acidity. Organic matter including sphagnum peatmoss or leafmold, reduce the ability of pH to affect the availability of nutrients. While in themselves usually acid, they act as a "buffer" in soils either too acid or too alkaline, absorbing excess acid or alkaline elements and tending to maintain the pH at a near constant level.

Characteristics of Organic Potting Media.

When vegetable matter decomposes it becomes humus. This organic medium comes in many forms, and its purpose is manifold. Humus enriches the soil far more than its chemical constituents might indicate. It "buffers" the bad effects either of too acid or too alkaline soils. Organic matter raises the water-holding capacity of mineral soils and brings about granulation for better aeration and porosity. Humus is like a sponge for holding water and may absorb three to five times its own weight. Mixing organic matter with sandy soil increases its waterholding ability, or when mixed with clay it spreads the clay particles apart, allowing more air to circulate. In either case the effect of humus is beneficial for roots and growing plant, and in addition, organic matter stimulates development of the micro-organisms in the soil, and itself is teeming with numerous microscopic forms of plant and animal life. Ancient tillers of the soil have known this even without their fancy names for thousands of years, and 300 years before Christ, Theophrastus wrote that the Greeks used crops of beans to enrich the soil; the Romans used similar crops to turn under as green manure.

Kinds of Humus

There are many kinds of organic matter available for incorporation with soil. **BARN MANURE** has been a time-honored source of organic matter but is becoming scarcer each year in urban areas. Fresh manure contains much nitrogen whereas rotted manure is primarily valuable as soil conditioner and biologically for its high population of nitrifying bacteria.

PEAT has become the successor to manure as the universal source of organic matter. Sphagnum peatmoss, most of it found in Canada or Europe, is considered the best as it is fairly resistant to decomposition, can hold a large amount of water sponge-like, is uniform in body, and consistently acid in reaction—a favorable environment for roots. Peat from sedges breaks down quicker, is not as uniform, and its acidity is various.

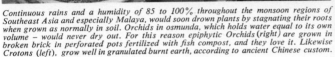

Continuous rains and a humidity of 85 to 100% throughout the monsoon regions of Southeast Asia and especially Malaya, would soon drown plants by stagnating their roots when grown as normally in soil. Orchids in osmunda, which holds water equal to its own volume – would never dry out. For this reason epiphytic Orchids (right) are grown in broken brick in perforated pots fertilized with fish compost, and they love it. Likewise Crotons (left), grow well in granulated burnt earth, according to ancient Chinese custom.

What is offered as **HUMUS** is generally well-decomposed, fluffy, fine-textured decayed vegetation often from lake bottoms that is so well broken down that its plant origin is no longer identifiable. Its particles are usually very fine and may actually be what is better known as muck. MUCK is black swamp earth and may be used on lawns, but it easily binds when wet, is poorly aerated and often saline or alkaline.

WOODS-SOIL, and known as **LEAF MOLD,** is partially decayed forest litter or decomposed leaves, and is an excellent natural humus compost where available. However, it should be flaky when used and not powdery. It is rich in bacterial life and organic fertilizer value where fresh sphagnum peat lacks both, although the latter is the best single substitute for leaf-mold.

For epiphytic plants various fibrous materials are used as growing media. Chunks of roots of the cinnamon fern, **OSMUNDA,** are of consistently good quality and long-lasting, and widely employed for orchids. Also root-masses of Polypodium fern and coarse treefern fibers or blocks (Hapu), even cocos fibers are used where available. Shredded fir-bark (Abies concolor) has been adopted commercially for epiphytic orchids and Anthurium. Redwood sawdust or shavings have become an important ingredient in Western mixes, but voraciously need nitrogen to offset their complete lack of it.

Soil Mixtures for potting.   All plants have preferences for certain potting media depending on the type of plant and the environment of its original habitat. However, the selection of a proper soil mix for any given plant is not as important as many specific recommendations would make it seem. Proof of this is found in the fact that based on University of California experiments, a universal mix could well be substituted and suited for nearly all plants, consisting only of equal parts of peatmoss and fine sand. Such a blend would of course have to be enriched with fertilizer because it would be lacking in basic plant food. But we must, for practical reasons, never be afraid to use whatever materials are available.

In areas with good garden soils fancy trimmings are not necessary, as friable loam mixed with some humus, will suit most any house plant satisfactorily. If, as in **Eastern States** such as Pennsylvania, we have the kind of rich loam that grows tall corn, it may be used as is for most flowering and bedding plants and, with only peatmoss added, for tropical foliage plants. Such loamy soils are not only easy to water, but encourage a healthy root system and grow sturdier stems and thick-textured leaves that will stand up longest. Eastern experiment stations recommend a basic mix of 1 loam, 1 peatmoss, and 1 pebbly perlite or sand. Where quick runoff is desired, include more coarse builders sand, especially for xerophytic plants as in desert-type cacti. Add a little agricultural lime to aggregate the clay particles in soil not only for good drainage and aeration, but also to sweeten it.

In sandy regions of the **South** more organic matter must be added to offset the lack of clay, and more fertilizer is required to sustain growth.

Large areas of the **Midwest** are covered with a deep layer of black-earth or gumbo which is often alkaline. This is predominantly mineral but dark because the soil particles are coated with ancient humus. When wet it is terribly sticky, but the addition of sand and of peatmoss will build aerated structure and permit the flushing out of alkalis. If alkali salts are not leached out of alkaline soil, they will stunt the growth and turn leaf margins brown.

In the **Southwest,** growers have to struggle with brick-hard Adobe, a non-glaciated clay, and toward the **California Coast** with decomposed granite—soil that forms solid rock when dry and undiluted. Many formulas have been devised to create growing media to resemble Eastern loams. Liberal additions of perlite granules and peatmoss to the decomposed granite builds good cohesive yet porous container mixes. Sometimes granitic soil is omitted altogether in favor of combinations of $\frac{1}{2}$ sand and $\frac{1}{2}$ redwood shavings, or $\frac{1}{2}$ perlite with $\frac{1}{2}$ peatmoss. However, it has been reported that while such blends grow good plants in greenhouses, they don't hold up well in

Some basic ingredients of good potting soils; peatmoss may substitute for leafmold, and perlite for sand and charcoal.

Fresh sphagnum moss and its partially decomposed, carbonized peatmoss form, are not only good soil conditioners, but are widely used in supports for climbers such as Philodendron and Scindapsus.

How fortunate that modern garden centers can provide us with such shortcuts to perfect and specialized potting soils in the same manner as we can buy endless varieties of prepared canned food!

the home; soil-grown plants keep best. A recommended California mixture is $\frac{1}{3}$ by volume sandy soil, $\frac{1}{3}$ peatmoss, and $\frac{1}{3}$ fine ground bark or redwood sawdust. Lately, subsoil of crushed rock has proven to give best drainage and aeration as a basis for soil mixes on the West Coast.

Europeans favor a Universal mix based on a clay and peat combination, with clay to form a structural skeleton for aeration even after the peat has gone.

Humusy soils for delicate plants

The more delicate-fragile exotics with hair-fine fibrous roots will want more humusy soil. Leafmold, peatmoss, sphagnum, shredded fir-bark or humus-compost are good basic ingredients, mixed with a small amount of loam or clay for structure, and coarse sand or perlite, broken brick, charcoal, or granite chips to improve drainage. Such soils have the advantage of not needing the addition of much plant food as the decomposing organic matter will provide for average needs.

Some sensitive tropic plants that depend on constant moisture yet resent stagnant soil conditions respond well to culture in still other materials. I have seen excellent Alocasias grown in Hapu (treefern fiber), and Dracaena goldieana in volcanic cinders in Hawaii; the same plants, and others, did very well in live sphagnum moss in Japan. In Malaya and Indonesia, with wet climate, and where plants are exposed to heavy Monsoon rains, I have watched tropical plants of all kinds quite content in broken brick or burnt clay, needed nutrients supplied by manure or fish fertilizer.

If in doubt with some of your fancy plants, rooted cuttings, or even Christmas cactus, and fresh sphagnum is not available, try dried sphagnum moss soaked in a weak complete fertilizer. Even a sick plant with poor roots may be saved if planted in such moss, but remove dead roots, and water from a saucer at first.

Soil Pasteurisation and Sterilization.

To eliminate undesirable disease-carrying organisms, fungi, nematodes, soil insects and weed seeds, and to prevent damp-off in seedlings or cuttings, pasteurization of the potting soil is desirable. This can be done with varying results chemically with fumigants, formaldehyde, methyl bromide and other gases; by heating with electricity, or by baking, boiling or steaming.

Pasteurization stops short of total sterilization (212° F and higher, under pressure) by using temperatures no higher than necessary to destroy harmful soil life. Dairies reduce bacterial life in milk by pasteurizing (heating) it to 140–150° F for 30 min. Exposure to moist heat at 140° F for 30 minutes will kill most disease-causing pathogens, insects and weeds and allowing beneficial organisms to survive.

Fungi are relatively sensitive to heat—Rhizoctomia is controlled at 125° F for 30 min. Most bacteria that cause plant diseases are killed at 140° F for 10 min. Nematodes are also quite susceptible to heat—the root-knot nematode succumbs at 118° F in 10 minutes. Insects and mites cannot long survive 140° to 160° F. Most weeds are destroyed at 158°–176° F for 15 min. Beneficial nitrifying bacteria perish with complete sterilization, and this is practiced only in spore cultures of ferns or laboratory seeding of orchids where foreign competition would be destructive to young embryos. The laboratory method of sterilization is by steam autoclave, 240° F at 15 lbs. pressure for an hour.

Where available, commercial growers use steam as most effective for 30 minutes after soil temperature reaches 180° F at the coolest corner. For small lots, in the home, a pan of moistened soil may be baked in a moderately hot oven for 45–60 minutes at 180 to 250° F depending on results intended. Place a medium-size potato in the middle; when it is baked the soil is cured. Soil placed in a pressure cooker, with a little water in the bottom will be adequately pasteurized under compressed heat for 15 minutes. Or simply boil soil in shallow water in a covered saucepan for 7–15 minutes.

Potting and Containers.

Repotting should be done only when there is a real reason, especially in the case of those decorative plants that are intended to maintain their original size. Even when the roots have completely filled the container it is not always necessary to repot. Some bulbous plants like Clivias and Hippeastrum can remain in the same pot year after year. Or woody plants like Citrus, Ficus or oleander; even aroids as Dieffenbachias or Philodendrons need only to be

Illustrating the wise rule never to OVERPOT! Even a geranium plant will root much quicker and bloom better if a container only slightly larger is used when repotting.

When repotting orchids with creeping rhizomes a swelling eye should be part of any division. Clean off decayed roots and distribute the live ones within a handful of shredded osmunda before placing the plant inside the pot.

Orchid roots run shallow, and for best drainage a 3/4 pot is often used. When leafless "back bulbs" are cut off, the plant may likely fit into the same size pot. Place the rhizome far back to allow for future growth, and ram the osmunda into the pot firmly.

trimmed in shape and left another year. Many well-established old-time decorative evergreens including Podocarpus and palms are better off if left alone. All that is necessary is to try and change the topsoil, adding fresh mulch including fertilizer once a year. Or remove some of the old soil around the root ball, and plant back into the same pot.

When transplanting is unavoidable, select a container only slightly larger. House plants generally keep best in relatively small pots or tubs. An ideal time to repot is when there are signs of new growth, or when the days are lengthening. Plastic pots are very popular and pretty too but must be watered much more carefully. Old-fashioned clay pots are still best for indoor plants as because of their porosity they do not tend to stagnate the soil. To best control their moisture they should stand in a saucer, or loosely as an inner pot inside an ornate jardiniere of stone ware, plastic, fiberglass, wood, or metal, but any excess water should not be allowed to remain for long, as plants dislike "wet feet". To plant directly into any water-tight container may be dangerous except if kept outdoors. To ensure that they don't remain wet too long, and contrary to general advice, it is best to fill glazed ceramic planters that come without a drainage hole, all the way to the bottom with the same soil without gravel drainage. Otherwise, the soil ball may be unable to lift upward, by capillarity, any surplus water that may form a stagnant pool in the bottom of the vessel.

Plant Nutrients and Feeding. If it is our purpose to contain a decorative plant near its original size, and merely to maintain it, plant nutrients and feeding should be held to a minimum. Undue stimulation is not desirable and may result in unattractive, ragged growth. Under the limited light conditions prevailing indoors, a slow-growing plant especially, should receive just enough fertilizer to keep it from starving, perhaps only once or twice a year. It is different with flowering plants and as we give them more light energy, they also require more food.

As mentioned earlier, green plants need a series of major and minor elements to exist. However, most good garden soils contain all the basics, and we are concerned mostly with those nutrients that are consumed in larger proportion, primarily Nitrogen (N), Phosphorus (P), and Potassium (K); also Calcium, or lime. Each of these favor the development and growth of particular aspects of the plant, although all work in unison—including the minor, trace elements—for its well-being. Nitrogen in abundance results in happy growth of shoots and leaves; Phosphorus builds strong roots and sturdy stems, and improves flower color. Potassium promotes good flowering and disease resistance.

We need fertilizers for two purposes: for mixing in soil, and for "feeding" afterwards. Slow-acting organic fertilizers are best with soil mixes but nevertheless should be used sparingly. Barn manure from cattle (9% N) or sheep (20% N) is always most beneficial, also dried blood (13% N), fish emulsion (6–10% N) and hornmeal (10–16% N) are high in nitrogen. Bonemeal (20% P) and tankage (8–30% P) are old sources of phosphorus. Rich in both nitrogen and phosphorus are Seabird guano (12–15% N, 12–25% P), and fishmeal (8–10% N, 12–14% P). Woodashes is a good natural source of potash (6% K) as well as calcium. Lime stone, chicken grit, or oyster shells, or crushed eggshells are good sources of calcium. The U.C. mixes of ½ peatmoss and ½ sand are fortified initially per bushel with 5 spoons dehydrated sheep manure, 3 spoons hoof and hornmeal, 3 spoons bonemeal, and two level teaspoons potassium nitrate (46% K).

For "follow up" feeding, when needed, many formulas of balanced, mostly synthetic, water-soluble plant foods are available—including tablets. Inorganic fertilizers are just as good and effective as organics, and cheaper, provided organic matter is present in the soil. General recommendations for their use are, subject to instructions on each label, a liquid feeding every 2 weeks of ½ teaspoon dissolved in 1 quart of lukewarm water, applied when the plant is moist—not dry. Less, or nothing should be given to a plant that is resting, cut back, or not wanted in active growth. Some foliage plants like aroids and bromeliads respond to foliar feeding by spraying the leaves with high nitrogen fertilizer such as Urea (46% N) or ammonium nitrate (33% N) 1 oz. to 1 gal. of water. But a pinch of any old-fashioned fertilizer as top-dressing will be just as gratefully received by a house plant. One should not worry so much about when and what to feed—but more not to over-feed. Excessive salts in the soil will by osmosis draw water from the roots causing them to dry out and subsequently rot—stunting growth and burning leaves.

Hydroponics answers the lazy gardener's wishful dream for exotic beauty without tedious work and worry. Here a glass brick, or any other container, is filled with proven plants doing well in water: Dracaena sanderiana, Aglaonema, Scindapsus ("Pothos"), and Philodendron oxycardium. Right, drawing of a European Hydro-vessel.

How to grow Plants without Soil

Hydroculture offers a fascinating new direction to plant-growing in the home for lazy gardeners and hobbyists. As if to challenge all horticultural knowledge, many plants actually will grow or at least keep very well in ordinary water. In Europe a regular cult has developed with techniques aimed at growing just about all indoor plants, foliage and flowering included, scientifically in perfected nutrient culture solutions. For this purpose, there are specially designed double Hydro-vessels, with perforated inner pots or nests of wire-mesh to hold the roots and support the plant; this is suspended in the nutrient solution of a decorative outer container. Extra oxygen is supplied through various devices, perhaps by air bubbled through the water daily with a bicycle pump, or using oxygenating, porous pebbles such as vermiculite. The culture solution is changed regularly every 3 to 6 weeks, in winter less often.

Our approach in this country is not to actually "grow" the widest variety of plants in hydro-culture, but confine their purpose to be a decorative but living and attractive object. There are a number of very pretty plants of smaller habit that are capable of forming roots and subsisting for long periods when their stems are kept in ordinary water from the faucet. Various Aglaonemas including Chinese evergreens; Dracaena sanderiana, Philodendron panduraeforme and sodiroi, Fatshedera; Piggy-backs (Tolmiea) and Coleus; the artistic Hawaiian treefern, even Dieffenbachias or Brassaias all do well in water alone.

In smaller vines, Philodendron oxycardium, Scindapsus and Syngoniums, Tradescantias, Cissus and ivies are popular. One can use either vigorous unrooted shoots, or if rooted, wash off all soil and arrange them into the container, the water covering only the basal or root area. New roots will adjust to the watery environment but growth will purposely be slowed much as we aim for mere maintenance of plants in soil by holding them on short rations. If the vessel is of clear glass, shield from bright sun, or use colored glass to prevent formation of algae; a piece of charcoal will help to keep the water clear.

Routinely, the liquid should be changed and the plants flushed every 2 to 4 weeks or so, but if the water looks clean there is no real reason why it should have to be. But fresh, cool water, warming to the temperature around it, will diffuse needed new oxygen to the roots. Some oxygen is replenished by surface absorption, but generally, the foliage also supplies oxygen to plants with roots submerged in water. Water itself contains about 9 ppm. of oxygen which is available to the roots and is constantly replenished.

Naturally, a little feeding will sustain plants so much longer. A bit of water-soluble plant food, complete with trace elements—say 1 teaspoon to 1 tablespoon for the equivalent of a 6-inch pot of water, and depending on recommendations on the label—is all that is needed to provide an adequate nutrient solution. Water should test between pH 5 and 6.5. If higher and too alkaline add a few drops of vinegar; if too acid a few drops of a solution of bicarbonate of soda will correct acidity. When working with nutrient salts, this can become an interesting study on plant behavior for those chemically inclined.

The great advantage of Hydro-culture is that with a minimum of care, plants kept in water can give pleasure for years, and without having to worry about bad soil nor soil-borne insect pests and diseases, no water spots on furniture, no going-away problems.

Beautiful

Tropicalia

to

live with . . .

on opposite page

TOP SHELF, LEFT TO RIGHT:
Cryptanthus bromelioides tricolor
Kalanchoe tomentosa
Acalypha hispida
Fittonia verschaffeltii argyroneura
Anthurium warocqueanum
Piper porphyrophyllum
Calathea makoyana
Hedera helix 'Maculata'
Vriesea splendens 'Major'
Anoectochilus roxburghii
Anthurium scherzerianum

SECOND SHELF:
Iresine lindenii formosa
Hedera canariensis fa. arb. 'Variegata'
Ficus radicans 'Variegata'
Begonia masoniana ('Iron Cross')
Homalomena wallisii
Syngonium wendlandii
Maranta leuconeura massangeana
Philodendron ilsemannii
Peperomia verschaffeltii
Ficus rubiginosa variegata

THIRD SHELF:
Pleomele reflexa variegata (Song of India)
Bougainvillea 'Harrisii'
Aphelandra fascinator
Aglaonema 'Pseudobracteatum'
Aphelandra aurantiaca
Chamaeranthemum igneum
Ananas comosus variegatus
Sinningia regina
Begonia boweri
Cordyline terminalis 'Baby Ti'

BOTTOM SHELF:
Osmanthus heterophyllus variegatus
Costus sanguineus
Scindapsus aureus 'Marble Queen'
Codiaeum (Croton) 'Duke of Windsor'
Episcia cupreata 'Frosty'
Dieffenbachia 'Exotica' ('Arvida')
Begonia rex 'Merry Christmas',
Dracaena sanderiana

Red spider mite on Maranta

Hard shell scale on Erythrina

Soft scale on Ficus

Mealybugs on Butterfly gardenia

Watch out for Insect Pests!

With the best of care—one day some troublesome insect will appear as if from nowhere and try to make a living from your houseplant. It will bore into the life stream of the plant and suck its juice, stunt tender growth, and sometimes even eat up the entire leaves. Growth will be deformed, and infested plants lose their vigor and look sick and ailing.

Insect pests feel invited, especially in winter, when temperatures indoors are artificially high while humidity runs low. Only a few of the sucking kinds of insects are frequently encountered on house plants.

Red Spider

Scale

Mealybug

Aphis

White Fly

Mite

Insects most common on House plants

Near hot radiators or in drafty locations, **RED SPIDER** will usually attack first. They are minute spider-like mites which can be seen with a hand lens, and thrive on dry warm air, usually living on the underside of leaves where they spin webs if allowed to remain. These tiny eight-legged creatures, usually red or brown, will introduce new generations rapidly from transparent eggs the size of a pin point. Red spiders when sucking the sap from the leaf, puncture the plant tissue causing speckling and discoloration to the leaf surface. They attack such plants as ivy, Araucarias, Aspidistras, Hibiscus, Marantas, and Red Cordylines. **Control:** Rotenone, oils, Malathion, Cygon.

SCALES are likely to build their turtle-like houses on leathery plants. They are small sucking insects protected under hard shells that may be tan, brown, black or white, and their shape oval, oblong or circular. Their shield-like appearance is nothing more than a waxy coat which covers the indistinct body of the insect. In the adult scale this shell is stationary and protects it from most contact insecticides. The young, however, are vulnerable while they move around. The Soft brown scale is more flat and can move sluggishly. Typical host of scale are shrubby house plants such as oleander, Citrus and Ficus; also palms, ferns, Aralia, ivy, Pandanus, cactus, orchids, and bromeliads. With plants having hard foliage, a soft sponge, brush, or rag dipped in an insecticide can kill the young, while dislodging the adult shell sheltering them. **Controls:** Alcohol, soap-nicotine, Rotenone-Pyrethrum w. oil, Malathion, Cygon.

Most distasteful and common on house plants are the messy **MEALYBUGS.** They are slow-moving pinkish-white soft-bodied crawlers equipped with many leg-like filaments covered with a powdery-waxy substance, which tends to prevent penetration of insecticides other than oils to their bodies. The young which emerge from cottony masses are easier to kill. Found usually in axils of leaves and alongside their ribs, they live by syphoning sap which causes sickly foliage, and bud-drop on flowering plants. Mealybugs persistently infest many indoor plants including Saintpaulias, Dieffenbachias, Dracaenas, Crotons, Cissus, Syngoniums, Philodendrons, Gardenias and ferns. **Controls:** Alcohol, Rotenone-Pyrethrum w. oil, Malathion Cygon.

APHIDS usually occur in clusters on tender growing shoots of plants and if the leaves are covered with sticky "honeydew" it is a sure sign of infestation. These Plant lice are small, soft-bodied, green or black sucking insects about $\frac{1}{8}$ inch in size and come in both winged and wingless forms. They move around freely on their six long legs, and appear to stand on their heads while sucking plant juices, causing stunting and curled, distorted growth. They multiply joyously on tender plants, especially on frail and weakened ones, but fortunately are easy to eradicate. Aphids are seen less on tropical house plants and they favor herbaceous blooming kinds and the cooler group of porch plants such as fast growing vines, ivy, Fatshedera, Chrysanthemums and Pelargoniums; occasionally also Dieffenbachias, Gardenias and ferns. **Controls:** soap and nicotine, Rotenone, Malathion, Cygon, Benzine hexachloride.

Aphids on Hibiscus   *Whitefly on Fuchsia*   *Cyclamen mite on Saintpaulia*   *Broad mite on ivy*

WHITE FLIES are tropical insects and can be persistently troublesome. Adults are moth-like, $\frac{1}{16}$ inch long and covered with white waxy powder; they take flight when disturbed. Immature stages resemble miniature pale greenish, semi-transparent scale-like insects. They suck juices from the underside of the leaves and excrete "honeydew", giving rise to sooty mold. **Controls:** Rotenone, Malathion, Cygon.

Broad and Cyclamen **MITES** are not true insects but microscopic eight-legged, oval arachnoids related to spiders, nearly transparent, and less than 1/100 inch in size. The Broad mite moves rapidly, the Cyclamen mite slowly. The Cyclamen mite prefers to suck in the newly forming leaves and buds of plants where they are difficult to reach, crippling the growing tips. It feeds preferably on ivy, Begonias and Cissus; in Saintpaulias, Episcias and other gesneriads extra effort must be made to force the insecticide through the shielding hairs of the plant. Broad mite is more easily controlled and is found on ivies, Vitis. **Controls:** Rotenone-Pyrethrum, and specific miticides.

CHEWING INSECTS such as beetles and grubs are less likely to bother indoor plants, but have a certain fondness for succulents and young seedlings, and tender cuttings. If they do start causing damage, they may be controlled by Pyrethrum, Chlordane, Rotenone, or Lead arsenate as a stomach poison; also Malathion, Cygon, Benzine hexachloride.

The so-called Garden **CENTIPEDES** or "Hundred-legs" (Symphylans) are soil pests eating off fine roots which can stunt the growth. Controlled by flooding, or Benzine hexachloride dust. **MILLIPEDES** or "Thousand-legs" are curling, hard-shelled wiry worms, scavenging in decaying organic matter and usually harmless to indoor plants. They are, however, apt to feed on small seedlings, but more often tunnel into vegetables or fruit. Spray with Malathion; or else spray or dust with Chlordane to control them.

SOWBUGS, and **PILLBUGS** or Roly-Poly, are Soil pests related to crayfish. They are seen in damp places on the surface of soil hiding under leaves or pots, scrambling along or rolling up when disturbed. They usually scavenge rotting plant parts, but may be quite injurious to seedlings—eating roots and girdling stems. Paris green and sugar or calcium arsenate was long used for control. We spray soil surface with Malathion, or dust with Chlordane, Benzine hexachloride.

SLUGS are slimy snails without shells; **SNAILS** form little calcareous houses into which they withdraw when disturbed. Both these mollusks are voracious feeders, chewing tender leaves and flowers. They are found hiding in damp places doing their destructive work at night. Try to pick them off when spotted; otherwise Metaldehyde baits (apple bait), or beer, are standard control.

Simple home control of Pests

House plants should be diligently observed for the possible appearance of insect enemies, and to prevent any population build-up that is difficult to control. Where plants are few, it is good practice to wipe the foliage with a damp cloth to remove possible pests, to keep the breathing pores open, and for cleaner and better appearance. When it becomes necessary, clean or spot-spray for specific problems as they arise. A light spray of rubbing alcohol half diluted with water in a hand atomizer causes any mealybugs present to turn brown. A touch with a cotton-tipped toothpick or artist's brush dipped in alcohol (including whisky) or ether (nailpolish remover) is sure death for a mealybug, scale or aphid, but care must be taken or these concentrates may burn the foliage.

General infestations on house plants can be discouraged by the simple hygienic means of forcefully syringing each plant every week or two with clear water—especially the undersides of leaves. This is best done in sink, bathtub, the shower, or outdoors, and should dislodge any unwanted guests. Washing keeps the foliage clean for breathing, and a plant generally more healthy.

Some of the old-fashioned home remedies are still useful today. A bath in warm soap suds (2 tablespoons soap flakes or octagon soap in 1 gal. water) is a mild old home remedy for foliage plants. If the plants are large and unwieldy, such a soap solution can be used for washing of the leaves with a sponge or a toothbrush.

Of course, washing with water alone is not always the complete answer if pests should entrench themselves in hiding places of leaf axils, growing tips or in dense foliage. If badly infested, the entire foliage of a plant may be immersed for 30 seconds in a bath of a warm soapy solution containing a teaspoon of nicotine sulphate to each gallon of water, and rinsed a few hours later; this should control green or black aphids and scale. Nicotine spray, if not available, can be home-made by soaking tobacco from cigars or from pipe tobacco in water for several days; dilute to the color of weak tea for use.

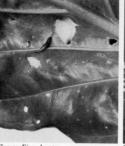

Cleaning of Philodendron by sponging   *Slug on Ficus lyrata*   *Mildew on Euonymus*   *Scrubbing off scale on Aechmea fasciata*

Commercial Insecticides

If more drastic measures of control must be taken, try and recognize the type of insect causing trouble, and choose insecticides specifically made for their control.

A number of manufactured leguminous insecticides, and inorganic poisons long used in house-holds are still adequate and relatively safe. Check the shelves of garden supply shops, neighborhood florists, hardware stores, or the garden sections of larger department stores and supermarkets for "bug-killers" currently available. They are usually supplied not only with old, or well-known remedies, but also the latest products of chemistry for plant protection. As some of these newer products are toxic and dangerous to health, read label instructions carefully, but to play safe, dilute to half-strength for indoor use.

As improved insecticides are constantly coming on the market, I am reluctant to enumerate in this Manual more than a few of the compounds that are presently and effectively used com-mercially. Some that are employed commercially or experimentally are not approved for public use by the U.S. Department of Agriculture. Scientific progress does not stand still, and what may be highly valued today may be superseded and obsolete tomorrow. Moreover, insect populations in greenhouse plant production tend to develop strains becoming resistant to certain chemicals; and alternate methods of control have to be found.

For household use there are reliable, all around INSECT SPRAYS combining Rotenone and Pyrethrum—two naturally occurring poisons, usually in emulsion with vegetable oils (pine oil or soya bean) and aromatic oils (petroleum derivative or mineral base).

ROTENONE is a contact and stomach insecticide and the principal toxic constituent of the leguminous "fish poison" plants Derris and Cube. Extracts from them and their ground roots are still the basis of our most dependable insecticides. PYRETHRUM, the ground flowers of a daisy-like plant, is a contact poison and has also long been used as a household or horticultural dust. In combination with oils. Rotenone and Pyrethrum are normally an effective yet safe control for aphids, mealybugs scales, exposed thrips, white fly, red spider and other mites; also beetles, cut worm, lace bugs and more. Depending on the concentration and formula, Rotenone sprays are diluted approx. 2 teaspoons per gal. of water. Such combination "bug killers" are available even in Ten-cent stores. Some are sold in form of AEROSOL pressure dispensers with push-button tops, which is the simplest modern method of insect control, but care must be taken not to hold the can closer than 10–12 inches to the plant or its propellant may actually freeze the foliage. Small hand atomizers or compressed-air sprayers are available for wet-sprays not in pressurized containers.

Refined emulsifiable, ammoniated Summer foliar OIL SPRAY, about 2–3 tablespoons mixed with each gal. of water help to control red spider, and are especially effective in penetrating the waxy coat of mealybugs, also to loosen the hard shells of scale and smothering their young crawling stage. The addition of a small amount of Rotenone or nicotine will increase its potency without harm to foliage. Use as a spray or dip, but with caution at weakest concentration recommended on label, and no oftener than once a month. Oil sprays are best suited to plants with leathery foliage which will not burn easily. Apply at relatively warm temperature 70–80° F and not in direct sunlight. After a few hours, syringe plant with water to remove any excess since saturation of foliage with oil, especially at low temperature will burn or cause leaf-drop. Not advisable on gesneriads or ferns.

Until something better will take its place the organic phosphate MALATHION, though poison-ous, is relatively safe if handled judiciously, and widely serves as an effective chemical control for many plant pests. It is used as a high pressure spray or dip, either in its 50% emulsion form diluted at the rate of $1\frac{1}{2}$–2 teaspoons per gal. of water, or as the 25% wettable powder, 2 teaspoons mixed with 1 gal. water. In the latter form it is considered safer for use on bromeliads, ferns, poinsettias, and other tender plants; not recommended on Anthuriums, Kalanchoe, Crassulas and certain delicate ferns. Malathion will control aphids, red spider, white flies, thrips and the younger stages of mealybugs and scales; also leaf rollers, ants and even cockroaches.

Ever since Leonardo da Vinci injected a fruit tree with arsenic, scientists have been intrigued with the idea of controlling insects by poison within the sap stream. After much research, it has been found that certain chemicals can be absorbed into a plant and these are known as systemics.

Systemic DIMETHOATE insecticides such as Cygon, used as a fly spray, have come into wide use on the West coast for control of aphids, mealybugs, scales, mites (red spider), white fly, ex-posed thrips and leafhoppers as a spray, diluted 1 teaspoon to 1 gallon water. Succulents in

California are often infested by root-mealybugs which suck juices from roots, and Cygon is used either to drench the ground or as a dip for infested roots.

MITICIDES have been a problem. Chlorinated hydrocarbons such as Kelthane or chlorobenzilate, and organophosphates like Malathion or Cygon can provide control. Alternate one group with the other if mites become resistant. The systemic fly-spray CYGON has been applied effectively for mites including red spider as well as other pests. Fine dusting sulfur is repellent to broad mite.

CHLORDANE, an organic hydrocarbon, is a widely used contact insecticide, and seems to kill anything that crawls, but is quite toxic by skin contact. Chlordane dusting is effective against ants, flies, spider mites, grasshoppers, beetle grubs, centipedes, and sowbugs by paralyzing their nervous system. Roaches that nip at night at juicy succulents such as Lithops, and even mice, can be found lying on their backs in the morning. As a liquid drench, Chlordane is also effective against root-mealybugs in cacti. Its use however is under government restriction.

DDT is a broad-spectrum hydrocarbon with long residual effect, the first of the organic insecticides to come into wide use, killing by contact or stomach action. DDT was first employed in Europe to get rid of beetles, and during World War II is credited with saving millions of lives by controlling malaria-carrying mosquitoes; flies and lice in war areas. DDT as a spray or dust will control chewing insects such as beetles, caterpillars and leaf miners; also juvenile mealybugs, scales, and thrips. It is effective against adult white fly. WARNING: A government study commission has found that this "miracle" pesticide, when used outdoors, may linger on for years in soil and water as a hazard to fish, wildlife and human health, and its use is restricted or prohibited.

The Carbaryl SEVIN, used with caution, is a promising stomatic insecticide with lasting residual effects, safer to mammals, and used to control leaf rollers, leaf miners, lace bugs, caterpillars, grass hoppers, weevils, beetles, centipedes, cut worms, grubs, sowbugs, earwigs, mosquitoes, moths, also aphids, and other surface insects, but is deadly to bees.

Refined BENZINE HEXACHLORIDE, another organic hydrocarbon, is a good contact insecticide, stomach poison and fumigant, but of unpleasant odor. When used as a spray 3 teaspoons per gal. water, it controls aphids, thrips, some beetles, and termites. For troublesome soil insects such as wireworms, grubs, weevils, pill or sowbugs, symphilids, use 25% wettable powder at rate of 1 oz. per 30 gal. of water, and apply as a soil drench.

Injury and disfiguration by insects, caused by their feeding on parts of the plant and its juices, is one of the problems of its maintenance. By periodic preventative spraying with combinations of a contact poison (as Malathion), a stomatic (try Rotenone or Sevin), and a miticide (now recommended: Kelthane), 3 – 4 weeks apart, insect invasions can be avoided altogether. Clean plants are healthy and contented plants and will reward you with their bright and happy faces.

Diseases and other Ailments

Diseases of house plants, generally are not a major problem due to the dry atmosphere in the average home. Powdery MILDEW is caused by fungi which may become a problem with sudden changes in temperature, cold drafts, or dampness. Use one of the spray mixtures containing Karathane with a soapy spreader added to penetrate the waxy powder; very satisfactory on Begonias, Kalanchoe, Hydrangeas, Roses, Saintpaulias, Asters, larkspur, snapdragon. Sulfur controls mildew or rust but requires high temperature to vaporize effectively. "DAMPING OFF" covers a number of fungi that will attack seedlings, but under proper conditions of pasteurization or soil sterilization, cleanliness, aeration, moisture and temperature, that is not particularly troublesome in the home. Damping off can be held in check by drenching the soil with one of the captan fungicides, used at the rate of 1 teaspoon of a 50% captan in 1 gal. of water. Apply at the rate of 1 pint per 2½ sq. ft. of soil surface. Under conditions of low humidity, the usual stem-rots and leaf-spots do not secure a foothold, but should they appear, the captan drench will help to control stem rot, and sprays with one of the zineb fungicides, a zinc carbonate, used at the rate of 1 tablespoon per gal. of water will check most leaf spot diseases.

Green algae, as found on orchid pots, may be checked by using yeast tablets and, as used in Germany, beer will also inhibit their growth.

In a book that is supposed to be useful also in future years, present trade products may eventually be outdated, but reputable insecticide and fungicide makers will continue to offer their best products on the market. If in need of specific advice, your local County Agricultural agent, or the entomologist of Agricultural colleges, Horticultural Stations or Botanical Gardens will be available to answer questions.

New biodegradable insecticides are increasingly introduced and their non-permanent effects will be safer for wildlife and our health.

A sad-looking Croton (left) and a partially defoliated Cissus (right) caused by an infestation of mealybugs – and the tools that can be used to control these and other problems. From left to right a rubber bulb for Fermate dusts or Sulphur to control mildew or plant diseases, also for use of insecticide dusts; a Bronx sprayer-mister with suction pump action; a rust-proof plastic handsprayer for mist, with lever action; a push-button-valve aerosol pressure sprayer in metal can filled with a general bug-killer combining pyrethrum, rotenone and petroleum distillate.

Propagation of Exotic Indoor Plants

Perpetuation of their kind is one of the wonderful processes with which nature has endowed all plant life. Each individual has been given this right and every method is exactly suited to whatever surroundings, favorable or unfavorable, may be characteristic of its habitat. In the difficult, dry climate of the high plateaux of Mexico a xerophytic agave, once having sent up its spiked inflorescence, may give birth to self-sufficient seedlings right inside its nest of flowers, nursed there to independence by moisture collected during the dewy night. The seed of a coconut palm, encased in a buoyant floating shell, travels for hundreds of miles on ocean currents until it is tossed up onto some far tropical beach, there to strike root and grow. Or in some humid forest, the feather-light, delicate seed of an epiphytic cattleya orchid will burst its capsule and float into the proximity of the root system of its elders and settle where, in symbiosis with friendly fungi present, it can spend the long months of its incubation into an embryo seedling. And why do we find ferns in moist places? Because a drop of water is needed during mating time to convey the swimming sperms from the male antheridia under the prothallus to the female archegonia, since this group of plants does not depend, like flowers, on birds and bees.

In these mysterious ways nature safeguards the reproduction of its own, each individual adapted entirely to its climatic surroundings. But if we anticipate to try and reproduce any of these species in our own habitations, we must know what methods they respond to, and provide a milieu that is aware of the climatic patterns prevailing where our captive plants originate.

Plant Propagation in the Home

Propagation is the miracle of rebirth—the creation of a new plant. This can be accomplished in numerous ways—some plants lend themselves better to one method, others to another technique.

Many plants can be reproduced at home with less effort than we fear, and with greater success than we dare hope. There is something so fascinating in seeing new life spring from a tiny seed and the miracle of a small plant emerging from its swelling. And to see a house plant grow up from a cutting which we ourselves have managed to root, adds immensely to the satisfaction we get from working with plants. I know, because — although I had the installations of our 150 greenhouses in New Jersey and California at my disposal—it is a thrill far greater when I can watch roots forming on a cutting of my own placed in a glass of water on my window sill.

How Plants are propagated

There are two primary methods of reproducing plants—seminal and vegetative propagation. Seminal propagation is known as the sexual method of reproduction because it involves sexual union to form the seeds of phanerogamic (flowering) plants, and the spores of cryptogamous plants such as ferns.

Flowers may exist for our enjoyment, but their true function is to produce seeds for their perpetuation. Within the calyx and colorful corolla of a normal flower are the male and female organs of reproduction. The female, usually in the center, consists of a swollen base or ovary, tipped by style and stigma. Around it are the stamens, bearing anthers filled with pollen, the male fertilizing sperms. By nature's various methods, usually insects, pollen is transferred to the stigma, and seed is initiated when the sperms, moving downward, fertilize the egg cell. Both sexes are sometimes on the same flower, sometimes on different flowers, and may even be on different plants. Since hybrid seedlings each have their own unique hereditary traits, results will vary according to the relative dominance of their parents.

Crassula rooting in water

In asexual, or vegetative propagation, however, the new plant will reproduce the same character as the one from which it was taken. Both methods of reproduction can be used for most plants. But although a Dracaena tree or a Rubber plant can be grown from seed, a near-mature young plant can be had so much sooner by taking a cutting or by mossing of the stem.

The Seed or 'Sexual' method

Short-lived herbaceous annuals are usually grown from seed. So are other herbaceous plants like Cinerarias or Calecolarias; fast-growing wax begonias or such indoor tropicals as Brassaias and Anthurium. Almost all palms must be grown from seed; and many ferns including Adiantum, Asplenium and Pteris from spores.

Rapid and vast increases in the number of plants can be brought about by seed. While highly bred seed is quite uniform in its offspring, hybrid cross-pollination in the home offers the possibilities of new cultivars. Though seedlings of woody house plants may take longer to mature, they present the fascinating opportunity of begetting interesting and colorful variations as for instance in Rex begonias or in Coleus. Many worthy new house plant hybrids, especially of the fast-maturing kind have been originated by design or accident by dedicated plant hobbyists. There is always the probability that at least one seedling will be an improvement over the parent plants, although the best features usually appear further along in the second generation of hybrids.

Unlike some hardy trees and shrubs from Temperate regions, seed of tropical origin must always be fresh and sown promptly when ripe as its viability may be extremely fleeting. Old seed is usually the reason why exotics, for instance aroids, Araucarias and Dizygotheca, and most other tropical trees and vines fail to germinate despite the best of care. In ferns a cause of non-germination is that they cannot survive if the spores were exposed to freezing as in a sense they are living, delicate, and unprotected embryos.

Seeds are best sown in sterile clay pans with drainage such as broken pots in the bottom, and sandy soil to near the top. Friable soil with sand and some peatmoss added should be pasteurized to prevent damp-off simply by baking it in a moderate hot oven in moist condition at 180–250° F for ¾ to 1 hr. For fine seed, sift the top layer of soil through a window screen and lightly pat it down level. A scattering of finely ground sphagnum is good as a surface layer as it is an excellent inhibitor of fungus. Then water thoroughly with a fine spray, if possible with boiling water, or immerse the bottom inch of a clay container in water until the moisture seeps to the top from below and barely appears on the surface. After some draining, scatter the seeds thinly, and sift fine soil or sand over them to a depth of once or twice their thickness. Dust-fine seed such as Begonias, Petunias or gloxinias should not be covered at all. Most seed want a warm environment of around 70° F or more. Cover with glass or plastic film and keep shaded until they begin to break ground. While most seeds are kept dark until germination takes place, there are exceptions. Primulas or bromeliads, for instance require light to germinate. Until seeds do, never allow the soil to dry out, and try to replenish their moisture need by setting the pan again in shallow warm water to soak up what it needs. Examine the seed pan daily and when the first seedlings are seen, remove the shade and gradually wean them to better light depending on whether they need sun or love shade. Transplant as soon as they are big enough to handle as this is always safer than to let them continue to grow in the seed pan. Then water overhead lightly with a fine sprinkler. When established, move cool or temperate plants to lower temperatures for sturdier growth.

screening of sphagnum

tamping and firming

moistening before seeding

sowing the seed

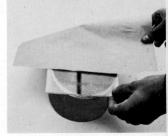

covering and shading

Orchids from Seed

Most orchids are propagated simply by division or offsets. Quantity production, however, is from seed, but this is a slow and tedious process. It takes Cattleyas 5 to 7 years to reach the flowering stage, Phalaenopsis 2 to 3 years. Orchid seeds are dust-like and of fleeting viability, containing highly vulnerable embryos that may germinate but do not have enough food to continue to develop and grow by themselves. In nature they do so in symbiosis or amicable association with a fungus present in roots of the parent plant, and which carries out the functions of a root and supplying the germinating seed with minerals and water, at the same time receiving the vital carbohydrates synthesized by the seedling.

This explains why orchid seeds are sometimes seen germinating on the surface of old orchid pots. Based on this principle, orchids can be sown in covered glass-jars filled with finely chopped sphagnum which is sterilized in a steampot and then inoculated with the root fungus from the species to which the seed plant belongs. For this purpose, portions of infected roots are cut into small pieces and scattered over the sphagnum seed bed. As soon as the root-fungus has grown through the sphagnum, the seeds should be sown in the jars.

In 1909, Dr. Burgeff in Germany developed a method of seed germination employing aseptic laboratory practices that could be depended upon to give constant results. Orchids were sown in test tubes on cultures of agar, a product of seaweed, permeated with root-fungus. In this environment they not only germinated successfully but the swelling embryos continued to grow and form roots and leaves. I remember back during my student days in Vienna inoculating jellied agar test tubes with cultivated root-fungus preparatory to sowing the seed.

Since then Dr. Knudson of Cornell has demonstrated that the fungus was not absolutely necessary for orchid seed to germinate and that the primary cause of the difficulty of germination was due to the inability of the orchid embryo to manufacture its own food. Under his asymbiotic system which I have used successfully as a grower of orchids for years, the embryo is supplied with mineral salts and sucrose while excluding all harmful microorganisms.

Knudson's nutritive formula "C" is basic for most orchids, especially Cattleyas. Per 1 liter (1000 cc or 1 qt. and 2 oz.) of distilled water, add the following ingredients in succession: 1 gram Calcium Nitrate, 0.25 g Monobasic Potassium Phosphate, 0.25 g Magnesium Sulfate, 0.5 g Ammonium Sulfate, 20 g Sucrose (kitchen sugar), 0.025 g Ferrous Sulfate, 0.0075 g Manganese Sulfate. Add 15 grams of plain agar, and warm in a double boiler until the agar is dissolved. The pH reaction must then be checked, as most orchids grow best at acid pH of 5.0 to 5.2. If too alkaline, add a drop or two of 1/10 normal hydrochloric acid and check again; if too acid, add 1/10 normal potassium hydroxide or sodium hydroxide to the solution. When the pH level reads satisfactory, the agar-nutrient mixture is poured into test-tubes, pyrex flasks or even milk bottles, plugged with non-absorbent cotton or a rubber stopper, or otherwise capped, and sterilized in a household pressure cooker for 20 minutes at 10 lbs. steam (225° F), then allowed to cool and jell.

The seed to be sown is decontaminated in a solution of bleaching powder (Calcium hypochlorite), 10 grams per 140 cc of distilled water; the platinum wire loop used for the transfer of the seed is also purified by heating over a flame. After the seed is introduced on the agar-slope, the glass vessel is stoppered again and kept at 70–75° F in a moist place with subdued light for several months to a year. The availability of the nutrients and sugar enables the embryos to develop and grow until they make leaves and roots after which they are ready to be transplanted into community pots.

Cattleya orchids usually take seven years from seed to bloom: left, mother plant with seed pod; in front some of my agar culture flasks and test tubes with seedlings several months old; right, seedlings transplanted into community pots on osmunda fiber; in back the mature blooming plants.

The sowing of Fern spores

Spores on non-flowering plants like ferns are the parallel of seeds in flowering plants. When the sori or spore masses turn brown on the fertile fronds of ferns they are cut and allowed to ripen in paper bags in a warm room. As the spores begin to fall out dust-like, they may be sown on pans with very moist, sterile peat or humus soil. Such soil is sterilized by steam autoclave 240° F at 15 lbs. pressure for an hour, or burn or bake in an oven set at 350° F for 2 or 3 hours. Don't water overhead after sowing, and cover with a pane of glass for two or three weeks until the cultures begin to turn green. Shield from light with dark paper, as germination takes place in near darkness under a light intensity of only 10 to 15 foot-candles. With green beginning to show the forming of prothallia is indicated, whereupon they are weaned to more light between 50 to 75 foot-candles. Fern embryos respond to blue or white light, while yellow light retards growth at the earliest stages. The optimum temperature for ferns in the prothallium stage is 77° F (25° C), with a possible range between 70 and 85° F. A higher temperature hinders the equal development of both sexes, consequently the forming of sporophytes. Also, bear in mind the needed presence of a film of water between prothallium and substratum to enable the male sperms to swim to the female to effect fertilization. Normal condensation, forming on the underside of the covering glass as the air cools off at night, should provide the necessary water conduit. After about 3 weeks, a little air is admitted but care must be taken that the cultures stay constantly moist, yet never soaking wet. Water where needed with rain or distilled water, or water boiled 3 times, preferably from the sides or the bottom where capillary action will pull the moisture where it is needed. After the formation of the male and female reproductive organs and subsequent development of the characteristic plant or sporophyte and its first true leaves, watering may be done over the surface. When the leaves are big enough, transplanting may begin, best in small clusters for mutual support. The rate of growth from then on varies—Pteris may develop into a $2\frac{1}{4}$ in. pot plant in 3 months, while Asplenium and Platycerium will take a year or more.

fertile fronds of Pteris and Cyrtomium showing spore-cases

Factors to Remember

Fern spores are living organisms and must never be exposed to freezing. Sowings should be made well ahead of cold winter weather, especially the slow-growing kinds to avoid possible chilling while germinating. All containers must be sterile, even the covering glass is best steam-sterilized or dipped in Formaldehyde solution. During the process of germination containers must not be turned, so light will always strike from the same direction as the prothallia form, with their faces to the light. Several fungi may attack young prothallia from the time they first emerge causing them to dampen off; these are mostly soil-borne but Pythium may also be carried by the water. Touching up with lime dust or powdered brick, also proper light and circulating air, have a drying and healing effect.

transplating young sporophytes in the greenhouse

harvesting of fertile fronds of Polypodium, for drying

The Vegetative or 'Asexual' method

While almost all plants would grow from seeds, there are many that are with advantage, or must be, reproduced vegetatively. Forest trees that never become mature enough to reach the flowering stage in our clime, others that are valuable but sterile and won't bear seeds at all, must be propagated asexually. For home needs many bulbs and rhizomes such as orchids are divided much more easily than to wait for seedlings to mature. Depending on their makeup, plants may be reproduced asexually using various vegetative parts: Softwood or hardwood stem cuttings, un-rooted tips, joints, or leaves, root-cuttings, basal separations, runners, offsets or suckers; also through air-layering, cane sections, rhizomes, division of rootstocks and tubers, or by bulblets or scales from bulbs; or by grafting.

There is a difference in the reaction and response to vegetative propagation by the two great classes of flowering plants or angiosperms. In monocotyledons the embryos and seedlings have a single seedleaf, the veins in their character leaves are parallel, their vascular system is dispersed and growth in diameter occurs throughout the stem. Orchids, Amaryllis, palms, grasses, lilies and Musas are such monocots, typically difficult to propagate from cuttings. On the other hand, dico-tyledons, germinating with two seedleaves, and which include the great majority of flowering plants, reproduce more readily by cuttings. In dicots the vascular system encircles the plant in the cambium region, the area where bark and wood meet. Growth in diameter is by adding tissue to the outside of the stem, resulting in growth rings. The concentration of vascular tissue in a cambium area lends itself to root production in moist media better than where vascular tissues are scattered through the pith as in monocots.

Cuttings should be taken from plants in active growth, most likely in the spring after awakening from winter rest. The green growth used for soft-wood cuttings should be half-ripened and crisp so that it would snap between fingers, and cut about 3 to 5 inches long. When propagating tropical plants from cuttings keep in mind that these are not usually the hardwood type so characteristic of hardy shrubs and trees but often softer with herbaceous leaves. Growth thus deprived of roots has extremely limited ability and those with thin, membranous leaves may quickly wilt. It means winning half the battle toward root formation if any wilting is prevented from the start by high humidity around their foliage to keep it turgid. Although cuttings must be kept steadily moist, watering is not enough. Wilting can be prevented in several ways, easiest by covering such a cutting with a tent of transparent polyethylene film, a glass jar, plastic cake cover, or in a glass covered aquarium tank. Cuttings root much faster when given a close, humid atmosphere. Constant mist will do the same but while ideal in the greenhouse, is not practical in the home. Cuttings with more leathery foliage, or waxy begonias, geraniums or Peperomias do not necessarily require any glass or glass-like shelter, but would benefit from it just the same. For one thing, such shelters help to keep the temperature more warm and even, and a warm temperature of at least 65–70° F is vital to the rooting of most exotic plants. Tropicals such as Croton and Ficus root best in 75° F but not over 80° F. Moisture-holding canopies may be removed after a cutting "heals" its cut, forming callus, and as the first roots begin to show.

Much scientific advice has recommended and proven the advantage of using root-inducing hormones, usually indolebutyric or naphthalene-acetic acid, or "Rooting powders". This is hardly needed with quick-rooting cuttings, but to dip the base into powder or solution can assist harder, slower-rooting cuttings in speedier rooting; in hardwood the butt ends may be wounded for more roots to form from the wounds. It has been our experience that hormone compounds are most effective if the bottom temperature is at least 60° F or more.

Light is also a factor in the rooting of cuttings. While in the beginning it is prudent to shield the propagating container from bright sun with a layer of newspaper, cheesecloth, or Venetian blinds,

light intensity for most cuttings except the shade lovers should be as high as possible without causing the plants to wilt. In commercial greenhouse and nursery practice best and quickest rooting is achieved in full sun without cover under constant mist and strangely, with a minimum of fungus and disease trouble due probably to the presence of freshening oxygen in the water vapor. In sun, cuttings continue to manufacture food without drawing on the stored supply and exhaust it; constant mist retards loss of water from the leaf, and air and leaf temperatures are kept lower.

A sandbox in the window if covered with polyethylene film provides a comfortable propagating atmosphere for the rooting of cuttings.

Vegetative Meristem Propagation

A revolutionary technique of clonal multiplication has been developed in France promising new dimensions in the propagation of rare or sterile clones of horticultural importance. From the growing tip or apical meristem of a plant, where the cells are dividing and have not yet differentiated into leaves, stems and roots, microscopic tissues are sectioned onto agar test tube cultures under sterile laboratory conditions. Further quartered from time to time and kept under regulated light and temperature these embryonic tissues develop into plants theoretically with identic characteristics. This process was originally designed to eliminate virus by heat therapy, producing for instance virus-free potatoes. Soon its application was extended aiming not only for disease-free plants but also to mass-reproduce outstanding cultivars such as of carnations and Chrysanthemums. By this means of vegetative propagation thousands of identical plantlets can be produced from a single cultivar within one year, and meritorious traits or superior flowers may be perpetuated in this manner. This is particularly significant to the future availability of fine orchids, promising to make prize-winning rarities available to all.

Rooting Materials

The oldest rooting medium is good clean, sharp builders SAND mined from pits. Beware of salty sand from ocean beaches. A bit of sterile sphagnum peat if added to the sand will hold moisture longer and aid in rooting woody cuttings.

Milled SPHAGNUM moss is widely used since it retains moisture well, and it is a natural fungus inhibitor without discouraging root formation. Some plant fanciers have rooted cuttings simply by placing them in moist sphagnum inside a plastic bag. The exploded volcanic glass-like PERLITE is most promising. This is a white, granular material used as a plaster aggregate. It is light, clean and sterile, does not break down, and holds more than 10 times its weight or 60% by volume of water, without becoming water-logged, and plants root readily in it. This medium may be used alone, but preferably in mixture with peatmoss. Perlite lets air into this mixture and prevents the peatmoss from becoming too wet.

Horticultural grade VERMICULITE, light-weight, spongelike heat-popped mica mineral is used as medium for the rapid rooting of Chrysanthemums, Coleus and similar soft or succulent plants, but its harmonica-like kernels must not be compressed or over-watered. However, for plants slow to root, including Azaleas, peat with sand or perlite is considered better, and conifers still do well in plain sand.

Cuttings root easily in water but seem to have more difficulty adjusting themselves to soil. Roots grown in water are thicker and more brittle and succulent than those in soil and sand media. Such cuttings should be potted as roots are forming and before they become too long and tangled.

Containers for Cuttings and Seed sowing

Various containers are suitable for cuttings or seeds. Clean flower pots, preferably shallow, with drainage in the bottom are good, especially as they can be set onto a dish to hold some moisture. Tin cans with holes punched in the bottom will do. Plant flats or little wood boxes are also used, all provided with a hood to hold moisture. A glass tank, a plastic food or refrigerator box makes a miniature greenhouse which is translucent, sterile and moisture-holding. Glass-covered Pyrex casseroles also are ideal for seed or soft cuttings. Even a plastic polyethylene bag may be used for the rooting of cuttings. Gentle bottom heat will accelerate the rooting of cuttings or the germination of seed. While electric cables are available, a simple alternative is to place an electric light bulb under an inverted clay pot on top of which the container is set.

On the following pages, most typical propagating methods are demonstrated in photos. Patterned after these examples, it should be relatively easy to select one or other alternate possibilities to propagate, divide or rejuvenate a plant getting overly spreading, too tall, or unsightly and lanky.

And, how often have we collected some interesting exotic plant in a far tropical habitat and then have it fail to survive, when a cutting successfully rooted in time would have saved this species as a new introduction for horticulture to enjoy.

Plastic breadbox with leaf cuttings of rhizomatous Begonias in porous perlite.

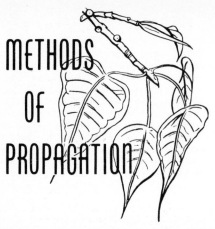

METHODS OF PROPAGATION

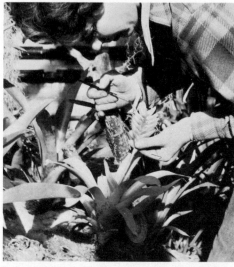

Many plants can be reproduced at home with little effort. There is great satisfaction in seeing a small plant emerge from a swelling seed, or a root to start from a stem which has been standing in a glass of water. Most house plants are propagated by division, offsets, or cuttings. Seedlings take longer to mature, but offer the fascinating opportunity to produce new hybrids. Short-lived annuals, and most palms, however, are grown from seed.

For best germination of seedlings, the seed used should be fresh. Seed may often be obtained right at home if the principles involved are understood and peculiarities of parent plants observed. When flowers are pollinated, seeds are initiated. Both sexes are usually present in perfect flowers, but seed is more likely to be fertile when pollen from one flower is transferred to the stigma of another. And if a different plant exhibits features that would add to the qualities of the maternal plant, the anthers of the mother flower must be removed previous to fertilizing its stigma with pollen from the paternal plant. Above, a Vriesea is pollinized with a dry artist's brush.

Propagation: Plants from Seed

A good general soil mixture for sowing seed is leafmold or peatmoss, with equal parts of loam and sand. To prevent damp-off, pasteurize a pan of moistened soil in a moderately hot oven for 45 to 60 min. at 180 to 250 deg., or use a pressure cooker at 10 lbs. for 15 minutes, or boil it wet for an hour. Placed in the seed pan, sift a layer of fine soil on the surface, tamp down smooth, and sow seed. If this seed is dust fine such as in Begonias, don't cover; with coarser seed sift a thin layer of sandy soil over them, and set pan in saucer with water to seep up even moisture. Keep moist, shaded, and evenly warm 65–70° F until germination. On germinating, embryonic seedlings of flowering plants first show small seedleaves, a single one in the monocotyledons, and a pair in the dicotyledons, followed by the first true leaves which are characteristic of the plant (Brassaia seedlings, above).

The best time for the transplanting of the more robust type of plants is when the first character leaves appear. Delicate seedlings from dust-fine seed however, are best picked off as soon as they are up, using a forked label to handle them. If left in the seed dish, there is a tendency for damping-off and they are easily destroyed by fungus and therefore are much safer once in another pot or flat. Soil for transplants should be sandy, with friable loam and humus, firmed down around the outer edges. Take a pencil to make a hole big enough to get the roots straight down, firm gently, and water well to settle the soil around the roots. Little water will be required thereafter until growth commences.

Germination of most seed, and seedling development may be accelerated 10 to 50 times under high light intensity 2000 to 4000 fc of fluorescent with incandescent illumination for a 16 to 24 hour day during the first 10 to 18 days, at 75–85° F. and using 60% rel. humidity.

Orchids bearing pseudo-bulbs can be divided by cutting the surface rhizome. Cattleya ('Enid' left) is best cut when the front eye is swelling and new roots begin to appear. Oncidium flexuosum (right) can be cut below each joint. Pot placing cut to back of pot, in osmunda fiber or shredded firbark, or similar fibrous media.

Dendrobium nobile orchids (left) form young plantlets, complete with roots, toward the tip of their long pseudo-bulbs; these plants can be cut off and potted. Phalaenopsis may form plantlets at the end of flower spikes from adventitious buds. Paphiopedilum (Cypripedium) (right), can be pulled apart into individual growths.

For large scale production, Cattleyas, and other orchids, are grown from tiny seed in a warm, 70° F minimum, shaded place in sterile culture flasks in nutrient-enriched agar-jelly. Seed is sterilized in calcium hypochlorite and transferred to the flask with a platinum loop. After several months, when they are large enough to face the rigors of this world, seedlings are transplanted into small pots with finely chopped osmunda fiber or other fibrous media. Then follow years of size increase before flowering-strength is reached.

Most PALMS are grown from seed. Howeias (Kentia) (left) are sown in summer on at least 2 inches of loam, covered with peat-moss, and kept moist at 65° F. The dwarf Chamaedorea palm (bottom), also Phoenix, are covered with sphagnum and want a warm 75–80° F to germinate. Syagrus (Cocos) weddelliana (top) require even more heat, and are sown right into pots to avoid breaking the taproots later; tip of the seed embryo is left exposed to prevent rotting. Rhapis (right) is one of the few palms that are satisfactorily propagated by division. Palm seed must be well-ripened and fresh. Palms are monocotyledons and show only a single seed-leaf when germinating.

Most tender plants can be propagated from brittle 4-inch tips, or portions of the stem with eyes. Certain waxy plants like fibrous Begonias, Scindapsus, Peperomia or Ivy do not need glass cover. Exploded volcanic Perlite rock or $\frac{1}{8}$ to $\frac{1}{4}$ inch crushed granite is often safer as a rooting medium than sand because of better drainage. Ivies, Pileas and Aeschynanthus form roots along the stem. Bottom heat of 65 to 75° F aids rooting.

Soft cuttings, having no roots, give off more water than they can take up, and it is most important that their leaves and stems are kept from wilting until a callus has been formed at the base. A glass enclosure, or hood of plastic film for the first week or two prevents excessive dehydration as it will maintain a high humidity. Typical examples Peristophe, Chrysanthemum, Hydrangea, Strobilanthes, Gardenia (above).

Since even moisture is important for the rooting of most cuttings, this method will provide it. A small pot, with drainage hole closed and kept filled with water, is plugged into the center of a larger pan with clean sharp mason sand. Even moisture is maintained by water seeping through the porous clay. Hormone dips will help on woody cuttings provided bottom heat is over 60° F.

Tip cuttings or stem sections of the moist tropical-pitcher plant, Nepenthes (above), root best in a sweaty glass enclosure at 75–80° F. Wrap a ball of fresh sphagnum moss around base of cutting, or place it into an inverted pot stuffed with moist moss. Semi-hardened young growth is far more satisfactory than woody stems, and will produce pitchers more willingly.

Fresh Anthurium seed is germinated at 75–80° F (right) on moist, shredded sphagnum moss, covered by glass, in humid-warm house. Sphagnum moss is a good fungus inhibitor without discouraging root formation, and is used successfully for other seed as well. To obtain seed, the characteristic spadix which is bisexual bearing both male and female organs, must be properly pollinated (lower left, seed developing). As the stigmas are moist and receptive a week or two before the pollen on the same spadix, pollen is usually transferred when it is dusty, from another spike. When the berries are ripe, the kernels are squeezed out and pressed firmly onto the surface of the medium. Anthurium seed must be sown absolutely fresh. Mature plants can be cut off with aerial roots (left) and repotted into sphagnum, peatmoss, or osmunda or other fibers.

To retain the best of character leaves in climbing aroids such as Philodendron (P. elegans, above) cut off the tip of older or straggly plants, with 3 to 5 leaves and some aerial roots attached. Pot into rough mixtures of peatmoss, or humus, sawdust, sand, coarse rotted manure and soil, and train against a slab of bark or a mossed pole drenched with nutrients to maintain healthy development of large leaves.

Joint cuttings of the sometimes heavy, fleshy stems of Philodendron and Dieffenbachia will easily break from eyes in humid-warm temperature, as will most Araceae. Place in a tray or pan with loose peatmoss or sphagnum and hold moist but not too wet. A tent of plastic film will help to set up an ideal tropical-humid environment —a big inducement for the sleeping eyes to swell and sprout. If space permits, any live aerial roots should remain on Philodendron or Monstera sections or a cut-off top, as these roots, where exposed to air are covered with spongy tissues capable of absorbing atmospheric moisture, and where they strike the ground act as true root.

CANE sections, freshly cut and ends powdered with charcoal; if too thick, the cane may also be split lengthwise. Dracaena warneckei (left), Cordyline terminalis (center), or Dieffenbachia (right), sprout from hidden eyes continuously, producing cuttings. Or, the sprouting cane can be planted upright and used as a single plant.

The rooting of cuttings in water is a favorite method of amateurs and it is a time-honored practice out in the country to root "slips" of woody plants as well as softer kinds in a glass partially filled with water. In a glass jar or drinking cup, the leaves, especially the herbaceous kind benefit from the humidity of partial glass enclosure. I myself have rooted many precious exotics successfully on the window shelf in my office. Proven subjects are Coleus, Saintpaulias, geraniums, Begonias, Fuchsias, Impatiens, Tradescantias, Cissus, Cyperus, ivy, Philodendron, Pilea, Tolmiea, Aucuba, Dracaena, Gardenia, Hibiscus, oleander, myrtle; even succulents such as Crassulas can be rooted in this manner.

Propagation by LAYERING is the rooting of branches or the top of woody plants while still attached to the parent shrub or tree. Layering often works on plants which are difficult or impossible to root by other methods. Ground-layering is a simple method outdoors in which branches are notched and hooked down into the soil to make them take root. Air-layering, also known as "Marcottage" or "MOSSING" is an ancient technique adaptable for larger woody plants indoors. For decorator plants that have become too large it is safe because the top stays on the plant while rooting. Cut upward about one third into the stem, or remove a narrow ring of bark including the cambium, wrap with damp sphagnum inserting some into the incision, and cover with poly-thene film which permits air to pass but confines humidity. Examples: Ficus (F. lyrata, above); Ardisia, Croton, Coccoloba, Dracaena, Medinilla.

HARDWOOD CUTTINGS 6 to 10 in. long made from ripened cane of heat-loving shrubs such as (left to right) Roses, Poinsettias, Allamanda, Croton, Acalypha, Hibiscus (at bottom); others: Citrus, Bougainvillea, Jasmine. Insert a deep two-thirds into sandy soil or peat with perlite to keep the woody stem moist. Much practiced in the South where softwood cuttings are apt to rot. This method also works with many cool shrubs and trees, using ripened cane of last season's growth, provided the temperature is kept down. With two 40 watt tubes of cool-white fluorescent light 12 in. above the rooting medium for 12 hours or more daily a cutting box may also be kept in the basement. Rooting hormones are especially effective on woody plants particularly if temperatures of 60 or 65° F can be maintained.

Woody evergreens from warmer regions sometimes require special methods for best propagation. Normally weak-growing Azalea simsii ("indica") do better if cleft-grafted on more robust under-stock in a glass-covered case. Semi-woody tips of variegated Osmanthus (above) are side-grafted on hardy privet. Remember that the crucial point of grafting is to align a freshly cut section of the cambium layer (the soft green inner tissue between bark and wood) of the scion with the cambium layer of the understock, binding them tightly together in close contact until union takes place. Araucaria tip or leader cuttings (right) are placed in sandy leafmold under glass jar at 65–70° F and kept evenly moist. Other conifers such as junipers, cypress, Chamaecyparis, Taxus are best propagated when the spring growth is completed but before it hardens; this also applies to broadleafed evergreens such as Erica and holly. Constant misting gives excellent results where practical.

Root-promoting hormones, if properly used, will induce cuttings to root earlier and stimulate them to produce roots in greater numbers and less time. Known as plant hormones, these chemicals are actually minute concentrations of indolebutyric or indoleacetic acid and known under several trade names such as Hormodin or Rootone. The bases of cuttings are either dipped in solution or powder, as the geranium above, or the bases are dusted. Rooting preparations bring about remarkable response on difficult hardwood species, especially if the base is wounded, but to be effective the temperature of the rooting medium should be above 60° F.

Since the advent of plastic film, entirely new methods of propagation have become possible. A simple Polythene bag makes an ideal little greenhouse suitable for the rooting of cuttings that need a close, humid atmosphere, including the woody kinds that take 6 to 10 weeks to form roots such as Camellias and Azaleas (above). Dip crisp cuttings of semi-hardened growth in Hormone rooting powder and insert into mixture of peatmoss or sphagnum and sand or perlite, well wetted down, seal the bag and set into light window. Plastic film allows for the exchange of oxygen and carbon dioxide without which roots do not develop properly.

Propagation by ROOT-CUTTINGS is sometimes possible. Of tender plants heavy roots of Bouvardia ternifolia (above), Plumbago, Pelargonium, Ligularia, Clivia, Echites, Breynia, Bananas, Oreopanax, or Clerodendrum are cut into 1–2 inch sections in early spring and shallowly laid into peatmoss with sand. Adventitious buds will soon break, developing into a young plant. If Pelargonium plants are buried upside down, plantlets will sprout from the roots. Many hardy plants are propagated in this manner, such as Phlox, Anemone, Aralia, Daphne, Gypsophila, Oriental poppy, Paeonia, Primula, Wisteria, raspberries, Roses and Yucca are increased by root-cuttings. Oddly, Selaginellas, and ferns such as Ophioglossum, Diplazium and Platycerium will also regenerate from root portions or root tips.

Mild, comfortable bottom heat around 70° F is of great benefit to the rooting of cuttings and the germination of seeds. In the absence of warm-water radiators, electric heat is the cleanest and easiest to operate. In propagating beds, insulated heating cables (above), are most efficient for even distribution of bottom heat. A 30 ft. loop uses about 180 watt. Temperature can be automatically controlled by a thermostat set between 65 and 75° F; higher temperature is not advisable. Rubberoid heating mats may be used under flats or pans; socket heaters or simply a 25 watt light bulb placed under a box will also furnish gentle heat.

LEAF SECTIONS of Begonia rex (above), and some other large-leaved rhizomatous begonias such as masoniana and the Beefsteak begonia develop young plants from their prominent primary veins, if these arteries or ribs are cut. Mature leaves can be laid on sand, in a warm place, the ribs cut at junctions, and fastened down with pegs or hairpins. The rooting medium should be porous and not too wet; sand or peatmoss or both is commonly used. A bell jar placed over the leaf sections will provide a moist atmosphere without having to wet the leaves which often leads to rotting. After the callus has formed and roots develop, young plantlets will soon sprout from adventitious buds. This method of propagation is important where rapid reproduction is wanted and yet preserve the character of highly complex hybrids which may not breed true from seed.

Plants with fleshy leaves or thick petioles can often be multiplied by LEAF-PETIOLE cuttings which will form several growing points at or near the cut. Above, Tolmiea, Saintpaulia, Peperomia, Rex begonia, Aphelandra, and (right) Cyperus; also most gesneriads, Cotyledon, Hoya. Leaves of many plants will root but fail to produce growth unless a leaf-bud or eye in the axil of the leaf with a fraction of the stem where removed, remains attached. Such LEAF-BUD cuttings can be made from Hydrangeas, Christmas begonias (Begonia x cheimantha), Camellias, geraniums, Citrus, Chrysanthemums, Dahlias, Ficus, Poinsettias, Rhododendron, and most other plants with fleshy or woody stems or branches. On opposite-leafed plants the stem may be split. Treating the cuts with rooting powders hastens rooting. Coarse sand is a good rooting medium, sometimes with peatmoss added, but for watery cuttings which rot easily, use perlite or crushed stone for best aeration and drainage.

Certain house plants will produce offsets or small plantlets on runners (stolons) which, if allowed to strike soil, will root easily, offering a simple means of propagating. RUNNERS are slender, usually prostrate shoots which start from the base of the mother plant, forming roots and plantlets of their own accord at their joints or tips. Preferably, allow the young plant to root well before cutting it off. Runners are formed by Chlorophytum (right), Episcia (left), and Saxifraga sarmentosa (center), also Neomarica, strawberries, Boston ferns (Nephrolepis). OFFSETS are a quick source of new plants, and are similar to runners but make a shorter stem. They may be pulled off whenever big enough, usually with rudimentary roots. Examples are Sempervivum, Haworthia, Dendrobium orchids, some bromeliads, and many bulbs including Amaryllis, Eucharis, Hymenocallis, Oxalis, Zantedeschia.

Tropical plants possessing thick-succulent leaves or fleshy veins are frequently propagated by LEAF CUTTINGS. A leaf-cutting may consist of a leaf or a piece of a leaf. Peperomias (above) root ideally from leaves; others are Saintpaulias, many rhizomatous Begonias including Rex, Gloxinia, Hoya, Streptocarpus, Zamioculcas, and such succulents as Cotyledon, Crassulas, Echeverias, Kalanchoe, Sedum and Sansevieria. However, variegated Peperomias and Sansevierias will come green from leaves, because such chimeric plants may have outer layers of tissue or skin carrying variegation and the inner layers and core green like the parent. Place the cutting in a porous rooting medium kept moist but not wet, in a uniform humid-warm atmosphere.

Pollination of bromeliads varies with different species. It is important to know that their flowers last only a few hours, and close attention is required to the ripening of the stamens and to the readiness of the pistil during the short period it may receive pollen. Remove anthers with pincers to keep them from rotting, usually early in the morning, if necessary forcing open the petals, and allow the pollen masses to become powdery in a clean dish. The critical hours may also occur in the evening, or during the middle of the night. When the stigma becomes moist, usually in a few hours, this indicates its being receptive and it may then be pollinated. As some bromeliads are self-sterile, pollen is usually transferred from one plant to another, best done with a dry brush as in the Vriesea above.

Bromeliad seed can be germinated remarkably well on 4-fold, sterile kleenex tissue soaked in water, in a shallow dish. Cover with a pane of glass and keep in light place at 65–70° F. Shade at first with green wax paper and add water as needed to keep evenly moist. Seed should be very fresh and if possible no more than 2–3 days old. Snip off feathery parachutes such as found on Vriesea seeds as these may cause fungus. This method of sowing seed on paper tissue is practical for other plants also, although fresh sphagnum moss is equally good and just as sterile.

Pineapple (Ananas) belongs to the Bromeliad family. It can be propagated not only from suckers at the base of the plant but by cutting off the leafy top with part of the fruit. Clean off the flesh and leave it lie 10 days to dry (heal) the cut , before placing on sand to root.

A SUCKER is a secondary shoot that starts from below the ground from a root or underground stem, so typical in raspberries which come up alongside the main row. They can be dug up and transplanted. An OFF-SHOOT is an unexpected growth that starts from the base of the plant. Typical house plants producing suckers are some of the Agaves, most bromeliads (Billbergia nutans, above), Echeveria, Musa and Platycerium, the Staghorn fern. Pandanus typifies a true offshoot. Break or cut from mother plant and allow to root in a small pot in peatmoss and perlite. Bromeliads should be planted deep to induce suckers and initiate a root system of their own.

Cactus propagation by joint cuttings, or grafting. For cuttings, use a sharp knife, dip wound in charcoal, and allow to dry for several days, after which place shallowly on dry sand to root (above left Opuntia vulgaris joint, right Cereus peruvianus cutting). Fancy cacti are often slow-growing, affected by nematodes, or otherwise sick and may be cut and grafted on vigorous understock. Above (R) a globular Mammillaria on a column Cereus, with both the vascular centers matched and tightly held by string or rubber bands; the edges or ribs are cut at a bevel to keep the growing tissue from pushing off the top. To the left a leafy Schlumbergera in cleft graft on Selenicereus pteranthus, the scion held by a cactus spine until new cambium forms to heal the union.

Most cactus seed is very fertile and easy to germinate provided it is fresh. Germination may vary from within a day to as long as a year. Use porous mixture of equal parts of leafmold, loam, and coarse sand, with perlite added, and sterilize, and the container also. Seed is sown on the surface and is covered coarsely with small crushed limestone grit which keeps the surface open and reflects the sun to keep from burning shallow roots. Maintain constant moisture by drawing up the water by capillarity from a saucer. Keep hot at 80 to 90° F to germinate—a 10 watt bulb under a tin pan supplies good bottom heat—and place near full sun. For quickest growth, transplant often but close together and as soon as seedlings touch. Watering may then be overhead. (transplanted Coryphantha above).

SUCCULENTS can be multiplied in various ways. Fast growing, branching species are propagated most easily from stem cuttings. The fresh cut should be allowed to dry, then insert lightly into dry sand, and if conditions are not very hot, don't water, unless they shrivel, until callus is formed or growth begins. Give plenty of air and light, and no cover is needed or they may rot. Most members of the Crassula family including Echeveria, Kalanchoe, and Sedum will also sprout from single leaves. Do not water, and place on top of dry sand. Stone-mimicry plants such as Lithops, and similar compact bodies are best grown from seed; many rosettes such as Aloe and Haworthia are propagated by suckers; clustering species by division. Examples in photo above: Top left, Crassula argentea tip cutting and leaf cuttings; top center, Echeveria from suckers and leaves; top right, Sedum tip and leaf cutting. Lower left, Haworthia suckers and leaf cuttings; center, Kalanchoe (Bryophyllum) leaf cutting, and plantlets forming at edge of leaf; right, Sansevieria leaf section forming plantlet.

The flowers of this night-blooming cactus, Harrisia tortuosa, have been pollinized, a showy red fruit is forming. With the skin peeled off it is edible, and its fragrant, juicy flesh houses numerous tiny seeds When the fruit is soft and ripe, it usually splits open. To harvest the seed, squash the pulp in a bowl of tepid water to wash the seed free. Strain off in a fine sieve and spread on absorbent paper. After drying, rub clean or wash again to remove any remaining mucilage as this is a prime cause for fungus. The seeds from smaller seed pods such as Mammillarias are scraped off onto a newspaper to dry awhile, then agitating them in water in test tubes to wash, and pouring on a paper towel. When dry, rub them off and store in a capped flask.

Many plants such as perennial herbs produce rhizomes that are usually swollen rootstocks or underground or surface stems which are not roots but storage organs for food and are provided with prominent leaf buds or eyes from which the new plant will sprout, and with roots attached as food gatherers. Typical are Iris, and it is well known how easily they may be divided. Typical surface rhizomes are found in many ferns, such as the Rabbit's-foot fern (Davallia). Many gesneriads, principally Achimenes, true Gloxinias, Kohleria and Smithianthas produce pinecone-like scaly rhizomes, and propagate both from rhizomes as well as the individual scales. The photo above shows (right) the cutting apart of a rootstock of Acorus gramineus into fan-shaped sections with feeder roots attached to each; left a shoot of Chrysanthemum growing from a slender rhizome. Other typical plants with rhizomes: Aspidistra, Astilbe, Sansevieria. In East Africa, I have collected Saintpaulia rupicola that formed thick rhizome 8″ long.

One of the simplest methods of propagation is DIVISION. All plants that develop—with age—multiple crowns, may be divided into several individual plants by separating them by pulling apart, splitting, or cutting the plant at the base into as many divisions as it has growth centers. Many house plants can be multiplied in this manner, and care must be taken that each section is worked loose so that it retains all the original roots that are attached to it for prompt potting in individual pots. Typical subjects are (above, left to right) ginger, Calathea, Cyperus. Easily divided also are Spathiphyllum, Aspidistras, Agapanthus, Clivias, Crinums, terrestrial orchids, and many ferns. Saintpaulias when they grow old and form multiple crowns, are best pulled apart, developing into better flowering plants.

TUBERS usually are modified, swollen underground stems, or thickened and fat subterranean branches similar to a rhizome but shorter, beset with buds. Tubers are storing reserve food for the plant to use during dormancy. For multiplication, they may be divided into as many pieces as they have eyes. Some tubers may be borne in the air, such as Dioscorea and Ceropegia. The potato is a typical underground tuber; they develop from underground stems (stolons). Dahlias, tuberous begonias and Cyclamen form their tubers from swollen roots. Above (left) Zantedeschia (Calla), Caladium (right). Alocasias (bottom) form baby tubers on root tips. Other tuberous plants are Cannas, Gloriosa, and Ranunculus, also Sinningias and Rechsteinerias. CORMS are also short, fat, underground stems similar in appearance to bulbs, Technically a corm differs from a bulb in that food is stored in its solid centre, not in the scales, and with buds eventually forming on the top; typical are Gladiolus, Crocus, Freesia. New corms or cormels form on top of the old one by the end of the season's growth, and the original corm dries up.

BULBS are extremely short stems reduced to disks with closely packed fleshy layers, like tulips, hyacinths and Narcissus, or leaf scales as in lilies attached to a basal plate from which roots are produced. A true bulb is an entire blooming plant enclosing buds. Eucharis (right), Hippeastrum (Amaryllis), Clivia can be divided after flowering by breaking the old bulbs and bulblets apart. Scaly bulbs such as lilies (left), will also form bulblets above the mother bulb and even up the stem, but for highest multiplication can be propagated from scales as well. Inserted into light, humusy soil, they will form a new bulblet from their base. Some bulbs such as Hippeastrum, Crinum and Lachenalias may be multiplied by cutting the mature bulb into longitudinal wedges each with a part of the base. But many bulbs propagate themselves naturally by the production of tiny bulbils between the scales eventually breaking the mother bulb apart, or bulblets may form on the outside of the old bulb.

Many ferns form surface RHIZOMES such as Davallia (right), and which may be cut into pieces, and planted on top of peatmoss with sand or spaghnum and hooked down with a hairpin to root. Staghorn fern (Platycerium, left) propagates easiest from pups which appear with age from adventitious buds on slender root-like stolons or on their root tips if kept moist enough; to raise them from spores is a slow process. Mother fern (Asplenium viviparum, center) produce plantlets on their fronds, which can be removed and planted in sandy leafmold. The fern allies Selaginella emmeliana and others are propagated by laying older fronds on moist peatmoss and partially covering them; kept warm and humid, numerous rosettes will form on the branches. Almost any piece of growth will sprout when cut into sections. Selaginella kraussiana and brownii form cushions which can be split apart.

Boston FERNS are commonly propagated by RUNNERS, long stringlike growths produced by older plants. Runners form most easily in good light during summer and warm autumn, forming plantlets if allowed to ramble over loose peatmoss mixed with fertilizer. Plants also sprout from root-like underground stolons. Bushy plants may be divided by cutting them or pulling them apart. Cultivated Boston ferns and their feathery varieties do not make spores. However, the similar Nephrolepis cordifolia 'Plumosa', known in California nurseries as 'Dwarf Whitmanii fern', bears fertile sori and may be propagated sown from spores.

FERNS are cryptogamous and unlike angiosperms do not flower nor bear seed. Instead, they form spores on the underside of their fertile fronds, usually brown and looking like scaly pests. When they become dusty and ready to fall out of their spore cases, they are harvested and air-dried. Sow on sterile peat or leafmold, preferably in a humid-warm greenhouse under fairly good though diffused light. Cover dish with glass and keep in saucer which permits keeping the cultures constantly moist from beneath. Shield-like prothallia will form, and by means of a drop of water through condensation, male antheridia and female archegonia can mate, developing into the sporophyte or, true fern plant.

Spores of rare and valuable ferns may be sown on sterile, jellied agar cultures in test tubes similar to the non-symbiotic method used for orchids. The South American staghorn, Platycerium andinum above is forming sporophytes in the laboratory of Dr. Morel at Versailles, France. Ferns need a fair amount of diffused light to germinate into prothallia The agar-nutrient solution for ferns is a modified orchid formula and omits sucrose. Terrestrial species require higher acidity than air plants, and acidity can be increased by the addition of Citric acid.

Dieffenbachia x bausei

Xanthosoma lindenii 'Magnificum'

Anthurium andraeanum rubrum

Spathiphyllum phryniifolium

Caladium 'Roehrs Dawn'

Scindapsus aureus 'Marble Queen'

Aglaonema commut. 'Pseudobracteatum'
also known as 'White Rajah'

Alocasia x amazonica

Anthurium bakeri

Monstera deliciosa (inflor.)

Philodendron domesticum (inflor.)

Zantedeschia (Calla) elliottiana

Exotic Plants Illustrated

Pages 66 to 545

The Indispensable AROID Family

The Aroid family, botanically Araceae, has given us more good "long life" house plants than any other group and, not surprisingly, has become quite indispensable. Most of its members, the best coming from Tropical America and the West Indies, the others from Southeast Asia, are of tropical origin and lend themselves ideally to the environment of the living room or modern office. As decorative subjects to accent architecture, or as "pets' they require a minimum of care and present few problems. Moreover, almost all Aroids have proven themselves tolerant of adverse and extreme conditions so often prevalent in the average home. Even the most delicate-looking, such as Caladiums or Dieffenbachias, stand up surprisingly long. Certainly Aglaonema and most cultivated Philodendrons, for instance, can give pleasure as long as any artificially-made plant and, in so doing, will offer us life and vitality. Prototypes made of plastics will often bleach and become dead-looking within less than a year if exposed to strong light. What could be more

Typical inflorescence of Monstera pertusa

intriguing than to know that the plants that surround us come to us from far exotic regions all over the tropical world. Yet they adapt themselves easily to our dry-warm surroundings, or even air-conditioning, perhaps after an initial loss of a leaf or two.

The word Aroid means "Arum-like" When we refer to Arums our mental picture is usually the "Jack-in-the-pulpit", its inflorescence bringing to mind a prompter in his box on the theater stage. This feature forms the unique characteristic of the family. An elongated central axis, called the spadix, usually upright, is densely crowded with minute flowers, devoid of petals. In Anthuriums and Monsteras male and female elements form perfect flowers distributed over the whole spadix. In Philodendrons and Dieffenbachias the staminate, male flowers are separate, toward the apex of the column, while the pistillate, female sex organs are grouped at the lower, basal part.

Attached near the base of the spadix is a leaf-like bract or funnel-shaped spathe, often highly colored and showy, popularly considered the "flower". The fruit is a fleshy berry, sometimes arranged cone-like as in Monstera, and may be edible.

The watery or milky, bitter sap in stems, tubers or rhizomes is often poisonous due to the presence of oxalic acid which forms calcium oxalate crystals. The Dieffenbachia for such reason is known as the "Dumbcane" because its juice if tasted will irritate and severely swell the inside of the mouth. The "Taro", Colocasia esculenta, is also dangerous, with a pungent taste, but the poison is dissipated by heating and the crushed, starchy rootstock is a widely used food in the tropics, particularly in Melanesia and Polynesia.

In habit Aroids range from herbs, often with tubes or rhizomes and large leaves, to climbing epiphytes to water-loving Zantedeschia and Colocasias and even aquatics, as Pistia and Cryptocoryne.

The climbing epiphytes have two kinds of roots, one to cling with, such as ivy has, and the other aerial and absorbing moisture from the air, but eventually reaching down into the ground. I remember the normally trunk-forming Philodendron selloum taking to the forest trees, in the area above the mighty cataracts of the Rio Iguacu near the border of Paraguay, in western Brazil, where clouds of mist saturate the air and this Philodendron luxuriates epiphytically on branches sending down its cord-like aerial roots drinking up the moisture.

Philodendron selloum growing above Iguassu Falls, in Western Brazil

Most Aroids are very handsome, with showy foliage in interesting forms. Palmate or pinnate leaves are common, and series of holes or perforations are typical in Monsteras. It must be noted, however, that the character of the leaf in many genera changes from a "juvenile" to the "adult" or fully grown stage. For this reason leaves of seedlings or developing from joint cuttings may look very different from the typical size and form when reaching maturity and preparing to flower. This "aging" process can be sustained if top cuttings are used in propagation; the growing tip or meristem will then develop toward increasingly divided margins and perforations.

Most of our cultivated aroids are easily reproduced. Vining types are propagated from top or joint cuttings; tuberous or rhizomatous kinds by division. The others are easily raised from seed if a few basic conditions are observed. Male and female flowers are both on the same spadix. But female receptiveness precedes pollen formation and this prevents self-pollination. Pollen can be saved when powdery, to apply to pistils when they are sticky with the sugar solution of stigmatic fluid and receptive, or fresh pollen can be transferred from staminate flowers of one spadix and dusted onto receptive female flowers on another each sunny day. In Philodendrons their receptiveness is indicated by the presence of viscid fluid and a rising temperature and this may occur even during the middle of the night. Aroid seed loses its viability quickly and must be sown fresh, preferably cleaned and best on top of moist, shredded or live, fungus-inhibiting sphagnum moss, under a plastic cover or a glass jar, at a humid-warm 75 to 80° F.

Propagation of Anthurium *Sectioning of Philodendron cane*

An estimate of the number of genera in the Araceae would be about 115, with a conservative estimation of some 1500 species known. While some 60 genera have been taken into cultivation, including about 1,000 species, only about a dozen genera have become the backbone of today's favorite family of indoor decorative plants, at least in our high-temperature American homes.

PHILODENDRONS:

These are foremost in importance and offer a wide range of suitable house plants and decorators. The genus includes some of the world's most strikingly handsome foliage plants. Contrary to appearance, all are climbers, or "lovers of trees" as the Greek root of their name indicates. Those that grow fast enough to be classified as vining should have some support, preferably a slab or porous bark, where roots can anchor and derive a measure

Assortment of various types of Philodendron by a window

of nourishment and moisture. But though I could never quite understand why, I once saw in a home in Holland a "Philodendron pertusum" with several vining stems reaching up to the ceiling and down again, supported by nothing but bamboo sticks. However, I noted that from every internode a long aerial root had grown, swinging free in space and presumably absorbing moisture whenever available.

Those species that grow extremely slowly, or imperceptibly, are known as "self headers"; some of these will actually form stately trees with thick trunks, and I have observed them in Brazil, in tropical climate, 15 feet tall. Such arborescent types require lots of room, and a highly satisfactory species is P. selloum, with large, deeply slashed and frilled leaves. This and similar varieties are used much for outdoor decoration in the warmer South where space is no problem.

Philodendron oxycardium benefits from the light of a table lamp.

The vining types vary greatly in size, requirements, and appearance. The most popular of the small trailing kind is P. oxycardium, commonly known in the trade as "cordatum". This will survive under quite unfavorable conditions, even with a light intensity as low as 10 foot-candles. Usually trained to a support are the medium-sized hastatum, mandaianum, panduraeforme, and the Swiss Cheese plant, with cut leaves, commonly known as "P. pertusum". Actually this is but the juvenile form of the "Mexican Breadfruit", Monstera deliciosa, which, in maturity, possesses the grandest perforated leaf pattern. All Philodendron want it warm and, for best growth, like to be kept moist though most species tolerate neglect and dry conditions, too. Those with velvety or highly variegated foliage would prefer a moist-warm greenhouse.

Soil in containers should be kept open and porous, and the addition of acid peat to loam will make a good, friable mixture. Sunlight must be filtered or the leaves may scorch and bleach.

MONSTERA:

These are very close to Philodendron and most of those we know are characterized by their interesting perforated leaves. The best known decorator in this genus is the "Swiss Cheese plant", M. deliciosa, with huge, green-enameled leaves marvelously stencilled with series of holes and slashings. But most of the other species, while interesting, are rather tender and soon become stringy, and new leaves become smaller and insignificant.

RHAPHIDOPHORA and EPIPREMNUM:

These are Asiatic counterparts of our Philodendrons and Monsteras but, as a rule, while quite attractive, seem to have more difficulty in adapting themselves to our rather dry living rooms.

DIEFFENBACHIA:

These have become indispensable as individual table plants since most of them are of compact size and offer a variety of shades of green, variegated with silver and white. Their color patterns are most attractive and nearly all varieties demand the least of care as long as they are in warm surroundings of between 65-75° F. Though most originate in moist tropical areas, from Central America to Northern Brazil, they tolerate low humidity quite well and even prefer to dry out well between waterings. Poor light may make them lanky, but the canes can be cut back. New breaks will easily develop from the cut-off base, and the top can be rooted in a jar of water. Best keepers are: the old D. picta, splashed with white, and its gorgeous nile-green, cream and gold mutant, D. 'Rudolf Roehrs'. The compact hoffmanii cultivar 'Exotica', and the bold D. amoena have also found lasting places in the decorator's schemes.

ANTHURIUM:

This genus is best known through its "flowering" species, the "Wax-flower", A. andraeanum, and the "Flamingo flower", A. scherzerianum, both with spectacular waxen spathes in burning reds. The latter is somewhat easier to keep in the home but both prefer an atmosphere of high humidity; A. andraeanum is difficult without a greenhouse. Some Anthuriums are very decorative due to their large velvety leaves covered with a net-work of white veins. The best keeper in this group is A. clarinervium, from Mexico. Anthuriums need a lighter, very porous soil, preferably mixed with coarse leaf mold, fern fiber, or pure peatmoss.

SPATHIPHYLLUM:

Allied to Anthurium but with white or greenish "flowers". They bloom much more freely and their shiny, dark-green foliage adapts wonderfully to poor light conditions. Nor are they so particular in their requirements and they like rich, loamy soil. The best known commercial variety is C. 'Clevelandii', of medium height; good types of S. 'Mauna Loa' do have larger spathes. S. phryniifolium grows into larger specimen, and S. floribundum is a free-flowering miniature.

CALADIUM:

These tuberous plants come in a great variety of fancy leaves, very often translucent, and their gay colors provide a bright splash wherever they are shown. Being of Brazilian origin, they thrive in warmth but will want a resting period during mid-winter. Cold drafts must be avoided, and they hate wet feet. With care, their fragile-looking foliage will keep surprisingly well.

AGLAONEMA:

Mostly of small growth habit, suitable for plant arrangements or planted in combination for effect. This genus includes the famous "Chinese Evergreen", A. modestum, plain green but known for its durability—it will even live in a jar of water. Its variegated cousins, especially A. commutatum and commutatum elegans, are difficult to hurt and I remember one such plant in a department store in New York in a neglected corner, dry and dark, yet still alive in its original container, after many years. A larger species with showy, silver-painted leaves, and tough and leathery, is A. crispum, commonly known as "Schismato-glottis". Most stunning, splattered all over with cream, and with ivory-white leaf stems, is A. 'White Rajah', in horticulture as 'Pseudobracteatum', an ideal house plant provided it is kept warm and dry.

SCINDAPSUS:
The so-called "Devil's Ivy" with small, leathery, heart-shaped leaves, is a good vine for the home. Best are the various, more-or-less variegated forms of S. aureus, the "Pothos" of florists. But they do require heat—in fact, can't have too much of it, and thrive best in a superheated room. Albinos, like S. 'Marble Queen', must be kept quite dry to keep from losing leaves and rotting away.

SYNGONIUM:
These dwarf climbers, primarily for novelty planting or trained to cling to slabs of bark are quite thrifty. As they advance in age they go through several interesting leaf changes, usually from the juvenile arrow shape into divided, compound-palmate forms of nine leaflets.

ALOCASIA, COLOCASIA and XANTHOSOMA:
"Exquisite" is the word for some of these "Elephant ears", such as Alocasia sanderiana, a tuberous plant with silver-green, waxy, deeply-scalloped leaves, strongly margined and veined with white. Generally, however, they are not good house plants, requiring too much warmth with humidity. Xanthosoma lindenii is highly patterned with cream but also does best as a greenhouse subject. All prefer a fibrous peaty soil with perfect drainage or wood shavings with charcoal. The Colocasias are water-loving and are best known as the "Taro", a chief source of food in the South Pacific.

ZANTEDESCHIA:
Most familiar as "Calla lilies", they sprout from tuberous rhizomes, with both attractive, arrow-head foliage and showy, trumpet-like "flowers", snow white, pink or golden yellow. Rich, stiff soil and medium-warm temperature is all they ask, and all look very exotic as cut blooms.

Philodendron selloum, a tree philodendron　　　　　　*Philodendron bipinnatifidum*

Monstera deliciosa, with its picturesque slashed leaves is variously known as "Ceriman", "Mexican Breadfruit", and "Hurricane plant". Where a show plant with exotic accent is required, its giant, dark green foliage, coated with glossy enamel, will always create the most unusual effect.

Monstera obliqua expilata (leichtlinii)
"Window-leaf"

Monstera pertusa
formerly known as "Marcgravia paradoxa"

Known horticulturally and erroneously as "Philodendron pertusum", this is but the juvenile, vining-stage form of Monstera deliciosa. This type is often referred to as "Split-leaf", and "Fruitsalad plant" and its foliage averages smaller than in the mature stage. Cord-like aerial roots either cling to support, or swing free to absorb moisture from the atmosphere.

Philodendron mello-barretoanum

Philodendron x evansii

Philodendron eichleri
"King of Tree-philodendron"

Philodendron lundii 'Sao Paulo'

Philodendron x barryi

Philodendron sellowianum

Philodendron speciosum
"Imperial philodendron"

Philodendron williamsii
(syn. 'Espiritu Santo')

Philodendron giganteum
"Giant philodendron"

Philodendron tweedianum
from Argentina

Philodendron oxycardium
(cordatum hort.)
"Heartleaf philodendron"

Philodendron panduraeforme
*"Fiddle-leaf", "Horsehead",
or "Panda"*

Philodendron squamiferum
"Red-bristle philodendron"

Philodendron erubescens
"Blushing philodendron"

Philodendron x mandaianum
"Red-leaf philodendron"

Philodendron domesticum
(hastatum hort.)
"Elephant's-ear"

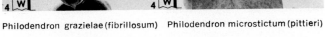

Philodendron grazielae (fibrillosum)   Philodendron microstictum (pittieri)   Philodendron asperatum

Philodendron verrucosum (lindenii)
"Velvet-leaf"

Philodendron gloriosum
"Satin-leaf"

Philodendron mamei
"Quilted silver leaf"

Philodendron
'Burle-Marx's Fantasy'

Philodendron sodiroi (juvenile)
"Silver leaf philodendron"

Philodendron micans
"Velvet-leaf vine"

Philodendron ilsemannii

Philodendron andreanum
"Velour philodendron"

Philodendron melanochrysum
"Black Gold"

Philodendron x corsinianum
"Bronze shield"

Philodendron melinonii
"Red birdsnest"

Philodendron longistilum

Philodendron wendlandii
"Birdsnest philodendron"

Philodendron martianum

Philodendron cannifolium
"Flask philodendron"

Philodendron x 'Florida compacta'

Philodendron x 'Florida'

Philodendron x 'Burgundy'

Philodendron (hastatum x imbe) x wendlandii

Philodendron x 'Lynette'
"Quilted birdsnest"

Philodendron x 'Emerald Queen'

Philodendron x 'New Yorker'

Philodendron x 'Wend-imbe'

Philodendron lacerum

Philodendron radiatum (dubium)

Philodendron elegans

Philodendron polytomum

Philode.1dron laciniatum

Philodendron warscewiczii

Philodendron trisectum

Philodendron tripartitum
"Trileaf philodendron"

Philodendron trifoliatum

Philodendron cruentum
"Red-leaf"

Philodendron oxycardium
"Florists cordatum"

Philodendron karstenianum
Philo. "Mexico" or "Mex."

Philodendron lingulatum

Philodendron auriculatum

Philodendron guttiferum

Philodendron imbe

Philodendron 'Angra dos Reis'

Rhaphidophora "laciniosa"
(Monstera subpinnata)

Rhaphidophora celatocaulis
"Shingle plant"

Epipremnopsis media

Rhaphidophora decursiva

Epipremnum pinnatum
("Monstera nechodomii")
"Taro vine"

Philodendron pinnatilobum
"Fernleaf"

Syngonium podophyllum
'Trileaf Wonder'

Syngonium podophyllum
xanthophilum
"Green Gold"

Syngonium podophyllum
albolineatum
(Nephthytis triphylla)

Syngonium auritum
(Philodendron auritum)
"Five fingers"

Syngonium podophyllum
(Nephthytis liberica)
"African evergreen"

Syng. pod. 'Emerald Gem'
"Arrowhead plant"

Syngonium hoffmannii
"Goose foot"

Syngonium podophyllum
'Atrovirens'

Syngonium podophyllum
albolineatum (juv.)
(Nephthytis triphylla)

Syngonium podophyllum
'Albo-virens'

Syngonium podophyllum alb.
'Ruth Fraser'

Syngonium wendlandii

Syngonium macrophyllum
"Big leaf syngonium"

Syngonium erythrophyllum
"Copper syngonium"

Syngonium triphyllum 'Lancetilla'

Scindapsus aureus 'Marble Queen'
"Taro-vine", *"Variegated Philodendron"*

Scindapsus aureus
"Devil's ivy", *"Hunter's robe"*, *"Ivy arum"*,
commercially known as "Pothos"

Scindapsus pictus

Scindapsus pictus argyraeus
"Satin pothos"

Monstera standleyana
("guttiferyum" hort.)

Pothos jambea

Pothos hermaphroditus
"True pothos"

Anthurium andraeanum rubrum
"Wax flower"

Anthurium x ferrierense
"Oilcloth-flower"

Anthurium scherzerianum rothschildianum
"Variegated Pigtail plant"

Anthurium scherzerianum
"Flamingo flower", "Flame plant"

Anthurium x mortefontanense

Anthurium veitchii
"King anthurium"

Anthurium warocqueanum
"Queen anthurium"

Anthurium magnificum

Anthurium crystallinum
"Crystal anthurium"

Anthurium clarinervium
"Hoja de Corazon"

Anthurium lindenianum
("Spathiphyllum cordata")

Anthurium ornatum

Anthurium andraeanum
rhodochlorum

Anthurium pentaphyllum

Anthurium polyschistum
"Aztec anthurium"

Anthurium aemulum

Anthurium podophyllum
"Footleaf anthurium"

Anthurium holtonianum

Anthurium kalbreyeri (araliifolium)

Anthurium pedato-radiatum

Anthurium fortunatum (elegans)

Anthurium digitatum

Anthurium gladifolium

Anthurium scandens
"Pearl anthurium"

Anthurium bakeri

Anthurium harrisii
"Strap flower"

Anthurium hookeri (huegelii)
"Birdsnest anthurium"

Anthurium tetragonum

Anthurium corrugatum
(papilionense)

Anthurium imperiale

Anthurium crassinervium

Spathiphyllum cochlearispathum

Spathiphyllum 'Clevelandii'
"White flag"

Spathiphyllum phryniifolium
"Bold Peace lily"

Spathiphyllum x 'McCoy'

Spathiphyllum 'Mauna Loa'
"White anthurium"

Spathiphyllum patinii
"White sails"

Spathiphyllum cannaefolium
(dechardii)

Spathiphyllum floribundum
(multiflorum)
"Spathe flower"

Spathiphyllum blandum (gardneri)

Aglaonema commutatum
'Treubii'
"Ribbon aglaonema"

Aglaonema commut
'Pseudobracteatum'
(syn. 'White Rajah')
"Golden evergreen"

Aglaonema 'Malay Beauty'
('Pewter')

Aglaonema simplex 'Angustifolium'

Aglaonema modestum (sinensis)
"Chinese evergreen"

Aglaonema simplex

Aglaonema commutatum elegans
(marantifolium hort.)
"Variegated evergreen"

Aglaonema commutatum
maculatum
"Silver evergreen"

Aglaonema 'Fransher'

Aglaonema nitidum 'Curtisii'

Aglaonema crispum
(Schismatoglottis
roebelinii hort.)
"Painted droptongue"

Aglaonema nitidum
(Curmeria oblongifolia)

Aglaonema costatum
"Spotted evergreen"

Aglaonema pictum

Aglaonema rotundum
"Red aglaonema"

Aglaonema siamense

Aglaonema brevispathum
hospitum

Aglaonema costatum foxii

Dieffenbachia picta (maculata)
"Spotted dumbcane"

Dieffenbachia picta 'Rud. Roehrs'
"Gold dieffenbachia"

Dieffenbachia x bausei

Dieffenbachia leoniae

Dieffenbachia amoena
"Giant dumbcane"

Dieffenbachia bowmannii

Dieffenbachia picta barraquiniana

Dieffenbachia picta 'Superba'

Dieffenbachia imperialis

Dieffenbachia seguina irrorata

Dieffenbachia seguina
"Dumbcane"

Dieffenbachia parlatorei

Dieffenbachia wallisii

Dieffenbachia 'Exotica Perfection'

Dieffenbachia picta angustior
angustifolia

Dieffenbachia 'Exotica' ('Arvida')
Dieffenbachia hoffmannii

Dieffenbachia picta angustior angustifolia
Dieffenbachia picta jenmanii

Dieffenbachia fournieri
Dieffenbachia picta barraquiniana

Dieffenbachia leopoldii
Dieffenbachia x splendens

Dieffenbachia daguense
Dieffenbachia parlatorei marmorea

Dieffenbachia x memoria-corsii
Dieffenbachia pittieri

Dieffenbachia oerstedii
Dieffenbachia seguina lineata

Dieffenbachia oerstedii variegata
Dieffenbachia picta shuttleworthii
Dieffenbachia fosteri

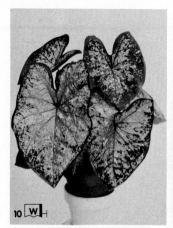

Caladium 'Mrs. W. B. Halderman'
"Coloradium"

Caladium 'Candidum'
"White fancy-leaved caladium"

Caladium 'Frieda Hemple'
"Red elephant-ear"

Caladium 'Lord Derby'
"Transparent caladium"

Caladium 'Miss Muffet'
"Baby elephant-ear"

Caladium 'Marie Moir'
"Blood spots"

Caladium 'Ace of Spades'
"Lance leaf"

Caladium humboldtii (argyrites)
"Miniature caladium"

Caladium 'Flame Beauty'
"Strap leaf"

Alocasia watsoniana

Alocasia sanderiana
"Kris plant"

Alocasia x amazonica

Alocasia cucullata
"Chinese taro"

Alocasia cuprea

Alocasia x sedenii

Colocasia antiquorum fontanesii
"Asiatic taro"

Colocasia antiquorum illustris
"Black caladium"

Cyrtosperma johnstonii

Alocasia macrorhiza
"Elephant's ear"
at Bien Hoa, Vietnam

Colocasia esculenta
"Taro", "Elephant's ear"

Zantedeschia aethiopica
"White calla", "White arum-lily", "Pig lily"

Zantedeschia elliottiana
"Yellow calla"

Xanthosoma violaceum
"Blue taro", "Yautia"

Xanthosoma lindenii 'Magnificum'
"Indian kale"

Spathicarpa sagittifolia
"Fruit-sheath plant"

Zamioculcas zamiifolia

Nephthytis gravenreuthii

Hydrosme rivieri (Amorphophallus)
"Devil's tongue"

Arisarum proboscideum
"Mouse plant"

Alocasia indica metallica (plumbea)

Rhektophyllum mirabile
(Nephthytis picturata)

Callopsis volkensis
"Miniature calla"

Stenospermation popayanense

Homalomena rubescens

Homalomena wallisii
"Silver shield"

Schismatoglottis neo-guineensis
variegata

Pistia stratiotes
"Water lettuce", *"Shell flower"*

Rhodospatha picta

Acorus gramineus variegatus
"Miniature Sweet-flag"

Cryptocoryne willisii
"Underwater aquatic"

Cryptocoryne griffithii

Cryptocoryne cordata

Introduction to the BEGONIA Family

Few plant families include the infinite variety in size, growth habit, foliage, and flower that is found in the Begonia family. Size ranges from the two-inch miniature with leaves smaller than a penny, to a gigantic six feet tall, or leaves two feet long. Some begonias creep and crawl, some grow like a tree; others are viney or bushy. Many are apt mimics—with leaves so like the elm, palm, maple, ivy, and other plants it's easy to be fooled on first sight.

Aside from at least 1200 species of begonias, there are countless hybrids, mutations and chance seedlings. Their rather succulent foliage is usually lopsided and often resembles other plants or trees. Characteristic of the genus is the presence of separate male and female flowers on the same plant; the male flowers have two petals, the female ones 3 to 5 petals, generally with 3-angled ovary.

In a family so diversified, clear-cut classifications are difficult to establish. But popular usage divides all begonias into 3 basic groups:

1. Fibrous-rooted begonias. This includes the favorite house plant group of "cane-stemmed", or "Angel wing begonias", with wing-shaped leaves, bamboo-like stems, and showy, drooping flower clusters; the "hirsute" or "hairy begonias", where flowers or foliage, or both, are covered with bristle-like hair; the "semperflorens" or "Wax begonias", of bushy and branching habit, and almost everblooming. The "cane-stems" usually make the best house plants.

2. Rhizomatous begonias. With thick, root-like rhizomes either creeping above the soil, or erect. Included are the "Rex begonias", distinguished by their showy tapestry-textured leaves, richly decorated with zones of color.

3. Tuberous-rooted begonias. Tubers are formed to allow such plants a resting period. Typical are the spring and summer-flowering "Tuberous hybrids" of gardens, and the semi-tuberous "hiemalis" and "cheimantha" groups for winter-blooming in pots.

Begonias are chiefly of tropical origin. The first species were discovered in Santo Domingo in 1690 by Charles Plumier, who named them for his patron, Michel Begon. Since then, new begonias have been collected and brought into cultivation from all parts of the tropical world. In the Andes in the 1860's were found the first tuberous species, ancestors of today's summer-and winter-blooming hybrids. From Africa have come the semi-tuberous maple leaf begonias; from Brazil, the original Begonia semperflorens; from Mexico, the first rhizomatous species. And today, along with still more newly discovered species, hybridists are constantly adding new plants to the list.

Begonia rex 'Merry Christmas' as a table plant.

Basically, begonias are "shade plants" and respond to tropical moist-warm growing conditions, from about 60 or 65°F. at night to 80°F. by day. Semperflorens and Tuberhybridas prefer the cooler range; Rex begonias and delicate fibrous-rooted and rhizomatous types require the most warmth but laced with humidity. Most begonias, especially the semperflorens, the cane-stemmed, and the hardier rhizomatous kinds need fresh and even cool air and the good light of an East or West window. Rex begonias do best away from sun.

Begonia serratipetala, a frilled angelwing begonia known as "Pink Spot," in an 8 in. hanging pot.

Because begonias have a rather delicate root-system, soil should be loose, well-drained and rich in humus. A good mixture would consist of 2 parts garden loam, 1 part humus, leaf-mold or peatmoss, 1 part sharp sand, gravel or perlite, adding a bit of organic fertilizer. Pots should be kept evenly moist, but not water-soaked, as the fine roots are subject to rot if kept constantly too wet. Rather give them a chance to become somewhat dry occasionally, especially if the temperature is low, or when not actively growing. Half-strength feeding with water-soluble fertilizer during active growth will support and nourish a plant once it is well rootbound.

PROPAGATION

Most begonias are fairly easily germinated and grown from their powder-fine seed. Sow without soil covering on pasteurized soil and protect with plastic or a pane of glass, taking care to keep soil evenly moist through subirrigation. Cane-stemmed varieties propagate readily from half-mature tip or stemcuttings in sand. Rhizomatous kinds are usually increased by taking tips or rhizome sections and half-bury them in sphagnum or peat mixed with sand or perlite, kept barely moist. Rexes, rhizomatous, winter-blooming tuberous, and some fibrous begonias can also be propagated from leaf cuttings, inserting the short petiole into a perlite-peat mixture, sand or other propagating medium.

Inserting rhizome cuttings in peat moss and sand mix, covered by plastic bag.

Begonia rex propagation by leaf-sections.

Begonia serratipetala
"Pink spot angelwing"

Begonia x'Corallina de Lucerna'

Begonia x argenteo-guttata
"Trout-begonia"

Begonia coccinea
"Angelwing begonia"

Begonia 'Picta rosea'

Begonia 'Alzasco'

Begonia 'Sachsen'

Begonia 'Pink Parade'

Begonia x'Pink Spot Lucerna'

Begonia x'Pres. Carnot'

Begonia 'Medora'
"Troutleaf begonia"

Begonia angularis

Begonia 'Dancing Girl'

Begonia echinosepala

Begonia luxurians
"Palm leaf begonia"

Begonia incarnata sandersii

Begonia odorata (nitida odorata)

Begonia isoptera hirsuta

Begonia x'Tingley Mallet'

Begonia x margaritacea

Begonia nitida rosea

Begonia sceptrum (aconitifolia 'Hild. Schneider')

Begonia eminii (mannii)
"Roseleaf begonia"

Begonia egregia (quadrilocularis)

Begonia x'Veitch's Carmine'

Begonia valdensium
"Philodendron-leaved begonia"

Begonia x'Tea Rose'

Begonia evansiana
"Hardy begonia"

Begonia barkeri

Begonia parilis
"Zig-zag begonia"

Begonia 'Preussen'

Begonia dregei
"Miniature mapleleaf begonia"

Begonia olbia
"Bronze mapleleaf"

Begonia x weltonensis
"Grapevine begonia"

Begonia vitifolia
"Grapeleaf begonia"

Begonia richardsiana
"Serrated mapleleaf begonia"

Begonia x'Orange-Rubra'

Begonia x'Catalina'
(Lady Waterlow)

Begonia dichroa

Begonia 'Richard Robinson'

Begonia fuchsioides floribunda
"Fuchsia begonia"

Begonia foliosa
"Fernleaf begonia"

Begonia cubensis
"Holly-leaf begonia"

Begonia 'Marie B. Holley'

Begonia x'Frutescans'

Begonia 'Phyllomaniaca'
"Crazy leaf begonia"

Begonia x'Rosea gigantea'

Begonia 'China Boy'

Begonia hispida cucullifera
"Piggyback begonia"

Begonia scharffii (haageana)
"Elephant-ear begonia"

Begonia x'Viaudii'

Begonia schmidtiana

Begonia hydrocotylifolia
"Miniature pond-lily"

Begonia bartonea
"Winter jewel"

Begonia fernandoi-costa

Begonia x'Alto-Scharff'

Begonia scharffiana

Begonia scabrida

Begonia bradei

Begonia leptotricha
"Woolly bear"

Begonia x'Braemar'

Begonia venosa

Begonia kellermannii

Begonia incana

Begonia x'Thurstonii'

Begonia metallica
"Metallic-leaf begonia"

Begonia x margaritae

Begonia 'Mrs. Fred Scripps'

Begonia x credneri

Begonia 'Nelly Bly'

Begonia epipsila

Begonia ulmifolia
"Elm-leaf begonia"

Begonia involucrata

Begonia semp. albo-foliis 'Maine Variety'
"Calla-lily begonia"

Begònia semp. fl. pl. 'Lady Frances'
"Rose-begonia"

Begonia semp. 'Luminosa'
"Wax bedding begonia"

Begonia semp. fl. pl. 'Curlilocks'
"Thimble begonia"

Begonia x cheimantha 'Marina'
"Scandinavian winter begonia"

Begonia x cheimantha 'Lady Mac'
"Christmas begonia"

Begonia x hiemalis (elatior)
'Rieger's Schwabenland'

Begonia x hiemalis 'The President'
"Dutch Christmas begonia"

Begonia x tub. multiflora fl. pl. 'Maxima'
"Dwarf tuberous begonia"

Begonia x tuberhybrida 'Camellia type'
"Tuberous begonia"

Begonia x tub. mult. fl. pl. 'Helene Harms'
"Lilliput tub. begonia"

Begonia x tub. pendula fl. pl. 'Orange-flowered'
"Hanging basket tub. begonia"

Begonia x'Elsie M. Frey'

Begonia limmingheiana (glaucophylla)
"Shrimp begonia"

Begonia secreta (macrocarpa)

Begonia 'Shippy's Garland'

Begonia x tuberhybrida pendula fl. pl. 'Sunset'
"Hanging tuberous begonia"

Begonia sutherlandii

Begonia x erythrophylla (feastii)
"Beefsteak begonia"

Begonia x heracleicotyle
"Mrs. Townsend begonia"

Begonia manicata 'Aureo-maculata'
"Leopard begonia"

Begonia kenworthyi

Begonia goegoensis
"Fire king begonia"

Begonia masoniana
"Iron cross begonia"

Begonia versicolor
"Fairy carpet begonia"

Begonia pustulata
"Blister begonia"

Begonia strigillosa in hort.
botanically B. tacanana

Begonia x'Cleopatra'
"Mapleleaf begonia"

Begonia x'Spaulding'

Begonia boweri
"Miniature eyelash begonia"

Begonia x'Skeezar'

Begonia imperialis smaragdina
"Green carpet begonia"

Begonia imperialis
"Carpet begonia"

Begonia mazae
"Miniature begonia"

Begonia mazae viridis 'Stitchleaf'
"Stitchleaf begonia"

Begonia x'Bow-Arriola'

Begonia x erythrophylla 'Bunchii'
"Curly kidney begonia"

Begonia manicata
'Aureo-maculata crispa'

Begonia x erythrophylla helix
"Whirlpool begonia"

Begonia x fuscomaculata

Begonia x ricinifolia
"Bronze leaf begonia"

Begonia heracleifolia nigricans

Begonia 'Immense'

Begonia heracleifolia
"Star begonia", "Parsnip begonia"

Begonia x verschaffeltii

Begonia 'Templinii'
"Crazy leaf begonia"

Begonia x'Silver Star'

Begonia x sunderbruchii
"Finger leaf begonia"

Begonia x lettonica

Begonia x pearlii

Begonia popenoei

Begonia nelumbiifolia
"Pond lily"

Begonia acida

Begonia "purpurea" hort.

Begonia x crestabruchii
"Lettuce-leaf begonia"

Begonia 'Randy'

Begonia 'Ricky Loving'

Begonia 'Leslie Lynn'

Begonia 'Fischer's ricinifolia'

Begonia x'Mac-Alice'

Begonia 'Paul Bruant'

Begonia x'Kumwha'

Begonia macdougallii

Begonia olsoniae
(vellozoana hort.)

Begonia paulensis

Begonia x'Joe Hayden'
"Black begonia"

Begonia x'Silver Jewel'

Begonia 'Whirly-Curly'

Begonia conchaefolia 'Zip'
"Zip begonia"

Begonia listida

Begonia x'Edith M'

Begonia x'Beatrice Haddrell'

Begonia x'Zee Bowman'

Begonia rex 'Merry Christmas'

Begonia rex 'Frosty Dwarf'

Begonia rex 'Silver Dollar'
"Baby rex-begonia"

Begonia rex 'Helen Teupel'
"Diadema hybrid"

Begonia rex 'President' ('Pres. Carnot')

Begonia rex 'Comtesse Louise Erdoedy'
"Corkscrew begonia"

Begonia rex 'Fairy'

Begonia rex 'Her Majesty'

Begonia rex 'Curly Silversweet'

Begonia rex 'Autumn Glow'

Begonia rex 'Thrush'
"Upright miniature rex"

Begonia rex 'Mikado'
"Purple fanleaf"

Begonia rex 'Robin'

Begonia rex 'Purple Petticoats'

"Spanish moss," Tillandsia usneoides, draped epiphytically from a Swamp cypress (Taxodium distichum) in Florida.

Companions of orchids and aroids from the mountains and rain forests to the rocky seasides of Central and South America, "bromeliads", peculiar to the Western hemisphere, are a distinguished group of plants dwelling as epiphytes on trees and rocks, or as terrestrials on the forest floor, and to which belong some of our most fascinating and decorative ornamentals. Their shape is usually in form of a rosette of leathery, concave leaves, sometimes plain, and others with bizarre design, or strikingly variegated. Their flowers may be hidden deep in the center surrounded by brilliant inner leaves, or carried high on showy spikes, feathery racemes, or brush-like heads between highly colored bracts, or on panicles of bright, long lasting berries, and even bring forth delicious fruit, such as our pineapple. Dr. Lyman Smith, of the National Herbarium, estimates that there are now known more than 2,000 species, in about 46 valid genera, in the Bromeliad family, against 14 species recorded by Linnaeus in 'Species Plantarum' in 1753.

Notwithstanding their beautiful coloring and leaf designs, bromeliads are easy to maintain. They can be classed as succulents since they also store an emergency supply of water, not inside fleshy leaves but within a natural vase-like center formed by their durable foliage. Their root sytem, particularly in the epiphytic varieties, serves largely as a means of attaching themselves to trees, rocks, or other convenient hosts. In fact, it has been found that as long as bromeliads receive their needed moisture through their center funnel, they can get along for a considerable length of time without any roots at all. It is a most important characteristic of this family that these plants absorb their food through the scales found at the base and the surfaces of the leaf. This is also why some oil sprays can easily cause damage. Bromeliads are therefore ideally suited for home decoration, even in unfavorable corners, in hanging pots, wall pieces, dishgardens, on driftwood, and for table adornment. In Europe, these interesting plants have been used in this manner for over a hundred years.

Bromeliads are grouped by characteristics into the following sub-families:

1. Pitcairnioideae: Includes the most primitive of species, in appearance more curious than beautiful, in habitat growing on the ground or on rocks. Typical are the whorled rosettes of Puya, Hechtia and Dyckia, with their spiny, succulent-stiff leaves, and the Pitcairnias with foliage more grass-like or oblanceolate, having needle-like barbs at their bases. The tubular flowers come in a wide range of color, primarily yellow or red, and seeds are borne in dry pods.

2. Tillandsioideae: The most important genera are Tillandsia, Vriesea, Guzmania and Catopsis. Mostly epiphytic, they have green, striped, or mottled, smooth-edged leaves with bases forming water-holding cups. Most Tillandsias and Vrieseas form single or branched

BROMELIADS

flattened, colorful flower spikes. Guzmania usually have globular heads of durable red bracts with white or yellow flowers. Seeds are air-borne by featherlight little parachutes.

3. Bromelioideae: Here belong the well-known Ananas, Aechmea, Billbergia, Cryptanthus, Neoregelia, Nidularium, Wittrockia; also Canistrum, Orthophytum, Portea, Quesnelia and Streptocalyx. Their interesting leaves are usually toothed; some form rosettes, others are tubular. The spikes or heads of Aechmea are colorful and long-lasting, featuring colored leathery bracts or bright berries. Billbergias are tubular and their inflorescence inclines to be pendant. Rosettes of Neoregelias and Nidularium turn brilliant red in center at blooming time; their flower heads are hidden deep in a center cup of water. The more terrestial members such as Cryptanthus, the "Earth Stars", and Ananas, the "Pineapple", grow well in soil and their rosette-forms have no need to hold a reserve of water. Fruits are berry-like either separate in clusters or heads, or welded together as in pineapple.

"Bromeliad tree," exotic-looking yet easy to maintain, in a Park Avenue apartment, New York City.

Bromeliads in the Home

Bromeliads tolerate a wide range of climatic conditions, from near freezing to high room temperature. Most of them prefer filtered sunlight, particularly the types with highly variegated leaves. As a growing medium, almost any light and porous material rich in humus, will be found satisfactory; best is peatmoss, also leafmold or shredded fir bark, with broken pots, sand, charcoal, Perlite, with some organic fertilizer added. Orchid fibre (Osmunda) is also recommended but not essential. The roots should be kept moist but don't like continuously wet feet, however, fresh water must stand in their funnels. Occasional mild feeding, perhaps at monthly intervals, with some organic, or water soluble complete synthetic plant food, will result in stronger growth. Chemical fertilizer should be used more diluted than recommended, and may be sprayed or poured freely over the foliage, as the leaves are capable of absorbing and utilizing it direct.

As for insect pests, soft brown, or hard black scale may develop, but sponging with soap or scale-oil with Pyrethrum-Rotenone in solution, or Malathion spray followed by forceful syringing with water, will keep them clean. Malathion is primarily effective as a control of scale when in their young, crawling stage, but the shells of the black scale are very hard and must be scraped to dislodge, as with a fingernail, before insecticides can act effectively.

An assortment of bromeliads will live in a minimum of root-space, as here planted into a ceramic log (l. to r.: Cryptanthus bivittatus minor,· Billbergia nutans, Vriesea splendens, Cryptanthus zonatus zebrinus).

BROMELIADS

Colorful, tree-dwelling Bromeliads face fashionable Fifth Avenue at Rockefeller Center in mid-town Manhattan during New York's tropical season in July and August, when temperature and humidity can be every bit as tropical as in Brazil, home of many of these beautiful yet enduring epiphytes.

Premature flowering in advance of normal blooming time can be initiated practically at will by pouring a solution of calcium carbide (Ca C2) (8-10 pea-size pellets, 5 gram, or approx. ¼ ounce in 1 quart of water) into the center of the plant. Acetylene gas is released in the reaction between this compound and water, which interferes with the normal plant growth regulator metabolism by competing with the natural plant hormone indoleacetic acid at the stem tip and initiating or shocking the plant into flower. Calcium carbide is flammable in solution and an accidental spark or match could cause the gas to momentarily explode into flame. Ethylene gas in various forms such as used in commercial pineapple plantations will accomplish the same result, and lends itself to larger operations outdoors, but may be explosive in closed areas. Beta-hydroxyethyl-hydrazine (B.O.H.), an additive to liquid oxygen for making rocket fuel, has been successfully used as a growth retardant by commercial growers but is not generally available. Some greenhouses employ compressed acetylene gas in tanks, covering a bench of plants with plastic. A method practical in the home is to place a mature bromeliad, emptied of water, into an airtight plastic bag together with a fragrant apple for 5 days, exposing the plant to the ethylene gas given off in the ripening process of the fruit. This works most efficiently if temperature is fairly warm.

Home propagation is done easiest from off-shoots, when suckers are large enough to handle, rooting them in moist sphagnum peat. Variegated pineapple (Ananas comosus variegatus) can also be propagated by cutting off the top growth forming on the fruit, and by cutting the fruit stalk into sections. Raising seedlings is more exacting but fresh seed will germinate readily on wet, sterile Kleenex or surgical tissue, also synthetic sponge in a shallow dish covered with a pane of glass, in a light place, at 65-70°F. Seeds are produced easily in most varieties. The pollen can only be applied when it is dry, onto the stigma which must be damp and thus receptive. As the pollen usually stays moist while in the flower, it should be removed with tweezers as soon as the flower opens, and let it dry—in a clean, open dish—for two hours, at room temperature. By then the stigma will have become viscid or sticky, and the pollen masses are then transferred by means of a fine brush. Seed will develop in from 3 weeks for Billbergias to 5 months for Vrieseas. The plumpest seed gives best promise of germination.   (See page 61).

(See page 61).

Cutting taken off the top of a pineapple fruit (Ananas comosus).

Aechmea chantinii seed germinates readily on moist layer of sterile Kleenex tissue.

Removing a basal off-shoot or sucker from mother plant of Billbergia distachya.

Aechmea caudata variegata
"Billbergia forgetii" hort.

Aechmea chantinii
"Amazonian zebra plant"

Ananas bracteatus 'Striatus'
"Variegated wild pineapple"

Ananas comosus variegatus
"Variegated pineapple"

Aechmea x 'Royal Wine'

Aechmea fasciata
"Silver vase"

Aechmea weilbachii leodiensis
"Tropical lilac"

Aechmea miniata discolor
"Purplish Coral-berry"

Aechmea fulgens
"Coral berry"

Aechmea mariae-reginae
"Queen aechmea"

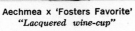

Quesnelia marmorata
"Grecian vase"

Aechmea x 'Fosters Favorite'
"Lacquered wine-cup"

Aechmea x'Bert'

Aechmea fulgens discolor
"Large purplish coral-berry"

Aechmea x maginalii

Ananas comosus 'Nanus'
"Dwarf pineapple"

Aechmea x 'Red Wing'

Aechmea x 'Foster's
Favorite Favorite'

Aechmea ramosa

Aechmea nudicaulis cuspidata
"Living vase plant"

Aechmea mertensii
"Chinaberry"

Aechmea lueddemanniana

Bromelia balansae

Ananas bracteatus 'Striatus'

Guzmania lingulata 'Major'

Neoregelia carolinae

Tillandsia cyanea

Aechmea fasciata

Vriesea x mariae

Billbergia euphemiae

Guzmania berteroniana

Guzmania musaica

Cryptanthus bromelioides tricolor

Aechmea caudata variegata

Aechmea bromeliifolia
"Wax torch"

Aechmea fasciata variegata
"Variegated silver vase"

Aechmea filicaulis
"Weeping living vase"

Araeococcus flagellifolius

Acanthostachys strobilacea
"Pine-cone bromel"

Bromelia serra variegata
"Heart of flame"

Bromelia balansae
"Heart of fire", "Pinuela"

Billbergia x 'Fantasia'
"Marbled rainbow plant"

Billbergia pyramidalis concolor
"Summer torch"

Billbergia nutans
"Queen's tears", "Indoor oats"

Billbergia macrocalyx
"Fluted urn"

Billbergia saundersii
"Rainbow plant"

Billbergia pyramidalis 'Striata'
"Striped urn plant"

Billbergia x 'Albertii'
"Friendship plant"

Billbergia x 'Santa Barbara'
"Banded urn plant"

Billbergia pyramidalis
var. pyramidalis

Billbergia amoena

Billbergia vittata
"Showy billbergia"

Billbergia horrida tigrina
"Tigered urn plant"

Billbergia venezuelana
"Giant urn plant"

Billbergia zebrina
"Zebra urn"

Cottendorfia guianensis

Canistrum lindenii roseum
"Wine star"

Catopsis morreniana

Cryptanthus zonatus zebrinus
"Pheasant leaf"

Cryptanthus diversifolius
"Vary-leaf star"

Cryptanthus x 'It'
"Color-band"

Cryptanthus beuckeri
"Marbled spoon"

Cryptanthus fosterianus
"Stiff pheasant-leaf"

Cryptanthus bromelioides tricolor
"Rainbow star"

Cryptanthus zonatus
"Zebra plant"

Cryptanthus acaulis ruber
"Miniature red earth star"

Cryptanthus bivittatus minor
(roseus pictus)
"Dwarf rose-stripe star"

Cryptanthus lacerdae
"Silver star"

Cryptanthus x rubescens
"Brown earth star"

Cryptanthus acaulis
"Green earth star"

Cryptanthus x osyanus
"Mottled earth-star"

Cryptanthus bivittatus
lueddemannii
"Large rose-stripe star"

x Cryptbergia rubra
"Cryptanthus-Billbergia hybrid"

Navia nubicola
"Cloud plant"

Fascicularia bicolor

Dyckia brevifolia
"Miniature agave"

Dyckia fosteriana
"Silver and gold dyckia"

Ochagavia carnea
(lindleyana hort.)

Orthophytum navioides

Puya alpestris 'Marginata'
"Banded puya"

Hechtia argentea
"Vicious hechtia"

Fosterella penduliflora

Deuterocohnia meziana

Guzmania berteroniana
"Flaming torch"

Guzmania lingulata 'Major'
"Scarlet star"

Guzmania monostachia
"Striped torch"

Guzmania lingulata 'Minor'
"Orange star"

Guzmania zahnii
"Pencil-stripe ribbon plant"

Guzmania lindenii
"Large snake vase"

Guzmania musaica
"Mosaic vase"

Guzmania vittata
"Snake vase"

Neoregelia carolinae 'Marechalii'
"Blushing bromeliad"

Neoregelia carolinae 'Tricolor'
"Striped blushing bromeliad"

Neoregelia farinosa
"Crimson cup"

Neoregelia x'Mar-Con'
(marmorata x concentrica)
"Marbled fingernail"

Neoregelia spectabilis
"Fingernail plant"

Neoregelia mooreana
"Ossifragi vase"

Neoregelia tristis
"Miniature marble plant"

Neoreglia x'Marmorata hybrid'
"Marbled fingernail plant"

Neoregelia sarmentosa chlorosticta

Neoregelia zonata

x Neomea 'Popcorn'
(Neoregelia x Aechmea)

Neoregelia ampullacea

Hohenbergia stellata

Nidularium fulgens
"Blushing cup"

Nidularium innocentii nana
"Miniature birdsnest"

Nidularium innocentii 'Striatum'
"Striped birdsnest"

Nidularium innocentii var. innocentii
"Black Amazonian birdsnest"

Pitcairnia corallina
"Palm bromeliad"

Pitcairnia xanthocalyx
as growing in Leningrad

Ronnbergia morreniana
"Giant spoon"

Quesnelia liboniana

Quesnelia testudo

Portea petropolitana extensa

Pitcairnia tabulaeformis

Streptocalyx holmesii

Pitcairnia andreana

Tillandsia flexuosa
"Spiralled airplant"

Tillandsia wagneriana
"Flying bird"

Tillandsia streptophylla
"Twist plant"

Tillandsia stricta
"Hanging torch"

Tillandsia recurvata
"Ball moss"

Tillandsia araujei

Tillandsia usneoides
"Spanish moss"

Tillandsia bulbosa
"Dancing bulb"

Tillandsia cyanea
"Pink quill"

Tillandsia lindenii
"Blue-flowered torch"

Tillandsia flabellata
"Red fan"

Tillandsia imperialis
"Christmas candle"

Tillandsia ionantha
"Skyplant"

Tillandsia balbisiana

Tillandsia fasciculata
"Wild pine"

Tillandsia utriculata
"Big wild pine"

Vriesea x mariae
"Painted feather"

Vriesea splendens 'Major'
"Flaming sword"

Vriesea imperialis
"Giant vriesea"
in the Organ Mountains, Brazil

Vriesea hieroglyphica
"King of bromeliads"

Vriesea carinata
"Lobster claws"

Vriesea x erecta
"Red feather"

Vriesea incurvata
"Sidewinder vriesea"

Vriesea saundersii
"Silver vriesea"

Vriesea fenestralis
Netted vriesea"

Vriesea barilletii

Vriesea psittacina
"Dwarf painted feather"

Vriesea philippo-coburgii vagans
"Vagabond plant"

Vriesea rodigasiana
"Wax-shells"

Vriesea guttata
"Dusted feather"

Vriesea heliconioides
"Heliconia bromel"

Vriesea scalaris

Vriesea x'Polonia'

Vriesea x'Favorite'

Vriesea x poelmannii

Vriesea x retroflexa

Wittrockia superba

Aechmea x 'Electrica'

Vriesea glutinosa

Vriesea altodaserrae

Aechmea nidularioides

Aechmea x 'Black Magic'

Aechmea mooreana

Streptocalyx poeppigii

Billbergia tessmanniana
"Showy billbergia"

Streptocalyx poitaei

Tillandsia "mooreana"

Vriesea splendens longibracteata

Vriesea platynema variegata

The Cactus Family

Out of privation evolves strange beauty

Like other succulents, cacti are plants that have developed a special capacity to store water in thick, fleshy leaves or stems. They are mostly children of the sun, and love light. Because they need a minimum of care and, except when actively growing, very little water, many cacti have become popular as thrifty house plants on a sunny window sill or in patio plantings in drier regions.

The family is native to the Americas where they are found ranging from Canada in the north upwards toward the Arctic Circle, down through Central America and the West Indies, south to the frigid regions of Chile and Patagonia. There are a great many, often curious, forms and close to 2000 species have been recorded. While the dry-sunny Mexican plateaux possess the richest concentration of varieties, many cacti persist on our western deserts, on the dry side of Caribbean isles, and high in the Cordilleras of Peru, Bolivia, and Argentina. They may be tiny pebbles, hiding in the gravel and scrub of the wastelands, or giant trees raising their ghost-like arms to the blazing sun. In the wet forests of tropical Brazil they live as epiphytes, together with orchids and bromeliads, where their roots get scant moisture from what little foothold they have on moss or bark, anchored high in the trees.

Carnegiea gigantea, the "Giant Saguaro", 35 ft. high, and 250 yrs. old, in Southern Arizona

During the prehistoric Eocene period, 50 million years ago, the American continent was covered with many lakes and moist, steaming jungle teeming with vegetation. Then gradually the earth and its climate began to change, and with lack of rains came the great hostile deserts. The once abundant plant life had to struggle for survival, and only by adjustment could it live. Plants slowed their growth; to prevent quick evaporation, the lush leaves which transpired much precious water into the air, became smaller or dropped off; the sap thickened or became milky; the stems greatly enlarged to store more water against long periods of drought, their ribbed or flattened bodies never turning a full side toward the sun for any extended period. The succulent, green stems themselves took over, for the missing leaves, the process of manufacturing food by photosynthesis. To keep from being eaten up, they armed themselves with spines in their defense. By such wondrous ways has Nature given us one of our most interesting and unique plant families, the Cacti, which, after long and bitter contest, learned to live in difficult environment.

A characteristic common to all Cactaceae are the usually cushion-like areoles on the surface of the plant body which correspond to internodes on other plants, and from which all growth takes place. From these growing points arise branches, leaves if any, flowers, and also wool, bristles or spines.

Pereskia grandifolia, a primitive leafy cactus; the "Lemon-vine" or "Rose-cactus"

The Cactus Family

Cacti as House Plants

The home care of cactus plants is governed, like that of other house plants, by the climatic conditions which prevail in their native habitat. Because of their succulent bodies most of them readily adapt themselves to varying conditions of environment and tolerate a good deal of neglect. If a xerophytic barrel cactus is not watered for months, at worst it will become lean and stop growing; if it is denied sun and fresh air it will lose its beautiful metallic tinting or snowy dress, develop an abnormal growth, and refuse to bloom—but it will hang on to life. Given a little attention to their wants however, they will reveal a capacity to grow much more than one would think possible of dwellers of the deserts and, provided they have sunlight at least several hours each day, bring forth some of the showiest flowers in the plant kingdom. But cacti don't grow in pure sand or in areas without vegetation, prefer to be amongst desert shrub.

A collection of cacti and other succulents growing under ideal conditions next to the good light of a large living room window, on shelving of glass and wood, in New York City. Next to window glass, on southern exposures, there is a great variation of temperatures—lower at night and generally higher in daytime—which simulates the climate of terrestrial cacti in their native habitats.

The active growing season for most cacti begins in early spring, when the plants should get fresh air on a window sill, or may be set outdoors. As the sun becomes stronger, healthy plants will absorb more water until, in midsummer, it may be necessary to water daily on clear days, preferably in the early morning. This is also the period for increased feeding, favoring fertilizers rich in phosphorus and potassium, in weak solutions.

Ideal requirements for Cacti are difficult to cover adequately in our code to care. It should be understood that plants marked T (temperate) or even C (cool) are probably from high altitudes with cool or even chilly nights, but during daytime may be exposed to temperatures of torrid heat.

In warm climates, where the absence of winter cold does not interrupt plant growth, it is generally the rainless season that brings active growth to a halt and starts off a resting period which, in turn, initiates buds and bloom. Under cultivation, water should be gradually withheld at beginning of winter, and a cool place between 45-50° F (7-10° C) is ideal for resting. Cacti of the deserts, and high Andean types should actually be chilled, kept quite dry, and watered no more than once in two weeks, on sunny days.

The soils of arid regions are generally high in mineral foods, lime, and even decaying organic matter, and these elements must be supplied for healthy growth. To enable their roots to absorb food, the soil must be kept moist during the active growing season, but good drainage is essential. For this reason a **porous cactus soil** for **terrestrial genera** should be of a type that falls apart. A third coarse builders sand (not saline beach sand), a third of garden loam, and some decayed leaf-mold and humus for organic food, is a good basic mixture; older plants benefit from a little old rotted manure in addition. For white-hairy cacti, Mammillaria and Astrophytum, add a handful of agricultural lime or old mortar to each bushel of soil.

The Cactus Family

Cacti with leaves, or flat, leaf-like stems are denizens of the humid-moist tropical forest where they are likely to live epiphytic on trees and will appreciate more warmth at night, an **open, humusy soil** and partial shade. When growing actively they will take an abundance of moisture, with frequent feeding to produce their lovely, showy blossoms.

Winter-blooming Zygocactus, and Schlumbergera bridgesii, the **Christmas cactus,** grow well in shredded firbark or peatmoss enriched with tankage, or sphagnum moss soaked in fertilizer. When well rooted, feed every two weeks and keep damp always. During the summertime they like an airy bedroom, or a screened porch, or can be moved outdoors under shade until the cool nights begin. The growth will ripen best from September on at 60-65° F. Christmas cactus are short-day plants and buds will be initiated in autumn when the daylength is not over 11 hours, or end of October in the New York area. All inci-

Authentic Mexican baking dish of red Jalisco clay, handpainted and glazed—the fireproof kitchenware of the villagers; planted with Euphorbia lactea and E. grandicornis, Aloe arborescens, Agave sisalina, Golden barrel, and Rainbow cactus.

dental light at night must then be shielded or avoided. Buds will form if temperature is kept within 55-70° deg. F. High temperatures result in sporadic flowering; cool 50° F. temperatures may stop development altogether.

Insect Control:

Troublesome **mealybugs** are occasionally encountered, for which we found a satisfactory control in Malathion; applied as a dilute spray at high pressure, so that it will penetrate to their bodies and hiding places. Malathion will also control **scale** if the hard protective shell cover of the old scale is loosened with a toothbrush so that the spray can get to work beneath it. Dormant oil sprays do eliminate scale but have a tendency to damage and scar the plant body. **Red spider** and the tiny **thrip** discolor skin portions of the cactus stem. In California, Cygon, as a systemic has given good control of root-mealy bugs. As these chemicals are poisonous, great care in handling is necessary.

Propagation: See pg. 62; also under genera in descriptive text.

Cactus propagation by cuttings; weak growers by grafting

Propagation by seed: germinating cactus seedlings in clay pan

Cereus "peruvianus" hort.
popular "Column-cactus"

Myrtillocactus geometrizans
"Blue myrtle" tree cactus

Trichocereus pachanoi
"Night-blooming San Pedro"

Cereus hexagonus
"South American blue column"

Escontria chiotilla
"Little-star big-spine"

Lemaireocereus marginatus
(Pachycereus)
"Organ-pipe"

Lemaireocereus thurberi
"Arizona organ-pipe"

Lophocereus schottii
"Totem-pole"

Lemaireocereus stellatus
known as "L. treleasei" in hort.

Cephalocereus russellianus

Trichocereus spachianus
"Torch cactus"

Haageocereus chosicensis
"Fox tail"

Eulychnia floresii
"White fluff-post"

Espostoa lanata
"Peruvian old man"

Cephalocereus senilis
"Old man cactus"

Oreocereus celsianus
"Old man of the Andes"

Cleistocactus straussii
"Silver torch"

Lemaireocereus beneckei
"Silver tip", "Chalk candle"

Neobuxbaumia   polylopha
(Cephalocereus)
"Aztec column"

Pachycereus pringlei
"Mexican giant"

Erdisia maxima

Trichocereus pasacana
"Torch cactus"

Nyctocereus serpentinus
"Queen of the Night"

Heliocereus speciosus
"Sun-cactus"

Machaerocereus eruca
"Creeping devil"

Borzicactus humboldtii
(Binghamia)
"Peruvian candle"

Rathbunia alamosensis
"Rambling Ranchero"

Harrisia tortuosa
*"Red-tipped
dog-tail"*

Cleistocactus smaragdiflorus
"Firecracker-cactus"

Nyctocereus serpentinus
"Snake-cactus"

Acanthocereus
pentagonus
"Big-needle vine"

Selenicereus grandiflorus
"Queen of the Night"

Hylocereus undatus
"Honolulu queen"

Selenicereus urbanianus
"Moon-cereus"

Cryptocereus anthonyanus
"Anthony's rick-rack"

Selenicereus hamatus
"Big-hook cactus climber"

Selenicereus wercklei
"Werckle's moon-goddess"

Aporocactus flagelliformis
"Rat-tail cactus"

Echinopsis multiplex
"Easter-lily cactus"

Rebutia kupperiana
"Red crown"

Echinocereus purpureus
"Purple hedgehog"

Echinocereus dasyacanthus
"Rainbow-cactus"

Chamaecereus silvestri
"Peanut-cactus"

Echinocereus viridiflorus
"Green-flower hedgehog"

x Chamaelopsis 'Firechief'
"Firechief hybrid"

Echinocereus reichenbachii
"Lace-cactus"

Lobivia aurea (Pseudolobivia)
"Golden lily-cactus"

Notocactus leninghausii
"Golden ball"

Echinocactus grusonii
"Golden barrel"

Echinocactus ingens
"Blue barrel"

Notocactus haselbergii
"White-web ball"

Ferocactus covillei
"Coville's barrel"

Hamatocactus setispinus
"Strawberry-cactus"

Ferocactus acanthodes
"Fire barrel"

Ferocactus latispinus
"Fish-hook barrel"

Toumeya papyracantha
"Paper-spine pee-wee"

Notocactus scopa
"Silver ball"

Notocactus mammulosus
"Lemon ball"

Notocactus rutilans
"Pink ball"

Epithelantha micromeris
"Button-cactus" or "Golfballs"

Lophophora williamsii
"Peyote" or "Sacred mushroom"

Neoporteria neteracantha
"Unkempt boy"

Echinofossulocactus
zacatecasensis
"Brain-cactus"

Echinomastus macdowellii

Malacocarpus vorwerkianus
"Colombian ball"

Blossfeldia liliputana
"Liliput-cactus"

Mila caespitosa

Obregonia denegrii
"Artichoke-cactus"

Arequipa leucotricha

Melocactus bahiensis
a *"Turk's-cap"*

Parodia aureispina
"Golden Tom thumb"

Parodia sanguiniflora violacea
"Crimson Tom thumb"

Parodia maassii
"Tom thumb cactus"

Gymnocalycium mihanovichii
friederickii
"Rose-plaid cactus"

Gymnocalycium mihanovichii
"Plain chin-cactus"

Gymnocalycium leeanum
"Yellow chin-cactus"

Astrophytum capricorne
"Goat's-horn"

Leuchtenbergia principis
"Prism-cactus", "Agave cactus"

Ariocarpus fissuratus
"Living rock cactus"

Astrophytum ornatum
"Star-cactus" or "Monk's hood"

Astrophytum asterias
"Sand dollar"

Astrophytum myriostigma
"Bishop's cap"

Mammillaria compressa
"Mother of hundreds"

Mammillaria camptotricha
"Birdsnest-cactus"

Mammillaria confusa
"Variable cushion"

Mammillaria elongata
"Golden stars"

Mammillaria multiceps
"Grape-cactus"

Mammillaria tetracantha
galeottii
"Ruby dumpling"

Mammillaria sempervivi

Mammillaria vaupelii
"Vaupel's pincushion"

Mammillaria "dolichocentra" hort.

Mammillaria microhelia
"Little suns"

Mammillaria fragilis
"Thimble-cactus"

Mammillaria echinaria
"Needle fingers"

Mammillaria columbiana
"South American pincushion"

Mammillaria geminispina (bicolor)
"Whitey"

Mammillaria bocasana
"Powder puff"

Mammillaria celsiana
"Snowy pincushion"

Mammillaria magnimamma
"Mexican pincushion"

Mammillaria hahniana
"Old lady cactus"

Mammillaria rhodantha
"Rainbow pincushion

Mammillaria parkinsonii
"Owl's eyes"

Mammillaria nejapensis

Mammillaria potosina

Mammillaria candida
"Snowball pincushion"

Mammillaria aurihamata
"Yellow-hook cushion"

Coryphantha bumamma
"Starry ball"

Coryphantha elephantidens
"Elephant tooth cactus"

Coryphantha macromeris

Solisia pectinata
"Lace-bugs"

Thelocactus bicolor
"Glory of Texas"

Cochemiea poselgeri
"Long hooks"

Escobaria shaferi

Pelecyphora aselliformis
"Hatchet-cactus"

Dolichothele longimamma
"Finger-mound"

Opuntia tomentosa (velutina)
"Velvet opuntia"

Opuntia vulgaris variegata
"Joseph's coat"

Opuntia ficus-indica 'Burbank'
"Spineless Indian fig"

Opuntia schickendantzii
"Lion's tongue"

Opuntia vulgaris (monacantha)
"Irish mittens"

Opuntia brasiliensis
"Tropical tree-opuntia"

Opuntia strobiliformis
"Spruce cones"

Opuntia erinacea ursina
"Grizzly bear"

Opuntia bigelovii
"Cholla cactus" or *"Teddy bear"*

Cryptocereus anthonyanus

Astrophytum asterias

Opuntia elata elongata

Rhipsalidopsis x graeseri 'Rosea'

Hamatocactus setispinus

Schlumbergera bridgesii

Nopalxochia ackermannii

Pereskia grandifolia

Mammillaria bocasana

Rhipsalis quellenbambensis

Trichocereus pachanoi

Rebutia kupperiana

Opuntia rufida
"Cinnamon-cactus"

Opuntia microdasys
"Bunny ears"

Opuntia microdasys
albispina
"Polka-dots"

Opuntia basilaris
"Beaver tail"

Opuntia subulata
"Eve's-pin cactus"

Opuntia vilis
"Little tree-cactus"

Opuntia cylindrica
"Emerald idol"

Opuntia leucotricha
"White-hair tree-cactus"

Opuntia verschaffeltii

Opuntia leptocaulis
"Toothpick-bush"

Consolea falcata
"Tree opuntia"

Opuntia vestita
"Cotton-pole"

Nopalea cochenillifera
"Cochineal plant"

Opuntia erectoclada
"Dominoes"

Opuntia turpinii (glomerata)
"Paperspine cactus"

Mammillaria vaupelii cristata
grafted on Trichocereus spachianus

Cereus peruvianus monstrosus
"Curiosity plant"

Notocactus scopa cristata
(on Trichocereus)
"Spiralled silver-ball"

Opuntia clavarioides
"Black fingers"

Cereus peruvianus monstrosus
(juvenile form)
"Curiosity plant"

Opuntia linguiformis 'Maverick'
"Maverick cactus"

Lophocereus schottii monstrosus
"Monstrose totem"

Opuntia fulgida
mamillata monstrosa
"Boxing-glove"

Epiphyllum hybrid 'Elegantissimum'
"Dwarf orchid cactus"

Epiphyllum x hybridus
"Orchid cactus"

Nopalxochia ackermannii
x 'Fire Glory'

Epiphyllum oxypetalum
"Queen of the night"

Nopalxochia phyllanthoides
"German Empress"

x Aporophyllum 'Star Fire'
(Aporocactus x Epiphyllum)

Rhipsalidopsis rosea
"Dwarf Easter-cactus"

Schlumbergera bridgesii
"Christmas cactus"

Rhipsalidopsis x graeseri 'Rosea'
(Epiphyllopsis)

Rhipsalidopsis gaertneri
"Easter cactus"

Rhipsalidopsis x graeseri
(Epiphyllopsis)

Zygocactus truncatus delicatus
"White crab-cactus"

Zygocactus truncatus
"Thanksgiving cactus", "Crab-cactus"

Schlumbergera russelliana
"Shrimp cactus"

Rhipsalis capilliformis
"Old man's head"

Rhipsalis cassutha
"Mistletoe-cactus"

Rhipsalis mesembryanthemoides (grafted)
"Clumpy mistletoe cactus"

Rhipsalis quellebambensis
"Red mistletoe"

Lepismium cruciforme
"Tree-and-rock cactus"

Pseudorhipsalis macrantha
"Fragrant moondrops"

Acanthorhipsalis monacantha
"Spiny rhipsalis"

Rhipsalis houlletiana
"Snowdrop cactus"

Erythrorhipsalis pilocarpa
"Bristle-tufted twig-cactus"

Rhipsalis paradoxa
"Chain cactus"

Rhipsalis rhombea
"Copper-branch"

Hatiora salicornioides
"Drunkard's dream"

Rhipsalis cereuscula
"Coral cactus"

Color and Glamour in Foliage

Nature has endowed many plants with special glamour, for the same purpose that she has presented us with beautiful birds. Because of a coddling climate, the daintiest exotics with the most colorful leaves or bizarre and beautiful markings, are found in moist-tropical regions. Such delicacy generally does not survive in colder or temperate zones which produce hardier pioneers mostly with plain-looking foliage. It follows that many of our beautiful-leaved plants are not amongst the toughest and most dependable decorators, as desirable as this would otherwise be.

Charming attractions: Pilea cadierei, the silvery "Aluminum plant" or "Watermelon pilea" (left), and Geogenanthus undatus, the "Seersucker plant" with quilted leaves (right).

It should be noted that variegated foliage forms usually need a little more light, less water and less feeding than the plain green leaf type of the same species, or the variegation may be lost. They also have less lee-way in culture, needing more exact attention to their needs; over-watering or dryness, temperature extremes and light extremes are more harmful to them than to the green-leaved plant.

Particularly beautiful is the Maranta family where nearly every species shows off with a leaf design and color pattern more artistic than the next. Fortunately, most of them do pretty well as house plants, under lamplight or in off-sun windows, but they require attention as to soil moisture and an atmosphere not too dry. We know the Marantas as "Prayer plants" because of their mysterious habit of folding pairs of leaves together in the dark, like hands in prayer.

Another large group of fancy foliage in a riot of colors are the delicate-looking yet remarkably sturdy and long-season Imperial Caladiums, belonging to the Aroid family. Originating in the alluvial, low-level tropical forests of Brazil and the Guianas, they are ideally adapted to warm living rooms, though requiring good light. For long life they will do best if planted out in boxes, where they ask for little care. Over-watering can be as harmful to them as too much dryness. But eventually their tubers should be allowed a winter dormant-period of dryness and rest to restore their vigor for new growth.

Included in the Aroid family are many other showy plants, such as Dieffenbachias and Alocasias, both also quite tropical. Of these the Dieffenbachias have proven to be exceptional decorators, with ease of care; they don't mind a little neglect since they prefer to be kept on the dry side anyway as do also the Aglaonemas. But the silvery Alocasias are more delicate, and for best success are recommended for the warm-tropical greenhouse or plastic enclosure. Some Anthuriums have velvety foliage with attractive ivory veins. Silver veining or ivory veining, or silver overlay, is also found in some of the Scindapsus, Syngoniums, Xanthosomas and Philodendrons. Variegated sports of normally green plants, occasionally found in Philodendron or Monstera, may lose their ivory-yellow leaf areas when a plant is growing too lustily and kept in much darkness. As mentioned earlier, plants with such variegated foliage should be kept more on the dry side.

Crotons, botanically Codiaeum, are known to everyone in the South for their gaily colored foliage painted in greens, yellows, reds, and blacks. They require strong sun, by the south window or on the patio for best color, as with insufficient light, and too high temperatures, they easily turn plain green. As with most woody plants, liberal watering at the roots is necessary to avoid the dropping of leaves. Other members of the Euphorbia family with showy foliage are the Acalyphas, but they do prefer the open patio or garden in the summer.

On the more delicate side are the Episcias of the Gesneriad family, and the Hoffmannias, Sonerilas and Bertolonias of the Melastome family. Their foliage is a carpet-like velveteen, often of quilted texture, and they thrive best in the warm humidity of a greenhouse or terrarium. Similar in texture but larger and showier are the big-leaf Miconias.

Some of the tropical Hibiscus have foliage strikingly variegated, although they are best known for their large, showy flowers. In the Lily-family we have some gorgeously colored Cordylines and Dracaenas. The Cordyline terminalis complex, commonly known as "Red Dracaenas", show their most intensely brilliant colors with advent of winter, and they respond to heavy water saturation to bring out their best in reds and yellows at that time. Dracaena goldieana with its zebra cross-banding might be called the Queen of Exotics.

The large Begonia genus has for generations been a favorite of the house plant lover, who has adopted the begonia as a pet as one would a cat. All are attractive, whether clothed in wax, colored hair or velvet fabric, and, with the help of a little fresh air, moderate light and a bit of attention to watering well deserve their place in the home. A few dainties require relatively high humidity combined with warmth, as do the Rex-begonias.

The long-lived Bromeliads rival Cacti in being able to stand neglect. Those with the fanciest leaf color and design need the most shade. Strikingly colorful center cups develop at blooming time in Neoregelia and Nidularium, and hieroglyphic cross-banding is characteristic of many Vrieseas.

Orchids in foliage are seldom beautiful, but the varied group of Jewel orchids, such as Haemaria and Anoectochilus, can rival any foliage plant with their striking velvety leaves, veined or striped gold or silver.

Numerous miniature plants with attractive foliage can be combined into very exotic looking arrangements. These include Fittonias, Hoyas, Gynura, Ligularia, Chamaeranthemum, Strobilanthes, Kaempferia, Pileas, many Peperomias and even Coleus, and they will create a significant representation of the beauty of the tropics, in all its romantic aspects, in any home.

The real tropical vegetation apparent to the casual observer shows very little foliage color other than shades of green, but in bloom is most likely very striking. Nature is frugal: species with spectacular blossoms seldom have colorful foliage, while plants with richly colored leaves often have insignificant flowers though quite often very fragrant. We like to envision a tropical scene in gorgeous design and color and thus nourish the dream picture that insists on living in the mind of the dweller of colder zones.

This planter-box of tropical beauties, consisting of an assortment of marantas and calatheas, dieffenbachia, alocasia, anthurium and platycerium, will bring a touch of the exotic to any warm room. Such Exotics will thrive best planted in moist peat moss or a leafmold mixture, or the pots can be plunged in peat, to facilitate the changing of plants whenever desired.

Calathea insignis
"Rattlesnake plant"

Calathea veitchiana

Calathea kegeliana

Calathea makoyana
"Peacock plant"

Ctenanthe oppenheimiana

Ctenanthe oppenheimiana tricolor
"Never-never plant"

Calathea warscewiczii

Calathea zebrina
"Zebra plant"

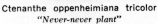

Calathea princeps

Maranta leuconeura kerchoveana
"Prayer plant"

Maranta leuconeura massangeana
"Rabbit's foot"

Calathea undulata

Maranta leuconeura erythroneura
"Red-veined prayer plant"

Calathea argyraea
"Silver calathea"

Maranta arundinacea variegata
"Variegated arrow-root"

Ctenanthe
lubbersiana

Ctenanthe pilosa (compressa)
"Giant bamburanta"

Calathea micans
"Miniature maranta"

Calathea roseo-picta

Calathea lindeniana

Calathea ornata 'Roseo-lineata'

Calathea ornata 'Sanderiana'

Calathea leopardina

Calathea bachemiana

Calathea picturata 'Argentea'

Calathea picturata vandenheckei

Calathea wiotii

Calathea louisae

Calathea lietzei

Dieffenbachia x bausei

Dieffenbachia picta
'Rudolph Roehrs'
"Gold dieffenbachia"

Dieffenbachia picta 'Superba'
"Roehrs superba"

Aglaonema crispum
(Schismatoglottis roebelinii)
"Painted droptongue"

Aglaonema commutatum
'Pseudobracteatum'
"Golden evergreen"

Syngonium podophyllum
xanthophilum
"Green Gold" or
"African evergreen"

Aglaonema pictum

Aglaonema commutatum 'Treubii'
"Ribbon aglaonema"

Aglaonema rotundum
"Red aglaonema"

Monstera deliciosa variegata
"*Variegated monstera*"

Philodendron verrucosum
"*Velvet-leaf philodendron*"

Philodendron gloriosum
"*Satin-leaf*"

Philodendron ilsemannii
"*Variegated philodendron*"

Philodendron mamei
"*Quilted silverleaf*"

Philodendron sodiroi
"*Silverleaf*"

Homalomena wallisii
"*Silver shield*"

Scindapsus aureus 'Tricolor'
"*Devil's-ivy pothos*"

Scindapsus pictus argyraeus
"*Satin pothos*"

Anthurium warocqueanum
"Queen anthurium"

Anthurium clarinervium
"Hoja de corazon"

Anthurium veitchii
"King anthurium"

Alocasia x amazonica

Alocasia sanderiana
"Kris plant"

Xanthosoma lindenii
'Magnificum'

Rhektophyllum mirabile
(Nephthytis picturata)

Caladium 'White Queen'
"White Fancy-leaved caladium"

Caladium 'Frieda Hemple'
"Red elephant-ear"

Codiaeum 'Mona Lisa'
"White croton"

Codiaeum 'L. M. Rutherford'
"Giant croton"

Codiaeum 'Elaine'
"Lance-leaf croton"

Codiaeum 'Johanna Coppinger'
"Strap-leaf croton"

Codiaeum 'Gloriosum superbum'
"Autumn croton"

Codiaeum 'Imperialis'
"Appleleaf croton"

Codiaeum 'Norwood Beauty'
"Oakleaf-croton"

Codiaeum 'Aucubaefolium'
"Aucuba-leaf"

Codiaeum 'Punctatum aureum'
"Miniature croton"

Abutilon striatum
'Aureo-maculatum'
"Spotted Chinese lantern"

Abutilon x hybridum
'Souvenir de Bonn'
"Variegated flowering maple"

Abutilon striatum thompsonii
"Spotted flowering maple"

Hibiscus rosa-sinensis cooperi
"Checkered hibiscus"

Acalypha wilkesiana macafeana
"Copper-leaf"
"Match-me-if-you-can"

Hibiscus rosa-sinensis 'Matensis'
"Snowflake hibiscus"

Acalypha wilkesiana obovata
"Heart copperleaf"

Abutilon x hybridum 'Savitzii'
"White parlor-maple"

Acalypha godseffiana
"Lance copperleaf"

Begonia imperialis
"Imperial begonia", "Carpet begonia"

Begonia serratipetala
"Pink-spot angel wing"

Begonia rex 'Helen Teupel'
"Diadem begonia"

Begonia rex 'Mikado'
"Purple fanleaf"

Begonia pustulata argentea
"Silver pustulata"

Begonia x 'Cleopatra'
"Mapleleaf begonia"

Begonia goegoensis
"Fire king begonia"

Begonia masoniana
"Iron cross begonia"

Episcia cupreata
'Chocolate Soldier'
"Carpet plant"

Episcia reptans (fulgida)
"Flame violet"

Stenandrium lindenii

Chamaeranthemum igneum

Crossandra pungens

Ruellia makoyana
"Monkey plant"

Pseuderanthemum alatum
"Chocolate plant"

Ligularia tussilaginea
aureo-maculata
"Leopard plant"

Ligularia tussilaginea argentea
"Silver farfugium"

Bertolonia 'Mosaica'

Hoffmannia roezlii
"Quilted taffeta plant"

Hoffmannia ghiesbreghtii
"Tall taffeta plant"

Bertolonia marmorata aenea
"Bronze jewel plant"

Bertolonia hirsuta
"Jewel plant"

Bertolonia sanguinea

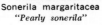

Sonerila margaritacea
'Mme. Baextele'
"Frosted sonerila"

Sonerila margaritacea
"Pearly sonerila"

Gynura aurantiaca
"Velvet plant"

Graptophyllum pictum
"Caricature plant"

Pseuderanthemum reticulatum

Pseuderanthemum atropurpureum
tricolor

Chamaeranthemum venosum

Graptophyllum pictum
albo-marginatum

Chamaeranthemum gaudichaudii

Kaempferia roscoeana
"Peacock plant"

Ruellia blumei

Hemigraphis "Exotica" hort.

Scilla violacea
"Silver squill"

Justicia extensa

Drimiopsis kirkii

Sanchezia nobilis glaucophylla

Ixora borbonica

Strobilanthes dyerianus
"Persian shield"

Myrtus communis variegata
"Variegated myrtle"

Manihot esculenta variegata
"Cassava" or "Tapioca plant"

Hosta undulata 'Medio-picta'
"Waxy-leaved funkia"

Aphelandra squarrosa 'Dania'
"Zebra plant"

Acanthus montanus
"Mountain thistle"

Aphelandra squarrosa
'Fritz Prinsler'
"Saffron spike"

Ficus rubiginosa variegata
"Miniature rubber plant"

Ficus elastica 'Doescheri'
"Variegated rubber plant"

Ficus parcellii
"Clown fig"

Coccoloba uvifera aurea
"Variegated sea-grape"

Clusia rosea aureo-variegata
"Monkey-apple tree"

Nicodemia diversifolia
"Indoor oak"

Amphiblemma cymosum

Miconia magnifica
"Velvet tree"

Myriocarpa stipitata
(Boehmeria argentea)

Pilea cadierei
"Aluminum plant" or
"Watermelon pilea"

Peperomia verschaffeltii
"Sweetheart peperomia"

Pilea 'Silver Tree' hort.
"Silver and bronze"

Peristrophe angustifolia
aureo-variegata
"Marble-leaf"

Saxifraga sarmentosa 'Tricolor'
"Magic carpet"

Hypoestes sanguinolenta
"Freckle-face"

Fittonia verschaffeltii
"Mosaic plant"

Fittonia verschaffeltii pearcei
"Snake-skin plant"

Fittonia verschaffeltii
argyroneura
"Nerve plant"

Pelargonium x hortorum
'Mrs. Pollock double'
"Double tricolor"

Senecio cineraria 'Diamond'
"Dusty miller"

Pelargonium x hortorum 'Velma'
"Tricolor geranium"

Coleus blumei 'Candidum'
"White coleus"

Coleus rehneltianus
'Lord Falmouth'
"Red trailing nettle"

Coleus blumei 'Defiance'
"Painted nettle"

Iresine herbstii
'Aureo-reticulata'
"Chicken-gizzard"

Impatiens marianae
"Silvery patience plant"

Impatiens hawkeri 'Exotica'
from New Guinea

Pleomele reflexa variegata
"Song of India"

Dracaena goldieana
"Queen of dracaenas"

Aspidistra elatior 'Variegata'
"Variegated cast-iron plant"

Pandanus veitchii
"Variegated screw pine"

Cordyline terminalis 'Tricolor'
"Tricolored dracaena"

Rhoeo spathacea (discolor)
"Moses in the cradle"

Heimerliodendron brunonianum
'Variegatum' (Pisonia)
"Bird-catcher-tree"

Siderasis fuscata
"Brown spiderwort"

Anoectochilus roxburghii
"Golden jewel orchid"

Macodes petola
"Gold-net orchid"

Haemaria discolor dawsoniana
"Golden lace orchid"

Neoregelia carolinae 'Tricolor'
"Variegated crimson cup"

Vanilla fragrans 'Marginata'
"Variegated vanilla"

Billbergia x 'Bob Manda'
"Manda's urn plant"

Nidularium innocentii lineatum
"Striped birdsnest"

Guzmania vittata
"Snake vase"

Vriesea hieroglyphica zebrina
"Zebra king"

Aechmea caudata variegata
(Billbergia forgetii)

Ananas bracteatus 'Striatus'
"Variegated wild pineapple"

Aechmea chantinii
"Amazonian zebra plant"

Neoregelia x 'Mar-Con'
"Marbled fingernail"

Vriesea fenestralis
"Netted vriesea"

Aechmea fasciata variegata
"Variegated silver vase"

Guzmania musaica
"Mosaic vase"

Aechmea 'Foster's
Favorite Favorite'

Cryptanthus zonatus zebrinus
"Zebra plant"

Polystichum aristatum variegatum
"East Indian holly-fern"

Pityrogramma pulchella
"Silver-fern"

Polypodium aureum 'Undulatum'
"Blue fern"

Pteris quadriaurita 'Argyraea'
"Silver bracken"

Pteris ensiformis 'Victoriae'
"Victoria table fern"

Selaginella uncinata
"Rainbow fern"

Bougainvillea 'Harrisii'
"Variegated paper-flower"

Athyrium goeringianum pictum
"Miniature silver-fern"

Breynia nivosa roseo-picta
"Leaf flower"

Polyscias balfouriana 'Pennockii'
"White aralia"

Polyscias guilfoylei victoriae
"Lace aralia"

Polyscias balfouriana marginata
"Variegated Balfour aralia"

Polyscias paniculata 'Variegata'
"Variegated rose-leaf panax"

x Fatshedera lizei variegata
"Variegated miracle plant"

Dizygotheca elegantissima
"Spider aralia", "Splitleaf maple"

Hedera canariensis arb. 'Variegata'
"Ghost tree ivy"

Hedera helix 'Silver King'
"Small variegated ivy"

Hedera canariensis 'Variegata'
"Variegated Algerian ivy"

Yucca aloifolia 'Marginata'
"Variegated Spanish bayonet"

Sansevieria trifasciata laurentii
"Variegated snake plant"

Agave angustifolia marginata
"Variegated Caribbean agave"

Haworthia papillosa
"Pearly dots"

Sansevieria trifasciata
'Golden hahnii'
"Golden birdsnest"

Aloe variegata
"Partridge-breast"

Crassula 'Tricolor Jade'
"Tricolored jade plant"

Kalanchoe tomentosa
"Panda plant"

Agave victoriae-reginae
"Queen agave"

DRACAENAS and Dracaena-like plants

Dracaenas are woody members of the Lily family, and include some of the most attractive, shapely, and durable decorative plants we have at our disposal. What we loosely call "Dracaenas" may actually belong to one of three genera, Dracaena proper, Pleomele, and Cordyline.

Dracaenas are endemic primarily to Africa, with but Dracaena americana native in Central America. Pleomele dominate in Southeast Asia and on Pacific Islands. Cordylines are primarily from the South Pacific, Indonesia, New Guinea, Australia and New Zealand.

DRACAENAS — Dracaenas come in many sizes, from small bushes to big trees, and they hold their handsome foliage well. Their leathery leaves are in shades of green often striped or marbled with white or cream. Flowers are carried in panicles; are typically small and whitish with 6 divided segments that differ in this respect from the tubular flowers of the Cordylines. Their roots are orange-colored, or yellow, often fragrant but not suitable for propagation. Being tropical most Dracaenas require warmth and plentiful moisture with good drainage; a sudden drop in temperature may precipitate flowering, which will cause the growing top of the tree-like species to split and branch.

Dracaena draco, the "Dragon tree", 20 ft. high, at home on Tenerife, Canary Islands. Reddish resin appearing in the cracks of its bark gave rise to the legendary belief that this thickened sap is dragon's blood.

Of the better-known Dracaenas, the "Madagascar dragon tree", Dracaena marginata has rapidly become top favorite of modern architects. Its oddly crooked stems, tipped by linear, leathery leaves edged in red, and long-lasting qualities even if neglected makes this species highly desirable as a decorator. Next in popularity is the white-striped Dracaena deremensis warneckei, an ideal room-divider plant and good keeper. They have been used for years in the United Nations building in New York in a very poor and dark location. Dracaena massangeana needs more attention to warmth and moisture but is an old-time favorite as house plant. The pretty, variegated Dracaena sanderiana and D. godseffiana are much used in small dish-gardens, and practically last forever. The Queen of dracaenas, Drac. goldieana, is of all species the most beautiful but also the most difficult. Essential to it is warmth, moisture and rapid drainage, best obtained in a gravelly soil with peat. I have seen this luxuriate in Hawaii planted in volcanic ash.

PLEOMELES—By some botanists, Pleomele are classified under Dracaena, but their flowers are tubular as in Cordyline. In habit they are tree-like or often scandent, such as Pleomele reflexa and its varieties. These branch freely, and the willowy stems are closely set with durable leaves, creating an unusual effect. Pleomeles are very popular as ornamental plants in India. With us they have proven to be long lasting and intriguing in architectural decor.

CORDYLINES—The Cordylines, passing in horticulture as "Dracaenas", are trees and shrubs with leathery leaves, often clustered and palm-like in appearance. They are distinguished from Dracaenas by their branched clusters of flowers that are tubular, not showy and usually white or pinkish. Their rootstock is swollen and can be used for propagation; their roots are white.

Cordyline fall into two main groups: The "Palm-lilies" such as Cordyline australis and indivisa, and the colorful "Red dracaenas" mainly belonging to the Cordyline terminalis complex. The Palm-lily we know best is Cordyline australis known to florists as "Dracaena indivisa" which is from New Zealand where it is known as "Cabbage tree" and it is very decorative and frugal as a container plant for cooler locations. It is also used widely for bedding as a smaller plant.

The "Red dracaenas" of the Cordyline terminalis group have spread far and wide in the tropics and are naturalized in Malaysia and South America where they have been introduced by mythical Polynesian seafarers. These Cordyline have highly colored foliage in shades of red or yellow and there is hardly a village under the tropic sun that ever is without them. These terminalis cultivars are somewhat seasonal in behavior. Toward middle of the winter the topmost leaves will burst into a brilliant show of color, in shades from purplish-crimson through pink, bright green, cream or yellow. But after that, the lower foliage is apt to drop if not kept wet sufficiently. At all times will the colored Cordyline require warmth, but unlike Dracaenas all prefer a location of bright light. There is one important subspecies, Cordyline terminalis 'Ti', widely known as the "Hawaiian Good luck plant" which is plain-green but is advertised because it lends itself well to being cut into sections of cane which then sprout readily in water or soil as the "Miracle plant".

DRACAENA-LIKE ASSOCIATES—Following the law of parallel evolution, there are other plants in different families that are dracaena-like, and all have long found their place in a decorator's scheme—Pandanus, Aspidistra, Chlorophytum and Yuccas all are old time favorites of the house plant fancier. Yucca elephantipes is an easy to please character that as a desert dweller is satisfied with almost any corner cool or warm, where an individual succulent plant is to create visions of the West. Variegated Pandanus need plenty of water to keep their beautiful white banding. Chlorophytum, the "Spider plants" are good in baskets; and the old "Cast iron plant" Aspidistra will ever persist where nothing else ever would.

Dracaena hookeriana
"Leather dracaena"

Dracaena marginata
"Madagascar dragon tree"
(as floor plant)

Dracaena draco
"Dragon tree"

Pleomele reflexa
"Malaysian dracaena"

Dracaena arborea
"Tree dracaena"

Dracaena x 'Gertrude Manda'
"Broad leather dracaena"

Pleomele thalioides
"Lance dracaena"

Dracaena marginata
"Madagascar dragon tree"
(as table plant)

Dracaena deremensis 'Warneckei'
"Striped dracaena"

Dracaena fragrans massangeana
"Cornstalk plant"

Pleomele reflexa variegata
"Song of India"

Dracaena godseffiana
"Gold-dust dracaena"

Dracaena sanderiana
"Ribbon plant"

Dracaena godseffiana
'Florida Beauty'

Dracaena fragrans 'Victoriae'
"Painted dragon-lily"

Dracaena goldieana
"Queen of dracaenas"

Yucca aloifolia 'Marginata'
"Variegated Spanish bayonet"

Yucca elephantipes
"Spineless yucca"

Yucca gloriosa
"Palm-lily"

Dasylirion acrotriche
"Bear-grass"

Cordyline terminalis 'Negri'
"Black dracaena"

Hesperaloe parviflora
"Western aloe" or "Red yucca"

Dianella revoluta (longifolia)
"Flax lily"

Cordyline indivisa 'Rubra'
"Red palm-lily"

Beaucarnea recurvata
"Bottle palm" or "Pony-tail"

Cordyline indivisa
"*Mountain cabbage-tree*"

Cordyline australis (juv.)
"*Cabbage-tree*"

Cordyline terminalis var. 'Ti'
"*Good luck plant*" or "*Miracle plant*"

Cordyline terminalis 'Firebrand'
"*Red dracaena*"

Cordyline terminalis
"*Tree of kings*"

Cordyline terminalis
'Mad. Eugene Andre'
"*Flaming dragon tree*"

Cordyline stricta
(Dracaena congesta)

Cordyline terminalis 'Tricolor'
"*Tricolored dracaena*"

Cordyline terminalis 'Amabilis'
"*Pink dracaena*"

Pandanus utilis
"Screw-pine"

Pandanus baptistii
"Blue screw-pine"

Pandanus veitchii
"Variegated screw-pine"

Rohdea japonica
"Sacred lily of China"

Aspidistra elatior (lurida)
"Cast-iron plant"

Arthropodium cirrhatum
"Rock-lily"

Chlorophytum comosum
'Variegatum'
"Large spider plant" or "Green-lily"

Chlorophytum comosum
'Vittatum'
"Spider plant"

Chlorophytum bichetii
"St. Bernard's lily"

Bananas and Gingers:

Musa nana (cavendishii), "Chinese dwarf banana", with edible, delicious fruit

The Banana (Musaceae) and the Ginger families (Zingiberaceae) with their showy leaves and strangely constructed inflorescence magically give us visions of the Exotic and the Tropics—balmy breezes, gaily colored birds, and smiling South Sea maidens wearing ginger blossoms. True enough, most members of these families are tropical, huge herbs with gigantic leaves, and often too large to be considered house plants, although many of them allow themselves to be confined to containers, will stand up surprisingly well, and even bloom. A greenhouse or the Southern patio would of course suit their nature better considering where they originate.

The term "Banana family" is perhaps too loose a definition of the Musaceae when it is considered that this includes such apparently different plants as the Strelitzia or "Bird of Paradise"; Heliconia, the spectacular "Lobster claws"; as well as the true bananas. What they have in common is their showy inflorescence with flowers mostly under sheathing or in brightly colored, boat-shaped bracts. The Banana family comprises the largest herbs in the world with huge leaves on long stalks, often tree-like but true herbs, except for the extraordinary Traveler's tree of Madagascar (Ravenala) and in some Strelitzia.

BANANAS—Bananas are best known for their flavorful "Fruit of Paradise" produced from clustering stalks the year round. Each stalk bears fruit only once and then disappears, and new stems arising as suckers from the stolon give a succession of ripening fruit. The fleshy pseudo-trunks are formed of the tightly packed sheaths of leaf bases. The relatively dwarf and compact "Chinese banana", Musa nana is most easily adapted for house culture in tubs, as it can be kept small enough for decoration indoors. This dwarf banana will still require good light and steady moisture, but to bear its sweet yellow fruit it would need a large container or better still be planted out in an area set aside for this purpose, such as in a covered atrium.

Some Bananas are purely ornamental. The "Blood banana", Musa zebrina has satiny leaves splashed with deep red; the "Variegated-leaved banana", beautifully feathered white and green, is unfortunately temperamental and difficult to grow being much troubled by nematodes. As a merely ornamental banana, Musa ensete, botanically Ensete ventricosum, has been planted since the times of ancient Egypt, and is used today as a center plant in formal flower beds because of its elegant, symmetrical shape. It is not grown for fruit and since it is not stoloniferous, must be propagated from seed. A striking new addition to ornamental bananas is the "Black banana", Ensete maurelii, a large rosette of dark-complexioned, blackish leaves that tend however to fade with age.

HELICONIAS—The tropical Heliconias, "Lobster claws" or "False Bird of paradise" are banana-like herbs to 15 ft. high, with unique, mostly very spectacular inflorescence of striking waxy, red or orange boat-shaped bracts aligned in two rows, from which peek creamy flowers. Some species such as the "Parrot flower" are of smaller habit with erect inflorescence and more suitable for limited space if used in tubs. The Heliconia illustris group stays small also, with highly colored foliage, but is quite delicate and requires greenhouse care and warmth.

STRELITZIA—The South African Strelitzia reginae, or "Bird-of-Paradise flower", with hard leaves and long-lasting inflorescence are much more tolerant of abuse and will do best not quite so warm; they require lots of sunlight and should be set outdoors in summer. Once old enough and well established in a tub they should be left undisturbed to bring forth regularly the exotic bird-like blue and orange flowers for which they are so famous, from winter into spring.

THE GINGERS—Members of the Ginger family are rhizomatous herbs mostly with clumps of cane-like stems and alternate elongated leaves; their irregular flowers are borne within large bracts, in long panicles or spikes. They need rich, well watered soil and high humidity to produce their showy flowers.

Probably the best-known of the larger species are the gorgeous "Shell ginger", Alpinia speciosa, with waxy flower clusters one foot long; and the fragrant "Kahili ginger", Hedychium gardnerianum. Alpinia sanderae is a beautiful, highly variegated, dwarf ornamental plant; unfortunately for the warm greenhouse only.

Amomum cardamon is a tough houseplant, its leathery leaves with aroma spicy to the touch. The various Costus or "Spiral gingers" are attractive and curious foliage plants. Smallest gingers of all are the Kaempferias with their fleeting flowers; the "Peacock plant" is aptly named for its iridescent leaves, but like the other species requires dormancy in winter. The "Bitter ginger", Zingiber zerumbet, is grown for its aromatic knobbed roots; the ginger of commerce is the similar Zingiber officinale, but it is not particularly attractive as an ornamental.

Nicolaia elatior (Phaeomeria magnifica), the magnificent "Torch ginger"

Heliconia illustris
rubricaulis

Heliconia illustris
aureo-striata

Heliconia velutina

Heliconia bihai
"Firebird"

Heliconia caribaea
"Wild plantain"

Heliconia humilis (wagneriana)
"Lobster claw"

Heliconia psittacorum
"Parrot flower"

Heliconia platystachys

Heliconia marantifolia
(syn. metallica)

Musa zebrina
"*Blood banana*"

Musa x paradisiaca
"*Common banana*"

Musa nana
"*Chinese dwarf banana*"

Ensete maurelii
"*Black banana*"

Ensete ventricosum (Musa ensete)
"*Abyssinian banana*"

Musa x paradisiaca 'Koae'
"*Variegated banana*"

Globba winitii

Roscoea alpina

Zingiber zerumbet
"*Bitter ginger*"

Brachychilum horsfieldii
in fruiting stage

Amomum cardamon
"Cardamon ginger"

Alpinia sanderae
"Variegated ginger"

Alpinia speciosa 'Variegata'
"Variegated shell ginger"

Hedychium gardnerianum
"Kahili ginger"

Curcuma roscoeana
"Hidden lily"

Alpinia calcarata
"Indian ginger"

Hedychium coronarium
"White ginger"

Cautleya spicata

Strelitzia alba (augusta)
"Great white strelitzia"
palm-like trees 30 ft. high, in Natal

Ravenala madagascariensis
"Travelers tree"
gigantic tree, holding water; from Madagascar

Strelitzia reginae
"Bird-of-Paradise"
strikingly exotic, from South Africa

Alpinia speciosa
"Shell ginger", from China
with fragrant, porcelain flowers

Costus sanguineus
"Spiral flag"

Costus speciosus
"Stepladder plant"

Costus malortieanus
"Emerald spiral ginger"

Costus igneus
"Fiery costus"

Canna coccinea
"Indian shot"

Orchidantha maxillarioides
"Orchid-flower"

Kaempferia roscoeana
"Peacock plant"

Kaempferia masonii

Kaempferia pulchra
"Pretty resurrection lily"

Kaempferia decora
"Dwarf ginger-lily"

Kaempferia galanga

Kaempferia gilbertii
"Variegated ginger-lily"

FIG TREES and other Broadleafs for warm locations

The Moraceae, known as the Fig or Mulberry family, includes about 1000 species, from woody trees to little vines. Best known to us for their economic importance are the Fig, Mulberry, Breadfruit, Hemp, and Hops. What they have in common is a peculiar inflorescence and compound "fruit": minute flowers covering the inner surface of a hollow receptacle as the fig of the fig tree, or covering the outside of globular (Mulberry) or flat receptacles (Dorstenia).

For ornamental use, the genus Ficus is the most important, and its best known representative, the age-old "Rubber plant", Ficus elastica, can be seen almost everywhere a reliable decorator plant is needed. Today there are many other species used indoors. What they share in common is their milky latex, a source of rubber, and the characteristic hollow "fruit", actually a fleshy receptacle that bears scores of minute male and female flowers on its inner walls, which thus bloom and mature wholly in the dark interior.

FICUS FOR DECORATION—Ficus as a rule are evergreen with glossy green, leathery foliage different in each species. In habit of growth and character there is a choice from large-leaved, sparry types like Ficus lyrata and decora to those of weeping habit with gracefully arching branches and little, pendulous leaves, as Ficus benjamina. Most Ficus are easy to maintain in containers and are extremely durable. Should they become too big they can be cut and shaped to any size desired and this will help to make them grow more dense. Such Ficus trees can be employed in most decorative schemes, and they lend themselves for use in imitating other trees. Ficus benjamina could well substitute for an indoor birch; Ficus exotica or stricta for a weeping willow. Ficus

Ficus benjamina 'Exotica' tree; this "Java fig" creates outdoor atmosphere at a cafeteria on Park Avenue, midtown New York.

nitida, especially if sheared, resembles a formal baytree, as used in the Metropolitan Museum. I once saw a florist on Beacon Street in Boston using plastic oranges on a standard Ficus nitida and the result looked cannily like a real citrus tree. Ficus elastica of course represents the "India Rubber tree" or "Banyan fig". Ficus are sunlovers and need 50 fc of light; more to hold most foliage.

THE FIDDLE-LEAF FIG—The monarch of all fig trees is the large-leaved Ficus lyrata, commonly known as pandurata, the "Fiddle leaf fig". This magnificent African tree is a bit more particular to care for and likes a warm location with good light. It is best to grow it in a relatively small container with holes for drainage, and set onto a saucer or loosely into a jardiniere, or plunge into a recess filled with peat. This will allow the roots to drain off excess water, breathe and stay healthy. While the "Fiddle leaf" enjoys moisture, it wants neither too wet nor too dry. If too dry, foliage may burn, turn grayish and drop. But if kept excessively wet, and drainage is poor, the roots will rot, the plant can't transport sufficient sap to the large leaves, and they will develop those unsightly black spots or marginal areas of brown. It is always best to let the soil get on the dry side and then soak it thoroughly. Once established in a good location, Ficus lyrata will keep remarkably well; an example are the large specimen trees in New York's Philharmonic Hall.

DIVERSE FICUS AND NEAR-FIGS—In small-size ornamental figs most popular is still Ficus elastica 'Decora' which is a broadleaved cultivar of the long-known "Rubber plant". It is tough and stands neglect though it might lose a leaf from time to time; should it grow too tall then cut it back and it will branch. These Ficus would enjoy being set outdoors in summer, for stocky growth, to rebuild weak tissues, and a glossy new look will be the result. The variegated forms of this species are of slower growth and prefer slight shade.

Ficus rubiginosa variegata is very pretty as a small potplant; also Ficus diversifolia, the "Mistletoe fig"; full of yellow berries even while small. The "Clown fig" needs the humidity of a greenhouse or it will drop its variegated leaves; the "Blue Mexican fig" is equally difficult but very beautiful with its metallic leaves blood red beneath.

In ficus-like plants both Clusia rosea with its thick-fleshy leaves, and the Sea grape, Coccoloba uvifera, at home along the shores of the Caribbean, are both as tough as leather though somewhat sparry in their habit. The eye-catching Coccoloba rugosa with its corrugated leaves made as if of tin appears to be more temperamental away from the coddling climate of its native tropical Puerto Rico.

To create the effect of large size trees indoors, Ficus in their wide selection of types offer the greatest opportunity to the interior decorator, considering ease of care coupled with their excellent keeping quality even in air-conditioned buildings.

Ficus retusa nitida tree, the "Indian laurel", in a bright corner on the 60th floor of a Wall street bank, downtown New York.

Ficus lyrata
commonly known as Ficus pandurata,
the "Fiddleleaf fig"

Ficus cyathistipula
fruiting as a young plant

Ficus retusa nitida
the "Indian laurel" as standard,
Santa Barbara, California

Ficus carica
the "Common fig"
bearing edible fruit

Ficus lyrata (pandurata)
"Fiddleleaf fig"
as tubbed combination

Ficus benjamina
graceful "Weeping fig"
in redwood container

Ficus elastica variegata
"Variegated rubber plant"
branched specimen

Ficus elastica 'Decora'
"Wideleaf rubber plant"
combination in tub

Ficus benjamina 'Exotica'
the "Java fig"
with gracefully pendant branches

Ficus stricta
in Florida as "philippinense";
larger foliage than benjamina

Ficus benjamina
the "Weeping fig"
in pyramid form

Ficus benjamina nuda
in Florida nurseries as
the "Small-leaved philippinense"

Ficus benjamina
"Weeping fig", specimen pyramids
in Balboa Park, San Diego, California

Ficus lyrata (pandurata)
"Fiddleleaf fig"
in terracotta eggshell container

Ficus retusa nitida
"Indian laurel"
in pyramid form

Ficus retusa, *the "Chinese banyan"*
as standard trees; very durable
as decorators and lending grand decor

Ficus nekbudu (utilis)
"Zulu fig tree"

Ficus macrophylla
"Australian banyan"

Ficus microphylla hort.
"Little-leaf fig"

Ficus rubiginosa
"Rusty fig"

Ficus wildemanniana
(panduriformis hort.)

Ficus jacquinifolia
"West Indian laurel fig"

Ficus roxburghii
"Ornamental fig"

Ficus altissima
"Lofty fig"

Ficus afzelii
(eriobotryoides)

Ficus religiosa
"Sacred Bo-tree" or "Peepul"

Ficus krishnae
"Sacred fig tree"

Ficus sycomorus
"Sycamore fig"

Ficus rubiginosa variegata
"Miniature rubber plant"

Ficus elastica 'Doescheri'
"Variegated rubber plant"

Ficus parcellii
"Clown fig"

Ficus dryepondtiana
"Congo fig"

Ficus diversifolia (lutescens)
"Mistletoe ficus"

Ficus petiolaris
"Blue Mexican fig"

Clusia rosea
"Autograph tree" or "Fat pork tree"
with thick-leathery leaves

Coccoloba uvifera
the "Sea-grape", from tropical shores;
red veins in stiff-leathery foliage

Bischofia javanica (trifoliata)
the very decorative "Toog tree"

Coccoloba rugosa, from Puerto Rico,
the interesting "Corrugated tin-tree"

PALMS and Palm-like Plants

Cocos nucifera, the "Coconut palm" and "Tree of Life" on tropical shores, swaying in the balmy breezes of the Indian Ocean, near Mombasa, East Africa.

What could convey better the image of the romantic South Seas than the view of graceful Coconut palms bending from the shores, their waving fronds giving with the balmy breeze. Thus, Cocos nucifera is truly the expression of the real tropics, although seldom seen confined to small containers.

Palms have been named "Principes", the "Princes of Plants" in an order of the Monocotyledons set apart by Linnaeus, exclusively for the family Palmae. Their majesty of stature in every location where they are seen makes them distinct from all other, more lowly trees, and only the regal orchid is a worthy enough companion in their society. Palms are evergreen and the most distinctive and noble foliage plants of the world's warm climates.

DISTRIBUTION: — The ancient Palm family includes between 2600 and possibly 4000 species. They are chiefly tropical, evergrowing woody plants, mostly tree-like, but some are climbing, others bushes.

Equatorial America, especially Brazil, and Tropical Asia share in having the greatest number of species of these amazing plants; there is less variety in Central America and the Caribbean, while Africa is poorest in types of palms. Some species have spilled over to the edge of the temperate zone; in the extreme south at latitude 42 deg. in New Zealand I found Rhopalostylis sapida; in the northern hemisphere I remember a "Windmill palm", Trachycarpus for-

Clusters of Phoenix reclinata, the slender "Senegal date palm," in romantic Balboa Park, San Diego, California.

PALMS

tunei, probably the hardiest of palms; at Edinburgh, Scotland at 56 deg. latitude north; its natural limit in Japan is 35 deg. N. The sole palm native in Europe is Chamaerops humilis, which one can observe wild from southern Spain over into Morocco and eastward, where it grows as small clusters right in the open desert.

ECONOMIC IMPORTANCE—While we are concerned with the decorative aspect of the Palms, their economic importance to humanity is by far the greatest. The Date palm, Phoenix dactylifera, has been the Tree of life of the Bible and the ancient people of the cradle of man from Asia Minor to Egypt for over 5000 years. It was cultivated in Mesopotamia since Sumerian times in 3500 B.C., and is seen chiseled in stone on a temple dating from 1370 B.C. at Karnak in Upper Egypt. With the increase in world population, another Tree of life has become even more important, the ubiquitous Coconut tree, the most important of all palms today. It is the plant of a thousand uses, its fruit yielding milk and oil; the fronds are

Hyphaene thebaica, the "Dhoum palm" or "Gingerbread palm", with typically forked trunk, in Kenya, E. Africa.

used for thatching. Other useful palms are the Palmyra palm, Sago palm, Betelnut palm, African Oil palm, Rattan palm, Raffia palm, and Sugar palm, all of which are of great value in their respective areas.

CHARACATERISTICS OF PALMS—Palms are unique amongst trees. Belonging to the Monocots, alongside aroids, orchids, lilies, and grasses, palms have evolved substantially into tree-types. The trunk of a palm is quite unlike that of our familiar trees. The latter have a hard central core of wood which regularly increases not only in length but also in diameter by its cambium layers. Most palms have only a single point of growth activity, the terminal bud. If this is removed the tree will die. Since palms have no cambium layers their trunks cannot increase in diameter, and for this reason they remain slender and graceful. Instead of a central mass of hard tissues, palms have a soft spongy heart surrounded by a hard casing of strong protective fibers arranged in vertical bundles which serve as the tree's conductive system for the transportation of water and mineral elements. This outer rind is extremely tough to cut.

Palm flowers are not showy and are usually borne in pendant clusters which rise from boat-shaped spathes from the leafy crown. From the flowers develop the fruits or nuts, ranging from the size of a pea in Euterpe to the great coconut. As some species, such as the Date palm, carry male and female flowers on separate trees, the female trees have to be artificially pollinated to produce their fruit. This was known and already practiced by the Babylonians 5000 years ago; in Egypt since 2000 B.C.

Palms are easily divided into two main groups by their leaves—the "Feather palms" with pinnate fronds looking like enormous feathers, to 65 ft. long in the Raffia palm—and the "Fan palms" with palmate, fan-like leaf-blades resembling giant hands. Important Feather palms are Aiphanes, Areca, Arenga, Arecastrum, Calamus, Caryota, Chamaedorea, Chrysalidocarpus, Cocos, Euterpe, Howeia, Phoenix, Pinanga, Ptychosperma, Raphia, Roystonia, Syagrus, Veitchia. Leading Fan palms include Borassus, Chamaerops, Coccothrinax, Erythea, Hyphaene, Latania, Licuala, Livistona, Pritchardia, Rhapis, Sabal, Thrinax, Trachycarpus and Washingtonia.

Palms as Graceful Decorators

PALMS FOR INTERIORS—Because of their majestic beauty and decorative value as foliage plants there is no limit to the number of types of palms that lend themselves for culture indoors. 50 years ago collections of such palms were in style in every hall and parlor, flowing over into contemporary conservatories, and were seen in all hotels worthy of a good name. The 1913 catalog of Roehrs Exotic Nurseries listed 98 different palms. These were used in company of Boston ferns, Dracaenas, Pandanus, the Rubber plant, and Aspidistra, as well as seasonable flowering plants, but the assortment of other decorative house plants was then small. Through the years some palms have proved by their good behavior to be of special merit, being tolerant of dark locations, and their leathery leaves resist drought and will take much abuse.

INDOOR FAVORITES TODAY—For the home or office a number of the graceful Feather palms have become known as reliable container plants. Best for general decoration is the sturdy "Paradise palm", Howeia forsteriana, better known to florists as the "Kentia palm". This long-suffering species from the South Pacific is relatively new as an indoor palm having become popular only since Victorian times after 1871, when it was first imported from Norfolk Island into Europe. With a minimum of care these Kentias have been kept alive in many a difficult location for many years on end, and combinations of several per tub are still the florist's favorite at weddings and other social functions.

Though in collections for at least 70 years, the charming little "Parlor palm" or "Mexican Mountain palm", Chamaedorea elegans, has become very popular of late. Commonly known as Neanthe bella, or even Collinia, it is essentially a dwarf palm eventually forming clusters. It is easy to grow and as a small plant it is the best keeper yet for use in dish gardens or small pots, having replaced Syagrus, better known as Cocos weddelliana, which is much more cold-sensitive and requires tropical warmth. There are several other Chamaedoreas also quite satisfactory for decoration, especially clump-forming C. erumpens, a "Bamboo palm". Its reed-like stems and feathery leaves remind us of bamboo which,

PALMS

as a decorative motif, is much sought after. Chamaedorea costaricana is also recommended and will stand considerable neglect in watering, coming from regions of prolonged dry seasons.

The elegant "Butterfly palm", Chrysalidocarpus lutescens, and better known as the "Areca palm", is very dainty with feathery fronds and slender yellow stems, but not as tough as Howeias if used for decoration, and preferring a warmer clime. Most popular of the Phoenix palms is the "Pygmy date", Phoenix roebelenii, a compact plant with gracefully arching leaves, and an excellent house palm if kept warm and moist. In larger decorating schemes, Veitchia merrillii, the "Manila palm", and Ptychosperma elegans, the "Solitair palm", are good and sturdy and give the impression of a full-sized palm tree in small scale. Curious for its fan-shaped leaflets is Caryota mitis, the suckering "Fishtail palm". Unfortunately, some of the more exotic featherpalms such as the Coconut palm, Royal palm, Cocos plumosa, and others, do not stand up well indoors but may be used temporarily for decoration or as small specimen.

Chamaedorea erumpens, the suckering "Bamboo palm" with slender reed-like stems, an excellent decorator for a New York office.

The Fan palms are less popular in homes today. Formerly, that old parlor palm, Livistona chinensis, better known as Latania borbonica or "Chinese fan palm", was very popular but as it grows, its big leaves require too much space. A more modest substitute would be one of the Thrinax group, also Chamaerops excelsa, the "European fan palm", but our homes are generally too warm for the latter. A clump-type cane palm with fan-shaped leaves that can be recommended is the slender "Lady palm", the durable Rhapis humilis. It is slow to grow and keeps its shape. This species will tolerate more warmth than the more robust Rhapis excelsa, a favorite potted palm in Japan and China.

MAINTENANCE AND CARE—Palms do well as container plants beacuse they put up with crowding and have no tap roots. Most prefer shade, all tolerate it. They can be maintained best in pots or tubs that are small in proportion to the size of the plant, and this to some extent controls their growth. Potted palms prefer to be left undisturbed in their containers as long as possible even if the soil becomes filled with roots. Periodic feeding during the warmer and growing season, preferably with mild soluble or organic plant food will maintain them adequately. What is important is sufficient watering. Palms enjoy a plentiful supply of water and two or three thorough soakings per week is not too much provided the container is small, the plants well established, and the drainage good. No plant likes to remain standing for long with "wet feet". Water requirements may change with the seasons and type of container used. On the other hand, if a potted palm is kept too dry the small feeding roots dry up and die, the plant cannot feed itself, and damage results. This can happen if on repotting, the new soil is not compacted with great firmness. The roots do not like to penetrate into new soil if loose, and prefer instead to remain in and around the old root ball. Besides, the old ball would not receive sufficient water as all of it would pass the easy way through the more porous soil outside it. For potting we prefer a fairly heavy, good loam combined with enriched peat or humus material. If palms are used in indoor plantings, boxes or jardinieres, it is always best to leave them in their original containers and to plunge them into peat. This allows them to be moved, turned around, or taken out for cleaning.

PEST CONTROL—Among the few pests that may afflict palms indoors are mites, known as Red spiders; Mealybugs and Scale, encouraged by dry atmosphere. Aside from using chemical sprays or weak oil emulsion, simple remedies are still a periodic sponging or scrubbing with lukewarm soap solution, or soap with nicotine, and forceful syringing with water to dislodge the "insects".

PROPAGATION—Most palms are reproduced from seed which must, to germinate, be fresh and not overheated in transit. Sow in peat or leaf-mold at a warm 70°F and high humidity. A few palms, such as Rhapis, can be propagated by division.

Howeia forsteriana, the popular "Kentia palm" or "Paradise palm" in decorative metal container, in a living room in Los Angeles. Kentia palms have been fashionable and very satisfactory since Victorian times, seed first having been imported into Europe in 1871.

Archontophoenix cunninghamiana
"Seaforthia" or "Piccabeen"
in San Marino, California

Veitchia merrillii
"Manila palm" or "Christmas palm"
at Miami, Florida

Ptychosperma elegans
the shorter "Solitair" or "Princess palm" (to 20 ft.)
Commonly called "Alexander palm"

Archontophoenix alexandrae
the tall-growing "King palm"
or "Alexandra palm" (to 70 ft.)

Ptychosperma macarthurii
"Hurricane palm" or *"MacArthur palm"*
leaflets as if cut off at tips

Veitchia merrillii
"Christmas palm" or *"Manila palm"*

Chrysalidocarpus lutescens
"Areca palm" or *"Butterfly palm"*
in Miami shopping center, Florida

Roystonea regia
the "Cuban Royal palm"
in the Climatron, St. Louis, Missouri

Cocos nucifera 'Golden Malay'
"Dwarf coconut palm"
in North Miami, Florida

Arecastrum romanzoffianum
usually known as Cocos plumosa
the "Queen palm", in Santa Barbara, California

Phoenix canariensis
the ornamental "Canary Islands date"
Los Angeles, California

Howeia forsteriana
"Kentia palm" or "Sentry palm"
bearing seed, in the South Pacific

Trachycarpus fortunei, *the "Windmill palm", in Edinburgh, Scotland, northernmost at latitude 56 degrees, only 700 miles from the Arctic Circle.*

Rhopalostylis sapida, *the "Nikau palm" or "Shavingbrush palm" at southerly limit of palms, 42 degrees south, South Island, New Zealand.*

Chamaerops humilis, *an interesting cluster of the arborescent "European fan palm" from Mediterranean Spain and North Africa, on the landscaped grounds of the Mormon Temple, in Los Angeles, California.*

Phoenix roebelenii
the graceful "Pigmy date" or "Dwarf date palm"

Phoenix reclinata
clustering "Senegal date palm"

Phoenix canariensis
the fresh-green "Canary Islands date palm"

Phoenix dactylifera
the true Arabian "Date palm"

Howeia forsteriana
popularly known as "Kentia palm"

Howeia belmoreana
the "Curly sentry palm"

Hedyscepe canterburyana
"Umbrella palm" or sometimes "Kentia"

Chrysalidocarpus lutescens
widely known as "Areca palm"

Phoenix roebelenii
"Pigmy date palm", with slender trunk

Licuala grandis
"Ruffled fan palm"

Aiphanes caryotaefolia
"Spine palm" or "Martinezia"

Caryota mitis
suckering "Dwarf fishtail palm"

Rhapis excelsa (flabelliformis)
the cane-stem "Lady palm" as cluster

Rhapis humilis
graceful "Slender lady palm"

Washingtonia filifera
"Desert fan palm" or "Petticoat palm"
in Goleta nursery, California

Washingtonia robusta
"Mexican fan palm"
planted in Manhattan Beach, California

Sabal minor
"Dwarf palmetto" or
"Scrub palmetto"

Sabal palmetto
"Cabbage palm", in Florida

Serenoa repens
bushy *"Saw palmetto"*

Coccothrinax argentata
"Florida silver palm"

Thrinax parviflora
"Florida thatch palm"

Thrinax microcarpa
"Key palm"

Trithrinax acanthocoma
"Webbed trithrinax"

Coccothrinax argentea
"Silver palm", in Leningrad

Trachycarpus fortunei
"Chamaerops excelsa" of horticulture
the "Windmill palm"

Chamaerops humilis
"European fan palm", bushy type

Latania lontaroides
syn. L. borbonica, commersonii
"Red Latan"

Livistona chinensis
"Latania borbonica"
of horticulture
"Chinese fan palm"

Livistona australis
"Australian fountain palm"

Erythea edulis
"Guadalupe palm"

Livistona rotundifolia
"Roundleaf livistona"

Copernicia macroglossa
"Cuban petticoat palm"

Rhopalostylis baueri
"Norfolk Island shaving brush"

Reinhardtia gracilis gracilior
"Little window palm"

Linospadix monostachya
"Walking-stick palm"

Licuala spinosa
"Spiny licuala palm"

Rhapis excelsa 'Variegata'
"Variegated lady palm"

Arenga engleri
"Dwarf sugar palm"

Reinhardtia simplex
"Fanleaf reinhardtia"

Syagrus weddelliana
in hort. as "Cocos weddelliana"
the *"Terrarium palm"*

Linospadix minor
"Miniature walking-stick palm"

Chamaedorea elegans 'bella'
"Neanthe bella"

Chamaedorea elegans
"Parlor palm"

Chamaedorea metallica
(tenella in hort.)
"Miniature fishtail"

Chamaedorea seifrizii
(graminifolia)
"Reed palm"

Chamaedorea erumpens
"Bamboo palm"

Chamaedorea elatior
"Mexican rattan palm"

Chamaedorea costaricana
"Showy bamboo palm"

Chamaedorea stolonifera
"Climbing fishtail palm"

Chamaedorea glaucifolia
"Blueleaf chamaedorea"

Calamus siphonospathus,
a furniture "Rattan" with canes to 200 ft. long,
climbing by means of hooks

Calamus ciliaris
"Lawyer canes",
an ornamental "Rattan palm"

Curculigo capitulata
"Palm grass"
(Amaryllis family)

Cyclanthus bipartitus
"Splitleaf cyclanthus"
(Cyclanthus family)

Carludovica palmata
"Panama hat plant"
(Cyclanthus family)

FERNS and Fern-Allies

For decorating your home indoors; for making it feel cool, woodsy and alive; for softening stark, harsh architectural lines or unifying a decorative scheme—few plants rival the beauty of graceful, green ferns. Here are plants with such purity of form, one can't go wrong however they are used. In addition to decorative value, ferns give a great deal of growing satisfaction, they stay green and lush the year around. Ferns are always cool, calm and refreshing.

DISTRIBUTION and HISTORY—Ferns, botanically Filices, belong to the series of non-flowering plants collectively known as Cryptogams. They are of all plant groups, the most widely distributed, being found from the Arctic to near the Antarctic, from sea-level to cool timberline at 10,000 feet or more on the highest mountain ranges in the tropics. Fern history goes back 400 million years to the Paleozoic era, before dinosaurs roamed the earth, when seed-bearing Pteridosperms with fern-like leaves were the first real land plants to develop from their seaweed ancestors. Fossils of ferns like those which live today first appeared 240 million years ago, 100 million years before the first flowering plants.

TYPES OF FERNS—Ferns come in a wide variety of forms, gracefully bold to daintily fine, in close to 12,000 species. They include plants varying in size from a hair-like creeping stem with moss-like leaves to tall, palm-like treeferns over 80 feet in height. Some inhabit trees as epiphytes of the tropical rain forest, others grow in shaded spots of moist, cool or tropical woods. Treeferns live on the misty, cooler elevations of the mountains of the tropics and subtropics; a few have wandered into the temperate zone as in the mild oceanic climate of New Zealand in sight of glacier ice. There are even aquatic ferns that float or live entirely in water.

Nephrolepis exaltata bostoniensis, the old-fashioned "Boston fern", is quite happy near a window, shielded from burning sun by Venetian blinds. Moisture condenses on the glass, a situation most ferns enjoy.

FERNS AS HOUSE PLANTS AND FOR DECORATION—Treeferns of the tropics are the giants of the great fern group, and some species raise their slender trunks to the height of a 6 story building. Most graceful are the West Indian treeferns, Cyathea arborea, and for exotic decor they always excite the imagination of architects. Container-grown they are unfortunately difficult to keep indoors for more than several weeks, unless in a covered atrium with relatively high humidity. Their roots must never be allowed to dry. We have found it a wise precaution to place such tubs inside an outer container with peat, or wrapping the tub in plastic to protect the delicate roots from the danger of desiccation.

The best of the Hawaiian treeferns is the "Man fern", Cibotium chamissoi, in the trade as C. menziesii, and of great appeal to decorators. More stout of habit, they are usually sold as sawed-off tops, and can be grown in a dish of shallow water. Lumps of charcoal may be used for support if necessary. But we have been more successful growing them in soil. Brown-hairy coils unfold successively into the friendly green and shining, leathery fronds.

The "Mexican treefern", Cibotium schiedei, is a favorite for decoration at weddings and other formal occasions because of the grace of its light green arching fronds combined with wonderful keeping qualities. Once tub-bound it is not difficult to maintain, so dainty and yet remarkably durable even under dry-room conditions. Other ferns such as Alsophila and Dicksonia are too water-thirsty, requiring high humidity, and are best kept in a greenhouse, or a shade house in the south.

What we know as Parlor ferns were those in vogue at the turn of the century. Before central heating, dwellings were kept cooler, the humidity as a result was higher, and ferns were thriving on plant-stands or in jardinieres. Think how they grow in the wild—in shaded spots of moist woods with a cool and airy feeling—this is the secret of keeping them happy. To succeed with ferns indoors, temperatures should go down to 60° or 55°F at night.

The best indoor fern always was and is today the "Boston fern". It enjoys good light and is fairly tolerant of dry air. It was found in Boston in 1894 as a sport of a Sword fern, Nephrolepis exaltata, shipped north from Florida. From it originated many excellent mutations frilled and lacy; the "Lace fern", Nephrolepis 'Whitmanii' is still a favorite. These ferns keep best if root-bound which allows them to be watered freely without danger of stagnation. Any repotting necessary is best done in the spring when active growth begins, using porous soil.

Another old-fashioned friend is the "Holly fern", Cyrtomium 'Rochfordianum' with glossy-dark leaves as tough as leather. Also leathery but lacy is the "Leather fern", Rumohra adiantiformis, whose long-lasting fronds are cut and used in flower arrangements. The "Birdsnest fern", Asplenium nidus, is in a class by itself, a rosette of fan-like light green leaves. If kept shady, warm and moist it will prove a good and interesting houseplant. Nice also on the windowsill is the "Fishtail fern", Polypodium polycarpon 'Grandiceps'. With wax green, leathery fronds forked at the tips it is as easy to keep as it is curious.

There are many Polypodiums, and we know them as "Hare's foot fern" because of their paw-like woolly rhizomes creeping along the surface. Probably the best for indoor use is the "Blue fern", Polypodium aureum, in its crisped form 'Mandaianum'. The wavy, shimmering blue fronds develop best in warmth and shade, but otherwise they are not difficult at all.

The Davallias, or "Rabbit's-foot ferns" likewise have slender creeping, furry rhizomes which bear on wiry stalks their lacy but leathery fronds, small enough for a windowsill. They take much neglect and tolerate both light or shade and are often used in hanging baskets.

The Culture and
Care of FERNS

Amongst our best-loved ferns are the dainty "Maidenhairs", but they are not easy under house conditions and do best in a terrarium or behind plastic shelter. Most satisfactory in this genus are Adiantum raddianum, better known as A. cuneatum, and the larger-leaved Adiantum tenerum 'Wrightii', but even these need shade, warmth and moisture.

The so-called "Table ferns" are a varied group of mainly Pteris and Pellaea. Some are frilly, others variegated, and in their small stage are ideal for terrariums. But they are also widely used in pottery bowls, or together with other plants as fillers. One of the most attractive is Pteris 'Victoriae', a dainty-looking but sturdy little fern of white and silver.

"Staghorn ferns" always have aroused the greatest curiosity because of their unusual shapes. Growing as epiphytes on trees, their sterile fronds cling snugly against branch or trunk, while their much-divided fertile leaves resemble the antlers of deer or elk. Platycerium are not difficult if grown on fibrous cushions fastened to a board or basket, and their root system is submerged in water whenever it is beginning to get dry. Platycerium bifurcatum and its cultivar, P. 'Netherlands' are less tropical, and quite tolerant even if abused.

10 |W|

Platycerium "diversifolium" hort., an "Elkshorn fern", growing erect in a glazed pot with attached saucer.

FERN-ALLIES and FERN-LIKE PLANTS—Fern-Allies are plants that in public usage are associated with ferns, although they belong to other families. Best-known as simple house plants are the so-called Asparagus ferns, Asp. densiflorus sprengeri and setaceus (plumosus), members of the Lily family, producing a steady supply of plumy cut "fern" used with bouquets.

The pretty Selaginellas, ominously called "Sweat plants", are survivors of the extinct club mosses which existed as trees in the coal age. These are strictly warm terrarium plants and their delicate fronds greedily soak up moisture from the atmosphere, as without it they will shrivel and cannot survive. Only the "Resurrection plant", which coils into a ball when dry, will again unfold fresh-green if placed in water.

LIFE CYCLE OF A FERN, AND PROPAGATION—Ferns are not like flowering plants, they do not flower and don't bear seed. Instead, their life cycle alternates between two different generations, represented by two separate and unlike plants. The leafy fern plant as we know it represents the sexless or sporophyte phase of growth, producing seed-like spores which are usually borne on the back of fertile fronds. When spores are ejected and come to rest on moist soil, they will germinate into a flattened, shield-like prothallus which produces male and female organs on its lower surface. By way of a drop of dew or water these will unite, effecting fertilization. Thus the embryo of a new plant is formed, the fern-plant as we know it, with the fronds that ultimately bear the spore cases. See directions on the sowing of fern spores on page 51.

Many ferns can be propagated other than by spores. Ferns that form several crowns may be increased by division; others send out runners. Rhizomes can be cut apart and rooted. Platyceriums are propagated from adventitious suckers. On Motherferns, the little bulbils or plantlets carried on their fronds may be hooked down to a bed of humus to let the young brood form roots. To sow the spores, scatter them over most, sterilized peat, or humusy soil, and keep at an optimum temperature of 77°F (25°C), cover with glass, and shade. Keep evenly moist by setting dish into a saucer of water when necessary.

INSECT PROBLEMS—Where the temperature is high and atmosphere too dry, ferns are easily afflicted by mealy bugs and scale. Where practical, these pests can be dislodged by syringing the underside of the fronds with water. Otherwise, mealy bugs could be killed by touching them with a cotton-tipped toothpick dipped in alcohol. Scale may be scratched off by hand, or scrubbed with a soft brush and soapy water; rinse off afterwards. Severe infestations may be controlled with Malathion spray, using an atomizer, at 20% of recommended strength. Somewhat safer for sensitive ferns is a Rotenone-Pyrethrum insect spray combined with a dilute foliar oil emulsion. Hopelessly infested fronds should be cut off altogether.

Cibotium schiedei, the "Mexican treefern", originally from the mountains of Chiapas, a favorite decorator because of its graceful, fresh-green fronds, so dainty yet so durable.

10 |W|

Alsophila cooperi
long known in horticulture as "A. australis"
"Australian treefern"

Cibotium schiedei
"Mexican treefern"
in popular 8 inch tub

Dicksonia squarrosa
"Rough New Zealand treefern"

Cyathea arborea
the slender "West Indian treefern"

Dicksonia fibrosa
"Golden treefern"

Dicksonia antarctica
"Woolly treefern" or "Tasmanian treefern"

Sadleria cyatheoides
"Pigmy cyathea"

Diplazium esculentum
"Edible dwarf treefern"

Blechnum moorei
(Lomaria ciliata)

Blechnum brasiliense
"Ribfern"

Cibotium chamissoi
known as "menziesii" in Hawaii,
the "Man tree fern", with leathery fronds

Cibotium glaucum
in horticulture as "chamissoi",
the "Female" or "Blonde Hawaiian treefern"

Cyathea medullaris
"Black treefern" or "Sago fern"
in fern-rich New Zealand

Alsophila cooperi 'Robusta'
"A. australis" of horticulture
with a charming gardener, in Los Angeles

Asplenium nidus
"Birdsnest fern"

Polystichum tsus-simense
"Tsus-sima holly fern"

Rumohra adiantiformis
(Polystichum coriaceum in hort.)
"Leather fern"

Cyrtomium falcatum
"Fishtail fern"

Cyrtomium falc. 'Rochfordianum'
"Holly fern"

Polystichum aristatum variegatum
"East Indian holly-fern"

Pellaea viridis macrophylla
"Pteris adiantoides" in hort.

Pellaea rotundifolia
"Button fern"

Pellaea viridis (hastata hort.)
"Green cliff-brake"

Nephrolepis exaltata 'Norwoodii'
"Norwood lace fern"

Nephrolepis ex. bostoniensis
compacta
"Dwarf Boston fern"

Nephrolepis exaltata 'Whitmanii'
"Feather fern" or "Lace fern"

Nephrolepis exaltata
'Fluffy Ruffles'
"Dwarf feather fern"

Nephrolepis cordifolia 'Plumosa'
"Dwarf whitmanii" in hort.

Nephrolepis exaltata 'Verona'
"Verona lace fern"

Nephrolepis exaltata 'Hillii'
"Crisped featherfern"

Nephrolepis exaltata
'Rooseveltii plumosa'
"Tall feather-fern"

Nephrolepis exaltata
the ubiquitous "Swordfern"

Polypodium aureum glaucum
"Hare's-foot fern"

Polypodium aureum 'Mandaianum'
"Crisped blue fern"

Drynaria quercifolia
"Oakleaf-fern"

Polypodium vulgare
"Adder's fern"

Polypodium vulgare virginianum
"Wall fern"

Polypodium polypodioides
"Resurrection fern"

Nephrolepis biserrata
(ensifolia hort.)
"Bold sword fern"

Nephrolepis biserrata furcans
"Boston fishtail fern"

Nephrolepis cordifolia 'Duffii'
"Pigmy swordfern"

Davallia fejeensis plumosa
"Dainty rabbit's-foot"

Davallia fejeensis
"Fiji rabbit's-foot fern"

Davallia bullata mariesii
"Ball fern"

Davallia trichomanoides
"canariensis" of horticulture
"Carrot fern"

Humata tyermanii
"Bear's foot fern"

Davallia griffithiana
"Squirrel's foot"

Davallia solida (lucida)
"Rabbit's-foot fern"

Davallia pentaphylla
"Dwarf rabbit's-foot"

Adiantum raddianum
'Pacific Maid'

Adiantum tenerum 'Wrightii'
"Fan maidenhair"

Adiantum raddianum
"cuneatum" of horticulture
"Delta maidenhair"

Adiantum hispidulum
"Australian maidenhair"

Adiantum caudatum
"Walking fern"

Adiantum tenerum 'Farleyense'
"Magnificent maidenhair"

Adiantum pedatum
"Five-fingered maidenhair" or "American maidenhair"

Adiantum trapeziforme
"Giant maidenhair"

Pteris quadriaurita 'Flabellata'
"Leather table fern"

Pteris tremula
"Trembling brake fern"

Pteris cretica 'Rivertoniana'
"Lacy table fern"

Pteris quadriaurita 'Argyraea'
"Silver bracken"

Pteris ensiformis 'Victoriae'
"Silver table fern"

Pteris cretica 'Albo-lineata'
"Variegated table fern"

Pteris multifida
"Chinese brake"

Pteris cretica 'Wilsonii'
"Fan table fern"

Pteris cretica 'Wimsettii'
"Skeleton table fern"

Pteris dentata
"Sleepy fern"

Pteris vittata
in hort. as "longifolia"

Pteris semipinnata
"Angel-wing fern"

Platycerium andinum
"American staghorn"

Platycerium x lemoinei

Platycerium angolense
"Elephant's ear fern"

Platycerium quadridichotomum

Platycerium hillii
"Stiff staghorn"

Platycerium veitchii

Platycerium bifurcatum cv. 'Netherlands'
"Regina Wilhelmina staghorn"

Platycerium grande
"Regal elkhorn"

Platycerium bifurcatum
"Common staghorn fern"

Platycerium willinckii
"Silver staghorn fern"

Platycerium wilhelminae-reginae
"Queen elkhorn"

Platycerium coronarium
"Crown staghorn"

Platycerium vassei
"Antelope ears"

Platycerium bifurcatum majus
"Greater staghorn"

Platycerium "alcicorne" hort.
"Elkhorn fern"

Platycerium "diversifolium" hort.
"Erect elkhorn"

Platycerium ellisii

Platycerium wallichii

Platycerium "ridleyii" hort.

Platycerium sumbawense

Platycerium stemaria
"Triangle staghorn"

Platycerium madagascariense

Aglaomorpha coronans
"Crowning polypodium"

Aglaomorpha meyeniana
"Bear's-paw fern"

Drynaria quercifolia
"Oakleaf fern" in Fiji

Aglaomorpha heraclea
"Hercules polypodium"

Drynaria rigidula
"Feather polypodium"

Doryopteris pedata
"Hand fern"

74 T

Polypodium subauriculatum
"Jointed pine fern"

62 W

Polypodium subauriculatum 'Knightiae'
"Lacy pine fern"

66 T

Lygodium japonicum
"Climbing fern"   (fertile fronds)

74 T

Stenochlaena palustris
"Liane fern"

54 W

Gleichenia linearis
"Savannah fern"

72 T

Microlepia hirta
Microlepia "speluncae" in horticulture

72 T

Microlepia speluncae
"Grotto lace fern"

Tectaria cicutaria
"Button fern"

Diplazium proliferum
"Double-dot fern"

Woodwardia radicans
"Chain fern"

Asplenium bulbiferum
"Mother spleenwort" or *"Hen and chicken fern"*

Polystichum setiferum proliferum
"Viviparous filigree fern"

Onoclea sensibilis obtusilobata
"Sensitive fern"

Asplenium viviparum
"Mother fern"

Pityrogramma chrysophylla
"Gold fern"

Pityrogramma caiomelanos
"Silver fern"

Hypolepis punctata
"Poor man's cibotium"

Pityrogramma triangularis
"California gold fern"

Polystichum acrostichoides
"Christmas fern"

Athyrium filix-foemina
"Lady fern"

Polystichum setosum
"Bristle fern"

Pyrrosia lingua
"Tongue fern"

Phyllitis scolopendrium 'Crispum'
"Crisped hart's-tongue"

Phyllitis scolopendrium
"Hart's-tongue fern"

Polypodium phyllitidis
"Strap fern"

Polypodium polycarpon
'Grandiceps'
"Fishtail"

Polypodium polycarpon
(P. irioides or punctatum in hort.)

Polypodium musifolium
"Bananaleaf fern"

Polypodium angustifolium
"Ribbon fern"

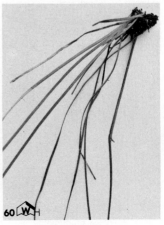

Vittaria lineata
"Shoestring fern"

Aneimia phyllitidis
"Flowering fern"

Osmunda cinnamomea
"Cinnamon fern"

Angiopteris evecta
"Turnip fern"

Marsilea fimbriata
"Water clover"

Acrostichum aureum
"Swamp fern", or "Leather fern"

Marsilea drummondii
"Hairy water clover"

Azolla caroliniana
"Mosquito plant" or "Floating moss"

Ceratopteris thalictroides
"Water fern"

Asparagus falcatus
"Sickle-thorn"

Asparagus densiflorus 'Sprengeri'
"Emerald feather" or
"Sprengeri fern"

Asparagus setaceus
(plumosus in hort.)
"Asparagus fern"

Asparagus asparagoides myrtifolius
"Medeola" or *"Smilax"*

Asparagus densiflorus myriocladus
"Zigzag-asparagus"

Salvinia auriculata
"Floating fern"

Bugula species, a moss-animal
known as "Air-Fern"

Asparagus densiflorus cv. 'Meyers'
"Plume asparagus"

Lycopodium phlegmaria
"Queensland tassel fern"

Selaginella kraussiana
(denticulata)
"Club moss" or
"Trailing Irish moss"

Selaginella kraussiana brownii
"Cushion moss"

Selaginella emmeliana
"Sweat plant"

Selaginella willdenovii
climbing "Peacock fern"

Selaginella caulescens
"Stalked selaginella"

Selaginella martensii divaricata
"Zigzag selaginella"

Selaginella uncinata
"Rainbow fern" or *"Blue selaginella"*

Selaginella martensii 'Watsoniana'
"Variegated selaginella"

Selaginella lepidophylla
"Resurrection plant"
"Rose of Jericho"

Entering a room where a vine softly frames a window with green leaf tracery, you feel an effect of gracious living and the pleasant kinship of the room with its outdoor view. Vines are exuberant, and both artistic as well as functional in container gardening. Their relaxed, flowing lines soften harsh geometric patterns of bare areas, they are mobile and full of vitality.

Vines are available in endless variety of character, size, color and form, and most can be trained to any shape, line, or curve desired. For dangling from the edge of a decorative pot, or climbing a piece of driftwood at the back, there are dainties like the "Creeping fig", the succulent "Lover's chain", or the variegated Hoya; and more luxuriously Scindapsus or a Philodendron oxycardium. To create a big, bold masculine effect a great number of astonishing Philodendron and Monstera would answer. For a tidy little pot spilling over with color there are a number of good Tradescantias and Zebrinas, the "Wandering Jew". Many of the smaller vines can be hung from wall brackets, or planted into hanging baskets. In a cooler window, ivies are ideal; its many varieties, especially the cute miniature types, would make collecting them an interesting hobby. If a window sill is warmer and with shade, Peperomias and Pileas don't ask much space; mainly spreading, or lightly scandent, they are easily kept in bounds.

Trailers for every Purpose

As the assemblage of plants in this chapter plainly illustrates, trailers can be found in almost all plant families, in great variety of habit, offering unlimited opportunities for decorating and individual taste. Just what can be utilized best depends on the location intented or available. Those scandent shrubs and woody clamberers that are known for their beautiful flowers need strong light and ample soil to bloom at all, and a greenhouse in the winter, or a patio in summer would be a necessity. This includes such favorites as Allamandas and the Passion vines, although Bougainvilleas, some Jasmines, Clerodendrums and Dipladenias might be coaxed into bloom on a bright window sill.

Hanging pots or baskets are popular for many trailers but can be used only where dripping after watering cannot harm the floor. There are self-contained hanging pots with built-in reservoirs or attached saucers, less apt to drip, or decorative jardinieres on chains to place a pot inside. Hanging baskets made of wire, redwood slats, or of plastic, are laid out with burlap, moss, or with polyethylene to hold the soil; watering is best done by submerging, for thorough saturation. Ingenuity can bring to interesting use such objects as wooden buckets, tea kettles, strawberry jars, lanterns or gourds. When plants are well established, a little plantfood is appreciated especially by those that are meant to bloom. Beautiful flowered hanging plants are the Columneas, though somewhat delicate, requiring lighter soil, warmth and shade.

True vines such as Stephanotis and Hoya do well trained onto trellis, or something slim to curl themselves around. Hoyas when old enough will bloom again and again from the old spur which must never be cut off. These "Wax vines" actually resent pampering, and do best if left undisturbed once well established. Tendril climbers such as the Cissus group, the Passifloras and Gloriosas will grasp at anything for climbing up. Cissus rhombifolia and antarctica with their glossy foliage are very satisfactory for room dividers or individual pots, and their branches become gracefully pendant if allowed to do so. Some plants spread by runners such as the Chlorophytum, Episcia, Saxifraga, and Nephrolepis fern. Climbers by aerial roots include the popular Philodendron, and they do best on a slab of bark, milled tree-fern pillar, or on a mossed "totem" pole. Ivies also creep or climb by aerial rootlets.

Vines and trailers are actually weeping plants with stems too weak to support themselves. They try to grow but do not have sufficient strength. Their lengthened stems reach out for any possible support that they can reach. They attach themselves to trees or walls by twining, clambering, clinging by aerial roots, by tendrils or by suction cups, and even fishhooks in certain climbing palms, competing for their very life with other stronger, and self-sufficient plants. But in doing so they have aroused in us a very special sympathy, beloved by all, as they catch our imagination because of their unlimited possibilities.

A practical way to grow flowering vines in the home. Ipomoea tricolor, the lovely "Heavenly-blue morning glory", is a tropical American perennial which can be grown as an annual. Climbing on wire support from an ample box of soil alongside a bright window, its cheerful flowers are red when in bud, bright blue with white tube when open, and fade to purplish red.

52

Macropiper excelsum
"Lofty pepper"

Cissus discolor
"Rex begonia vine"

Piper crocatum
"Peruvian ornamental pepper"

Piper porphyrophyllum
"Velvet cissus"

Piper magnificum
"Lacquered pepper tree"

Piper ornatum
"Celebes pepper"

Piper nigrum
"Black pepper"

Piper betle
"Betel-leaf pepper"

Piper futokadsura
"Japanese pepper"

Cissus rhombifolia
in hort. as "Vitis" or "Rhoicissus"
"Grape-ivy"

Cissus rhombifolia 'Mandaiana'
"Bold grape-ivy"

Cissus antarctica
"Kangaroo vine"

Ampelopsis brevipedunculata
elegans
"Colored grape-leaf"

Rhoicissus capensis
"Evergreen grape"

Cissus rotundifolia
"Arabian wax cissus"

Cissus adenopoda
"Pink cissus"

Cissus striata
"Miniature grape-ivy"

Tetrastigma voinierianum
"Lizard plant"
"Chestnut vine"

Cissus sicyoides
"Princess vine"

Parthenocissus henryana

Cissus gongylodes
in Frankfurt, Germany

Cissus erosa

Cissus guadrangularis
"Veld grape"

Cissus albo-nitens
"Silver princess vine"

Cucumis anguria
"West Indian gherkin"

Piper sylvaticum
"Silver cissus"

Parthenocissus
quinquefolia vitacea
"Virginia creeper"

Syngonium podophyllum
"Arrowhead plant"

Scindapsus aureus 'Wilcoxii'
"Golden pothos"

Rhaphidophora celatocaulis
"Shingle vine"

Syngonium erythrophyllum
"Copper syngonium"

Philodendron oxycardium
"cordatum" of florists
"Heart leaf philodendron"

Pothos jambea
"True pothos"

Ficus radicans 'Variegata'
"Variegated rooting fig"

Ficus pumila
"Creeping fig"

Ficus tikoua
"Waupahu fig"

Marcgravia rectiflora
"Shingle plant"

Ficus radicans
"Rooting fig"

Ficus quercifolia
"Oakleaf ficus"

Callisia elegans
"Striped inch plant"

Cyanotis somaliensis
"Pussy-ears"

Cyanotis kewensis
"Teddy-bear vine"

Setcreasea pallida

Tradescantia navicularis
"Chain plant"

Setcreasea purpurea
"Purple heart"

Commelina benghalensis variegata
"Indian dayflower"

Tradescantia velutina
"Velvet tradescantia"

Commelina communis aureo-striata
"Widow's tears"

Dichorisandra thyrsiflora
"Blue ginger"

Callisia fragrans 'Melnikoff'
(Spironema in hort.)

Gibasis geniculata
(Tradescantia multiflora in hort.)
"Tahitian bridal veil"

Zebrina pendula
"Silvery wandering Jew"

Tradescantia fluminensis
'Variegata'
"Variegated wandering Jew"

Tradescantia albiflora
'Albo-vittata'
"Giant white inch plant"

Zebrina purpusii
"Bronze wandering Jew"

Zebrina pendula 'Discolor'
"Tricolor wandering Jew"

Zebrina pendula
'Quadricolor'
"Gay wandering Jew"

Tradescantia albiflora
'Laekenensis'
"Rainbow inch plant"

Tradescantia fluminensis
"Wandering Jew"

Zebrina pendula
'Discolor multicolor'

Tradescantia sillamontana
"White velvet"
"White gossamer"

Tradescantia blossfeldiana
"Flowering inch plant"

Tradescantia blossfeldiana
'Variegata'

Siderasis fuscata
"Brown spiderwort"

Callisia fragrans
(Tradescantia dracaenoides)

Hadrodemas warszewicziana
(Tripogandra or Spironema)

Dichorisandra thyrsiflora
variegata

Campelia zanonia 'Mexican Flag'
(Dichorisandra albo-lineata)

Dichorisandra reginae
"Queen's spiderwort"

Palisota barteri

Geogenanthus undatus
"Seersucker plant"

Palisota elizabethae

Geogenanthus ciliatus
(Dichorisandra 'Blackie' in hort.)

Rhoeo spathacea (discolor)
"Moses-in-the-cradle"

Rhoeo spathacea 'Vittata'
"Variegated boat lily"

Plectranthus purpuratus
"Moth king"

Plectranthus oertendahlii
"Prostrate coleus"

Plectranthus australis
"Swedish ivy"

Pellionia daveauana
"Trailing watermelon begonia"

Plectranthus coleoides
'Marginatus'
"Candle plant"

Pellionia pulchra
"Satin pellionia"

Senecio mikanioides
"German ivy"

Fittonia verschaffeltii
argyroneura
"Nerve plant"

Manettia inflata (bicolor)
"Firecracker plant"

Lotus berthelotii
"Winged pea"

Hemigraphis colorata
"Red ivy"

Saxifraga sarmentosa
"Strawberry geranium"

Wedelia trilobata

Mikania ternata
"Plush vine"

Senecio confusus
"Orange-glow vine"

Campanula fragilis
"Star of Bethlehem"

Campanula isophylla mayii
"Italian bellflower"

Schizocentron elegans
"Spanish shawl"

Campanula elatines alba plena
"Double white Bethlehem star"

Browallia speciosa major
"Sapphire flower"

Browallia speciosa alba
"White bush-violet"

Lamium galeobdolon variegatum
"Silver nettle vine" or *"Archangel"*

Coleus rehneltianus
'Trailing Queen'
"Trailing coleus"

Vinca major variegata
"Band plant"

Impatiens repens
"Creeping impatiens"

Vinca minor
"Periwinkle" or *"Virgin flower"*

Sibthorpia europaea 'Variegata'
"Cornish moneywort"

Helxine soleirolii
"Baby's tears"

Dichondra repens
"Lawn leaf"

Mentha requienii
"Corsican mint"

Glecoma hederacea variegata
"Variegated ground-ivy"

Cymbalaria muralis
"Kenilworth ivy"

Ardisiandra sibthorpoides

Bignonia argyraea
"Silvery trumpet vine"

Smilax ornata
"Sarsaparilla"

Serjania glabrata

Dioscorea multicolor argyraea
"Silverleaf-yam"

Rubus reflexus
"Trailing velvet plant"

Dioscorea discolor
"Ornamental yam"

Ipomoea batatas 'Blackie'
"Black-leaf sweet-potato"

Gynura bicolor
"Oakleaved velvet plant"

Gynura 'Sarmentosa' hort.
"Purple passion vine"

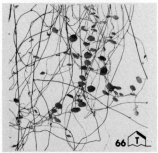

Muehlenbeckia complexa
"Maidenhair vine"

Echites rubrovenosa

Aristolochia leuconeura
"Ornamental birthwort"

Peperomia clusiaefolia
"Red-edged peperomia"

Peperomia caperata
"Emerald ripple"

Peperomia maculosa
"Radiator plant"

Peperomia marmorata
"Silver heart"

Peperomia sandersii
"Watermelon peperomia"

Peperomia verschaffeltii
"Sweetheart peperomia"

Peperomia glabella 'Variegata'
"Variegated wax privet"

Peperomia obtusifolia 'Variegata'
"Variegated peperomia"

Peperomia scandens 'Variegata'
"Variegated philodendron leaf"

Peperomia puteolata
"Parallel peperomia"

Peperomia polybotrya
"Coin leaf peperomia"

Peperomia ornata

Peperomia griseo-argentea
"Ivy peperomia"

Peperomia obtusifolia
"Baby rubber plant"

Peperomia magnoliaefolia
"Desert privet"

Peperomia rubella
"Yerba Linda"

Peperomia crassifolia
"Leather peperomia"

Peperomia rotundifolia
(syn. nummularifolia)

Peperomia tithymaloides
"Pepper-face"

Peperomia resedaeflora
"Mignonette peperomia"

Peperomia fosteri
"Vining peperomia"

Peperomia velutina
"Velvet peperomia"

Peperomia caperata 'Tricolor'
"Tricolor ripple"

Peperomia bicolor
"Silvery velvet peperomia"

Peperomia incana
"Felted pepperface"

Peperomia blanda

Peperomia pulchella
"Whorled peperomia"

Peperomia pereskiaefolia
"Leaf-cactus peperomia"

Peperomia dolabriformis
"Prayer peperomia"

Peperomia orba
"Princess Astrid peperomia"

Peperomia scandens
"Philodendron peperomia"

Peperomia glabella
"Wax privet peperomia"

Peperomia acuminata

Pilea cadierei 'Minima'
"Miniature aluminum plant"

Pilea involucrata
"Friendship plant"

Pilea repens
"Black leaf Panamiga"

Pilea nummulariifolia
"Creeping Charlie"

Pilea depressa
"Miniature peperomia"

Pilea sp. 'Black Magic'

Pilea sp. 'Silver Tree'
"Silver and bronze"

Pilea microphylla (muscosa)
"Artillery plant"

Pilea pubescens 'Argentea'
"Silver Panamiga"

Dischidia imbricata
"Thruppence urn plant"

Senecio macroglossus variegatus
"Variegated jade vine"

Senecio jacobsenii (petraeus)
"Weeping notonia"

Sedum x orpetii
"Giant burro tail", *"Lamb's tail"*

Lampranthus emarginatus
an *"Ice plant"*

Sedum morganianum
"Burro tail"

Crassula sarmentosa

Plectranthus tomentosus
"Succulent coleus"

Senecio tropaeolifolius
"Succulent nasturtium"

Hoya cinnamonifolia
"Cinnamon wax vine"

Hoya keysii

Hoya imperialis
"Honey plant"

Ceropegia haygarthii
"Wine-glass vine"

Ceropegia caffrorum
"Lamp flower"

Ceropegia radicans
"Umbrella flower"

Ceropegia sandersonii
"Parachute plant"

Ceropegia woodii
"String of hearts"

Ceropegia debilis
"Needle vine"

Hoya carnosa variegata
"Variegated wax plant"

Hoya carnosa
"Wax plant"

Hoya bella
"Miniature wax plant"

Hoya motoskei
"Spotted hoya"

Hoya australis
"Porcelain flower"

Hoya purpureo-fusca
"Silver-pink vine"

Hoya sp. "minima"
(Dischidia minima hort.)

Hoya lacunosa

Hoya kerrii
"Sweetheart hoya"

Gloriosa verschuuri
a "Climbing lily"

Gloriosa rothschildiana
"Glory-lily"

Gloriosa carsonii

Gloriosa superba
"Crisped glory-lily"

Gloriosa simplex
"Dwarf glory-lily"

Pelargonium peltatum
'L 'Elegante'
"Sunset ivy-geranium"

Episcia lilacina 'Lilacina'
"Blue flowered teddy-bear"

Vanilla fragrans 'Marginata'
"Variegated vanilla"

Begonia limmingheiana
"Shrimp begonia"

Phaseolus caracalla
"Snail flower"

Dioscorea bulbifera
"Air-potato" or *"True yam"*

Tolmiea menziesii
"Piggy-back plant"

Bowiea volubilis
"Climbing onion"

Freycinetia multiflora
"Climbing pandanus"

Chlorophytum comosum 'Vittatum'
"Spider plant"

Chlorophytum comosum
'Variegatum'
"Green-lily","Large spider plant"

Aristolochia elegans
"Calico flower"

Wisteria floribunda
"Japanese wisteria"

Bougainvillea glabra
"Paper flower"

Bougainvillea x buttiana
"Crimson Lake"

Campsis radicans (Tecoma)
"Trumpet bush"

Tecomaria capensis
"Cape honey-suckle"

Stigmaphyllon ciliatum
"Brazilian gold vine"

Russelia equisetiformis
"Fountain bush"

Albizzia julibrissin
"Silk tree" or "Pink mimosa"

Calliandra haematocephala
(inaequilatera)
"Red powderpuff"

Dipladenia boliviensis
"White dipladenia"

Dipladenia x amoena
"Pleasing dipladenia"

Dipladenia sanderi
"Rose dipladenia"

Hibbertia volubilis
"Guinea goldvine"

Petrea volubilis
"Purple wreath"

Lapageria rosea
"Chilean bell-flower"

Beaumontia grandiflora
"Herald's trumpet"

Allamanda cathartica hendersonii
"Golden trumpet"

Solandra longiflora
"Chalice vine"

Antigonon leptopus
"Coral vine"

Clerodendrum ugandense
"Blue glory-bower"

Swainsona galegifolia
"Winter sweet pea"

Bougainvillea glabra 'Sanderiana variegata'
"Variegated flowering bougainvillea"

Streptosolen jamesonii
"Marmalade bush" or *"Firebush"*

Clerodendrum thomsonae
"Bleeding-heart vine"

Mahernia verticillata
"Honey bells"

Kennedia rubicunda
"Coral-pea"

Jasminum simplicifolium
"Little star jasmine"

Jasminum rex
"King jasmine"

Jasminum nitidum
"Angelwing jasmine"

Jasminum mesnyi
"Primrose jasmine"

Jasminum polyanthum
"Pink jasmine"

Jasminum grandiflorum
"Spanish jasmine"

Trachelospermum jasminoides
"Confederate jasmine"

Stephanotis floribunda
"Madagascar jasmine"

Araujia sericofera
"Bladder flower"

Clitorea ternatea
"Butterfly pea"

Tropaeolum majus florepleno
"Double English nasturtium"

Lathyrus odoratus
"Sweet pea"

Mazus reptans
"Wart flower"

Asarina erubescens
"Creeping gloxinia"

Chaenostoma grandiflorum
"Purple glory-plant"

Tropaeolum tricolor
"Tricolored Indian cress"

Clematis lawsoniana 'Ramona'
"Virgin's bower"

Viola hederacea
"Trailing violet"

Ipomoea nil
"Imperial morning-glory"

Thunbergia alata
"Black-eyed Susan"

Calonyction aculeatum
"Moon flower"

Passiflora racemosa
"Red passion flower"

Passiflora x alato-caerulea
"Showy passion flower"

Passiflora caerulea
"Passion flower"

Passiflora coriacea
"Bat-leaf"

Passiflora trifasciata
"Three-banded passion vine"

Passiflora maculifolia
"Blotched-leaf passion vine"

Passiflora coccinea
"Scarlet passion flower"

Passiflora incarnata
"Wild passion flower"

Passiflora quadrangularis
"Granadilla"

Hedera helix 'Abundance'
large leaved

Hedera helix poetica
in adult, fruiting form

Hedera helix 'Albany'
"Albany ivy"

Hedera helix 'Fluffy Ruffles'
crested and self-branching

Hedera helix 'Fan'
"Fan ivy"

Hedera helix 'Star'
"Star ivy"

x Fatshedera lizei
"Miracle plant" in flower

Hedera helix
'Manda's Needlepoint'
"Needle ivy"

Hedera helix 'Arrow'
"Arrow ivy"

Hedera helix
'California Needlepoint'
"California parsley"

Hedera helix 'Pixie'
"Pixie miniature"

Hedera helix
"English ivy"

x Fatshedera lizei
"Ivy tree" or *"Botanical wonder"*

Hedera helix 'Pittsburgh'
"Pittsburgh ivy"

Hedera helix 'Patricia'
a dense, selfbranching ivy

Hedera canariensis 'Variegata'
"Variegated Algerian ivy"

Hedera helix 'Harald'
"White and green ivy"
"Improved Chicago variegata"

Hedera colchica
'Dentato-variegata'
"Variegated Persian ivy"

Hedera can. 'Gloire de Marengo'
in Boskoop, Holland

H. h. 'Green Ripples', 'Sylvanian'   'Maple Queen'

H. h. 'Shamrock',   'Ivalace',   'Green Spear'

H. h. 'Minima',   'Smithii',   'Denticulata'

H. h. poetica,   'Pin Oak',   'Pedata'
"Birdsfoot ivy"

H. h. 'Erecta',   'Conglomerata'   'Meagheri'
"Japanese ivy"   *"Green feather"*

H. h. 'Glymii',   'Scutifolia',   'Sagittaefolia'
"Heart-leaf ivy"   *"Taurian ivy"*

H. helix hibernica,   H. helix '238th Street'
"Irish ivy"   *"Bronx ivy"*

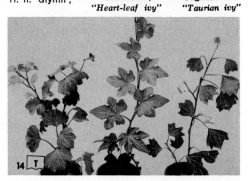

H. h. 'Weber's fan', 'Weber's Californian', 'Digitata'

H. h. 'Walthamensis',   baltica,   Hedera helix
    "Baby ivy"   *"Baltic ivy"*   *"English ivy"*

H. h. 'Hahn's   'Merion Beauty',   'Pittsburgh'
Selfbranching',   *"Pittsburgh ivy"*

H. h. 'Discolor'   'Marginata',   "Silver Emblem'
    "Marmorata"   *"Silver garland"*

H. canariensis   H. canariensis   H. helix
'Margino-maculata'   'Variegata'   'Maculata'

H. h. 'Hahn's Variegated',   'Glacier'   'Golddust'

H. h. 'Silver King',   'Jubilee',   'Williamsiana'

H. h. 'Manda's crested',   'Parsley',   'Curlilocks'

H. h. 'Helvetica'   'Goldheart'   'Chicago'

Selaginella willdenovii
"Peacock fern"

Nephrolepis exaltata bostoniensis
"Boston fern" on wall

Asparagus asparagoides
myrtifolius
"Baby smilax"

Asparagus densiflorus 'Sprengeri'
"Emerald feather" in basket

Lygodium japonicum
(sterile fronds)
"Climbing fern"

Polypodium subauriculatum
'Knightiae'
"Lacy pine fern"

Polypodium aureum 'Mandaianum'
"Blue fern" with crisped fronds

Davallia fejeensis
"Fiji rabbit's-foot fern"

BROADLEAVED EVERGREENS
mainly for cooler locations

Interior decorators have long recognized that modern buildings with their severe and simple lines must introduce some softening effect if architectural design is not to appear stark, cold and uninhabited. This is why we find murals or paintings on bare walls, fountains or futuristic art in entrance halls; period furniture and statuary in corridors. Here is where specimen foliage plants can add warmth and a breath of nature alive, thereby enriching the empty vastness of such edifices, making employees and visitors alike feel welcome and more comfortable.

Along the limits of the subtropic zone in our northern hemisphere, specifically Japan, South China, the slopes and valleys of the Himalayas, and around the Mediterranean, and south of the Tropic of Capricorn, in New Zealand, Australia and Southern Africa, there are regions of cool nights and warm days, safe from devastating freezes. Here live sheltered plants remaining evergreen and yet not tropical, shrubs and trees with leathery leaves which are ideally groomed for decorative purpose. But even under the equator, where elevations are high enough to moderate the climate, such evergreens exist, and that is where we find plants of similar nature in tropical Central and South America, on mountains in Central Africa and even in Malaysia.

Brassaia actinophylla, known in horticulture as "Schefflera" or the "Queensland Umbrella tree", in the corner of a living room. This is a warmth - loving species, whereas true Schefflera come from cooler regions.

Broadleafed evergreens of this type are usually rather plain and because they were not spectacular, were often overlooked by early plant collectors. But by their very nature they are built to last, they are accustomed to the change in temperature and tolerate neglect and much abuse. For this same reason they will be at their best where temperatures are lowered to 60° or even 50°F at night. As floor plants in lobbies with good light or other exposed places, to decorate the outside of a building, or high up on a roofgarden, they are both attractive and dependable. The most formal of all decorator trees is the noble "Baytree", but unfortunately difficult to obtain in our country. A fair substitute for cooler levels is the "Wax privet", Ligustrum lucidum. Ficus retusa nitida is more pliable and superior, but this "Indian laurel" while tolerant, prefers to be a little warmer (see under Fig trees, chapter 8). There are a number of decorative and good araliads which are recommended primarily for cooler locations. Fatsia japonica with its glossy, hand-like leaves, long holds its shape and rates as one of the best. The Panax group including Pseudopanax, Nothopanax, Oreopanax; Meryta, Tupidanthus and the "Snowflake plant", Trevesia, with its peculiar duck's-foot leaves all offer a wide variety of leaf patterns together with good keeping quality. Fatshedera are nice in boxes if kept short, but have a tendency to become scandant especially if warm. An outstanding beauty in this family and great favorite with decorators is the graceful "False aralia" from the South Pacifc, Dizygotheca elegantissima; most exotic-looking with its hands of ribbon-like scalloped leaflets in coppery-brown. The one araliad requiring warmth is the lovely "Queensland umbrella tree", Brassaia actinophylla, better known in the trade as "Schefflera". Symmetrical and of compact habit is the dense, glossy Pittosporum, and the Japanese boxwood, usually grown as small pyramid or formal globe. Euonymus are old-time decorators too but easily catch mildew. The "Chinese horned holly", Ilex cornuta assists the holiday spirit especially if decorated with artificial red berries. The golden-spotted Aucubas are always good for underplanting. All these plants would greatly benefit and remain sturdier if they could be placed outdoors in summer, and care must be taken that they always receive an adequate supply of water.

Tubbed oleanders are good roof garden subjects, flowering freely in spring and summer. Eugenias are elegant in pyramid form and much in favor on warm patios, but are apt to drop their little leaves if once neglected. Prunus caroliniana is similar and may prove perhaps a little sturdier.

The South Seas have given us several pretty Polyscias, often called "Aralias", and which are good window subjects as small plants with their lacy or variegated foliage. Clustering older specimen of Polyscias fruticosa, with their flexible, curiously crooked stems bearing bouquets of feathery foliage, look so much like a tropical replica of the Chinese ming tree, that decorators have enthusiastically come to use them with great effect as the "Ming aralia".

Polyscias fruticosa is known as the "Ming aralia" because of its twisted stems and lacy foliage. This specimen lends unusual decor on the 17th floor of·a Wall Street bank in downtown New York.

Casuarina equisetifolia,
the "Horsetail tree" or "Australian pine"
sheared and trained as giant Bonsai,
in Miami, Florida

Eugenia myrtifolia,
the "Australian brush-cherry",
tall sheared columns in Mediterranean terracotta
jars; Huntington Library, San Marino, California

Laurus nobilis, the
"Grecian Laurel" or "Bay tree",
formal shaped specimen in front of the Orangerie,
at Longwood Gardens, Kennett Square, Pennsylvania.

Prunus caroliniana cv. 'Bright'n Tight',
a compact "American cherry-laurel",
in pyramid form, in the Spanish patio
of Monrovia Nurseries, Azusa, California

Polyscias fruticosa
the "Ming aralia"
sometimes known as P. filicifolia,
because of its lacy, fern-like foliage.

Fatsia japonica,
the "Japanese aralia"
in horticulture as Aralia sieboldii;
excellent for cooler locations.

Trevesia palmata sanderi,
the "Vietnam snowflake plant";
its curious duck-foot leaves are covered
with silvery dots.

Brassaia actinophylla, the decorative
"Queensland umbrella tree"
and "Australian schefflera"
of florists, is an effective decorator plant
needing a warm and light location,
and should be kept fairly dry.

Oreopanax peltatus
"Mountain aralia"

Tetraplasandra meiandra
"Hawaiian Ohe tree"

Cussonia spicata
"Spiked cabbage tree"

Oreopanax xalapensis

Meryta sinclairi
"Puka tree"

Dendropanax chevalieri
"Tree aralia"

Schefflera digitata
"Seven fingers"

Tupidanthus calyptratus
"Mallet flower"

Schefflera venulosa
(Heptapleurum in hort.)

Polyscias filicifolia
"Fernleaf aralia"

Polyscias guilfoylei victoriae
"Lace aralia"

Polyscias fruticosa
"Ming aralia"

Polyscias balfouriana 'Pennockii'
"White aralia"

Dizygotheca elegantissima
"Spider aralia", "Splitleaf maple"

Polyscias balfouriana marginata
"Variegated Balfour aralia"

Pseudopanax lessonii
"False panax"

Tetrapanax papyriferus
"Rice paper plant"

Polyscias balfouriana
"Dinner plate aralia"

**Polyscias guilfoylei
'Quinquefolia'**
"Celery-leaved panax"

Polyscias fruticasa 'Elegans'

Polyscias paniculata 'Variegata'
"Variegated rose-leaf panax"

Feijoa sellowiana
"Pineapple guava"

Eugenia myrtifolia
"Australian brush cherry"

Melaleuca leucadendron
"Paperbark tree"

Prunus laurocerasus
"English laurel"

Euonymus japonicus
"Japanese spindle tree"

Eriobotrya japonica
"Chinese loquat"

Aucuba japonica serratifolia
"Sawtooth laurel"

Buxus microphylla japonica
"California boxwood"

Myrtus communis boetica
"Citrus myrtle"

Ligustrum lucidum
"Glossy privet" or "White-wax tree"

Pittosporum tobira
"Mock-orange" or "Australian laurel"

Ligustrum lucidum 'Texanum'
"Wax-leaf privet" in pyramid form

Callistemon lanceolatus
in hort. as Metrosideros floribunda,
the "Bottle-brush"

Cocculus laurifolius
"Platter-leaf"

x Fatshedera lizei
"Ivy tree" or "Botanical wonder"

Olea europaea
"Olive tree"

Ligustrum japonicum
"Japanese privet"

Viburnum odoratissimum
"Sweet viburnum"

Pittosporum viridiflorum
"Cape pittosporum"

Viburnum suspensum
"Sandankwa viburnum"

Coprosma baueri
"Mirror plant"

Ulmus parvifolia
"Evergreen elm"

Heimerliodendron brunonianum
'Variegatum' (Pisonia)
"Bird-catcher-tree"

Hebe speciosa 'Variegata'
"Variegated evergreen veronica"

Aucuba japonica 'Picturata'
"Golden laurel"

Coprosma baueri marginata
"Variegated mirror plant"

Aucuba japonica variegata
"Golddust plant"

Pieris japonica variegata
"Lily-of-the-valley bush"
in hort. as "Andromeda"

Pittosporum tobira 'Variegatum'
"Variegated mock-orange"

Osmanthus heterophyllus
'Variegatus'
"Variegated false holly"

Elaeagnus pungens 'Maculata'
"Variegated Russian olive"

Nerium oleander 'Album'
"White oleander"

Nerium oleander
"Oleander" or "Rose bay"

Nerium oleander
'Carneum florepleno'
"Mrs. Roeding oleander"
double oleander

Elaeagnus pungens
"Silver berry"

Osmanthus fragrans
"Sweet olive" or "Fragrant olive"

Daphne odora 'Marginata'
"Variegated winter daphne"

Griselinia lucida
"Shining broadleaf" or "Kapook"

Dovyalis hebecarpa cv. 'Kandy'
"Ceylon gooseberry"

Ilex cornuta 'Burfordii'
"Burford holly"

Ilex aquifolium albo-marginata
"Variegated English holly"

Ilex cornuta
"Chinese horned holly"

Ilex vomitaria
"Yaupon holly"

Ilex opaca
"American holly"

Ilex cornuta 'Rotunda'
"Dwarf Chinese holly"

Osmanthus heterophyllus
"Holly osmanthus"

Mahonia lomariifolia
"Chinse holly grape"

Mahonia beallii
"Leatherleaf mahonia"

Cunonia capensis
"African red alder"

Myoporum laetum
"Mouse-hole tree" or "Ngaio"

Stenocarpus sinuatus
"Fire-wheel tree"

Leucadendron argenteum
"Silver-tree"

Eugenia myrtifolia globulus
"Myrtle-leaf eugenia"

Corokia cotoneaster
"Golden Korokio"

Sarcococca ruscifolia
"Sweet box"

Semele androgyna
"Climbing butcher's broom"

Melianthus major
"Honey-bush"

Homalocladium platycladum
"Tapeworm plant" or "Ribbon bush"

Acanthus mollis
"Greek akanthos" or "Grecian pattern plant"

Euonymus japonicus medio-pictus
"Goldspot euonymus"

Euonymus jap. aureo-variegatus
"Yellow Queen"

Euonymus jap. argenteo-variegatus
"Silver Queen"

Euonymus fortunei
radicans gracilis
"Creeping euonymus"

Euonymus jap.
microphyllus variegatus
"Variegated box-leaf"

Euonymus jap. albo-marginatus
"Silver leaf euonymus"

Myrtus communis variegata
"Variegated myrtle"

Eugenia myriophylla
"Needle eugenia"

Hosta sieboldiana 'Variegata'
"Seersucker plantain-lily"

Pachysandra terminalis variegata
"Japanese spurge"

Hedera canariensis arb. 'Variegata'
"Ghost tree ivy"

Ligustrum sinense 'Variegatum'
"Chinese silver privet"

Leucothoe catesbaei
"Sweet bells"

Myrtus communis microphylla
"German myrtle"

Ardisia japonica
"Marlberry"

Galax aphylla
"Beetle-weed" or *"Florists galax"*

Asarum shuttleworthii
"Wild ginger"

Ruscus hypoglossum
"Mouse-thorn"

Ruscus aculeatus
"Butchers broom"

Vaccinium ovatum
florists *"Huckleberry greens"*

Gaultheria shallon
"Salal" or *"Florists greens"*

Cuphea hyssopifolia
"False heather"

Malpighia coccigera
"Miniature holly"

Myrsine africana
"African boxwood"

Eucalyptus globulus
"Blue gum" planted 1941
in southern California, 1966

Eucalyptus cinerea
"Spiral silver-dollar" of florists

Eucalyptus polyanthemos (mature)
"Silver-dollar tree"
or "Red box gum"

Grevillea robusta
"Silk oak"

Sparmannia africana
"Indoor-linden"
in Oberammergau, Bavaria

Jacaranda acutifolia
"Mimosa-leaved ebony"

Schinus terebinthifolius
"Brazilian pepper-tree"

Ricinus communis coccineus
"Red castor-bean"

Eugenia myrtifolia
shaped in topiary form,
in Los Angeles

Bambusa multiplex
"Oriental hedge bamboo"
a clump-forming type

Pseudosasa japonica
"Hardy Metake" or "Female arrow bamboo"
slowly running and fairly hardy

Phyllostachys aurea
the attractive "Golden bamboo" or "Fish pole bamboo"
a running type, medium hardy

Phyllostachys bambusoides
"Japanese timber-bamboo"
a "Running bamboo" with green to yellow culms

BAMBOO and GRASS-like plants

The renaissance of oriental landscaping has brought bamboos into prominence as a plant group distinctive in character from any other. Bamboo, wonderfully compatible with the glass walls of modern homes and office buildings, softens the glare of the sun with a lilting play of flickering leaves. Their slender, arching stems and feathery, handsome foliage form patterns of great beauty. There is no other ornamental plant that will grow so tall in such a small area.

Bamboos are long-lived woody grasses ranging from dwarfs of several inches to giants 100 feet tall, native to every continent except Europe. They fall into two main groups, the clump forms and the running types. The clump forms are tropical and subtropical and derive their name from the fact that each new rhizome promptly turns upward, growing into a culm and resulting in a dense clump. In the running bamboo the underground rhizome continues its horizontal development, and culms come up at intervals from lateral buds. If such species are confined into a tub however they will form attractive clusters also. Bamboos are the fastest growing plant known; their shoots develop very rapidly, and growths of 18 in. per day have been recorded. The hollow canes are widely utilized in their native countries for building, fishing, furniture, flutes, baskets, flower pots, fencing, rafts, fuel, papermaking and to pipe water; locally we may use them for plant stakes, trellis, or fishing poles.

Unfortunately, bamboo have the bad habit of constantly to shed leaves when used as container plants. Plentiful watering will help to minimize this problem. Neatest in behavior is the Metake or "Arrow bamboo", Pseudosasa japonica, with large, leathery, deep green leaves; it is also one of the hardiest outdoors.

The graceful hedge bamboos of the Bambusa multiplex group are much used where compactness is required and their willowy canes, bearing an abundance of feathery leaves, look charming wherever they are placed. Where space permits, the larger species of Bambusa, and Phyllostachys bambusoides, with their bold, bright yellow culms best convey the real spirit of the Orient.

Bambusa beecheyana, the "Beechey bamboo," sometimes known as Sinocalamus, developing thick culms to 40 ft. high and a source of edible bamboo shoots in China. I photographed this tubbed, tall specimen at the Art Institute on Michigan Boulevard, Chicago, Illinois.

By using a method seen in Hawaii, timber bamboo will leaf out and grow branches even after the trunks are sawed off at ground level. The partitions between the hollow internodes are rammed through with a metal rod and the culm filled with water. Standing in shallow buckets of water they will soon sprout foliage giving the effect of bamboo in full growth, remaining alive and attractive for as long as a year.

The favorite subject of artists in Japan and China, bamboo is esteemed there more as a symbol than a thing of beauty. It stands for purity of mind, and such virtues as fidelity, humility, wisdom, and gentleness, and because it becomes stronger as it grows — also promises long life. In Japanese allegory, rather than to resist, bamboo will bend with the winds of adversity but does not break.

Dendrocalamus giganteus, the "Giant bamboo," photographed with the author in tropical Ceylon. This giant of the Bamboo tribe grows to 100 ft. high, with stems 10 inches in diameter.

Bambusa vulgaris, the "Feathery bamboo," is most common in the tropics and very showy with yellowish stems. This tubbed plant lends oriental decor on the street floor of Macy's Department Store, New York.

Carex foliosissima albo-mediana
"Miniature variegated sedge"

Acorus gramineus variegatus
"Miniature sweet flag"

Scirpus cernuus (Isolepis)
"Miniature bulrush"

Rohdea japonica marginata
"Sacred Manchu lily"

Oplismenus hirtellus variegatus
"Panicum" or "Basket grass"

**Stenotaphrum
secundatum variegatum**
"St. Augustine grass"

Cyperus alternifolius
"Umbrella plant"

Reineckia carnea
"Fan grass"

Ophiopogon jaburan 'Variegatus'
"Variegated Mondo"

Liriope muscari 'Variegata'
"Variegated blue lily-turf"

Ophiopogon jaburan
"White lily-turf"

Ophiopogon japonicus
"Snake's beard"

Bambusa multiplex riviereorum
"Chinese goddess bamboo"

Bambusa tuldoides
"Punting-pole bamboo"

Bambusa multiplex 'Stripestem'
"Fernleaf bamboo"

Sasa pygmaea
"Pigmy bamboo"

Sasa fortunei
"Miniature variegated arundinaria"

Bambusa ventricosa
"Buddha's belly bamboo"

Phyllostachys sulphurea
"Yellow running bamboo"

Phyllostachys nigra
"Black bamboo"

Bambusa oldhamii
in hort. as B. "falcata"

Maranta arundinacea
"Arrow-root"

Ctenanthe pilosa (compressa)
in hort. as "Bamburanta"

Phormium tenax 'Variegatum'
"Variegated New Zealand flax"

Nandina domestica
"Heavenly bamboo"

Oryza clandestina
"Wild rice"

Cortaderia selloana pumila
"Pampas grass"

Arundo donax versicolor
"Variegated giant reed"

Saccharum officinarum
"Sugar cane"

Setaria palmifolia
"Palm-grass"

CONIFERS for containers, and Dwarf BONSAI Trees

We are aware of Conifers as the "Needle trees", evergreen and of striking, regular habit, and though lacking showy flowers, they present an ornamental quality quite their own. Coniferae means "Cone-bearers", but in a technical sense this group name covers the evergreen trees and shrubs having needle-like foliage and the balsamic resin typified by fir or spruce.

The history of Conifers extends back 200 million years to the appearance of the early dinosaurs. The Coniferae are a distinct order in the Gymnosperms, plants bearing free or unprotected seeds, thus differing from the great Angiosperms or true flowering plants which carry their seed enclosed in ovaries. These two divisions comprise the phaneroganic or seed-bearing plants. All gymnosperms bear "flowers", but without petals and sepals, better described as cones, both pollen cones and seed cones, although the latter are so greatly reduced in some species that the cone structure is not always obvious. Typical male cones consist of a central axis bearing the pollen either in cones or on catkins. The female, or seed cones form their seed on scales attached to an axis, as in the cones of the Pine family; in some kinds such as podocarpus and yews, the scales become fleshy and the fruit will be berry-like.

Conifers are overwhelmingly represented in the temperate and cold zones of the northern hemisphere. We are familiar with woody cones of the Pine family, and such leading genera as Abies, Picea, Pinus, and Tsuga, which dominate our northern forests in their majesty. Conifers not quite so hardy include cedars, Chamaecyparis, Cryptomeria, cypress, junipers, Taxus and Thuja, and the broad-leaved Ginkgo, but these "Evergreens" are primarily outdoor subjects and could be used indoors only as temporary decorators. When planted in containers, most kinds will be found useful in pyramidal form gracing a building entrance or a cold, light foyer. Taxus, junipers and cypress are most often seen in this manner.

INDOOR CONIFERS—Fortunately, there are coniferous evergreens from along the subtropical zone, that can be used with confidence indoors, provided there is sufficient light and the temperature is not too warm. As a smaller house plant the "Norfolk Island pine", Araucaria heterophylla, better known as excelsa, has long been known to be a thrifty, easy-to-care-for indoor conifer; one that can also be used as miniature Christmas tree. Lately, several Podocarpus have been introduced as floor plants. These "Buddhist pines" in their leisurely yet elegant stature are reminiscent of shady Mediterranean gardens clothed as they are in their deep green, somber needles, and they should keep like iron once they are properly seasoned and well established in their tubs.

The Art of Bonsai

Shaped Bonsai, or dwarfed trees, grown in shallow earthenware containers, are objects of loving care and admiration in every home in Japan. Though this cult had evolved out of China early in the 12th century, yet here in America it is as new as tomorrow. The ultimate aim of the subtle art of Bonsai is to control the growth of a normal plant by constant and skillful pruning and shaping to attain the stately shape of ancient big trees in miniature. The trunk of the tree, the spread of the roots, the distribution of branches are all used to give an aged appearance to the tree. The Japanese white pine, Pinus parviflora, is widely used in Japan as an apt subject for bonsai work. Other pines can similarly be used.

While coniferous evergreens have been the popular favorites, other woody plants, with a little imagination, can be given the look of an ancient tree. Promising subjects are ginkgo, cedar, juniper, pomegranate, Camellia; even deciduous trees such as maple, apricot or beech are used. Much professional skill also is lavished on dwarfed azaleas.

Promising young plants or stunted collected trees are groomed according to a conceived plan by first thinning out 75% of the inner twigs and needles. Desired effects are created by coiling stiff wire around branches to hold them in position until they are set. The growth in pines can be controlled by pinching out new candles in the spring. Soil must be kept moist by daily sprinkling. Mild feeding with fish emulsion twice a year should be sufficient. Should transplanting become necessary, in conifers about every 5 years, it is best to do so before the bonsai tree starts to grow. Remove two-thirds of the old soil, trim broken or decayed roots and replant firmly. Equal parts of loam and sand, with peat or leaf humus added, is a good soil for the little that is needed. Remember that the hardy type of Bonsai trees are outdoor plants and, should not be kept indoors except on special occasions, but they are ideal for the open patio or terrace.

As true art all these Bonsai are created with an interest to span generations and centuries, and I have, in Japan, observed ancients only 2 feet high yet more than 300 years old. In Japanese belief, the cultivation of these Lilliputian trees develops patience and soothes nerves.

While the purpose of Bonsai is to achieve the dwarfing of evergreens in the extreme, it is well to remember that in contrast the conifers also include the greatest of trees. The Giant Sequoia is without question the world's most massive living organism and probably the oldest; the related Sequoia sempervirens, with a maximum height of 360 ft., the tallest tree known. This species has the unusal ability to reproduce itself from basal shoots or cut-off burls that will sprout when placed in water.

Araucaria heterophylla, long known in horticulture as A. excelsa, the "Norfolk Island pine" or "Christmas tree plant"—very satisfactory as a house plant, bringing a breath of cool forest into office and home.

22

Podocarpus elongatus
"Weeping fern-podocarpus" or *"African yellowwood"*

Araucaria bidwillii
"Monkey-puzzle" or *"Bunya-Bunya"*

Podocarpus gracilior
"African fern pine"

Podocarpus elatus
"Weeping podocarpus", in Los Angeles

Pinus pinea
the "Italian stone pine" of the Mediterranean

Juniperus chinensis 'Torulosa'
the artistic "Hollywood juniper" or
"Twisted Chinese juniper"

Podocarpus macrophyllus
"Buddhist pine"

Podocarpus macrophyllus 'Maki'
"Southern yew" or *"Bamboo-juniper"*

Podocarpus Nagi
"Broadleaf podocarpus"

Araucaria heterophylla tree, better known as excelsa,
in Vista, San Diego, California

Araucaria araucana
in hort. as A. imbricata, the so-called
"Hardy monkey puzzle", in British Columbia

Cupressus arizonica pyramidalis
"Blue pyramid Arizona cypress"

Tsuga canadensis
"Hemlock-spruce"

Cupressus sempervirens
the classical "Italian cypress"

x Cupressocyparis leylandii
bigeneric hybrid with flat branches

Ginkgo biloba (as Bonsai)
"Maidenhair-tree"

Juniperus scopulorum 'Chandleri'
"Chandler's silver juniper"

Cryptomeria japonica nana
"Dwarf Japanese cedar"

Sequoia sempervirens
the "California redwood",
a knotty burl sprouting in water

Chamaecyparis obtusa nana
"Dwarf Hinoki cypress"

Taxus cuspidata 'Capitata'
"Upright Japanese yew"
in typical sheared pyramidal form

Thuja occidentalis
the "American arbor-vitae" or "White cedar"
its dense pyramids are often used in Flower Shows

Picea abies, in horticulture as Picea excelsa,
the nordic "Norway spruce", widely used
as "Living Christmas tree"

Metasequoia glyptostroboides
the "Dawn redwood",
a fossil-age conifer from Central China

Juniperus chinensis
a dwarfed "Chinese juniper"

Juniperus chinensis sargentii
as bonsai, 60 yrs. old

Larix leptolepis var. minor
the deciduous "Golden larch"

Myrtus communis compacta
a bonsai-shaped "Compact myrtle" in California

Chamaecyparis obtusa
a dwarfed "Hinoki cypress" 175 years old;
"False cypress"

Rhododendron indicum (lateritium)
"Satsuki azalea" as bonsai, 75 yrs. old

Pinus parviflora (pentaphylla)
the popular "Japanese white pine".
annealed copper wire is used
for initial shaping.

Pinus nigra,
the "Austrian pine",
an early stage bonsai
with mat-forming Arenaria verna, the "Irish moss"

Pinus thunbergii
the "Japanese black pine"

Cedrus atlantica glauca
the "Blue Atlas cedar" 15 yrs. old

Pinus densiflora
"Japanese red pine", 100 yrs. old

Pinus mugo mughus
"Dwarf Swiss mountain pine"

Acer buergerianum, as shapely bonsai;
bright new leaves on a "Trident maple"
80 years old

Carpinus japonica
deciduous "Japanese hornbeam" 40 years old

Punica granatum
a dwarfed "Pomegranate" 65 years old

Elaeagnus pungens
"Thorny silver-berry" 60 years old

Enkianthus perulatus, with white urn-flowers,
a "White enkianthus" 40 years old

Camellia japonica   *as evergreen bonsai,*
the wired branches bearing red flowers

CYCADS, the prehistoric

The Cycadaceae are living fossils, representing the most primitive or lowest living form of seed-bearing plants. They occupy a place intermediate between true flowering plants, and the cryptogams or flower-less plants like the ferns. Though generally known as "Sago palms", as gymnosperms they are more closely related to the conifers such as the pines than to the palms. Like conifers, the cycads carry male and female cones but on separate trees; they differ in the palm-like appearance of their trunks and the large pinnate leaves. Also, the cones are borne at the apex of the stem and not axillary as in conifers. The staminate cones are columnar; the pistillate, female cones as in Cycas revoluta resemble a nest and sheltering the nut-like, vivid red seed. The trunk contains starch, yielding a sort of sago used by natives, but the true Sago palm which produces the sago of commerce is the palm Metroxylon rumphii.

These ancient "Sago palms", Cycas revoluta, well over 100 years old, are an impressive sight fronting the entrance to Huntington Library, San Marino, California.

ORIGIN—Great forests of the palm-like cycads covered the earth during the age of the dinosaurs 200 million years ago. Extinct cycads form a large part of that important material which we know as coal. Only 9 genera remain of the large group that once existed, and these interesting plants are on the verge of extinction. The family is essentially subtropical and tropical, and colonies are found from Central Florida and Mexico down to northern South America; the American genera are Zamia, Ceratozamia. Dioon and Microcycas. In tropical and southern Africa we find the spectacular Encephalartos, and the odd Stangeria from Natal. Australia has two major groups, Bowenia and Macrozamia, as well as some Cycas, which are at home from Polynesia through Southeast Asia to southern Japan.

CYCADS AS DECORATORS—The most handsome of the ornamental cycads are the arborescent kinds with tall trunks such as the Cycas and most of the Encephalartos. The low, tuberous species have subterranean stems; typical are Zamia, Stangeria and Bowenia. Though cycads are regarded as woody plants they really have only a scanty zone of wood surrounding the very large pith. The trunk is topped by a luxuriant rosette of rigid, long persistent leaves; a new crown is produced annually in some species, every 2 years in others. Cycads of the columnar type attain their maximum height only after many centuries of growth.

Cycas revoluta, with its lacquered fronds, the best-known "Sago palm", is much used for decoration, and though not very graceful because of its hulking trunk, it stands out as an individual, prefers cool temperature, and requires little care. Cycas media is more tropical, tall and slender, and truly palm-like; Cycas circinalis is most graceful with arching, feathery and leathery fronds. The Encephalartos can be quite gigantic, resembling tree ferns, and specimen I have observed in Africa were more than 30 feet in height. Dioon also require a tremendous amout of space. Some of the smaller cycads have lately become popular and although Cycas revoluta can be kept dwarf, Zamia furfuracea, the "Jamaica Sago tree", is unique as a small ornamental with its broad leaflets as if made of iron.

PROPAGATION—Viable seed is necessary for good germination; cuttings may be obtained from branching types. Old plants send up suckers which may be taken off when in dormant state and rooted, care being taken to remove the leaves to guard against excessive transpiration. Aged Cycas revoluta and some other cycads form woody knobs along their stem which can easily be removed, dusted with charcoal, and rooted with success. Even slanting sections of Cycas 2-3 in. thick may be used for propagation. Bulb-like buds will develop on the trunk when it is scarred or wounded, and these may be detached while the tree is dormant, and used for propagation — the method employed by nurseries in Japan.

A miniature combination of three Cycas revoluta, the "Sago palm," kept dwarfed in stoneware jar, in Japan.

A female Dioon edule from Mexico cradling its fruiting cone. This stately, palm-like plant is known as "Virgin's palm", and also as "Chestnut dioon" because of its nut-like, edible seed.

Cycas revoluta
"Sago palm", young plant

Zamia furfuracea
"Jamaica sago-tree"

Cycas circinalis
"Fern palm" or "Crozier cycad"

Zamia floridana
"Coontie" or "Seminole bread"

**Macrozamia peroffskyana
(denisonii)**
a "Palm-fern" showing male cone

Ceratozamia sp. 'Hilda'
a "Bamboo zamia" from Mexico

Encephalartos horridus
"Ferocious blue cycad"

Ceratozamia spiralis
"Spiral horn-cone"

Dioon spinulosum
"Giant dioon"

Cycas media
"Australian nut palm"
slender, graceful tree, at Roehrs greenhouses
in New Jersey

Cycas revoluta
"Sago palm" or "Funeral palm"
a massive specimen in Florida

Encephalartos latifrons
"Spiny Kaffir bread"
with pollen-bearing cone

Stangeria eriopus
in horticulture as S. paradoxa
the fern-like "Hottentot's head"

Elizabethan Knot garden showing herbs with foliage varying from green to silver, form-ing intertwining borders and alternating with colored pebbles to create patterns. The Tudor garden was in vogue in England throughout the 16th century, combining utility with beauty. Author photo at Brooklyn Botanic Garden.

A collection of fragrant herbs on a sunny windowsill or, in summertime, on a little plot outdoors, are more than any other group of plants both useful and ornamental, and a center of interest and enjoyment for their flavor, variety and romantic charm. An aura of mystery has surrounded the use of herbs since ancient times; early medical texts go back to 1500 B.C., when coriander was men-tioned as an herb in Sanskrit writings. Hippocrates, the Greek physician, taught the use of herbal medicines in the 5th century B.C.; Theophrastus, a student of Aristotle, wrote about herbs in his great botanical works. Herb culture was encouraged by Charlemagne early in the 9th century, but their use flourished most in later medieval .times, protected in monasteries and castle gardens. Numerous herbals were printed in the 16th century, mainly in England and Germany, and in my library is an illustrated volume printed in Venice in 1650. Early Botanic gardens were created primarily as physic or apothecary gardens, as for instance the one in Leningrad. Even in early American history, herbs played an important role, and many superstitious colonists believed in love potions and witches brews.

WHAT ARE HERBS?—"Herbs" as popularly known are a special gathering of herbaceous annuals as well as perennial, and persistent shrubby plants, selected for their practical usefulness and pur-pose, and the aromatic properties of their roots, stems, leaves, flowers, or seeds. Most of them are used for seasoning as condiments, some for perfumes or sachets, or their medicinal properties, and others are grown only for their sentimental worth.

Herbs differ from vegetables in that herbs are used to flavor other dishes while most vegetables are a main source of food. A great many plants have been termed "herbs", and a recent English herbal lists some 1000 plants that have medicinal, culinary, cosmetic or other uses.

HERBS IN THE WINDOW—A small indoor culinary garden of a practical nature can be kept close at hand in a sunny kitchen window or on the sun porch, in pots, pottery or even tin cans; or several can be planted together in a redwood "Spice box". A basic beginning could be made with Mints, Basil, Chives, Parsley, Thyme, Marjoram, and Rosemary. Cool fresh air at the window will promote sturdy growth; Parsley may welcome a little shade. But as practically all herbs are normally outdoor subjects, they would do best on the patio or the garden patch, at least during the frost-free season. Use liberal containers and well drained good garden loam, not too rich, with some decayed manure or humus added to hold moisture, and water regularly. To control aphids, spray with nicotine insecticide and soap spreader, then syringe with clear water after a few hours.

THE USE OF HERBS—A cook will first think of the seasoning herbs, to flavor and garnish food or drinks. The housekeeper can use herbs of sweet smell that will perfume and help protect linens and clothes. The main reason for growing your own culinary herbs is to have them handy for use fresh from the plant. But why not dry some of them too? To prepare them for drying cut the leafy herbs before flowers are open. Snip or cut early in the day because the foliage then will be permeated with that volatile oil—the essence for which we grow them. Bunch loosely and hang to dry in an airy room or in paper bags. They may also be dehydrated in a barely warm oven. When leaves are crispy dry, strip from stems and store in air-tight jars.

PROPAGATION—Most of the common herbs can be propagated easily by one or more of the usual methods, from seed, cuttings, division or layering. Annuals and biennials such as Basil, Borage, Hys-sop, Sweet marjoram, and Parsley are grown from seed; the bulbous Allium from seed or bulbels. The most practical method in clump-forming perennials like the Mints, Dictamnus, Galium, Hyssop, Marrubium, Melissa, Nepeta, Origanum, Ruta, Stachys, Tanacetum, Tarragon, Teucrium, also Chives, is by division. Shrubby plants such as Artemisia, Chrysanthemums, Laurus, Lavandula, Lemon-verbena, Pelargonium, Rosemary, Salvia, Santolina, and Thyme are best propagated from half-hard cuttings.

Mentha citrata
"Orange mint" or "Bergamot mint"

Lavandula stoechas
"French lavender"

Mentha spicata
"Spearmint"

Artemisia abrotanum
"Southernwood"

Laurus nobilis
"Sweet bay" or "Laurel"

Petroselinum crispum
"Parsley"

Mentha rotundifolia variegata
"Pineapple mint"

Majorana hortensis
"Sweet marjoram"

Artemisia absinthium
"Wormwood"

Pelargonium graveolens
'Variegatum'
"Mint-scented rose geranium"

Artemisia dracunculus
"Tarragon" or *"Estragon"*

Origanum dictamnus
"Dittany of Crete" or
"Hop marjoram"

Thymus vulgaris aureus
"Golden lemon-thyme"

Teucrium flavum
"Yellow germander"

Aloysia triphylla
in hort. as Lippia citriodora
"Lemon verbena"

Rosmarinus officinalis
"Rosemary"

Teucrium fruticans
"Tree germander"

Borago officinalis
"Common borage"

Mentha rotundifolia
"Woolly apple-mint"

Tanacetum vulgare crispum
"Crisped tansy" or *"Jesus-wort"*

Nepeta mussinii
"Catnip" or *"Cat-mint"*

Tanacetum vulgare
"Common tansy"

Lavandula officinalis
"English lavender"

Thymus vulgaris
"French thyme" or *"Common thyme"*

Artemisia purshiana
"Cudweed"

Hyssopus officinalis
pungent *"Hyssop"*

Galium verum
"Lady's bedstraw"

Ruta graveolens
"Rue" or *"Herb-of-grace"*

Marrubium candidissimum
"Horehound"

Stachys officinalis
"Wood betony" or *"Wound-wort"*

Dictamnus albus
"Gas plant" or *"Burning bush"*

Melissa officinalis
"Lemon-balm" or *"Sweet Mary"*

Ocimum basilicum
"Sweet basil"

Allium schoenoprasum
"Chives" or "Schnittlauch"

Thymus hirsutus
"Woolly thyme"

Salvia officinalis
"Garden sage"

Mentha piperita
"Peppermint"

Salvia sclarea
"Clary sage"

Allium porrum
"Swiss leek"

Chrysanthemum ptarmicaeflorum
"Silver lace"

Allium moly
"Lily leek"

Salvia leucantha
"Mexican bush-sage"

Mentha requienii
"Corsican mint"

Santolina chamaecyparissus
"Lavender cotton"

OXALIS and SHAMROCKS

The characteristic trifoliate leaves so typical of the true clover, or Trifolium, belonging to the Leguminosae family, are common also within the genus Oxalis of the Oxalidaceae, and for this reason both these are popularly known as "Shamrock", adapted from the Irish word for clover. Since both Trifolium and Oxalis are present in Ireland there is no certainty as to the exact identity of which of these Shamrocks was picked by St. Patrick as the symbol of Trinity when he landed as a missionary in Wicklow in 432.

Trifolium repens, the "White clover", or Trifolium dubium, the "Yellow clover", are not indoor plants and would soon become leggy. But both are grown as a specialty in little pots for the celebration of St. Patrick's Day, March 17; repens as the "Irish shamrock" in America, dubium in Ireland, now the Irish national emblem.

OXALIS AS HOUSE PLANTS—In the Oxalis, however, there are many wonderful little house plants for the sunny window, their pretty flowers of white, yellow, pink or rosy-red blooming tirelessly, opening only to the sun. The tender Oxalis in cultivation are primarily from the Andes of Chile and Peru to Brazil and north to Mexico, and are well represented in South Africa; the hardier kind are spread over northern Europe and North America. Most are small herbs, often producing tuberous rhizomes or bulbs; some species have branching, leaf-bearing stems, others are scapose with the flower stalks arising directly from the crown; a few are succulent or even subshrubs. The leaves are clover-like digitate but sometimes with more than 3 leaflets, and which close umbrella-like at night or in darkness as if going to sleep.

A commonly encountered creeping Oxalis is O. acetosella, the "European Sheep sorrel", also known as "Sleeping beauty". It is being cultivated for sentimental reasons and many students of Irish tradition convincingly reason that this is the true "Irish shamrock". Bulbous Oxalis bloom either in summer or winter, but should be given a resting period after blooming. Amongst the best of these is O. purpurea, the "Cape oxalis", in the trade as "Grand Duchess", with showy bright rose flowers. The popular 4-leaf "Lucky clover", Oxalis deppei, produces tubers that are edible. The succulent kinds are intriguing, particularly Oxalis herrerae with bright yellow flowers. The tree-like "Fishtail oxalis", Oxalis ortgiesii, is another curiosity with bronzy leaflets notched like gold-fish tails. Most eye-catching is the yellow-flowered, shrubby "Fire-fern", Oxalis hedysaroides rubra, with glowing wine-red, fern-like foliage, its leaflets sensitive to the touch.

Oxalis are not particular as to soil, and prefer temperatures not too warm. They like to be watered liberally but not kept constantly wet. Propagation of the fibrous and spreading types is by division or breaking up the roots which are frequently formed into strings of bulblike, scaly masses. The tuberous or bulbous species are often very liberal in the production of new bulbils. Shrubby varieties propagate from stem cuttings. Seeds develop in a capsule which when ripe, turns inside out, flinging seeds in all directions.

Oxalis deppei has 4 leaflets; the "Lucky clover" or "Good luck plant"

Trifolium repens minus, in 2 inch pot, the "Irish shamrock" or "Little white clover" as sold in America

Oxalis martiana 'Aureo-reticulata', "Gold-net sour clover"

Wearing the green shamrock emblem, Irish lassies proudly parade on 5th Avenue, an annual event on St. Patrick's Day, March 17, in New York City

Trifolium dubium as used in Dublin, also an "Irish shamrock", the "Yellow clover" of Ireland

Oxalis braziliensis
with rosy flowers

Oxalis acetosella
"European wood-sorrel" and "Irish shamrock"

Oxalis purpurea (variabilis)
"Grand duchess oxalis"

Oxalis pes-caprae (cernua)
"Bermuda buttercup", with yellow flowers

Oxalis bowiei
"Giant pink clover"

Oxalis incarnata

Oxalis carnosa

Oxalis melanosticta

Oxalis regnellii
in hort. as "rubra alba"

Oxalis vulcanicola
(siliquosa in hort.)

Oxalis peduncularis

Oxalis rubra alba
also known as "crassipes alba"

Oxalis hedysaroides rubra
the "Firefern"

Oxalis adenophylla

Oxalis flava

Oxalis rubra
"crassipes" or "rosea" in hort.

Oxalis herrerae
"Succulent oxalis"

Oxalis ortgiesii
"Tree oxalis"

SUCCULENTS
other than CACTI

In arid regions of the world many totally unrelated plants have been forced to adopt some method to exist through long periods of deficient rainfall. Plants that were once delicate and with abundant foliage have, in their struggle for survival in areas that have turned arid, evolved smaller leaves and developed stems or leaves capable of storing water, becoming succulent in the process and changing character and appearance.

In stem-succulents as in Cacti and Euphorbias all life process is carried on in the stem itself; in leaf-succulents as represented by Aloe, Echeveria, or Lithops the whole plant may be reduced to a few highly modified fat leaves, and roots. Some desert plants, instead of building water storage tissues, have become drought-resistant by reducing their leaf area, by covering leaves or stems with insulating scales, hairs, wax, varnish or resin, or a protective woody bark; these are properly designated as Xerophytes, and typified by Dyckia, Fouquieria, Yucca, and some Agave.

Glazed ceramic planter featuring the heart-warming Disney fawn "Bambi", as planted with an assortment of miniature succulent plants such as kalanchoe, sedum, crassula and aloe. This modest arrangement enhances a corner of the living room, requiring a minimum of attention.

Some succulents protect themselves in other ways as well, particularly from desert animals in search of juicy food and water. Simulating the coloration and the form of rocks among which they grow, they hide and pull themselves into the surface of the soil from which they can be distinguished only with difficulty. Amongst the most characteristic of these "Mimicry plants" are the Lithops, or "Flowering stones" of the rain-starved deserts of Southern Africa; and the Pleiospilos, popularly known as "Cleft rock". Other plants such as Fenestraria, Frithia, some Haworthia, and many more are known as "Windowed plants", because they have their flat tops surfaced by a translucent window which permits entry of the sun's rays and filters and dilutes them before reaching the green tissues lying below. Mesembryanthemum crystalinum, the true "Iceplant", covers itself with crystalline pebbles which glisten in the sunlight like the quartz sand in which it grows.

Succulents as House Plants

Because of their variety of form, beauty of coloring, and oddity of behavior, succulents are highly prized by hobbyists. They are durable and need a minimum of attention, and interesting collections in small pots can be kept close together in a very small space. As "hot weather plants" most of them prefer a place in the sun which brings out the best of their inborn coloring. Tones of glowing-red or copper, glaucous blue or silver will attract the eye wherever on display. When planted into glazed pottery as "Dishgardens", artistically arranged by character and color, or as individuals in novelty containers, such planters are perfect for use on furniture or the office desk, as these are quite attractive in their compactness, yet requiring little watering and no special care.

MINIMUM OF CARE—Succulent plants are from areas of wide temperature variation, native to regions that are arid or semi-arid for at least part of the year. It is in the study of their habitat therefore that gives the cue to the best cultivation and handling of succulents. Although from tropical and subtropical areas of intense daytime heat, temperatures in the tropics usually drop to a pleasant coolness or even chilliness at night. While rains may be absent, morning mist or fog is common, refreshing to shallow-rooted leaf succulents such as Aeoniums or Echeverias. It is fortunate that we can approximate climatic conditions of their natural habitats right on our window sill, the sun-porch, or the open patio. At a sunny south window it is warm and bright during day and cooler at night—ideal for Aloe, Crassulas, Echeverias, Kalanchoe, Sedums, Senecio; also the various Mesembryanthema, such as Lithops, although the latter need to be kept particularly dry. Haworthias and Gasterias prefer an East window, and perhaps a little warmer, as do Euphorbias, Stapelias and Sansevierias.

Succulents need a porous, gritty soil that readily drains superfluous water and yet is sufficiently compact and retentive to hold the moisture the plant requires. A good mixture is two-thirds loam and one-third sand, with some leafmold or peatmoss added.

A collector of lithops and other small, light-hungry succulents could not provide a more ideal place in his home than this window-greenhouse extending out into the bright light of day, with all its benefits, so necessary especially to South African plants.

SUCCULENTS

As desert soils are rich in mineral content, it follows that succulents need fertilizer during their growing season. Extra lime is added to the soil especially if species to be potted have white spines, white bodies, whitened leaves or white incrustation. Roots of succulents are apt to harbor nematodes which impair their growth and which can be detected by the little knobs forming on the root system. To check this hazard, these roots should be cut off when replanting, and as an additional precaution, the new soil ought to be pasteurized.

Succulents need less water than most other plants. Leafy kinds can take more than stem succulents or those with leaves modified into thick-fleshy bodies. Water requirements are greater in the summer than in the dark days of winter unless much artificial heat should dry them out prematurely. A good rule is to water thoroughly twice a week during the growing season, but not to water on cloudy or rainy days. During winter months or when the plants are dormant, watering once every two weeks would be sufficient.

The stone-mimicry and windowed plants from the driest areas of southern Africa have two growing periods, in the spring and in the autumn; rest periods from June through August and November to March. They require rapid drainage, and 10% perlite added to a mixture of equal parts of loam, sand, and old leafmold will wonderfully improve aeration and growth response. Feeding will produce exuberant growth but impair keeping quality and hardiness. No more than a token watering is needed when plants such as Lithops are dormant, or they may rot overnight.

The author's patio in Manhattan Beach, California, is typical of an outdoor living room in warmer areas, and except for the filtering foliage of an eucalyptus tree, is open to blue sky and sun. A sheltering wall with view windows protects the windward side from the prevailing breeze of the Pacific ocean. This is an ideal place enjoyed by plants, especially the easily pleased succulents.

During the winter rest period it is advisable not to water mature plants at all especially if kept cool; a little water once or twice a week will be about right during the hot summer period; sub-irrigation from a saucer underneath is safest. Baby seedlings require more coddling; moistening the soil lightly and more frequently may be necessary to keep them from shriveling. To expose these sun-lovers as much as possible close to the light hardens their bodies and brings out their best in coloring. Mimicry plants mostly belong to the family Aizoaceae.

Should a few mealybugs or soft scales infest any succulent, they may be destroyed by touching the insects with rubbing alcohol, applied with a small water-color paintbrush or a pipe-cleaner. Malathion spray could normally be recommended, but members of the Crassula family as well as Lithops are sensitive to it and subject to damage. If root-mealybugs occur, roots may be thoroughly washed, or dipped in alcohol; if severe it is best to cut off the roots altogether and syringe the plant body; a pot drench with Chlordane liquid has also proven satisfactory in their control.

Propagation of fast growing succulents is easiest by stem cuttings, division or suckers. Most members of the Crassula family will also sprout from individual leaves. Stone-mimicry plants and other "Mid-day flowers" germinate easily from seed at warm 75°F. The various species of this Mesembryanthema family are self sterile and two individually separate plants are needed for fertilization to obtain seed.

One of the miracles of nature: the fleshy leaves of the "Panda plant," Kalanchoe tomentosa, sprouting young plantlets from their base — a means of propagation characteristic of the Crassulaceae family of succulents.

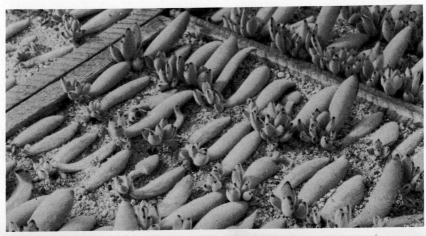

Aloe variegata, "Tiger aloe"
Stapelia hirsuta, "Hairy toad plant"
Gasteria verrucosa, "Wart gasteria"

Agave decipiens, "False sisal"
Agave striata 'Nana', "Lilliput agave"
Aloe arborescens, "Octopus plant"

Kalanchoe marmorata, "Pen wiper"
Senecio serpens, "Blue chalk sticks"
Kalanchoe 'Roseleaf', (beharensis x tomentosa)

Euphorbia lactea, "Candelabra plant"
Sansevieria 'Hahnii', "Birdsnest"
Euphorbia splendens, "Crown of thorns"

Aeonium decorum, "Copper pinwheel"
Kalanchoe tomentosa, "Panda plant"
Echeveria multicaulis, "Copper roses"

Kalanchoe fedtschenkoi, "Purple scallops"
Cotyledon barbeyi, "Hoary navelwort"
Crassula tetragona, "Miniature pine tree"

Gibbaeum petrense, "Flowering quartz"
Lithops marmorata, "Living stone"
Dinteranthus wilmotianus, "Split rock"

Pleiospilos nelii, "Cleft stone"
Titanopsis calcarea, "Limestone mimicry"
Pleiospilos bolusii, "Living rock cactus"

Crassula argentea, "Jade plant"
Crassula arborescens, "Silver dollar"
Crassula rupestris, "Rosary vine"

Crassula 'Tricolor Jade', "Tricolored jade plant"
Portulacaria afra variegata, "Rainbow bush"
Crassula dubia; "obvallata" hort.

Pachyphytum 'Cornelius', "Moonstones"
Sedum adolphii, "Golden sedum"
Pachyphytum compactum, "Thick plant"

Kalanchoe fedtschenkoi 'Marginata', "Aurora Borealis"
Sedum rubrotinctum, "Christmas cheers"
Graptopetalum paraguayense, "Ghost plant"

Sempervivum tectorum calcareum, "Houseleek"
Euphorbia mammillaris, "Corncob cactus"
Faucaria tigrina, "Tiger jaws"

Echeveria elegans, "Mexican snowball"
Echeveria derenbergii, "Painted lady"
Echeveria 'Pulv-oliver', "Plush plant"

Aloe nobilis, "Gold-tooth-aloe"
Aloe humilis 'Globosa', "Spider aloe"
Aloe brevifolia depressa, "Crocodile jaws"

x Gasterhaworthia 'Royal Highness'
Haworthia subfasciata, "Little zebra plant"
x Gastrolea 'Spotted Beauty', "Gasteria-aloe hybrid"

Aeonium 'Pseudo-tabulaeforme'
"Green platters"

Aeonium canariense
"Giant velvet rose"

Aeonium tabulaeforme
"Saucer plant"

Aeonium haworthii
"Pin-wheel"

Aeonium burchardii

**Aeonium arboreum
atropurpureum**
"Black tree aeonium"

Aeonium sedifolium
"Tiny-leaved aeonium"

Aeonium domesticum variegatum
"Youth and old age"

Aeonium decorum, in flower
"Copper pinwheel"

Adromischus festivus
"Plover eggs"

Cotyledon undulata
"Silver crown"

Adromischus maculatus
"Calico hearts"

Cotyledon orbiculata oophylla

Lenophyllum guttatum (in hort.)
"Basin leaf"

Adromischus cristatus
"Crinkle-leaf plant"

Cotyledon ladismithiensis
"Cub's paws"

Adromischus tricolor
"Spotted spindle"

Cotyledon paniculata
"Botterboom" of the Boers

Spectacular Aloe bainesii tree,
60 feet high, in view of Table Mountain,
Cape Town, South Africa

Aloe dichotoma, the "Dragon tree aloe"
later branching in pairs;
Desert of Namaqualand, South Africa

Aloe arborescens
the "Candelabra aloe" or "Octopus plant"
in brassbound redwood tub,
a picturesque table plant

Aloe marlothii, bold blue rosette
heavily armed and becoming tree-like;
showy salmon-yellow inflorescence;
Veld of Northern Transvaal

Aloe saponaria
"Soap aloe"

Aloe polyphylla
"Spiral aloe"

Aloe striata
"Coral aloe"

Aloe vera chinensis
"Indian medicine aloe"

Aloe vera
"Medicine plant" or "Bitter aloe"

Aloe suprafoliata
"Propeller aloe"

Aloe x humvir
"Needle-aloe"

Aloe aristata
"Lace aloe"

Aloe x'Walmsley's Bronze'
freely branching rosette

Aloe brevifolia
A. "humilis" of the trade

Aloe brevifolia depressa
"Crocodile jaws"

Aloe humilis echinata
"Hedge hog aloe"

Aloe africana
"Spiny aloe"

Aloe x spinosissima
"Torch plant"

Aloe ciliaris
"Climbing aloe"

Aloe virens
"Green aloe"

Aloe variegata ausana
"Partridge breast"

Aloe ferox
"Ferocious aloe"

Aloe plicatilis
"Fan aloe"

Aloe lineata
"Lined aloe"

Aloe eru maculata
"Nubian aloe"

Crassula x portulacea
"Baby jade"

Crassula argentea
in the trade as C. arborescens
"Jade plant"

Crassula argentea variegata
"Variegated jade plant"

Crassula cornuta
"Horned crassula"

Crassula dubia
"obvallata" or
"tomentosa" in hort.

Crassula hemisphaerica
"Arabs turban"

Rochea coccinea
in hort. as Crassula rubicunda

Crassula falcata
"Scarlet paintbrush"

Crassula schmidtii
"Red flowering crassula"

Crassula rupestris
"Rosary vine"

Crassula perforata 'Pagoda'
"Pagoda plant"

Crassula perfossa
"String o' buttons"

Crassula perforata
"Necklace vine"

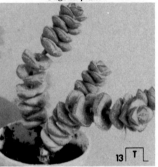

Crassula 'Marnieriana hybrid'
"Jade necklace"

Crassula teres
"Rattlesnake tail"

Crassula lactea
"Tailor's patch"

Crassula 'Lactea hybrid'

Crassula cultrata
"Propeller plant"

Crassula x imperialis
"Giant's watch chain"

Crassula pseudolycopodioides
"Princess pine"

Crassula lycopodioides
"Skinny fingers" or *"Toy cypress"*

Echeveria peacockii var. subsessilis
desmetiana hort.

Echeveria elegans
"Mexican snowball" or "Mexican gem"

Echeveria x'Haageana'
"Fruit cups"

Echeveria simulans
"True Mexican rose"

Echeveria gilva
"Wax rosette"

Echeveria x perbella

Echeveria glauca
"Blue echeveria"

Echeveria x glauco-metallica
"Hen and chicken"

Echeveria secunda
"Blue hen and chickens"

Echeveria pulvinata
"Chenile plant"

Echeveria x'Pulv-oliver'
"Plush plant"

Echeveria x'Set-oliver'
"Maroon chenile plant"

Echeveria x'Pulvicox'
"Red plush plant"

Echeveria affinis
"Black echeveria"

Echeveria sprucei
"spruceana" in hort.

x Sedeveria derenbergii
"Baby echeveria"

Echeveria derenbergii
"Painted lady"

Echeveria expatriata
"Pearl echeveria"

Echeveria agavoides
"Molded wax"

x Pachyveria 'Curtis'
"White cloud"

Echeveria multicaulis
"Copper rose"

x Pachyveria scheideckeri
"Powdered jewel plant"

Dudleya virens
(Echeveria insularis)

x Pachyveria 'E. O. Orpet'
"Large jewel plant"

Echeveria x'Meridian'
crenulata hyb. 15 in. dia.

Echeveria crenulata 'Roseo-grandis'
showy 10 inch rosette

Echeveria x'Can-Can'
edged with red lace

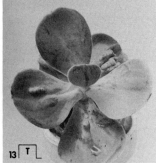

Echeveria gibbiflora carunculata
"Blister echeveria"

Echeveria crenulata
"Scallop echeveria"

Echeveria x'Ballerina'

Echeveria gibbiflora metallica

Echeveria purpusorum
(Urbinia purpusii in hort.)

Echeveria x kewensis

Echeveria runyonii

Echeveria pallida
"Argentine echeveria"

Echeveria x'Doris Taylor'
"Woolly rose" or "Plush rose"

Euphorbia abyssinica
"Imperial Ethiopian euphorbia"
majestic plant becoming tree-like

Euphorbia coerulescens
the "Blue euphorbia"
branched specimen 25 years old

Euphorbia ingens
the cactus-like "Candelabra tree"
on the Veld in northern Transvaal

Euphorbia grandicornis
the ferociously thorny "Cow horn euphorbia"
with irregularly constricted stems

Euphorbia ingens
"Candelabra tree"

Euphorbia lactea
"Candelabra cactus"

Euphorbia trigona
in hort. as "hermentiana"
"African milk tree"

Euphorbia polyacantha
"Fish bone cactus"

Euphorbia mammillaris
"Corncob cactus"

Euphorbia mammillaris 'Variegata'
"Indian corn cob"

Euphorbia ramipressa
"alcicornis" in hort.

Euphorbia tirucalli
"Milk bush" or *"Pencil cactus"*

Euphorbia grandidens
"Big tooth euphorbia"

Euphorbia flanaganii
"Green crown"

Euphorbia obesa
"Baseball plant" or
"Turkish temple"

Euphorbia valida,
similar to meloformis

Euphorbia heptagona

Euphorbia stenoclada
"Silver thicket"

Euphorbia pseudocactus
"Yellow euphorbia"

Euphorbia caput-medusae
"Medusa's head"

Euphorbia lophogona
"Pink-flowering euphorbia"

Euphorbia pulvinata
"Cushion euphorbia"

Euphorbia grandicornis cristata
"Crested cowhorn"

Euphorbia x'Manda's Cowhorn'
grandicornis hybrid

Euphorbia ingens monstrosa
"Totem pole"

Euphorbia neriifolia
cristata variegata
"Variegated oleander cactus"

Euphorbia lactea cristata
"Elkhorn" or "Frilled fan"

Euphorbia flanaganii cristata
"Crested green crown"

Euphorbia neriifolia cristata
"Crested oleander cactus"

Euphorbia mauritanica cristata
"African fan"

Euphorbia x'Zigzag'
"Zigzag cactus"

Euphorbia x keysii
"Dark Florida Christ-thorn"

Euphorbia splendens bojeri
"Dwarf crown-of-thorns"

Euphorbia x'Giant Christ thorn'
"Giant California Christ-thorn"

Euphorbia splendens
"Crown-of-thorns"

Euphorbia x'Flamingo'
"Pink Florida Christ-thorn"

Euphorbia canariensis
"Hercules club" or *"African cereus"*

Monadenium lugardae
"Green desert rose"

Synadenium grantii rubra
"Red milk bush"

SUCCULENTS: *Haworthias and Gasterias*

Haworthia margaritifera
"Pearl plant"

Haworthia fasciata
"Zebra haworthia"

Haworthia subfasciata
"Little zebra plant"

x Gastrolea beguinii
"Pearl aloe" or "Lizard-tail"

Haworthia papillosa
"Pearly dots"

Gasteria x'Hybrida'
"Oxtongue" or "Bowtie-plant"

Haworthia cymbiformis
"Windowed boats"

Haworthia cooperi
"Window haworthia"

Haworthia cuspidata
"Star window plant"

Haworthia coarctata
"Cowhorn haworthia"

Haworthia truncata
"Clipped window plant"

Haworthia reinwardtii
"Wart plant"

Kalanchoe x'Grandiflora hybrid'
large flowered "Surprise"

Kalanchoe x'Yellow Darling'
"Yellow blossfeldiana hybrid"

**Kalanchoe
blossfeldiana 'Compacta'**
vivid "Flaming Katy"

Kalanchoe pumila
"Dwarf purple kalanchoe"

Kalanchoe x kewensis
"Spindle kalanchoe"

**Kalanchoe fedtschenkoi
'Marginata'**
"Rainbow kalanchoe"

Kalanchoe x'Fernleaf'
"Fernleaf felt bush"

Kalanchoe 'Oakleaf'
"Oakleaf felt bush"

Kalanchoe beharensis
"Velvet leaf" or "Elephant ear"

Kalanchoe pinnata (Bryophyllum)
"*Airplant*" *or* "*Miracle leaf*"
sprouting plantlets on leaves

Kalanchoe gastonis-bonnieri
"*Life plant*" *with young plants at leaf tips*

Kalanchoe synsepala
"*Cup kalanchoe*"

Kalanchoe daigremontiana
"*Devil's backbone*"

Kalanchoe longiflora
"*somaliensis*" *of the trade*

Kalanchoe nyikae
"*Shovel kalanchoe*"

Kalanchoe
'daigremontiana x tubiflora'
"*Good luck plant*"

Kalanchoe tomentosa
"*Panda plant*"

Sedum 'Adolphii hybrid'
"Golden Glow" or *"Honey sedum"*

Graptopetalum
'Paraguayense hybrid'
"Giant Indian rock"

Sedum adolphii hybrid
'Peach blossom'

Sedum treleasei
"Silver sedum"

Sedum pachyphyllum
"Jelly beans"

Sedum 'Weinbergii' hort.
"Mother of pearl plant"

Pachyphytum oviferum
"Pearly moonstones"

x Sedeveria 'Green Rose'
(Sedum x Echeveria)

Pachyphytum brevifolium
"Sticky moonstones"

Sedum lucidum
"tortuosum" of the trade

Sedum platyphyllum
"Powdered jade plant"

Sedum multiceps
"Pigmy Joshua tree"

Sedum morganianum
"Burro tail" in basket

Sedum x orpetii
"Giant burro tail", "Lamb's tail"

Sedum sieboldii
hardy "October plant"

Sedum lineare 'Variegatum'
"Carpet sedum"

Sedum praealtum cristatum
"Green coxcomb"

Sedum mexicanum
"Mexican stone crop"

Sedum spectabile 'Variegatum'
"Variegated live-forever"

Sedum spectabile
hardy "Live foreever"

Sedum telephium
hardy "European ice plant"

Senecio herreianus
"Green marble vine"

Sempervivum arachnoideum
"Cobweb house leek"

Sempervivum arachnoideum
glabrescens

Chamaealoe 'Africana hybrid'
(Chamaealoe x Aloe)

Pedilanthus
tithymaloides variegatus
"Ribbon cactus" or *"Red bird"*

Dyckia brevifolia
"Miniature agave"

Chamaealoe africana
"False aloe"

Adenium obesum
"Desert rose"

Plectranthus tomentosus
"Succulent coleus"

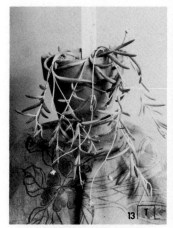

Othonna capensis
"Ragwort vine"

Senecio macroglossus variegatus
"Variegated wax-ivy"

Senecio jacobsenii (petraeus)
"Weeping notonia"

Senecio articulatus
"Candle plant"

Senecio scaposus
"Silver coral kleinia"

Senecio stapeliaeformis
"Candy stick"

Senecio haworthii
"Coccoon plant"

Senecio fulgens
"Scarlet kleinia"

Senecio kleiniaeformis
"Halberd kleinia"

Stapelia variegata
"Carrion flower" or *"Spotted toad cactus"*

Stapelia gigantea
"Giant toad plant" or *"Zulu giants"*

Stapelia hirsuta
"Hairy star fish flower"

Stapelia nobilis
"Noble star flower"

Stapelia clavicorona
"African crown"

Huernia pillansii
"Cockle burs"

Tavaresia grandiflora
"Thimble flower"

Huernia zebrina
"Owl eyes"

Caralluma lutea

Caralluma nebrownii
"Spiked clubs"

Trichocaulon cactiforme

Trichodiadema stellatum
as bonsai, 25 years old

Hoya carnosa
'Hummel's compacta'
"Hindu rope plant"

Hoodia rosea
"African hat plant"

Pelargonium crithmifolium
"Succulent geranium"

Didierea madagascariensis
"Madagascar cactus"

Pachypodium lamieri
"Club foot"

Portulaca grandiflora
"Rose moss" or "Sun plant"

Anacampseros buderiana
"Silver worms"

Aptenia cordifolia variegata
"Baby sun rose"

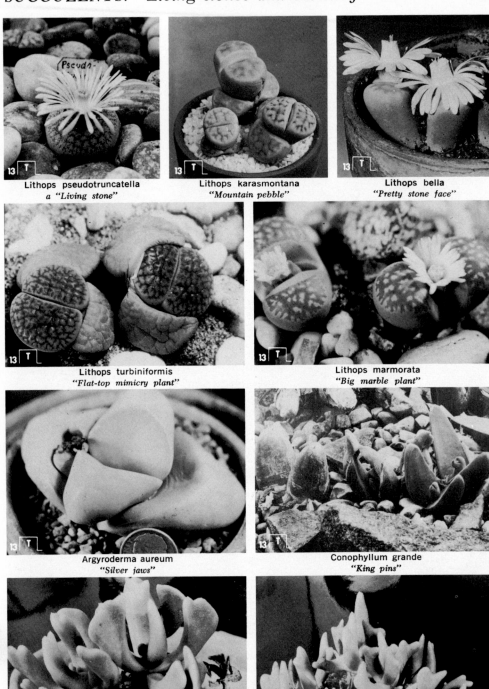

Lithops pseudotruncatella
a "Living stone"

Lithops karasmontana
"Mountain pebble"

Lithops bella
"Pretty stone face"

Lithops turbiniformis
"Flat-top mimicry plant"

Lithops marmorata
"Big marble plant"

Argyroderma aureum
"Silver jaws"

Conophyllum grande
"King pins"

Rhombophyllum dolabriforme
"Hatchet plant" (Collection Marnier-Lapostolle)

Rhombophyllum nelii (Hereroa hort.)
"Elkhorns", (Marnier-Lapostolle, France)

Titanopsis calcarea
"Limestone mimicry"

Dinteranthus inexpectatus
"Surprise split rock"

Nananthus malherbei
"Miniature blue fan"

Trichodiadema densum
"Miniature desert rose"

Faucaria tigrina
"Tiger-jaws"

Faucaria tuberculosa
"Pebbled tiger jaws"

Pleiospilos simulans
"African living rock"

Fenestraria aurantiaca
"Baby toes"

Dinteranthus puberulus
"Downy split rock"

Anacampseros rufescens
"Love plant"

Conophytum simile
"Cone plant"

Cheiridopsis candidissima
"Victory plant"

Glottiphyllum linguiforme
"Green tongue leaf"

Corpuscularia algoense
"Grayhorns"

Frithia pulchra
"Purple baby toes"

Oscularia deltoides
"Pink fig marigold"

Fenestraria rhopalophylla
"Baby toes"

Mesembryanthemum crystallinum
the true "Ice plant", with ice-like crystals,
on the dunes of Manhattan Beach, California

Dorotheanthus gramineus, the "Tricolor
mesembryanthemum" or "Buck Bay daisy", an
annual of the Cape of Good Hope, in varied colors
opening to the sun, its crystalline foliage glittering.

Lampranthus roseus
the "Pink ice plant"
on Table Mountain, South Africa

Lampranthus spectabilis
the beautiful "Red ice plant"
widely planted in Southern California

Lampranthus aureus
the "Golden ice plant"
at Mission San Luis Rey, California

Carpobrotus edulis
"Hottentot-fig", with silky flowers,
in the Karroo desert, Cape Province

Agave ferdinandi-regis
the shapely, very beautiful "King agave"

Agave victoriae-reginae
"Queen Victoria agave",
a very formal, small rosette

Agave potatorum
"Drunkard agave"

Agave parviflora
"Little princess agave"

Agave filifera
"Thread agave"

Agave sisalana
"Sisal-hemp"

Agave decipiens
"False sisal"

Agave miradorensis
"Mirador agave"

Agave stricta
"Hedgehog-agave"

Agave striata 'Nana'
"Lilliput agave"

Agave striata
"Dark-striped agave"

Agave attenuata
very artistic "Dragon tree agave"
with arching stem, in Italian stone container
at Huntington Botanic Garden,
San Marino, California

Agave angustifolia marginata
"Variegated Caribbean agave"
a very decorative, densely formal rosette,
and freely suckering

Agave americana
the large "Century plant" or "Maguey"

Agave americana 'Marginata'
"Variegated century plant"

Sansevieria trifasciata 'Hahnii'
"Birdsnest sansevieria"

Sans. trifasciata 'Golden hahnii'
"Golden birdsnest"

Sans. trif. 'Silver hahnii marginata'
"Gilt-edge silver birdsnest"

Sansevieria intermedia
"Pygmy bowstring"

Sansevieria guineensis
"Bowstring hemp"

Sansevieria cylindrica
"Spear sansevieria"

Sansevieria ehrenbergii
"Blue sansevieria" or "Seleb"

Sansevieria senegambica
"cornui" in hort.

Sansevieria grandis
"Grand Somali hemp"

**Sansevieria trifasc.
'Bantel's Sensation'**
"White sansevieria"

Sansevieria "Kirkii" hort.
"Star sansevieria"

Sansevieria zeylanica
"Devil's tongue"
true species from Ceylon

Sansevieria trifasciata laurentii
"Gold-band sansevieria" or "Variegated snake plant"
combination in 9 inch glazed jardiniere
widely used because of its enduring qualities

Sansevieria trifasciata
"Snake plant", "Mother-in-law-plant" or "Lucky plant"
an old house plant favorite;
combination in glazed pot.

Sansevieria parva
the "Flowering Parva sansevieria"
as a hanging plant,
cascading runners developing with age

Sansevieria suffruticosa
"Spiral snake plant",
young plants forming at ends of thick stolons,
in an oak grove at Arcadia, California

Yucca elephantipes (gigantea)
"Spineless yucca" or "Bulb stem palm-lily"

Yucca aloifolia 'Marginata'
"Variegated Spanish bayonet"

Yucca gloriosa 'Variegata'
"Variegated mound-lily"

Yucca elephantipes 'Variegata' (gigantea var.)
"Giant variegated palm-lily"

Furcraea gigantea
"Giant false agave"

Furcraea selloa marginata
"Variegated false agave"

POPULAR BLOOMING PLANTS for festive occasions

Flowering plants are joyful envoys of nature to brighten your home, and particularly so during the drab days of winter. To "say it with flowers" is not an empty phrase; flowers tender like no other gift, the deepest sentiments of romance and feelings of good will that we as friends or neighbors are capable of expressing. There are countless occasions in our lives where we would like to convey to someone our innermost feelings, and never once would the gift of a flowering plant, because of their intrinsic value, be weighed in monetary terms whether lowly and demure, or costly and magnificent. Through flowers we can speak of admiration and of love, of gladness and of heartfelt sympathy, or extend congratulations and best wishes, or merely express good will.

As the year advances, certain blooming plants have become associated with sentimental or religious holidays as symbolic of the season. We look for the first primroses and azaleas as the heralds of spring, we must have hyacinths and tulips, lilies and hydrangeas to decorate our homes and churches for the Easter season; roses, gloxinias and gardenias follow up to Mother's Day; geraniums and spring bedding plants on the windowsill lead into summer; in the fall

An "Easter basket" breathing spring, and gay with pastel colors of seasonable potted plants: fragrant Easter-lilies; showy heads of sky-blue Hydrangeas; salmon-pink Azalea 'Coral Bells' and red 'Hexe'; and a deep salmon Polyantha Rose 'Margo Koster'.

what could be more natural than chrysanthemums in their autumn shades, and cyclamen, begonias and fruited plants take over with winter; and a Christmas without poinsettias is now almost unthinkable.

FLOWERING PLANTS INDOORS—Holiday blooming plants in pots require much more water and strong light to keep in good condition, than the average foliage plant. We Americans like nice warm homes, but unfortunately our popular flowering plants don't like our comfortable temperatures; for long life and extended bloom our rooms are much too warm unless wanted for temporary display. Most holiday plants keep better down around 60°F during the day and 50° at night. Find a location by the window where it is sunny and yet cooler, at least at night, or on the sunporch; also keep away from heating units.

Azaleas, cyclamen, primulas, chrysanthemums, cinerarias, hydrangeas, Christmas cherries, kalanchoe, pyracantha, heathers, lilies, hyacinths, tulips, daffodils, daisies, geraniums, roses, fuchsias, calceolarias, astilbe, begonias all keep better, and for a reasonably long time, at cooler temperatures, and unfold their buds in a normal manner. There is no harm in using them where needed temporarily for display on a table or other piece of furniture, but real dividends are earned if these plants are carried back to a sunny window and a nice cool place for the night. On the other hand, saintpaulia, gloxinias, poinsettias and gardenias should not be exposed to temperatures much below 65°F, as their flowers and buds, as well as their leaves are sentive to cold and may be severely checked. But stuffy heat is resented by all flowering plants, and with drafts and drying out, is the cause of most complaints.

Because their pots are likely to be full of roots, take special care that the pot is watered, and thoroughly, as soon as the soil "feels" dry. If practical, immerse the pot at least once a week into a bucket of water until the soil is thoroughly soaked, and be sure to place a saucer under it that can hold water. Setting a smaller pot into a larger one with soil or peat around it, also is good practice. A porous clay pot can be wrapped in waxed paper, tinfoil or plastic cloth to retard quick loss of moisture in a dry room, a method that can well be used when going away for a few days. Placing a pot plant inside a glazed jardiniere will look more attractive, but don't allow accumulated water to stand in the bottom for more than a couple of days or the roots will become waterlogged and subject to rot, causing leaf drop.

Attractive Azalea 'Haerens Beauty', a large shell pink of the "Belgian indica" (simsii) group of tender azaleas, grown in tree form. Grafted high on a stem of Azalea concinna, it takes 6-8 years to develop this showy standard.

The usual seasonal flowering gift plant would grow with difficulty in the house. But it is a challenge to many a housewife with a "green thumb", once a plant has been received, to try and keep it over for another year and make it bloom again. With some understanding of the plant's requirements this is not impossible, and certainly most rewarding. Most holiday plants are of a seasonal nature and require a decided rest period after bloom, and almost all of them should be plunged out into the garden for the summer period for recuperation and preparing for another indoor season.

Of the more important plants, **Azaleas** would be worth trying to keep over. While the hardy kind can simply be planted into the garden and left there for the future, the large-flowered indica type and other tender varieties should be left in pots on your terrace or buried to their rims in the semi-shade of the garden and left there until after the first light frost, protecting the developing soft buds from sunshine should the temperature go to 28°F of frost. When brought inside in fall, keep them cool until buds begin to swell. **Hydrangeas** are similar to Azaleas in that they also need a cold period in the fall to develop the flower buds. Prune them back to several inches above the pot and set them outside in the spring, keeping them well watered. In the fall keep them in a cool basement until January.

Chysanthemums can also be planted in the garden in the spring; cut back, and keep them growing in the garden until they are potted and brought inside in September before frost. **Cyclamen** after blooming should be allowed to rest, and later be plunged outside in a shady, moist and cool place under a tree, replacing the top soil with fresh compost and fertilizer added. **Gardenias** also are best moved outdoors in summer to a lightly shaded area, keeping them moist and fertilized regularly before moving back inside before heating is needed in the fall. **Roses** need too much sun to succeed indoors and should receive a permanent place in the garden, or a large tub on the patio. **Poinsettias** will ever be a challenge but by attention to a few essentials can be carried over for another Christmas. After flowering reduce watering and store in an airy place until May. Cut back to six inches and when the outside temperature stays above 60°F, plunge outdoors and feed, pinch to shape in August and bring back into a sunny window as nights become chilly. Poinsettias are "short-day" plants, and no light whatsoever, not even from a street lamp, must reach them at night at bud initiation time in September and October as this would "lengthen" the day and inhibit the setting of flowers. Also the night temperatures during this period must be kept below 70°F. but not less than 60°F.

Seasonal Lists of Popular Blooming Plants

By long established usage, the following plants have become popular and are generally grown commercially—

For Sweetheart's Day, February 14, and **early spring:** Azaleas, African Violets, Wax begonias, Cyclamen, Gloxinias, Amaryllis, Hyacinths, Tulips, Spathiphyllum, and arrangements of these plants in pottery.

For the Easter season: Hydrangeas with giant heads in red, white or blue; Azaleas in gay colors, Roses in pots, timed Chrysanthemums, Cinerarias, Calceolarias occasionally, fragrant Gardenias, Rose begonias, Marguerites, early Geraniums, Gloxinias and Saintpaulias; in bulbous stock, Easter lilies of course, then Tulips and Hyacinths, Grape hyacinths, potted Daffodils, yellow Callas, Lily-of-the-Valley. These bright-colored seasonal plants are very charming planted into Easter baskets.

For Mother's Day in May: Hardy Azaleas, late Hydrangeas, Gloxinias, Fuchsias, Tuberous and Wax begonias, Petunias, Hybrid and Polyantha roses, Cinerarias, Chrysanthemums.

For Memorial Day, May 30: Beginning with Mother's Day, bedding plants in four inch pots such as Geraniums, Fuchsias, Petunias single and double, Ageratum, Heliotrope, Wax begonias, Lantanas and Marigolds—all are nice little gift plants and ideal for a gayly colored window box. Combinations planted into large white pots add a bright spot of spring color on the patio or to decorate a grave.

For the Autumn season and Thanksgiving: Long-lasting Chrysanthemums in various shades are by far the natural leaders. Early Cyclamen, Kalanchoe, Saintpaulias, Heathers and Begonias usher in the indoor season when outdoor blooms begin to fade.

For Winter and Christmas time: More than any other plant the bright "Christmas star", or "Poinsettia", has been adopted as the symbol of the season. New types have been developed that are less sensitive to draft and change in temperature, and I have kept such new varieties in my office window in bright bloom for six months and more. Next in popularity are the so-called Indica Azaleas, pastel-shade Cyclamen, winter-flowering Erica; red Kalanchoe and African violets; long-flowering Christmas begonias of the cheimantha group are well adapted to our warmer homes but the otherwise so spectacular Dutch hiemalis type unfortunately drops its flowers unless it can be kept quite cool. Popular berried plants are the Calamondin orange with little golden fruit, Jerusalem cherries and gay Christmas peppers, and scarlet-berried Pyracantha, the Firethorn.

A large-flowered pot-grown Chrysanthemum morifolium hybrid of the decorative "Princess Ann" group, known as the "Florists chrysanthemum" and popularly nick-named "Mum plant". Year-round flowering can be scheduled following regulation of day-length through short-day treatment.

63

"Easter lily",
Lilium longiflorum 'Croft',
a herald of fragrant spring

"Gloxinia"
Sinningia speciosa 'Emperor Frederick'
in season spring and summer

"Dwarf Azalea indica",
Azalea simsii 'Hexe', a compact, free-flowering
tender variety for spring.

"Hortensia", or "Snowballs":
Hydrangea macrophylla 'Merveille',
a showy Easter and Mothers Day plant.

Florists "Mum plant":
Chrysanthemum morifolium 'Yellow Delaware';
normally blooms early November

"Poinsettia" or "Christmas star":
Euphorbia pulcherrima 'Barbara Ecke Supreme',
its starry bracts a holiday symbol

Azalea 'Constance' in Easter pottery;
cerise-pink rutherfordiana azalea

Azalea 'Alaska',
white rutherfordiana, winter flowering

Azalea 'Redwing',
large cerise-red (Pericat)

Azalea 'Pride of Detroit',
vivid red Pericat hybrid

Azalea 'Southern Charm',
"Southern Sun azalea", carmine rose

Azalea 'Sweetheart Supreme',
"Sweetheart rose azalea" (Pericat)

Azalea 'Ambrosiana' (grafted),
superb glowing red for Christmas

Azalea 'Leopold-Astrid',
white with crisped cherry-red margin

Azalea 'Triomphe',
double, frilled crimson-red (own root)

Azalea 'Mad. John Haerens',
deep rose, double flowers; vigorous

Azalea 'Lentegroet',
excellent crimson-red, medium flower

Azalea 'Hexe', dwarf indica,
free-blooming, glowing crimson

Azalea 'Hinodegiri', hardy kurume,
carmine-red, an old garden favorite

Azalea 'Hino-Crimson' (hardy)
single flowers clear crimson

Azalea 'Linwood Pink' (hardy)
frilled hose-in-hose, soft rose flower

Azalea 'Coral Bells' (kurume Kirin),
dainty coral pink, early

Azalea 'Peggy Ann', hardy,
lovely appleblossom white, edged rose

Azalea 'Roehrs Tradition',
deep carmine-pink, fairly hardy

Azalea 'Delaware Valley White',
single pure white, partially deciduous

Azalea 'Hershey Red', rosy crimson;
resembling indica but fairly hardy

Azalea 'Polar Bear', U.S.D.A. hardy,
pure white, hose-in-hose

Azalea 'Elizabeth Gable', hardy,
light purple; partially deciduous

Rhododendron 'Pink Pearl',
tender "Rose bay" for early forcing

Rhododendron 'Jean Marie de Montague',
bright crimson, for Easter forcing

Primula obconica
"German primrose", for winter-spring

Primula malacoides,
"Fairy primrose", flowers from January

Primula sinensis,
"Chinese primrose"

Primula x polyantha, for spring,
"Polyanthus primrose", in many colors

Chrysanthemum frutescens
"White Marguerite" or "Paris daisy"

Chrysanthemum frutescens chrysaster,
"Boston yellow daisy"

Senecio cruentus 'Grandiflora nana',
large flowered "Cineraria", for spring

Senecio cruentus 'Multiflora nana',
small-flowered "Cineraria" of florists

Calceolaria herbeohybrida 'Grandiflora',
the gorgeous "Giant pocketbook plant"

Calceolaria integrifolia (rugosa),
the "Bedding calceolaria"

Calceolaria herbeohybrida 'Multiflora nana',
"Lady's pocketbook", a cool spring plant

Schizanthus x wisetonensis,
"Butterfly flower" or "Poor man's orchid"

Cytisus canariensis, a spring bloomer,
bright yellow potted "Genista"

Astilbe japonica 'Gladstone',
perennial "Spiraea" forced for Easter

Gardenia jasminoides 'Veitchii',
fragrant "Everblooming gardenia"

Dianthus caryophyllus 'Cardinal Sim',
"American Remontant carnation" of florists

Celosia argentea cristata,
crested "Cockscomb" for summer bloom

Celosia argentea plumosa,
"Feather celosia" or "Burnt plume"

Convallaria majalis, a perennial,
forced "Lily-of-the-Valley", sweetly fragrant

Narcissus cyclamineus 'February Gold',
a compact "Pot narcissus"

Crocus moesicus,
"Dutch yellow crocus", in spring

Lachenalia aloides,
"Tricolor Cape cowslip"

Muscari botryoides album,
"Pearls of Spain"

Muscari armeniacum 'Heavenly Blue',
spring-blooming "Grape hyacinth"

Lilium 'Enchantment' (tigrinum hybrid)
bright orange-red "Mid-century lily"
for forcing or for garden planting

Lilium longiflorum 'Ace',
bushy American "Easter lily" with waxy-white
flowers medium-large but numerous

Zantedeschia elliottiana,
the "Yellow calla" of horticulture; cupped golden
spathe with leaves marked translucent silver

Hippeastrum x 'Leopoldii hybrid'
superb type of "Dutch amaryllis" with large,
open-faced trumpet flowers usually red

Tulipa x gesneriana 'Robinea', 6 inch comb. pan;
a stocky, late-blooming "Triumph tulip", a good
keeper and ideal for Easter.

Lilium longiflorum 'Croft',
an elegant "Easter lily" with long white bells
of firm texture, from the Pacific Northwest

Hyacinthus orientalis 'Delft Blue',
fragrant, porcelain-blue "Dutch hyacinth",
5½ inch pan comb., for Easter

Narcissus pseudo-narcissus 'Gold Medal',
large flowered, golden-yellow "Pot daffodil",
stocky and compact in 6 inch pan

Hydrangea macr. 'Strafford'
clear rosy-red, firm

Hydrangea macr. 'Enziandom'
for coloring gentian-blue

Hydrangea macr. 'Soeur Therese'
noble heads of pure white

Rosa x polyantha 'Margo Koster'
"Baby-rose", soft orange-red

Rosa x polyantha 'Mothers-day'
deep crimson "Baby rose"

Rosa x floribunda 'Spice'
very double salmon-red

Rosa 'Magna Charta'
carmine-rose "Hybrid Perpetual"

Rosa x floribunda 'Fashionette'
exquisite shell pink, fragrant

Rosa x grandiflora
'Queen Elizabeth'
free flowering soft rose-pink

Hydrangea macrophylla 'Chaperon Rouge'
(Red Ridinghood), also known as "Red Cap".
Superb "French hortensia" with vivid, pure
rosy-crimson flower heads.

Hydrangea macrophylla 'Merveille',
a bushy "French hortensia" of robust habit, and
good keeping qualities; large heads of carmine-rose
flowers, forced for Easter and Mother's Day.

Rosa 'Bonfire', a climbing "Rambler rose",
the long canes trained into basket shape
and forced for Easter; crimson-red.

Rosa odorata 'Mrs. W. C. Miller'
a dependable old "Hybrid Tea" rose of robust
habit flowering "monthly"; dainty pink flushed
with rose and intensely fragrant.

Fuchsia x hybrida 'Winston Churchill'
"Lady's eardrops"; red sepals, blue skirt

Begonia semperflorens 'Pink Pearl'
shapely 'Wax begonia", rich pink

Heliotropium arborescens
old-fashioned "Heliotrope", intensely fragrant

Lantana camara
"Shrub verbena", yellow to orange

Petunia x hybrida 'Pink Magic'
"Single petunia", ideal for bedding

Petunia x hybrida 'California Giant'
showy "Giant pot petunia"

Petunia x hybrida florepleno 'Caprice' all-American frilled "Double petunia" in lovely rose pink, as a 4 inch pot plant for Mother's Day or Summer

"Spring combination pan" of colorful bedding plants (clockwise): Coleus, Pigmy marigold, Giant petunia, Geranium, Dusty miller, Double petunia, French marigold, Ageratum

Pelargonium x hortorum 'Olympic Red', a good "Zonal geranium"; free bloomer with fiery red flowers; an old-fashioned window plant all summer

Pelargonium x domesticum "Springtime', noble "Lady Washington geranium" gay with large white flowers prettily zoned rose; best where cool.

Sinningia speciosa 'Emperor Frederick',
florists "Gloxinia', velvety ruby, edged white

Sinningia speciosa florepleno 'Chicago',
"Double gloxinia", crimson with rows of
frilled white lobes

Saintpaulia 'Kenya violet-blue',
vigorous "Kenya double violet", retains flowers well

Saintpaulia 'Blushing Bride',
"Pink African violet" daintily frilled

Begonia semperflorens fl. pl. 'Lady Frances'
everblooming "Rose begonia" with coppery foliage

Begonia tuberhybrida multiflora fl. pl. 'Maxima'
"Miniature tuberous begonia" for pots or bedding

Chrysanthemum morifolium 'Golden Lace'
"Spider mum" or "Fuji" type

Chrysanthemum morifolium 'Princess Ann'
excellent "Pot mum", of the "decorative" class

Chrysanthemum morifolium 'Beautiful Lady'
bushy with large "Anemone" type flowers

Chrysanthemum morifolium 'Handsome',
full "Pompon mum", purplish rose

Chrysanthemum morifolium 'Delaware',
"incurved" flowers deep red with yellow reverse

Chrysanthemum morifolium 'Golden Cascade'
"Cascade mum" as bush, with daisy-like flowers

Cyclamen persicum giganteum flore pleno,
"Double cyclamen", blooms with many petals above
bluish leaves patterned with silver

Cyclamen persicum 'Perle von Zehlendorf',
"Florists cyclamen" or "Alpine violet"
winter-flowering vivid salmon, for cool locations

Erica gracilis, the "Rose heath"
covered with tiny rosy bells from autumn on

Erica melanthera, the "Christmas heather",
feathery shrub snowy with pinkish flowers
in winter

Euphorbia pulcherrima 'Paul Mikkelsen',
sturdy "Poinsettia" with bright crimson bracts on
compact plants holding foliage well

Euphorbia pulcherrima 'Eckespoint' Cl Red,
stocky diploid "Poinsettia" intense scarlet red
crinkled 12 inch bracts perfect for Christmas

Euphorbia pulcherrima 'Ecke's White',
snowy "White poinsettia" with stars of glistening
bracts lasting a long time

Euphorbia pulcherrima plenissima 'Henrietta Ecke',
this showy "Double poinsettia" carries an extra
crown of red above its normal, vermilion bracts

Kalanchoe blossfeldiana 'Tom Thumb'
a dwarf, very compact "Flaming Katy"
with masses of red flowers in late winter

Kalanchoe' Yellow Darling' (Gelber Liebling),
"White kalanchoe" in 4 inch pot, unusual
and attractive with clusters of creamy flowers

Kalanchoe 'Grob's Rotglut',
"Red glow kalanchoe" in 5½ inch pot
the fiery red flowers won't "go to sleep"

Kalanchoe 'Orange Triumph', very robust,
in the trade as "Gravenmoer's Glorie"
large coral-red flowers which will stay open

Begonia x hiemalis 'Apricot Beauty'
"Apricot winter begonia", best where cool

Begonia x cheimantha 'Lady Mac'
"Christmas begonia" or "Busy Lizzie begonia"

**Begonia x cheimantha
'Dark Lady Mac'**
"Dark Christmas begonia"

Begonia x hiem. 'The President'
"Dutch Christmas begonia"

Begonia x hiemalis 'Emita'
"Rosy winter begonia"

Begonia x ch. 'White Christmas'
"Improved Turnford Hall begonia"

Begonia x cheimantha 'Marina'
"Scandinavian winter begonia"

Begonia x cheim. 'White Marina'
"White Scandinavian begonia"

Fruit usually grows on trees or shrubs in orchards or our gardens. But since the time of ancient Egypt in 1530 B.C. and of Babylon in 1137 B.C. man has transplanted fruit trees into containers to bring into gardens or his home. Miniature fruit trees, or plants resembling fruit trees will continue this tradition ideally even in limited space.

Solanum pseudo-capsicum, the "Jerusalem cherry" or "Christmas cherry" loaded in winter with orange-scarlet, cherry-like fruit

Citrus taitensis, the "Otaheite orange" or "Dwarf orange", an orange tree in miniature, potgrown for its fragrant flowers and ornamental golden fruit

Capsicum annuum 'Birdseye', the "Christmas pepper", a cheerful Christmas plant covered candle-like with waxy, berry-like fruits scarlet red.

Citrus mitis, the "Calamondin" orange, loaded at Christmas time with numerous small golden-yellow fruits

Pyracantha koidzumii 'Victory', "Red Fire-thorn", a robust shrub with long-lasting scarlet berries in winter

BULBOUS Flowering Plants

In your living room, forced Narcissus x poetaz 'Geranium', a "Cluster narcissus", its clay pot dressed in foil to hold moisture—heralds spring with sweet fragrance and cheerful color while a blanket of snow may still be covering the bulbs waiting outdoors.

Bulbs indoors are undoubtedly among the most rewarding kinds of plants and their beauty and versatility is unlimited; they may be had in bloom throughout the year, from January through December. We may divide our best known bulbous subjects into two groups: the hardy garden types that are planted into pots in autumn and forced for winter bloom as temporary house plants, and the tender, tropical species that may be grown indoors the year round—except for a time of dormancy and rest —and which bloom at their normal season.

TYPES OF BULBS—What we know as "bulbs" is a horticultural term and may properly include true bulbs, corms, tubers, rhizomes or tuberous root-stocks, and that may be dried off and cured and stored over winter. Such "bulbs" are actually highly developed plants, whose dormant life awaits the call to life, to break out of their tunic prison and unfold with leaves and showy flowers to full beauty. What bulbs have in common is an underground food store, and the fact that they can be preserved out of the ground in a dormant state.

A true **Bulb** consists of a central growth bud sheathed in graded layers of modified white leaves known as scales. Such bulb plants include Amaryllis, Eucharis, Haemanthus, Hippeastrum, Hyacinths, Lilies, Narcissus and Tulips. **Corms** are very bulb-like but solid, consisting of a swollen, fleshy base of the stem, which stores up the nourishment gathered by the leaves during the period of vegetation. Cormous plants are Crocus, Freesia, Gladiolus, Ixia, Sparaxis, Tritoma. The **Tuber** is a thickened subterranean stem looking like an irregular sphere; it has no tunic and no basal plate, but a rough leathery skin and forms widespread roots. On its surface are the latent buds, so well known in the potato. Some of the Begonias are tuberous; so are Caladiums, Colocasias, Cyclamen, Gloriosa, many Oxalis, Rechsteinerias, and Sinningias. Sometimes tubers form in the air, as in Dioscorea. The **Rhizome** is thinner and more elongated in shape than the tuber and is formed by the stem of the plant. It is in fact a rootstock or underground stem, bearing buds. Better known rhizomatous plants are Anemone, Cannas, Iris, Zantedeschia, in horticulture as Calla; and the Gesneriads Achimenes, Kohleria and Smithiantha with their scaly rhizomes. **Tuberous Roots** are true, swollen roots which have no buds on the root itself but congregate round the base of the stem, so typical in Dahlias and Ranunculus.

HISTORY AND GEOGRAPHY—Asia Minor and the countries around the Mediterranean are the cradles of the classic bulbous plants so popular for winter forcing—Tulips, Hyacinths, Crocus and Narcissus. Lilium candidum, the fragrant Madonna lily, has been recorded in Egypt in 2500 B.C.; King Solomon caused the columns of his Temple to be decorated with lilies. Tulips have been cited in literature by Omar Khayyam back in the 12th century. The flora of South Africa has given us many non-hardy bulbous plants, such as Agapanthus, Ixia, the fragrant Freesia, and the popular Gladiolus. From tropical Africa come the climbing Gloriosas. North America has few spectacular species other than Lilies, but Central and South America are the home of the Dahlia, Sprekelia and Tigridia, and the more tropical Caladium, Canna, Hippeastrum, Hymenocallis, and Sinningia, our "Gloxinia". Most of our splendid Lilies have originated in Eastern Asia.

BULBS IN THE HOME—To force bulbs for flowering indoors is one of the easiest and most certain of all types of flower growing. All hardy bulbs such as Tulips, Hyacinths, Narcissus or Muscari are planted into pots late in the fall in a sandy, friable soil, thoroughly soaked with water, and stored in a cool, dark cellar, not over 48° F, for 6-8 weeks, and watered every two weeks thereafter, or better still, buried outside until well rooted. When wanted, bring them inside into some light, keeping the pots not to exceed 60°F. for the first 8-10 days, after which they may be given full light and kept at regular room temperature, taking care to water daily. Tender bulbs in pots that cannot endure freezing make satisfactory house plants the year around, or placed outdoors during summer. Best known of these are Amaryllis, or Hippeastrum, with their giant trumpet blooms. Many other tropicals give a joyful succession of bloom, such as Lachenalia, Tuberous begonias, Hymenocallis, Veltheimia, Clivia, Crinum, Eucharis, Nerine, Vallota, Tuberoses, Gloriosas, Haemanthus, Neomarica, Freesia, Oxalis, Ornithogalum and "Callas"; also the Gesneriads Achimenes and Gloxinia. The possession of a small greenhouse increases tremendously the pleasure which may be had from bulbous plants. With "glass" one can enjoy many kinds that cannot be handled in the house. Also, plants wanted for room decoration can be cared for in the greenhouse until they come into flower, and returned to it afterward for resting, to be used another year. Thus such window space or other locations in the home can constantly be replenished with plants always in bloom.

"Hardy" bulbs can be planted outdoors in the spring, where many of them, if not too exhausted, will proliferate in the garden. Tender bulbous plants will renew themselves, or can be increased through scales, offsets, bulbils or by division if kept growing after flowering until their foliage is fully developed and going into dormancy.

At Eastertime, passersby on Fifth Avenue in midtown New York like to stop in front of the show windows of a beauty-conscious bank to admire their seasonal display of white lilies and blue and pink hyacinths, against a background of yellow acacias and the vivid-green of Hawaiian treeferns.

Hippeastrum vittatum
"Striped amaryllis"

Hippeastrum (leopoldii) 'Claret'
crimson "Dutch amaryllis"

Amaryllis belladonna
the "Cape belladonna"

Hippeastrum striatum fulgidum
"Everblooming amaryllis"

Hippeastrum puniceum
"American belladonna"

**Hippeastrum reticulatum
striatifolium**
"Stripe-leaf amaryllis"

Hippeastrum procerum
"Blue amaryllis"

**x Crindonna memoria-corsii
(x Amarcrinum howardii)**
"Amaryllis-Crinum hybrid"

Hippeastrum psittacinum
"Parrot amaryllis"

Crinum 'Ellen Bosanquet'
"Red angel lily"

Crinum x powellii
"Powell's swamp lily"

Crinum 'Roozenianum'
a white Swamp-lily

Brunsvigia 'Josephine hybrid', carmine rose
the "Josephine's lily"

Crinum asiaticum
"Poison bulb" or "Grand crinum"

x Brunsdonna parkeri
"Naked lady lily"

Crinum giganteum hort.
"Giant spider lily"

Crinum moorei
"Longneck swamp lily"

Alstroemeria ligtu
"Inca lily"

Eucharis grandiflora
"Amazon lily"

Alstroemria pelegrina
"Peruvian-lily"

Clivia nobilis
"Greentip kafir lily"

Clivia miniata 'Grandiflora'
"Scarlet kafir lily"

Vallota speciosa
"Scarborough lily"

Nerine curvifolia fothergillii
"Curveleaf Guernsey lily"

Polianthes tuberosa
sweetly scented "Tuberose"

Nerine masonorum
"Dwarf pink Guernsey lily"

Hymenocallis caribaea
"Caribbean spider lily"

Hymenocallis americana
"Crown beauty"

Hymenocallis speciosa
"Winter spice hymenocallis"

Hymenocallis narcissiflora
(Ismene calathina)
"Basket flower"

Pamianthe peruviana
"Peruvian daffodil"

Haemanthus albiflos
"White paint brush"

Haemanthus katherinae
"Blood flower"

Haemanthus multiflorus
"Salmon blood lily"

Sprekelia formosissima
"Jacobean lily"

Hypoxis hygrometrica
"Star-grass"

Ixiolirion montanum
"Siberian lily"

Zephyranthes grandiflora
"Zephyr-lily"

Galanthus elwesii
"Giant snowdrop"

Habranthus andersonii texanus
"Texas amaryllis"

Phaedranassa carmiolii
"Queen-lily"

Cyrtanthus mackenii
"Bow-lily"

Stenomesson incarnatum
"Crimson stenomesson"

Ipheion uniflorum
"Spring-star flower"

Lycoris aurea
"Golden spider lily"

Leucojum aestivum
"Summer snowflake"

Neomarica gracilis
"Apostle plant"

Neomarica northiana
"Walking iris"

Trimeza caribaea
"Long House iris"

Moraea iridioides
"African iris"

Moraea glaucopsis
"Peacock iris"

Tigridia pavonia
"Tiger flower"

Crocosmia masonorum
"Golden swan tritonia"

Freesia x hybrida
intensely fragrant

Ferraria undulata
"Orchid iris"

Babiana stricta rubrocyanea
red and blue "Baboon flower"

Chionodoxa sardensis
"Glory of the snow"

Streptanthera cuprea
"Kaleidoscope flower"

Hesperantha stanfordiae
"Evening flower"

Sisyrinchium striatum
"Satin flower"

Ixia x 'Rose Queen'
"Rosy corn-lily"

Homeria collina ochroleuca
"Cape tulip"

Crocus 'Vernus hybrid'
"Alpine spring crocus"

Tritonia deusta
"Blazing stars"

Iris x 'Wedgwood'
"Dutch forcing iris"

Iris histrio
"Syrian iris"

Iris pallida 'Argentea'
"Variegated sweet iris"

Acidanthera bicolor murieliae
"Darkeye gladixia"

Gladiolus x hortulanus 'Valeria'
"Florists glads" or
"Painted lady"

Watsonia humilis maculata
"Southern bugle-lily"

Homeria lilacina
"Lilac shower"

Hermodactylus tuberosus
"Snake's-head iris"

Sparaxis tricolor alba
"Harlequin flower"

Lachenalia aloides nelsonii (syn. tricolor)
"Yellow Cape cowslip"

Lachenalia bachmanii
"White Cape cowslip"

Lachenalia lilacina
"Lavender Cape cowslip"

Agapanthus africanus
"Blue African lily"

Lachenalia purpureo-caerulea
"Purple Cape cowslip"

Veltheimia glauca 'Rosalba
"Pinkspot veltheimia"

Veltheimia viridifolia
"Forest-lily"

Kniphofia uvaria
"Torch lily"

Lilium 'Imperial Crimson'
"Empress hybrid lily"
with originator Jan de Graaff in Oregon

Lilium longiflorum 'Croft'
firm-textured elegant "Easter lily" for pots
from the Pacific Northwest

Lilium tigrinum
"Tiger lily"

Lilium tigrinum splendens
"Red tiger-lily"

Lilium regale
fragrant "Regal lily"

Gloriosa rothschildiana
climbing "Glory lily"

Lilium speciosum
"Japanese lily"

Lilium 'Harmony'
orange "Mid-century lily"

Muscari armeniacum 'Heavenly Blue'
spring-flowering "Grape hyacinth"

Muscari racemosum
"Starched grape-hyacinth"

Fritillaria meleagris
"Checkered lily" or "Snake's head"

Chionodoxa luciliae
"Glory of the snow"

Miniature narcissus, left to right:
'W. P. Milner', odorus 'Orange Queen', bulbocodium,
triandrus 'Silver Chimes', tazetta canaliculatus

Narcissus tazetta papyraceus
the fragrant "Polyanthus narcissus" or "Paper-white"

Tulipa 'Makassar'
yellow Triumph tulip

Tulipa 'Kees Nelis'
Triumph tulip red, edged yellow

Tulipa 'Blizzard'
white Triumph tulip

Narçissus 'King Alfred'
Yellow trumpet daffodil

Tulipa 'Red Giant'
scarlet Triumph tulip

Tulipa 'Livingstone'
double red "Peony-flowered tulip"

Hyacinthus 'Ostara'
violet blue Dutch hyacinth

Hyacinthus 'Carnegie'
Dutch hyacinth pure white

Hyacinthus 'Pink Pearl'
rose pink Dutch hyacinth

Ornithogalum umbellatum
"Star of Bethlehem"

Ornithogalum thyrsoides
"Chincherinchee"
or *"African wonder flower"*

Eremurus elwesii
"Desert candle"

Urginea maritima, the "Sea onion" or "Squills", from the Mediterranean, and colonized in Normandy;
left to right: bulb flowering before leaves appear; close-up of inflorescence; pot plant with foliage after blooming.

Eucomis zambesiaca
"Pineapple lily"

Ornithogalum caudatum
"False sea-onion"· or "Healing onion"

Allium triquetrum
ornamental *"Triangle onion"*

Scilla sibirica 'Spring Beauty'
"Siberian hyacinth"

Scilla peruviana
"Cuban lily" from Madeira

Scilla violacea
"Silver squill"

Hemerocallis lilio-asphodelus
"Tall yellow day-lily"

Scilla tubergeniana
"Persian squill"

Tulbaghia fragrans
"Pink agapanthus"

Convallaria majalis 'Rosea'
"Pink lily-of-the-valley"

the species Cyclamen persicum *(left) and its development through
horticultural effort into the present large-flowered Cyclamen persicum
giganteum (right), the "Florists cyclamen"*

Anemone coronaria
"Poppy anemone" or "Lily of the field"

Sinningia speciosa florepleno
"Double gloxinia" of horticulture

Astilbe japonica 'Deutschland'
"Florists spiraea"

Ranunculus asiaticus
"Persian buttercup"

Paeonia albiflora festiva
"Double-flowered peony"

Dahlia pinnata 'Unwin's Dwarf'
"Dwarf Aztec dahlia"

Begonia x tuberhybrida
'Camellia type'
"Scarlet tuberous begonia"

Dahlia pinnata 'Pompon strain'
"Pompon garden dahlia"

Zantedeschia aethiopica
'Compacta'
"Dwarf white calla"

Zantedeschia rehmannii
"Pink calla"

Zantedeschia albo-maculata
"Spotted calla"

Zantedeschia elliottiana
"Yellow calla" of hort.

Zantedeschia aethiopica
"White calla-lily"

Canna coccinea
"Scarlet Indian shot"

Canna 'Striped Beauty'
"Variegated Bangkok canna"

Canna generalis 'President'
red *"Garden canna"*

Canna generalis 'King Humbert'
"Bronze garden canna"

BEDDING or GARDEN PLANTS for the Window and outside it

An operating Flower clock of bedding plants of contrasting colors in Golden Gate Park, San Francisco. Favorite silver-grays are Echeverias, Dusty Millers and Santolinas; Wax begonias, Alternanthera and Iresine for red and yellow. But I have also watched gardeners there to paint some of the plants if certain bright colors were needed for their scheme.

A bed of flowers, artistically planted, is the gardener's way of expressing his art, in the same way as the painter expresses himself on canvas. Although the interest in gardening goes back thousands of years to the culture of the Assyrians and Egyptians, these were primarily interested in exotic trees producing fruit and spice. From medieval times, starting with Charlemagne in 812, emphasis was on gardens of healing and culinary herbs with their rather somber colors. Annuals and bi-ennials for ornament as grown today came into use in cottage gardens only with the 16th century, and in America the gardens of the Governor of Virginia about 1720 showed a good example of decorative bedding. But the wide assortment of semi-hardy and tender plants as we know it today was made possible only with the spontaneous evolution of the glass planthouse in the 18th century. An example and the most spacious greenhouse in Europe at the time was built at Schoenbrunn in Vienna in 1753, followed amongst others by Paxton's house at Chatsworth in 1836. With the unfolding era of the prosperous and peaceful Victorian times, bedding plant gardening came into highest bloom. Having greenhouses to winter the propagating stock of tender plants, cuttings and seedlings could be started in late winter, for "bedding out" in spring with the advent of warm weather. This made possible the tremendous popularity of such flowering bedding plants as Geraniums, Fuchsias, Heliotrope and Lantanas; mini-Calceolarias, Cuphea and Marguerites, all of which are propagated from cuttings, together with the colorful foliage of Coleus, Alternantheras and Iresine. Usually raised from seed is a long list of smaller annuals and biennials that may be sown between January and March. For germination a warm temperature between 65° and 75° F is best, then keeping cooler between 50-58° F to harden growth. The most popular of annuals are Ageratum, Alyssum, Amaranthus, dwarf Aster, Balsam, Begonia semperflorens, Celosia, hybrid Coleus, Dianthus, Dusty Miller, Gomphrena, Impatiens, Lobelia, Marigold, Mignonette, Myosotis, Petunias, dwarf annual Phlox, Salvia, Snapdragon, dwarf Stocks, Torenia, Verbenas, Vinca rosea (Catharanthus), and Zinnias. An early planting of Pansies and Bellis may precede these summer-flowering annuals in bloom.

BEDDING PLANTS FOR POTS—While we think of bedding plants as subjects to beautify the summer garden, it is fun and a source of pride to grow them in the open window, in pots or boxes, on the sunporch or the patio. Nothing will be as profuse with gaily-colored bloom all summer long as a selection of bright Geraniums, in scarlet or pink; sun-loving, multi-colored Petunias; Ageratum, Verbenas and Lantanas, Marigold and Phlox; Wax begonias, Heliotrope or Fuchsia. We cannot help but think nostalgically of the charming window boxes spilling over with gay color, and so popular in all Alpine countries. A combination of all kinds of annuals can also be grown in jardinieres, large stoneware urns, and porchboxes—a Cordyline australis in the center and Vinca vines planted round the outside. Such sun-lovers will attractively frame a doorway, or enhance the patio, where even as a hanging basket they would find an ideal summer spot.

The planting and maintenance of a formal, European style "Carpet bed" requires high professional skill to create intricate and symmetrical designs of lasting attraction through the summer. The Channel Gardens at Rockefeller Center in New York provide an excellent opportunity to observe and admire such colorful patterns primarily of vari-colored Alternantheras with Echeverias.

Most Bedding Plants can easily become Window plants. As a rule they are lovers of sun, abundantly found near a good window. A geranium in bloom, miniature roses and wax begonias appear to be happy on this broad window sill. Most bedding plants require sufficient water to bloom, and trays are a necessity. Normally a window should and would be open to let in fresh air to please these garden subjects, but the plants in this Madison Avenue office in New York have adjusted themselves quite well to modern air-conditioning.

Most treasured in this widely varied group of plants are members of the Geranium family in their many species and cultivated forms. Easily the most popular bedding plant is the zonal geranium, Pelargonium x hortorum, with large heads of flowers in fiery red, soft pink or salmon, pure white, or even purple. Some geraniums are known for their highly variegated foliage, especially the "Tri-colors", and these as well as some of the miniature geraniums are good house plant subjects the year around. The Ivy geraniums are especially favored for window boxes and baskets, some with white edged leaves, or delicate pastel-shade flowers.

Pelargonium x domesticum, the fancy "Martha Washington" geraniums, have large and strikingly variegated flowers but they must have cool surroundings to succeed around the home, and their flowering period is brief.

Fuchsias, the "Lady's eardrops", in hundreds of named varieties, are beloved by many fanciers, but need humidity with coolness and are not good subjects for the warm living room unless by a semi-shaded window with fresh air, or better still, on a glassed-in sunporch. Fuchsias generally are good hanging basket subjects where the place is suitable, and dripping is no problem. Begonia semperflorens, as "Wax begonias", or better still, the double-flowered "Rose begonias" such as 'Lady Frances' are quite satisfactory in warm rooms, but again, are better near the window or under light. Warm rooms will encourage Red spider and White fly on Verbenas, Lantanas and the fragrant Heliotrope. As a foliage plant, Coleus, in variety, will do remarkably well in the home but may lose some of their brilliant color if away from good light. Most other plants in this group of bedding "Annuals" do not feel well in warm rooms except in the strong light at the window or under artificial light. There they can give us, with little care but sufficient watering, a succession of flowers beginning with late spring until the sun gives way to the dark days of winter.

CARPET-BEDDING—Formerly the art of "Carpet planting" was very much in favor in formal gardens, and the greatest technical skill was required not only for artful design but in constant care and trimming. Fanciful geometric patterns, insignia or scrolls can be created with striking effect, composed of plants of diminutive habit such as many-hued Alternantheras, red or yellow Iresines and Coleus, blue Lobelias, and Wax begonias, outlined by silvery Santolinas, Dusty Millers and succulent Echeverias, a sight in public gardens which never fails to impress the traveler to Europe or the Canadian West.

Celosia argentea pyramidalis
"Plume celosia"

Celosia argentea cristata
crested "Coxcomb"

Amaranthus hypochondriacus
"Princess feather"

Impatiens oliveri
"Giant touch-me-not"

Impatiens balsamina
"Garden balsam"

Amaranthus tricolor
"Joseph's coat"

Impatiens walleriana var.
sultanii 'Variegata'
"Variegated patient Lucy"

Impatiens platypetala aurantiaca
"Tangerine impatiens"

Impatiens walleriana var. holstii
"Busy Lizzie"

Iresine herbstii 'Acuminata'
"Painted bloodleaf"

Iresine herbstii 'Aureo-reticulata'
"Chicken gizzard"

Iresine lindenii formosa
"Yellow bloodleaf"

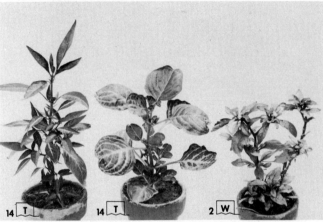

Iresine lindenii, "Bloodleaf";
Iresine herbstii, "Beefsteak plant";
Alternanthera amoena, "Parrot leaf"

Alternanthera bettzickiana
"Red calico plant"

Alternanthera amoena 'Brilliantissima', "Red parrot-leaf":
Alt. bettzickiana 'Aurea', "Yellow calico plant";
Alternanthera versicolor, "Copper leaf".

Alternanthera dentata 'Ruby'
"False globe amaranth"

Coleus blumei 'Brilliancy'
"Flame nettle"

Coleus blumei verschaffeltii
"Painted nettle"

Coleus blumei 'Pride of Autumn'
"Autumn nettle"

**Coleus renneltianus
'Lord Falmouth'**
"Red trailing nettle"

Coleus thyrsoideus
"Flowering bush coleus"

Coleus blumei 'Klondyke'
"Butterfly coleus"

Coleus blumei 'Candidum'
"White coleus"

Coleus blumei 'Pink Rainbow'
"Showy coleus"

Coleus blumei 'Defiance'
"Painted leaf"

Coleus rehn. 'Trailing Queen'
"Trailing coleus"

**Coleus blumei
'Frilled Beauty'**

Coleus pumilus
"Dwarf creeping coleus"

Ageratum houstonianum
'Midget blue'
"Floss-flower"

Begonia semperflorens 'Pink Pearl'
"Rosy wax begonia"

Heliotropium arborescens
"Heliotrope"

Phlox drummondii
"Dwarf annual phlox"

Callistephus chinensis
'Dwarf Queen'
"Dwarf pot aster"

Calendula officinalis
'Apricot Queen'
"Pot-marigold"

Myosotis sylvatica 'Compacta'
dwarf *"Forget-me-not"*

Tagetes patula 'Spry'
"Dwarf French marigold"

Tagetes patula 'Naughty Marietta'
"Single French marigold"

Lantana camara
"Shrub verbena"

Verbena peruviana 'Chiquita'
"Peppermint-stick verbena"

Senecio (Centaurea) leucostachys
"Tall dusty miller"

Echeveria peacockii
"desmetiana" of hort.

Santolina chamaecyparissus
"Lavender cotton"

Centaurea cineraria
"Dusty miller"

Senecio cineraria
(Cineraria maritima 'Diamond')
"Silver groundsel"

Centaurea candidissima
"Woolly-white miller"

Zinnia elegans 'Pumila'
"Lilliput zinnia"

Lobularia maritima
"Sweet alyssum"

Salvia splendens
"Scarlet sage"

Gomphrena globosa
"Globe amaranth", "Immortelle"

Bellis perennis florepleno
"English daisy"

Verbena x hortensis
"Garden verbena"

Exacum affine
"Persian violet"

Viola tricolor hortensis
"Garden pansy"

Viola cornuta 'Bluette'
"Horned violet"

Fuchsia x hybrida 'Black Prince'

Fuchsia x hybrida
'Winston Churchill'

Fuchsia x hybrida 'Rollo'

Fuchsia x hybrida 'Chickadee'

Fuchsia x hybrida 'Lace Petticoats'

Fuchsia x hybrida 'Pride of Orion'

Fuchsia magellanica variegata
"Variegated hardy fuchsia"

Fuchsia triphylla
'Gartenmeister Bohnstedt'
"Honeysuckle fuchsia"

Fuchsia x hybrida
'Swanley Yellow'

Fuchsia x hybrida 'Sunray'
foliage with cream margins

Fuchsia x hybrida 'Golden Marinka'
leaves variegated golden yellow

Fuchsia x hybrida 'Ting-a-ling'
bell-flowers all-white

Fuchsia x hybrida 'Jubilee'
skirts dark cerise, sepals flushed rose

Fuchsia x hybrida 'Fluorescent'

Fuchsia x hybrida 'Swingtime'

Fuchsia x hybrida 'Brigadoon'

Linaria maroccana
"Toadflax"

Delphinium ajacis
annual "Larkspur"

Chrysanthemum parthenium
(Matricaria capensis)
"Feverfew"

Nierembergia caerulea
blue "Cup flower", with yellow eyes

Nemesia strumosa
"Cape jewels" or "Pouch nemesia"

Euphorbia marginata
"Snow on the mountain"

Helichrysum bracteatum
"Straw flower"

Kochia scoparia trichophylla
"Summer cypress"

Gypsophila elegans
"Baby's breath"

Godetia amoena plena
"Farewell to-spring"

Clarkia elegans
"Rose clarkia"

Reseda odorata
sweet-fragrant "Mignonette"

Schizanthus x wisetonensis
"Poor man's orchid"

Torenia fournieri
"Wish-bone plant"

Linum grandiflorum
"Flowering flax"

Catharanthus roseus
(Vinca rosea in hort.)
"Madagascar periwinkle"

Helianthus annuus florepleno
"Double sunflower"

Mathiola incana
"Stocks" or "Gilliflower"

Antirrhinum majus
"Snapdragon"

Cheiranthus cheiri
"English wallflower"

Nicotiana alata grandiflora
"Jasmine tobacco"

Convolvulus tricolor
"Dwarf morning glory"

Nicotiana x sanderae
"Flowering tobacco"

Moluccella laevis
"Bells of Ireland"

Penstemon x gloxinioides
"Beard-tongue"

Digitalis purpurea
"Foxglove"

Dianthus x 'Mitt's Pinks'
"Baby carnation"

Dianthus barbatus
"Sweet William"

Limonium sinuatum
"Statice" or *"Sea-lavender"*

Gazania rigens hybrid
"Treasure flower"

Phlox paniculata
"Summer perennial phlox"

Coreopsis auriculata 'Nana'
"Dwarf tickseed"

Bergenia ciliata ligulata 'Rosea'
rose-flowered "Winter begonia"

Leontopodium alpinum
the woolly, silvery-white "Edelweiss"

Canna generalis 'President'
red "Garden canna"

Tulipa fosteriana princeps
"Scarlet rockery tulip"

**Begonia tuberhybr. multiflora fl. pl.
'Helene Harms'**
"Lilliput tuber-begonia"

Dahlia pinnata 'Unwin's Dwarf'
"Dwarf Aztec dahlia"

Helleborus niger
"Christmas rose"

Hosta plantaginea
"Fragrant plantain lily"

Astilbe japonica rubens
"Rosy florists spiraea"

Aquilegia 'Long-spurred hyb.'
"American columbine"

Dicentra spectabilis
"Bleeding heart"

Petunia x hyb. grandiflora
'Crusader'
"Bicolored petunia"

Petunia x hyb. fl. pl. **'Sonata'**
"Double white petunia"

Petunia x hyb. **'California Giant'**
giant patterned single

Petunia x hyb. grandiflora
'White Cascade'
hanging "Balcony petunia"

Petunia x hyb. grandiflora
'Popcorn'
single white ruffled

Petunia x hyb. multiflora
'Celestial Rose'
"Dwarf bedding petunia"

Petunia x hyb. multiflora
'Satellite'
small-flowered bicolor

Petunia x hyb. grandiflora
'Elk's Pride'
"Blue balcony petunia"

Petunia x hybrida fl. pl.
'Caprice'
frilled double rose

Pelargonium x hortorum
'Salmon Supreme'
prolific salmon pink

Pelargonium x hortorum 'Irene'
popular pot geranium

Pelargonium x hortorum
'Improved Ricard'
brick red geranium

Pelargonium x hortorum
'Beaute' Poitevine'
salmon French geranium

Pelargonium x hortorum 'Carefree Deep Salmon'; one of several
outstanding color forms of a new race of garden geraniums grown directly
from seed, photographed in bloom during July (Pan American Seed Co.)

Pelargonium x hortorum 'Appleblossom Rosebud'
"Rosebud geranium", white fl. edged pink

Pelargonium x hortorum 'S. A. Nutt'
old fashioned zonal geranium, crimson

"Geraniums" have always been one of our most loved flowers not only in gardens, but in the sunny window, on the porch or a special place in the roofgarden. However, the ordinary zonal geraniums with their brightly colored cluster flowers we see about us everywhere are not in cultivation for as long a time as their importance today might indicate. It was a true Geranium from Greece that was first mentioned in 64 A.D. by Dioscorides, a genus having regular flowers with 5 petals and without spurs. When geraniums from South Africa were first introduced in the 17th century, it was discovered that these species had irregular flowers with the two uppermost petals being larger, and a spur behind resembling a storks-bill and now classified as Pelargoniums.

Pelargoniums, the cultivated 'Geraniums'

Although we still know them as "Geraniums" today, it is these Pelargoniums that have increasingly come into public favor so that today we know some 300 species and 6000 cultivars in this genus. At first, it was the scented geraniums that were introduced from the Cape of Good Hope and loved by the Dutch and the English people because of their foliage which exudes aromatic oils, having pungent or fruity flavors like those of roses, apple, orange or lemon, strawberry and apricot; nutmeg, peppermint, ginger and spice, or pine. These old-fashioned plants are most easy to grow in the home, preferring a cool temperature and flowering in spring, but may need to be severely pinched to prevent them from growing out of bounds in pots.

Enthusiasm went wild in England with the appearance of fancy-leaved geraniums in 1845, and to this day nothing can equal in attractiveness the tricolored foliage of 'Miss Burdett Coutts' which originated in 1850 after decades of hybridizing with the zonal geraniums from South Africa. Colored leaf geraniums like it warm and fairly dry, also it is best to hold fertilizer to a minimum.

Where space permits, our most popular geraniums today are the large-flowered, brilliantly colored Pelargonium x hortorum (zonale), offered in many good commercial varieties and colors. They tolerate a wide range of conditions but are happiest in a sunny spot and kept on the dry side. Roots are best confined for if they are allowed to spread too freely they may go into luxuriant foliage at the expense of prolific bloom. With sunshine, geraniums will flower the year round, best in summer.

This is where the present trend to miniature geraniums has received a tremendous boost in popularity because they easily retain their dwarfness in 2¼ or 3 inch pots, ranging in height between 3 and 8 inches. Despite this diminutive size, their flowers are relatively large, bloom boldly if kept sunny and can be kept in the same pot for several years; quite a number will fit in a limited area or on shelves in the window.

The ivy-leaved geraniums first made their appearance in 1698. They are of trailing habit, so useful in window boxes and hanging baskets. One of the most attractive is Pelargonium peltatum 'L'Elegante', its foliage edged in white and tinted pink.

The "Lady Washington" pelargoniums are spectacular in flower but their blooming period is short and they grow best only under cool conditions. They are not too satisfactory as a house plant, refusing to bloom if kept much over 60°F.

There are a number of odd geraniums such as the Rosebud or Poinsettia-flowered, and Succulent or Cactus-type Pelargoniums, all of which will certainly intrigue the collector. All geraniums are easily propagated from stem cuttings, best taken with beginning of fall, leaving the cut to dry for a day, and rooting them in clean sand.

Charming all-dwarf or miniature geraniums ideally suited for a bright window which because of their small size can accommodate twice the number of "regulars" on sill and glass shelves.

Pelargonium x hortorum 'Alpha',
*small bronze and
gold leaf geranium*

Pelargonium x hortorum
'Variegated Kleiner Liebling',
gray, small leaves bordered white

Pelargonium x hortorum
'Distinction'
blackish zone on light green

Pelargonium x hortorum
'Mme. Salleron'
non-flowering variegated

Pelargonium peltatum 'L'Elegante'
"Sunset ivy geranium"

Pelargonium peltatum
"Ivy geranium"

Pelargonium peltatum 'New Dawn',
double flowers vivid rose cerise

Pelargonium x hortorum 'Velma'
bold tricolor-leaf

Pelargonium x hortorum
'Skies of Italy'
"Maple-leaf tricolor geranium"

Pelargonium x hortorum
'Miss Burdett Coutts'
beautiful "Tricolor geranium"

Pelargonium x hortorum 'Pastel'
flowers salmon pink

Pelargonium x hortorum
'Happy Thought'
"Butterfly geranium"

Pelargonium x hortorum
'Wilhelm Langguth'
flowers cherry red

Pelargonium x hortorum
'Mountain of Snow'
silver green bordered white

Pelargonium x hortorum
'Mrs. Pollock double'
double-flowered tricolor

Pelargonium x hortorum
'Mrs. Parker'
soft pink blooms

Pelargonium x hortorum 'Tangerine',
miniature with salmon flowers

Pelargonium x hortorum 'Brownie',
dwarf with gold leaves zoned chocolate

Pelargonium x hortorum 'Sirius',
dwarf with salmon-pink stars

Pelargonium x hortorum 'Tiny Tim Pink',
miniature with tiny foliage prettily zoned

Pelargonium x hortorum 'Rober's Lavender',
semi-dwarf flowering lavender-pink

Pelargonium x hortorum 'Fairy Tales',
miniature white with pink eye

Pelargonium x hortorum 'Meteor miniature'
P. 'Black Vesuvius'
P. 'Salmon Comet'

Pelarg. 'Pride' (semi-dwarf)
P. 'Jaunty' (strawberry-red)

Pelargonium x hortorum 'Sparkle',
semi-dwarf, bright red

**Pelargonium x hortorum
'Robin Hood',**
fiery red, semi-dwarf

Pelargonium x hortorum 'Tu-tone',
pale pink, streaked white

Pelargonium x hortorum 'Pygmy',
miniature, vivid red

Pelargonium x hortorum 'Antares',
burning dark scarlet

**Pelargonium x hortorum
'Prince Valiant',**
purplish crimson flushed orange

Pelargonium x hortorum 'Tempter',
purplish crimson semi-dwarf

**Pelargonium x hortorum
'Brooks Barnes',**
soft salmon pink

**Pelargonium x hortorum
'North Star',**
flowers white with pink veins

Pelargonium x domesticum 'Chicago Market'
lavender "Martha Washington geranium"

Pelargonium x domesticum
'Firedancer'
vivid crimson with white edge

Pelargonium x domesticum
'South American Bronze'
bronzy red with orchid border

Pelargonium x domesticum
'Carlton's Pansy'
large-flowered pansy geranium

Pelargonium x domesticum
'Mad. Layal'
French "Pansy geranium"

Pelargonium x domesticum
'Earliana'
a bushy, free-blooming "Pansy geranium"

Pelargonium x domesticum
'Edith North'
free-blooming salmon pink

Pelargonium x domesticum
'Springtime'
white flowers zoned rose

Pelargonium x domesticum
'Jessie Jarrett'
violet rose to purple

Pelargonium x domesticum
'Josephine'
white flowers blotched rose

Pelargonium x domesticum
'MacKay No. 12'
lavender with crimson

Pelargonium x domesticum
'Easter Greeting'
popular rosy-carmine

Pelargonium x domesticum
'Grand Slam'
striking flame-red

Pelargonium x domesticum
'Mrs. Marie Vogel'
salmon-rose with crimson

Pelargonium x domesticum
'Mrs. Mary Bard'
pure white, striped purple

**Pelargonium x hortorum
'Maxime Kovalevski'**
bright orange

**Pelargonium x hortorum
'Dr. Margaret Sturgis'**
rosy stripes over white

**Pelargonium x hortorum
'Appleblossom'**
soft pink shading to white

**Pelargonium x hortorum
'Souvenir de Mirande'**
rosy with white eye

**Pelargonium x hortorum
'Olivia Kuser'**
free-blooming pure white

**Pelargonium x hortorum
'Honeymoon'**
apricot with pale margin

**Pelargonium x hortorum
'Bronze Beauty'**
light salmon blooms

Pelargonium 'Formosa'
salmon "Fingered flowers"

**Pelargonium x hortorum
'Royal Fiat'**
fringed petals pink

Pelargonium x limoneum ("Lemon geranium")
P. 'Prince Rupert' (lemon-scented)
P. 'Prince Rupert variegated'

P. grossularioides hort. ("Gooseberry")
P. graveolens 'Minor' ("Little-leaf rose")
P. 'Torento' ("Ginger geranium")

P. graveolens ("Rose geranium")
P. capitatum 'Attar of Roses'
P. graveolens 'Variegatum' ("Mint-scented rose")

P. x fragrans ("Nutmeg geranium")
P. denticulatum ("Pine geranium")
P. dent. filicifolium ("Fernleaf geranium")

P. 'M. Ninon' ("Apricot geranium")
P. tomentosum ("Peppermint geranium")
P. quercifolium 'Fair Ellen' ("Oakleaf ger.")

P. x nervosum ("Lime geranium")
P. 'Prince of Orange' ("Orange geranium")
P. odoratissimum ("Apple geranium")

Pelargonium x ardens,
"Glowing storksbill" with
knobby stems, red flowers

Pelargonium x fragrans
'Variegatum',
"Variegated nutmeg geranium"

Pelargonium acetosum
"Sorrel cranesbill"

Pelargonium graveolens
'Lady Plymouth',
"Variegated rose geranium"

Pelargonium violarium
"Viola geranium"

Pelargonium dasycaulon
"Succulent geranium"

Pelargonium echinatum
"Cactus geranium"

Pelargonium gibbosum
"Knotted geranium"

GESNERIADS –
the African violet family

The Gesneriaceae family, with few exceptions, is chiefly warm-climate, found in tropical areas from sea-level on up into rainforest where they may luxuriate as epiphytes; they are at home on the high veldt and again in cool mountains to rocky heights above 7000 feet. This giant family of some 120 different genera, and approximately 1800 species, includes hundreds of handsome, often velvety-leaved plants with showy wheel-shaped, tubular, or bell-like flowers. The gesneriads are so variable in foliage, texture, growth, and flowering habits that one would have to be a specialist to realize that all were members of the same plant group.

The Western hemisphere—Mexico, the West Indies, Central and South America—offer the greatest wealth in species of gesneriads in cultivation today. Most important of this tropical American group are the Sinningias, better known as "Gloxinias"; then Columneas with their brilliant flowers; Episcias with carpeted leaves; Achimenes, the "Magic flower"; Smithiantha, the "Temple bells"; Kohlerias, looking like little gloxinia trees; and Rechsteinerias which include the scarlet "Cardinal flower" and the silver-hairy "Brazilian Edelweiss".

Many curious gesneriads come from Asia, amongst them are Aeschynanthus, Boea, Lysionotus, Petrocosmea, Titanotrichum, Chirita, and the hardy Rehmannia. Further south to New Guinea we find Cyrtandra; the shrubby Rhabdothamnus is the sole genus in New Zealand. Africa is not so rich on species, but from a rather small area in East Africa originate the various Saintpaulia, which as the "African violet" is easily today the most admired little flowering plant in the American home. Endemic in Africa also is Streptocarpus, the "Cape primrose". Even Europe has a few mountain species in Ramonda, Jankaea and Haberlea, sometimes buried under snow.

A Saintpaulia hybrid, the well beloved "African violet" is one of our best little flowering plants for the home provided it is given sufficient light, steady moisture and food as needed, rewarding such attention with continuous blooms.

Culture in the Home

Countless hybrids, mutations and other cultivars of gesneriads exist. It is therefore small wonder that this interesting family has caught the fancy of many collectors, especially since most species adapt themselves fairly easily to cultivation in our homes. Combine ample humidity with the warm temperatures between 70 and 75°F. we are accustomed to live with, and one has a suitable place to grow them. At the window with filtered light, or on tables under fluorescent tubes, even in the basement, gesneriads respond willingly to a little extra care, and in their variety a display of flowers is possible all year. Some are relatively easy, others are rather exacting in their demands, requiring resting periods. It is important to remember to maintain an even moisture at the roots, preferably watering from a saucer beneath the pot. A tray with peatmoss, sand or pebbles and kept wet, will raise humidity which is an important key to their success. Relative humidity in the rain forest home of some species is as high as 95%, but I have noted that even in the humid Usambara Mountains of East Africa, the habitat of many of our Saintpaulias, humidity occasionally drops to 50%. This would be an ideal condition to try to maintain for most gesneriads in the living room, since it is conceded that a relative humidity of about 50% is best also for our own good health and comfort. For especially delicate plants a converted aquarium or glass bowl may be used with good success to confine and maintain high humidity.

Pests: Cyclamen mite is the most crippling gesneriad enemy. Mite damage causes plants to grow distorted or oddly twisted. Repeated applications of rotenone or malathion are recommended controls.

All gesneriads do best if provided moisture in the atmosphere. But this delicate lilliputian Sinningia pusilla finds an ideal home planted in a 5 inch brandy snifter where high humidity is easily maintained.

American housewives are our most ardent collectors of Saintpaulia hybrids, the "African violet", now grown in thousands of varieties and a wide range of color. Perhaps unwittingly, they choose the best place in the house, a kitchen window, where temperature is right and moisture abundant. Glass dishes under each pot help to maintain an even moisture.

Columnea microphylla, the "Small-leaved goldfish vine", sends down long strands of showy flowers burnt-red with yellow along tube.

It is natural that most gesneriads like a humusy, yet well-drained soil, and steady feeding during active growth will help produce best blooms. When nothing else will help shake out the old soil, divide or thin out the plant, and replant in new soil, removing bad or excessive roots. A good mix is 60% peatmoss and humus, 30% enriched loam and 10% cinders, sand or coarse perlite mineral for aeration, and a little lime. Sufficient light also is important and directly influences bud initiation; Saintpaulias for instance would like at least 500-600 foot-candles for 16 hours daily to flower most successfully. As a practical rule, this is light diffused to show fairly well the shadow of your hand if passed above the foliage.

PROPAGATION—Excepting the most difficult, gesneriads flower in less than a year from dust-fine seeds. But most of the fibrous rooted kinds are propagated vegetatively by leaf or stem cuttings. Leaf propagation is practiced primarily in Saintpaulias, but can also be used with Sinningias, Streptocarpus, Smithiantha and Episcia; this often leads to worthy new mutations. Tuberous species such as Sinningia and Rechsteineria need a rest period, and dormant tubers may be divided like a potato. Then there are genera that form scaly rhizomes that are broken apart, even to the extent of individual scales; this includes Achimenes, Kohleria and Smithiantha; these also need a dormant period, but should not be stored below 60°F. Some gesneriads such as Episcia produce runners, or stolons which root easily; still others can be multiplied by division.

Delicate Gesneriads, mainly Episcias, are good subjects for culture under fluorescent light in the home, at the same time high-lighting the colorful patterns of their carpet-like foliage.

Achimenes longiflora 'Major'
lavender "Orchid pansy"

Achimenes candida
white "Mother's tears"

Achimenes antirrhina
"Scarlet magic flower"

Achimenes flava
bright orange-yellow

Achimenes patens
purple "Kimono plant"

Achimenes erecta (coccinea)
scarlet "Cupid's bower"

Achimenes ehrenbergii
orchid-colored

Achimenes longiflora 'Andersonii'
purplish "Trumpet achimenes"

Achimenes andrieuxii
violet with white throat

Saintpaulia amaniensis, a creeper

Saintpaulia grandifolia, of large habit

Saintpaulia orbicularis, a small species

Saintpaulia shumensis, a miniature

Saintpaulia velutina, a dwarf species

Saintpaulia nitida, violet-blue

Saintpaulia pendula, robust grower

Saintpaulia grotei, a creeper

Saintpaulia ionantha
original "African violet", violet-blue

Saintpaulia rupicola
"Kenya violet", wisteria-blue

Saintpaulia diplotricha
pale violet flowers

Saintpaulia confusa (kewensis hort.)
"Usambara violet", pale blue

Saintpaulia magungensis
creeper with flowers medium blue

Saintpaulia difficilis
pretty violet blue

Saintpaulia tongwensis
robust; pleasing pale amethyst

Saintpaulia intermedia
medium-blue with violet eye

Saintpaulia 'Spring Sky'
"Single pale blue"

Saintpaulia 'Azure Beauty'
"Double white with violet center"

Saintpaulia 'Fair Elaine'
"Double pink with white edge"

Saintpaulia 'Red Spark'
"Semi-double red"

Saintpaulia 'Icefloe'
"Large double white"

Saintpaulia 'Pink Amour'
"Single frilled pink with dark edge"

Saintpaulia 'Star Girl'
"Girl-leaf single"

S. 'Blue Boy-in-the-Snow'
"Leaves mottled cream"

Saintpaulia 'Rhapsodie Elfriede'
European triploid

Saintpaulia 'Kenya Violet'
"Long-lasting double violet-blue"

Saintpaulia 'Diana'
"European single violet"

Saintpaulia 'My Flame'
"Double pinkish with violet margin"

Saintpaulia 'Blue Caprice'
"Double light blue"

Saintpaulia 'Sea Foam'
"Semi-double rose with white edge"

Saintpaulia 'Calypso'
"Single purple with white edge"

Saintpaulia 'Savannah Sweetheart'
"Holly-leaf; double bright rose"

Saintpaulia 'Pooty-cat'
"Star-type wine purple, pale border"

Aeschynanthus x splendidus
"Orange lipstick vine"

Aeschynanthus speciosus
(Trichosporum splendens hort.)

Aeschynanthus pulcher
"Royal red bugler"

Aeschynanthus marmoratus
"Zebra basket vine"

Aeschynanthus lobbianus
"Lipstick vine"

Alloplectus schlimii
shimmering light green

Alloplectus capitatus
"Velvet alloplectus"

Alloplectus vittatus
"Striped alloplectus"

Columnea teuscheri
(Trichantha minor hort.)

Chirita sinensis
"Silver chirita"

Codonanthe macradenia
Central American bellflower"

Nautilocalyx forgetii
"Peruvian foliage plant"

Nautilocalyx lynchii
"Black alloplectus"

Nautilocalyx bullatus
(Episcia tessellata hort.)

Besleria lutea
clamberer from the West Indies

Boea hygroscopica
"Oriental streptocarpus"

Columnea crassifolia
orange-red flowers

Columnea lepidocaula
orange with yellow

Columnea verecunda
bright yellow

Columnea sanguinea
"False alloplectus"

Columnea nicaraguensis
brilliant red

Columnea illepida
yellow flowers striped red

Columnea microphylla
"Small-leaved goldfish vine"

Columnea linearis
rose pink flowers

Columnea x banksii
vermilion-orange

Columnea x 'Vega'
"Goldfish bush", flowers coxcomb-red

Columnea allenii
slender trailer flowering red-orange

Columnea arguta
salmon red, marked yellow

Columnea gloriosa
fiery red with yellow throat

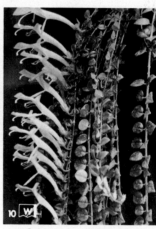

Columnea x 'Stavanger'
"Norse fire plant"

Columnea x 'Cornellian'
orange with yellow underside

Columnea x 'Cayugan'
large flame-red flowers

Columnea x 'Canary'
bright canary-yellow

Episcia punctata
creamy-white spotted purple

Episcia melittifolia
magenta flowers

Episcia lineata
cream with purple spots

Episcia dianthiflora
white "Lace flower vine"

Episcia hirsuta
flowers pale lilac

Episcia lilacina 'Viridis'
large light blue flowers

Episcia 'Moss-Agate'
crimson, very floriferous

Ep. x wilsonii 'Pinkiscia'
lovely rose-pink

Episcia cupreata 'Tropical Topaz'
"Canal Zone yellow"

Episcia reptans (fulgida)
blood-red flowers

Episcia lilacina 'Lilacina'
"Blue-flowered teddy-bear"

Episcia cupreata 'Acajou'
blooms orange-scarlet

Episcia cupreata 'Silver Sheen'

Episcia cupreata 'Metallica'
"Kitty episcia"

Episcia cupreata 'Frosty'
splashed silver

Episcia cupreata 'Musaica'
coppery with silver veins

Episcia reptans 'Lady Lou'
variegated pink and white

Episcia cupreata
'Chocolate Soldier'
chocolate brown with silver

Diastema quinquevulnerum
white flowers dotted purple

Chirita lavandulacea
blue "Hindustan gentian"

Chirita micromusa
flowers orange-yellow

Chrysothemis pulchella
yellow tube in red calyx

Dichiloboea birmanica
white-flowered bush

Gloxinia perennis (maculata)
"Canterbury bell gloxinia"

Didymocarpus kerrii
flowers dark purple

x Gloxinera 'Rosebells'
dainty bells of deep rose

x Eucodonopsis 'Kuan Yin'
yellowish flowers streaked purple

Kohleria lindeniana
vivid green with silver veins

Kohleria tubiflora
"Painted kohleria"

Kohleria 'Eriantha hybrid'
known as Isoloma hirsutum

Kohleria digitaliflora
"Foxglove kohleria"

Kohleria strigosa
blooms cinnabar-red

Kohleria lanata
"Woolly kohleria"

Kohleria "allenii" hort.
inflated red tubes tipped yellow

Kohleria 'Amabilis hybrid'
known as Isoloma ceciliae

Hypocyrta strigillosa
"Pouch flower"

Hypocyrta teuscheri
leaves olive green with silver

Hypocyrta nummularia
"Miniature pouch flower"

Nematanthus fluminensis
"Thread-flower"
rosy pouches suspended

Petrocosmea kerrii
"Hidden violets"

Klugia ceylanica
violet with yellow eye

Lysionotus serratus
pale lilac funnels

Gesneria cuneifolia
(Pentarhaphia reticulata)
"Fire cracker"

Koellikeria erinoides
"Dwarf bellflower"

Rechsteineria cyclophylla
"Roundleaf helmet flower", scarlet red

Rechsteineria leucotricha
"Rainha do Abismo" or "Brazilian edelweiss"

Rechsteineria verticillata
"Double-decker plant"

Rechsteineria cardinalis
"Cardinal flower"

Rechsteineria macropoda
"Vermilion helmet flower"

Rehmannia angulata
"Foxglove gloxinia"

Ramonda myconii (pyrenaica)
"Alpine gesneria"

Seemannia latifolia
tubular flowers brick-red

Sinningia eumorpha
milky white and lilac

Sinningia regina
"Cinderella slippers"

Sinningia speciosa
"Violet slipper gloxinia"

Sinningia barbata
bearded pouches white

x Streptogloxinia 'Lorna'
bigeneric "Stroxinia"

Sinningia schiffneri
white with purple dots

x Paschia hybrida
"Gloxiloma"

Sinningia pusilla
"Miniature slipper plant"

Sinningia 'Doll Baby' (3 in. pot)
pretty, unceasing bloomer

Sinningia speciosa 'Emperor Frederick'
popular "Florists gloxinia", crimson, bordered white

Sinningia speciosa 'Switzerland'
ruffled grandiflora gloxinia

Sinningia speciosa florepleno 'Chicago'
frilled "Double gloxinia", crimson with white

Sinningia speciosa florepleno 'Monte Cristo'
"Double gloxinia", velvety crimson

Sinningia 'Tom Thumb'
Fischer's true "Miniature gloxinia";
3 inch pot in English teacup

**Sinningia speciosa
'Princess Elizabeth'**
*"Blue bell gloxinia",
light blue on white*

Sinningia speciosa 'Tigrina'
"Tigered bell-gloxinia"

Streptocarpus x hybridus
"Hybrid cape primrose", in many colors

Streptocarpus rexii
"Cape primrose"; pale lavender, lined purple

Titanotrichum oldhamii
golden bells brown-red inside

Streptocarpus holstii
flowers blue-purple

Streptocarpus caulescens
"Violet nodding bells"

Streptocarpus polyanthus
"Sky-blue fountain"

Streptocarpus wendlandii
"Royal nodding bells"

Streptocarpus saxorum
"False African violet"

Smithiantha zebrina (Naegelia)
bells scarlet with yellow

Smithiantha 'Orange King'
velvet-leaves patterned red

Smithiantha multiflora
flowers creamy-white

Smithiantha 'Exoniensis'
orange-yellow nodding bells

Smithiantha cinnabarina
"Red velvet-leaf gesneriad"

Smithiantha 'Rose Queen'
charming rose-pink

Smithiantha x hybrida 'Compacta'
lemon-yellow "Temple bells" above velvet leaves

Smithiantha 'Zebrina discolor'
beautiful plush leaves green and red

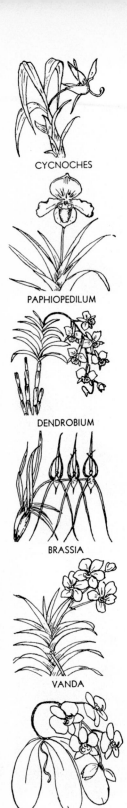

CYCNOCHES

PAPHIOPEDILUM

DENDROBIUM

BRASSIA

VANDA

PHALAENOPSIS

Orchid growing is an interesting, rewarding, and relaxing hobby. A family of aristocrats world-wide, to know of their origins conjures up visions of tropical islands or humid mountain forests, but also gives some indication of their requirements in cultivation.

35,000 species of orchids in 800 genera, plus countless hybrids have been recorded, of which 85% dwell within the tropics and the subtropical belt. Only 250 varieties may be shown on these pages, these however are carefully chosen to include a wide and varied collection of all types of orchids found in collections and proven in cultivation, at the same time illustrating the beauty, imagination and artistry of nature found in this regal family.

Probably the richest representation of orchids are found in Tropical Asia, amongst the most popular being Coelogyne, Cymbidium, Dendrobium, Paphiopedilum, Phalaenopsis, and Vanda. Tropical America is distinguished by an isolated orchid flora including the important Cattleya, Epidendrum, Laelia, Lycaste, Masdevallia, Miltonia, Oncidium, Odontoglossum, and Stanhopea. In Africa are many endemic genera including Lissochilus, Polystachya, Ansellia, and the mountain-dwelling Disa. Australia has the least number of orchids in the world, but we find Bulbophyllum, Cymbidium, Dendrobium, Eria, Spathoglottis, and others. But adjacent New Guinea is extremely rich in species, estimated at over 2500; the variety of forms is especially numerous in Dendrobium and Bulbophyllum. The true Cypripedium is at home in temperate North America, Europe, and Asia.

In the temperate zones orchids grow terrestrial. In the tropics more of them are epiphytes, growing attached to other plants, generally trees. But altitude directly determines a climatic zone, and high mountain species even near the equator exist under cool, frequently moist conditions. In the Cordilleras of Colombia, Oncidiums are found at an elevation of 14,000 feet.

Orchids belong to the Monocotyledons, like palms, lilies, bromeliads and grasses. Their showy or interesting, often fragrant flowers cannot easily be mistaken, having an outer whorl of sepals which are almost always colored as in iris or tulips; the corolla, or inner whorl of 3 petals, has two laterals that are identical, while the third, lower petal always differs in shape. It may be modified into a trumpet-shaped tube broadening into a conspicuous labellum or "lip" like in Cattleya; resemble a pouch or "slipper", as in Paphiopedilum; or it can assume various other shapes, with fringes or tassels, and be decorated with color in an interesting manner. Orchids are unique in having stamen and style fused to form a prominent column; a majority of species have swollen stems known as pseudobulbs, which store food and water to enable the plant to withstand seasonal periods of drought.

Success with Orchids in the Home.

All orchids prefer the hospitable and controlled conditions of the greenhouse, but many can be grown and flowered with success and pleasure in the home if given the environment that meets their needs. A sunny, well ventilated room and a place, if possible, next to the window where the temperature is normally lowest, plus a relatively high humidity, preferably between 50 and 70%, is ideal. As the habitats of various orchids are found in different climatic zones, specialists' greenhouses aim to provide similitudes through several ranges of temperature for their cultivation under glass: **Warm** (62-80°F.), **Temperate** or **Intermediate** (50-65°F.), and **Cool** (40-60° and up).

A lovely Cattleya hybrid, the "Royal orchid", brings glamour into any sober business office, if only for temporary adornment.

A small lean-to greenhouse adjoining suburban residence, can be ideal for orchids.

ORCHIDS *in the Home*

TEMPERATURE:

While orchids in general are rather tolerant and even tough, which allows them to adjust themselves to many conditions, the species of the "Intermediate" group will normally prove the most rewarding for growing in the home. Actually, the important point with a given temperature range is not so much how warm, but to keep it cool enough, particularly at night, for the same reason that we sleep best in a cool room. The plant is enabled to reduce respiration, and to store food manufactured during the day, the maximum period of light. Modern airconditioning is ideally suitable to many cool orchids.

LIGHT:

Most orchids, particularly the species with hard pseudobulbs, such as Cattleyas, want considerable sunlight, if possible 4-5 hours per day —and need at least 1500, to 3000 foot-candles— to insure good flowering. To place them outdoors during the summer time, hung under a tree or arbor, or set onto a slatted bench, will tend to rebuild jaded tissues, and favor the development of new growth and bud initiation. The pseudobulbless, or softer-growthed kinds will prefer a location protected from direct rays of the sun, and an east window would be best during the summer season. If grown entirely under artificial light, 650 foot-candles for 16 hours have proven sufficient for many light-hungry species.

WATERING AND MOISTURE:

A shallow tray of fiber glass or metal with gravel, charcoal, perlite, vermiculite, peatmoss, and partially filled with water, and slats or even upside-down pots on which to set the plants, will help to maintain moisture in the pots as well as increase humidity around the plants. When they are in full growth and with active roots, orchids should be watered copiously, usually once or twice a week but only when the pot is becoming dry, especially the varieties with pseudobulbs. Epiphytes with their spongy roots hold a great deal of water, and if in a constant state of saturation these roots will rot. If potted in bark, a good drenching may be needed every day; Osmunda keeps moist longer. Most orchids require a period of rest with reduced moisture immediately after flowering.

A relative humidity of 50%, which can easily be checked with a hygrometer, is fairly satisfactory for most orchids, but difficult to maintain in winter when it often drops to 30%. Frequent misting of the foliage with a hand-sprayer will materially increase it; so would a plastic enclosure, with the top left open; or a space humidifier. Gentle movement of fresh air is always good, either by opening a window or with the aid of a small fan.

POTTING AND FEEDING:

Diced roots of the Osmunda fern are an old, reliable growing medium for epiphytic orchids. But many growers pot them into coarsely shredded firbark or bark chips, frequently with addition of some peat-humus or perlite for better moisture retention. Deficient in nitrogen, porous bark materials need additional feeding with dilute liquid high nitrogen fertilizer or fish-emulsion monthly. Epiphytes generally grow well on tree fern slabs, provided they can be dunked daily, but terrestrials often prefer a humusy compost, even sphagnum moss. In the South, treefern fiber lasts better than bark.

PROPAGATION

Reproduction from seed is difficult for the amateur. Terrestrial orchids bearing pseudobulbs can be divided by cutting the surface rhizome. Those without pseudobulbs, and after clustering, may be pulled apart and separated. Some genera such as Dendrobium produce plantlets directly on their stems. Orchids from seed see page 50.

MINIATURE ORCHIDS

Once known as "Botanical orchids", the smaller growing species take up much less room and allow space for a wide collection in a small area, with the good chance of having orchids in bloom at all times. Their cute little flowers are every bit as beautiful and intriguing as the larger, more regal species, and in addition, are usually sweetly fragrant.

Orchid miniatures to live with, at a large bay window facing east, in a fancier's home in Madison, New Jersey.

Anoectochilus sikkimensis
"King of the forest"
(Sikkim)

Haemaria discolor dawsoniana
"Golden lace orchid"
(Malaya)

Erythrodes nobilis
"Silver orchid"
(Brazil)

Malaxis calophylla
"Plaited jewel orchid"
(Malaya)

Cyclopogon variegatum
leaves marbled with pink
(Brazil)

Anoectochilus roxburghii
"Golden jewel orchid"
(Himalayas)

Pelexia maculata
bronze foliage spotted white
(West Indies)

Macodes petola
"Gold-net orchid"
(Borneo)

Sarcoglottis metallica 'Variegata'
leaves banded with silver
(Costa Rica)

Cymbidium 'Doris'

Sophronitis coccinea

Cycnoches ventricosum

Vanda tricolor

Coelogyne pandurata

Stanhopea tigrina

x Laeliocattleya canhamiana alba

Paphiopedilum fairrieanum

Odontoglossum crispum

Oncidium kramerianum

Phalaenopsis amabilis

Epidendrum prismatocarpum

Angraecum eichlerianum
(West Africa)

Angraecum sesquipedale
"Star of Bethlehem orchid"
(Madagascar)

Angraecum eburneum
(Madagascar)

Aspasia variegata
(Panama)

Neofinetia falcata
(Angraecum)
(Japan)

Aganisia cyanea
"Blue orchid"
(Brazil)

Ascocentrum miniatum
(Java)

Arundina graminifolia
(Burma)

Ansellia gigantea nilotica
"Leopard orchid"
(East Africa)

Aerides fieldingii
"Foxbrush orchid"
(Assam)

Acineta superba
(Colombia)

Aerides crassifolium
"King of Aerides"
(Burma)

Ascotainia viridifusca
(Assam)

Anguloa virginalis
"Cradle orchid"
(Colombia)

Arachnis lowii
"Scorpion orchid"
(Borneo)

Aerangis thomsonii
(Kenya)

Ada aurantiaca
(Colombia)

Acampe papillosa
(India)

Brassavola cordata
(West Indies)

Brassavola nodosa
"Lady of the night"
(Costa Rica)

Brassavola digbyana
(Honduras)

Brassia gireoudiana
"Spider orchid"
(Costa Rica)

Chysis aurea
(Mexico)

Brassia verrucosa
"Queen's umbrella"
(Central America)

x Brassolaelia 'Charles J. Fay'
(Brassavola x Laelia)

x Brassocattleya veitchii
(Brassavola x Cattleya)

Bulbophyllum grandiflorum
(New Guinea)

Bulbophyllum pulchellum
"Walnut orchid"
(New Guinea)

Bulbophyllum virescens
"Windmill orchid"
(Java)

Broughtonia sanguinea
(Cuba, Jamaica)

Bletilla striata
(Vietnam, China)

Barbosella anaristella
(Costa Rica)

Bifrenaria harrisoniae
(Brazil)

Restrepia xanthophthalma
(Guatemala)

x Brapasia 'Panama'
(Brassia x Aspasia)

Cattleya intermedia alba
a "Cocktail orchid"
(Brazil)

Cattleya dowiana aurea
"Queen cattleya"
(Colombia)

Cattleya mossiae
"Easter orchid"
(Venezuela)

Cattleya skinneri
(Guatemala)

Cattleya lueddemanniana
(Brazil)

Cattleya forbesii
another "Cocktail orchid"
(Brazil)

Cattleya bowringiana
a "Cluster cattleya"
(Honduras)

Cattleya trianaei
"Christmas orchid"
(Colombia)

Coryanthes bungerothii
"Helmet orchid"
(Venezuela)

Calanthe x bella
East Indian terrestrial

Calanthe striata
(Japan)

Comparettia speciosa
(Ecuador)

Chloraea membranacea
(Argentina)

Spiranthes longibracteata hort.
(Brazil)

Cycnoches egertonianum
(Mexico), *female flowers*

Cycnoches egertonianum
male inflorescence

Cycnoches loddigesii
"Swan orchid"
(Surinam)

Catasetum warscewiczii (scurra)
(Panama, Brazil)

Catasetum bicolor (Panama)
female flowers on left; male flowers right

Catasetum saccatum christyanum
(Brazil: Amazon)

Cymbidium 'Flirtation'
"Miniature hybrid cymbidium"

Cymbidium lancifolium
(Himalayas)

Cymbidium virescens angustifolium
(Japan)

Cymbidium x alexanderi
"Corsage cymbidium"

Cymbidium aloifolium
(Sikkim Himalayas)

Coelogyne pandurata
"Black Orchid"
(Sumatra)

Coelogyne lawrenceana
(Vietnam)

Coelogyne massangeana
(Assam to Java)

Coelogyne sparsa
(Philippines)

Coelogyne cristata
(Nepal Himalayas)

Coelogyne dayana
"Neck-lace orchid"
(Sumatra)

Cyrtopodium punctatum
"Cigar orchid"
(Florida, W. Indies)

Cypripedium macranthum
(Tibet, Siberia)

Cypripedium acaule
"Pink Ladyslipper"
(Newfoundland to Carolina)

Dendrobium falconeri
(India)

Dendrobium phalaenopsis
(Queensland, New Guinea)

Dendrobium thyrsiflorum
(Burma)

Dendrobium johnsoniae
(New Guinea)

Dendrobium densiflorum
(Himalayas, Burma)

Dendrobium aggregatum
(lindleyii) (Yunnan)

Dendrobium nobile
(Himalayas to Yunnan)

Dendrobium primulinum
(Himalayas, Burma)

Dendrobium moschatum
(Himalayas, Burma)

Dendrochilum cobbianum
(syn. Platyclinis) (Philippines)

Dendrochilum glumaceum (syn. Platyclinis)
"Chain orchid"
(Philippines)

x Diacattleya 'Chastity'
(Diacrium x Cattleya)

Disa uniflora (So. Africa)
"Pride-of-Table Mountain"

Diacrium bicornutum
"Virgin orchid"
(Trinidad, Guyana)

Dendrobium formosum
(Himalayas)

Dendrobium distichum
(Philippines)

Dendrobium linguaeforme
"Tongue orchid"
(Queensland)

Epidendrum stamfordianum
(Guatemala to Colombia)

Epidendrum brassavolae
(Guatemala)

Epidendrum prismatocarpum
"Rainbow orchid"
(Costa Rica)

Epidendrum pentotis
(C. America to Brazil)

Epidendrum mariae
(So. Mexico)

Epidendrum radiatum
(Mexico)

Epidendrum ibaguense (radicans)
"Fiery reed orchid"
(Mexico to Peru)

Epidendrum cochleatum
"Cockle-shelled orchid"
(W. Indies, C. America)

Epidendrum atropurpureum
"Spice orchid"
(Mexico to Brazil)

x Epiphronitis veitchii
(Epidendrum x Sophronitis)

Eria ferruginea
(Himalayas)

Elleanthus capitatus
(Mexico to Peru)

Eulophidium mackenii
(South Africa: Natal)

Epidendrum polybulbon
(West Indies)

x Epicattleya 'Gaiety'
(Epidendrum x Cattleya)

Epidendrum fragrans
(West Indies to Brazil)

Epidendrum medusae (syn. Nanodes)
fantastic and sinister (Andes of Ecuador)

Galeandra devoniana
(Amazonas, Guyana)

Gastrochilus dasypogon
(Himalayas)

Gomesa crispa
(Brazil)

Gomesa planifolia
"Little man orchid"
(Brazil)

Gongora galeata
"Punch and Judy orchid"
(Mexico)

Habenaria radiata
"Egret flower"
(Japan)

Grammatophyllum speciosum
"Queen of orchids"
(Vietnam to New Guinea)

Hexisea bidentata
(Costa Rica to Colombia)

Huntleya meleagris
(Brazil)

Ionopsis utricularioides
"Violet orchid"
(C. America to Brazil)

Kefersteinia tolimensis
(Andes of Ecuador)

Liparis longipes
(India to Philippines)

Lockhartia lunifera
"Braided orchid"
(Brazil)

Lycaste gigantea
(Colombia, Ecuador)

Lycaste cruenta
(Guatemala)

Lycaste deppei
(So. Mexico)

Lycaste virginalis alba
"White Nun"
(Guatemala)

Lycaste aromatica
(Mexico)

Laelia purpurata
"Queen of orchids" in So. Brazil

Laelia perrinii
(Brazil)

Laelia tenebrosa
(Brazil: Bahia)

Laelia anceps
(Mexico)

Laelia cinnabarina
(Brazil)

x Laeliocattleya hassalii alba 'Majestica'
(Laeliocattleya x Cattleya)

Laelia crispa
(Brazil)

Mesospinidium warscewiczii
(Brazil)

Mormodes igneum
"Goblin orchid"
(Central America)

Mystacidium distichum
syn. Angraecum (W. Africa)

Mormodes lineata
(Guatemala)

Meiracyllium trinasutum
(Mexico)

Maxillaria lepidota
(Colombia)

Maxillaria tenuifolia
(Mexico)

Maxillaria picta
(Brazil)

Masdevallia bella
(Colombia)

Masdevallia veitchiana
(Peru)

Masdevallia chimaera
(Colombia)

Masdevallia grandiflora
(Ecuador)

Masdevallia liliputana
miniature (Brazil)

Masdevallia houtteana
(Colombia)

Masdevallia infracta purpurea
(Brazil, Peru)

Miltonia laevis
(Mexico, Guatemala)

Miltonia roezlii alba
white "Pansy orchid"
(Colombia)

Miltonia spectabilis
(Brazil)

Miltonia vexillaria 'Volunteer'
(Ecuador)

Miltonia clowesii
(Brazil)

Miltonia phalaenopsis
"Pansy-face"
(Colombia)

Megaclinium purpureorhachis
"Cobra orchid"
(Congo)

Neobenthamia gracilis
(Zanzibar)

Notylia sagittifera
(Brazil)

Oncidium flexuosum
"Dancing doll orchid"
(Brazil, Paraguay)

Oncidium splendidum
(Guatemala)

Oncidium sarcodes
(Brazil)

Oncidium stipitatum
(Panama)

Oncidium lanceanum
"Leopard orchid"
(Trinidad, Guyana)

Oncidium sphacelatum
"Golden shower"
(Mexico to Honduras)

Oncidium ornithorhynchum
(Mexico to Salvador)

Oncidium kramerianum
(Ecuador, Colombia)

Oncidium papilio
"Butterfly orchid"
(Trinidad to Peru)

Odontoglossum crispum
"Lace orchid"
(Colombia)

Odontoglossum grande
"Tiger orchid"
(Mexico)

Odontoglossum pulchellum
"Lily of the valley orchid"
(Guatemala)

Ornithochilus fuscus
(Nepal to South China)

Maxillaria densa
syn. Ornithidium (Mexico)

Orchis spectabilis
"Showy woodland orchid"
(Eastern North America)

Ornithocephalus dolabratus
"Mealybug orchid"
(Ecuador)

x Odontioda 'Wey'
(Odontoglossum x Cochlioda)

Otochilus fusca
(Nepal, Burma)

Paphiopedilum x harrisianum
deep red "Lady slipper"

Paphiopedilum x maudiae
exquisite hybrid, green and white

Paphiopedilum insigne
"Lady slipper"
(Himalayas)

Paph. callosum splendens
(Thailand, Vietnam)

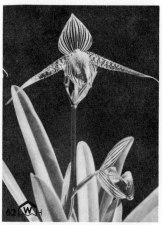

Paphiopedilum rothschildianum
(Papua, Borneo)

Paphiopedilum lowii
(North Borneo)

Paphiopedilum venustum
(Himalayas, Nepal)

Paphiopedilum godefroyae
(Vietnam, Thailand)

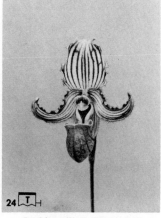

Paphiopedilum fairrieanum
(Bhutan, Assam)

Pleurothallis tuerckheimii
(Guatemala)

Phaius wallichii
"Nun orchid"
(Sikkim)

Ponthieva maculata
(Venezuela)

Plectrophora cultrifolia
(Ecuador)

Pleurothallis cardiothallis
flowers from leaf (Nicaragua)

Peristeria guttata
(Guyana: Demerara)

Polystachya cucullata
(West Africa, Nigeria)

Pleione formosana alba
"Indian crocus"
(Taiwan)

Pholidota imbricata
"Rattlesnake orchid"
(Malaysia)

Phalaenopsis amabilis 'Summit Snow'
glistening white "Moth orchid"
(Malaya, Indonesia)

Phragmipedium caudatum (syn. Selenipedium)
"Mandarin orchid"
(Peru, Ecuador)

Phalaenopsis schilleriana
"Rosy moth orchid"
(Philippines)

Phalaenopsis lueddemanniana
(Philippines)

Phragmipedium x grande
"Spiralled lady slipper"

Rodriguezia batemanii
(Brazil, Peru)

Renanthera monachica
"Fire orchid"
(Burma)

Rhynchostylis coelestis
"Foxtail orchid"
(Thailand)

Dendrobium acuminatum
svn. Sarcopodium (Philippines)

Stenoglottis longifolia
(South Africa: Natal)

Stenorrhynchus navarrensis
(Costa Rica)

Stelis ciliaris
(Mexico)

Sobralia decora
(Mexico, Guatemala)

Spathoglottis plicata
(Malaysia)

Thrixspermum arachnites
(Indonesia: Java)

Trichocentrum orthoplectron
(Ecuador)

Spiranthes elata
"Ladies tresses"
(West Indies to Brazil)

Stanhopea tigrina (Mexico)
tigered "Horned orchid" powerfully scented

Stanhopea wardii (Guatemala, Venezuela)
*waxy flowers appear through bottom of basket;
known as "El Toro"*

Rhynchostylis gigantea
"Foxtail orchid"
syn. Saccolabium (Vietnam)

Sarcanthus lorifolius
(India)

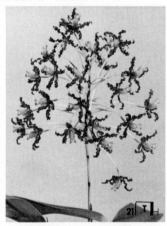

Schomburgkia undulata
(Trinidad, Venezuela)

Sophronitis coccinea
(Serra do Mar of Brazil)

Trichopilia suavis
(Costa Rica)

Vanda tricolor
"Tricolor strap vanda"
(Java)

Vanda coerulea
"Blue orchid"
(Himalayas)

Vanda x 'Miss Agnes Joaquim'
"Hawaiian corsage orchid"
"Moon orchid"

Vanilla fragrans
"Vanilla"
(E. Mexico)

Vanilla fragrans 'Marginata'
"Variegated vanilla"

Vanilla pompona (lutescens)
(Mexico to Venezuela)

Xylobium powellii
(Costa Rica, Panama)

Zygopetalum mackayi
(Brazil: Serra do Mar)

Scuticaria steelei
(Guyana)

FANCY FLOWERING Plants
Shrubs and Trees for the hobbyist

Flowering plants for pots or tubs could easily separate into two groups: the kind we popularly associate with holiday giving and produced and timed by commercial growers as seasonal blooming plants; and another category, less common but of a type that can be grown as a hobby and cared for at home. To bring these into flower may present a challenge but would bring a wonderful feeling of satisfaction and sense of pride to a real plant lover. A great and abundant class, the seed-bearing angiosperms or true flowering plants through the process of millions of years of evolution now represent the bulk of the present world vegetation; their seeds are typically protected by a fruit. It is estimated that there are now recognized some 350,000 species of flowering plants in 12,500 genera. An overwhelming wealth of material has been collected in all parts of the world and brought into cultivation, especially since the times of the great seafarers. But we can go far back in history and read of the early plant expeditions of Queen Hatshepsut of Egypt in 1500 B.C.; we know of the exotic collections of container plants on the terraces of the Hanging Gardens at the time of Nebuchadnezzar II of Babylon in 605 B.C. And so through every age and culture there has developed an increasing interest to live with plants, to try and see if such dwellers of the tropics could be brought successfully into bloom especially in colder climates. Nature offers such a wide selection that unlimited possibilities exist, to experiment with most any flowering plant, whether small and of a herbaceous nature, or exotic shrubs and even trees.

We do not always realize that what we know as "flowers", sometimes refers to their floral display only figuratively. There are plants, or shrubs, having flowers typical with a colored corolla as in Hibiscus, but what catches the eye in

Small standard tree of Hibiscus rosa-sinensis 'Regius Maximus', a "Chinese hibiscus", or "Rose-of-China", with blazing scarlet flowers. Also known as "Flower-of-a-day" because its large blossoms remain open for one day only, but will be almost everblooming wherever there is sufficient sunlight.

others as their bloom may also be manifested in various other forms. We know the showy bracts surrounding the insignificant flowers of Poinsettias or Bougainvilleas; the enlarged sepal of Mussaenda; the often spectacular spathe of Anthuriums, "Callas", or other Aroids; the pipe-shaped calyx of an Aristolochia simulating a corolla; or the floral effects of the bright conspicuous stamens of Acacia, Callistemon and Metrosideros; even leaves that masquerade as flowers as in Breynia. Many small warm-climate shrubs can be successfully flowered in the house if given a place at a bright window or the sunporch—in summer the patio, or garden border. But tropical trees rarely develop their blossoms while small enough to fit indoors, and a large conservatory would be required to house a tree of sufficient maturity to reach the flowering stage. We cannot easily duplicate in the temperate zones the glory of exotic flowering trees of the warmer zones, many of which bloom while bare of foliage, and which causes the effect of their display of color to be that much more striking. Though tempted to include more of these spectacularly showy trees as found in tropical gardens, this section must limit itself to those subjects that are small enough and willing to flower with reasonable assurance near a window or in the sun-room, or else grown in a greenhouse and used indoors for temporary decoration when in bloom.

Since the possible choice amongst flowering shrubs from the tropics or subtropics is so great, the photos assembled on the following pages may well speak for themselves, in giving a glimpse of their variety, and the beauty not only of their flowers but often their foliage as well. They come from many families but as grouped here are no more subject to botanical classification than are paintings in an art gallery. All, however, have been tested in cultivation, being representative of those flowering exotics which can be successfully grown indoors or under glass.

Shrubby plants usually, with some exception, prefer an intermediate or even cooler temperature at night, and enjoy being set outdoors for the summer. Such woody subjects are not difficult to keep alive but require thorough and more than ordinary watering when grown in containers. The floral results we can ultimately achieve will bring delight, and their own reward for patient care, in letting the wondrous loveliness of nature be revealed in our daily lives, for beauty is where man finds it.

Camellia japonica 'Debutante', also known as 'Sara Hastie', in glazed pot with saucer. Not all camellias lend themselves to pot culture, but this South Carolina cultivar is a willing bloomer, producing exquisite, clear pink flowers of peony form in late winter, breathing all the romance and loveliness of our South.

Brunfelsia latifolia
"Kiss-me-quick"

Brunfelsia calycina floribunda
"Yesterday, today and tomorrow"

Brunfelsia undulata
"White raintree"

Brunfelsia americana
"Lady of the night"

Viburnum tinus
"Laurestinus"

Viburnum suspensum
"Sandankwa viburnum"

Hebe salicifolia
"Evergreen veronica"

Daphne odora 'Marginata'
"Variegated Winter daphne"

Carissa acokanthera
"Bushman's poison"

Pittosporum tobira
"Mock-orange"

Raphiolepis indica 'Enchantress'
compact *"Indian hawthorn"*

Choisya ternata
"Mexican orange"

Ochrosia elliptica
in horticulture as Kopsia arborea

Thryallis glauca (Galphimia)
"Gold shower thryallis"

Gelsemium sempervirens
"Carolina yellow jessamine"

Camellia japonica 'Elegans'
"Peony camellia"

Camellia sasanqua
'Showa-no-sakae'
"Pink sasanqua"

Gardenia jasminoides 'Veitchii'
"Everblooming gardenia"

Gardenia jasminoides 'Fortuniana'
"Cape jasmine"

Gardenia jas. radicans fl. pl.
"Miniature gardenia"

Jasminum sambac 'Grand Duke'
"Gardenia jasmine"

Jasminum sambac
"Arabian jasmine"

Ervatamia coronaria
"Crape jasmine"

Cestrum nocturnum
"Night jessamine"

Carissa grandiflora
"Natal plum"

Osmanthus fragrans
"Fragrant" or "Sweet olive"

Plumeria rubra acutifolia
*the sweetly fragrant "West Indian jasmine",
waxy white; sacred as the "Temple tree" in India*

Plumeria rubra (acuminata)
the rosy "Frangipani tree"

Monodora grandiflora
"African orchid nutmeg" with orchid-like flowers

Poinciana pulcherrima
in hort. as Caesalpinia
*"Dwarf poinciana" or
"Barbados Pride"*

Brownea coccinea
"Scarlet flame bean"

Calliandra emarginata
*"Powder puff"
with bright red stamens*

Serissa foetida variegata
"Yellow-rim serissa"

Punica granatum legrellei
"Double-flowering pomegranate"

Coffea arabica
"Arabian coffee" in bloom

Rosa odorata 'Golden Rapture'
*yellow "Hybrid tea rose"
deliciously fragrant*

Rosa chinensis minima
"Pigmy rose" (R. roulettii)

Rosa odorata 'Happiness'
*red "Hybrid tea rose"
a fancy greenhouse type*

Solanum jasminoides
"Potato-vine"

Clianthus formosus (dampieri)
"Glory pea"

Solanum rantonnetii
"Blue potato tree"

Strophanthus gratus
"Climbing oleander"

Nerium oleander
"Oleander" or "Rose bay"

Thevetia peruviana
"Yellow oleander"
or *"Be-still tree"*

Ixora javanica
"Jungle geranium"

Ixora chinensis rosea
"Pink flower of the woods"

Ixora coccinea
"Flame of the woods"

Protea barbigera
"Giant woolly-beard"
or *"Frilled panties"*

Protea cynaroides
"King protea"
or *"Honeypot sugarbush"*

Telopea speciosissima
"Crimson waratah"

Dombeya x cayeuxii
"Pink ball"
or "Mexican rose"

Datura suaveolens
"Angel's trumpet"

Datura candida
in hort. as arborea
"Floripondio tree"

Jacaranda chelonia
"Blue jacaranda"

Cassia splendida
"Golden wonder senna"

Quisqualis indica
"Rangoon creeper"

Allamanda neriifolia
"Golden trumpet bush"

Bauhinia variegata
"Orchid tree"
or "Mountain ebony"

Erythrina crista-galli
"Coral tree"
or "Cry-baby tree"

Medinilla magnifica,
the magnificent "Rose grape"
with showy pink inflorescence

Acalypha hispida (sanderi)
the "Chenille plant" or "Foxtails",
in pendulous cat-tails of bright red flowers

Jacobinia velutina,
the "Brazilian plume", with fountain-like
heads of arched rosy blooms

Ixora macrothyrsa (duffii in hort.)
the "King ixora" or "Jungle flame"
with striking clusters of rosy-red flowers

Clerodendrum bungei
"Kashmir bouquet"

Clerodendrum speciosissimum (fallax)
"Java glorybean"

Clerodendrum x speciosum
"Glory-bower"

Plumbago indica coccinea
"Scarlet leadwort"

Clerodendrum fragrans pleniflorum
double "Glory-tree"

Plumbago capensis
"Blue Cape plumbago"

Mimulus aurantiacus
syn. Diplacus glutinosus
"Monkey-flower"

Strobilanthes isophyllus
"Bedding cone-head"

Barleria lupulina
"Hop-headed barleria"

Jacobinia carnea
(syn. Justicia magnifica)
"Flamingo plant"

Scutellaria mociniana
"Scarlet skullcap"

Pachystachys coccinea
"Cardinal's guard"

Centratherum intermedium
"Brazilian button flower"

Tibouchina semidecandra
"Glory bush"
or *"Princess flower"*

Pycnostachys dawei
"Blue porcupine"

Ruellia macrantha
"Christmas pride"

Ruellia amoena
"Redspray ruellia"

Ruellia strepens (ciliosa)
bluish purple

Cuphea platycentra
"Cigar flower"

Tacca chantrieri
"Bat flower"
or "Cat's whiskers"

Arbutus unedo
"Strawberry-tree"

Lopezia lineata
"Mosquito-flower"

Allophyton mexicanum
"Mexican foxglove"
or "Mexican violet"

Malpighia coccigera
"Singapore holly"
or "Miniature holly"

Crossandra pungens
light orange flowers

Crossandra infundibuliformis
"Firecracker flower"

Beloperone guttata
"Shrimp plant"

Hibiscus schizopetalus
"Japanese lantern"

Hibiscus rosa-sinensis
'Regius Maximus'
"Scarlet Chinese hibiscus"

Hibiscus rosa-sinensis plenus
"Double rose of China"

Hibiscus x 'White Wings'
white with crimson eye

Hibiscus syriacus coelestis
a "Shrub-althea", "Rose-of-Sharon"

Hibiscus mutabilis
"Confederate rose"
or "Cotton-rose"

Malvaviscus
penduliflorus (conzatii)
"Turk's cap"
or "Sleepy mallow"

Hibiscus rosa-sinensis 'Matensis'
"Snowflake hibiscus"

Pavonia "intermedia rosea"
rosy flowers with blue anthers

Bouvardia ternifolia 'Rosea'
"Giant pink bouvardia"

Bouvardia longiflora (humboldtii)
"Sweet bouvardia"

Bouvardia ternifolia 'Alba'
"White Joy bouvardia"

Eranthemum nervosum
"Blue sage"

Bouvardia ternifolia
"Scarlet trompetilla"

Bouvardia versicolor
orange flowers tipped yellow

Correa x harrisii
"Australian fuchsia"

Iochroma tubulosum
"Violet bush"

Lisianthus nigrescens
"Black gentian"

Abutilon striatum thompsonii
"Spotted flowering maple"

Abutilon x hybridum
"Flowering maple"
or *"Parlor maple"*

Abutilon striatum
'Aureo-maculatum'
"Spotted Chinese lantern"

Abutilon megapotamicum
variegatum
"Weeping Chinese lantern"

Abutilon x hybridum
'Souvenir de Bonn'
"Variegated flowering maple"

Abutilon x hybridum
'Golden Fleece'
"Golden parlor maple"

Dimorphotheca ecklonis
"Cape marigold"

Gerbera jamesonii
"African daisy"
or *"Transvaal daisy"*

Felicia amelloides
"Blue daisy"

Feijoa sellowiana
"Pineapple guava"

Pentas lanceolata
"Egyptian star-cluster"

Hypoestes aristata
"Ribbon bush"

Diosma reevesii
"Breath-of-heaven"

Corokia cotoneaster
"Golden korokio"

Ardisia wallichii
"India spice berry"

Ochna serrulata
"Mickey-mouse plant"

Turnera ulmifolia angustifolia
"West Indian holly"
or *"Sage rose"*

Reinwardtia indica
"Yellow flax"

Aphelandra aurantiaca
"Fiery spike"

Aphelandra squarrosa 'Louisae'
a compact "Zebra plant"

Aphelandra squarrosa
"Saffron spike zebra"

Aphelandra squarrosa
'Uniflora Beauty'

Aphelandra sinclairiana
"Coral aphelandra"

Aphelandra chamissoniana
"Yellow pagoda"

Aphelandra squarrosa leopoldii

Aphelandra aurantiaca roezlii
"Scarlet spike"

Aphelandra squarrosa
'Louisae compacta'

Euphorbia fulgens
"Scarlet plume"

Euphorbia pulcherrima
"Poinsettia"
or *"Christmas star"*

Euphorbia splendens (milii)
"Crown of thorns"

Dalechampia roezliana
with rosy bracts

Adenium obesum
"Desert rose"
or *"Impala lily"*

Mussaenda luteola (flava)
"Buddha's lamp"
creamy bracts

Asclepias currassavica
"Blood-flower"

Jatropha hastata
"Peregrina"

Asclepias tuberosa
"Butterfly-weed"

Chorizema cordatum
"Australian flame pea"

Boronia elatior
"Scented tall boronia"

Holmskioldia sanguinea
"Chinese hat plant"

Kopsia fruticosa
red-eyed *"Shrub vinca"*

Leptospermum
scoparium 'Keatleyi'
"Pink tea tree"

Leptospermum scoparium
'Ruby Glow'
"Double red manuka"

Stenocarpus sinuatus
"Wheel of fire"

Callistemon lanceolatus
"Crimson bottle-brush"

Epacris longiflora
"Australian heath"

Azalea 'Concinna' (phoenicea)
"Queen Elizabeth azalea"

Rhododendron 'Roseum elegans', *a rosy-lilac "Rose-bay"*
a hardy catawbiense hybrid, forced for Easter

Calluna vulgaris
"Scotch heather"

Erica 'Wilmorei'
"French heather"

Erica gracilis
"Rose heath"

Rhododendron x 'Bow Bells'
dainty "Miniature rhododendron"

Rhododendron charitopes
charming pink miniature

Rhododendron racemosum
small-flowered soft pink

Acacia baileyana, *a spreading tree with fragrant flowers,*
the "Golden mimosa" or "Fernleaf acacia", widely
planted in California and on the Riviera.

Acacia retinodes
commonly known as "floribunda"
"Everblooming acacia"

Acacia armata (paradoxa)
"Kangaroo thorn", for pots

Acacia longifolia
"Sydney golden wattle"

Acacia armata pendula
"Weeping kangaroo thorn"

Acacia longifolia latifolia
"Bush acacia"

Acacia podalyriaefolia
"Pearl acacia"

Acacia longifolia mucronata
"Narrow Sydney wattle"

Spartium junceum (Genista)
"Spanish broom"

Cytisus x racemosus
(canariensis hybrid)
"Genista of florists"

Acacia cultriformis
"Knife acacia"

Piqueria trinervia
fragrant "Stevia" of florists

Astilbe japonica 'Gladstone'
white "Spiraea of florists"

Buddleja crispa
"Curly butterfly bush"

Malus x astrosanguinea
"Flowering crab-apple"

Syringa vulgaris 'Marie Legraye'
"White lilac"

Prunus triloba plena
double "Flowering almond"

Spiraea x vanhouttei
"Bridal wreath"

Weigela praecox
"Early rose weigela"

Deutzia gracilis
pure white "Slender deutzia"

Pieris japonica (Andromeda)
"Lily-of-the-valley bush"

Lagerstroemia indica
"Crape myrtle"

Abelia x grandiflora
"Glossy abelia", fragrant fl.

FRUITED and BERRIED PLANTS in Containers

The fruit of the Tree of Knowledge grew in the Garden of Eden since the creation of man and woman. Tropical and subtropical fruit trees have been planted in the centuries to follow, at first for subsistence, later for ornamentation, since the beginnings of civilization. We know that date palms were cultivated in ancient Babylon in 3500 B.C., and olives in Minoan Crete since 3000 B.C. For the ancient Hebrews the fig tree was a symbol of abundance. To the Chinese, the symbols of the Three Greatest Blessings were the pomegranate, brought to China in 126 B.C., and standing for a hopeful future; the peach, as an emblem of longevity; and the citrus, a symbol of happiness. But it was the Citron tree, first reported in China in 2200 B.C., as a fruit, that took the fancy of Royal gardeners and wealthy aristocrats in Europe, causing them to grow Lemon trees and Oranges in containers, housed in winter

Ancient olive trees, Olea europaea, on the Mount of Olives overlooking the ramparts and walls of the old city of Jerusalem. Olive trees are as old as history having been cultivated on the island of Crete as early as 3000 B.C., and pickled olives were found amongst the ruins of Pompeii. Its fruit is vastly important to the Mediterranean people, olive oil substituting to them for butter. The olive branch is the emblem of peace through victory, and the winners at the Olympic Games were crowned with an olive wreath since 776 B.C.

in orangeries, and displayed outside in formal gardens during summer. One of the noblest, and best classic example, is the great orangerie at Versailles, built in 1685 for King Louis XIV and which can winter store more than 1200 orange trees. Smaller oranges in terracotta jars were popular in every Italian garden, and, inspired by the Moors, in Spain since the 11th Century.

Little container trees bearing fruit or berries are always ornamental and intriguing if grown in the home or on the patio particularly if the anticipated fruit is edible as well. To grow your "own oranges" has been the ambition of almost everyone that has seen our South or West—or—"why throw away the seed of an avocado or the sweet mango fruit?" Put it into a pot with soil or sand, and soon, a little avocado or mango tree will have sprouted. You will probably never see any fruit on it but it is thrilling to hope. That is precisely how the many venerable and appreciated old mango trees were spread through Eastern Africa because the Arab traders were commanded by Mohammed never to throw away a mango seed but to plant it every time. It is encouraging to know that the little trees we receive for decoration, and loaded with fruits or berries, will keep surprisingly well, holding their decorative fruit for weeks or even months, especially if not kept too warm, well watered, and in a reasonably bright location. Some fruit is more decorative and ornamental than edible. These include the "Jerusalem Cherry" and the "Christmas pepper"; the Ardisia or "Coral berry", Pernettya and Pyracantha, long-lasting Skimmia, and female plants of the popular holly, or Ilex.

EDIBLE FRUIT—Of tropical or subtropical trees producing edible fruit, many are ornamental but usually too large for decoration indoors unless used in high lobbies, the patio, or in exhibition buildings. This includes coconut and date palms, breadfruit and the mango, peach of the tropics. Not quite so large are avocado with vegetable-like fruit rich in fat and vitamins, and of high nutritional value; the freely-bearing olive tree; common figs with fruit-like fleshy receptacles which are very sweet. Figs are best kept cool, and easily propagated from cuttings. The papaya is a giant herb, maturing most rapidly, its fruit resembling a melon; guavas make wonderful jelly.

Fortunately, there are a number of plants that bear fruit small enough to be grown in the home. Dwarf Ananas plants in pots will easily produce an aromatic pineapple; little cherry tomatoes could be grown in the kitchen ready for plucking. Pomegranates are both attractive and of pleasing flavor. We can even eat the fruit of such ornamentals as Monstera or of various Cactus if we look for the unusual or exotic. The dwarf banana does well in tubs but to bear its delicious fruit would have to have a large container. But the most favored of container fruits are the Citrus in their many forms, known already in Greek mythology as the Golden apples of the Hesperides. The larger-fruited kinds of oranges, the American Wonder-lemon, and the grapefruit, easily bear fruit, but unless the tree is large, they will seldom be able to support and hold more than one or several fruits. The smaller fruited Citrus and Fortunellas are much better for home or decoration. A little Meyer-lemon bush, bearing numerous bright yellow, flavorful lemons, is a joy to see. Likewise the Calamondin and the Kumquat are usually loaded with little orange-yellow fruits all year. The golden fruit of the cute Tahiti orange is charming but primarily ornamental, however, it rewards us, as all citrus, with the sweet fragrance of its waxy flowers. Of all the plants we grow as a hobby what could be more exciting than to produce your own fruit at home, and there is great promise for all that have the courage, the understanding, and the patience to try. According to Oriental belief, fruit trees in containers, symbolize those qualities which do enrich our life.

Fruiting female "Date palms", Phoenix dactylifera,
in the dry-hot but irrigated desert of the
Coachella Valley, Southern California

Cocos nucifera 'Dwarf Samoan',
an early fruiting "Coconut palm" with short trunk

Mangifera indica, the tropical "Mango tree"
bearing aromatic, sweet-fleshy fruit

Artocarpus altilis (communis),
the "Breadfruit tree", with ornamental incised
leaves and large yellow, prickly fruit; Capt. Bly's
ill-fated cargo on the "Bounty" in 1789.

Persea americana,
the "Avocado tree" or "Alligator pear",
bearing large edible, blackish fruit
with buttery flesh

Ficus carica,
the "Common fig tree" of the Mediterranean region,
much planted for its sweet, pear-shaped "fruit"

Musa nana (cavendishii),
the "Chinese dwarf banana" or "Dwarf Jamaica",
of compact habit, and bearing edible deliciously
fragrant, yellow fruit.

Musa x paradisiaca
the tall "Common banana" or "Plantain", producing
the well-known large fruit shipped to our markets.

Coffea arabica,
the "Arabian coffee tree", in fruit; fragrant white
flowers are followed by crimson berries, each
enclosing two coffee beans.

A female tree of Carica papaya, the tropical
"Papaya" or "Melon tree", bearing melon-like
aromatic fruit.

Vitis vinifera 'Black Alicante', a black "Tokay wine
grape", grown under glass in Nelson, New Zealand.

Passiflora edulis,
the "Purple granadilla", a climber with striking
flowers and edible purple passion fruit.

Feijoa sellowiana
"Pineapple guava"
fruit with guava flavor

Psidium guajava
"Common guava"
source of guava jelly

Psidium cattleianum
"Strawberry guava"
fruit with strawberry flavor

Eriobotrya japonica
"Chinese loquat" or "Japan plum",
pear-shaped edible fruit

Eugenia uniflora
"Surinam cherry",
spicy-flavored crimson fruit

Garcinia xanthochymus
"Camboge tree",
fruit yields cathartic

Carissa grandiflora
the "Natal plum"
fragrant flowers, red fruit

Punica granatum nana
"Dwarf pomegranate"
showy orange-red fruit

Olea europaea
the "Olive tree"
green fruits ripen to black

Aechmea fulgens discolor
"Large purplish coral berry"
long-lasting berried spike

Ananas comosus (sativus)
the cultivated "Pineapple",
as a fruited pot plant

Aechmea miniata discolor
"Purplish coral-berry"
or "Living vase flower"

Monstera deliciosa
"Mexican breadfruit"
fruit with pineapple aroma

Cereus peruvianus monstrosus
"Peruvian apple cactus"
with large, ruby-red fruit

Anthurium scandens
"Pearl anthurium"
strands of exquisite pearls

Opuntia ficus-indica, *a large flat-jointed cactus,*
the "Indian fig", widely cultivated in
Italy and Spain and elsewhere for its
juicy orange-red fruit

Rhipsalis quellebambensis
"Red mistletoe"
red-berried epiphytic cactus

Citrus x 'Meyeri', a hybrid of lemon with Sweet orange;
the *"Dwarf Chinese lemon"* or *"Meyer lemon"*,
with bright yellow lemons for ornament, and delicious for juice

Fortunella margarita
"Nagami kumquat"
or "Oval kumquat"

Fortunella hindsii
"Dwarf" or *"Hongkong kumquat"*
tiny orange-red fruit

Poncirus trifoliata
"Trifoliate orange" "Hardy orange"
spiny deciduous tree

Citrus limon 'Ponderosa'
"American wonder-lemon"
or "Giant lemon"

Citrus aurantium
"Sour Seville orange"
bearing freely in containers

Citrus aurantium myrtifolia
"Myrtleleaf orange"
small leaves and showy fruit

Citrus sinensis 'Washington Navel'
winter-fruiting "Sweet orange"
waxy flowers sweetly fragrant

Citrus paradisi, originally from the West Indies, "Grapefruit" is now planted in extensive groves for its juice, but will grow well in large containers also

Citrus taitensis, the "Dwarf Tahiti orange", has been a popular ornamental potted plant during winter, as an orange tree in miniature, laden with pretty ornamental golden fruit

Fortunella crassifolia "Meiwa kumquat", the Chinese "Golden bullet", a compact bush bearing rounded little orange-fruits with sweet and good-flavored peel

Citrus mitis, the everbearing "Calamondin orange", also known as "Orangequat", is an attractive potted plant when covered with its small but numerous long lasting orange-yellow fruit.

Skimmia japonica 'Nana'
"Dwarf Japanese skimmia"
female plant with red berries

Ardisia crispa (crenulata)
"Coral berry"
scarlet berries long persisting

Pernettya mucronata
"Chilean pernettya"
female plant with berries

Bucida buceras
tropical "Black olive"
branches symmetrically tiered

Hedera helix poetica
"Poet's ivy"
yellow-fruited adult form

Murraya exotica
"Orange jessamine"
with vivid-red berries

Cotoneaster horizontalis
"Rock clusterberry"
covered with coral-red fruit

Nertera granadensis (depressa)
"Coral-bead plant"
or "Hardy baby tears"

Rivina humilis
"Rouge plant"
or "Baby pepper"

Ilex aquifolium
"English holly"
berries on female trees

Ilex crenata
"Japanese holly"
small black berries

Ilex cornuta 'Burfordii'
"Globose Chinese holly"
few spines, scarlet berries

Pyracantha coccinea 'Lalandii'
hardy "Firethorn"
orange-red fruit, robust

Pyracantha crenulata flava
"Yellow Nepal firethorn"
fruit shining yellow

Ilex x altaclarensis
'James G. Esson'
big-berried hybrid

Glycyrrhiza glabra
"Licorice" or "Sweetwood"
sweet roots produce licorice

Pyracantha x duvalii
"Red firethorn"
dense with berries

Pyracantha koidzumii 'Victory'
"Christmas firethorn"
scarlet berries for Christmas

Lycopersicon esculentum 'Tiny Tim'
a miniature "Cherry tomato", with palatable scarlet
fruit; handy in the kitchen window

Solanum pseudo-capsicum
the "Jerusalem cherry" or "Christmas cherry", a
popular ornamental fruited plant with orange-scarlet
cherry-like fruit

Capsicum annuum conoides 'Red Chile' (frutescens),
a "Tabasco pepper", its bright red conical fruit
ornamental as well as edible and sweet-spicy

Capsicum annuum conoides 'Christmas Candle',
a "Christmas pepper"; cheerful holiday plant
with ornamental fruit waxy yellow turning to red

Jatropha curcas
"Barbados nut" or "Physic nut"

Palisota barteri
orange-red berries at base

Malpighia glabra
"Barbados cherry"
with edible red fruit

Water plants may be living under water, float on its surface, or merely grow along the edge of a pool or pond. Literally, the term Aquatics should be restricted to plants that spend their life in deep water. Some are with roots anchored in the bottom, with leaves and flowers floating at the surface as do most water lilies; others swim completely free as true aquatics such as some water-ferns, floating moss, duckweed or water-lettuce. Then there are the underwater aquatics that are most appreciated in tropical aquariums and that act as oxygenators to keep the water clean and healthy for the fishes. Amongst the best of these are Cryptocoryne or "Tape grass", and Echinodorus, the "Amazon sword plant". Of bog plants we are most familiar with Cyperus, the "Umbrella plant", and the giant "Papyrus" of Eastern Africa, which has been cultivated for paper making in the Nile delta since ancient times, and which was even known by the Assyrians.

Water gardening is the art of grouping various aquatic and waterside plants for best effect, into all sizes of aquatic plantings, from small tub gardens and artificial or natural brooks to large informal or formal pools. The great inducement that water gardens offer is not only their inviting restfulness and refreshing visions of coolness, but also their ease of care. However, sunlight is important to waterlilies as they are shy to bloom if under the shade of trees. Nymphaeas fall into two general divisions: tropical or tender, and winter-hardy. Tropicals, with their large, fragrant blooms well above the water, form tubers; generally their foliage and roots die at the approach of winter and the tuber may be dug and kept in moist sand until spring. Some day-blooming tropicals are viviparous and will grow continuously by forming new plantlets at the junction of petiole and leaf. One of these, the dwarf N. x daubenfana, potgrown and placed into a pail of water will, in a sunny window, keep blooming with its dainty sky-blue flowers all winter long. The hardy type of pondlilies have rhizomatous rootstocks and may be taken up, divided, and replanted, in rich soil, whenever it is not freezing; their lovely blooms seem to float gently on the water.

Water Gardening Indoors

Outdoor water gardens are a constant delight throughout the summer. But for year-round pleasure a small pool would enhance a winter garden or in the home, a glass aquarium, bowl or other glass container.

A 2 to 3 inch layer of medium coarse sand or washed aquarium gravel is all that is needed at the bottom of the water tank for the planting of a few aquatics needed to arrange a well balanced aquarium. While plants aren't absolutely necessary to support the life of fish when kept inside, they do add oxygen, absorb impurities, and reduce the green growth of algae, thus keeping the water fresh and clear. Two hours of good sunlight are considered essential and ideal for oxygenating aquarium plants to do a good job to keep the sand sweet, without making the water too warm. No airpump is needed unless the aquarium is over-crowded with fish. Plants use the energy from the sun to take in carbon dioxide and give off the oxygen which fish must have to breathe. After a night-long darkness, in strong morning sunlight, bubbles of oxygen will be seen to rise from the leaves. The most popular aerators are Sagittaria, Echinodorus, Vallisneria, Myriophyllum, Ludwigia and Cabomba; in poor light Cryptocoryne is useful. The floating plants will also help to clear the water. Temperatures for tropical fish should be within a range of 70-80° F; goldfish between 60-70° F. All new water should be allowed to stand for a day for any chlorine to dissipate and the temperature to adjust before using it for fish or plants.

An aquarium can be a very decorative and artistic asset in any home, a miniature world of aquatic life, with plants in their mellow greens enhancing the sparkling color of the fishes.

Aquatics indoors: A tropical Aqua-Terrarium for the living room, with strong electric light above benefiting both aquatic plants and fish. Underwater aquatics, like the "Amazon sword plant", Echinodorus tenellus (center), and Ludwigia mulertii (left), spreads easily in sand-loam mix, with the water a warm 70°F. Suitable companion plants would be Cryptocoryne, Aponogeton, Vallisneria, Myriophyllum, and Cabomba, and the floating Salvinia, Azolla, Lemna, Pistia and Ceratopteris.

Nymphaea x 'Pink Pearl'
long-stemmed tropical day bloomer
delicate pearl-pink and fragrant

Nymphaea x 'Golden West',
floriferous day-blooming tropical water lily in
autumn shades yellow and pink; foliage mottled red

Nymphaea x marliacea rosea,
a fragrant hardy water lily,
cup-shaped deep pink shading to rose

Nymphaea x 'Somptuosa',
a large, fragrant hardy water-lily
rose pink, deepening toward center

Nymphaea x helvola (tetragona x mexicana),
smallest hardy "Pygmy water-lily"
with reddish leaves and canary-yellow flowers

Nymphaea x 'Missouri'
beautiful tropical night-blooming water lily;
huge flowers creamy-white; foliage tinted red

Euryale ferox, at Nymphenburg Botanic Garden,
a giant aquatic from India, with violet flowers
and floating puckered, spiny leaves

Victoria cruziana on the French Riviera,
the "Santa Cruz water lily"
with floating leaves to six feet across

Nymphaea x daubeniana
tropical day-blooming "Pygmy water lily",
free-blooming lavender-blue and viviparous

Nelumbium nelumbo, the sacred "East Indian lotus",
symbol of perpetual life in Buddhism;
delicate pink flowers of haunting fragrance

Eichhornia azurea
"Peacock hyacinth" or "Creeping water hyacinth"

Eichhornia crassipes
the "Floating water hyacinth"

Cyperus diffusus var. elegans
"Broad umbrella palm"

Cyperus alternifolius
"Umbrella plant"

Cyperus diffusus variegatus
"Striped umbrella palm"

Cyperus alternifolius gracilis
"Dwarf umbrella plant"

Cyperus alternifolius variegatus
"Variegated umbrella plant"

Cyperus papyrus
"Egyptian paper plant" or *"Papyrus"*

Sagittaria montevidensis
"Giant arrowhead"

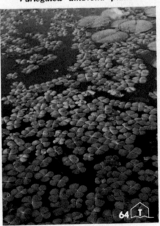

Marsilea quadrifolia
four-leaf *"Water clover"*

Hydrocleys commersonii
"Water poppy" or *"Water-keg"*

Cryptocoryne ciliata
"Fragrant tape grass"

Aponogeton fenestralis (Ouvirandra),
the skeleton-like "Madagascar lace-leaf" or "Lattice leaf"

Colony of Salvinia auriculata, a fern-ally,
known as "Floating fern", with boat-shaped leaves

Myriophyllum proserpinacoides
oxygenating "Parrot's feather"

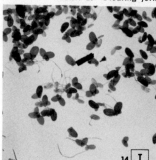

Lemna minor
floating "Duckweed"

Pistia stratiotes
"Water lettuce", "Shell flower"

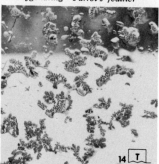

Azolla caroliniana
"Floating moss", "Mosquito plant"

Ceratopteris thalictroides
*"Water fern"
or "Water staghorn"*

Aponogeton distachyus
*"Cape pondweed"
or "Water-hawthorn"*

Hydrocotyle rotundifolia
shiny "Water-pennywort"

Darlingtonia californica, from California and Oregon,
the weird "Cobra plant" or "California pitcher plant";
sweet glands in pitchers entice insects to crawl inside

Sarracenia drummondii
the showy "Lace trumpets",
apex marbled white

Sarracenia psittacina,
"Parrot pitcher plant"
green with red veining

Sarracenia flava,
"Yellow pitcher plant" or "Umbrella trumpets",
insects are attracted into funnels and drown

Sarracenia purpurea, in flower,
"Northern pitcher plant" or "Side-saddle flower",
in bogs from Labrador to the Rocky Mountains

Sarracenia minor,
"Hooded pitcher plant" or "Rainhat trumpet",
yellow hood protects the throat

CURIOSITIES of the Plant World
Carnivorous Plants

One of the great miracles and mysteries of plant-life is the often unexplained ability of some plants to act like animals by setting in motion parts of their bodies, or thinking up means of trapping moving or flying insects and small animals. It is not known for sure all that prompts such "flesh-eaters" to react to the proximity or touch of insects or animals, which sense of intelligence it is that closes their traps or causes them to secrete deadly juices or reach out with sensitive sticky hairs or grabbing tentacles to attract and ensnare unwary insects. Insectivorous plants are a physiological assemblage of some 450 species belonging to a number of distinctive families. They agree in having the extraordinary habit of adding to the supplies of nitrogenous material and salts, by capturing and digesting the protein of ensnared insects and other small animal forms; they will even absorb bits of meat or of eggs. The various and curious mechanical arrangements by which these supplies of animal food are obtained and utilized differ with the various genera.

According to their methods of catching insects, the plants fall into three main categories: those that drown them in pitchers or urns as do Nepenthes, Sarracenia and Darlingtonia; others have sweet-sticky hairs in which the insect becomes enmeshed as in Drosera, or the gluey fly-paper leaves of Pinguicula; lastly, those that have traps or mobile tentacles that catch the insect as demonstrated by Dionaea and Utricularia.

Fascinating Nepenthes rafflesiana, and other species of tropical pitcher plants collected by the author in humid-warm Malaya. Honey glands at the mouth of the hollow flasks attract flies and other insects inside where they drown in pepsin liquid which digests them.

DISTRIBUTION — Carnivorous plants are widely distributed throughout the world. The tropical Nepenthes are chiefly Malaysian and are the most impressive of this group and their pendulous, decorated pitchers are an amazing spectacle. While they may be made to subsist under home conditions, Nepenthes need a high humidity greenhouse, or some form of sweatbox, and a temperature between 75-80° F. to form and develop their strange pitchers which attract not only insects but also small animals including birds and mice. Enticed by nectar glands to crawl inside, opposing hairs will keep entrapped insects from getting out again, and the pepsin that fills the tanks eventually digests them. More popularly known are the worldwide Sundews, which include Drosera, Drosophyllum and Dionaea. Of these, the "Venus fly-trap", Dionaea muscipula, endemic only to Carolina, is the most vicious looking—its little bear trap snapping shut on bugs that come within its reach. Heliamphoras are at home on the table mountains of Guyana and Venezuela. Prominent among the terrestrial pitcher plants are the Sarracenias, which range from Labrador to Louisiana, and the Darlingtonias of our Pacific Coast. They both attract their victims into elongated tubes ingeniously equipped as tank traps. Widely distributed are the Pinguiculas, found from the north temperate zone down to Patagonia, and the epiphytic or aquatic Utricularias which inhabit cold Greenland, tropical America, as well as South Africa. The curious Cephalotus is confined to Western Australia.

CULTURAL REQUIREMENTS—With the exception of the xerophytic Drosophyllum and, strangely, a number of Nepenthes, all the carnivorous plants agree in inhabiting damp heaths, bogs, swamps and stagnant soils with abundant water but a characteristic dearth of nitrogenous material. Except for Nepenthes most carnivorous plants prefer to live in cool or intermediate temperatures—but if grown indoors, all should be provided with an enclosure of plastic or of glass which favors condensation of moisture and a high degree of humidity; aquarium tanks, fishbowls or even brandy snifters are ideal for smaller plantings. Their best growing medium is probably fresh sphagnum moss, at least on top, with sandy composts of sphagnum peat or humus underneath. For watering, mineral-free rain or distilled water is best; allow chlorine-treated tap water to stand for a day. Feeding should be omitted altogether. Some species, such as the Venus fly-trap, are not at all difficult to grow and it is most interesting to tease this little plant into closing its ferocious jaws.

An "insect trap" with its highly developed, almost intelligent sense of perception, or of touch, activating a variety of mechanisms or secretions through curious devices, will always be a fascinating conversation piece, since it is the nearest a living, yet passive plant has come to animal life which is active and aggressive. Paradoxically, a carnivorous plant turns the tables on the very animals which normally are the ones that eat them.

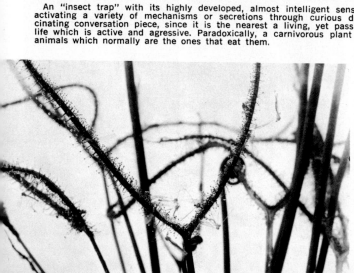

Forked leaves of Drosera binata, the "Twin-leaved sundew" from Australia and New Zealand, enlarged to show insects caught by sensitive glandular hairs discharging sticky fluid, to hold and finally digest them.

Cephalotus follicularis
"West Australian Pitcher plant",
quaint pitchers marked maroon

Dionaea muscipula
"Venus fly-trap"
jaws snap shut on contact

Drosera rotundifolia
glistening red "Sun dew plant",
sticky hairs embrace victim

Heliamphora nutans
"Sun pitcher"
pitfall insect trap

Pinguicula caudata
"Tailed butterwort"
feet get stuck on leaves

Utricularia alpina
"Mountain bladderwort",
bugs sucked through trap door

Drosophyllum lusitanicum
Mediterranean "Dew leaf",
glands secrete sticky drops

Drosera binata
"Twin-leaved sundew"
sticky tendrils hold insects

Drosera filiformis
"Threadleaf sundew"
hairs curve inward to entrap

Nepenthes x coccinea (distillatoria x mirabilis), forms its crimson and yellow pitchers willingly; lids keep out the rain.

Nepenthes maxima (curtisii), sturdy "Pitcher plant" tolerating lower temperature; firm flasks pale green marbled blood-red

Nepenthes x dicksoniana (rafflesiana x veitchii), large cylindric pitchers marbled crimson, and containing the pepsin fluid that traps and eats overly curious flies.

Nepenthes 'Superba', a vigorous tropical rambler, with pendulous urns yellow-green blotched with wine-red, formed at the end of tendrils from the leaf tips at high humidity.

CURIOSITIES
of the Plant World

Nature is full of wonders and never ceases to remind us that there must be forces beyond our understanding which can only lead us more to the belief in an Almighty and All-Knowing Hand, that is capable of creating this wondrous universe and all that lives within it. There are plants that are sensitive to touch; plants that mimic their surroundings or build defenses against them; plants that move to trees to reach the light they crave, and others that climb up them to strangle their hosts in their own selfish battle to exist, until they themselves become a tree. There are plants with flowers insignificant but rich in fragrance, or equipped with showy bracts to attract the birds they need to help them with their reproductive functions. There are stems that resemble leaves, and leaves that look like flowers. Science has determined plants whose flowering period is regulated or inhibited by length of day or night. Plants from xerophytic regions surprise us with their stubbornness and will to live, but others are so delicate that without the presence of saturated humidity they would quickly perish. Succulents or cacti may be deformed or fasciated and yet be strangely beautiful. We know plants that are nothing but leaves and stems and manage to live through pores or absorptive scales; others carry on their life process through their roots only, lacking leaves entirely.

Who has not marvelled at plants which fold their hands and go to sleep; flowers that open their blooms only to the midday sun and others that unfold only at night. We have trees that have lived for thousands of years, and Arctic annuals that must complete their life cycle and process of regeneration in just a few weeks.

Bombax malabaricum, the "Red silk-cotton" or "Kapok tree", its powerful roots struggling for centuries with the massive temples of Angkor built without mortar by the Khmers 1000 years ago, deep in the tropical forests of Cambodia.

Some ferns and other plants bear young plantlets on older leaves, or old bulbs produce tiny bulblets sitting on their body. Many plants have healing powers and others are narcotic or poisonous. Plants parasitic that drink up the lifeblood of other life, and plants merely appearing parasitic but in fact care for themselves. Some plants or their flowers resemble animals or birds but, conversely, true animals such as the "Air fern" have long been mistaken for plants. The very existence of a group of "insect-eating" plants creates the image of a bridge from botany to zoology.

The illustrations in this section are a sampling only of the many peculiarities found in plant life —though often aided by the imagination and skill of horticulture. If we take the time and learn to attune ourselves to nature, we become absorbed in the study of its patterns, its beauty and its mysterious workings. This will make the habit of living with plants one of the richest, most restful, and enjoyable experiences in our life.

Seedbearing, tree-like fossil fern, of the extinct genus Sphenophyllum, imbedded in red shale 260 million years ago and now extinct, down the tortuous Kaibab trail at Cedar Ridge, Grand Canyon, Arizona.

Xanthorrhoea arborea, a curious "Grass tree", centuries old, in northern New South Wales, Australia; long-lived perennial plant with palm-like woody trunk and narrow linear leaves.

Adansonia digitata, the monstrous "Baobab", also known as the "Dead rat tree", in the Transvaal, South Africa. One of the largest trees in the world, its pulpy trunk reaches 30 feet diameter, and is said to become 5000 years old; this is the tree that "God turned upside down".

This "Giant Redwood" or "Big tree" of the High Sierra in California, Sequoiadendron giganteum, 5000 years old, has grown to a height of 280 feet, and its diameter measures 38 feet.

The most unique creation of the flora of the Canary Islands is the "Dragon tree", Dracaena draco. World famous is this ancient specimen, 15 ft. thick and thought more than 2000 years old, at Icod on Tenerife; photographed in 1965.

The evergreen "Cork oak", Quercus suber; in Andalucia, southern Spain. Its elastic, thick outer bark is stripped every 8 years and is the cork-bark of commerce.

An ancient "Olive tree", Olea europaea, believed to stand since biblical times, its weathered trunk 4 feet in diameter, in the Garden of Gethsemane outside Jerusalem.

Fuchsia excorticata, the unbelievable "Tree fuchsia" in the south of New Zealand forms woody trunks to 2 feet thick.

Ginkgo biloba, the "Maidenhair tree" or "Living fossil" is perhaps our most ancient existing flowering plant, from the age of dinosaurs 80 million years ago.

Festooned with Spanish moss, Taxodium distichum, the "Bald cypress" of swamps in Southeastern U.S., produces "cypress knees" from the roots showing above water with which to "breathe".

The remarkable Beaucarnea recurvata, or "Bottle palm" planted along Sepulveda Boulevard in Los Angeles. The swollen base will store a supply of water against periods of drought.

Yucca brevifolia, the "Joshua tree", a grotesquely branching liliaceous succulent tree 40 feet high, in the storied Mojave of California—a ghostly silhouette on the high desert at 4000 ft.

Carnegiea gigantea, the "Saguaro bouquet" in Arizona, one of the world's largest cacti. These candelabra-like giants grow 50 feet high and may be 250 years old, enduring torrid temperatures reaching 120°F.

The extra-ordinary "Sausage tree", Kigelia pinnata, seen in Uganda and elsewhere in tropical Africa, with sausage-shaped fruit swinging from long cords.

Brachychiton rupestris, the peculiar "Bottle tree" of Queensland, has a trunk 11 feet thick, built like a thermos-bottle, an outer compartment holds water; the inner reservoir contains edible, sweet jelly; enabling the tree to withstand long periods without rain.

Couroupita guinanensis, the "Cannonball-tree", in Panama; enormous round fruits hang in clusters direct from trunk

Ficus glomerata, known as the "Cluster fig", with edible reddish fruit pods produced from the trunk

Theobroma cacao, the "Chocolate tree" in Central America, bearing pods containing cacao beans; "Food of the gods" of the Mayas

Artocarpus heterophyllus, the "Jackfruit", one of largest edible fruits 2 feet long and weighing 40 lbs.; grows on trunk

Phaleria capitata, a small tree in Java, with beautiful waxy-white flowers, sweet-scented like daphne, blooming from trunk

Ruscus hypoglossum, the "Mousethorn", a small "Butcher's broom" with leaf-like branches, bearing tiny flowers on their center

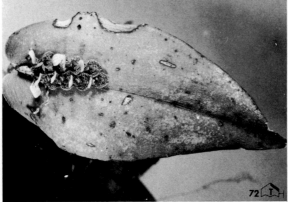

Pleurothallis cogniauxiana, an epiphytic orchid from Central America, its malodorous red-brown flowers emerging from the axil of a solitary leaf and resting on it

Phyllanthus speciosus (Xylophylla), "Woody-leaf flower"; tiny flowers at edge of leaf-like branchlets

Pantropic Mimosa pudica, the "Sensitive plant", also known as "Humble plant", "Live and die", "Touch-me-not"; in Vietnam as "Shame plant"; remarkable because its leaflets close and the petioles drop at the slightest touch.

Biophytum zenkeri, the "Sensitive life plant" from the Congo; the leaflets will fold back when touched

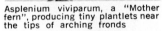

Asplenium viviparum, a "Mother fern", producing tiny plantlets near the tips of arching fronds

Maranta leuconeura kerchoveana, the "Prayer plant" in prayer, its leaves folding upward when night approaches

Tolmiea menziesii, the "Piggyback plant" from California to Alaska; young plantlets develop from mature leaves

Mimosa speggazzinii, "Sensitive-leaf tree" from Argentina, spiny shrub with forked, sensitive leaves that fold when touched

Begonia hispida cucullifera, the "Piggy-back" or "Crazy-leaf begonia" produces adventitious leaflets on the surface of its leaves

Ornithogalum caudatum, the "False sea-onion" or "Healing onion"; the mother bulb carries bulblets on its side.

Kalanchoe pinnata (Bryophyllum), "Airplant" or "Curtain-leaf"; plantlets sprout from leaves even if broken off.

hybrid Kalanchoe 'daigremontiana x tubiflora', "Mother of thousands", develops masses of little plants from leaf margins

Agave attenuata, the "Dragon-tree agave", carrying reproductive bulbils and plantlets on its arching flower spike.

male plant of Welwitschia mirabilis, the "Tree tumbo" or "Mr. Big" of Southwest Africa, with cone-like inflorescence. This curious desert plant grows monstrous leaves to 20 ft. long.

Kalanchoe gastonis-bonnieri from Madagascar, known as "Life plant"; its glaucous-white leaves produce perfect young plants at tips.

Sempervivum arachnoideum, the "Cobweb hen-and-chicks"; tiny clustering rosette covered by a cobweb of white hairs.

Aloe vera, the "Medicine plant" or "Bitter aloe" of the Africans; the jelly-like pulp is used as a poultice to heal burns and cuts; also taken internally

Aeonium tabulaeforme, the "Saucer plant", grows flat as a plate, clinging to rocks on Tenerife

Mesembryanthemum crystallinum (Cryophytum), one of the mysteries of plant distribution; the true "Ice plant", found in coastal California and distinguished by its glistening ice-like crystals covering the fatty leaves from other so-called ice plants whose home is South Africa.

Haworthia truncata, the unusual "Clipped window plant", with windowed apex of each leaf flat as if cut off

Lophophora williamsii, the "Sacred mushroom" or "Peyote", having narcotic properties, and revered by Indians

Lithops karasmontana, the "Mountain stone-face", a "Living stone" of Southwest Africa, mimics the pebbles of the desert

by a miracle of nature, white-felted leaves of Kalanchoe tomentosa, the beautiful "Panda plant", develop new plantlets from their leaf bases, though merely lying on the sand

branches from a "Candelabra tree", Euphorbia candelabrum, planted as a living fence in Central Africa, resembling the cactus fences of the pueblos of Mexico

grafted Notocactus scopa cristata, the "Spiralled silver ball"; beautifully contorted into shape of a coiled white snake

Euphorbia lactea cristata, the "Crested euphorbia", "Elkhorn", or "Frilled fan", an intricate fasciation into little fans and paws

Cereus peruvianus monstrosus, the "Curiosity plant" or "Giant club", a column cactus gone freakishly monstrose

Aeonium arboreum cristatum, resembling a bird; the fasciated ridge supports a chain of tiny leaf rosettes

Opuntia clavarioides, the strange "Black fingers", "Sea-coral" or "Fairy castles"; from fasciated fans rise little fingers

Hoya carnosa 'Hummel's compacta', a wax vine densely compacted with folded and cupped leaves oddly twisted

Opuntia linguiformis 'Maverick', the "Maverick cactus"; an attractive mutation, the flat joint forming monstrose fingers

grafted Mammillaria vaupelii cristata, the "Silver brain"; fantastic fasciations of coiled crests form symmetrical globe

x Pachyveria clavata cristata, silver-blue miniature rosettes tightly clustered and fused on a malformed stem

Dischidia rafflesiana, the remarkable "Malayan urn vine", with modified fleshy leaves formed into hollow pitchers, a refuge for ants

Tillandsia usneoides, the "Spanish moss", a root-less true epiphyte; lichen-like airplants with silvery scales capable of absorbing atmospheric moisture

Rhaphidophora celatocaulis, a "Shingle vine" from Borneo, with juvenile stage leaves clinging flat to walls or trees

Stanhopea eburnea, an epiphytic "Horned orchid", the peculiar waxy-white, strongly perfumed pendulous flowers push out through their basket

Habenaria radiata, the pretty "Egret orchid", a terrestrial from Japan with green and white flowers, the fringed lip resembling a bird in flight

the leafless Taeniophyllum fasciola in Fiji, one of tiniest of orchids, has spidery roots functioning for absent foliage

Phanera williamsii, a "Chain liane", leguminous climber with chain-like stems, photographed in the jungles of New Guinea

Bowiea volubilis, the "Climbing onion", or "Zulu potato", a drought-resistant curiosity, the large green bulb anually sending up a twining stem

Ceropegia fusca in flower, the "Carnival lantern" or "Stick plant"; erect succulent with curious lamp-like flowers.

Amorphophallus campanulatus, a giant aroid in New Guinea, the huge spadix bearing a multitude of golden-yellow fruits

Ravenala madagascariensis, the "Travelers tree" in Madagascar; its leaf fan supposedly aligned East and West guides travelers, and supplies them water.

Hydrosme rivieri, the "Devil's tongue" or "Voodoo plant", its dark, sinister, carrion-scented inflorescence appears before leaves

Dorstenia yambuyaensis, from the Congo; what looks like a flower is really a plate-like receptacle imbedded with many male and female flowers

Ficus krishnae, the "Sacred fig tree" from India; its leaves folded into cups to collect drops of dew when the god Krishna was thirsty.

Spathicarpa sagittifolia, the "Fruit sheath plant", bearing male and female flowers attached along recurved green spathe

Strelitzia parvifolia juncea, the "Rush-like strelitzia" seen in South Africa, a "Bird of paradise" with spear-shaped stalks but without leaves

Nepenthes mirabilis, in New Guinea; this pitcher plant lives on the ground in hard volcanic soil entrapping unwary beetles and ants.

Anigozanthus manglesii, the "Kangaroo paw" of Western Australia; striking thick-felted flowers red, green and blue

Zantedeschia elliottiana, the "Yellow calla"; a flower-like golden spathe surrounds the actual inconspicuous flowers on the column

the showy red bracts of a "Poinsettia", Euphorbia pulcherrima, are not flowers; true flowers are the little yellow-green knobs in the center of the star

in Guzmania lingulata 'Major', the "Scarlet star bromelia", a burst of bracts, brilliant red tipped yellow, subtend the white, true flowers

the "Flamingo flower", Anthurium scherzerianum, looks like a bird— the showy "flower" is the spathe; the tiny, real flowers line the twisted spadix

Helichrysum bracteatum, an everlasting "Strawflower", where gay petal-like, long-lasting bracts surround the inconspicuous floral disk

Ipomoea tuberosa, known as "Wood-rose" because of the wooden appearance of its "flower", actually enlarged sepals and seed pod

Pereskia grandifolia, a "Rose cactus", with true leaves, a living image of the ancestors of the cactus family that lived millions of years ago

Consolea falcata, a "Tree opuntia" from Haiti; "leaves" are actually flattened joints of branches, but true leaves are absent

Breynia nivosa roseo-picta, the "Leaf-flower" or "Snow bush" from Polynesia; gaily colored foliage lavishly masquerades as "flowers"

Brassica oleracea acephala crispa, the "Flowering kale"; in autumn the leaves turn ivory, rose and purple, looking like a giant flower

Oxalis deppei, bulbous plant from Mexico, sentimentally known as "Lucky clover" or "Good luck plant" because of its four leaflets

Cordyline terminalis 'Ti', advertised as the Hawaiian "Miracle plant"; sections of its stem will sprout new plants

The amazing "Santa Cruz water lily," Victoria cruziana, has big floating leaves with airfilled, buoyant ribs, a temptation to young people to use these giant platters as a boat.

A semi-circular marble bench considerably accommodates one of the Aleppo pines in a shady grove within spectacular view of the ramparts and temples of the Acropolis in Athens, Greece

gnarled specimen of Hedera helix hibernica, the "Irish ivy", has grown a woody trunk 12 in. thick, since planted, 90 years ago, at Princeton University, in New Jersey in 1870.

Sequoia sempervirens, the "Coast redwood", can reproduce itself from knotty burls, and these will sprout in dish of water

"Mata de Rosa Blanca" or "White Rose tree", a custom in Puerto Rico, where eggs speared on Yucca supposedly ward off misfortune

"Sugar cane", Saccharum officinarum, also produces rum, basic ingredient of "Rumandia cocacoliensis" listed in Exotica 1

As seen in Hawaii, Golden bamboo, Bambusa vulgaris aureo-variegata, is growing in buckets of water after the canes were sawed off, "rodded" out inside and filled with water.

water is piped into irrigation channels for the shade trees of a boulevard near the Prado Museum in arid Madrid, Spain

to create an Ivy-tree, Hedera helix 'Curlilocks' (or others) can be side-grafted on the erect stem of x Fatshedera lizei

the "Sweet-potato vine", Ipomoea batatas, with its edible tuberous root can easily be grown as a curiosity in a bowl of water

"Potomato", the result of my grafting a tomato scion on a sprouted potato tuber, in Sioux Falls, South Dakota

Eugenia myrtifolia, the ornamental "Australian brush cherry", clipped and shaped into fanciful topiary in Los Angeles

a rooted piece of cane of Dracaena fragrans massangeana is host to the climbing Philodendron oxycardium (cordatum of florists)

the "Vegetable sponge" or "Dish-cloth gourd" is produced by the tropical climber Luffa cylindrica. Its fibrous skeleton is widely used as sponge for scrubbing and bathing.

Rafflesia arnoldii, a fleshy parasite without leaves, known as the largest flower in the world, 3 feet across; in **Sumatra**

the parasitic "Mistletoe", Phoradendron flavescens, from New Jersey to New Mexico, a traditional invitation to a kiss at Christmas time

another "Mistletoe", Phoradendron juniperum, a true parasite, mimicking juniper foliage along the Grand Canyon rim in Arizona

Psilotum triquetrum, the "Whisk fern", primitive vascular plant; first step of evolution from liverworts toward higher plants

Phrygilanthus aphyllus, a hostile parasite which fastened itself on a Cereus cactus stealing its life blood

looking prehistoric, a female, flowerless "Liverwort", Marchantia polymorpha, stands botanically between mosses and the algae

the Love tree in Florida, Ficus aurea, a "Strangler fig"; embraces and strangles a palmetto palm which in the end must die

Tillandsia flexuosa, the "Spiralled bromelia" in Florida, clings to a twig as epiphyte, but is not a parasite

Dendrobium d'Albertisii, the "Rabbit's ear orchid", is not parasitic, though it perches like one on a tree in New Guinea

Polypodium bifrons, a creeping fern forming curious hollow, chestnut-like vessels separate from its fronds (Palmengarten, Frankfurt, Germany)

a "Ball fern", Davallia bullata mariesii, its slender rhizomes trained in form of a monkey, or Shinobu style, in Japan

the four-leaf "Water clover", Marsilea drummondii from Australia, a true fern but normally growing as an aquatic

the "Walking fern" or "Trailing maidenhair", Adiantum caudatum, its elongate fronds rooting at tip and forming new plant

"Cinnamon fern", Osmunda cinnamomea, appears to have "flowers"; actually these are furled fertile fronds bearing spore

the "Resurrection plant", Selaginella lepidophylla, curls into ball when dry, but unfolds fresh green if placed in water

the "Queensland tasselfern", Lycopodium phlegmaria, is a tropical epiphytic clubmoss, pendulous, needle-sharp and tasseled

the "Mystery plant" is Lycopodium obscurum, a "Groundpine" tied together like a tree, studded with Eriocaulon or "Skyrockets"

the "Air-fern" of dime-store fame, is actually a moss-animal invertebrate polyp of the genus Bugula, and dyed green

Unique in the palm family, Hyphaene thebaica, the "Gingerbread" or "Dhoum palm" is a botanical wonder because its trunk is forked. Its beautiful orange, edible fruit tastes like ginger bread. Photographed in Tanzania, East Africa.

Livistona chinensis, known in horticulture as Latania borbonica, the "Chinese fan palm", properly has a stout single trunk. But this tree is unusual because it has developed a crestate fan resulting in numerous trunks; at Icod, on Tenerife.

the "Siamese date palm", Phoenix paludosa, has the queer habit of sending out trunks forming loops, with the tips subsequently growing erect. At Huntington Botanical Gardens, San Marino, California.

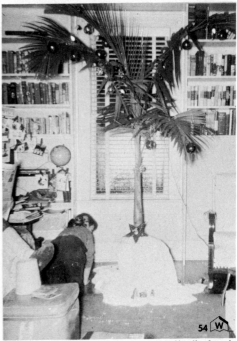

In tropical regions and far from the Nordic forests it is frequently the custom to decorate other than spruce for the holidays, but most unusual is this "Piccabeen palm", Archontophoenix cunninghamiana used as a Christmas tree in southern Florida.

a mesozoic Dinosaur come to life, Clerodendrum inerme, the "Indian privet"; carefully trained over wire frames in the art of topiary and sheared every two weeks; at Hanging Garden, Bombay, India.

this "Japanese white pine", Pinus parviflora, trained and dwarfed as a miniature Bonsai tree, is especially remarkable because this old specimen manages to exist on a slab of lime stone rock.

a grouping of Ficus retusa nitida, the "Indian laurel", united through grafting and shaped into a basket, in China.

the "Chinese banyan", Ficus retusa, growing in porcelain jar, and trained as a Ming tree with snake-like twisted stem, in Hong Kong.

a potted "Buddhist pine", Podocarpus macrophyllus, in the shape of a Ming tree, its stem skillfully snaked and looped by in-grafting.

Rooster-shaped topiary, with several varieties of English ivy, Hedera helix, densely planted into wire frame stuffed with moss.

decorative standard "Baytrees" or "Laurel", Laurus nobilis in tubs, some most unusual and trained with great patience with spiralled and looped stems; on exhibit at National Flower Show in Essen, Germany.

this diminutive "Monstera" in my office is a cute replica of its big brother, and never a wet spot on my desk (3 inch pot)

Pseudo "Splitleaf Philodendron" on real bark, as costly as a living plant; never loses a leaf but doesn't make new ones either

this imitation "Crinkly Ivy" may be useful in a place where insufficient light does not allow the use of real ivy, but in sunlight, it would bleach

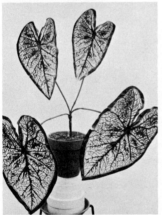

a "Fancy-leaved Caladium" breathes tropic charm from its delicate, highly colored leaves, but this awkward plastic copy is unaffected by cold drafts or lack of water

warmth-loving Dieffenbachias are amongst the easiest of house plants to maintain, requiring very little water. So why this unnatural, stiff creation?

A stark Hong Kong "Araucaria", lacking elegance and looking like a brush, and easy to catch dust, is so unnecessary because a live tree is one of our most modest keepers

this luscious pink "Saintpaulia" looks almost real and will bloom forever. But is there any fun in a plant never changing with the seasons although it may fool a visitor with its near-perfect imitation?

Chinese "Ming tree" molded of polyethylene, imitating the traditional dwarfed Pine trees of the Ming period and of Korea

These plastic Boston ferns look real enough, ideal in a furniture store because they never require watering, but are out of place on a sun porch at Nature's own Lake Placid, in New York's Adirondack Mountains

The current trend toward man-made imitations masquerading as real plants is out of harmony with true fondness for plants and flowers that are alive and breathing in all their delicate and natural beauty. Even so, artificial blooms and foliage have their place in our decorating schemes in places where watering is difficult or impossible. Their presence can soften stark architectural lines in locations with insufficient light where live plants could not survive for long without the aid of potent reflector lamps. However, there are now available categories of living plants which, with only fair reading light (15-20 foot-candles) can be maintained indefinitely and will, in the long run, not cost any more than artificial plants.

FLOWERS IN HISTORIC ART—The Chinese, to whom flowers have long been symbolic of all human virtues, have reproduced flowers in bronze or semi-precious stones 2000 years ago. The emperor Wu Ti of the Han dynasty, a century before the Christian era, caused a Taoist temple to be built with a garden of trees of jade, the branches of red coral and leaves of green jade, the blue and red flowers carved of precious gems. After the discovery of kaolin by the Mongols, flowers were painted on porcelain, or boldly enamelled on vessels of bisque. Miracles of beauty were created with the beginning of the Ming dynasty in 1368, to the Manchus under poet-emperor Ch'ien Lung who brilliantly ruled from 1736-1795, and encouraged the arts. During these periods Chinese jade in mysterious colors, always treated with reverence by the people, began to be carved and fashioned with great patience, together with other semi-precious stones, into Jewel-flowers and so-called Jade-trees, rich and glowing in their polished, eternal loveliness. Later, adaptations of these "Ming trees" were, and still are being made of the less costly "Peking glass". In Europe, following the Middle Ages, flowers were cut from precious gems and semi-precious stones, arranged in bouquets; masterpieces of this art can be seen at the Imperial Treasury in Vienna, and Russia's Hermitage in Leningrad.

The several thousand "glass-flowers" now on display in the Botanical Museum of Harvard University, Cambridge, Massachusetts, skillfully and patiently made by Bohemian artist-naturalists between 1890 and 1936 are breathtakingly life-like and famous for their scientific detail and fragile beauty.

From England came the diminutive colonial-type nosegays of flowers made of bone china, and widely loved for their endearing and enduring charm.

ARTIFICIALS IN DAILY LIFE—Fresh and natural-looking waxed paper-flowers were popular with the beginning of the 20th century and can still be found, particularly in country villages and farms of Europe and Latin America. Flowers of silk fabric from France and Germany are 3-dimensional in subtly changing hues, but are expensive and soon become dusty, losing their vivid colors. Still other flowers are fashioned of cloth, coated with clear chemical finish, with charming effect.

With the advent of the age of plastics, permanent blooms of vinyl made their appearance about 1958. The following year flowers and plant-foliage began to be copied from nature and cast in superior tough yet flexible polyethylene, half of them manufactured in the Orient. Through improved techniques today's plastic creations are almost life-like, more so the blossoms than the foliage. When, inevitably, dust accumulates, they may be washed and cleaned with detergent and water, but eventually, the polyesters will break down and discoloration sets in, particularly if exposed to sunlight.

In the business world—in building lobbies, offices and showrooms; conference rooms rarely used; hotels and motels; restaurants and cocktail lounges; department stores and window displays; hospitals and reception rooms—artificial foliage is frequently seen, often introduced through an interior decorator. Here the location of the planters, usually dark, drafty or super-heated areas, may dictate the use of plastics. Sanitary requirements in a dining area may also limit the use of live plants where there is fear of contamination by insect sprays. In itself, this would indicate that the use of plants in buildings is considered desirable to enhance a setting in the general decor. Happily, reputable firms, with discerning good taste, increasingly favor the installation and maintenance of long-lasting, reliable live specimen.

In the home, the housewife would most always prefer "the real thing"—fresh flowers and growing plants, buying plastic substitutes as a poor second. The muted colors of dried branches; berries, pods, cones and thistles; grasses and other "everlastings", are especially appropriate during winter—for a bit of nature that can long be enjoyed.

This redwood planter box containing an assortment of plastic "plants" including Schefflera, Ficus lyrata, and Philodendron, would find a place where due to lack of light a live plant could scarcely live. However, a spot-light would do much to accent any such exhibit also and thereby allow real plants to be used instead.

Thoughts On ARTIFICIAL PLANTS

CRITICAL COMMENTS—Ardent plant lovers are in decided opposition to plastic flowers and vehemently voice their disdain of plants artificial as expressed here by Marlene Robinson, a perceptive interior decorator: "Artificial plants when used to substitute for live ones are an insult to human beings. They offend the senses and affront the intelligence. The joy that arises in one on seeing a lovely, healthy plant is shattered upon discovering the plant not only is fake but is plastic—a symbol of all that is easy-come-quickly-gone in our modern society. The compliment that was about to be extended to the owner turns to sour frustration with the shock of having been fooled. For what the eye first discerns in a plant, the other senses must then help bring into focus the full character of the plant: the tactile depth, the fragrance, the imperfections of growth. Character is what plastic plants lack. There is not a withered leaf nor a bare portion of stem—owing to some moment of human frailty. There is no tropism, or turning to the light, evident in the stretch of the leaves. The sun has not blotched, bleached nor reddened them. Indeed, the non-plant shows no involvement with sun, soil nor human nature. One gets tired of looking at its ever-sameness and while outwardly brightly painted it is inwardly dead. It sits there a plague to the air around it, a collector of dust. An inorganic non-growth that gradually fades as its chemical polyesters break down. An expensive investment in a half-slice of reality. A frightening expression of the "I can't" part of our society."

"Whether or not it is the basic commitment to life that determines whether a person buys a real or non-real plant, it would indicate that everyone wants plants. Is it possible that some people prefer plastic ones over the real ones? Does it have something to do with a person's capacity for trial and error? Or is it more a matter of time and lack of knowledge of what type plant to choose for what dim corner or cold window sill? When one sees so many windows in both high and low income areas of a crowded metropolis burgeoning with obviously growing plants, the observer is left with the distinct feeling that the more removed from nature a person is, the more he tries to recapture it. Artificial plants may be a fad. Through owning them and watching them discolor more people may see their grotesqueries and return to gardening." When they do, they will find that there are available real plants that have proven near permanent in keeping quality, and through their active metabolism, able to repair tissues to prolong life which a plastic plant cannot.

FLOWERS FOR SENTIMENT—Unlike the creations of nature, simulated flowers and plants lack warmth and character, and elegance of form; they don't have the individuality, exquisite charm, and meaningful esthetic values of live ones. They fail to rouse within us a mental uplift. A live plant is something we could look at daily; an artificial one is quickly ignored and speedily forgotten. If this were not so, the growing of real plants would very likely stop.

The famous architect Eero Saarinen once remarked unhappily that there exist enough fake things in this world, without creating more in trying to mimic nature which embodies that which is pure and beautiful in cardinal virtues. When we think of flowers, we think of something that is living, delicate and fragrant as the gift of nature that they were meant to be. This is especially true where sentiment is involved. To "Say it with Flowers"—or—"Flowers bring Joy" are slogans only too valid when, as nothing else can, they bear a message of understanding and affection on happy or on sad occasions. Flowers and plants that are alive have a meaning all their own, for enchantment in the home or when presented to a loved one.

A delicate, fragrant yellow Tea rose decorates the tables in the dining car of a Santa Fe streamliner en route from Chicago to Los Angeles. What could be more exquisite, touching and inviting than a flower that is alive and speaks of hope and romance and life itself?

Basket of plastic Carnations and Roses, mingled with Philodendron and Boston fern, hanging from lamp post along busy Wabash Avenue in Chicago. Poinsettias are used at Christmas.

"Wreath of Happiness" in Tokyo, Japan. Such wreaths are made with chrysanthemums and lilies of gaily painted tin or paper and displayed at store openings, weddings and anniversaries.

Three Roystoneas, or Royal palms of Andean stone and bronze, united bear the figure of 'La India', commemorating the battle of Carabobo in 1821, and the liberty of Gran Colombia, at Caracas, Venezuela.

Echinocereus engelmanii, a hedgehog cactus; painstakingly reproduced in colored glass near Dresden, Germany in 1895 by father and son Blaschka from Bohemia; now at Harvard University.

Rose-bush characteristic of the flower composites so elaborately modelled during the reign of Manchu emperor Ch'ien Lung in China: the dish of green nephrite; leaves of mottled green jade, flowers of coral, quartz, and pearls.

"Blumen-vase" — precious antique vase of tulips and other flowers carved of gem stones, onyx and jade, by Miseroni; in the Schatzkammer of the Imperial Palace, Vienna, Austria.

Chinese Chrysanthemums, the "Golden flower of the East", and symbol of Long Life and Happiness, here reproduced of fine porcelain and adorning an ancient temple in Kwangtung, China.

An exquisite and excellent ivory carving of a cattleya orchid in miniature, 4 inches high, daintily tinted in natural color as if alive; in Japan.

Mingtree in the Chinese style made of gnarled California manzanita wood and tree moss as found in the high Sierras, and dyed (8 in. high).

Glossary of Botanical and Scientific Terms, as used in this Manual

Words difficult to understand are kept to a minimum in this Manual, using descriptive English equivalents wherever possible. Those that could not be avoided because of their specific meaning, are listed below. The Botanical Chart on opposite page also illustrates many technical terms.

acuminate — tapering to a point

acute — sharply pointed, but not drawn out

adventitious — other than usual place

alternate — arranged along a stem at different levels

anther — pollen-bearing top of stamen

apex — the tip of an organ (as a leaf)

appressed — flatly pressed against

areole — cushion-like structure out of which can arise spines, branches, and flowers, a characteristic confined to cacti

asexual — propagates without benefit of sex

attenuate — becoming narrow, tapered

axil — the point just above the leaf where it rises from the stem

bifurcate — forked

bilabiate — divided into two or equal lips

bipinnate — both primary and secondary divisions with separate leaflets

bisexual — possessing perfect (hermaphrodite) flowers having both stamens and pistils

blade — the expanded portion of a leaf

bract — modified leaves intermediate between flower and the normal leaves, frequently colored

bulb — a growth bud with fleshy scales, usually underground

calyx — outer circle or cup of floral parts (usually green)

caudex — upright root stock or trunk

caulescent — becoming stalked

cephalium — woolly cap at the apex of cacti

chromosomes — microscopic rodlike bodies in the plant cell, bearing the hereditary material

ciliate — fringed with eyelash hairs

cladode — branch simulating a leaf

clasping — leaf surrounding stem

column — combined stamens and style into one body (as in orchids)

concave — hollowed out

connate — united

convex — umbrella-like

cordate — heart-shaped

corm — bulb-like solid; enlarged fleshy base of a stem

corolla — complete circle of petals

crenate — with teeth rounded, scalloped

crested — with elevated and irregular ridge

culm — the hollow stem or stalk of grasses and bamboo

cultigen — plant originated in cultivation

cultivar — special form originating in cultivation

cuneate — wedge-shaped, triangular

cuspidate — tipped with a sharp and stiff point

deciduous — with leaves falling

deltoid — triangular

dentate — with coarse teeth, usually directed outward

dichotomous — forked equally

dioecious — unisexual; the male and female reproductive organs in different plants

diploid — having 2 sets of chromosomes

distichous — in two ranks, arranged in two rows

diurnal — day blooming

dorsal — back; in orchids usually a top sepal

endemic — native to a restricted region

ensiform — swordshaped

epiphyte — air-plant; a plant growing on another, but not taking food from its host

F 1 hybrid — first generation hybrid obtained by artificial cross-pollination between two dissimilar parents (F meaning filial), each from a pure line or race, and each bearing hereditary factors (genes), which characteristics will be transmitted, according to their dominant genes, to an F 1 hybrid; as a rule imparting also greater vigor (heterosis).

fertile — spore bearing or seed-bearing

fimbriate — fringed

frond — leaf of fern

glabrous — smooth, not hairy nor rough

glaucous — covered with a white powder that rubs off

glochid — barbed hair, or bristle, as in cacti

herbaceous — non-woody

hirsute — hairy, with long rather stiff hairs

imbricated — shingled

irregular flower — a flower which cannot be halved in any plane, or in one plane only

laciniate — slashed into narrow irregular pointed lobes

lanceolate — lance-shaped; tapering toward the tip

limb — the border or expanded part of corolla (or spathe) above the throat

lip — the principal lobes of a bilabiate corolla; in orchids a much modified petal

membranous — thin, semi-transparent

monoecious — the stamens and pistils in separate flowers but borne on the same plant

mutant — form derived by sudden change from a species

nocturnal — night-blooming

oblique — slanting; unequal-sided

obovate — inverted ovate, the broad end upward

orbicular — leaf with circular outline

panicle — an open and branched flower cluster

peltate — leaf-blade attached to stalk inside its margin

perianth — the calyx, or corolla, or both

petiole — the supporting stalk of a leaf; leaf stem

photosensitive — responding to daylength

phyllodia — leaf-like stems and no blades (as in Acacia or Epiphyllum)

pinnae — primary division of a pinnate leaf, its leaflets

pinnate — feather formed; separate leaflets arranged along side of leaf stalk; separation complete

pistil — the female organ of a flower, consisting of ovary, style and stigma

pollen — the fertilizing powder contained in the anther

procumbent — lying along the ground; leaning

pseudobulb — thickened and bulb-like portion of stem in epiphytic orchids

pubescent — covered with short, soft hairs, downy

raceme — elongated simple inflorescence with stalked flowers

rachis — axis bearing flowers or leaflets

reniform — kidney-shaped

rhizome — creeping rootstock, on or under the ground

rhombic — irregularly slanting rectangle

rugose — covered with wrinkles

runner — a slender prostrate shoot, rooting at the end or at joints

salverform — slender tube abruptly expanded into disk-like flat limb

saxicolous — rock-dwelling

scandent — climbing, in whatever manner

scapose — bearing leafless stems

sepal — each segment of a calyx, or outer floral envelopes

serrate — notched like saw; finely toothed

sessile — sitting close, without stalk

sinuate — with a deep wavy margin, curved

sinus — the curve between two lobes of a leaf

sori — spore masses (in ferns)

spadix — a fleshy spike bearing tiny flowers as in aroids

spathe — a flower-like bract often colored or showy partly surrounding the inflorescence

spatulate — oblong, broadly rounded at tip but tapering to narrow base

spike — elongated flower stem, with flowers not stalked

spore — in ferns a reproductive cell, somewhat corresponding to seed in flowering plants

stamen — the pollen-bearing or "male" organ

stellate — star-form; stellate hairs have radiating branches

stigma — that part of the pistil or style which receives the pollen

stipule — a leaf-like appendage at base of petiole

stolon — rooting branch

stoloniferous — sending out, or propagating itself by stolons

subtend — to extend under, or be opposite to

synonym — a name rejected in favor of another

terete — circular, rounded in cross section; cylindric and usually tapering

terrestrial — plants growing in the ground

tomentose — densely covered with matted wool

trifoliate — three-leaved

trifoliolate — with three leaflets, as in clover—commonly, but incorrectly, termed "trifoliate"

triploid — having 3 times the haploid chromosome number

truncate — as if cut off at the end

tuber — modified underground stem; the thickened portion of subterranean stem, provided with "eyes"

tubercle — a wart-like or knobby projection

umbel — inflorescence in which flower stalks or cluster arise from same point

unisexual — of one sex; staminate (male), or pistillate (female) only (see dioecious)

vaginate — sheathed; surrounded by a sheath, usually of leaf stems

viable — capable of living or germinating

viviparous — producing young, while attached to parent

whorled — leaves in circle around stem

xerophytic — growing in dry situation, subsisting with little moisture

zygomorphic — can be divided into two symmetrical halves by cutting down the middle across the face one side reflecting the other side like a mirror

Descriptive Botanical Terms Illustrated

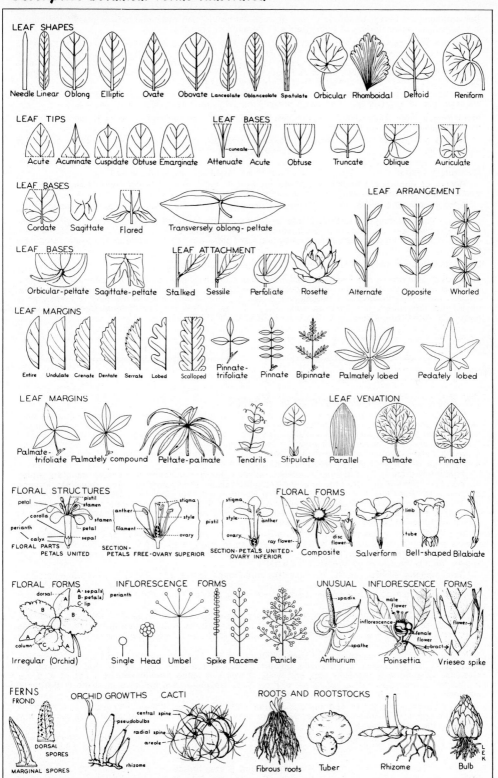

LEAF SHAPES
Needle · Linear · Oblong · Elliptic · Ovate · Obovate · Lanceolate · Oblanceolate · Spatulate · Orbicular · Rhomboidal · Deltoid · Reniform

LEAF TIPS
Acute · Acuminate · Cuspidate · Obtuse · Emarginate

LEAF BASES
cuneate · Attenuate · Acute · Obtuse · Truncate · Oblique · Auriculate

LEAF BASES
Cordate · Sagittate · Flared · Transversely oblong- peltate

LEAF ARRANGEMENT
Alternate · Opposite · Whorled

LEAF BASES
Orbicular-peltate · Sagittate-peltate

LEAF ATTACHMENT
Stalked · Sessile · Perfoliate · Rosette

LEAF MARGINS
Entire · Undulate · Crenate · Dentate · Serrate · Lobed · Scalloped · Pinnate-trifoliate · Pinnate · Bipinnate · Palmately lobed · Pedately lobed

LEAF MARGINS
Palmate-trifoliate · Palmately compound · Peltate-palmate

LEAF VENATION
Tendrils · Stipulate · Parallel · Palmate · Pinnate

FLORAL STRUCTURES
petal · pistil · stamen · corolla · stamen · perianth · petal · calyx · sepal
FLORAL PARTS - PETALS UNITED

anther · stigma · style · filament · ovary
SECTION - PETALS FREE · OVARY SUPERIOR

stigma · style · anther · pistil · ovary · ray flower
SECTION - PETALS UNITED - OVARY INFERIOR

FLORAL FORMS
disc flower · Composite · limb · tube · Salverform · Bell-shaped · Bilabiate

FLORAL FORMS
dorsal · A · A: sepals B: petals C: lip · B · B · A · A · column · C · A
Irregular (Orchid)

INFLORESCENCE FORMS
perianth · Single · Head · Umbel · Spike · Raceme · Panicle

UNUSUAL INFLORESCENCE FORMS
spadix · male flower · inflorescence · spathe · female flower · bract · flower
Anthurium · Poinsettia · Vriesea spike

FERNS
FROND · DORSAL SPORES · MARGINAL SPORES

ORCHID GROWTHS
pseudobulbs · rhizome

CACTI
central spine · radial spine · areole

ROOTS AND ROOTSTOCKS
Fibrous roots · Tuber · Rhizome · Bulb

L E K

Combination Text-Index: Numerals at end of each listing indicate page numbers to plant illustrations, a short-cut to easier finding of photographs (* denotes in color).

C: Culture Code numbers to Pictograph symbols and Key to Plant Care shown on pages 18-19; listing the elementary growing conditions for each. Further comments on the culture, maintenance and use of important subjects, or of those requiring special care, are given with the descriptive text; also methods of propagation unless this is self-evident by the habit of a plant. (See also chapter on propagation, pages 48-64). Quick reference Key to Care – inside back cover.

Recommendations on care are principally for container plants indoors. For outdoor conditions in warm or tropical regions see 'Gardening in the Tropics' in the Plant Geography Section.

Plant Names other than species, in conformance with the International Code of Nomenclature, are generally distinguished as follows:

Names of Hybrids: Generic names of bigeneric hybrids, and Latinized specific names of hybrids derived from two species, are as a rule preceded by the x mark;

Cultivar names (horticultural sports or varieties, hybrids with fancy names, and clonal selections) start with a capital initial letter, and are enclosed within single quotation marks (');

Names of uncertain standing, or where incorrectly used in horticulture, are shown in double quotes ("), and/or are followed by the abbreviation "hort".

Synonyms: Active synonyms, or important names used in horticulture, are cross-referenced.

Blooming times of long or short-day plants, as indicated, are for Northern latitude day-lengths; Southern hemisphere seasons are as a rule reversed. Plants originating from within the Tropical belt are usually unaffected though many such exotics react as "short-day" plants when brought into Temperate climate.

Light intensities are given in footcandles (fc.), explained on page 23.

Terms of measurement, and Temperatures: Following long-established custom, measurements of length cited are usually in inches and feet, based on the English system; these are convertible into metric terms, 1 inch (in.) equalling 2.54 centimeters; 1 foot (ft.) = 30.4 cm. Likewise, we read temperatures in degrees of Fahrenheit, with freezing point at 32° F., as against the international Centigrade, water freezing at zero degrees Celsius (0° C). Water boils at 212°F (100° Celsius).

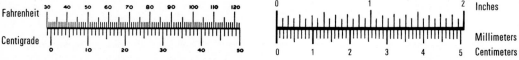

On Pronunciation of Botanical Names

Botanical nomenclature is basically Latin, or words adopted from other languages, with Latin endings, and conceived to be understandable internationally. Correctly pronounced and clearly enunciated, the recital of botanical names has a stately and noble sound. However, in the words of L. H. Bailey, there is no standard agreement on rules for the pronunciation of botanical binomials. Many English-speaking people pronounce generic and descriptive specific names simply as if the words were English, in what is known as the Traditional English system.

Alternately there is the Restored Academic, or phonetic pronunciation of classical scholars, which comes close to the manner of speech of the ancient Romans. Their idiom has been conserved through the centuries. The Florentine vernacular Latin of Dante in the 14th Century is the touchstone of modern Italian. Castilian Spanish also is an enhanced but faithful perpetuation of spoken Latin practically unchanged since the 5th Century. Anyone conversant with these languages will have no difficulty to articulate Botanical Latin in the classical tradition.

In spoken Latin, much the same as in Italian and Spanish, also in German or even Japanese. the vowels are pronounced precisely and uniformly: a as in apart; e as in pet; i as in pin; o as in note; u as in full; y in phyllus as in the French rue, the German or Chinese ü. Typical of the clear sound of spoken Latin is the Spanish expression "Te amo!".

Combinations of two vowels, or imperfect diphthongs found in Latin or Greek are enunciated separately as two syllables: aë (ah-eh = Gr. aer, Aërides); ai (ah-ee = eye); au (house); ei (eight); eo (areole); eu (eh-oo = aureus); ie (ee-eh = variegata); oi (oh-ee = deltoides); iu (ee-uu = folius); ue (uu-eh = cruentus); ui (ruin).

Exceptions are the perfect diphthongs or inseparable ligatures æ or ae (in caeruleus, Linnaeus, Caesarea), sounded as one vowel halfway between ah and eh, as in hat or fair, in French père, or the German "Umlaut" ä; œ or oe (in Coelogyne, coelestis) as in the French heureux, also the German or Swedish "Umlaut" ö.

Consonants: In the classical Latin used by Cicero in the first century B.C., the Romans never pronounced C like an English s, or G as j, but always like k and g (in get). By 180 A.D. however, the classical standard was gradually lost, and while C was still being pronounced as k before a, o, u— it changed to sound as z or s before ae, e, i, oe, y. G remained hard before a, o, u—as in Gardenia, but became a soft j as in joy before e and i (Geranium).

Many botanical names or epithets are derived from foreign-root personal or geographical names with Latin endings. To be recognizable, these are best pronounced in the idioms of their source, with accent on the preferred syllable.

The accent for nomenclatural Latin names with two syllables is on the first syllable; in words with several syllables the stress is usually on the next-to-the-last. If in doubt, pronounce all syllables with equal emphasis.

With a background of European schooling, I find it appropriate to use the pleasing phonetics of Continental Latin; those employing the English inflections may have difficulty being understood in non-English speaking areas. However, since English is increasingly a universally understood world language, it may well be employed, wherever found to be more convenient, in the pronunciation of Botanical names. I feel that a language should be our servant, not our master.

ABELIA (*Caprifoliaceae*)

x grandiflora (chinensis x uniflora), "Glossy abelia"; attractive half-evergreen hybrid shrub from Chinese parents, with gracefully arching branches and opposite ovate, glossy leaves 1 to $2\frac{1}{2}$ in. long; terminal clusters of small bellshaped fragrant white flowers flushed with pink; freely blooming from June to October. An excellent patio container plant. Prop. from softwood or fall cuttings. **C**: 14 p. 504

ABUTILON (*Malvaceae*); the "Flowering maples" or "Chinese lanterns" are favorite shrubby house plants of coarse and rangy, rather rapid growth, with handsome alternate mostly lobed and maple-like herbaceous foliage, and colorful bell or lantern-like axillary flowers bright red, salmon, yellow, white or two-tone; blooming best in good sunlight. Long cultivated as easy-to-care-for pot plants, they are also well suited for containers on the summer patio. Propagated by cuttings of young wood in spring or fall, or from seed.

x hybridum, "Flowering maple" or "Parlor maple"; an old fashioned house plant of robust, sparry habit; typical of the numerous hybrids of the tropical American species striatum, pictum and darwinii, resulting in this floriferous group of herbaceous shrubs with pubescent, soft green foliage varying from lobed to not lobed, 4 to 8 in. long, often resembling maple leaves, and with bell-like flowers more or less 2 in. long, in shades from white through yellow and salmon to red; in bloom the year round, the main season from April to June. **C**: 14 p. 497

x hybridum 'Golden Fleece' (A. darwinii x' Boule de Neige'), a "Golden parlor maple"; with showy bell-like flowers of golden yellow, and crenate, lightly lobed, soft-pubescent, green leaves; flowering over a long period throughout the summer and into winter. **C**: 14 p. 497

x hybridum 'Savitzii', "White parlor-maple"; colorful old cultivar of bushy habit, with its little maple-like lobed and toothed leaves grayish-green, and highly and irregularly variegated white from the margins in. **C**: 16 p. 182

x hybridum 'Souvenir de Bonn', "Variegated flowering maple"; herbaceous shrub and an old houseplant, with soft, long-stalked, slightly hairy maple-like leaves grayish-green bordered in creamy-white; bell-shaped flowers salmon, veined with crimson. Grows to twice the size as 'Savitzii'. **C**: 16. p. 182, 497

megapotamicum variegatum (Brazil), "Weeping Chinese lantern"; a pretty vine-shrub of lax, graceful habit, the slender, drooping branches with small, arrow-shaped, crenate leaves, $1\frac{1}{2}$ to 3 in. long, fresh-green with ivory to yellow variegation; gaily decorated by small 2 in. pendulous flowers lemon-yellow with lantern-like red calyx; from May to Oct. Excellent for hanging pots. **C**: 14 p. 497

striatum 'Aureo-maculatum', "Spotted Chinese lantern"; attractive herbaceous shrub with pointedly 5-lobed and toothed foliage, highly blotched and marbled with yellow on rich green, the leaves being pubescent and smaller than thompsonii; flowers spread bell-shaped with corolla light reddish-orange with red veins. The variegation in the foliage is caused by a transmissible virus, and if a shoot of it is grafted on a green-leaved understock, this stock later becomes variegated also. **C**: 14 p.182, 497

striatum thompsonii (Guatemala), "Spotted flowering maple"; shrubby plant with slender branches and soft, maple-shaped leaves deeply 5–7 lobed, the middle lobe narrowed at base, dark green with chartreuse-yellow mottling; bell-like orange-salmon flowers; leaves not pubescent. An undemanding houseplant. **C**: 14 p. 182, 497

ACACIA (*Leguminosae*); the "Mimosa trees" or "Wattles" are showy tropical and subtropical flowering shrubs and trees of the Pea family, mostly Australian, with bipinnate foliage or leaves reduced to leaf-like, flattened petioles or phyllodia and resembling a simple leaf; small usually yellow, and often fragrant, flowers conspicuous by their long stamens, in stalked heads or spikes blooming in winter and early spring: July to October in Australia; January to April in the Northern hemisphere. Acacias are sun-lovers and must be wintered in a cool (40 to 50° F.) and light greenhouse or closed-in sunporch, with ventilation; and where they will bloom like yellow fountains, one species after another. After flowering they should be cut back severely and plunged outdoors in late spring; Acacias prefer to remain well established in smallish containers. Propagated from cuttings of half-ripened wood in May or June under glass; also from seed but this should be soaked in water for a day or two to soften the seed coat.

armata (paradoxa) (Australia), "Kangaroo thorn" or "Hedge wattle"; dark green, densely branched erect shrub, to 10 ft. high, but usually much smaller, with stems ribbed and bristly-spiny, dense phyllodia half-ovate, to 1 in. long; flowers rich yellow in $\frac{1}{3}$ in. globose heads and close to stem; March-April blooming. A valuable flowering pot plant, willingly bloom for Easter, even in small pots. **C**: 25 p. 503

armata pendula (Australia), "Weeping kangaroo thorn"; shrubby bush of straggling habit, with phyllodia longer and narrower than the upright species, the small yellow ball-shaped fl. on longer stalks; April-May. **C**: 75 p. 503

baileyana (New South Wales), "Golden Mimosa" or "Fernleaf acacia"; handsome spreading tree, to 30 ft. high, with branches having a tendency to be pendulous, very leafy with fern-like, bipinnate, glaucous, bluish-silvery leaves; the beautiful fluffy bright yellow $\frac{1}{4}$ in. globose flower heads massed in great sprays of clustering racemes very fragrant, blooming from January into April. Widely planted in California and on the Riviera and much shipped and used as cut flowers. **C**: 75 p. 503

cultriformis (South Australia, New South Wales, Queensland), "Knife acacia"; bushy shrub to 6 ft. or more, with stiff, spirally set, triangular 1 in. hatchet-like phyllodia silvery-gray and glaucous, one edge hugging the stems; small fluffy, ball-shaped yellow flowers in long, arching clusters forming a terminal spray much beyond the "leaves"; between February and April. **C**: 75 p. 503

longifolia (Eastern Australia to Tasmania), "Sydney golden wattle"; rapid-growing shrub or tree 15 to 30 ft., of willowy, spreading habit, the arching branches with 6 in. long linear, leathery dark green phyllodia; bright yellow, small globose flowers in thin finger-like spikes $2\frac{1}{2}$ in. long from a long stem during February-March and into summer. Popular subtropic street tree, or shrub, of roundish, billowy habit, but rather short-lived. **C**.: 75 p. 503

longifolia latifolia (Australia), "Bush acacia"; small flowering tree or large bush with long-linear glaucous, leaf-like phyllodia to 6 in. long and 2 in. wide; the cylindrical flower heads in masses of loose, fluffy spikes to 2 in. long, clear yellow in spring. **C**: 75 p. 503

longifolia mucronata (S. E. Australia), "Narrow Sydney wattle"; small spreading tree 8 to 12 ft. high with lightly drooping branches having very narrow linear, stiffly thick phyllodia usually only 2 to 3 in. but occasionally longer, and the lemon-yellow flowers in $1\frac{1}{2}$ to $2\frac{1}{2}$ in. cylindrical, finger-like fluffy spikes from the stem. March. **C**: 75 p. 503

podalyriaefolia (N.S.W., Queensland), "Pearl acacia"; tall silvery-gray winter-blooming pubescent shrub or tree 8 to 20 ft. high, set with ovate, silvery glaucous 1 to $1\frac{1}{2}$ in. phyllodia; flowers in axillary clusters 2 to 4 in. long, carrying up to 20 globose $\frac{1}{2}$ in. balls of golden yellow stamenlike little powder puffs; November to March, one of the earliest to flower. Lovely alone for its silvery white "foliage"; also known as the "Queensland silver wattle". **C**: 75 p. 503

retinodes, in California nurseries as longifolia floribunda, and commonly planted on the Riviera as A. "floribunda"; (South Australia), the "Everblooming acacia"; dense, upright tree to 25 ft., having most of its foliage toward the ends of its branches, with narrow linear, dark green phyllodia to 5 in. long, and small $\frac{1}{4}$ in. globular light yellow, fragrant flowers in loose, clustered racemes 6 to 9 in. long, blooming constantly from February nearly all year. **C**: 75 p. 503

ACALYPHA (*Euphorbiaceae*); tropical and subtropical shrubs found in both hemispheres, with large, often showy and highly colored foliage as in the "Copperleaf", and small, imperfect and inconspicuous flowers but these may be assembled in long, showy red spikes as in the "Chenille plant", A. hispida. Both kinds like warmth and sunlight, liberal watering and feeding, and if possible a moist atmosphere to help keep the red spiders under control. Soil should be rich friable loam with humus added. The colored-leaf species are much planted as hedges in the tropics; they make good container plants during summertime in our patios. Propagated by cuttings, best from December to April.

godseffiana (New Guinea), "Lance copperleaf"; profuse shrub, bushy and dense, with ovate and obovate bright green leaves having a broad cream serrate margin; flowers are greenish-yellow. **C**: 2   p. 182

hispida, better known as A. sanderi (India), the "Chenille plant" or "Foxtails"; striking tropical shrub to 15 ft. high, with broad ovate, bright green hairy leaves 5 to 8 in. long having crenate margins; bright crimson flowers lacking petals, in long slender, pendant spikes to 1½ ft. or more long, resembling tassels of red chenille. Blooming most heavily in summer; warmth with humidity and rich soil are required for best development of tassels. Easily subject to red spider in dry atmosphere. Cuttings during late winter or spring. **C**: 4
p. 43*, 491

wilkesiana macafeana (New Hebrides), "Copperleaf" or "Match-me-if-you-can"; robust branching shrub dense with fairly large 5 to 8 in., ovate, red leaves marbled crimson and bronze, the margins serrate; slender insignificant male and female reddish flowers. **C**: 2   p. 182

wilkesiana obovata (Polynesia), "Heart copperleaf"; large obovate leaves emarginate or notched at apex, green edged cream-white when young, later changing to copper with orange-rose margins. **C**: 2   p. 182

ACAMPE (Orchidaceae)
papillosa (Saccolabium) (Himalayas, India to Burma); strong epiphyte with fleshy 2-ranked leaves on stout stem to 10 in. high, and relatively small ½ in. waxy flowers sweetly fragrant, the sepals and petals deep yellow spotted with brown; lip white marked with rose, in small dense clusters; blooming mostly in autumn. Requires tropical conditions, similar to Vandas, and of interest mainly to connoisseurs; best grown in porous baskets. **C**: 60   p. 458

ACANTHOCEREUS (Cactaceae)
pentagonus (horridus) (Florida, Louisiana, W. Indies, So. America), "Big-needle vine"; nightblooming slender clamberer to 20 ft.; 3–4 in. dia., 3–4–5 angled, with short spines, forest green to deep green, long jointed; large white flowers 8 in. long, outer petals brownish; deep red, spiny fruit. **C**: 1   p. 155

ACANTHORHIPSALIS (Cactaceae)
monacantha (S.E. Peru, Bolivia, Argentina) "Spiny rhipsalis"; erect stocky plant to 3 ft. high with flat or mostly 3-cornered, waxy green joints, the wings notched and usually with stiff brown spines; flowers waxy golden orange; fruit red. Sometimes in horticulture as A. micrantha. **C**: 16 p.172

ACANTHOSTACHYS (Bromeliaceae)
strobilacea (So. Brazil, Paraguay, Argentina), "Pine-cone bromeliad"; epiphytic plant with long pendant, very narrow, succulent and channeled leaves deep green with gray scurf and spiny; inflorescence on reed-thin stems bearing red cone-like fruit. **C**: 20   p. 133

ACANTHUS (Acanthaceae)
mollis (So. Europe), "Greek akanthos", "Bear's breech" or "Grecian pattern plant"; handsome perennial herb with basal, large leaves to 2 ft. high, dark glossy green with attractive pale veining, sinuately deeply lobed and toothed but not spiny, cordate at base and faintly white-hairy; rigid spikes of pinkish flowers with purplish spiny bracts 2 ft. high. Fairly winter-hardy. The acanthus leaf is much sculptured in classical Greece. A truly ornamental plant for shade or sun. From division. **C**: 16   p. 298

montanus (Trop. West Africa), "Mountain thistle"; handsome shrub with decorative but formidable hard, black-green, pinnatifid or lobed, puckered, very spiny leaves to 12 in. long; white flowers tinted purple, in a showy terminal spike to 10 in. long. Best suited for a conservatory. **C**: 16
p. 187

ACER (Aceraceae)
buergerianum (China, Japan), "Trident maple"; deciduous small tree to 25 ft. high, of low, spreading growth, with very attractive 3-lobed glossy, rich green leaves 3 in. wide, glaucous beneath. Beautiful in autumn colors varying from yellow to orange and red. Decorative patio tree and favorite Bonsai subject. Needs constant supply of water. Propagated by seed, cuttings, or grafting on A. palmatum. **C**: 76   p. 314

ACHIMENES (Gesneriaceae); the "Magic flowers" are generally low, tropical American herbs with peculiar scaly rhizomes, slender stems that trail gracefully over the side of a pot or basket then bend and grow upright; showy axillary tubular to flat-faced flowers from ½ to 3 in. across over the plant in summertime and fall, in great profusion. Many good, free-blooming hybrids have been produced since 1840. Achimenes make wonderful hanging plants but need warmth with humidity for growth until budded; after blooming they require a period of dormancy through fall and winter. Their curious catkin or cone-like rhizomes are dried and stored at 60° F.; they are used for propagation either whole or by separating the scales. Also from cuttings or seed.

andrieuxii (So. Mexico: Chiapas); compact, hairy little plant forming rosette of corrugated, ovate leaves; from the leaf axils rise numerous 2 in. stalks bearing the small violet flowers, their pure white throat lined with purple dots, giving a bicolor effect, and looking like miniature gloxinias. With its pretty flowers and dwarf, shapely habit this promises to be a worthy plant for blooming in 3 or 4 in. pots alongside Saintpaulia. **C**: 60   p. 433

antirrhina (So. Mexico, Guatemala), "Scarlet magic flower"; erect herbaceous plant with hairy stems to 1 ft. high or more; opposite, ovate leaves to 1½ in. long, green and slightly hairy above, reddish beneath, corrugated and with serrate margins; the tubular flowers to 1½ in. long, straw-yellow with red-brown lines outside, yellow with red lines inside, the flaring lobes bright scarlet. **C**: 60   p. 433

candida (Dicyrta) (Mountains of Guatemala), "Mother's tears"; low growing hairy herb 6 to 18 in. high with scaly rhizome; red-brown wiry stems and rugose, serrate, oblique 2½ in. leaves; small but pretty, nodding curved funnel-shaped flowers under 1 in. long, creamy-white, buff or reddish outside, the throat spotted purple; bl. in summer. **C**: 60 p. 433

coccinea: see erecta

ehrenbergii (lanata) (Mountains of Mexico); erect bushy white-woolly plant to 1½ ft. high, with pairs of hairy, light green, ovate leaves 4 to 6 in. long, scalloped at margins and white-hairy beneath; stalked, solitary, nodding slipper-type flowers 1¾ in. long, orchid-colored with white throat marked purple, and blooming most of the year. **C**: 60 p.433

erecta, better known in horticulture as A. coccinea (Mexico, Jamaica, Panama), a scarlet "Cupid's bower"; first Achimenes imported to England 1778; herbaceous plant of neat miniature habit, with thickened scaly underground stolons; trailing stems 4 to 18 in. long, with 2 to 3 crenate, hairy 1 in. leaves in a whorl, and small axillary flowers under 1 in. across, intense bright scarlet, produced in profusion from late August into winter. **C**: 60   p. 433

flava (So. Mexico); attractive branching, straggling herb with slender beige stems, and opposite, dark metallic green, ovate leaves, puckered between veins, lightly hairy, toothed at margins, and paler beneath; small, oblique flowers bright orange-yellow. **C**: 60   p. 433

longiflora 'Andersonii', a "Trumpet achimenes"; purple-flowered cultivar of A. longiflora, with violet-blue blooms; the species is from Mexico and Guatemala; a variable plant with slender branches to 1 ft. or more long, forming pear-shaped rhizomes at the roots; small elliptic 1 to 3 in. stiff-hairy, toothed green leaves; dipping salver-shaped flowers 2 to 2½ in. long, blooming in July and August. **C**: 60
p. 433

longiflora 'Major', "Orchid pansy"; slender cultivar with metallic green leaves, larger than the species and in greater number, and with 1 to 3 in. long flowers lavender to pale purple, golden throat and yellow tube. The largest bloom in achimenes. **C**: 60   p. 433

patens (Mexico: Michoacan), "Kimono plant"; branching herb to 1½ ft. high, with slender brown-red stems; small oblique-elliptic dentate leaves, dark lustrous green with pale veins and short-hairy, 1¾ in. long; small 1¼ in. flaring oblique, trumpet flowers deep violet-purple marked yellow in throat, and spurred like an aquilegia; bl. in summer. **C**: 60 p.433

ACHYRANTHES: see Iresine

ACIDANTHERA (Iridaceae)
bicolor murieliae (Gladiolus murieliae), (from the highlands of Ethiopia), "Darkeye gladixia", or "Abyssinian sword lily"; beautiful cormous plant with flat, linear leaves, in appearance between Gladiolus and Ixia; stalk about 2–3 ft. high, with clusters of nodding fragrant star-like flowers 2–3 in. across, having long tubes and spreading petals,

creamy-white with crimson star in center. The variety murieliae is taller growing than the species, and heavily scented, blooming in late summer and fall. For blooming in pots, plant corms in March; when well rooted must feed. Rest at 60° F., not cooler, during winter. From cormlets as offsets. **C**: 64   p. 395

ACINETA (*Orchidaceae*)
   superba (Venezuela, Colombia, Ecuador); fantastic, strong growing epiphyte with stout pseudobulbs bearing 3 leaves; big cup-like spicily fragrant, waxy 3 in. flowers in pendulous racemes, reddish-brown spotted red inside, the lobed lip yellow or brown-red spotted with purple, not opening fully; blooming mostly in spring. Because of the pendulous inflorescence often pushing through the bottom of the growing medium, these plants should be grown in baskets or on treefern slabs. Requiring a rest period during winter. Prop. by separating the pseudobulbs. **C**: 60   p. 458

ACOKANTHERA: see Carissa

ACORUS (*Araceae*)
   gramineus variegatus (Japan), "Miniature sweet flag"; water-loving dwarf perennial 4 to 12 in. high with creeping rhizomes and iris-like linear, flat, leathery leaves, arranged as in a fan, light-green and variegated white. Wonderful as little miniature plant for dish-gardens, terrariums or in the shallow water of aquarium pools. Prop. by division. **C**: 4   p. 98, 304

ACROSTICHUM (*Filices: Polypodiaceae*)
   aureum (Old and New World Tropics), coarse "Swamp fern" or "Leather fern", generally from tropical sweet water swamps; with stout rootstock sending up dark green, thick-leathery pinnate fronds 2 to 9 ft. high, only the upper pinnae fertile, smaller than the barren pinnae, and covered with the deep reddish-brown coating of the spore-cases. A leather-fern growing luxuriantly on the Florida Keys, next to mangrove swamps and salt marshes, forming dense masses. If in pots, set into saucers with water. Propagation from young plants appearing at base of older specimen. **C**: 4   p. 252

ACTINOPHLOEUS: see Ptychosperma

ADA (*Orchidaceae*)
   aurantiaca (Andes of Colombia at 9,000 ft.); handsome epiphyte about 1 ft. high, with compressed pseudobulbs bearing 1–3 tapering leaves to 9 in. long; arching racemes of showy 1 to 1½ in. somewhat bell-shaped flowers bright cinnabar red, often spotted black; petals and sepals narrow, with shorter lip; blooming winter to early spring. Attractive for the spectacularly colorful flowers of a shade scarce in orchids. Culture cool as high altitude Odontoglossums. Propagated by separating the pseudobulbs. **C**: 84   p. 458

ADANSONIA (*Bombacaceae*)
   digitata (across Trop. Africa), the "Baobab" or "Dead rat-tree"; one of the largest trees in the world, thought to become more than 5,000 years old and while only 60 ft. tall, the swollen trunk attains a diameter of more than 30 ft., is of pulpous wood without growth rings. Leaves deciduous digitately compound with leaflets 5 in. long; large 6 in. solitary, scented, pendulous white flowers with purplish stamens; oblong woody, hairy fruit 1 ft. long. A landmark on grassy savannahs across tropical Africa, it is surrounded by legend. Because of its horizontally spreading branches, the Africans believe that when God created the world he turned this tree upside down in anger, with its roots into the air. Water is stored in its huge trunk maintaining the life of the tree through the driest season. Old hollow trunks are used as houses. **C**: 51   p. 525

ADENIUM (*Apocynaceae*)
   obesum (Trop. E. Africa: Mozambique, Tanzania, Kenya, Uganda), the "Desert rose" or "Impala lily"; spreading succulent bush to 6 ft. high with globose, twisted base and thick stem with short branches; clustered toward their tips the deciduous obovate, fleshy leaves glossy dark green with pink midrib, 3 in. long, when young sometimes with minute hairs; numerous showy 2 in. flowers with spreading petals pinkish, edged with carmine or all carmine-rose; early summer blooming. Slow growing and beautiful in flower; reduce watering when the plant is resting; its milky juice is poisonous. Propagate from cuttings; allow cut wound to heal. **C**: 51   p. 350, 500

ADIANTUM (*Filices*); dainty shade-loving mostly tropical ferns of the Polypodiaceae family with short rhizomes, and thin, delicate fronds on wire-like blackish stalks. They all like a moist atmosphere, and dislike sun, and generally do best in the greenhouse or terrarium, and should receive a rest period into winter, but do not keep cold nor let the soil become too dry. Older plants may be divided.
   caudatum (South Africa, India, Indonesia, China), the "Walking fern"; small tufted plant with creeping rhizome, and long, simply pinnate, grayish-green, hairy fronds to 10 in. long, growing almost horizontally, and rooting at tip when it touches the ground, from whip-like extensions, and from which new plants will arise. **C**: 74   p. 242, 539
   cuneatum: see raddianum
   hispidulum (Australia, New Zealand), "Australian maidenhair"; handsome and different species with 2 to 3-pinnate palmate fronds, forked at base; borne on long wiry, hairy stalks; the leaflets almost stalkless, thin leathery, arranged along axis; veins running into teeth. **C**: 62   p. 242
   pedatum (North America), "American maidenhair" or "Five-fingered maidenhair"; hardy outdoors, and equally useful in the greenhouse, with fronds to 2-ft. high; the purplish stalks forked fan-like, each pinnate branch set with neat rows of papery, peagreen leaflets. **C**: 74   p. 242
   raddianum, long known in horticulture as cuneatum (Brazil), "Delta maidenhair"; an old greenhouse favorite because of its tolerance, sturdiness, and the simple elegance of its dark green fronds 8 to 15 in. long, with many small, firm ¼ in. leaflets having a wedgeshaped base, the veins running into sinus between lobes. A fairly sturdy tropical species that may be tried as a house plant. Never let the soil ball become dry. **C**: 10   p. 242
   raddianum 'Pacific Maid' (Peru); an attractive compact variety developed in San Francisco; with 2 to 3-pinnate fronds, the pinnae stiffly set in stages above each other; satiny green leaflets large, with veins running into sinuses. Much used for its stiffish fronds in flower arrangements. **C**: 62   p. 242
   tenerum 'Farleyense' (Barbados), "Magnificent maidenhair"; strikingly beautiful feathery, but delicate fronds that may become 2 ft. high; gracefully drooping, rose-tinted when young, pea-green later, the leaflets large and lacily cut and crisped, base jointed; veins run into teeth. Cold-sensitive tropical fern with infertile fronds; must be propagated by division or sections of growth under warm, moist-tropical conditions. **C**: 62   p. 242
   tenerum "Wrightii" (W. Indies, Mexico, Venezuela), the "Fan maidenhair"; good pot-plant with graceful medium-size fronds of good texture, pink when young, later fresh-green; the fan-shaped leaflets occasionally growing large and lacily lobed. Probably the best of the maidenhairs for pot growing, because of its medium-large, friendly green pinnae and shapely habit; to attempt as a house plant the pot should stand in a saucer with water, and try to mist the foliage occasionally. **C**: 12   p. 242
   trapeziforme (Mexico and W. Indies south to Brazil), delicate looking, yet bold-growing "Giant maidenhair"; with slowly creeping rhizome and large 2-pinnate fronds on black stems, the stalked trapezoid leaflets to 2 in. long and brilliant green. **C**: 62   p. 242

ADONIDIA: see Veitchia

ADROMISCHUS (*Crassulaceae*); fascinating South African succulents of the stonecrop family, forming small rosettes with curious and colorful leaves thick and fleshy, usually painted silver with red or purple; small white or pinkish tubular flowers in spike-like raceme. To maintain their attractive foliage colors, they should grow slowly with good light and fresh air; winter cool and dry. Propagation by cuttings or the easily dropping leaves which root easily. Collection plants.
   cristatus (South Africa: Cape Prov.), "Crinkleleaf plant"; very pretty succulent rosette with stout stem covered with red hair-like aerial roots, the wedge-shaped fleshy 1–1½ in. leaves with crested apex, light green and soft pubescent with white hairs; tubular flowers reddish-white. **C**: 15   p. 331
   festivus (South Africa: Cape Prov.), "Plover eggs"; colorful clustering succulent with cylindric or thick spatulate or club-shaped leaves flattened toward the crested apex, silvery green marbled maroon; small tubular reddish flowers, in tall raceme. **C**: 15   p. 331

maculatus (South Africa: Cape Prov.), "Calico hearts"; attractive dwarf succulent rosette with few flat leaves almost round, 1–1½ in. long, on both sides convex, gray-green and heavily flecked red-brown except at corky edge; flowers tipped red-white. A favorite collector's plant. **C**: 15   p. 331

tricolor (South Africa: Cape Prov.), "Spotted spindle"; very attractive succulent with stout stem covered by curled aerial roots and spindle-shaped leaves about 2 in. long, pale green and richly blotched maroon and marked with silver; flowers pale reddish purple. Leaves don't fall off as easily as in some other species. **C**: 15   p. 331

AECHMEA (*Bromeliaceae*); probably the most popular of all bromeliads; generally with foliage so attractive that growers are content to raise them for this alone, though the shapes and colorings of the inflorescences are spectacular. Most of the species are epiphytic forming shapely rosettes of stiff leaves excitingly patterned or richly colored, the edges are spiny and most have deep cups to hold water. The inflorescence is carried well above the foliage on erect or arching stalks in spikes or clusters with flowers small and which rarely open but often with persistent colored bracts; it is however their vivid long-lasting berry-clusters that makes this genus outstanding, together with their other charms that makes them most spectacular amongst decorative plants. Their care is easy, Aechmeas prefer light for best coloring but they are quite tolerant and in Europe are widely used in department stores and windows particularly the summer blooming. A. fasciata with its leathery pink bracts remaining in beauty for months. Aechmeas generally will bloom when mature especially following a dry period in summer or any other time of year. Propagated by offsets or seed.

x'Bert' (orlandiana x fosteriana); stocky rosette of short leathery leaves matte green marked with irregular purplish-brown crossbands, along margins with prominent dark spines; arching inflorescence with dense head of red bracts and pale flowers. **C**: 10   p. 129

x'Black Magic' (victoriana discolor x racinae); Hummel hybrid similar to 'Foster's Favorite' but more black; shiny black wine-red leaves 2 in. wide; outstanding for its sinister coloring as a foliage plant; inflorescence with red berries. **C**: 10   p. 148

bromeliifolia (B. Honduras, Guatemala to N.E. Argentina), "Wax torch"; large tubular rosette of variable leaves 2–3½ ft. long, green with white scaly coating, a few brown teeth toward apex, with tips curled under; erect, stout cylindric, long-lasting inflorescence, densely white-woolly with leathery, broad floral bracts; the flower petals greenish-yellow soon turning black. **C**: 10   p. 133

caudata variegata (Brazil) (Billbergia forgetii in hort.); big sparry rosette of rich green stiff leaves broadly banded creamy-yellow; tall inflorescence borne on white-mealy stalk topped by a bold panicle of yellow bracts sheathing golden-yellow flowers. **C**: 10   p. 128, 131*, 193

chantinii (Venezuela, Amazonas, Amazonian Peru), "Amazonian zebra plant"; spectacular open rosette of hard olive-green leaves with pronounced pinkish-gray crossbands; inflorescence on branched spike with tight red bracts tipped yellow, subtended by red floral bracts; flower petals canary-yellow. **C**: 10   p. 128, 193

x'Electrica'; elegant Hummel hybrid involving Aechmea dealbata x miniata x fasciata; large rosette with leaves dominated by wine and red; the bold inflorescence a compound head dense with orange bracts and lavender flowers. **C**: 9   p. 148

fasciata (Billbergia rhodocyanea), (Brazil: Guanabara), "Silver vase"; stocky rosette of leathery green leaves covered by gray scales and richly tigered silver-white; blackish spines along margins; durable inflorescence of rose-colored leathery bracts with blue flowers in globose heads. An important decorator plant. **C**: 9.   p. 129, 131*

fasciata variegata, "Variegated silver vase"; variety with center of the channeled leaves attractively striped and banded ivory-white lengthwise through the regular green, and with silver cross-banding. **C**: 9   p. 133, 193

filicaulis (Venezuela), "Weeping living vase"; open rosette with grass-green, thin-leathery, strap-shaped, oblanceolate leaves glossy on both sides and with dark mottling; tiny soft, marginal spines; long pendulous flowering panicles on snaky, string-like axis, with distant, red bract leaves and white flowers. **C**: 10   p. 133

x'Foster's Favorite' (victoriana x racinae), "Lacquered

wine-cup"; upright rosette of strikingly lacquered wine-red leaves; pendulous spike bearing coral red pear-shaped berries tipped with midnight-blue flowers. **C**: 60   p. 129

x'Foster's Favorite Favorite'; a most beautiful plant; variegated sport of A. 'Foster's Favorite'; rosette with soft-leathery, glossy, strap-like leaves bordered by broad cream margins, the upper leaves tinted all over glowing coppery rose to wine-red or maroon according to prevailing light conditions. **C**: 10 or 60   p. 130, 193

fulgens (Brazil: Pernambuco), "Coral berry"; loose rosette of stiff-leathery green leaves dusted gray: inflorescence in showy panicles with the oblong red berries tipped with purple flowers. **C**: 10   p. 129

fulgens discolor (Brazil: Pernambuco), "Large purplish coral-berry"; free-growing rosette of soft-leathery, dark olive-green leaves purple beneath, covered on both sides with glaucous gray crossbands; produces showy spikes with oval red berries tipped by violet flowers. **C**: 10   p. 130, 509

lueddemanniana (coerulescens) (Mexico, C. America); stiff rosette of metallic green leaves mottled dark green, and with bronze base; flower spike a panicle of white berries turning a beautiful bright purple after flowering; petals lavender. **C**: 10   p. 130

x maginalii (miniata discolor x fulgens discolor); open rosette with broad soft-leathery olive-green glaucous leaves red-purple beneath; erect spike bearing inflorescence with oblong berry-like salmon-red bracts tipped by deep blue flowers. **C**: 10   p. 130

mariae-reginae (Costa Rica), "Queen aechmea"; robust rosette of broad, gray-green, leathery leaves, recurved and with toothed edge; stout spike subtended with pendant, delicate pink bract leaves, topped by a cylindrical head of red-tipped berries and violet flowers. **C**: 10   p. 129

marmorata: see Quesnelia marmorata

mertensii (Trinidad, Guyana, Venezuela, Colombia, Peru, North Brazil), "China-berry"; epiphytic, open rosettes with few green leaves to 2 ft. long, covered with white scales especially beneath, and having marginal spines; stalks with large rose bracts, the slender inflorescence many-flowered, with yellow or red petals; fruit blue. **C**: 10   p. 130

miniata discolor (Brazil), "Purplish Coral-berry"; open rosette of soft-leathery olive-green leaves, pale red reverse; inflorescence a panicled spike of orange-red rounded berries tipped by pale blue flowers. Attractive and long-lasting. **C**: 10   p. 129, 509

mooreana (Amazonian Peru); very showy bromeliad with branched inflorescence similar to chantinii but with lower bracts carmine-rose, the upper, flattened bracts lime green tipped flame orange. Foliage shining bronzy green. **C**: 9   p. 148

nidularioides (Colombia); epiphytic rosette of open habit, with strap-shaped, white-scaly, green leaves 2 ft. long, turning intense lacquer wine-red in strong sun, and with broad marginal spines; globular inflorescence on short stalk, red bracted and with white flowers. **C**: 9   p. 148

nudicaulis cuspidata (South Brazil), "Living vase plant"; stiff rosette, of glossy, deep green leaves, tubular at base and edged with black spines; inflorescence bracts rosy-carmine; yellow flowers arranged cylindrically on red axis. **C**: 9   p. 130

ramosa (South Brazil); large symmetrical rosette composed of many leathery, medium green leaves coated with gray scurf; inflorescence a vermilion-red spike with loose panicle of greenish-yellow berries and yellow flowers. **C**: 10   p. 130

x'Red Wing'; stunning Hummel hybrid involving Aechmea penduliflora x mutica; handsome rosette with large coppery leaves, dark purple beneath. The wine-red stalk bears an inflorescence of many berries first pink then purple, in heavy clusters; the flowers are straw-colored. **C**: 10   p. 130

x'Royal Wine' (miniata discolor x victoriana discolor); open rosette of broad, soft-leathery, highly glossed apple-green leaves beautifully lacquered burgundy above, and more so beneath; inflorescence a heavy panicle with orange bracts and pointed berries tipped by blue flowers, on leaning stalk wine-red. **C**: 10   p. 129

weilbachii leodiensis (Brazil: Rio), "Tropical lilac"; attractive rosette of oblanceolate coppery-green leaves wine-red beneath, and spined; inflorescence on panicle with glowing crimson bracts and orchid-colored ovaries and flowers. **C**: 10   p. 129

AEONIUM (*Crassulaceae*); attractive, lush succulent plants of Madeira, the Canary Islands and Northern Africa, intermediate between Sedum and Sempervivum; usually small or forming diminutive stems resembling little trees, topped by plate-like rosettes of spatulate leaves; the flowers in branched clusters, yellow, white or red, in spring. Very easily grown and excellent durable house plants; light and cool in winter; the patio in summer with liberal moisture during the warm season. Propagated from seed, leaf cuttings or side branches; these can be sprouted by cutting off the head.

arboreum atropurpureum, "Black tree aeonium"; striking decorative variety with coppery to deep purple leaves. The species Ae. arboreum is Mediterranean from Morocco and Portugal east to Crete; an erect bold succulent to 3 ft. high, little branching, topped by a flaring 6–8 in. rosette of spatulate light green, fleshy leaves, fringed white at margins; flowers golden yellow. **C**: 15   p. 330

arboreum cristatum; interesting fasciation of the gray-brown fleshy stem broadened obliquely fan-like, the curved apex supporting like a comb, a dense chain of tiny, deep lacquer-red rosettes, the little oblanceolate leaves with ciliate margin; the whole plant resembling a sitting bird. **C**: 15   p. 532

burchardii (Sempervivum burchardii) (Canary Islands: Tenerife); erect succulent stem 8 to 12 in. high, eventually branching, bearing toward the top a flat fleshy rosette of obovate pointed leaves, grayish green and metallic bronze, 4 in. across, the margins ciliate; flowers soft ochre yellow. Beautiful, but slow-growing species. **C**: 15   p. 330

canariense (Canaries: Tenerife), known as the "Giant velvet rose"; this loose succulent rosette reaching out with foliage 8 to 18 in. or more across, forming a thick short stem; light green waxy, limp, soft-fleshy, spatulate leaves undulate along the sides and covered with white velour, tip reddish; flowers pale yellow. Propagated from the suckers appearing at base. **C**: 16   p. 330

decorum (cooperi) (Canary Islands), "Copper pin-wheel"; much branched shrublet topped by several open rosettes 1¾ to 2½ in. dia.; small spatulate succulent green leaves flushed glowing coppery red and with deep red margins; slender sprays with white flowers lined with pink. Freely clustering even as small plant, a pretty dishgarden subject. From suckers. **C**: 13   p. 328, 330

domesticum variegatum (Aichryson) (Canary Islands), "Youth and old age"; freely branching succulent rosettes with round-spatulate, thin fleshy finely-hairy, 1 in. leaves light green with white margins; flowers yellow. Very gay old house plant. **C**: 15   p. 330

haworthii (Tenerife), "Pin-wheel"; bushy plant to 2 ft. high with short woody branches topped by rosettes 2–3 in. across, of thick, obovate-acute gray-green leaves with red margins having horny teeth; flowers pale yellow flushed rose. A cute tree-like miniature and a favorite in dish-gardens. **C**: 13   p. 330

'Pseudo-tabulaeforme' (tabulaeforme hybrid), "Green platters"; shrubby bush with distinct, thick, brown stem, the branches topped by flat open succulent rosettes, waxy green, leaves broad fan-like, ciliate at margins; golden flowers. **C**: 15   p. 330

sedifolium (Canary Islands), "Tiny-leaved aeonium"; twiggy shrub 4 to 6 in. high with, woody branches, tipped with rosettes of tiny succulent, obovate ¼ in. leaves usually brownish marked with red, shining and sticky; flowers bright yellow. Forms dense cushions, and likes humidity. **C**: 13   p. 330

tabulaeforme (Canary Islands: Tenerife), "Saucer plant"; circular, plate-like succulent rosette to 20 in. across, of fresh waxy green spatulate leaves arranged flattened like a shingled roof, the margins ciliate; flowers yellow. As seen in habitat, the rosettes usually grow almost vertically flattened against moist rock walls, their roots clinging to the stone or into clefts. Propagation by seed or leaf cuttings. **C**: 16   p. 330, 530

AERANGIS (*Orchidaceae*)

thomsonii (from the Highlands of Kenya, Usambaras of Tanzania, Uganda); attractive vandaceous epiphyte with stout stem 1½ to 2 ft. high, sending out aerial roots from between the ranks of alternating, strap-shaped leaves which are notched at tip; axillary pendant racemes set with erect, firm-textured flowers 1½ to 2 in. across, glistening white, having long 6 in. spurs light bronze; delicately scented,

more intensively after dusk. Fantastic with the blooms resembling birds in flight, a long tail floating behind them. Flowering irregularly throughout the year. Propagated by cutting the stem with roots. **C**: 22   p. 458

AERIDES (*Orchidaceae*); the "Fox tail orchids" are noble epiphytes of Tropical South-east Asia, to New Guinea, of vandaceous habit, with two-ranked succulent leaves on elongate stems without pseudobulbs; the inflorescence an elongate pendulous cylindrical racemes dense with exquisite, waxy flowers distinguished by an upturned spur, and possessing a powerful fragrance. They need to be grown moist and warm all year, especially during summer, less so in winter. Aerides delight in basket culture to give their rampant roots access to the outside; in chunks of treefern fiber, or osmunda with sphagnum. Propagated by offshoots with roots attached, or cutting the stem into pieces.

crassifolium (Burma), so beautiful that it has been called the "King of Aerides"; delightful epiphyte of compact habit, with erect stem about 10 in. high, clothed by thick-leathery two-ranked leaves sheathing at base; the wax-like, heavy-textured, very fragrant 1½ in. flowers amethyst-purple, with lip an upturned greenish spur, in pendulous racemes; blooming May to summer. **C**: 10   p. 458

fieldingii (Himalayas: Sikkim and Assam), the "Fox-brush orchid"; a magnificent free-blooming epiphyte of bold and striking character with stems to 3 ft. high set with fleshy leaves; the cylindrical, brush-like inflorescence 1 to 2 ft. long, crowded with waxy 1 to 1½ in. fragrant flowers white and beautifully dotted and suffused with bright rose, the triangular lip rosy purple; blooming from May into summer. **C**: 22   p. 458

AESCHYNANTHUS (*Gesneriaceae*); fibrous rooted creepers, or climbers, with weak rooting stems or somewhat woody; straggling epiphyte from tropical Asia, for a time known as Trichosporum, with opposite fleshy or leathery leaves and showy, 2-lipped tubular long-persistent flowers in shades of red to yellow with green; smaller plants make good pot plants, the larger ones very effective in hanging pots or baskets, on treefern blocks, or on treebranches like Orchids. When growing Aeschynanthus thrive best with warmth and high humidity, and should be grown in porous materials rich in humus, coarse peat, sphagnum or osmunda. To flower at their best in spring, they should be kept cooler and rather dry in winter giving them a rest. From stem cuttings.

lobbianus (Trichosporum) (Java), epiphytic trailer with slender, wiry stems, and small elliptic, fleshy, dark green leaves; tubular, two-lipped flowers with hairy calyx cup soot-red glistening like silk, the downy corolla fiery red, creamy-yellow in throat, only twice as long as calyx; summer-blooming. **C**: 60   p. 438

marmoratus (zebrinus) (Burma, Thailand), "Zebra basket vine"; epiphytic trailer with flexuous stems, and pairs of beautiful waxy leaves to 4 in. long, dark green with a reticulated network of contrasting yellow-green, and maroon underneath; tubular green flowers blotched chocolate brown, 1¼ in. long. Cultivated mainly for its ornamental foliage. **C**: 10   p. 438

pulcher (Trichosporum) (Java), "Royal red bugler" or "Scarlet basket vine"; straggling epiphytic plant with pendant, flexuous branches, small opposite ovate, waxy leathery light green 2½ in. leaves, and showy tubular flowers axillary or in terminal clusters; calyx green and smooth, the bilabiate corolla 3 times longer, 2 to 2½ in. long, vermilion red with yellow throat, blooming in summer or winter. A lovely houseplant easy to bloom. **C**: 4   p. 438

speciosus (Trichosporum splendens in horticulture), (Java, Borneo, Malaya); strong straggler with wiry, 4-angled stems to 2 ft. long, and large lanceolate, waxy-green 3 to 4 in. leaves; showy tubular 4 in. flowers in terminal clusters, corolla flame-orange, yellow at base and beneath; throat marked brown-red on yellow. **C**: 60   p. 438

x splendidus (parasiticus x speciosus), "Orange lipstick vine"; robust trailer with ovate, tapering leaves 4 in. or more long; the fleshy erect, curving tubular flowers 3 in. long in large clusters, orange-red blotched with deep maroon. A spectacular hybrid when in bloom. **C**: 60   p. 438

AGANISIA (*Orchidaceae*)

cyanea (Acacallis), (Brazil: Amazonas), a "Blue orchid" from the Rio Negro; small epiphyte with creeping rhizome, one-leaved ovate pseudobulbs; plaited 8 in. leaves tapering into a furrowed petiole; basal stalks with racemes of light blue fragrant flowers 1½ to 2½ in. across, the clawed, lobed lip bluish-purple with pale veins; column streaked with red; summer-blooming, starting in May. A charming plant, unique in color, requiring a damp and warm atmosphere. Increased by division. **C**: 60   p. 457

AGAPANTHUS (*Liliaceae*)

africanus (South Africa: Cape Prov.), the "Blue African lily" or "Lily-of-the-Nile"; strikingly beautiful plant to 3 ft. high, with tuberous root stock and fleshy roots; basal strap-like, rich green leaves; and stout, stiff-erect stalks with great trusses of funnel-shaped, showy flowers 2 in. long, pale porcelain-blue with darker center and margins, forming mid-summer to early fall. A favorite, and very effective in July and August when grown in large containers or tubs and enjoying the sun on the patio or by the side of the house, requiring copious watering; keep cool and moderately dry during winter. No repotting necessary for several years. Propagate by division of clumps. **C**: 14   p. 396

AGATHEA: see Felicia

AGAVE (*Amaryllidaceae*); counterparts of the African aloe and because of their resemblance known as "American Aloe"; also popularly known as "Century plants"; noble succulents of the Amaryllis family; mostly large or even gigantic, forming regular rosettes of fleshy, strap-shaped leaves with or without spines. The flower clusters are big but not colorful, the tall inflorescence often bears adventitious reproductive bulbils or plantlets. In warm climate flowering takes place, according to species, and contrary to popular belief, in from 10 to 15 years, sometimes as much as 60 years or more, after which the clump of foliage dies, leaving behind suckers which may be used for propagation. Agaves are sun-lovers, and tolerate abuse; they are excellent in containers but larger specimens require much space; distinctive as accent plants, ideal for the summer outdoors. Indoors they take airconditioning well, also drafts and other abuse, and tolerate fair light to 100 fc. Some species such as A. victoriae-reginae are propagated from seed.

americana (Mexico; but widely naturalized in South Europe and around the Mediterranean), "Century plant" or "Maguey"; large stoloniferous, trunkless rosette of wide-spreading, broad and thick-succulent, glaucous gray-green leaves 6 to 7 ft. long and 8 in. wide, recurving or sharply bent downward above the middle, with sharp brown hooks at lightly sinuate margins, and ending in a wicked spiny point; when 10 years old or more, the tall erect, rather slender, branched inflorescence shoots up to 15 or 30 ft. high, bearing clusters of yellowish flowers. Easily propagated and grown from offsets. **C**: 13   p. 359

americana 'Marginata', "Variegated century plant"; large, loose rosette with the broad laxly recurved, glaucous gray leaves displaying showy broad, intense golden-yellow or creamy bands along the margins. A beautiful house plant when small, and striking as a larger show plant on the patio or where space permits. **C**: 13   p. 359

angustifolia marginata (caribaea), (West Indies), "Variegated Caribbean agave"; beautiful, freely suckering densely formal rosette with stiff-erect, short, sword-shaped leaves to 2 ft. long, bluish-gray with broad white marginal bands and diminutive brown spines. Because of its compact, regular shape one of the most attractive of decorative century plants. **C**: 1   p. 196, 359

attenuata (Mexico), "Dragon tree agave"; spectacular species, with age forming thick trunk to 5 ft. high, sometimes leaning; suckering from the base; unarmed rosettes with spreading broad-elliptic fleshy leaves 2½ ft. long and 8 in. wide in the middle, soft glaucous green or almost whitish coated; its crowning beauty is the fantastic inflorescence spike forming a complete arch 10 to 14 ft. long, under the weight of as many as 10,000 greenish flowers, often followed by reproductive plantlets. A beauty in a tub or a show specimen in a landscape scheme. **C**: 13   p. 359, 530

decipiens (Mexico: Yucatan), "False sisal"; stocky stoloniferous plant, arborescent and bearing bulbils on the 15 ft. inflorescence, with trunk eventually 3 to 8 ft., rosette with concave thick-fleshy dagger-like and recurving, glossy-green leaves 1 to 5 ft. long with triangular red-brown marginal spines and black-brown terminal spine. A nice agave in miniature, for dishgardens and novelties. From suckers or bulbils. **C**: 1   p. 328, 358

ferdinandi-regis (victoriae-reginae nickelsii) (N.E. Mexico), "King agave"; solitary, globose rosette similar to A. victoriae-reginae but fewer-leaved and more open; the very stiff, pointed leaves 5 to 9 in. long, deeply concave on the upper surface, almost folded, keeled beneath, dark green with converging white lines, usually terminated by 2 to 3 spines. Very striking small species. Produces few suckers. **C**: 1   p. 358

filifera (Mexico: Pachuca), "Thread agave"; stemless, many-leaved rosette of narrow, stiff, incurving bright green leaves 1 or 2 ft. long, with white lines and horny margins along edges which split into loose filaments. Inflorescence 8 to 18 ft. high, with maroon flowers. Forms runners producing offsets. **C**: 13   p. 358

miradorensis (desmetiana), (Mexico: Veracruz), "Mirador agave"; large trunkless rosette 5 to 6 ft. in dia., the spreading, leathery-fleshy lanceolate leaves to 3 ft. long and 5 in. wide, pale gray-green lightly glaucescent; when growing luxuriously or under cultivation in pots, the leaves are gracefully recurving; some plants with terminal spines; others with finely toothed edges. The species is short-lived, blooming in Florida in 6 to 8 years, the 10 ft. inflorescence bearing numerous adventitious plantlets; also freely suckering from the base. A handsome pot or tub plant. **C**: 13 p. 358

parviflora (Mexico: Sonora, Chihuahua; U.S.: Arizona), "Little princess agave"; striking small stemless rosette 6 to 8 in. dia., dense with stiff leaves 4 in. long, dark green to olive-gray with white lines above, the margins with white threads toward tip, lower part dentate; gray-brown terminal spike. Inflorescence 3 to 5 ft. high. A pigmy agave for pots in limited space. From seeds; offsets only seldom. **C**: 13   p. 358

potatorum (C. Mexico), "Drunkard agave"; trunkless, formal, very attractive rosette with stiff fleshy, obovate leaves 8 to 12 in. long, attractively lobed and scalloped at margins; the inflorescence to almost 15 ft. high with yellowish green flowers. **C**: 1   p. 358

sisalana (Mexico: Yucatan), "Sisal-hemp"; large rosette, forming 3 ft. trunks, developing offsets; with straight and stiff, sword-shaped leaves to 5 ft. long and 4 in. wide, matte gray-green, or green, with horny edge and occasional deformed teeth, keeled beneath, and with small black-brown terminal spine; branched inflorescence to 20 ft. high, with greenish, odoriferous flowers. Young plants will form as suckers on the ground as well as on the flowering stalk, and plantlets or bulbils rise vegetatively from the floral bracts, from observations I made on sisal plantations in Tanzania; the leaves of A. sisalana furnish the sisal fiber. **C**: 1 p. 358

striata (Mexico: Hidalgo), "Dark-striped agave"; short-stemmed rosette with flexible-leathery, long linear, angled leaves 18 in. long and ½ to 1 in. wide, glaucous gray striped with dark green, and with brown terminal spine; branching when old. Inflorescence to 12 ft. high with green flowers. The form with longer, elegantly outward curving leaves is var. recurva. **C**: 13   p. 358

striata 'Nana' (Mexico), "Lilliput agave"; miniature rosette of flat, pale yellowish or gray-green, linear, short leaves 2–3 in. long, characteristically suckering as a young plant. Very slow growing; leaves will become to 12 in. long with age. The cutest little agave for 2 or 3 inch pots or novelty planting. From suckers. **C**: 1   p. 328, 358

stricta (Mexico: Tehuacan), "Hedgehog agave"; hard, roundish ball-like rosette dense with stiff, spreading then incurving, narrow linear green leaves to about 15 in. long and only ½ in. wide, slightly keeled on both sides, with horny edges and faint gray lines, on the margins tiny teeth often much branched from a short thick stem when old. Inflorescence a dense raceme 7 ft. high. **C**: 13   p. 358

victoriae-reginae (North Mexico), "Queen agave"; shapely small, many-leaved symmetrical rosette with numerous keeled, dark-green leaves about 6 in. long, with white-horny margin and an abrupt point tipped by a short, blunt, terminal spine. Forms no or few offsets. Inflorescence to 15 ft. high with greenish flowers. Beautiful slow-growing species. Generally from seeds only. **C**: 1   p. 196, 358

AGERATUM (*Compositae*)

houstonianum 'Midget blue', "Floss-flower"; the popular garden bedding plant, 12 in. high, a compact horticultural form of Mexican origin; bushy herbaceous perennial grown as an annual, with tassel-like heads of clear blue tubular florets, carried nicely above the fresh green, small foliage. From seed, or occasionally cuttings. **C**: 64
p. 409

AGLAOMORPHA (Filices: *Polypodiaceae*)

coronans (Polypodium) (India, Malaya, South China, Taiwan, Philippines), "Crowning polypodium"; striking epiphytic fern with thick rhizomatous base densely matted with light brown scales, and from which rise large, hard-leathery, fresh green fronds 2–4 ft. long and 1–1½ ft. broad, arranged in a circle, their axis with reddish midrib, at the base winged, passing gradually into a deeply lobed frond cut nearly to axis; the glossy surface having a pebbled effect, and in places parchment-like translucent. **C**: 60   p. 247

heraclea (Polypodium, Drynaria), (Java, Philippines, New Guinea), "Hercules polypodium"; giant epiphytic fern with fronds 8 ft. long carried on thick rhizomes, covered with silky light brown scales, and climbing spirally around big trees; stalks with broad wings along lower part, higher up broadening into leaves 2 feet wide; deeply lobed, light green, thin-leathery, with puckered surface, the pointed tips curled. **C**: 60   p. 247

meyeniana (Polypodium), (Philippines), "Bear's-paw fern"; distinctive epiphyte with thick, paw-like rhizome covered thickly with brown hair, the long pinnate, glossy 2–3 ft. fronds have broad barren segments on the lower part, while the upper third bears narrow, fertile pinnae with prominent sori on their pearly margins. **C**: 62   p. 247

AGLAONEMA (*Araceae*); the "Chinese evergreens" are smallish tropical Asiatic rhizomatous herbs, slowly spreading, or with fleshy stems becoming lax with age; very durable, leathery leaves often painted with attractive silvery or colorful patterns; the inflorescence is unisexual, with the spadix carrying the tiny petal-less female flowers near its base, the males toward the apex, and subtended by a soon withering green or whitish spathe; and provided it has been pollinized followed by bright red, fleshy berries, each containing a single seed. Aglaonemas are mostly of small growth habit and ideal for planting into small containers for table decoration, surviving long under neglect and poor light conditions down to 10 foot-candles. Their growth is very slow and with their diminutive shape ideal on furniture away from windows; also doing well under airconditioning. Though getting along on only moderate watering if pot grown, Aglaonemas also take admirably to water culture. Easily propagated by division, tip or joint cuttings; also fresh seed.

brevispathum hospitum (Thailand); spear-shaped, leathery, glossy, ovate, dark-green leaves with cream-white spots, on long, wiry petioles. Relatively tall plant. **C**: 56 p. 90

commutatum elegans (marantifolium hort.) (Moluccas), "Variegated evergreen"; robust-growing plant with long, lanceolate leaves deep green with greenish-gray feather design. One of the most durable and long-lived of houseplants, in small size ideal for dishgardens. Tolerates the darkest locations down to 10 fc., also airconditioning. **C**: 5   p. 89

commutatum maculatum (Philippines, Ceylon), in the trade as "commutatum", the "Silver evergreen"; durable plant with oblong-lanceolate 20 cm leaves deep green with markings of silver-gray, waxy-white spathe; berries yellow to red. Good for hydroculture; tolerates locations of poor light down to 10 footcandles. **C**: 5   p. 89

commutatum 'Pseudo-bracteatum' "Golden evergreen"; possibly a hybrid cultivar; durable plant with leathery, oblong-lanceolate leaves deep green with markings of silvery-gray; waxy-white spathe; berries yellow to red. A beautiful foliage plant but requires warmth. **C**: 5
Also known as 'White Rajah'.   p. 43*, 66* 89, 178

commutatum 'Treubii' (Celebes), "Ribbon aglaonema"; slender plant with narrow, leathery, bluish-green leaves attractively marked with silvery-gray; petioles marbled. **C**: 5   p. 89, 178

costatum (Perak), "Spotted evergreen"; very decorative, low growing plant with stiff, broad-ovate, glossy, dark-green leathery leaves spotted white, a band of white through center. Slow but beautiful. **C**: 56   p. cover*, 90

costatum foxii (Malaya); short plant with leathery, shiny, cordate-pointed leaves, prettily spotted and a bold, broad white center band; slow-growing. **C**: 56   p. 90

crispum (Schismatoglottis roebelinii hort.), (Malaya, Borneo), "Painted droptongue"; large and showy, ovate-pointed, leathery leaves medium grayish-green, largely variegated with silver. Very satisfactory in dark locations to 10 fc.; handsome. **C**: 5   p. 90, 178

'Fransher' (A. 'Treubii' x commutatum tricolor); colorful cultivar named after Francis Scherr, originator of the

Parrot Jungle in South Florida; slender plant with fleshy lanceolate leaves green to milky green variegated with cream, on white or ivory petioles. **C**: 5   p. 89

'Malay Beauty' ('Pewter'); a free growing horticultural variation of A. 'White Rajah', less highly variegated than the type but larger in all parts; with lanceolate leafblades to a foot or more long, matte grass or deep green marbled with milky green and some cream. This form is no doubt a reversal toward the type, hybrid or species as encouraged by ideal growing conditions, which seems to throw off the albino variegation retarding normal growth. **C**: 5   p. 89

marantifolium: see commutatum elegans

modestum (sinensis) (So. China: Kwangtung), the popular "Chinese evergreen"; with durable, leathery, waxy-green leaves, ovate-acuminate and, to some extent, pendant, on a slender cane. Spathe green, spadix cream. Easily adapted for water culture, as are most Aglaonemas. **C**: 17   p. 89

nitidum (A. oblongifolium; Curmeria oblongifolia) (Malaya); slow-growing, stiff aroid with plain-green, thick leathery, long elliptic leaves. **C**: 56   p. 90

nitidum 'Curtisii' (oblongifolium curtisii) (Malaya); large but slow-growing aroid with thick stem; densely set, long, elliptic, bluish-green leaves and silver feather design. Tough as leather. **C**: 5   p. 90

pictum (Malaya); dainty plant with broad-ovate, velvety leaves bluish-green with irregular patches of silver gray. **C**: 56   p. 90, 178

roebelinii: see crispum

rotundum (No. Thailand), "Red aglaonema"; compact, slowly growing beauty which I found in Thailand, with stout stem covered by sheathing short petioles; leaves broad, ovate or rhombic 4–6 in. or more long, thick-leathery, dark metallic glossy green to coppery with pronounced pink midrib and lateral veins, later paling; the reverse side glowing wine-red with rosy veins. **C**: 56   p. 90, 178

siamense (Thailand); long, leathery, deep-green leaves with white center vein, on slender petioles. **C**: 56   p. 90

simplex (Java), a more tropical, faster-growing species similar to A. modestum but not as durable; leaf is more oblong and narrow, with a twist; texture more thin and papery; deep green and glossy; veins depressed. **C**: 5 p. 89

simplex 'Angustifolium' (Malaya: Penang); erect plant with long, slender, pointed, dark-green shining leaves, with depressed lateral veins giving it a corrugated appearance. **C**: 5   p.89

sinensis: see modestum

AICHRYSON: see Aeonium

AIPHANES (*Palmae*)

caryotaefolia (Martinezia), (Colombia, Venezuela, Ecuador), a "Coyure" or "Spine palm"; very unusual tropical palm with slender trunk to 40 ft., which is ringed with long black spines, and topped by a crown of light green pinnate leaves resembling a Caryota, 3 to 6 ft. long, the short, irregular pinnae suddenly expanded like a fishtail, squared off and lobed at tip, clothed with long black spines beneath; yellow-red fruits. This "Chonta palm" is very extraordinary for tropical collections and flower shows. **C**: 54   p. 226

"AIR-FERN", as sold in department stores, is a curious fern-like plume to 1 ft. long, by early naturalists (Rondelet 1558) considered as of the nature of plants and grouped as a Zoophyte or "Sea plant", at a time when sea coral was regarded as a "Stony plant". Now recognized to be a lowly moss animal of the genus Bugula, it is found growing under water on rocks or floating pieces of wood off the English or Norman coast. The plume-like structure is produced by the labor of invertebrate polyps which build colonies, skeletal cases of gelatinous material of their own secretions; originally a translucent brownish-gray, but dyed a luminous green and sold for decoration like lacy ferns, never requiring water or soil.   p. 253, 539

ALBIZZIA (*Leguminosae*)

julibrissin (Subtropical Iran to Japan), "Pink mimosa" or "Silk tree"; deciduous tree to 40-ft. high, with large, fern-like, bipinnate leaves to 18 in. long, the leaflets finely feathered; and terminal clusters of flowers with long, conspicuous, pink stamens arranged like fluffy tassels 1 in. or more across, reminding of acacias. Fairly winter-hardy. From seed. **C**: 26   p. 276

ALLAMANDA (*Apocynaceae*); the "Golden trumpets" are handsome and vigorously growing, robust tropical American woody clamberers or bushes, with leathery leaves, and large funnel-shaped flowers mostly golden yellow, some purplish, profusely blooming in summer and fall best in full sun. Watering is reduced during winter, and old branches should be pruned back in spring. Whether in pots or planted out, the main stems are tied to wire or staked support, ideal against walls or under ridges of greenhouses. Propagated by spring cuttings.

cathartica hendersonii (So. America: Guyana), "Golden trumpet"; robust shrubby, tropical clamberer with whorled, long, leathery, glossy leaves to 6 in. long; extra large wax-like, funnel-shaped golden flowers 4 in. across. With its showy flowers one of the most satisfactory summer bloomers in a sunny location, in rich soil, with some support. From cuttings. **C**: 52   p. 277

neriifolia (Brazil), the "Golden trumpet bush" or "Oleander allamanda"; half-climbing or erect shrub to 3 ft. high; oleander-like dark green leaves in whorls of 2 to 5, 3 to 5 in. long; smallish flowers 1½ in. across, swollen at base, light yellow and streaked orange-red, blooming in summer and fall. Not as vigorous and golden as the cathartica group, but may be kept in smaller bush form in pots. **C**: 2   p. 490

ALLIUM (*Liliaceae*)

moly (So. Europe), "Lily leek"; attractive small garden allium to 1 ft. high, with clustered bulbs, flat blue-green leaves 1 in. wide; showy flowers star-like, bright yellow, in 3 in. umbels. A favorite for its decorative and showy flowers. Resembles garlic with the same odor, but with broader foliage; hardy. By division. **C**: 64   p. 322

porrum (of Eurasian origin), "Swiss leek", or cultivated "Leek", well-known in the kitchen garden; a stout succulent onion-like herb 2 ft. or more high, with long fleshy stalks unfolding into flat or keeled leaves not hollow, about 2 in. wide; white or pinkish flowers in dense globular heads. Bulb is simple and scarcely more than an enlargement of the stalk. Both the soft bulb and the leaves are used in cooking, the lower part is blanched and delicate to eat; the aroma is strong but the flavor is sweet and milder than onion. The whole herb is prepared as a vegetable, or parts may be used in flavoring. Fairly hardy. From seed. **C**: 76   p. 322

schoenoprasum (Northern hemisphere: Siberia to North America), "Chives" or "Schnittlauch"; small, clump-forming perennial onion relative; tufted, clustered bulbs with grass-like but awl-shaped round and hollow green leaves, 10 to 18 in. high, tipped by round heads of clove-like rose-purple flowers. Use: snip fresh leaves into soups, salads, cream-cheese, egg dishes, omelette, hamburgers; for a delicious mild onion flavor or as garnish. Hardy perennial, but often kept in pots on the kitchen window sill in winter. By division. **C**: 26   p. 322

triquetrum (S.E. Europe), the ornamental "Triangle onion"; small egg-shaped bulb producing narrow-linear, channeled and keeled leaves ½ to 1 in. wide, and a 3-angled stalk 12 to 18 in. high bearing a cluster of attractive bell-shaped, strong-scented ½ in. flowers, each pointed segment white with keeled green line, blooming from March to May. Relative of the edible onion, and suitable for naturalizing in the garden; excellent cut flowers fresh or dried; also as cool-house container plant. Prop. by offsets, division or seed. **C**: 78   p. 400

ALLOPHYTON (*Scrophulariaceae*)

mexicanum (Tetranema mexicanum) (Mexico: Vera-cruz), "Mexican foxglove" or "Mexican violet"; darling little plant with very short stem, topped by a rosette of long obovate, dark-green leathery, flexible leaves to 6 in. long, glaucous beneath; from the base angled purplish stalks topped by clusters of little nodding, trumpet-shaped, pretty, ½ in. flowers, orchid-colored with large, lobed, whitish lip and purple-violet throat, blooming mostly in summer but almost continuously the year round. Excellent house plant for window planters or the winter garden. Propagation easy from seed, also division of older plants. **C**: 16   p. 494

ALLOPLECTUS (*Gesneriaceae*); handsome fibrous-rooted tropical American plants grown for their decorative foliage; the pairs of leaves unequal in size, one not quite so large as the other. Alloplectus develop their regal foliage best in the warm-humid greenhouse. Easy from cuttings; as older plants become unsightly; they should be replaced with new plants yearly.

capitatus (Andes of Colombia, Venezuela), "Velvet alloplectus"; erect large plant becoming 2 to 3 ft. high, the 4-angled succulent stem handsomely red-downy; large ovate, olive green leaves 6 to 8 in. long, covered with a rich velvety nap, reddish beneath; dense clusters of bright yellow flowers with blood-red calyx; fall-blooming. **C**: 60   p. 438

lynchii: see Nautilocalyx
sanguineus: see Columnea

schlimii (Andes of Colombia); handsome herb with erect fleshy green stem becoming woody; large, broad-lanceolate leaves to 4 in. long, bristly but shimmering light green, scalloped and fringed with red hairs, purplish beneath; small golden-yellow axillary flowers with crimson lines. **C**: 60   p. 438

vittatus (Eastern Peru: Moyobamba), "Striped alloplectus"; beautiful, fleshy plant to 2 ft. high, with stout stems and ovate, quilted, crenate 6 in. leaves bronzy moss-green, strikingly patterned with silver-white feathering along the midrib and lateral veins, and covered with white hair; glowing red-purple beneath; the clustered flowers yellow, with orange-red calyx and red bracts. A lovely foliage plant best in the warm-humid greenhouse. **C**: 60   p. 438

ALOCASIA (*Araceae*); exotic tropical Asiatic herbs growing from a tuberous rhizome or thick rootstock, with large and showy thick-fleshy or soft-leathery leaves, peltate when young, carried nobly on long, sheathing stalks; their inflorescence unisexual with the minute petal-less male flowers along the top part of the spadix, the female organs down at its lower portion, inside a boat-shaped fleshy spathe. The colored-leaf Alocasias belong to our most beautiful foliage plants, but need warmth and humidity during growing time; a decided rest at 60 to 70°F. with less watering in winter. They grow well in pure sphagnum moss, or very porous pasteurized humus compost. Propagated by division of the tubers, or baby tubers forming on root tips. (See p. 63.)

x amazonica (sanderiana x lowii grandis); robust and bushy grower with leaves very dark lacquered green, veins contrasting white, the scalloped margin white. One of the easiest to grow. **C**: 62   p. 66*, 95, 180

cucullata (Bengal, Burma), "Chinese taro" or "Chinese ape"; shapely plant with fleshy peltate, pointed leaves deep green and with prominent veins, the base drawn together to form a spoon. Will tolerate somewhat cooler conditions. **C**: 54   p. 95

cuprea (Borneo), "Giant caladium"; tropical Asian herb with large, 10 to 12 in., peltate-cordate fleshy leaves with deeply depressed veins; the blade dark metallic, shining green; purple beneath. Attractive in collections. **C**: 60 p. 95

indica metallica (plumbea) (Malaya); big plant with waxy-stiff, sagittate leaves deep olive-green, with metallic purple overcast, veins prominent; reverse of leaf and stalks reddish-purple. **C**: 60   p. 97

macrorhiza (Malaya, Ceylon), "Giant elephant's ear", "Giant alocasia"; becoming tree-like to 15 ft. high, with thick trunk; large 2-ft. or more, broadly arrow-shaped, fleshy leaves waxy green, with prominent ribs and wavy margin; very showy. **C**: 54   p. 96

sanderiana (Philippines), "Kris plant"; tuberous plant with graceful, sagittate leaves 12 to 15 in. long, shining metallic silver-green, with grayish-white ribs; margins deeply lobed and outlined in white; reverse purple. Most attractive of alocasias, does best in warm greenhouse. **C**: 62   p. 95

x sedenii (cuprea x lowii); stocky plant with large, peltate-cordate leaves deep metallic-green with raised veins and wavy, gray margin; purple beneath; pale green stem. **C**: 62   p. 95

watsoniana (Sumatra); large, corrugated, leathery leaf blue-green with silver-white veining and margin; purple beneath. **C**: 62   p. 95

ALOE (*Liliaceae*); handsome and decorative succulent plants of the lily family, usually forming rosettes, some small, many becoming treelike; these are the Eastern hemisphere counterparts of the Western hemisphere Agaves; with fleshy, often stiff, toothed leaves, and attractive, striking inflorescence with tubular flowers in vivid reds to yellows. Many species have beautifully patterned leaves and nearly all make excellent house plants or decorators with a minimum of care. Most Aloes are sun-lovers from the African veldt and love to be outdoors on the patio or roofgarden in the summertime. A number of species have been prized as medicine plants since ancient times, in Africa and India, and even now, the juicy pulp of such species as A. vera, the "Bitter aloe", is used as a poultice for burns and cuts, or taken internally. Aloes are easily propagated from suckers or top cuttings; A. aristata will root from leaves.

africana (Transvaal), "Spiny aloe"; a very pretty tree aloe; good compact rosette when young, though reaching 12 ft. or more in its habitat; the glaucous bluish leaves are hard, and brown spined at the margins; flowers yellow, tipped green. **C**: 1   p. 334

arborescens (perfoliata arborescens) (So. Africa), "Tree aloe", "Octopus plant" or "Candelabra aloe"; becoming tree-like to 18 ft. high; branching stems carry spreading rosettes of sword-like fleshy, tapering leaves to 2 ft. long, glaucous pale bluish-green, edged with yellow horny teeth; winter-blooming with tubular vermilion to yellow flowers in long, spiky clusters. An easy growing unassuming and popular house plant full of character. From top cuttings or suckers. **C**: 13   p. 328, 332

aristata (So. Africa), "Lace aloe"; small stemless clustering 4 to 6 in. rosette of densely set, slender, dark green leaves, covered with soft white wart-like spines, especially on the back; the margins with white horny edge and horny teeth; each leaf ending in a long pale bristle; flowers orange-red. Greatly admired little house plant with its as many as 100 to 150 leaves almost forming a ball. Propagation possible by leaf cuttings. **C**: 15   p. 333

bainesii (So. Africa, Mozambique), tree to 60 ft. high with age, trunk 5 ft. or more thick at base, and crown to 20 ft. across, branching dichotomously (in pairs); one of the largest of the genus; sword-shaped green leaves 2–3 ft. long, leathery and concave, with small marginal prickles; flowers salmon-pink tipped with green. When young, a dense rosette with recurved-spreading leathery leaves. **C**: 13   p. 332

beguinii: see x Gastrolea

brevifolia (Aloe "humilis" of the California trade) (So. Africa: Cape Prov.); robust 4 to 5 in. rosette with triangular-oblong leaves twice as broad as humilis, flat on top, rounded beneath to keeled, glaucous grayish-green and with white horny teeth at margins. Forms suckers much more freely than var. depressa, and when fully grown rarely exceeds 8 in. in dia., with narrower leaves in proportion to brevifolia depressa, it is also more spiny, especially on the keel below. Flowers red, in dense racemes 20 in. high. Excellent house plant. **C**: 13   p. 334

brevifolia depressa (Aloe brevifolia of the trade), (So. Africa: Cape: Caledon), "Crocodile jaws"; open bluish-green rosette of broad fat, blunt leaves having a white cartilage edge and only very few spines on back of leaf toward tip only. Normally 5 to 6 in. dia., but with age will grow to about 12 to 15 in., with leaves 3–4 in. wide at the base. Clump-forming, producing many suckers that can be used for propagation. Very popular in dishgardens. Tall clusters of flame-scarlet flowers on older plants, intermittent all year. **C**: 13   p. 329, 334

ciliaris (So. Africa), "Climbing aloe"; a small succulent with slender stem later scandent and scrambling to more than 30 ft. into trees; the soft fleshy, 3 to 6 in. leaves in open spirals, plain dark green with pale teeth along the edge; flowers scarlet with green tips. Also known as one of the "Medicine plants"; a good, easy house plant. **C**: 15   p. 334

dichotoma (Cape Prov. to S.W. Africa), "Dragon tree aloe"; bold succulent rosette with broad fleshy leaves 10 in. or more long, glaucous green edged in brownish-yellow and with triangular teeth; growing into monstrous trees to 30 ft. high, with branches forking, on a smooth trunk, with age to 3 ft. thick; flowers canary yellow. **C**: 13   p. 332

eru maculata (Eritrea, Nubia), "Nubian aloe"; very attractive variety with leaves smaller than the species and with more pronounced blotching. A branching succulent rosette with stem becoming 1½ ft. high, the very fleshy arching leaves in the type to 2½ ft. long, concave above, glossy dark green with creamy-white oblong spots on both sides, and with prominent marginal spines; salmon red flowers in branched racemes. **C**: 1   p. 334

ferox (So. Africa), "Ferocious aloe"; bold rosette, developing a stem to 15 ft. high, with broad, fleshy leaves bronzy-green, hollow above, curved and warty beneath, margins with brown-red teeth; flowers orange-red. Very distinctive as a small dishgarden plant. The Zulus use this species as an unguent, and for stomach troubles. **C**: 13 p. 334

humilis echinata (So. Africa: Cape Prov.), "Hedgehog aloe"; pretty rosette with pearly foliage; narrower, more concave leaves than humilis, not over ½ in. wide; pale prickly, fleshy spines on the upper surface and pearly warts below. Very charming. **C**: 3   p. 334

humilis 'Globosa', "Spider aloe"; a cultivar in the California nursery trade; small suckering rosette with glaucous bluish leaves more incurved and more numerous than humilis, the tips turning purplish in the sun, prominent horny teeth cream to pinkish. Much used in dishgardens for its pretty blue color. **C**: 13   p. 329

x humvir (humilis x virens), "Needle-aloe"; neat little, freely suckering rosette, offered by California growers; with slender tapering leaves only about 2 in. long, rich green and dotted pale green, the margins with soft pale spines. A favorite dishgarden plant long remaining small. **C**: 15 p. 333

lineata (So. Africa), "Lined aloe"; large rosette on short stems to 2½ ft. high; with broad, lightly concave leaves to 18 in. long, pale glaucous blue-green, both sides with vertical green lines, the margins with reddish horny prickles; nodding red flowers. **C**: 15   p. 334

marlothii (Botswana, Transvaal, Natal), attractive tree with single trunk, to 18 ft. high; leaves to 2½ ft. long and 7 in. wide, in dense rosette, pale glaucous blue or greenish, concave and very spiny with short thick purplish-brown thorns; flowers red-orange. I did however, in the northern Transvaal, discover an old tree having two heads. Bold succulent rosette for the patio. **C**: 13   p. 332

nobilis (So. Africa: Cape Prov.), "Goldtooth-aloe"; an attractive, robust, short-stemmed rosette usually 4 to 5 in., but with age to 10 in. across; the thick leaves friendly yellowish-green, rounded below, and with pale horny teeth at margins; clusters of flowers orange-yellow on tall stalks. From offsets. **C**: 13   p. 329

plicatilis (So. Africa: Cape Prov.), "Fan aloe"; small branching tree to 15 ft. high, with closely packed foliage repeatedly forking in two ranks; the linear, fleshy 1 ft. leaves are of strap-like, even width, and rounded at the apex, of a pale glaucous blue with translucent yellow margins; large scarlet flowers in loose spike. A distinctive succulent, each cluster of leaves resembling an open fan. **C**: 13   p. 334

polyphylla (So. Africa: Drakensberg; Basutoland), "Spiral aloe"; striking low rosette 2 to 2½ ft. dia., dense with fleshy, broad leaves relatively short, but becoming 8 to 12 in. long, pale glaucous green and with a strong, red-brown tip; the foliage spirally arranged and overlapping; branched inflorescence bearing salmon-pink flowers. A succulent of unique beauty whether large or small, with its leaves regularly arranged as in shingles on a roof, in clockwise or counter-clockwise coils. Somewhat difficult in cultivation; in habitat at 7,800 ft. they are often under snow. They do not increase from suckers or offshoots; from seed only. **C**: 13   p. 333

saponaria (Natal, Transvaal), "Soap aloe"; short-stemmed, suckering rosette with fleshy, broad lanceolate leaves 6–8 in. long, bluish-green marked with elongate yellow-green blotches arranged in transverse bands; margins with large cream to brown thorns; flowers red-yellow. **C**: 15   p. 334

x spinosissima (humilis echinata x arborescens), "Torch plant"; tall spidery rosette with blue-gray concave warted leaves to 1 ft. long, spirally arranged, the margins set with horny teeth; flowers brilliant red in dense racemes. **C**: 3   p. 334

striata (S.W. Africa, So. Africa), "Coral aloe"; handsome large rosette, nearly stemless, of broad fleshy leaves becoming almost 2 ft. long, pale glaucous gray-green and faintly striped, the hard margins pale pink to red and spineless; brilliant pendulous coral-red flowers in erect branched clusters. One of the few aloes without spines. **C**: 13   p. 333

suprafoliata (So. Africa: Swaziland), "Propeller aloe"; very attractive and interesting succulent, especially as a young plant when its thick, elegantly recurved, tapering leaves are arranged in two opposing series; glaucous blue, but bronzy-red toward margins and dressed with reddish marginal spines; with age forming rosettes with thick-fleshy leaves to one foot long; flowers in tall inflorescence, coral-red with green tips. **C**: 15   p. 333

variegata (So. Africa: Cape Prov.), "Tiger aloe" or "Partridge-breast"; strikingly beautiful succulent to 12 in. high with triangular blue-green leaves arranged in 3 ranks and painted with oblong white spots in irregular crossbands, the margins horny and white-warty; tubular flowers salmon red, in loose racemes. From seed, or suckers that eventually form on old plants. **C**: 15   p. 196, 328

variegata ausana (S.W. Africa), "Partridge breast"; a succulent rosette similar to A. variegata but with less and broader foliage, not marked as distinctly as the species; the blue-green leaves deeply channeled, and the white markings not clearly forming a crossband. A very charming collection plant. Propagated from seed. **C**: 5   p. 334

vera (Cape Verde, Canary Isl., Madeira), the "Bitter aloe", also popularly known as "Medicine plant", "True aloe", "Barbados aloe" or "First-aid plant"; short-stemmed, freely suckering rosette with fleshy, dagger-shaped, channeled leaves 1 to 2 ft. long, bluish or gray-green and glaucous, spotted when young; edged with soft pale spines; nodding, cylindrical yellow flowers on usually unbranched inflorescence. Since ancient times the thickened sap was employed as bitter aloe; even now, especially in Africa and Asia, the juicy pulp is used as a poultice to heal sore burns and cuts, or is taken internally. A good plant to keep on the window sill. **C**: 13   p. 333, 530

vera chinensis (India, Vietnam, Taiwan), "Indian medicine aloe"; smaller Asiatic form; the fleshy, lanceolate leaves to 12 in. long and curving at tips, blue-green with white markings and whitish teeth, rounded beneath; flowers yellow to orange. Used medicinally in India and Southeast Asia. **C**: 15   p. 333

virens (So. Africa), "Green aloe"; stemless, freely tufting rosette of numerous lanceolate-fleshy upright tapering leaves to 8 in. long, light green with scattered pale spots and pale marginal teeth; flowers red. **C**: 15   p. 334

x'Walmsley's Bronze'; commercial California cultivar; small branching rosette with broad-stubby, hard fleshy, channeled leaves glaucous bronzy green. Very willing to grow and sucker, and forming robust clusters turning coppery in good light. **C**: 13   p. 333

ALOYSIA (Verbenaceae)

triphylla (Lippia citriodora in horticulture) (Argentina, Chile), the "Lemon verbena" or "Citronalis"; scented branching, tender herb-shrub to 6 ft. or more, with angular woody shoots; narrow lanceolate leaves to 3 in. long, arranged in whorls of 3 or 4; shiny yellowish-green; the small pinkish-white flowers in graceful pyramidal spikes and very fragrant; an old-fashioned house plant. The fresh leaves are used for their fragrance in finger bowls, and to give a lemony taste to teas, iced drinks, jellies, custards, appetizers and fruit salads. When dried, their aroma in sachets is long-lasting. From cuttings. **C**: 14   **p. 320**

ALPINIA (Zingiberaceae)

; the "Giant gingers" are tropical herbs with leafy stems and fleshy aromatic rhizomes; their inflorescence with showy flowers in terminal clusters sometimes pendulous and very striking. They would require large containers and abundant watering and feeding to develop their exotic-looking flowers; equally interesting for their colored capsular fruit which springs open into 3 gaping wings. Propagated by division.

calcarata (India), "Indian ginger"; attractive tropical plant with aromatic rhizomes, bearing leafy reed-stems 3 to 5 ft. tall, the smooth, spicy foliage 1 ft. long; at the apex a terminal panicle of beautiful waxy flowers greenish-white, the large lip yellow marked with pink and maroon. Seed capsule red, springing open when ripe. **C**: 54   p. 206

sanderae (New Guinea), "Variegated ginger"; very beautiful tropical ornamental ginger of dwarf habit, having creeping rhizome and clustered, leafy stems about 18 in. high, the lanceolate leaves arranged somewhat in 2 ranks, pale green, attractively edged and obliquely feathered outward from the center to the margins with pure white variegation. Requires the high humidity and warmth of a greenhouse as with lack of it the leaves will curl. Prop. by division. **C**: 54   p. 206

speciosa (syn. Renealmia nutans, Catimbium speciosum), (China, Japan), the famed "Shell ginger" so popular in Hawaii; majestic rhizomatous plant forming dense clumps of leafy, arching canes to 12 ft. high, with smooth, long-bladed, leathery leaves arranged in 2 rows; the striking, fragrant, porcelain-textured flowers with bell-shaped, waxy-white calyx, corolla white flushed pink, and tipped with red and yellow lip, in dense pendulous racemes to 1 ft. long. **C**: 54   p. 207

speciosa 'Variegata', "Variegated shell ginger"; a variety seen in Hawaii, with the broadly lanceolate, thin-leathery dark green leaves prettily variegated in feather design, with stripes and bands of creamy yellow. Prop. by division. **C**: 54   p. 206

ALSOPHILA (Filices: Dicksoniaceae)

cooperi, in horticulture generally as A. australis (Tasmania, Australia), the "Australian treefern"; a noble treefern with heavy trunk to 20 ft. high, with well proportioned, spreading crown even when small; the arching fronds finely divided, metallic green, on rough stalks covered with small pale brown, hair-like scales. Wonderful for planting out in ferneries in subtropical climate, but as a tub plant would require a copious supply of water, best left standing in a saucer under the container. Once dried out the fronds desiccate badly, with permanent damage. **C**: 60   p.235

cooperi 'Robusta', also known as A. australis or excelsa in horticulture; cultivated in California as a more vigorous form of the well-known A. cooperi, a handsome treefern with greater, long arching fresh green fronds, and more quickly forming trunk to 20 ft. high. Much used in ferneries and patio gardens of our southwest; the fibrous trunks must be kept moist in dry areas. Prop. from spores. **C**: 60   p. 237

ALSTROEMERIA (Amaryllidaceae)

; the "Peruvian-lilies" are herbaceous plants with fibrous roots from egg-shaped tubers which are attached to a common stem, and treated as bulbs; they are showy, free-flowering plants blooming in summer on tall leafy stems in loose clusters, the blooms Azalea-like in beautiful colors from orange and yellow to shades of rose and lilac, also white, many streaked with darker colors. Roots are planted in fall in large containers and kept growing until flowering, then tapering off to rest them; shake out the roots and repot. Propagated by division.

ligtu (Chile), the "Inca lily"; herbaceous perennial plant with thick fibrous roots, the 18–24 in. stem, with linear leaves, topped by large clusters of flowers $1\frac{1}{2}$ in. long, the spreading segments pinkish or lilac, streaked with purple. Also a long-lasting cut flower. **C**: 76   p. 390

pelegrina (Chile), "Lily of Lima"; low-growing "Peruvian-lily" only 18 in. high, with glossy green, lanceolate, succulent leaves along the stems, the numerous 3 to 4 in. azalea or lily-like flowers with spreading segments lilac, spotted and lined with maroon inside, blooming in summer. **C**: 76   p. 390

ALTERNANTHERA (Amaranthaceae)

; the "Calico plants" or "Parrot-leaves" are low growing tropical perennial herbs primarily grown, and greatly admired, for their colorful foliage as small border plants ideally suited for designs in carpet bedding, and kept in shape by frequent pinching of the growing tips or shearing. Their various colors are brought out most brilliantly at warm temperature in full sunlight. Their tiny, chaffy flowers are not showy and hide in the leaf axils. Propagated by tearing apart the mother plants in fall and kept at warm 60–65° F. in winter, or by August, or spring cuttings.

amoena (Telanthera) (Brazil), a "Parrot leaf"; small, densely branching herb of robust growing habit for its size, with broad elliptic 1 in. leaves brownish-red and carmine, and changing into orange in calico patterns; tiny whitish flowers. Much used for carpet-bedding because of its deep colors, and kept in shape by shearing. **C**: 2   p. 407

amoena 'Brilliantissima', "Red parrot leaf"; vigorous bushy bedding plant with ovate-elliptic leaves to 2 in. long, coppery green overlaid with red; irregular variegated areas mainly in center; intense carmine pink to red and with purplish veins if grown in full sun; if grown in the shade variegated yellow to pinkish. Kept in shape by pinching or shearing. **C**: 2   p. 407

bettzickiana (Brazil), "Red calico plant"; dainty little clustering tropical herb 2 to 6 in. high, dense with twisted leaves narrow and spoon-shaped, $\frac{1}{2}$ to $1\frac{1}{2}$ in. long, blotched and colored salmon-red to cream-yellow; the dwarfest for carpet-bedding and the red counterpart for the variety 'Aurea nana'. The assortment of gay colors is brought out best by a combination of heat, moisture, and sunshine. **C**: 2   p. 407

bettzickiana 'Aurea', "Yellow calico plant"; a pretty horticultural variety of slightly more open habit; bushy little plant with its small 1 to $1\frac{1}{2}$ in. spoon-shaped leaves prettily variegated pale yellow and fresh green. The green counterpart in carpet-bedding to the red A. bettzickiana, and kept in shape by shearing. **C**: 2   p. 407

dentata 'Ruby' (Brazil), "False globe amaranth"; a bushy herbaceous plant very much resembling Gomphrena, one of the "Everlastings"; redleaf form of A. dentata, with elliptic, metallic wine-red 3 in. leaves rather flat, purple beneath; the inflorescence strawflower-like greenish white globes. A pretty, somewhat tall plant for bedding or pots that can be kept short by trimming. **C**: 2   p. 407

versicolor (Brazil), "Copper leaf"; low, compact bedding plant having opposite, broad spoon-shaped 1 to 1½ in. leaves nearly round and with a short cusp at apex, crisped and corrugated around the margins, coppery or crimson in good light; bronzy green in the shade; with purplish-carmine veining and edged in pink and white. Needs to be pinched or sheared to keep low for edging. **C**: 2   p. 407

AMARANTHUS (*Amaranthaceae*)

hypochondriacus (hybridus var. hypoch.) (Trop. America), "Princess feather"; colorful tropical annual 2 to 5 ft. tall, the stout furrowed reddish stem with lance-shaped leaves purplish green tinged deep red; the inflorescence an erect branched spire with spikes of densely packed deep crimson or brownish-red flowers without petals tassel-like, blooming from July to September, and requiring full sunshine. Grand show in containers on the patio. From seed. **C**: 52   p. 406

tricolor (gangeticus) (East Indies), "Joseph's coat" or "Fountain plant", known in the South as "Summer poinsettia"; vigorous, erect, herbaceous tropical annual 1 to 3 ft. high, with small red flowers not showy, hidden in the leaf axils, but important for its rosettes of 4 to 8 in. leaves brilliantly colored green, red and yellow during summertime. A favorite garden annual relishing the sun, and spectacular when grown in patio containers for its gaily colored foliage. From seed. **C**: 52   p. 406

x AMARCRINUM: see x Crindonna

AMARYLLIS (*Amaryllidaceae*)

belladonna (syn. Brunsvigia rosea) (So. Africa), the "Cape belladonna" or "Belladonna lily", the "true" Amaryllis; a bulbous plant also known as "Naked lady" because the lovely trumpet-shaped, lily-like clear rosy-pink fragrant flowers, 3½ in. long, appear after the leaves die down, in clusters on solid, reddish 2 ft. stalks, bare and ahead of the new strap-like leaves, in August; forming clumps. Replant right after bloom and only if necessary as bulbs may not bloom for several years if disturbed at wrong time. Slightly hardy. In containers they are good patio bloomers; after leaf growth is completed, let them go to rest for the winter. Propagated by division; also from bulblets which take 3–4 years to bloom, or cutting bulb into sections. **C**: 13   p.388

AMARYLLIS: see also Hippeastrum, x Brunsdonna

AMOMUM (*Zingiberaceae*)

cardamon (Java), "Cardamon ginger"; perennial herb with creeping rootstock, and clustering leafy stems, which can attain 10 ft., but usually grown as a durable foliage plant less than 2 ft. high; linear-lanceolate, leathery leaves deep dull green, and finely hairy, emitting, when rubbed, a spicy aroma; yellow flowers in cone-like spikes beneath the foliage; produces cardamon seeds. **C**: 4   p. 206

AMORPHOPHALLUS (*Araceae*)

campanulatus (So. India, Ceylon to New Guinea and Fiji); a giant aroid with tuber 10 in. thick; dark green solitary lobed leaves pinnately cut; inflorescence with ovate spathe 8 x 10 in., fleshy and funnel-shaped below, green spotted white, purple inside; the expanded part green with wavy margin, and purplish outside; the huge conical, purple spadix 18 in. long, later developing a phenomenal cone bearing a multitude of golden yellow fruits. In cultivation loses its foliage, to rest during winter. **C**: 54   p. 534

AMORPHOPHALLUS: see also Hydrosme

AMPELOPSIS (*Vitaceae*)

brevipedunculata elegans (heterophylla variegata), (Manchuria, Japan), "Colored grape-leaf"; attractive climber with blood-red, hairy branches, and thin, forking vines; variable leaves simple, 3-lobed, or 5-lobed, bluish green and prettily variegated with milky green, creamy white and rose. Colorful grapevine for cool windows or winter-gardens. **C**: 66   p. 257

AMPELOPSIS: see also Parthenocissus

AMPHIBLEMMA (*Melastomaceae*)

cymosum (W. Africa: Sierra Leone); showy tropical shrub with large bristly, heart-shaped, rich satiny green leaves marked by a network of depressed veins, pale beneath and tinted red; flowers bright carmine-red. **C**: 54   p. 188

ANACAMPSEROS (*Portulacaceae*)

buderiana (So. Africa: Cape Prov.), "Silver worms"; queer, fascinating little succulent branching with prostrate, twisting worm-like stems of ¼ in. dia., spirally covered with minute leaves, but hidden by silvery white membranous scales; small flowers greenish-white. From cuttings. **C**: 13   p. 353

rufescens (arachnoides grandiflora) (So. Africa: Cape Prov.), "Love plant"; small clustering succulent to 3 in. high with roots tuberously thickened; stem erect or creeping with spiral rosettes of thick ovate-pointed ½ in. leaves, green above, purplish-brown beneath, and white cobwebs in the axils; 1½ in. flowers deep pink, opening in the sun, and lasting only for the afternoon. Slow growing, they love sun; keep dry in winter. From stem or leaf-cuttings, or seed. **C**: 13   p. 356

ANANAS (*Bromeliaceae*); the "Pineapples" are tropical American stiff terrestrial rosettes, usually grown for their delicious edible fruit-heads, some also for their decorative effect. The fruit which ripens in summer is botanically very interesting, being really a thickened stem in which the sterile ovaries forming the berries are imbedded producing the collective fruit; actually the stem passes right through the fruit and bearing on top a tuft of scaly leaves. This crown of leaves if cut off will grow into a new plant. Suckers from the base are also formed, as well as slips or stem-suckers along the stalk below the fruit; the base of cuttings should be allowed to dry for a day or two before being placed on sand with bottom heat to root.

bracteatus 'Striatus', "Variegated wild pineapple"; colorful variegated form of the "Red pineapple", a large terrestrial rosette; flat leathery leaves coppery green in center, cream-yellow toward margin; spines larger and more widely spaced than comosus. Produces medium-large brown-red, edible fruit. Flowers lavender. **C**: 59
                                                          p. 128, 131*, 193

comosus (sativus) (Brazil: Bahia, Mato Grosso), the cultivated "Pineapple"; large formal terrestrial rosette 2 to 3 ft. high, of 30 to 50 stiff, tapering and spiny-edged leaves grayish to bronze-green, gracefully arching; violet flowers borne in dense heads crowned by tufts of leaves, and producing the spiny greenish to light brown, aromatic edible pineapple, with its delicious yellow flesh, 6 to 8 in. long. The main season of fruit maturity is from May to July, but plants can be induced to produce at other times of the year by the use of chemicals or gas. (p. 127). A conversation piece as a decorative plant; if space permits even in the living room. **C**: 7   p. 509

comosus 'Nanus', "Dwarf pineapple"; a miniature form ideally suited for pots as everything is smaller in proportion: the leaves are shorter, with small spines, and the fruit is diminutive and cute. **C**: 7   p. 130

comosus variegatus, "Variegated pineapple"; beautiful variety with the channeled leaves having broad ivory bands along the margins, and red spines; center of rosette tinted rosy-red. **C**: 59   p. 43*, 128

ANDROMEDA: see Pieris, Leucothoe

ANEIMIA (*Filices: Schizaeaceae*)

phyllitidis (Trop. America: Cuba, Mexico, to Brazil), known as the "Flowering fern"; a xerophytic fern with fronds to 2 ft. long, bearing leaflets on the lower part; the stalk continuing bare and carrying at apex a panicle bearing the sporangia, but looking like flowers; the barren pinnate section 4–12 in. long, with firm, ovate 1 to 6 in. leaflets. Slow-growing, and does not require high humidity. **C**: 60
                                                          p. 252

ANEMONE (*Ranunculaceae*)

coronaria (So. Europe to Central Asia), "Poppy anemone", or "Lily of the field" of the bible; attractive perennial 1 ft. or more high with blackish tuberous roots; fresh green, finely divided, lacy leaves both basal and sessile along the erect, hairy stalk which bears the large solitary, poppy-like flowers 2 to 4 in. across, the showy sepals in vivid shades and combinations of red, blue, yellow, or white; blooming from March to May. A favorite cut flower; or a pot plant often forced to flower from February on. The knobby tubers are planted in September-October and kept 40–50° F. nights. Planted in April they will bloom in June-July; potted in June for September flowering. After the growing period they are taken up and dried in a cool spot. By division of the rhizomes. **C**: 64 p. 402

ANGIOPTERIS (*Filices: Marattiaceae*)

evecta (Japan to Australia and Madagascar) "Turnip fern"; giant, robust fern from swampy places, developing a stout stem to 2 ft. thick; the bipinnate fleshy leaves on swollen stalk may grow to 18 ft. long; the linear-oblong pinnae are succulent, dark green, and finely toothed. If grown in pots, place in saucer with water. Propagation by basal leaf scales removed from the bulbous stem, in peat-sand. **C**: 16 p. 252

ANGRAECUM (*Orchidaceae*); beautiful vandaceous

epiphytes from tropical and Southern Africa, Madagascar to the Philippines, with climbing stems clothed by 2-ranked very thick leaves; handsome, fragrant, star-shaped long-lived waxy flowers in shades of white to green, extending back into a curious long hollow spur. Angraecum like to be grown warm and humid with fresh air and in filtered shade; constant moisture while growing, thereafter moderate watering but never dry. Propagated by cutting the stems into pieces each with roots attached; or any rooted young plants from the base.

distichum: see Mystacidium

eburneum (Mascarenes, Madagascar); robust epiphyte with thick leaves neatly arranged in 2 ranks on stems that may become 4 ft. or more high; very fragrant 3 to 4 in. long-lived flowers upside-down on stout stalk dense in one-sided raceme, the waxy sepals and petals green, the broad lip ivory-white with green spur; blooming mostly in winter to March; flowers will last 5 to 6 weeks in perfection. **C**: 10 p. 457

eichlerianum (Trop. West Africa), epiphyte with scandent, leafy stems to 5 ft. or more long, developing aerial clinging roots; the thick light green leaves in 2 ranks, heavy-textured fragrant 3 in. flowers with yellow-green sepals and petals, white lip emerald green near base and with funnel-shaped greenish spur, blooming mostly in summer. **C**: 10 p. 457

falcatum: see Neofinetia

sesquipedale (Madagascar), because of its Christmas blooming habit known as the "Star of Bethlehem orchid"; spectacular epiphyte with usually solitary stem to 3 ft. tall, the leaves densely 2-ranked; magnificent, thick-fleshly, fragrant flowers to 6–7 in. across, ivory-white, and with a long greenish, flexuous spur that may grow to 1 ft. long; largest blooms in the genus, flowering mostly in winter from November to March. **C**: 10 p. 457

ANGULOA (*Orchidaceae*)

virginalis (uniflora), (Andes of Colombia and Ecuador), the "Cradle orchid", or "El Torito"; a delightful terrestrial of bold habit, with thin plaited leaves; the solitary curious, cupped fleshy long-lasting 4 in. flowers pure white, spotted and flushed with pink; blooming spring-summer. Easy to grow with humidity and good but diffused light; willing bloomer provided a strict resting period of several weeks is given after new pseudobulbs are matured. Prop. by division. **C**: 74 p. 458

ANIGOZANTHUS (*Amaryllidaceae*)

manglesii (Western Australia), the spectacular "Kangaroo paw" or "Australian sword-lily"; a fascinating rosette of onion-like leaves from which rises a striking inflorescence 3 ft. tall with velvety stem of intense red topped by a paw-like cluster of emerald green, woolly tubular flowers 3 in. long, red at calyx, with dark green lines, the lip reflexed and tipped blue, blooming from May to July. Kept cool and dry to rest in winter; fairly sunny, warm and well watered in summer; ideal for the patio. Easily propagated by division of the rhizome in spring. **C**: 69 p. 534

ANOECTOCHILUS (*Orchidaceae*); the charming "Jewel

orchids" are small tropical epiphytes of great beauty growing on trees, or terrestrials beneath trees or on mossy rocks under moist-warm conditions, and are cultivated primarily for their gorgeous velvety, often bronzy foliage with tracery in muted colors or in gold or silver; the flowers being very small and not showy; lacking pseudobulbs, they require constant humidity and even temperature, and need to be grown in greenhouses; in our super-heated homes in a terrarium or under a plastic hood or in bell-jar, in very humusy soil with sphagnum or fern fiber, broken pots, charcoal, or similar. Less moisture needed in winter. Removing the flowers on weak shoots will conserve the strength of the plant. Propagated by cutting the fleshy rhizomes into pieces preferably with a root attached. To my surprise I have seen Jewel orchids successfully grown as cut-off branches simply by placing them in a glass of water near the window where they have been blooming regularly for several years.

roxburghii (Himalayas, Java), "Golden jewel orchid"; dwarf terrestrial, with gorgeous fleshy, broad ovate leaves on succulent stems, the blade beautifully reticulated in red over dark bronzy green, with golden center; reddish flowers with white lip, on a spike, blooming variously between January and August. As other jewel orchids, grow in osmunda and sphagnum with crushed potsherds and charcoal, in small pots. **C**: 60 p. 43*, 192, 454

sikkimensis (Sikkim Himalayas), dwarf "King of the forest"; terrestrial orchid, with succulent stems and broad ovate, dark bronzy red 2½ in. leaves with velvety sheen, reticulated with a network of orange red veins; the 9 in. inflorescence with small ¾ in. flowers green and white, blooming in autumn. **C**: 62 p. 454

ANSELLIA (*Orchidaceae*)

gigantea nilotica (Kenya, No. Tanzania), the "Leopard orchid"; in habit usually smaller than africana and with pendulous inflorescence, the flowers 2½ in. across, larger and brighter yellow with chocolate flecks, golden lip marked red-brown in throat, blooming in spring. It is a strong-growing epiphyte 1½ ft. high with ribbed, slightly swollen pseudobulbs, bearing linear leaves, and an erect terminal cluster of long-lasting fleshy flowers, with spreading linear sepals and petals, faintly fragrant during the heat of day. Strangely, in East Africa it is the curiously branching Hyphaene palm which is host to this Ansellia, perching on which it forms large clumps, freely flowering with clusters of a hundred flowers on a single spike. By division. **C**: 7 p. 457

ANTHURIUM (*Araceae*); the "Tail flowers" are tropical

American ornamentals highly exotic, and cultivated for their decorative, often velvety or silver-patterned foliage, or their strikingly curious inflorescence distinguished by a slender, often tail-like colored spadix and more or less showy, waxy or leathery spathe impersonating a flower. The actual minute, petal-less, perfect flowers are arranged densely along the column. Anthuriums generally are not good house plants, unless humidity is kept relatively high; they are shade-loving and unfold their beauty best in a humid-warm greenhouse. Grown in porous fibrous or humus composts, best in fern fiber such as osmunda with sphagnum or in peatmoss. Propagated by top cuttings, stem sections, or division; also from seed absolutely fresh. (See p. 57.)

aemulum (heptophyllum) (Mexico to Costa Rica); climber in Central American forests with soft, membranous leaves palmately compound, glossy green with depressed veins; spathe and spadix green. **C**: 60 p. 86

andraeanum (S.W. Colombia, at altitude 3,000 to 5,500 ft.), the "Oilcloth flower" or "Tailflower"; an aristo-cratic tropical epiphyte perching on tree branches in wet forests at 58° to 86° F., exposed to almost continuous rains or mist, but with good drain-off at the roots. Erect plant with long-lobed, heart-shaped green, leathery leaves; the showy spathe glossy, waxy coral red and quilted, 4 to 5 in. long, and typically a pendant spadix tipped yellow, with white band marking the zone where stigmas are receptive. As may be surmised from its climatic background, A. andraeanum, or its hybrids known under the group name A. x cultorum, with spathes to 6 in. or more long, needs constantly high humidity, warm sections and in the case of A. scherzerianum, not less than 65 to 70° F., East or West light, and grown in very porous compost such as peatmoss or coarse humus, with osmunda, sphagnum moss, charcoal or perlite, and a little organic fertilizer added, over crocks. Mist the foliage daily. A. andraeanum blooms the year round, and the spectacular spathe is very long-lasting. **C**: 62   Cover*

andraeanum album; a beautiful, chaste color variant, with round-cordate, quilted, waxy-white spathe; spadix dipping, with white base, center area purplish-rose and tip yellow: glossy leaves. **C**: 62   Cover*

andraeanum rhodochlorum; a robust form featuring giant, deeply cordate, salmon-red spathe to 8 in. long, in which lobes or tip are green; curiously beautiful. **C**: 62 p. 85

andraeanum rubrum (Colombia), "Wax flower"; a striking color form of the normally lacquer-red type species, with large, waxy, quilted spathe of dark crimson; spadix white, yellow at tip. One of the most desirable types, but not a good house plant. **C**: 62   p. 65*, 84

araliifolium: see kalbreyeri

bakeri (Costa Rica); from a short stem, the pendent strap-like leathery, elliptic, long ribbon-like deep green leaves, with stout midrib; spathe and spadix green; the little charming, berry-like fruits turning a beautiful scarlet. **C**: 60   p. 65*, 87

clarinervium (Mexico), "Hoja de Corazon"; exquisite dwarf, ornamental species found growing in the clay of the Chiapas mist forest; of compact habit, the dark green, velvety, heart-shaped leaves smaller and more leathery than crystallinum, with clear, silver-gray ivory veins borne on stiff-erect petioles; the narrow spathe reddish green. Very promising as a house plant. **C**: 11   p. 85, 180

corrugatum (papilionense) (Colombia); ornamental leaf plant from Putumayo; the sagittate foliage light green, leathery, corrugated, with pointed basal lobes; green spathe, spadix purplish-black. **C**: 62   p. 87

crassinervium (Venezuela, Panama); bold rosette with fleshy, shining-green, elliptic leaves to 3 ft. long; thick stalk 3-ridged beneath. **C**: 60   p. 87

crystallinum (Colombia, Peru), "Crystal anthurium"; strikingly beautiful 10–15 in. heart-shaped leaves, velvety green with contrasting white veins; petioles round; spathe green and linear. **C**: 62   p. 85

digitatum (Venezuela); climber with palmately-compound, dark green leaves; stalked segments obovate, dark green, leathery; spathe green and purple, spadix purple. **C**: 60   p. 86

elegans: see fortunatum

x ferrierense (andraeanum x ornatum), "Oilcloth-flower"; robust climber with lobed, heart-shaped leaves; ovate-cordate, rosy spathe waxy-smooth, carried upright, and erect spadix white to rose; willing bloomer and easier to grow than others. **C**: 60   p. 84

fortunatum (elegans) (Colombia); showy plant from the Rio Dagua, with large, glossy-green, lobed leaves; the undulate, oblanceolate lobes arranged along the curving base of the sinus; reddish veins, spathe green, spadix brown. **C**: 60   p.86

gladifolium (Venezuela, Brazil); short-stemmed epiphyte, with long thick-leathery, strap-like, pendent leaves to about 5 ft. long with prominent veins, on pale, stippled petioles; narrow spathe. **C**: 62   p. 87

harrisii (Brazil), "Strap flower"; leathery, glossy green, and narrow-lanceolate leaves borne on slender, flat petioles; narrow spathe green or rose; spadix purple-brown. **C**: 60   p. 87

holtonianum (Panama, Colombia); striking plant with large, palmately compound, glossy green leaves 2 ft. across; the segments pinnately lobed and undulate. **C**: 60   p. 86

hookeri (huegelii) (Guyana, W. Indies), "Birdsnest anthurium"; robust, symmetrical rosette resembling bird-nest, with broad, obovate, grass-green leaves 2 ft. long; spathe green. **C**: 60   p. 87

imperiale (Trop. America); birdnest-like rosette with short-stalked leathery leaves lanceolate oblong, corrugated and wavy, on thick, angled petioles; inflorescence nearly hidden, with spathe chocolate-colored inside, green outside; long spadix chocolate brown, on very short stems. **C**: 60   p. 87

kalbreyeri (araliifolium) (C. America); climbing plant with shining-green, palmate leaves, the 5–7 broad and sinuate segments attached to base by short petioles; spathe purplish green. **C**: 60   p. 86

lindenianum ("Spathiphyllum cordata") (Colombia); attractive ornamental, broad, fleshy sagittate-pointed leaves with open sinus; slightly fragrant inflorescence on slender

stalks with broadly ovate white spathe tipped green, similar to Spathiphyllum, and whitish spadix. Known in South California nurseries as "Spathiphyllum cordata". **C**: 62 p. 85

magnificum (Colombia); large and showy heart-shaped, velvety, olive-green leaves with prominent white veins; the petioles 4-angled; spathe and spadix green. **C**: 62   p. 85

x mortefontanense (veitchii x andraeanum); glossy, sagittate leaves with large sinus, surface as if quilted; cream-white spathe; upright, pink spadix. **C**: 62   p. 85

ornatum (Venezuela); noble leaf ovate-cordate, bright green; cupped spathe white turning purplish-rose toward tip; upright spadix red-purple. **C**: 60   p. 85

pedato-radiatum (Mexico); elegant plant with large, graceful, glossy-green leaves palmately lobed; segments slender, oblanceolate; basal lobes broader; spathe reddish, spadix green. **C**: 60   p. 86

pentaphyllum (Colombia to Brazil); climber with large, palmately fingered, glossy-green, leathery leaves, with pale veins, carried on grooved petioles; spathe green, spadix dark green. **C**: 60   p. 86

podophyllum (Mexico), "Footleaf anthurium"; showy, self-heading plant from Vera Cruz; large leathery leaves digitately lobed with stiff, finger-like segments, carried on long stalks from trunk. **C**: 60   p. 86

polyschistum (Colombia: Putumayo), "Aztec anthurium"; scandent species with slender stem and distantly spaced, palmately compound, graceful leaves, 7–9 leaflets 4 in. long x ¾ in., matte dark green, thin-turgid, narrow, with crisped margins, on slender channeled petioles: small 1 in. green spathe, spadix brownish. **C**: 60   p. 86

scandens (W. Indies, to Ecuador, at 8,000–10,000 ft.), "Pearl anthurium"; dwarf climber with small, leathery, glossy, dark green elliptic leaves; spathe green; purple or white fruit, like exquisite pearls. **C**: 60   p. 87, 509

scherzerianum (Costa Rica), "Flamingo flower", "Flame plant"; fascinating small plant with lanceolate, leathery green leaves; the curious inflorescence a sub-tended broadly ovate spathe, brilliant in shades of scarlet and crimson to soft salmon; the slender spadix coiling and spirally twisted, flamingo-like, golden-yellow; blooming most of the year, but mainly from February to July. If conditions can be kept sufficiently moist, this will be a durable and ever interesting house plant. **C**: 74   p. 43*, 84, 535

scherzerianum rothschildianum, "Variegated pigtail plant"; spathe more or less red with white spots and specks. **C**: 74   p. 84

tetragonum (Costa Rica); loose rosette of 2 to 4 ft. oblanceolate, fresh-green leaves with bold midrib and wavy margin, on 4-angled, short petiole; spathe green. **C**: 60   p. 87

veitchii (Colombia), "King anthurium"; unusual plant with pendant, showy leaves to 3 ft. long, cordate at base, rich metallic green; curved lateral veins sunken, giving a quilted look; pale midrib. Inflorescence with narrow green spathe. Beautiful but difficult. **C**: 74   p. 85, 180

warocqueanum (Colombia), "Queen anthurium"; climbing species with showy, long tapering, velvety leaves to 3 ft. long; deep green with ivory veins; small spathe green to yellowish. **C**: 62   p. 43*, 85, 180

ANTIGONON (*Polygonaceae*)

leptopus (Mexico), "Coral vine"; showy tendril climber, with tuberous roots; slender zigzag stems bearing arrow-shaped, light green 3 in. leaves, and axillary racemes of small flowers a lovely shade of bright rose-pink with deeper center. One of the sweetest tropical vines, trained to support, requiring lots of sunlight to bloom. From cuttings. **C**: 52   p. 278

ANTIRRHINUM (*Scrophulariaceae*)

majus (So. Europe, No. Africa), the popular "Snap-dragon"; charming herbaceous perennial but usually grown as a biennial or annual; stiff erect, leafy stems about 1½ to 2½ ft. high topped by terminal spikes of very showy, curiously sac-shaped, two-lipped 1½ to 2 in. flowers looking like dragon's jaws; in many colors and shades, velvety red to pink, copper, yellow or white, the yellow mouth closed but is forced open by the bees or by hand. Among the best flowers for the garden or for bouquets, they reach their greatest perfection in spring or early summer, and when cut will continue to bloom. However, since "snaps" react to daylength and light intensity, groups have been developed that make it possible to have them in flower the year around. Easy from seed; also from cuttings. **C**: 75   p. 416

APHELANDRA (*Acanthaceae*); the "Zebra plants" are showy tropical American evergreen woody herbs, with opposite simple but magnificent leaves beautifully painted with silver or ivory, and 4-sided shingled spires of colored bracts, with brilliant somewhat 2-lipped flowers primarily blooming in autumn, combining both ornamental foliage with a striking inflorescence in one plant. As a group Aphelandras are amongst our most colorful indoor plants, easy to grow in a humid warm greenhouse, but requiring careful attention in maintaining steady continuous moisture at the roots without ever drying out, to grow as a house plant. Commercially most attractive are the squarrosa cultivars with their white-veined foliage and golden pyramids of bracts; they are summer-autumn-blooming and will flower when the plant has become pot-bound; by exposing it to a cool shock of 40–50° F. for 7 weeks in fall, flowering may be advanced to spring; from October to January these plants are somewhat rested and cut back.

aurantiaca (Mexico) "Fiery spike"; striking tropical plant of erect, shrubby habit to 3 ft. high, with stiff stems and broad ovate, smooth green leaves 4 to 6 in. long, silvery-gray in the netted vein areas, and more or less wavy-margined; showy bracted spikes with brilliant scarlet-red flowers, orange in throat and along tube, blooming from October to December. Easily reproduced from seed. **C**: 10
        p. 43*, 499

aurantiaca roezlii (Mexico), "Scarlet spike"; a lower growing variety with somewhat twisted, dark green leaves, silvery between main veins; flowers orange scarlet in 4-ranked bracted spike; fall and winter-bl. **C**: 10   p. 499

chamissoniana (Brazil), "Yellow pagoda"; erect plant with wiry stems and closely set, thin, slender pointed 3 to 4 in. green leaves, painted and spotted silver-white along mid-rib and main lateral veins; the terminal inflorescence a slender spike of shingled, pointed yellow bracts with recurved, spiny green tips; the protruding flowers clear yellow; autumn to winter-blooming. **C**: 60   p. 499

fascinator (Colombia), emerald-green, satiny leaves marked with silver, veins a silvery amethyst, purple beneath; large, scarlet flowers. **C**: 10   p. 43*

sinclairiana (Central America, to Darien Prov., Panama), "Coral aphelandra"; exceptional tropical flowering plant of compact habit with limp, ovate leaves rich glossy-green, and with pale veins depressed; the gorgeous inflorescence branching and clustered like a bouquet, with imbricated cupped herbaceous bracts a striking orange-coral, the corolla of the protruding flowers rosy-pink. **C**: 60   p. 499

squarrosa (Brazil), "Saffron spike zebra"; striking tropical plant of robust habit, with stiff erect, reddish stems, and glossy dark green leaves 6 to 10 in. long beautifully veined with white; in summer with terminal pyramids of shingled long-lasting waxy bracts of vivid golden yellow, and which remain beautiful long after the 1½ in. green-tipped canary-yellow flowers have passed; fall-blooming.
C: 10   p. 499

squarrosa 'Dania', a "Zebra plant"; very stocky Danish clone selected from A. sq. 'Fritz Prinsler', A. 'Dania' is a handsome foliage plant of compact habit, with short inter-nodes between the attractive white or creamy-veined leathery, deep glossy green leaves, on a red brown stem; the inflorescence a spike with orange-yellow bracts and yellow flowers; every shoot pinched should form 2–5 flowering heads. A. squarrosa comes from Brazil. **C**: 10
        p. 187

squarrosa 'Fritz Prinsler', (squarrosa leopoldii x sq. louisae), "Saffron spike"; valuable cultivar combining the remarkable white to straw-yellow leaf-veining on dark olive-green of louisae with the leaf form and willingness of leopoldii, but remaining smaller; floral bracts deep yellow, marked green, corolla canary yellow. Colorful both as an attractive foliage plant with its waxy, ovate, variegated leaves and its bold, spiked inflorescence. Requires attention to moisture or it will drop leaves in the home. **C**: 10   p. 187

squarrosa leopoldii (Brazil); most robust variety with fleshy red stems, dense with large elliptic 8 to 12 in. glossy leaves olive green, the lateral veins silvery white; short-lived, pale yellow 1½ in. flowers stand out one-third from the 4-ranked spike of red, shingled bracts, distinctly toothed, and appearing singly or several at the top of the plant. **C**: 10   p. 499

squarrosa 'Louisae' (Brazil), a compact "Zebra plant"; low growing, exquisite variety with smaller, shiny, emerald-green, elliptic leaves only 3 to 5 in. long and with equally prominent white veins; bright yellow flowers tipped green, on fleshy reddish terminal spikes out of long-lasting, waxy, golden-bracted pyramids 3 to 6 in. long; fall blooming. One of the best as a pot plant. **C**: 10   p. 499

squarrosa 'Louisae compacta'; a selection of European growers for its most compact, low habit; similar to the Danish cultivar 'Dania', but having foliage darker green, and with boldly contrasting veins of greenish ivory; in habit even shorter and more closely jointed; most attractive as a small foliage plant with its leaves in effect arranged like a rosette. **C**: 10   p. 499

squarrosa 'Uniflora Beauty'; a slow, compact grow-ing cultivar distinguished by its small 3 to 4 in. leaves which are a deep green with contrasting greenish-ivory veins. **C**: 10   p. 499

APONOGETON (*Aponogetonaceae*)

distachyus (So. Africa), "Cape pondweed" or "Water-hawthorn"; a beautiful aquatic herb with floating linear-oblong, bright green to coppery, fleshy leaves rising from a tuberous rootstock. The hawthorn-scented flowers are arranged in 2 ranks on the forked stalk, each flower with a waxy-white single petal, blooming from April to October. A good pond plant fairly hardy, but also suitable for aquaria at 70 to 80° F. provided there is good light; a rest period at 55° F. or less is advisable. Propagate by division or offsets. **C**: 64   p. 519

fenestralis (Madagascar), known in horticulture as Ouvirandra, the skeleton-like "Madagascar laceleaf" or "Lattice leaf"; an extraordinary submerged tropical aquarium plant, in habitat in slowly running streams; from a tuberous rootstock looking like a small potato, ascend the curious leaves to float just beneath the water surface; 6 to 16 in. long, these leaves are but a skeleton network of pale yellow to dark olive veins; the purplish flowers rise a little above water, on long stalk, dividing into two curved spikes. This difficult but beautiful plant may be successfully grown in water temperature of 65 to 70° F. (not over 75° F.) in summer; in winter 60 to 65° F. for some rest; in soft (rain or filtered) water with diffused light. Stagnant water with algae kills the plants, and it is best to change half of it weekly. Plant in lime-free sand or sandy loam, at least 8 in. below the water surface. Propagated by division of the rhizome, or offsets. **C**: 54   p. 519

APOROCACTUS (*Cactaceae*)

flagelliformis (Mexico), the popular "Rat-tail cactus"; slender creeping or pendant stems to 1 in. thick and 6 ft. long, close-ribbed and covered with small reddish spines; small but pretty day-blooming crimson flowers; the fruit a small red berry. **C**: 1   p. 156

x APOROPHYLLUM (*Cactaceae*)

'Star Fire' (Aporocactus x Epiphyllum); beautiful Johnson bigeneric hybrid between Orchid cacti and Rat-tail cacti, normally epiphytic and ideally suited for baskets. This and similar hybrids are free-flowering, earlier than orchid cactus and bloom at odd times. The slender stems are ribbed and pendant but may be trained on a trellis. The large rose-red flowers are flushed scarlet outside. **C**: 21
        p. 169

APTENIA (*Aizoaceae*)

cordifolia variegata, known in horticulture as Mesem-bryanthemum cordifolium (Eastern coastal deserts of South Africa), "Baby sun rose"; soft, rambling succulent with prostrate rooting branches to 2 ft. long, dense with small fleshy, heart-shaped leaves ½ to 1 in. long, grayish or fresh green, prettily variegated with cream margins, and shimmering in the sun; tiny starry, purple flowers. Exquisite for carpet-beds or for hanging pots. From cuttings or seed. **C**: 14   p. 353

AQUILEGIA (*Ranunculaceae*)

'Long-spurred hybrid', the beautiful "American columbine" or "Garden aquilegia", derived from such species as the European vulgaris and chrysantha, and caerulea from the U.S. Rocky Mountains; attractive hardy perennial with fibrous roots and stalked maidenhair-fern-like compound, glaucous leaves; exquisitely colorful flowers 2 to 3 in. across consisting of 5 petals each with a long straight, hollow, nectar-bearing spur attracting hummingbirds. These hybrids come in pastel and deeper shades of pink, yellow, copper and blue, often with sepals and petals in contrasting colors, blooming in spring and early summer. For forcing, pot up clumps and store in cold house to February, then gradually increase to 50° F. Prop. by division of clumps, or seed. **C**: 78 p. 418

ARACHNIS (*Orchidaceae*)
lowii (Vandopsis), the "Scorpion orchid"; a giant vandaceous spider orchid growing high on trees in the humid forests of Borneo; magnificent epiphyte climbing with fleshy roots on stout stems to 6 ft. long, clothed with ranks of glossy green 2 to 3 ft. leaves; the remarkable 3 in. flowers regularly spaced on pendulous racemes that may stretch from 6 to 12 ft. long; the 1 or 2 lower, basal blooms orange-yellow finely spotted red, the succeeding flowers greenish-yellow with large brown blotches, blooming August to November. This robust plant needs large containers to accommodate the rampant roots, with good drainage; shade only to keep from burning; watering is somewhat reduced during winter. Prop. from stem cuttings. **C**: 60 p.458

ARAEOCOCCUS (*Bromeliaceae*)
flagellifolius (North Brazil, Surinam, Venezuela, Colombia); cluster-forming slender rosettes of few narrow, stiff, channeled, brown-scaly, toothed leaves 2–4 ft. long; the inflorescence in erect racemes with light red bracts, and subsequent greenish berries. **C**: 19 p. 133

ARALIA: see Dizygotheca, Fatsia, Polyscias, Tetrapanax

ARAUCARIA (*Araucariaceae*); tropical and subtropical evergreen trees of great height belonging to the Conifers, having symmetrical whorled branches and scale or awl-like stiff leaves often resembling needles, and large woody cones. Generally known as "Star-pine" or "Monkey-puzzle" they are very durable and tolerant to draft and abuse as exceptional container plants for decoration indoors or the patio, especially the popular A. heterophylla. They prefer good light but adjust down to a mere 15 fc. but should be turned around to maintain form. Propagation by very fresh and viable seed; or by top cuttings in winter, glass-covered moist-warm between 65–70° F.

araucana, in horticulture as imbricata (So. America: Mts. of So. Chile, No. Patagonia), the so-called "Hardy Monkey-puzzle"; an oddity in trees, becoming 50 to 100 ft. high, with resinous bark and heavy, spreading branches in whorls; tangled, rope-like branchlets closely shingled with sharp-pointed, ovate, thick-leathery, dark green leaves 1–2 in. long, uniformly arranged around shoot. Slow growing and the hardiest of the Araucarias, growing as far north as British Columbia, where the temperature may drop to zero ° F. A conversation piece that may be grown in containers. **C**: 22 p. 309

bidwillii (Australia: Queensland), "Bunya-Bunya" or "Monkey-puzzle"; slow-growing coniferous tree becoming 150 ft. high with age; with wide-spreading branches; long, glossy sharp-pointed needles with depressed parallel veining, to 2 in. long, arranged in two flat rows in the juvenile stage; in later stage oval, ½ in. long and spirally arranged. A vicious-looking, fearsome tree supposed to keep the monkeys puzzled because they can't manage to climb over the spiked leaves. Of interesting appearance, also grown in containers but sparry, requiring much space, and passers-by risk tearing their clothes. **C**: 16 p. 308

excelsa: see heterophylla

heterophylla, best known in horticulture as A. excelsa (Norfolk Island), the "Norfolk Island pine", "Christmas tree plant" or "Starpine"; a noble evergreen pyramidal tree from the South Pacific, to 220 ft. high. Usually grown in the juvenile stage 1 to 3 ft. high, when they are very formal with tiers of 5 to 7 pinnate branches parallel to the ground, closely set with overlapping soft, awl-shaped needles to ½ in. long; one of the best and easy-to-care-for house plants, tolerant to shade or sun, chills or heat, and even occasional drying out, and lasting for years, although with abuse they may loose their symmetry. **C**: 20 or 22 p. 307, 309

imbricata: see araucana

ARAUJIA (*Asclepiadaceae*)
sericofera (So. Brazil), "Bladder flower" or "Cruel plant"; prodigious woody twiner resembling Stephanotis, with several forms of leaves, mostly oblong, 2 to 4 in. long, green above, white and minutely felted beneath; salver-shaped flowers 1 in. across, white or pinkish-tawny, fast-fading, and fragrant. The seed pods explode with characteristic silk of asclepiads. Biologically interesting because flowers are able to trap insects. For the cool wintergarden. Propagate by seeds, or cuttings of ripe wood. **C**: 64 p. 279

ARBUTUS (*Ericaceae*)
unedo (So. Europe from Dalmatian coast to Portugal; also found in Ireland), "Strawberry tree"; small, slow-growing evergreen shrubby tree to 8 ft. and even more than 20 ft. high with rough red-brown shreddy bark, and sticky-hairy branches; handsome oblong, shiny, toothed 2 to 4 in. leaves, and small waxy white or pinkish urn-shaped ½ in. flowers in drooping clusters; strawberry-like, orange-red ¾ in. fruit edible but without flavor, and appearing at the same time as the blooms from fall into December. A fine decorative plant for the summer patio, cool in winter. Artificial pollination is advisable for best fruit set. From seed, or cuttings of partially ripened wood in fall; also layers. **C**: 63 p. 494

ARCHONTOPHOENIX (*Palmae*)
alexandrae (Australia: Queensland), the tall-growing "King palm" or "Alexandra palm", with erect ringed trunk 6 in. thick and to 70 or even 100 ft. high, and bulging at base; bearing a majestic crown of arching pinnate fronds, the narrow leaflets 1½–2 in. wide, green above but prominently grayish-white beneath; flowers white or creamy; attractive red fruit. **C**: 54 p. 220

cunninghamiana (Australia: Queensland, N.S.W.), "Seaforthia palm" or "Piccabeen palm"; tall feathery palm with slender trunk not enlarging below except at surface of ground; the gracefully arching fronds to 10 ft. long; broad leaflets dark green on both sides, 3–4 in. wide; pendulous inflorescence with lilac flowers; the coral fruit less than 1 inch. Sometimes in the trade as Seaforthia elegans. **C**: 54 p. 220, 540

ARDISIA (*Myrsinaceae*)
crispa (crenulata) (China and Japan to Malaya), "Coral berry"; graceful little tree slow growing to 4 ft., with long elliptic, thick-leathery, shining dark green leaves 2½ to 5 in. long, prettily crisped and thickened at margins; the fragrant white or reddish flowers in axillary stalked clusters, followed in winter by a tier of bright scarlet, waxy berries, often persisting for 6 months and until the next crop of flowers. A long-enduring, handsome potplant for the house, both for its waxy foliage and beckoning fruit; the patio in summer. Washed seed is sown in January; for stocky plants the tops of the seedlings are cut off and rooted as cuttings. Old, lanky plants may be rejuvenated by mossing of the top. **C**: 16 p. 512

japonica (Japan, China), "Marlberry"; small broad-leaved evergreen shrub 1½ ft. high, with elliptic leaves 2–3 in. long, glossy dark green and sharply toothed and crowded at ends of branches; small white flowers in clusters, followed by ¼ in. red berries. Rather slow-growing, for cool corners. Propagated from fresh seed or half-hard cuttings. **C**: 16 p. 300

wallichii (sanguineolenta) (India), "India spice berry"; a durable small evergreen bush about 2 ft. high, with branches smooth except when young, obovate leaves minutely toothed, 3 to 5 in. long; in summer with loose axillary clusters of small red flowers having pointed lobes, followed by black berries. Propagated by cuttings of half-ripe wood, or from seed. **C**: 16 p. 498

ARDISIANDRA (*Primulaceae*)
sibthorpoides (Africa: Cameroons Mts., Fernando Po); miniature creeping herbaceous plant, with meandering, round pink stems; alternate doubly-crenate, membranous, roundish, rugose nettle-like, small leaves, 1¼ in. dia., with cordate base, flesh-green, covered with white hairs; small axillary flowers with white corolla. **C**: 78 p. 265

ARECA: see also Chrysalidocarpus

ARECASTRUM (*Palmae*)
romanzoffianum (So. Brazil to W. Argentina and Bolivia), widely known as Cocos plumosa, the "Queen palm"; very handsome with straight smooth trunk to 40 ft. high and a widely spreading crown of graceful, long-arching

plumy fronds 8–15 ft. long, the narrow soft, shining dark green leaf segments densely set in clusters; suddenly pendulous at the middle, and giving this palm its characteristic habit. Gorgeous under moist and warm conditions, and now widely planted in So. California and So. Florida. Edible orange fruit. A tolerable tub plant for tropical effect but then not holding more than a few fronds; requires much water; shade tolerant to 20 fc. **C**: 4   p. 222

ARENGA (*Palmae*)
   engleri (Taiwan), "Dwarf sugar palm"; handsome dwarf but showy palm 5 to 10 ft. high, suckering with rather short stems covered by black fibers, the spreading pinnate leaves to 4 ft. long; dense leaflets dark green above, silvery-tomentose beneath; irregularly notched on each side, broadening toward apex and with ends jagged as if chewed off. The sap of the related A. pinnata yields arenga sugar. **C**: 54   p. 230

AREQUIPA (*Cactaceae*)
   leucotricha (Chile, Peru: Arequipa); globular to cylindric cactus to 2 ft. long, becoming prostrate; grayish-green covered with long yellow spines, centrals much longer than radials, on tubercled ribs, and topped by yellow wool; long and slender funnel-shaped, carmine-red flowers. **C**: 13   p. 159

ARGYRODERMA (*Aizoaceae*)
   aureum (So. Africa: Cape Prov.), "Silver jaws"; sun-loving, solitary, extremely succulent little body looking like smooth silvery pebbles, with a thick-fleshy pair of nearly white leaves resembling a gaping jaw 1 in. across, each leaf chin-like and keeled; the stalkless 1 in. flowers golden-yellow from between the leaves, in habitat amongst white quartz appearing as if silver-white stones were blooming. They grow in summer and rest in winter. From seed. **C**: 13   p. 354

ARIOCARPUS (*Cactaceae*)
   fissuratus (W. Texas, Mexico), known locally as "Living rock cactus"; star-shaped flat succulent rosette to 6 in. across, with weird, horny gray surface, and turnip-like root; lovely pink flowers. **C**: 13   p. 160

ARISARUM (*Araceae*)
   proboscideum (Italy: Apennines), "Mouse plant"; small tuberous herb with long rhizome forming arrow-shaped leaves; near ground level appears the curious inflorescence resembling a mouse, a swollen tube grayish-white below and olive green above, the tip narrowed to a long curved tail; dormant in summer. **C**: 72   p. 97

ARISTOLOCHIA (*Aristolochiaceae*)
   elegans (Brazil), "Calico flower"; graceful tropical climber with corky stem and kidney-shaped 2–3 in. leaves glossy green, glaucous beneath; the curious 5 in. flowers a yellowish, inflated tube and expanded cup rich purplish-brown inside with white markings. This very satisfactory, free flowering species may be grown in pots with sufficient nourishment, in sunny winter gàrden or greenhouse; cut back in winter; oldest plants bloom best with their very remarkable flowers. Propagation by cuttings. **C**: 54   p. 275
   leuconeura (Colombia), "Ornamental birthwort"; decorative climber of the tropics, with glossy yellow-green, heart-shaped leaves with wide sinus, attractively veined cream-white. **C**: 54   p. 266

ARTEMISIA (*Compositae*)
   abrotanum (So. Europe), "Southernwood" or "Old man"; deciduous fragrant semi-shrubby plant 2–4 ft. tall, with finely divided, grayish-green, pungent smelling leaves 1–2½ in. long, at first gray downy; flower heads yellowish-white. Hardy. A favorite for its sweet, charmingly fragrant aroma; dried stems were once thought to keep moths away. By division. **C**: 25   p. 319
   absinthium (Europe), "Wormwood"; handsome hardy herbaceous perennial with a woody rootstock, to 3 ft. tall, with silky lobed leaves 1½–3 in. long, green and nearly glabrous above, white with cottony down beneath; yellowish flower heads. Best known as an ingredient of absinthe liqueur, and a good accent in plantings. From seed or cuttings. **C**: 25   p. 319
   dracunculus (So. Europe), "Tarragon" or "Estragon"; hardy herbaceous perennial 2 ft. tall, with fresh green linear leaves 1–3 in. long, scented like anise; panicles of whitish flowers. Use: young leaves and stemtips in salads, steaks, chicken, sauces and other cookery; yields flavor to vinegar in pickles. By division or root cuttings. **C**: 28   p. 320
   purshiana (Nebraska to the Pacific, California and British Columbia), "Cudweed" or "Lady's Tobacco"; hardy herbaceous perennial with horizontal rootstock and stems to 3 ft., oblong leaves 1–3 in. long, covered alike both sides of the leaf with white wool, the lower ones toothed or parted; overtopped by spiked panicles of white flower heads. Attractive with their silvery-white foliage for edging in the herb garden. Considered a "cattle-fattener", and given its popular name as we may assume that it is chewed over and over again. **C**: 25   p. 321

ARTHROPODIUM (*Liliaceae*)
   cirrhatum (New Zealand), "Rock-lily"; tufted perennial herb to 3 ft. high, with fleshy roots, numerous spreading flexible, lanceolate, light green, clasping leaves, grayish beneath, with a narrow translucent edge and parallel veins; cluster of white flowers. **C**: 14   p. 202

ARTOCARPUS (*Moraceae*)
   altilis (communis, incisa) (Malaysia to Tahiti), the "Breadfruit"; a strikingly handsome tropical tree 30 to 60 ft. high, widely planted in tropical countries because of the large, yellow 5 to 8 in. dia. prickly fruit with milky juice, containing sweet, starchy pulp, and which is made edible and resembling bread when baked; beautiful and decorative huge, deeply lobed leathery leaves, 1 to 2 ft. long luxuriantly green with yellowish veins; the small male and female tubular flowers in different clusters on the same plant, the male in catkin-like spike, the female in clusters. Breadfruit comes from moist lowlands, and doesn't like to be moved; if established in large containers, they require copious watering. As the best fruits are seedless, propagation is from cuttings of side shoots. **C**: 2.   p. 506
   heterophyllus (integrifolia) (India to Malaya), the "Jackfruit"; interesting tropical tree to 50 or 70 ft. high related to the breadfruit, with milky juice; glossy oblong leaves, lobed on younger branches; remarkable for its enormous fruit dangling directly from the trunk and biggest branches of the tree. This fruit is one of the largest in the world, 1 to 3 ft. long and weighing up to 40 pounds; a green knobby rind encloses a soft sweet or acid pulp of unpleasant odor, and is eaten raw or cooked, the seed roasted, in tropical Southeast Asia. In the greenhouse for botanical collections. Prop. by sideshoots or seed. **C**: 52   p. 528

ARUNDINA (*Orchidaceae*)
   graminifolia (Himalayas through Malaysia, Indonesia to Tahiti), tall-growing terrestrial orchid with reedy stems to 6 ft. tall, evenly set with series of plaited, grass-like leaves; topped by terminal clusters of showy, delicately fragrant cattleya-like flowers 2–3 in. across, rosy-lilac with purple lip and pale disk; opening in succession one or two at a time, each lasting about 3 days, but blooming throughout the year. Flowering best if planted in large containers with loam, leafmold and fern-fiber plus fertilizer; ideal for the patio in fairly sunny spot short of burning. Propagated by division, or by taking off young plants produced on stems. **C**: 60 p. 457

ARUNDINARIA: see Sasa

ARUNDO (*Gramineae*)
   donax versicolor (So. Europe), the "Variegated giant reed"; majestic perennial grass, the species becoming 10 to 15 or more feet high, and the variety versicolor only 6 ft., with knotty rootstock and stout stems almost woody; arching ribbon-like leaves gray-green striped creamy white, to 2 ft. long, alternately arranged on canes, and topped by showy plume-like panicles at first reddish, then white. Very impressive with its bold, bamboo-like leafy canes; quite ornamental in containers, requiring lots of water. By division. **C**: 64   p. 306

ASARINA (*Scrophulariaceae*)
   erubescens (Maurandia) (Mexico), "Creeping gloxinia"; strongly vining perennial herb, hairy with alternate, tri-angular, toothed, downy leaves, on twining flower-stalks, bearing large, 3 in. trumpet-shaped blossoms, having broad green sepals, and carmine-rose corolla with pale throat, spotted rose; blooming from summer into winter in a cool greenhouse or wintergarden. Easily grown as annuals from seed, but also from cuttings. **C**: 66   p. 280

ASARUM (*Aristolochiaceae*)

shuttleworthii (U.S.: Virginia to Alabama), "Wild ginger"; a stemless, rhizomatous woodland perennial with showy kidney-shaped, soft-leathery 3 in. leaves, dark green with silvery-green marbling; resin-scented brownish flowers urn-shaped, mottled violet inside, near ground-level. Called wild ginger because of their strong scent and flavor. Requires lots of moisture. Increased by division of their creeping rootstocks. **C**: 78   p. 300

ASCLEPIAS (*Asclepiadaceae*); the "Milkweeds" are perennial summer and fall-blooming herbs usually with milky juice; the inflorescence in clusters, with the 5 petals of the blossoms strongly reflexed showing a crown of 5 horned hoods. Attractive for their showy flowers and of easy cultivation. Propagated by division or seeds.

currassavica (Trop. America), "Blood-flower"; a showy tropical sub-shrub to 3 ft. high, with woody base, stems with milky sap, oblanceolate leaves 2 to 6 in. long, bluish beneath; the flowers in clusters with reflexed, 5-parted corolla brilliant red-purple, exposing the crown of 5 horned orange hoods; June to September blooming, and very showy for its rare coloring as a pot plant. **C**: 63   p. 500

tuberosa (U.S.: Maine to Florida and Arizona), the "Butterfly-weed"; hardy perennial 2 to 3 ft. high, without milky juice; from creeping rhizomes the wiry hairy stems, with alternate rough, green, narrow ovate leaves 3 to 5 in. long, topped by a brilliant cluster of small $\frac{1}{4}$ in. vivid orange flowers, blooming from June to September; the showiest of our native milkweeds; excellent for the sunny patio. By division. **C**: 51   p. 500

ASCOCENTRUM (*Orchidaceae*)

miniatum (Saccolabium) (Java, Borneo, Philippines); pretty, small epiphyte with short erect stem clothed with ranks of fleshy, linear leaves, and short cylindric racemes 5 in. high of small but gay red-orange flowers nearly 1 in. across; blooming late winter to June and again October. A charming miniature for the warm window collection, or hung near the glass in the greenhouse; dependable, willing bloomer. Propagated from side shoots or by cutting the top, and stem sections with roots attached. **C**: 10   p. 457

ASCOTAINIA (*Orchidaceae*)

viridifusca (Ania or Tainia) (Assam, Burma); robust terrestrial orchid, with flask-shaped pseudobulbs each bearing a single plaited leaf; from the base an erect stalk to 3 ft. high with long-lasting flowers to 2 in. across, the sepals and petals brownish olive-green, lip yellowish-white. February-March blooming. Growing in fibrous compost with some loam, the plant needs abundant water while growing but after completion of growth, rest and keep cooler for a month without letting it shrivel. By division. **C**: 8   p. 458

ASPARAGUS (*Liliaceae*); these fern-like plants are mostly woody vines or spiny shrubs, some perennial herbs of dry Africa and Asia, including such unlike plants as the "Smilax" of florists', the common garden asparagus and the feathery so-called "Asparagus fern" widely used with bouquets of cut flowers, and being leafless, quite resistant to heat. All have tuberous or fleshy roots and no distinct true leaves, but what looks like green leaves are branchlets or needle-like branchlets called cladodes; the small flowers are white or greenish. The decorative Asparagus are of easy care though they prefer moderate temperatures; some are effectively used in hanging baskets. Propagated by divisions of older plants; or from seed. Germination is faster from seed not fully ripe, or soak seed in tepid water.

asparagoides myrtifolius (Medeola) (So. Africa: Cape of Good Hope); the "Baby smilax" or "Medeola", used where available by florists for weddings and other decorations; graceful twining vine with thread-like stems and dainty, fresh glossy-green, little ovate leaves to 1 in. long, usually allowed to climb on strings for cutting. When needed, cut with string. **C**: 53   p. 253, 286

densiflorus cv. 'Meyers'; "Plume asparagus"; a very charming compact-growing form of A. densiflorus from So. Africa; plants send up many stiffly erect wiry stems 1 to 2 ft. long, densely clothed as in a spindle with needle-like deep green cladodes. With their fluffy look they are exceptionally attractive in pots or baskets; very durable. **C**: 15
            p. 253

densiflorus 'Sprengeri' (S. Africa: Natal), "Feather" or "Sprengeri fern"; much branched plant with tuberous roots,

scarcely climbing; the arching stalks woody and wire-like, carrying the fluffy branchlets set with soft, fresh green, needle-like cladodes; true leaves reduced to thorns; small fragrant flowers white, followed by bright red berries. Durable, pretty pot plant or for baskets; cut branches used for flower arrangements. Propagation by seed, although division is possible. **C**: 15   p. 253, 286

falcatus (Ceylon, So. Africa), the "Sickle-thorn"; widely climbing on woody stems to 40 ft. long; branches straw-colored, slender and with rigid spines, the clustered, firm, bright-green 3 in. leaves narrow and sickle-shaped; flowers white. **C**: 15   p. 253

myriocladus (So. Africa: Natal), "Zigzag-asparagus"; erect, much branched sinuous shrub to 6 ft. high, from swollen roots; gray stems with zigzag branches, the cladodes (branches simulating leaves) thread-like $\frac{1}{4}-\frac{3}{4}$ in. long, in dense clusters, bright green. **C**: 15   p. 253

plumosus: see setaceus

setaceus; plumosus in hort.; (So. Africa), "Asparagus fern"; ornamental sub-shrub with lacy fronds bearing branchlets set with rich green needle-like cladodes, and arranged in a horizontal plane, on thin wiry stems with sharp prickles; in time climbing or straggling upward. Neat little, durable potplant tolerating much abuse. The fern-like strands are cut for use in flower arrangements. **C**: 3 p. 253

ASPASIA (*Orchidaceae*)

variegata (Odontoglossum variegatum) (Panama, Trinidad, Guianas, Brazil); choice epiphyte of compact habit, with 1 to 2-leaved oblong 2-edged pseudobulbs bearing 6 in. leaves; the inflorescence a cluster of handsome, thick-textured and long-lived $2\frac{1}{2}$ in. flowers, their sepals and petals green with transverse purple-brown bars, the 3-lobed lip dotted white with violet; blooming late winter to spring. Easily grown and a willing bloomer recommended for the hobbyist; deliciously sweet-scented in the morning. By division. **C**: 72   p. 457

ASPIDISTRA (*Liliaceae*)

elatior (lurida) (China), "Cast-iron plant" or "Parlor palm"; old-fashioned tough-leathery foliage plant with thick roots and blackish-green, shining oblong basal leaves to $2\frac{1}{2}$ ft. long, narrowed to a channeled stalk; purple bell-shaped flowers at the surface of the ground. Ideal for cool, unfavorable locations, very tolerant of neglect. By division. **C**: 18   p. 202

elatior 'Variegata', "Variegated cast-iron plant"; attractive variegated form having leaves alternately striped and banded green and white in varied widths. **C**: 17   p. 191

ASPIDIUM: see Polystichum, Rumohra

ASPLENIUM (*Filices*); the "Spleenworts" are a large group of widely distributed ferns belonging to the family Polypodiaceae, both from temperate regions and therefore hardy, and others tropical and grown indoors. They are of small or medium size, with short rhizome, and fronds that may be undivided as in the Birdsnest fern or feathery 2 or 3 pinnate, but their texture is leathery and usually glossy green, and of great decorative value as indoor plants. Furthermore, there are some that delight the plant lover by being viviparous, literally giving birth to young plantlets on the older leaves as in the Motherferns. Aspleniums are not difficult, but of course wish to receive adequate moisture and warmth. Propagated from spores, or division of older plants; in viviparous kinds by pegging the fronds carrying adventitious buds or bulbils down onto moist peatmoss, to allow them to form roots and develop into plantlets.

bulbiferum (New Zealand, Australia, Malaya), "Mother spleenwort" or "Hen and chicken fern"; with arching bipinnate or tripinnate fronds to 3 ft. long, borne on wiry, grooved black stalks, the pinnae fresh green, and much larger than viviparum, and not as finely and deeply lacily cut, the segments becoming linear only when spore-bearing; bulbils or plantlets are produced on upper surface of frond. Best in ferneries. **C**: 6   p. 249

nidus (nidus-avis) (India to Queensland and Japan), the "Birdsnest fern"; great epiphytic rosette of simple oblanceolate, stiffly spreading, friendly green fronds anywhere from 1 to 3 ft. long, of thin leathery texture with prominent blackish midrib and wavy margins, rising from a crown densely clothed with black scales. One of the most interesting and attractive of ferns for pots, keeping surprisingly well but should be kept steadily moist and warm, or new fronds may become deformed; avoid drafts; tolerates poor light to 25 fc. Propagated from spores. **C**: 6   p. 238

viviparum (Mauritius, Réunion), the popular "Mother fern"; tufted plant with dark green, finely lacy, arching fronds 1–2 ft. long, on firm stems, the little thread-like linear segments giving rise to tiny bulblets from which develop little plants. A good house plant if given sufficient moisture, and always an interesting conversation piece. The plantlets may be propagated in most-warm enclosure, in peatmoss and sand, cutting off with part of the frond when strong enough, and placing the older part below ground until rooted. **C**: 56   p. 249, 529

ASTER: see Callistephus

ASTILBE (*Saxifragaceae*); the japonica dwarf hybrids known by florists as "Spiraea" or "False spiraea", "Silver sheath", "Meadow sweet" and "Goat's beard"; are winter-hardy perennials not to be confused with the woody deciduous Spiraea and well known in northern gardens. Astilbes are valued for the airy quality of their plumy spires of flowers and attractive ferny foliage, blooming according to variety with tiny white, pink, or red flowers from May to July. They are often grown in tubs; or forced in pots for Easter, but in order to respond the clumps of roots must be big enough and have been exposed to a period of frost or cold storage before potting, or they may become blind. When in growth Astilbes are very thirsty and might be kept standing in saucers with water. Propagated by division.

japonica 'Deutschland'; an excellent, robust florists' "Spiraea" to 2 ft. high, normally blooming in June-July with long graceful plumes of pure white flowers, and much favored for Easter forcing. The species japonica from which these florists' varieties are derived is at home in Japan. **C**: 76   p. 402

japonica 'Gladstone', an early-blooming (June), compact florists' "Spiraea"; robust, winter-hardy, hybrid perennial, about 12–15 in. high, very showy and attractive with its fluffy and plumy pyramids of white flowers carried on slender, wiry stalks well above the fern-like, pinnate foliage; a charming potplant for the Easter season, and willing forcer. **C**: 76   p. 372, 504

japonica rubens (Japan), "Rosy florists' spiraea"; a rose-colored variety of the normally white-flowering species; winter-hardy perennial with dense clumps of fibrous roots, sprouting lacy, much divided fresh green foliage, and stiff-wiry stalks with branched spires 1½ to 2 ft. high, bearing a profusion of tiny rosy-crimson flowers in June-July, a colored counterpart to the white "Silver sheath", A. japonica. Not as easy or safe to force as 'Gladstone', and unless strong enough apt to come blind. **C**: 76   p. 418

ASTROPHYTUM (*Cactaceae*); the "Star cacti" or "Bishop's caps" are unique and very attractive Mexican cacti of small habit, with more or less flattened ribs and pebbled or flecked surface, with a few weak, or no spines at all; during summer large and showy flowers appear from the top, gaily colored yellow with reddish center. Astrophytum are of easy culture; they need soil containing limestone; moist and warm in summer but with some protection from burning sun; during winter cool at 45° F. and dry. From seed.

asterias (No. Mexico), "Sand dollar"; interesting low, dome-shaped body to 3 in. across, green tinged coppery, and covered with white scales; grooved on top in shape of a star, the areoles tufted with white wool; flowers yellow with red throat. **C**: 1   p. 160, 165*

capricorne (No. Mexico), "Goat's horn"; hard green globe stretching to 10 in. high, attractive with some silver marking, the 7 to 8 high ribs with contorted spines; spreading yellow flower with red throat. **C**: 1   p. 160

myriostigma (C. Mexico), "Bishop's cap" or "Star cactus"; very attractive small globe with usually 5 prominent ribs without spines, growing to 8 in. dia. with age; entirely covered with small white spots giving it a stone-like appearance; large yellow flowers. **C**: 1   p. 160

ornatum (Mexico), "Star-cactus" or "Monk's hood"; small and hard plant subglobose, becoming cylindric with age to 1 ft. high, or even taller, and having 8 prominent spiral folds green, beautifully studded with silvery spots; sharp, talon-like spines; beautiful clear lemon-yellow flowers. **C**: 1   p. 160

ATHYRIUM (*Filices: Polypodiaceae*)

filix-foemina (Arctic and Temperate zones of Europe,

Asia, No. America and Southern So. America), "Lady fern"; variable hardy fern of the humusy forest floors, with tufted, graceful feathery, herbaceous fronds 2 to 3-pinnate with brownish stalks, 1½ to 3 ft. high, scaly below; the bright green leaflets deeply toothed and with rolled-in margins. Cultivated in many named horticultural forms, especially in England. Propagated by division. **C**: 74   p. 250

goeringianum pictum (Japan), "Miniature silver-fern"; little tufted fern with pretty, variegated fronds, which are spearshaped and pinnate, the segments toothed, the stalks wine-red, and a band of gray down through each lateral pinnae. Winter-hardy, and deciduous in temperate regions. **C**: 84   p. 194

AUCUBA (*Cornaceae*)

japonica 'Picturata', also known in horticulture as "Goldiana", the "Golden laurel"; mutant with large leaves almost entirely golden yellow and with only the margins light green, and occasionally dotted yellow. A truly showy colorform although more easily subject to leafscorch or spotting from wetness, due to the absence of chlorophyll. **C**: 27   p. 295

japonica serratifolia, "Sawtooth laurel"; vigorous bushy variety distinguished by its saw-edged foliage.; the willowy stems densely covered with rather long and narrow, thin-leathery lanceolate leaves to 9 in. long, waxy, plain blackish-green; the margins coarsely toothed; female plants are charmingly decorated with clusters of glossy red fruit; propagated from cuttings. **C**: 27   p. 292

japonica variegata (Himalayas to Japan), the "Gold-dust plant"; evergreen shrub of rounded, bushy shape to 6 and 12 ft. high, with willowy branches, and opposite elliptic thick-leathery leaves 6 to 8 in. long, shining dark green more or less blotched, spotted, or variegated yellow, toothed above the middle; purple flowers near base; the female plants with scarlet berries, but the flowers must be pollinized from male plants. Fairly winter-hardy. Widely used decorator plant tolerant of neglect, but preferring cool locations and protection from sun, and ample watering. Easy from cuttings which may be rooted in water. **C**: 27   p. 295

AZALEA (*Ericaceae*); members of the Heath family, Azaleas are amongst the most beautiful and widely used flowering shrubs in the garden, and forced in pots for bloom indoors. Linnaeus himself established the genus AZALEA in 1735, and in addition in 1753 the genus RHODODENDRON. As then understood, Azaleas were deciduous in North America with funnel-shaped flowers having 5 stamens, Rhododendron evergreen, with 10 stamens. It has been found since, however, that there are Azaleas that are evergreen (in E. Asia) and with 10 stamens, and Salisbury in 1796 pointed out the technical similarity between the two genera and combined them both under Rhododendron, leaving Azaleas a sub group only. Yet, in horticultural usage and in literature the name Azalea is firmly fixed, and for practical reasons two main groups are so listed: the ever-green, mostly small-leaved kinds from Asia, highly hybri-dized and including the tender Belgian indica, flowered in pots from December to May; and the hardy deciduous types blooming in our gardens, or forced for flower shows, in a riot of bright colors including yellow and orange. All Azaleas like a humusy, acid soil, moisture, and cool en-vironment. Propagated from soft cuttings, or grafting. The evergreen Azalea hybrids, grown in pots, fall into several groups according to special characters, size of flowers or hardiness. Foremost are the "Belgian Indica" descendants of Rhododendron simsii, native in moist-warm subtropical China, Yunnan and along the Yangtse where they grow to 12 ft. high, with deep rosy-red flowers. These simsii hybrids were developed primarily for their large and showy double flowers, principally in Belgium for forcing in pots; usually grafted on understock of phoenicea concinna for better growth and longer life. Not winter-hardy. Some of these hybrids were selected and acclimated for planting in our southern states for vigor, resistance to sun, and occasional cold down to 20° F. and are known as "Southern Indica", mostly with single flowers. Widely planted in gardens and grown in pots are the so-called "Kurume hybrids" which are compact, small-flowered evergreen azaleas originating on Kyushu Island in southern Japan during the Meiji period (1868–1912) from species involving Rhododendron obtusum, kaempferi, kiusianum and sataense. They are fairly winter-hardy to about 5 or 10° F.

The "Pericat hybrids" were introduced from Pennyslvania in 1931 and are intermediate between Belgian indica and Kurume. Not as hardy as Kurume, but lovely in pots, with flowers opening like rosebuds. The "Rutherfordiana" group of hybrids was introduced in 1937; they are of vigorous habit with long-lasting flowers borne in trusses, a character derived from their parentage of Rhododendron proper, indica alba, Belgian indica, macrantha and kurume.

'Alaska' (Rutherfordiana hyb. 1937) ; (Vervaeneana alba x Snow, with Rhododendron), very satisfactory hybrid of vigorous habit; dark elliptic foliage, and trusses of medium-large 2½ in. semi-double (hose-in-hose) pure white flowers; early forcer, and after a period of cooling can be flowered for Christmas with supplementary light. Not hardy. **C**: 70 p. 366

'Ambrosiana' (simsii hyb.) ('Mad. Petrick x Reinhold Ambrosius); a superb early flowering, fairly new German cultivar easily forced for Christmas; vigorous grower with obovate glossy leaves and 3 in. double flowers like little roses, glowing crimson red. **C**: 70 p. 367

'Concinna' (phoenicea), the "Queen Elizabeth azalea"; according to Haerens Tuinbouw probably a strong growing erect cultivar of 'Phoenicea' which is a variety of Rhododendron pulchrum originally introduced from China. This 'Concinna' has replaced 'Phoenicea' as understock for grafting of tender Rhododendron simsii hybrids, commonly known as "Belgian Indica azaleas"; a prolific bush with hairy branches having elliptic or oblanceolate leaves 2¼ in. or over, matte dark green above, glossy beneath, and thinly furnished with appressed brown hairs on both sides; flowers large when grown under glass, 2 to 3 in. diameter, wide-open, single, purplish crimson rose, spotted crimson on upper petals, the margins lightly wavy. **C**: 70 p. 502

'Constance', (Rutherfordiana hyb. 1937); vigorous variety with happy green foliage and Rhododendron-like, large trusses of lovely, medium-large 2 to 2¼ in. single flowers, vivid cerise-pink with pale throat and frilled margins; early and mid-season. An excellent gift plant for Sweetheart's Day, Feb. 14. **C**: 70 p. 366

'Coral Bells' (Kirin) (Kurume hyb.); low, freely branching and early blooming plant with shiny, fresh green leaves, covered with a multitude of 1⅛ in. bell-shaped hose-in-hose flowers of dainty silver-pink, shading to coral-pink in center. Very charming as a small pot plant and of a color that appeals to ladies. Fairy hardy. **C**: 82 p. 368

'Delaware Valley White', form of Rhod. mucronatum; a woody semi-evergreen plant of robust, somewhat sparry, tall habit, with shiny obovate light green leaves and large single, pure white 2½ to 3 in. flowers, late blooming; may be held for Mother's Day. Partially decidouus; hardy. **C**: 82 p. 369

'Elizabeth Gable' (Gable hybrid), Rhod. indicum with poukhanense and kaempferi; very hardy spreading bush, with medium size hose-in-hose flowers light purple, and wavy petals spotted carmine; late blooming. Deciduous in gardens. **C**: 82 p. 369

'Haerens Beauty' (simsii hyb. 1950), very attractive 'Paul Schaeme' seedling, similar to 'Roberta', with 3 in. double flowers of good substance and frilled petals soft shell pink with outer areas silver pink; mid-season. **C**: 70 p. 363

'Hershey Red' (Hershey Nurs. about 1945); result of hybridizing with seedlings of Kurume varieties including Hinodegiri; shapely evergreen plant with glossy olive green foliage, and beautiful full round, flaring 2 in. hose-in-hose flowers soft clear rosy-crimson; blooming late and ideal for Mother's Day mid-May. Resembling a small indica it has withstood temperatures to 5° F. but is apparently not bud-hardy in Pittsburgh, and foliage may "burn" if lower than 15° F. **C**: 82 p. 369

'Hexe' (simsii x obtusum amoenum, 1888), a "dwarf indica"; excellent late season potplant, very free-blooming and of small but even habit, with dark green leaves, and long lasting 1¾ in. flowers a glowing crimson, coming alive under light, the hose-in-hose petals recurved and frilled. **C**: 70 p. 365,* 367

'Hino-Crimson' (Kurume hyb.), bushy evergreen similar to Hinodegiri but while its growth is more compact, it is an improvement over Hinodegiri in that its single flowers are a clear crimson red without the bluish overtones of the latter; late bloomer; fairly hardy. **C**: 82 p. 369

'Hinodegiri' (Mist-of-the-Rising-Sun), (Kurume hyb.); an old favorite, bushy evergreen much planted in gardens,

fairly hardy with glossy foliage and small 1½ in. single flowers vivid carmine-red; late blooming, and lends itself to wintering in coldframes without heat under glass for blooming at Mother's Day in mid-May. Fairly winter-hardy to about zero or even −10° F. in the vicinity of New York City, but not hardy in up-state New York. **C**: 82 p. 368

'Lentegroet' (Easter Greetings) (simsii hyb. 1909) (Camille Vervaene x Hexe), similar to Hexe but with larger 2 to 2½ in. flowers, crimson-red semi-double; midseason. **C**: 70 p. 367

'Leopold-Astrid' (Picotee) (simsii hybrid 1933); sport of 'Vervaeneana'; a lovely improvement over 'Albert-Elizabeth', with very pretty 3 in. double flowers white with pink and bordered in a rich salmon-red, the margins prettily frilled; midseason. **C**: 70 p. 367

'Linwood Pink' (Fischer hybrid); vigorous evergreen exceptionally good, but of limited hardiness; medium-large 2 in. hose-in-hose flowers soft rose, similar to 'Roehrs Tradition' but with daintily frilled margins. Especially effective as larger specimens **C**: 82 p. 368

'Madame John Haerens' (simsii hyb. 1907); in the trade as 'Jean Haerens'; an excellent and free-blooming variety of vigorous habit and even growth, the large 3 in. double flowers are elegant and firm, of a pleasing shade of deep rose (midseason or late). **C**: 70 p. 367

'Peggy Ann' (Roehrs-Bauman hyb.); one of the loveliest creations the bees have brought about, between Kurume hybrids and kaempferi; small 1½ in. wide open hose-in-hose flowers, with two rows of white petals edged in rosy-pink like apple blossoms; late blooming; fairly hardy. **C**: 82 p. 368

'Polar Bear' (Beltsville hyb.) (Firefly x Snow); a floriferous hardy type azalea suitable for late blooming; with small 1½ in. flowers pure white, hose-in-hose with flaring petals; shining green foliage **C**: 82 p. 369

'Pride of Detroit' (Pericat hyb.) ; sport of 'Mad. Pericat'; an outstanding variety noted for its color and keeping quality, of spreading growth habit, with large round flaring flowers vivid fiery red, having a double row of petals; somewhat shy in blooming unless heavily fertilized; mid-season to late. **C**: 82 p. 366

'Redwing' (Pericat hyb.); rapidly growing evergreen requiring twice as much pinching as other Azaleas if grown for pot culture, or else they will grow loose, straggly and floppy; large, wide open showy flowers, 3 in. or more, with 2 rows of petals, ruffled at the margins, cerise-red. Flowers larger and more frilled than Lentengroet, but sets buds best only in older specimen. **C**: 82 p. 366

'Roehrs Tradition' (Roehrs-Bauman hyb.); evergreen Kurume-kaempferi cultivar of low, shapely, compact habit, with small shiny leaves which hold well through winter; hose-in-hose 1¼–1½ in. flowers deep carmine-pink, lighter than Roehrs Mother's Day; fairly hardy. An excellent Mother's Day plant. **C**: 82 p. 368

'Southern Charm' (Southern indica azalea), sport of 'Formosa', a phoeniceum hybrid; tall evergreen shrub with large 3½ in. single flowers carmine rose with red spots, the petals rounded and half free; rough hairy, 2½ in. elliptic leaves. Sun tolerant and vigorous. **C**: 70 p. 366

'Sweetheart Supreme' (Pericat hyb. 1931), probably the best liked of the Pericat hybrids, of strong spreading, irregular growth; the buds are rosy-pink unfolding like a dainty sweetheart rose, opening flat to starlike, 2 in. flowers with successively smaller circles of delicate pink petals toward center; midseason to late. **C**: 82 p. 366

'Triomphe' (simsii hyb. 1923), (Mad. Aug. Haerens x Lentegroet); strong grower, budding readily and early-blooming; the 3 in. double flowers are crimson red and beautifully frilled, holding their fresh color for a long time. **C**: 70 p. 367

AZOLLA (Filices: Salviniaceae)
caroliniana (United States to Argentina), "Floating moss" or "Mosquito plant"; mosslike aquatic fern, the floating bodies scarcely 1 inch long; pinnately divided leaf-like stems, with root fibers on the underside of the stems, and minute lobed, densely overlapping leaves pale green to reddish and bearing spores; forms attractive floating patches in pools or aquaria. Best kept in winter in shallow vessels in a light, moderately warm place; in summer growing quickly in an indoor aquarium, or outdoors where it becomes reddish-purple, but may become too invasive there. **C**: 14 p. 252, 519

BABIANA (*Iridaceae*)
   stricta rubrocyanea (So. Africa: Cape), "Baboon flower"; low cormous herb to 1 ft. high, the stems bearing several downy, pleated leaves and clusters of colorful fragrant flowers 2 in. across, with 6 spreading lobes, their basal half rich crimson, brilliant blue toward tip, and normally blooming in May and June. The species B. stricta has flowers blue and white. Allow to go dormant after leaves turn yellow, and keep dry from September to January; corms may be stored like gladiolus. By offsets. **C**: 64   p. 394

BAMBURANTA: see Ctenanthe pilosa

BAMBUSA (*Gramineae*); the true, typical "Bamboos" are tall, woody, hollow-stemmed grasses of imposing, often gigantic appearance, native in the Eastern hemisphere, planted both for ornament, as hedges, and their numerous construction uses in the tropics and subtropics. The polished pipe-like stems are interrupted by a partition at the joints which conspicuously ring the culms. Their leaves are short-stalked, with parallel veins; their small flowers in clustered spikelets but rarely seen in cultivation. Like few other plants, bamboos can best create a truly oriental effect in decor, and they are excellent container plants in bright and cool locations, or on the patio, but easily shed foliage if away from light and fresh air for any extended period; best to rotate plants by taking them outdoors to revive them. Bamboos require a large supply of water if in tubs. Prop. by division of the rhizome with canes. Also classed as bamboos are a number of other genera, such as Pseudosasa, Phyllostachys, Sasa, Arundinaria, Dendrocalamus.
   beecheyana, sometimes known as Sinocalamus beecheyanus (Southeastern China), "Beechey bamboo"; a large running, tender timber bamboo developing thick, bright green culms to 40 ft. high and 4 in. thick, with age and weight gracefully arching; 6 to 10 heart-shaped leaves grouped on a twig, 4 to 7 in. long. This bamboo is one of the important sources of edible bamboo sprouts, cooked and eaten in China and elsewhere. **C**: 14   p. 303
   multiplex (Vietnam), "Oriental hedge bamboo"; a variable woody grass usually 4 to 5 ft., but becoming 12 ft. high, the green reedlike hollow stems bearing graceful twigs with many very small leaves, fern-like, 1 to 6 in. long, glabrous deep green, silver-blue beneath, and with a ring of hair at base of leaf; the older culms become 1 to 1½ in. thick. Keep wet to hold foliage; however may be used for extended periods indoors in cool, bright rooms. Propagate by division. Temperature tolerant to 15° F. **C**: 14   p. 302
   multiplex riviereorum (China), "Chinese goddess bamboo"; graceful clump-type bamboo of smaller habit much grown as hedges in southern China, with willowy culms ½ in. thick and 4 to 6 ft. high, with age becoming 10 ft. high; the canes are entirely green and always solid and less stiffly erect than multiplex; small dainty fern-like leaves 1½ to 2½ in. long, in two ranks on slender branchlets or twigs which curl down toward the tips, and with silvery green reverse. Hardiness to 15° F. **C**: 14   p. 305
   multiplex 'Stripestem', known as the "Fernleaf bamboo", because of its usually small, fern-like foliage; the leaves are occasionally banded with white; yellowish or pinkish solid canes ½ to 1 in. dia., striped green, and forming clumps 10 to 15 ft. high. Hardy to 15° F. **C**: 14   p. 305
   oldhamii, in horticulture sometimes as B. "falcata", or Dendrocalamus latiflorus (So. China, Taiwan), "Clump giant timber bamboo"; handsome ornamental giant bamboo, forming open clumps, with hollow green to yellow culms to 60 ft. high, 3¼ in. thick, successively branching at nodes, the sheaths with deciduous hairs; broad leaves 7 in. × 1½ in., dull green above, almost shiny green beneath. Densely foliaged for good screen effect. Hardy to 20° F. **C**: 14 p. 305
   tuldoides (S.E. China), known as "Punting-pole bamboo", because the canes are used to push junks in Chinese rivers; handsome semi-hardy clump bamboo with straight culms 15 to 20 ft. and on to 50 ft. high, with 2 to 3 in. dia., the broad, rich green leaves to 6 in. long, and green reverse. Prolific producer of slender poles. Endures 20° F. **C**: 14   p. 305
   ventricosa (China), "Buddha's belly bamboo"; a clump bamboo with 7 in. leaves, distinctive with its dark olive culms characteristically swollen between internodes, especially so if potgrown or in poor dry soil and the growth remains stunted 3–6 ft. high; when planted out it will grow

from 15 to 50 ft. high, and the internodes stretch and become straight. Endures 20° F. **C**: 14   p. 305
   vulgaris (Java, and wild in tropical regions of E. and W. Indies, Africa, C. and So. America), the "Feathery Bamboo"; widely grown ornamental bamboo because of its attractiveness and large size; forms open clump with spreading rhizomes, arching culms 20 to 80 ft. high and 4–5 in. thick, hollow with thin walls, usually green, or green at first later yellowish, the joints covered with deciduous brown hairs; dark green leaves 6–9 in. long borne on graceful branches on the upper part of the stem. Adaptable for planting in tubs for at least extended periods for decoration indoors, for that oriental touch. **C**: 14   p. 303
   vulgaris aureo-variegata, the "Golden feathery bamboo"; giant canes of a rich golden yellow color, penciled or banded lengthwise with green; a beautiful sight in the tropical landscape, and very artistic with but a few culms in a tub. **C**: 14   p. 537

BAMBUSA: see also Pseudosasa, Phyllostachys

BARBOSELLA (*Orchidaceae*)
   anaristella (Pleurothallis) (Costa Rica); mat-forming small epiphyte, resembling a Pleurothallis; from June to August literally covered with charming single flowers over 1 in. long, gold-bronze with maroon lip; only the lateral sepals are united to near their tips; the petals and dorsal sepal are free; petals not club-shaped at their tips as in Restrepia. By division. **C**: 72   p. 460

BARLERIA (*Acanthaceae*)
   lupulina (Mauritius), "Hop-headed barleria"; evergreen spiny shrub 2 ft. high, with narrow lanceolate leathery green leaves having prominent pink or red midrib; two pairs of downward-turned thorns at each node; the 5-lobed tubular flowers with 1 in. peach-yellow corolla, in terminal hops-like spikes; blooming in summer. Curious for its inflorescence and handsome foliage; likes fresh air and needs to be pinched for bushy shape. From cuttings, or seed. **C**: 64
                                                              p. 492

BAUHINIA (*Leguminosae*)
   variegata; in horticulture sometimes as B. purpurea (India, China), the purple "Orchid tree" or "Mountain ebony"; small semi-deciduous tropical tree 10 to 25 ft. high, also grown as a shrub with multiple stems, with curious foliage; the thin-leathery leaves connate or cleft beyond the middle, 3 to 4 in. long, dull green; gorgeous 3½ in. axillary flowers like a cattleya orchid, carmine-rose, the petals with dark purple center stripe, the broad lip lined with crimson, fan-like; blooming January to May. An excellent patio tree attractive even for its "twin" leaves, but slow-growing in the greenhouse and difficult to bloom indoors unless kept very sunny. Propagated from seed or semi-ripened cuttings under glass. **C**: 52   p. 490

BEAUCARNEA (*Liliaceae*)
   recurvata (Nolina) (Mexico), the curious "Bottle palm" or "Pony-tail"; tree-like plant with tall trunks to 30 ft. high, swollen balloon-like at base, and carrying a dense rosette of narrow linear pendulous, concave, green leaves to 6-ft. long, rough and thin but not spiny-toothed; panicles of small whitish flowers. An imposing and durable succulent which can store a year-long water supply; related to Yuccas. **C**: 4 or 13   p. 200, 526

BEAUMONTIA (*Apocynaceae*)
   grandiflora (India: Himalayas), "Herald's trumpet"; woody twiner or small tree with opposite, ovate leaves to 8 in. long; young shoots tinted rose and rusty-haired; large, fragrant, showy trumpet-shaped, 5 in. flowers with 5 twisted lobes, white, with dark throat, in terminal clusters on wood from the previous year. Blooms best in winter garden or conservatory. The long shoots are cut back after bloom to form bud-bearing side branches. From soft cuttings. **C**: 54
                                                              p. 277

BEGONIA (*Begoniaceae*); many species and their hybrids in a great tropical genus of foliage and flowering herbs, sappy or succulent, with fibrous, rhizomatous or tuberous roots, around the world. Foliage nearly always lop-sided; flowers slightly irregular, the male and female separate. Most begonias are more or less shade-loving, and enjoy humid surroundings with fresh air. The annual fibrous rooted kinds are grown from seed; the stem-forming fibrous rooted from stem or tip cuttings; the rhizomatous species by division of the rootstocks; the heavy-leaved kinds, as B. rex, from leaf cuttings; the tuberous varieties from division of tubers.

acida (Brazil); rhizomatous species with large, roundish leaves to 1 ft. across, bright apple green, roughly puckered; tall inflorescence with white flowers. **C**: 22.   p. 119

aconitifolia: see sceptrum

x'Alto-Scharff' (bradei x scharffiana), beautiful hirsute, bushy plant with white-hairy, wing-shaped leaves with red, depressed veins, and red beneath; hairy pink blossoms. **C**: 10   p. 109

'Alzasco' (Lucerna seedling); tall, cane-stemmed angel-wing type with blackish-green, satin, ruffled leaves with silver spots, deep purple beneath; bronzy-red flowers in drooping clusters. **C**: 66   p. 101

angularis (Brazil); tall, cane-stemmed plant with drooping branches, long, ovate-pointed, glossy green leaves with grayish-white veins, margin undulate; small white flowers. **C**: 10   p. 102

x argenteo-guttata (albo-picta x olbia), "Trout-begonia"; bushy canestem with angel-wing leaves growing at an angle, waxy, olive green heavily spotted silver, and red beneath; flowers cream with pink. Well-known house-plant. **C**: 22   p. 101

barkeri (sparsipila) (Mexico, Central America), fibrous-rooted plant becoming tremendous when grown outdoors in California, with foliage to 3 ft. across. Completely covered with brown scurf, the lobed, toothed, bright green leaves edged in red and usually depressed in center; large, pink flowers in winter. **C**: 72   p. 105

bartonea (Puerto Rico), "Winter jewel"; delicate, low-growing jewel, with green leaves, overcast with gem-like sheen; tiniest pink blossoms like an ethereal shower. **C**: 10   p. 108

x'Beatrice Haddrell' (boweri x sunderbruchii); magnificent small rhizomatous plant, with star-shaped leaves, boldly marked chartreuse on brown; flowers in tall pink showers. **C**: 10   p.122

x'Bow-Arriola' (boweri x C42); rhizomatous miniature starleaf; satiny green with eyelash edging of intermittent purple markings at edge of leaf or along veins; flowers blush pink. **C**: 72   p. 117

boweri (So. Mexico: Chiapas and Oaxaca), "Miniature eyelash begonia"; rhizomatous plant, bushy with small waxy ovate leaves vivid green, and blackish-brown patches and erect hairs along margin; flowers shell pink. May be temperamental; grows well in sphagnum moss in terrarium. **C**: 60   p. 43*, 117

bradei (laetevirides, laetevirens, macrocarpa pubescens) (Brazil: Sao Paulo); a species formerly known as Alto da Serra' with branches soft-hairy and more upright than secreta, velvety olive green angelwing leaves with red margins, reddish underneath; flowers white with crimson hairs on the outside of petals. **C**: 59   p. 109

x'Braemar' (scharffii x metallica); leggy plant with white-hairy stems and large, orbicular-pointed, dark, lustrous green leaves, glossy red beneath; flowers white with bright pink beard. Resembles scharffiana but without fur coat. **C**: 16   p. 109

x'Catalina' ('Lady Waterlow') (odorata x fuchsioides); slender, spreading, fibrous plant with ovate, toothed leaves glossy green, silvery with pink veins beneath, petioles red; flowers pinkish. **C**: 22   p. 106

x cheimantha, the "Lorraine begonias"; a group of prolific winter-blooming hybrids evolved from B. socotrana x dregei; the first hybrid was introduced as 'Gloire de Lorraine' in France in 1891. The present "Christmas begonias" are much improved and have the advantage of holding their masses of lovely flowers under the trying conditions of heated rooms, especially 'Lady Mac' type, and with some care continue to bloom until Easter. Propagated from young leaf petiole cuttings in winter or from tip cuttings in summer.

x cheimantha 'Dark Lady Mac', "Dark Christmas begonia"; a welcome variant of the old-reliable Lady Mac'; bushy plant, with the rounded, satiny foliage overcast with burnt copper, and the 1½ in. flowers a deep rose, in a large bouquet, in November-December. **C**: 60   p. 385

x cheimantha 'Lady Mac', "Christmas begonia" or "Busy Lizzie begonia"; bushy plant covered in winter with masses of clear pink smallish, but long-lasting flowers.

Cultivar of the old 'Gloire de Lorraine', the first French socotrana x dregei hybrid 1891. **C**: 60   p. 112, 385

x cheimantha 'Marina', "Scandinavian winter begonia"; Danish sport of Solbakken, with deep rose, durable flowers larger than 'Solfheim' but not quite as dark, and contrasting nicely with the yellow anthers; of medium growth habit with rich green, medium leaves; flowers October on to December and later. **C**: 72   p. 112, 385

x cheimantha 'White Christmas', "Improved Turnford Hall begonia"; a freer growing form of 'Turnford Hall', which was a rather delicate mutant of 'Lorraine'; during winter, the foliage is literally snowed under by buds flushed apple blossom and opening to pure white. **C**: 60   p.385

x cheimantha 'White Marina' (Snow Princess); "White Scandinavian begonia"; cultivar of 'Marina' a color form of the winter-begonias from Sweden, vigorous plant bursting with large glistening white flowers, the edges of the petals flushed delicate pink, to 2 in. across, over large fresh green, somewhat brittle, waxy foliage. The Scandinavian winter-begonias grow and feel better at intermediate temperatures and under these conditions remain in bloom a long time. **C**: 60   p. 385

'China Boy'; low-growing dwarf plant; the reddish petioles carrying smooth, crinkled, dark green leaves with red veins above, red with darker veins beneath; flowers in pink and white. **C**: 60   p. 107

x'Cleopatra' (Maphil x Black Beauty), "Mapleleaf begonia"; lovely hybrid, with translucent maple-leaf foliage nile-green with pronounced chocolate-red areas toward margin; rhizomatous; clusters of perfumed, pink flowers. Very attractive; more vigorous and rugged than 'Maphil'; leaves become larger 4 to 5 in. across. **C**: 10   p. 116, 183

coccinea (rubra) (Orgaos, Brazil), "Angelwing begonia"; tall, bamboo-like canestem to 15 ft. high, with ovate leaves glossy green spotted silver, and edged red; drooping clusters of coral-red flowers blooming constantly. **C**: 10   p. 101

conchaefolia 'Zip' (Brazil), "Zip begonia", miniature, rhizomatous species, with petioles brown-fuzzy; waxy green, rounded, cupped peltate leaves, having red sinus spot in center; the margins slightly scalloped; veins brownish beneath; clusters of small whitish flowers. **C**: 10   p. 122

x'Corallina de Lucerna' (Lucerna) (teuscheri x coccinea); tall, branching canestem with large olive green angelwing leaves silver spotted, wine-red beneath; flowers droop in giant clusters. Old house plant. **C**: 16   p. 101

x credneri (scharffiana x metallica), hirsute, bushy plant with thick, broad-ovate pointed, softly white-hairy leaves olive-green, red beneath; larger than scharffiana; flowers pink and bearded. **C**: 21   p. 110

x crestabruchii (manicata crispa x sunderbruchii), "Lettuce-leaf begonia"; curious rhizomatous plant with large, heavy, roundish leaves, glossy bronzy to yellow-green occasionally blotched with cream and very much twisted and frilled; rosy petioles and nerves; flowers pink. Rests after winter-flowering. **C**: 60   p. 119

cubensis (acuminata hort.) (West Indies), "Holly-leaf begonia"; widely known as the "Cuban species"; an attractive miniature fibrous plant with small crinkled, holly-like shiny leaves of darkest metallic green without hairs, narrow and sharply toothed, 2½–3 in. long; snowy white flowers in winter. Nice for baskets. **C**: 10   p. 107

'Dancing Girl'; low, fibrous, angelwing variety with whirling growth of twisted leaves no two alike, olive green with silver spotting, and metallic pink along veins and margin; red flowers. **C**: 59   p. 102

dichroa (Brazil); low, fibrous-rooted species inclined to be lax; ovate-pointed, glossy green leaves silver-spotted when young; wavy, dentate margins; orange flowers. **C**: 10   p. 106

dregei (So. Africa), "Miniature mapleleaf begonia"; semi-tuberous species with succulent, red stems annually; small angled, thin leaves shallowly lobed, bronzy-green with purple veins; flowers white. **C**: 60   p. 106

echinosepala (Brazil); species of medium growth with red cane stems dotted with bright green spots, carrying small, obliquely lance-shaped leaves, dull satiny green on top, younger leaves bright green, shiny, reddish maroon and covered with short hairs below, irregularly crenate to serrate margins. **C**: 9   p. 103

x'Edith M' (boweri x reichenheimii); miniature plant with creeping rhizome, carrying star-shaped leaves in chocolate brown and chartreuse, on red-spotted petioles. **C**: 10   p. 122

egregia (quadrilocularis) (Brazil); fibrous-rooted species with long lanceolate, light green, puckered leaves somewhat cupped and brittle; white flowers with 4-winged seed pods. **C**: 10   p. 104

x**'Elsie M. Frey'** (baumannii x limmingheiana); basket type plant with short stems, freely branching, and metallic green, red-lined leaves hugging the pot; fragrant pink flowers in late winter. **C**: 60   p. 114

eminii (mannii) (W. Africa), "Roseleaf begonia"; fibrous-rooted begonia with toothed, ovate leaves glossy green, pale beneath and tinged red, on long arching stems; flowers white, streaked with red, the ovaries without wings. **C**: 10   p. 104

epipsila (Brazil); low-growing, fibrous-rooted plant with fleshy, roundish, enamel-green leaves, covered with reddish felt beneath; white flowers. **C**: 60   p. 110

x **erythrophylla** (feastii) (manicata x hydrocotylifolia), "Beefsteak begonia"; creeping rhizome, with tough leathery, rounded leaves polished to a high green gloss, violent red beneath; free-flowering pink in winter and spring; withstands most problems. **C**: 16   p. 115

x **ertyhrophylla 'Bunchii'**, "Curly kidney begonia"; rhizomatous; curious decorative mutant with fleshy leaves lighter green, red-tinged and ruffled and crested at the margins, the lobes rise and meet. **C**: 21   p. 118

x **erythrophylla helix** (feastii helix), "Whirlpool begonia"; interesting form with fleshy leaves spiralled like a corkscrew in center, with undulate margins. **C**: 22   p .118

evansiana (China, Japan), "Hardy begonia"; somewhat winter-hardy tuberous species, to 2 feet; tolerating some frost; olive green, ovate leaves, veined purple beneath; large, flesh-colored flowers. Attractive garden plant. Propagates by axillary bulblets. **C**: 34   p. 105

feastii: see erythrophylla

feastii helix: see erythrophylla helix

fernandoi-costa (Brazil); fleshy, low, fibrous-rooted plant with large, roundish, soft-hairy, bright green leaves, rose-pink beneath; erect clusters of white flowers, each forming a perfect cross. **C**: 72   p. 108

'Fischer's ricinifolia'; pink flowered ricinifolia seedling, with low, creeping rhizome; leaves somewhat smaller than 'Ricinifolia' and less deeply lobed, red-tinged underneath; habit bushier and more compact. **C**: 10   p. 120

foliosa (Colombia), "Fernleaf begonia"; willowy plant with drooping branches; fibrous; tiny, waxy, bronzy-green, oval leaves notched toward tip; small blush-white flowers. **C**: 16   p. 107

x**'Frutescans'** (fruticosa seedling); low, fibrous-rooted foliage begonia, with small, thin-leathery, olive-green leaves, wavy and cupped to show red underneath; flowers white. **C**: 22   p. 107

fuchsioides floribunda (Mexico), "Fuchsia begonia"; fibrous-rooted plant with slender arching stems 2–3 ft. high; very small oblique-ovate toothed leaves to 1½ in. long, waxy dark green; nodding fuchsia-like red flowers. Old-time houseplant. **C**: 16   p. 107

x **fuscomaculata** (rubella) (heracleifolia x daedalea); small rhizomatous, hairy plant with lobed leaves smooth; bronze-green with pale veins and chocolate spots; flowers pink. **C**: 72   p. 118

glaucophylla: see limmingheiana

goegoensis (Sumatra), "Fire king begonia"; low-growing gem on creeping rhizome, silky leaves round peltate like water-lily pads, puckered, dark bronze green with lighter veins, reddish beneath; pink flowers. A beauty that needs warmth and humidity. **C**: 60   p. 116, 183

haageana: see scharffii

x **heracleicotyle** (heracleifolia x hydrocotylifolia), "Mrs. Townsend begonia"; rhizomatous; fleshy leaves a 7-pointed star of fresh, clean waxy-green, dark edged; beautiful deep pink flowers. Good house plant. **C**:10 p.115

heracleifolia (Mexico), "Star begonia" or "Parsnip begonia"; creeping rhizome with large, palmately lobed leaves bristly-hairy, bronzy green with irregular green patches along veins, flowers pink. **C**: 4   p. 118

heracleifolia nigricans (Mexico); handsome, rhizomatous plant smaller than sunderbruchii, the palmately lobed leaves blackish-green with contrasting pale green areas along main veins. **C**: 10   p. 118

x **hiemalis**; a class of autumn and early winter-blooming hybrids of B. socotrana with summer-flowering tuberous types, originated by Veitch (1883) and Clibrans (1908) in England, followed by Dutch specialists in Holland about 1946 as the 'Elatior' begonia; semi-tuberous plants pre-terring moderate warmth. Unfortunately quickly dropping flowers in superheated rooms. Prop. from summer tip or winter leaf-cuttings.

x **hiemalis 'Apricot Beauty'**, "Apricot winter begonia"; early winter-blooming, semi-tuberous plant, free grower large and spready, and with semi-double, salmon-pink flowers flushed rosy-red. **C**: 72   p. 385

x **hiemalis 'Emita'**, (Veitch 1910), "Rosy winter begonia"; vigorous, winter-flowering variety with very large, 3 in. coppery-orange single flowers. **C**: 72   p. 385

x **hiemalis 'The President'**, "Dutch Christmas begonia"; one of the most desirable commercial socotrana x tuberous hybrids, a good keeper, but subject to mildew; double flowers brilliant crimson-red to 2 in. across, a beauty for Christmas. Loves fresh air. **C**: 72.   p. 112, 385

x **hiemalis 'Rieger's Schwabenland'** (1964, Pat.). Robust plant with pointed, oblique leaves metallic deep green, and large, glowing scarlet flowers to 5 cm. dia., with contrasting yellow stamens. The new German Rieger "Elatior" strain is distinguished by its bushy, free-growing and floriferous habit, with flowers in vivid new colors of superior keeping quality, blooming for 8–10 weeks even in the home, normally in fall-winter. Prop. by top or leaf cuttings. **C**: 72   p. 112

hispida cucullifera (Sao Paulo), "Piggyback begonia"; interesting, fibrous-rooted species with large, velvety, maplelike, pale green leaves producing adventitious leaflets on top surface of the leaf along the veins; red petioles covered with white hairs; white flowers. Conversation piece. **C**: 10   p. 108, 529

hydrocotylifolia (Mexico), "Miniature pond-lily"; hairy, rhizomatous plant with small, rounded, thick, waxy 2–2½ in. leaves, light olive green with dark veins; small rosy flowers. **C**: 22   p. 108

'Immense', rhizomatous ricinifolia seedling; with large star-like, acutely lobed, light green leaves, waxy and flat, some short bristles on surface, edged red; leaf-stalks with red scale-like hairs; flowers pink. **C**: 10   p. 118

imperialis (Mexico), "Carpet begonia" or "Imperial begonia"; choice, dwarf rhizomatous, hairy plant, with small heart-shaped, plushy, pebbly leaves in dark green to bronze, with lighter areas along veins; small white flowers. Needs warmth and humidity. **C**: 62   p. 117, 183

imperialis smaragdina (Mexico), "Green carpet begonia"; differs from imperialis in its larger leaf with lighter, emerald-green leaf color, with only occasional slight variegation. **C**: 56   p. 117

incana (peltata) (Mexico); thick-stemmed plant woolly white with hoary scurf; fleshy peltate, white-tomentose leaves; drooping white flowers. **C**: 7   p. 109

incarnata sandersii (subpeltata, insignis) (Mexico); delightful, bushy, medium tall fibrous rooted plant with apple-green leaves lobed and crinkled at margins; big, graceful pink flowers. **C**: 19   p. 103

involucrata (Costa Rica); bushy plant with brown stems; fibrous rooted; velvety green leaves lobed and toothed, lightly white-tomentose; fl. white and fragrant. **C**: 10 p. 110

isoptera hirsuta (E. Indies); fibrous-rooted plant with erect stems but less robust than isoptera, the shining green leaves smaller, very oblique-ovate, with toothed red margins on hairy petioles; flowers greenish with pink. **C**: 10   p. 103

x**'Joe Hayden'** (mazae x reichenheimii), "Black begonia"; upright rhizomatous hybrid, with large star-like, satiny, bronzy, black-green lobed leaves, with cream spot in center, red beneath; winter flowering red. **C**: 10   p. 121

kellermannii (Guatemala); dainty, fibrous-rooted species with peltate, cupped, oval leaves yellow-green to rose-colored, and felt-like white-scurfy beneath; flowers pink. **C**: 7   p. 109

kenworthyi (Mexico: Chiapas); distinctive, slow-growing, upright rhizomatous species with smooth, fleshy, slate-colored or bluish green, 5 lobed leaves with lighter veins, looking very much like an ivy leaf, up to 10 in. in size; flowers faintly pink in winter. **C**: 9   p. 115

x**'Kumwha'** (kenworthyi x reichenheimii); huge, vigorous rhizomatous plant, with large, dull dark green, star-shaped leaves; free-flowering habit with clusters of pink flowers above foliage. **C**: 10   p. 121

leptotricha (Paraguay), "Woolly bear"; dwarf plant with stout, felted, succulent stems and dark green, oval, thickish leaves, the underside thickly felted with brown; flowers ivory; everblooming. **C**: 15   p. 109

'Leslie Lynn' (Lexington x dayii); sturdy rhizomatous hybrid, with large dark lustrous satiny, palmately lobed, star-like leaves cut into 5–7 pointed fingers, the edges serrate and ruffled. **C**: 10   p. 120

x lettonica (heracleifolia x nelumbiifolia); massive-looking plant with thick, erect rhizome bearing large ovate, light green leaves, with red-toothed and lobed margin, red beneath and brown-pubescent; flowers pink. **C**: 22   p. 119

limmingheiana (glaucophylla) (Brazil), "Shrimp begonia"; trailing with slender stems, glossy light green leaves with wavy margin; coral red flowers at end of branches; ideal plant for baskets. **C**: 16   p. 114, 274

listida (So. Brazil); attractive fibrous-rooted miniature angelwing, with narrow oblique-ovate pointed waxy leaves to 6 in. long; olive green with bronzy cast; chartreuse with cream along center; red reverse, on hairy leaf stalks. **C**: 59   p. 122

'Lucerna': see 'Corallina de Lucerna'

luxurians (Chile), "Palm leaf begonia"; ornamental, tall fibrous-rooted plant sparsely rough-hairy; leaves peltate-palmate, with up to 17 narrow, serrate leaflets, reddish above; small cream, hirsute flowers. **C**: 12   p. 103

x'Mac-Alice' (imperialis x macdougallii); handsome plant with procumbent rhizome; rough, pimpled star-leaves with 4–5 points, bluish-green overlaid with silver. **C**: 10   p. 120

macdougallii (Mexico: Chiapas); creeping rhizomatous stem with long, reddish petioles bearing the enormous, palmately compound, waxy leaves, consisting of 7–10 stalked segments like a waving palm-tree; the outer ones sickle-shaped, bronzy green; red beneath, and with toothed margins. **C**: 10   p. 121

macrocarpa: see secreta

manicata 'Aureo-maculata', "Leopard begonia"; robust plant with stout, ascending rhizome; the large fleshy green leaves are smooth and blotched yellow or ivory, and occasionally rose-red; reddish beneath. Characteristic collar of red bristles at top of leaf-stalk. Flowers pink. The species is from Mexico. **C**: 16   p. 115

manicata 'Aureo-maculata crispa'; difficult but beautiful form; the waxy, fresh-green leaves blotched yellow and occasionally rose, the margin thickly crested. Water carefully. **C**: 53   p. 118

mannii: see eminii

x margaritacea (Arthur Mallet x coccinea); distinctive, fibrous-rooted, delicate hybrid with metallic, purplish-pink leaves overlaid with silver; flowers pink. **C**: 62   p. 104

x margaritae (echinosepala x metallica), bushy fibrous plant with soft white-hairy, small, ovate-pointed and toothed, quilted leaves faded bronze green, veins purple beneath; flowers pink with red whiskers. **C**: 21   p. 110

'Marie B. Holley'; branching fibrous-rooted plant with thinnish maple-shaped leaves of glossy bright green; large waxy-white flowers. **C**: 10   p. 107

masoniana (Vietnam or China; introduced from Singapore), the "Iron cross begonia"; one of the most beautiful begonias in cultivation; spectacular rhizomatous plant of robust habit, with white-hairy, reddish stems and large roundish, firm, puckered leaves nile-green, marked with contrasting, bold pattern of brown-red, the older leaves overlaid with silver, and covered with bristly red hair and red-ciliate; waxy flowers greenish-white with maroon bristles on back; flowering March-May. Propagates from leaf sections. **C**: 16   p. 43*, 116, 183

mazae (Mexico: Chiapas) "Miniature begonia"; small rhizomatous plant, with red and white creeping stalk; roundish satiny leaves, shaded in various greens to bronzy; paler veins in center; flowers pink, spotted red. Keep fairly dry, with perfect drainage. **C**: 9   p. 117

mazae viridis 'Stitchleaf', the lovely "Stitchleaf begonia"; small rhizomatous plant with satiny, heart-shaped, light green leaves, with purple marks like stitches along margin; flowers pink. **C**: 9   p. 117

'Medora', the miniature "Troutleaf begonia"; slender cane-stemmed, leaves triangular lance-shaped, glossy green, spotted silver, margins wavy; large pink flowers. **C**: 16   p. 102

'Mrs. Fred Scripps' (scharffiana x luxurians); fibrous-rooted, vigorous, hairy; leaves velvety forest-green with red veins, deeply lobed, red beneath; basal lobe sometimes forms separate leaflet, like a finger; small pinkish, hirsute flowers. Distinctive. **C**: 22   p. 110

metallica (Brazil: Bahia), "Metallic-leaf begonia";

bushy, fibrous-rooted, silver-hairy plant with small broad-ovate pointed leaves metallic olive-green, and depressed purple veins, red-veined beneath; showy pink bristly flowers. **C**: 16   p. 110

'Nelly Bly' (metallica cypraea seedling); bushy fibrous plant, hairy; quilted leaves ovate, dark green and toothed, red beneath; flowers pink, red-bearded. **C**: 13   p. 110

nelumbiifolia (macrophylla) (Mexico), "Pond lily" or "Lily pad begonia"; rhizomatous species with large, peltate leaves attached to stem in lotus-leaf fashion, green and rounded, to 1 ft. long, hairy on nerves beneath; flowers pinkish. **C**: 10   p. 119

nitida rosea (odorata rosea); fine form of minor (syn: nitida) from Jamaica; erect stems to 3 ft. high, leaves smooth and rather fleshy, oblique heart-shaped, medium green, to 7 in. long; flowers clear pink. **C**: 10   p. 104

odorata (nitida odorata) (Jamaica); fibrous-rooted, bushy plant to 5 ft. tall, with fleshy, reniform, glossy green leaves; rosy-pink flower clusters, fragrant. **C**: 20   p. 103

olbia (Brazil), "Bronze mapleleaf"; bushy, fibrous-rooted plant with thin, lobed leaves bronzy green, and with black-green veins; scattered white-hairy; red beneath; the surface shimmering in light; flowers greenish-white. **C**: 60   p. 106

olsoniae (vellozoana hort.) (Brazil); lush, low rhizomatous species, with large fleshy leaves in changeable colors bronze to green, contrasting with striking ivory veins; flowers whitish, outer edge sometimes rosy. **C**: 10   p. 121

x'Orange-Rubra' (dichroa x Coral Rubra); cane-stemmed; large leaves obliquely ovate, of the angelwing type, smooth green; silver spotted when young; lacquered orange-red flowers. **C**: 4   p. 106

palmifolia: see vitifolia

parilis (Brazil), "Zig-zag begonia"; stiff, upright, well branched grower to about 3 ft., branches zig-zagging between nodes, old branches covered with brown scurf; leaves long, narrow, shining green on top, red under; long, arching stems with flowers in white with predominant yellow stamens; when in bud, so tight that the cluster resembles a snow ball. **C**: 10   p. 105

'Paul Bruant' (heracleifolia longipila x Frutescans); interesting old hybrid with stout stem; pointed leaves lobed and toothed, glossy green, on red petioles; pink flowers in winter. Produces adventitious leaves and new plants on the stem, but these are difficult to grow on. **C**: 10   p. 120

paulensis (Brazil); showy rhizomatous species, with large round, peltate, quilted leaves, waxy and bristly, fresh-green with pale veins which are red beneath; the bright red, creamy flowers bearded with wine-colored hairs. Grows well in sphagnum. **C**: 60   p. 121

x pearlii; rhizomatous imperialis hybrid; low, bushy plant, with coarsely bristly, roundish leaves fresh jade-green and pebbled, suffused with a pearly sheen. Hybrid with B. rex. **C**: 60   p. 117

'Phyllomaniaca' (incarnata x manicata), "Crazy-leaf begonia"; fibrous-rooted, branching plant with narrowly oblique-ovate, twisted leaves to 7 in., side-lobed and toothed, green on both sides; producing from stems and all over leaves many buds and leafy growths; flowers pale pink. **C**: 60   p. 107

'Picta rosea'; probably coccinea form; cane-stem; narrow waxy, angelwing leaves light olive green with silver spots; pendant clusters of salmon-rose flowers. **C**: 16 p. 101

'Pink Parade' (dichroa seedling); angelwing begonia of upright, branching habit with oval-pointed waxy, bronze leaves thickly silver-spotted; shrimp-pink fl. **C**:10 p. 102

x'Pink Spot Lucerna' (Lucerna x dichroa); robust cane-stemmed plant with dark angelwing leaves, accented by pink polka dots; enormous clusters of crimson flowers. **C**: 10   p. 102

popenoei (Honduras); thick ascending rhizome carrying large, rounded, bright green leaves, sparsely hairy, the margins reddish and toothed; large, white fl. **C**: 10   p. 119

x'President Carnot' (corallina x olbia); vigorous cane-stem, with ovate, green leaves sometimes silver-spotted, slightly lobed and toothed, red-tinged; trusses of large carmine flowers in constant bloom. **C**: 9   p. 102

'Preussen' (German seedling); sturdy, bushy plant with small ovate-pointed, coppery olive green leaves, spotted with silver while young; large delicate pink flowers, blooming nearly year round. **C**: 22   p. 105

"purpurea" hort. (Brazil). also as "McDougallii var. Brazil" in the trade; impressive, large, rhizomatous plant with many fleshy, beet-red petioles carrying palmately compound leaves of up to eight leaflets, sickle-shaped, shiny deep green above and light maroon below, with red veins, and frilly, dentate margins; long, erect flower stems with very large clusters of flowers in chartreuse, edged with pink. **C**: 60 p. 119

pustulata (Mexico), "Blister begonia"; beautiful rhizomatous species of rambling, hanging tendencies, with pebbly leaves off-emerald green with silver veins, underside deep rose. Similar to imperialis, with fewer hairs. Likes moss-lined wire-basket, with warmth and humidity. **C**: 60 p. 116

pustulata argentea (Mexico), "Silver pustulata"; beautiful small, low-growing rhizomatous plant with plush nile-green, puckered leaves richly variegated with silver; flowers greenish. **C**: 62 p. 183

quadrilocularis: see egregia

'Randy' (Verde Grande x Missouri); rhizomatous begonia with large, fresh green seven-pointed, star-shaped leaves, edges stitched in brown, entire leaf often marbled brown; leaf edge and petioles hairy; pink fl. in winter. **C**: 10 p. 120

Rex cultorum hybrids: this aristocratic class of "Painted leaf" or "Fan begonias" is grown and cherished for its gorgeous brocaded foliage, and numberless named cultivars have been produced, vying with one another for the ever different combinations of iridescent and shimmering shades and patterns, from emerald green and bronze to deep maroon, strikingly overlaid or zoned with silver on their large, tapestry-textured oblique-ovate leaves, usually growing basally from a horizontal rhizome; a few grow upright and are branching. These hybrids are very complicated, offsprings of the original B. rex, discovered in Assam in 1856, and other related species; some are crossed with B. diadema, with deeply parted leaves. Flowers are rose, autumn-blooming, and seeds are easily raised. Rex begonias are strictly "shade plants" and while enjoying good daylight, cannot take bright sun; ideal in the north window. Once accustomed to a certain location they should not be moved and turned around. They require warmth and a high degree of humidity. Rex begonias should have a period of rest during winter; receive occasional fertilizer while growing in summertime. Prop. by pieces of the rhizome, or leaf, or leaf section cuttings.

rex 'Autumn Glow'; stem-forming hybrid; with erect, stem-like branches bearing smallish, rex-type bronze foliage overlaid with silver and spotted red. **C**: 60 p. 124

rex 'Comtesse Louise Erdoedy' (rex Alex. von Humboldt x argenteo-cupreata), "Corkscrew begonia"; first hybrid with spiralled leaves; large oblique-ovate, red hairy leaves light olive green and zoned with silvery rose; one or both basal lobes spirally curled. Unusual. **C**: 60 p.123

rex 'Curly Silversweet'; elaborately ruffled and curled mutant; upright grower; the silvery leaves accented with garnet veins, on wine-red petioles. **C**: 60 p. 124

rex 'Fairy'; a silvery, and strong diadema hybrid of shapely, upright habit; lightly lobed, pointed leaf with silver dominant over the puckered area, and tinted pink; dark gray-green along veins. **C**: 60 p. 124

rex 'Frosty Dwarf' (Curly King Edward x Silver Queen); bushy and sturdy Brooklyn Botanic Garden hybrid of dwarf habit, with firm, oblique cordate leaves silvery, and outer edge olive green; pinkish when young; brown-red veins beneath. **C**: 60 p. 123

rex 'Helen Teupel' (rex x diadema), "Diadema hybrid"; bushy plant with long leaves sharply lobed, center and margin gaily patterned dark fuchsia-red, metallic green along veins, silvery-pink areas between. Probably the best of the lacy diadema type. **C**: 10 p. 123, 183

rex 'Her Majesty'; broad, tapering satiny leaves in oriental tones of blackish and reddish purple, with olive green zone frosted with rosy silver blotching. One of the darkest; a collector plant. **C**: 56 p. 124

rex 'Merry Christmas' ('Ruhrthal' or 'Gundi Busch', and similar to 'Reiga'); beautiful smooth leaf with well defined color zones; center velvety blackish-red, broad adjoining area shimmering ruby-red, narrow zone of silver and pink, outer zone forest-green, edged fuchsia-red. Most attractive, and durable. **C**: 60 p. 43*, 99, 123

rex 'Mikado', "Purple fanleaf"; good American hybrid (1916), of bushy habit and enduring the heat of summer; center of leaf purple spilling over into metallic silver, and edged silvery purple. **C**: 62 p. 124, 183

rex 'President' ('Pres. Carnot'); sturdy old hybrid and an easy grower; quilted leaf with base deep moss-green,

and the raised area a clear silver. **C**: 60 p. 123

rex 'Purple Petticoats'; sturdy plant, having leaves covered with silver, except for purple areas along veins, and purplish toward margin; foliage on short petioles forming cups and spiralling, with ruffled margins curling in on both sides. **C**: 60 p. 124

rex 'Robin'; upright and branching; leaves with center area colored ruby, and speckled silver pink, with a clearly outlined outer zone of silver green; maroon beneath. **C**: 60 p. 124

rex 'Silver Dollar', "Baby rex-begonia"; pretty plant of compact habit with many small, roundish leaves to 5½ in. long, the oblique point broader and more blunt than on most rexes: basically silver on top, with dark olive green markings outlining the veins; reverse maroon. **C**: 60 p. 123

rex 'Thrush' (rex x dregei), "Upright miniature rex"; different because it forms upright, stem-like branches; the small maple-shaped leaves deep crimson-red sprinkled all over with fine silver and without any definite zones; prolific bloomer. **C**: 60 p. 124

'Richard Robinson' (macbethii seedling); small, low plant with silvered leaves lobed and finely toothed, center green, on short petioles; flowers white; deciduous in winter. **C**: 60 p. 106

richardsiana (So. Africa), "Serrated mapleleaf begonia"; semi-tuberous, small, succulent plant with thin, light green leaves digitately lobed, and a red spot at the juncture of the long lobes; free-flowering white. **C**: 10 p. 106

x ricinifolia (heracleifolia x peponifolia), "Bronze leaf begonia"; robust old hybrid with erect rhizome and large roundish leaves to 1 foot long, deeply lobed like a maple leaf, fresh moss green with a taffeta sheen on top; reverse pale green, the pale veins with red scales; petioles ringed by red, broad-scaly hairs, especially near apex; pink flowers. Durable old-time parlor begonia. **C**: 22 p. 118

'Ricky Loving' (Billy x sunderbruchii); rhizomatous plant with large, 7-pointed, blackish velvet leaves, laced with silvery-gray along the green veins; underleaf and petioles densely hairy; rose-colored flowers in fall and winter. **C**: 10 p. 120

x'Rosea gigantea' (semperflorens x roezlii); vigorous fibrous-rooted hybrid with roundish, bright glossy green leaves; flowers coral-red, during winter. **C**: 72 p. 107

'Sachsen' (German seedling); low, spreading, fibrous-rooted; small, angelwing leaves dark bronzy green, underside red; flowers rosy red; a free-blooming houseplant. **C**: 16 p. 102

scabrida (Venezuela); decorative, fibrous-rooted, erect bushy plant, with ovate bright green, rough-hairy leaves irregularly toothed; small white hirsute flowers like pearls. **C**: 22 p. 109

sceptrum (aconitifolia 'Hild. Schneider') (Brazil); fibrous-rooted species with erect stems, and large leaves digitately lobed, the segments long and finger-like, green streaked with silver; flowers pinkish-white. **C**: 60 p. 104

scharffiana (Brazil); compact, spreading, densely white-hairy, fibrous-rooted species with red stems, broad ovate leaves, olive-green, red beneath; pale pink hirsute flowers. Of smaller habit than scharffii. **C**: 9 p. 109

scharffii (haageana) (Brazil), "Elephant-ear begonia"; lovely, rugged, white-hairy plant with fibrous roots, and larger 10 in. ovate leaves brownish yellow-green with red veins, and red beneath; large pinkish flowers with beards (hirsute) in clusters. Parent of many velvety beauties. **C**: 16 p. 108

schmidtiana (Brazil); small, fibrous-rooted begonia with reddish, hairy stems and oblique heart-shaped, olive-green, hairy toothed leaves; beneath reddish with green margin; hirsute flowers in pale pink. **C**: 9 p. 108

secreta (macrocarpa) (Trop. W. Africa); fibrous-rooted species with arching or drooping branches, elliptic pointed leaves glossy olive-green, sparsely hairy and toothed; flowers white, tinged pink. **C**: 21 p. 114

semperflorens albo-foliis 'Maine Variety', "Calla-lily begonia"; a sturdier, heavy-leaved cultivar of the calla begonia; succulent plant with glossy, light green leaves; marbled white and flushed bronzy-red toward margins; the terminal growth glistening white suggesting miniature calla lilies; single pink flowers. A dainty-looking beauty. **C**: 11 p. 111

semperflorens flore pleno 'Curlilocks', "Thimble begonia"; eye-catching "thimble-type" wax begonia with deep coppery-bronze leaves and somewhat pendant clusters of large flowers, the pronounced back petals rosy red, and the center with a high crest of stamen turned into petals, first green then bright yellow. **C**: 21   p. 111

semperflorens flore pleno 'Lady Frances', "Rose-begonia"; endearing and enduring plant, with waxy, mahogany-red leaves, and a profusion of ruffly camellia-type flowers fully double bright pink, with white ovary but without prominent back petals, in constant bloom. **C**: 21
p. 111, 380

semperflorens 'Luminosa', "Wax bedding begonia"; compact variety with succulent stems, and flowers soft scarlet, and light green waxy leaves with ciliate red border, turning deep red-brown if grown in full sun. Popular old fibrous begonia much used for bedding. **C**: 19   p. 111

semperflorens 'Pink Pearl', a "Wax begonia"; compact growing F1 hybrid and a good "bedding" variety of the gracilis type, 5 to 8 in. high, with fresh-green waxy foliage and covered with single flowers bright rose-pink. Excellent also in the sunny window, even for winter-blooming. Don't cover the very fine seed with soil after sowing. **C**: 19
p. 378, 409

serratipetala (New Guinea), "Pink spot angelwing"; beautiful, ornamental species, fibrous-rooted, freely branching, with arching stems carrying small 3 to 4 in. shiny, frilled, pleated, deeply lobed and doubly toothed leaves in dark olive green, with iridescent, deep pink to red spots; deep pink flowers. **C**: 21   p. 100, 101, 183

'Shippy's Garland'; freely branching fibrous-rooted trailer with glossy, grass-green, lanceolate leaves to 4 in. long, slightly undulate and scalloped; abundant flowers cherry rose **C**: 10   p. 114

x'Silver Jewel' (imperialis x pustulata); glamorous hybrid, luxurious as pustulata argentea but sturdier; cordate leaves with silver blisters, streaked with emerald green. **C**: 60   p. 121

x'Silver Star' (caroliniaefolia x liebmannii); outstanding rhizomatous hybrid; star-shaped leaves, deeply lobed, olive green, largely overlaid with iridescent rosy silver and puckered, margins toothed; red beneath; flowers pinkish white. **C**: 22   p. 119

x'Skeezar' (dayii x liebmannii); plant rhizomatous; a favorite because of its strong bushy growth, and attractive, smallish rippled, silver-splashed leaves; flowers greenish-white. **C**: 10   p. 117

sparsipila: see barkeri

x'Spaulding' (boweri x hydrocotylifolia); extra bushy rhizomatous dwarf, with rounded leaves, shading from grass to moss green into velvety black lines, underneath oxblood red, and edged with eyelashes. Pink flowers. **C**: 10   p. 117

strigillosa in hort. (bot. tacanana) (Mexico); small plant, slightly hairy, with rounded waxy, oblique-ovate light green leaves, covered by a network of darker markings, and ciliate; flowers pink. **C**: 60   p. 116

x sunderbruchii (ricinifolia x heracleifolia), "Finger leaf begonia"; a rhizomatous old-fashioned Star begonia, with palmately lobed, bright and bronze-green leaves, and silver bands along nerves; the stems ringed with a collar of red hairs; flowers pinkish. **C**: 21   p. 119

sutherlandii (So. Africa: Natal); slender tuberous plant with drooping branches, lance-shaped, serrate, crisped leaves bright green and with red veins; flowers orange. **C**: 70   p. 114

x'Tea Rose' (odorata alba x dichroa), fibrous-rooted plant with waxy, glossy bright-green leaves; flowers dainty, delicate pink and fragrant. Grows bushy if pinched. **C**: 10   p. 105

'Templinii', a variegated "Crazy leaf begonia"; mutant of phyllomaniaca; fibrous-rooted, self-branching; leaves ovate-pointed, glossy green and blotched yellow, margins ruffled; forms adventitious leaves all over foliage; flowers pink. Baby plants on the leaves have been rooted but often revert to plain green. **C**: 60   p. 119

x'Thurstonii' (metallica x sanguinea); fibrous-rooted, bushy plant with white-haired red stems, and cupped ovate, angelwing leaves glossy, bronzy green, red beneath; hairy flowers pink. Old favorite in shopwindows. **C**: 20   p. 110

x'Tingley Mallet' (rex Eldorado x incarnata purpurea); handsome erect, shrubby plant, brighter than 'Arthur Mallet' and less difficult; oblique-ovate leaves metallic maroon and red-hairy, margins toothed; large pink blooms. Beautiful. **C**: 60   p. 104

x tuberhybrida 'Camellia type', or "Tuberous begonia"; sturdy hybrids of Andean species, with watery stems and brittle, pointed leaves; large, single and double, waxy flowers, white, yellow, orange or red, during summer. 'Camellia type' has giant double, formal flowers. For cool porch or outdoor bedding. **C**: 83   p. 113, 402

x tub. multiflora flore pleno 'Helene Harms', "Lilliput tuberous begonia"; miniature tuberous begonia of compact bushy habit, with smaller, canary yellow double flowers, blooming profusely through the summer; originated from the dwarf Andean species B. davisii, and the velvety marbled foliage of B. pearcei. **C**: 83   p. 113, 418

x tub. multiflora flore pleno 'Maxima', "Dwarf tuberous begonia"; strain developed by crossing the large camellia-flowered type with multiflora; compact plants with rigid stems and dark foliage, covered with many medium-sized double blooms, red, rose, yellow, white. Ideal for bedding. **C**: 72   p. 113, 380

x tub. pendula fl. pl. 'Orange-flowered', "Hanging basket tuberous begonia"; tuberous begonia with drooping stems carrying triangular leaves with frilled, dentate margins, and double, orange flowers with many narrow petals. Good hanging basket begonia. Derives its pendant habit from B. boliviensis. **C**: 72   p. 113

x tub. pendula fl. pl. 'Sunset' (Lloydii), a "Basket begonia"; drooping stems with narrow leaves, and numerous small double pendulous flowers, produced from a single tuber. 'Sunset' is orange, but other colors are red, rose, salmon, white, yellow; derives its pendant habit from B. boliviensis. **C**: 70   p. 114

ulmifolia (Colombia, Venezuela), "Elm-leaf begonia"; tall, rapid-growing, fibrous-rooted, odd species with elm-like ovate leaves rough brown-hairy, double-serrate; small white flowers. **C**: 16   p. 110

valdensium (Brazil), "Philodendron-leaved begonia"; rounded, pretty, satiny, green leaves with prominent light veins, blade flat with undulate ruffled edge. A beautiful cane-stemmed foliage plant. **C**: 60   p. 105

x'Veitch's Carmine' (dregei x coccinea); cane-stemmed, bushy plant with ovate, pointed, green leaves shallowly lobed, red margins; drooping red flower clusters. **C**: 10   p. 104

vellozoana: see olsoniae

venosa (Brazil); fibrous-rooted plant covered with white scurf; stem thick; succulent cupped leaves reniform and appearing frosted; flowers white. **C**: 7   p. 109

x verschaffeltii (manicata x caroliniaefolia), tall and stately like a palm tree; thick erect rhizome; glossy bright green, fleshy leaves shallowly lobed, red-edged and toothed; drooping pink flowers. **C**: 22   p. 119

versicolor (China), the beautiful "Fairy carpet begonia"; with thick corrugated, roundish leaves radiating a design of silver, emerald-green and bronze, and covered with red hairs, small, rhizomatous plant. Needs high humidity, but good drainage; does best in terrarium. **C**: 72   p. 116

x'Viaudii' (duchartrei x pictaviensis); fibrous-rooted, hairy plant with thick, ovate leaves olive green, toothed and cupped; large white flowers. **C**: 9   p. 108

vitifolia (dichotoma, lobulata, palmifolia) (Brazil), "Grapeleaf begonia"; vigorous, handsome begonia with upright, branching stems, hairy petioles carrying broad-ovate, somewhat leathery, pointed to rounded, shallowly lobed leaves, somewhat downy, margins finely toothed; small white and pinkish flowers. **C**. 10   p. 106

x weltonensis (sutherlandii x dregei), "Grapevine begonia"; semi-tuberous bushy plant with light green leaves shallowly lobed and toothed, with purple veins, on wine-red petioles and stems; flowers pink. There is also a white form, 'Alba'. **C**: 60   p. 106

'Whirly-Curly'; heracleifolia nigricans sport; small plant with twisted foliage, leaves grow fan-like, copper with green vein area, red beneath; dwarf habit. **C**: 10   p. 122

x'Zee Bowman' (liebmannii x boweri); dwarf rhizomatous plant, with many small lobed and quilted leaves silvered, and pointing in all directions; flowers pink. **C**: 10
p. 122

BELLIS (*Compositae*)
perennis florepleno (W. Europe, Asia Minor), "English daisy"; low perennial meadow plant grown as a biennial, in the original single form the hardy "Common daisy" of European meadows and lawns; a clump-forming, spreading herbaceous rosette, with creeping rootstock, to 6 in. high, obovate fleshy fresh green basal leaves and more or less double flowers 2 in. across with several layers of white or

rosy florets closing at night; blooming during the cool spring months from March to June. Its cheerful, early bloom made it a favorite bedding plant or for window boxes often planted in fall like pansies, from seed sown in August, easy by division. **C**: 76   p. 411

BELOPERONE (*Acanthaceae*)
   guttata (Mexico), "Shrimp plant"; ever-blooming somewhat shrubby plant 1 to 3 ft. high, with soft-wiry stems; opposite ovate, hairy 2½ in. leaves; curiously the 2-lipped white flowers peek out from beneath showy, overlapping, reddish-brown bracts, in drooping terminal hops-like spikes; blooming nearly continuously throughout the year. A very satisfactory house plant, and a spot of bronzy color in larger containers on roof garden or patio. Remove flower buds until desired size is obtained. From cuttings anytime. **C**: 13
                                                                                        p. 494

BERGENIA (*Saxifragaceae*)
   ciliata ligulata 'Rosea' (in hort. as Saxifraga ligulata), (Nepal Himalayas), the "Winter begonia"; an elegant perennial with thick rootstocks forming clumps of glossy green, leathery, broad-obovate leaves to 12 in. long, ciliate at margins and pale beneath; thick stalks 12 to 18 in. high bear graceful nodding clusters of small pink, begonia-like flowers, deep rose in the variety 'Rosea', blooming in early spring. Somewhat winter-hardy with protection. Prop. by division of rootstocks and stem-sections. **C**: 76   p. 417

BERTOLONIA (*Melastomaceae*); the "Jewel plants" are beautiful tropical South American dwarf herbaceous plants or creepers, grown for their gorgeous brocaded or iridescent foliage, often splashed with silver, and freely blooming with small white or purple flowers. These beauties are strictly delicate tropical conservatory plants, requiring warmth, high humidity and shade, but they are small enough to grow them in a glass terrarium or under a bell jar. Best in small pots, they need open and porous, humusy soil. Propagated by seed, or stem cuttings; also by leaves cut across the veins.
   hirsuta (Trop. America), "Jewel plant"; small herb to 8 in. high, with oval, pubescent leaves vivid green, red-brown along center; flowers white. **C**: 62   p. 185
   marmorata aenea, "Bronze jewel plant"; beautiful little herbaceous plant with quilted ovate leaves velvety moss-green with reddish sheen, shimmering bronze without the silvery-white markings of the species from Ecuador. Underside purple; purple flowers. Very delicate. **C**: 62 p. 185
   'Mosaica'; colorful German cultivar of low habit with beautiful velvety green leaves, distinguished by white longitudinal bands broadest along the midrib, iridescent in several colors; fairly durable. **C**: 62   p. 185
   sanguinea (Trop. South America); rather vigorous, small ornamental plant with hairy, bronzy leaves, prettily silver-banded along middle, and red beneath. **C**: 62   p. 185

BESLERIA (*Gesneriaceae*)
   lutea (West Indies), warmth-loving herbaceous or semi-woody clamberer to 6–10 ft. high, having opposite, thickish smooth, large elliptic leaves with ribs especially prominent underneath; small axillary, tubular swollen, golden yellow flowers, blooming in July. Propagated by cuttings. **C**: 60
                                                                                        p. 439

BIFRENARIA (*Orchidaceae*)
   harrisoniae (Brazil); lovely epiphyte of sturdy habit; the 4-angled pseudobulbs with a solitary plaited leathery leaf to 1 ft. long; basal stalks with 1 or 2 large fleshy, waxy-looking 3 in. flowers, their sepals and petals ivory white or yellowish, and tinged with red, the lip violet-red with yellow hairy callus, blooming from spring to early summer. An exquisite orchid, long-lasting in the living room, easily grown with good light, and freely developing while actively growing; after new pseudobulbs are completed rest for several weeks in cooler spot to aid next season's blooming. By division after flowering. **C**: 22   p. 460

BIGNONIA (*Bignoniaceae*)
   argyraea (So. America: Amazon), "Silvery trumpet vine"; greenhouse vine with tendril climbing wiry stems, ovate-lanceolate leaves deep green with vein-areas silvery gray; toothed at the margins and with a velvety look; purplish beneath. Very attractive for its foliage. **C**: 54 p. 266

BILLBERGIA (*Bromeliaceae*): the popular "Queen's tears" are ornamental prolific epiphytes with graceful mostly

tubular or urn-shaped leaf rosettes, and free-blooming with usually arching stalk carrying a weeping inflorescence very showy with typically bright rose bracts; the slender flowers are large, with petals coiled back, and stamens and pistils extended; their berries are mostly ribbed and lack the brilliant coloring of Aechmea. The inflorescence is short-lived but the foliage is generally quite attractive, barred with silver, splotched or variegated with cream, pink or bronze or covered with silvery scales, or white, waxy powder, and have small spines along their margins. Billbergias are easy-to-care for, undemanding durable house-plants, tolerating all kinds of light and dry air indoors; blooming from summer into winter. Easily propagated from rooted off-shoots.
   x'Albertii' (distachia x nutans), "Friendship plant"; tall clustering tubular hybrid, the offshoots with narrow tapering leaves, dark green and gray-scurfy; inflorescence rosy-red with pendulous green flowers edged blue. Tough and tolerant. **C**: 21   p. 134
   amoena (So. Brazil); fluted rosette consisting of stiff gray-green leaves with pronounced silver cross bands; inflorescence an arching spike with rose bracts and green flowers edged blue. **C**: 9   p. 135
   x'Bob Manda', "Manda's urn plant"; a colorful, compact pyramidalis hybrid with broad leathery leaves in a bold rosette, highly variegated and blotched cream over copper to olive and even purple. **C**: 7   p. 192
   euphemiae (So. Brazil), stiff tubular plant with gray-scurfy, green leaves and gray crossbands; rosy bracts and pendant blue flowers. **C**: 7   p. 131*
   x'Fantasia' (saundersii x pyramidalis), "Marbled rainbow plant"; colorful urn-shaped hybrid with coppery green leaves highly blotched and variegated creamy-white and pink; rose bracts and blue flowers. **C**: 7   p. 134
   forgetii: see Aechmea caudata
   horrida tigrina, "Tigered urn plant"; stocky fluted plant with brown leaves richly tiger-banded white beneath; upright inflorescence with red bract leaves and spidery blue flowers with narrow petals. **C**: 7   p. 135
   macrocalyx (Brazil; Minas Geraes), "Fluted urn"; clustering stiff green tubes with scattered silver bands; inflorescence with showy rosy bracts, and large pale blue flowers on mealy white stalk. **C**: 21   p. 134
   nutans (So. Brazil, Uruguay, Argentina), "Queen's tears" or "Indoor oats"; slender rosette of narrow, silvery bronze foliage, forming clusters. An arching flowerstalk bears the nodding inflorescence of rosy bracts and green flowers edged violet—a tear-drop forming on the stigma. Very tolerant. **C**: 21   p. 134
   pyramidalis concolor (thyrsoidea) (Brazil), "Summer torch"; rosette shaped like a birdsnest: broad, glossy, apple-green leaves; showy but short-lived inflorescence, on mealy-white stem, the head consisting of large red bracts and crimson red flowers tipped purplish, with blue stigma. Very showy in summer bloom. **C**: 9   p. 134
   pyramidalis var. pyramidalis (Brazil); vase-shaped rosette of thin-leathery, glaucous, dark green leaves, and faint gray banding beneath; inflorescence an erect cluster of scarlet flowers tipped blue; bract leaves red. **C**: 9   p. 134
   pyramidalis 'Striata', "Striped urn plant"; an attractive seedling clone raised by M. Foster 1950 from the species collected in Brazil; has broad tomentose leaves blue-green, not glabrous yellow-green as in the type, and in addition striated and variegated at margins with cream; flowering in winter not summer as B. pyramidalis var. pyramidalis. **C**: 7
                                                                                        p. 134

   quintitissima: see Quesnelia liboniana

   rhodocyanea: see Aechmea fasciata
   x'Santa Barbara', "Banded urn plant"; colorful yet tough California hybrid of probably B. distachya, nutans and pyramidalis; cluster-forming tubular rosette of compact habit, scaly gray-green branded ivory suffused with pink. **C**: 9   p. 134
   saundersii (Bahia), "Rainbow plant"; attractive smallish, tubular species with stiff olive to bronzy foliage blotched ivory and tinted pink; inflorescence bracts red, and pendant blue flowers. **C**: 7   p. 134
   tessmanniana (Peru), "Showy billbergia"; long tubular-shaped plant with leaves mottled gray and white; possibly the showiest of all of this genus for its inflorescence which is a giant pendulous spike with red and pink flowers and bracts flaring out into a fabulous show. **C**: 7   p. 148

venezuelana (Venezuela), "Giant urn plant"; bold tubular plant with attractive coppery leaves marked with pronounced crossbands of silver; margins toothed; pendant inflorescence with rosy bracts, sepals white farinose, and petals yellow-green. **C**: 7   p. 135

vittata (S. E. Brazil), "Showy billbergia"; clustering, fluted species with leathery, olive to purplish-brown leaves silver-banded cross-wise; leaning inflorescence with dark flowers and glowing red bract leaves. Formerly known as B. calophylla. **C**: 7   p. 135

zebrina (Brazil: Minas Geraes to Rio Grande do Sul), "Zebra urn"; attractive species with long fluted leaves purplish bronze in strong light, heavily crossbanded silvery white, and armed with thorns; inflorescence with large rosy-red bract leaves and nodding violet flowers, the salmon calyx leaves tipped violet. **C**: 9   p. 135

BIOPHYTUM (Oxalidaceae)

zenkeri (Congo), the "Sensitive life plant"; a small sensitive plant with slender stem 2 to 4 in. high, topped by an umbrella of pinnate leaves 2–3 in. long, occasionally forking, the pinnae short and roundish, dark glossy-green, and which fold back and downward when touched; small yellow flowers. Easy to grow as annuals; a pretty little plant to play with. From seed. **C**: 54   p. 529

BISCHOFIA (Euphorbiaceae)

javanica (trifoliata) (Trop. Asia: Java), the very decorative "Toog tree"; ornamental evergreen tree to 75 ft. high, somewhat deciduous, with large alternate compound leaves of 3 ovate, fleshy leaflets deep green to bronzy, minutely toothed; small greenish flowers without petals, and reddish or blue pea-size fruit on female flowers. Requires warmth. **C**: 1   p. 216

BLECHNUM (Filices: Polypodiaceae)

brasiliense (Brazil, Peru), "Ribfern"; coarse rosette, with age developing a scaly trunk to 3 ft. high, the leathery green fronds to 3 ft. long deeply pinnatifid, widest in the upper third, the midrib broad, with pinnae overlapping and wavy, and coppery when young. Sterile and fertile fronds of similar appearance. Very decorative as potplant but difficult as house plant and must never be allowed to dry out. Prop. from spores. **C**: 62   p. 236

moorei (New Caledonia), known as Lomaria ciliata in horticulture; open rosette of simply-pinnate light green, thin-leathery fronds 8–12 in. long, on blackish hairy stalks, with age forming short trunk; the broad sterile pinnae are ovate-oblong, toothed or lobed, the segments on the fertile fronds almost thread-like narrow linear. Very sensitive to drying out at the rootball when in pots. **C**: 62   p. 236

BLETIA: see Bletilla

BLETILLA (Orchidaceae)

striata, better known in horticulture as Bletia hyacinthina (Vietnam, So. China, Japan); handsome terrestrial orchid 1 to 2 ft. high, growing leafy stems from tuberous rhizomes, bearing 3 to 5 rather thin plaited leaves; light purple flowers with trilobed lip lined by deep purple ridges; the blooms are 1 to 2 in. across, usually not fully opening, in terminal clusters on erect leafless stalk, rising from the center of the new shoots, mostly in June-July. A good orchid for summer on the patio. Easy to grow and fairly winter-hardy with protection. Bletillas bloom best well-established in pots, with fertilizer and plenty of water while growing; when the foliage withers they should be rested until new growth begins in spring. Propagated by division during dormant period. **C**: 71   p. 460

BLOSSFELDIA (Cactaceae)

liliputana (No. Argentina: Salta 6,000 ft.), "Lilliput-cactus"; the tiniest cactus genus; clustering miniature heads only $\frac{1}{2}$ in. dia., dull grayish-green with ribs barely indicated, and spineless; tiny areoles in spirals; flowers very tiny, creamy-white to buttercup-yellow. Should be grafted for better growth. (See photo.) **C**: 13   p. 159

BOEA (Gesneriaceae)

hygroscopica (East Asia: Malaysia), "Oriental streptocarpus"; lovely small fibrous-rooted perennial herb about 6 in. high, close to Streptocarpus and resembling Saintpaulia; fleshy oval, quilted, fresh green leaves 1½–3 in. long, deeply veined and ribbed and covered with short white hairs, the margins pronounced crenate; small $\frac{1}{2}$ in. 5-lobed flowers purplish-violet and wheel-like, showing yellow pollen, and held in clusters above the foliage; if dry, leaves will shrink and curl but spread again when watered. From seed, or crowns developing from base or leaf axils; leaves root easily in vermiculite or peat-perlite. **C**: 60   p. 439

BOEHMERIA: see Myriocarpa

BOMBAX (Bombacaceae)

malabaricum (India to Vietnam), the "Red silk-cotton" or "Kapok tree"; one of the best-known of Indian flowering trees; an imposing big soft wooded tropical tree 75 to 100 ft. high, with buttressed roots, and spiny trunk 8 ft. in dia.; palmate leaves with 5–7 leaflets; brilliant crimson or orange 4 in. flowers, clustered near ends of the whorled branches. The fruit-pod is 4 to 7 in. long, splitting open and disgorging quantities of silky cotton, like kapok but reddish and of inferior quality, used in India as pillow stuffing. The tree makes a magnificent splash of bright red color in spring. Its powerful, voracious roots are able to move stones, as may be seen amongst the ruins of Angkor in Cambodia. **C**: 51   p. 524

BORAGO (Boraginaceae)

officinalis (Europa, No. Africa), "Borage" or "Common borage"; sturdy, luxuriant annual to 2 ft. high, with coarse, oblong leaves 4–6 in. long, stiff-hairy, gray green; beautiful sky-blue nodding starry flowers, attractive to bees. The leaves are edible and taste like cucumbers, they may be used for iced drinks, salads, pickling; or cooked to flavour soups and stews, and may take the place of parsley. From seed. **C**: 25   p. 320

BORONIA (Rutaceae)

elatior (Western Australia), "Scented tall boronia"; leafy evergreen dense shrub to 4 or 5 ft. high, the wiry shoots thickly furnished with spreading hairs, lacy, fresh-green pinnate foliage divided into needle-like leaflets; along the branches the numerous $\frac{1}{4}$ in. nodding globose, very attractive, fragrant flowers carmine-red, and often of different colors on the same plant, blooming March to May. One of the greatest providers of fragrance, and with long lasting flowers, also used for cut. From ripened cuttings, or seed. **C**: 34   p. 501

BORZICACTUS (Cactaceae)

humboldtii (Binghamia) (Peru), "Peruvian candle"; erect stems to 3 ft. long and 2 in. thick, dull green dotted white, with 12–13 ribs notched between the areoles which are dressed with yellow wool and needle spines; night-blooming pink to crimson. **C**: 13   p. 155

BOUGAINVILLEA (Nyctaginaceae)

x buttiana (peruviana x glabra), the clone 'Mrs. Butt' is widely popular as 'Crimson Lake'; a vigorous, pubescent woody clamberer with large recurved spines, broader and thicker ovate leaves than glabra; blooming in big panicles, with the insignificant flowers subtended by corolla-like, showy cordate bracts of bright crimson, cascading in masses of color from late winter to summer. Requiring sun; tolerates drought. Keep cool in winter. Propagate by cuttings from ripened wood. **C**: 1   p. 276

glabra (Brazil), "Paper flower"; strong woody rambler with bright green, smooth leaves slender pointed and with narrow base; flower clusters smaller than spectabilis, the branchlets with showy acute purplish-pink bracts in threes, and imitating flower petals; blooming in summer; more compact than spectabilis. **C**: 1   p. 276

glabra 'Sanderiana variegata', "Variegated flowering bougainvillea"; colorful cultivar with the smooth foliage attractively bordered or variegated with creamy white. However, the variegated foliage seems to detract from the beauty of the flowering bracts. **C**: 1   p. 278

'Harrisii', "Variegated paper-flower"; a strikingly beautiful, though delicate variegated foliage plant, possibly a form of glabra but with smaller foliage and rarely seen to flower; of low bushy habit freely branching, dense with small ovate, thin leaves, friendly green with cream when young, grayish green with age, liberally splashed with glistening white inward from the margins. Flowers white, subtended by purple bracts. **C**: 53   p. 43*, 194

BOUVARDIA (*Rubiaceae*); subtropical American flowering shrubs with wiry-woody branches, opposite leaves, and bearing showy terminal clusters of slender tubular 4-lobed flowers in vivid colors from orange to scarlet, also pink and white, blooming variedly through the year, heaviest in fall and winter, and at one time much used as cut flowers, the white forms for wedding bouquets. Grown in pots, the plants may be plunged outdoors in summer but need a sunny place indoors in October; intermediate temperature of 55–60° F. favors bud initiation, at higher temperature only during short days. After their winter-blooming, the bushes are generally cut back and rested until spring. Propagation of B. longiflora best by cuttings from soft young shoots in January-February; the smaller flowered ternifolia cultivars usually from 1 in. root pieces between February and April.

humboldtii: see longiflora

longiflora, better known amongst florists as humboldtii (Mexico), the "Sweet bouvardia"; beautiful fall-to-winter-flowering shrub 2 to 3 ft. high with woody and flexible branches, opposite, glossy fresh-green leaves, and bearing toward the tip several waxy, salver-form sweet-scented flowers having slender tubes 2 to 4 in. long, opening into spreading lobes of purest white. These glistening snowy blooms are great favorites for weddings as corsages and bouquets and normally bloom beginning in May-June; then prune back for a second crop for fall and winter blooming to February; the flowers will fill a room with the fragrance of orange-blossoms. Prop. from tip cuttings. **C**: 13   p. 496

ternifolia (syn. triphylla) (Mexico; Texas), "Scarlet trompetilla"; subtropical flowering shrub 2 to 3 ft. high of dense but somewhat straggling habit with thin, woody branches, whorled ovate hairy 2 in. leaves, opposite on the branchlets which are terminated by clusters of 1¼ in. tubular flowers of fiery scarlet red, blooming naturally from June to February, but more or less most of the year. Horticultural cultivars are 'Christmas Red' and 'Firechief', but these may be hybrids involving B. longiflora, leiantha as well as ternifolia. Commercially these are mostly bloomed from October to February. Best propagated from root cuttings. **C**: 13 p. 496

ternifolia 'Alba', "White Joy bouvardia"; dainty albino cultivar with clusters of numerous small, glistening white flowers, and with pubescent foliage. Pinched to bloom commercially mainly from October to February, but will flower from June on. **C**: 13   p. 496

ternifolia 'Rosea', "Giant pink bouvardia", a favorite horticultural cultivar grown by florists for cut flowers, with clusters of flowers larger than the species, an exquisite rose-pink with salmon sheen. Bloomed commercially, the first cut from June, second cut end of August, the third from November to February. According to Dr. Encke, this plant may be a hybrid involving B. longiflora, leiantha and ternifolia. **C**: 13   p. 496

versicolor (So. America); small shrub 2 to 3 ft. high, with thin-wiry branches becoming pendant; small narrow ovate dark green leaves 1½ in. long; terminal clusters of charming slender tubular flowers 1 in. long, orange-red and prettily tipped yellow; blooming from July to September. **C**: 14   p. 496

BOWIEA (*Liliaceae*)
volubilis (Schizobasopsis) (So. Africa), "Climbing onion"; succulent, light green bulb 5 to 8 in. dia., growing partially above ground, each year sending up a twining, fresh-green, stringy branched stem with few linear deciduous leaves; small greenish-white flowers. Grown as a curiosity; dry off to rest in summer to October. Propagation by seed, division of bulb, or offsets. **C**: 15   p. 275, 533

BRACHYCHILUM (*Zingiberaceae*)
horsfieldii, sometimes found in horticulture as Alpinia calcarata (India, China); a slender ginger with leafy stems 3–5 ft. high, furnished with sessile, long-sheathed, lanceolate leaves to 1 ft. long, glossy green and leathery; flowers with greenish-white corolla, and white lip; the fleshy fruit a 3-celled red capsule, showing orange when open, with crimson-red seed masses. **C**: 54   p. 206

BRACHYCHITON (*Sterculiaceae*)
rupestris (Sterculia) (E. Australia), the "Bottle tree" or "Barrel tree"; a remarkable semi-deciduous tree to 50 ft. high, inhabiting the dry interior of Queensland; characteristic for its huge bottle-shaped trunk that may swell to 11 ft. in dia.; the spreading branches with variable blackish-green

leaves lanceolate or finger-like divided; hairy bell-shaped flowers; nourishing and edible fruit. The tree is built like a thermos-bottle, an outer compartment holds water, the inner reservoir contains a great quantity of sweet edible, juice-like jelly which is wholesome and nutritious, but also enables the Bottle tree to withstand long periods without rain. **C**: 63   p. 527

x BRAPASIA (*Orchidaceae*)
'Panama' (Brassia caudata x Aspasia principissa); a lovely small bigeneric epiphyte with 2-leaved pseudobulbs, and flowers with long-pointed sepals and petals green and yellowish-white, and broad lip; bl. in autumn. **C**: 72 p. 460

BRASSAIA (*Araliaceae*)
actinophylla (Schefflera macrostachya) (Queensland, New Guinea, Java), the "Queensland umbrella tree", a grand ornamental tub plant in U.S.A. where it is generally known as 'Schefflera actinophylla'; in Queensland becoming a tall tree 50 to 100 ft. high and with buttressed trunks 18 in. or more dia.; in New Guinea I have seen them growing as epiphytes 100 ft. up on rainforest trees, 15 ft. high; the sparry, willowy brownish, later woody branches are each topped with a rosette of successively larger, palmately compound leaves forming umbrella-like symmetrical heads; soft-leathery lacquered-green leaves, when young with 3 to 5, in older plants between 7 and 16 oblong leaflets to 1 ft. long; with pronounced teeth irregularly spaced, especially in mature stage leaves. Terminal inflorescence of several 3–4 ft. straight spikes set with sessile round, dense clusters of honey-laden flowers with fleshy wine-red petals, and which form into the purple fruit. Very satisfactory decorator if kept warm and dry, dropping foliage when cold and wet, never let stand in water. A plant that should not be pampered; hold back on watering if in big container, and allow to become thoroughly dry before watering again. Prefers full sunlight but tolerates shade down to 50 fc. Feed for growth if in sun. Quickly growing from fresh seed which loses viability rapidly. **C**: 3   p. 287, 289

BRASSAVOLA (*Orchidaceae*); showy tropical American epiphytes close to Laelia, with cylindrical, stem-like pseudobulbs topped by a solitary fleshy leaf; the striking flowers are typically green, yellowish or white; sepals and petals are narrow, the showy lip sometimes conspicuously bearded. Brassavolas are easily grown in intermediate temperature with good light and abundant moisture during active growth, followed by a short 2-week rest. The pendulous kinds are best planted on treefern slabs or rafts. Propagated by division, but blooming is best if allowed to develop into specimen plants. Parent of many intergeneric hybrids with Cattleya or Laelia.

cordata (West Indies, especially Jamaica); small epiphyte of pendulous habit with long linear, channeled leaves on cane-like, prostrate pseudobulbs; the long-lasting, fragrant flowers 1¾ in. across white, the sepals and petals pale green, and porcelain-white heart-shaped lip with its tube-like base marked purple; blooming summer to autumn. Should be grown as a hanging plant. **C**: 21   p. 459

digbyana (Laelia) (So. Mexico, Honduras, Guatemala); spectacular white-mealy epiphyte, in habit resembling Cattleya, the pseudobulb bearing a solitary, rigid, glaucous leaf; large and showy, fragrant flowers 4 to 6 in. across, with the narrow petals and sepals greenish-white, tinged purple; the rounded lip cream-white with green throat, strikingly fringed at the margin with a long beard; blooming in spring and summer. Famous parent of "bearded" hybrids with Cattleya and Laelia. **C**: 21   p. 459

nodosa (Mexico, Costa Rica, Jamaica, Colombia to Surinam), the exquisite "Lady of the night"; compact epiphyte with stemlike pseudobulbs continuing into solitary, very fleshy channeled leaf; the charming flowers solitary or in clusters; linear greenish-yellow sepals and petals, and broad, pointed, white lip, very fragrant at night and blooming mostly from September to December but may continue throughout the year. **C**: 21   p. 459

BRASSIA (*Orchidaceae*); these "Spider orchids" are easy-to-grow epiphytes of the American tropics, with 1 to 3-leaved pseudobulbs, and lateral clusters of fantastic flowers often with long tail or thread-like sepals and petals, predominating in yellows, greens and browns, and with a pleasant fragrance. A wonderful, free-blooming house plant even for the amateur. They require abundant moisture, though in smallish pots, during active growth, with circulating air, followed by a 2-week rest. Propagated by division. The long lasting flowers should not be left too long on the plant to prevent shrinking of the bulbs.

gireoudiana (Costa Rica, Panama), a magnificent "Spider orchid"; robust epiphyte with flat pseudobulbs bearing 2 oblong leaves; long arching racemes to 2 ft., with numerous, very spidery flowers, the 6 in. sepals and 2½ in. petals yellow tinted green, spotted brown-red, the pale lemon tip marked purple, blooming from December through spring into fall. A fascinating houseplant with flowers very fragrant and lasting for several weeks. **C**: 22   p. 459

verrucosa (Mexico, Guatemala, Honduras, Venezuela), known as the "Queen's umbrella"; vigorous epiphyte with pseudobulbs each bearing 2 oblong leaves; arching, wiry floral stem with spidery flowers sitting along it like a row of birds; the long, threadlike, 5 in. sepals greenish-yellow, spotted at base with purple, as are the petals; the finely brown-warty lip white. A very willing house plant, with fragrant, long-lasting waxy flowers abundantly from April to June. **C**: 22   p. 459

BRASSICA (Cruciferae)

oleracea acephala crispa, the "Flowering kale"; thick-leaved, glaucous perennial, a form descended from the European species, grown as an ornamental, reported to be a hybrid between 'Flowering cabbage" and ordinary Kale; forms a big rosette of large frilled and fringed foliage, glaucous-green in the younger stage, but in autumn as the season advances and they are kept cool, remarkably beautiful colors begin to appear with the center blooming forth in shades of ivory, yellow, rose and purple. These plants are used as Christmas and New Year's pot plants in Japan, and have equally been used in flower arrangements in New York. From seed. **C**: 76   p. 536

x BRASSOCATTLEYA (Orchidaceae)

veitchii (Brassavola digbyana x Cattleya mossiae); the first bigeneric hybrid between these genera; very beautiful with large delicate rose-pink flowers having a wide open, broad bearded lip, crimson in the throat changing to yellow and fading out to cream. A happy cross combining the huge lacily fringed lip of Brassavola digbyana with the vigor and floriferous habit of the Easter orchid.   **C**: 71   p. 459

x BRASSOLAELIA (Orchidaceae)

'Charles J. Fay' (Brassavola nodosa x Laelia crispa), exquisite yet showy bigeneric hybrid with its short stem-like pseudobulb bearing a fleshy leaf, and flowers with long slender white sepals and petals; the broad lip white prettily spotted with purple in radiating lines, yellow in throat; blooming in winter. Combines the compact habit of Brassavola with the grace of a Laelia. **C**: 71   p. 459

BREYNIA (Euphorbiaceae)

nivosa roseo-picta (Phyllanthus), a color form known as "Leaf flower" because the little oval, papery leaves in their coloring look like flowers, being mottled and variegated green, white and pink, with red stems and petioles. The species B. nivosa, the "Snow bush" is at home in the South Sea islands, where it grows as a loose growing bush to 4 ft. high, its pendant branches with alternate fern-like leaves 2 in. long, and variegated white. **C**: 54   p. 194, 535

BRODIAEA: see Ipheion

BROMELIA (Bromeliaceae); large tropical American terrestrial rosettes with stiff sword-shaped wide-spreading leaves armed with vicious hook spines; approaching flowering time, in summer, the center leaves on their base turn a brilliant and glowing red. The beauty of the inflorescence of the true Bromelia lies in its large lustrous red bracts. Not practical as a container plant unless it may be placed a safe distance from passers-by, as its needle-sharp spines are very dangerous and may tear clothes and skin. Propagated by offsets.

balansae (Brazil, Argentina), "Pinuela" or "Heart of Fire"; large and vicious terrestrial rosette used for fencing; the stiff green leaves with dangerous hook spines facing both directions; center turning red anticipating blooming time; flowers white, in paniculate inflorescence forming branches of small, ovoid orange-yellow fruit with pineapple flavor. Easily confused with B. pinguin, but balansae has

broad sepal tips, pinguin needle-like tips and loose inflorescence. **C**: 7   p. 131*, 133

serra variegata, "Heart of flame"; large showy plant but dangerously spiny; the spreading, recurving leaves grayish-green with broad ivory margins; at flowering time center of rosette turns billiant red. Inflorescence globose (unlike the elongate inflorescence of balansae), with red bract-leaves and maroon flowers. Fruit orange-colored. The green-leafed species B. serra is a native of N.W. Argentina. **C**: 9   p. 133

BROUGHTONIA (Orchidaceae)

sanguinea (Cuba, Jamaica); small epiphyte in habitat growing on sunny rocks, with flat pseudobulbs crowded and bearing 1 to 2 rigid leaves; the roundish, flattened flowers 1 in. across, usually vivid crimson, sometimes yellow, in terminal cluster on slender, wiry stalks to 18 in. high, blooming from autumn to spring, and into summer; the brilliant flowers lasting a long time. Grows best on blocks of wood or treefern. Propagated by separating the pseudobulbs. **C**: 22   p. 460

BROWALLIA (Solanaceae)

speciosa alba, "White bush-violet"; profusely blooming herb of South American origin, and while becoming shrubby at base, is grown as an annual, easy to flower at almost any time of year, depending on the time of sowing the seed. The variety alba has large 1½ in. salverform white flowers, and lively green foliage. **C**: 72   p. 264

speciosa major (Colombia), "Sapphire flower"; straggling herb grown as an annual, 8 to 12 in. high or more, dark green leaves, and large, solitary, salverform flowers having slender tube and spreading limb to 2 in. across, violet-blue with white edge, pale lilac beneath; blooming almost continuously. While cuttings may be used, propagation is usually from seed, and when sown in summer, they start to flower end of winter in a cool, light location. **C**: 72   p. 264

BROWNEA (Leguminosae)

coccinea (Venezuela), "Scarlet flame bean"; beautiful tropical evergreen shrub 6 to 10 ft. high with vining or clambering branches, oblong leathery pinnate leaves having 2 or 3 pairs of undulate leaflets, and clustering scarlet trumpet-shaped flowers with protruding stamens tipped strikingly by vivid-yellow pollen, and blooming in summer. For the moist-warm greenhouse, and particularly spectacular when the foliage on young shoots hangs laxly like a fantastic, brightly colored coppery mass. Propagated from fresh seed, semi-ripened cuttings moist-warm under bell-jar, or mossing. **C**: 53   p. 487

BRUNFELSIA (Solanaceae); handsome tropical American evergreen shrubs and trees, formerly known as Franciscea, free-flowering and usually winter-blooming; with alternate somewhat leathery leaves, and showy often fragrant flowers abundantly between January and April, but frequently continuing through the year Brunfelsia perform best if undisturbed and at even temperatures in intermediate or warm greenhouses or winter gardens with adequate humidity. After flowering a slight rest will ripen wood for next season's prolific bloom. Propagated from semi-hard cuttings under glass jar or plastic hood

americana (West Indies), "Lady of the night"; evergreen shrub to 8 ft. high, the leathery leaves oval to obovate, 3 to 4 in. long; exquisite flowers white, fading to pale lemon yellow with age, usually solitary, with a slender tube to 4 in. long and spreading limb to 2¼ in. across; very fragrant especially at night; bl. in spring and summer. **C**:52 p. 484

calycina floribunda (Brazil), "Yesterday, today and tomorrow"; floriferous evergreen shrub with spreading branches, dark green, long elliptic 2 to 4 in. leathery leaves, and large flowers 2 in. across, rich violet to lavender with small white eye, quickly fading to white, lavishly blooming from January to July. **C**: 64   p. 484

latifolia (Franciscea) (Trop. America), "Kiss-me-quick"; low evergreen shrub 2–3 ft. high, with broad, leathery, obovate leaves to 4 in. long, slightly pubescent beneath, flowers 1½ in. across, very fragrant in daytime, pale violet with white eye at first, changing in a day or so to white, and blooming freely in winter and early spring. **C**: 64

undulata (Jamaica), the "White raintree"; magnificent strong-growing evergreen shrub, or small tree, beginning to flower freely while quite small; light green, elliptic leaves, 2 to 6 in. long, the pendant, salverform flowers with long slender tubes and wavy limb, deliciously fragrant, white, but changing to creamy-yellow with age, blooming in terminal clusters, best in October; requires warmth. **C**: 52   p. 484

x BRUNSDONNA (Amaryllidaceae)

parkeri (Brunsvigia or Amaryllis x parkeri), the "Naked lady lily"; hybrid of the So.African Amaryllis belladonna, the Cape Belladonna; and Brunsvigia josephinae; handsome bulbous flowering plant 18 in. high with strap-shaped basal leaves and a stalked cluster of funnel-form flowers with entrancing fragrance, the color an exquisite pink and with yellow throat, blooming during summer. By offsets. **C**: 64
p. 389

BRUNSVIGIA (Amaryllidaceae)

'Josephinae hybrid' (Amaryllis), a South Australian hybrid of "Josephine's lily" from Natal where it blooms in March; in the Northern zone in late summer; a gigantic 6–8 in. bulb bears a slender, tall stalk 2 to 3 ft. high crowned by a large cluster of glowing carmine-rose, irregular flowers 3 in. long, with pretty purplish filaments showing, and appearing before the some 18 broad, strap-like leaves. Brunsvigia need a complete resting period after the leaves have dried down. From lateral bulbs. **C**: 64   p. 389

BRUNSVIGIA: see also Amaryllis, x Brunsdonna

BRYOPHYLLUM: see Kalanchoe

BUCIDA (Combretaceae)

buceras (Florida, West Indies, to Panama), the "Black olive"; handsome tropical evergreen tree becoming 50 ft. high, as a younger shrub of very symmetrical growth habit, with branches laterally in tiers, small leathery obovate leaves 2 to 3 in. long, at the end of twigs; small, insignificant greenish flowers without petals displaying long stamens, followed by small $\frac{1}{3}$ in. black ovoid berries. A decorative ornamental for containers where space allows for its spreading, sculptured shape. From seed or cuttings. **C**: 2
p. 512

BUDDLEJA (Loganiaceae)

crispa (No. India: Himalayas), "Curly butterfly bush"; charming deciduous shrub 6 to 15 ft. high, robust with woolly branches bearing lanceolate, opposite leaves 2 to 5 in. long, coarsely toothed, woolly above, white or yellow tomentose beneath; fragrant flowers lilac with white eye, in showy clusters; summer blooming. **C**: 76   p. 504

BUGULA: see "Air-fern"

BULBOPHYLLUM (Orchidaceae); a large genus of varied epiphytes, often more curious than beautiful, generally of small habit and including some of the smallest orchids known, others with flowers more showy; distributed widely from Australia and New Guinea through Asia, Africa and America. The creeping rhizomes bear pseudobulbs usually with single leaves; remarkable is the wide variety of flowers characterized by having the lateral sepals joined to the column at their broadened base; the small lip is delicately hinged to the base so that it trembles at the slightest touch. Ideally grown on treefern blocks, or osmunda, in moist atmosphere, some followed by a short rest depending on their area of origin; they are rather intolerant of disturbance, and should only be transplanted if the growing medium has become stale. Propagated by division.

grandiflorum (New Guinea); probably the most remarkable orchid it has been my privilege to see in New Guinea; the large 4 to 6 in. flowers solitary, olive-green with pale chartreuse spots and netted with brown, the sepals 4–5 in. long, strongly arching over at base with the dorsal hanging down in front hood-like, petals and lip very small and almost hidden, the lip greenish dotted with red-brown; flowering in autumn in the north, in New Guinea in May-June; short 4-angled pseudobulbs with solitary leaves, on creeping rhizome. **C**: 10   p. 460

pulchellum (Cirrhopetalum) (Malaya Peninsula), "Walnut orchid"; small epiphyte with pseudobulbs bearing single solitary leaves; a basal stalk carries fan-like a semi-circle of very remarkable flowers having tiny dark-striped petals and a small dorsal sepal; the two rosy lower sepals, however, are comparatively large, long lanceolate and folded over in front, each ending in a long tail; the dark lip a small pouch; blooming in early spring. **C**: 60   p. 460

virescens (Indonesia: Amboina, Java), "Windmill orchid"; very striking epiphyte, with flattened, one-leaved pseudobulbs carried at intervals on creeping rhizome; the flowers strikingly arranged in a full circle but with objectionable odor; long-tapering 5 in., greenish sepals with their slender points forming a star; the fleshy lip is yellow with red-purple at the base and a white waxy center; blooming in summer. **C**: 60   p. 460

BUXUS (Buxaceae)

microphylla japonica (Japan), "California boxwood" or "Japanese little-leaf boxwood"; evergreen densely branching shrub to about 6 ft. high, the wiry shoots with small $\frac{1}{2}$ to 1 in. obovate leathery leaves closely set, and a glossy bright green to deep green. Clipped into hedges, or globes, pyramids or other topiary shapes in containers. Tolerates the dry heat of a California patio, and alkaline soil, but not fully frost-resistant. From spring cuttings. **C**: 26
p. 292

CAESALPINIA: see Poinciana

CALADIUM (Araceae); tropical American tuberous herbs with showy foliage, of which the most striking are the many named "Fancy-leaved caladiums", hybrid cultivars and mutations known under the group name C. hortulanum; the larger peltate heart-shaped leaves have C. bicolor blood, the lanceolate strap-leaved hybrids go back to C. picturatum. The hortulanum hybrids carry their gorgeously colorful, membranous often translucent foliage on slender petioles, producing their gay leaves almost the year round if kept growing in pots or planted out. Caladiums are very cold-sensitive and must not be exposed to drafts; they are not satisfactory under airconditioning. They may be maintained remarkably well, despite their fragile appearance, in a warm, well-lighted place between 75 to 200 fc., but away from direct sun which may scorch them. Being seasonal, a short rest is advisable, allowing their foliage to wither, keeping the tubers warm (60–65° F.) in their pots or laid in flats in dry peatmoss until wanted. Propagated by separating the tubers.

hort. 'Ace of Spades', "Lance Leaf"; a glorious lance leaf, with broad cordate base, on long petioles; the dominant red veins over rose and white marbling, the margin dark green. **C**: 10   p. 94

hort. 'Candidum', "White fancy-leaved caladium"; exquisitely beautiful glistening white leaves 8 to 12 in. long, traced with dark green veins and green border; delicate, yet a good keeper. One of the best commercially, especially for Easter. **C**: 12   p. 94

hort. 'Flame Beauty', "Strap leaf"; excellent strap variety with firm leaves to 8 in. long, ruffled at margins, red-veined, translucent between; edge deep green; stems pink. **C**: 10   p. 94

hort. 'Frieda Hemple', "Red elephant-ear"; leading red variety; bushy, compact and sturdy; medium size leaves clear bright red with primary ribs scarlet, the outer edge deep green. **C**: 10   p. 94, 180

hort. 'Lord Derby', "Transparent caladium"; bushy plant with translucent, quilted leaf having wavy edge; delicate rose-pink with network of thin, green veins and green margin. Very cold-sensitive. **C**: 12   p. 94

hort. 'Marie Moir', "Blood spots"; a beautiful variety because of its simplicity; pure white with a network of green tracings, dark green ribs, and a sprinkling of showy red blotches like drops of blood. **C**: 10   p. 94

hort. 'Miss Muffet' ('Midget Princess'), "Baby elephant-ear"; wonderful little miniature with intriguing leaf: sturdy, compact, nile green or chartreuse, with white primary ribs, and deep blood-red blotches and heart. **C**: 10
p. 94

hort. 'Mrs. W. B. Halderman', "Coloradium"; bushy variety with medium-size pointed leaves transparent rose, with zone of white marbling and dark green border. Well-known commercial hybrid. **C**: 10   p. 94

hort. 'Roehrs Dawn', a great caladium with large, dainty quilted, transparent leaves of shimmering creamy-white, contrasting with the rosy red veins and a narrow green edge. **C**: 10   p. 65*

hort. 'White Queen', "White Fancy-leaved caladium"; stunning introduction with pure white, lightly crinkly leaves contrasted by clear crimson primary veins, a tracing of green laterals, and a fine deep green edge. **C**: 10   p. 180

humboldtii (argyrites) (Pará, Brazil), "Miniature caladium"; smallest and daintiest of genus; the tiny leaves light green with transparent white areas between veins. Best in moist-warm terrarium or greenhouse. **C**: 62   p. 94

CALAMUS (*Palmae*)
ciliaris (Borneo, Java, Sumatra), "Lawyer canes"; a tropical "Rattan palm" with very long, thin and tough, lightly spiny stems having vivid green pinnate leaves to $2\frac{1}{2}$ ft. long, closely set with numerous narrow, hairy leaflets; with age climbing by long separate leafless branches with hooked spines, which appear in the axils of each leaf. Very attractive as a pot plant when young, requiring warmth and moisture; best to set pot into a saucer kept filled with water. **C**: 54   p. 232

siphonospathus, or more correctly C. maximus (Old World Tropics: Malaysia, Philippines), a "Rattan"; climbing feather-leaf palm whose flexuous stems scramble 100–200 ft. high into the treetops by means of hooks at the ends of the pinnate leaves, these hooks being modified pinnae on an extended, whiplike midrib; the bamboo-like, tough stems are densely covered with sharp black spines. Commercially used as furniture rattans and for canes, but ornamentally only for the warm, moist conservatory mainly because of its special interest. **C**: 54   p. 232

CALANTHE (*Orchidaceae*); known as the "Beautiful flowers", these are mostly terrestrial orchids, some with deciduous leaves, others evergreen. The deciduous species have angular conical pseudobulbs with heavily ribbed leaves which fall before or at the time of flowering; the evergreen species are without significant pseudobulbs and leaves grow directly from the creeping rhizome; clusters of showy flowers are carried on tall erect or arching stalks. Terrestrial calanthe grow well in a mixture of sandy loam, shredded peatmoss or osmunda and, as they are heavy feeders, rotted cow-manure or other organic fertilizers should be added, and perlite for drainage.

x bella (turneri x veitchii); a very handsome winter-blooming English hybrid by Veitch 1880; squarish, silvery-gray pseudobulbs with broad-lanceolate plaited leaves developing after flowering; the inflorescence in a showy cluster carried on long arching stalk 2 ft. or more tall, the large flowers $1\frac{1}{2}$ to 2 in. or more across having white sepals, the petals blush, the four-lobed lip blush-pink with crimson blotch edged in white. One of many good, named deciduous Calanthe hybrids, largely used as cut flowers. Keep warm to grow; cooler when in bloom and when resting. **C**: 4   p. 462

striata (sieboldii) (Korea, China, Japan); elegant dwarf, evergreen species 15 in. high, with broad, dark green plaited leaves, and erect raceme of large yellow flowers $1\frac{1}{2}$ in. across, tinted coppery orange, the lip dark yellow marked crimson; blooming spring or summer. **C**: 66   p. 462

CALATHEA (*Marantaceae*); beautiful tropical American foliage plants with highly decorative, thin-leathery leaves springing from underground rhizomes; the foliage wonderfully patterned, barred, mottled or striped, and gorgeously colored in shades of yellowish to bluish green, silver and white, chocolate and maroon; similar in foliar habit to Maranta, but with an inflorescence in compact clusters or cones, the flowers spirally in basket-like bracts, instead of branched as in Maranta. Formerly considered exclusively warm conservatory subjects, we have found that Calatheas adapt very well to growing in the East or West window, provided they are kept constantly moist and they might be kept standing in a saucer. They grow in porous soil rich in humus and should not be disturbed more than necessary. When they have developed into a well-branched plant, Calatheas may be divided in February-March at the end of their off-season and as new growth begins.

argyraea (Brazil), "Silver calathea"; sturdy plant with foliage borne on stiff-erect, winged stalks; leaves oblanceolate, unequal-sided, glossy, the feathered vein areas grass-green, silvery-gray between; reverse wine-red; from the collection at Paris Museum. **C**: 60   p. 176

bachemiana (trifasciata) (Brazil); small very pretty plant with slender stalks and oblique, narrow tapering, leathery leaves pale greenish-gray; dark green, almond-shaped blotches alternately fanning out from midrib, and a narrow border of the same deep color; green beneath. **C**: 60   p. 177

insignis (Brazil), "Rattlesnake plant"; very pretty, bushy species with narrow leathery, tapering, almost linear, stiffly erect foliage, wavy at the margins, yellow-green with lateral ovals in dark green, alternately large and small; underside a showy maroon red; while slow and low growing at first, I have seen it measure 2 ft. tall or more, as growing in Brazil. **C**: 60   p. 175

kegeliana (Brazil); robust, durable plant to 18 in. high with wiry stalks bearing heavy, thick-leathery oblique-ovate leaves, having cordate base, laxly pendant; the surface silvery gray, attractively patterned with lateral veins and lance-shaped bands of yellowish to dark green; grayish green beneath with midrib showing yellow. **C**: 60   p. 175

leopardina (Brazil); attractive species to 15 in. high, with beautiful oblique-lance-shaped, waxy leaves on slender stalks, the blade is nile-green, marked with bold dark green triangles alternately on both sides of the midrib, and not reaching to the margin; underside tinted bronzy purple. **C**: 60   p. 177

lietzei (Brazil); very pretty plant with narrow-ovate, thin-leathery leaves deep green with sharply contrasting, alternate lateral bands in silvery green to margin, and purple beneath; sending up erect runners bearing young plantlets. **C**: 10   p. 177

lindeniana (Brazil); vigorous plant to 3 ft. high; broad oval leaves deep green, with a feathery, pale olive zone either side of midrib and near the border, a darker zone between; purple underneath except near midrib. Needs humid warm conditions. **C**: 60   p. 176

louisae (albertii) (Brazil); tufted plant somewhat similar to lietzei but not making runners; the 5 in. leaves are larger and more broadly ovate, dark green, with light yellowish-green, irregular feathering along the midrib; purple underneath; fl. white out of pale green bract head. **C**: 60   p. 177

makoyana (Brazil: Minas Geraes); the "Peacock plant"; bushy, and stunningly beautiful on both sides of its oval leaves, the surface with a feathery design of opaque, olive green lines and ovals alternately short and long, in a translucent field of pale yellow-green; this pattern of lines and ovals is purplish-red beneath. A better house plant than its exquisite beauty would indicate. **C**: 10   p. 43*, 175

micans (Brazil, Peru), "Miniature maranta"; perhaps the tiniest of Calatheas; narrow, pointed leaves 3 to 4 in. long hugging the ground, in shining medium-green with a flameshaped silvery-white center band above, gray-green beneath. **C**: 60   p. 176

ornata 'Roseo-lineata'; attractive tropical plant featuring long narrow-ovate, unequal-sided, metallic olive-green leaves, in the juvenile stage prettily marked with closely set pairs of rosy-red lateral stripes, these later turning white in older leaves; purple beneath. This species is from Guyana, Colombia and Ecuador, a variable plant growing from 18 in. to 8 ft. tall, in adult stage plain metallic coppery-green. **C**: 60   p. 177

ornata 'Sanderiana'; a showy cultivar with erect leathery, rather broader leaves glossy dark olive green, with boldly contrasting pink or white stripes curving across the blade in groups of two or more; reverse rich purple-red. **C**: 60   p. 177

picturata 'Argentea' (Venezuela); dwarf plant similar in habit to vandenheckei but with short-stalked leaves almost entirely shining silver except for a border of dark green; wine-red beneath. Cv. 'Wendlinger' may be the same plant. **C**: 10   p. 177

picturata vandenheckei (Brazil: Amazonas; Colombia), juvenile form of metallica (picturata); dwarf plant with oblique, glossy leathery leaves, deep olive to blackish green, with feathered center band of silver white, and another one on both sides two-thirds toward margin; wine-red beneath. **C**: 10   p. 177

princeps (Brazil: Amazonas); large oblong, showy leaves of yellow-green, and a feathery center band of blackish-green, with dark veins running to the dark border, violet-purple beneath. I was surprised to see this species grow in the rock-hard floor of the primeval forest north of Manaus, in Brazil. According to Kyburz, princeps is the juvenile form of altissima, which grows to 7 ft. tall and becomes plain green with age. **C**: 60   p. 175

roseo-picta (Brazil); low plant with short-stalked, large 9 in. rounded, unequal-sided leaves dark olive green above with red midrib and narrow zone of bright red near margin, changing to silver pink when old; underside purple. Slow growing. **C**: 60   p. 176

undulata (Peru): tufted small plant 8 in. high with thin-leathery, oblique-ovate leaves on short petioles, with waxy undulate surface following lateral veins to margins; metallic dark green with bluish sheen when young, and a jagged band of greenish silver along midrib, purple beneath over green; flowers white with green bracts margined and dotted white. **C**: 60   p. 176

veitchiana (Ecuador, Peru); strikingly beautiful plant from the Jivaro country with large, stiff-leathery, glossy leaves to 1 ft. long, obliquely broad-ovate, in four different shades of green, with a peacock-feather design outlined in yellow-green, encircling brownish-green half moons which adjoin the pale bluish green feathered center zone; marginal area bright green; the peacock-feather design outlined in red over bluish green underneath also. **C**: 60   p. 175

warscewiczii (picta) (Costa Rica); robust plant to 2½ ft. high, with short-stalked, fairly large oblong leaves undulate at margins, velvety deep green except for yellow-green midrib, feathering into pale green lateral veins; underside wine-red. Inflorescence a short spike with tubular, leathery white bracts, and white flowers. **C**: 60   p. 175

wiotii (Phrynium brachystachyum) (Brazil); dwarf plant with thin-leathery, short-ovate leaves to 4 in. long, glossy light green with elevated oval blotches of dark green on either side of midrib; reverse mostly gray-green, tinted purple. **C**: 60   p. 177

zebrina (Brazil: Rio, Sao Paulo) "Zebra plant"; a magnificent vigorous warmhouse plant to 3 ft. high, with gorgeous, large thin leaves deep velvety green, the midrib and lateral veins pale or yellow-green, purplish beneath. **C**: 60   p. 175

CALCEOLARIA (*Scrophulariaceae*); the "Slipperworts" are annual herbaceous or somewhat shrubby perennial plants from the cool Chilean Andes, with curious, fascinating pouch-like flowers, and inflated lower lip, in yellows and bronzes and spotted with red. They do best at cool temperatures and especially the spring-flowering herbaceous kinds quickly collapse in summer heat, while the shrubby types tolerate more heat and sun. Grown from seed or cuttings.

herbeohybrida 'Grandiflora', "Giant pocketbook plant"; herbaceous cultigens derived principally from C. crenatiflora which grows in the cool Andes of Chile and north; developed into beautiful annual potplants with thin, fresh-green leaves and clusters of showy, membranous 2 to 3 in. flowers, distinguished by a large inflated lower lip resembling a lady's handbag, in shades of yellow or red, often brilliantly spotted or tigered orange-red to maroon, blooming in spring. From seed. **C**: 35   p. 371

herbeohybrida 'Multiflora nana', "Lady's pocketbook", herbaceous plant grown as an annual; crenatiflora form of dwarf, bushy habit, ideal as a potplant, with thin ovate, toothed and quilted leaves and clusters of numerous, smallish 1 to 2 in. pouch-like flowers, most often yellow or orange to red, usually spotted or tigered with crimson. From seed. **C**: 85   p. 371

integrifolia (known in horticulture as rugosa) (Chile), "Bedding calceolaria"; shrubby perennial of bushy habit, with woody, hairy stems, small, wrinkled, toothed leaves, and clusters with masses of small ½ in. pouch-like flowers yellow to red-brown. Blooming from spring to fall and used as a small pot plant or for bedding, tolerates sun and some heat or cold. From seed, or from cuttings taken in fall. **C**: 83   p. 371

CALENDULA (*Compositae*)

officinalis 'Apricot Queen', a highly-bred "Pot-marigold"; popular annual herb 12 to 18 in. high originating in So. Europe; with fleshy entire, brittle and clammy leaves not scented, on sticky stalks; the large double flowers 3 in. or more across, closing at night, in apricot, also shades of orange, lemon or golden yellow in other named varieties. Excellent, long-lasting cutflower developing largest blooms if disbudded and grown cool. Their normal outdoor season is from June to frost, however, with a cool greenhouse or wintergarden, pot-marigolds may be had in bloom in the dark of winter; for Christmas from August sowings. By seed only. **C**: 76   p. 409

CALLIANDRA (*Leguminosae*)

emarginata (So. Mexico and C. America); "Powder puff" or "Flame bush"; exotic-looking clambering shrub

with smooth bipinnate leaves, the rich green leaflets large and oblique-ovate; showy flower heads conspicuous by their profusion of 1 in. long bright red silky stamens arranged hemispherically as in a fantastic brush. Willing bloomer with warmth and moisture in sunny locations. Should be cut back yearly to keep it from getting too sparry. From fresh seed, or cuttings of firm young wood, in heat, under glass. **C**: 2   p. 487

haematocephala (Ecuador, Bolivia), "Red powderpuff"; tropical small tree having bipinnate foliage, the pinnae and leaflets in equal pairs; beautiful, very showy globose inflorescence, larger and more dense than tweedii, with long, conspicuous silky, bright red stamens 1½ in. long; should be cut back yearly to keep compact. Cuttings or seed. (Syn. inaequilatera). **C**: 2   p. 276

CALLISIA (*Commelinaceae*)

elegans (introduced as Setcreasea striata) (Mexico: Oaxaca), "Striped inch plant"; vigorous little creeper hugging the ground and resembling Tradescantia, with succulent triangular, clasping leaves olive green and lined with white stripes, deep purple beneath, and arranged like shingles; flowers white. **C**: 4   p. 260

fragrans (Spironema, or better known as Tradescantia dracaenoides), (Mexico: Oaxaca); robust fleshy rosette almost dracaena-like, becoming several feet tall; sending out long runners with young plants; broadly lanceolate, recurving clasping leaves, 8 to 12 in. long, fresh glossy green and turning reddish-purple in strong light, margins smooth and pale green; long inflorescence of clusters of waxy-white flowers fragrant as hyacinths. Very effective in hanging baskets with pendant plantlets on long runners. **C**: 4   p. 262

fragrans 'Melnikoff' (Spironema in horticulture); fleshy creeper with glossy, clasping ovate leaves to 8 in. long, fresh green, striped lengthwise with yellowish bands; the foliage is as closely sheathed as shingles on a roof, and much space is needed to allow runners to develop. Terminal inflor. with small white fl. fragrant like hyacinths.   **C**: 16   p. 260

CALLISTEMON (*Myrtaceae*)

lanceolatus of horticulture; botanically citrinus (syn. Metrosideros floribunda) (Australia: Victoria, N.S.W., Queensland), the crimson "Bottlebrush"; tree-like shrub 10 to 15 ft. high or more, of sparry, bare habit, with hard heavy wood, sun-loving and drought-resistant; the silky twigs with rigid, long-linear, 3 in. leaves, coppery in new growth, vivid green when mature; cylindrical flower spikes with brush-like masses of brilliant crimson, thread-like stamens tipped by dark yellow anthers, out of a grayish, felted calyx; the 6 in. brushes appear in cycles throughout the year, especially in summer. A most satisfactory and durable container plant for the sunny patio or roofgarden. Blooming willingly even as a small pot plant. Propagated by ripped-off cuttings of ripened wood in fall to spring, or spring-sown seeds. **C**: 13   p. 293, 501

CALLISTEPHUS (*Compositae*)

chinensis 'Dwarf Queen', "Dwarf pot aster" or "China aster"; an herbaceous self-branching annual 10 in. high, of Chinese origin, with stiff, leafy branches terminated by 2½ in. floral heads densely double with florets in various colors from violet, purple and blue to rose and white but no true yellow. The flowers somewhat resembling 'Mums' are long-lasting and normally bloom from July until frost, but by early sowing may be had in bloom for Easter and Mother's Day. Being long-day plants, Asters can be brought into bloom the year round provided the seedlings receive additional daylight by lighting during short days. From seed only. **C**: 63   p. 409

CALLOPSIS (*Araceae*)

volkensis (Tanzania), "Miniature calla"; short, branching little plant with underground rhizome, resembling a calla-lily but only 4 in. high, from the Usambara Mountains near the Sigi River; cute, cupped spathe waxy-white, with an attractive yellow spadix; quilted, leathery, heart-shaped leaves; winter flowering. **C**: 60   p .98

CALLUNA (*Ericaceae*)

vulgaris (No. Europe), "Scotch heather"; densely bushy small evergreen shrub, 1 to 2 ft. high, the thin woody branchlets with scale-like, overlapping leaves, and a profusion of tiny bells typically purplish pink, or white, the colored $\frac{1}{4}$ in. calyx concealing the corolla; this longer calyx distinguishing Calluna from Erica. With branches spreading and covered with dense masses of usually purplish flowers from June to November, these heathers are characteristic of the heath and moor landscapes of Europe and Asia Minor; attractive in roof and rockgardens, or in pots in a light, cool greenhouse. Best grown in lime-free, sandy peat; may be cut back in spring. From cuttings taken before bloom. **C**: 82 p. 502

CALONYCTION (*Convolvulaceae*)
aculeatum (American and other Tropics), the legendary tropical "Moon-flower"; perennial twiner related to morning-glories (Ipomaea), but blooming at night instead of in sunshine; stems more or less prickly, with milky juice, climbing to 15 ft. high; cordate, smooth leaves 6 in. long, and large salver-shaped flowers 5 to 6 in. wide, pure white, and fragrant, opening toward evening and closing in the morning. For a light, warm yet airy greenhouse, or as a porch climber; best as annual from seed; notched to hasten germination. **C**: 52 p. 280

CAMELLIA (*Theaceae*); ornamental evergreen shrubs or trees of the Tea family, native to eastern Asia, having alternate simple finely toothed, leathery and usually very ornamental shining deep green leaves, and showy, white and pink to red sessile flowers with deciduous sepals; popular in containers, for the cool greenhouse or winter-garden, in the summer on the patio or terrace, or under trees. Most widely cultivated for their exquisite flowers and handsome leathery foliage is Camellia japonica of which there are known over 3000 named kinds, and widely planted in Southern and Western gardens, and where they are synonymous with romance and history. These have flowers from 2 to 5 in. across, borne on woody branches. Originating in Northern China, and the mountains of Japan and Korea where they grow into trees 40 ft. high; camellias may bloom according to variety and climate from October to May. The large-flowered kinds are much used as cut flowers for which purpose they are twisted or snipped off, or at most cut with one or two leaves; they keep longest if floated in water. Best for pots or tubs are the smaller-bloomed varieties, but the holding of flowers or buds of pot-grown plants is always risky and depends on evenly moderate temperature and steady moisture supply at the roots. The sasanqua varieties are small-flowered, but these are even more sensitive to sudden changes in temperature resulting in bud-drop; their flowers are more flimsy and short-lived. Camellias require soil rich in organic material; want to be kept light and moist but not wet. Kept a cool 45 to 50° F. in autumn and winter, the temperature is increased after flowering to 60° F. or more for new growth and to form the new buds. Camellias do best if long undisturbed, even in small pots or tubs. Propagation is occasionally by grafting but mainly by semi-hard tip cuttings of nearly ripe wood, preferably in January and August, treated with rooting hormone, in peat-sand on warm 65–70°F. bottom heat.
japonica (Mountains of Japan and Korea), the "Common camellia", a fine ornamental evergreen tree with great age to 40 ft. high; well-shaped, with woody branches and dark green, glossy, leathery, ovate leaves to 4 in. long, finely toothed; very showy, exquisite axillary flowers to 5 in. across, variably single to double; white, pink, red, or variegated. This is the species that is most highly hybridized in many named cultivars, and much grown in pot or tubs; for decoration or for the cut flowers which must not be cut with a stem longer than a fraction of an inch. **C**: 34 p. 314
japonica 'Debutante', also known as 'Sara Hastie'; originating in South Carolina, this is a shapely fast-growing beauty of upright habit and ideal pot or tub plant; early and free-blooming with exquisite bud forms, full double peony-form flowers of clear soft pink, from October to January and later. **C**: 34 p. 483
japonica 'Elegans', often found as 'Chandler's elegans'; "Peony camellia"; one of the first English hybrids (1824) and still most important today, of wide-spreading habit with glossy foliage and very large flowers of the peony type, 4 in. across, cherry-red and carmine-rose variegated with white areas. A mid-season variety blooming in profusion in late January. Slow grower, and great for training on espaliers. **C**: 34 p. 485
sasanqua 'Showa-no-sakae', a hiemalis type "Pink Sasanqua"; graceful evergreen shrub, lower in general habit than japonica and somewhat frost-hardy; dense with willowy branches and small thin leathery shining dark green, flat foliage 2 to 2½ in. long and with serrate margins; lavish with semi-double 2½ in. rather flattened flowers soft carmine rose, blooming prolifically from October to April; the flowers shed easily but are quickly followed by others. The species is at home in Kyushu, Southern Japan, and bears mostly single flowers; good subject for containers as bush or trained on espaliers, against walls, also hanging baskets; in sun or shade. **C**: 22 p. 485

CAMPANULA (*Campanulaceae*)
elatines alba plena, the "Double white Bethlehem star"; a slender, sprawling, nearly vine-like herb that makes a pretty hanging basket plant, with glossy green leaves, and frilly double white flowers blooming from summer to fall; a selected clone of the C. elatines complex which is native in S.E. Europe and the Adriatic region. **C**: 78 p. 264
fragilis (limestone rocks of Southern Italy), "Star of Bethlehem"; small prostrate tufted herbaceous trailer, with little spoon or kidney-shaped crenate leaves glossy green; large 1½ in. star-like flowers not as flat as isophylla, petals grooved, pale purplish blue with white center. Not hardy, but loves coolness and fresh air; in regions of cool summers they may be grown in pots, flowering from summer through autumn. **C**: 76 p. 264
isophylla mayii, "Ligurian" or "Italian bellflower"; herbaceous trailer with small ovate leaves, white-hairy in the variety mayii, and saucer-shaped lavender blue flowers to 1½ in. across, larger than isophylla. The species is from the Ligurian Alps of Northern Italy, and is not winter-hardy. But in areas with cool summers like New England or Oregon they are ideal basket plants, showered with pretty blossoms from summer to fall. They like it cool and airy, with limestone soil. **C**: 76 p. 264

CAMPELIA (*Commelinaceae*)
zanonia 'Mexican Flag', (long known as Dichorisandra albo-lineata), a striking cultivar of the green species (Mexico to Brazil); attractive and vigorous foliage plant, with erect, thick-fleshy stalk to 3 ft. high, and large thin-fleshy, lance-shaped, light green leaves 8–12 in. long in a rosette, striped white lengthwise throughout center and with broad, cream-white marginal bands edged red; flowers purple and white, in dense cluster, from leaf axils. **C**: 53 p. 262

CAMPSIS (*Bignoniaceae*)
radicans (Tecoma), (U.S.: New Jersey to Florida and Texas); "Trumpet bush" or "Trumpet vine"; deciduous woody climber with large glossy, toothed, pinnate leaves; from summer to autumn with large trumpet flowers 2 in. across, orange with scarlet limb. Winter-hardy, the branches cling by means of aerial roots like ivy; ideal against a warm wall or other support. Propagation from root sections and half-hard cuttings. **C**: 63 p. 276

CANISTRUM (*Bromeliaceae*)
lindenii roseum (So. Brazil), "Wine star"; stocky, open rosette with broad leaves waxy yellow-green on both sides, and with darker mottling, the underside with rosy wine-red bands and lines, soft scattered teeth at margins; the erect-stalked inflorescence in a dense head like a basket with rose red bracts and white and green flowers. **C**: 10 p. 135

CANNA (*Cannaceae*); striking tropical or near-tropical sun-loving perennial herbs 3 to 6 ft. high with thick branching tuberous rhizomes, each live eye likely to sprout stiff erect unbranched canes with large, showy, often colored foliage, and in the Garden cannas (C. generalis) crowned with gorgeous flowers in vivid pink, scarlet, salmon, orange, or yellow, presenting a distinct tropical effect through summer until frost. Very handsome when grown in large containers on terrace or the patio, but such plants require plentiful watering. After flowering cut back the canes and store rhizomes with all the soil possible upside-down in cool cellar. Divide in spring.
coccinea (W. Indies to S. America), "Scarlet Indian shot"; tropical plant 4 ft. or more high, growing from tuberous rhizome, slender stalks with large oblong pointed leaves; the inflorescence with clusters of flowers subtended by small green bracts, sepals green, and 1½ in. narrow petals scarlet, the lip spotted yellow. Attractive though not showy;

but with C. indica and others one of the parents of the spectacular Garden cannas. **C**: 52   p. 208, 403

generalis 'King Humbert', a popular "Bronze garden canna"; striking perennial tropical herb 4 to 5 ft. high, with thick branching rootstocks; handsome bronze-red leaves on stout, almost woody canes bearing a showy truss of large orange-red blooms of the orchid-flowered type with recurved petals, blooming through summer until frost. There is also a yellow-flowered variety, which nicely complements the colors of C. 'The President'. **C**: 52   p. 403

generalis 'President', one of the leading "Garden cannas"; 3–4 ft. high, with large vivid green leaves, each cane topped by a magnificent terminal cluster of enormous scarlet flowers 5 or 6 in. across. This variety is very free-blooming, and planted in nice contrast to the bronze foliage of 'King Humbert'. **C**: 52   p. 403, 418

'Striped Beauty', "Variegated Bangkok Canna"; a stunning beauty of stocky habit found in Thailand, probably a mutation in the C. generalis group, with striking foliage brilliant green having broad midrib and following lateral veins in contrasting yellow to ivory-white; the large truss of deep scarlet flowers. A show plant for the exotic greenhouse or the summer patio. **C**: 52   p. 403

CAPSICUM (*Solanaceae*); the "Red-peppers" are tropical shrubs of the potato family grown as herbaceous annuals for their edible or ornamental fruit, yielding red pepper, Tabasco and Cayenne pepper, as well as the milder garden pepper, commonly grown for seasoning but adaptable as pot plants for their colorful and pungent fruit, of particular interest to housewives. Peppers in pots need warmth and even moisture; checks and chills may cause them to drop their foliage. From seed.

annuum 'Birdseye', a "Christmas pepper"; small tropical shrub originally from South America, mostly grown as an herbaceous annual ornamental potplant; much branching, the twigs dense with small lanceolate, fresh-green thin leaves bearing a multitude of small white flowers, followed by rounded $\frac{1}{2}$ in. berry-like, waxy fruit first green then scarlet red, beginning to color in autumn and keeping well for Christmas. **C**: 52   p. 386

annuum conoides 'Christmas Candle', a "Christmas pepper"; cheerful holiday plant with ornamental erect, slender conical fruit waxy yellow turning to red, 1$\frac{1}{2}$ to 2 in. long. A Craig (Philadelphia) clone of robust, broadly spreading habit, with its little peppers long remaining firm and showy well through the Christmas season. **C**: 2   p. 514

annuum conoides 'Red Chile' (frutescens), a "Tabasco pepper"; actually an edible vegetable with its sweet-spicy, large conical, 2 in. fruit, but also an attractive annual pot-plant with white star-like flowers and fruits beautifully lacquered cardinal red, borne stiffly above the rich green, scattered foliage on few, wiry branches; keeps well to Christmas. **C**: 2   p. 514

CARALLUMA (*Asclepiadaceae*)

lutea (So. Africa, Botswana), clustering leafless succulent branching from the base with thick, 4-angled sinuate stems sharply toothed, 2 to 4 in. high, gray and mottled with purple; the interesting flowers 2 to 3 in. across, are star-like with long narrow lobes, striking but with unpleasant odor; canary-yellow with a fringe of red hairs along the edges. These ciliate hairs are sensitive and vibrant and start trembling at the least breeze, probably to attract passing insects. From cuttings or division. **C**: 13   p. 352

nebrownii (Tropical S.W. Africa), "Spiked clubs"; mat-forming leafless succulent 6 to 8 in. high, with fat, 4-angled, branching stems having prominent teeth distantly along ridges, gray-green and marbled dull-red; the fleshy flowers star-like, 4 in. across, beautiful dark red-brown with purple hairs; malodorous. **C**: 13   p. 352

CAREX (*Cyperaceae*)

foliosissima albo-mediana (elegantissima) (Sakhalin to Japan), "Miniature variegated sedge"; grass-like tufted plant with elegant, narrow-linear flat leaves $\frac{1}{4}$ in. wide, bright green with a white stripe near each margin, of stiff-erect habit, toward apex gracefully recurving. By division. **C**: 16   p. 304

CARICA (*Caricaceae*)

papaya (Colombia), "Papaya" or "Melon tree"; fast-growing dioecious tropical tree 10 to 25 ft. high, with

straight succulent, palm-like trunk containing milky juice, topped by a cluster of immense leaves 2 ft. across, deeply 7-lobed and lobed again on long hollow stalks; male and female greenish-yellow flowers on separate trees, both of which are usually needed to produce the melon-like or oblong fruit, 6 to 12 in. or more long, with delicious edible, aromatic flesh. An interesting container plant. To obtain fruit, the flowers are hand-pollinated. Propagated from seed in late winter, but these result in mostly male plants; female plants may be recognized by their less lobed leaves. Cuttings of ripe shoots are possible. **C**: 52   p. 507

CARISSA (*Apocynaceae*)

acokanthera (Acokanthera venenata), (So. Africa: Cape), "Bushman's poison"; evergreen shrub becoming 12 ft. high with large 4 in. leathery, long elliptic, glossy green to purple leaves; especially prized for its very fragrant, shining white flowers, the corolla tubes almost 1 in. long, with star-like spreading lobes, lightly hairy outside, and borne in axillary clusters close between foliage; blooming throughout the year but most prodigiously in late winter-spring. The blackish fruit is very poisonous and so is the sap of the plant. Propagated by cuttings under cover. **C**: 14   p. 484

grandiflora (So. Africa, Natal), "Natal plum"; spiny spreading, evergreen shrub to 5 and 7 ft. or more high, with milky juice; with leathery ovate rich green lustrous 2 to 3 in. leaves without marginal teeth, the forked spines along branches and at the end of each twig; fragrant white, star-shaped flowers to 2 in. across, appearing throughout the year; followed by plum-like, red 2 in. fruit, the edible pulp with the flavor of rather sweet cranberry. May be pruned into shapes including hedges; for sun or shade. Very handsome with flowers and showy fruit. Propagated from seed or ripened cuttings in late summer. **C**: 14   p. 486, 508

CARLUDOVICA (*Cyclanthaceae*)

palmata (from W. Brazil through Ecuador, Peru to C. America), the "Panama hat plant"; ornamental, stemless plant with friendly green, fan-shaped, palm-like leaves 2–3 ft. across, usually cut into 4 parts, the plaited segments cut again; on channeled leafstalks that may become 5 to 8 ft. tall. Inflorescence aroid-like with spadix and spathes. Leaves cut into strips, and bleached, provide the material for panama hats. Requiring abundant water and good drainage. Propagated by division. **C**: 4   p. 232

CARNEGIEA (*Cactaceae*)

gigantea (Cereus) (So. America, No. Mexico), the spectacular, massive "Giant Saguaro"; post-like columns 20–60 ft. high, sometimes 2 ft. in diameter, tapering to either end and branching with age; ribs 18–21 in mature plants; spines different on fertile and sterile areoles; flowers white, diurnal (day-flowering), 4 in. long or somewhat more; fruit 2 to 3 in. long, pale red, edible. **C**: 1   p. 149, 527

CARPINUS (*Betulaceae*)

japonica (Japan), "Japanese hornbeam"; hard-wooded deciduous small tree of the birch family, to 35 ft. high, with smooth gray bark, spreading branches with lanceolate leaves 2 to 4 in. long and more or less 2-ranked; the male flowers in pendant catkins, blooming before the leaves unfold; ribbed nutlets sheltered by leaf-like bracts in attractive, drooping clusters. Subject for dwarfed Bonsai with contorted trunks in Japan. From seed. **C**: 76   p. 314

CARPOBROTUS (*Aizoaceae*)

edulis (So. Africa), "Hottentot-fig", so known because the fruit capsules resemble figs which are edible; creeping, freely growing, robust branching succulent, often planted as a sand-binder along subtropical seashores; angled rooting branches to 3 ft. long, with coarse, very fleshy long linear, sharply 3-angled leaves 3 to 5 in. long, grass green or glaucous; large showy flowers 3 to 4 in. across with narrow, silky rays opening with the warming sun toward noon, yellow or varying purple. In South Africa, I have observed the purple form primarily on the cool Cape Peninsula, the yellow form in the glaring hot Karroo. Young plants from cuttings 1 year old bloom better than older specimens. **C**: 13   p. 357

CARYOTA (*Palmae*)

mitis (Burma, Malaya, Indonesia), "Dwarf fishtail palm"; spectacular cluster-forming palm with numerous suckers growing up from the base, the green-gray trunks to 25 ft. high and topped by dense tufts of irregularly bipinnate dull-green leaves, the pinnae fan-shaped, jagged at apex and many-veined, nodding at the tips; fruit blackish-red. Caryotas are the only palms that have bipinnate foliage; the species urens has similar leaves but is taller and not suckering. Flowering is from leaf axils from the top down, and when the lowest is reached, the stem dies. As a decorative specimen this palm creates interesting effect but is temperamental, requiring more than average light, 100 fc. up; warmth and humidity; not for air-conditioning. **C**: 4   p. 226

CASSIA (Leguminosae)
splendida (Brazil), "Golden wonder senna"; a winter-blooming floriferous evergreen shrub to 10 ft. high; the pinnate leaves with oval leaflets to 3 in. long; from November to January the bush covers itself from the branch ends with loose clusters of large 1½ in. orange-yellow flowers, a spectacular sight against blue sky. Give some rest after flowering and prune back to size desired. Sun-loving, tolerant container plant for patio or roof garden outdoors where it may bloom at odd times. Propagated from seeds or cuttings of semi-ripe wood but not too warm. **C**: 64
p. 490

CASUARINA (Casuarinaceae)
equisetifolia (No. Australia, Queensland), the "Horse-tail tree" or "Australian pine"; hardwooded tree to 75 ft. high, with pendulous branches swaying in the breeze; wirelike branchlets with apparently leafless twigs, the leaves reduced to scales and suggesting the horsetail (Equisetum). Adaptable as a patio tree or the sunny, airy greenhouse, when sheared into compact shapes or imaginative topiary forms. Cuttings or seed. **C**: 52   p. 288

CATASETUM (Orchidaceae)
most remarkable epiphytes of the American tropics including some of the most unusual and handsome of orchids, with spindle-shaped pseudobulbs clothed with large membranous plaited foliage; some species with flowers that are perfect, and others that have a separate male and female inflorescence, the male or sterile flowers hornless; the large and fleshy female or seed-bearing flowers are furnished with horns or feelers, and either may bloom at different seasons. Catasetums are easily grown, best in baskets, and adaptable as house plants, they benefit from ample moisture while growing, but need a few weeks' rest when new pseudobulbs have matured. By division.
bicolor (Panama, Colombia); distinctive epiphyte with swollen pseudobulb carrying the plaited leaves; the two kinds of inflorescence fantastic with greenish female flowers on short spikes with upright, helmet-shaped lip greyish-green, while the 3 in. male flowers are on long pendant racemes, having chocolate-brown sepals, greenish-brown petals and a white, purple-spotted and purple-margined lip; blooming from summer to October, and very fragrant. **C**: 10   p. 463
saccatum christyanum (Brazil: Amazon); curious epiphyte with spindle-shaped stems clothed with plaited leaves, and arching stalks with scattered, large fragrant flowers about 4 in. across, the sepals reddish-brown, petals lighter brown pale-spotted at base, lip green marked purplish at base, the lateral sac-like lobes with long purple fringes; blooming variously from late summer into winter.
C: 10   p. 463
warscewiczii (scurra) (Costa Rica, Panama to Colombia, Brazil); rather dwarf, compact species with egg-shaped pseudobulbs becoming ridged with age; several lively green leaves to 12 in. and more long; the inflorescence drooping, 5 flowers or more together, each to 1½ in. across and perfect; very fragrant and long-lived, the sepals and petals greenish-white striped with pale green, the dorsals forming a hood above them, lip with bearded midlobe, waxy white and lined with green; blooming mostly in summer to fall, or various. **C**: 10   p. 463

CATHARANTHUS (Apocynaceae)
roseus (better known as Vinca rosea) (Pantropic from Java to Brazil), "Madagascar periwinkle"; tropical fleshy perennial often grown as an annual, 1 to 2 ft. high, with opposite 1 to 3 in. oblong leaves glossy green with pretty white midrib; showy phlox-like 1½ in. flowers rosy-red with purple throat; in variety also white with rose eye, blush pink or bright rose, blooming from spring to fall. Invaluable

for color and very showy in dry summer heat; tolerant of poor, stony soils as seen in South India. From seed or cuttings. **C**: 52   p. 415

CATOPSIS (Bromeliaceae)
morreniana (So. Mexico, Guatemala, and C. America); small epiphytic or terrestrial rosette with strap-like glossy green leaves 4–8 in. long; the inflorescence tall, erect and branched, with pale yellow bracts and tiny white flowers. **C**: 71   p. 135

CATTLEYA (Orchidaceae); the "Queen of orchids"; tree-perching, true epiphytic orchids peculiar to tropical America, slowly creeping on woody rhizomes; the growing tip developing periodically into succeeding more or less erect pseudobulbs, thickened club-shaped stems carrying a reserve of food and water; these are topped usually by one, sometimes 2 or 3 massive fleshy leaves, the magnificent flowers with flaring sepals and petals in many lovely tints and colors, make it the favorite of florists as probably the showiest of orchids with blooms that may measure to 8 in. across. Cattleyas like the genial, moist atmosphere of a greenhouse, and if attempted in the home would need a shelter of plastic, and frequent misting, to maintain 40 to 70% humidity; they require maximum light, but not burning sun for best development of flower buds, ideally 2000 to 3500 fc., under normal daylength. When in active growth, Cattleyas need quantities of water but over porous potting medium, such as osmunda, shredded fernfiber, or bark chips. As soon as the new pseudobulbs are mature and have expanded their flowers, watering should be reduced until the new "eye" begins to swell and new rootlets form. Propagated most easily from divisions.
bowringiana (Honduras, Guatemala), a beautiful autumn-flowering "Cluster cattleya"; strong-growing, prolific epiphyte with long cane-like, 2-leaved pseudobulbs, that may become 3 ft. tall, topped by large clusters of numerous small but firm, 3 in. flowers, with sepals and petals rosy-purple, the lip much darker with maroon, and blooming freely and willingly from October to December. Easy to grow as a house plant, and sometimes producing as many as 20 flowers on a stem; in care it requires more watering than most Cattleyas, growing in habitat on rocks and near streams. **C**: 19   p. 461
dowiana aurea (Colombia), "Queen cattleya"; a superb epiphyte with stout pseudobulbs bearing solitary leaves, and beautifully colored, highly fragrant flowers to 6 in. across, the crisped petals and sepals golden yellow, and a great, frilled lip deep crimson-red distinctly streaked or banded with old gold; blooming from May to Autumn. The species dowiana is from Costa Rica and has sepals shaded with red underneath and a lip more completely crimson except for its golden yellow veining. This species is the most gorgeous of cattleyas, and widely used for hybridizing, but not as free-blooming nor easy to grow, also requiring more warmth than others. **C**: 59   p. 461
forbesii (South Brazil), another "Cocktail orchid"; a lovely clustering epiphyte 12 to 15 in. tall, and a sight to see it nesting in small trees in its jungle habitat, loaded with medium-size 3 to 4 in. flowers, the sepals and petals yellow-green, lip white outside, yellow streaked with red inside, borne on stem-like, 2-leaved pseudobulbs, blooming between May and October, and resting in winter. The waxy flowers are quite fragrant and long-lived; a charming species different in its greenish hues. **C**: 19   p. 461
intermedia alba (intermedia parthenia) (Brazil), a "Cocktail orchid"; variety with heavy-textured, pure white flowers with crisped lip, on cane-like pseudobulbs much longer than the species; blooming in spring-summer. C. intermedia is a 2-leaved epiphyte with slender pale rose, fragrant flowers to 5 in. across and having a purple lip; a beautiful, neat plant with the charming blooms usually 3 to 5 or more in a cluster, long-lasting for 3 to 4 weeks if kept cool. **C**: 21   p. 461
lueddemanniana (speciosissima) (Brazil, Venezuela); handsome epiphyte with cylindrical pseudobulbs bearing a rigidly erect solitary leaf; large fragrant, very variable flowers 6 in. or more across, 2 to 5 together, with broad petals a delicate purplish-rose suffused with white, the front lobe of the lip amethyst-purple and crisped; throat yellow, lined purple, blooming July into autumn. **C**: 21   p. 461
mossiae (labiata var. mossiae) (Venezuela), the "Easter orchid" or "Flor de Mayo"; magnificent, free-blooming

epiphyte with pseudobulbs bearing solitary leaves; the handsome fragrant flowers are large, and 5 to 6 in. across, 3 to 5 or more in a cluster, variably blush or light rose, the big, beautiful frilled lip crimson and rose with golden yellow markings often on a suffused white ground, blooming from early March into June. Imported from South America since 1836, it is one of the first Cattleyas introduced to cultivation, and at one time was very popular and regularly timed as an Easter cut flower and for Mother's Day in May, in northern greenhouses, and worn in holiday corsages. C. mossiae is a parent of many excellent hybrids; it is still a reliable beginner's plant, and if kept cool the flowers last 3 to 4 weeks. **C**: 21   p. 461

skinneri (Guatemala to Colombia); robust, free-flowering epiphyte with long, almost cane-like, twin-leaved pseudobulbs to 12 in. or more high; the comparatively small 2–3 in. flowers in clusters of 4 to 10, sepals and petals rose-purple, with lightly pointed dark lip yellowish-white in throat, blooming between March and June or July. Recommended for beginners as a willing house plant. **C**: 19   p. 461

trianaei (Andes of Colombia), the "Christmas orchid"; a lovely winter-blooming epiphyte; 1-leaved pseudobulbs with snowy flowers to 7 in. dia., 2–3 together, sepals and broad wavy petals rose, the frilled lip purple-crimson with yellow throat, normally blooming from December to March. Very valuable because of its flowering time, introduced in 1856, and long imported; grown in northern greenhouses before the advent of modern hybrids, timed for Christmas and late winter. A charming Cattleya adaptable to culture in the home. New growth starts after flowering and continues until August; flower buds develop from October on. **C**: 21
p. 461

CAUTLEYA (*Zingiberaceae*)

spicata (Simla Himalaya); growing in habitat at an elevation of 5000–8000 ft.; rhizomatous herb 2½ ft. high, with dark leafy stems, broad-lanceolate, sessile leaves, and terminal spikes with crimson-maroon bracts, appended to light yellow flowers with tubular corolla, the upper lobe erect, concave, lateral lobes broader and reflexed; calyx red, slit on one side; capsule red, when ripe the 3 valves are reflexed exposing their red lining and numerous small seeds. **C**: 65   p. 206

CEDRUS (*Pinaceae*)

atlantica glauca (No. Africa: Algeria), "Blue Atlas cedar"; magnificent coniferous tree to 60 ft. high or more, a true cedar of open habit with wide spreading branches, evergreen with erect leading shoots, the lateral branchlets with needles in tufted clusters. Needles a beautiful silvery-blue, under 1 in. long; softer than those of the Lebanon cedar, which are green and very stiff. Propagation from fresh seed, or by side-grafting, best on C. deodara. **C**: 26   p. 313

CELOSIA (*Amaranthaceae*);

the cultivated Celosias are striking herbaceous sun-loving annuals from Tropical Asia, with succulent stems and branches terminated by dense chaffy spikes highly colored and often in fantastic shapes; the flowers are small and not individually showy. They thrive in humusy soil with ample moisture and fertilizer, blooming best during the hottest weather, from July until frost. Good pot subjects provided they are never stunted by cold or by being too long in small pots, as they must be allowed to grow rapidly by transplanting into larger container as soon as the roots show, or they will produce small flowers prematurely. Propagated by seed from mother plants producing the fewest seeds.

argentea cristata (E. Indies) "Cockscomb", one of the most spectacular summer-flowering annuals; a tropical fleshy herb 1 ft. or more high with succulent stem and big fresh-green or sometimes bronzy lanceolate leaves; the inflorescence in showy stiff, cockscomb-like velvety fans 6 to 10 in. wide, often very contorted and fluted; most effective in glowing crimson, but available also in yellow or orange color forms. Large flower heads develop best during hot summer. **C**: 58   p. 372, 406

argentea plumosa, the "Feather celosia" or "Burnt plume"; herbaceous annual about 1 ft. high in bushy types, with fleshy stem and fresh green ovate leaves; the branches terminated by dense erect spikes of feathery plumes, highly colored fiery red as in 'Fiery Feather', or yellow such as in 'Golden Feather'; developing best in high heat with plenty of sunshine. **C**: 58   p. 372

argentea pyramidalis (Trop. Asia), "Plume celosia"; a taller growing annual 2 to 4 ft. high with erect stems bearing

spectacular pyramidal plumes with silky threads, and looking like the feathery inflorescence of Pampas grass, but in brilliant colors of luminous crimson-red, bronze, copper or golden yellow, carried well above the fresh-green foliage. Useful cut flowers in late summer; may also be dried for winter bouquets. **C**: 58   p. 406

CENTAUREA (*Compositae*)

candidissima (So. Italy: Sicily), "Woolly-white miller"; small herbaceous perennial 6 in. or more high, densely covered with white matted wool, the foliage at first spoon-shaped and lightly crenate, later pinnately parted and lobed in maturity; inflorescence golden yellow. Attractive velvety all-white plant for pots or bedding. Best from seed. **C**: 13
p. 410

cineraria (gymnocarpa) (So. Italy: Capri, Sicily; No. Africa), "Dusty miller"; herbaceous perennial to 1 ft. or more high, used for bedding; densely white-felty, leaves two pinnate cut into narrow, irregular lobes; the inflorescence a large head of thistle-like lavender flowers with purple style, sometimes yellow, blooming in summer. Marvelous container plant for the open dry patio, its white foliage glistening in the sun. From seed. **C**: 13   p. 410

CENTRATHERUM (*Compositae*)

intermedium (Brazil), "Brazilian button flower"; herbaceous bush with serrate leaves, bearing at the ends of branches long-lasting, bluish-lavender, fluffy button flowers 1 in. across, over a long blooming period. Needs to be pinched for bushy shape. Easy from soft cuttings. **C**: 14
p. 493

CEPHALOCEREUS (*Cactaceae*)

russellianus (Colombia, Venezuela); attractive column growing into trees 25 ft. high, branching at top, satiny moss-green or glaucous, 4–6 ribs, with channels curving back from the notched rib, the areoles set with white hair and short brown spines; flowers cream. **C**: 1   p. 153

senilis (Mexico), "Old man cactus" or "Beatle"; slender gray-green column closely ribbed, and with translucent white needle spines, covered with long silver-white matted hairs; nocturnal flowers rose-colored. With great age becoming 20 to 35 ft. high. **C**: 1   p. 154

CEPHALOTUS (*Cephalotaceae*)

follicularis (Southern W. Australia), the "West Australian pitcher plant"; a charming, curious little herbaceous perennial rosette only 2–3 in. high, with creeping rhizome; forming two types of leaves, one normal spoon-shaped flat, green and fleshy in the center of the rosette; later the other in form of a curious little Nepenthes-like pitcher 1–2 in. high on the outer margins of the plant and hugging the ground, with down-pointing teeth inside the trap mouth, holding water and carnivorous; pale green with maroon tinting, and shielded by a red-veined lid; tiny buff flowers on 2 ft. stalk. Grown in small pots in sandy humus with chopped sphagnum added, set inside larger pot with sand, the surface covered with fresh sphagnum moss; cool and light, under glass. Propagated by division, or leaf cuttings in spring. **C**: 72   p. 522

CERATOPTERIS (*Filices: Parkeriaceae*)

thalictroides (in quiet waters of the Tropics of Asia, Africa and America), the "Water fern", a true aquatic fern with rosettes of light green, succulent fronds; barren ones floating, and forming young plantlets along the lobed margins; the fertile fronds divided into linear segments. Variable species depending on place where they live; landforms, with staghorn-like fronds; half-submerged, with floating leaves; floating, with leaves swimming on the water. A favorite water fern for indoor aquaria and for ponds. Prefers good light and warmth, best in soft or rain water. Propagated by division. Winter in shallow dish in moist peatmoss, warm and light. **C**: 2 or 54   p. 252, 519

CERATOZAMIA (*Cycadaceae*)

sp. 'Hilda' (Mexico), "Bamboo zamia"; palm-like cycad with only short trunk, bearing stiff pinnate fronds 3 to 4 ft. long, in shade even 8 to 10 ft. tall; the bamboo-like axis beige and lightly armed with prickles; thin-leathery, shiny grass green leaflets 5–6 in. long, whorled in pairs nearly around stem. Quite unusual in its type. **C**: 60   p. 316

spiralis (Mexico), "Spiral horn-cone"; unusual looking Zamia-like cycad with short trunk carrying slender pinnate leaves about 3 ft. long, having axis without spines and scattered, very slender, narrow linear matte green leaflets 6 in. long by ¼ in. wide. Rather lacy introduction for collectors. **C**: 60   p. 316

CEREUS (*Cactaceae*); these fast-growing tropical "Column" or "Torch cactus" are usually massive and treelike, but leafless, with ribbed, fleshy stems of continuous growth, with age branching, and characteristic in drier regions of the subtropics and tropics, although similar in appearance to many separated genera of the same subtribe Cereanae; the areoles are more or less spiny; the white flowers are funnelform and night-blooming, but the popular "Night-blooming Cereus" belong to other genera. The fleshy fruit looks like apples and is edible in some species. Cereus are striking decorator plants for desert effect. They love sun and warmth; will, however, tolerate limited light reduction down to 50 fc. but then must be kept quite dry between only occasional waterings, especially in winter. Don't feed unless in good light. Requiring practically no care, and tough even when under airconditioning. Propagated by fresh seeds; cuttings are dipped in charcoal and allowed to heal for several days before placing them shallowly on warm sand to root.
  hexagonus (Colombia), "South American blue column"; tall tree type to 45 ft., branching near base; glaucous bluegreen columns, deeply 4–6 ribbed, few spines; flowers white. **C**: 1   p. 152
  "peruvianus" hort., a popular "Column-cactus"; the commonly cultivated tree-cereus in the trade, possibly a hybrid; fleshy columns 6–9 ribbed, glaucous bluish-green free branching; few brown spines; whitish flowers. **C**: 1   p. 152
  peruvianus monstrosus, "Curiosity plant"; an irregular growing monstrose form constantly developing new heads, which are a pretty glaucous bluish-green; slow growing but always retaining its character. No need to graft as this is a plant of robust habit. Sometimes a young seedling will develop many dense heads, retaining its brownish spines and remain dwarf, as illustrated in the "juvenile form". **C**: 1   p. 168, 509, 532
  triangularis: see Hylocereus undatus

CEROPEGIA (*Asclepiadaceae*); the "Lantern flowers" are succulent half-shrubs or perennial herbs; with fleshy stems and leafless, or twining and with opposite leaves, often with roots tuberous or corm-like. Lantern-like flowers from the nodes with corolla lobes often attached at the tips forming parachutes. Curious and easy house plants. Keep moderately moist and airy in summer, drier in winter; in well drained soil. Propagation by cuttings; C. woodii also by the tubers which form at the internodes of vines.
  caffrorum (Eastern So. Africa: Natal to Mozambique), "Lamp flower"; wiry vine with small fleshy, grass-green 1 in. ovate leaves in pairs, with red veins and petioles; ½ in. flowers green outside with purple lines, purplish black inside. Ideal as a hanging vine with string-like strands. **C**: 9   p. 272
  debilis (Rhodesia, Zambia, Malawi), the tropical "Needle vine"; threadlike vines, from corm-like; tuberous roots, pendant or twining, the small succulent leaves narrow-linear about 1 in. long, upper surface grooved and green, the flowers 1 in. long, tinged red, inside bright red, blooming almost all year. **C**: 9   p. 272
  fusca (Canary Is.), "Carnival lantern" or "Stick plant"; a remarkable succulent of shrubby habit with at first upright later spreading, forked, chalk-white to purplish cylindrical pencil-like columns ½ in. thick, constricted at joints; curious lantern-like flowers brown, and inside yellow, carried erect from the branches of the preceding year's growth. Keep very dry in winter. **C**: 19   p. 533
  haygarthii (Natal), "Wine-glass vine"; twining, succulent stem with small ovate 2 in. leaves; the flower like a fluted wine glass with bent stem, cream with specks of maroon, covered by maroon parachute to 2 in. across; from center rises a maroon stalk topped by red knob, sparsely covered by white hair. **C**: 69   p. 272
  radicans (So. Africa), "Umbrella flower"; succulent, prostrate creeper with rooting stems, and small ovate, dark green leaves to 1¾ in. long; flowers 3–4 in. long, a slender tube expanding into joining lobes zoned in purple, white and green. **C**: 69   p. 272
  sandersonii (So. Africa: Natal, Mozambique), "Parachute plant"; twining, quick-growing succulent with small

ovate, fresh green, fleshy leaves to 2 in. long, chiefly on younger tips; large flowers 2 in. long, formed like a parachute, light green and mottled dark green. One of the most curious lantern flowers; prefers moderate temperatures. **C**: 69   p. 272
  woodii (Rhodesia, Transvaal, Natal), "Rosary vine" or "Heart vine", also called "String of hearts", because of the trailing or hanging, threadlike purplish stems, set with pairs of succulent, heart-shaped or kidney-shaped, bluish leaves marbled with silver, less than 1 in. wide; gray-green to purple beneath; the vines grow from large corm-like roots; small purplish flowers. At the internodes on vines also form little tubers which may be used to start new plants. Charming house plant with long pendulous strands for basket or hanging pot. **C**: 9   p. 272

CESTRUM (*Solanaceae*)
  nocturnum (West Indies), "Night jessamine" or "Queen of the Night"; evergreen shrub to 6 to 10 ft. high, with wiry, brownish branches; shining, thin-leathery, 3 to 6 in. ovate leaves, and small, greenish or creamy-white, slender tubular flowers with pointed lobes, to 1 in. long, axillary in profuse clusters, intensely perfumed at night, and blooming mostly from July to September, but at intervals throughout the year. Propagated by cuttings of half-ripened wood. For powerful fragrance in the sunroom, or the patio outdoors. **C**: 2 p. 486

CHAENOSTOMA (*Scrophulariaceae*)
  grandiflorum (Sutera) (So. Africa), "Purple gloryplant"; sticky pubescent small subshrub to 4 ft. high, with 1½ in. ovate, toothed leaves, and showy terminal racemes of deep purple tubular flowers, with spreading petals 1¼ in. wide. Attractive for their compact habit and profuse blooms, for the sunny window. From seed or cuttings. **C**: 13   p. 280

CHAMAEALOE (*Liliaceae*)
  africana (So. Africa: Cape Prov.), "False aloe"; low, mat-forming stemless rosette with numerous fleshy leaves, broad at the lower end, together forming a bulb-like base; narrow linear to awl-shaped toward the apex, 4 in. long, the upper surface grooved, pale green with distinct lines lengthwise; margins with small white teeth, the underside with raised creamy spots; small greenish-white tubular flowers on 8 in. stalk. Cultivation like Haworthia; a rare, somewhat delicate collector's plant. **C**: 53   p. 350
  'Africana hybrid' (Chamaealoe africana x Aloe parvula); beautiful bigeneric California creation by Hummel; a dense succulent rosette 8 to 10 in. across, of slender, soft-fleshy erect spreading leaves about 6 in. long arranged spirally in ranks; leaves rounded on bottom, flat on top, glaucousgreen and decorated all over with long, soft creamy spines. A promising new house plant. **C**: 15   p. 350

CHAMAECEREUS (*Cactaceae*)
  silvestri (Argentina), the miniature "Peanut cactus"; little clusters of short cylindric, fresh green branches covered with rows of soft bristle-like white spines; flowers a beautiful orange-scarlet. **C**: 13   p. 157

CHAMAECYPARIS (*Cupressaceae*)
  obtusa (Japan), "Hinoki cypress" or "False cypress"; slow growing evergreen coniferous tree of irregular habit 40 to 50 ft. high or more, with horizontal branches, the branchlets flattened and frond-like, scale-like leaves densely clothing them glossy green; fruiting with small brown ½ in. cones. Excellent container plant with Oriental character, developing into contorted specimen when older; for cool locations. Also splendid Bonsai subject. Prune to shape. From summer cuttings. **C**: 26   p. 312
  obtusa nana (Japan), "Dwarf Hinoki cypress"; tough, slow growing, very dwarf and compact form of an otherwise large, spreading tree, growing to 2 ft. high and wide, with glossy dark green, scale-like leaves dense in flat, fan-like horizontal branchlets. A miniature which takes years to increase in size and a much-sought-after pot or container plant as diminutive evergreen, yet with all the features of a large tree. **C**: 26   p. 310

CHAMAEDOREA (*Palmae*); the "Dwarf mountain palms" or "Bamboo palms" are shade-loving, small and graceful feathery palms with slender green, single or clustered stems, and rich green pinnate fronds. They have only recently been discovered by decorators for their outstanding worth as durable and compact, long-lasting container plants for decor in offices and corridors of modern buildings; they require a minimum of space; take airconditioning well, and tolerate locations of poor light down to 40 fc.; in bright sun the foliage will bleach. These palms prefer small pots or tubs; should be watered regularly but don't resent an occasional drying out; when established occasional feeding will keep them in growth. Chamaedoreas will bloom while quite young; male and female flowers are usually carried on different plants (monoecious), and must be pollinated to set their clusters of red, orange-yellow or purplish berry-like fruit. Fresh seed germinates easily on moist warm peatmoss; the suckering species may be propagated by division.

costaricana (Costa Rica), "Showy bamboo palm"; a clustering palm of great beauty, with bamboo-like dark green canes to 15 and 18 ft. high, furnished from top to bottom with graceful fronds pinnate to the very tips, solid green not glaucous, longer and holding better than erumpens, and with some tolerance to cold. **C**: 4   p. 231

elatior (Mexico), "Mexican rattan palm"; a tree-like but actually climbing rattan-type palm reaching 30 ft. high; slender green cane ¾–1 in. thick; fronds pendant becoming 4–6 ft. long, the widely separated broad, ribbed, glossy green leaflets in almost opposite pairs, the lower ones to 10 in. long, toward the hanging tips only 5 in.; in seedlings simply two-lobed. Inflorescence near base; black berries in panicles. **C**: 4   p. 231

elegans (Collinia) (Mexico), "Dwarf mountain palm" or "Parlor palm"; small graceful, relatively fast grower with stiff, slender stem, to 8 ft. high, 3 times as high as Neanthe bella and eventually forming clusters; pinnate leaves, loosely spirally arranged, the thin-leathery, dark green leaflets broadly lanceolate; clusters of berry-like fruit yellow to white, larger and more round than 'Bella'. Excellent keeper in shady locations, from 40 fc. up; takes airconditioning well. Ideal for pottery planting and considered the most satisfactory small living room palm today, tolerating all kinds of abuse. Wonderful as seedlings for dishgardens and terrariums. **C**: 4   p. 231

elegans 'bella' (E. Guatemala), "Neanthe bella"; a tree palm in miniature, from the mountain forests of Mayaland, more dwarf in habit than Chamaedorea elegans and slow-growing; thin stems bearing near the top a graceful rosette of small, pinnate fronds, with narrow, hard-leathery dark green pinnae 4–5 in. long; will flower as a youthful plant in pots. **C**: 6   p. 231

erumpens (Honduras), well described as the "Bamboo palm"; suckering dwarf palm, forming bushy, erect clusters of thin, bamboo-like reed-stems, loosely furnished down to the base with short pinnate, drooping leaves; the segments are broad, almost papery, dark green and recurved. The "Bamboo palm" is a favorite of decorators because of its oriental appeal and good keeping qualities, even under airconditioning once well established; shade tolerant but should have 40 fc. up. **C**: 4   p. 219, 231

glaucifolia (Guatemala), "Blueleaf chamaedorea"; slowly climbing bamboo-like solitary stems to 22 ft. high, draped with slender pinnate fronds 4 to 6 ft. long, the many well separated long and narrow leaflets glaucous gray over dark green; inflorescence between the foliage. **C**: 4   p. 231

metallica, introduced in hort. as tenella (Mexico), the "Miniature fishtail"; one of the smallest of palms, slowly developing a stiff stem to 3 ft. high, and which bears a rosette of broad, leathery, shining green entire leaves forked only toward apex, and distantly toothed; blooming when quite young with simple, short spadix bearing minute yellow flowers. A tough and unusual small decorative plant. **C**: 4   p. 231

seifrizii (graminifolia) (Mexico: Yucatan), a "Reed palm"; small stoloniferous palm with clustering slender cane-like stems alternately furnished near the top with broadly spreading pinnate fronds, their pinnae long and narrow and spaced well apart, giving the plant a lacy appearance; withstands wind, sun, and some cold. **C**: 4   p. 231

stolonifera (Guatemala), "Climbing fishtail palm"; stoloniferous small palm to 6 ft. high, covering, in time, large areas by running stolons; the plaited leaves are satiny green and slightly glaucous and remain simple or bifid, on slender rattan-like cane ¼ in. thick. **C**: 6   p. 231

x CHAMAELOPSIS (*Cactaceae*)
  'Firechief' (Chamaecereus x Lobivia), striking Johnson

bigeneric "Firechief hybrid", forming dense clusters of stubby finger-like olive-green columns bearing needle spines and white radials, 9–10 ribbed; free-blooming brilliant flame-colored flowers to 3 in. across. **C**: 13   p. 157

CHAMAERANTHEMUM (*Acanthaceae*)
  gaudichaudii (Brazil); densely branching tropical herb with creeping wiry stems, thick with small 2–3 in. oval leaves dark green, painted silver-gray along center and into the primary veins; small lavender flowers with white star in center, in short raceme. A favorite old ground-cover in conservatories. **C**: 54   p. 186
  igneum (Xantheranthemum) (Peru); low, spreading herb with beautiful, dark brownish-green, oblanceolate, veloured leaves, and a vein pattern in red to autumn-yellow; bright yellow flowers. **C**: 60   p. 43*, 184
  venosum (Brazil); dwarf plant with wiry stems spreading close to the ground, its small, hard-leathery 3–4 in. ovate leaves grayish, and attractively marked with a network of silver along veins. **C**: 54   p. 186

CHAMAEROPS (*Palmae*)
  excelsa: see Trachycarpus
  humilis (Western Mediterranean: Spain, Morocco), the "European fan palm"; usually dwarf, cluster-forming, sometimes to 20 ft. high in arborescent forms, with rough trunks covered by old leaf bases, tough dull green leaves relatively small, 2–3 ft. across, stiff and folded, gray-scaly beneath, with many narrow segments nearly to the base, on very thorny, flat stalks; small red-brown fruit. As a low spreading tub plant for cold interiors, or the patio in summer outdoors; tough but not particularly attractive. **C**: 14   p. 223, 229

CHEIRANTHUS (*Cruciferae*)
  cheiri (So. Europe), "English wallflower"; charming old house plant especially in Europe, where it is cultivated since ancient times for the cool window or winter garden, blooming in spring; perennial becoming woody and usually grown as a biennial; erect bushy plant 1 to 1½ ft. high, with narrow, bright green leaves 3 in. long; the sweet-scented 1 in. flowers, often double, carried in terminal clusters, resembling stocks but in a different color range of velvety tones of yellow, orange, brown, red, pink and burgundy. Sow seeds in spring for bloom following year; flowering is best if seasoned by near frost winter exposure. Propagated also by cuttings. **C**: 75   p. 416

CHEIRIDOPSIS (*Aizoaceae*)
  candidissima (So. Africa: Cape Prov.), "Victory plant"; beautiful tufting, highly succulent plant with one or two pairs of long, prismatic or boat-shaped, silver-gray leaves 3 to 4 in. long, united at the base and spreading in the "V" for victory sign; an exciting plant in glistening white, and with lovely pink flowers 2½ in. across. From seed. **C**: 13
                                                                p. 356

CHIONODOXA (*Liliaceae*)
  luciliae (Turkey: mountains of Anatolia; Crete), "Glory of the snow"; blooming in early spring as soon as the snow has melted; charming, small bulbous plant to 8 in. high; related to Scilla and hardy, with narrow bright green basal leaves and attractive, funnel-shaped flowers expanded star-like, 1 in. across, intensely bright blue, paling to white in the center, in loose clusters on short stalk. Will naturalize and spread in the garden. For pots, plant bulbs in fall and keep cold and dark until good roots have formed, then bring into a light house near the glass; water freely from April to June, when leaves disappear keep dry. By offsets. **C**: 78
                                                                p. 398
  sardensis (Asia Minor: Turkey), "Glory of the snow"; charming early-spring bulbous herb to 8 in. high with narrow, channeled basal leaves, and a very short stalk with clusters of all-blue flowers having long spreading lobes, 1 in. across, on nodding stemlets. Coming from cool mountains, Chionodoxa are winter-hardy in gardens but may be grown in pans for early flowering, potting in September, and bury outdoors until January, then bring into cool window or greenhouse. By offsets. **C**· 76   p. 394

CHIRITA (*Gesneriaceae*); herbaceous somewhat succulent plants from tropical Asia, with handsome foliage and gloxinia-like funnel-shaped though small flowers, most often bluish to white, occasionally in shades of yellow, blooming from winter into summer. They like warmth with humidity, and a humusy soil as for gloxinias. Propagated from seeds; also by large leaves, cutting their main-ribs through and laying them flat on sandy soil, in summer.

lavandulacea (Malaya), "Hindustan gentian"; erect, charming, rather succulent plant 1 to 2 ft. high, with erect branching stems, supporting pairs of large ovate, soft-hairy, light green leaves with toothed margins, to 8 in. long; and whorls of axillary flowers with white pouchlike corolla tube 1 to 1¼ in. long, and a spreading limb of pale lavender blue, marked yellow in throat. Easy to grow, this species is influenced by day length, and flowers during the long days of summer; under light it may be bloomed at other times. **C**: 60   p. 444

micromusa (Thailand); rapid-growing tropical plant of succulent habit, with stout stems and large shiny, bright green cordate leaves softly hairy; small flowers about 1 in. long, brilliant chrome-yellow with dark orange in throat. Of curious appearance and deserving its name "Little banana". **C**: 60   p. 444

sinensis (So. China), "Silver chirita"; beautiful stemless plant 6 in. high, with thickened, persistent roots; large quilted, thick-fleshy ovate leaves 3 to 4 in. long, set cross-wise; white-hairy over deep green and silver variegation, with lightly toothed margins carried on flat petioles; lavender-lilac 1½ in. flowers gloxinia-like, in many-flowered clusters on red-hairy stalks, blooming in summer. **C**: 60
p. 439

CHLORAEA (Orchidaceae)

membranacea (Argentina); a little-known, choice terrestrial orchid with finger-like subterranean tubers producing a cluster of oblong leaves, and a shaft topped by an elongate cluster of creamy-white flowers slightly tinged green, 2 to 2½ ft. tall, blooming in Spring. Growing in pots in well-drained compost with sandy soil, the plant needs moisture and fresh air while actively growing; rested several months after blooming when the leaves die down. Divide and repot at least every two years. **C**: 82   p. 462

CHLOROPHYTUM (Liliaceae)

bichetii (Trop. W. Africa), "St. Bernard's lily"; small, pretty tropical herb with broad, but grass-like, thin-leathery leaves fresh-green with yellowish-white stripes, forming bushy tufts. By division. **C**: 4   p. 202

comosum 'Variegatum', "Large spider plant" or "Green lily" in Europe; large rosettes with fleshy roots and arching, fresh green, linear leaves 10 to 16 in. long, ¾ to 1 in. wide, having margins prettily edged in white; long arching stalks appearing from the center will first produce small white flowers, then develop tufts of leaves with aerial roots. A modest old house plant favorite. The green-leafed species is from South Africa. Propagated by division, or plantlets on stolons. **C**: 16   p. 202, 275

comosum 'Vittatum' (C. elatum vittatum) (So. Africa: Cape of Good Hope), "Spider plant"; attractive low, clustering rosette with fleshy roots and channeled narrow-linear, soft, thin-leathery, recurving leaves 4 to 8 in. long and ½ in. wide, dark green banded white along the middle; successive plantlets develop from the long, wiry flowering stalks which become pendant; a good old basket plant. Propagate from plantlets. **C**: 16   p. 202, 275

CHOISYA (Rutaceae)

ternata (Mexico), the "Mexican orange"; handsome evergreen shrub 3 to 6 ft. or more high, with aromatic foliage divided into 3 ovate, lustrous yellow-green leaflets to 3 in. long, and clusters of fragrant white 1 in. flowers somewhat like orange blossoms open in early spring and blooming into April and through summer. Durable, long-blooming decorator plant. Propagated by cuttings. **C**: 14   p. 485

CHORIZEMA (Leguminosae)

cordatum, better known in horticulture as ilicifolium; (Western Australia), the "Australian flame pea" or "Flowering oak"; sprawling shrub 3 to 8 ft. high, but usually much smaller, with thick, wiry branches and small leathery, glossy-green, ilex-like ovate or cordate 1 to 2 in. leaves sharply toothed and spine-tipped; small flowers a gaudy red-orange with a yellow basal blotch and purplish keel; blooming from December to June or later. Charming both as a miniature house plant or in hanging baskets, or larger containers for the sunny patio; may be pruned to shape. From cuttings of fairly firmed wood in March or in late summer. **C**: 13   p. 501

CHRYSALIDOCARPUS (Palmae)

lutescens (Areca) (Madagascar, Mauritius, Reunion), widely known as "Areca palm", also "Yellow butterfly palm"; growing in dense clusters with slender, graceful, yellowish stems, its suckers forming an attractive, compact bushy clump to 10—20 ft. tall, with pinnate foliage nearly to the base; narrow, papery pinnae glossy yellow-green and well spaced, on yellow, willowy, furrowed stalks; fruit violet-black. Very popular and grown in warmer climates in combinations as a container-grown decorator palm, but in colder climates not as durable as Howeia or Chamaedorea. Requires warmth, and when grown indoors, keep water in the saucer beneath the plant. **C**: 4   p. 221, 225

CHRYSANTHEMUM (Compositae);

mostly perennial herbs of the daisy family with strong-scented foliage, and flowers in showy terminal heads made up of individually small flowers, the florets modified by long cultivation into petal-like marginal ray flowers and small, usually tubular disk-flowers forming center cushions. Horticulturally most important are the showy florists' chrysanthemum. Popularly known under the nickname "Mums" they are cultigens of Chinese origin of the species morifolium cultivated in the Orient for 3000 years. Mums are "short-day" plants, setting buds when daylength is 12 hours or less, normally blooming in autumn but their flowering time may be advanced by long-night treatment of 12 to 15 hours beginning 10 to 12 weeks before blooming date wanted; cover with black cloth or place under a box or in a dark cabinet until buds show. Flowering may be delayed by short-night treatment giving light in the middle of the night. Potmums have justly become favorite decorators because they bring a wide range of color indoors lasting for weeks. Largest blooms can be obtained if sidebuds are removed while small. Mums need sufficient light, 100 fc. up to open their flowers, prefer coolness for best keeping, and an even moisture. Propagated from soft cuttings; the morifolium types also by division.

frutescens (Canary Islands), "White marguerite" or "Paris daisy"; a tender branching herbaceous plant of bushy habit, to 3 ft. high, becoming woody at base; with lacy, finely divided grayish-green, somewhat fleshy leaves, and numerous single white daisy flowers with yellow disk 1½ to 2½ in. across, on slender stalks. A pretty pot plant for spring blooming; from soft cuttings. **C**: 75   p. 370

frutescens chrysaster, the "Boston yellow daisy"; a form with fleshier stem, fresh green, fern-like divided leaves, and lemon-yellow ray-flowers in daisy-like heads heavier and not as willing or free-flowering as the white type. **C**: 75
p. 370

morifolium 'Beautiful Lady', large "Anemone mum" of bushy habit, with strong stems and good foliage; flowers semi-double lavender pink, lighter toward center, and a yellow cushion; blooms Nov. 25. **C**: 63   p. 381

morif. 'Delaware', a typical good "Incurved" type commercial, free-branching pot plant; with firm foliage, and semi-incurved double flowers, deep amaranth-red bleaching to bronze-red; reverse bronzy-yellow; normally blooming Nov. 8, this proven commercial hybrid is valuable because it may be grown and flowered during the heat of summer. **C**: 63   p. 381

morif. 'Golden Cascade', commercial "Cascade mum" of the single daisy type; very floriferous with light yellow ray-petals around a golden yellow center disk; earlier than 'Jane Harte'; its dense habit with long wiry branches is equally adaptable to training to forms or trellis as well as into broad bushy globes, blooms Oct. 30. **C**: 63   p. 381

morif. 'Golden Lace', the strange "Spider mum" or "Fuji mum"; morifolium cultivar of the spider type, branching with very slender stems, bearing large flowers with graceful threadlike, golden yellow tubular quilted florets, somewhat pendulous from the double center, the tips are open and curled; normal blooming date Oct. 25. For that oriental touch in flower decoration. **C**: 63   p. 381

morif. 'Handsome', full "Pompon" type mum, with large size flowers purplish rose, in sprays on stiff stems with closely internoded foliage, and of erect bushy habit; ideal for pots or cut flowers; blooms Nov. 1. **C**: 75   p. 381

morif. 'Princess Ann', an excellent "Pot mum" of the decorative type; popular pot plant of compact, shapely growth habit with firm foliage and substantial flowers 4 in. across, broad-petaled, peach-pink, slightly darker in center paling to light pink at edges; normal bloom Nov. 10. Very satisfactory keeper. **C**: 63   p. 364, 381

morif. 'Yellow Delaware'; intermediate "semi-incurved" sport of 'Delaware'; a shapely, bushy plant with rich yellow flowers, 4 in. across when grown in pots, larger if planted out; when ripe the lower petals turn outward exposing the center as if semi-double; normal blooming date Nov. 8. **C**: 63   p. 365*

parthenium (in horticulture as Matricaria capensis) (Asia Minor to Caucasus, naturalized in Europe), the florists "Feverfew"; bushy, strong-scented perennial herb 1 to 2 ft. high, with lacy leaves and masses of small white, daisy-like, usually double ¾ in. flower-heads in open clusters, blooming from June to September. The doubles are a favorite cut flower, as early as April-May if grown in the cool greenhouse, from January cuttings. Also good bedding plant; from seed only 90% double. **C**: 76   p. 414

ptarmicaeflorum (Canary Islands), "Silver lace"; white-tomentose, shrubby perennial 1 ft. and more high; lacy, silvery leaves bipinnately parted much as in certain achilleas, and topped by flat clusters of small flower heads with white ray petals. Very attractive with silvery foliage glistening in the sun. Primarily ornamental and of easy care. **C**: 13   p. 322

CHRYSOTHEMIS (*Gesneriaceae*)

pulchella (Trinidad), erect succulent herb producing tubers with age; large fleshy opposite, rough-bristly, puckered 6 in. leaves shiny bright green, with crenate margins; the flowers in the leaf axils are small, bristly ½ in. tubes buttercup-yellow with red stripes and markings inside, but curiously interesting for the cluster of large flame-red calyces which appear well before the blossoms and keep long afterwards, giving the plant a colorful appearance over a long period. Blooms in spring; allow to go dormant when too tall. Propagated from cuttings; from offsets in leaf axils; by seed or from tubers. **C**: 10   p. 444

CHYSIS (*Orchidaceae*)

aurea (Mexico, to Venezuela, Colombia and Peru); charming deciduous epiphyte with spindle-like stems bearing 3–5 soft-textured folded leaves; the thick-waxy, brightly colored 2 to 3 in. aromatic flowers closely clustered in short raceme, and appearing with the flush of new growth; petals reddish-buff with golden yellow edges and base, throat yellow marked with red. Mostly blooming from May to August, but the long-lasting flowers may appear twice a year. After growth is finished, keep cool and on the dry side to rest until the plant begins to grow again. By division. **C**: 21   p. 459

CIBOTIUM (*Filices*); small group of stout, slow-growing tropical treeferns of the family Dicksoniaceae, growing 30 ft. and more tall, their trunks covered with thick matted fiber, and bearing majestic crowns of light or fresh green fronds mostly tripinnate and very lacy. They are wonderful decorator plants for tropical effect; provided they are kept warm, shady and quite moist at all times. Trunks or trunk sections without foliage of the large Hawaiian treeferns are sent from the islands and start to grow readily if kept constantly moist. The smaller C. schiedei will tolerate more abuse, even occasional drying out; these ferns must be grown from spores.

chamissoi (Hawaii), formerly and still popularly known as C. menziesii in Hawaii, also locally as the "Man tree fern"; huge treefern to 35 ft. tall, with monstrously stout, fibrous trunk 8 to 12 in. thick, the upper parts covered with yellow-brown to black-brown hairs, bearing a magnificent crown of handsome, massive, tripinnate fronds in color a delicate, glossy light nile-green of a smooth leathery-hard texture; on stalks thickly covered with stiff blackish hairs. For its artistic decorative value firm top sections from one to several feet long are sawed off and can be grown in shallow bowls filled with rocks and water, or in soil in pots where they will root, and their uncoiling fronds, with several older ones, present exotic decor, and are known to keep well indoors. Larger tubbed specimen also last indefinitely in a moist-warm, light location, or the patio outdoors, but the fibrous trunks must be kept constantly wet. **C**: 10 p. 237

glaucum (Hawaii), formerly known as C. chamissoi and known in Hawaii as "Hapu'u", the "Blonde" or "Female Hawaiian treefern" of the rainforest; with stout, fibrous trunks to 16 ft., but usually less, consisting of a mass of aerial roots around a core of starch, capable of storing moisture, and covered, especially toward the apex with limp golden yellow-brown hairs; bearing heads of large, soft-textured, tripinnate crinkled fronds 3 to 8 ft. long, luxuriously green and smooth above, and grayish glaucous beneath; the stalks covered with silky, fawn-colored wool (pulu). (The similar C. splendens is different by having cobwebby hairs on the underside of the leaf). Top sections sawed off are used for their decorative effect, grown in bowls of water or coarse soil kept wet. **C**: 10   p. 237

schiedei (mountains of Chiapas in Southern Mexico; also Guatemala), "Mexican treefern"; with shapely crown of graceful, lacy, tripinnate fronds from 2 to 5 ft. long, thin-leathery with matte-satin surface, light apple green, glaucous beneath; will eventually, over many years, form a fibrous trunk to 15 ft. high. In cultivation, probably the most favored fern for indoor decoration, much used by florists, usually as a small tub plant with arching fronds spreading 3–5 ft. in dia.; in spite of its dainty appearance yet very durable, even taking some abuse. Soil may become somewhat dry but then should be soaked thoroughly with water. Main enemy mealybugs in dry air. Propagated from spores only. **C**: 10   p. 234, 235

CINERARIA: see Senecio, Centaurea

CIRRHOPETALUM: see Bulbophyllum

CISSUS (*Vitaceae*); except for some thick-fleshy African succulents, mostly tendril-climbing grape-like vines, some moderately, others quickly scrambling high on supports, or pendulous, many of them popularly grown. Similar to Cissus and often confused with them are Vitis, Piper, Ampelopsis, Parthenocissus, Rhoicissus, Tetrastigma. Cissus is from the Greek word for ivy. Prop. by cuttings.

adenopoda (njegerre), (Tanzania: Usambaras; Uganda), "Pink cissus"; rapid herbaceous climber with tuberous roots, the slender vines with large trifoliolate, green leaves having a coppery sheen and covered with purple hair, the stalked leaflets coarsely serrate and with sunken veins, wine-red beneath; flowers pale yellow. A good indoor vine in spite of its soft appearance. **C**: 4   p. 257

albo-nitens (Brazil), "Silver princess vine"; rapid tropical climber, with small, herbaceous, arrow-shaped leaves shining metallic silvery-gray on surface, the veins green, and green beneath. There is the possibility that this silvery vine is merely the juvenile stage of the green-leaved sicyoides. **C**: 10   p. 258

antarctica (Rhoicissus) (Australia: New South Wales), "Kangaroo ivy" or "Kangaroo vine"; attractive vine with shrubby base and flexuous, slowly climbing, tomentose branches; the elegant, firm leaves to 6 in. long, almost leathery, bright shining green to a deep metallic shade, and light brown veins, the base cordate, and margins sinuately toothed. A good keeper, and excellent indoor vine, tolerating warmth and air-conditioning; needs good light, 50 fc. up; in bright location add fertilizer. **C**: 15   p. 257

discolor (Java, Cambodia), "Rex begonia vine", beautiful tendril-climber, with thin, angled, dark-red vines and petioles; the strikingly colored, oblong-cordate, quilted leaves 4 to 6 in. long and with toothed margin, the sunken network of veins moss-green, with elevated ridges painted shimmering silver, and the center variegated violet and red-purple with velvet sheen; reverse glowing maroon. A gorgeous foliage vine, best in the warm greenhouse on wire support or trellis. **C**: 60   p. 256

erosa (Puerto Rico); rambling tropical tendril climber along the waysides over hedges and small trees; thin wiry vines bearing compound, stiff succulent, trifoliolate, glossy leaves waxy fresh-green, the obliquely ovate, stalked leaflets about 4 in. long, coarsely toothed at the margins, and pale beneath. **C**: 4   p. 258

gongylodes (Brazil, Paraguay, Peru); robust liane for the large conservatory, rapid climber with 4-angled stems, large trifoliolate rugose leaves to 8 in. long; hairy at the nerves and at the margins; the tendrils with suction cups. Most attractive are the long pendant, red aerial roots looking like a bead curtain. At end of growing season the ends of branches produce a terminal fleshy tuber which drops to the ground and develops into a new plant. **C**: 54   p. 258

quadrangularis (Vitis) (Trop. Africa, Arabia, India, Moluccas), "Veld grape"; odd succulent with thick, fleshy, rich green 4-winged stems much constricted at the nodes; climbing, mostly glabrous, often nearly leafless, looking like a spineless cactus; small herbaceous, 3-lobed, green leaves; green flowers, berries red. I have seen this plant clambering over euphorbias and various shrub on the steppes in Uganda, Tanzania and northern Transvaal. **C**: 51   p. 258

rhombifolia (in American horticulture as Vitis, in Europe under Rhoicissus) (West Indies, N. So. America), the "Grape-ivy"; scandent, herbaceous plant with vine-like, flexuous, brown-hairy branches having coiling tendrils, and compound leaves 3–4 in. long, of 3 rhombic-ovate, stalked, thin-fleshy leaflets wavy-toothed, the glabrous, somewhat quilted surface fresh green to metallic deep green and with brownish veins and petioles, pubescent underneath, the young growth covered with white hairs. Small flowers with 4 petals, unlike Rhoicissus which have 5–7 petals. One of the best ramblers for living room or office; also under air-conditioning. Needs fair light, at least 50 fc. up to hold 6 months or more; should be trimmed or pinched occasionally. **C**: 4   p. 257

rhombifolia 'Mandaiana' (Vitis), "Bold grape-ivy"; meritorious sport introduced by Manda in 1935, of more upright, compact habit, thicker stems and substantial, fleshy, deep green leaves, the leaflets broad and firm, almost leathery, lightly quilted and recurved, shining like polished wax; very attractive and a good keeper. **C**: 4   p. 257

rotundifolia (E. Africa: Tanzania, to Yemen), "Arabian wax cissus"; climbing plant with green, 4-angled stems having sharp corky edges in older specimen; rounded waxy-fleshy leaves 2½–3 in. dia., distantly toothed at margins, glossy fresh green to olive; inconspicuous green flowers, followed by red berries. Excellent house plant, very dry resistant. Ideal as a small dishgarden plant substituting for Peperomia obtusifolia, but must be pinched to keep in shape. **C**: 3   p. 257

sicyoides (Brazil), "Princess vine"; tall, prolific tendril climber of the tropical rainforest, for the conservatory where, with high humidity, it will send down a multitude of string-like aerial roots 10 ft. long or more, forming a veritable curtain; the light green, cordate leaves are somewhat fleshy with serrate margins, about 4 in. long. Ideal in large conservatories to create the impression of a rain forest full of swinging lianes. **C**: 54   p. 258

striata (listed in Europe as Vitis orientalis) (Chile), "Miniature grape-ivy"; cute little plant with thin reddish shoots, climbing by tiny tendrils; dense with small, palmately compound leaves to 1½ in. across, with 5 obovate leaflets toothed toward the apex, bronzy-green above with pale midrib, and wine-red beneath. **C**: 4   p. 257

CISSUS: see also Parthenocissus, Piper

CITRUS (*Rutaceae*); evergreen aromatic shrubs and small trees mostly subtropical and tropical, with handsome leathery foliage, sweetly fragrant flowers, and decorative usually edible fruit. Many Citrus are good container plants, especially if out on the patio, and have been used as such for centuries, creating visions of the sunny South; these may need daily watering to support their vigorous root system and fruit. A slight rest at cooler temperature should be allowed the trees after fruiting, best during winter, which will ripen the wood and induce budset and flowering. To hand-pollinate the blooms with a brush or by shaking them (a beehive nearby would allow bees to do this job), will ensure fruiting; when the swelling ovaries are very small, the young fruits are very sensitive to drop, if growth is checked by sudden temperature changes or dryness of the soil; constant moisture and adequate feeding at increased temperature is necessary to bring the fruit to maturity; keep cool to ripen. Spraying with modern hormones is supposed to help. The most charming of small potted oranges is the "Otaheite", but this is comparatively difficult while the smaller-fruited Calamondin may serve as a good substitute and is easier to grow and fruit. All Citrus fruit need good light and may be used for decoration where intensity is 100 fc. or more; they will drop foliage if kept dark. The small citrus-like Kumquats belong to Fortunella. Tree-like citrus are usually grafted by specialist nurseries; the Otaheite and Calamondin are raised from half-hard August cuttings. Citrus will also grow easily from seed.

aurantium (Vietnam), "Seville orange" or "Sour orange"; medium small spiny tree up to 30 ft. high, deep orange fruit of large size with glossy skin, and usually depressed at both top and base; with, however, rather acid, sour juice, too bitter for use as a fresh fruit; large fragrant, white-waxy flowers, the slender pointed dark green leaves with broadly winged petioles. Its fruit is mainly used in marmalade, also refreshing beverage. Often grown as Seville orange are hybrids with sinensis, the Sweet orange, and better known as "Bittersweet", which has fruit with sweet pulp. C. aurantium is very close to C. sinensis, the true Sweet orange now dominantly grown for table use; nurseries, also offer the similar looking sweet Temple orange as a form of aurantium; these, however, are actually "Tangors", hybrids of orange and tangerine. The "Sour orange" was introduced by the Arabs in the 11th century; there are great plantings in Southern Spain around Sevilla. This was also the orange of the Middle ages and later of the orangeries, maintained for ornament on wealthy estates, as it fruits freely as a small tree when grown in containers. **C**: 13   p. 510

aurantium myrtifolia (China), "Myrtleleaf orange" or "Citrus myrtle"; dwarfish, thornless tree with dense, deep green, stiff, little 1 in. myrtle-like leaves; waxy flowers, and small, sour, bright orange fruits; while blooming in April in the northern hemisphere, I remember the strikingly impressive sight of these trees in Uruguay and Argentina in October, when literally covered with fragrant white flowers. Excellent as a container plant. In southern Italy this orange is known as "Myrtus boetica", and easily confused in the leaf stage, but Myrtus has opposite leaf pairs, Citrus alternate leaves. **C**: 14   p. 510

limon 'Ponderosa' (Maryland hybrid of C. limon and medica 1887), the popular "American wonder-lemon" or "Giant lemon"; ornamental tree 8–10 ft. high, also known as "Ponderosa lemon", often grown in tubs, with irregular branches and short, stout spines, large oblong leaves on short petioles nearly wingless, the large waxy flowers white, the somewhat pear-shaped, big fruit lemon-yellow, almost 5 in. long, weighing 2¼ lbs., and while edible, it is of inferior grade for use, as its juice is sour. However, it is a well-loved container plant, although often of awkward shape; willingly bearing one or more of its large, showy lemons even as younger trees. Easily raised from a seedpit, though cuttings bear earlier. **C**: 13   p. 510

x'Meyeri', (hybrid of lemon with sweet orange), the "Meyer lemon" or "Dwarf Chinese lemon", found as a pot-plant near Peking 1908; semi-dwarf, spreading tree that may become 12 ft. high; nearly thornless, with small leaves, sweetly scented, lavender to white flowers, and bearing good-looking, light orange-colored, thick-skinned, oval lemons 2½ to 3½ in. long. A conversation plant, as it is easily grown in pots, long remaining not over 1 to 2 ft. high, and willingly fruiting at an early stage, furnishing lemons of table quality, and containing an excellent acid juice, nearly all year round; main season from December to April. **C**: 13   p. 510

mitis (originally from China, widely cultivated in the Philippines), the "Calamondin orange" or "Orangequat"; a small, spineless mandarin tree 8 to 10 ft. high, but usually grown as a small ornamental bush, with upright slender branches dense with broad-oval, leathery leaves on narrowly-winged petiole; white, fragrant flowers borne singly at the tips of twigs; small ¾ to 1½ in. deep orange-yellow fruit round or somewhat flattened, loose-skinned, the juicy flesh having a strongly acid, but pleasant flavor. Possibly a garden hybrid in China of mandarin (C. reticulata) with kumquat (Fortunella). Presently the most popular of miniature oranges grown in pots or tubs, with fruits easily produced, almost everbearing the year round, but with the main season in November and December for the Christmas holidays when even small plants are literally covered with lush foliage and masses of long-lasting, golden fruit; ideal for the sunny window or the wintergarden, also for decoration where light intensity is 100 fc. or more. The skin and flesh of the fruit makes good marmalade. **C**: 13   p. 386, 511

paradisi (West Indies), the "Grapefruit" or "Pomelo"; spreading tree to 30 ft. high and wide, with big leaves on broadly winged petioles, and large white flowers, the smooth branches bending under the weight of the large and heavy, pale lemon-colored, round fruit, averaging 4 to 5 in. in size, much grown for its somewhat tart, but refreshing juice. C. paradisi of the West Indies is a satellite species rising from the tropical C. grandis (maxima) of S.E. Asia. The variety most widely planted for commercial fruit production in subtropical regions is "Marsh seedless", with juicy pale yellow or pinkish pulp containing very few seeds, and ripening 18 months after flowering, from November to June. **C**: 13   p. 511

sinensis 'Washington Navel' or more properly 'Bahia', a "Sweet orange" from Brazil; excellent, large commercial, seedless orange for table use, 3 to 4 in. long with deliciously sweet fresh, early ripening from November to April (Valencia from May to Nov.); best adapted to California and Arizona climate; producing low yields and poor fruit in semi-tropical regions; it derives its name from the characteristic rudimentary secondary fruit imbedded in the apex of the primary fruit, and which serves as a trademark; deep orange in color, and easy peeling; flowers in large clusters, waxy-white and sweetly fragrant, on large round trees 20 to 25 ft. high. **C**: 13   p. 510

taitensis (otaitense), "Dwarf Tahiti orange" or "Otaheite orange"; dwarf tree only a few feet tall, with glossy-green, dense foliage and few spines, the waxy-white, very fragrant flowers tinged with pink appearing in January, followed by little fruits developing through the summer and ripening toward Christmas into colorful golden oranges from 1½ to 2 in. dia.; a miniature version of the sweet orange, except that the juice is tart, more resembling a lime. Probably a lemon (or lime) x mandarin (aurantifolium x reticulatum) hybrid, possibly of garden origin in South China, and naturalized in the South Pacific. This charming little plant lends itself beautifully to cultivation as a small ornamental plant in 6 to 7 in. pots, but it needs to be understood, have proper rest to initiate buds, hand pollinize the blooms, followed by even more careful watering and feeding to nurse the developing golden fruit. The Calamondin orange is much easier to grow and fruit. This may be the "Tahiti orange" carried on the 'Bounty' in 1789. **C**: 13 or 63   p. 386, 511

CLARKIA (Onagraceae)
elegans (California), showy annual herb 1½ to 3 ft. high with alternate narrow leaves and handsome purple, scarlet, salmon or rose-colored flowers 2 in. wide, often double,in few-flowered clusters, on erect, reddish, glaucous stems; blooming from July to October; in the greenhouse in winter and spring. Excellent cut flowers, easily grown. From seed. **C**: 63   p. 415

CLEISTOCACTUS (Cactaceae)
smaragdiflorus (Argentina, Uruguay, Paraguay), "Fire-cracker-cactus" or "Emerald-tipped cleisto"; slender, erect stems with 12–14 ribs, 2 in. thick, to 6 ft. high, and then leaning; closely set yellow areoles and numerous thin yellow to brown very sharp spines; a spectacular sight when in bloom with tubular flowers orange-red or scarlet, tipped emerald-green. **C**: 13   p. 155

straussii (Bolivia), the lovely "Silver torch"; slender, light green, clustering, many-ribbed column to 3 ft. high, covered with bristle-like white spines, central spine pale yellow; flowers red. **C**: 1   p. 154

CLEMATIS (Ranunculaceae)
lawsoniana 'Ramona'; a "Virgin's bower" vine of the 'lanuginosa' (China) x 'patens' (Japan) hybrid complex; shrubby deciduous winter-hardy climber with leaves of 3 leaflets, and striking flowers with spreading petals 6 in. or more across; in the clone 'Ramona' lavender blue with purple stamens, blooming during summer. One of the most beautiful of flowering hardy vines; widely planted outdoors, but may be grown in large pots on trellis or wire support in cool, light greenhouse or wintergarden; in limestone soil, never allowing them to become too dry. Cuttings from semi-ripened wood. **C**: 75   p. 280

CLERODENDRUM (Verbenaceae); the "Glory bowers" are chiefly tropical shrubs or climbers with curious white, violet or red flowers in brightly colored terminal clusters, the showy bell-shaped calyx often colored, the stamens long protruding. Primarily for the greenhouse or winter-garden; outdoors in summer. Propagated from cuttings of semi-ripe wood.
bungei (Sikkim, China, Okinawa), "Kashmir bouquet"; erect-growing shrub with wiry branches to 6 ft. high, spreading by suckers from the base; big, broadly ovate, quilted leaves to 1 ft. long and coarsely toothed, dark green on the surface, rusty-fuzzy beneath; sweetly fragrant 3 to 4 in. flowers rosy-red in a head-like cluster 4 to 8 in. across, blooming from June to September. Propagated by division in spring. **C**: 66   p. 492

fragrans pleniflorum; "Glory-tree" from Japan and China; double-flowered form of the shrubby plant to 6 ft. or more high, with stiff, angled stems covered by white hairs; broad ovate, opposite leaves, 3 to 8 in. long, light

green hairy on the surface, green beneath, the margins toothed; the fragrant 1 in. flowers in a dense terminal cluster, delicate peach-white and fully double, their calyx purplish-red. I photographed this plant in November, but it may bloom at various times during the year. A wonderful house plant if space permits and with fresh air, both for its long enduring bouquet of flowers and its penetrating perfume. Propagated from tip cuttings or root-sections. **C**: 66   p. 492

inerme (India: Bombay), the "Indian privet"; evergreen branching shrub with hard-leathery, small boxwood-like shiny deep green leaves, and little ½ in. white flowers with purple stamens, indigenous to mangrove areas near Bombay, and planted at the Hanging Gardens of Bombay, trained over wire frames, into topiary shapes of animals; sheared every 2 weeks. **C**: 52   p. 541

speciosissimum, grown in horticulture as C. fallax (Ceylon, Java, New Guinea, Polynesia), "Java glorybean"; a magnificent soft-wooded shrub I remember from the mountains of both Java and Ceylon, the showy flower clusters to 12 in. across a blaze of fiery scarlet above white-pubescent, heart-shaped foliage 6 to 10 in. long, on white-hairy, angled stems to 5 ft. high. Individual flowers with long stamens measure almost 2 in. across, blooming in our northern latitude from summer to autumn. Unfortunately, the flowers drop off easily in handling; for best results in airy conservatory or winter-garden. **C**: 54   p. 492

x speciosum (splendens x thomsonae), "Glory-bower"; shrubby plant of semi-erect, climbing habit with scandent soft-woody branches, dark green glossy leaves, and terminal 6 in. clusters of flowers having a pinkish calyx and deep crimson corolla shaded violet, blooming throughout summer. **C**: 4   p. 492

thomsonae (Trop. West Africa, Congo), "Bleeding-heart vine"; robust twining tropical shrub climbing to 15 ft., with ovate, quilted, glossy, papery leaves to 5 in. long deep green; exquisitely showy flowers in forking clusters, the inflated calyx pure white changing to pink, corolla deep crimson. A favorite winter garden and greenhouse flowering plant because of its striking inflorescence especially in spring; should be pinched to keep in shape; keep dry and cool in winter to rest for 2–3 months. **C**: 4   p. 278

ugandense (Uganda to Rhodesia), "Blue glory-bower"; climbing tropical shrub to 15 ft., with bright green, toothed obovate leaves to 4 in. long, and terminal panicles of bright blue flowers, the front lobe dark violet, and with blue stamens blooming from winter to spring. Propagated from root cuttings or stem cuttings. **C**: 54   p. 278

CLIANTHUS (Leguminosae)
formosus (dampieri) (Australia), "Glory pea"; a striking clambering shrub to 4 ft. high covered with long white, silky hairs; with glaucous-gray pinnate leaves feather-fashion, and pendant, showy clusters of scarlet, pea-shaped flowers to 3 in. long, and a parrot-bill-shaped keel-petal, blooming from March on. Young plants may flower from August into winter; silky-hairy pods follow blooms. **C**: 63
                                       p. 488

CLITOREA (Leguminosae)
ternatea (Pantropic: Panama to India and Moluccas), "Butterfly pea"; Tropical annual or biennial slender twiner with pinnate leaves, and pea-like solitary, showy flowers 2 in. long with broad, fan-like lip narrowing to the base, bright blue with beautiful markings in the throat, blooming all summer. For the warm, sunny greenhouse or outdoors. From seed. **C**: 52   p. 280

CLIVIA (Amaryllidaceae)
miniata 'Grandiflora', a highly developed commercial cultivar of the "Kafir lily" from Natal; an evergreen somewhat succulent plant with fleshy roots and broad, strap-like thick leathery deep green leaves to 2½ ft. long, arranged in two rows, with their bases expanded and forming an imperfect bulb; the scentless, more or less erect lily-like flowers flaming scarlet, and yellow in the throat, in clusters. Mostly blooming between March and May, they must be watered well all summer and fall, then slightly rested and kept cool. By division or from seed. An old house plant favorite because of their trusses of brilliant flowers, and demanding little care; leave undisturbed as long as possible. **C**: 15   p. 390

nobilis (So. Africa: Cape Prov.), "Greentip kafir lily"; a "different" clivia with strap-shaped leaves having roughish margins; the fleshy 1 to 1½ ft. stalk bearing clusters of 30 to

50 nodding, slender tubular flowers bright salmon red and tipped with green; bl. in summer and fall. **C**: 15   p. 390

CLUSIA (*Guttiferae*)
 rosea (W. Indies, Panama, Venezuela), known as the "Autograph tree" or "Fat pork tree"; evergreen tree 20 ft. or more high, growing naturally on rocks, or epiphytic on other trees; sparry branches with obovate thick-leathery, deep green, opposite leaves 8 in. long, and without lateral veins; large rosy flowers. With its horizontal branches and oddly fat leaves an artistic and durable decorator plant; needs light 50 fc. or more. **C**: 4   p. 216
 rosea aureo-variegata, "Monkey-apple tree"; rare and showy variety of this rugged plant having its broad obovate, thick-leathery leaves variously variegated and streaked dark green, light green and creamy to golden yellow. **C**: 4   p. 188

COCCOLOBA (*Polygonaceae*)
 rugosa (West Indies, Puerto Rico), the "Corrugated tin-tree"; a sparry but spectacular tropical tree to 30 or 40 ft. tall; the erect woody stems bearing at long intervals the well-spaced large, clasping, broad roundish leaves 1 ft. long and 8–10 in. wide, glabrous deep green and heavily corrugated and stiff as metal, appearing to be made of heavy aluminum; the showy inflorescence a long tail-like raceme crowded with tiny salmon-rose flowers. Best for a tropical conservatory, and tends to drop an occasional leaf when confined to tubs, and in dry atmosphere. **C**: 53   p. 216
 uvifera (South Florida, W. Indies), the familar "Sea-grape" along sandy tropical shores; a good decorative shrub or tree to 20 ft. high, with flexuous branches and stiff-leathery, rounded leaves to 8 in. across, glossy yellowish to olive green with prominent crimson-red veins on young growth, later changing to ivory; small fragrant white flowers, followed by purple fruits resembling bunches of grapes; used for jelly. A long-lasting tub plant for indoor decoration or the patio, preferring good light, minimum 50 fc., or full sun, as otherwise leaves become smaller in darkness. **C**: 2   p. 216
 uvifera aurea (Florida), "Variegated sea-grape"; spectacular sport with its large, almost sessile, leathery leaves largely variegated with creamy yellow. A collector's item. **C**: 4   p. 188

COCCOTHRINAX (*Palmae*)
 argentea (Haiti, Dominican Rep.), "Silver palm" or "Guano palm"; a slender unarmed fan palm to 40 ft. high, with medium-large palmate leaves, divided nearly to the base into slender arching dull green segments, pronounced silvery underneath but not as white as argentata; brown-black fruit. **C**: 66   p. 228
 argentata (So. Florida, Bahamas), the charming small "Florida silver palm" usually not more than 3 ft. tall, but occasionally to 20 ft. high, with small, very deeply divided star-shaped leaves which appear peltate, to 3 ft. across, the segments narrow and more or less pendant, glossy green above and decidedly silvery-white beneath. Begins fruiting when only 2 ft. high; fruit black. An attractive small palm in containers. **C**: 4   p. 228

COCCULUS (*Menispermaceae*)
 laurifolius (Himalayas), "Platter-leaf"; decorative though sparry shrub to 15 ft. high, with wiry green branches, and alternate stiff-leathery, ovate to narrow-elliptic leaves 5–6 in. long, shining forest green, concave with prominently raised yellow-green parallel veins. Can be kept bushy by pruning, or trained as espalier. For sun but tolerates shade. **C**: 13   p. 294

COCHEMIEA (*Cactaceae*)
 poselgeri (Baja California), "Long hooks"; cylindric plant with numerous sprawling or hanging stems to 6 ft. long and 1½ in. thick, bluish green, with conical warts, white-woolly in axils, the radials reddish and a solitary, hooked central spine; small glossy scarlet fl. **C**: 1   p. 163

COCOS (*Palmae*)
 nucifera; the majestic "Coconut palm" and "Tree of life"; thought to be a native to islands of the South Pacific Ocean, but now naturalized near most seashores within the tropics; characteristic with curving-erect, swaying slender trunks to 100 ft. high, leaning toward the water, and topped by graceful crowns of glossy, feathery fronds, a feature of tropical shores. The attractive curving, pinnate fronds may be

from 6 to 20 ft. long, with leathery, glossy yellowish-green, pendant leaflets set across from each other at regular intervals; its edible nuts inside a large yellow-brown husk. The flowers are insignificant but the juice or toddy that can be drawn from the unopened flower clusters is used in the manufacture of sugar, and when it is fermented turns into alcoholic arrack. The fruit is the economically important coconut, covered by a fibrous husk 8–12 in. long; its meat is dried into copra, which is used to manufacture palm-oil, fats and soap. The fronds are used for thatching. Fresh coconuts germinate easily in moist-warm propagating bed, but grown trees are difficult to maintain in containers. **C**: 54   p. 217
 nucifera 'Dwarf Golden Malay'; originated from seedlings of the dwarf King-coconut from the Andaman Islands; more diminutive and more slender-trunked than 'Dwarf Samoan', and smaller-fruited (yellow, green, or red); starts producing in about three years, in time 100 or more nuts per year as against 30 to 40 of the common coconut palm, and has a trunk so low that the clusters may need props. For containers only if supplied warmth and plenty of water and light; better planted in the ground. **C**: 54   p. 222
 nucifera 'Dwarf Samoan' (Marquesas), a "Dwarf coconut palm"; forming stout but short trunk, to 25 ft. high, and bearing fruits larger, and more rounded than that of C. nucifera typica; begins to bear at 5 years from planting seed while var. typica takes 8–10 years; husk color either green or red, with average 20 oz. of fresh meat per nut against the normal 12 oz. in the type coconut. **C**: 54   p. 506
 plumosa: see Arecastrum
 weddelliana: see Syagrus

CODIAEUM (*Euphorbiaceae*); the magnificent, multi-colored foliage plants long though erroneously known in horticulture as "Crotons", are beautiful hybrids derived primarily from such species as C. variegatum pictum, widely distributed in the monsoon territory from South India, Ceylon and Malaya to the paradise islands of Indonesia; tropical shrubs with highly ornamental, thick-leathery glossy leaves in many shapes and sizes, ovate to linear, entire or lobed, or spirally twisted; variegated in gay colors; the greens and yellows of the young leaves later often change to shades of orange and of red. Codiaeums are monoecious, bearing separate male and female flowers in elongate racemes on the same plant; these blooms are small, the males with white petals. For good coloring Crotons require strong sun, warmth with humidity, and liberal watering at least while growing; ideal in the south window; in summertime on the sunny patio. Without good light while kept warm, they will easily turn plain green. Crotons are not normally good house plants and will feel much happier in a tropical conservatory where light intensity and atmospheric humidity are high. They are at their best from early summer into winter. Cuttings are taken from January to June, rooting them in 80°F. bottom heat; mossing of tall plants is also practiced. Seed is obtained by pollinizing flowers; but resulting seedlings would probably not be true to name.
 var. pict. 'Aucubaefolium', "Aucuba-leaf"; bushy, sturdy cultivar of Polynesian ancestry, dense with small elliptic thin leathery 3 to 4 in. leaves bright glossy green, blotched and spotted yellow. Resembling a miniature Aucuba. **C**: 1   p. 181
 var. pict. 'Duke of Windsor', a good, robust, resistant, standard variety, always colorful, with wide linear leaves slightly broader near base, deep green, with gold to orange center, red midrib and red-orange margin. **C**: 2   p. 43*
 var. pict. 'Elaine', "Lance-leaf croton"; a good, old cultivar (1910), with stiff upright lanceolate leaves, shallowly lobed, and smaller than 'Reidii'; fresh green with bright yellow veins tinted into pink and red. **C**: 2   p. 181
 var. pict. 'Gloriosum superbum', an "Autumn croton"; dependable variety with long leathery leaves lobed near the apex, nicely variegated according to age of leaf with yellow veins and margins changing to autumn and crimson, the blade fresh green to almost red-black; good keeper. **C**: 2   p. 181
 var. pict. 'Imperialis' ('Mrs.Iceton'),"Appleleaf croton"; small compact plant and good keeper, with simple, elliptic 5 in. leaves almost entirely colored yellow, shading to peach and turning rose and red, green midrib and apex turning metallic purple. **C**: 2   p. 181
 var. pict. 'Johanna Coppinger', "Strap-leaf croton"; vigorous bushy plant with strap-like leaves dark green

variegated and blotched yellow to orange-red, the ribs red. **C**: 2   p. 181

var. pict. 'L.M. Rutherford', "Giant croton"; large showy plant, with heavy, undulate, large semi-oakleaf foliage consistently pink with gold and crimson on green. **C**: 2 p. 181

var. pict. 'Mona Lisa', "White croton"; outstanding Bachmann cultivar with large, broad ovate leaves lobed or constricted toward apex, the young growth almost entirely a waxy creamy-white edged in green; later the center area and ribs with reddish tints, the green turning into olive or coppery. **C**: 53   p. 181

var. pict. 'Norwood Beauty', "Oakleaf-croton"; an old (1924), formerly widely grown variety with smallish leaves and of compact bushy habit; tough 3-lobed 6 in. leaves of the oakleaf type, dark green with bronze to brown-red and rosy edge, yellow in vein areas. **C**: 2   p. 181

var. pict. 'Punctatum aureum', "Miniature croton"; densely bushy, free branching shrub with narrow linear, dark glossy green leathery leaves freely spotted yellow; good as miniature plant. **C**: 1   p. 181

CODONANTHE (*Gesneriaceae*)

macradenia (Central America), "Central American bell flower"; tropical vine, in habitat creeping over rocks and trees, the flexuous stems rooting below the nodes; small opposite ovate, thick shiny leaves similar to Aeschynanthus but with red spots scattered on the underside; axillary slipper-shaped waxy-white flowers with red markings inside the tube, and everblooming; followed by long-lasting white berries. Similar to C. crassifolia but which has red berries. Propagated from cuttings. **C**: 60   p. 439

COELOGYNE (*Orchidaceae*); large group of desirable tree-perching orchids chiefly Malaysian and from China to the South Pacific, having pseudobulbs bearing 1 to 2 leaves and showy flowers in clusters, often large, pendulous and chain-like and typically spicy-fragrant. Of easy culture if they receive ample moisture with perfect drainage; the warm-growing Coelogyne do not require a rest period. No repotting unless necessary as they dislike disturbance and may not bloom again for several years. Propagated by dividing the pseudobulbs.

cristata (Nepal Himalayas 5000 to 7000 ft.), a most beautiful and easy-growing small epiphyte with branching rhizome forming mats of fleshy apple-green pseudobulbs with thin twin leaves to 12 in. long; comparatively large 3 to 5 in., sweetly fragrant flowers in pendant clusters, glisten-ing pure white with undulate sepals and petals, the lip with fringed yellow keels. Winter flowering, it blooms most dependably between January and April if well established. An excellent orchid on the cool window sill; also grown in baskets in the cool to temperate greenhouse or winter garden. The snowy flowers may last 4 to 5 weeks if kept cool; also good as cut flowers. **C**: 34   p. 464

dayana (Malay Peninsula to Borneo, Sumatra and Java), "Necklace orchid"; a large and decorative epiphyte with long 5 to 10 in. pseudobulbs bearing 1 or 2 narrow leathery plaited leaves; the long pendulous inflorescence hangs down 3 ft. chain-like with $2\frac{1}{2}$ in. flowers loosely in ranks, the narrow sepals and petals are lemon-yellow, the broad, lobed lip with chocolate blotches and white keels from the base; blooming in spring and summer. Very striking when in flower but best grown in baskets in the warm greenhouse. **C**: 60   p. 464

lawrenceana (Vietnam); small epiphyte of compact habit, with ovoid pseudobulbs topped by 2 glossy green lanceolate leaves; 1 to 3 large and showy 4 in. waxy flowers on short arching stalk; broad sepals and linear petals buff-yellow, large lip with white center lobe and brown side-lobes, disk orange with brown keels; blooming spring and early summer; very fragrant. **C**: 22   p. 464

massangeana (Assam through Malaysia to Java); remarkable free-blooming epiphyte with pear-shaped, 2-leaved pseudobulbs; from their base appears the mag-nificent, pendant inflorescence $1\frac{1}{2}$ to 2 ft. long, with up to 25 or more flowers; these are $2\frac{1}{2}$ in. across, fragrant, pale yellow, the lip chocolate with bright yellow veins and light yellow keels on the mid-lobe, and edged in white; usually blooming twice a year, between January and May, and November. Very handsome and vigorous, best grown in baskets. **C**: 10   p. 464

pandurata (Borneo, Sumatra), one of the "Black orchids"; striking epiphyte of straggling habit with large

pseudobulbs and shining plaited leaves; the large and very fragrant flowers to 4 in. across, in arching sprays $2\frac{1}{2}$ ft. long, appearing from the base, the sepals and petals lovely green, lip greenish-yellow with black raised ridges; blooming variously between May and autumn. **C**: 10   p. 455*, 464

sparsa (Philippines); dwarf epiphyte with tapering pseudobulbs and two 3 to 4 in. glaucous leaves; arching short clusters of small 1 in. fragrant flowers white tinted green, and a white, lobed lip with sides dotted purple, and brown cross-bar on front lobe; summer blooming. **C**: 22   p. 464

COFFEA (*Rubiaceae*)

arabica (highlands of Ethiopia, Mozambique and Angola), the "Arabian coffee"; handsome evergreen shrub to 15 ft. high, with willowy branches bearing lustrous dark green rather decorative, thin-leathery elliptic foliage 3 to 6 in. long, wavy at the margins; the pure white, starry delightfully fragrant flowers massed in axillary clusters at the base of leaves, and blooming in late summer, are followed by brilliant crimson, pulpy, $\frac{1}{2}$ in. cherry-like berries containing the two seeds or "beans", which are roasted into coffee. This type of "Blue Mountain" coffee, grown at higher altitudes or cool climates, is considered superior to varieties of C. liberica which grow better in hot climates. Coffee was first used as a beverage by Arabs about 1450, and only since 1720 in South America. Very ornamental as a container plant, and an interesting conversation piece; the attractive fruits are produced on plants mature enough, and kept warm and fairly sunny, but take several months to ripen. Propagated from very fresh seeds only briefly viable; also by ripened cuttings from top growth. **C**: 4   p.488, 507

COLEUS (*Labiatae*); the "Painted nettles" or "Flame nettles" are gorgeous herbaceous foliage plants having succulent square stems, freely branching when pinched, and opposite ovate leaves more or less toothed, and usually beautifully painted or variegated so that they are popularly known as "The Foliage plant". The most decorative types with brilliantly colored foliage are cultivars of C. blumei from Java, they are willing, fast-growing container plants for the warm indoors-outdoors, as well as attractive bedding subjects that can be kept low by pinching. Easy from cuttings or seed.

blumei 'Brilliancy', horticultural cultivar of the "Flame nettle" from Java, a soft but showy-leaved perennial herb about 2 ft. high unless pinched; with square, fleshy stems and blue flowers; the leaves of 'Brilliancy' are boldly crimson-red, marked gold on finely crenate edge, 3 to 4 in. or more long. **C**: 2   p. 408

blumei 'Candidum', a "White coleus"; striking new hybrid with typically large, broad wavy leaves featuring a central area of creamy white, prettily framed with green, spreading into the outer zone of friendly light green. **C**: 2   p. 190, 408

blumei 'Defiance', a "Painted nettle", long used in bedding planting; showy standard variety with lanceolate leaves finely crenate, brownish-red and bordered in striking contrast with a broad yellow margin. **C**: 2   p. 190, 408

blumei 'Frilled Beauty'; a lace-leaf cultivar quite fantastic with the medium-small ovate-pointed leaves undulate and deeply lobed, coloring in various shades of red; blackish-maroon when young, turning into brighter crimson later; margins first green, later yellow; stems green. **C**: 2   p. 408

blumei 'Klondyke', the "Butterfly coleus"; robust variety with sturdy foliage wildly streaked and variegated brownish-red and yellow. A well known plant in the spring bedding trade. **C**: 2   p. 408

blumei 'Pink Rainbow', "Showy coleus"; large, showy new hybrid with broad, wavy leaves 4 to 6 in. or more long, beautifully painted in copper, rusty-red, and moss green; carmine-red along the veins; the crenate edge outlined in green. One of the newer hybrids that come 90% true raised from seed. **C**: 2   p. 408

blumei 'Pride of Autumn', the "Autumn nettle"; a most attractive cultivar with ivory-white stem and medium-sized ovate leaves having bright carmine center, the outer area chocolate brown, and bordered by a contrasting pale margin lightly crenate; the underside pale green with carmine center showing through. **C**: 2   p. 408

blumei verschaffeltii (collected in Java 1860); a most beautiful "Painted nettle", with relatively large, glowing crimson-red leaves 4 to 5 in. long, purple in center, and with

narrow, nile-green border, the underside purplish-red also; stems and petioles white to purple; flowers pale blue. Possibly the most exquisitely colorful of the older varieties of Coleus and still very much in favor by horticulturists. **C**: 2
p. 408

pumilus (Philippines), "Dwarf creeping coleus"; a rather weak growing species with lax stems, more or less reclining and rooting at lower joints; the leaves small, 1 to 1½ in. long, broad, blackish-red, lighter along veins, and marked nile-green around the crenate edge. **C**: 2   p. 408

rehneltianus 'Lord Falmouth', "Red trailing nettle"; beautifully colored tropical herbaceous plant with stems first erect, later creeping, the small ovate, long-stalked crenate leaves a glowing carmine red with contrasting rosy center. **C**: 2   p. 190, 408

rehneltianus 'Trailing Queen', "Trailing coleus"; a popular commercial cultivar densely branching with prostrate stems rooting along the ground; the foliage somewhat larger than the species from Ceylon; ovate leaves 1 to 2 in. long, chartreuse to emerald green along margins, prominently overlaid and marbled with brownish purple, a carmine center near base. Long used as a low creeper and ground cover in garden beds, the window box, hanging baskets or as a window-sill plant, and easily kept compact by pinching off the tips. **C**: 14   p. 265, 408

thyrsoideus (Central Africa), "Flowering bush coleus"; tender shrub 2–3 ft. high, with 4-angled pubescent stems dense with cordate-ovate, plain green leaves up to 7 in. long, and coarsely lobed; bright blue flowers in long, spike-like racemes. This otherwise plain plant is a nice house plant mainly for its showy flowers. **C**: 2   p. 408

COLLINIA: see Chamaedorea

COLOCASIA (*Araceae*); the "Elephant-ears" or "Taros" are big tropical fast-growing Asian and Polynesian herbs with rhizomes or starchy tuberous roots, some of them edible after boiling; showy, peltate ornamental leaves; the inflorescence with thick fleshy spathe enclosing a spadix with male flowers toward the apex, the females at the base, separated by a ring of neutral organs. Grow in fibrous peaty soil with good drainage. Colocasias are water-loving, requiring high humidity, and generally are not good house plants preferring the edge of pools. Propagated by division.

antiquorum fontanesii (Ceylon), "Asiatic taro"; tropical bog plant with soft, peltate-heart-shaped leaves, satiny, happy-green with purple edge, on succulent purplish brown stems; spathe pale yellow. **C**: 60   p. 95

antiquorum illustris (E. Indies), "Black caladium"; the soft, peltate, heart-shaped leaves are fresh spring-green, especially in vein areas, balance brownish-purple which shows through to grayish reverse; stems green or violet. **C**: 60   p. 95

esculenta (Hawaii and Fiji), "Elephant's ear", "Green taro" or "Dasheen"; soft, fleshy herb with edible tuber (after baking or boiling), and large, quilted leaf to 3 ft., bright satiny green. **C**: 54   p. 96

COLUMNEA (*Gesneriaceae*); the "Goldfish plants" are tropical American fibrous-rooted usually epiphytic, tree-perching plants, either vine-like with creeping or hanging branches, or shrubby, with stiffish, erect or spreading stems, which sometimes root at the nodes; leaves are in pairs along the stem, one blade mostly dissimilar in size or shape from the other; the resplendent flowers irregular or 2-lipped with tubes generally having a wide arching hood formed from two top lobes which with the lower lobe looks like the open mouth of a fish; very striking in vivid red, yellow, orange or pink. Columneas grow best in porous fast-draining soil with fern-roots or sphagnum and do well in hanging pots. Growing warm, most kinds bloom in spring and summer, but bud setting is aided by keeping them dry and a cool 55° F. (12° C.) in fall or winter. The hanging vines should not be pinched. Propagated through tip cuttings, stem-sections, or division; individual leaves will also root but develop slowly.

allenii (Panama), creeper from tropical mountain forests, with slender hanging, pale-hairy pliant stems, with elliptic ¾ in. leaves in equal pairs, shining dark green; flowers to 3 in. long with reddish calyx and 2-lipped corolla usually red, sometimes orange, solitary from the leaf axils, in late summer and fall. **C**: 10   p. 441

arguta (Panama, in mountain forests); epiphytic trailer with pendulous strands 3 to 5 ft. long covered with maroon bristles and densely set, almost overlapping with waxy ovate, very sharp pointed 1¼ in. leaves in equal pairs; large 2½ in. wide-flaring orange-red flowers marked yellow, during the leafless state in its native habitat; in cultivation blooming in autumn, on old wood. Baskets with long hanging garlands covered by a profusion of bright blooms are a spectacular sight. **C**: 10   p. 441

x banksii (hybrid of the pendulous oerstediana x the more erect schiedeana); rambling or creeping plant as showy as oerstediana or gloriosa, with small 1½ in. waxy ovate leaves red beneath; the downy, vermilion-orange or scarlet slender, 2-lipped flowers faintly lined yellow at mouth, blooming in spring. Very vigorous, and tolerant as a houseplant. **C**: 10   p. 440

x'Canary' (Cornell hybrid 1959 of yellow verecunda x moorei); semi-erect plant with thickish rambling stems, constricted at nodes; dark green, smooth leaves in pairs, one of each twice as large as the other, 2 and 1 in. long, wine-red with green veins beneath; the medium small 1¾ in. flowers bright canary yellow, covered with silky white hairs, produced from October to March. **C**: 10   p. 441

x'Cayugan'; a clone from the Cornell hybrid 1959 C. crassifolia x hirta; intermediate between the two parents, of erect-spreading, shrubby habit; stems with reddish hairs, set with unequal pairs of all-green narrow leaves 1½ and 2 in. long; from each joint the vivid reddish orange flowers, intensified by dense red hairs, 2½ to 3 in. long. Although in bloom only in winter and spring, the abundance of flame-colored flowers creates a striking effect; one of a group of college-bred columneas promising to be better, more vigorous houseplants than most species. **C**: 10   p. 441

x'Cornellian'; an outstanding hybrid clone produced at Cornell University 1959 from a yellow-flowered C. verecunda with the orange-red crassifolia; an erect-growing, somewhat shrubby plant with glossy dark green narrow-elliptic leaves, dark red underneath, 5 in. long for the larger of the pair, 3 in. for the smaller; profuse with 2 in. flowers dark reddish-orange with yellow throat. Tested as a good, vigorous houseplant, blooming throughout the year. **C**: 10
p. 441

crassifolia (Mexico, Guatemala); erect, succulent plant with long linear, fleshy, glossy-green leaves 2 to 4 in. long and nearly erect; orange-red, spread-open, tubular flowers 3 in. long, covered with erect red hairs, blooming from early in the year until June. **C**: 10   p. 440

gloriosa (Costa Rica); graceful epiphytic trailer with slender rooting stems becoming pendulous, the nearly equal pairs of small oblong leaves to 1¼ in. long, quite ornamental, covered with brown-red hairs; large solitary, bilabiate, fiery red flowers 2½ to 3 in. long, with wide open yellow throat, fanning into the broad, helmet-like upper lip; blooming best in winter and spring, but with some flowers almost throughout the year. Beautiful basket plant but somewhat delicate. **C**: 60   p. 441

illepida (Peru ?), robust, ungraceful herb with ascending stems, covered with brownish hairs; opposite hairy leaves in unequal pairs light green, prettily splotched red beneath; axillary 3 in. flowers yellow with dark maroon stripes lengthwise along tube and limb; almost always in bl. **C**:10 p. 440

lepidocaula (Costa Rica); erect shrubby herb with stout succulent stem 18 in. or more high, becoming woody with age; densely set with fleshy, glossy deep green elliptic leaves 3½ in. long; axillary flowers with green calyx, the 3 in. corolla orange, shading to yellow at throat and lower side of tube, and covered with long pale hairs; blooming all year. **C**: 10   p. 440

linearis (Costa Rica); attractive shrubby epiphyte, with branching jointed stems to 1½ ft. high; long narrow, equal-sized shiny dark green leaves to 3½ in. long; from the axils solitary bilabiate 1¾ in. flowers with flattened throat, uncommon because they are rose-pink, covered with silky white hair; spring-blooming. Good for both pots and baskets. **C**: 10   p. 440

microphylla (diminutifolia) (Costa Rica), the "Small-leaved goldfish vine"; trailing epiphytic vine with brown cord-like stem and long hanging branches, dense with tiny ½ in. rounded or broad-elliptic, coppery-hairy soft leaves; from the axils the relatively very large 2-lipped flowers ending in a spreading hood, 2½ to 3 in. long, scarlet-red with a yellow patch along bottom of tube. Similar to gloriosa but with smaller foliage. Very glorious in a basket when the long pendant red-velvety vines are afire with the striking blooms in spring. **C**: 10 .   p. 432, 440

nicaraguensis (Central America), profusely blooming, stiff-stemmed, epiphytic clamberer to $2\frac{1}{2}$ ft. high, with large lanceolate, corrugated, satiny green leaves red beneath, the larger ones 5 in., the opposite very small or missing altogether; axillary, brilliant red 3 in. flowers with a flaring, hood-like upper lip, often yellow at the throat. **C**: 10   p. 440

sanguinea (West Indies), in horticulture as Alloplectus sanguineus, the "False alloplectus"; sparry, densely hairy shrubby plant, with strong scandent stems to over 4 ft. long, and looking like an Alloplectus; short-stalked lanceolate leaves emerald green, blotched blood-red beneath, the long ones 8–10 in., the opposite short ones only 2 in. long; small 1 in. furry, pale yellow flowers in leaf axils. **C**: 10   p. 440

x'Stavanger' (microphylla x vedrariensis'), the "Norse fire plant"; (Haualand 1949), free-growing Norwegian diploid trailer with pendant branches; the small succulent $\frac{1}{2}$ in. leaves rounded or broad-elliptic, dark green, smooth and fairly shiny; willing bloomer with 2-lipped tubular, erect hairy flowers 3–4 in. long, fiery cardinal-red from orange base, with a little yellow in throat. A fantastic basket plant with long strands like a chain of beads and flowers rearing up like hungry dragons. **C**: 10   p. 441

teuscheri (Trichantha minor in horticulture) (Ecuador); rambling creeper with long thread-like vines set with lance-shaped blackish-green leaves shiny above and hairy beneath; from the leaf axils the very unusual tubular 2 in. flowers dark purple, with 4 stamens, the calyx a feathery cluster of carmine bristles, and the face of the corolla contrasting yellow and black, with 5 curious yellow horns at the mouth; blooming for 4 months, starting in summer. **C**: 60   p. 439

x'Vega' (C. 'Stavanger' x vedrariensis), "Goldfish bush"; 1956 Norwegian diploid, with erect or arching wiry branches first light green, later reddish; fresh green round-oval, smooth leaves 1 in. long; the erect bilabiate flowers $2\frac{1}{2}$–3 in. long and evenly coxcomb-red; of stiffer habit than C. 'Stavanger', flowering 3 weeks earlier. To initiate buds for spring-blooming, plants in pots are chilled in September, keeping them cool and dry for 6 weeks, then increase temperature to warm 60°–65° F. or more; lovely as a branching flowering bush. **C**: 10   p. 441

verecunda (Costa Rica); compact, shrubby, erect plant to $2\frac{1}{2}$ ft. high; the waxy, deep olive green, lanceolate, convex leaves burgundy red beneath, along the stems in pairs but so unequal that they appear to be alternate, the larger one about 4 in., the opposite shorter one only $\frac{1}{2}$ in. long; in the axils the $1\frac{1}{2}$ in. 2-lipped flaring flowers lemon to golden yellow from above, and beautifully wine-colored underneath; bright pink berries. A flowering plant that should be looked at from below. **C**: 10   p. 440

COMMELINA (*Commelinaceae*)

benghalensis variegata (Trop. Asia, Africa), "Indian dayflower"; delicate creeper with watery stems and soft fleshy, pale grayish green, narrow leaves 2 to 3 in. long striped and edged in white; flowers, with 2 kidney-shaped petals, sky blue. **C**: 51   p. 260

communis aureo-striata, the variegated "Widow's tears"; a white-striped form found in New Jersey; annual creeper with watery-succulent stems and lanceolate 2 to 4 in. leaves bright glossy green, banded and striped with cream; the flowers with 2 large blue petals and one smaller white petal. The green species is from East Asia, naturalized in N.E. U.S. **C**: 75   p. 260

COMPARETTIA (*Orchidaceae*)

speciosa (Andes of Ecuador); beautiful and distinct small epiphyte with one-leaved pseudobulbs, and pendulous racemes of numerous flowers, remarkable for their broad lip and long spur, the sepals and petals light orange, lip about $1\frac{1}{4}$ in. wide, cinnabar-red, orange at base, blooming in summer. Grown either in smallish pots or suspended on tree fern slabs; sphagnum added to fern fiber is a good growing medium; plants should not be permitted to dry out. Prop. by division. **C**: 72   p. 462

CONOPHYLLUM (*Aizoaceae*)

grande (So. Africa: Namaqualand), "King pins"; highly interesting sun-loving succulent from the driest part of South Africa; a pair of light green sterile leaves 8 in. high together forming a cone, later split apart to let the flowering growth emerge on elongate internodes, rising to 18 in. high

and bearing pairs of leaves joined during resting, splitting later to permit shining white flowers to emerge. From seed. **C**: 13   p. 354

CONOPHYTUM (*Aizoaceae*)

simile (So. Africa: Cape Prov.), "Cone plant"; very small succulent under 1 in. high, forming clumps, similar to Lithops but smaller; the fat little bodies heart shaped, with the notch on top, smooth and bluish green or gray-green dotted with dark green all over; large 1 in. yellow flowers. Conophytum belong to the pigmies amongst the mimicry plants; resting in summer; blooming into winter, a joy in the window. From seed and cuttings. **C**: 13   p. 356

CONSOLEA (*Cactaceae*)

falcata (Opuntia) (N.W. Haiti), a "Tree opuntia", robust tree 10 ft. or more high, with straight, massive brown trunk, and a heavy crown of thick-fleshy, oblong, flat joints sickle-shaped 8 to 12 in. long, glossy rich green, with raised knobs especially on margins, nearly spineless; flowers red turning orange. A tree appearing to be literally covered with lush foliage, but the "leaves" are actually flattened joints of branch stems and true leaves are absent. **C**: 1   p. 167, 535

CONVALLARIA (*Liliaceae*)

majalis (Europe, Temp. Asia. No. America), "Lily-of-the-Valley"; small perennial, winter-hardy herb 6 to 8 in. high with horizontal slender, creeping rootstock, the upright, detachable parts called pips and containing the buds, each sprouting 2 basal, fresh green, lanceolate leaves, and a stalk with nodding little bells of hauntingly fragrant, waxy-white flowers. Normally spring-blooming, large pre-chilled European forcing pips from cold storage and available during winter can be had in bloom indoors in about 3 weeks; pot in sphagnum or peatmoss and keep warm (70°F.). moist and dark for 10 days then accustom them to light. Propagated by division. **C**: 78   p. 373

majalis 'Rosea', "Pink Lily-of-the-valley"; dainty and more dwarf than the white-flowered species, the leaves slightly glaucous; nodding waxy, $\frac{1}{4}$ in. urn-shaped flowers delicate ivory flushed pink, and more reddish at base, in erect, one-sided racemes; normally blooming in May or June. Winter-hardy; seldom seen in pots because the white species is traditionally favored. **C**: 16   p. 401

CONVOLVULUS (*Convolvulaceae*)

tricolor (Sicily and Spain to No. Africa), "Dwarf morning glory"; branching annual somewhat trailing but only 8 to 16 in. high and 2 ft. wide, with obovate leaves and clustered showy, funnel-shaped flowers to 2 in. or more across, azure-blue with yellow throat and white tube, opening on nice days in the morning and closing in late afternoon; blooming most prolific from June to September. Color variations are all-white, vivid rose, lavender, blue and violet. Striking little pot plants for the bright window or for hanging baskets. Needs sun and warmth, and blooms best when kept on dry side. Nick the tough seed coats with a knife. **C**: 63   p. 416

COPERNICIA (*Palmae*)

macroglossa ("torreana" of hort.) (Cuba), the "Cuban Petticoat palm"; so called because of the dense cylindrical hanging growth of dead persistent leaves covering the trunk of this fan palm; growing to about 25 ft., the stiff, folded, lustrous deep green palmate leaves 5–6 ft. across on short thorny stalk. A curiosity plant for the palm fancier. **C**: 54   p. 229

COPROSMA (*Rubiaceae*)

baueri (New Zealand), the "Mirror plant"; flexible densely leafy, dioecious shrub, with male and female flowers on different plants, and growing 15 to 20 ft. high, with opposite, roundish, soft-leathery leaves 2 to 3 in. long, bright green and glossy as if varnished; small flowers greenish, the berries on female trees orange-yellow. A very salt-tolerant species, and I remember planting this evergreen in landscaping along the seashore in Southern California, because of its resistance to salt spray and ocean breezes. Also charming as a potted decorative plant, very durable and tolerant except for its need for water. From cuttings. **C**: 16   p. 294

baueri marginata, the "Variegated mirror plant"; with lacquered, vivid green to milky green leaves prettily edged in creamy-white. Very attractive as a low plant for the cool window. **C**: 16   p. 295

CORDYLINE (*Liliaceae*); known as "Dracaenas" in horticultural usage, these are tropical and subtropical foliage plants, many becoming tree-like; principally from Malaysia to New Zealand and Polynesia, their swollen root-stocks with white roots; the slender stems topped by clusters of leathery often beautifully colored leaves. They are distinguished from true Dracaenas by their branched clusters of flowers that are tubular, not showy and usually white or pinkish. The colored-leaf terminalis or "Red dracaena" group while very decorative, are not ideal house plants; they are somwhat seasonal in behavior, displaying their most brilliant colors toward mid-winter, but after that they tend to shed the lower leaves; they require lots of water when not resting to develop and hold their foliage. Red dracaenas also easily catch Red Spider indoors. They prefer good to bright light short of burning sun; and like a position on the patio with fresh air better than a darkish interior. Propagated by mossing the top; stem sections, pieces of roots; sprouts will root easily even in water; C. australis from seed.

    australis (New Zealand); popularly known as the "Cabbage-tree" or "Grass-palm" in New Zealand, but by our florists erroneously as "Dracaena indivisa"; with age a mighty branched tree to 40 ft. high and forming a thick trunk; until it flowers, a single erect slender stem crowned by a dense cluster of flat, and clasping, tough-leathery, bronzy-green, arching leaves with fresh-green midrib, to 2½ in. wide in a mature tree, but narrow-linear ½ to 1 in. in seedlings; fragrant white flowers in large terminal panicles, following which the head will fork; an ancient monocot. Young seedling plants are used as central plant in spring bedding plantings. **C**: 15   p. 201

    indivisa (New Zealand), "Mountain cabbage-tree"; slender tree to 45 ft. high, usually with single, flexible stem topped by a large head of sessile, flat, tough leaves, matte green with raised orange midrib, glaucous blue beneath, to 8 in. wide, or twice as broad as australis; the inflorescence a pendulous panicle of whitish flowers; not branching after flowering. **C**: 26   p. 201

    indivisa 'Rubra', "Red palm-lily" or "Red cabbage tree"; a popular broadleaved cultivar in California nurseries, with the flexible-leathery leaves entirely colored dark bronze; the inflorescence a beautiful erect panicle of fragrant white flowers. **C**: 4   p. 200

    stricta (Dracaena congesta) (Queensland, New South Wales); tree-like, with slender stem to 12 ft. high, occasionally branched; narrow clasping leaves swordshaped, leathery, matte green with rough edges, inconspicuously toothed, narrowing toward a constricted base; young growth reddish. **C**: 4   p. 201

    terminalis (India, Malaysia, to Polynesia), known as the "Tree of kings"; widely spread tropical palm-like plant with clustering canes to 10 ft. high, bearing rosettes of leathery, rather narrow, sword-shaped bronzy-red leaves carried in an elegant-erect manner. A selected color clone is commercially known as "Red dracaena" with foliage copper-green shading into red; the young winter-growth approaching flowering turns intense rosy-crimson, at which time the plant requires lots of water. Flowers lilac-tinted, in panicles; followed by red berries. **C**: 54   p. 201

    terminalis 'Amabilis', "Pink dracaena"; strong growing color form with the broad, shining green to bronze foliage prettily variegated or edged with cream, suffused with pink. **C**: 54   p. 201

    terminalis 'Baby Ti' (Hawaii); one of the tiniest of "Red dracaenas", miniature rosette with narrow, concave, gracefully recurved leaves deep metallic green suffused with copper, and a red border. **C**: 54   p. 43*

    terminalis 'Firebrand'; excellent commercial "Red dracaena", which in addition to being a compact rosette, also holds its foliage well; slender stiff, gracefully recurving leaves a satiny purplish-red with glaucous sheen, the younger foliage mahogany red with crimson midrib. **C**: 54   p. 201

    terminalis 'Mad. Eugene Andre', "Flaming dragon tree"; showy commercial cultivar of spreading habit, with broad, leathery leaves deep green to coppery, and red border; the new growth in winter a beautiful shocking pink, indicating the flowering season, after which the lower leaves tend to drop if disturbed. **C**: 54   p. 201

    terminalis 'Negri', known in horticulture as Dracaena Negri, the "Black dracaena"; a striking cultivar from Louisiana, outstanding for its big leathery, glossy coppery maroon leaves almost black, 2½ to 5 ft. long and 5 in. wide, in graceful rosette on slender canes. **C**: 4   p. 200

    terminalis var. 'Ti' (Taetsia fruticosa), (native in Southern Pacific Islands from Hawaii to New Guinea), "Good luck plant"; palm-like, robust, with slender cane to 12 ft. high, topped by a cluster of oblong leaves to 2 ft. long spirally arranged, smooth, flexible and plain yellowish green. This foliage is used for hula skirts; sections of cane will sprout young plants, and are sold as the "Miracle plant". A curiosity but not particularly attractive. **C**: 54   p. 201, 536

    terminalis 'Tricolor', "Tricolored dracaena"; very colorful rosette of broad leaves on top of slender canes, beautifully variegated red, pink, and cream over a base of fresh green; a striking beauty and good keeper. **C**: 54   p. 191, 201

COREOPSIS (*Compositae*)
    auriculata 'Nana' (U.S.: Virginia to Florida), "Dwarf tickseed"; low, branching perennial 3 to 6 in. high, spreading by stolons to 2 ft. or more, with 1 to 4 in. spoon-shaped leaves on ciliate petioles, and carrying a profusion of 1½ to 2½ in. bright orange-yellow flower heads, with florets grooved and apex jagged. Blooming from spring to fall, the blooming season may be lengthened by removing flowers as they fade. Propagated by division of root crown. **C**: 76   p. 417

COROKIA (*Cornaceae*)
    cotoneaster (New Zealand), "Golden Korokio" or "Zigzag bush"; slow-growing, stiff evergreen shrub to 10 ft. high, the numerous slim, nearly black branches contorted and so interlaced as to make an intricate, fantastic pattern; the little 1 in. leaves dark green above, white tomentose underneath; tiny, yellow, starry jasmine-like ½ in. flowers in spring to summer, followed by small bright yellow to red fruits. Thrives in containers, best in sun; night-lighting from beneath enhances bizarre but charming branch pattern. From seeds, or half-ripened cuttings under glass. **C**: 13   p. 298, 498

CORPUSCULARIA (*Aizoaceae*)
    algoense (Delosperma) (So. Africa: E. Cape Prov.), "Grayhorns"; dwarf stemless shrubby succulent with prostrate branches, the slender 3-angled keel-shaped, opposite leaves about 1 in. long and pretty gray-green; solitary whitish flowers to 2 in. across. **C**: 13   p. 356

CORREA (*Rutaceae*)
    x harrisii (speciosa hybrid), "Australian fuchsia"; handsome evergreen shrub with rusty-hairy twigs and opposite simple, downy leaves, the striking nodding tubular 1 in. flowers bright scarlet red with protruding stamens and yellow anthers. Very similar to but shapelier than, the species C. speciosa, the "Red Correa" from Eastern Australia, found from Tasmania to Queensland, with flowers often tipped with green or yellow, and flowering there in their spring, September-October; in our northern latitude in April-May. Propagate from cuttings; sometimes grafted. **C**: 32   p. 496

CORTADERIA (*Gramineae*)
    selloana pumila (Argentina), the "Pampas grass"; evergreen perennial giant ornamental grass, in the species growing to 20 ft. high, but the variety pumila forming white compact clumps only 4 ft. high; a fountain of saw-toothed, long grassy, narrow leaves several feet long, above which arise low stalks bearing masses of decorative white, silky 1 to 3 ft. flower plumes. Will grow in dry or wet soil; prefers sun, tolerates dry and hot or cool and moist. Magnificent decorator in large tubs; keep cool in winter. From young divisions. **C**: 64   p. 306

CORYANTHES (*Orchidaceae*)
    bungerothii (Venezuela), "Helmet orchid" or "Bucket orchid"; an extraordinarily fascinating epiphyte with small ovoid 2½ in. pseudobulbs; ribbed leaves narrow lanceolate, 1 ft. long; the drooping 1½ ft. floral stalks emerge from the base similar to Stanhopea, and bear 2 to 5 remarkable flowers with a pouched lip molded to resemble a bucket or helmet; sepals and 3 in. petals greenish-white covered with purple spots, the hooded lip brownish yellow, orange inside with red spots; blooming in May but flowers only

last 3 to 4 days. A rare curiosity, requiring warmth with moisture, in fairly good light. By division. **C**: 60   p. 462

CORYPHANTHA (*Cactaceae*); a wide-spread North American genus of "Pincushions" found from British Columbia to Southern Mexico; globular or cylindrical cacti of small and medium-sized habit, their bodies covered with large knobs or tubercles which are grooved through the middle and bearing spines. The flowers are mostly yellow and showy, sometimes red or purple, and blooming late summer into autumn. Coryphanthas are slow-growing and sensitive against wetness, but neither do they like to be too dry; for the collection in the sunny window; grow in sandy loam containing calcium; cool and dry at 45 to 50°F. during winter. From seed.

bumamma (Mexico), "Starry ball"; small globular plant to 6 in. high, in habitat arising only a little above ground, relatively few very large bluish-green tubercles carry recurved brownish radial spines; large reddish-yellow flowers. **C**: 13   p. 163

elephantidens (Mexico: Michoacan), "Elephant tooth cactus"; beautiful globose plant, to 5 in. high and somewhat broader; blue green tubercles obtuse and 2 in. long, densely white-woolly in axils; 6–8 brownish spines and all radial; large flowers about 4 in. across, rose-colored. When old, freely sprouting from the warts and extending. **C**: 13   p. 163

macromeris (Mexico, Texas, New Mexico); freely clustering globose plant pale grayish green, becoming 8 in. high, with very high, dense knobs to several inches long, tipped with woolly areoles, awl-shaped white radials and long black needle-like central spines; funnel-shaped flowers carmine-red. **C**: 13   p. 163

COSTUS (*Zingiberaceae*)
igneus (Brazil), "Fiery costus"; beautiful tropical herb of small size, with stout leafy stems maroon, carrying the 6 in. smooth leaves, green above, reddish beneath, spirally arranged; large 2½ in. fl. deep scarlet orange. **C**: 54   p. 208

malortieanus (zebrinus) (Costa Rica), "Emerald spiral ginger"; showy, suckering plant with stout stalks usually compact, but becoming 3 ft. tall, furnished in spiral order with broad, recurved, fleshy leaves to 1 ft. long, bright emerald green, banded lengthwise with dark zones and covered with glistening silky hair; flowers yellow marked with brownish red. **C**: 4   p. 208

sanguineus (C. America), beautiful "Spiral flag"; a small tropical herb with coppery green stems clasped by wine-red petioles, bearing in spiral order the gracefully recurving, oblique-elliptic, fleshy leaves of shimmering velvet bluish-green, marked with a central band of silver, thin gray lines, and a zone of yellow-green toward the margin; deep blood-red beneath. **C**: 56   p. 43*, 208

speciosus (India), "Spiral ginger" or "Stepladder plant"; perennial herb, and a conversation piece, with heavy rootstock, slender reed-like green, leafy stems usually short and clustering, but in habitat to 10 ft. long; twisting and growing upward in loose spirals, or drooping in graceful curves, set in spiral order with fresh-green, oblanceolate, glossy, succulent, slender-pointed leaves 6–8 in. long, silky beneath; flowers in dense spike, white with yellow center and red bracts. **C**: 4   p. 208

COTONEASTER (*Rosaceae*)
horizontalis (Western China), "Rock clusterberry"; charming, low growing semi-evergreen shrub forking and spreading horizontally over rocks, as far as 15 ft., branching herringbone-like, and set neatly with small ½ in. dark glossy green, leathery, rounded leaves, and parallel rows of small pinkish flowers in May-June and later, with shiny coral-red fruit. Very satisfactory in containers with branches spilling over the sides; also trained to espaliers. Hardy in my garden in New Jersey. From seed or cuttings. **C**: 76   p. 512

COTTENDORFIA (*Bromeliaceae*)
guianensis (Venezuela); a rather primitive species from the Guayana Highlands, living on rocks; long linear, stiffish leaves unarmed, 3–4 ft. long, with spiny apex, white-scaly beneath; the small inflorescence on long cane; flowers white. **C**: 9   p. 135

COTYLEDON (*Crassulaceae*); old-world succulents in the Stonecrop family having diverse sizes and appearances,

some with striking showy leaves, others rather monstrous but all with clusters of large, drooping tubular or urn-shaped flowers in many colors; the genus is characterized by a corolla that is much longer than the 5-parted calyx. They are good, easily grown house plants in a bright airy place; in winter at no more than 50°F. Propagation from cuttings or seed; leaf cuttings grow less freely.

barbeyi (Eritrea, Somaliland, Arabia), "Hoary navel-wort"; bold. branching succulent with large, opposite, thick, shovel-shaped leaves 2½–5 in. long, hoary-white on light green; inflorescence 2 ft. tall with clusters of nodding orange to reddish flowers. Outstanding for the window sill. From cuttings. **C**: 13   p. 328

ladismithiensis (So. Africa: Cape Prov.), "Cub's paws"; attractive branching succulent with woody stems and small and plump-fleshy 1–2 in. leaves yellowish green and covered with rough white hair, apex claw-like with little teeth and maroon; shiny flowers pale yellow tinged apricot, darker outside. **C**: 13   p. 331

orbiculata oophylla (S.W. Africa); elongate rosette of thick-fleshy, obovate gray-green leaves 1½ to 4 in. long, covered with silvery blue powder, the apex prettily edged in brown-purple; nodding bell-shaped flowers orange red. Very charming. **C**: 13   p. 331

paniculata (So. Africa: Karroo to Cape; S.W. Africa), the "Botterboom" of the Boers; tree-like, thick-stemmed succulent to more than 6 ft. high, with swollen trunk as thick as a man, covered with papery yellow-brown skin; few branches, with obovate fleshy leaves 2 to 4 in. long, gray-green with yellow margin, deciduous during resting time. Red flowers with green stripes. **C**: 13   p. 331

undulata (So. Africa: Cape Prov.), "Silver crown" or "Silver ruffles"; strikingly beautiful succulent to 18 in. high but usually much smaller; thick stem with opposite, broad wedge-shaped thick 3 to 5 in. leaves covered with silvery-gray bloom, the rounded pure white apex crimped and wavy; flowers orange. Overhead watering washes off the white powder. One of the noblest "hard" succulents and an eye-catcher in every collection. **C**: 15   p. 331

COUROUPITA (*Lecythidaceae*)
guianensis (Guyana), the "Cannonball-tree"; tall, soft-wooded, deciduous tree with armed branches; obovate, serrate 12 in. leaves, and fragrant 5 in. cup-shaped, waxy flowers rose-colored inside, orange-yellow outside; these spring curiously directly from the tree-trunk on tangled stems, later followed by the large round 6–8 in. brownish fruits, looking like medieval cannon balls. This is a striking example of cauliflory, common amongst woody plants in the tropics, where buds break through the bark and set fruit dangling from the trunk instead of its leafy branches. **C**: 51   p. 528

CRASSULA (*Crassulaceae*); the "Jade plants" belong to a large and interesting genus of succulents mostly concentrated in South Africa, ranging from diminutive mimicry forms to handsome, shrubby bushes; their small flowers are in large clusters, usually white or yellow, but sometimes a brilliant scarlet as in C. falcata. The most cultivated species are wonderful, unpretentious foliage plants, ideal for planting into novelty containers or for pots or tubs; there are also creeping kinds that may be used in hanging pots. Crassulas prefer a cool location, especially in winter, but tolerate the warmer interiors, they are not sensitive to drafts and respond well to air-conditioning; normally sun-lovers, they best maintain good foliage in a maximum of light and with fresh air, but hang on to life in poor lighting down to 25 fc. but stems will etiolate and foliage become smaller. Crassulas like small containers but appreciate occasional feeding. Propagated easily from cuttings, also from leaves which form roots and plantlets at their base.

arborescens (So. Africa: Namaqualand to Natal), known commercially as argentea or "Silver dollar" or "Silver jade plant"; heavy, branching succulent growing tree-like with thick trunk to 12 ft. high, the robust branches with boldly fleshy, broad obovate, opposite leaves 1½–3 in. long, united at the base, silver gray with reddish dotting and contrasting red margin; starry flowers white, turning pink. Slow growing and rarely blooming; excellent in pots. **C**: 13   p. 329

arborescens hort.: see argentea

argentea (So. Africa: Cape Prov., Natal), the "Jade plant", "Money-tree" or "Chinese rubber plant"; known as C. arborescens in horticulture, and so listed in Loudon's

Encyclopedia of Plants (England 1836); freely-branching or forking succulent shrub with enormously stout, fleshy trunk, growing to 10 ft. high, even the flexuous branches are thick, the leaves obovate, thick-fleshy pads, 1–2 in. long, the upper surface convex, the lower flat, glossy bright or jade-green, turning reddish in the sun, edged red; clusters of small pinkish-white starry flowers form in profusion from winter to spring, on older specimen. One of the best all-around house plants tolerant of neglect; loves sun but takes shade down to 25 fc.; stands up under draft and air-conditioning. An enduring favorite plant in dishgardens. Propagates easy from cuttings, even in a glass of water. Botanically correct probably C. obliqua. **C**: 13   p. 329, 335

argentea hort.: see arborescens

argentea variegata, "Variegated jade plant" or "Variegated crassula"; obovate leaves green and grayish, variegated with cream to orange-yellow and edged in orange-red and with rounded apex; tends to revert to green. Not as colorful and pretty as C. 'Tricolor jade' as it lacks the contrast of deep green with clear white. **C**: 15   p. 335

cornuta (S.W. Africa), "Horned crassula"; small gray column of shingled, snugly fitting, opposite stubby leaves only $\frac{1}{2}$ to $\frac{3}{4}$ in. long, short keel-shaped, glaucous silver gray and puckered; small white flowers. A pretty slow-growing miniature "hard" succulent. **C**: 13   p. 335

cultrata (So. Africa: Cape Prov.), "Propeller plant"; small branching succulent with well spaced pairs of obovate, oblong, light green leaves $1\frac{1}{2}$ to $2\frac{1}{2}$ in. long, curved and twisted in opposite directions like a propeller; clusters of small whitish flowers. Very interesting foliage. **C**: 13   p. 336

dubia, in the nursery trade as obvallata or tomentosa (South Africa); interesting succulent with fleshy, obliquely obovate leaves $1\frac{1}{2}$ in. or more long, flat on the surface, rounded beneath, and arranged diagonally like propeller blades, grayish green and covered with fine white felt, the margins red. **C**: 13 or 15   p. 329, 335

falcata (often known as Rochea falcata) (So. Africa), "Scarlet paintbrush" or "Propeller plant"; handsome succulent shrub to 3 ft. high, with wide, flattened and sickle-shaped, curved 3 to 4 in. leaves arranged like parallel shingles, rough gray-green, and sending up fleshy stalk with large and showy flat-topped cluster of bright crimson flowers. A spectacular sight when in bloom in late summer, silvery foliage glistening in the sun. The patio plant or for the sunny window. Propagated from seeds, or side cuttings from topped plants. **C**: 13   p. 335

hemisphaerica (S.W. Africa), "Arabs turban"; small round plant forming cushions, the rounded 1 in. leaves are opposite and united at the base, and closely imbricated or shingled, overlapping and curving downwards; dark gray-green and finely fringed at edges; flowers white. A collection miniature. **C**: 13   p. 335

x imperialis (pyramidalis x lycopodioides), "Giant's watch chain"; small branching succulent with semi-erect stems similar to lycopodioides but the little columns are more robust and thicker, almost $\frac{1}{2}$ in. dia.; the overlapping shingled leaves much larger and rich green. Fine for imitation ferneries. **C**: 15   p. 336

lactea (Natal, Transvaal), "Tailor's patch"; quick-growing succulent with thick stems leaning or lying on the ground; broad, obovate, flat, opposite fleshy leaves $1\frac{1}{2}$–2 in. long, smooth dark green and lightly covered by scurfy scale; sunken white dots along margins, starry flowers white and fragrant, in large clusters. Used in dishgardens; easy from cuttings. **C**: 15   p. 336

'Lactea hybrid'; robust cultivar with a small-leaved Crassula; an attractive open rosette of short-ovate, closely set flat leaves smooth glossy green without the typical marginal dots, and edged in red. **C**: 15   p. 336

lycopodioides (S.W. Africa), "Skinny fingers" or "Toy cypress"; spreading lycopodium-like, with stringy erect or scandent brittle branches, covered tightly with scale-type pointed leaves appressed in 4 ranks and hiding the stem, fresh yellowish-green; minute whitish flowers in the axils of the leaves. **C**: 15   p. 336

'Marnieriana hybrid' (marnieriana x falcata), "Jade necklace"; slender chains dense with plumb broad-ovate leaves $\frac{1}{2}$ in. long, opposite and united at the base, grayish with pronounced red edge. Easy grower. **C**: 13   p. 336

obvallata hort: see dubia

obliqua: see argentea

perforata (So. Africa), "Necklace vine"; thin wiry stems more scandent and creeping than perfossa, the opposite triangular $\frac{3}{4}$ in. leaves somewhat thin and recurved, dark green, finely dotted gray and with white ciliate edges; little yellow flowers. For hanging baskets. **C**: 13   p. 336

perforata 'Pagoda', the "Pagoda plant"; more stiffly upright and with fleshier, larger 1 to $1\frac{1}{2}$ in. leaves than perforata, which seems to be one of its parents as it has the same rich green color, and white ciliate margins. **C**: 13 p. 336

perfossa (S. Africa), "String o' buttons"; attractive succulent of erect habit with thin woody stem on which are, as if threaded, pairs of united thick, keeled $\frac{1}{2}$ in. leaves glaucous bluish-gray with purplish-red dotting and edging; small yellow flowers. When the branches become longer they will creep along the ground. **C**: 13   p. 336

x portulacea (argentea x lactea), "Baby jade"; type of Jade plant with fleshy obovate leaves somewhat smaller than argentea, entirely plain green, and pointed at the apex; the upper surface flat, rounded beneath. **C**: 15   p. 335

pseudolycopodioides (S.W. Africa), "Princess pine"; small shrubby succulent of sprawling habit, lycopodium-like, resembling lycopodioides but more vigorous and freely branching, the 4-ranked scale-like leaves rich green or grayish and spread away from stem which is lax and rambling. **C**: 13   p. 336

rubicunda: see Rochea coccinea

rupestris (monticola) (So. Africa: Cape Prov.) ,"Rosary vine" or "Bead vine"; shrubby, quite pretty succulent with scandent branches, the thin wiry stems set with opposite sessile, fat triangular leaves united at the base, $\frac{1}{2}$ to 1 in. long, glaucous light gray green, the margins tinged red; clusters of yellowish flowers. Good in baskets. **C**: 13   p. 329, 336

sarmentosa (S.E. Africa, Natal); succulent perennial shrub with slender, wire-stiff leaning branches; the ovate-pointed fleshy, shining green 2 in. leaves with greenish-cream center, coarsely toothed and cut at margins; tall panicles of starry pinkish flowers. Neat little variegated house plant. **C**: 13   p. 271

schmidtii (S.W. Africa, Transvaal, Natal), "Red flowering crassula"; small mat-forming floriferous succulent 3 in. high, sometimes grown as 'Crassula rubicunda', which is a taller plant (see Rochea coccinea); fleshy needle-like channeled leaves gray green with darker dots and with marginal hairs; reddish beneath; the leafy stems topped by clusters of carmine-red flowers. A miniature flowering plant. **C**: 13   p. 335

teres (S.W. Africa, So. Africa), "Rattlesnake tail"; dwarf clustering succulent plant forming slender columns to 4 in. high, of closely appressed, pale green stubby leaves $\frac{1}{2}$ to 1 in. broad, imbricated in 4 close rows, having a pale-translucent margin, which give it a glazed appearance; scented flowers white or yellowish. After flowering the plants dry up but form buds which root easily. **C**: 13 p. 336

tetragona (So. Africa: Cape Prov.), "Miniature pine tree" or "Chinese pine"; erect brittle shrub to 3 ft. high, with slender branches bearing opposite, tapering spindle-shaped, glossy green leaves 1–$1\frac{1}{2}$ in. long; small white flowers. Small size plants in dishgardens suggest miniature pine trees; however, if too dry, needles have a tendency to drop off. **C**: 13   p. 328

'Tricolor Jade' (argentea x lactea), "Tricolored jade plant"; beautiful succulent of hybrid origin; of upright habit, the glossy, somewhat pointed leaves are true green and strikingly variegated gray, white and pink, shading to purplish towards margin; surface flat, rounded beneath. Has the typical needle-work patch of lactea. A charming house plant. From cuttings or leaves. **C**: 15   p. 196, 329

x CRINDONNA (Amaryllidaceae)

memoria-corsii, known in horticulture as x Amarcrinum howardii (Amaryllis belladonna x Crinum moorei); this beautiful intergeneric hybrid was originated in California as x Amarcrinum, and at the same time (1921) in Italy as x Crindonna; having the persistent long leaves of a Crinum, and with better, larger blooms than A. belladonna; the very pretty pink funnelform, somewhat fragrant flowers with recurving segments, about 4 in. across, borne in clusters on $2\frac{1}{2}$ to 4 ft. stems. Autumn blooming in September-October. As a container plant they do best in larger tubs because of their size. **C**: 64   p. 388

CRINUM (Amaryllidaceae); the "Crinum-lilies", also known as "Swamp lily" or "Milk and Wine lily", or "Angel lily", are handsome tropical and subtropical amaryllis-like flowering bulbous plants mostly with persistent sword or strap-

shaped leaves; long-necked bulbs with fleshy roots; the long solid stalks supporting clusters of lily-shaped often fragrant blooms, distinguished from Hippeastrum by their long slender tube, in spring or summer. Because they are so massive, their fleshy roots would need large containers. Crinums are kept semi-dormant during winter in a light cool cellar or similar place; in summer best outdoors on the patio. Propagated from natural offsets or the large fleshy, bulb-like seeds.

asiaticum (Tropical South Asia), the "Grand Crinum", also known as "Asiatic poison bulb"; giant, showy rosette resembling an agave, with stem-like bulb to 1 ft. long, very decorative alone for its numerous broad, sword-like fleshy, light bluish green, persistent leaves 3–5 in. wide and 3 to 4 ft. long, spirally arranged and with broad, clasping base; center rib depressed. The inflorescence on 1½ to 2 ft. 2-edged stalks, in clusters of 20 or more deliciously fragrant, pure white flowers with red stamens, the tube greenish and the 3 in. linear petal segments drooping; blooming almost the year round even in winter when warm enough. The bulb is used medicinally as an emetic; juice from the leaves to treat inflammation. If in containers, move into frost free place for the winter. From offsets, not freely produced. **C**: 52
                                                                      p. 389

'Ellen Bosanquet', "Red angel lily"; a Florida cultivar by Louis Bosanquet, one of the best hybrid Crinums, and which can be grown in pots, flowering in September; stocky plant with broad, fleshy, spreading leaves, and an umbel of large amaryllis-like deep wine-red flowers; can be grown in pots at cool temperature, but does best planted out and taken up in October and planted in a tub. **C**: 64   p. 389

"giganteum" of hort., "Giant spider lily"; big bulbous plant with wide, strap-shaped leaves, and showy white flowers with linear segments, in large umbels. The true West African species C. giganteum has lily-like, white, bell-shaped flowers, strongly vanilla-scented. **C**: 64   p. 389

moorei (So. Africa: Natal), "Longneck swamp lily"; herbaceous plant from a large bulb with stem-like neck to 1 ft.; broad and thin, smooth-edged, somewhat wavy, bright green leaves 2–3 ft. long, and showy, lily-like, soft rose bell-shaped fragrant flowers 5 in. or more across, very attractive with pink filaments, 6 to 8 blooms to a cluster, on stout stalk, during summer; free bloomer; in the greenhouse or winter garden. Remarkable for producing young bulbs from stolons or runners. **C**: 64   p. 389

x powellii (bulbispermum x moorei), "Powell's swamp lily"; a spectacular old English hybrid (1732), with globose bulb, carrying abundant, decorative foliage, about 20 sword-shaped leaves 3 to 4 ft. long; a 2 to 3 ft. firm floral stalk appears in summer, crowned by a cluster of up to 10 trumpet-shaped flowers 4 in. long, generally deep rose, and opening in succession; blooming at a time when there are few other bulbous plants of its stature of beauty in flower. Slightly hardy. **C**: 64   p. 389

'Roozenianum', (Jamaica), a white swamp-lily; robust plant rather compact in habit with 3–4 in. bulb, strongly channelled deciduous leaves to 18 in. long; stiff floral stalks each with large clusters of 5 to 7 white flowers 5 in. long with slightly curved purple-crimson tubes, the back of the segments crimson and anthers cream-colored; blooming well in cool house in spring. **C**: 52   p. 389

CROCOSMIA (Iridaceae)

masonorum (Tritonia masonorum) (So. Africa: Transvaal), "Golden swan tritonia"; cormous herb forming clumps of sword-shaped basal leaves 2½ ft. long; and branched, arching stems bearing two-tiered spike-like clusters of starry flowers flaming orange to orange-scarlet, 1½ in. across, and blooming in July or August. Plant new corms into pots in October or March, water when growth commences. When foliage dies down keep dry; store corms out of soil like gladiolus. Flowers last 2 weeks when cut. From offsets. **C**: 75   p. 393

CROCUS (Iridaceae); popular winter-hardy low cormous, stemless herbs with grass-like leaves and relatively large flowers 3 to 6 in. long with stem-like tubes and cup-shaped or flaring petals mostly blooming in late winter or earliest spring, in many vivid colors. Often naturalized in gardens. Crocus may be flowered indoors during winter and are the easiest of bulbs to force. Plant several in a pot in autumn and bury outside for 8 weeks or more or keep in cool cellar. When wanted bring inside in a cool sunroom at 50°F. nights.

If kept too warm only leaves will grow and the flowers are checked. Flowered-out plants may be planted in the garden. Propagated by offsets from old corms in summer.

moesicus (aureus) (S.E. Europe: Hungary) in cultivation as the "Dutch crocus" or "Dutch yellow"; a popular early-spring flowering stemless herb with large corm, grass-like leaves with silver midrib, and bright yellow tubular flowers to 4 in. long. A hardy and beloved low garden plant. The Dutch crocus, which also come in mauve, may also be bloomed in pots indoors. **C**: 26   p. 373

'Vernus hybrid', "Alpine spring crocus"; spring-blooming hardy cormous herb with large, erect, funnel-shaped lilac or white flowers in the species; the tube 3 in. and the segments 1–2 in. long, the spreading lobes faintly striped purple; the hybrids in other colors from yellow to blue, but throat never yellow; much larger than the original So. European species, the wild crocus of the Alps and the Pyrenees. Normally blooming from February to April, depending on climate. **C**: 26   p. 394

CROSSANDRA (Acanthaceae)

infundibuliformis (India), "Firecracker flower"; attractive somewhat shrubby herbaceous tropical plant with glossy, ovate leaves 3 to 5 in. long; from the axils appear clusters of showy salmon-red tubular flowers with one-sided flattened limb, in angled, bracted spike. While grown in our greenhouses as a small 4 or 5 in. potplant, I have seen these in their habitat in India and Ceylon ever-blooming as a bush 3 ft. high. In our latitude profusely flowering for a long time during the warm summer months. A handsome and exotic-looking house plant; also as durable cutflower. From seed or soft cuttings. **C**: 10   p. 494

pungens (Tanzania: Usambara Mts.); attractive, bushy plant to 2 ft. high, with oblong leaves olive-green in beautiful contrast with creamy-white veins; flowers light orange, set in oval bracts hairy at margins. **C**: 10 p. 184, 494

CROTON: see Codiaeum

CRYPTANTHUS (Bromeliaceae); the charming "Earth-stars" are more or less flattened star-shaped rosettes with white flowers from Eastern Brazil growing terrestrial on the forest floor or on rocks; grown for their striking leaves that may be mottled, striped or cross-banded in silver over greens or bronzes, their margins rippled like pie-crust edges; they prove durable even in poor light, and are undemanding, ideal for small novelty containers, planting on roots, or into dishgardens. Offsets are mostly produced from in between the leaves; these may be pulled off and rooted in peatmoss; some species form offsets from the base or on stolons.

acaulis (Brazil: Guanabara), "Green earth star"; small flattened, suckering, terrestrial rosette 4 to 6 in. across with waxy leaves medium green and covered with pale gray scurf; the margins prickly; white fl. low in center. **C**:9 p. 137

acaulis ruber (Brazil), "Miniature red earth star"; small rosette 3 to 5 in. across with leaves green to purplish-bronze in center and along margins, covered over-all with beige scurf. **C**: 9   p. 137

beuckeri (Brazil), "Marbled spoon"; very attractive, irregular medium-size rosette of flat, thin-leathery spoon-shaped, slender-pointed leaves, prettily painted with rich green marbling on pale green; flowers whitish. **C**: 9   p. 136

bivittatus lueddemannii (Brazil), "Large rose-stripe star"; large, flattened rosette 10 to 15 in. from tip to tip, with narrow leaves much longer than the similarly colored biv. minor; strap-shaped and undulate-margined, thick and fleshy, with two pale green bands over coppery base. **C**: 9
                                                                      p. 137

bivittatus minor (in trade as roseus pictus) (Brazil), "Dwarf rose-stripe star"; flattened, small, star-like terrestrial rosette 3 to 6 in. across, satiny olive green with two pale bands, overcast with salmon rose, and turning coppery red in strong light; finely toothed margins. Long popular for use in novelty planting. **C**: 9   p. 137

bromelioides tricolor, "Rainbow star"; strikingly variegated sport of bromelioides (terminalis) from Brazil; a flaring rosette with long-elliptic tapering, smooth wavy-edged leaves, fresh-green to bronzy, edged and striped with ivory-white, the margins and base tinted carmine-rose in good light. **C**: 9   p. 43*, 131*, 136

diversifolius (Brazil) (acaulis diversifolius of hort.), "Vary-leaf star"; relatively large rosette with arching, wavy-margined leathery leaves to 10 in. long, grass green and thickly covered with silvery scurf. **C**: 9   p. 136

fosterianus (Brazil: Pernambuco), "Stiff pheasant-leaf"; large and beautiful terrestrial rosette to 32 in. across, of very flat habit; marked similar to zonatus, but leaves much thicker and very stiff, coppery green to purplish brown, with tan zebra cross-banding. **C**: 9   p. 136

x'lt'; aptly known as "Color-band"; spectacular mutant developed from a Foster hybrid probably involving C. bivittatus lueddemannii. Terrestrial rosette with strap-like leaves to 18 in. long and 1½ in. wide, coppery green with longitudinal striping and broad outer bands of ivory; the margins a glowing rosy-red intensified by good light. **C**: 9
p. 136

lacerdae (Brazil), "Silver star"; flat rosette of thin leathery leaves emerald green with silver borders, and broad, pale center band; finely toothed; whitish flower. **C**: 9 p. 137

x osyanus (beuckeri x acaulis), "Mottled earth-star"; medium size, irregular rosette of broad leathery leaves warm green, with pale mottling, tinted pink to coppery red; white flowers. **C**: 9   p. 137

roseus pictus: see bivittatus minor

x rubescens, "Brown earth star"; low, medium-large rosette of leathery, obovate, waxy leaves predominantly purplish-brown, the center covered with silvery scales becoming more scattered toward apex; whitish flowers. **C**: 9   p. 137

terminalis: see bromelioides

zonatus (Brazil: Pernambuco), "Zebra plant"; spectacularly attractive spreading rosette 8 to 18 in. across of waxy lance-shaped soft-leathery leaves, brownish green to copper, and with tan to light brown, irregular crossbands; silvery-scurfy beneath; translucent white flowers; growing terrestrial. **C**: 9   p. 136

zonatus zebrinus (fuscus), "Pheasant leaf"; strikingly beautiful form 12 to 18 in. across, with bronzy-purple, long wavy leaves, the pronounced silvery to beige crossbanding somewhat resembling the stripes of a zebra. **C**: 9 p. 136, 193

x CRYPTBERGIA (Bromeliaceae)

rubra (Crypt. bahianus x Billb. nutans), "Cryptanthus-Billbergia hybrid"; small open bigeneric rosette with thick, narrow, recurved leaves bronzy green, shading to wine-red in center; soft, spiny margins; raised cluster of whitish flowers. **C**: 9   p. 137

CRYPTOCEREUS (Cactaceae)

anthonyanus (Mexico: Chiapas), "Anthony's rick-rack"; fragrant night-bloomer; a spectacular climber of the rain forest, using aerial roots; the unusual deeply lobed stems look like fish bones; 4 in. flowers intensely fragrant, lasting but a single night, beautifully colored burning-red with cream-yellow petals. **C**: 60   p. 156, 165*

CRYPTOCORYNE (Araceae); aquatic Asian herbs with creeping rhizomes and stalked leaves which act as oxygenators in the water for fish, also creating, in their different shades of green, the effect of an underwater landscape. The inflorescence is typically aroid-like, with spathe enclosing the slender spadix bearing the male flowers on the upper part and female organs on lower portion. They grow best in tropical aquaria and prefer a water temperature of 68 to 76°F., in coarse, sandy loam; the water slightly acid, or rainwater if possible. Cryptocoryne belong to the best aquarium plants supplying oxygen and are appreciated especially where the light is poor. Propagated by separation of the rhizomes with offsets.

ciliata (Monsoon India: Bengal to Malaya), "Fragrant tape grass"; tropical aquatic herb or bog plant to 15 in. high, with creeping rhizome and stalked lance-shaped leaves of firm texture, usually growing under water, fresh to yellowish-green with undulate margins, the depressed midrib dark green. Inflorescence submerged with the enclosed spadix bearing female flowers on the lower part, male flowers on the upper, only the extended spathe above water surface. **C**: 4   p. 519

cordata (Malaya: Borneo; Java); aquatic herb with creeping rhizome, the 3 to 4 in. lanceolate leaves on 4 to 8 in. stalks, olive green, reddish-purple beneath; used in aquaria as oxygenators. **C**: 54   p. 98

griffithii (So. Malaya): aquarium plant with stalked ovate 3 to 5 in. leaves dark green, marked purple beneath; tubular part of spathe rosy; blade blood-red. **C**: 54   p. 98

willisii (Trop. S.E. Asia), "Underwater aquatic"; aquatic plant with narrow lanceolate leaves 3 to 5 in. long, when young reddish-brown, marked with greenish-black; curly margin. **C**: 54   p. 98

CRYPTOMERIA (Taxodiaceae)

japonica nana (China, Japan), "Dwarf Japanese cedar"; dwarf form, differing by its symmetrical habit with the normally large C. japonica, a magnificent evergreen tree more than 100 ft. high, with shreddy reddish bark; the small awl-shaped ½ to 1 in. needles emerald green, keeled and curved inward, and spirally arranged on stiffly upright branches; small, red-brown cones. Needs ample water; not dependably winter-hardy. By grafting or cuttings. **C**: 76   p. 310

CTENANTHE (Marantaceae)

lubbersiana (Phrynium) (Brazil); spreading herb with forking stems; the oblique, linear-oblong leaves are firm, green, variegated and mottled with yellow above, paler beneath. **C**: 54   p. 176

oppenheimiana (Brazil); strong, compact, branching plant forming a dense, broad bush eventually 6 ft. high; the narrow lanceolate leaves are leathery and firm, dark green with lateral bands of silver, and wine-red beneath, attached at an angle to downy stalks. **C**: 10   p. 175

oppenheimiana tricolor (Brazil), "Never-never plant"; very colorful, tufted variety with narrow leaves, highly variegated white over green and silver-gray, their wine-red underside in vivid contrast, and glowing through the surface. **C**: 60   p. 175

pilosa (compressa), in horticulture erroneously as Bamburanta arnoldiana (Brazil: Pernambuco), "Bamburanta" or "Giant Bamburanta"; decorative perennial herb with basal and with stem leaves; the foliage unequal-sided, oblong leathery, waxy green; grayish-green beneath; borne at an angle on wiry petioles; the plant is bushy, 1 to 3 ft. high, but on occasion throws up bare stalks bearing 2 to 4 small plants at the top forming a heavy tuft of foliage; the small flowers borne under flattened, hairy bracts. A sturdy foliage plant for indoors or the patio in pots or tubs. By division or stem cuttings. **C**: 4   p. 176, 306

CUCUMIS (Cucurbitaceae)

anguria (So. U.S.: Florida and Texas), "West Indian gherkin" or "Gooseberry gourd"; cucumber-like vine with slender angled, rough stems and palmately lobed 4 in. leaves; yellow flowers; the females producing spiny ovoid, yellowish-green fruits 2 to 3 in. long, and which can be used for pickles. Interesting for the conservatory, or as an annual for the hot bed. **C**: 54   p. 258

CUNONIA (Cunoniaceae)

capensis (So. Africa), "African red alder"; small evergreen tree to 40 ft. high, with pinnate, leathery glossy green leaves, the 3 in. leaflets toothed, on wine-red petioles and twigs; the new growth is bronzy red; dense feathery axillary racemes of small white flowers; somewhat hardy. Good large decorative tub plant, or smaller as a house plant. From semi-ripened cuttings or seed. **C**: 15   p. 298

CUPHEA (Lythraceae)

hyssopifolia (Mexico, Guatemala), "False heather" or "Elfinherb"; small hairy, woody shrublet to 2 ft. high, the wiry branches crowded with tiny linear, leathery fresh green leaves less than ½ or ¼ in. long; numerous small starry flowers, consisting of a green calyx and 6 purplish-rose petals, axillary amongst the fern-like foliage, make this a pretty little flowering plant for indoors, or outdoors in summer. From cuttings. **C**: 14   p. 300

platycentra, as known in horticulture; bot. ignea; (Mexico), "Cigar flower"; low herbaceous plant to 12 in. high with slender branching stems, fresh green lanceolate leaves, and solitary axillary flowers having a slender, bright scarlet tubular calyx ¾ to 1 in. long with white mouth and dark ring at end, without petals; spring-summer blooming. Charming little pot plant for the windowsill. Easy from cuttings. **C**: 14   p. 494

x CUPRESSOCYPARIS (Cupressaceae)

leylandii; bigeneric hybrid between Cupressus macrocarpa and Chamaecyparis nootkaensis, resembling the latter in habit and foliage; very robust and healthy-looking, and a fast grower to 20 ft. high; long slender upright branches of flattened, striking green foliage sprays, with scale-like leaves appressed along the branchlets. From cuttings. **C**: 64   p. 310

CUPRESSUS (*Cupressaceae*); characteristic in the landscape, these evergreen coniferous, mild climate trees of the cypress family are usually pyramidal, sometimes spreading, with aromatic foliage, the small scale-like leaves pressed close to the cord-like twigs, but often linear in very young plants; globular cones with woody scales. Propagated from seeds or cuttings.

arizonica pyramidalis (glabra 'Pyramidalis'), "Blue pyramid Arizona cypress"; very symmetrical compact form of Arizona cypress having an intense silver blue-green color. C. arizonica from Arizona and New Mexico is a fast growing pyramidal conifer with glaucous bluish-green scale-like needles. Drought resistant and tolerant to dry heat, it is a fairly good indoor tub plant where coniferous effect is wanted, also for patio; small plants in dishgardens. **C**: 1   p. 310

sempervirens (So. Europe, W. Asia, No. India), "Italian cypress"; the classic conical cypress of the Greek and Roman writers, with very short branches, forming a dense, narrow column slowly to 60 ft. or more high; the stout branchlets with scale-like leaves dark green with grayish cast in 4 ranks. Esteemed for formal effect because of its stiff, picturesque outline. Suitable in containers for warm, sunny places; cool in winter. **C**: 64   p. 310

CURCULIGO (*Amaryllidaceae*)
capitulata (recurvata) (Java), "Palm grass"; stemless tropical perennial plant with short, thick rootstock palm-like but belonging to the Amaryllis family; the handsome recurved, glossy-green leaves plaited like a fan, 3 ft. or more long; small yellow flowers near base. Attractive foliage plant, the leaves moving with the air; needs water copiously with good drainage. Propagate by division. **C**: 4   p. 232

CURCUMA (*Zingiberaceae*)
roscoeana (Burma), "Hidden lily"; robust perennial, tropical herb with tuberous roots, sending up 6—8 handsome, long-stalked, lanceolate, ribbed leaves with dark green nerves; the inflorescence a splendid spike about 8 in. long, with cone-like, showy bracts gradually changing from green to vivid scarlet-orange, corolla yellow, with rich golden lip. **C**: 54   p. 206

CURMERIA: see Aglaonema

CUSSONIA (*Araliaceae*)
spicata (Transvaal), "Spiked cabbage tree"; evergreen tree with handsome leathery leaves palmately compound, the smooth, grayish-green segments dentately cut or lobed again, and arranged in flat rosettes. Decorator with fascinating foliage, requiring lots of sun and plenty of water with fast drainage. **C**: 14   p. 290

CYANOTIS (*Commelinaceae*)
kewensis (Tradescantia) (India: Malabar), "Teddy-bear vine"; succulent creeper rather compact, with brown-woolly branching stems, densely set almost shingle-like with small fleshy triangular leaves about 1 in. long, olive green wholly covered with woolly brown hair, underside violet; small purple flowers. Cute plant for collections of miniatures. **C**: 13   p. 260

somaliensis (East Africa, Somaliland), "Pussy-ears" or "Fuzzy ears"; succulent branching little creeper related to Tradescantia, with linear-lanceolate, clasping, glossy green fleshy leaves 1½ in. long, the whole plant entirely covered with long, soft white hair; flowers blue-purple. **C**: 13 p. 260

CYATHEA (*Filices: Cyatheaceae*)
arborea (mountains of Puerto Rico to Jamaica), the graceful "West Indian treefern"; regal treefern to 50 ft. high, the slender trunk rather bare and brown, the upper part covered with pale brown scales, and bearing a crown of spreading bipinnate, fresh-green, soft-textured, finely toothed fronds 3—6 ft. long, paler below, and without spines. Stems dug in the Caribbean forests may be established in tubs; they are probably the most tropical-looking treeferns for decoration, but as most treeferns, are difficult to maintain

indoors more than a few months; the roots must practically stand in water, or the tubs may be wrapped in a plastic bag to hold moisture in. **C**: 10   p. 235

medullaris (New Zealand, S.E. Australia, Tasmania), "Sago fern", the "Black treefern"; tallest of N.Z. treeferns to 60 ft. or more, with slender black trunk, at base covered with matted aerial roots; on top a great crown of spreading, curving, feathery fronds 8 to 20 ft. long, firm-leathery, tripinnate, deep green above, paler beneath; the apex and leaf bases clothed with long black scales. In cultivation, trunks must be kept constantly moist. **C**: 60   p. 237

CYCAS (*Cycadaceae*); Bold palm-like or fern-like, slow growing members of the ancient Sago-palm family, with stout, trunk-like stem and a large crown of frond-like, stiff leaves divided feather-fashion into innumerable rigid segments without veins except midribs; male and female cones are carried on separate plants. Choice, long enduring container plants for tropical effect, indoors or on the patio, requiring a minimum of care. Propagated from suckers or bulbils from near the base of trunk, which are detached when the plant is dormant, between the periods when a new crop of leaves appears. Adventitious knobs, like growth buds or "Bulbs", also appear along older trunks from scarred leaf bases, especially where injured, and these are taken off and used for propagation by potting them in moist loam with sand and humus, best covered with a plastic bag until growth commences. In the summer they may be set out-doors. The pith of the trunk of C. revoluta furnishes a sort of coarse sago, but not the commercial kind.

circinalis (Madagascar, South India, Ceylon to the Philippines and New Guinea), "Fern palm" or "Crozier cycad"; tropical palm-like tree faster growing and more lush than C. revoluta, with heavy trunk normally to 12 ft., but may become 40 ft. high, topped by a graceful rosette of stiff glossy fronds to 8 ft. long, pinnately divided, with the long, shiny dark green leaflets flat on edges. I have seen colonies of this species in the Bulolo highlands in New Guinea, at 4000 ft., in tall Kunai grass, giving a tropical effect to the landscape. Wonderful and enduring indoor subject where space permits; one of the easiest to grow. **C**: 22
                                                                    p. 316

media (No. Australia, Queensland), "Australian nut palm"; trunk normally 12 to 18 ft. high, but reaching up to 30 ft., the arching fronds shorter than circinalis, to 5 ft. long, with its numerous leathery flat leaflets a rich green, and narrowed to a spine. The female cones bear nuts the size of a walnut, with a thin fleshy covering. A most graceful ornamental cycad, and a grand decorator, very tenacious even when neglected. The tree I photographed in our palm house at Roehrs in New Jersey has seen decades of exotic plant exhibits at New York flower shows, exposed to much abuse, yet serving faithfully year after year. **C**: 16   p. 317

revoluta (So. Japan to Java), "Sago palm"; palm-like tree slowly forming hulking trunk to 10 ft. or more high, mostly solitary but sometimes branched; stiff deep glossy green, pinnate fronds 2 to 7 ft. long, in a terminal crown, periodically appearing all at one time every year or two; the leathery segments are spine-tipped and rolled down at margins. This is the hardiest (15°F.) and most widely used stately cycad. Excellent as patio or winter-garden subject; prefers coolness in winter but is not too particular. Forms woody growth buds or knobs at the base and from leaf-bases along old trunks which are removed for propagation. **C**: 15   p. 315, 316, 317

CYCLAMEN (*Primulaceae*); exquisite autumn and winter-flowering plants with large, flat tubers, bearing a rosette of succulent heart-shaped bluish leaves beautifully marbled and traced in silver; elegant long-stemmed nodding flowers with 5 corolla lobes turned up like little flags in delicate colors. The florists' cyclamen (C. persicum) has been much improved and enlarged through the years, variously colored from red through salmon and rose to white and delicate pastel shades between, and has become a beloved flowering house plant, longest enduring however in a cool room or window. Cut flowers keep best if the base is cut and slit, and the blooms set overnight into a refrigerator to stiffen. After blooming, tubers may be rested and kept over, but while again producing flowers the foliage will not come back to perfection. Normally grown from seed which takes up to 18 months to bloom.

persicum (indicum) (Greece and Mediterranean Islands to Syria); charming low, fleshy herb with large, hard tuberous roots, heart-shaped basal leaves in a rosette, prettily patterned with silver; and long-stalked solitary, 1½ in. nodding, fragrant flowers; with purplish-rose flaring corolla lobes elegantly reflexed. **C**: 22   p. 401

persicum giganteum (indicum); the "Florist's cyclamen" or "Alpine violet"; highly developed horticultural form of the original C. persicum with small 1½ in. flowers; now an important and well-beloved cool winter-flowering house plant with attractive bluish leaves marked with silver, and flowers much enlarged, by horticultural effort since its introduction in the 17th century, with erect petals to 3 in. or more, and now available in many named varieties; pastel shades and colors from white through salmon, rose or glowing red, some with purple eye. **C**: 16   p. 401

persicum giganteum flore pleno, the "Double cyclamen"; beautiful double-flowering cultivar, during wintertime with heavy blooms of numerous petals principally in shades of salmon-pink, in a bouquet above the bluish leaves patterned with silver. **C**: 72   p. 382

persicum giganteum 'Perle von Zehlendorf' (1907); an old and still favorite color form of the "Florists' cyclamen" or "Alpine violet"; with elegant flowers glowing dark salmon, deeper red toward eye, shading to lighter salmon rose at margins; very free-blooming beginning from November; a great holiday plant for Christmas, and if kept cool, or from late sowings, blooming plants may be had into March. **C**: 72   p. 382

CYCLANTHUS (*Cyclanthaceae*)

bipartitus (Guyana), "Splitleaf cyclanthus"; stemless, palm-like perennial foliage plant with a cluster of milky-juiced, large leathery, quilted leaves with two depressed ribs, and forked at top, about 1 ft. long or more, on stalks that may become 6 ft. tall; scented flowers, aroid-like on spadix with yellow spathe. Best in moist-warm greenhouse. **C**: 54   p. 232

CYCLOPOGON (*Orchidaceae*)

variegatum (Brazil), terrestrial Jewel orchid with fleshy roots; very ornamental small ovate leaves marbled with pink; inflorescence an upright spike with pinkish bracts and small pale green flowers. **C**: 62   p. 454

CYCNOCHES (*Orchidaceae*); the peculiar "Swan orchids" are among the most handsome tropical American epiphytes, with elongated pseudobulbs bearing plaited, deciduous leaves; lateral clusters of peculiar, long-lived flowers having an arching column simulating the neck of a swan; the sexes are separate with either male or female flowers; the male ones usually in long pendulous inflorescences; the females are as a rule larger with fewer in a cluster. Cycnoches are of easy cultivation. While actively growing they enjoy moisture, humidity and warmth, but when the new pseudobulbs are matured, water should be withheld for some weeks for a rest. Propagated by division.

egertonianum (Mexico and Guatemala to Colombia, Peru, Brazil); singularly beautiful epiphyte with long, leafy pseudobulbs, and greenish flowers suffused with purple; male flowers 2 to 2½ in. across, in long pendulous racemes, sepals and petals spotted with red-brown, the lip with purple fringe-like filaments; the fleshy female flowers considerably larger, yellowish-green with creamy lip; blooming from late summer into winter. **C**: 10   p. 462

loddigesii (Guyana, Surinam to Colombia), a typical "Swan orchid"; curious epiphyte with pendant cluster of brownish green, fragrant flowers suffused and marked with purplish brown, and fleshy, rosy lip, together bearing some resemblance to the expanded wings of a swan; the male flowers 4 to 5 in. long; blooming variously from late summer into early winter. The plant often produces male and female flowers very distinct from each other; the long-lasting blooms may keep for 3 weeks. **C**: 10   p. 462

ventricosum (Mexico, Guatemala), a "Swan orchid"; striking epiphyte with fleshy stems to 1 ft. high; from near the apex of the stout pseudobulb the arching raceme of sweet-scented, waxy flowers 4 to 5 in. across, greenish-yellow, a white lip, and an arched slender column resembling a swan's neck; lasting about 3 weeks; male and female flowers almost identical; free-blooming, from July to November. **C**: 60   p. 455*

CYMBALARIA (*Scrophulariaceae*)

muralis (Linaria cymbalaria) (So. Europe, No. Africa, W. Asia; naturalized in C. Europe and E. No. America), "Kenilworth ivy"; creeping perennial, herbaceous ground cover, with thread-like stems rooting at the nodes, small kidney-shaped, fresh-green, waxy leaves irregularly lobed, purplish beneath; tiny lilac-blue flowers with yellow throat, like miniature snapdragons. An old-fashioned basket plant; winter-hardy. **C**: 78   p. 265

CYMBIDIUM (*Orchidaceae*); handsome terrestrial and epiphytic orchids with short pseudobulbs bearing long narrow leaves; the flower stalks laterally with clusters of long-lasting blooms. Cultivated today are mainly the numberless hybrids that have been developed; the large-flowered with arching sprays of 3 to 5 in. blooms very much in favor as a corsage flower; the smaller types as potplants. Most Cymbidiums are treated as semi-terrestrials; they like porous, rapid-draining growing mixtures made up of such as loam, peatmoss, shredded bark, chunks of fern-fiber, pebbles or perlite or fir bark chips, with fertilizer. The robust kinds need lots of water while in active growth, but after the bulbs mature, a partial rest period of several weeks should be allowed to assure development of flower buds. Most Cymbidium hybrids enjoy relatively cool and airy conditions and they are best placed out into the sun during the warm season. Requiring lots of light, 4000 to 8000 fc. in summer is not too much for them while in winter 3000 fc. is sufficient. The large-flowered hybrids are in the height of their blooming season in our northern latitudes between February and June; in the Southern hemisphere around September-October. Propagated from single-bulb divisions, but these may take 3 to 5 years to flower.

x alexanderi (eburneo-lowianum x insigne), a "Corsage cymbidium"; well-known commercial hybrid of the white with pink color group; terrestrial of robust habit, with bold, arching spray of waxy, long-lasting 4 to 4½ in. flowers flushed pink, with a purple horseshoe on the lip; tending to bloom in spring. **C**: 84   p. 463

aloifolium (widespread from the Himalayas, India to Ceylon and Burma); a small-flowered epiphyte which I have seen growing in the forking branches of hardwood trees at 8000 ft. in the Sikkim Himalayas; clustering plant with leathery, stiff-erect linear obtuse leaves nearly without pseudobulb; the inflorescence in elongate clusters flowering on stalks 12 to 18 in. long, becoming pendant, the fleshy flowers 1½ in. or more across, with spreading linear sepals and petals yellow with purple center stripe, the lip brownish-red with yellow center, blooming from May into autumn. Several very similar species are known in horticulture as C. aloifolium, and the plant pictured may be C. simulans from tropical Burma and Indonesia. **C**: 20   p. 463

x'Doris' (insigne x tracyanum); well known commercial primary hybrid ¶1912, with large 4 in., fleshy, long-lasting flowers; light yellow sepals, petals closely veined with red, the lip spotted with red-brown on front lobe and veined in same color on side lobes. **C**: 84   p. 455*

x'Flirtation' (pumilum x 'Zebra'), an excellent and typical "Miniature hybrid cymbidium"; attractive, variable semi-miniature recently developed, ideal as a shapely, compact flowering pot plant, blooming successively twice a year, usually November through April; arching racemes with flowers spreading 2 in., pale pink to greenish ivory, shaded to coppery purple and striped orchid, lip white in throat spotted and blotched maroon. **C**: 72   p. 463

lancifolium (Himalayas at 6000 ft.; India to Japan, south to Malaysia); charming, terrestrial forest dweller of low habit; spindle-shaped pseudobulbs clothed with 3 to 5 ribbed thin-leathery leaves 8 in. long; stiff-erect stalks bearing inflorescence 12 in. high of scattered, fragrant and long-lasting flowers 1½ to 2 in. across, greenish-cream with purple spots, the lip· white boldly marked with maroon, blooming in spring and early summer. **C**: 72   p. 463

virescens angustifolium (Japan); tufted, grass-like plant with linear leaves to 10 in. long; stalked, waxy flowers 2 in. dia., vivid green with faint purple stripes, two petals join over lip to form a hood, lip curved under, creamy-white with yellow throat and purple markings; winter-blooming. **C**: 34   p. 463

CYPERUS (*Cyperaceae*); the "Umbrella plants" or "Sedges" are perennial rush or grass-like partly aquatic herbs, all moisture loving, including the papermaking giant

"Papyrus" of the ancient Egyptians. From a mass of fibrous roots rise the usually 3-angled solid culms bearing small greenish, inconspicuous flowers in flat, bracted spikelets subtended by an umbrella-like rosette of leaves. Depending on size, these plants may be used at the edge of pools or partially submerged in aquaria. Propagation by division or suckers, from seed; in Cyp. alternifolius the crown of leaves may be cut off and rooted in moist sand or peatmoss, or in a glass of water as in hydroculture.

alternifolius (Madagascar, Réunion, Mauritius), the "Umbrella plant"; clustering perennial bogplant to 4 ft. or more high, but in cultivation usually short; with ribbed stalks bearing a crown of bright green, grass-like leaves spread in a rosette umbrella-like around a head of small green flowers. I have seen this also along the Athi River in Kenya, East Africa. Good house plant with a saucer of water under the pot, or in shallow pools or aquariums. To propagate, a leaf rosette with a piece of stem is cut off, and may be rooted in peatmoss or a glass of water. **C**: 16   p. 304, 518

alternifolius gracilis (Australia, New Caledonia); "Dwarf umbrella plant"; dwarf form of stiff, erect, stilted habit, in all parts smaller than the species, growing only 12 in. or so high, with wiry slender stems and narrower, abbreviated linear leaves. Best for the edge of the smaller aquarium, with that "Papyrus" look in miniature. **C**: 4 p. 518

alternifolius variegatus; "Variegated umbrella plant"; a fugitive variety with stems and leaves striped and banded lengthwise with creamy-white contrasting with shiny green. Unfortunately, the variegation is not stable, and unless the variegated portions of the plant are constantly reselected, it will, if too happy, revert back to the green species. On the other hand, this form needs more warmth and care to stay alive and evenly variegated; occasionally turning into a difficult all-white stage. **C**: 4   p. 518

diffusus var. elegans; laxus in horticulture; (Mauritius), the "Broadleaf umbrella palm"; vigorous, bushy perennial plant, 1 to 3 ft. high, with big basal leaves, sending out runners with suckers which are used for propagation; study, 3-angled culms topped by a crown of broadly sword-shaped, matte green, rather rough-edged leaves, 4 to 12 in. long; greenish or pale brown spikelets on thin wiry spreading stalks. A bold, handsome house plant of easy care. **C**: 4 p. 518

diffusus variegatus, a "Striped umbrella palm"; compact and bushy growing, attractive variety having its broad green leaflets more or less striped and banded pale yellow or cream, both near the base, or also crowning the wiry slender stems. **C**: 4   p. 518

papyrus (Tropical Central Africa, from there brought to Egypt in ancient times), the "Egyptian paper plant" or "Papyrus"; a stately aquatic sedge with stout dark green culms 4 to 10 ft. or more high, topped by a brush-like dense cluster of spreading and drooping flower spikelets with green rays that are thread-like, 6 to 18 in. long and extending out beyond the smaller leaves beneath them. A striking plant for greenhouse pools, growing in shallow water either planted out or grown in pots or tubs; these may be used in winter as house plants. Along the shores of Lake Victoria in Central Africa I have seen them 15 ft. tall; they are cut for roofing, etc. Used in Egypt for making papyrus paper since 2750 B.C. **C**: 52   p. 518

CYPRIPEDIUM (*Orchidaceae*); the "Lady-slippers" or "Moccasin flowers" are deciduous terrestrials mainly native in the North Temperate Zone, and are the most primitive of orchids; growing from fibrous-rooted underground rhizomes, producing the erect stem bearing sheathing, strongly ribbed leaves, and fantastic blooms characterized by their large lip inflated pouchlike, mostly in woodsy colors. These true Cypripediums are essentially for outdoor flowering and are fairly winterhardy planted in the ground; in cultivation they may be grown in well-drained pots filled with a compost of leafmold, gritty sand and sphagnum moss; while they are in growth they need water copiously, but as the shoots begin to wither a definite resting period is given from November to March until new sprouts begin to appear; they are best wintered in a cold, 45°F., greenhouse, or cold frame under sash. Propagated by division. True Cypripediums must not be confused with the tropical Asiatic Lady-slippers erroneously called "Cypripedium" in horticulture which are actually and botanically Paphiopedilum.

acaule (Canada and U.S.: Newfoundland west to Alberta, south to No. Carolina and Alabama), "Pink Lady-slipper"; deciduous terrestrial of the North temperate zone and hardy; with two light green plaited leaves silvery beneath, and large nodding solitary flowers 4 to 5 in. across, hairy sepals and petals greenish-brown, the divided, pouch-like lip rose-pink veined with crimson; blooming according to climate April to July. Very showy in the woodland garden, but needing attention in cultivation in pots, best kept over in covered cold frame. **C**: 84   p. 464

macranthum (S.W. Tibet, Siberia); handsome terrestrial to 15 in. high, with 3 to 4-leaved stems; the large foliage elliptic, bright green; attractive, solitary flowers 4 in. across, the large dorsal and narrow petals purple with darker netting, the full-inflated lip rose-pink; May-June blooming. **C**: 84   p. 464

CYRTANTHUS (*Amaryllidaceae*)
mackenii (So. Africa: Natal), "Bow-lily"; a charming bulbous herb with linear leaves 15 in. long, having prominent rib on back; the 1 ft. stalks carrying clusters of fragrant slender flowers 2 in. long with an ivory white curving tube and 6 spreading lobes, blooming through the winter and in spring; in Natal in July-August. Plants grow actively through most of the year, producing numerous offsets; rest when foliage turns yellow. Repot annually before new growth begins. **C**: 64   p. 392

CYRTOMIUM (*Filices*); small ferns of the family Polypodiaceae, sometimes known as Aspidium; these are of compact habit, with feather-fashion, pinnate fronds of leathery texture, rising on wiry stalks from crowns along the short rhizome, with age forming tufts. Known in cultivation as very tough and durable ferns, they are of easy care and tolerate much abuse, as a potted plant indoors, including occasional drying out at the roots. They will grow in ordinary garden soil with humus added but prefer not to be too warm and love fresh air. Propagated from spores or by division of older plants.

falcatum (Japan, China, India, Celebes, Hawaii), the "Fishtail fern"; a "Holly-fern" with handsome pinnate fronds on brown scaly stalks to 2 ft. long, the leathery, shining dark green leaflets ovate, slender pointed and very durable under adverse conditions. **C**: 18   p. 238

falcatum 'Rochfordianum', the popular form of "Holly fern" usually in nurseries; this cultivar is of more robust, compact habit, the fronds are broader and fuller, and the large leathery, glossy leaves are saw-toothed and wavy at the margins; rarely becoming more than 12 in. high. As a potfern very easy to maintain, tolerant of abuse, shade and neglect; prefers cool, but also grows in warmer interiors. **C**: 18   p. 238

CYRTOPODIUM (*Orchidaceae*)
punctatum (So. Florida and Mexico, West Indies to Brazil and Argentina), the "Cigar orchid", "Beeswarm" or "Cowhorn orchid"; noble rock-dwelling orchid with bold spindle-shaped pseudobulbs 2 to 4 ft. tall, the upper part with partially deciduous leaves; from the base, concurrently with new growth, an erect stalk, branching near the top, and bearing a showy display of clusters with colored bracts and greenish-yellow flowers 1½ in. across, having crisped petals, the 3-lobed lip yellow at base and thickly spotted with red-brown; February to early summer-blooming. These orchids bloom best in good light; need humus compost with osmunda and rich loam, plentiful watering while growing, but a rest and dry after blooming until new spikes arise. By division. **C**: 22   p. 464

CYRTOSPERMA (*Araceae*)
johnstonii (Solomon Islands); small plant with firm, sagittate leaves, with long, widely spread basal lobes, fresh green with crimson veins; spiny stem with dark-purple zebra bands. Reluctant and slow. **C**: 60   p. 95

CYTISUS (*Leguminosae*); the "Brooms" are bushy, sun-loving shrubs of the Mediterranean region and Asia Minor, admired for their profusion of usually golden yellow pea-like flowers in showy terminal clusters in spring. The C. canariensis group are better known in horticulture as "Genista", and they are usually shaped and sheared by pot plant growers into small formal globes or as semi-standards in 6 to 8 in. pots for Easter forcing and spring-blooming. They grow best in fast-draining soil somewhat on the dry side; in winter they are kept cool at 40°F., and dry, increasing temperature only slightly to force into profuse bloom for the Easter season. Cut back, or shear for shaping, after flowering with a last trimming in September; plunge outdoors for the summer. Propagated from half-hard cuttings in December-January; also from seed sown in January, after soaking it 48 hours in lukewarm water.

canariensis (Canary Islands), florists "Genista" or "Canary Island broom"; much branched shrub to 6 ft., but usually grown as smaller formal globe by shearing, for spring blooming; the small ferny foliage trifoliolate with little obovate, silky leaflets, and showy spike-like clusters of bright yellow, fragrant $\frac{1}{3}$ in. pea-like flowers at the end of each shoot. **C**: 75   p. 372

x racemosus, a hybrid of canariensis from the Canary Islands, and maderensis from Madeira; the "Genista of florists", a plant favored by growers because of its smaller darker green leaflets, and shorter, more numerous clusters of deep yellow flowers, set closely at the end of branches, the blooms longer lasting than canariensis. **C**: 25   p. 503

DAEDALACANTHUS: see Eranthemum

DAHLIA (*Compositae*); the Garden dahlias are amongst the greatest of summer-autumn flowers, having been cultivated since ancient times by the Aztecs in Mexico and presenting a gorgeous display of color and shapes in our cool northern gardens and as cut flowers. They are tender perennial herbs from 15 in. to 6 ft. high, with large underground tubers, and showy heads of flowers with florets arranged into "Pompon", "Single", "Anemone", flat "Decorative" or pointed-petalled "Cactus" forms with blooms 5 to 10 in. across. For pot culture only dwarf hybrids of D. pinnata are practical, and these are occasionally forced for Easter or Mother's Day. For forcing plant 3 strong tuber divisions into 7 in. pot end of January at warm 60–65°F.; when in growth and buds show set cool, to bloom for Easter. Propagation best by division, taking care that each tuber retains a stem bud; also from cuttings; the single dwarfs from seed.

pinnata 'Pompon strain'; exquisite usually small, very symmetrical, fully double round flower heads about 2 to 3 in. dia., ball-shaped or slightly flattened, the floral rays fluted or tubular and blunt at tip, tightly arranged spirally like honey-combs; generally of stiff erect, compact bushy habit and very free-flowering. **C**: 76   p. 402

pinnata 'Unwin's Dwarf', "Dwarf Aztec dahlia"; miniature strain suitable for growing as a compact pot plant, only 18 in. high, producing $2\frac{1}{2}$ to 3 in. double and semi-double blooms in colors red, salmon, yellow, and lavender, flowering over a long period; usually grown from seed but eventually forming small tubers; blooming within first year from seed in cool climate. Seed sown in February will bloom by fall; seed started in October produces blooming 4 in. pots by Mother's Day in May and onwards. **C**: 76   p. 402, 418

DALECHAMPIA (*Euphorbiaceae*)
roezliana (Mexico); branched sub-shrub to 4 ft. high, with alternate 6 in. oblanceolate dark green leaves, and small fragrant, yellow flowers without petals in dense clusters, subtended by two rose-pink bracts and a fringe of waxy threads on top. Attractive, sun-loving warmhouse plant blooming almost all year. From seed or cuttings. **C**: 51 p. 500

DAPHNE (*Thymelaeaceae*)
odora 'Marginata' (China, Japan), "Variegated winter daphne"; handsome ornamental evergreen, wide-spreading low shrub to 4 ft. high, with glossy dark green, leathery, 3 in. leaves margined with ivory; the flowers are creamy-white to pink and very fragrant, in dense terminal clusters, blooming from January to early spring. The variegated-leaf form is slow-growing, adaptable as a cool house plant, the little waxy flowers in nosegay clusters hauntingly with intense sweet perfume. Somewhat winter-hardy. Propagated by layers and cuttings. **C**: 16   p. 296, 484

DARLINGTONIA (*Sarraceniaceae*)
californica (U.S.: coast of California, Oregon), the weird "Cobra plant" or "California pitcher plant"; insectivorous perennial bog or marsh plant 1 to $2\frac{1}{2}$ ft. tall, with tubular, fluted leaves, yellowish green tinged with purple and usually containing liquid, enlarged upward into an arched hood with translucent, whitish windows, the opening partially concealed by a purple-spotted forked tongue; equipped inside with tangled hairs entrapping insects; the slender stalks with nodding yellowish-purple flowers from

April to July. In cultivation it will last longest in a cool, moist greenhouse grown in acid spagnum or peatmoss with sand (pH 4 to 5); watering if possible with soft rain water free of lime. Propagation by division, or seed. **C**: 84   p. 520

DASYLIRION (*Liliaceae*)
acrotriche (Mexico), "Bear-grass"; stiff tree-like desert plant with trunk 3 ft. or more, topped by a dense rosette of 3 ft. linear leaves only $1\frac{1}{2}$ in. wide, these tipped by a fibrous tuft and sharp yellow spines; white flowers on panicle to 15 ft. high. **C**: 13   p. 200

DATURA (*Solanaceae*); the "Angel's trumpets" are large shrubs or small trees, also annuals, of the potato family, with pulpy growth, large herbaceous leaves, and big trumpets of usually pendant flowers most often white—a spectacular sight in large containers, enchanting at night in the moonlight. Daturas in tubs should be set outside for the summer in a sheltered location; they are voracious feeders and, if kept in good light, sufficiently watered and supplied with fertilizer, could bloom almost continuously, although they like to take a rest in a cool place in winter, a time for cutting them back. The shrubby species are propagated from semi-ripened cuttings; the annuals mostly from seed.

candida, known in horticulture as D. arborea (Peruvian to Chilean Andes), the "Floripondio tree"; small pubescent tree to 10 ft. high, with dull green soft-hairy somewhat leathery, ovate leaves 8 to 12 in. long; the nodding, trumpet-shaped flowers usually not over 7 in. long, white with green nerves and recurving pointed lobes, the long green calyx spathe-like, tapering to one tip. Widely planted in tropical gardens, this is also a sturdy container plant, blooming in summer through autumn into December with a haunting fragrance, especially at night. **C**: 52   p. 490

suaveolens (So. Brazil), "Angel's trumpet"; robust, tree-like shrub, with pulpy canes to 15 ft. high; large lanceolate, 1 ft. leaves, oblique at the base, green and smooth, thin and quickly wilting; the large pendulous funnel-shaped, fragrant flowers with tubular 5-toothed calyx and a showy, white corolla to 12 in. long; blooming mainly in late summer and autumn, occasionally also at other times. The most widely grown species. **C**: 52   p. 490

DAVALLIA (*Filices*); mostly small slow-growing tropical ferns of the Polypodiaceae family, with finely divided foliage, in habitat growing epiphytically, on rhizomes creeping above the surface of the soil, on rocks or trees. Rhizomes may be divided, hooked down on wire hooks on peatmoss or sphagnum, under moist-warm plastic or glass cover. Much seen in hanging baskets planted in sphagnum moss or other fibers, and which must be dipped periodically in a bucket of water. Also used as cut sprays for flower arrangements.

bullata mariesii (Japan), the famous "Ball fern" of the Japanese fancier; long creeping, flexible slender, light brown, hairy rhizomes with uniformly small, 4-pinnate, finely lacy yet tough triangular fronds to 6 in. In Japan they are trained onto wired forms filled with sphagnum or tree-fern fiber into many shapes, such as balls, pillars, bells, animals, monkeys and dolls; deciduous in cool climates. Requires steady moisture when not dormant; best to set into saucer with water, or dip in bucket; don't keep too warm. **C**: 22   p. 241, 539

canariensis: see trichomanoides
fejeensis (Fiji Islands), "Fiji rabbit's-foot fern"; a name referring to its brown woolly, creeping rhizomes, from which rise graceful triangular, durable fronds 1–$1\frac{1}{2}$ ft. long on wiry stems, 4-pinnatifid but more finely cut than solida. Good for hanging baskets. **C**: 10   p. 241, 286
fejeensis plumosa (Polynesia), "Dainty rabbits-foot"; a dainty, dwarf variety with fresh-green, more finely cut, 4-pinnatifid, plume-like, fronds with very narrow segments, the tips gracefully arching or pendant. A durable stand-by pot plant for cutting graceful long-lasting fronds for cut flower arrangements. **C**: 10   p. 241
griffithiana (India, So. China), "Squirrel's foot"; epiphytic fern with creeping rhizomes covered by glistening white scales; wiry fronds 3 to 4-pinnatifid into well spaced segments, with the ultimate leaflets deeply toothed; not deciduous. **C**: 10   p. 241
pentaphylla (Java, Polynesia), "Dwarf rabbit's-foot"; a distinct dwarf growing species with black-haired creeping rhizomes, bearing wiry fronds of two or three pairs of lateral wavy-toothed, linear, glossy green leaflets and a terminal one. **C**: 12   p. 241

solida (lucida) (Malaya, Queensland), "Rabbit's-foot fern"; brown densely hairy or scaly rhizomes, bearing broadly massive, 3-pinnate triangular fronds to 2 ft. high, the stiff-leathery leaflets ovate-rhomboid and crenate. **C**: 10
p. 241

trichomanoides (Malaya), D. "canariensis" of horticulture, the "Carrot fern"; rather robust epiphyte with wide-creeping rhizome covered by pale brown scales; wiry gray stalks bearing rather leathery fronds 6–9 in. long, 4 times pinnate, with leaflets overlapping; the final pinnae cut into strap-shaped segments. More durable than fejeensis, seldom fertile. **C**: 10
p. 241

DELPHINIUM (*Ranunculaceae*)

ajacis (So. Europe, wide-spread from Switzerland to the Taurus), "Larkspur" or "Annual delphinium"; grand garden herb to 4 ft. high, with lacily divided fresh-green leaves along the stiff-erect stalks bearing columns of flowers, often double, each with a prominent spur, and sepals usually in lovely shades of blue or violet, sometimes rose or pink, to white, blooming from June through August. May be massed effectively in containers for the patio, whether all in the popular sky-blue, or in mixed colors. From seed. **C**: 76 p. 414

DENDROBIUM (*Orchidaceae*); a magnificent genus of

tropical and warm-climate epiphytes of great variety, partly with tall-jointed stems bearing flowers along the stems or from their apex; others have squat pseudobulbs topped by leaves and arching sprays of blooms. Nearly all are profuse and willing bloomers, and of easy care, even in the home, and this accounts for their wide popularity; the flowers are dainty and of a glistening texture, in colors from rose and violet to brilliant yellow and white. To bloom well, Dendrobiums need good light, but also appreciate warmth and humidity with fresh air; the cooler, partially deciduous kinds need about 3000 fc. of light and should be kept cooler and more dry after growth is completed, to ripen the stems for bud initiation. The evergreen species doing best at 4000 fc. of light, are kept somewhat more moist, and constantly warm even after bloom. They like small containers; with good drainage grow well in osmunda, fern fiber or various composts. Progagation is generally by division; some form plantlets on the old stems; in others the stems are cut from the plants and laid on moist sphagnum; they will break at joints. Any Dendrobium can be summered outdoors in direct sun once growth is well under way.

acuminatum (syn. Sarcopodium or Epigeneium) (Philippines); charmingly attractive epiphyte with creeping rhizome; squat, four-angled glossy green pseudobulbs terminated by a pair of rigid elliptic leaves, and an arching spray of many beautiful and fragrant carmine rose, star-like flowers $2\frac{1}{2}$ to $3\frac{1}{2}$ in. across, the base maroon and with whitish tips, blooming February to May, and long-lasting. **C**: 71
p. 479

aggregatum (lindleyii) (So. China: Yunnan, Burma, Malaya); charming dwarf evergreen epiphyte with small 2 in. pseudobulbs bearing solitary leathery leaves, the rounded $1\frac{1}{2}$ in. flowers opening flat, delicate golden yellow with orange center, from 10 to 25 in arching sprays, in spring from March to June. **C**: 9
p. 465

d'Albertisii (antennatum var.) (So. New Guinea: Papua); the "Rabbit's ear orchid", pretty epiphyte with square, slender tapering stems set alternately with narrow, fleshy leaves; axillary erect clusters of fragrant, spurred flowers about 3 in. long, with white sepals, the narrow greenish twisted petals standing erect looking like rabbit's ears, the lip whitish with purple veins; blooming in summer. **C**: 59
p. 538

densiflorum (Nepal Himalayas, Assam, Burma); beautiful, sturdy epiphyte with 15 in. tall, stout cane-like, 4-sided pseudobulbs bearing 3 to 5 leathery leaves; the $1\frac{1}{2}$ in. flowers golden yellow with velvety orange-yellow lip, in dense, pendant trusses, from near the top of either old or young growth; blooming in spring from March to May. Magnificent display of sunny, sparkling flowers, sometimes 50 to 100 completely encircling the canes. **C**: 21
p. 465

distichum (Philippines); curious, distinctive epiphytic species without pseudobulbs; the flat and fleshy sickle-shaped leaves arranged in two ranks along stem, their bases overlapping; small $\frac{1}{2}$ in. fleshy flowers in clusters at tip of stem, yellowish with conspicuous, red-brown stripes; blooming from March to May. **C**: 9
p. 466

falconeri (Bhutan, Himalayas to Burma and Thailand); a magnificent epiphyte of pendulous growth, with pseudobulbs stem-like, very slender, profusely branching, often rooting from the nodes, 2 to 3 ft. long, forming dense thickets of interwoven pseudobulbs; the few narrow leaves on the young shoots, are quickly deciduous; fragrant flowers solitary from the nodes of the older pseudobulbs, to $4\frac{1}{2}$ in. across, white or bluish-white, all segments tipped with rich purple, the orange lip blotched deep purple in the base, blooming spring to summer, and lasting about 10 days. **C**: 21
p. 465

formosum (Himalayas to Burma and Thailand); remarkably handsome evergreen epiphyte with stout erect pseudobulbs to $1\frac{1}{2}$ ft. tall, with thin leathery oblong leaves, and at the apex, several large, fragrant, snowy-white 3 to 5 in. flowers with orange throat, of good substance, blooming late autumn to early spring, and lasting 6 weeks or more to perfection. **C**: 21
p. 466

johnsoniae (macfarlanei) (Mountains of New Guinea); noble epiphyte with erect cylindric, 2 to 3-leaved pseudobulbs bearing terminal racemes of large 4 to 5 in. flowers pure white except for the 3-lobed lip marked purple inside; blooming variously from summer into winter; I photographed my plant in February. **C**: 59
p. 465

linguaeforme (Australia: N.S. Wales, Queensland), the "Tongue orchid"; curious dwarf but charming epiphyte with creeping rhizome, pseudobulbs almost absent; thick-fleshy 1 in. leaves, furrowed and bean-like, olive-green, matted and hugging tree trunk or slab; clusters of small greenish-white $\frac{3}{4}$ in. flowers having narrow greenish-white sepals and petals with light brown radial lines; the short lip with yellow tips; blooming season variable but usually spring. An ideal button-hole orchid. **C**: 3
p. 466

moschatum (Himalayas to Burma and Laos); handsome, mostly evergreen epiphyte with tall, cane-like pseudobulbs that may become 3 to 6 ft. long, and which occasionally produce young plants at their tip; loose clusters of lovely 3 to 4 in. heavy-textured flowers from the older wood, unusual with petals pale to orange-yellow tipped with rose, the lip is pouch-shaped with lines of fringe in the center, and a large black-purple blotch on each side inside surrounded by orange; blooming starting in winter, to spring and summer. **C**: 21
p. 465

nobile (Nepal Himalayas, Yunnan, Assam to Vietnam, Taiwan); popular, deciduous epiphyte free-blooming to 2 ft. tall, with large fragrant 3 in. flowers produced in twos and threes from nodes of 2 year pseudobulbs; sepals and petals white, tipped rosy-pink, lip white with rosy tip and deep velvet-crimson throat; blooming mostly between January and June. One of the most widely grown species, known to have many variations, with long-lasting flowers lending themselves to dainty corsages; also a good window plant, requiring 4 hours sun; moisture and warmth until growth is mature, but when the leaves fall into winter they are kept dry and cool. **C**: 21
p. 465

phalaenopsis (Queensland, New Guinea, Timor); a most beautiful epiphyte with stiffly slender canes 2 to 3 ft. tall, bearing leaves in 2 ranks, and long, arching chain-like sprays of large, neatly draped 3 in. flowers, the sepals pale magenta with reticulated darker nerves, petals rose, much larger, lip dark purplish-red, blooming variously, mainly in spring, sometimes also in fall to November. **C**: 9
p. 465

primulinum (China, Himalayas, Burma to Vietnam); beautiful free-blooming, deciduous epiphyte with cylindrical, fleshy stems to 12 in. long, erect or pendulous, with fragrant flowers in 2 rows, from nodes along the pseudobulb when it is leafless; large blooms to $2\frac{1}{2}$ in. across, with relatively small, almost equal sepals and petals pale rose, and shell-shaped lip primrose-yellow streaked red, blooming from February to May. Because of its pendant trend it does well in baskets, the flowers lasting 10 days. **C**: 21 p. 465

thyrsiflorum (Himalayas to Burma, Thailand); showy epiphyte with the habit of densiflorum but stems are more round and taller, to 30 in. high, the 2 in. fragrant flowers in longer pendant racemes, sepals and petals white, often flushed pink, the pubescent lip golden orange, blooming February to May. A stunning display with its pendant lantern-shaped spray of sparkling, crystalline blooms in winter and spring. **C**: 9
p. 465

DENDROCALAMUS (*Gramineae*)

giganteus (Burma, Malaya), "Giant bamboo"; this tropical giant of the tribe, largest of all bamboos, grows to

100-ft. high with stems 10 in. or more in diameter, thin-walled, joints to 16 in. apart; later developing branches with large leaves to 20 in. long borne in graceful masses toward the top. The young emerging culms, growing at the rate of 18 in. a day, were an instrument of torture and death to prisoners of war in Ceylon, pushing through their bodies when they were tied down to the ground in a bamboo grove. In temperate climate for the large conservatory only. Widely planted in Southeast Asia and used for building, water-spouting or plant pots. **C**: 52   p. 303

DENDROCHILUM (*Orchidaceae*); the "Chain orchids", long known by horticulturists as Platyclinis are tropical Asiatic epiphytes of compact, graceful habit, with small one-leaved pseudobulbs; from the top the arching, wiry stalks supporting exquisite ribbon-like slender sprays of small mostly whitish, very fragrant flowers. Plants grow best in small pots with sphagnum and fern-fiber, or mounted on treefern slabs; they are kept moist and growing almost year round, fairly humid-warm but with fresh air. Propagated by division as new growth starts.
   cobbianum (syn. Platyclinis) (Philippines); graceful dwarf epiphyte with small, one-leaved pseudobulbs, their soft-leathery leaves 6 in. long, and producing from the apex of the bulbs a wiry stalk extending into gracefully pendant ribbons of densely 2-ranked, tiny straw-yellow flowers with orange lip; strongly fragrant; blooming between April-October. **C**: 22   p. 466
   glumaceum (syn. Platyclinis) (Philippines), a popular, pretty "Chain orchid"; vigorous, clustering epiphyte with evergreen leaves to 10 in. long, from the pseudobulbs the lovely arching slender spikes to 12 in. long, of delicately scented, 2-ranked flowers, sepals and petals straw-white and with a greenish-yellow lip; blooming mostly from March to May; often at other times, continuing 3 to 4 weeks in perfection. **C**: 22   p. 466

DENDROPANAX (*Araliaceae*)
   chevalieri (D. arboreus; Gilibertia japonica) (Japan, Korea) "Tree aralia"; glabrous evergreen small tree to 20 ft. high with dark green, leathery, 3-lobed, 3 to 6 in. leaves with pale veins, variable becoming oval in later stage. **C**: 14   p. 290

DEUTEROCOHNIA (*Bromeliaceae*)
   meziana (So. America: Mato Grosso, Paraguay); xerophytic rosette with densely set, sharply recurved and tapering, flexible leaves light waxy green, with brown marginal spines in two directions; reverse with whitish scales; inflorescence a large 3 ft. panicle, to 7 ft. high, with red-orange petals. **C**: 1   p. 138

DEUTZIA (*Saxifragaceae*)
   gracilis (Japan), "Slender deutzia"; attractive floriferous winter-hardy deciduous shrub of low habit, about 3 ft. high or sometimes more, with slender branches, wide-spreading or arching; lanceolate leaves with star-like hairs above, serrate at margins, to 2½ in. long; masses of pure white ¾ in. flowers in open simple or compound clusters; blooming outdoors in May and June. Favored for early forcing; flowering off wood of previous summer's growth, such one year growth should be lightly tipped only, to remove any unripened wood if needed to flower for same season. From cuttings, hardwood, and layered branches. **C**: 76   p. 504

x DIACATTLEYA (*Orchidaceae*)
   'Chastity' (Diacrium bicornutum x Cattleya granulosa); bigeneric plant to 20 in., with stout pseudobulbs topped by fleshy leaves, and heavy-textured, full-shaped fragrant 4 to 5 in., green flowers with pink lip; blooming everytime a growth matures. **C**: 59   p. 466

DIACRIUM (*Orchidaceae*)
   bicornutum (Caularthron) (Trinidad, Colombia to Guyana and Amazonas), the magnificent "Virgin orchid"; handsome epiphyte with stout, hollow, furrowed pseudo-bulbs to 9 in. long, clothed by 3 to 5 leathery leaves; from the apex the long-stemmed clusters of 5 to 20 pure white flowers 2 in. across, with a few crimson spots on the spread-ing lip and a yellow, fleshy crest. One of the finest American orchids, most exquisite with its waxy, long lived pure

ivory-white, fragrant blooms in spring and summer carried high on erect stalks. Diacrum needs heat and moisture, and doesn't like to be disturbed or divided more than necessary; after flowering there should be a short rest. **C**: 59   p. 466

DIANELLA (*Liliaceae*)
   revoluta (longifolia) (Tasmania, New Caledonia), "Flax lily"; branching rootstock forms elongate leafy stems grass-like, the sheathing, keeled leaves thin-leathery and dark green, in ranks, 1½ to 3 ft. long, purplish at edge; greenish-blue flowers followed by light blue berries. **C**: 15   p. 200

DIANTHUS (*Caryophyllaceae*)
   barbatus (So. and Southeastern Europe to So. Russia and East), "Sweet William" or "Bunch pink"; popular garden perennial better treated as a biennial; from some seed also annual; smooth herb 18 to 24 in. high with green flat and broader leaves than most Pinks; the sturdy stem with 1 in. flowers in large bracted heads, the spreading petals fringed and calyx bearded, in colors from white, pink, rose, red, purplish to bi-colored, blooming from June to August; winter-hardy. From seed. **C**: 76   p. 417
   caryophyllus 'Cardinal Sim', a well-known florists' carnation typical of the "American Remontant carnation", diploid form of D. caryophyllus, the "Clove pink"; branching glaucous perennial of stiff habit, commercially grown under glass for their cut flowers because of their perpetual, year round flowering; bluish linear leaves and stiff-wiry, leafy stems with showy double flowers in many colors, often spicy fragrant. The cultivar 'Cardinal Sim' is a vigorous and productive variety, with flowers 3½ in. dia. if disbudded, vivid scarlet. Best at cool to intermediate temperature, in full sun. From soft cuttings. **C**: 75   p. 372
   x'Mitt's Pinks', the "Baby carnation"; a true miniature hybrid carnation derived from D. caryphyllus, plumarius, deltoides and latifolius; very cute as a diminutive flowering plant when grown in 2¼ in. pot 8 to 10 in. high; dense little bush with narrow linear, glaucous foliage and long-stemmed semi-double, fringed flowers 1½ in. across, in several pretty colors, pink to blood red with white eye, or red with pink border; blooming in spring. **C**: 25   p. 417

DIASTEMA (*Gesneriaceae*)
   quinquevulnerum (Colombia, Venezuela), dwarf branching, stiff-hairy herbaceous plant 4 to 6 in. high, with creeping scaly rhizome; ovate, rugose, serrate, bright green 3 to 4 in. leaves; loose clusters of small white, dainty tubular 1 in. flowers, the lobes each with a purple spot; blooming all summer long. With its lush green foliage and pretty miniature flowers an attractive plant in pots. Going dormant, it needs a rest period in winter. Propagated from cuttings or the scaly rhizomes. **C**: 60   p. 444

DICENTRA (*Fumariaceae*)
   spectabilis (Manchuria, Korea, Yunnan, Japan), "Bleeding heart" or "Dutchman's breeches"; popular somewhat watery-juiced perennial herb 1½ to 2 ft. and more high, with fleshy rootstocks, feathery divided leaves; the very pretty, exquisitely heart-shaped, irregular rosy 1 in. flowers with inner, white petals or spurs protruding, all in a row pendulous along the arching reddish, leafy stalks. Normally May blooming, Bleeding hearts are of swiftly passing splendor. For forcing in pots, lift clumps in the fall, pot and store cold until February, then start in cool 45° F. greenhouse increasing to 50—55°F.; it takes 4 weeks to bloom. Propagated by division, suckers or root-cuttings. **C**: 78   p. 418

DICHILOBOEA (*Gesneriaceae*)
   birmanica (Burma; China: Yunnan), erect herbaceous bush 1½ ft. high, with white or beige felted woody stem; well-spaced oblong or elliptic, deep green, corrugated, thickish leaves to 5 in. long; the inflorescence in axillary, elongate racemes of bracted small white flowers with corollas less than ½ in. long. From seed. **C**: 71   p. 444

DICHONDRA (*Convolvulaceae*)
   repens (West Indies); known as "Lawn leaf"; low herb creeping close to the ground and rooting, with close-matting stolons; small, rounded or kidney-shaped, silky leaves to ¾ in. across, tightly overlapping, fresh green above, white-hairy beneath when young; the little flowers greenish-yellow; seeds freely. Used primarily as a lawn substitute in

California and sunny desert regions where lawn-grasses do poorly; will stand mowing. Propagation from seed or division. **C**: 64   p. 265

DICHORISANDRA (Commelinaceae)

albo-lineata: see Campelia

'Blackie': see Geogenanthus ciliatus

reginae, also known as Tradescantia reginae (Peru), the "Queen's spiderwort"; stiffly erect, rather slow growing plant, the waxy, long pointed, dark green leaves 6 in. long, banded and spotted on both sides with sparkling silver, the center metallic red-violet and spilling over into deep purple beneath, flowers lavender. Beautiful foliage plant but appears to need warm greenhouse. **C**: 53   p. 262

thyrsiflora (Brazil), the "Blue ginger"; herbaceous perennial; stout canes to 4 ft. high, bearing rosettes of broad lance-shaped, glossy green leaves 6 to 10 in. long with purplish reverse, umbrella-like; topped by a showy huge raceme of brilliant, deep electric-blue flowers with yellow anthers. Goes into a rest period in winter, when it should be kept dry until new growth begins in spring. **C**: 54   p. 260

thyrsiflora variegata; very attractive variety with the long green leaves about 6 in. long, having two silver bands down the length of the leaf, a reddish midrib; stem is mottled purple. Resembling Zebrina but the foliage is much longer. **C**: 4   p. 262

DICKSONIA (Filices); stately treeferns of the family

Dicksoniaceae, at home in the Southern hemisphere; slow-growing to almost 50 ft. high, their fronds large, 2 to 3 pinnate and of leathery texture, and with trunks covered by matted fibrous aerial rootlets. These great ferns are very durable and are particularly suited for cool locations, the sheltered patio or shade house in summer, or all year in frost-free climate; however, wherever they are they need constant moisture especially by misting their absorbent trunks. Formerly fern trunks without leaves used to be imported for starting in a moist and shady greenhouse; growth from spores takes years.

antarctica (Australia, Tasmania), "Woolly treefern" or "Tasmanian treefern"; a woody trunk eventually 30–40 ft. high, covered with matted aerial rootlets, bearing a rather flat crown of large fronds to 6 ft. long, 3-pinnate, dark green and leathery, with its lanceolate, toothed segments rather hard and showing a network of straw-colored veins with the margins turned down. One of the most useful of the treeferns for decoration; very slow-growing; popular in sub-tropical ferneries and in cool winter-gardens both planted out and as a tub plant; fairly resistant; stands unfavorable conditions, even few degrees frost. **C**: 72 p. 236

fibrosa (New Zealand), the "Golden treefern"; shade and cool-loving tree fern of 20 ft. height, the stout trunk covered with brown, fibrous aerial rootlets; large fronds to 8 ft. long, 2–3 pinnate, vivid fresh green and fairly stiff. **C**: 72   p. 236

squarrosa (New Zealand), "Rough New Zealand treefern"; a medium-sized treefern to 20 ft. high, the slender black trunk studded with leafbases; crown with fronds nearly horizontal, 2 to 3-pinnate, to 8 ft., long, dull, dark green and stiff-leathery, harsh to the touch, on black-brown stalks clothed when young by long brown, stiff hairs. **C**: 22   p. 235

DICTAMNUS (Rutaceae)

albus (So. Europe to No. China), "Fraxinella"; also known as "Gas plant" or "Burning bush", because the volatile oil exuding from the foliage and flowers may be ignited by a match at high temperature; hardy woody perennial 2–3 ft. high, with strong-smelling glossy olive green leathery leaves, pinnate with 1–3 toothed leaflets, scented of lemonpeel when rubbed, of balsam when bruised; spike-like clusters of white flowers with green stamens. Attractive ornamental and for cut flowers. Propagate by seed, root cuttings or division. **C**: 63   p. 321

DIDIEREA (Didiereaceae)

madagascariensis (S.W. Madagascar), "Madagascar cactus"; erect tree-like succulent 12 to 18 ft. high, in habitat with the appearance of a cactus-like euphorbia; thick gray stems set with short, thorn-like branches which at their ends produce one long spine and several shorter ones and soft, succulent, narrow-cylindric green leaves. A rarity, cultivated like euphorbias. **C**: 1   p. 353

DIDYMOCARPUS (Gesneriaceae)

kerrii (S.E. Himalayas); tropical herbaceous, fibrous-rooted, low branching plant to 6 in. high, with fleshy stems, and pairs of slightly hairy, saintpaulia-like green leaves crenate at margins; from the leaf axils branched clusters with small 1 to 1¼ in. dark purple flowers, the corolla a swelling trumpet with flaring lobes. From cuttings or seed. **C**: 60   p. 444

DIEFFENBACHIA (Araceae); the "Dumbcanes" are

handsome tropical American foliage plants with fleshy or woody stems usually with beautiful variegated leaves; the inflorescence having long spathe partially encircling the slender spadix which is unisexual carrying the petal-less staminate or male flowers toward the apex and the pistillate, female elements separately at the lower, basal part. Dieffenbachias are excellent decorative plants suitable for superheated interiors because they thrive in warmth, and tolerate the dry air of the living room and reduced light away from the window down to 15–25 fc.; requiring only moderate watering. The sap of Dieffenbachia when brought in contact with delicate skin is harmful, containing crystals of calcium oxalate, which get under the skin, and due to mechanical irritation, not poison, causing intense pain and swellings. Rinsing the mouth or skin with vinegar dissolves the crystals and reduces the swelling. Propagated by mossing off the top; from side suckers, or sections of the cane laid in moist moss or sand.

amoena (Colombia 4000–5000 ft.), "Giant dumbcane". "Mother-in-law plant", or "Tuft root"; sturdy and handsome, thick-stemmed species to 5 ft. or more tall, with large oblong-pointed, glossy, thin-leathery leaves 12 to 18 in. long, deep green and feathered with cream-white bands and blotches along lateral veins. Large table plant very decorative and durable with low-light tolerance (15–100 fc.). **C**: 3   p. 91

x bausei (picta x weirii); beautiful, compact plant; pointed leaf delicate yellowish-green with dark green and white spots and dark green at edge. Striking but not as tough as others. **C**: 3   p. 66*, 91, 178

bowmannii (Colombia); some resemblance in markings to D. amoena, but the leaves are much broader, and the lateral markings of cream are less noticeable; reverse freely spotted white. The picture shown may be D. macrophylla. **C**: 53   p. 91

daguense (Colombia); thick-stemmed plant with heavy, leathery, pointed leaves; thick, elevated midrib, lateral veins depressed; rich shiny green. **C**: 53   p. 93

'Exotica' ('Arvida') (Costa Rica); attractive mutant, probably of hoffmannii, from the Rio Ysidro region; quite compact and shapely; of slender, narrow habit; smallish ovate pointed, firm leaf 8–10 in. long, deep matte-green and highly variegated cream-white, and of good texture. **C**: 3
p. 43*, 93

'Exotica Perfection'; a Florida selection of compact habit, the smallish leaves 6–8 in. long, broadly bordered with deep matte green and of leathery feel; highly variegated greenish-ivory. **C**: 3   p. 92

fosteri (Costa Rica); dwarf species with small ovate-acuminate, deep green, leathery leaves having satiny sheen, closely set on slender cane. **C**: 53   p. 93

fournieri (Colombia); upright, large, shiny, leathery, black-green leaves with white spots; slender and very elegant. A collector's plant. **C**: 53   p. 93

hoffmannii (Costa Rica, 11,000 ft.); oblong pointed leaves deep, satiny green, marbled and spotted white as are the petioles; midrib white. **C**: 3   p. 93

imperialis (Peru); tall plant with large leathery, dark green leaves, and silvery-gray feathering along midrib, pale green and pale yellow blotching toward apex. **C**: 53   p. 92

leoniae (Colombia); attractive plant with satiny, somewhat reflexed, oblong leaves of firm texture, friendly green with dominant yellow variegation; faintly cream along midrib. **C**: 55   p. 91

leopoldii (Costa Rica); velvety, corrugated, emerald-green leaf with white midrib and pale band alongside it; brownish stem and petioles, spotted white. **C**: 53   p. 93

x memoria-corsii (picta x wallisii); broadly lanceolate

leaves rather gray, with dark, gray-green veining, interrupted by solid green blotches and occasional ivory spots. **C**: 53
p. 93

oerstedii (Guatemala, Costa Rica); small, graceful plant; oblong-ovate, leathery, matte blackish-green leaves on slender petioles. **C**: 3   p. 93

oerstedii variegata (Costa Rica), the slender, hard and smooth, dark-green leaf with a contrasting ivory-white midrib. Quite attractive, smallish species. **C**: 3   p. 93

parlatorei (Colombia); tall tropical foliage plant of robust habit, slowly forming stout stem, leaning with age and weight; long clasping petioles winged halfways, carry the leathery oblong leaves, 18 in. long, shiny deep green, with prominently broad, thick midrib, lightly channeled higher up. I photographed this specimen at the Brooklyn Botanic Garden. **C**: 53   p. 92

parlatorei marmorea (Colombia); long oblong, tough-leathery, dark lustrous leaves with fleshy midrib partially white; the blade blotched with silver, especially when young; strong, winged petioles. **C**: 53   p. 93

picta (maculata, brasiliensis) (Brazil: Amazonas); "Spotted dumbcane"; handsome aroid of robust habit, slowly developing a green cane at fist erect, later stout and leaning, bearing highly ornamental glossy and firm, grass-green ovate-cordate leaves 10 in. or more long, with beautiful ivory-white marbling and blotching; the petioles dotted pale green, spathe greenish. Found growing wild by the writer at the confluence of the Solimoes and Rio Negro in Amazonas. An excellent, very decorative houseplant, with very little care. Brazilians believe that a tall cane should always be tied to a stake to keep from toppling, as this would bring bad luck. **C**: 3   p. 91

picta angustior angustifolia (So. America); elegant plant with narrow-lanceolate leaves of almost leathery texture, glossy, grass-green, richly spotted with ivory-white; stem and winged petioles olive brown with pink bands lengthwise; spathe pale green. **C**: 53   p. 92, 93

picta barraquiniana (Brazil); slender plant with leaves oblong-pointed, bright green with bold, white midrib and occasional white spots; white leaf stalks. **C**: 53   p. 92, 93

picta jenmannii (Guyana); slender plant with long-oblong, rather narrow leaves glossy fresh green, with distinct ivory bars in feather design. **C**: 3   p. 93

picta 'Rudolph Roehrs' (Roehrs cultivar, named by the author 1937), "Gold dieffenbachia"; a striking mutant having oblong pointed 10–12 in. leaves almost entirely yellow or chartreuse, with ivory-white blotches, and only the midrib and border dark green. Both beautiful and showy, and proven an excellent house plant. Thrives in dry room conditions but needs good light for best variegation (25 to 100 fc.). **C**: 3   p. 91, 178

picta shuttleworthii (Colombia); long-lanceolate, deep green leaves with feathery, gray band along midrib and pea-green blotches on blade. **C**: 53   p. 93

picta 'Superba' (Roehrs 1950); very attractive, compact mutant of picta, with thicker, more durable foliage dominated by a high degree of creamy-white variegation, over the glossy green leafblade. **C**: 3   p. 92, 178

pittieri (Costa Rica); long, satiny, emerald-green leaf with depressed veins, and marbled pale green or ivory; base of petioles surrounding stem. **C**: 53   p. 93

seguina (Puerto Rico, W. Indies), "Dumbcane"; robust, variable species with long elliptic, dark green fleshy leaves, base acute, prominent midrib and depressed veins; pleasing green beneath; spathe green. Mostly of botanical interest. **C**: 53   p. 92

seguina irrorata (Brazil: Pará); oblong, pointed, rather thin leaf, predominantly yellow-green, with dark green blotching and edging; petioles whitish. **C**: 53   p. 92

seguina lineata (marmorata) (Venezuela, Colombia); similar to the type but with petioles striated white. **C**: 53
p. 93

x splendens (leopoldii x picta); velvet, bronze-green leaf with nile-green and ivory spots and ivory midrib; slender. **C**: 55   p. 93

wallisii (No. S. America); medium tall, slender plant with oblique lanceolate leaves having long tapering apex, matte to semi-glossy grass-green, a broad feathery band of light gray to creamy silver along midrib, on winged petioles. **C**: 53   p. 92

DIGITALIS (*Scrophulariaceae*)
purpurea (Western Europe), "Foxglove"; stately her-

baceous perennial usually grown as a biennial, with hairy wrinkled light green leaves in basal rosettes and diminishing in size along the boldly erect 2 to 4 ft. stalk, topped by an elegant, dense spike of bell-shaped, slightly 2-lipped, nodding flowers to 3 in. long, purple, varying to yellow or white, with dark purple spots, and edged white inside, blooming in summer. Winter-hardy, and one of the handsomest garden flowers long cultivated; best in cool, moist areas. Its leaves yield an important heart remedy. From seed. **C**: 78   p. 416

DIMORPHOTHECA (*Compositae*)
ecklonis, as known in horticulture; botanically probably Osteospermum ecklonis, (So. Africa), "Cape marigold"; shrubby perennial spreading and mound forming, with narrow green toothed leaves and large daisy-like 3 in. flowers with ray-florets white above, purplish beneath, and a blackish-blue disk in center; often grown as an annual, these charming daisies are undemanding and bloom profusely in spring if indoors, July to fall outdoors; the flowers opening with the warming sunshine until afternoon, closing by night. Prop. from seeds, cuttings, or division. **C**: 14. p. 497

DINTERANTHUS (*Aizoaceae*)
inexpectatus (Southwest Africa), "Surprise split rock"; normally compact, attractive stemless succulent about 2 in. high with pairs of thick-fleshy, keeled, whitish leaves united halfway, and with translucent greenish dots; daisy-like 1 in. bright orange-yellow flower opening in late afternoon. Growing time is summer; needing sunny-dry conditions. From seed. **C**: 1   p. 355

puberulus (So. Africa: Cape Prov.), "Dawny split rock"; cluster-forming highly succulent little plant to 1½ in. high, with 1–2 pairs of basally united, swollen, rounded leaves grayish-brown dotted with dark green, the upper surface convex, rounded beneath, and with a velvety feel from microscopic hairs; 1 in. flowers golden yellow. **C**: 1
p. 355

wilmotianus (So. Africa: Cape Prov.), "Split rock"; tiny clump-forming succulent ¾ in. in dia., with paired fat leaves resembling a round cleft stone, outside and top keeled, the surface wrinkled, grayish green, tinged with pink, and covered with violet dots; flowers golden-yellow. Must keep quite dry when resting after flowering. From seed. **C**: 13   p. 328

DIONAEA (*Droseraceae*)
muscipula (E. U.S.: No. and So. Carolina), the popular "Venus fly-trap"; a remarkable carnivorous perennial of damp mossy places; cultivated for its amazing method of catching and digesting insects; a flat rosette, with tiny scaly, lily-like bulbs; 4 to 8 leaves 1 to 5 in. long, the lower part a flattened winged petiole, the upper part an oblong yellow-green blade of which the two hinged halves are bent upward, the margins with long teeth like the jaws of a shark; the inside surface covered with reddish digestive glands and on each side also 3 long hairs, the sensitive trigger hairs. When insects such as flies touch 3 of the sensitive bristles, the trap springs shut in a fraction of a second and the unwary victims are subsequently digested. White flowers in July-August. In cultivation the plants grow well in acid peatmoss or sphagnum compost; pots may be set in a dish with water to keep roots steadily moist; use mineral-free water. We grow Dionaeas successfully in small glass brandy snifters or glass jars. During winter keep quite cool in order to give them a complete rest, still keeping them moist. Propagated from seed, division, or leaf cuttings under glass; also by bulblets forming from base of mother bulb. **C**: 82   p. 522

DIOON (*Cycadaceae*)
edule (Mexico), "Chestnut dioon" or "Virgin's palm"; rather tender palm-like plant from the hot open country, closest to the fossil cycads; slowly forming stocky trunks to 10 in. thick and 6 ft. high, topped by a beautiful spreading crown of stiff, dusty blue-green pinnate leaves 3–6 ft. long, the leaflets having spiny tips; the young petioles are covered with white wool; flowers in cones to 1 ft. long. **C**: 1 p. 315

spinulosum (Mexico: Yucatan), "Giant dioon"; tree-fern-like, handsome cycad from the rain forest, with tall and slender trunks to 45 ft. or more high, the regal spreading dark green pinnate fronds to 6 ft. long in magnificent rosette, the leaflets with spiny teeth at margins. Very slow growing, yet probably the tallest of cycads; for warm locations. **C**: 54   p. 316

DIOSCOREA (*Dioscoreaceae*)
bulbifera (Trop. Asia, Philippines), "Air-potato" or "True yam", also known as "Bitter yam"; twining plant with small or no underground tuber like most yams, but forming aerial, axillary, angular gray-brown tubers to 1 ft. long, weighing several pounds; alternate leaves heart-shaped, and small unisexual greenish flowers. The tubers of this species are potato-like in flavor but must be properly cooked. Primarily grown as an oddity, best on supports in a greenhouse. Propagation by division of tubers or planting of the aerial tubers. **C**: 54   p. 275
discolor (Surinam), "Ornamental yam"; beautiful tropical herbaceous twining plant with large tuberous roots and thin-wiry stems, bearing handsome ornamental, large, papery, cordate leaves 5–6 in. long, and with long basal lobes, velvety dark olive green marbled light green and silver-gray, veins carmine to silver, reverse purple. Gorgeous foliage vine in the tropical greenhouse trained on wire supports. **C**: 54   p. 266
multicolor argyraea (Colombia), "Silver-leaf yam"; tuberous-rooted twiner with showy ornamental heart-shaped leaves 3 to 5 in. long, emerald green with the areas between veins overlaid with silvery gray. During winter the plant goes into a resting period when the tubers are kept dry and cool. Propagation by division of tubers while at rest. **C**: 54   p. 266

DIOSMA (*Rutaceae*)
reevesii in horticulture (botanically Coleonema album) (So. Africa), "Breath-of-heaven"; evergreen heather-like shrub to 5 ft. high, with slender branches and feathery, very small needle-like, channeled spiny-tipped leaves pleasingly scented like pine-needles, the little white flowers like Baby's breath (Gypsophila), freely carried over a long season in winter and spring, with scattered bloom the year round. Usually grown as a small pot plant in the sunny window; shear lightly after main bloom. Propagated by cuttings. **C**: 13   p. 498

DIPLACUS glutinosus: see Mimulus aurantiacus

DIPLADENIA (*Apocynaceae*)
x amoena (amabilis x splendens), "Pleasing dipladenia"; climbing shrub with oblong, rugose leaves 4–8 in. long; beautiful rose-pink, funnel-shaped 3½ in. flowers with deep rose eye, and yellow throat. **C**: 53   p. 277
boliviensis (Bolivia), "White dipladenia"; free-blooming shrubby climber with slender branches; shining green oblong, slender pointed 2–3 in. leaves; and 2 in. funnel-form flowers white, with orange-yellow throat, in axillary or terminal racemes. **C**: 53   p 277
sanderi (Brazil), "Rose dipladenia"; woody vine with wiry stems having milky sap; small leathery 1½–2 in. ovate leaves glossy green, bronzy beneath; the 2½–3 in. flowers rose-pink with pure yellow throat. A beautiful climber, blooming throughout the year in good light, even as a smaller, shrublike plant; requires copious watering when growing, with good drainage. Propagated from cuttings. **C**: 3   p. 277

DIPLAZIUM (*Filices: Polypodiaceae*)
esculentum (India to Polynesia), "Edible dwarf tree-fern"; lovely semi-dwarf treefern, with short trunk 1–2 ft. long, the fronds 4–6 ft., pinnate or bipinnate, the segments notched or crenate, fresh green. Young shoots edible. **C**: 62   p. 236
proliferum (asperum) (Polynesia, Malaya, Trop. Africa), "Double-dot fern"; tufted fern with spreading pinnate fronds 2 to 4 ft. long, the leathery leaflets 1–2 in. broad, often bearing young plantlets in axils. **C**: 62   p. 249

DIPLAZIUM: see also Pyrrosia

DISA (*Orchidaceae*)
uniflora (grandiflora) (So. Africa), "Pride-of-Table Mountain"; a stunning terrestrial rosette of shining, lanceolate leaves with tuberous rootstock, and producing an erect, leafy spike from 6 to 18 in. high bearing a lax cluster of brilliantly fiery scarlet, showy, waxy flowers 3 to 4 in. across, the helmet-shaped dorsal sepal red outside, inside lighter, veined with crimson over yellow, the large lower sepals a vivid scarlet, but the lip itself is small; blooming from January to June, lasting a long time in perfection. Disas are probably the grandest, most showy terrestrial

orchids, but their needs must be understood to succeed away from the fogs of Table Mountain. I have seen them grown successfully in fibrous sandy loam, dressed with sphagnum, kept cool and steadily moist or misted while in active growth; when flowering-stems die down, the subterranean tubers welcome some rest, perhaps in the shady garden. Propagated from suckers, or seed. **C**: 72   p. 466

DISCHIDIA (*Asclepiadaceae*)
imbricata (Malaysia), "Thruppence urn plant"; tropical succulent high-climbing epiphytic vine, with thin-wiry twining stems bearing opposite fleshy, small rounded leaves convex like an umbrella, and growing against their support shingle-like; small urnshaped, hoya-like flowers. Grown in a moist-warm greenhouse, these epiphytes appear to do best growing on branches of locust (Robinia) and possibly other trees. **C**: 53   p. 271
rafflesiana (Malaya, New Guinea), "Malayan urn vine"; remarkable epiphytic climber with rambling wiry stems rooting at the joints, bearing opposite, oddly thick-fleshy, oval or rounded leaves in the juvenile period, formed into pitchers in the maturity stage 3 to 4 in. long, grayish to dark green, pear-like and hollow, purplish inside, the cavity filled with roots, in habitat frequented as a refuge by ants; yellowish fleshy flowers in clusters. Cultivated in the warm greenhouse as a curiosity, these epiphytes grow best fastened to logs with bark, or branches of locust (Robinia) under moist-warm conditions. Propagated by seed or stem sections with leaves. **C**: 53   p. 533

DIZYGOTHECA (*Araliaceae*)
elegantissima (Aralia) (New Hebrides), "Spider aralia" or "Splitleaf maple"; graceful tropical shrub, branching with slender, flexuous stems mottled cream, and long-stalked, palmately compound, leathery leaves, the threadlike, narrow segments metallic red-brown and lobed; grows into a small tree to 25 ft. high, forming a woody trunk; with the maturity stage, the leaf character changes, the leaflets becoming much broader, lanceolate and lobed. A good decorator plant with fair light 50 fc. up. Propagated by mossing or fresh seeds. **C**: 4   p. 195, 291

DOLICHOTHELE (*Cactaceae*)
longimamma (C. Mexico) "Finger-mound"; interesting, clustering plant of globular shape to 6 in. high but consisting almost entirely of cylindrical knobs to 2 in. high, deep green and soft-fleshy, tipped with scattered soft, pale yellow spines; showy yellow flowers borne in the axils; free-blooming. **C**: 13   p. 163

DOMBEYA (*Sterculiaceae*)
x cayeuxii (D. mastersii and wallichii), "Mexican rose" or "Pink ball"; a hybrid shrub of tropical African parents, with soft, rough-hairy cordate-kidney-shaped and palmately veined and netted, rich green leaves 6 to 8 in. across, dentate at margins; the lovely inflorescence a pendant 3 to 4 in. ball-like cluster of fragrant 1½ in. flowers rosy-pink with pale center, and which bloomed for me in January. Propagated from spring cuttings of firm young shoots. **C**: 3   p. 490

DOROTHEANTHUS (*Aizoaceae*)
gramineus (S. Africa: Cape of Good Hope), the "Tricolor mesembryanthemum" or "Buck Bay daisy"; annual spreading succulent herb forming clumps to 8 in across; opposite fleshy fresh green cylindrical leaves 2 to 3 in. long, and covered with crystalline blisters, the medium-large 1½ in. flowers normally bright carmine with a dark center, but cultivated in South African gardens in many lustrous colors in pink, rose, red, orange, yellow to white, sometimes with blue or red centers, the most beautiful of all annual Mesems, and blooming when the sun warms the sleeping buds. Make charming pot plants. From seed. **C**: 13   p. 357

DORSTENIA (*Moraceae*)
yambuyaensis (Congo); interesting erect bristly herb related to figs., 12 to 18 in. high with creeping rootstock; oblanceolate leaves to 6 in. long, shining green above, pale and dull beneath, and irregularly toothed, the inflorescence a curious green receptacle angularly rounded, almost 1 in. across, the winged margins with rays to 4 in. long, a study object for schools, demonstrating that what looks like a handsome blossom is actually a plate-like receptacle im-

bedded with many, insignificant greenish, not colored male and female flowers. Grown best in a moist-warm greenhouse. Propagated by division or seed. **C**: 54   p. 534

DORYOPTERIS (*Filices: Polypodiaceae*)
   pedata (W. Indies to Argentina), the "Hand fern"; an excellent, curious dwarf fern with attractive thin-leathery, bright green, palmate fronds, 6 in. across, carried horizontally on black wiry stalks, the 3 main divisions having black veins, parted again into linear segments and pointed lobes which are edged by a line of sori. **C**: 60   p. 247

DOVYALIS (*Flacourtiaceae*)
   hebecarpa cv. 'Kandy' (Aberia gardneri) (Ceylon, India), "Ceylon gooseberry"; decorative evergreen growing into a small dioecious tree to 20 ft. high; the alternate ovate lanceolate, wavy-edged, 3–4 in. papery leaves glossy green with metallic sheen, and rosy midrib, grayish beneath, arranged as if pinnate, in two ranks on the thin wiry branches, set with scattered $\frac{1}{2}$ in. needle-like thorns; small greenish flowers, followed on female plants by 1 in. maroon-purple gooseberry-like edible berries; the female flowers must be pollinated from male tree. A tough, long-lived decorator plant very handsome with its feather-like undulate foliage. **C**: 15   p. 296

DRACAENA (*Liliaceae*); the "Dragon lilies" are handsome tropical foliage plants, some becoming tree-like, principally with African affinities, forming woody stems topped by a crown of leathery leaves suggesting palms. Flowers are carried in clusters, they are typically small and whitish with sepals and petals together forming 6 divided segments that differ from the tubular flowers of the Cordylines which are often misnamed "Dracaena" in horticultural practice. Their wiry roots are orange-colored or yellow but not suitable for propagation. Being tropical most Dracaenas require warmth and constant moisture at the roots. They are generally well adapted as house plants, tolerating low light conditions to 30 fc., depending on the species. Most kinds are excellent keepers in water culture, even as unrooted cuttings. Propagated mostly by cuttings, stem sections (p. 57) or mossing; Dracaena draco from seed.
   arborea (Africa: North Guinea), "Tree dracaena"; tree-like, to 40 ft. high, forming trunk with dense head of broad, sword-shaped, evenly fresh-green, sessile, wavy leaves to 3 ft. long, with sunken veins and prominently raised midrib. **C**: 4   p. 198
   deremensis 'Warneckei' (Trop. Africa), "Striped dracaena"; handsome, and symmetrical rosette, formed by long sword-shaped, leathery leaves, sessile and dense along a stout cane, growing to 15 ft. high and eventually branching; the foliage to 40 cm., green, streaked milky in center, and bordered by a translucent white band on each side inside the narrow bright green edge. Attractive as house plant and very satisfactory for decoration, suitable for locations with poor light down to 15 fc.; also doing well under air-conditioning. **C**: 4   p. 199
   draco (Canary Islands), the curious "Dragon tree"; strangely unique tree to 60 ft. high with monstrous trunk to 15 ft. dia.; decorative as a young plant with its crowded rosette of sessile, sword-shaped, thick fleshy leaves to 2 ft. long, smooth to glaucous green, and with translucent edges, outlined in red if in strong sun; flowers greenish, in panicles, followed by orange berries. Ancient superstition believes that the dark red, resinous substance exuding from leaves and trunk is "dragon's blood". **C**: 4   p. 197, 198, 525
   fragrans massangeana, "Cornstalk plant" or "Variegated dragon-lily"; an old-fashioned robust house plant with its rosette of rich green, laxly arching leaves broadly striped and banded light green and yellow down the center. Requires warmth and moisture as chills cause brown blotches, and stop the growth, inducing it to bloom; the yellowish flowers appear in clusters and are fragrant. The species fragrans is plain green and grows into a branched tree 20 ft. high, at home in tropical West and East Africa. Propagates best from cane sections. **C**: 4   p. 199, 537
   fragrans 'Victoriae', "Painted dragon-lily"; stunningly beautiful, slow-growing tropical conservatory plant, with gracefully pendant, wide, soft-leathery leaves green streaked silvery-gray in center, bordered by contrasting, very broad margins of cream to clear golden yellow. Not recommended as a house plant as even with ideal conditions the foliage, lacking chlorophyll, tends to develop unsightly brown areas. **C**: 53   p. 199

x'Gertrude Manda' (hookeriana x grandis), "Broad leather dracaena"; very decorative Manda hybrid, intermediate between grandis and hookeriana, and characterized by very thick leaves 4 in. or more wide, translucent margins; very durable. **C**: 4   p. 198
   godseffiana (Congo, Guinea), "Gold-dust dracaena"; a charming small, shrubby plant with spreading, wiry stems bearing thin-leathery, elliptic leaves in pairs or whorls of three, glossy deep green, irregularly spotted yellow, maturing to white; greenish-yellow flowers followed by red berries; an ideal miniature for novelty planting. **C**: 4   p. 199
   godseffiana 'Florida Beauty', a striking seedling of kelleri, with the same characteristic thick-leathery leaf as the latter, but more richly variegated, being almost entirely covered with creamy-white blotching over the glossy green. **C**: 4   p. 199
   goldieana (So. Nigeria, 3000 ft.), "Queen of dracaenas"; the most spectacular dracaena in cultivation; branching foliage plant with slender canes to 10 ft. high, furnished with leathery foliage from bottom to top; the broad-ovate, glossy deep green, long-stalked leaves in a spiralling rosette, and strikingly marked with crossbands of pale green, maturing to almost white. Flowers in a dense spiral head, fragrant, white, opening at night. Difficult to maintain unless given humid-warm conditions and a very porous growing medium; I have seen them thrive in crushed pure lava in Hawaii. **C**: 56   p. cover*, 191, 199
   hookeriana (grown in California nurseries as "Rothiana") (So. Africa: Cape of Good Hope, Natal), the "Leather dracaena"; heavy plant with woody trunk to 6 ft. high, occasionally branching, topped by a crowded rosette of spreading sword-shaped, very thick leathery leaves dark green and glossy, to 1$\frac{1}{2}$ to 2 in. wide and with translucent edges; slow-growing, and with greenish flowers in large panicles. In the coastal bush of Natal I have measured leaves 4$\frac{3}{8}$ in. wide. Tough as leather and an excellent house plant. **C**: 4   p. 198
   marginata (gracilis) (Madagascar), "Madagascar dragon tree"; a favored decorator; tree-like, with branching slender trunk, growing to 12 ft. high, each cane topped by a dense terminal rosette of thick-fleshy, narrow linear, clasping leaves 15 in. long, rigidly spreading horizontally, shiny deep olive green prettily edged in red. The flexuous canes have the tendency to grow snaky and twisted, giving a branched specimen very artistic appearance. Slow and durable; best in small containers, and not too wet, if in cool location. Tolerant to poor light down to 30 fc. **C**: 4   p. 198
   rothiana: see hookeriana
   sanderiana (Cameroons, Congo), "Ribbon plant"; neat and attractive, very durable little rosette, erect on slender cane until becoming too heavy; with narrow lanceolate, thin-leathery, relaxed 6 to 8 in. leaves, deep green somewhat milky, and with broad marginal bands of white. Very tolerant to all conditions except cold. Excellent for small pottery dishes. **C**: 4   p. cover*, 43*, 199

DRACAENA: see also Cordyline, Pleomele

DRIMIOPSIS (*Liliaceae*)
   kirkii (Zanzibar); small bulbous plant with lax, strap-shaped. keeled, fleshy leaves narrowing toward base, pale blue-green with dark blotches; white flowers on short spike in July. **C**: 54   p. 186

DROSERA (*Droseraceae*); the "Sundews" are interesting insectivorous plants that received their vernacular name because of their glistening foliage. Some grow in sphagnum bogs, or their sandy shores, in North America, others are tender and are at home in Africa and Australia. They are small perennial herbs with thread-like to spoon-shaped leaves covered with glistening, sticky hairs. When an insect comes in contact with the leaf, the sensitive hairs curve inward and entrap it; the glandular hairs discharge from their tips a sticky fluid which retains and digests these small creatures, the digested remains of which are absorbed by the leaves. Sundews require sun and air; they grow best in moist sphagnum peat in a cool greenhouse or terrarium; pots should stand in a dish with water; this should be lime-free or rain water. Propagated by seed, or root division; D. binata also from root cuttings kept warm under glass cover.
   binata (Eastern Australia, New Zealand), "Twin-leaved sundew"; interesting insectivorous plant called "Sundew" for

the glistening foliage; a small tender perennial 6 to 18 in. high with long-stalked linear leaves deeply divided into two thread-like lobes covered with reddish, sensitive glandular hairs, discharging from their tips a sticky fluid which holds and digests insects and other small creatures. Propagation of this species from root cuttings, as well as division. **C**: 70
p. 521, 522

filiformis (Atlantic North America from Mass. to Florida), "Threadleaf sundew", the common "Sundew"; insectivorous perennial with threadlike, linear green leaves 4 to 15 in. long equipped with fine purplish red grandular hairs which curve inward to entrap any unwary and curious insects. Th¹s pretty species grows in habitat on moist sandy soils near the coast; not safely winter-hardy and best wintered in cool greenhouse or cold frame. **C**: 82   p. 522

rotundifolia (No. America, No. Europe, No. Asia), "Sundew plant"; insectivorous herb with small white flowers; flat, hairy petioles and light green round, spoon-shaped leaves to 1½ in. long, the ¼ to ½ in. cups equipped at the margins with long threads of red tentacles like the comparative strength of an octopus in ability to hold its victims. Insects are lured by the odor of the sticky hairs, they become trapped as they move around and fight to free themselves, other tentacles bend secreting more of the digesting enzymes. These are the smallest and most fragile of carnivorous plants. I collected the plant photographed at Barnegat, N.J. and planted it into a 1½ in. pot in sphagnum moss. **C**: 82   p. 522

DROSOPHYLLUM (Droseraceae)

lusitanicum (Portugal, S.W. Spain, N. Morocco), Mediterranean "Dew leaf"; low shrubby deep-rooted carnivorous plant from gravelly locations in pine forests; with linear green leaves to 10 in. long, densely set with immovable glandular hairs which in the sun exude a honey odor from a dewlike sticky drop by which insects are captured; clusters of handsome sulphur-yellow flowers in summer. Drosophyllum requires less humidity than Drosera, grown in sandy peat or sandy loam, and kept drier and warmer especially in fall and winter. Propagated from fresh seed, **C**: 65   p. 522

DRYNARIA (Filices: Polypodiaceae)

quercifolia (Polypodium) (So. India, Queensland, Fiji), "Oakleaf fern"; growing on trees and rocks, with a thick, brown, woody rhizome, and stalkless brown, bluntly lobed barren fronds, above which rise long-stalked green, deeply pinnate rigid fertile fronds 2–3 ft. high, with brown veins, the pinnae cut to center. May be planted into baskets or on blocks. **C**: 62   p. 240, 247

rigidula (Malaysia, Queensland, Fiji), "Feather polypodium"; epiphytic fern with thick creeping rhizome covered by long brown hair-like scales and looking like a rabbit's foot; the fronds in two forms; the sterile ones stalkless to 9 in. long and merely lobed, the fertile fronds pinnate, 2 to 4 ft. long, leathery and rich green, on brown rachis first erect, later pendant. Propagated by division. Good basket plant. **C**: 60   p. 247

DUDLEYA (Crassulaceae)

virens (Echeveria insularis) (California: Catalina and other offshore islands); large flattened rosette 4 to 10 in. dia., with obovate linear, strap-like leaves bluish gray and densely covered with silver bloom, reddish toward tips; branching with offsets from the base and forming clumps; flowers reddish-yellow. Enjoys cool-humid air. **C**: 15 p. 338

DYCKIA (Bromeliaceae);

xerophytic terrestrial rosettes of thick-leathery or succulent more or less rigid, dagger-like or narrow-tapering leaves, spiny and sharp-pointed at tip and margins; at home on the South American campos or growing on rocks, ranging from miniatures to plants 5 ft. or more tall. The underside of the leaves is covered by silvery scales often in parallel lines; their tall, branching inflorescence is showy with bright, usually orange flowers. With age, Dyckias form dense clumps not particularly attractive, but as small single plants are nice in novelties or dishgardens resembling aloe or agave. Propagated from suckers, or seed.

brevifolia (sulphurea) (So. Brazil), "Miniature agave"; small clustering, star-like rosette of thick, stiff-succulent, glossy green, sharp-pointed leaves, with silver lines beneath; inflorescence off center, a long stalk with bright orange flowers. **C**: 22   p. 138, 350

fosteriana (Brazil: Paraná), "Silver and gold dyckia"; very ornamental robust rosette, in a dense whorl of narrow stiff silvery-purple arched leaves, to 8 in. or more long, the margins with prominent silver spines; inflorescence a spike of rich orange flowers. **C**: 10   p. 138

ECHEVERIA (Crassulaceae);

large genus of mostly small, typical leaf succulents in the shape of charmingly elegant rosettes, chiefly Mexican, and popularly known as "Hen-and-chickens"; with broad, fleshy leaves having waxy, velvety or glaucous surface, often shimmering in iridescent glow, and edged in red; the stalked inflorescence with clusters of pretty flowers in shades from white to flame orange and brilliant red. They are outstanding in collections, for planting into novelty pottery or strawberry jars, and good light brings out their best in coloring. For the winter they are best kept cool and dry; in summer some species are used in carpet-bedding, the larger forms are good individual accent plants requiring very little care. Grown from offsets (chicks), or more slowly from leaf cuttings or seed; the giant types like crenulata from the leafy bloom-stalks.

affinis (nigra) (Mexico), "Black echeveria"; stiff rosette of fleshy dark green, waxy, pointed and keeled leaves to 2 in. long, turning almost completely black in full sun; flat-topped clusters of scarlet-red flowers on leafy stalk, in winter. Distinctive for its black coloring. **C**: 13   p. 338

agavoides (in the trade as Urbina agavoides) (Mexico: San Luis Potosi), "Molded wax"; solid starlike rosette with rigid, fleshy leaves triangular-pointed, 1½– 3in. long, glossy pale apple-green; the margins frequently reddish, tipped yellow; tall pink stalk with reddish to yellow flowers. Sometimes branching from the base and forming clumps. One of the "hard" succulents used in dishgardens. **C**: 13
p. 338

x'Ballerina'; a giant San Francisco hybrid rosette 18 in. across, with branching leafy stems carrying the inflorescence; glaucous leaves with wavy edges, the young foliage violet and with very red border; flowers pink and glaucous, on pinkish purple branches. **C**: 13   p. 339

x'Can-Can'; magnificent large California hybrid rosette which I photographed at Johnson's; 14 in. or more across, with thick-fleshy, stout leaves arranged in a perfect circle, glaucous copper-green to brown-violet when older; the margins are doubly crenulate and the red edge is fringed and ruffled as a French petticoat. Flowers pink. **C**: 13 p. 339

crenulata (Mexico: Morelos), "Scallop echeveria"; showy, stem-forming large rosette 15 or more in. across, with broad, obovate leaves to 1 ft. long, and tapering to narrow petiole; pale green and glaucous bluish-gray, the margins undulate and red; flowers yellowish red on stout stalk. **C**: 13   p. 339

crenulata 'Roseo-grandis'; a very showy and decorative large rosette 10 in. across, with broad, always rich green, waxy leaves beautiful maroon-red at the undulate margins; some light bluish glaucescence on young inside leaves. These large Echeverias may be propagated from inflorescence stalks. **C**: 13   p. 339

curtisii: see x Pachyveria 'Curtis'

derenbergii (Mexico: Oaxaca), "Painted lady" or "Baby echeveria"; a miniature globe-shaped clustering rosette of 1½ to 2½ in. dia., the numerous thick leaves pale green and glaucous with waxy silvery blue bloom and tipped red; blooming freely on short stalks from between the foliage, with flowers golden yellow with orange. Very charming in dishgardens. From offsets. **C**: 13   p. 329, 338

x'Doris Taylor' (setosa x pulvinata), "Woolly rose" or "Plush rose"; charming rosette of robust habit, to 5 in. dia., with glossy deep green, obovate leaves beautifully covered with white plush, the apex tipped red. Brilliant red and yellow blossoms in spring. A stunning succulent, also for partial shade. **C**: 13   p. 339

elegans (Mexico: Hidalgo), "Mexican snowball", "Mexican gem" or "White Mexican rose"; exquisite clustering incurved rosette 2 to 4 in. across, of spoon-shaped succulent leaves, waxy glaucous pale blue, with white translucent margins; clusters of coral pink flowers on long pink stalk, lasting for weeks. One of the most beautiful small echeverias, like a small ball of ice about to break open; for carpet beds, in dishgardens or small pots. Propagated from offsets or leaves. **C**: 13   p. 329, 337

expatriata (Mexico), "Pearl echeveria"; short branching stems, topped by small rosettes of thick, oblanceolate, glaucous gray leaves 1 to 1½ in. long, rounded on both sides and with pointed tip; flowers pinkish yellow on horizontal stalks. **C**: 13   p. 338

gibbiflora carunculata, "Blister echeveria"; curious large rosette 10 or more in. across, with the large spoon-shaped coppery, red-edged leaves covered with metallic pink to amethyst bloom; having raised blister-like warty growth on the surface. The species E. gibbiflora is from So. Mexico, forming a heavy stem crowned with a large rosette of glaucous leaves 5 to 12 in. long. Large clusters of flowers red outside, yellow within. A strange curiosity plant. New leaf rosettes sprout at the base of old plants. **C**: 13   p. 339

gibbiflora metallica; a rather smooth rosette to a foot or more across with large spoon-shaped leaves neatly rounded toward the tip; bronzy amethyst with metallic luster and glaucous purple, the margins translucent and reddish; branching from the base. A clean-cut big succulent very decorative in containers. **C**: 13   p. 339

gilva (Mexico), "Wax rosette"; shapely open rosette of broad obovate, spoon-like swollen leaves 2–3 in. long, pea green and slightly glaucous, with tip flushed orange red. Very friendly color, and durable in dishgardens. **C**: 13   p. 337

glauca (Valley of Mexico), "Blue echeveria"; open rosette about 4 in. across; with leaves broad spatulate, almost round, and somewhat thin, glaucous gray-green tinted purple, and tipped with red; flowers on slender coiled stems, reddish outside, yellow within. Makes offsets freely. Much used for carpet bedding. **C**: 25   p. 337

x glauco-metallica (glauca x gibbiflora metallica), one of several known as "Hen and chickens"; fairly large colorful rosette about 6 in. across, with broad spatulate, flat leaves, olive-green covered with gray bloom, and tinted amethyst toward edge; flowers dull orange-red. Forms offsets which cluster like chicks around mother hen. **C**: 13   p. 337

x'Haageana' (derenbergii x pulvinata), "Fruit cups"; robust open rosette of obovate leaves, rather broad, fleshy leaves keeled beneath, light green with a rough rugose white surface, tips slightly edged and flushed carmine. There is another E. 'Haageana' of record, hybrid of agavoides x pulchella, but this is apparently not in cultivation. **C**: 13   p. 337

x kewensis; large open rosette about 4 to 5 in. across, of broad obovate leaves light olive green with silvery blue bloom; margins tinted pink, and the blunt, spine-tipped apex red; pinkish-red flowers tipped with yellow, on pink stems from between the leaves. **C**: 13   p. 339

x'Meridian' (crenulata hybrid); a handsome vigorous giant rosette 15 to 18 in. across, with thick-fleshy fan-like, wavy, glaucous green leaves crenulated toward apex, the double-curly leaf margins light red. Flowers glaucous pink on long leafy stalks. A show plant. Harry Johnson recommends feeding such large Echeveria hybrids. **C**: 13   p. 339

multicaulis (Mexico: Guerrero), "Copper rose"; attractive small succulent with branching stems to 8 in. high, each with a small, loose rosette of waxy fresh green spatulate leaves 1—1½ in. long, coppery toward apex and edged brown; flowers on short red-bracted stalks red, yellowish inside. **C**: 13   p. 328, 338

pallida (Argentina), "Argentine echeveria"; large stem-forming South American succulent rosette of open, loose habit, with large obovate, spoon-shaped, concave 6 in. leaves pale green, with edge lightly glaucous and reddish, tapering to a stalk-like petiole, the back surface keeled; bell-shaped pink flowers on erect stalk 20 in. high. A specimen for solitary display. **C**: 15   p. 339

peacockii (Mexico), well known in horticulture as E. desmetiana; lovely symmetrical, stemless rosette 2½ to 5 in. dia., of numerous cupped, long obovate fleshy leaves richly covered with silver-blue glaucescence, the margins and tips outlined in red; intense red flowers, glaucous outside, on tall coiling stalks. A favorite for carpet bedding in Europe. From seed or suckers. **C**: 25   p. 410

peacockii var. subsessilis, according to Eric Walther, (subsessilis or desmetiana hort.) (Mexico: Tehuacan); beautiful rosette more compact than the species, with obovate leaves shorter, 1½–2 in. long, covered with waxy bluish-white bloom, the margins and short-pointed tip red, and forming a tight round circle; numerous bright red flowers in one-sided tall clusters, freely blooming. **C**: 13   p. 337

x perbella, nice dense rosette of thick, concave, obovate, waxy green leaves tinted copper and with contrasting reddish edge. Very stiff and sturdy. **C**: 13   p. 337

x'Pulvicox' (pulvinata x coccinea), "Red plush plant"; attractive hybrid forming red-hairy leafy stem, topped by a rosette of broad gray-green, recurved leaves, densely covered with white velvet, turning reddish in strong light; flowers yellow and vermilion. **C**: 13   p. 338

pulvinata (Mexico: Oaxaca), "Chenile plant"; beautiful velvety rosette of thick fleshy, broad obovate, ashy-green, red edged leaves 2 to 3 in. long, thickly covered with silver-white felty hair, and turning reddish in good light; in time forming a shaggy, rust-colored stem; red or yellow flowers on a leafy stalk. Very pretty in collections; best in cool weather. **C**: 13   p. 337

x'Pulv-oliver' (pulvinate x harmsii), "Plush plant"; very fine plushy hybrid with oblanceolate, broad, medium green leaves edged maroon toward apex only, and covered with glistening white velvety hairs; stem-forming; very free-blooming with flowers orange and red on leafy stalk. **C**: 13   p. 329, 337

purpusorum, known in horticulture as Urbinia purpusii (So. Mexico); charming stemless rosette of nearly triangular pointed, keeled fleshy leaves only 1½ in. long, gray green dotted and mottled with brown; flowers vivid red outside, yellow green inside and at tips. Very pretty in its regular, starry habit, resembling a miniature agave. **C**: 13   p. 339

roseo-grandis: see crenulata 'Roseo-grandis'

runyonii (Mexico); handsome, formal succulent rosette of numerous rather flat, obovate 3 in. leaves upcurving, very glaucous-blue; flowers pink in forked racemes. Clustering from basal off-shoots. **C**: 13   p. 339

secunda (So. Mexico), "Blue hen and chickens"; small stemless, saucer-shaped 3–4 in. rosette of obovate oblong leaves which stay fairly narrow, always bluish-green, not metallic; the apex with short tip red; flowers reddish. Used for bedding work because of its blue coloring, tinted red in good light. Forms many basal offsets, or chickens. **C**: 13   p. 337

x'Set-oliver' (setosa x harmsii), "Maroon chenile plant"; a robust Echeveria x Oliveranthus plush-covered rosette, of many close-set olive-green oblanceolate leaves, white hairy beneath, flushed or edged maroon; flowers orange, yellow and red on leafy stalk. **C**: 13   p. 337

simulans (Northern Mexico), "True Mexican rose"; dense rosette of numerous comparatively narrow pointed, fleshy, concave leaves 1½–2½ in. long, rather incurving in center, light green and slightly glaucescent, the upper half of leaf rich crimson red, particularly the short cusps; flowers reddish-yellow. A favorite dishgarden plant. **C**: 13   p. 337

sprucei (quitensis sprucei) (Andes of Ecuador), "spruceana" in horticulture; a small but dense rosette from the Andes, crowded with many slender oblanceolate, fleshy leaves round on both sides, fresh-green with red apex; the flowers orange-red. The foliage is smaller than quitensis, to which it is related, and which forms a brown stem 3 ft. high. **C**: 13   p. 338

ECHINOCACTUS (Cactaceae)

grusonii (C. Mexico), "Golden barrel"; striking species growing into a giant globe of 3 ft. dia., light green, closely ribbed, covered with golden spines; yellow flowers imbedded around top, opening in sun. Grows well in cultivation. **C**: 13   p. 158

ingens (Mexico), "Blue barrel"; large barrel type, to 5 ft. high, glaucous blue, woolly at top; few ribs as seedling, later densely ribbed, with rigid straight brown spines; flowers yellow. Slow-growing. **C**: 13   p. 158

ECHINOCEREUS (Cactaceae); the "Hedgehogs" are spiny or bristly columnar, sometimes cylindric cluster-forming desert cacti of small size but with comparatively large flowers, which bloom from April to July during the daytime but don't always close at night. They are amongst the easiest to bring into flower, producing them at an early stage. Hedgehogs are easy to grow, but generally are kept very much on the dry side, especially if plants are desert-collected. They require sunlight and fresh air to bring out the best coloring in their often vari-colored spines; keep quite cool and dry in winter. Propagated by rooting offsets, or from seed.

dasyacanthus (W. Texas; Mexico), "Rainbow-cactus"; small cylindric, hard, and closely ribbed xerophyte to 8 in. high, the stiff spines very attractive with colors in zones of straw-yellow, purple, tan, red-brown and amethyst; spines pointing out and downward; funnel-shaped flowers yellow. **C**: 13   p. 157

purpureus (Oklahoma), "Purple hedgehog"; cylindrical plant to 5 in. tall, olive-green with shallow ridges densely set with flattened, all-radial white spines tipped purple; flowers deep magenta-purple. **C**: 13   p. 157

reichenbachii (Texas; Mexico), "Lace-cactus"; small globe to 8 in., close-ribbed, attractively covered by flat rosettes of stiff spines in a range of colors from white to red-brown; flowers iridescent light purple. **C**: 13   p. 157

viridiflorus (So. Dakota, Wyoming to Texas), "Green-flower hedgehog"; small rigid, nearly globular, dark green plant to 8 in. high, with 14 ribs; long stiff areoles and spines, in zones of white or brownish; flowers greenish. Winter-hardy. **C**: 13   p. 157

ECHINODORUS (*Alismaceae*)

tenellus (widespread from temperate No. America, Mass. and Michigan, to Florida; in So. America from Colombia to So. Brazil), a dwarf "Amazon sword plant"; tuft-forming aquatic, existing both in hardy perennial and persistent tropical forms; the latter used in aquaria as under-water oxygenator, with bright green lanceolate or linear leaves 2 to 6 in. high, growing from a rosette, sending out runners from the crown creeping above the soil and forming thickets, appreciated by fish in spawning; a pretty sight with the light shining through the translucent leaves. Growing best planted shallowly, in soft warm water at 70 to 80° F. in fairly good to subdued light. Rapidly propagated from rooted runners. **C**: 4   p. 515

ECHINOFOSSULOCACTUS (*Cactaceae*)

zacatecasensis (Stenocactus) (Mexico: Zacatecas), "Brain-cactus"; weird-looking depressed globe to 3½ in. wide with pale green, 50–60 sharp ribs closely compressed into undulate folds, and set with broad hooked central spines and smaller radials; flowers white, tipped lavender. **C**: 1   p. 159

ECHINOMASTUS (*Cactaceae*)

macdowellii (No. Mexico); small globular plant 3 in. high, with 20–25 ribs, very beautiful with white radials and long white needle spines; showy flowers lively rose-pink. **C**: 1   p. 159

ECHINOPSIS (*Cactaceae*)

multiplex (So. Brazil), "Easter-lily cactus"; small barrel to 6 in. high, forming clusters, dark green, close-ribbed, with sharp brown spines; long, erect funnel-formed very showy rosy flowers, intensely fragrant. **C**: 13   p. 157

ECHITES (*Apocynaceae*)

rubrovenosa (Prestonia), (So. America: Rio Negro); tropical wiry climber with striking foliage on slender twining stems; the large elliptic emerald green leaves, to 6 in. long, netted with red veins. The beautiful coloring of the leaves is brought out to perfection only under moist-warm conditions. **C**: 53   p. 266

EICHHORNIA (*Pontederiaceae*)

azurea (Brazil), "Peacock hyacinth" or "Creeping water hyacinth"; widely spread in ponds and streams of Tropical and Sub-tropical America often choking the waters. Aquatic herb rooting in shallow water, and sending long runners across its surface, with fleshy stalks not inflated; fresh green, waxy leaves 3 to 6 in. across, rounded or obovate; terminal spikes of showy flowers in summer, lilac-blue with purple center, the two lower petals fringed. Propagated by division; plants may be wintered by planting in loamy soil in shallow pan, warm and in good light. **C**: 54   p. 517

crassipes (speciosa) (Trop. and Sub-trop. America), the "Floating water hyacinth"; small and pretty aquatic plant usually floating, with bluish, feathery roots; runners forming clumps, the roundish glossy, green leaves in rosettes, on petioles much inflated at base causing them to stay afloat; clusters of large 1½ in. lilac flowers with yellow center sur-rounded by blue blotch and purple stripes. Very attractive with their waxy foliage and exquisite flowers on the surface of aquariums or ponds, also in tubs of water. Requires good light for wintering, not too warm; with their long plumy roots the finest plant to use for spawning fish. By division. **C**: 54   p. 517

ELAEAGNUS (*Elaeagnaceae*)

pungens (Japan, China), the "Silver berry"; slow-grow-ing brown-stemmed evergreen shrub of rather rigid, angular sprawling habit of growth, to 6–15 ft. high; leathery oval grayish green leaves with rusty-brown tinting, 1 to 3 in. long and with wavy edges and silvery beneath; the surface is covered with silvery scales reflecting sunlight which give the plant a special sparkle; white, fragrant flowers. Tough container plant in reflected heat or wind; very frugal; sometimes grown as dwarfed bonsai in Japan. From cuttings. **C**: 14   p. 296, 314

pungens 'Maculata' (pung. aureo-variegata), "Varie-gated Russian olive" or "Golden elaeagnus"; sparry ever-green shrub with woody brown branchlets, and silver-scaled 3–4 in. green leaves boldly and irregularly painted rich yellow in center, sometimes all yellow. Resists drought; requires sun; somewhat winter-hardy. **C**: 14   p. 295

ELLEANTHUS (*Orchidaceae*)

capitatus (West Indies, Mexico to Peru and Brazil); terrestrial of the habit of Sobralia, with leafy, slender reed-stems 1 ft. or more high; thin but rigid leaves to 9 in. long, prominently length-ribbed ending in two short points; terminal at the apex of the wiry stalks a showy, round head 2 in. in dia. of many rose-purple 3¼ in. tubular flowers peeking out from purplish bracts; blooming April-June. **C**: 72   p. 468

ENCEPHALARTOS (*Cycadaceae*)

horridus (So. Africa: Karroo), "Ferocious blue cycad"; wide-spreading ornamental cycad with short stout trunk crowned by a cluster of stiff pinnate fronds, rich glaucous green, 2 ft. long, the leaflets oblique lanceolate and spinily lobed; male cone cylindric and stalked, 1 ft. long, female cone broad ovoid 15 in. long. A remarkable and unusual xerophyte and durable tub plant commanding attention by its pronounced blue coloring and dense, dangerous-looking crown of tangled, spiny leaves. **C**: 15   p. 316

latifrons (horridus latifrons) (So. Africa: Port Elizabeth), "Spiny Kaffir bread"; a monstrous cycad, with stout globular trunk, to 8 ft. high by 4 ft. around, the recurving vicious looking pinnate leaves 2 to 3 ft. long, reflexed toward apex; the broad, thick leaflets overlapping and with coarse spine-tipped lobes; the cones brownish-yellow. **C**: 16   p. 317

ENKIANTHUS (*Ericaceae*)

perulatus (Japan), "White enkianthus"; good-looking deciduous shrub to 5 ft. high with tiers of nearly horizontal branches, the smooth, reddish young shoots with oval, serrate leaves to 2 in. long, whorled at branch ends; nodding bell-shaped flowers with ⅓ in. white corolla in terminal clusters, blooming before foliage. Very pretty in autumn when the foliage turns orange or scarlet. Acid-loving; if necessary add peatmoss, and keep moist. From cuttings. **C**: 26   p. 314

ENSETE (*Musaceae*)

maurelii (Musa) (Ethiopia), the "Black banana"; symmetrical plant of almost sinister effect; a compact rosette of sturdy, broad, dark-complexioned leaves, deep green suffused with blackish-red above, brown-purple midrib and margins, purplish-brown beneath. With the decorative habit of Ensete ventricosum, of which species it may be a variety; excellent as a patio plant; good light brings out the darkest coloring. Not stoloniferous but can be propagated by shaving heads from old bases, cover with moss or moist burlap, and young plantlets will sprout under moist-warm conditions. Take off as cuttings when 3 to 4 in. high. **C**: 54   p. 205

ventricosum, well known in applied horticulture as Musa ensete (moist Central and E. Africa), the "Abyssinian banana"; a showy horticultural subject of formal habit; stout solitary pseudostems, conspicuously swollen at the base, bearing a full rosette of erect, broad banana-like leaves 5 ft. or more long, in a cluster; dying after fruiting because the plant is not stoloniferous; leaves bright green with red midrib and purple edge, on short red stalks. Erect inflores-cence with dark red bracts and whitish flowers. Used as shapely centerpiece in formal summer beds where it is usually 5 to 8 ft. high. Have seen this at alt. 5,000 ft. on Mt. Kilimanjaro (Tanzania), and at 7,700 ft. in Kikuyu country of northern Kenya where it reaches height of 40 ft. Propagated from seed. **C**: 66   p. 205

EPACRIS (Epacridaceae)

longiflora (autumnale) (Australia: New South Wales), "Australian heath" or "Fuchsia heath"; a beautiful shrub 3 to 5 ft. high, with closely set, sharp-pointed $\frac{1}{4}$ in. ovate leaves, and thin tubular 1 in. flowers rosy crimson excepting the 5-lobed, spreading limb which is creamy-white, blooming mainly in spring, and scattered at other times also. Epacris belong to the most attractive flowering shrubs of the cool house, blooming prolifically once well established in lime-free soil, in good light and with fresh air, with careful watering not too much and not too little; outside in summer. Old plants are cut back after blooming. Propagated like Ericas from cuttings in late summer or late winter. **C**: 69
p. 501

x EPICATTLEYA (Orchidaceae)

'Gaiety' (Epidendrum plicatum x Cattleya 'Rowena Prowe'); flowers $3\frac{1}{2}$ in. across, white and pale lavender, in spring. A bigeneric hybrid of vigorous habit, showing characters of Epidendrum in the growth, and of Cattleya in the large size and color of the blossom. **C**: 21   p. 468

EPIDENDRUM (Orchidaceae); tropical and subtropical

very diverse American epiphytes at home from sea level to the chilly high Cordilleras; well loved by amateur growers for their ease of cultivation and free-blooming habit, some producing their beautiful clusters of brightly colored flowers almost all year round. Epidendrums fall into two main groups: the species with pseudobulbs bearing one or a few leathery leaves, and the "reed-stems" with long slender stems bearing alternate, usually fleshy leaves; rambling or climbing by aerial roots, and topped by a terminal cluster of smaller or larger, long-lasting flowers. Epidendrums like good light, preferably 2000 to 3000 fc.; the pseudobulbous types tolerate some shade, while the "reed stems" will stand practically full exposure to morning sun and bloom the most profusely and almost constantly. The bulbous kinds should be given a short rest after new growth is matured; the reedy varieties require moist conditions at the roots at all times with frequent feedings. Propagated by division of the pseudo-bulbs, or sectional cuttings of the leafy stems.

atropurpureum (W. Indies. Mexico to Panama, Peru and Brazil), the "Spice orchid"; a very lovely evergreen epiphyte with pear-shaped pseudobulbs bearing 2–3 long leaves; stout stalks to 2 ft. long, with clusters of 2 to 3 in. heavy-textured, fragrant flowers, having sepals and petals a beautiful chocolate with incurved green tips, often tinged with purple; large white lip with crimson center; blooming between late winter into summer, and lasting 4 to 5 weeks in good condition. **C**: 21   p. 467

brassavolae (Mexico, Guatemala to Panama); graceful epiphyte with small pear-shaped 2-leaved pseudobulbs, the waxy, rather fleshy spidery flowers to 4 in. across in erect-stalked clusters $1\frac{1}{2}$ ft. tall, narrow sepals and petals yellowish brown, the heart-shaped lip white, narrowing to a purple point; sweet-scented in the evening, blooming from April to September. **C**: 21   p. 467

cochleatum (Cuba to Central America and Brazil), the "Cockle-shelled orchid"; a distinctive epiphyte with pear-shaped, usually 2-leaved pseudobulbs, and an erect cluster of curious spidery, long-lived upside-down flowers to $3\frac{1}{2}$ in. long, the narrow greenish-white petals and sepals grouped together, the shell-like lip black-violet striped with yellow lines, the whole reminding of an octopus; blooming almost all year, mostly from November to February. Well-known and suitable as a house plant, practically ever-blooming. **C**: 21   p. 467

fragrans (W. Indies, C. America, Peru and Brazil); stocky epiphyte with compressed pseudobulbs bearing a solitary leathery leaf; the waxy, very fragrant flowers in short clusters, inverted upside-down, the narrowly lanceolate sepals and petals creamy-white, and curled; the heart-shaped pointed lip white and candy-striped red-purple, blooming from late winter through summer. A safe and easy plant for the amateur, willing to bloom in the window if kept moist enough, or even outdoors under a tree in summer; always attracts attention by means of its pervading perfume. **C**: 21   p. 468

ibaguense (known in horticulture as E. radicans) (Peru, Colombia, Guatemala, Mexico), the "Fiery reed orchid"; a very handsome and useful reed-type terrestrial with slender thin-wiry, rambling, leafy stems extending to 4 ft. and even 20 ft., with long white aerial roots developing from nodes; the widely spaced 3 to 5 in. leaves very fleshy; the extreme

end of the stem thin-wiry and leafless, terminating in a dense, almost round head of waxy orange-scarlet or cinna-bar red small 1 in. flowers, with yellow in fringed lip, in good light blooming throughout the year, mostly in spring and summer, and lasting a long time. Very popular because it grows so easily and blooms willingly on the windowsill, or on the sunny patio in summer, with adequate moisture at the roots. Larger plants need support such as stakes, on trellis or over globe, also against walls. Propagated by cutting the plants apart. **C**: 19   p. 467

mariae (So. Mexico); a very lovely little epiphyte, with small 2 in. pear-shaped pseudobulbs carrying 1 or 2 four-in. leaves, topped by one to five large and beautiful waxy, long-lasting flowers to 3 in. long, having greenish yellow sepals and petals, and a huge, wavy white lip lined green in the throat, blooming in spring or summer. **C**: 21   p. 467

medusae, also known as Nanodes medusae, (Ecuador); sinister looking epiphyte without pseudobulbs, with densely tufted, branching pendant stems 4 to 12 in. long, and crowded with two-ranked, glaucous green, fleshy leaves; the showy flat, fleshy 2 to 3 in. flowers from the axils of the terminal leaves, with narrow sepals and petals yellowish green tinged with brown, and a very large round, 2 in. lip deep maroon purple with green in center, deeply fringed all around; blooming during summer. Suited best to growing on treefern slab or in basket, and kept in a cool greenhouse with moist atmosphere—with its lurid purple flowers out-standing in collections. **C**: 83   p. 468

pentotis (C. America, Brazil); a compact-growing, stiff epiphyte with long spindle-shaped pseudobulbs bearing 2 leaves; the large waxy, very fragrant flowers to 3 in. long, inverted and carried back to back in pairs, on short stalk; their tapering, fleshy sepals and petals greenish-white, lip white with purple stripes; blooming March into summer. Of easy and safe cultivation in the house. **C**: 21   p. 467

polybulbon (West Indies, C. America); pretty, dwarf epiphyte not over 3 in. high, on creeping rhizomes with tiny pseudobulbs bearing two leaves; solitary $\frac{3}{4}$ in. long-lived and fragrant flowers brownish with yellow margins, the lip pure white; blooming winter and spring. In habitat, forming flat masses on trees and rocks, in season literally covered with bloom. **C**: 21   p. 468

prismatocarpum (Costa Rica, Panama), the strikingly handsome "Rainbow orchid"; robust epiphyte with flask-shaped pseudobulbs bearing 2 leaves; projecting from its apex an erect spike to 15 in. tall of beautiful, brightly colored, waxy, long-lasting flowers 2 in. across, sulphur-yellow blotched chocolate-maroon, the pointed lip marked rosy-red, and fragrant; blooming from May to autumn, lasting several weeks in perfection. One of the showiest of Epidendrums, suitable for cult. in home. **C**: 21 p. 455*, 467

radiatum (Mexico to Costa Rica, Venezuela); compact-growing epiphyte with short spindle-shaped pseudobulbs to 5 in. high, bearing 2–3 stiff leaves; very fragrant, heavy-textured fleshy, cream-white blossoms, having a broad white lip with radiating purple lines, carried upside-down in few to many-flowered, dense clusters, in spring and summer. Of easy culture and a dependable bloomer. **C**: 21   p. 467

stamfordianum (Mexico to Panama, Venezuela, Colombia); beautiful and distinctive epiphyte with spindle-shaped pseudobulbs 10 to 12 in. long, bearing 2 to 4 leaves, peculiar with the inflorescence appearing from the base of the bulbs; the fragrant 1 to $1\frac{1}{2}$ in., long-lived flowers in large arching or pendant clusters to 2 ft. long, sepals and petals yellow or greenish-yellow strikingly spotted with red, white lip 3-lobed, the fringed mid-lobe with yellow, together looking like a bird with wings spread. Winter-spring bloom-ing. **C**: 21   p. 467

x EPIPHRONITIS (Orchidaceae)

veitchii (Epid. ibaguense x Sophronitis grandiflora); an attractive small hybrid with reed-type stems and alternate leaves, topped by clusters of striking, fiery red 1 in. flowers, golden yellow in the center of the lip, blooming spring to fall. One of the earliest bigeneric crosses (1890), lower than Epid. ibaguense, with flowers intermediate between parents, and darker than Soph. grandiflora. **C**: 21   p. 468

EPIPHYLLOPSIS: see Rhipsalidopsis

EPIPHYLLUM (Cactaceae); the mostly day-blooming

"Orchid cacti", sometimes known as Phyllocactus, de-servedly take the spotlight for the distinctive, breath-taking beauty of their magnificent blooms, ranging from purest white to flashy shimmering shades of carmine, cerise and purple, often blending with creamy yellows and soft pinks; in size they vary from clusters of 3 in. blossoms to giant 10 in. flowers, primarily blooming in spring and summer. Epiphyllums are tropical near-epiphytes of vigorous habit with flattened, fresh green leaflike stems, rarely spiny; their parents originating in luxuriant forests, growing in rich humus on the ground or perching in the crotches of trees. They may be grown in hanging baskets, or on stakes or trellis in pots, in the window, wintergarden or glassed-in porch; in summertime the shaded patio or under a tree. Orchid cacti need rich but porous soil of loam, leafmold or peat moss, and sand, grit or perlite, with organic fertilizer added. They are rested and kept only moderately moist from November to February, cool at near 50° F.; when growth begins in March they are kept warm with liberal watering and bi-weekly feeding, in a light location that benefits from the morning sun; mist the stems as often as possible to raise the humidity; flowers develop on 1 and 2 year old branches. Propagated by rooting cuttings in moist peatmoss during spring and summer.

x hybridus, a typical "Orchid cactus"; there are more than a thousand named "Orchid cacti", all with large flowers of great beauty in iridescent colors from nearly blue through red, pink, yellowish to white. Recorded genera which were used in breeding were such epiphytes as Heliocereus, Nopalxochia, Nyctocereus, Hylocereus, Selenicereus. These plants will flower more willingly if given occasional plant food, liquid or dry. During summer and early autumn the plants should, if possible, be kept in the open under tall trees or shadehouse. **C**:.9   p. 169

hybrid 'Elegantissimum', "Dwarf orchid cactus"; superb, dwarf basket type of "Orchid cactus," large, perfect flower with broad petals bright crimson, the outer petals flushed scarlet, and green throat, on glossy green flattened branches. One of the many showy named orchid cacti. **C**: 9   p. 169

oxypetalum (latifrons) (Mexico to Brazil), another "Queen of the night"; probably the best night-blooming cactus for the home; a large epiphytic plant to 6 ft. high with branches usually feather-like flat and thin, waxy green and deeply crenate; large fragrant white flowers, reddish outside. **C**: 10   p. 169

EPIPHYLLUM: see also Nopalxochia, Schlumbergera

EPIPREMNOPSIS (*Araceae*)
media (Epipremnum medium) (Borneo); wiry tree climber with palmately 3-lobed parchment-like leaves deep green with faint mottling of yellow-green on puckered surface. Baileya (March 62) puts this under Epipremnum. **C**: 54   p. 81

EPIPREMNUM (*Araceae*)
pinnatum ("Monstera nechodomii") (Malaya through Java to New Guinea), "Taro vine"; prolific tropical climber with large, oblong leaves pinnately parted into regular segments, and tiny pinholes appearing as silvery dots along midrib. The juvenile leaves are entire, oblique-ovate. The Eastern hemisphere parallel to our Western Philodendron. Recently referred to Rhaphidophora. **C**: 60   p. 81

EPISCIA (*Gesneriaceae*); the "Peacock gesneriads" or "Flame violets" are low, fibrous-rooted tropical American herbs generally with a short stoutish central stem and producing stolons which creep along the surface, rooting at the joints, or cascade in rosettes of leaves which are often very beautiful, quilted, like gorgeous brocade, or velvety and painted with patterns of silver or metallic pink over shades of green or copper; the small tubular or bell-shaped axillary flowers with 5 flaring lobes in bright colors of scarlet, pink, purple, blue or yellow, also waxy white, blooming at any time of year; but best in spring and summer. The creeping Episcias are ideal basket plants, or planted as ground cover in the warm conservatory or wintergarden; they thrive luxuriously with high relative humidity of at least 40 to 75% at a warm 65° to 75° F. or more; even the water should be tepid. Episcias are most cold-sensitive and start suffering below 65°, defoliate and "freeze" at 50° F., going dormant; however will come back again when conditions are warm enough. They grow well in comfortably air-conditioned rooms. Easily propagated from cuttings or runners; each rosette roots readily; leaves may also be rooted but this method takes longer.

cupreata 'Acajou', an exquisite hybrid of cupreata x cup. 'Variegata'; 3 to 4 in. leaves dark mahogany, except for the contrasting network of veins, and the central area of bright metallic silver-green; orange-red flowers 1 in. long. Very excellent and a good keeper; more pubescent than 'Harlequin'. **C**: 10   p. 443

cupreata 'Chocolate Soldier', "Carpet plant"; a robust hybrid of cupreata x cup. 'Variegata' having leaves of a dark chocolate color with a silver-gray center band; orange-scarlet flowers. **C**: 10   p. 184, 443

cupreata 'Frosty'; lustrous, robust growing cultivar, with downy emerald-green leaves tinted copper at margins, the center area silvery white and splashing over into the lateral veins; flowers orange-red. **C**: 10   p. 43*, 443

cupreata 'Metallica', "Kitty episcia"; variety with large 3 to 4 in. quilted coppery leaves having a band of pale silvery green through the center, margined with metallic pink; orange-scarlet flowers $\frac{7}{8}$ in. across. Similar to the commercial cupreata cultivar 'Kitty.' **C**: 60   p. 443

cupreata 'Musaica' (Panama); seedling from the Canal Zone, with coppery foliage attractively traced with an even network of silver veins, as if the leafblade were laid out with mosaic. **C**: 60   p. 443

cupreata 'Silver Sheen'; a most attractive branch mutant of cupreata with hairy, wrinkled foliage, most of the leaf and outer veins a bright silver-gray leaving only the marginal areas a coppery green; flowers orange-red. **C**: 10
                                                                p. 443

cupreata 'Tropical Topaz'; a mutant form of E. cupreata once known in horticulture as 'Canal Zone Yellow', and widely cultivated there; this clone is distinguished by having bright yellow flowers, deeper inside, instead of scarlet blooms as in cupreata 'Viridifolia' of which this appears to be a sport; plain pale green leaves with erect hairs instead of being appressed. **C**: 60   p. 442

dianthiflora (C. Mexico), "Lace-flower vine"; charming miniature clustering rosette, prolifically sending out rooting branches; small oval, velvety dark green leaves 1 to 1$\frac{1}{2}$ in. long, with purple midrib and scalloped margins; free-blooming with glistening white, deeply fringed, feathery flowers 1$\frac{1}{4}$ in. long, appearing singly in the leaf axils, and blooming in summer. **C**: 60   p. 442

fulgida: see reptans

hirsuta (Trop. America), beautiful low-growing rosette, with large wrinkled leaves green or bronzy, with pale gray-green central area; tubular flowers 2 in. long, pale lilac. **C**: 10   p. 442

lilacina 'Lilacina' (Costa Rica), "Blue flowered teddy-bear"; a beautiful blue-flowered species also known as "Fannie Haage" and 'Variegata'; pubescent creeper with dark coppery, rough-puckered 2 to 4 in. leaves decorated with a prominent fishbone pattern of silvery green; large flowers of lavender-blue, 1 to 1$\frac{1}{2}$ in. long, good grower, but sensitive to cold. **C**: 10   p. 274, 443

lilacina 'Viridis' (fendleriana) (Costa Rica); attractive variety having soft-hairy bright emerald-green 3 to 4 in. leaves, faintly traced with glistening silver from the midrib; the foliage feels plushy; flowers light blue. **C**: 60   p. 442

lineata (Panama); a curiously tufted, fibrous-rooted plant, with long oblanceolate leaves plain green above, grayish beneath and with pronounced reddish veins on succulent stalks; the tubular flowers cream, with red-purple spots on the expanded limb, low between the foliage. Propagated from leaves. **C**: 60   p. 442

melittifolia (Dominica, Brazil), sturdy upright grower with stout, angled stem to 1 ft. high; large elliptic, glossy dark green 6 in. leaves with wrinkled surface and crenate margins, pale beneath; small axillary flowers toward top, with down-curving tube carmine-purple; blooming more or less continuously but best in spring. Propagated from cuttings; forms no runners. **C**: 10   p. 442

'Moss-Agate' (syn. 'Mosaic'), floriferous seedling with cupreata background from Ancon, Panama; freely branching with stolons; the corrugated and puckered leaves vivid to blackish-green, overlaid with a network of silver; large bright red flowers of greater than average size. **C**: 10 p. 442

punctata (Guatemala, Mexico); with thickish central stem to 6 in. high, developing numerous rank creeping, brown stolons, with leathery ovate, crenate leaves to 3 in. long, green except for red-purple midrib, on lightly erect branches; tubular flowers 1 to 1$\frac{1}{4}$ in. long, solitary with spreading fringed lobes creamy-white, and spotted purple into the throat. **C**: 10   p. 442

reptans (fulgida) (Brazil, Guyana, Surinam, Colombia), a "Flame violet"; tropical pubescent creeper with broad ovate, quilted, brown-green 2 to 5 in. leaves and bright silvery-green veins, margins crenate; flowers with long blood-red 1½ in. corolla tube having fringed edges, and pink inside. **C**: 10   p. 184, 443

reptans 'Lady Lou'; a variegated leaf sport 1952, very striking for its beautiful foliage; the heavily embossed, somewhat hairy, large leaves 3 to 5 in. long, basically bronzy-green with silver veining, entire sections irregularly splashed or colored in clear rose-pink, also occasional white or gray. This fancy coloring however is not constant, and unless care is taken that only the best variegation is allowed to remain, new growth may revert to the species. ¾ to 1 in. flowers are red. Thrives best in warmth with high humidity, in greenhouse or terrarium. **C**: 60   p. 443

tessellata: see Nautilocalyx

x wilsonii 'Pinkiscia' (cupreata x lilacina), lovely hybrid with 3 to 4 in. leaves of thick texture, metallic chocolate-bronze with a touch of silver along the green veins, and covered with pink hair; flowers with broad rose-pink limb, and orange-yellow dotted rose inside the 1¼ in. tube. The green-leaf type of this hybrid is known as E. x wilsonii 'Catherine'. **C**: 10   p. 442

EPITHELANTHA (Cactaceae)

micromeris (Texas, N. Mexico), "Button-cactus" or "Golf balls"; tiny globes usually ½ to ¾ in. dia. but may grow to over 2 in. high, depressed on top; small tubercles in spirals, almost entirely covered by flattened white spines; pinkish flowers; forms small groups; slow-growing, small but charming. **C**:.13   p. 159

ERANTHEMUM (Acanthaceae)

nervosum (Daedalacanthus) (India), the tropical "Blue sage"; charming winter-flowering plant of shrubby habit 2 to 4 ft. high, opposite rough, ovate, dark-green leaves with veins depressed, 4 to 8 in. long; the pretty ¾ in. flowers gentian-blue with purple eye, from under pointed bracts. Desirable for its fall and winter bloom; best grown from new cuttings annually. **C**: 54   p. 496

ERDISIA (Cactaceae)

maxima (Cereus) (Andes of Central Peru); bushy plant about 3 ft. high with slender angular branches fresh green with 5 ribs bearing depressed areolas well furnished with white radials and 1 cream central spine; small but many-petalled pretty red flowers, gold outside, and red stamens. **C**: 1   p.   154

EREMURUS (Liliaceae)

elwesii (robustus) (Turkestan), "Desert candle" or "Foxtail lily"; stately perennial from the steppes of Central Asia, with narrow basal leaves in rosette, the striking floral stalk 6 ft. or more high, bearing a 2½ ft. spire dense with numerous stalked flowers having spreading segments 1 in. across, pink with deeper center stripe; blooming in late spring. Somewhat hardy with protection. After blooming the foliage fades away. If in containers, copious water is required during growth, dry side when resting. Propagated by division of rhizomes. **C**: 75   p. 400

ERIA (Orchidaceae)

ferruginea (Himalayas); handsome small epiphyte with creeping rhizome, and cylindric pseudobulbs topped by 2 to 4 leathery leaves, the small 1½ in. flowers loosely in a lateral arching cluster, hairy outside, sepals olive-brown with darker stripes, the petals white flushed pink, and a 3-lobed lip white with pale purple throat, blooming March to May. Erias require liberal quantities of water while in active growth, but on completion of the new growth a rest should be allowed. Propagated by division of rhizomes. **C**: 71 p. 468

ERICA (Ericaceae); the true heaths are densely shrubby

evergreens with tiny, needle-like leaves and masses of little urn-shaped or tubular flowers in tall plumes that differ from Calluna, the true heathers, in having a short calyx while Callunas have a colored calyx longer than the corolla and concealing it. Many ericas are not winter-hardy. After blooming the leading branches are headed back slightly and plunged outside during summer; new buds must be developed by fall and the plants are kept as cool and light as possible into winter. Rain water is preferable for watering.

Propagation in winter or summer from little side shoots at 65° F. in peat-sand under a bell jar or plastic.

gracilis (So. Africa), "Rose heath"; bushy little shrub to 1 ft. high with small light green needle-like leaves in 4's, the side shoots loaded with terminal clusters of tiny globose, rosy flowers from autumn on into winter. A long-lasting decorator for late-season window boxes or cemetery graves; in Europe they are often sprayed with red dye which makes them look alive and in flower the year round even if dried up. **C**: 69   p. 382, 502

melanthera (canaliculata) (So. Africa), the "Christmas heather"; an attractive and shapely shrubby heath seldom over 2 ft. high, with downy shoots and tiny linear needles in threes; snowy with small globular rosy flowers displaying prominent black anthers, blooming in winter and spring, literally covering the feathery plant. A charming Christmas plant. **C**: 69   p. 382

'Wilmorei', "French heather"; a grand old, well-known large-flowered hybrid, close to perspicua, known since 1835; strong growing shapely erect bush with showy spikes, having stiff linear leaves in threes, and very large tubular inflated rosy flowers 1 in. long, prettily tipped with white; blooming winter to spring; very showy and with long-lasting flowers. **C**: 69   p. 502

ERIOBOTRYA (Rosaceae)

japonica (China), "Chinese loquat" or "Japan plum"; symmetrical evergreen tree to 20 ft. high, with noble decorative, thick obovate foliage, 6-12 in. long, glossy-green on the surface, strongly ribbed, and toothed, the underside and new branches rusty-woolly; fragrant white flowers; and pear-shaped yellow to orange edible fruit 1 to 3 in. long, of pleasant, sprightly flavor. Plant in well-drained soil but must be kept moist to hold its crisp, heavy leaves; if in tubs, best for cool locations. **C**: 14 or 64
                                                                      p. 292, 508

ERVATAMIA (Apocynaceae)

coronaria, better known as Tabernaemontana (India), the "Crape jasmine", "Clavel de la India", or "East Indian rose-bay"; a handsome shrub 5 to 8 ft. high with glossy green, elliptic, thinnish leaves 3 to 5 in. long, and waxy white, undulate flowers 1½ to 2 in. wide, the lobes prettily crisped, very fragrant at night; blooming in summer. Widely cultivated in tropical gardens. Propagated by semi-ripened cuttings, with bottom heat, under glass or plastic cover. **C**: 52   p. 486

ERYTHEA (Palmae)

edulis (Mexico: Guadalupe Is., Baja California), "Guadalupe palm"; robust, squat-growing fan palm, becoming about 30–40 ft. high; trunk to 15 in. thick and ringed with scars; crown of few palmate, rigid leaves 7 ft. across, green on both sides, the numerous segments deeply cut and cleft or torn at apex. The black fruit is eaten in Mexico. **C**: 1   p. 229

ERYTHRINA (Leguminosae)

crista-galli (Brazil, Argentina), "Coral tree" or "Cry-baby tree"; handsome free-flowering small tree or shrub to 15 ft. high with thorny stems eventually rough-barked, leathery leaves having 3 leaflets, on spiny petioles; and showy butterfly or bird-like waxy flowers deep scarlet-red, the flaring flag 2 in. long in dense spike-like clusters usually produced before the leaves. Very striking during their main blooming period in summer into fall, profusely flowering provided they are completely rested, and kept cool from October to April. The old branches should be cut back severely in late fall, or at least the old flower stems pruned out. Ideal for the sunny patio or roof garden. Propagated best in spring from young shoots with a piece of bark of the main stem. **C**: 51   p. 490

ERYTHRODES (Orchidaceae)

nobilis (Physurus) (Brazil), "Silver orchid"; terrestrial Jewel orchid with creeping stem, grown for its beautiful foliage; the clasping ovate leaves about 1 to 2 in. long, light grayish-green or moss-green with a network of silver veins, becoming more dense over the grayish center area in the variety argyrocentrum; shimmering gray underneath; erect spike with ½ in. waxy white flowers yellow at throat and marked purple; February blooming. Culture like Anoectochilus. **C**: 62   p. 454

ERYTHRORHIPSALIS (Cactaceae)

pilocarpa (Brazil: Rio, Sao Paulo), "Bristle-tufted twig-cactus"; a pretty epiphytic cactus, at first erect, later pendant, with its thin cylindrical, grooved, purplish-green stems mostly branching in whorls, and closely set with areoles bearing whitish hair-like bristles; flowers white, fruit wine-red. **C**: 10   p. 172

ESCOBARIA (Cactaceae)

shaferi (Mexico); small globe with spiralling tubercles hidden under bristle-like spines; flowers similar to Mammillaria, bronzy-pink. **C**: 13   p. 163

ESCONTRIA (Cactaceae)

chiotilla (So. Mexico), "Little star big-spine"; slender columns branching tree-like, sparry-branched to 20 ft., seedlings stocky, bright green, 7–8 sharp ribs, long light colored spines; yellow flowers; edible purple fruit. **C**: 13   p. 153

ESPOSTOA (Cactaceae)

lanata (Peru, Ecuador), "Peruvian old man"; small column, 20–25 ribs, beautifully covered with cottony, snow-white hair; flowers pinkish. Becoming 3 ft. high with age. **C**: 1   p. 154

EUCALYPTUS (Myrtaceae); the Gum-trees are evergreen, mostly subtropical great trees belonging to the Myrtle-family, mainly Australian, with aromatic foliage, planted world-wide in semi-arid areas for reforestation, windbreaks, timber and oil, but equally for their decorative character. Most Eucalyptus have two kinds of foliage: soft, variously shaped juvenile leaves found on seedlings, saplings, and new branches that grow from stumps; and the tougher adult (usually elongate leaves). The trees are very drought-tolerant, and require sun, are free from insect pests, and fast-growing, some 10–15 ft. a year in early stages. Very voracious and greedy, they are difficult to maintain in containers for long. Easy from seed. The fruits and leafy aromatic branches are often dried, sometimes colored and used as winter decorations.

cinerea (cephalophora) (N.S. Wales, Victoria), "Spiral silver-dollar" of florists; small glaucous tree 20 to 50 ft. high, with brown-red willowy branches bearing opposite pairs of sessile leaves rigid, stiff-leathery and silvery glaucous, rounded like a silver coin, 1 to 3 in. across in the juvenile stage; later when 6 ft. or more high the leaves become ovate to lanceolate, with yellow midrib; the juvenile branches are used by florists for decorative arrangements. **C**: 14   p. 301

globulus (Tasmania, Victoria), "Blue gum"; a pretty plant when young, with shoots and clasping oval leaves a glaucous blue, but rapidly growing into a gigantic tree to 300 ft. high and shedding bark; with advancing age the leaves become elongate, dark green and sickle-shaped 6 to 10 in. long; flowers creamy-white in winter-spring; the glaucous 1 in. fruit capsules are quite attractive on cut branches. Much used for windbreaks in California, but very greedy and messy with constantly dropping leaves. Its foliage contains aromatic oil used for medicinal purposes. **C**: 14   p. 301

polyanthemos (Victoria, New South Wales), "Red box gum" or "Silver-dollar tree"; tree 20 to 60 ft. and more high with persistent bark, the slender, sparry, glaucous branches with distantly spaced attractive leathery leaves almost round, 2–3 in. across, glaucous bluish gray and edged all around with purple. Mature leaves lance-shaped. Excellent cut foliage. **C**: 14   p. 301

EUCHARIS (Amaryllidaceae)

grandiflora (amazonica) (Colombia), the exquisite "Amazon lily", also known as "Eucharist lily"; first collected along the Rio Magdalena, a low bulbous plant with broad, glossy green basal 6–10 in. leaves narrowed into petioles; delightfully fragrant, glistening white flowers starlike, with a central corona and 6 spreading segments 2½ to 3 in. or more across, in clusters on fleshy stalk. A tropical plant enjoying moist-warm conditions, this "Star-of-Bethlehem" blooms normally in late winter and spring, but may be flowered several times a year by alternate resting and growing periods. By drying off to some extent, for a few weeks, a crop of flowers may be had at almost any season, providing the bulbs are strong enough and the foliage is not completely lost. Rested in March or August, they should bloom in May or autumn. Propagation by bulblets. **C**: 54   p. 390

x EUCODONOPSIS (Gesneriaceae)

'Kuan Yin'; intergeneric hybrid between Achimenes and Smithiantha; in habit somewhat resembling Kohleria, with herbaceous stems, and ovate dark green smallish leaves with scalloped margins; large Achimenes-like flowers with oblique limb, creamy-yellow streaked with purple, blooming in late summer and fall. Propagated from leaf or stem-cuttings, or the scaly rhizomes. **C**: 60   p. 444

EUCOMIS (Liliaceae)

zambesiaca (E. Tropical Africa), the "Pineapple lily"; exotic-looking bulbous plant with wrinkled conical bulb, topped by a basal rosette of unusual, broad lance-shaped, firm channeled leaves to 2 ft. long, bright shining green, gracefully spreading, and spotted purple beneath; the inflorescence a stout stalk bearing a dense cylindrical truss of sweet-scented, waxy-looking starry ½ in. flowers greenish with violet ovary, from July to September, and topped by a tuft of red-edged bracts. A remarkable pot plant resembling a bromeliad, long in bloom. Bulbs are planted in fall or February in fresh soil. Propagated by offsets. **C**: 1   p. 400

EUGENIA (Myrtaceae)

myriophylla (So. Cent. Brazil), "Needle eugenia"; slow growing, densely branched, compact evergreen bush, with thin brown, drooping branches, soft pubescent when young; closely set with narrow, needle-like leathery foliage 1 to 2 in. long and rich green; flowers white. Most unusual and charming small decorative plant; requires warmth and moisture. Cuttings moist-warm and under cover. **C**: 52 p. 299

myrtifolia (Syzygium paniculatum) (Australia), "Australian brush-cherry"; vigorous evergreen, compact shrub to 20 ft. high, which lends itself to shearing into pyramids; slender branches dense with small elliptic foliage to 2½ in. long, vivid red when young, later shining green; fluffy white flowers followed by edible red berries. Very ornamental trained into pyramid or other topiary forms but easily drops foliage unless it has sunlight, fresh air, and does not dry out too much. From cuttings. **C**: 64   p. 288, 292, 301, 537

myrtifolia globulus (Syzygium), the "Myrtle-leaf eugenia"; very compact evergreen bush of globular shape and dense habit, with shiny, stiff leathery vivid amber to deep green 1 in. leaves, densely clothing the numerous branchlets. Attractive indoors, and for hedging. **C**: 14 p. 298

uniflora (Brazil, Guyana), "Surinam cherry" or "Pitanga"; evergreen compact shrub or tree 6 to 20 ft. high, with glossy, coppery green ovate, 2 in. leaves, and fragrant white flowers bursting with stamens like little brushes; producing distinctive round fleshy deep crimson fruit with 8 deep grooves, to 1 in. across; edible and of spicy flavor. Attractive with its numerous red cherries as a large pot plant. From cuttings. **C**: 52   p. 508

EULOPHIDIUM (Orchidaceae)

mackenii (So. Africa: Natal); small evergreen terrestrial orchid with short pseudobulbs and solitary leathery, oblanceolate dark-spotted leaf, and basal erect cluster of light brown flowers with creamy lip veined purple. Propagated by division. **C**: 69   p. 468

EULYCHNIA (Cactaceae)

floresii (North Chile), "White fluff-post"; very pretty column, olive green, with many close ribs, attractively and regularly dressed with large white-woolly tufts from the areoles, long brown-tipped central spines. According to Backeberg, this species is close to E. ritteri, freely branching, to 3 in. dia. and 9 ft. high; small pinkish flowers. **C**: 1 p. 153

EUONYMUS (Celastraceae); bushy shrubs with willowy branches, and stem-rooting creepers; many evergreen, sub-tropical or somewhat hardy, and which are cultivated for their often variegated, tough-leathery dense foliage, as decorators for cooler locations; suitable for entrance halls, wintergarden or the patio; in small sizes also for dishgarden planting or the sunny window. They are not particular as to soil, but need fresh air; if exposed to close-damp environment are susceptible to mildew. Propagated from cuttings; the dwarf and creeping forms from division.

fortunei radicans gracilis (W. China), variegated "Creeping euonymus"; a pretty evergreen, trailing or climbing by rootlets, with small leathery leaves grayish green and variously variegated with white or yellow, and tinted pink.

Very attractive as a ground or wall cover, also for terrariums; winter-hardy. Prop. by division or cuttings. **C**: 26   p. 299

japonicus (Southern Japan), "Japanese spindle tree"; dense evergreen shrub to 15 ft. high, with erect willowy branches, closely set with opposite small leathery oval leaves glossy dark green and 1–2½ in. long, obscurely toothed; small greenish-white flowers. **C**: 28   p. 292

japonicus albo-marginatus, "Silver-leaf euonymus"; attractive form with somewhat smaller, narrower leaves silver gray on milky green, bordered with a narrow white edge; very bushy. **C**: 14   p. 299

japonicus argenteo-variegatus (Japan), in horticulture as "Silver Queen"; leaves broadly oval, glossy fresh green boldly variegated and edged with white; a pretty dish-garden plant. **C**: 16   p. 299

japonicus aureo-variegatus, in horticulture as "Yellow Queen"; a colorful evergreen of bushy habit with waxy-green, oval leaves richly variegated and margined yellow toward edge. Particularly attractive as a small plant, and a favorite for dishgarden planting; fairly resistant to mildew. **C**: 16   p. 299

japonicus medio-pictus, "Goldspot euonymus"; waxy oval leaves fresh green at margin, golden yellow in center and down the petiole and stem. Mildew-prone. **C**: 14   p. 299

japonicus microphyllus variegatus (pulchellus var.) (Japan), "Variegated box-leaf"; charming little, stiff erect very bushy evergreen 4 in. high or more, with a dense mass of tiny oblong, toothed leaves ½ to 1 in. long, glossy deep green and circled by a white border, closely arranged on the erect little branches. Very attractive dwarf pot or dishgarden plant, also for low hedging. From division or cuttings. **C**: 14   p. 299

EUPHORBIA (*Euphorbiaceae*); the "Spurges" are a large and varied group of subtropical herbs, shrubs and leafless trees, many of them succulents, mainly of the Eastern hemisphere, where these resemble many members of the Western hemisphere Cactus family, a remarkable example of converging or parallel evolution. The tree-like Euphorbia in their bizarre candelabra silhouettes are a characteristic feature of the African landscape, the same as many column Cereus are found in arid American regions, but differ by having a milk-like juice, often poisonous, and flowers not showy as in cacti. Their inflorescence is curious in having flowers reduced to a small, inconspicuous cyathium which consists of many male flowers or stamen surrounding one central female flower, subtended to which are the petal-like, usually showy bracts as in Poinsettia. What appear to be spines in some species are the converted, persistent leaf bases of their more or less deciduous foliage. The succulent Euphorbias are fantastic and excellent, undemanding, slow-growing house and decorative plants, needing little care in watering and are generally fond of warmer night temperatures unlike cacti of similar habit. Propagation mostly by cuttings during the warm season; as a precaution care should be taken to wash off the milky latex of the wound in lukewarm water, or dip the base in charcoal dust; and allow the cut to dry or heal a while before inserting shallowly into porous rooting material such as gritty sand or peat-perlite mixes, on warm bottom.

abyssinica (Ethiopia, Eritrea), "Imperial Ethiopian euphorbia"; majestic plant, becoming a large tree to 30 ft. high; young succulent stems 5 to 8-angled, olive green marked milky and dark green, the undulate, sharp notched ridges densely set with corky knobs, and downward pointed pairs of light brown spines. Slow-growing, and a fantastic show plant out of reach of its multitude of spines. **C**: 3 p. 340

canariensis (Canary Islands), "Hercules club" or "African cereus"; large cactus-like succulent densely branching from the base, to 35 ft. high; the stems 5, 4 or 6-angled and 2 to 3 in. thick, brownish fresh green, the sinuate angles set with black spine-pairs; small yellow flowers; poisonous. In appearance an old world parallel to the American column cereus of the cactus family, especially when seen growing in clumps in the desolate lava beds of the volcano Teide on Tenerife. **C**: 1   p. 344

candelabrum (Trop. Africa: Sudan, Uganda); a "Candelabra tree"; massive candelabra, tree-like to 30 ft. high, at home in the Nile basin, with 4-angled yellowish green branches about 3 in. thick and constricted to joints, the angles wavy-toothed and set with spines; commonly used as living fences around kraals in Central Africa resembling

the cactus fences of many villages in Mexico. This species is claimed by some authorities to be identical with Euph. ingens of southern Africa. **C**: 1   p. 531

caput-medusae (So. Africa: Cape Prov.), "Medusa's head"; short thick or globose stem to 8 in. dia., bearing numerous wide-spreading serpentine, gray-green branches 1 to 2 in. thick and to 2 ft. long, densely knobbed and with tiny deciduous leaves at the growing tips. A favorite in collections. From seed, as cuttings from lateral branches grow one-sided. **C**: 13   p. 342

coerulescens (So. Africa: Cape Prov.), "Blue euphorbia"; thorny shrub, branching underground and forming dense clusters of upright bluish green columns 2 in. thick, 4 to 6 angled, to 5 ft. high, constricted to joints and with hollow sides, the margins toothed and horny. Wonderful exhibition plant because of shape and blue color. **C**: 3 p. 340

x'Flamingo', "Pink Florida Christ-thorn"; a robust, woody shrub with stout, angled branches densely set with thorns, long obovate leaves and long-stalked clusters of flesh-pink bracted flowers. Probably only a coral pink colorform of E. x keysii, an almost constant bloomer, especially all winter and spring; a great improvement over the regular Crown of thorns. **C**: 2   p. 344

flanaganii (South Africa: Cape Prov.), "Green crown"; low, spineless dwarf succulent, with short, thick stem only 2 in. high, bearing a crown of 3 or 4 rows of slender rich green branches around the tubercled head; branches semi-erect, older ones spreading wide, tiny linear leaves deciduous. A pleasure for the collector. **C**: 13   p. 342

flanaganii cristata, "Crested green crown"; a cushion-like fasciation dense with multitudes of rich green little crests fanning out from the short conical but invisible stem. A collector's plant photographed in California. **C**: 13   p. 343

fulgens (jacquinaeflora) (Mexico), "Scarlet plume"; beautiful leafy shrub to 4 ft. high, with slender thin-wiry branches gracefully arching, the small 3 to 5 in. leaves narrow lanceolate, and the flowers axillary in terminal sprays of small petal-like brilliant orange-scarlet bracts about ½ in. across, blooming in cultivation normally end of December to January. Charming when several young plants are planted together in a shallow pan. Responds to short-day treatment in under 12 hours; if placed under a carton or into a closet from about 5 p.m. to 8 a.m. for six weeks beginning 20 September, buds will be initiated to flower for Christmas. Cut back after blooming and rest to March; propagated from soft or hard cuttings. **C**: 1   p. 500

x'Giant Christ-thorn', "Giant California Christ-thorn"; a phenomenal flowering plant; the result of 25 years of hybridizing by Ed Hummel of Carlsbad, California, involving E. splendens bojeri, breonii and lophogona; a swollen, stout grayish stem to 2¼ in. thick, with long brown thorns; toward the apex with bluish-green leathery leaves 8 to 10 in. long and, arranged as in a giant bouquet, strong stalks bearing the clustered inflorescence with firm, large bracts 1 to 2 in. across, larger than a silver dollar, in glowing cerise pink with salmon sheen. Slow-growing. A superb container plant for the sunny patio or sunporch. **C**: 1   p. 344

grandicornis (Natal to Kenya), "Cow horn euphorbia"; thorny branching succulent to 6 ft., of interesting shape, the green to gray-green branches 3-angled to 6 in. thick, irregularly constricted, the angles winglike with horny edge, set with pairs of stout light brown spines. Most unusual as solitary container plant, out of reach of its vicious 2 in. needle spines. **C**: 3   p. 340

grandicornis cristata, "Crested cowhorn"; formidable, crestate form growing into a contorted body of angles and long spines in all directions; very slow to grow, but a collector's conversation piece; quite durable and easy to maintain if warm and not too moist. **C**: 3   p. 343

grandidens (So. Africa), "Big tooth euphorbia"; spiny succulent tree to 40 ft. high, with 6-angled trunk and 3–4-angled branches in tiers, often spirally twisted, fresh green with silvery markings, the ridges prominently toothed and set with brown spines. Free-growing. **C**: 13   p. 341

heptagona (So. Africa: Cape Prov.); thorny, erect succulent shrub to 3ft. high with main stem 7 to 10-angled and 1¾ in. thick, the sparry, angled branches grass-green turning silvery when old, with tiny linear leaves and light brown needle spines. **C**: 13   p. 342

hermentiana: see trigona

ingens (So. Africa: Natal, Transvaal, Mozambique, Rhodesia, Zambia), the "Candelabra tree"; a giant spiny, succulent tree to 30 ft. high, the trunk branched and rebranched into dense crown; stems to 5-angled; the leafless branches succulent, constricted into joints, dark green, angled into 4 to 5 wavy wings. In habitat in the bush-veld of northern Transvaal. I noted that growing seedlings in their juvenile stage up to 5 ft. have stems beautifully marked with silver. **C**: 1   p. 340, 341

ingens monstrosa, popularly known as the "Totem pole"; interesting slender column, later branching, with contorted stems dark green with pale markings; the angles are dissolved and scrambled, the prominent teeth point in all directions and are set with brown spines. Slow growing. **C**: 3   p. 343

x keysii (lophogona x splendens), "Dark Florida Christ-thorn"; a spiny xerophytic shrub with woody, rambling branches resembling splendens in habit but with stems much more stout, and larger, fleshy obovate leaves 6 in. long that turn coppery-red in the sun; and stalked clusters of flowers with large showy rosy-carmine bracts, or red in saline soil. **C**: 2   p. 344

lactea (India, Ceylon), "Candelabra plant", "Candelabra cactus" or "Hatrack cactus", known in Viet-Nam as "Dragon-bones"; cactus-like plant slowly growing into form resembling a candelabra, branching and treelike to 20 ft. high; the sparry branches 3–4 angled $1\frac{1}{8}$–2 in. dia., distantly but deeply scalloped and black-spined, dark green with greenish-white marbling down the center. Curious but easy succulent requiring warmth. Used in India medicinally, as a hot jam, for rheumatism. An excellent solitair plant for the house. Cuttings must lay several days to heal wound before inserting in sand. **C**: 3   p. 328, 341

lactea cristata, "Elkhorn" or "Frilled fan"; an intricately monstrose crested Euphorbia with fan-shaped branches forming a snaky ridge or a crowded cluster; friendly green marked and variegated with silver-gray. The best form for growing as a house plant. With little care will slowly increase in size; regular branches are discouraged. Propagate from cut heads, but wound must be allowed to heal before setting lightly in dryish sand with perlite or charcoal to root. **C**: 3   p. 343, 532

lophogona (Madagascar). "Pink-flowering euphorbia"; branching shrub to 2 ft. high with clubshaped 4 to 5-angled branches, the ridges reddish and fuzzy-hairy; a species with pronounced oblong leaves, fresh green with pale veins, 4 to 8 in. long; stalked inflorescence with rounded bracts a lovely light pink. Easy from seed. **C**: 4   p. 342

mammillaris (So. Africa: Cape Prov.), popularly known as the "Corncob cactus"; small succulent to about 8 in. high, with a reduced mainstem which branches in a clustered manner; erect olive green cylindrical 8–17-angled columns to $1\frac{3}{4}$ in. thick, the angles tubercled, looking like a corncob, occasionally set with $\frac{1}{2}$ in. buff needle spines. In cultivation indoors the mainstem elongates and develops outstretched arms of club-like lateral branches. **C**: 13   p. 329, 341

mammillaris 'Variegata', "Indian corn cob"; a beautiful California cultivar of the Corncob cactus, the notched columns largely greenish-white marked fresh-green and tinted pink, with occasional buff spines. A most attractive variegated succulent for individual pots or dishgardens. **C**: 3

x'Manda's Cowhorn' (pseudocactus x grandicornis); a fasciated New Jersey hybrid with deformed growths in a series of ascending fan-like crests, the largest and upper-most with long extended spine-tipped, horn-like tubercles, bright green with brown edging. A curiosity like a landscape on the moon. **C**: 3   p. 343

marginata (in horticulture as E. variegata), (Prairies of Minnesota, Dakota to Texas), "Snow on the mountain" or "Ghost-weed"; hardy pubescent annual to 2 ft. high; with numerous branches, having glaucous gray-green, ovate to oval, soft-fleshy leaves 1–3 in. long containing milky sap, the upper, terminal ones white-margined; small white flowers in clusters subtended by showy petal-like variegated bract-leaves having attractive white border, from July to October. Popular in the flowergarden; long-lasting when cut, but stems should be scorched. From seed. **C**: 63   p. 414

mauritanica cristata, "African fan"; interesting crested form of the normally much branched spineless shrub with pencil branches native in So. Africa. In this fasciated shape the cylindric stems are transformed into spread-out, thin flattened fans, dull grayish green, with tiny leaves attached along apex, and an occasional normal, pencil-like stem. **C**: 13   p. 343

milii: see splendens

neriifolia cristata (India: Deccan), "Crested oleander cactus"; monstrose form of the normally cylindric, leafy stems of the species; in its fasciated shape, this deformed succulent is growing into repeatedly forking fan-like crests, and the ribs are dissolved. Strangely attractive in light green, later gray, with short black spines and obovate, leathery-fleshy deciduous leaves to 5 in. Slow-growing. **C**: 3   p. 343

neriifolia cristata variegata, "Variegated oleander cactus"; weird fasciation with stems crested fan-like; the contorted body vivid grass-green with milky-white variega-tion; fleshy light green obovate leaves and short brown spines. **C**: 3   p. 343

obesa (So. Africa), "Turkish temple" or "Baseball plant"; handsome small spineless globe to 5 in. or more high, in male plants somewhat pear-shaped, female plant more globular; the body gray-green marked with reddish length and cross stripes, and rows of small knobs along ridges that resemble stitching of a baseball. A precious, slow-growing succulent more often under 3 in. high; requires good drain-age and light. Keep dryish in winter. By seed only. Female mother plants must be pollinated from male plants; cover seed capsules with net to prevent premature loss of seed. **C**: 3   p. 342

polyacantha (Ethiopia, Eritrea), "Fish bone cactus"; sparry branching succulent plant to 5 ft. high, with 4 to 5-angled leafless stems to $1\frac{1}{2}$ in. dia. grayish green; the horny angles crenately toothed, and wide-spreading black-tipped spines. **C**: 1   p. 341

pulcherrima (So. Mexico: Guerrero, Oaxaca, Chiapas), the "Poinsettia", "Christmas star" or "Flor de Noche Buena" of the Christmas season; a branching shrub to 10 ft. high, with woody trunk and milky juice; ovate leaves deciduous when disturbed or resting; the terminal shoots forming the spectacular scarlet bract leaves which surround the tiny yellowish actual flowers in the center of the star. Poinsettias need warmth and maximum light to develop their showy bracts; the soil should become dry before watering again. To bloom kept-over plants for another year, they are cut back and best set outdoors or in the sunny window with beginning of the warm season and kept pinched to shape until early August. Poinsettias are "short-day" plants and buds must be initiated starting at latitude 40 to 45° N. (New York) about September 21 to 25 provided the night temperature is below 65° F. For Christmas timing the day-length must be limited to less than 12 hours from October 1 until bracts form, and as little as $1\frac{1}{2}$ foot candles of light during this time will inhibit budset; cover with opaque black cloth, a box, a heavy paper bag, or set into a closet for the night. Poinsettias may be brought into bloom at other times of year from cuttings produced under light, then subjected to short-day treatment; flowering will occur 10 to 12 weeks later. Propagated from softwood or hardwood cuttings. (Photo p. 500 shows cv. 'Oakleaf'.) **C**: 51
                                                                    p. 365*, 500, 535

pulcherrima 'Barbara Ecke Supreme' (Ruth Ecke sport, 1945), a tetraploid of superior quality, with heavy stems freely branching; good fleshy, rich green lobed leaves, and broad ovate flexible bracts of bright cardinal red, closely surrounding the yellow-green central flowers in perfect disks; measured from tip to tip, an inflorescence may spread its bracts 10 to 15 or even 22 inches. (56 Chromosomes) **C**: 51   p. 365*

pulcherrima 'Eckespoint' C1 Red; a new strain of California poinsettia hybrids by Paul Ecke; diploids developed under a series of numbers, of which C1 has been judged excellent for commercial blooming in pots for Christmas, and also timed for summer-blooming in bedding plant use by lighting, followed by short day treatment. Of vigorous, stocky growth and medium height, the foliage exudes vigor, and the somewhat crinkled bracts from a perfect circle 12 to 14 in. across, their coloring is a vivid scarlet red with rosy sheen. The sturdy stems require no staking; the bracts are long-lasting and the central flowers just right without dropping out for Dec. 25. For their first commercial test in 1968 growers report enthusiastic customer satisfaction because of their beautiful bracts and bright color though not as tough for leaf-holding quality as 'Paul Mikkelsen' if dried out unreasonably. During active growth, responds best at warmer temperature 68° to 72° F., with steady moisture and extra feeding; perfect branching plants may be had for Christmas as a result of soft pinching in July. **C**: 51   p. 383

pulcherrima 'Ecke's White' (E. pulcherrima alba cultivar), a "White poinsettia"; shapely, freely branching and late flowering seedling with bright green, ovate leaves, the flexible, wiry stems producing perfect close disks of creamy-white, membranous bracts; long-lasting, and if kept cool, will keep both leaves and bracts from Christmas to Easter. **C**: 51   p. 383

pulcherrima 'Paul Mikkelsen' (U.S. Pat.), a poinsettia seedling of 'Barbara Ecke Supreme' and 'Ecke's White'; a happy break-through in potgrown poinsettias for Christmas; compact plant with strong root system and sturdy close-jointed, wiry stems leafy with smallish, durable foliage and topped by an inflorescence with a circle of sturdy bright crimson bracts spreading 8 in. and up to 12 in. across from tip to tip. The small greenish flowers in the center of the star unfortunately begin to fall out somewhat prematurely unless lighted from end of September to October 5. Growing plants develop best if kept warmer than older varieties; they require more feeding but stand up stiffly without staking. This poinsettia, in increasing commercial production since 1964, holds up exceedingly well under adverse, even chilly conditions, and also in shipping, with its bracts and importantly, the foliage retained on the stems even into Easter and longer. **C**: 51   p. 383

pulcherrima plenissima 'Henrietta Ecke', a "Double poinsettia"; sported in 1927 in Hollywood, blooming early December, with two kinds of bracts, the broad, lanceolate horizontal bracts of glowing vermilion, crowned by a cluster of smaller bracts of various shapes and sizes, giving the inflorescence their characteristic "double" effect. This double form develops to perfection only in areas with intense autumn light such as in southern California, Florida or in tropical latitudes. **C**: 51   p. 383

pseudocactus (So. Africa: Natal), "Yellow euphorbia"; thorny, succulent shrub with sparry branches from near the ground, to 2 in. thick, forming clumps to 3 ft. high; usually 4- or 5-angled, irregularly constricted into segments, bright green with yellowish fan-shaped markings, the angles coarsely toothed and with horny edge; small yellow flowers. **C**: 13   p. 342

pulvinata (So. Africa), "Cushion euphorbia"; low, branching succulent with many angled stems only 4 to 8 in. high, forming convex clusters moundlike; mostly 7-ribbed, fresh green and lightly notched; small 1 in. leaves and numerous persistent wine-red, spine-like flower stalks. **C**: 13   p. 342

ramipressa, in horticulture erroneously as alcicornis (Madagascar); leafless succulent shrub to 10 ft. high; fleshy stems to 3 in. thick, obscurely 5-angled, lateral shoots forming narrow green branches, cylindrical at base, then 3-angled, and flat and straplike 2-angled toward the apex; sinuately scalloped at the margins, and with brown spines. **C**: 1   p. 341

splendens (milii) (W. Madagascar), "Crown of thorns"; xerophytic spiny shrub with slender scandent woody stems to 6 ft. long, the spreading branches about ½ in. dia., grooved and armed with spines; obovate 1½ in. leaves dull green, deciduous and soon falling if disturbed or too dry; flower bracts soft salmon-red with pale center. May be trained against trellis or wire frame; very cheerful when in bloom at Easter time; good house plant for warm location. From stem cuttings. **C**: 3   p. 328, 344, 500

splendens bojeri (W. Madagascar), "Dwarf crown-of-thorns"; bushy gray-spined plant of compact habit with upright gray main stem, freely branching, the stubby branches with slender thorns and small persistent dark green leaves about 1 in. long, pale gray-green beneath, and with terminal clusters of flowers with pretty, rounded cardinal-red bracts. Much cultivated as a "Christ-thorn" for Easter; also planted in dishgardens as it holds foliage better than splendens. **C**: 3   p. 344

stenoclada (Madagascar), "Silver thicket"; beautiful leafless but forbidding woody shrub 6 ft. or more high, densely bushy, and with flattened silvery gray limbs forking into vicious long-pointed spikes and thorns. Very slow and requiring good light; a sight to see it glistening silvery under the African sun. A collectors item for pots. **C**: 1   p. 342

tirucalli (Tropical Eastern Africa: Congo, Uganda, Tanzania), the ubiquitous "Milk bush" or "Pencil cactus" of the African steppe; succulent shrub growing into dense tree to 30 ft. high; the branches slender cylindrical and pencil-thick, glossy green and bursting with poisonous milk; narrow deciduous leaves. As a young plant with willowy stem and branches an excellent decorative plant tolerant of much neglect. **C**: 3   p. 341

trigona, in the horticultural trade as E. hermentiana (Trop. West Africa: Gabon), "African milk tree"; attractive succulent, growing in candelabra-form, the branches usually 3 to 4-angled, winglike, 1 to 1½ in. dia., more slender than lactea, and the arms more erect, rich green and prettily marbled pale green to white, the ridges closely crenate and with deciduous, obovate leaves and small, reddish spines. Much used in dishgardens and very durable. **C**: 3   p. 341

valida (So. Africa: Cape Prov.); low hemispheric succulent plant to 5 in. thick and with 8 ribs; later becoming more cylindric to 12 in. high and forming basal clusters; coppery green with light green transverse lines; numerous flowering stalks which become woody and spine-like. Similar to meloformis but without turnip root. **C**: 13   p. 342

x'Zigzag' (grandicornis x pseudocactus) x pseudocactus; "Zigzag cactus"; erect 3-angled, rather compressed column growing more or less upright in hither-thither style; 1¼ in. thick, yellow-green and prettily painted dark green in herringbone pattern, the wavy edges meandering upward in zigzag fashion, corky gray-brown and set with pairs of sharp spines. A worthy California (Hummel) hybrid involving ¾ E. pseudocactus and ¼ grandicornis; quite tough and durable; slow-growing. **C**: 3   p. 343

EURYALE (Nymphaeaceae)

ferox (Tropical and Subtropical India to China and Japan); a widespread aquatic perennial with large circular floating bright glossy green leaves 3 to 5 ft. across, very spiny and with inflated bubbles, and spiny-ribbed beneath, small 2 in. fl. green outside, purplish red within, barely above the water and short-lived; the seeds are edible when roasted. Much at home in rice paddies in the Orient; a stunning exotic in the conservatory pool. Treated as an annual easily grown from fresh seed. **C**: 52   p. 517

EXACUM (Gentianaceae)

affine (Arabian Sea: Socotra Is.), "Persian violet" or "Arabian gentian", also known as "Blue Lizzie"; charming little herbaceous, free-flowering tropical perennial grown as biennial; about 6 in. high with waxy, stalked, ovate leaves, and tiny wide open, star-like, bluish-lilac, fragrant ½ in. flowers with pretty eye formed by the deep yellow stamen. For August bloom, seed should be sown in March, or they may be grown in pots as greenhouse perennials, flowering from June to October. **C**: 60   p. 411

FARFUGIUM: see Ligularia

FASCICULARIA (Bromeliaceae)

bicolor (Bromelia bicolor) (Chile); clustering rosette of numerous gray-green, stiff-recurved, channeled linear leaves to 18 in. long, brownish and with gray ridges beneath; the margins with occasional spines; center of rosette turns carmine-red. Flowers pale blue, enclosed by ivory bracts. **C**: 19   p. 138

x FATSHEDERA (Araliaceae)

lizei, "Ivy tree". "Miracle plant" or "Botanical wonder"; bigeneric hybrid between Fatsia japonica 'Moseri' and Hedera helix hibernica, the "Irish ivy". Evergreen shrub combining characters of both parents, growing erect over 6 ft. high if the woody stem has support; leathery leaves 5-lobed, similar to English ivy but larger, often 5 to 8 in. wide, dark lustrous green. Small light green flowers in dense clusters. Does best in cool or intermediate locations, requiring relatively good light 50 fc. up; tolerates chills but needs moisture. Propagated from joint cuttings. **C**: 16
                                                        p. 282, 283, 294

lizei variegata, "Variegated pagoda tree"; horticultural form with the fresh green leaves prettily variegated and edged cream-white. Prefers moist-temperate climate. **C**: 66
                                                                  p. 195

FATSIA (Araliaceae)

japonica (Japan), the "Japanese Aralia"; handsome, evergreen, sparingly branching shrub that may grow to 5 to 8 ft. high, with flexible stem and big glossy green, thin-leathery leaves 10 to 15 in. wide, deeply lobed like the fingers of a hand, the broad lobes pointed and toothed; small, milky white flowers in roundish clusters, followed by little shiny black fruits. An excellent decorative plant for cool locations, requiring elbow room; it can take variations in temperature but needs to be kept moist; tolerates poor lighting to 50 fc. but keeps better appearance with good light; in summer on the patio. Propagated from fresh seed in late winter-spring; cuttings or mossing are also possible. **C**: 28   p. 289

FAUCARIA (Aizoaceae)

tigrina (So. Africa: Cape Prov.), "Tiger jaws"; ferocious-looking small succulent with opposite, thick, keeled or boat-shaped, gray-green leaves 1–2 in. long, and marked with numerous white dots; margins armed with stout recurved teeth so that the leaf-pairs resemble a gaping jaw. Large golden yellow sessile flowers. From suckers or seed. **C**: 13   p. 329, 355

tuberculosa (So. Africa: Cape Prov.), "Pebbled tiger jaws"; striking little succulent with wide-open jaws of dark green thick leaves about 1 in. long, upper side covered with white tubercles, the edges armed with stout teeth, underside keeled; 1½ in. flowers yellow, opening in the afternoon. Very attractive little house or dishgarden plant, easy and durable. **C**: 13   p. 355

FEIJOA (Myrtaceae)

sellowiana (So. Brazil, Paraguay, Uruguay, No. Argentina), the "Pineapple guava"; tall evergreen shrub or small tree to 18 ft. high, dense with whitish-felted branches and thick opposite, oval leaves 2 to 3 in. long, waxy dark green above with white midrib, silvery white to brown woolly beneath; waxy-white inch-wide flowers with fleshy, edible petals, purplish inside, and a tuft of showy dark red stamens, blooming in spring; followed by green fruit tinged red, edible and tasty with guava flavored pulp. From seed; cuttings and layers are both very slow. **C**: 13
                                                           p. 292, 498, 508

FELICIA (Compositae)

amelloides, sometimes grown as Agathaea coelestis (So. Africa), the "Blue Daisy" or "Blue marguerite"; a popular shrubby perennial 1½ ft. high and spreading, with opposite obovate roughish rather aromatic leaves, and daisy-like 1¼ in. flowers consisting of sky-blue florets and yellow disk on long stalks. Blooming almost continuously, especially in late fall, if dead flowers are picked off. A charming house plant for the sunny window, especially in winter, or as a container plant, in the balcony box, on the patio or roof garden. From seed, cuttings or division. **C**: 14   p. 497

FENESTRARIA (Aizoaceae)

aurantiaca (So. Africa: Cape Prov.), "Baby toes"; charming, highly succulent plant forming attractive little groups, with club-shaped glaucous gray-green leaves 1 to 1½ in. high, the pearly top whitish and with glossy, translucent windows; large 2 in. flowers golden yellow. From the sand dunes of South-western Africa, where they grow buried to their windowed tops; in cultivation they are better grown high to avoid rotting. Water moderately during growing time in summer, hardly at all in winter. From seed. **C**: 1   p. 355

rhopalophylla (S.W. Africa), "Baby toes"; tufts of cylindrical, succulent club-like leaves 1 to 1½ in. long, thickened above and bearing a translucent window at top to filter the sun down to the chlorophyll; large white flowers 1½ in. across. Wants sunny and only moderate watering even in summer while growing; quite dry in winter. Prefers sandy soil with lots of lime. Best from seed. **C**: 1   p. 356

FEROCACTUS (Cactaceae)

acanthodes (S. Nevada, California, Baja California), "Fire barrel"; globular plant glaucous green, becoming cylindric to 9 ft. tall, ribs up to 27 covered entirely with big curving or awl-shaped spines to 5 in. long, pinkish to red; flowers yellow to orange. **C**: 13   p. 158

covillei (Arizona, Sonora), "Coville's barrel" or "Bird cage"; globular to barrel type to 5 ft.; dull bluish green, 22–32 ribs, deeply notched; with vicious hook spines variable in color, red to white; flowers red, tipped yellow. **C**: 13   p. 158

latispinus (Mexico), "Fish-hook barrel"; depressed globe to 16 in. high, dull grayish-green, with many dented folds, the central spines very broad, curving and crimson-red; bell-shaped flowers whitish to rose or purple. **C**: 13
                                                                        p. 158

FERRARIA (Iridaceae)

undulata (So. Africa: Cape), the "Orchid iris"; strangely beautiful cormous plant 1½ ft. high with flat, sword-like

glaucous leaves from the base and along the stout branched stalk, passing into ovate bracts and topped by cup-shaped flowers 2 in. across with six spreading, crisped lobes and appearing orchid-like; and although lasting only from morning to afternoon, and somewhat ill-scented, their peculiar coloring of greenish brown, spotted and blotched with purple, makes them very attractive when in bloom from March into June. Keep dry and dormant when foliage begins to die. By offsets. **C**: 63   p. 393

FICUS (Moraceae);

the "Rubber plants", "Banyans" or "Fig trees" are tropical and subtropical decorative evergreens, trees, also a few root-clinging vines, of the Mulberry family, having milky sap; densely branching with leathery usually very handsome leaves; their inflorescence not showy, but distinct with minute flowers and seeds covering the inside of a hollow fleshy receptacle as typified by the edible fig. Ficus are very ornamental for every decorative use and are very durable and satisfactory provided they receive sufficient light of preferably 50 to 500 fc. to hold their foliage well. They grow easily in good fibrous loam, and need an even moisture at their roots; allow the surface to become dry then water thoroughly; other than that they tolerate the dry atmosphere usually found indoors; also perform quite satisfactorily under air-conditioning. Propagated most safely by mossing.

afzelii (eribotryoides) (Ghana, Nigeria, Congo, Mozambique); large tropical tree with foliage at ends of branches; stems bronzy with light brown hairs; closely-set long-oblanceolate leaves 16 in. long and 5 in. wide, dark olive green and smooth above, brown tomentose beneath; limply soft when young, stiff-leathery when mature. **C**: 53 p. 214

altissima (India), "Lofty fig"; large spreading tree with few aerial roots; leaves thick-leathery, broadly oval, to 8 in. long and with short tip, shining deep green with ivory veins, on slender stem; attractive clustering orange-red fruits. **C**: 3
                                                                        p. 214

aurea (So. Florida), a "Strangler fig"; epiphytal at first, later growing into a tree 60 ft. high, with rough bark exuding whitish, sticky sap; thin-leathery, green, oblong 4 to 6 in. leaves on buff branches with age forming aerial roots; small ⅓ in. fruits orange. It is quite a common occurrence in the tropics for a bird to drop a Ficus seed into some humusy recess of a proud forest tree or palm, germinating and growing there with roots reaching downward along the trunk, thickening with age, girdling the host tree and finally choking it, the Ficus itself becoming a mighty tree. **C**: 64   p. 538

benjamina (India, Malaya), the graceful "Weeping fig"; a beautiful tropical tree of dense growth, forming aerial roots, and with branches of somewhat pendant habit; the pendulous, shining deep green leaves long ovate, slender pointed, 3 to 4 in. long; small round fruit blood-red when ripe. One of the most attractive tubbed decorators for tropical effect, preferring smallish containers, and rejoicing in warmth and good light, 40 fc. and up to 500 fc., the more light the more leaves. Allow surface to become dry, then water thoroughly; tolerates air-conditioning. **C**: 1
                                                           p. 211, 212, 213

benjamina 'Exotica' (Java, Bali), "Java fig"; especially graceful, weeping form with slender, arching branches, and smaller, rather narrow and pendulous glossy green, leathery, oblique 3½ to 4½ in. leaves, which are given special charm by the coquette twist of their slender tip; small red berries; thrip-proof. I have seen large trees of this variety quite common in Bali. Good indoor decorator tub plant but needs at least 50 fc. of light; some leaves may drop at first when temperature changes, or a tree is moved to air-conditioning. **C**: 1   p. 209, 212

benjamina nuda (Philippines), the "Small-leaved philippinense"; small-leaved variety planted in South Florida, having grown into large banyan trees, and where it has become known horticulturally as "philippinense"; the shiny leaves are inclined to be tufted toward the ends of zigzag branches, 1½–2 in. long, somewhat narrower than benjamina, more tapering toward base and slender pointed; a distinct pair of basal veins run for some distance up the side margins before uniting with the most basal side-veins. The tiny figs are greenish-yellow whose basal bracts fall off early denuding the fruit. **C**: 1   p. 212

carica (Mediterranean region, North Africa, Asia Minor), the "Common fig"; broad, irregular, subtropical tree-like shrub to 30 ft. high, much planted and economically important for its sweet, pear-shaped fruit ripening in summer;

deciduous in climates with occasional frost: leaves thick, rough above, hairy beneath, deeply 3–5 lobed and with palmate veining. Not a living room plant; best in a glassed-in porch, or the patio in summer. Somewhat winter-hardy with protection. Prop. by hardwood cuttings. **C**: 63   p. 210, 506

cyathistipula (Trop. Africa); bushy tree with thin woody branches dense with long-oblanceolate leathery, dark green leaves 4 to 10 in. long, suddenly tapering to a slender point at apex; fruiting as a young plant with globular fig 1½ in. dia. **C**: 3   p. 210

diversifolia (lutescens) (India, Malaya, Java), "Mistletoe ficus"; small woody shrub to 3 ft. high, mostly growing epiphytic in its tropical homeland; with small stalked, obovate, hard leaves only 2 in. long, dark green with brown specks above, pale beneath; liberally bearing small yellowish fruit lined with gray. A miniature fruiting Ficus, and good little houseplant. **C**: 1   p. 215

dryepondtiana (Congo), "Congo fig"; tropical foliage plant of rare beauty; woody, brown stems carry stiff, long-ovate, quilted wavy leaves 8–12 in. long, metallic deep olive green with the underside strikingly colored red-purple. Very attractive but needs warm greenhouse. **C**: 54   p. 215

elastica 'Decora', "Wideleaf rubber plant"; superb commercial seedling of the "India rubber tree" from India and Malaya, a mighty tropical banyan tree to 100 ft. high, yielding latex-bearing milky sap; a good decorator plant of bold habit, with leaves larger and much broader and heavier in texture, 10–12 in. long, deep glossy green, with prominent lateral depressed veins, nearly at right angles to the ivory midrib which is red beneath, the sheath at the growing tips is red. Very light-hungry, preferably full sun, but tolerates more darkness, to 20 fc., than other Ficus; the more light the more robust the growth, foliage luxurious and shining. **C**: 1   p. 211

elastica 'Doescheri' (New Orleans 1925), "Variegated rubber plant"; outstanding variegated cultivar of Ficus elastica, with beautiful leathery leaves very colorful in a striking range of colors, from green with gray into white and cream-yellow, while midrib and leafstalks are pink; coloring is stable and not reverting to green, nor its variegation becoming dull as does F. elastica variegata. **C**: 3   p. 188, 215

elastica variegata, "Variegated rubber plant"; the common Rubber plant in variegated form; the long, leathery leaves are usually variegated gray and edged in creamy-yellow. Foliage has a tendency to turn rusty if kept too cold. Not as good as 'Doescheri'. **C**: 1   p. 211

glomerata (racemosa) (India, Burma, Malaya, Australia), the "Cluster fig" is a strangely curious thick-topped tree that oddly chooses to bear its large clusters of edible reddish fruit directly on the sometimes large trunk and naked thickset branches. This production of flowers and fruit from trunks and branches is known as cauliflorous, a character also found on a few other species of Ficus. The ovate leaves are 4 to 7 in. long and are handsome with their metallic luster. **C**: 2   p. 528

jacquinifolia (West Indies, Bahamas), "West Indian laurel fig"; small-leaved tree to 30 ft. high, forming thick trunk, dense with oval or oblong, convex foliage 2–3 in. long, glossy-green with veins not prominent. As a decorative plant it resembles Pittosporum. **C**: 3   p. 214

krishnae (India, Pakistan), the "Sacred fig tree", probably a form of benghalensis; small tree sacred to the Hindu god Krishna, said to have folded its leaves into cups to collect drops of dew, when the god was thirsty in the desert; leathery, irregularly cupped, deep green leaves with raised ivory ribs, finely pubescent inside, on grayish branches. Very interesting collection plant because of its curious leaves. **C**: 51   p. 215, 534

lyrata (Trop. West Africa), commonly known as Ficus pandurata, the "Fiddleleaf fig"; close headed tree to 40 ft. high, with woody stems nobly carrying large obovate, hard-leathery leaves 1 to 2 ft. long, resembling a violin with wide rounded apex, deep waxy green, quilted and wavy, with attractive yellow-green veins; fruits with white dots. This magnificent decorator tree does best in relatively small containers; allow surface of soil to become dry before watering. Needs good, filtered light but avoid full sun or foliage may scorch; tolerance 40–50 fc., optimum 200 fc. **C**: 3   p. 210, 211, 213

macrophylla (Queensland, New South Wales), the "Australian banyan" or "Moreton Bay fig"; large tree with ovate to broad oblong leathery leaves blunt at apex, 6 to 10 in. long and 4 in. wide, cordate at base, glossy green with pronounced ivory veins, shiny light green underneath, and netted. Popular in Australia and New Zealand as a dec. pot plant, but not as durable as F. elastica. **C**: 4   p. 214

microphylla hort., "Little-leaf fig"; rapid-growing, gray-barked evergreen tree, with broad ovate to obovate, glossy, intense dark green, soft-leathery leaves 2 to 6 in. long, glossy green beneath. This plant grown in California nurseries as microphylla may actually be F. rubiginosa or form. **C**: 1   p. 214

nekbudu, better known in horticulture as F. utilis (Trop. Africa), "Zulu fig tree"; large tropical forest tree; the young growth pubescent; fresh green thick-leathery leaves oval to obovate, 6–15 in. long, rounded at both ends and with pretty, yellow ribs; good examples of their shapely form are the old specimen in front of the Capitol in San Juan, Puerto Rico. Not very common as a decorator plant. **C**: 3   P. 214

nitida: see retusa nitida

pandurata: see lyrata

panduriformis: see wildemanniana

parcellii (South Pacific Islands), the "Clown fig"; strikingly colorful small shrub with sparry branches, bearing large oblong, toothed, rough hairy, pointed leaves to 8 in. long, grass-green and milky-green, wildly variegated and marbled ivory white; even the pear-like fruits are variegated white. Needs warm greenhouse as otherwise it drops leaves; catches Red spider easily. **C**: 54   p. 188, 215

petiolaris (jaliscana) (Mexico), "Blue Mexican fig"; beautiful small tree to 6 ft. high, developing with age a very wide, swollen base, crowned by a rosette of heart-shaped leathery, wavy leaves with long pointed tip, metallic blue-green with veins and petioles ivory pink to showy red. An outstanding collector's plant. **C**: 53   p. 215

"Philippinense" hort.: see stricta and benjamina nuda

pumila (widely known in horticulture as F. repens), (China, Japan, Australia), "Climbing fig" or "Creeping fig"; freely branching creeper with small, obliquely cordate, dark green leaves less than 1 in. long, clinging to walls by roots like ivy and then flattened; fruiting branches erect or growing outward, with stiff, much larger, elliptic or oblong leaves to 4-in. long, and bearing heart-shaped brown-purple figs. Handsome, small-leaved plant attaching itself to wood or walls by roots like ivy; almost hardy, and tolerating a variation of temperatures. **C**: 4   p. 259

quercifolia (Burma, Malaya), "Oakleaf ficus"; shrub with prostrate, stiff-hairy branches creeping or erect, elliptic or lobed leathery deep green leaves 5–8 in. long, shaped like those of the common oak; produces small green figs. Rather sparry in habit, and interesting mainly because of its oak-like foliage. **C**: 4   p. 259

radicans (rostrata) (E. Indies), "Rooting fig"; robust trailing species rather coarse-looking with rough-wiry stems rooting at the leaf axils; small lanceolate, slender-pointed, quilted leathery leaves 2 in. long, dark green and glabrous above, rough beneath, and becoming larger, to 4 in. long. Good basket plant, or on bark. **C**: 4   p. 259

radicans 'Variegata', "Variegated rooting fig"; attractive variegated creeper grown under glass, with lanceolate, grayish green leaves irregularly marked with creamy-white, the variegation beginning at the margin. Requires warmth with moisture; when chilled the variegated leaves develop brown areas and drop. **C**: 54   p. cover *, 43*, 259

religiosa (India, Ceylon), "Sacred Bo-tree" or "Peepul"; sacred to Hindus and Buddhists, and under which Buddha, according to scriptures, received enlightenment in the 6th century B.C.; large tropical tree to 100 ft. high, with slender grayish, willowy branches carrying the gracefully pendant, smooth and thin, bluish-green leaves, with pronounced ivory or pinkish veins, heart-shaped with a remarkable long, tail-like drip-tip; fruit purple. An attractive and curious ornamental plant, requiring light and air. **C**: 4   p. 215

repens: see pumila

retusa (microcarpa) (Malaya, So. China, Hongkong, Philippines), the "Chinese banyan"; shapely large tree with dense foliage, much planted for shade in American tropics; the branches first ascending but becoming pendant, with small leathery leaves 3 to 4 in. long, broadly obovate and waxy. Very decorative grown as shaped standards, with the dark green, shiny foliage eventually gracefully pendulous giving it a "weeping" effect. Light-hungry the same as its more erect form nitida. Very durable. **C**: 1   p. 213, 541

retusa nitida (Malaya), "Indian laurel"; attractive, glabrous thick-topped banyan tree forming aerial roots and buttressed trunk in habitat; with erect branches and small, rubbery, elliptic leaves 3–4 in. long, waxy, nice green and very smooth; can be shaped into pyramids and standards resembling a true laurel. Good tub plant; also widely used as a containerized street tree in our Southwest. In the interior of buildings this sun-lover requires 50 fc. to 500 fc.; the more light, the more leaves it can hold. Allow soil surface only to become dry, then water. **C**: 1

p. 209, 210 ,213, 541

roxburghii (auriculata) (India), "Ornamental fig"; low, spreading tree to 20 ft. high, very showy for its large foliage on slender woody branches, the light brown petioles covered with white hairs; big papery, rounded and toothed, slightly glossy leaves to 15 in. long, their surface covered with a fine pubescence, and with depressed veins: when young an attractive mahogany-red. Big flat-globose figs are borne on the stems. Not recommended as a house plant as foliage easily catches red spider and drops. **C**: 53   p. 214

rubiginosa (australis) (New South Wales, Queensland), the "Rusty fig"; low shrub 6 to 12 ft. high, spreading by means of rooting branches; in habit compact and dense, and usually grown as a little bush, with small shining leathery 3 to 6 in. oval or elliptic leaves, rusty-pubescent beneath when young. Slow-growing, and will tolerate cooler temperatures. **C**: 3   p. 214

rubiginosa variegata, the variegated "Miniature rubber plant"; a pretty, freely branching, graceful miniature foliage plant with its small eggshaped, deep green leathery 3 in. leaves richly marbled and edged yellowish-cream; tomentose in juvenile stage only, as noted in Queensland. Slow-growing; does best in a greenhouse. Safest propagation by mossing. **C**: 54   p. 43*, 188, 215

stricta (Guam), known in some South Florida nurseries as "philippinense", but with generally smaller foliage than the true species philippinensis; starting as a climbing woody epiphyte sending down aerial roots and developing into trees; triangular branches with broad-ovate, leathery deep green, smooth leaves 6 in. or more long, equipped with water pores scattered along margins; fruits are axillary. A very decorative tropical ornamental with branches broadly arching and its medium-large foliage pendant; requires good light 75 fc. up; water when surface of soil shows dry. **C**: 1

p. 212

sycomorus (Syria, Egypt, Sudan, south to the Transvaal), "Sycamore fig"; round-headed, freely branching tree to 40 ft. high, partially deciduous. In Africa, one sees them often along rivers, very attractive with their pale fawn-yellowish trunk and shapely crown; ovate, rugose leaves with hairy ribs, 3–4 in. long and with blunt apex, soft when young, then almost bluish-green; when mature with leathery "feel" but quickly dessicating, and fresh green to deep olive; greenish flowers and small, abundant, edible fruit produced from the trunk and branches. **C**: 2   p. 215

tikoua (China: Yunnan 3000 ft., Tibet, Assam, No. Indonesia), "Waupahu fig"; scandent shrub fast spreading with woody, flexible brown branches; leathery obovate, rough leaves to 5 in. long, shallowly toothed, matte dark green above, pale green beneath; Pear-shaped figs are borne on prostrate branches, ⅜ in. dia., brown to black. Promising ground cover. **C**: 54   p. 259

utilis: see nekbudu

wildemanniana (Trop. Africa); introduced by Voster's in Florida as panduriformis hort.; sturdy, compact plant small enough to look good in a 4 in. pot; leaves are glossy dark forest green and very leathery, oblique elliptic, 4 to 6 in. or more long, and with their midrib and lateral veins ivory, looking very attractive. **C**: 4   p. 214

FITTONIA (Acanthaceae)

verschaffeltii (Peru), "Mosaic plant"; a pretty tropical ground cover; low, creeping herb with hairy, rooting stems and colorful oval 3–4 in. leaves dark olive green and entirely covered with a network of deep-red veins. Although the leaves are somewhat papery they are fairly sturdy. Small yellow, green-bracted flowers. **C**: 54   p. 189

verschaffeltii argyroneura (Peru), "Nerve plant"; charming tropical herb, spreading and hugging the ground, with flat, papery, oval leaves vivid green and beautifully netted with white veins. Likes warmth and high humidity. **C**: 54   p 43*, 189, 263

verschaffeltii pearcei, "Snake-skin plant"; a variety with larger foliage than the species, the leaves glaucous light olive-green with rose-pink veins; thin-papery and not as sturdy. **C**: 54   p. 189

FORTUNELLA (Rutaceae)

crassifolia, "Meiwa kumquat", known in China as the "Golden bullet"; freely branching shrub of compact habit widely grown and with nearly thornless twigs, rather thickish leaves, and nearly round to broadly oval, small 1 to 1½ in. fruit, having thick, sweetly flavored peel which is tasty to eat. As a container plant this free-fruiting kumquat is quite ornamental during fall and winter. From cuttings, or grafted. This may be a Chinese garden hybrid. **C**: 14   p. 511

hindsii (Hong Kong, Kwangtung), the "Dwarf kumquat" or "Hongkong kumquat"; small spiny, ornamental tree, with leathery green, oval-obovate leaves, and small nearly round, ½ in. bright orange-red fruit of 3–4 cells, almost without juice. Somewhat sparry in habit, it is esteemed in South China for its brilliant flame-scarlet berries which are preserved in jars, and used as a spicy flavoring. Grown in pots as a curiosity plant since the days of the Sung dynasty in the 12th century. **C**: 14   p. 510

margarita (China: Kwangtung), "Nagami" or "Oval kumquat"; small and shapely vigorous citrus, generally to 5 ft. and up to 10 ft. high, with slender, erect, angled branches, long-pointed, shining dark green 3 in. leaves, white flowers with orange-blossom perfume; the small, rather persistent, oblong golden-orange fruit is 1½ in. long, having a finely-flavored pulp; it is used for preserves or candied; the edible rind is sweet, the flesh tart. Often grown as a standard in tubs, admirably suited for the patio; generally flowering in spring to summer, and ripening their prolific miniature oranges from October to January. **C**: 14   p. 510

FOSTERELLA (Bromeliaceae)

penduliflora (Lindmania) (N.W. Argentina); terrestrial rosette of spreading linear-lanceolate, soft leaves with smooth edges, satiny grayish green and a faint yellow-green stripe down the middle; tall 1 ft. wiry stalk carrying the branched inflorescence, bearing to one side small white flowers. **C**: 21   p. 138

FRANCISCEA: see Brunfelsia

FREESIA (Iridaceae)

x hybrida; well-loved small cormous flowering plant derived from such South African species as F. refracta and armstrongii, resulting in this large-flowering strain; bulb-like corms with branched, wiry stems and linear leaves, with very fragrant, waxy-firm, funnel-form flowers 2 to 2½ in. long in one-sided spikes, now in many colors from creamy-white and yellow through orange to lilac-blue; August-blooming unless forced. Prized for the rich fragrance and assortment of gay colors of their long-lasting flowers. Freesias have become great favorites for extended periods of blooming by timing and forcing, both in pots or as cut flowers. Corms may be planted all year round to flower 3 to 5 months later, at a cool 50° F.; no cool storage is necessary after potting. When ripened and rested shake out the soil and keep dry; later repot. Propagate by offsets. **C**: 77   p. 393

FREYCINETIA (Pandanaceae)

multiflora (Philippines), "Climbing pandanus"; climbing tropical shrub with slender rooting stems becoming pendant from trees; sessile leathery, black-green, linear lanceolate keeled leaves, and showy terminal inflorescence with glowing salmon-rose bracts; and with rosy spadix. **C**: 52   p. 275

FRITHIA (Aizoaceae)

pulchra (So. Africa: Transvaal), "Purple baby toes"; dainty little windowed plant; clustering succulent rosette similar to Fenestraria, with 6 to 9 club-like gray-green leaves under 1 in. high, the flat top with translucent windows; large carmine-purple flowers with white center lasting 2 to 3 weeks. Keep dry during summer rest period. Desirable but difficult. From seed. **C**: 1   p. 356

FRITILLARIA (Liliaceae)

meleagris (Britain, Norway, C. Europe, to Caucasus), "Checkered-lily" or "Snake's Head", also known as "Guinea-hen tulip" or "Toad-lily"; beautiful and unusual bulbous perennial, to 1½ ft. high, with linear to oblanceolate glaucous leaves scattered along middle of stem; usually solitary, bell-shaped nodding flowers 1½–2 in. long, purple with white checkering, or white with green network of veins; blooming in April-May; winter-hardy. For cool and damp locations. To grow in pots plant in fall and keep cold until growth begins, and moist while growing; rest after blooming. By offsets. **C**: 84   p. 398

FUCHSIA (*Onagraceae*); handsome, popular flowering shrubs and even trees native in cooler areas of the Americas and New Zealand, most of them adaptable for growing in pots, with flexible stems becoming woody; simple leaves, and charming, usually pendant blossoms. The flowers are characteristic with tubular or bell-shaped calyx, flaring sepals or calyx lobes usually recurved, and petals a skirt-like corona, often double. The cultivated "Garden fuchsias" are grouped as F. x hybrida (speciosa), nicknamed "Lady's eardrops", and derived from such species as fulgens, magellanica and corymbiflora, having large gay flowers 2 to 3 in. or more long, in a wide range of color combinations, in bloom the year round, but most profusely during the season of longer days. Fuchsias grow easily in cooler, semi-shady, moist locations, and are ideal as house plants in the window or in window boxes, in hanging baskets; on the patio where they may be trained into little trees, or for bedding outdoors. In the greenhouse they can be had in bloom from March on, without the aid of artificial light, and may be flowered until November. In the home it is best to bring fuchsias to rest from September on, by reducing watering and exposing them to cool fresh air for hardening; when almost denuded bring into cold greenhouse, frost-free garage or cool basement until growth should again begin in April. Propagated from cuttings.

excorticata (New Zealand), the unbelievable "Tree fuchsia"; in habitat becoming a tree to 45 ft. high, with woody trunk to 2 ft. thick, having loose, papery bark and brittle branches, thin dark green lanceolate leaves 2–5 in. long, silvery beneath; small waxy flowers with spreading calyx lobes first yellowish then deep dull red, the petals deep purple with blue stamens. It is not only difficult to realize that our fragile, blooming bedding plant Fuchsias have cousins that grow into veritable trees, but it is interesting to note that this establishes the fact of an ancient landbridge between our American species and the South Pacific, across Antarctica. **C**: 78   p. 526

x hybrida 'Black Prince', popular slender free-growing commercial plant for early bloom; the strong shoots with pendulous axillary and terminal single flowers having red calyx lobes or sepals, and blue-purple, skirt-like petals. **C**: 66   p. 412

x hybrida 'Brigadoon'; huge flowers with double, purple corolla; sepals flushed and marbled rosy-pink. Excellent for hanging baskets. **C**: 66   p. 413

x hybrida 'Chickadee'; distinctive, profuse bloomer of slender habit, with slim flowers, the spreading, narrow spur-like sepals around a dainty orchid-colored corolla. **C**: 66   p. 412

x hybrida 'Fluorescent'; willowy grower with long medium-small flowers, the tube and broad, recurving sepals white, faintly marked pink, the semi-double corolla in a lovely shade of orchid. Recommended for hanging baskets. **C**: 66   p. 413

x hybrida 'Golden Marinka' (1955); trailing variety easily branching, with green leaves splashed and edged with golden-yellow, later turning cream; semi-double flowers with tube and sepals rich red, the corolla deeper red. Not very vigorous but beautiful with its variegated foliage; good for baskets. **C**: 66   p. 413

x hybrida 'Jubilee' (1953); spreading, heavy plant with light green foliage and large, extremely attractive double flowers, the bicolored corolla cerise edged with dark magenta-red, the tube and spreading calyx lobes waxy-white flushed with deep rose. Recommended for hanging pots or baskets. **C**: 66   p. 413

x hybrida 'Lace Petticoats' (1952), fancy variety with lovely, intriguing large flowers having broad white recurving sepals, and double white, fluffy corolla with center petals serrated and fringed like lace, and with an undertone of blush pink; free-blooming. Growth is upright but inclined to straggle with the weight of blossoms. **C**: 66   p. 412

x hybrida 'Pride of Orion' (Veitch); an old variety; bushy, upright plant with lovely, large double white petals veined cerise against bright crimson tube, and spreading narrow calyx lobes. Fairly free-blooming. **C**: 66   p. 412

x hybrida 'Rollo' (Lemoine 1913); floriferous variety of vigorous growth habit having flowers 3 in. long, with large white double corolla, and shell pink sepal falls; foliage light

green. For large bushes as well as training into standards. Rollo Duke of Normandy (860–930) was a 9 ft. giant, according to legend. **C**: 66   p. 412

x hybrida 'Sunray' (1872); an old, bushy variety of upright habit until the branches are pulled down by the weight of the blooms; very attractive for its colorful foliage, silvery light green suffused cerise, and variegated with cream to yellow margins; small single flowers with fuchsia-red corolla and rosy-red sepals. Mainly grown as a pretty foliage plant; especially graceful in baskets. **C**: 66   p. 413

x hybrida 'Swanley Yellow', an old English hybrid (1900), of the continuously-blooming honeysuckle type; vigorous plant with pendant branches with bronzy foliage, and single flowers unusually colored having a long orange pink tube with pinkish white sepals and small orange-vermilion petals. Excellent basket plant. **C**: 66   p. 412

x hybrida 'Swingtime' (1950); good trailer and early, free-blooming variety; large fine-shaped double flowers having a milky-white corolla with slight pink veining, and shiny red crepe-textured, short upturned sepals; free-blooming; suitable for hanging baskets but also for bush or standard forms. **C**: 66   p. 413

x hybrida 'Ting-a-Ling' (1958); vigorous upright grower with large clusters of bell-like, all white flowers, the single corolla resembling a flaring skirt, blooming over a long period. Suitable for both bush and standard form. **C**: 66   p. 413

x hybrida 'Winston Churchill'; excellent commercial potplant of compact habit and a willing bloomer, with dark stiff stems and green leaves, the large round flowers with salmon-red sepals, and deep purplish-blue, fully double petals, best in spring. **C**: 66   p. 378, 412

magellanica variegata, the "Variegated hardy fuchsia"; a variegated leaf form of the "Hardy fuchsia" from Peru and Chile to Tierra del Fuego, and which is a bushy, herbaceous plant growing into a shrub to 6 ft. high, and with support, on walls to 20 ft.; arching branches with small ovate leaves to 2 in. long, wavy-toothed at margins, and pendulous slender flowers 1 to 2 in. long with purplish red calyx lobes longer than the purple blue petals. The variegated form with leaves green, white and red is a choice hanging basket plant, blooming best from May to fall. Roots are fairly hardy and withstand some frost. **C**: 66   p. 412

triphylla 'Gartenmeister Bohnstedt' (1905), the "Honeysuckle fuchsia"; a pretty descendant of this little West Indian shrub, with dark metallic bronze, purple-veined leaves, red-purple beneath, and lustrous, slender tubular flowers of salmon-rose with inner petals orange-scarlet, on purple stems, in large terminal nodding clusters, a continuous bloomer during winter and over a long period of time; a popular old house plant. **C**: 16   p. 412

FUNKIA: see Hosta

FURCRAEA (*Amaryllidaceae*)

gigantea (foetida) (So. Brazil), "Giant false agave"; also known as "Mauritius hemp" for which it is grown commercially; huge rosette of up to 50 fleshy, stiffly erect, flat, sword-shaped leaves 4–8 ft. long, shining green and with negligible spines, in time forming 3–4 ft. stems; the branched inflorescence on a slender stalk 20 or more feet high with flowers milk-white inside, greenish outside, accompanied by bulbils or plantlets which may be used for propagation. **C**: 1   p. 362

selloa marginata (Colombia), "Variegated false agave"; strikingly beautiful succulent rosette, forming stout trunk to 4 ft.; the broad leaves 3 to 4 ft. long, sword-like, narrowed toward base, thin-fleshy and flexible, glossy green with broad cream margins, and armed with vicious curved, brown teeth. The laxly branched inflorescence 15 to 20 ft. tall, with greenish flowers often in company of small bulbils freely produced. Very handsome for decoration where space permits, similar to agave but more sensitive to cold. **C**: 3
                                                                        p. 362

GALANTHUS (*Amaryllidaceae*)

elwesii (Mountains of the Aegean Is., Turkey, Asia Minor), the "Giant snowdrop", in Britain "Fair-maids-of-February"; in Germany "Little snow-bells"; pretty little spring-blooming bulbous herb 9–12 in. high, with glaucous green basal, narrow-linear channeled leaves 8 in. long; solitary nodding, bell-like flowers 1¼ in. long, the 3 long outer segments white and spreading, the 3 short inner segments green at base and tips; winter-hardy. Universal favorites in temperate-climate gardens where they colonize and are the first bulbous flowers to make their appearance between January and April according to climate. Bulbs potted in fall are kept buried outdoors or in cool cellar until growth begins. Increased by offsets or dividing the clumps. **C**: 78   p. 392

GALAX (*Diapensiaceae*)
aphylla (U.S.: Virginia to Alabama), the "Beetle-weed" or "Florists galax"; evergreen perennial stemless, tufted herb spreading with basal, thin-leathery ornate leaves to 5 in. across, heart-shaped to nearly round, waxy green with crenate red margins and prominent netted veins, on thin-wiry petioles, and turning bronze in autumn; small white flowers. The leaves are widely used for indoor arrangements, also by florists for wreaths; a good ground cover; winter-hardy. Propagated by division. **C**: 84   p. 300

GALEANDRA (*Orchidaceae*)
devoniana (Amazonas, Venezuela, Colombia, Guyana); a beautiful slender epiphytic orchid with erect leavy stems 2 ft. or more tall, topped by terminal clusters of fragrant, heavy-textured flowers having brownish-purple narrow sepals and petals margined yellow, the large white lip pencilled violet; blooming at different times of year, usually summer or fall, lasting a long time. Essentially a warm orchid, it needs plentiful water while growing, but cooler and some rest after growth is completed. Propagated by division. **C**: 10   p. 469

GALIUM (*Rubiaceae*)
verum (Europe; naturalized in E. No. America), "Yellow bedstraw" or "Our Lady's bedstraw"; dainty perennial of filmy effect 1–2 ft. high, with narrow-elliptic, bristle-tipped leaves about an inch long, pubescent beneath, in starry whorls on weak, partly prostrate stems; small yellow flowers in dense, elongate inflorescence; hardy. By seed or division. The flowers are sometimes used to curdle milk for cheese-making. Dried plants are used to stuff mattresses. **C**: 75   p. 321

GARCINIA (*Guttiferae*)
xanthochymus (India), the "Camboge tree"; tall tropical evergreen shrub or tree to 40 ft. high, with handsome dark green leathery lanceolate leaves 9 to 18 in. long; small $\frac{3}{4}$ in. white flowers in summer, followed by lemon-yellow, pointed fruit $2\frac{1}{4}$ in. long, the thick rind containing pith-like flesh and one large seed. The slightly acid fruit yields cathartic. From fresh seed or cuttings. **C**: 52   p. 508

GARDENIA (*Rubiaceae*); the "Cape jasmines" are evergreen warm-region shrubs or trees of Asia and Africa, with handsome leaves and mostly white, thick-textured flowers treasured for their haunting jasmine-like fragrance; the corolla lobes twisted in the bud then spreading wide. Depending on variety flowers vary in size from 1 to 5 in., blooming mostly during the long-day periods from spring to summer. Gardenias are light-hungry and require both good, intense sunlight and moist-warm surroundings to feel comfortable; temperatures much below 60° F. may cause chlorosis, or yellowing of the foliage; this may be treated with iron sulphate. Optimum temperature in day and for summer growth is between 65 and 85° F.; during light-poor winter 60 to 64° F. or the forming buds will drop unless supplementary light is added; buds are initiated at temperatures below 65° F. and this will determine time of flowering. When buds are developing adequate feeding is important. Root-knot nematodes are a serious problem, and plants should be planted in steamed or otherwise pasteurized soil which should be rich in humus and slightly acid. Propagated by half-hardened cuttings from new growth, winter or summer, in peat-sand under glass or plastic cover, with warm 75° F. bottom temperature.
jasminoides 'Fortuniana' (grandiflora or florida) (So. China), "Cape jasmine"; robust rather sparry shrub with strong woody branches, large shining dark green, leathery, quilted leaves, and big 4 in., fragrant waxy-white double flowers of heavy substance, turning creamy-yellow with age; spring and summer flowering. The yellow fruit is eaten in China. A good patio plant with warmth, good drainage and occasional feeding. The Opera gardenias cv. 'Belmont' or 'Hadley', are mutations of this variety, grown for cut flower production in greenhouses, with flowers 4 to 5 in.

across, and popular as a corsage flower for festive occasions; flowers keep better covered by moist cotton or floating on water. **C**: 58   p. 486
jasminoides radicans fl. pl. (Japan), "Miniature gardenia"; dwarf evergreen shrub 8 to 12 in. high, with spreading, rooting branches; narrow 2–3 in., pointed leaves and small, solitary, irregularly double, very fragrant white flowers 1 to 2 in. across, blooming most in summer. **C**: 8   p. 486
jasminoides 'Veitchii', "Everblooming gardenia"; double-flowering form of the Chinese cape jasmine; an evergreen shrub with willowy branches, of compact, bushy habit, dense with handsome elliptic shining green 3 in. leaves, and intensely fragrant, extra double, pure white flowers averaging $2\frac{1}{2}$ to 3 in. in dia., blooming from January to May in great profusion. These pot gardenias require high light intensity, warmth and moisture, and if grown in the house need the sunniest place. If possible set outdoors during summer; bring indoors when the nights get chilly, then keep 65° F. nights until December when the night temperature is lowered to 58°–60° F. for a cool period to initiate buds; then they are kept warmer again; frequent feeding is necessary to nourish the developing flowers which will fill any room with sweet perfume for the Easter season. From soft wood cuttings. **C**: 58   p. 372, 486

x GASTERHAWORTHIA (*Liliaceae*)
'Royal Highness' (Gasteria verrucosa x Haworthia margaritifera), attractive California bi-generic hybrid; small shapely, easy-growing rosette 3–4 in. across with short, tapering, stiff leaves concave on top, dark green and densely covered with white warts. A beautiful "hard" succulent. May be propagated from leaves. **C**: 15   p. 329

GASTERIA (*Liliaceae*); the so-called "Ox-tongue cacti" are small Aloe-like South African succulents with long fleshy leaves 2-ranked or opposite and slowly spiralling into rosettes; the tongue-like foliage often prettily specked or tubercled; the curved-tubular flowers in lax clusters, greenish but tipped with red. Propagated by offsets, or leaf sections.
x 'Hybrida', "Oxtongue" or "Bowtie-plant"; representative of the numerous unnamed hybrids in California nurseries; succulents from South African parents with pairs of thick, tongue-shaped leaves in two ranks, 6 in. or more long; succeeding pairs spirally arranged, green turning purplish, and with large greenish-white tubercles. A popular house plant, shade tolerant and easy growing. Forms offsets. Propagation from seed would not be true to type. **C**: 15
                                                                          p. 345
verrucosa (So. Africa), "Wart gasteria" or "Warty aloe"; attractive succulent suckering from the base, with 2-ranked, fleshy, tongue-like, concave leaves pointed at tip, dull deep green and closely covered with raised white warts or tubercles. Propagation easy from leaf cuttings. **C**: 15 p. 328

GASTROCHILUS (*Orchidaceae*)
dasypogon (Saccolabium calceolare) (Himalayas, Burma, to Java); small, pretty epiphyte with stout leafy stem, the leaves to 7 in. long; short purple-spotted axillary stalks bearing a round cluster of small waxy, slightly fragrant $\frac{3}{4}$ in. flowers, their sepals and petals yellowish overlaid with brown, the white lip with orange blotch and tiny purple spots. Blooming in autumn, the fleshy flowers last a long time. **C**: 72   p. 469

x GASTROLEA (*Liliaceae*)
beguinii (Aloe aristata x Gasteria verrucosa) known in horticulture as Aloe beguinii, the "Pearl aloe" or "Lizard-tail"; dense rosette of rather narrow, leathery, short-pointed, keeled leaves, to 8 in. across, dark green marked with pale tubercles, and with warty teeth at margins. Combines the rosette shape of Aloe with the pretty coloring of Gasteria. Adventitious plantlets are produced on the red-green inflorescence which may serve for propagation. Formation of plantlets is stimulated by breaking off the flowers. **C**: 15
                                                                          p. 345
'Spotted Beauty' (Gasteria x Aloe), "Gasteria-aloe-hybrid"; a California hybrid combining the rosette-shape of Aloe with the hard nature of Gasteria in an attractive, trim miniature to $3\frac{1}{2}$ in. across; deep green leaves, tinged bronze, are covered with white tubercles and have translucent spines on keel and margins. **C**: 15   p. 329

GAULTHERIA (*Ericaceae*)
    shallon (W. No. America: Alaska to California), "Salal" or "Lemon-leaf", a "Florist's greens"; low evergreen shrub to 2 ft. high in sun, in shade to 10 ft., densely spreading and forming thickets, as in the Redwood forest; the ornamental foliage oval, 3 to 4 in. long, leathery bright green and with depressed ribs. From cuttings or layers. **C**: 78   p. 300

GAZANIA (*Compositae*)
    rigens hybrid, "Treasure flower"; spectacular, branching herbaceous perennial, forming clumps, dense with narrow obovate dark green leaves white-woolly beneath, the pretty, daisy-like flower heads golden yellow in the South African species, and 1½ in. across and with a brown-black center, on 6 to 8 in. stalks; the gorgeous hybrids with flowers 2 to 3 in. or more across, in various colors red, cream, orange, buff, painted black at the base of each floret. These African sun-loving daisies give a dazzling color display during peak of bloom in late spring and through the summer, and inter-mittently throughout the year, but the flowers are open only in sunshine, closing on dark days and at night. Suitable for patio containers, window boxes or on roof gardens. By division, fall cuttings or seed. **C**: 63   p. 417

GELSEMIUM (*Loganiaceae*)
    sempervirens (Virginia to Florida, Texas and Central America), "Carolina yellow jessamine" or "False jasmine"; sparry evergreen shrub with scandent or clambering brown branches reaching to 18 ft.; leathery glossy-green ovate-lanceolate, opposite leaves 1½ to 4 in. long; small flaring trumpet flowers 1¼ in. dia., yellow and sweetly fragrant like a tea rose blooming from December into summer and later. The thin slender branches may be trained on wire or over trellis and porches for a neat screen; resembles miniature Allamanda. All parts of the plant are poisonous. Propagated by half-ripened cuttings or seed. **C**: 14   p. 485

GENISTA: see Cytisus, Spartium

GEOGENANTHUS (*Commelinaceae*)
    ciliatus (Dichorisandra 'Blackie' in hort.) (Colombia); curious low, tropical plant with thick stems, spirally bearing fleshy-succulent, clasping, rounded leaves to 20 cm, sinister in glossy purplish black and contrasting red veins; tiny lavender 3-petaled fringed flowers near base. **C**: 53   p. 262
    undatus (Peru), the "Seersucker plant", conspicuous plant, low-growing and suckering to 10 in high, with stiff-fleshy, broad-ovate quilted leaves 3 to 5 in. long, dark metallic green with several parallel bands of pale gray; wine-red beneath; light blue flowers. With its curiously corrugated, strangely colored foliage a deserving exotic plant, best in a warm green house. Prop. by cuttings or division. **C**: 60
                                                                 p. 173, 262

GERBERA (*Compositae*)
    jamesonii (So. Africa: Transvaal), "African daisy" or "Transvaal daisy"; striking herbaceous tufted perennial with root-crowns forming clumps; pinnately lobed 8 to 10 in. leaves very woolly beneath, and brilliantly colored long-lasting solitary flowers daisy-like and about 4 in. across, with slender, usually orange-flame colored florets, but varieties or hybrids have been developed also in shades ranging from cream-white through yellow to coral and red, even pink and violet, also duplex and double strains; some of these measure 5 to 6 in. across. In South Africa they bloom from November to May; in our northern latitudes from May to December, with peaks in early summer, and a second period in September-October. Gerberas require lots of sun-light and to be well established in hearty soil to bloom well, best in a cool greenhouse, or in the bright window, unless climate permits planting outdoors. Propagated from fresh seed in October, by division February–April. As elegant cut flowers they will last longer if the stalk is slit at the bottom before placing in water. **C**: 25   p. 497

GESNERIA (*Gesneriaceae*)
    cuneifolia (Pentarhaphia reticulata) (Cuba, Puerto Rico, Hispaniola), "Fire cracker"; low-growing fibrous-rooted, herbaceous rosette of thin-leathery glossy grass-green, long wedge-shaped 4 in. leaves with toothed margins; and striking little 1 in. tubular, somewhat bottle-shaped flowers burning red with yellow inside, borne singly on short wiry axillary stalklets, blooming constantly all year. From seed or division also cuttings if long enough. **C**: 10   p. 446

GIBASIS (*Commelinaceae*)
    geniculata, in horticulture as Tradescantia multiflora (Trop. America), the "Tahitian bridal veil"; small free-branching herbaceous creeper with string-like rambling stems dense with tiny ovate, shining olive-green leaves only 1 in. long, purplish on back; tiny white flowers. When well established and bushy in baskets, and the branches are literally covered with white bloom. one could indeed think of a bridal veil. **C**: 13   p. 260

GIBBAEUM (*Aizoaceae*)
    petrense (So. Africa: Cape Prov.), "Flowering quartz" or "Shark's heads"; minute, nearly stemless succulent with fleshy heads forming clumps; growths with 1 or 2 pairs of modified fat leaves, united at base and only ½ in. long, whitish gray-green, flat on top, keeled below; small reddish flowers. As in Lithops, water must be withheld when growth is completed and the plant rests. From seed. **C**: 13   p. 328

GILIBERTIA: see Dendropanax

GINKGO (*Ginkgoaceae*)
    biloba (China), the "Maidenhair-tree"; perhaps the most ancient existing flowering plant, sole survivor of an extinct race of the Carbon age. A hardy deciduous resinous tree with woody trunk to 100 ft. high; having gray, corky bark; related to the conifers but differs in having broad, fan-shaped leathery bright green leaves 2 to 3 in. wide, with parallel veining, rather than the usual narrow needles of conifers. Flowers without petals in catkins; male and female on separate trees. Grown in pots as dwarfed bonsai in Japan. Very artistic, with leaves tinted yellow in autumn. Best from stratified seeds. **C**: 13   p. 310, 526

GLADIOLUS (*Iridaceae*)
    x hortulanus 'Valeria', "Garden gladiolus", "Florists glads" or "Painted lady"; with medium-large soft salmon-red flowers about 3 in. across similar to 'Halley', in shape of spreading, curved funnels with rounded petals, on one-sided, stiff fleshy spike 2 to 3 ft. or more long, with flattened, sword-shaped leaves in ranks, produced on bulb-like corm; descendant of numerous So. African species and lately much influenced by the tropical G. primulinus, resulting in the highly developed primulinus grandiflorus type, very much in favor in gardens and by florists as cut flowers and for decoration; about 10,000 named cultivars have been introduced. Garden glads have become so popular because of their superb keeping qualities, and even spikes cut before petals have unfolded will open flower after flower when placed in water. In addition, in areas with sufficient light intensity as Southern California, gladiolus may be had in bloom all year round, by planting successive stages of corms stored cool at 40° F. for at least 60 to 90 days; large size corms will keep a year or more; to flower takes from 60 days in good light to 90 or even 110 days in winter. In the north temperate zone where the light intensity is weak, forcing before February may lead to blindness; best forcing tempera-ture nights is at 50° to 60° F. A number of bulbs may be planted in large containers for summer bloom on the patio. After flowering, the remaining foliage is kept growing until it yellows and new corms and cormlets are developed which are then stored at 40–50° F. with ventilation. **C**: 64   p. 395

GLECOMA (*Labiatae*)
    hederacea variegata, better known as Nepeta hederacea (Temperate Europe and Asia; naturalized in No. America), "Variegated ground-ivy", or "Gill-over-the-ground", small, lively creeper resembling Sibthorpia but with leaves opposite; thin creeping or hanging stems with hairy, kidney-shaped leaves 1 to 1½ in. across, light green, variegated white at the crenate margins; flowers light blue. Used as a winter-hardy ground cover, or for baskets. **C**: 16
                                                                 p. 265

GLEICHENIA (*Filices: Gleicheniaceae*)
    linearis (dichotoma) (Polynesia, New Zealand, Japan, Java, Malaya, India, Sikkim, other Tropics), "Savannah fern"; handsome, variable fern widely spreading by shallow-rooted, wiry, creeping, shiny rhizomes; fronds successively branching, on zigzag stalk with pairs of pinnately cut leaflets, 6 to 18 in. long, bright green, glaucous beneath and at times tomentose. Requires lots of space for its shallow root system, and though tropical, sometimes does better in a cool-moist location. **C**: 54   p. 248

GLOBBA (*Zingiberaceae*)
winitii (Thailand); curious tropical perennial herb to 3 ft. high, with slender rhizome and fibrous roots; large lanceolate leaves, the base sheathing the stem, and hairy underneath; strange inflorescence with flowers in a loose, drooping terminal panicle to 6 in. long, with rose-purple bracts, the corolla yellow with a curved tube. **C**: 54   p. 205

GLORIOSA (*Liliaceae*); the "Glory-lilies" are weak-stemmed showy tropical vines with angled tuberous roots of the lily-family, and fresh green, glossy leaves which are prolonged into tendrils by which they climb. The beautiful flowers are lily-like in upper leaf axils, mostly strikingly red and/or yellow, with separated segments. Requiring a warm-temperature, sunny house, dormant tubers are planted soon after January, and will then bloom on supports from summer to fall, after which keep dry to rest. Propagated by offsets or division of the tuber; care must be taken that tubers are not damaged; their hidden "eye" usually sprouts at the outer bend of the tuber.
carsonii (Central Africa); tuberous climber with shiny wiry stems set with clasping 4–5 in. thin leaves ending in a grasping tendril; magnificent flowers with broad recurving 2½ in. segments wine-purple, edged with lemon-yellow especially toward base. **C**: 51   p. 274
rothschildiana (Kenya, Uganda), "Glory-lily"; a climbing "lily" with tuberous roots, the fresh-green, lanceolate leaves prolonged into tendrils; striking flowers with broad recurved petals 4 in. long, crimson-scarlet, golden-yellow toward base; blooms early spring on, but can be flowered any time. I remember seeing this species, both going north from Mombasa (Kenya), along the Malindi road, and south on the way to Tanga (Tanzania), brightly clambering over low bush not far from the coast of the Arabian Sea. **C**: 51   p. 274, 397
simplex (virescens, plantii, greenei), (No. Uganda, Mozambique, Congo, Fernando Po); a "Dwarf glory-lily", with broader petals than superba, not crisped, and which I first saw in Surinam where, in the shade of trees, it is a clear yellow, while in sunlight, the petals turn orange; summer-blooming. **C**: 51   p. 274
superba (India, Ceylon), "Crisped glory-lily"; tall vining herb; flowers smaller than rothschildiana, with narrow but crisped petals 3 in. long, first green, then yellow, changing to orange red; blooming late summer to fall. **C**: 51   p. 274
verschuuri (East Africa?), a "Climbing lily"; tuberous vine rather compact-growing to 5 ft., leaves broader than rothschildiana, shorter flower stalks, 3¼ in. flowers with broad, reflexed segments of substantial texture deep vivid crimson, and buttercup-yellow at base and margins. **C**: 51   p. 274

GLOTTIPHYLLUM (*Aizoaceae*)
linguiforme (So. Africa), "Green tongue leaf"; free-growing branching, very succulent creeper with tongue-shaped, soft-fleshy, glossy green leaves 2½ in. long, in two opposite rows. Recommended for home growing sunny and fairly dry, in small containers to bring out its large 2½ in. golden yellow flowers. Safest from cuttings. **C**: 13   p. 356

x GLOXINERA (*Gesneriaceae*)
'Rosebells'; freely blooming intergeneric hybrid between Sinningia eumorpha x Rechsteineria macropoda, resembling a slipper-type gloxinia; with shiny, dark green, heart-shaped leaves to 6 in. long, and a stalk bearing several flowers 1¼ in. long, the dainty bells are deep rose with a lavender throat. Like Sinningia, the plants go dormant after blooming, and the tubers should be rested with their pots in a warm 60 to 65° F. From seed, young shoots or leaves. A lovely new type of blooming house plant. **C**: 10   p. 444

GLOXINIA (*Gesneriaceae*)
perennis (maculata) (Colombia, Brazil to Peru) "Canterbury bell gloxinia"; this true Gloxinia grows with single or branching fleshy, red-spotted stem to 2 ft. high, from a knotted or scaly rhizome, not a tuber as in the florists' "Gloxinia" (Sinningia speciosa); bearing a spire of large downy, bell-shaped flowers of 1¼ in. dia., purplish blue with darker throat, fragrant of peppermint; basal 4 in. leaves heart-shaped, crenate, waxy above and reddish beneath. Fall-blooming, it is grown in pots indoors or outdoors in the tropics for its resemblance to lilac Campanulaš, becoming sprawling with age. The rhizomes are potted in February and kept a warm 70–75° F. until after bloom; when the stem goes dormant keep dry for the winter. **C**: 60   p. 444

GLOXINIA: see also Sinningia

GLYCYRRHIZA (*Leguminosae*)
glabra (C. Europe, Spain, Mediterranean), "Licorice" or "Sweetwood"; economically interesting, hardy perennial forming runners, erect stem to 5 ft. high with vivid-green pinnate leaves, small pale blue pea flowers in axillary clusters, in summer-fall; fruits in flattish pods 3 to 4 in. long. The long sweet, woody roots, yellow inside, furnish licorice, a long-known cough remedy. Propagated by division or seeds. **C**: 64   p. 513

GODETIA (*Onagraceae*)
amoena plena (California to British Columbia), double "Farewell-to-spring" or "Satin flower"; slender annual branching herb 1½ to 3 ft. high, with narrow leaves, often with smaller ones in the axils; wiry stalks profuse with double flowers 1½ to 2½ in. wide, the satiny petals lilac-crimson or reddish pink, opening in the day time and closing at night. June to October blooming, depending on time of sowing. Showy garden plant and good cut flower. **C**: 76
        p. 415

GOMESA (*Orchidaceae*); floriferous, rather dwarf epiphytes with compressed pseudobulbs, and arching to pendulous, densely many-flowered clusters of unusually shaped, fragrant flowers, all their segments attractively waxy, the lateral sepals joined behind the lip, their shape giving the name "Little man orchids". Grown similar to Oncidiums, they are of easy cultivation, delight in bright light and copious watering with good drainage. Planted on treefern slabs or in pots with osmunda, they grow throughout the year. Propagated by division.
crispa (Brazil); a floriferous epiphytic orchid, with oblong, compressed, tightly clustered pseudobulbs to 4 in. tall, with 2 soft-leathery leaves to 8 in. long; the inflorescence gracefully arching, densely many-flowered to 8 in. long; with fragrant small ¾ in. yellow-green waxy flowers; all segments very crisped and undulate, blooming spring into late summer. **C**: 21   p. 469
planifolia (Brazil), a "Little man orchid"; small epiphyte with 2-leaved pseudobulbs and gracefully pendant spray of small waxy, ¾ in. sweetly fragrant, greenish-yellow flowers; blooming in June-July. Similar to G. crispa but the sepals and petals less wavy, the lateral petals only partially joined together. **C**: 22   p. 469

GOMPHRENA (*Amaranthaceae*)
globosa (India and pantropic), "Globe amaranth" or "Immortelle"; tropical annual, stiffly branching, heat-resisting herb about 1 ft. high with elliptic green leaves ciliate at edge, and strawy bell-shaped clover-like 1 in. flower heads usually white to purple, blooming in gardens from July to frost and which if cut just before maturity may be dried as "Everlastings" for winter arrangements, retaining their color for a long time. Atrractive in smaller pots or for bedding. From seed. **C**: 63   p. 411

GONGORA (*Orchidaceae*)
galeata (Mexico), "Punch and Judy orchid"; extraordinarily curious small epiphyte with little 2 in. clustered pseudobulbs bearing 2 plaited leaves; from the base a graceful, pendulous group of strongly scented small flowers 1 to 2 in. long on gracefully coiling stalks; their sepals and petals brownish-yellow, the upper sepal hooded, fleshy lip wine-red; blooming May to September. Adapted for growing in baskets or hanging pots; needs humid, moist conditions at all times, a little cooler when buds are ready to open. By division. **C**: 73   p. 469

GRAMMATOPHYLLUM (*Orchidaceae*)
speciosum (widespread from Burma to Vietnam, Philippines and Java to New Guinea); a magnificent tropical epiphyte, called "Queen of orchids" or "Giant orchid" because of its noble stature and beauty; stout stemlike pseudobulbs 6 to 10 ft., in habitat even 25 ft. long, clothed by numerous 2-ranked strap-leaves; basal floral stalks like lamp-posts, to 6 and 10 ft. long, carry giant clusters with as many as 100 showy, heavy-textured flowers measuring 4 to 6 in. across; the undulated sepals and petals rich yellow, blotched with reddish-brown; old, mature plants blooming in autumn or winter, and lasting a long time. Because of its huge size, the Queen orchid is difficult for the average greenhouse, and is more ideally suited for places like

Hawaii where I have seen it flourish in bloom on a sunny patio in a large tub with tree fern fiber. During growth it needs much water, and sticky heat. **C**: 60						p. 469

GRAPTOPETALUM (*Crassulaceae*)
paraguayense, also known by its synonym Sedum weinbergii (Mexico), "Ghost plant" or "Mother-of-pearl plant"; branching succulent loose rosette forming thick stem and fleshy, broad 2–3 in. leaves flat and somewhat recurved on surface, keeled beneath, amethyst-gray with silvery bloom; flowers white. The foliage has a subtle opalescent blending of colors, but the leaves are brittle and drop easily when handled; good in hanging pots. **C**: 13			p. 329
'Paraguayense hybrid', "Giant Indian rock"; a magnificent California cultivar, forming clustering rosettes to 8 in. across, composed of broad, fleshy leaves practically spoon-shaped and glaucous silvery gray and with an opalescent glow. **C**: 13						p. 348

GRAPTOPHYLLUM (*Acanthaceae*)
pictum (New Guinea), "Caricature plant"; gay tropical shrub with oval-pointed, leathery leaves, purplish green splashed with yellow, the markings sometimes resembling faces; center vein and stem pink; flowers crimson. **C**: 4 p. 186
pictum albo-marginatum; a variety of the Caricature plant with leaves variously dark green, milky green, and gray, splashed and margined with white. **C**: 2		p. 186

GREVILLEA (*Proteaceae*)
robusta (Queensland, New South Wales), "Silk oak"; a daintily lacy, ornamental plant while small, usually 50–60 ft. but growing into a mighty evergreen tree 150 ft. high; silvery downy shoots with fern-like, green leaves 6–8 in. long, 2-pinnate into finely lobed silky-haired segments, giving them a grayish appearance; flowers golden yellow in one-sided racemes, in older trees only. A favorite fern-like house plant easily pleased; the patio in summer. From fresh seed. **C**: 13						p. 301

GRISELINIA (*Cornaceae*)
lucida (New Zealand), "Shining broadleaf" or "Kapook"; handsome evergreen shrub or tree to 30 ft., in habitat epiphytic and forming aerial roots; leathery fleshy, lustrous green, oblong or broad-ovate leaves 4 to 8 in. long, oblique at base; minute male and female flowers in axillary clusters on separate plants. Excellent foliage plant with a well-groomed look. Good near saltwater exposures. From cuttings. **C**: 13						p. 296

GUZMANIA (*Bromeliaceae*); handsome usually epiphytic rosettes from Andean rain forests, the glossy leaves are thin-leathery, strap-like to broad and with smooth edges; often prettily patterned or figured with tracery, cross bars or pencil lines; their spike-like or nested inflorescence appearing in winter is varied but always conspicuous or even spectacular, especially their leathery, brilliantly colored and long-lasting bracts; the white, yellow or red flowers peeking out from behind them; a floral bouquet rising well above the foliage, perhaps the most beautiful of all bromeliads. Guzmanias require more humidity, shade, and warmth than most other bromels. Propagated by offsets forming after bloom; also fresh seed.
berteroniana (Puerto Rico), "Flaming torch"; formal rosette of thin-leathery leaves fresh green to glowing wine-red depending on the light, and a spectacular inflorescence in form of a tight cylindrical head of scarlet bracts with yellow flowers. **C**: 10				p. 131*, 139
lindenii (Northeastern Peru), "Large snake vase"; showy large rosette of glossy-leathery leaves beautifully mottled moss-green with jagged ivory or pinkish cross-bands, made up of parallel lines across the leaf; these lines show brownish beneath. Inflorescence with numerous white flowers. **C**: 60							p. 139
lingulata 'Major' (Broadview), "Scarlet star"; magnificent clone of a plant collected in Ecuador, epiphytic rosette of smooth, metallic green leaves, from the center of which rises the bold inflorescence of recurving, glossy-leathery bracts in vivid scarlet red, its center with short waxy, incurved red bracts tipped yellow, and white flowers; the scape leaves typically red at base. **C**: 10 p. 131*, 139, 535
lingulata 'Minor', "Orange star"; small, clustering rosette of strap-like, thin-leathery, yellowish green leaves, with maroon pencil lines starting at base and diminishing

toward tip; long floral bracts bright orange-red, and small white flowers. **C**: 10						p. 139
monostachia (tricolor) (W. Indies, C. America, to Brazil), "Striped torch"; formal rosette of thin-leathery bayonet-shaped yellow-green leaves; inflorescence a stiff spike with bracts salmon-red striped brown, and white flowers. **C**: 60						p. 139
musaica (Colombia), "Mosaic vase"; bold, very showy rosette of broad, pea-green leaves marked with crossbands consisting of multiple waxy cross-lines dark green to red-brown; underneath purplish with lines much darker; flower spike with red-lined bract leaves, the head with orange-red bracts, golden flowers tipped white. **C**: 60 p.131*, 139, 193
vittata (Colombia, Brazil), "Snake vase"; a most handsome epiphyte with bizarre, stiff-leathery narrow leaves 1½ in. wide and 2 ft. long, arranged more or less erect and partially funnel-shaped, glossy dark green beautifully cross-banded by means of small silvery scales; the inflorescence an erect spike with rather insignificant bracts, and white flowers. **C**: 60						p. 139, 192
zahnii (Colombia, Panama), "Pencil-stripe ribbon plant"; very ornamental plant with strap-like, soft-papery, olive-green leaves pencil-striped maroon-red, the center tinted pink to coppery red; strong branched inflorescence with pink to yellow bracts and white flowers. **C**: 60 p. 139

GYMNOCALYCIUM (*Cactaceae*); the attractive "Chin cacti" are small and medium-sized globular cacti much in favor by fanciers because of their interesting bodies, having their ridges notched, and raised into prominent chin-like knobs beneath the areoles, or spine bundles; and also for the ease with which they are brought into bloom even when the plants are still quite small. An excellent window sill subject with long-lasting reddish or yellow flowers, over a long season from April to October; of easy culture in porous soil and small pots; in summer warmer, airy and in good light short of burning, with moderate watering; for the winter cooler at 50 to 55° F. From seed.
leeanum (Argentina, Uruguay), "Yellow chin-cactus"; small depressed globe, glaucous green, with 11–14 indistinct ribs, covered with slender spines; large flowers yellow, the outer perianth segments purplish. **C**: 13   p. 160
mihanovichii (Paraguay), "Plain chin-cactus"; grayish green, depressed little globe to 2½ in. dia. with, 8 triangular, notched ribs and banded with maroon; straw-colored spines; free-flowering with outer petals deep red, the inner greenish white. A favorite house plant. **C**: 13		p. 160
mihanovichii friederickii (Argentina, Uruguay), "Rose-plaid cactus"; little depressed globe, coppery green with triangular ribs, the spines banded cream and yellow; flowers freely at a young stage, in a lovely shade of pink. **C**: 13							p. 160

GYNURA (*Compositae*)
aurantiaca (Java), the popular "Velvet plant"; a beautiful tropical herbaceous plant with stout stems and fleshy, broad-ovate, serrate, 4 to 5 in. leaves densely velvety with violet or purple hairs and deeper purple veins; orange disk-flowers. The plant pictured may be ovalis. **C**: 4		p. 185
bicolor (Moluccas), "Oakleaved velvet plant"; tropical branching foliage plant with heavy, fleshy, slightly clammy leaves, 5 to 6 in. long, deeply and irregularly lobed on sides, metallic green with purple cast and short-hairy on top, midrib light purple, and glossy, rich purple beneath; terminal flower heads with orange florets. **C**: 2		p. 266
'Sarmentosa' hort., the "Purple passion vine"; creeping and rooting, angled stems with soft-fleshy, lobed and serrated leaves 10 cm or more long, dark metallic green with purple veins, covered with purple, bristly hairs; purple beneath. **C**: 2						p. 266

GYPSOPHILA (*Caryophyllaceae*)
elegans (Asia Minor to Caucasus), "Baby's breath"; dainty annual herb 1 to 1½ ft. high with bluish green narrow leaves and slender forking stems carrying a profusion of small single white or rosy flowers ½ in. or more across, blooming during summer from June to October, by a succession of sowings. The graceful sprays are favored cut flowers for trimming in bouquets. Sow seeds directly in pots or where wanted. **C**: 75					p. 415

HAAGEOCEREUS (*Cactaceae*)
chosicensis (Peru), "Foxtail"; group-forming fresh-green columns to 5 ft. high, with as many as 21 low ridges

completely hidden by the masses of bristle-like golden-
yellow spines; night blooming with rosy-red flowers. **C**: 1
                                                              p. 153

HABENARIA (*Orchidaceae*)
  radiata (Japan), the pretty "Egret flower"; terrestrial
orchid with tuberous roots, producing a slender, erect leafy
stem to 2 ft. high, topped by solitary or clustered small
flowers with green sepals and white petals, and a white lip
1 in. wide, with spreading side lobes beautifully fringed and
with a long green spur; these blooms appearing like a bird
in flight. In flower between July and September. Grown in
rich, porous compost in pots, they do well set onto the
patio during summer, requiring liberal watering and feeding.
When the leafy stalks die back, plants are then rested to let
the tubers ripen; when the foliage has gone the tubers may
be divided and repotted. **C**: 82   p. 469, 533

HABRANTHUS (*Amaryllidaceae*)
  andersonii texanus (Texas), "Texas amaryllis"; a form
of the South American species from Uruguay and Argentina,
apparently escaped into Texas; small bulbous plant 8 in.
high, close to Hippeastrum and Zephyrantes, and by which
names it is sometimes known; with narrow linear leaves, and
reddish stalks bearing solitary lily-like yellow flowers 1½ in.
long with petals recurved at tip, coppery and striped with
purple outside, usually blooming in summer. Showy in
pots in the cool window or on the patio but flowers are
fleeting. Rest after blooming but not too dry. By offsets.
C: 63   p. 392

HADRODEMAS (*Commelinaceae*)
  warscewicziana (Tripogandra), (Guatemala); very
thick-fleshy rosette with stout stem, resembling Dracaena or
Agave; the clasping leaves broad, long pointed, to 12 in.
long, pale green, ciliate at edge and recurved; clusters of
small pale purple flowers on long stalk. **C**: 3   p. 262

HAEMANTHUS (*Amaryllidaceae*); the "Blood-lilies" are
showy, low bulbous African plants grown for their peculiar,
handsome globular flower heads with protruding stamens,
looking like brilliantly colored or white shaving brushes for
which they are known in So. Africa as "Paint brush",
"April fools" and "Torch lilies". Easily cultivated in pots, the
foliage must be kept growing after flowering until the leaves
show signs of dying off; the plants may then be rested
through the winter; in some species, such as katherinae, the
foliage remains green, but watering must be reduced. Do
not repot more than necessary. By offsets removed at
potting time.
  albiflos (So. Africa), "White paint brush"; low bulbous,
succulent plant to 1 ft. high, with thick-fleshy leaves to 4 in.
wide and with ciliate margins, evergreen or appearing with
the flowers; the inflorescence a stout stalk carrying the
brush-like white inflorescence consisting of a 2 in. head of
greenish-white flowers, dense with protruding white
stamens seated in a cup of pale greenish bracts, in summer
and fall. **C**: 14   p. 391
  katherinae (Natal), the "Blood flower"; vigorous and
striking bulbous plant 1 to 1½ ft. high, branching from offsets; the thin, soft-fleshy fresh green sword-shaped leaves
with depressed midrib running into channeled petiole, a
separate solid stalk bears an umbrella-shaped or round head
of about 6 in. across, of star-like flowers with salmon petals
and long coral-red stamens, blooming with the foliage fully
developed, in July-August. After bloom, gradually reduce
watering. A beautiful and most frequently cultivated species
for the sunny window or patio. **C**: 14   p. 391
  multiflorus (Trop. Africa), "Salmon blood lily"; magnificent tropical bulbous plant with broad leaves nearly
1 ft. long on short, red spotted petioles; the showy inflorescence forming a perfect 6 in. ball with up to 100 coral-pink
flowers on a stalk separate from the foliage, coral-pink,
crimson-red at base of the narrow petals, the showy long
extended red stamens red and tipped by the yellow anthers;
blooming in spring into June, just before the foliage. The
floral head is smaller than the popular hybrid 'King Albert',
but produces an equally dense globe of dense filaments.
C: 14   p. 391

HAEMARIA (*Orchidaceae*)
  discolor dawsoniana (Anoectochilus) (Malaya),
"Golden lace orchid"; vigorous terrestrial with creeping,
branching rootstock and fleshy, ovate gorgeous leaves of

blackish red-green velvet with a network of coppery-red
veins, wine-red beneath; small, waxy-white flowers, with
yellow center, in terminal raceme, blooming Oct.-May.
Although they will grow in porous soil, osmunda-sphagnum
mixes are recommended, adding broken pots and charcoal.
C: 62   p. 192, 454

HAMATOCACTUS (*Cactaceae*)
  setispinus (No. Mexico, So. Texas), "Strawberry-
cactus"; globose plant to 6 in. high, with 13 thick ribs
arranged spirally, grayish matte green, the areoles with tufts
of white wool, radials gray-brown, the long central spines
with fish hook; flowers yellow with red throat. **C**: 13
                                                 p. 158, 165*

HARRISIA (*Cactaceae*)
  tortuosa (Argentina), "Red-tipped dogtail"; night-
bloomer with slender arching stems erect or sprawling, to
3 ft. long, bright green, round with indistinct ribs and knobby,
awl-like spines; white funnel-form flowers. Large round
and spiny, edible red fruit. **C**: 1   p. 155

HATIORA (*Cactaceae*)
  salicornioides (S.E. Brazil), the "Drunkard's dream";
known under this name in Brazil because of the bottle-
shaped joints or branchlets of this epiphytic tree dweller;
an erect shrubby plant much branched, the thin stems green
or purplish; flowers salmon outside, yellow inside; white
translucent fruit tipped red. **C**: 10   p. 172

HAWORTHIA (*Liliaceae*); large genus of charming, small
succulent jewels from South Africa; stemless rosettes, sometimes elongated, of thick-fleshy leaves in many curious
forms, from those forming perfect stars covered with silvery
tubercles looking like bead-pearls, to others with translucent
tips, or the pointed tips cut off altogether; the inflorescence
not showy, the tiny white or greenish flowers in a sparse
raceme. The "windowed" Haworthias in their desert
habitat are buried in the ground to avoid loss of water, and
sunlight is admitted inside through the windows to reach
the chlorophyll cells of the inner body. Haworthias are about
the sturdiest, and most durable little house plants even if
neglected; they are both pretty and long-lasting planted into
novelty containers or dishgardens. They are sensitive to
strong sun, and do not want to be too dry except for a short
resting period in summer; keep 55° F. or less in winter if
possible. Propagated from offsets, leaves, or seed.
  coarctata (So. Africa: Cape Prov.), "Cowhorn haworthia"; succulent rosette elongated column-like to 8 in. high,
suckering at the base; 2½ in. leaves in a close spiral, incurved,
short and wide, keeled, dark green with a few greenish white
tubercles. **C**: 15   p. 345
  cooperi (So. Africa), "Window haworthia"; interesting
small succulent rosette, with fleshy, boat-shaped 1¾ in.
leaves green to purplish-brown, the ciliate margins and the
pointed tip translucent and window-like. In habitat these
window plants are buried in the soil to avoid burning up, and
only the "windows" allow solar energy into the plant body.
C: 15   p. 345
  cuspidata (So. Africa: Cape Prov), "Star window
plant"; charming small, stocky, tufting, 2½ in. starry rosette
with short, fat, wedge-shaped and keeled leaves; pale
grayish-green, with abruptly shortened tips having translucent windows marked with green lines. Highly interesting
for collectors. **C**: 13   p. 345
  cymbiformis (So. Africa: Cape Prov.), "Windowed
boats"; prolific, stemless, suckering rosette 3 to 4 in. across,
of thick, soft-fleshy, broad-obovate, pointed leaves, boat-
like hollowed above, keeled beneath, smooth pale grayish
green, with darker green veins running into the translucent
windowed apex. **C**: 3   p. 345
  fasciata (So. Africa), "Zebra haworthia"; very charming
small succulent rosette 2 to 3 in. across, of erect or somewhat incurved slender tapering dark green leaves, with large
white wart-like tubercles in neatly connected cross bands.
A favorite long enduring and a pretty plant even in deep
shade; popular in dishgardens. Develops offsets for propagation. **C**: 15   p. 345
  margaritifera (So. Africa: Cape Prov. north to Karroo),
"Pearl plant"; attractive stemless succulent rosette to 6 in.
dia.; leaves at first bent inward, later spreading, to 3 in. long,
thick and firm-fleshy, upper surface flat or convex, underneath keeled toward apex, blackish-green, on both sides
with prominent pearl-like, large creamy tubercles; suckering

with age. Tiny whitish flowers in clusters on wiry stalks to 2 ft. high. **C**: 15 p. 345

papillosa (So. Africa: Cape Prov.), "Pearly dots"; beautiful, robust small 4 inch rosette with erect, lanceolate fleshy leaves rich deep green, adorned with rows of greenish-white raised, often hollow warts; rounded keel beneath. From leaf cuttings or suckers. **C**: 15 p. 196, 345

reinwardtii (So. Africa), "Wart plant"; attractive slender, elongate succulent rosette, with stem to 6 in. high, making basal offsets; the slender lanceolate leaves in closely shingled spiral, and keeled beneath, marked with cross rows of white tubercles. **C**: 13 p. 345

subfasciata (So. Africa: Cape Prov.), "Little zebra plant"; shapely rosette with erect spreading, tapering leaves in dense ranks, 2 to 4 in. long, dark green and cross-banded with regular, thin rows of translucent white tubercles, especially on the lower, convex side; lightly keeled beneath. This species has longer leaves than fasciata, and thinner rows of warts. Very popular in dishgardens for its lasting quality. **C**: 15 p. 329, 345

truncata (So. Africa: Karroo), the "Clipped window plant"; stemless succulent with fleshy leaves dark green-brownish and rough-warty, arranged in 2 ranks, and most unusual with the apex of each leaf flat as if cut off with a knife; the "cut" surfaces have translucent windows admitting sunlight. **C**: 15 p. 345, 531

HEBE (*Scrophulariaceae*)

salicifolia (New Zealand), "Evergreen veronica"; charming evergreen shrub of symmetrical form freely branching; dense with shining green sessile, always opposite, lanceolate thin-leathery leaves 2 to 4 in. long; and flowers densely packed in cylindrical spike-like clusters, variously lilac-tinged white to mauve or purple; summer-blooming. Attractive both as foliage or flowering plant in the cooler greenhouse or on the patio; may be pruned after blooming or clipped into informal hedges. Propagated by cuttings of soft wood in late summer. **C**: 64 p. 484

speciosa 'Variegata', long known in horticulture as Veronica, (New Zealand) "Variegated evergreen veronica"; pretty evergreen shrub 2–5 ft. high, the spreading angled branches dense with opposite thick, oblong leaves 2-4 in. long, more or less set in ranks, glossy dark green in center and downy on the midrib above, the margins broadly variegated creamy yellow; summer-blooming in dense axillary racemes of small purple-crimson flowers, opposite near tips of branches. Very attractive as a low plant for cool locations, but the flowers shed quickly; needs water and feeding. From cuttings. **C**: 14 p. 295

HECHTIA (*Bromeliaceae*)

argentea (Mexico), "Vicious hechtia"; attractive terrestrial rosette of stiff dagger-shaped, dark glossy green leaves 12 in. or more long, with prominent sawtooth spines; gray pencil lines beneath; tall inflorescence with orange flowers. **C**: 21 p. 138

HEDERA (*Araliaceae*). Hedera is the classical name for the "Ivies", evergreen woody vines of the Araliad family; widely spread as ground covers throughout the temperate and subtropical zones of the world. They are spreading and climbing by aerial rootlets which cling easily to masonry or the bark of trees. The leathery leaves are alternate on flexuous vines, later woody. With age or when ceasing to climb, or if young wood is constantly cut off, and finding no further support, even when only 10 or 15 years old, habit of growth and the foliage change character; the branches reach out and become semi-erect tree-like and start to bear their small flowers and berry-like fruit, and the leaves change to become ovate rather than lobed, in the maturity stage. The winter-hardy H. helix varieties generally like shade and cool-moist surroundings; the subtropical species prefer it warmer and sunny, are easily adapted as house plants. Propagated from cuttings.

canariensis arborescens 'Variegata', the "Ghost tree ivy"; so called because of the ghostly trembling of the pendant foliage in a breeze; arborescent or fruiting form of the variegated Algerian ivy, its ovate, hard leaves with cream variegation on light green or gray; fruit black. An excellent, very durable, but slow-growing decorative plant in containers, for the sunny window or the patio, the beautiful variegated foliage stiff as if varnished. From half-hard cuttings. **C**: 15 p. 43*, 195, 299

canariensis 'Gloire de Marengo'; a clone selected in Europe for its beautiful markings; this is probably identical with or a colorful form of H. canariensis 'Variegata'. The relatively large leaves are usually 3-lobed, green in the center changing to milky-gray, then white or cream to yellowish in irregularly variegated areas, mostly along the margins; stems and petioles red. **C**: 14 p. 283

canariensis 'Margino-maculata', "Canary Island ivy"; differs from variegata by its creamy yellow variegation speckled and marbled with green, running over into fresh green and dark green areas; stems and petioles red. **C**: 14 p. 285

canariensis 'Variegata' (maderensis), known as "Variegated Algerian ivy", and "Hagenburger's ivy" commercially, very colorful with reddish vines and petioles, and leathery leaves 3 to 6 in. long, in the center fresh-green to slate-green, joined by a zone of blue or gray-green, and marginal variegation of creamy white, even tinted pink. The favorite indoor ivy for larger pots as well as for small dishgarden planting; tolerates warmth or coolness, dry conditions, and sun or shade. Widely used also on wire fences or in patios in California. The green species canariensis, with 6 in. leaves is at home along the Western Mediterranean in North Africa, the Azores, Canary Islands and Madeira. **C**: 14 p. 195, 283, 285

colchica 'Dentato-variegata', (Caucasus, Iran), "Variegated Persian Ivy"; robust climber with stiff pea-green twigs and large variegated leaves to 8 in. long, elongate heart-shaped, and of a thick, hard quality, the margins lightly lobed and remotely toothed; young leaves green with nile green, old leaves gray-green and scaly, with cream-white areas more or less along margins. Somewhat bold and unwieldy in appearance. **C**: 14 p. 283

'Gloire de Marengo': see canariensis

helix (Europe, Asia, N. Africa), the "English ivy"; root-climbing vine with juvenile stage leaves 5-lobed, leathery glossy forest green, with creamy veins, fairly large, 3 to 5 in. wide and somewhat cupped; this species has 4 to 12 stellate hairs covering the foliage, mostly underneath, seen through a magnifying lens; in the arborescent or fruiting stage leaves are unlobed, the berry-like fruit black. The favorite ivy used as a ground-cover outdoors, or staked on bamboo canes as pyramids for cool area decorations. Not the best as an indoor pot plant except where fairly cool, as it easily catches red spider in dry-warm rooms. Needs fair light 50 fc. up; constant moisture to prevent leaf drop, but tolerates draft near doors, and air-conditioning. **C**: 26 p. 283, 285, 541

helix 'Abundance'; slow trailer and bushy, with broad and large 4, 5 to 7-lobed variable leaves dark green with pale veins; some are wavy in the sinus, some have 2-pointed apex. **C**: 26 p. 282

helix 'Albany', "Albany ivy"; self-branching from strongly flattened (fasciated) twigs with leaves sharply 5-lobed, to 3½ in. long, rich green to purplish green with pale veins; purplish beneath. **C**: 26 p. 282

helix 'Arrow', "Arrow ivy"; a sturdy California cultivar with leathery foliage, in variety from densely shingled small leaves 1 in. long, barely lobed, or with 3 lobes, each rather ovate and overlapping, to large leaves 3 to 5-lobed and 2–2½ in. wide; dark green, smooth to shiny with pale network of veins. **C**: 26 p. 282

helix baltica (Latvia), "Baltic ivy"; clinging vine similar to English ivy but the leathery foliage not so large and more cut; whitish veins; very hardy. **C**: 26 p. 285

helix 'California Needlepoint', the "California parsley ivy", in California nurseries as "Needlepoint"; a very attractive bushy little cultivar, of especially compact growth, the branches, while short, neatly erect, self-branching from nodes, and covered with tiny leaves ¾ to 1¼ in. long, rich green, tri-lobed and in all resembling parsley. **C**: 16 p. 282

helix 'Chicago', ('Typ Schaefer') small "Emerald ivy", of classical leaf-shape, much grown in Europe for its elegantly lobed, medium green, intermediate-size leaves of good texture, with raised, light veins, willing, fast growing habit, and good keeping qualities; like 'Pittsburgh' but more short-jointed, slower growing and somewhat self-branching, the 3 to 5-lobed leaves brighter green. 1½ by 2 in. **C**: 14 p. 285

helix 'Conglomerata', the "Japanese ivy"; slow growing shrubby, contorted stems closely crowded with small undulate hard-leathery leaves dark green and stiff, with green veins; very artistic with its shingled foliage, also for dishgardens, but of extreme slow growth. **C**: 13 p. 284

helix 'Curlilocks', self-branching sport of 'Parsley', with young growths sprouting from every axil of the crested leaves, resulting in densely bushy vines and plants. Wonderful indoor ivy. **C**: 16   p. 285, 537

helix 'Denticulata'; robust growing vine with medium-small, roundish, thick leaves, somewhat cupped, 5-lobed, with basal lobes full-round. **C**: 26   p. 284

helix 'Digitata' (palmata); an old, freely vining variety; small, dark, leathery leaves shaped like a fan, the whitish veins also palmate, and usually with 5, almost equal, toothlike lobes. **C**: 14   p. 284

helix 'Discolor' (marmorata minor), "Marmorata ivy"; robust, old-time variety with wiry twigs and the small, scattered, leathery leaves deep green mottled or spotted with white. **C**: 16   p. 285

helix 'Erecta' (conglomerata erecta) (Japan); similar to conglomerata but the stout twigs grow more erect, the leaves are more broadly triangular pointed and not undulate, dark green with whitish veins, and rigidly arranged opposite each other in two ranks. **C**: 13   p. 284

helix 'Fan' (crenata), "Fan ivy"; vining variety with attractive broad, lightly 5 to 7-lobed leaves rich to dark green, palmately veined pale green, the lobes wavy or pleated at sinus, and fairly even. **C**: 14   p. 282

helix 'Fluffy Ruffles'; unusual ivy with scattered roundish leaves very much undulate and crested at the margins; quite self-branching. **C**: 14   p. 282

helix 'Glacier'; good vining growth combined with nicely variegated, small, triangular, leathery leaves 1½ to 2 in. long, of several shades of green down to gray, with white marginal areas and pink edge. An excellent small-leaved variegated ivy for indoors and dishgarden planting. **C**: 16   p. 285

helix 'Glymii'; creeper hugging the surface with horizontal, wiry green branches making aerial rootlets; the small, medium green, ovate-cordate leaves spoonlike cupped. **C**: 26   p. 284

helix 'Golddust'; trilobed leaves varying in width and shape; variegated and mottled green and yellow, especially on new growth. **C**: 16   p. 285

helix 'Goldheart', a charming small-leaved ivy from Italy where it is known as "d'Oro di Bogliosco"; with slender vines, and neatly pointed, 3 to 5-lobed, leathery green leaves with its middle painted golden yellow and cream, on reddish stems. **C**: 14   p. 285

helix 'Green Ripples'; simple bold leaves large and small, bright green and dark, often deformed and frilly but produced in abundance, and clothing the stem in a pleasing way. **C**: 14   p. 284

helix 'Green Spear'; free growing form with willowy foliage, the spear-shaped, medium green leaves with light green veins, long pointed and laciniate. **C**: 26   p. 284

helix 'Hahn's Selfbranching', 'Hahn's self-branching ivy'; a bushy variant of "Pittsburgh ivy", branching near the growing tip, making a close mat of stems and leaves; constant reselection necessary to maintain self-branching character. An excellent indoor ivy, also for dishgarden work. **C**: 14   p. 285

helix 'Hahn's Variegated'; probably the best grower of the small albinos; long vines with silver-gray, leathery leaves, and narrow white edge turning reddish; not self-branching. **C**: 16   p. 285

helix 'Harald', "White and green ivy" or "Improved Chicago variegata"; a medium small-leaved yet robust, variegated clone favored in Europe because its 1½ to 3 in. 3 to 5-shallowly lobed leaves, somewhat rounded, are mostly green not gray in the center, broadly margined creamy white; also quite durable as a house ivy. **C**: 16 p. 283

helix 'Helvetica'; shield-like, deep matte-green leaves, cordate to 3-lobed with contrasting white veins, on strong, stiff, woody vines. **C**: 26   p. 285

helix hibernica (Ireland); the "Irish ivy" is the largest of the helix varieties, growing more dense and vigorous than the "English ivy" with roundish leaves to 5 in. across, bright green with pale green veining. **C**: 26   p. 284, 536

helix 'Ivalace'; vining plant with self-branching characteristics, resulting in close growth; the 5-lobed, dark green, leathery leaves with upcurled margins creating a lacy appearance. **C**: 26   p. 284

helix 'Jubilee'; tiniest of the variegated leaf forms, self-branching; the little snubnosed leaves friendly light green, gray and white, and quite irregular. **C**: 16   p. 285

helix 'Maculata'; may be a slow-growing, variegated form of the Irish ivy; the leaves are roundish, shallowly 5-lobed, rather flat and fleshy, and the yellow-green mixed with dark green, beautifully variegated white or cream. **C**: 14   p. 43*, 285

helix 'Manda's crested'; charming plant with star-shaped, jade-green leaves bordered by a rosy edge, the long lobes fluted and undulate, on straight, upright, reddish stalks. Excellent indoor ivy. **C**: 16   p. 285

helix, 'Manda's Needlepoint,' "Needle ivy"; very graceful little bushy, self-branching cultivar full of small leaves 1 in. long, with 3 very slender, needle-narrow lobes; some of the older leaves develop to 5 lobes. Ideal as a small house plant. **C**: 16   p. 282

helix 'Maple Queen'; very satisfactory variety indoors; although branching freely, its habit is even and constrained, the small, leathery leaves are fresh to dark green with pale veins; the keeping quality excellent. **C**: 14   p. 284

helix 'Marginata', "Silver garland ivy"; an old, variegated-leaf variety with small, leathery, 3 to 5-lobed leaves 1½ to 2 in. long, and which will grow progressively larger, green with gray, and margined white, edge turning reddish. **C**: 16   p. 285

helix 'Meagheri'; its name "Green feather" describes the feathery look of the narrow laciniate, leathery leaves, with the side lobes spoon-like. **C**: 13   p. 284

helix 'Merion Beauty' (procumbens); dense little plant with weak twigs weighted down by little auxiliary shoots, breaking out of all axils; the small leaves of varied shape. Unfortunately not a good keeper indoors; unless in a light, cool, and moist location, it loses many of its leaflets. **C**: 14   p. 285

helix 'Minima'; form with flexuous twigs, green petioles and small, 3 to 5-lobed, thin leathery leaves, somewhat undulate, base cordate. **C**: 26   p. 284

helix 'Parsley' (cristata); long vining variety called "Parsley ivy", because the small, leathery, medium green, 5-lobed, roundish leaves are minutely frilled and crimped on margins. The long-lasting cut branches are used by New York florists in flower arrangements. **C**: 16   p. 285

helix 'Patricia'; sport of 'Pittsburgh', an excellent Philadelphia cultivar; dense, self-branching ivy with medium-sized leathery leaves 1 to 2½ in. wide and long, usually remotely 5-lobed, and prettily curled in at sinuses. A good keeper and favorite in a 5 in. pot, with its neatly draped reddish branches. Better keeper indoors than the "Pittsburgh ivy". **C**: 14   p. 283

helix 'Pedata', the "Birdsfoot ivy"; dainty leaves cut into narrow lobes, apple-green becoming much darker, with contrasting whitish veins; long vining. **C**: 26   p. 284

helix 'Pin Oak'; weak, selfbranching creeper with thin-leathery leaves gracefully cut into narrow curving lobes, on reddish stem. **C**: 26   p. 284

helix 'Pittsburgh', "Pittsburgh ivy"; strong vining and branching, bushy plant with leaves smaller than the English ivy, more long-pointed, not as black-green, venation less pronounced and green. One of the cultivars of the "ramosa complex", probably a bud-sport of H. h. gracilis but less spindly; quite winter hardy. Fairly good indoor plant but naturally preferring the cooler environment. **C**: 14 p. 283, 285

helix 'Pixie'. "Pixie miniature"; fine miniature variety of very compact, almost stunted growth, with densely shingled small, very variable leathery leaves deeply trilobed, to more shallowly 5 to 7-lobed. **C**: 16   p. 282

helix poetica (chrysocarpa) (Italy, Greece, Asia Minor); yellow fruited variety, with flowers greenish, and adult leaves unlobed; in juvenile stage twigs green and with aerial rootlets, leaves broadly 5-lobed with prominent palmate veins. **C**: 26   p. 282, 284, 512

helix 'Sagittaefolia' (taurica) (Turkey), "Taurian ivy"; flat growing variety with narrow-sagittate, small to medium leaves, 1¾ to 4 in. long, very heavy, thick-leathery, deep grayish green with whitish vining, on spreading, purplish vines. Slow to start, but a very frost-resistant ground cover. **C**: 26   p. 284

helix 'Scutifolia' (cordata), "Heart-leaf ivy"; small heart-shaped leaves, thick leathery and a deep, matte green with pale veining, on reddish petioles and wiry vines. **C** :26   p. 284

helix 'Shamrock'; wiry plant with red stems and tiny, bright to rich green, leathery leaves having 3 lobes more or less of same size; side lobes folded forward alongside the center segment. **C**: 26   p. 284

helix 'Silver Emblem'; wiry grower lightly self-branching, with leathery, medium-small, pointed leaves bright green and gray-green with narrow cream border. **C**: 16
p. 285

helix 'Silver King', a "Small variegated ivy"; tiny, self-branching variety with leaves smaller than 'Glacier', weaker and softer, pointed, closely set and neatly arranged, light green to gray with white edge; slow and particular, but a good little houseplant if kept on the warm side and with fair light. **C**: 16
p. 195, 285

helix 'Smithii'; medium size, star-like, 5-lobed, thin-leathery leaves, medium to dark green, with lobes somewhat fluted and tips curving downward. **C**: 26
p. 284

helix 'Star', "Star ivy"; shapely self-branching, bushy variety, leaves 5-pointed, star-shaped with slender finger-like lobes. **C**: 26
p. 282

helix 'Sylvanian'; densely clothed twigs with leaves overlapping each other shingle-like and self-branching, pointed blade leathery, unequally lobed, forest green. **C**: 14
p. 284

helix taurica: see helix 'Sagittaefolia'

helix '238th Street', "Bronx ivy"; this ivy produces, in addition to flowering twigs, stiff, viny, green shoots with un-lobed, waxy green leaves, spreading horizontally, and which stay green in winter; flower greenish-cream, fruit green to black. Excellent for low hedges. **C**: 26   p. 284

helix 'Walthamensis' (minor); the "Baby ivy", because it is the smallest variant of the typical English ivy group; leaves tiny, deep green and leathery, veins whitish. **C**: 26
p. 285

helix 'Weber's Californian'; a really satisfactory indoor plant of bushy habit; medium green, 5-lobed leaves with light green veins, notched deeply, and wavy in the sinus, lobes rounded off; reddish stem. **C**: 14   p. 284

helix 'Weber's Fan'; perhaps a fasciated form of crenata, its broad rounded leaves with 5 to 8 small, rounded lobes and undulate margin, and light green palmate veining. **C**: 14   p. 284

helix 'Williamsiana'; shapely and vigorous variety with 3 to 5-lobed leaves, the long tips curled downward while edges of leaves are wavy; greenish ivory border around apple-green or gray-green center. Exquisite variegated little-leaf plant for the fancier. **C**: 16   p. 285

HEDYCHIUM (*Zingiberaceae*)

coronarium (Himalayas into China), "Butterfly lily", also known as "Garland flower"; this "White ginger" is most popular in Hawaii and used for leis because of the sweet perfume of its broad-petaled, pure white flowers, showing a yellow heart on their lip, and appearing from behind a green, waxen bulb of scale-like bracts in terminal clusters, on robust, leafy canes to 6 ft. long, the leaves silvery-haired beneath. **C**: 54   p. 206

gardnerianum (Nepal, Sikkim), "Kahili ginger"; beautiful, desirable species growing high in the Himalayas to 8000 ft. altitude, of stiff habit, canes to 6 ft. high, leaves to 1½ ft. and powdery-white beneath when young; the delightfully fragrant flowers in elongate, open terminal spikes to 1½ ft. long, and from a cylindrical cone of green bracts, appear the yellow flowers with long, conspicuous bright red filaments. **C**: 54   p. 206

HEDYSCEPE (*Palmae*)

canterburyana (Kentia, Veitchia) (Lord Howe Island in the South Pacific), the "Umbrella palm"; sometimes grown as Kentia, which it resembles; tall, robust pinnate palm to 35 ft. high, with a dense crown of recurving, sturdy leaves to 6 ft. long, arching downward like an umbrella; pinnae numerous, green on both sides, narrow-lanceolate and acuminate, drooping at ends; leaflets short, and only about 8–10 on each side of axis. In limited cultivation in California and Florida, and as satisfactory as a houseplant as Howeia. **C**: 16   p. 225

HEIMERLIODENDRON (*Nyctaginaceae*)

brunonianum 'Variegatum', better known by its synonym Pisonia (New Zealand), "Bird-catcher-tree"; showy sport of the Para-Para tree 20–50 ft. high and native from Tahiti and the Marquesas to New Zealand and Australia, with oblong foliage to 15 in. long, having very sticky ribs, on short, robust petioles, and slightly angled stem; the glossy leaves are marbled in two shades of green: edged with warm cream to almost white; unfolding young growth has a tinge

of red at edge and midrib; clusters of inconspicuous greenish flowers. The seed pods are covered with a sweet gum which attracts birds, hence the common name. Has a tendency to die off from disease. Prop. by mossing. **C**: 16   p. 191, 295

HELIAMPHORA (*Sarraceniaceae*)

nutans (So. America: Guyana), the "Sun pitcher"; carnivorous perennial rosette from the moist Roraima savannah, 4 to 12 in. high, with funnel-shaped pitchers hairy inside, forming an insect trap of the pitfall type; the green pitcher leaves are edged in red and red-veined; the delicate nodding flowers white. A rarity mostly cultivated in botanic gardens, grown in sandy humus together with sphagnum moss, best under glass cover to maintain humidity. Prop. by single crowns taken from the rhizome. **C**: 70 p. 522

HELIANTHUS (*Compositae*)

annuus florepleno, the "Double sunflower" or "Cut-and-come-again"; an ornamental dwarf, double-flowered form of the common sunflower, at home in the prairies in Western U.S.; annual, vigorous herb 4 to 5 ft. high with erect, rough hairy stems and ovate leaves; fully double yellow flower heads 4 to 6 in. across in varieties such as 'Sungold' shown; the rays very dense and largely serrated at tips. With their golden-yellow chrysanthemum-like flowers a picturesque cut flower, also very striking in containers on the sunny patio, blooming in late summer and fall. From seed. The well-known sunflowers, as painted by Vincent van Gogh, that become giants to 15 ft. tall with immense flower disks to 20 in. across, are not only showy but produce edible seeds, bird-feed, and furnish cooking oil. **C**: 64   p. 415

HELICHRYSUM (*Compositae*)

bracteatum (Australia), the everlasting "Strawflower" or "Immortelle"; a tender annual herb 2 to 3 ft. high, usually branched, with narrow, roughish leaves, the erect wiry stems each bearing an inflorescence consisting of flower disks which are enclosed by colored bracts, in shades from white through yellow and orange to red, coming into full bloom between July and September. Highly developed into quite double-flowered form in the cultivar 'Monstrosum', flowers 2 to 3 in. across are available in many vivid, glossy colors, which when cut and hung head-downward in bunches to dry in the shade, preserve their coloring wonderfully and are much treasured for winter bouquets as "Ever-lastings". From seed sown in early spring. **C**: 76 p. 414, 535

HELICONIA (*Musaceae*); the "Lobster claws" are tropical American herbs to 15 ft. high, with banana-like foliage, cradling in the base of their leaves the very curious and spectacular erect or pendant inflorescence of waxy red or orange boat-shaped bracts arranged in two ranks, from which peek the creamy flowers. For best blooming, the larger species require large containers with plenty of water, or should be planted out. Propagated by seed or division.

bihai (W. Indies, Brazil, New Caledonia, Samoa), "Firebird" or "Wild plantain"; large perennial tropical herb to 18 ft. high, widespread from Tropical America to the Pacific; long-stalked, oblong, smooth-textured, pointed, banana-like green leaves having a pale midrib and raised lateral veins; spectacular greenish-yellow flowers clustered in the axils of large, stiff boat-shaped, burnt-red bracts ending in a slender, twisted, yellow tip, the edges apple-green to yellow; inflorescence erect; often seen as a cut flower. **C**: 54   p. 204

caribaea (West Indies: Martinique), the yellow "Wild plantain"; huge perennial with large leaves 3 ft. or more long and rounded at base resembling bananas, but arranged in two ranks, the striking inflorescence is carried erect between the foliage, being a series of large, fleshy boat-shaped pointed, stiff bracts holding water, and compacted shingle-like in two alternate series, waxy golden-yellow with keel and tip greenish. **C**: 54   p. 204

humilis (wagneriana) (Trinidad, Brazil), "Lobster claw"; related to H. bihai, but with leaves shiny green; as well as smaller, salmon-red, boat-shaped bracts changing into green toward tip, and ridge of greenish-yellow; flowers yellowish-white; inflorescence erect. **C**: 54   p. 204

illustris aureo-striata, striking tropical foliage plant with broad, recurving leaves shining fresh-green, with contrasting ivory-yellow and pink midrib, and closely parallel lateral ivory veins showing on both surfaces;

sheathing petioles pinkish, striped or mottled green. The green, pink-veined species is from the South Sea Islands; possibly a form of indica. **C**: 54   p. 204

illustris rubricaulis, has beautiful foliage light green with midrib and the dense lateral veins clear rose pink, on clasping petioles vermilion red, and red underneath; rather delicate and unfolds best in humid-warm greenhouse, if possible planted out. **C**: 54   p. 204

marantifolia (syn. metallica), (Colombia); slender tropical perennial to 10 ft. high, with long red stalks and banana-like oblong leaves shimmering velvety emerald green, and with pearl-white midrib, purplish-red beneath; waxy inflorescence erect, rosy-red flowers with pale green tips, in green, widely separated boat-shaped bracts. **C**: 54
p. 204

platystachys (Colombia, Guatemala); large banana-like plant with big leaves green on both sides, and a long pendulous inflorescence with folded bracts widely separated on snaky axis, the flower bracts slender to 7 in. long, the terminal ones shorter; vivid red edged with light green or yellow, and downy near base. **C**: 54   p. 204

psittacorum (West Indies; coast of Guyana; Brazil), "Parrot flower"; tufted tropical perennial 4 ft. or less, with long-stemmed, slender, narrow lanceolate, leathery, rich green leaves, and a stalked inflorescence of 3 in., shining orange, long-pointed boat-shaped bracts tipped red; greenish-yellow flowers with black spots near apex. **C**: 4
p. 204

velutina (Trop. America); compact plant with beautiful, satiny, grass-green, pointed, rather short leaves having prominent pale midrib; lateral veins dense and raised; shining green beneath; petioles and stem green. **C**: 54 p. 204

HELIOCEREUS (Cactaceae)

speciosus (Central Mexico), aptly named the "Sun-cactus", because of its large and showy, day-flowering, bright scarlet flowers with a lovely steel-blue sheen; 4-angled stems erect or clambering, freely branching; the spines all alike. **C**: 4   p. 155

HELIOTROPIUM (Boraginaceae)

arborescens, the old-fashioned "Heliotrope"; (Peru), tender, fleshy herb becoming shrubby to 3 ft. high, small ovate wrinkled leaves and rounded clusters of $\frac{1}{8}$ in. purple flowers; very fragrant like vanilla. The sweet scent of a potted heliotrope in the window awakens feelings of nostalgia whenever one is near it. Usually grown as an annual about 8 in. high from cuttings, or by layering of long branches. **C**: 66   p. 378, 409

HELLEBORUS (Ranunculaceae)

niger (Europe: Alps, Apennines, Yugoslavia, Carpathians), the "Christmas rose"; cultivated in European gardens since the 16th century; a wintergreen hardy perennial herb 6 to 12 in. high, with thick blackish roots, palmately divided leathery, lustrous olive green leaves, pubescent beneath; charming solitary long-lasting flowers with 5 large petal-like sepals white or greenish-white, becoming purplish with age, to $2\frac{1}{2}$ in. across, the green petals very small, and mostly hidden by the numerous stamens, the nectaries yellow; if undisturbed blooming from December to early spring, sometimes under cover of snow. Strong plants in pots may be forced for Christmas but must stand cool until buds are formed, then warm. Good cut flowers; sear end of stems. Propagated by division of roots August-September or seed. Hellebores yield drugs. **C**: 78   p. 418

HELXINE (Urticaceae)

soleirolii (Corsica, Sardinia), popularly known as "Baby's tears", also as "Mind-your-own-business", "Irish moss", "Corsican curse" and "Japanese moss"; low moss-like creeping herb hugging the ground and forming dense mats or cushions as ground cover, in subtropical plantings and terrariums, or spilling over the edges of pots; tiny roundish lush-green leaves $\frac{1}{4}$ in. or less across on threadlike intertwining branches, and often pinkish in color, with minute greenish flowers in leaf-axils. **C**: 72   p. 265

HEMEROCALLIS (Liliaceae); the "Day-lilies" are strikingly beautiful and free-blooming winter-hardy, herbaceous perennials with short rhizome and fleshy roots, linear basal leaves, and clusters of showy trumpet-shaped flowers lasting only a day, but others follow on the branched stem.

Highly hybridized, the period of bloom may be extended from late spring into fall, in colors from yellow through orange to rusty reds. Of easy culture, they would require large containers if grown for the patio or sunny room. Propagated by division.

lilio-asphodelus (flava) (Siberia to Japan), "Tall yellow day-lily"; vigorous perennial herb with spreading rhizomes; linear basal leaves 2 ft. long, and weak, arching, branched 2 to 3 ft. stalks carrying beautiful lily-like lemon-yellow, fragrant flowers spreading $3\frac{1}{2}$ in.; blooming from May to July. **C**: 76   p. 401

HEMIGRAPHIS (Acanthaceae)

colorata (Java), "Red ivy"; prostrate tropical herb with stringy, rooting branches and opposite, broad-cordate, puckered and toothed leaves $2\frac{1}{2}$ to 4 in. long, shimmering metallic violet, underneath red-purple; terminal heads of small white flowers between large bracts, conspicuous trailer for the warm shady greenhouse. **C**: 56   p. 263

sp. "Exotica" hort. (New Guinea); trailing plant of robust habit, with flexible, pubescent, reddish branches, opposite oval to ovate 3 in. leaves irregularly depressed-puckered; metallic purplish-green surface, margins crenate; reverse wine-red. Smaller foliage and more bushy than H. colorata. **C**: 56   p. 186

HEPTAPLEURUM: see Schefflera

HERMODACTYLUS (Iridaceae)

tuberosus (So. France to the mountains of Greece), "Snake's head iris"; sun-loving plant close to Iris, with hand-like tubers, on creeping rhizome, 2–3 four-angled glaucous leaves 2 ft. long; solitary peculiarly attractive flowers of somber coloring, with green velvety "falls" painted black-purple, inner segments or standards bright yellowish-green, 2 in. across on rather weak stalks 1 ft. high; April-May blooming. Grown in pots and kept indoors, this iris can flower in February-March depending on temperature. Somewhat hardy. Propagation by division of the rhizomes. **C**: 64
p. 395

HESPERALOE (Liliaceae)

parviflora (Texas to Mexico), "Western aloe" or "Red yucca"; stemless, clustering rosette of hard, recurving, channeled leaves deep green to grayish, to 4 ft. long and less than 1 in. wide, the margins corky and with coiling threads; showy nodding, bell-like, day-blooming 1 in. flowers salmon-rosy, on tall spikes. **C**: 13   p. 200

HESPERANTHA (Iridaceae)

stanfordiae (So. Africa), the "Evening flower"; cormous plant to 15 in. high, with narrow-linear sickle-shaped basal leaves, and erect leafy, slender stalks with clusters of sweetly scented bright yellow flowers with flaring lobes, to 2 in. across, and which open from noon to sunset; blooming in April-May. For pots, plant in November and keep cool until growth begins; water moderately till flowers fade then keep quite dry from Sept. to January. Propagate by cormlets. **C**: 63   p. 394

HEXISEA (Orchidaceae)

bidentata (Mexico to Costa Rica, Panama, and Colombia); floriferous epiphyte of curious growth habit; with slender spindle-shaped two-leaved pseudobulbs set above each other like a jointed stem, bearing 2 leaves at each node and eventually $1\frac{1}{2}$ ft. tall; a burst of small bright scarlet red flowers to 1 in. long, appear from the nodes along the stem and near the top mostly during summer, but often blooming at intervals throughout the year. A species easy to grow and flower by the amateur, it does not require a rest period, likes good light, and does well on treefern slabs or in pots. **C**: 22
p. 469

HIBBERTIA (Dilleniaceae)

volubilis (Queensland, N.S. Wales), "Guinea goldvine" or "Snake vine"; twining shrub, reaching 30 ft. high, with obovate 3 in. leaves, glabrous above and clasping at base, silky-hairy beneath; showy 2 in. flowers rich yellow with numerous stamens, unpleasantly scented. May be kept compact and bushy by pruning. From cuttings. **C**: 65   p. 277

HIBISCUS (Malvaceae); the magnificent "Rose mallows" are principally shrubs or small trees of the tropics, a few in temperate regions; of sparry habit, with alternate leaves,

and mostly very large 5-petaled flowers from white to yellow and red to magenta. Hibiscus are among the most showy flowers of the tropics, especially the Chinese hibiscus, which is the state flower of Hawaii where some 5000 varieties are known. Raw flowers are eaten there to aid digestion. In Tahiti, a flower worn over the right ear shows that one is looking for a mate. Blossoms of most varieties remain open for one day only, unfolding early in the morning and dying after closing near sunset, but in good sunlight are almost everblooming, especially if given liberal watering and nourishment. Much used as flowering shrubs and hedges in tropical gardens. Chinese hibiscus are also wonderful container plants for pots or tubs indoors, or on the summer patio, preferring some rest, and cutting back in winter; need lots of food and water when growing, to support adequate foliage and prodigious bloom. In the sunny window or the winter garden container plants in bloom should not be turned or buds will drop. The H. rosa-sinensis types are propagated by soft wood cuttings in spring under plastic tent, or hardwood cuttings with bottom heat in fall; also grafting on seedling stocks.

mutabilis (China), "Confederate rose" or "Cotton-rose"; fast-growing shrubby bush becoming tree-like where planted in the tropics and subtropics; green stems becoming woody, the large 3 to 5-lobed leaves 4 to 8 in. wide, dull green and rough pubescent; toward branch ends the showy axillary flowers 4 to 5 in. across, opening white or rose in the morning with crimson center and a divided maroon column; by evening the flower becomes deep red. In colder areas the plant is deciduous, growing flowering branches from a woody base. Spring and summer blooming. **C**: 64   p. 495

rosa-sinensis cooperi (E. Indies), "Checkered hibiscus"; ornamental shrub mainly grown for its colorful foliage; the narrow lanceolate leaves are metallic green and brightly variegated and marbled with dark olive, white, pink and crimson; small scarlet flowers. **C** 2   p. 182

rosa-sinensis 'Matensis', "Snowflake hibiscus"; probably a clone or color form of H. rosa-sinensis cooperi; vigorously branching shrub with willowy reddish stems dense with rough ovate toothed leaves grayish-green, variegated mainly toward margins with creamy-white; single 3 in. flowers carmine-red with crimson veins and center. **C**: 2   p. 182, 495

rosa-sinensis plenus (India to China), "Double Rose of China"; magnificent robust flowering bush of somewhat sparry habit, with fresh green foliage and large showy blooms, 4 to 6 in. across, typically carmine-rose with dark center, but with flowers fully double; in bloom whenever the sunlight is intense enough, best during the warm summer season but also in winter. Other cultivars are in shades of yellow to deep red. **C**: 2   p. 495

rosa-sinensis 'Regius Maximus', (E. Indies to China), "Flower-of-a-day", "Chinese hibiscus", "Rose-of-China"; vigorous, free-flowering, spreading, tropical shrub 5 to 20 ft. high, with sparry, willowy branches and large ovate, toothed leaves deep glossy-green, and showy 5 to 7 in. single flowers with recurved wavy petals a deep glowing scarlet and a blackish center; the long staminal column red with golden anthers. Chinese hibiscus lend themselves beautifully to training as semi-standard trees by encouraging a straight central stem, cut off at the height desired; and are available in many other colors as well as bicolors. **C**: 2   p. 483, 495

rosa-sinensis 'White Wings'; unusual variety with attractive large flowers having flaring, narrow linear petals 3½ in. long, freely separated from each other, pure white but with a contrasting feathery crimson design at the base, from which rises a striking red staminal column. **C**: 2   p. 495

schizopetalus (Trop. E. Africa), "Japanese lantern"; exciting tropical flowering shrub with slender drooping branches, smooth ovate-elliptic, toothed leaves; the showy orange-red flowers 3 in. across, hanging from slender stalks, petals deeply slit and recurved, from their center a fantastic, long projecting pendulous staminal column; blooming from May to November. Wonderful on the sunny patio, especially if trained into little trees. **C**: 52   p. 495

syriacus coelestis (China, Japan), "Rose of Sharon" or the blue "Shrub-althea"; hardy deciduous flowering shrub to 10 ft. high, with lush-green rhomboid-ovate or 3-lobed leaves, crenate toward apex; widely bell-shaped, axillary 3 in. single flowers in a shade of delicate mauve-blue with darker eye, and with prominent ivory staminal column; blooming in late summer to October. The species is violet,

but there are other color forms including red and white, also doubles. For summer and fall-blooming outdoors; may be cut back in winter. **C**: 76   p. 495

HIPPEASTRUM (*Amaryllidaceae*); tropical American bulbous plants popularly known as "Amaryllis" but differing from the true Amaryllis by having hollow flower stalks and the presence of scales between the filaments in the flower. The Amaryllis belladonna, on the other hand, is African, with solid stalks and without scales between the stamens. With their striking, lily-like trumpets, in shades from red and rose to white, amaryllis are amongst the showiest of all bulbous flowers which appear in long-stalked clusters of 2 to 6 before or with their basal, strap-shaped leaves. Hybrids with gorgeous blooms 8 to 12 in. across have been developed, and all are excellent blooming plants for indoors. Bulbs that have been rested with their perennial roots intact are easiest to bring into bloom year after year. Bare bulbs of the large Dutch amaryllis, rested from August on and potted in October should start flowering from end of December. To force early bloom indoors, keep in warm, dark place 70 to 80° F. until rooted, and flower-stalk is 6 to 8 in. high, then put in warm light shade, and feed occasionally. After flowering, cut the stalks off and continue to grow in a sunny window, or if possible plunge the pots outdoors for the summer and feed; from August or September reduce watering and keep indoors at about 50° F. until new growth begins. Propagated from side bulblets, by cut fractions of bulbs, or by separating bulb scales.

'Leopoldii hybrid' (Amaryllis), a Dutch "Amaryllis"; the leopoldii hybrids are considered the finest class of fancy flowering amaryllis in pots, the result of breeding H. leopoldii from Peru with reginae and other species; large flat, open-faced flowers generally 8 in. but even to 12 in. across, with roundish, overlapping segments, and short tubes. Hybrid clones come in a wide range of colors, from deep glowing red, clear scarlet, gleaming orange-red, light rose, and white with stripes or finally, all white. Such clones are propagated vegetatively. **C**: 13   p. 374

x leopoldii hybrid 'Claret'; a crimson "Dutch amaryllis" by Warmenhoven of Holland; a perfect open-faced giant trumpet glowing crimson on wine-red, with blackish-red lines, 8 in. and more across. **C**: 13   p. 388

procerum (Worsleya rayneri) (Brazil), the "Blue amaryllis"; distinctive bulb with a stem-like neck to 3 ft. long, from which extend the drooping, sickle-shaped leaves 2 in. wide and becoming 3 ft. long, seamed with brown; the flowers with pointed segments. The false stem carries up to 6 large, funnel-shaped flowers with long pointed petals 5 to 6 in. long and with wavy edge, heliotrope blue or mauve, with paler lavender base shading to white, blooming July-August. This tropical beauty is difficult to flower; some growers recommend the addition of osmunda fiber to loam; keep constantly moist. **C**: 82   p. 388

psittacinum (So. Amer.), "Parrot amaryllis"; beautiful flowering plant with bulb having a stem-like long neck to 6 in. long; 6—8 strap-like, glaucescent leaves at the same time as the flowers, 2 to 4 together on a stalk 2 to 3 ft. long; the trumpet-shaped blooms 4—6 in. long, the narrow, spreading segments greenish and with thickened green keel in the center, crimson stripes radiating from it to the crimson-red edges. Requires warmth during winter, and kept dry. **C**: 13   p. 388

puniceum (H. equestre) (West Indies and Mexico to Brazil), popularly known as the "American belladonna" or "Barbados lily"; a smaller-flowered amaryllis with bulb having brown scales and short neck, narrow strap-shaped, pointed waxy green leaves ¾ to 1¼ in. wide, and long round stalk bearing 2 or more obliquely trumpet-shaped flowers of 5 in. dia. and 4 in. long, salmon-red, with center creamy-whitish with greenish bands, an oblique feathery corona in base; stamens when straight larger than stigma. An old-fashioned flowering houseplant almost continuously in bloom, preferring cooler location. From offsets freely produced. **C**: 13   p. 388

reticulatum striatifolium, "Stripe-leaf amaryllis"; interesting bulbous South Brazilian plant with strap-shaped dark green leaves to 2 in. wide having a prominent ivory-white midrib; the lily-like, fragrant flowers are 4 in. long, rose-pink lined and netted darker rosy-red, without corona in the throat; blooming in autumn. Keep warm and moist in summer, cool in winter. A good house plant. From offsets. **C**: 13   p. 388

striatum fulgidum (So. Brazil), the "Ever-blooming amaryllis"; a vigorous bulbous plant with broad obovate, fleshy green leaves 1½ to 2 in. wide, in time forming bushy clumps from lateral bulbs, the trumpet-like blooms clustered on stout stalk, salmon-rose with lemon-yellow center and base; the flowers 4 in. long, with 3-parted slender tube, without corona inside. Recurrent blooming, and an ideal flowering house plant. Propagated from bulbs forming alongside the old one. **C**: 13   p. 388

vittatum (Peruvian Andes, and Brazil), "Striped amaryllis"; large bulb producing trumpet-shaped flowers white with purple stripes and 5–6 in. across, with fairly broad, pointed petals, several on a hollow stalk; the strap-shaped leaves appearing with or after flowers. February-blooming. These plants may be grown year after year in the same pot, but require a rest in summer after the leaf-growth is completed. This species has become parent to many beautiful hybrids. **C**: 13   p. 388

HOFFMANNIA (*Rubiaceae*)

ghiesbreghtii (Higginsia) (Mexico), "Tall taffeta plant"; erect, herbaceous ornamental with 4-cornered, winged, green stems eventually to 4 ft. tall; the long lanceolate leaves soft and shimmering-velvety, bronzy to moss-green, quilted between the silvery green and pink ribs, rosy-red beneath; small, tubular flowers yellow with red spot. Requires high humidity and warmth. **C**: 62   p. cover*, 185

roezlii (Mexico), "Quilted taffeta plant"; very ornamental plant possibly a variety of H. refulgens; with 4-angled stem and roundish ovate leaves 4 to 8 in. long, scanty-hairy, wine-red in the center, puckered all over, and with a satiny lustre caused by the reflected light playing on the shades of green and rose-purple on the surface; underside purple; small, dark red flowers. Quite sensitive to chills. **C**: 62 p.185

HOHENBERGIA (*Bromeliaceae*)

stellata (Brazil, Venezuela); large, loose rosette of green leathery leaves to 3 ft. long, finely toothed; inflorescence a long leaning spike with alternate clusters of reddish bracts with purple flowers. **C**: 9   p. 141

HOLMSKIOLDIA (*Verbenaceae*)

sanguinea (Himalayas, Malaysia), "Chinese hat plant" or "Coolie's cap"; subtropical fast-growing straggling shrub attaining 30 ft. but in cultivation usually 3 to 5 ft. high, with slender-pointed ovate leaves to 4 in. long; and clusters of curious 1 in. flowers, their tubular, scarlet-red corolla rising from a nearly circular, widely flaring brick red calyx base, blooming all summer and beyond in good light; an excellent flowering patio plant and on trellis or against wall. From seed or cuttings. **C**: 52   p. 501

HOMALOCLADIUM (*Polygonaceae*)

platycladum (Muehlenbeckia) (Solomon Islands), "Ribbon bush" or "Tapeworm plant"; odd curiosity plant that may become 2 to 4 ft. high; weird, perfectly flat and jointed, ribbon-like fresh-green leathery stems, bearing small greenish flowers at alternate joints; the true leaves ½ to 2½ in. long appearing mostly in the vegetative stage. Usually from 1 to 5 ft. high, but in the tropics making round canes to 12 ft. Both interesting and a good old house plant. From cuttings or division. **C**: 2   p. 298

HOMALOMENA (*Araceae*)

rubescens (Sikkim to Java); durable plant, with heart-shaped leaves reddish-green having red-brown edge and sunken veins; the slender, brown-red petioles on thick stem. **C**: 10   p. 98

wallisii (Curmeria) (Colombia), "Silver shield"; beautiful low, compact, leathery plant with broad, oval, reflexed leaves dark olive-green and prettily blotched with yellowish-silver, translucent-silvery edge. A treasure in the exotic collection. **C**: 60   p. 43*, 98, 179

HOMERIA (*Iridaceae*); the "Cape tulips" are attractive South African early summer-blooming herbs with a coated corm and linear leaves basal and along the stalk; the flowers cup or bell-shaped at first then opening flat into 6 separated segments, closing in the absence of sunshine; the filament united in a tube. Homerias may be had in bloom earlier by planting them in pots in fall, keeping them cool and dark until growth begins, then set near the glass. Rest when foliage yellows, and repot. Propagated by cormels.

collina ochroleuca (So. Africa: Cape Prov.), the "Cape tulip"; pretty cormous herb 18 in. high, and allied to Moraea, with usually one linear basal leaf longer than the leafy flowering stalk; clusters of conspicuous yellow flowers 1½ in. long, with orange markings, bell-shaped at first, then opening nearly flat into 6 separate segments spreading 2¼ to 3 in. across, but closing on dull days. The type species collina has vivid red blooms with yellow center. Flowering in So. Africa from August to October, in the Northern hemisphere in early summer. **C**: 63   p. 394

lilacina (So. Africa), "Lilac shower"; small, pretty summer-flowering plant with 3 basal leaves narrowly linear, to 12 in. long; slender stalk to 9 in., with starry lilac flowers veined purple and with yellow-speckled purple blotch, the long lobes spreading wide and nearly flat 2½ in. across. **C**: 63   p. 395

HOODIA (*Asclepiadaceae*)

rosea (So. Africa, Botswana), "African hat plant"; clustering cactus-like succulent to 12 in. high with leafless circular grayish-green stems branching from the base; ribs about 14, dense with tubercles armed with pale brown spines; the spectacular flowers have a saucer-shaped flat corolla 3 in. across, rosy to cinnamon and covered with thin hairs. Keep dry and warm in winter; best from fresh seed. **C**: 1   p. 353

HOSTA (*Liliaceae*)

plantaginea, formerly known as Funkia (Japan, China), "Fragrant plantain lily"; bold perennial herb with thick durable rootstocks forming large clumps of big, ovate-cordate green leaves 10 in. long, having 7 to 9 sunken veins on either side of midrib, the petiole with incurved wings; a stout stalk 2½ ft. high topped by a cluster of large white waxy, 4 in. trumpet-shaped ascending, fragrant flowers carried well above foliage, blooming in late summer and fall. Good in containers both for their showy foliage and lovely flowers, for sunny or shady locations; foliage goes dormant in winter. By division of the clumps. **C**: 76   p. 418

sieboldiana 'Variegata' (Funkia glauca) (Japan), the "Seersucker plantain-lily" in variegated form; winter-hardy perennial with thick durable roots forming clumps of basal, handsome foliage; the giant soft-fleshy roundish to cordate, 10 to 15 in. leaves with many deeply depressed veins, deep green on surface, variegated with white toward margins, glaucous-blue beneath; nodding 2 in. flowers white tinged lilac, in raceme above or half-hidden between foliage. Attractive foliage clumps for moist sunny or shady places, also good in containers. Dormant in winter. Propagated by division. **C**: 76   p. 299

undulata 'Medio-picta' (Funkia) (Japan), "Waxy-leaved funkia" or "Plantain-lily"; stout clustering herb with ovate basal leaves 6 in. long, narrowed toward base, sharp-pointed and curved, margins undulate, largely variegated cream in the center areas; pale lavender flowers on stalk 30 in. high. Winter-hardy in cold temperate zones. **C**: 28   p. 187

HOWEIA (*Palmae*), the popular "Paradise palms" or "Sentry palms" are commonly known as "Kentias" in the horticultural trade; they are regally handsome monoecious feather palms 30 to 60 ft. high. Seed was first imported into Europe in 1871, during Victorian times, from Lord Howe Island, east of Australia's New South Wales, and soon becoming widely used in parlors and hotel lobbies; later exported to the United States. Today they are also grown in Southern California shade houses where the climate agrees with them. Tubbed "Kentias" are usually grown in combinations of several plants at least 5 or 6 years old; over the years they have proven ideal decorators for locations with moderate temperature, also under air-conditioning; they tolerate much abuse and poorer light than other palms, from 20 fc. to preferably 200 fc. They are maintained best in smallish tubs and must be kept evenly moist; if containers are too large, keep more on dry side. However, dryness in the extreme, and also when too wet while cold, and with poor drainage, will be the cause of brown tips on the leaves.

belmoreana, better known in horticulture as Kentia (Lord Howe Island in the South Pacific), the "Curly sentry palm"; handsome solitary feather palm to 25 ft. high with pinnate leaves erect and then arching downward on short petioles; the segments closely placed, narrower than forsteriana, first curving upward then gracefully pendant and slender-pointed, usually on reddish stalks; fruit yellow green, on single spike, The overall habit of this species is more compact and the fronds spreading, requiring more room as a younger plant than the more erect forsteriana, which spreads its fronds only with age. **C**: 16			p. 225

forsteriana (Kentia) (Lord Howe Island east of Australia), commonly known in horticulture as "Kentia palm" or the "Paradise palm"; elegant, sturdy decorator, relatively new to interior use, seed having been brought to Europe only since 1871 during Victorian times, but now widely used by florists because of its tropical effect and consistently good keeping qualities; compact yet graceful with pinnate fronds growing successively larger on long slender stalks, the well-spaced, waxy deep green pinnae leathery and spreading flat. With age, this palm will develop a robust trunk 60 ft. high with fronds to 9 ft. long; the 1 to 2 in. olive-shaped yellow-green fruit, hanging in heavy clusters in successive 4-strand branched racemes; the seed is ripe and viable only about 6 years after the flowers from which it developed. **C**: 16			p. 219, 222, 225

HOYA (*Asclepiadaceae*); tropical and subtropical vines of the Milkweed family; high-climbing and twining with wiry to woody, flexible stems and opposite, fleshy, leathery leaves, and stiff-waxy, long-lasting, wheelshaped flowers in showy axillary clusters, and known as the "Wax plants". Popular and durable house or greenhouse plants trained to trellis or wire support; use soil with good drainage; pots should not be disturbed or even turned around when in bud as changed light direction may result in the loss of buds; keep drier and cooler in winter, more moist and warm but airy in summer; feed when growing. Flowers must not be cut off as the new buds will set on the same old spur year after year. Propagation from shoots, or joints, of the preceding summer; also by layering.

australis (Australia: N.S.W., Queensland; also Fiji, Samoa), "Porcelain flower"; robust vine with broad and large oval, pointed leaves 3 to 5 in. long, thin-fleshy, waxy, light glossy green on both sides; small and dainty ⅝ in. flowers waxy greenish white with purple dots at base of petals and with white crown, in small clusters axillary from the leaf joints. **C**: 15			p. 273

bella (paxtonii) (India), "Miniature wax plant"; dwarf, shrubby plant with flexuous branches first upright, later drooping, and not climbing; the small thick leaves ovate, to 1 in. long, deep green, and a brown band along the midrib; flowers waxy-white with purple center. Charming little plant ideal for baskets. **C**: 59			p. 273

carnosa (Queensland, So. China), called "Wax plant" because of the waxy, wheelshaped, fragrant pinkish-white flowers, with a red, star-shaped crown in the center, in pendant umbels in summer; root climbing fleshy vine with elliptic thick-leathery waxy leaves 2 to 3 in. long. This old-time house plant favorite prefers the cooler or moderate temperature of 50 to 60° F. in a window, sun porch, or wintergarden, with fresh air, growing best and undisturbed in porous, well-drained soil; decidedly dry in winter. **C**: 15			p. 273

carnosa 'Hummel's compacta', "Hindu rope plant"; contorted cultivar of Hummel's H.c. 'Exotica' in California; a tortuous wax plant vine densely compacted with folded and cupped foliage oddly twisted, the 1½ to 2 in. leaves waxy green with occasional silver spots, grayish beneath; the waxy flowers in a dense cluster, old ivory tinted pink petals and starry crown, on brown base. Slow-growing and ideal for training on trellis for near the window. **C**: 13   p. 353, 532

carnosa variegata, "Variegated wax plant"; very ornamental variety with the fresh green to bluish, leathery leaves bordered creamy-white, and often shaded and edged red. A beautiful little house plant which tolerates some neglect; used as a long-lasting substitute for variegated Peperomia in dishgardens. **C**: 3			p. 273

cinnamonifolia (Java), "Cinnamon wax vine"; long branching stems twining to 10 ft. high, with ovate, thick-fleshy leaves 3 in. long, light olive with conspicuous pale veins; the young foliage with bronze cast; large yellow-green flowers, corona segments deep red. **C**: 9			p. 272

imperialis (Malakka, Borneo), "Honey plant"; tall, robust climber with felty stems and elliptic, leathery, shiny leaves 6 to 9 in. long lightly downy, margins wavy; waxy flowers reddish-brown, about 3 in. across, the crown creamy-white in pendant clusters. Needs rich soil and high temperature. **C**: 59			p. 272

kerrii (Thailand, Indochina, Fiji), "Sweetheart hoya"; robust tropical climber with curious, thick succulent, leathery leaves in form of an inverted heart, the obcordate foliage notched at the apex, 5 by 4 in. large, dull green, on stout green, rooting stems; clusters of small axillary flowers ½ in. across, the petals turned back, greenish tinted pink, and covered with white velvet; the crown dark purplish crimson. **C**: 59			p. 273

keysii (Australia: Queensland), close-jointed creeper with thick stem, later climbing, and pale, gray-green, obovate, thick leaves covered with gray felt. Starry flowers in large clusters all white. **C**: 1			p. 272

lacunosa (Malaysia); tall twiner with thin, rooting stems; long-elliptic, shiny, leathery leaves 1½ in. long, waxy olive green; wheel-shaped, fragrant small flowers greenish-yellow, velvety hairy inside, in flat, axillary clusters. Nice as a hanging plant. **C**: 59			p. 273

sp. "minima" (Thailand) (Dischidia minima hort.), introduced in California nurseries as "Dischidia minima"; slender vining plant with small, thick-fleshy round-oval leaves of ⅝ in. dia., surface dull green and rough above, silvery gray beneath; the pendant floral clusters of ½ in. flowers with pale pink petals covered by silky white scales, the crown white, center crimson. This may be H. serpens. **C**: 59			p. 273

motoskei (Indonesia: Borneo, Sumatra), "Spotted hoya"; robust climber with waxy broad elliptic leaves 3–4 in. long, irregularly spotted silver; thinner than australis; waxy, star-like somewhat cupped flowers in umbel, pinkish cream with maroon center; a prolific bloomer. May be a natural hybrid of H. carnosa. Very tolerant houseplant. **C**: 3   p. 273

purpureo-fusca, "Silver-pink vine", found in Queensland and probably from Java; introduced from Hawaii as 'Silver Pink'; a beautiful vining plant with leathery long-ovate leaves to 4½ in. long, distinguished by raised pinkish-silver blotching on the waxy green, and supported on red petioles; waxy flowers consisting of rusty-red, white-hairy stars with waxy, pinkish-white inner stars or crown, and with purple center, in large axillary umbels. **C**: 3   p. 273

HUERNIA (*Asclepiadaceae*)

pillansii (So. Africa: Cape Prov.), "Cockle burs"; dainty succulent with tiny cactus-like, 20–24 ribbed green to purplish columns only 1½ to 2 in. high, set with spiral rows of soft, maroon spines; densely clustering and creeping; the small star-like 1¼ in. flowers yellow, spotted crimson. The little fingers clothed with bristles give the plant a delicate soft-hairy aspect; free-blooming. Sensitive to over-watering. **C**: 1			p. 352

zebrina (So. Africa, S.W. Africa, Botswana), "Owl-eyes" or "Zebra flower"; dense clusters of little, fleshy, 5-angled stems 3 in. high, marked reddish, with spreading stout teeth; small, star-like 1¾ in. flowers with a thick ring around mouth, yellow with transverse purple bands. **C**: 13			p. 352

HUMATA (*Filices: Polypodiaceae*)

tyermanii (Cent. China), "Bear's foot fern"; related to Davallia; small epiphytic fern with creeping light brown rhizome covered by silvery-white scales, and looking like a rabbit's foot; brown, channeled petiole bears sturdy dark green, leathery, tripinnate 6 in. frond, always with spores; slow growing. **C**: 72			p. 241

HUNTLEYA (*Orchidaceae*)

meleagris (Costa Rica, Panama, Colombia, Brazil); a beautiful and interesting epiphyte without pseudobulbs, bluish-green leaves to 15 in. long, arranged fan-like in 2 ranks; from near the bases the large solitary 3 to 4 in. very waxy flowers on long stalks, with sepals and petals whitish at base, passing into yellow, heavily flushed and marked red-brown, the fringed lip white, with front yellowish-brown; blooming in summer and into fall. With their sweetly fragrant, long-lasting star-shaped flowers one of the finest American orchids; but coming from high elevations needs attention to adequate moisture with perfect drainage and fresh air. Propagated by division. **C**: 72			p. 469

HYACINTHUS (*Liliaceae*); historically important spring-flowering bulbs whose introduction in form of the "Garden hyacinths" (H. orientalis) from Constantinople in 1560 set fanciers in Renaissance Europe wild with speculation. Greatly improved since then, hyacinths in pots, 8 to 10 in. high, with their heavy, cylindrical clusters of waxy, intensely fragrant flowers in many colors, carried on watery-fleshy stalks and with fresh green foliage, are always beloved as a blooming plant indoors. After potting in fall and kept in a cool cellar or buried outdoors to form roots, hyacinths are easily forced, beginning mid-December, to bloom in 2 to 3 weeks. Hyacinths are best obtained and bought as bulbs of blooming size. Propagation may be from offsets but these need 3 to 4, or more, years to bloom.

'Carnegie' (orientalis cv.); an excellent late-blooming Dutch hyacinth of robust, compact habit with large and full heavy heads of pure white, waxy flowers; a favorite for flowering in pots especially if Easter is late, while the ivory-white 'L'Innocence' is recommended for an early Easter as the easier forcer. **C**: 26   p. 399

'Delft Blue' (orientalis cv.); a magnificent, medium-early Dutch hyacinth, excellent for Easter flowering in pans; large 3 to 4 year bulbs will produce a full spike of giant bright porcelain-blue bells, with rich, over-powering fragrance. **C**: 26   p. 375

'Ostara' (orientalis cv.); robust Dutch garden hyacinth with large, heavy spikes of medium sized, 1 in. porcelain-blue or violet-purple waxy flowers on pale center, suitable for both early and Easter flowering in pots; an improvement of the old favorite, pale blue 'Bismarck'. **C**: 26   p. 399

'Pink Pearl' (orientalis cv.); an excellent, early blooming hyacinth, usually with two spikes if the bulb is large and old enough; the trumpet-shaped flowers are somewhat loosely set and a lovely, vivid pink; large heads are borne on stout stalks which are brittle as in all hyacinths. As this cultivar also holds for later blooming, it is a great and dependable favorite for Easter and throughout spring in pots; when grown cool the petal-tips turn green. **C**: 26   p. 399

HYDRANGEA (*Saxifragaceae*); the "Snowballs" are ornamental woody plants mostly deciduous, with showy, long-lasting terminal clusters of individually small flowers, in summer; the outer flowers of the cluster often without stamens or pistils and with larger petals than the inner, fertile flowers. The florists' hydrangeas, also known as "Hortensias" or "Bigleaf hydrangeas", are cultivars of H. hortensis macrophylla, with huge globe-shaped clusters of rosy, white or blue flowers practically all of which are sterile. They are forced for Easter or Mother's Day following a cool period of dormancy with frost to 28° F. Hardy from New York southward. Good pink flowers are the result of available calcium in the soil at about pH 6.5, if more acid add ground limestone to maintain a level near neutral. Blue flowers occur naturally in acid soils at about pH 5 which makes the aluminum present available; if higher, add alum crystals, or bits of iron, to the soil, or aluminum sulphate in periodic waterings well ahead of bloom; only good deep rose-colored varieties will turn clear blue. The pot hydrangeas are propagated from softwood cuttings.

macrophylla 'Chaperon Rouge' (Red Ridinghood), also known as 'Chapeau Rouge' (Red Cap); superb "French hortensia" (Mouillère 1951) of compact habit; the stiff erect stem with smallish, dark green foliage bearing a large firm, rounded head dense with flowers in vivid, pure rosy-crimson to clear carmine, depending on degree of temperature and pH of soil; ideal pot plant, flowering medium-early. **C**: 64   p. 377

macrophylla 'Enziandom' (1949); excellent German hybrid; free-branching, tallish potplant for early forcing, with dark green, glossy leaves and large, dense heads of flowers, by nature vivid red but by the use of aluminum sulphate in an acid soil, of pH5 or below, easily coloring a beautiful, deep gentian-blue. **C**: 64   p. 376

macrophylla 'Merveille' (1927), a "French hortensia"; robust, vigorous midseason variety of excellent keeping qualities, with stout stems not requiring staking, firm foliage and large heads about 8 in. across of big round, carmine-rose flowers of good texture; lending itself also to coloring into blue or lilac by the use of alum; the showy calyces hold well after forcing. **C**: 64   p. 365*, 377

macrophylla 'Soeur Therese', a French hybrid (1954); vigorous, willing grower and early forcer, freely branching, with strong stems and dark foliage which sets off beautifully the noble flower heads of pure white; a great improvement over and more compact than the taller, often straggly, but very early Mad. E. Mouillère (1909), and with larger 2½ in. flowers. White hydrangeas are subject to sunburn if grown too soft; and if chilled during forcing brown spots may blemish the flowers. **C**: 66   p. 376

macrophylla 'Strafford' (1937) (originated in France as Mad. Cayeux); favorite commercial, well-proportioned variety with strong erect stem, small, dark green foliage, and medium-large, firm heads of clear rosy-red flowers—actually the enlarged calyx lobes, which are first pale yellow-green, then colored; mid-season, free to bud, but florets have a tendency to drop after hard forcing. Somewhat slower growing but an excellent keeper, of richer clear red without purple overtones. **C**: 64   p. 376

HYDROCLEYS (*Butomaceae*)
commersonii (nymphoides) (Brazil, Venezuela), "Water poppy" or "Water-keg"; floriferous tropical aquatic herb with creeping stems rooting in the mud; basal 2 to 3 in. leathery glossy-green oval, floating leaves with spongy midrib to make them buoyant on the water surface; from joints on the stems, the large 2½ in. poppy-like solitary yellow flowers. One of the most beautiful aquatics with floating stems and leaves for the indoor aquarium, blooming continuously with good light, especially in summer, and while each flower only lasts one day, they are replaced constantly by new ones. To keep from spreading too rapidly, best planted in sand, at 65 to 75° F.; may be wintered cooler. Propagated by division of rooting stems. **C**: 2   p. 518

HYDROCOTYLE (*Umbelliferae*)
rotundifolia (Trop. Asia, Africa); the shiny "Water-penny wort"; a dwarf waterside amphibious herb; low and attractive, creeping with slender rooting branches, the small ¼ to 1 in. shield-like round, lightly lobed, shining green leaves hugging the ground; very minute white, axillary flowers. Forms a dense, pretty carpet along the shore and when the water rises, the leaves swim on its surface. Propagate by sections of the branches. **C**: 66   p. 519

HYDROSME (*Araceae*)
rivieri, often known in horticulture as Amorphophallus (Vietnam) the sinister "Devil's tongue" or "Voodoo plant", also known as "Snake palm"; a curious tropical herb with tuber to 10 in. thick grown for its impressive 3 ft. flower spike which carries the large, reddish spadix and calla-like green and purple spathe 8 to 12 in. long, spreading a foul, carrion-scented odor; the stately foliage nearly 4 ft. wide, appearing after flowering, umbrella-like on a single, rose-marbled stalk with 3 main divisions, each consisting of numerous finger-like segments. The tubers are rested during winter by keeping them dry and not too cold, at about 50° F. Toward spring, tubers that are large enough, develop even without watering, their stately flowering shaft. Later in the season, with foliage growth and feeding, a new tuber will form. Propagated from bulblets. **C**: 54   p. 97, 534

HYLOCEREUS (*Cactaceae*)
undatus, known in horticulture as Cereus triangularis (Brazil), "Honolulu queen"; one of the largest night-blooming cereus; epiphytic, deep green 3-angled clamberer of 2 in. dia.; white flowers nearly 1 ft. long, blooming one night. Edible red fruit. Used as understock for grafting Zygocactus or Rhipsalis. **C**: 10   p. 156

HYMENOCALLIS (*Amaryllidaceae*); the "Spider-lilies" are beautiful American summer-flowering bulbous plants with sword or strap-shaped basal leaves; the stout, solid stalk bearing clusters of sweet-scented, spidery, mostly white flowers with long tube and spreading narrow segments, the stamens united into a conspicuous cup-like crown. Spider lilies in sufficiently large containers are good warm house subjects, or for the summer patio; they must be watered abundantly during growth and bloom; kept on the dry side when foliage begins to yellow. Propagation by offsets, bulb cuttings, or seeds.

americana (littoralis) (Trop. America), "Crown beauty"; magnificent bulbous plant with basal, sword-shaped leaves 2½ ft. long by 2½ in. wide; the inflorescence in umbels on top of solid stalk; white flowers with 4 in. tube and long linear segments, in the center a cupped, toothed crown formed by united stamens; in summer. **C**: 51   p. 391

caribaea (West Indies: Lesser Antilles), tropical summer-flowering "Caribbean spider lily"; with globose bulb, a dozen leaves 2–3 ft. long, not 2-ranked, narrowing at base; a solid stalk bearing a cluster of elegant, fragrant pure white flowers 6 in. long with toothed crown and long linear segments, green outside at base. This may be H. pedalis from South America, naturalized in Florida and the West Indies. **C**: 51 p. 391

narcissiflora, better known as Ismene calathina (Andes of Peru and Bolivia), the "Basket flower" or "Chalice-crowned sea daffodil"; long cultivated bulbous plant with strap-shaped leaves 2 ft. long, almost in 2 ranks; a two-edged floral stalk 2 to 3 ft. high, bearing 2 to 6 large white, very fragrant flowers with 4 in. tube, and spreading lanceolate segments 3 to 4 in. long; the large cup-like base of the stamens funnel-shaped with fringed edge and striped with green. Spring and summer-blooming, and easily cultivated in pots indoors. The Basket flower has to be kept growing after flowering, and watered freely from April to September until the foliage is matured, then gradually rested and kept cool for the winter. **C**: 63 p. 391

speciosa (W. Indies), "Winter spice hymenocallis"; evergreen tropical flowering plant with big bulb, the thick-fleshy, dark green obovate leaves on tapering, channelled petioles; from the center rises a flattened glaucous stalk crowned by a cluster of fragrant spidery, pure white flowers with 3 in. greenish tube and linear segments 2 in. long, a distinctive inner cup bearing the long anthers; one of the most beautiful and best known in cultivation; usually fall-blooming but may flower at other times. **C**: 51 p. 391

HYPHAENE (*Palmae*)

thebaica (Upper Egypt, Sudan, Kenya, Tanzania), "Dhoum palm" or "Gingerbread palm" of the African steppe; a botanical wonder because, unique in palms, it grows with repeatedly forked trunks, to 50 ft. high; the slender branches smooth like a Cordyline, each branch end tipped by a rosette of smallish, stiff, green fan-leaves, the blade 2–2½ ft. long, deeply cut to the middle, on spiny petiole. The orange edible fruit tastes like gingerbread. Although palms normally are not hosts to epiphytic orchids, I have often seen Ansellias growing in the forks of their branches in East Africa. **C**: 52 p. 218, 540

HYPOCYRTA (*Gesneriaceae*); tropical American fibrous-rooted plants either somewhat shrubby, or epiphytes with scandent, creeping stems, bearing opposite, unequal leaves, and curiously shaped pouch-like flowers, with corolla a swollen tube; spring and summer-blooming. Prune back after flowering as blooms appear on new growth. Propagated from soft tip cuttings, kept humid by placing inside plastic bag; also from seed.

nummularia (S. Mexico, C. America), "Miniature pouch flower"; dainty fibrous-rooted creeper epiphytic on mossy tree trunks, with branching, red-hairy stringy stems and oval, crenate ¾ to 2½ in. hairy leaves; the ½ to ¾ in. flowers a vermilion-red corolla tube with inflated pouch tipped by small yellow lobes. An odd but attractive little plant ideally suited for hanging baskets, blooming in summer; sometimes a period of dormancy follows, the plant losing leaves, but with new growth will bloom again in fall and winter. **C**: 60 p. 446

strigillosa (growing on tree trunks in tropical Brazil), "Pouch flower" or "Rough-leaved hypocyrta"; semi-shrubby, spreading plant having pendant branches with pairs of lanceolate bristly-hairy dark green 2½ in. leaves tipped by spines; interesting solitary 1 in. flowers in leaf axils, with coppery calyx, and a corolla inflated on the lower side like a pouch, crimson red, edged with yellow and covered with soft hairs; blooming in spring. **C**: 60 p.446

teuscheri (Ecuador)· attractive herbaceous plant becoming shrubby and straggling with red stems; the lanceolate rough stiff-hairy leaves to 6 in. long, olive-green painted with silver-green along the midrib and lateral veins, wine-red beneath; axillary lemon-yellow flowers with crimson lobes, bracts and calyx orange-red; blooming late spring and summer. **C**: 60 p. 446

HYPOESTES (*Acanthaceae*)

aristata (So. Africa), "Ribbon bush"; a perennial herb of erect habit to 3 ft. high, with soft-hairy, light green ovate leaves 3 in. long; downy 1 in. tubular flowers rose-purple with short lobes striped and spotted purple or white; looking like little orchids, and blooming in winter to February. A nice and long-lasting cut flower. Propagated by young cuttings in spring. **C**: 64 p. 498

sanguinolenta (phyllostachia) (Madagascar), "Freckle-face"; rather delicate tropical herb to 12 in. high; with soft, downy, small roundish leaves green splashed and spotted with rosy-red; small lilac flowers. **C**: 60 p. 189

HYPOLEPIS (*Filices: Polypodiaceae*)

punctata (helenensis; Dryopteris) (Japan, Trop. Asia, Australia, Polynesia, Hawaii, St. Helena, Africa, Trop. America to Chile), "Poor man's cibotium"; from thick, hairy, spreading rhizome, purplish stalks bear bipinnate, triangular fronds to 3 ft. long, soft and papery in texture, a friendly light green. Could serve as a decorator fern in tubs for Cibotium schiedei, but unlike the latter, requires constant supply of water, if possible from a saucer under the container, or the firm-looking, but actually only turgid fronds will suffer and collapse. **C**: 72 p. 250

HYPOXIS (*Amaryllidaceae*)

hygrometrica (Australia), "Star-grass" or "Golden weather-glass"; pretty little bulbous plant to 6 in. high, with narrow, grass-like channeled leaves to 10 in. long and somewhat soft-hairy, the starry six-parted flowers yellow, singly or in clusters; the petals closing in cloudy weather; normally blooming in April. Increased by division. **C**: 63 p. 392

HYSSOPUS (*Labiatae*)

officinalis (So. Europe to Black Sea and Siberia), "Hyssop"; hardy, somewhat woody perennial of the mint family, 1½ to 2 ft. high, with square stems and opposite narrow-elliptic or linear dark green leaves to 2 in. long; small blue variable flowers in one-sided spikes; evergreen in mild climate. By seed, division or cuttings. An oil distilled from the pungent foliage is used in toilet waters and liqueurs; tender young leaves to flavor fruit cocktails, rich fish, game, meats, stews, pies, soups, salads. **C**: 75 p. 321

ILEX (*Aquifoliaceae*); the "Hollies" are mostly aristocratic, evergreen shrubs and trees characterized by their handsome leathery foliage, having leaf margins usually indented and with sharp teeth. Most of them are slow growing, bushy and shapely, and fairly winter-hardy. Most holly plants are either male or female and both parents must be present for the female to bear fruit. There are exceptions; some female trees will set fruit without pollination. Holly prefers a rich, slightly acid soil; it tolerates sun and shade, but needs ample water with good drainage. For decoration in containers, in cool locations we favor Ilex cornuta and its varieties. Cut branches are usually of varieties of I. aquifolium. Propagation by cuttings of young ripe wood aided by rooting hormone.

x altaclarensis 'James G. Esson', hybrid of the English holly (aquifolium) x perado from the Canaries; a favorite, thrifty evergreen usually 6 to 8 ft. or more, with waxy-glossy leaves to 5 in. long, flatter than aquifolium, with more numerous but smaller teeth: distinguished on female plants by large clusters of bright red berries, bigger than the English holly. One of the best hollies outdoors, taking sun, shade and wind, a heavy producer of berries. **C**: 14 p. 513

aquifolium (C. and So. Europe, No. Africa, W. Asia), "English holly" or "Christmas holly"; handsome slow-growing dense evergreen to 45 ft. or more, stiff branches with the alternate leathery, ovate leaves 1½ to 2½ in. long, lustrous dark green, and with coarse spiny teeth along wavy margins; small unisexual flowers whitish, followed by scarlet-red berries on female trees, remaining on the tree during winter. Not safely hardy from New York north; an ideal decorative container plant during the winter season for the cool indoors. **C**: 14 p. 513

aquifolum 'Albo-marginata', "Variegated English holly" or "Christmas holly"; attractive horticultural form of the handsome all-green English holly I. aquifolium, native in W. and So. Europe, No. Africa and W. Asia to China; a pyramidal evergreen tree to 45 ft. high, with alternate ovate, leathery leaves 2–3 in. long, lustrous black green edged with ivory, silvery in places, and coarse spiny teeth along the margins; small whitish flowers followed by scarlet-red berries on female trees. Precariously hardy. Somewhat stiff-sparry in containers, and prefers large root area, cool and moist conditions. **C**: 16 p. 297

cornuta (No. China), "Horned ilex" or "Chinese horned holly"; evergreen shrub of dense bushy growth, to 10 ft. high; recurved leathery leaves 3 to 4 in. long, shining green, nearly rectangular, with pronounced spines at the 4 corners and at the tip of each leaf; large bright red, long lasting berries. Not quite winter-hardy. Good decorator plant for its character. **C**: 14 p. 297

cornuta 'Burfordii', "Globose Chinese holly", a rounded-headed, bushy form of the Chinese holly, with shining, 2 in. convex vivid green leaves, having few spines, prolific with scarlet berries on female plants. The species cornuta is from North China, known as the "Horned holly", an evergreen shrub of dense bushy growth to 10 ft. high, with very characteristic dark green 2 to 4 in. leaves quadrangular oblong, with spines at the 4 corners and the tip; its berries are large pea-size. These two are our most favored of all hollies for interior decoration, they are handsome, tolerant to some neglect especially if well established, growing in sun or shade. **C**: 14   p. 297, 513

cornuta 'Rotunda', "Dwarf Chinese holly"; compact low-growing holly with dense branching habit, usually broader than tall; leaves dark shining green, 2 to 4 in. long, somewhat rectangular, with a large strong spine at each corner, one at the end and 2 smaller ones, much recurved, in the middle; flowers small, dull white; produces no berries, but its low, globular appearance sets it apart. Limited hardy. **C**: 14   p. 297

crenata (Japan), "Japanese holly"; stiff evergreen shrub usually 3 to 4 ft., with age 15 ft. or more high, densely branched, with smooth, dark green, oval leaves 1 to 2 in. long, sharply pointed and sparsely toothed; looking more like a boxwood than a holly; small black berries. Fairly winter-hardy. **C**: 26   p. 513

opaca (U.S.: Massachusetts to Florida and Texas), "American holly"; evergreen, slow-growing spreading tree to 50 ft., with stiff, elliptic or obovate leaves dull or sometimes glossy green above, yellowish beneath, 2 to 4 in. long, and with large spiny teeth; small red fruit, usually solitary. Not as handsome as I. aquifolium, but more winter-hardy; rather somber looking. **C**: 14   p. 297

vomitaria (U.S.: Virginia to Florida and Texas), "Yaupon holly"; evergreen monoecious shrub or small tree 15–20 ft. high, with narrow shiny dark green, oval leaves to 1½ in. long, wavy-toothed; separate male and female flowers on same plant, on branches of previous year; tiny scarlet berries in profuse clusters. Often sheared into columnar form. Tolerates alkaline soils. **C**: 14   p. 297

IMPATIENS (*Balsaminaceae*); the "Balsams" or "Touch-me-nots" are more or less succulent annual and perennial herbs, mostly from mountains of Africa and Asia to New Guinea, with simple leaves and axillary, irregular spurred flowers. The seed pod bursts easily, scattering its seeds. Almost all species are easy to grow and some are old-time house plants such as the "Busy Lizzie". I. walleriana, beloved on the window sill for its almost constant blooms even in poor light, others have lightly variegated, colorful foliage and are quite ornamental. Propagated from seed or cuttings anytime.

balsamina (subtropical India to China), the "Garden balsam"; an old flowergarden annual herb 1½ to 2 ft. tall, with succulent knotty stem and soft-fleshy often reddish branches, with lanceolate leaves; the charming 2 in. flowers in the leaf axils close to the stem, in the species mostly rosy-pink, but in the present cultivars in double-flowering "Rose" or "Camellia" types in various colors from white and pale yellow to purple, lilac, bluish and red, sometimes spotted, blooming from June to fall. From seed. **C**: 14   p. 406

hawkeri 'Exotica' (New Guinea); beautiful herbaceous plant which I found in the Highlands of Eastern New Guinea; densely branching bush with succulent, channeled red stalk and gaily colored stalked, long ovate or obovate waxy leaves, ciliate, 3–4 in. long, dark green to bronzy in the outer area, cream to golden yellow in the center, the bold midrib and petiole crimson-red; large flowers carmine-rose with white throat. **C**: 72   p. 190

marianae (Assam), "Silvery patience plant"; herbaceous foliage plant with creeping, fleshy light green, hairy stems bearing thin oval crenate leaves, dark green with silver bands between the veins. **C**: 60   p. 190

oliveri (Trop. E. Africa), "Giant touch-me-nots" or "Poor-man's rhododendron"; large-growing, fleshy perennial herb becoming shrubby and 4 ft. or more tall with long, oblanceolate, succulent leaves 6 to 8 in. long, olive-green, having a prominent, pale midrib, and edged with coarse bristles; large slender-spurred flowers 1½ to 2½ in. across a delicate lilac-pink. An excellent container plant, blooming even in the shade. **C**: 20   p. 406

platypetala aurantiaca (Celebes), "Tangerine impatiens"; charming tropical herb with fleshy stems and fresh-green, ovate, corrugated leaves having a prominent pink midrib; spurred, attractive flowers in an uncommon shade of orange-yellow very striking with crimson eye. A welcome addition to the color range of the "Sultanas", and a good houseplant. **C**: 8   p. 406

repens (Ceylon), "Creeping impatiens"; small trailer with creeping, fleshy red branches, and alternate small round or kidney-shaped, ciliate leaves ¾ in. wide, waxy deep green, purplish beneath, on long red petioles; golden yellow hooded flowers with brownish net-like striping and curved spur. **C**: 59   p. 265

walleriana var. holstii (Trop. E. Africa), "Busy Lizzie" or "Sultana"; watery-succulent herb 12 to 18 in. high with fleshy stem striped red; and small coppery, ovate leaves; the fiery vermilion-scarlet 1 to 1½ in. flowers with spurs pointing downward. In horticulture in compact, bushy forms only 6 to 8 in. high, in several pastel colors from pink to rose, salmon, orange, lavender, also white, purple and red. Lovely window or bedding plants blooming summer and winter, easily grown from cuttings or seed. **C**: 14   p. 406

walleriana sultanii 'Variegata' (Zanzibar), the "Variegated patient Lucy"; a charming old-fashioned perennial houseplant 8 to 12 in. high, with watery-succulent branching stems, and whorls of waxy, long-tapering, crenate leaves milky-green irregularly bordered with white; with flat 1½ in. carmine-red flowers having an upturned spur, blooming continuously, and hence popularly known as "Busy Lizzie". From cuttings; the green species also from seed. **C**: 20   p. 406

IOCHROMA (*Solanaceae*)
   tubulosum (Colombia), the "Violet bush"; hairy shrub 4 to 8 ft. high, with soft herbaceous, thinly downy, long elliptic 5 in. green leaves; axillary pendulous clusters of lilac-purple cylindric-tubular flowers 1¾ in. long, pale lavender at the toothed mouth, and blooming during summer from June on. Ideal for the summer patio or roof garden, as they love fresh air. From cuttings or seed. **C**: 13   p. 496

IONOPSIS (*Orchidaceae*)
   utricularioides (Cent. America to Brazil), a "Violet orchid"; small tufted, pretty epiphyte of vandaceous habit with rigid, keeled leaves to 5 in. long and with yellowish raised veins; a branched inflorescence with clusters of ½ to ¾ in. flowers white flushed with rose, and marked with purple at base of large hairy forked lip, and somewhat resembling violets; blooming winter and spring. Charmingly floriferous on boards, fernslabs, branches, or in small pots where the roots like to cling to the support and thereafter don't like to be disturbed. **C**: 22   p. 470

IPHEION (*Amaryllidaceae*)
   uniflorum (in horticulture Brodiaea or Leucocoryne) (Argentina, Peru), "Spring-star flower"; charming small herb 8 in. high with corm-like bulb and having onion-like odor, and forming thick clumps; basal grass-like leaves, and solitary, star-shaped flowers with spreading petals 2 in. across, white tinged with blue, opening in bright weather; blooming in spring or early summer. Somewhat hardy. Attractive as pot plant; going dormant after bloom. Before new growth begins the bulbs should be planted in new soil. Increased by offsets. **C**: 76   p. 392

IPOMOEA (*Convolvulaceae*); the "Morning-glories" are mostly annual or perennial herbs more or less of climbing or twining nature, beloved for their large and showy funnel-shaped flowers with 5-angled border and which generally open in the morning and fade in the afternoon, but are quickly followed by others, blooming almost continuously where the sunlight is bright enough; they relish fresh air and moderate moisture, but not wet feet. Container-grown, Ipomoeas are best trained against support of trellis or wire and they grow very rapidly and rank, and can quickly enhance a sunporch or roofgarden. Their seeds are very hard and should be notched with a file, or soaked in warm water for several hours before sowing.

batatas (E. Indies), the "Sweet potato vine"; an economic as well as a hanging basket plant; trailing perennial having a rootstock with deep spindle-shaped or fist-size tuberous edible roots, which may be orange, white or purple; the long vines are stem-rooting, containing milky juice, the variable leaves ovate, angular or digitately lobed; large funnel-shaped 2 in. flowers purple, reddish or white. The tubers are often grown in water as a curiosity and for demonstration in schools. The Sweet potato is economically very important and is cultivated, cooked and eaten widely in all warm-climate regions as a valuable food holding a high number of calories. Propagated by division of tubers, and cuttings. **C**: 13   p. 537

batatas 'Blackie', "Black-leaf sweet-potato"; an ornamental variety grown for its decorative effect, as the whole plant, vines and foliage is in shades of blackish red; stems are purplish brown, and the deeply lobed leaves at first greenish with blackish veins, later almost all black-purple, and wine-red beneath. An easy vine for the warm window. **C**: 1   p. 266

nil (Old World Tropics), "Imperial morning-glory"; tropical perennial hairy, rank vine tendril-climbing, with yellow-green 3-lobed leaves 4–6 in. wide; and showy axillary funnel-form flowers about 4 in. wide, blue, purple or rose. The cultivar 'Scarlett O'Hara', shown in photo, is a spectacular rosy-red veined scarlet, with white tube. Can be grown as a bushy pot plant for the cool, sunny window, as in Japan, where plants are pinched back after every second or third leaf, to prevent climbing, resulting in large blooms all summer and fall; flowers open in the morning and fade in the afternoon, but in good light will produce new ones continuously. Grown as annual from seed. **C**: 63   p. 280

tricolor (rubro-coerulea) (Trop. America), "Heavenly-blue morning glory"; perennial grown as an annual climber to 10 ft. high, with cordate leaves, and large, beautiful blue, disk-shaped flowers, to 4 in. dia.; grown warm in large pots or boxes, with nourishing soil, on trellis; exposed to full sun and fresh air it will bloom freely from summer to autumn. **C**: 52   p. 255

tuberosa (India), the "Wood rose"; tropical perennial herb rising from an underground tuber, vining, with 8 in. leaves deeply parted into 5 to 7 lobes; funnel-form yellow-orange flowers 2 in. long; the globular fruit ultimately a woody pod, when ripe it opens and with its large leathery, persistent, rounded sepals forms the golden brown "Wooden rose", 4 in. across; widely used for decoration, especially in Hawaii, and in dry flower arrangements. **C**: 52   p. 535

IRESINE (*Amaranthaceae*); the "Bloodleafs", also known in horticulture as Achyranthes, or "Chicken gizzards", are chiefly tropical ornamental herbaceous perennial plants grown primarily for their gaily colored foliage; the flowers are woolly and not showy. Much used in summer bedding for borders but also attractive as a small window plant for boxes or pots and easily formed into various shapes by frequent pinching of the tips, or shearing. Easy from cuttings in fall or spring.

herbstii (Achyranthes verschaffeltii) (So. Brazil), a "Beafsteak plant"; bushy tropical herb with waxy $\frac{1}{2}$ to $2\frac{1}{2}$ in. leaves almost round, and notched at tip; glowing purplish-red and traced with light red veins. The ornate foliage coloring is brought out best in good sunlight. A bedding or border plant where red is required. **C**: 14   p. 407

herbstii 'Acuminata' (versicolor), "Painted bloodleaf"; a horticultural variety with showy red stems and ovate, sharply-pointed leaves to 3 in. long, deep red and with light carmine, contrasting veins. **C**: 14   p. 407

herbstii 'Aureo-reticulata', "Chicken gizzard"; erect tropical herb with small rounded leaves curiously notched at apex, brightly colored fresh green with yellow veins; fleshy stem and petioles red. A favorite bedding plant. **C**: 14
p. 190, 407

lindenii (Ecuador), the typical "Bloodleaf"; rather slender tropical herb with narrow lanceolate and sharp-pointed leaves 2 to 3 in. long, deep blood-red; most intense foliage color is brought out in bright sunlight. May be grown into bushy shapes by frequent pinching or shearing. **C**: 14   p. 407

lindenii formosa (reticulata), the "Yellow bloodleaf"; very attractive, colorful form with broader, pointed 2 to 3 in. leaves yellow with light green area between veins; stems and petioles red. Charming for carpet bedding and as a window plant. **C**: 14   p. 43*, 407

IRIS (*Iridaceae*); beautiful summer-flowering perennial herbs with stout rhizomes or bulbous rootstocks, and narrow, or sword-shaped leaves; the flowers are in 6 segments and arise from spathe-like bracts; the 3 outer segments or "falls" are reflexed while the inner 3 are usually smaller and erect and called standards. Iris are like captive butterflies;

most are hardy, especially the rhizomatous kinds and may be grown in large containers and flowered on the patio; propagated by divisions. Bulbous kinds may be potted in October and kept cold until growth begins then gently forced not over 60° F. for early blooming. Dormant bulbs may be stored until time to plant in fall, these are propagated from bulblets.

histrio (Lebanon, Syria, Turkey: Taurus Mts.), "Syrian iris"; dwarf bulbous iris 6 to 8 in. high, the linear leaves 4-angled, deeply grooved on each face and to 1 ft. long, usually appearing before the flowers; the short stalk bearing a solitary large 3 in. blossom in early February, the outer segments or "falls" violet-blue with a yellow line in center, bordered with creamy white and spotted with blue, the standards lilac. Off-shoots bursting into tufts after blooming in late winter. **C**: 64   p. 395

pallida 'Argentea', "Variegated sweet iris"; attractive variegated leaf-form of the species pallida from the South Tyrolean Alps; a perennial iris of the Germanica type, growing from rhizomes; the sword-like, milky-green leaves 12–18 in. long, very striking with upper edge variegated white; flowers sky-blue and scented like orange blossoms, on stout stalks 2–3 ft. long, the "falls" $3\frac{1}{2}$ in. long, with a yellow beard near the base which is veined with bright lilac on a white ground, in May and June. For planting near the edge of aquatic pools. Propagate by separating the tufts with roots. **C**: 76   p. 395

x'Wedgwood', "Dutch forcing iris", bulbous hybrid of the North African I. tingitana and the Spanish Iris, xiphium; sword-like leaves and stiff spike 18–24 in. tall with striking violet-blue flowers, the outer segments or "falls" lavender and marked with yellow. Much grown under glass, and gently forced at not over 55–60° F., to flower from December to May; bulbs planted into tubs in September-October take 10–12 weeks to flower; the earliest from bulbs precooled about 4 weeks at 40° F. **C**: 76   p. 395

ISMENE calathina: see Hymenocallis narcissiflora

ISOLEPIS: see Scirpus

ISOLOMA: see Kohleria

IXIA (*Iridaceae*)
x'Rose Queen', "Rosy corn-lily"; with starry flowers uniform soft pink. One of the many charming spring-blooming horticultural hybrids that have been developed for 150 years; of unknown parentage but probably involving the South African I. maculata and columellaris. Cormous plants with about 6 grass-like leaves in two ranks, and slender, wiry stems with terminal spikes or clusters of flowers $1\frac{1}{2}$ to 2 in. across and opening with the warming sun from cups; a little wider on succeeding days until wide open and like stars; available in gay colors from cream to yellow, orange and red usually with dark centers, and lasting 1–2 weeks. Normally blooming in May-June, Ixias may be flowered from April on by growing them in pots, planted in November; keeping them cool until growth begins and after, as they may only be forced gently in a winter garden or cool greenhouse at 50° F. After the foliage yellows, the corms are rested and stored dry. Propagate by corms. **C**: 13   p. 394

IXIOLIRION (*Amaryllidaceae*)
montanum (tataricum), (from the steppes of So. Russia and Siberia, to Afghanistan), "Siberian lily" or "Altai lily"; a pretty bulbous herb 1–1$\frac{1}{2}$ ft. high, with 3 to 8 broadly linear persistent, basal leaves, and clusters of long-lasting 1$\frac{1}{2}$ in. lilac flowers with spreading segments; filaments, style and stigma violet, the anthers white, on wiry, leafy stalks; blooming in spring to June. Somewhat winter-hardy. After the bulbs become dormant in autumn they should be kept dry. Propagated by offsets in spring, or from seed. **C**: 63   p. 392

IXORA (*Rubiaceae*); handsome tropical, mostly Asiatic evergreen shrubs with opposite or whorled leaves, grown for their striking, compact clusters of starry flowers, most showy in shades of brilliant red, beginning to bloom when quite young and grown in pots, but also planted in patio containers or trimmed into hedges in the tropics. By periodic pinching terminal branchlets may be had in bloom throughout the year. Propagated by moderately firm cuttings warm under bell jar or plastic at about 80° F.

borbonica (bot. Enterospermum), a beautiful tropical foliage plant somewhat resembling a croton, from the Indian Ocean Isle of Réunion; branching shrub with stiff-leathery narrow-lanceolate leaves about 9 in. long, in bluish or mossy-green, mottled with pale green; the midvein a bold salmon-red; small whitish flowers. **C**: 54   p. 187

chinensis rosea (India: Bengal), "Pink flower of the woods"; natural variety of taller habit (4 ft,.) than the species with dark shining green leaves; the flowers a lovely soft, clear pink, becoming darker with age, in large clusters. Ixora chinensis comes from South China to Malaya, has firm-leathery 3 to 4 in. leaves, and waxy tubular 1 to $1\frac{1}{2}$ in. flowers with spreading lobes, first orange-yellow, changing to brick red; blooming in summer. **C**: 52   p. 489

coccinea (India), "Flame-of-the-woods"; tropical flowering shrub of dense growth habit, with sparry branches to 4 ft. high; opposite, sessile, rather short, leathery leaves, and smallish clusters of vivid deep scarlet tubular flowers 1 to $1\frac{1}{4}$ in. long, with spreading lobes, summer-blooming. Less effective than others, but with flowers of deepest coloring. **C**: 52   p. 489

javanica (Java), "Jungle geranium"; showy flowering shrub 3 to 4 ft. high with forking, willowy, red branches; opposite long slender-pointed, smooth, leathery 4 to 7 in. leaves; and large terminal clusters of waxy, soft salmon-red flowers having long thin tubes $1\frac{1}{2}$ in. long, spreading into expanded lobes, and blooming in summer. A good flowering pot plant though blooms soon drop. **C**: 2 p. 489

macrothyrsa, formerly grown in horticulture as duffii (Sumatra), the "King ixora" or "Jungle flame"; large tropical flowering shrub 3 to 9 ft. high freely branching with wiry stems, oblong lanceolate deep green leaves 6 to 8 in. long and slender-pointed; the 1 in. sparry flowers rosy-red, becoming tinged crimson with age, in large striking clusters. **C**: 2   p: 491

JACARANDA (Bignoniaceae)

acutifolia (mimosifolia) (Brazil), "Mimosa-leaved ebony"; semi-evergreen tropical tree 25 to 60 ft. high, with lacy, bright green, bipinnate foliage to 18 in. long, lightly downy. In early spring the leaves drop, following which appear the big 8–12 in. erect clusters of hanging 2 in. lavender blue flowers, with silky, inflated tubular corolla. In the juvenile stage an attractive houseplant because of its fern-like foliage. Sow warm-moist from fresh seed. **C**: 1   p. 301

chelonia (Paraguay, Argentina), "Blue jacaranda"; deciduous or semi-evergreen hard-wooded flowering tree in its homeland becoming 90 ft. high, but coming into bloom when only several feet high; the ornamental foliage is bipinnate feather-like, held outdoors until late winter; the branches may make new leaves quickly or remain bare until flowering time in spring; the large inflated bell-shaped flowers $1\frac{1}{2}$ to 2 in. long, purplish-blue in showy terminal elongate clusters 1 ft. long. Best planted out, or grown in a large container, such a tree in bloom is an unforgettable sight. Smaller pots are very decorative foliage plants. Propagated from half-ripe shoots in early summer, or from only fresh seeds. **C**: 64   p. 490

JACOBINIA (Acanthaceae); the "Brazilian plumes" are summer-blooming tropical American soft-wooded shrubs often grown in horticulture as Justicia; admired for their showy, fountain-like terminial inflorescence of two-lipped flowers in dense clusters. From cuttings.

carnea, widely known in horticulture as Justicia magnifica (Brazil), the "Flamingo plant"; erect herbaceous shrub to 5 ft. high, with ovate, grayish-green, satiny leaves 6 to 8 in. long, on reddish stem, and a terminal head of arched clear rose to flesh-colored, sticky 2 in. flowers. Blooming in summer. **C**: 4   p. 493

velutina (Brazil), "Brazilian plume"; somewhat shrubby plant with long, olive-green leaves, soft hairy on both sides. The beautiful inflorescence a dense, fountain-like terminal head of arched rosy-pink blooms; from early summer. **C**: 4   p. 491

JACOBINIA: see also Pachystachys

JASMINUM (Oleaceae); the "Jasmines" are tropical and semi-tropical shrubs of the Olive family, attractive for their starry, usually fragrant flowers white, yellow or pink. They are of easy cultivation requiring sun; best for cooler winter-

gardens or the patio. Propagated by half-ripened cuttings or layers.

grandiflorum (Kashmir, Himalayas), "Spanish jasmine" or "Poet's jasmine"; straggling tender bush with slender, angled branches; opposite, pinnate leaves of usually 7 small leaflets, and showy white $1\frac{1}{2}$ in. flowers, reddish beneath, commonly in clusters and fragrant. Resembling J. officinale but with larger flowers and more tender. For winter-garden, blooming June to October. **C**: 14   p. 279

ilicifolium: see nitidum

magnificum: see nitidum

mesnyi (primulinum) (China), "Primrose jasmine"; free-flowering, evergreen, rambling shrub up to 15 ft. if supported; with 4-angled, glabrous branches, opposite leaves with 3 small thick, lanceolate shining green leaflets, and showy, single or semi-double, solitary yellow flowers with darker center, to 2 in. across, on wood of the previous season, in spring; somewhat winter-hardy. For the cool winter-garden or sheltered outdoors. **C**: 14   p. 279

nitidum (So. Pacific: Admiralty Isl.); known in Florida hort. as J. ilicifolium, in California hort. as magnificum; the "Angelwing jasmine"; semi-vining small evergreen shrub with shiny dark green leaves ovate with tapering tip, $2\frac{1}{2}$ in. long; large $1\frac{1}{2}$ in. windmill-like, glistening white flowers with lanceolate petals, purplish in bud, and sweetly fragrant. **C**: 14   p. 279

polyanthum (China: Yunnan), "Pink jasmine"; freely blooming, shrubby plant rapidly climbing, and reaching up to 20 ft., the reddish, glabrous branches with lacy 3 to 5 in. pinnate leaves of 5–7 lanceolate leaflets; early in the year it will be adorned by masses of deliciously scented $\frac{3}{4}$ in. flowers, white inside and rosy outside. Somewhat frost-resistant. For the cool wintergarden. **C**: 14   p. 279

rex (Southwest Thailand), "King jasmine"; glabrous climber with young branches green, round and wiry; simple, rigidly hard opposite leaves broad-ovate, dark green, 4 to 8 in. long; large pure white, salver-shaped flowers without scent, 2 in. or more across, during winter, usually in 2 to 3-flowered clusters. Very showy in bloom. **C**: 52   p. 279

sambac (Arabia, India), "Arabian jasmine"; woody shrub clambering 5 to 8 ft. high, with stiff-erect angled branches and firm, broad elliptic, dark green leaves, opposite or in threes, to 3 in. long; 1 in. flowers in clusters, gardenia-white but turning purple as they fade, and powerfully fragrant; a perpetual bloomer from early spring to late fall. Excellent in containers and grown on trellis. In Hawaii perfume is extracted from the flowers; in the Orient the blooms are added to tea to make jasmine tea. **C**: 2   p. 486

sambac 'Grand Duke', also known as 'Grand Duke of Tuscany'; the "Gardenia jasmine"; a button-flowered form from Italy with large, tightly double, gardenia-white 1 to $1\frac{1}{2}$ in. ball-shaped blooms that look like small roses and won't drop off, with an intense sweet fragrance; waxy, quilted, oval leaves in whorls, on stiff-woody pubescent stems; flowers more fully double than the semi-double 'Maid of Orleans' and slower growing. Almost constantly in bloom, the scent of a single flower is strong enough to penetrate throughout living room or winter garden. **C**: 2   p. 486

simplicifolium (South Sea Islands), "Little star jasmine"; rambling evergreen shrub, with leaves privet-like, ovate, 2 in. or less long, dark green and glossy, the sweetly fragrant, tiny star-shaped flowers white, $\frac{5}{8}$ in. across, with narrow-linear petals, in terminal clusters very free-blooming. Attractive as a little flowering pot plant. **C**: 14   p. 279

JATROPHA (Euphorbiaceae)

curcas (Trop. America), "Barbados nut" or "Physic nut"; interesting shrub or small tree to 15 ft. high, with milky juice and soft spongy wood; dark green leaves broadly heart-shaped and lightly 3 to 5-lobed resembling ivy, 3 to 7 in. long; small yellowish-green flowers are borne in clusters, and from the central female ones develop inch-long yellow, somewhat fleshy capsules, each containing 2 to 3 black seeds; containing oil; they are pleasant-tasting but reputedly poisonous. The oil is used as a purgative, for cooking and soap-making. Propagated from cuttings of firm wood, also seed. **C**: 52   p. 514

hastata (Cuba), "Peregrina"; small shrub with stiff-woody branches and milky sap; oblong-obovate leaves constricted below into fiddle-shape, and with tapering apex; showy scarlet flowers with fleshy petals. **C**: 52 p. 500

JUNIPERUS (*Cupressaceae*); the "Junipers" are coniferous evergreens of the northern hemisphere varying in type from low prostrate shrubs to tall slender, rather formal trees; leaves are either needle-like or scale-like pressed close to the twigs; the cones look like fleshy berries. Propagated by seeds, cuttings or grafting.

chinensis (China, Mongolia, Japan), "Chinese juniper"; handsome. erect pyramidal evergreen tree to 60 ft. high or sometimes shrubby, dense with short branches and leaves of two kinds on the same tree: juvenile type in threes, awl-shaped, glaucous above; adult type scale-like in fours, in opposite pairs; glaucous brown fruit. Winter-hardy in New York and New England. **C**: 26   p. 312

chinensis sargentii (Sakhalin, Japan, Korea), prostrate, ground-hugging, feathery variety to 2 ft. high, the creeping stems and ascending branchlets with whorled, scale-like grayish green leaves in adult plants; on young plants linear with 2 white bands below; fruit bluish. Widely used for Bonsai-culture in Japan. **C**: 26   p. 312

chinensis 'Torulosa', the "Hollywood juniper" or "Twisted Chinese juniper"; very artistic irregular growing juniper to 15 ft. high, branching with an appealing twisted effect; the tiny imbricated scaly leaves rich green on brown stem, the cord-like branchlets somewhat flattened and contorted. A relatively fast and vigorous grower, and the bigger the more interesting in its decorative silhouette. Excellent plant for containers, quite durable if it receives adequate light, fresh air and sufficent moisture. From cuttings. **C**: 64   p. 308

scopulorum 'Chandleri', "Chandler's silver juniper"; a beautiful silvery form of the Rocky Mountain red cedar, dense with outward arching, scaly branchlets and of compact, pyramidal habit; J. scopulorum is from Western Canada south to the mountains of Arizona and hardy in Eastern states; an evergreen tree to 30 ft. high, with red-brown bark and branching from the ground; leaves coarsely scale-like and glaucous when young, later yellowish green. **C**: 76   p. 310

JUSTICIA (*Acanthaceae*)

extensa (Congo); attractive shrubby plant with slender, flexible branches and small, long-ovate somewhat papery leaves fresh-green richly marked with silver. **C**: 60   p. 186

JUSTICIA: see also Jacobinia, Pachystachys

KAEMPFERIA (*Zingiberaceae*); warmhouse plants with attractive leaves, and pretty but short-lived flowers. Being tuberous they go into dormancy in autumn, and are kept dry and warm until growth begins again. Propagated by division.

decora (Mozambique), "Dwarf ginger-lily"; a fine, showy ginger growing along the road from Beira to Salisbury; tuberous plant with large leathery, lanceolate leaves glossy dark green, the underside grayish, depressed veins nearly parallel; the channeled petioles lined deep green and clasping; a separate basal stalk bears a raceme of large 3 in., bright canary-yellow funnel-shaped flowers, sweetly fragrant. **C**: 54   p. 208

galanga (India, Malaysia, Vietnam); stemless, low tropical herb with aromatic tuberous rhizome bearing two horizontal broad, fleshy, plain green leaves flat to the ground; fleeting fragrant flowers white with broad petals and narrow sepals, banded violet on lip. **C**: 54   p. 208

gilbertii (So. Burma), "Variegated ginger-lily"; stemless, fleshy-rooted herb with tufted, oblong-lanceolate soft-fleshy leaves grass-green, the marginal areas prettily variegated with milky-white and gray beneath; the flowers on separate stalks with white corolla and long white lip with violet stripes, in summer. **C**: 54   p. 208

masonii (elegans) (Trop. Asia); robust rhizomatous herb similar to K. pulchra, with fleshy leaves more upright, blackish coppery with silvery-green peacock markings; the fleeting flowers light purple with white spot in center. **C**: 54   p. 208

pulchra (E. Tropical Asia: Burma), "Pretty resurrection lily"; attractive tropical rhizomatous herb with broad, corrugated leaves flat to the ground, a gray band in peacock design over the bronze blade; large 1½ in. light purple flowers with broad petals and narrow translucent lip, white eye in center. **C**: 54   p. 208

roscoeana (pulchra) (Burma), "Peacock plant"; stemless plant with fleshy rhizome and wide, succulent leaves spreading horizontally; the foliage beautiful as shining bronzy-chocolate taffeta, iridescently veined and zoned pale green like a peacock tail; purplish and shining gray beneath; fleeting flowers pale purple with white eye, appearing day after day in summer. **C**: 54   p. 186, 208

KALANCHOE (*Crassulaceae*); handsome, usually robust, sometimes shrubby succulents of Tropical Africa and Madagascar, with fleshy, opposite leaves in many forms, entire, scalloped, lobed or arrow-shaped, smooth and patterned, or densely felted; the long-stalked inflorescence carrying showy, usually pendulous balloon or urn-shaped flowers white, yellow to red and purple. Kalanchoes are easily cared-for, excellent decorator house plants, both for their profusion of flowers and often very interesting or beautiful foliage; they like sunshine and fresh air. They are easily propagated from tips, branches, or leaf cuttings. In the section Bryophyllum, the "Life plants" such as Kal. pinnata, by some taxonomists held as a separate genus, the separated leaves have the curious ability to sprout young plantlets from the notches along their margin or at the tip, more so if pegged down on moist sand.

beharensis (S. Madagascar), "Velvet leaf" or "Elephant ear"; woody succulent shrub usually 3 to 5 ft., but to 10 ft. high, with opposite pairs of large, broadly arrow-shaped, lobed leaves 4 to 8 and up to 18 in. long and of thick-fleshy texture, rich green but densely rusty-haired above, silvery-felted beneath and on leaf stalks; tall inflorescence with clusters of yellowish urn-shaped flowers violet inside. With plush-like foliage attractively waved and crimped, this is a beautiful solitair plant for the sunny patio, but tolerates some shade indoors. Easily propagated from leaf cuttings or the small plants that are produced from the roots. **C**: 13   p. 346

blossfeldiana 'Compacta', the "Christmas kalanchoe", or "Flaming Katy" as known in Europe; a dwarf cultivar of K. blossfeldiana (globulifera coccinea) from Madagascar. Excellent compact flowering plant 6 to 8 in. high, the waxy fresh green foliage literally covered with clusters of vivid scarlet-red flowers, normally blooming in January-February. Being short-day plants and photosensitive, plants sown in January may be had in bloom for Christmas if shaded with black cloth or black paper, or put into a dark cabinet overnight for about 4 weeks beginning September, limiting day-length to 9 hours. Propagated from seed or cuttings. **C**: 13   p. 346

blossfeldiana 'Tom Thumb', a dwarf and very compact "Christmas kalanchoe"; with bronzy foliage, and covering itself during late winter with masses of bright red flower clusters. To encourage budset, place the plant outside for the summer. If flowers are wanted earlier than usual, daylight should be restricted to between 9 or 10 hours during July and August by covering the plant with black paper or a box; for Christmas flowering short-day treatment is given from early September until buds show. **C**: 13   p. 384

daigremontiana (Bryophyllum daigremontianum) (Madagascar), "Devil's backbone"; easy-growing, robust, erect succulent plant to 3 ft. high, with fleshy, long-triangular, tapering brownish-green concave leaves 6 to 8 in. long, reverse gray, flecked purple; flowers gray-violet. Freely producing adventitious buds from the crenelated margins, and which grow into young plants. **C**: 13   p. 347

'daigremontiana x tubiflora', the "Good-luck plant"; prolific and popular hybrid with succulent stem and fleshy, narrow pinkish-brown leaves about 2 in. long, channeled on the surface, keeled beneath, and spotted and lined purple. Young plantlets freely appearing on the toothed margins from where they drop off forming masses of growing plants wherever they touch the ground. **C**: 13   p. 347, 530

fedtschenkoi (Madagascar), "Purple scallops"; bushy succulent with wiry branches erect or creeping, small fleshy leaves notched at apex, about 1 in. long, metallic green and delicately glaucous amethyst, edged with purple; nodding inflated tubular flowers brownish rose on leafy stalk. **C**: 13   p. 328

fedtschenkoi 'Marginata' (California cultivar), "Rainbow kalanchoe" or "Aurora Borealis plant"; a charming sport which I first photographed in Mr. Orpet's garden in Santa Barbara; branching plant with flexuous stems and pale glaucous bluish-gray scalloped leaves beautifully margined creamy-white. **C**: 13   p. 329 346

x'Fernleaf', (beharensis x (possibly) tomentosa), "Fernleaf felt bush"; an attractive California hybrid; fleshy succulent entirely covered with short stiff hairs, with features of the giant K. beharensis but much smaller and more graceful, the thick, triangular lobed leaves doubly dentate, gray-felty with brown edge. C: 13   p. 346

gastonis-bonnieri (Madagascar), a "Life plant"; fleshy plant with robust stem and large, wide-spreading ovate-lanceolate soft-succulent, toothed leaves 5 to 7 in. long, pale to coppery green with darker spots, covered with white-mealy powder especially the young growth; brownish margins appearing as if stitched; flowers in tall clusters with corolla tube pale pink. A good house plant and a conversation piece because of the young plants produced on the foliage. C: 13   p. 347, 530

x'Grandiflora hybrid'; a striking cultivar with large clusters of $\frac{5}{8}$ in. flame-scarlet blooms of a type that won't close overnight as the blossfeldiana varieties tend to do. A number of good such hybrids have been created involving blood of K. grandiflora, blossfeldiana, flammea and others. The plant illustrated is grown under the name "Surprise" in U.S., but appears identical with the late blooming Swiss "Grob's Triomphe du Chef", a large flowered, stocky plant of vigorous habit, with shiny rich green, crenate leaves, and long-lasting leathery blooms especially developed for growing in small pots as single-stemmed mini-plants. Normally spring-blooming but responding to short-day treatment and special pinching, to flower during mid-winter. Propagated from leaf cuttings. C: 13   p. 346

'Grob's Rotglut' (translated "Red Glow"); a Swiss cultivar that comes true from seed; compact succulent plant with long stalked flower clusters 8 to 10 in. high above pot when in bloom; the broad, heavy foliage rich green and crenate, the $\frac{1}{2}$ in. flowers a fiery scarlet, staying open at night. Normally late winter flowering, this variety may be timed for Christmas, from January sowing, by short-day treatment beginning July 28 and light pinching August 18. C: 13   p. 384

x kewensis (teretifolia x flammea), "Spindle kalanchoe"; robust, branching succulent with fleshy stems, and grayish-green to coppery, $5\frac{1}{2}$ in. long spindle-shaped leaves having a hollow ridge on top, margins with an occasional tooth, the slender apex recurved; terminal clusters of tubular spreading flowers soft rosy pink. C: 13   p. 346

longiflora (petitiana salmonea), erroneously in the horticultural trade as K. somaliensis (So. Africa: Natal); robust leafy succulent with 4-angled stem and stalked oval 2 to 3 in. leaves scalloped toward apex, pale bluish or gray-green turning coppery, and edged with orange; light orange flowers with spicy fragrance. C: 13   p. 347

marmorata (Ethiopia), the "Pen wiper" plant; stout succulent plant branching from base, thick stems with soft fleshy, flaccid, broad obovate leaves $2\frac{1}{2}$–4 in. long, pinkish to bluish green, dusted with glaucous blue waxy coating, and blotched purple on both sides, margins scalloped; flowers white. C: 13   p. 328

nyikae (East Africa), "Shovel kalanchoe"; erect succulent with opposite, peltate leaves 2 in. long, bluish with red edge; very curious with the basal ends turned up and looking shovel-like with a long handle; stout terminal inflorescence with yellow-green flowers tipped rose. C: 13   p. 347

'Oakleaf', "Oakleaf felt bush"; fascinating large tomentose succulent, a bud sport of K. beharensis, and resembling it in habit but the gray-felted thick, giant 8–12 in. leaves are deeply lobed and laciniately cut. Very attractive sun-plant but requires elbow-room. C: 13   p. 346

'Orange Triumph' (Grob 1957), in the horticultural trade as "Gravenmoer's Glorie"; a Swiss 'Grandiflora hybrid'; very robust, stocky succulent plant 9 to 12 in. high, with thick ovate, shiny deep green $2\frac{1}{2}$ in. leaves deeply double-serrate at margins and with faint red edge; the showy $\frac{5}{8}$ in. flowers in 2 long-stalked clusters a rosy coral or orange-red, and remaining open at night. Normally blooming March-April, this cultivar reacts to short-day treatment willingly; for Christmas bloom, reduce daylength to less than 12 hours beginning in September, until budding. Propagate from tip of leaf cuttings. C: 13   p. 384

pinnata, in horticulture usually as Bryophyllum pinnatum (Tropical Africa and tropics around the world), known as the "Air plant" or "Miracle leaf", also "Floppers" or "Goodluck leaf"; remarkable because of the offspring produced on its foliage; erect succulent plant with fleshy stem 2 to 3 ft. high, the leathery-fleshy leaves green tinged with red, at first undivided oval and notched, in later stages divided into 3 to 5 scalloped leaflets, and 3 to 8 in. long;

inflorescence with pretty nodding inflated tubular greenish flowers purple-tinted to 3 in. long. Produces many plantlets in the notches of scallops. Leaves can be removed and pinned to curtain where they will produce plantlets until they dry up. C: 13   p. 347, 530

pumila (C. Madagascar), "Dwarf purple kalanchoe"; bushy little plant 4-8 in. high with closely set 1 in. obovate leaves notched along upper margin, purplish-brown and covered with white bloom; pitcher-shaped red-violet flowers. C: 13   p. 346

x'Roseleaf' (beharensis x tomentosa); an attractive Hummel hybrid with stocky stem; triangular, thick, spatulate-pointed, toothed leaves about 2–3 in. long symmetrically arranged as in a cross, with brown felt above, and silver felt beneath; teeth brown. C: 13   p. 328

somaliensis: see longiflora

synsepala (C. Madagascar), "Cup kalanchoe"; attractive short-stemmed succulent with cupped, broad oval, fleshy leaves 6 in. long, pale green with purple band inside of marginal teeth, long inflorescence with rosy flowers. Unique for its sending out young plants on runners from the leaf axils, and forming plantlets on roots. C: 13   p. 347

tomentosa (pilosa) (C. Madagascar), the "Panda plant"; strikingly beautiful succulent with erect branching stem and soft fleshy spoon-shaped leaves about 3 in. long and entirely clothed in dense white felt, apex dentate and the teeth marked brown; whitish flowers with light brown stripes. From cuttings, or leaves laid on the surface of sand. C: 1   p. 43*, 196, 327, 328, 347, 531

x'Yellow Darling' (blossfeldiana x schumacheri), "Yellow blossfeldiana hybrid" or "White Kalanchoe"; compact, bushy winter-flowering plant with waxy light green, crenate leaves topped by dense clusters of lovely creamy-white flowers having deeper lemon-yellow center. Normally late blooming in February-March. this cultivar may be flowered for Christmas if given short-day treatment early September. C: 13   p. 346, 384

KEFERSTEINIA (Orchidaceae)

tolimensis (So. Ecuador at 7000 ft.); lovely little epiphyte with tufted, narrow lanceolate leaves and short-stalked basal $1\frac{1}{4}$ in. very pretty flowers, yellowish-green and densely covered with small purple spots; broad, fringed whitish lip marked yellow and spotted maroon, darker in center, blooming at various times, January or June. Best grown on blocks or treefern slabs. Propagated by division. C: 22   p. 470

KENNEDIA (Leguminosae)

rubicunda (Australia: Victoria, N.S.W., Queensland), "Coral-pea"; strong-growing shrubby bean-like twiner, young growth covered with silky brown fur; variable palmate-trifoliate leaves, the 2 to 3 in. leaflets ovate; red pea-shaped, pendulous axillary $1\frac{1}{2}$ in. flowers in pairs, a long standard, or upstanding petal, reflexed upward. Best in winter garden with good light and fresh air, trained to support. From firm cuttings or seed. C: 63   p. 278

KENTIA

belmoreana: see Howeia
canterburyana: see Hedyscepe
forsteriana: see Howeia

KIGELIA (Bignoniaceae)

pinnata (Trop. Africa: Sudan, Uganda, Kenya, Rhodesia, Transvaal, Mozambique), the unbelievable "Sausage tree"; a humorous conversation piece wherever seen; spreading tree to 50 ft. high, with pinnate leaves, and pendulous clusters of large velvety, showy 3 in. flowers with curved tubular-flaring corolla orange-yellow at base, the corrugated lobes blood-red; having an unpleasant odor, the blooms are pollinated by bats at night. The tree is a curiosity mainly because of the somewhat conical, cylindric, pale brownish-gray hard-shelled fruit 12–18 in. long, containing solid pulp and dangling for many months on cords several feet long. In Africa the fruit is used medicinally; the Kikuyus of Kenya make beer from this fruit, brewed with sugar water and honey. C: 51   p. 527

KLEINIA: see Senecio

KLUGIA (Gesneriaceae)

ceylanica (Ceylon, Eastern India), branching herbaceous plant of rangy habit, 1 to $1\frac{1}{2}$ ft. high, with fleshy stems, alternate oblique-ovate leaves to 8 in. long, soft green above, pale beneath, and numerous nodding 1 in. flowers in one-sided clusters, the tubular corolla with 2 unequal lips, the short upper lip white, broader lower lip rich violet with yellow spot at base; blooming from spring to fall. Propagated from seed or cuttings. **C**: 60   p. 446

KNIPHOFIA (Liliaceae)

uvaria (Tritoma) (So. Africa), "Torch lily" or "Poker plant"; spectacular, robust perennial plant with thick fleshy roots, forming clumps of long grass-like basal leaves 1 in. wide; and a showy poker-like, stout 3 to 5 ft. spike of nodding tubular flowers to 2 in. long, the upper ones scarlet-red, the lower ones yellow, looking like a flaming torch, blooming all summer from June to fall, and attracting hummingbirds. Excellent as striking summer-bloomers in large containers; may be wintered in the cellar. Propagated by division, suckers, or seed. **C**: 76   p. 396

KOCHIA (Chenopodiaceae)

scoparia trichophylla (S.E. Europe, Temperate Asia), "Summer cypress", "Belvedere cypress" or "Fire bush"; showy, densely branched, ornamental annual, with its formal globe or columnar shape resembling Cupressus, 1½ to 3 ft. high; a subshrub with numerous narrow, partly almost threadlike leaves 2 in. long, fresh green; the foliage in the species remaining green; the form trichophylla coloring purplish-red in autumn hence the name "Burning bush". The tiny green axillary flowers are insignificant. May be used in patio container as an imitation evergreen. From seed. **C**: 63   p. 414

KOELLIKERIA (Gesneriaceae)

erinoides (Costa Rica to Venezuela, Colombia and Argentina), "Dwarf bellflower"; delicate herbaceous, low clustering miniature from scaly rhizomes; downy, quilted elliptic leaves 1 to 4 in. long, velvety green spotted silver, and numerous sprays 10 in. high of tiny, 2-lipped flowers less than ⅜ in., flushed purplish-red, with white lip and throat blotched yellow, and lined red outside; blooming in summer. Propagated from scaly rhizomes, cuttings or division. **C**: 10   p. 446

KOHLERIA (Gesneriaceae); the "Tree gloxinias" were formerly known as Isoloma or Tydaea; they are showy tropical American herbaceous plants with creeping scaly rhizomes, opposite ornate leaves, and flowers with inflated tube spreading into 5 lobes in many colors and combinations, usually blooming in summer and fall, but may continue to grow and flower almost the year round if old growth is removed. Kohlerias welcome a resting period, however, but the stems should not die down completely in winter, and require some moisture then. Most are good indoor plants in pots or baskets. Propagated by breaking up the scaly rhizomes; or from seed. Rooting tip cuttings successively will give continuous bloom.

"allenii" hort. (Trop. America), as cultivated at the Montreal Botanic Garden; stout stems with large tomentose, lanceolate leaves, and hairy flowers with long inflated red tube, the small spreading lobes yellow with red spots; fall to winter-blooming. **C**: 10   p. 445

'Amabilis hybrid', known in horticulture as Isoloma ceciliae; lovely herbaceous plant of low habit with weak, spreading branches, lanceolate hairy green leaves to 4 in. long, bronzy along the veins; exquisite tubular rosy-red 1 in. flowers, creamy yellow inside and barred and spotted with red; blooming almost all year. The species amabilis is from Colombia having deep rose, long-lasting flowers with throat almost white; the 'Amabilis hybrids' have added yellow, white and deep red to the color range. **C**: 10   p. 445

digitaliflora (Colombia), "Foxglove kohleria"; erect, robust, hairy herbaceous plant with large ovate, dark green leaves to nearly 8 in.; big bell-shaped flowers bulging in lower part, the 1 to 1¼ in. tube rosy-lavender with white hairs, white beneath, the showy lobes greenish with purple spots; blooming summer to fall. **C**: 10   p. 445

'Eriantha hybrid', widely known in horticulture as Isoloma hirsutum; robust plant 2 to 3 ft. high, more floriferous than the species from Colombia, and with more prominently inflated orange-red corolla tube, 1½ to 2 in. long, and large, spreading limb, throat and lower lobes yellow and all marked red. Soft-hairy stem from scaly

rhizome, with deep green, ovate, toothed leaves to 5 in. long, having a conspicuous border of reddish hairs; in bloom from late summer into fall, and a good subject for baskets. **C**: 10   p. 445

lanata (Mexico: Guerrero, Chiapas) the "Woolly kohleria"; stout herbaceous plant with stem and foliage entirely covered with pale hairs; the smallish, ovate leaves with crenate margins; lovely tubular flowers orange, shaded with red, the narrow spreading lobes spotted and edged with bright crimson; tubes are covered with white hairs, and before opening the buds look like a soft mass of yellow fuzz. **C**: 10   p. 445

lindeniana (Ecuador); compact, pretty plant, with reddish herbaceous erect stem, and with beautiful velvety, ovate leaves to 3 in. long, vivid green changing to copper toward the crenate margin; striking with a pattern of silvery veins; small white bell-shaped flowers with purple throat; blooming late fall. **C**: 60   p. 445

strigosa (Costa Rica); slender herbaceous plant to 18 in. tall with velvety-hairy foliage dark green, and prettily contrasting pale green veins; reddish beneath; the tubular trumpet-shaped flowers 1¼ in. long, cinnabar-red, yellow inside with throat heavily spotted red; ¾ in. across mouth. **C**: 10   p. 445

tubiflora (Costa Rica, Colombia), "Painted kohleria"; robust, tall-growing, hairy plant to 1½–2 ft. high, with large ovate, toothed, burnished dark green, corrugated leaves red beneath, and slender velvety-textured swollen-cylindric 1 in. flowers orange, constricted at the mouth which is yellow, the small lobes neatly marked with red. **C**: 10 p. 445

KOPSIA (Apocynaceae)

arborea hort.: see Ochrosia elliptica

fruticosa (India, Malaysia to Philippines), the red-eyed "Shrub vinca"; large tropical evergreen shrub, with opposite elliptic-lanceolate, somewhat leathery leaves 4 to 8 in. long; the vinca-like flowers in clusters, white with crimson center, 2 in. across, blooming in late spring. From cuttings of firm young wood. For Kopsia of horticulture see Ochrosia. **C**: 13   p. 501

LACHENALIA (Liliaceae); the "Cape-cowslips" are subtropical, very pretty small South African bulbous plants related to the hyacinths, with usually a pair of soft-fleshy basal leaves and short succulent stalk bearing waxy tubular or bell-shaped flowers which are long-lasting, sometimes up to 2 months in a cool location. Mostly blooming in late winter and spring, bulbs are planted in pots early September and kept in a cool room at 45 to 50° F. until buds show. After flowering keep watering until foliage turns yellow, then keep dry and in full sun to ripen the bulb; repot in August or September. Lachenalias should not be forced. Propagated by offsets or seed.

aloides (tricolor) (So. Africa), "Tricolor Cape cowslip"; dainty hyacinth-like bulbous plant to 12 in. high, with two broad linear lanceolate spreading, fleshy leaves dark green and spotted purple; bright, colorful nodding waxy tubular flowers 1 in. long, the outer segments yellow tipped green, inner ones scarlet-red at tip, much exceeding the outer, on erect fleshy stalk; spring blooming. **C**: 14   p. 373

aloides nelsonii, "Yellow Cape cowslip"; a pretty variety with many pendulous bells on a long raceme; bright yellow, slightly tinged green and with green tips—one of the most beautiful. **C**: 14   p. 396

bachmannii (So. Africa), "White Cape cowslip"; small bulbous herb to 8 in. high, with thick linear, channeled, recurving basal leaves, and short bell-shaped flowers with white segments keeled red, in short spikes; spring-blooming. **C**: 25   p. 396

lilacina (So. Africa), "Lavender Cape cowslip"; small bulbous plant 4–5 in. high, with pairs of sickle-shaped lanceolate 4 in. basal leaves spotted with purple; greenish-red stalks, prettily mottled red-brown, with numerous, about 20, erect tubular, lavender flowers in an oblong spike, their outer segments with blue base, the inner lilac, and spreading outward. **C**: 25   p. 396

purpureo-caerulea (So. Africa), "Purple Cape cowslip"; bulbous plant, 1 ft. high with lanceolate, blistered leaves, tiny ¼ in. bell-shaped, fragrant flowers purplish-blue with sky-blue base, the inner segments slightly longer than the outer, with protruding stamens, 40 to 100 on long fleshy stalk; April-blooming. **C**: 25   p. 396

LAELIA (*Orchidaceae*); the elegant "Amalias" are a lovely genus of epiphytic orchids, closely resembling Cattleyas but generally more slender, and with colors more brilliant which includes yellow, coppery-bronze, red-orange and scarlet as well as deep violet. Laelias have been widely used in breeding with Cattleyas, with the object to combine Laelia coloring with better Cattleya shape. Their culture is similar to Cattleya; if grown in the home they require the maximum of light of a sunny window, at least 2000 to 3000 fc., or they won't bloom. While growing a high percentage of moisture should be provided, best under a plastic curtain if no greenhouse is available; also liberal watering. Planted in osmunda, smallish pots are best. After flowering the water is reduced for a rest until the new eye begins to swell. Propagated easily by division.

anceps (Mexico, Honduras); a remarkably handsome epiphyte of robust, prolific habit with egg-shaped, angled pseudobulbs to 5 in., usually single-leaved, topped by long wiry stems to 3 ft. high, bearing small clusters of slender 3 to 4 in. flowers, their sepals mostly lilac rose, the petals slightly darker, and a contrasting lip purplish crimson with yellow throat; mostly winter-blooming, especially December-January; very useful at that time of year. A popular species easy to grow, delighting in an abundance of light and air, with copious watering while growing. Very variable in color forms from deep purple to white. **C**: 21 p. 471

cinnabarina (Brazil); charming epiphyte of graceful, compact habit, with cylindric tapered 5 to 10 in. pseudobulbs usually single leaved; the inflorescence slender with stalks 12 to 24 in. high, with solitary or clustered 2 to 3 in. flowers of leathery texture, uniformly bright orange-red. Blooming mostly between February and May, flowers last 6 weeks in continuous beauty. An excellent pot plant. **C**: 21 p. 471

crispa (Brazil); a splendid, free-growing epiphyte with tall club-shaped pseudobulbs 8 to 12 in. long, topped by a solitary leaf; the exquisite, fragrant 4 to 6 in. flowers in clusters, their narrow, undulate sepals and petals white, the crisped lip purple and prettily edged in white, and yellow throat; summer-blooming, and lasting on the plant for 2–3 weeks. **C**: 71 p. 471

perrinii (Brazil); a truly beautiful species, resembling a Cattleya in growth and flower; club-shaped pseudobulbs with a solitary leaf; the inflorescence rising from a compressed sheath, with small clusters of flowers to 5 in. across, the sepals and petals light rose, the lip around the front portion intensely rich purplish-crimson, and white throat; blooming autumn-winter. **C**: 21 p. 471

purpurata (Brazil), known as the "Queen of orchids" in So. Brazil, and the national flower of Brazil; probably the finest of the genus; magnificent epiphyte with tall, club-shaped pseudobulbs a foot or more tall, carrying a long, dark green, solitary leaf; the elegant flowers, 6 in. or more across, in showy clusters; narrow sepals and slender petals glistening white, sometimes flushed pink, the large lip a rich velvety crimson-purple, with pale yellow throat striped crimson, blooming mostly between May and July; ideal for June weddings. **C**: 71 p. 471

tenebrosa (Brazil: Bahia to Rio), a distinct and easily grown epiphyte, similar in habit to L. purpurata, but less robust; flowers few, but fragrant, often not lasting well, but very showy, 6 in. across, somewhat variable in color, but usually brownish yellow or purplish brown, the lip large, deep purple inside, lighter toward the margin, blooming May-June. **C**: 21 p. 471

x LAELIOCATTLEYA (*Orchidaceae*)

canhamiana alba (L. purpurata x C. mossiae reineckeana); excellent intergeneric hybrid of robust habit, free blooming with clusters of large, elegant flowers; sepals and petals ivory or pure white, with deep violet-purple frilled lip edged white, and a golden throat; blooming February to May or June; Lc. canhamiana is an early hybrid of 1885. A wonderful cut flower for spring weddings. **C**: 71 p. 455*

hassalii alba 'Majestica' (Laeliocattleya Britannia x Cattleya warscewiczii); magnificent bigeneric hybrid with showy flowers, 6 in. or more across, the sepals and broad wavy petals glistening white, the frilled lip a velvety, bright purplish-violet, yellow in the throat; blooming in spring. This hybrid genus has successfully brought together the brilliant color found in Laelia, with the more rounded shape and substance of Cattleya. **C**: 71 p. 471

LAGERSTROEMIA (*Lythraceae*)

indica (Japan, Korea, China), the "Crape myrtle"; handsome flowering tree to 30 ft. high, foliage falling annually, the elliptic leaves 1 to 2 in. long; at branch tips the gorgeous clusters of frilled, though scentless flowers, pink or purple and resembling crepe-paper, blooming profusely all summer, and fall from August to October; best in hottest or sunniest season. In older trees the bark flakes off to reveal a smooth pinkish inner bark. Keep cool in winter and moderately dry; cut back hard in March or April as flowers develop on current year's new growth. Best outdoors. From slow-germinating seed, or from ripened cuttings. **C**: 64 p. 504

LAMIUM (*Labiatae*)

galeobdolon variegatum (Canada: Quebec; C. and E. Europe, to Urals), "Archangel" or "Silver nettle vine"; rampant creeper with square thread-like, rooting stems covered with appressed pale hairs; nettle-like herbaceous, opposite, crenate leaves 1¼ to 2 in. long, deep green and rugose, prettily zoned and painted with silver; two-lipped yellow flowers, the lower lip marked red. While winter-hardy outdoors, also a willing house plant for the cooler window. **C**: 27 p. 265

LAMPRANTHUS (*Aizoaceae*)

aureus (Mesembryanthemum) (So. Africa: Cape Prov.), "Golden ice plant"; succulent creeper with brown-barked branches, the fleshy leaves 3-angled, rounded at the sides, 2 in. long and united at the base, fresh green and with transparent dots; very floriferous in the sun with 2 in. flowers, deep shining orange with yellow center. Naturalized in So. California. From cuttings. **C**: 13 p. 357

emarginatus (So. Africa), a so-called "Ice plant"; branched shrubby, succulent plant with spreading, thin woody stems to 12 in. long, with numerous short side shoots; furnished with semi-cylindrical, almost linear fleshy gray-green leaves slightly keeled or 3-angled; numerous large flowers purplish-red. For the sunny window, baskets or the patio. **C**: 13 p. 271

roseus (So. Africa: Cape Prov.). "Pink ice plant"; vigorous spreading succulent, erect or trailing, woody at the base, with compressed 3-angled, linear leaves 1 to 1½ in. long, and covered with prominent translucent spots; large 1½ to 2 in. soft pink flowers deeper rose toward base of the petals, and with yellow stamens in center. Used as blooming plant in the sunny, airy window or on the patio. From cuttings. Plants in the trade under the name L. roseus such as in the photograph shown appear to be highly intermixed and may be hybrids that would not come true from seed. **C**: 13 p. 357

spectabilis (Mesembryanthemum spectabile) (So. Africa: Cape Prov.), "Red ice plant"; somewhat woody perennial, branching with long prostrate reddish flowering stems; succulent keeled leaves ½ to 2 in. long, glaucous gray olive; the 1½–2 in. flowers gleaming in brilliant color, normally purplish, and with longer leaves, the color forms in the California trade as L. spectabilis are probably hybrids such as the pictured "Red ice plant" in shining crimson; other colors available are pink or rose. Beautiful flowering plants in spring and summer. From cuttings. **C**: 13 p. 357

LANTANA (*Verbenaceae*)

camara (West Indies), the "Shrub verbena" or "Yellow sage"; hairy shrub that grows to 4 ft. high in the tropics, but usually grown as smaller plants in pots, with thin-woody, angled branches sometimes prickly; ovate, toothed, rough-bristly leaves having a pungent odor; very floriferous with stiff-erect, small but showy 2 in. heads of verbena-like flowers in changeable colors, usually opening pink or yellow, becoming red or orange, and several color combinations may be found on the same plant; summer-blooming. Easily subject to red spider and white fly in dry air. From softwood cuttings. **C**: 63 p. 378, 410

LAPAGERIA (*Liliaceae*)

rosea (Chile), "Chile bells" or "Chilean bell-flower"; shrubby vine with alternate ovate, leathery leaves, and many pendulous, axillary, large bell-shaped flowers 3 in. long, rich rosy crimson, spotted white inside; summer blooming. For the cool winter garden, requiring copious watering when growing, moderately in winter to ripen wood. Lapageria rejoices in humid-cool fresh air. Propagated by layering, or fresh seed. **C**: 84 p. 277

LARIX (*Pinaceae*)
  leptolepis minor (Japan), "Golden larch", a dwarf form of the "Japanese larch"; attractive deciduous coniferous tree fast growing to 60 ft. or more, the horizontal branches with branchlets bearing fluffy tufts of needles, soft to the touch, about 1 in. long and clustered on short spurs, light or bluish green. Similar to Cedrus but differing in dropping its needles in winter; in spring the needle tufts are a soft yellow-green; in autumn they turn brilliant yellow. Can be dwarfed in containers. For cool locations. From seed. **C**: 26   p. 312

LATANIA (*Palmae*)
  "borbonica" hort.: see Livistona
  lontaroides (borbonica, commersonii) (Mauritius), "Red Latan"; robust, rapid-growing fan palm to 50 ft., with gray trunk swollen at base, bearing a large crown of numerous handsome thick leaves palmate fan-shaped, gray-green, and deeply cut, 6–8 ft. across; the segments edged with tiny sawteeth, veins and margins tinged with red, fuzzy beneath; the stalks colored orange, thorny when young, the rachis extending into the leaf for 18 in. or more. Highly ornamental. Loves warmth and moisture with good drainage. **C**: 54   p. 229

LATHYRUS (*Leguminosae*)
  odoratus (So. Italy, Sicily), "Sweet pea", annual, tendril-climbing herb with brittle, winged stems, paired oval leaves, and large, sweetly fragrant, butterfly-like flowers originally purple, but now hybridized into many horticultural varieties in delicate pastel tints, 2 in. wide. Requiring bright sunlight with cool, airy conditions; grown on support. Best in a light greenhouse planted out, or the winter garden; lack of light or temperature changes may cause bud drop. From seed. **C**: 75   p. 280

LAUROCERASUS: see Prunus

LAURUS (*Lauraceae*)
  nobilis (Asia Minor, naturalized in So. Europe), the "Grecian laurel" or "Bay tree"; the true laurel of history is an evergreen, pyramidal, aromatic, tree-like shrub to 40 ft. high, very leafy, with elliptic, stiff-leathery leaves 3 to 4 in. long, dark green and lightly crimped; the small, inconspicuous greenish-yellow flowers are followed by berry-like black fruits. The dense foliage lends itself to shaping by shearing, and tub-grown specimen clipped into formal pyramids, globes or standards with straight, or playfully with twisted trunks, are grown especially in Europe and displayed as impressive and noble decorators. Easy to maintain, but requiring sun, coolness and fresh air. The "bay-leaves" are used for seasoning in cooking. From cuttings. **C**: 26
                                    p. 288, 319, 541

LAVANDULA (*Labiatae*)
  officinalis (W. Mediterranean to Greece), "English lavender"; aromatic hardy semi-shrub with square stems and pleasing gray-green narrow 2 in. linear leaves, white tomentose when young; the fragrant small lavender-blue flowers are borne on long slender spikes in late summer. Good also indoors. A famous herb for the fragrance of its dried flowers and the oil distilled from them; used in perfumes, soaps and sachets; also medicinal and for moth prevention. From seed or cuttings. **C**: 13   p. 321
  stoechas (So. France), "French lavender"; gray-tomentose, attractive-looking smaller perennial shrub about 1 ft. high with hoary gray-green linear leaves and spikes of dark purple, scented flowers. Not winter-hardy. Use: flowering tips for perfumes, soaps, herbal tobacco, and scented sachets; fresh young tips for flavoring in jellies and beverages. Thrives in warmer climates, or indoors. From seed or cuttings. **C**: 13   p. 319

LEMAIREOCEREUS (*Cactaceae*); the "Organ pipes" are handsome, large and tree-like columnar and branching cacti of the Cereus group, with deep ribs, mostly furnished with many stout spines; medium-size white or pink flowers mostly day-blooming somewhat funnel-form; the juicy fruit is often edible. These are attractive, warmth-loving house plants but tolerate temperatures down to 50° F. By cuttings or seed.
  beneckei (C. Mexico), "Silver tip" or "Chalk candle"; slender cylindrical column to 10 ft., usually unbranched, heavily coated with white waxy powder; ribs 5–9, notched into knobs, and having few spines; flowers brownish outside, white inside. **C**: 1   p. 154
  marginatus (Pachycereus) (Mexico), "Organ pipe"; slender 5 to 6-ribbed columns, branching from base, dark green, with gray spines; flowers red outside, greenish-white inside. **C**: 1   p. 153
  stellatus (So. Mexico), in the trade as "treleasei"; pretty, slender columns to 9 ft. high, branching; pale bluish-green, 8–12 ribs, short white spined; flowers red, borne at the apex; edible fruit known in Mexico as "Joconostle", **C**: 1   p. 153
  thurberi (Arizona, Mexico), "Arizona organ-pipe"; forbidding-looking column with 12–17 acute ribs, dark green to grayish, the dense areoles with black cushions and clusters of stiff spines gray to black; branching from the base and becoming 20 ft. high; day-blooming purplish flowers with white margin. **C**: 1   p. 153
  "treleasei" hort.: see stellatus

LEMNA (*Lemnaceae*)
  minor (Temperate and Subtropical Northern hemisphere), "Duckweed" or "Duck's meat"; small floating perennial aquatic herb without stems and without true leaves, and consisting of one tiny floating frond-like organ (thallus) to $\frac{1}{4}$ in. long and one unbranched hair-like root; they cohere by their edges in twos and threes, and new fronds will grow out of the edge of the old ones; in autumn falling to the bottom of a pool and rising in the spring. Flowers are microscopic; Lemna belong to the smallest known flowering plants. Grown in ponds as food for fishes and ducks and they may completely cover the surface of still water. Much prized in the aquarium. Propagated vegetatively. **C**: 14   p. 519

LENOPHYLLUM (*Crassulaceae*)
  guttatum of the trade (Sedum guttatum) (Mexico: Saltillo), "Basin leaf"; pretty, suckering succulent with thick stem and opposite spatulate $1\frac{1}{2}$ in. leaves olive green-gray and scurfy, and covered with fine dots and stripes of black-brown, the margins folded upward to form a concave trough; flowers yellow. Very attractive as a window sill plant. **C**: 15   p. 331

LEONTOPODIUM (*Compositae*)
  alpinum (Alps of Europe, Pyrenees, Himalayas and Japan), the treasured "Edelweiss"; low tufted, white-woolly, perennial alpine herb, widely distributed in high mountains but rare, which I saw even in Central Asia in view of Mt. Everest; leafy stalks 4 to 8 in. high, topped by the star-shaped inflorescence with a series of dry silvery bracts radiating around the small yellow disk flowers; very pronounced glistening white when growing in the clear sun at high altitudes. For successful cultivation requiring deep gritty loam rich in calcium such as limestone. Occasionally grown in pots, as cool and light as possible, blooming normally from June to August. Propagated by division or seed **C**: 75   p. 417

LEPISMIUM (*Cactaceae*)
  cruciforme (Pfeiffera) (Brazil, Paraguay, Argentina), "Tree-and-rock cactus"; branching cactus growing on trees or rocks, with its angled stem waxy green and tinted purple, the areoles sunken into notched margins set with tufts of white hair-like spines; small white flowers followed by purplish fruit. **C**: 10   p. 172

LEPTOSPERMUM (*Myrtaceae*); the "Manukas" or "Tea-trees" are attractive sub-tropical flowering shrubs from New Zealand and Australia, with needle-like leaves and numerous small axillary flowers looking like miniature single roses, with flaring petals arranged around a hard central cup; and blooming in spring; when the petals fall they leave goblet-shaped seed capsules. Leptospermum are widely used in California, and one of the most popular is the white flowered L. scoparium or its forms. This species grows to 12 and more feet high, but will bloom as a small plant, mostly in spring and summer. They love sun and fresh air with only moderate watering. Propagated by seeds in spring; soft cuttings in May, under glass, or mature cuttings in autumn.
  scoparium 'Keatleyi' (New Zealand: North Island), "Pink tea tree" or "Manuka"; an attractive flowering shrub of picturesque habit which may grow to 6 ft. high, the young

growth silky, dense with tiny rigid, sharp pointed leaves under ½ in., and dotted with fragrant oil glands; numerous small pale pink, axillary ¾ in. flowers amongst the foliage, blooming in early summer. Charming as a small pot plant. Largest flowers of the manukas. **C**: 63   p. 501

scoparium 'Ruby Glow', the "Double red manuka", California cultivar 6 to 8 ft. high, with full double crimson flowers ½ in. across, long retaining their color; the leaves are bronzy-green on deep purple branches; the young growth white-silky. An excellent bloomer with blood-red flowers covering the entire plant in spring. **C**: 63   p. 501

LEUCACENDRON (Proteaceae)

argenteum (So. Africa), the "Silver-tree"; beautiful, eye-catching tree with gray-barked trunk to 40 ft. high, the branches dense with silky, clasping, lanceolate leaves 3 to 6 in. long, thickly covered with silvery pubescence which glistens like shining silver in the sun; globular flower heads yellow. Needs sunlight, humid air, and fast-draining soil not alkaline. Small plants are picturesque container subjects. Best from imported, fresh seed, or grafted. **C**: 64 p. 298

LEUCHTENBERGIA (Cactaceae)

principis (Mexico), the "Prism-cactus" or "Agave cactus"; a very different and unusual type of small cactus to 8 in. high, with a parsnip-like root, and the elongated tubercles looking like triangular fleshy grayish-green leaves as in Agave, but with the tips cut off, these bearing grayish wool and angular papery spines; fragrant yellow flowers near the center, borne at the tip of new tubercles. As a curiosity, the tubercles placed in sand will sprout new plants. **C**: 15   p. 160

LEUCOJUM (Amaryllidaceae)

aestivum (C. and So. Europe, Asia Minor, Caucasus), "Summer snowflake" or "Giant snowflake"; charming small bulbous herb 1 ft. high with 2–3 narrow basal leaves; the nodding little ¾ in. bell-shaped, fragrant flowers white and tipped with green, all 6 segments of even length, unlike Galanthus; late spring blooming, the Snowflakes follow the Snowdrops in flowering. Perfectly winter-hardy and well-loved in gardens where they form clumps. When planted in pots in early fall and buried outside until rooted, Snowflakes may be brought into a cool place indoors in December and flowered 6 weeks ahead of normal. Propagate by offsets or division. **C**: 76   p. 392

LEUCOTHOE (Ericaceae)

catesbaei (Andromeda catesbaei) (Eastern U.S.: Virginia to Georgia), "Sweet bells"; handsome broad-leaved evergreen shrub of low habit, slowly reaching to 6 ft.; with graceful, arching branches and glossy dark green, thick-leathery lanceolate leaves to 6 in. long, the young branches red; small urn-shaped white flowers in crowded racemes. Fairly hardy, and when exposed to cold the foliage turns bronze or red in winter. Also likes shade, and humus in soil. From cuttings. **C**: 26   p. 300

LICUALA (Palmae)

grandis (New Britain Isl., N.E. of New Guinea), "Ruffled fan palm"; very elegant small fan palm, with slim solitary 8 ft. trunk topped by plaited bright green fan leaves almost round, lobed and toothed along the continuous margin, on slender, thorny petioles; glossy crimson fruit. Best in the tropical conservatory, but does well in the warm living room if kept in a water-filled saucer. **C**: 54   p. 226

spinosa (Malaya to Vietnam and Java), "Spiny licuala palm"; densely suckering and clustering fan palm, forming compact tufts, with a mass of foliage from top to bottom, often short but may grow to 10 ft. high, glossy-green leaves parted to the center into plaited segments 12 in. long and ending abruptly in a toothed apex as if cut off, the rigid petioles armed with curved black thorns; fruit lustrous red. **C**: 54   p. 230

LIGULARIA (Compositae)

tussilaginea argentea, "Silver farfugium"; solitary cordate-orbicular toothed, soft-leathery leaves, on fleshy stems rising from rhizomatous basal rosette, glaucous fresh to grayish-green, beautifully variegated cream or white especially at margin. **C**: 66   p. 184

tussilaginea aureo-maculata (L. kaempferi or often known as Farfugium japonicum) (Japan), the "Leopard plant"; a colorful rhizomatous perennial with large rounded, green, somewhat leathery, smooth leaves blotched yellow and cream; light yellow daisy-like flowers. **C**: 16   p. 184

LIGUSTRUM (Oleaceae), the "Privets"; are ornamental

shrubs of the Olive family, some evergreen, others deciduous, having opposite entire, often thick leaves and small funnel-form white flowers in terminal clusters. The evergreen species L. lucidum and japonicum and their forms may be somewhat confused as their leaves are variable, and I have used the names common for horticulture but they may be subject to correction. Privets are voracious and of easiest culture, very tolerant to varying conditions in unfavorable sites, but generally prefer good light, coolness, and sufficient soil moisture. Propagated from soft or hardwood cuttings and division.

japonicum (Japan, Korea), the "Japanese privet"; fast growing evergreen bush to 10 ft. high, with leathery, rich dark glossy green, ovate short-pointed leaves 1½ to 4 in. long, on minutely downy twigs; deciduous during cold winters; small white flowers in terminal clusters, followed by black berries. Fairly winter hardy. **C**: 25   p. 294

lucidum (China, Korea, Japan), "Glossy privet" of the South, or "White-wax tree" in China; planted widely in the Southern states; dense leafy evergreen shrub to 30 ft. high, with glossy, dark green, thick-leathery, ovate leaves 3–6 in. long, on flexible, smooth branches; white flowers in long panicles, followed by black berries. Good tub plant for drafty doorways and entrance halls occasionally cold, also air-conditioning; very durable, tolerates dark places to 20 fc., but preferring sun. Water liberally, sufficient for its prolific root-system, also feeding. From cuttings. **C**: 13   p. 293

lucidum 'Texanum', the "Wax-leaf privet"; a cultivar of compact habit becoming 6 to 9 ft. high, with shiny, thick-leathery, lush dark green, ovate-acuminate leaves 2½ to 3½ in. long, somewhat wavy; ideally suited to shaping by trimming, and an enduring decorator under unfavorable conditions, but best in cooler locations. Some botanical authorities place texanum under L. japonicum, but the young shoots of this species are downy-hairy, while lucidum is not. **C**: 13   p. 293

sinense 'Variegatum', (China), "Chinese silver privet"; shapely variegated bush; the green species in China to 12 ft. high and deciduous; densely branched with thin twiggy, woody stems, the small ovate leaves ½ to ¾ in. long and faintly dentate, to 3 in. long in older plants, milky green with white borders irregularly outlined; pubescent on midrib beneath; fragrant whitish flowers. Not hardy North. Pretty though somewhat sparry house or decorative plant. **C**: 13   p. 299

LILIUM (Liliaceae); hardy, leafy-stemmed herbs with scaly

bulbs, parallel-veined leaves, and regular, 6-parted flowers having petals and sepals often colored alike, and with ovary inside the flower. Sometimes fragrant, lilies are amongst the most dignified, varied and showy of bulbous plants in a wide range of colors except blue, and have been highly hybridized. For container growing, bulbs may usually be potted from September to March, and kept cool until well rooted, and growth starts, then move to light window or sun room; when warm enough out to the patio. After flowering, do not cut down the foliage, to help mature the new bulbs before going into their dormancy period; the old bulbs usually split or give rise to young offsets or bulblets; or bulbils may form along the old stem; these planted out may take 2–3 years before they reach flowering size. Individual scales may also be rooted for propagation.

'Enchantment' (tigrinum hybrid), a "Mid-century lily"; outstanding clone of a robust group of floriferous, colorful Oregon hybrids ideally adapted to warmer climate and pot culture; of short, stocky habit, densely leafy stems 1 to 2 ft. high with dark green foliage 2½ in. long, and topped by a showy cluster of wide open flowers 4 to 6 in. across, bright orange with dark brown spots. Easy to force at moderate temperature of about 60° F. for spring flower shows; normally July-blooming; hardy with some protection with good drainage. From bulblets. **C**: 25   p. 374

'Harmony' (tigrinum hybrid); a "Mid-century lily", one of the finest, early-flowering in rich orange, the wide-spread blooms 5 in. across and spotted with brown; numerous flowers and buds on a stiff-erect leafy stem about 20 in. high or more; normally blooming in June. One of the best for forcing in pots; may be planted in the garden after flowering, but leave all the foliage possible to help mature the new bulb. **C**: 25   p. 397

'Imperial Crimson', an "Empress hybrid lily"; fantastic strain of magnificent, vigorous lilies raised by Oregon Bulb Farms, involving blood of L. auratum, speciosum, japonicum and rubellum; tall, leafy, wiry stalks 5 to 7 ft. high carrying giant rather flat, firm flowers 8 in. across, the petals gently curling back, basically white but toward the middle changing to pink with red spotting, the center of each segment intense crimson; blooming in August. From seed or bulblets. **C**: 75
                                                                                    p. 397

longiflorum 'Ace', or "Slocum's Ace"; a favorite commercial "Easter lily" for pot growing; Oregon cultivar (1935) of the 'Croft' Easter lily complex; prolific bloomer of rather short, husky habit, dark shining green rather erect foliage on green stem about 18 in. above pot, toward the apex of which appear buds and flowers one above the other, unlike Croft which are all in one cluster; the firm-textured white trumpets are generally shorter (5½ in. long) and force well from a slow start. To force this type of lily for Easter, bulbs are pre-cooled at 33 to 35° F. for at least 4–6 weeks, potted in November and kept warm at 60–65° F. nights. Should flower in 120 days **C**: 1   p. 374

longiflorum 'Croft', a superior American "Easter lily", developed in Oregon 1928 from a hybrid of the Japanese L. longiflorum 'Giganteum' x 'Erabu', and became commercially important when World War II in 1941 cut off supplies of L. longiflorum 'Giganteum'; large and elegant, firm-textured white trumpet-flowers 6 in. long, carried horizontally on stiff, medium-tall green stems, densely furnished with glossy-green leaves. Bulbs pre-cooled at 33 to 35° F. a minimum of 5 weeks, respond quickly to forcing at 60° F. or more, take about 120 days to bloom. **C**: 13   p. 365*, 375, 397

regale (W. China: Szechwan), the ubiquitous and well-loved "Regal lily" or "Royal lily"; a lovely bulbous plant with wiry, purplish stem anywhere from 1½ to 5 ft. high, dense along upper part with numerous, short, narrow leaves and topped by a whorl of 3 to 15 fragrant, trumpet-shaped, white flowers to 6 in. long, with recurved tips, with yellow throat and golden anthers; pale purple outside; blooming July-August. Winter-hardy and widely planted in gardens, and a good container plant for cool location, but superseded in quality by today's hybrid trumpet lilies. Quickly growing from seed and flowering the second year. **C**: 76   p. 397

speciosum (Southern Japan), the showy "Japanese lily"; well-known, attractive somewhat sparry species sometimes described as the Queen of lilies; with stem-rooting, scaly bulb and wiry stem 1 to 4 ft. high, with scattered, rather broad, lanceolate leaves, bearing 3–10 stalked, fragrant, wide-spread flowers with wavy, reflexed petals, 4–5 in. across, white suffused with rose, and with raised crimson spots, blooming in July-September. Good pot plant, lending itself to forcing for early bloom in the cool greenhouse or sun-room. Not quite winter-hardy. **C**: 63   p. 397

tigrinum (China, Japan; escaped to the roadsides of Eastern U.S.), the "Tiger lily", a handsome, strong-growing winter-hardy bulbous plant with slender, leafy stem 2–4 ft. high, covered with white, woolly hairs; the scattered leaves rather short, dark glossy green topped by a large cluster of drooping flowers to 5 in. across, the reflexed segments glowing orange-red and spotted with purplish-black; blooming from July to September. Small black bulbils are generally present in the axils of the foliage and which may be detached and used for propagation. The Chinese Tiger lily bulbs are cooked and eaten in the Orient, tasting like artichoke. **C**: 75   p. 397

tigrinum splendens, "Red tiger-lily"; a showy, robust variety of Tiger lily with broad, glossy leaves, and larger flowers fiery orange scarlet, densely covered with large glossy black spots, more numerous to 25 blooms on the strong, leafy stem. **C**: 75   p. 397

LIMONIUM (Plumbaginaceae)

sinuatum (Mediterranean from Portugal to Asia Minor), the widely planted "Statice" or "Sea-lavender"; perennial in habitat and somewhat hardy with protection, usually grown as biennial or annual 1 to 1½ ft. high; with deeply lobed leathery basal leaves 6 or more inches long; winged, branching, but almost leafless stems carrying one-sided dense clusters of numerous small flowers having papery ½ in. calyx usually blue, sometimes lavender or rose, and a tiny

yellowish-white corolla hidden inside. The flowers are amongst the best and most important for cutting as "everlastings", keeping color even when dried. Normally summer-blooming in the garden, statice is flowered in winter in mild regions, from seed sown in July. **C**: 63   p. 417

LINARIA (Scrophulariaceae)

maroccana (Morocco), "Toadflax" or "Baby snapdragon"; very pretty fast-blooming cool-weather annual to 18 in. high, with slender, branching stems bearing narrow-pointed flax-like leaves and long, dense spikes of showy flowers faintly resembling a miniature snapdragon; normally violet-purple and each with a long pointed spur and a yellow blotch on the lower lip, blooming in June to September. There are a number of color varieties in soft and unobtrusive lavender, violet, purple, pink, red and yellow, also white, and they are very effective if planted together in large containers. From seed. **C**: 76   p. 414

LINOSPADIX (Palmae)

minor (Bacularia) (Queensland), "Miniature walking-stick palm"; a delicate, attractive cluster palm, sprouting several stems ½ in. thick and 2–5 ft. long from underground rhizome; the leaves are oblong and entire in the juvenile stage, becoming divided and to 3 ft. long. **C**: 60   p. 230

monostachya (New South Wales; Queensland), the "Walking-stick palm"; clustering small palm (5–8 ft.), with cane-like smooth green stems less than 1 in. thick, and partly covered by remnants of leaf-petioles; few pinnate fronds 2–3 ft. long, the leaflets dark green and variable in width. Not very ornamental; reminding of a walking-cane. **C**: 66   p. 230

LINUM (Linaceae)

grandiflorum (Northern Africa), "Flowering flax"; showy annual 1 to 2 ft. high, with slender branching leafy stems, narrow grayish-green leaves, and abundant, shallow-cupped, 5-petaled flowers to 1½ in. across, in loose clusters, in shades from bright scarlet to rose and bluish purple, blooming from late spring into fall. Each bloom lasts only a day, but others keep coming on. Sun-loving and ideal for patio containers, sown right where wanted; for quick, easy color effects. **C**: 63   p. 415

LIPARIS (Orchidaceae)

longipes (viridiflora) (Mountains in India, China, Malaya, Vietnam, Philippines); attractive rock-dwelling orchid, sometimes on trees, of compact habit, with densely clustered elongate pseudobulbs thickened at base, bearing 2 leaves to 8 in. long; the erect inflorescence spike-like, very crowded with tiny greenish-cream flowers less than ¼ in. across, blooming during winter, or various. By division. **C**: 22   p. 470

LIPPIA: see Aloysia

LIRIOPE (Liliaceae)

muscari 'Variegata' (exiliflora) (Japan, China), "Variegated blue lily-turf"; tufting, grass-like perennial spreading by underground stems, with firm linear leaves 8 to 16 in. long, ⅜ in. wide, rich green with yellow margins and bands; the little lilac flowers carried erect on purple spikes. By division. **C**: 16   p. 304

LISIANTHUS (Gentianaceae)

nigrescens (Eustoma) (U.S.: Nebraska to Texas), the "Black gentian"; annual or biennial erect herb 1 to 2 ft. high; with glaucous, ovate to lanceolate-oblong leaves and tubular-flaring lavender-purple flowers 1½ in. long, blooming during summer time, a lovely bloomer for the window-sill or outdoor planter. From seed sown in June and kept cool at 50° F. over winter; final potting in February. **C**: 54 p. 496

LITHOPS (Aizoaceae); the "Stone-faces" or "Living stones" are the most fascinating group of the Mesembryanthemum family of South Africa; they are tiny succulents with pairs of fat, modified leaves tightly held together except for a cleft across the flattened top; when beginning to grow in spring, these bodies are burst apart by the succeeding pair of leaves and persist beside them for nearly a year; in autumn from the cleft appear the showy white or yellow, daisy-like flowers often bigger than the plant beneath. The translucent windows on the top surface of the plant are

vividly and prettily patterned, and serve to allow the needed solar energy to penetrate deep into the plant body to reach the cells containing chlorophyll that line the inner plant body, to carry on the process of food manufacturing normally handled by conventional green leaves. Lithops are grown in gritty soil that drains rapidly; they rest in midsummer and midwinter when they are kept fairly cool and dry; at other times they need careful watering once or twice a week. See culture p. 327. Propagated by seed.

bella (S.W. Africa), "Pretty stone face"; small succulent body about 1 in. dia., consisting of two thick leaves united except for the fissure across the slightly rounded top, brownish-yellow with darker markings or in some forms silver-gray, and resembling the granite pebbles of their natural habitat; pure white flowers 1¼ in. across. Very beautiful species. **C**: 13   p. 354

karasmontana (Southwest Africa), "Mountain pebble" or "Mountain stone-face", from quartzite hills; turf-forming; club-shaped cleft body almost circular, ¾ to 1 in. dia., pearl-gray, with wrinkled top brownish-ochre and with obscure dark markings; flowers white with yellow anthers. Very sensitive to excess moisture. **C**: 13   p. 354, 531

marmorata (So. Africa: Cape Prov.), "Living stone" or "Big marble plant"; pale buff succulent body ¾ in. thick, the top surface slightly convex, and patterned with a bold marble design of contrasting translucent brown; prominent fissure separating the pair of modified fat leaves; scented flowers white. **C**: 13   p. 328, 354

pseudotruncatella (S.W. Africa), a "Living stone" of the arid highlands around Windhoek, in red loam with quartzite gravel; tufting split rocks 1 in. across, with cleft running right across, the surface smooth, pale brownish-gray with network of brown lines; however, color of body changes according to soil color, brown over yellowish to chalky gray-white; 1¼ in. flowers golden yellow. **C**: 13
                                                            p. 354

turbiniformis (So. Africa: Cape Prov.), "Flat-top mimicry plant"; elegant round, split body to 1½ in. high and broad, the sides gray, the top brownish and wrinkled, with dark brown branched grooves between the pebbly surface. The succeeding pair of leaves splits open the older body. Large 1½ in. yellow flowers. **C**: 13   p. 354

LIVISTONA (*Palmae*)

australis (Queensland, New South Wales, Victoria), "Australian fountain palm" or "Australian cabbage palm"; forming slender, ringed trunk to 80 ft. or more, which is covered with brown leaf bases and brown fiber while young; dense crown of soft leathery palmate fronds, rounded in outline, 3–5 ft. across, divided to middle into narrow glossy dark green segments with yellow central nerve, extending straight and without threads between; the petiole with stout curved spines; the rib extends several inches into base of leaf. **C**: 16   p. 229

chinensis (South China), the formerly widely grown "Latania borbonica" of horticulture, "Chinese fan palm"; spectacular, large fan palm with thick trunk that may grow to 30 ft. high, but extending in spread sideways to a diameter of 25 ft., with gigantic, glossy, fresh green, plaited leaves more broad than long, to 6 ft. wide, cut halfway into many narrow, one-ribbed segments which are split again, the tips will hang like a fringe; petioles armed with small spines when the palms are young, usually disappear later; fruit metallic blue. Long popular in Europe and America in parlors, hotels and winter gardens before the advent of Howeias. Satisfied with medium-warm conditions but requiring lots of water and big space. **C**: 16   p. 229, 540

rotundifolia (Malaya, Indonesia), "Round-leaf livistona"; tropical fan palm becoming 50 ft. or more high, brown trunk topped by a dense crown of glossy green palmate fronds 2 to 5 ft. across, with broad lobes cut only 1/3 to base when young, slightly notched at tip, and arranged almost in a circle, on thorny red stalks. The younger plants are particularly attractive as tub plants where growth can be limited to their small, perfectly round leaves looking like little bar stools. Probably the best of the Livistonas for house plant use as it enjoys higher temperature. **C**: 10 p. 229

LOBIVIA (*Cactaceae*); the "Cob cacti" are small-bodied xerophytes from high deserts of South America, very floriferous and prized for their large colorful, funnel-shaped, beautiful flowers in profusion from June to September; these are day-blooming but frequently last only a few hours.

Because of their small size Lobivias are favored window sill plants, where they receive sunlight and fresh air and may be watered freely, provided their soil is gritty and porous. To ensure good budding for the next season they must have a rest period beginning in October, keeping them cold at near 40° F. and quite dry. Propagated by offsets or from seed.

aurea (Pseudolobivia) (No. Argentina: Cordoba), "Golden lily cactus"; lovely globular to cylindrical small plant similar to Echinopsis, and clustering; dark green, to 4 in. high, with 14–15 acute ribs, spines yellowish brown; short-funnelform flowers freely borne, glossy lemon-yellow. **C**: 13   p. 157

LOBULARIA (*Cruciferae*)

maritima (Mediterranean region), "Sweet alyssum"; a much branched little herbaceous plant grown as an annual, 4 to 6 in. high but spreading 6 to 12 in., with lanceolate to linear leaves 1 to 3 in. long, and many small pure white 4-petalled flowers in showy clusters blooming a long time from May on; if sheared back new flowers will cover the plants like a carpet until fall. Their intense fragrance of honey, and tolerance of dry-hot summers makes alyssum a favored bedding plant or for edging, for window boxes, or in pots on the window sill. From seed. **C**: 64   p. 411

LOCKHARTIA (*Orchidaceae*)

lunifera (Brazil), "Braided orchid"; curious little epiphyte scarcely looking like an orchid, having erect, later pendulous ribbon-like stems to 8 in. long, clothed with triangular leaves folded flat, out of which appear the tiny ½ in., golden yellow, long-lasting flowers with red-spotted lip; blooming summer to autumn. Easy to grow, and uniquely attractive even without the not very showy flowers; they must be warm and moist all the time. By division of the root-stock, or by sections of the stem. **C**: 10   p. 470

LOMARIA: see Blechnum

LOPEZIA (*Onagraceae*)

lineata (coccinea) (Mexico), "Mosquito-flower"; shrubby plant 1 to 3 ft. across, with hairy stems and serrate ovate leaves, the curious, insect-like, ¼ to ½ in. winged red flowers with 5 petals, the 2 upper ones bent away from center disclosing an apparent drop of honey, actually a glossy piece of hard tissue, or gland, which deceives flies. Blooming late spring to next winter. Charming small flowering house plant for the window sill; in summer outdoors. Cut back in late winter, the flowers appear on the young shoots. Propagated by cuttings. **C**: 14   p. 494

LOPHOCEREUS (*Cactaceae*)

schottii (Sonora, Arizona, Baja California), "Totem-pole"; moss-green column 3 in. thick, with 5–9 acute ribs, the areoles set with small clusters of white wool, and short black spines, night-flowering red. **C**: 1   p. 153

schottii monstrosus; "Monstrose totem"; very peculiar yet attractive monstrose form consisting of a short column entirely composed of large, smooth knobs or remnants of ribs with spineless areoles, waxy moss-green. The species is a columnar plant from Northern Mexico and Arizona growing 15 ft. high, and night-blooming with red flowers. **C**: 1 p. 168

LOPHOPHORA (*Cactaceae*)

williamsii (So. Texas, Mexico), the famous "Peyote", "Mescal", or "Sacred mushroom" of the ancient Mexicans who used to eat it fresh or dried, because of its powerful exhilarating but non-narcotic properties, similar to hashish; small depressed globe to 3 in. dia. freely sprouting laterally; bluish-green, with 5–13 low and wide ribs, the tubercles white-tufted; flowers pink to white. Revered by the Aztecs in their religious rites; contains Mescaline. **C**: 13 p. 159, 531

LOTUS (*Leguminosae*)

berthelotii (peliorhynchus) (Cape Verde, Canary Islands), "Winged pea" or "Coral gem"; remarkable silvery-haired shrub spreading with straggling branches; ferny, pinnate foliage of threadlike leaflets; pea-type boat-shaped, orange scarlet flowers, toward the ends of branches in the spring. Nice for training in baskets or boxes; needs careful watering, avoid severe drying out. **C**: 13   p. 263

LUDWIGIA (*Onagraceae*)

mulertii, hort. (So. America); the "Swamp spoon"; a small bog or aquatic perennial herb with opposite spoon-shaped leaves glossy above and crimson-purple beneath, on slender stems rooting at nodes; solitary yellow inconspicuous flowers in the axils. Does well, as an aerator, submerged in aquarium, but likes to break the surface, turning coppery in strong sun. Possibly a form of palustris. Best rooted in earth, in good light at 65 to 70° F. Propagated best by tip cuttings, on trays of saturated sand. **C**: 4   p. 515

LUFFA (Cucurbitaceae)

cylindrica (Eastern hemisphere Tropics), the odd "Vegetable sponge", "Dishcloth gourd" or "Sauna sponge", is produced by a fast growing tropical climber of the cucumber family; 4-angled stems pull themselves up by coiling tendrils, the rough-hairy herbaceous angled or lobed green leaves cucumber-like; 4 to 6 in. wide; separate male and female 3 in. flowers golden yellow; when pollinated the pistillate flowers produce the swollen cylindric fruit to 2 ft. long, which may be eaten when young, but the ripe yellow fruit contains the sponge-like fiber that is washed and dried, and marketed for scrubbing, stimulating massage and skin care. **C**: 58   p. 537

LYCASTE (Orchidaceae); charming tropical American epiphytes of easy culture, with short, thick pseudobulbs and ribbed or shining green leaves which last only a year or two; the large waxy, quaint and very showy flowers are remarkably durable, and are produced singly on short stalks from the base of mature pseudobulbs or concurrently with the flush of new growth, sometimes several at one time, and a plant may flower for months. Lycaste grow best in a little shade with free-moving air; watered generously during growth and flowering; fairly dry through winter. Propagate by division as new growth starts, separating the bulbs in groups of two or three.

aromatica (Mexico, Guatemala, Honduras); attractive dwarf epiphyte with oval pseudobulbs holding 1 or 2 dark green, deciduous leaves to 1½ ft. long; clustered flower stalks to 6 in. long, each bearing a solitary waxy blossom to 3 in. across, with sepals greenish-orange, petals golden-yellow, the concave lip golden-yellow dotted red, blooming in winter and spring. Prized for its strong, pleasant fragrance, free-flowering habit; recommended for the beginner. **C**: 21   p. 470

cruenta (Mexico, Guatemala, El Salvador); a desirable epiphyte with 2 to 3-leaved pseudobulbs and medium-sized long-lasting, individually stalked, waxy and very fragrant flowers, 2 to 4 in. across; sepals greenish-yellow, petals deep golden-yellow, lip orange-yellow with blood-red in throat; blooming March-April, and longer. Deciduous. **C**: 71   p. 470

deppei (So. Mexico, Guatemala); interesting and ornamental epiphyte with compressed pseudobulbs bearing 3–4 plaited, deciduous leaves; fragrant curiously colored, waxy, 4 in. variable flowers, the sepals greenish-yellow, spotted with red; in striking contrast, the petals are ivory-white, and the little lip white with golden crest and spotted with red; blooming nearly all year round, mostly in spring and autumn. A well-known species very satisfactory as a blooming house plant, flowers lasting 6 to 8 weeks on the plant. **C**: 21   p. 470

gigantea (Colombia, Ecuador, Peru); stately, robust species with big pseudobulbs and deciduous leaves to 2½ ft. high; large waxy flowers 6 in. across and sweetly fragrant, with greenish-yellow sepals and petals inclined forward, the toothed lip orange, center purplish-brown, the whole flower appearing like a laughing face; blooming spring or summer. **C**: 71   p. 470

virginalis alba, better known as L. skinneri alba (Guatemala), the chaste "White Nun" orchid; beautiful albino form of L. virginalis with large waxy flowers to 6 in. across, pure white, except for a crest of light yellow on the lip, and very fragrant; blooming between November and May. The species is a handsome, profusely flowering epiphyte with large pseudobulbs, each with 2 to 3 broad, plaited leaves to 2 ft. long and more persistent than in other species; flowers vary from rose-pink to carmine rose. One of the easiest of orchids to grow and its flowers can be enjoyed for a long time. **C**: 71   p. 470

LYCOPERSICON (Solanaceae)

esculentum 'Tiny Tim', a miniature "Cherry tomato"; a cute little decorative plant of dwarf habit, about 1 ft. high, and suitable for pot culture; freely bearing small round scarlet, edible tomato fruit to ¾ in. dia., with palatable flavor, in winter or summer; handy in the sunny kitchen window. Easy to grow but needs warmth and good light. Derived from the common tomato, a vegetable originating in the Andes of South America. Quickly from seed, which may be washed from your own fruit. **C**: 2   p. 514

LYCOPODIUM (Lycopodiaceae)

obscurum, at home in moist woods and along bogs in acid soil from Newfoundland to Alaska, south to Alabama, west to So. Dakota and Washington, the "Ground pine", also loosely known as "Princess pine"; evergreen moss-like herb of the clubmoss family, allies to the ferns; the main stem creeping horizontally. Sending up aerial branches 4 to 10 in. high, like lilliputian pine trees with branches covered by dark green scale-like leaves, and topped instead of flowers with club-shaped spikes producing spores resembling miniature cones. Ground pines are collected annually for the making of long-lasting Christmas wreaths. Similarly bound together into a tight little evergreen tree, 8 in. high, and sealed into small pots, the "Mystery plant" as sold by florists is created, studded with tiny dyed Skyrockets, immortelle flowers of the genus Eriocaulon. Such Mystery trees last a long time especially if the "plant" is occasionally sprinkled with water, but the uninitiated don't quite understand why the little "tree" won't grow.   p. 539

phlegmaria (Queensland, and Tropics of Eastern hemisphere), "Queensland tassel fern"; handsome tropical epiphyte; slender, wiry stems first erect then pendant, the pale rope-like, forking branches set closely with stiff spirals or whorls, light green beneath, the fertile tasseled fronds thin, like catkins, and pendulous. Slow-growing; can be grown in baskets or against the wall in sphagnum and osmunda or other fiber; avoid calcium; needs high humidity. **C**: 60   p. 253, 539

LYCORIS (Amaryllidaceae)

aurea (China), "Golden spider lily"; a pretty, bulbous plant about 1 ft. high, with narrow, strap-shaped bluish leaves which appear in the spring, ripen, then die down before bloom stalk develops; with clusters of spidery looking bright yellow 3 in. flowers having narrow segments; anytime from May to August or September. After blooming give plants a rest until new buds show but don't keep pots too dry. Lycoris flower best if pots are left undisturbed and well-rooted. Propagate by offsets. **C**: 63   p. 392

LYGODIUM (Filices: Schizaeaceae)

japonicum (Japan to Himalayas to No. Australia), "Climbing fern"; with twining, thread-like stems bearing pretty, pleasing green, papery triangular leaves pinnate into lobed segments, the sterile pinnae toothed and lobed, the fertile pinnae with tiny lobes along margins, often much reduced. The twining stem is really petiole and climbing axis of the elongate frond, growing unlimited in length. Best planted out in the greenhouse or in boxes with headroom and a permanent wire or nylon thread support; but will also grow in pots. Needs humidity. **C**: 66   p. 248, 286

LYSIONOTUS (Gesneriaceae)

serratus (India: Himalayas), small fibrous-rooted shrubby plant about 12 in. high, with dark green, waxy, elliptic leaves with pale veins, and light green, shiny reverse, margins lightly crenate; scandent clusters of bracted funnel-shaped, bilabiate flowers 1½ in. long, inflated in the middle, pale lilac with deeper veins, blooming in winter. From seed or by division. **C**: 66   p. 446

MACHAEROCEREUS (Cactaceae)

eruca (Baja California), "Creeping devil"; a ferocious creeper; prostrate, thick stems dark green with coppery tint, many-ribbed, ribs low and very spiny, the central spine flattened, dagger-like and directed downward; day-blooming, creamy flowers. **C**: 1   p. 155

MACODES (Orchidaceae)

petola (Anoectochilus) (Java, Borneo), "Gold-net orchid"; small, handsome terrestrial growing in caves, with broad ovate, 3 in. leaves a beautiful velvety olive-green, and a contrasting network of shimmering gold bronze; small flowers on long spike, brown with white tips; should be removed to conserve the strength of the plant; blooming January-February. **C**: 62   p. 192, 454

MACROPIPER (*Piperaceae*)
excelsum (New Zealand, Tahiti), "Lofty pepper"; small tree growing to 20 ft. high, having aromatic, cordate rounded, leathery, 5 in. leaves, olive-green with prominent pale veins and undulate margin, pale gray beneath; flowers in spikes very small. Attractive house plant. **C**: 16　　p. 256

MACROZAMIA (*Cycadaceae*)
peroffskyana (denisonii) (Queensland, New So. Wales), a giant "Palm-fern", with trunk to 20 ft. high and 1½ ft. dia., carrying a great crown of majestic arched pinnate fronds to 10 ft. long, the numerous leaflets sickle-shaped, bright shiny green above, paler beneath; the male and female cones exceptionally large. Needs plenty of water during growth, and a distinct rest in winter. For the large conservatory. **C**: 16　　p. 316

MAHERNIA (*Sterculiaceae*)
verticillata (So. Africa), "Honey bells"; rambling bush with fern-like, lacy foliage, and small, ¾ in., nodding, golden, bell-shaped flowers, very fragrant like lily-of-the-valley. Grown as a cool house plant or hanging basket, long-blooming from winter to spring; summer outdoors. From seed or soft cuttings. **C**: 70　　p. 278

MAHONIA (*Berberidaceae*)
bealii (China), "Leatherleaf mahonia"; distinctive shrub with stout, erect woody stems, reaching to 12 ft. high, carrying horizontal leaves a foot long, divided into 7–15 thick-leathery, broad leaflets as much as 5 in. long, bluish green with yellow at base and veins glaucous green underneath, the margins with a few large teeth, on red petioles. Fragrant yellow flowers followed by powdery blue berries. Excellent and effective as solitary cool decorator plant. Fairly winter-hardy. **C**: 14　　p. 297
lomariifolia (China: Yunnan), "Chinese holly grape"; showy evergreen plant with erect woody stem, later branching and 6 to 10 ft. high, symmetrically bearing clusters of pinnate leaves to 2 ft. long and held horizontally; the wiry axis set with pairs of hard-leathery, holly-like leaflets 2 to 3½ in. long, olive green with lighter veins, the margins undulate and with pointed spiny lobes; yellow flowers in late winter followed by powdery blue berries. This beautiful Mahonia with its fern-like appearance creates dramatic effect as an unusual decorator, best for a cool location. **C**: 14　　p. 297

MAJORANA (*Labiatae*)
hortensis (Origanum majorana) (So. Europe, No. Africa), "Sweet marjoram" or "Annual marjoram"; aromatic bushy, half-hardy spreading perennial 8 to 18 in. high, treated as an annual in cold winter climate; with tiny oval ½–1 in. furry gray-green foliage; small clusters of white flowers with green bracts on top. Very sweet-scented, and a favorite herb for seasoning; the fragrant leaves or dried sprigs are used in cooking for flavoring vegetables, salads, beef, lamb, poultry, stuffings, sausage, stews and other casserole dishes. Oil is distilled from flowering tips, and used medicinally, and as a scent for soaps and perfumes. From seed or cuttings. **C**: 13　　p. 319

MALACOCARPUS (*Cactaceae*)
vorwerkianus (Colombia, near Bogota), "Colombian ball"; small, flat-topped, deeply 20-ridged, shining bright green flattened globe to 4 in. dia., spines straw-colored, white woolly at top; flowers sulphur-yellow. **C**: 13　p. 159

MALAXIS (*Orchidaceae*)
calophylla (Microstylis) (Thailand, Malaya, Java), "Plaited jewel orchid"; handsome terrestrial with ornamental foliage membranaceous, on fleshy base; the oblique ovate leaves plaited or grooved, 4 to 6 in. long, greenish-brown in center, pale grayish-green in outer area, dark spotted, the margin undulate, reddish beneath; small ¼ in. yellowish flowers, with spreading sepals and narrower petals, on erect stalk to 10 in. tall, blooming mostly in spring. Culture as Anoectochilus. **C**: 60　　p. 454

MALPIGHIA (*Malpighiaceae*)
coccigera (W. Indies), "Miniature holly" or "Singapore-holly"; bushy evergreen shrub, densely covered with tiny, ¾ in., holly-like, stiff leaves glossy dark green with coarse spiny teeth; small starry pink, lovely flowers. One of the prettiest miniatures with cute foliage and blooming as a

small plant, during winter and summer. From cuttings. **C**: 13 or 63　　p. 300, 494
glabra (Southern Texas to Northern So. America), "Barbados cherry"; tropical shrub or small tree 5 to 10 ft. high, with slender branches, shining green ovate, leathery 3 in. leaves; small ½ in. carmine-rose flowers with fimbriate petals, followed by cherry-like edible red fruit, with thin skin, and pulp of acid flavor, high in Vitamin C; used in preserves. From cuttings or seed. **C**: 13　　p. 514

MALUS (*Rosaceae*)
x atrosanguinea (halliana x sieboldii), (China); a handsome carmine "Flowering crab-apple"; small tree to 18 ft. high, grown as an ornamental, with purplish somewhat drooping twigs, and ovate, deciduous, glossy leaves finely serrate, covered in spring with masses of fragrant rosy carmine flowers with narrow petals, not fading to white, normally late April-May blooming; followed by small, yellow and dark red apples, hanging on through winter. Propagated by budding or grafting on seedling stock; seed could not be expected to come true. Hardy. **C**: 76　p. 504

MALVAVISCUS (*Malvaceae*)
penduliflorus (conzatii) (Mexico and Guatemala), "Turk's cap" or "Sleepy mallow"; herbaceous shrub to 10 ft. high with large ovate hibiscus-like leaves, doubly crenate having roundish marginal teeth; very showy with pendulous bell-like flowers brilliant red, their petals narrowly folded, never quite opening, and about 2 in. long, blooming mostly from summer into late autumn. A good, colorful house plant for the sunny window, requiring little care. By cuttings. **C**: 14　　p. 495

MAMMILLARIA (*Cactaceae*); the many "Pin-cushion" or "Wart cacti", principally from Mexico, include some of the most delightful midget succulents for growing in small pots, and a fancier may well specialize, and cultivate a collection of varied species within this genus alone and they will be of never ending interest. These "Pin-cushions" are usually small globes, some solitary, others clustering, many covered with silky white hairs or bristles, others with starry clusters of spines; their surface is broken into many prominent knobs or tubercles spirally arranged. The day-flowers are small but often appearing in circles around the plant, appearing in spring and summer, and followed by bright shiny red fruits. Mammillarias grow easily in porous, gritty soil; those with white spines, with calcium in form of gypsum added; warm and moderately moist in summer; during winter as light as possible, cooler at 50–55° F., and dry. Propagated from seed easily produced; or offsets where formed.
aurihamata (Central Mexico), "Yellow-hook cushion"; pretty little globe 2½ in. high with glossy green tubercles bearing white, silky-hairy radials, and long yellow to brown central spines; yellow flowers in a ring around the top. **C**: 13　　p. 162
bocasana (Mexico), "Powder puff", as it looks like a bursting cotton ball; the lovely little, mound-forming globes, to 2 in. thick, are covered with snow-white silky hair, as well as brown fish-hook spines; blooms creamy-white, tipped red. A nice species easy to flower as small plant. **C**: 13　　p. 162, 165*
camptotricha (Mexico), "Birdsnest-cactus"; small globes to 2 in. thick and forming large clusters; fresh green, with extended nipples and long yellow bristle-like spines often twisted; flowers white, greenish outside, hidden in axils. Of easy cultivation. **C**: 13　　p. 161
candida (C. Mexico), "Snowball pincushion"; an exquisite little globe somewhat flat on top, to 3 in. dia., bluish-green, closely tubercled and covered with a multitude of pure white radial spines and very woolly; clustering flowers rose-colored. **C**: 13　　p. 162
celsiana (So. Mexico), "Snowy pincushion"; attractive small deep green globe becoming cylindric to 5 in. high, and forming clusters with age; tubercles neatly arranged and white-woolly in axils; radials white, central spines pale yellow; small flowers red in a ring around top. **C**: 13　p. 162
columbiana (bogotensis) (Colombia), "South American pincushion"; solitary, small club-shaped gray-green plant 4 to 10 in. high with dense nipples set with 20–30 white radials and reddish-brown central spines having dark tip; flowers deep pink with yellow throat. **C**: 13　　p. 162
compressa (C. Mexico), "Mother of hundreds"; small globes, pale bluish-green, later forming clumps and be-

coming 8 in. high; prominent knobs set with long white spines, the axils woolly; flowers pinkish. Very resistant. **C**: 13																												p. 161

confusa (Mexico: Oaxaca), "Variable cushion"; elongate globe to 6 in. high, close to pyrrhocephala, occurring in various forms; pyramidal green tubercles set with curving spines; the depressed center with white wool; flowers greenish straw-colored, midline of petals pink. **C**: 13		p. 161

"dolichocentra" as known in horticulture; possibly M. kewensis var. craigiana (Mexico: Queretaro); small globular to short-cylindric plant bluish gray-green, to 5 in. high; the broad knobs set with 4 to 6 short but strong white spines, the areoles at first woolly, the axils with curled white wool; small flowers rosy to red. **C**: 13					p. 161

echinaria (gracilis) (Mexico), "Needle fingers"; short cylindric plant to 3 in. high, forming clusters, similar to elongata but usually possessing central spines, yellowish and glossy, with the golden yellow radials; flowers yellow with salmon outer petals. **C**: 13				p. 161

elongata (Mexico), "Golden stars"; pretty species with small, clustering cylinders to 1 in. thick, abundantly clustering; light green, with the tubercles in spirals and covered with yellow interlacing radial spines, central spines directed outward; flowers white. Keep very dry in winter. **C**: 13 p. 161

fragilis (gracilis var. fragilis) (Mexico), "Thimble-cactus"; little cylindrical stem freely clustering and branching toward top, to 4 in. high, bright green, the knobs with white radial spines, central spines nearly absent; flowers cream, pinkish outside. **C**: 13				p. 161

geminispina (bicolor) (C. Mexico), "Whitey"; beautiful small, club-shaped plant becoming cylindric to 7 in. high, and clustering; bluish-glaucous, its prominent knobs topped by white radials and needle-like white central spines tipped black; flowers red. One of the finest for indoor culture. **C**: 13						p. 162

hahniana (Mexico), "Old lady cactus"; one of the nicest white-haired species; attractive small globe, to 4 in. across, rich green with long and curly snowy-white, hair-like bristles and red-tipped spines; flowers violet-red, starting in 3 year olds. **C**: 13					p. 162

magnimamma (centricirrha) (C. Mexico), "Mexican pincushion"; clustering globe, to 4 in., dark green, with large conic tubercles topped by abundant white wool in axils and areoles; 3–5 recurved horn-colored spines, creamy flowers. Easy to cultivate; exudes milky sap when wounded. **C**: 13						p. 162

microhelia (Mexico: Queretaro) "Little suns"; short cylindric plant 6 in. high, with low tubercles tipped by regularly radiating, bristle-like but stiff spines white to yellow, and 1–2 blackish red central spines toward apex, increasing in number in older plants; small silky flowers canary yellow to whitish. **C**: 13				p. 161

multiceps (So. Texas, No. Mexico), "Grape-cactus"; possibly a variety of prolifera, a native of Cuba; small globes ¾ in. thick and forming large groups; dark green and covered by white hair-like radials and needle-like yellow central spines; flowers salmon. **C**: 13			p. 161

nejapensis (Mexico: Oaxaca); beautiful elongate, clustering globe to 6 in. high having fresh green, prominent nipples tipped with white-woolly tufts, and with sets of stout, glistening white needle spines red-brown at the extreme tip, the axils also filled with white wool especially near top of plant; yellow flowers shaded with red-purple. **C**: 13						p. 162

parkinsonii (C. Mexico), "Owl's eyes"; branching and clustering little cylinder to 6 in. high, glaucous green; tubercles neatly arranged and richly furnished with white wool, white radials and prominent central spine; flowers yellowish, in a ring at the top; this handsome species exudes milky sap when bruised. **C**: 13				p. 162

potosina (Mexico: Queretaro); elongate globe to 4 in. high, with closely-set knobs which bleed milky sap if bruised; dark green but covered with white hair-like radials and prominent, long reddish yellow needle spines; flowers yellowish white. **C**: 13					p. 162

rhodantha (Mexico), "Rainbow pincushion"; seedlings globose, later cylindric to 1 ft. high, and in time forming clumps; dull green, with round tubercles, the radials white, and central spines reddish-brown; rose-colored flowers numerous in a ring near the top. **C**: 13			p. 162

sempervivi (Cent. Mexico), the true species is an attractive globular plant of 3 in. dia., dense with long angular green nipples tipped by very short, hard spines, the axils toward the apex filled with wool; the numerous flowers appearing near top, carmine rose. **C**: 13				p. 161

tetracantha galeottii (Mexico), "Ruby dumpling"; attractive little globe to 3 in. dia. with conic tubercles lighter green than the species, axils naked; long central spines yellow with brown tips, recurved and flexuous; radial spines usually none except for short bristles; small carmine-red flowers in a ring around the center. **C**: 13			p. 161

vaupelii (Coryphantha) (Mexico), "Vaupel's pincushion"; small solitary bluish green globe 2½ in. across, closely tubercled and covered with yellowish radials and long shiny central spines; flowers pinkish; inside white. **C**: 13						p. 161

vaupelii cristata, "Silver brain"; a fantastic fasciation of the normally globular Mexican species; coiled and wavy crests growing in time into a large and tight symmetrical, compound globe; fine brown spines silvery at crest over the fresh green body. Being of extremely slow growth, such cristate forms grow better and look better if grafted on Trichocereus or similar kind. **C**: 13			p. 168, 532

MANETTIA *(Rubiaceae)*

inflata (in horticulture as M. bicolor), (Paraguay, Uruguay), "Firecracker plant"; twining herb with thread-like stems; thin-fleshy, green, ovate leaves and attractive, solitary, ¾ in. tubular, waxy flowers from the axils, flask-like and vivid yellow, the lower part of the tube densely covered with bright scarlet bristles, giving the appearance of a red corolla tipped yellow. Graceful twiner freely blooming with charming little flowers, easily trained to wire, thread or trellis in boxes or pots; should succeed in warm window. **C**: 54						p. 263

MANGIFERA *(Anacardiaceae)*

indica (No. India, Burma, Malaya), the "Mango tree"; with large spreading, evergreen crown, 60 to 90 ft. high, and grown for its delicious fruit all over the tropics; leathery, lanceolate leaves to 16 in. long; small pinkish flowers in terminal panicles, followed by large, variably yellow to reddish sweet-fleshy aromatic fruit averaging 5 in. long, containing the large adhering stone. Prop. by fresh stony seed at high temperature; also semi-ripened cuttings at 90° F. **C**: 2						p. 506

MANIHOT *(Euphorbiaceae)*

esculenta variegata (Brazil), "Cassava" or "Tapioca plant"; widely grown in tropical regions for the starchy tubers which yield tapioca; digitate, fresh green leaves in this form are beautifully variegated yellow along veins. **C**: 53 p. 187

MARANTA *(Marantaceae)*; the "Prayer plants" are attractive perennial herbs of Tropical America, having rhizomatous or swollen, starchy tuberous roots; the foliage basal or along a stem; inflorescence scattered in branched racemes; the more or less tubular flowers often white and spotted or striped with purple. Long grown as small potted plants in greenhouses or in the home for their beautiful foliage, which is variously colored and blotched, and patterned with fantastic and bizarre designs; at night the leaves fold upward like hands in prayer, always one of the marvels of nature. Marantas need warmth and constant moisture, grow well in loam and humus mixed. They will take a slight rest from December to February, when they may be allowed to get quite dry before watering. As new growth begins, large clusters may be divided.

arundinacea (Mexico to So. America), "Arrowroot"; slender, erect herb to 3 ft. high in cultivation; in habitat to 6 ft.; having starchy roots and forking, zigzag branches bearing scattered, thin, long-arrowshaped leaves to 1 ft. long with rounded base, and tapering upward to a slender tip, fresh green when young, glaucous grayish in older leaves; flowers white. Quickly curls its leaves if too dry. **C**: 54 p. 306

arundinacea variegata (Phrynium micholitzii), "Variegated arrow-root"; slender, erect herb, to 3 ft. high, having starchy roots and forking, zigzag branches bearing scattered, thin, long-lanceshaped papery leaves with rounded base, and tapering upward to a slender tip; the light green foliage is prettily variegated or margined with white, the variegation showing underneath. Flowers white. The green, later glaucous grayish species is wide-spread from Mexico to South America. In dry air the leaves will quickly curl. **C**: 54						p. 176

leuconeura erythroneura (Brazil: Estado do Rio), the beautiful "Red-veined prayer plant" from the Organ Mts.

near Petropolis; a low-growing herbaceous plant, the foliage more or less horizontal with the ground; 4 to 5 inch leaves obovate, on short winged petioles, patterned with a herringbone design of carmine-red veins over light yellow-green to dark velvety olive green, jagged silvery green along center; reddish beneath except green along center; flowers lilac to whitish with purple eye. With all its beauty a surprisingly virtuous house plant and good keeper. **C**: 10 p. 176

leuconeura kerchoveana (Brazil), a well-known "Prayer plant"; low-growing with 6 in. oval leaves mostly hugging the ground, and folding upward in the evening as if hands in prayer; the surface is vivid to pale-grayish green, more pronounced along the midrib and feathering along the veins, with a row of chocolate, later dark green blotches on either side; blotched red beneath; small flowers white, striped purple, in a raceme. **C**: 4 p. 176, 529

leuconeura massangeana, (Brazil), "Rabbit's foot"; low, strikingly beautiful plant having satiny bluish-green leaves, with feathery centerband of silver and thin stripes of pink, later silver, extending out toward the margin, after first passing a broad zone of reddish brown; red-purple beneath. **C**: 54 p. 43*, 176

MARCGRAVIA *(Marcgraviaceae)*
rectiflora (Cuba, Hispaniola, Puerto Rico), "Shingle plant"; climbing epiphytic shrub seen in the rain forest of El Yunque in Puerto Rico; the stem-rooting sterile shoots with small ½ in., sessile leaves clinging shingle-like to trunks of trees; the mature, freely pendant fruiting branches have larger slender, 5 in. dark green leaves, and are tipped by a spreading candelabra-like, greenish inflorescence tinged red. Somewhat similar to Ficus pumila but more tropical. **C**: 4 p. 259

MARCHANTIA *(Marchantiaceae)*
polymorpha (E. United States), a "Liverwort" or "Prehistoric moss plant"; interesting primitive plant that looks like a prehistoric creation, botanically between the mosses and the algae; useful for colonizing in rockgardens, also in conservatories and moist dishgardens; leaf-like flat, scale-like plant body or thallus is 4 to 5 in. long and 1 in. wide; the fruiting body on female plants looks like a tiny umbrella. May be grown in shallow pans wherever the atmosphere is cool and damp. **C**: 86 p. 538

MARICA: see Neomarica

MARRUBIUM *(Labiatae)*
candidissimum (So. Europe), "Hore-hound"; tender aromatic, white-woolly low, woody perennial 1–3 ft. high, small ovate, wrinkled 2 in. leaves with crenate margins; white, mint-like flowers in whorls on foot-long, slender branching stalks. Horehound is the source of the old-fashioned horehound candy; dried stems and flowers used for tea; also medicinally for coughs and colds. From seed, cuttings, or division. **C**: 13 p. 321

MARSILEA *(Filices: Marsileaceae)*
drummondii (W. Australia), "Hairy water clover"; aquatic perennial tufted herb with creeping rhizome rooting at nodes, and floating 4-parted clover-like leaves 3 in. dia., the fan-like leaflets covered with whitish hairs, and with wavy margins, on long slender stalks; the bean-like spore cases or fruiting bodies at the base of leaf stalks. Planted out in aquarium, or if in pot, set in saucer or plunged in water. **C**: 64 p. 252, 539

fimbriata (West Africa: Ghana, Guinea); a tropical "Water clover", with clover-like leaves on petioles 6–12 in. long, the fan-like leaflets about 1 in. long, smooth above, stiff-hairy beneath, outer edge entire, often floating in water; spore case globose, solitary and hairy. An aquatic swamp fern, and pots should be partially plunged in water. Propagated by division. **C**: 60 p. 252

quadrifolia (North Temperate Zone: Europe, No. India, Japan, E. U.S.); the four-leaved "Water clover" or "Clover fern"; a spore-bearing hardy, wide-creeping perennial aquatic with floating clover-like 4-parted leaves which arise from long runners on thin petioles 3 to 6 in. long, the smooth deltoid ¾ in. leaflets having a rounded outer edge. Very attractive planted in aquaria or pools although it may choke the surface of the water with its creeping stems unless kept in check. May also be grown, in turfy loam, in pots partially plunged in water, or the plants can be submerged completely to live under water for a time. Propagated by pieces of the running rhizomes. **C**: 64 p. 518

MARTINEZIA: see Aiphanes

MASDEVALLIA *(Orchidaceae)*; the "Tailed orchids" are most unique epiphytic and American terrestrials, many from high Andean cloud forests, distinguished by the weird and grotesque shapes of their long-lasting flowers, to which some add brilliant coloring and fragrance, to make them spectacular and exciting in every collection. Without pseudobulbs, the stems grow from the rhizome in a tufted manner, typically with small thick leaves tapering at the base; the flowers hardly look like orchids, and the most conspicuous part are the sepals which are united at the base with spreading lobes prolonged into horns or tails. Plants are grown in pots or baskets, and generally kept on the cool side, with liberal watering all year, preferably using rain water, and good air circulation and high humidity are necessary for their success. Propagated by division.

bella (Colombia); a weird but beautiful epiphyte from 8000 ft. elevation, with tufted, oblanceolate leaves to 9 in. long; slender stalks, suspending large fragrant triangular flowers pale yellow, spotted with brownish-crimson, ending in long red-brown tails 4 in. long; oscillating, white, shell-like lip; blooming between autumn and June. **C**: 84 p. 473

chimaera (Colombia); grotesque but beautiful, very variable orchid of tufted habit with short-stalked leaves 6 to 10 in. long; the basal stalks with clusters of large, somewhat malodorous flowers about 8 in. long, brownish-yellow spotted with brownish-purple, the petals and lip small but the united sepals large and showy and prolonged into spreading reddish tails from 4 in. to nearly 1 foot long; blooming variously from November to February and into summer. Because of their pendant flowers this plant should be grown in a basket as the floral stalks sometimes grow down through the osmunda fiber. **C**: 84 p. 473

grandiflora (Ecuador; Peru); small but very floriferous rosette, dense with short, narrow oblanceolate leaves; the numerous spidery flowers 1 in. across, white on pale lemon-yellow, extending into thread-like curving spurs; the slender floral stalks bursting out in all directions like an exotic bouquet; blooming May-June. **C**: 84 p. 473

houtteana (Andes of Colombia and Venezuela at 8000 ft.); charming, free-flowering species of a very pleasing character, forming dense tufts of linear-lanceolate, thin-leathery, light green leaves, 10 to 12 in. long; the floral stalks half as long, each bearing a small solitary, triangular flower 1 in. across, creamy-white and profusely spotted with blood-red; the sepals extending into red-purple tails another 2 to 3 in. long; blooming from May to July. **C**: 84 p. 473

infracta purpurea (Brazil, Peru); although small and not very showy, this is a fantastic and very pretty species, 5 to 7 in. high, from the Organ Mountains; with spoon-shaped, ovate, leathery leaves tapering to a slender base; wiry stalks bearing solitary, curiously bell-shaped flowers 2½ in. long, the spreading yellowish-green tails extending another 2 in.; inside of sepals silky maroon purple with yellow base; spring to summer blooming. Darker form of M. infracta. **C**: 84 p. 473

liliputana (Brazil); low and beautiful plant with leathery, folded leaves; the relatively large flowers greenish and boldly overlaid with blackish tubercles; showy sepals broad at base and extending about 1 in. into long slender point; small blackish lip; blooming all year. **C**: 84 p. 473

veitchiana (Andes of Peru at 12,000 ft.); a most striking species with densely tufted leathery, dark green leaves linear oblong and shining, and a wiry stalk bearing 1 or 2 resplendent wide-open, brilliant orange-scarlet flowers to 6 in. or more long, and with iridescent sheen, the dorsal sepal with long tail, the tailed lateral sepals partly grown together, closely studded with purple hairs; blooming January to July, sometimes also in fall; their flowers lasting long in perfection. Requires very cool treatment. This lovely species is found growing on the cliffs of the ancient Inca fortress of Machu Picchu. **C**: 84 p. 473

MATHIOLA *(Cruciferae)*
incana, "Stocks" or "Gilliflower"; well-known for their spicy-sweet fragrance, by origin a semi-shrubby plant, usually grown as a biennial and lately as annual (10 to 14 weeks); at home around the Mediterranean from the Canary Is. to Asia Minor; cool-temperature plant highly developed, with brittle stems and grayish pubescent leaves, the popular branching stocks in the trade as 'Giants of California' or 'Bismarck', 1½ to 2 ft. high, with spike-like clusters of mostly

double, quite fragrant flowers 1 in. wide, in many pastel or vivid colors from rose, apricot and red to blue, violet, lavender, even canary-yellow and white. Normally blooming from April to fall in cool climate; stocks may also be flowered from January on until June in regions of hot summers, growing them at 45° F nights and not over 65° F day. Tolerant to some frost. From seed. **C**: 25   p. 416

MATRICARIA: see Chrysanthemum parthenium

MAXILLARIA (Orchidaceae); evergreen American orchids usually epiphytic or growing on rocks; of habit small, and falling into one group having short pseudobulbs in clusters or along the rhizome; the others elongated canes with 2-ranked foliage or with clusters of leaves fan-like. The flowers are generally small and not very showy, but plants bloom profusely even when still small; they are of easy care for the amateur, and adapted as house plants provided they receive good light, 2400 to 3600 fc., and moisture while in active growth, followed by a rest period of several weeks after flowering. Growing media osmunda, treefern, barks. Propagated by division.

densa (also known as Ornithidium) (Mexico to Honduras); robust epiphyte with small compressed pseudobulbs bearing solitary leaves to 15 in. long; curious little ½ in. hooded flowers borne singly on thin 1 in. stemlets in dense bundles, sepals and petals greenish or creamy-white, often tinted with rose or green, lip white with red center, blooming winter and spring. **C**: 22   p. 476

lepidota (Colombia, Ecuador); a free-blooming, ornametal epiphyte, with ovoid pseudobulbs clustered, 1 to 1½ in. long; bearing a single 9 to 12 in. glossy-leathery leaf; the floral stalks shorter than the foliage, carrying solitary yellow, spidery flowers almost 5 in. across, the lip spotted with brown; long lasting, and blooming mostly in spring and summer. **C**: 34   p. 472

picta (Brazil: Organ Mts.); bushy epiphyte with small 2 to 3 in. pseudobulbs, each with 1 or 2 thick, strap-shaped leaves 10 to 15 in. long; the individual, fragrant 2 to 2½ in. flowers yellow-streaked and dotted purple, chocolate inside, the petals incurved, lip white spotted purple; winter-spring blooming, and various. An easily grown species that blooms so profusely that many dozen flowers may be found on even small plants. **C**: 22   p. 472

tenuifolia (Mexico, through C. America); one of the oldest species, grown since 1839; a small densely clustering epiphyte, its flattened pseudobulbs at intervals on ascending rhizomes, topped by solitary, leathery linear leaves 1 ft. or more long, almost hiding the small, individually stalked, strongly scented flowers 1½ to 2 in. across, their sepals and petals dark rusty-red, lip spotted blood-red on yellow base, blooming willingly at various times, summer, fall, and winter. Not showy but easy to keep in a small space, their heavy-textured highly fragrant flowers lasting a long time. **C**: 22
                                                                    p. 472

MAZUS (Scrophulariaceae)
reptans (Himalayas), "Wart flower"; tiny mat-forming, tufted perennial herb to 2 in. high, rooting at nodes, with toothed obovate 1 in. leaves, and small ¾ in. purplish-blue two-lipped flowers, the lower lip spotted white, yellow and purple. A curious miniature for the cool, sunny window; propagated by division or seed. **C**: 64   p. 280

MEDINILLA (Malastomaceae)
magnifica (Philippines, Java), the "Rose grape"; gorgeous slow-growing tropical shrub, with age becoming to 5 ft. high; forking with 4-angled woody branches and large opposite, thick-leathery leaves to 1 ft. long, prettily patterned with ivory midrib; the striking inflorescence in a pendulous panicle to 1 ft. long, of carmine-red flowers with purple anthers and great showy, pink bracts. One of the most magnificent warm conservatory plants, blooming from late winter into summer. Medinilla bloom best if well established; they may be kept in smallish containers for years. Shift or cut after flowering. Give slight rest from November to February at 60° F. to ripen the new wood. Propagated from semi-ripened cuttings in late winter moist-warm at 75 to 85° F. under glass case or plastic hood; also by mossing. **C**: 60
                                                                    p. 491

MEGACLINIUM (Orchidaceae)
purpureorhachis (Bulbophyllum) (Congo Kinshasa), the "Cobra orchid"; a sinister-looking, fantastic epiphyte from inner Africa, with clustered pseudobulbs and paired,

rigid leaves to 6 in. long; the curious inflorescence 12 in. long, stalk a singular flattened green axis, densely spotted and overlaid with deep red, the tiny ½ in. deep brown flowers, borne in a single row on the flat sides; blooming August into winter. Snake-like in appearance, this is a striking curiosity plant in collections. May be grown on fern slabs or small pots, warm and moist. Propagated by division. **C**: 60   p. 474

MEIRACYLLIUM (Orchidaceae)
trinasutum (Guatemala, Mexico); charming small epiphyte on trees or rocks, with creeping rhizome, and secondary, thickened stems bearing a little solitary, ovate leaf only 1 to 2 in. long; inflorescence short, the pretty, tiny flowers to 1 in. across, creamy at base to red-purple toward tips and on the concave lip; blooming in summer, and singularly showy for their size. An easy plant to grow and freely flowering; best on slabs of treefern fiber so the rhizomes can spread in all directions. Propagate by division but plants won't like to be disturbed. **C**: 34   p. 472

MELALEUCA (Myrtaceae)
leucadendron (Australia, New Caledonia, Malaya), the "Paperbark tree" or "Swamp tea tree"; slender tree to 30 ft. high, the undulate trunk with white bark that peels in thick paper-like layers; the leaves 2 to 4 in. long, softly downy when young, later rich green with prominent veining and of firm texture, crowding the branches; bearing masses of honey-laden cream, bottlebrush-like flowers, the stamens variably whitish, greenish-yellow, pink, or purple. Tolerant to wind, poor soil and exposure but requiring sun. Shearing makes them dense. **C**: 1   p. 292

MELIANTHUS (Melianthaceae)
major (South Africa), "Honey bush"; evergreen semi-woody shrub with widely creeping roots and herb-like stems erect to 10 ft. high or sprawling, with striking pinnate foliage 8 to 12 in. long, the deeply serrate, soft-fleshy leaflets in pairs, glaucous green above, grayish beneath; strong-scented when bruised. Reddish-brown 1 in. flowers in dense racemes secreting honey. Attractive as solitary accent plant. From division, suckers or cuttings. **C**: 14   p. 298

MELISSA (Labiatae)
officinalis (So. Europe, No. Africa and East), "Lemon-balm" or "Sweet Mary"; aromatic herbaceous hardy perennial to 2 ft. high and spreading on the ground, with ovate, round-toothed leaves 3 in. long, fragrant of lemon when bruised; clusters of pale yellow 2-lipped flowers in July. Fresh sprigs or leaves are used in cool fruit drinks to add a lemony taste; as a delicate culinary seasoning, balm leaves add their flavor to many foods such as meats, sauces, soups, salads and stuffings; also to herb teas. From seed or root divisions. **C**: 25   p. 321

MELOCACTUS (Cactaceae)
bahiensis (Brazil: Bahia), a "Turk's-cap"; globose body to 6 in. dia., later becoming stretched, dark green with 10–12 deep ribs and clusters of straw-colored stiff spines; the white-woolly cap (cephalium) full of cinnamon bristles; flowers pink but hidden; seed pods red. **C**: 1   p. 159

MENTHA (Labiatae)
citrata (Europe), "Orange mint" or "Bergamot mint"; hardy perennial from square leafy stolons; branches declining but with ends erect 1–2 ft. long; bronzy green, thin and smooth egg-shaped leaves; purple flowers in upper axils on short spikes. Use: to flavor lamb, fish, sauces, apple-sauce, beverages, jelly, teas. The fragrant lemon-scented oil is distilled for use in making perfumes. Naturalized in Eastern U.S. **C**: 14   p. 319

piperita (Great Britain; naturalized in New York), "Peppermint" or "American mint"; hardy perennial to 3 ft. high, the spreading root-stocks and runners with reddish stems and numerous branches; dark green, toothed leaves 1 to 3 in. long, containing strong pungent or pepper-like oil; purple flowers in thick terminal spikes. Use: leaves make refreshing tea; also for flavoring liqueurs; oil in menthol. One of the most important of all plants in the production of essential oils. **C**: 26   p. 322

requienii (Corsica), "Corsican mint"; miniature creeping and spreading herb used for ground cover, with thread-like square stems, and very tiny round, stalked, bright green leaves, strongly peppermint-scented; flowers pale purple. Leaves used for flavoring. Not winter-hardy. An attractive, spicy trailer on the window sill, or a hanging basket. **C**: 16 or 84   p. 265, 322

rotundifolia (Europe), "Woolly apple-mint" or "Round-leaved mint"; hardy, pubescent perennial from leafy stolon; erect, slender stems 20–30 in. high; sessile, oval toothed leaves to 2 in. long; flowers purple in dense or interrupted spikes to 4 in. long. Apple-scented; sometimes used as substitute for peppermint or spearmint. Naturalized in No. America. **C**: 26   p. 320

rotundifolia variegata (Europe),· "Pineapple mint"; hardy perennial with soft-woolly oval leaves variegated light green and creamy-white; purple flowers. Use: to flavor fruit drinks, juleps, iced beverages, teas, jellies, soups and meats. **C**: 14   p. 319

spicata (Europe and Asia), "Spearmint"; hardy perennial from leafy stolons, stems erect with ascending branches 1–2 ft. high; smooth, light green leaves sharply serrate, to 2½ in. long; and purple flowers in dense terminal spikes. Use: leaves to flavor cold drinks and tea, chewing gum, jelly. Cut sprigs are used in making "mint julep". Also an ingredient in "mint sauce" to eat with lamb. Distilled as medicinal oil. Widely naturalized about old gardens in Eastern U.S. Propagated from perennial rootstocks. **C**: 26   p. 319

MERYTA (*Araliaceae*)
sinclairi (South Pacific Isl., New Zealand), "Puka tree"; small evergreen tree to 25 ft., with oblong, entire, leathery leaves to 2 ft. long crowded at ends of branches; blade glossy-green with irregular margin, pale prominent veins; petioles striped brown. **C**: 14   p. 290

MESEMBRYANTHEMUM (*Aizoaceae*); the "Fig-mari-golds" or "Mid-day flowers" comprise a supergenus of more than 1000 species of small xerophytic succulents, mostly from arid Southern Africa, all highly interesting, many of them grotesque or mimicking their desolate surroundings. Because of their marked differences and wide extent, this genus has been, with few exceptions, re-studied and divided into many separate genera; what they have in common is that their daisy-like flowers, in most species, open only to the warming sun, and close again in the afternoon. Their cultural requirements also are somewhat similar. The more succulent types should have a very sunny position and be given little or no water when resting and not too much even when growth is active. The resting period does not occur at the same time of year for all species; thus Lithops are rested about April-May; Conophytums in June and July; the state of the plant will show when water should be withheld; it is always safe to discontinue watering when growth appears to be completed. New growth will begin without water, which should be given again when new leaves begin to appear. Any stagnant moisture during the resting period can be fatal. Propagation from seed is usually easy and the seedlings require continuous but moderate watering during their first year. Some Mesembryanthemums can be propagated by cuttings.

crystallinum (Cryophytum) (So. Africa, S.W. Africa; carried off to the shores of the Mediterranean, Canary Islands, and also naturalized on the coasts and off-shore islands of California), the true or California "Ice plant"; annual branching succulent with shallow, fibrous roots and soft-fleshy stems creeping close to the ground, breaking and forking at every joint. Irregularly spatulate soft-fleshy leaves ½ to 2 in. long, grayish light green, bronzy at apex; both stems and leaves are thickly covered with crystal-clear round bubbles filled with watery fluid and looking like ice-crystals, turning jewel-like amethyst toward tips; the starry flowers ½ to 1¼ in. across, translucent white tinted lavender, in July-August, and opening only to the sun. Fascinating in the window or on the patio with its glittering foliage. From seed; self-seeding in California. **C**: 13   p. 357, 531

MESEMBRYANTHEMUM: see also Aptenia, Carpobrotus, Dorotheanthus, Glottiphyllum, Lampranthus, Oscularia, and other Aizoaceae.

MESOSPINIDIUM (*Orchidaceae*)
warscewiczii (Brazil); small epiphyte 4 to 6 in. high with flattened pseudobulbs bearing an oblanceolate leaf, and arching panicles of small greenish-yellow flowers to ½ in. across, these spotted or barred chestnut brown; the

petals with brown margins, with lateral sepals united, and spoon-shaped, concave yellow lip spotted with brown; the column with maroon-colored wings. A rare and difficult species requiring greenhouse care. **C**: 60   p. 472

METASEQUOIA (*Taxodiaceae*)
glyptostroboides (Central China: Hupeh), the "Dawn redwood"; this fossil age conifer has been found growing to 100 ft. high near Chungking in 1946 after having been thought extinct; handsome coniferous moisture-loving and fast-growing tree with symmetric branches, resembling Sequoia sempervirens, but deciduous, with light green, opposite needles ¾ in. long, and soft to the touch. A young tree with soft green lacy foliage is like a breath of spring; well adapted to containers. From cuttings or seed. **C**: 64   p. 311

METROSIDEROS floribunda: see Callistemon

MICONIA (*Melastomaceae*)
magnifica (Mexico), "Velvet-tree"; gorgeous, tree-like tropical foliage plant with woody stem and beautiful, long ovate, thin leaves to 2½ ft. long, velvety green with the sunken primary ribs ivory, and a network of pale green secondary veins; reddish purple beneath. Flowers insignificant. A show piece for the humid greenhouse. **C**: 60 Earliest name: M. calvescens (Wurdack 1971).   p. 188

MICROLEPIA (*Filices: Polypodiaceae*)
hirta (pilulosa; speluncae var. hirta), (So. India, Ceylon, Malaysia, Polynesia), grown in European horticulture as "speluncae"; strong growing, variable fern, with creeping rhizomes, bushy when small but developing handsome fronds 3 to 6 ft. long, 2 or 3 times pinnate, the ultimate segments deeply lobed, and toothed, rich green almost glossy, matte-green beneath, on brown-purple stalk. Easy to grow but not as elegant and durable as other ferns. **C**: 72   p. 248

speluncae (Davallia polypodioides) (East Indies, Polynesia) ,"Grotto lace fern"; prolific grower with strong stalks and herbaceous, soft and papery, pale green fronds 3–6 ft. high and covered with sparse white hair, 3–4 pinnate, the lower pinnae 6–12 in. long, and with toothed lobes. Excellent for decorative effect and for ferneries but not a lasting house plant because of soft, quick growth habit. **C**: 72   p. 248

MIKANIA (*Compositae*)
ternata (Brazil), "Plush vine" or "Purple haze vine"; a most attractive and promising, rapid-growing trailer with brown stems, covered with lighter, felt-like hairs; opposite small herbaceous, palmately compound leaves 1½ in. long, consisting of 5 lobed leaflets dark coppery green, purple beneath, densely covered all over with whitish hairs. Very graceful and eye-catching. **C**: 10   p. 264

MILA (*Cactaceae*)
caespitosa (Peru: Lima); low plant branching from the base with cylindric stems to 6 in. high, forming small clusters; about 10 ribs almost hidden by the needle-like yellowish spines; wheel-shaped yellow flowers. **C**: 1 p. 159

MILTONIA (*Orchidaceae*); the "Pansy orchids" belong to a group of very charming American epiphytes, mainly of small habit with prominent pseudobulbs and flexible, thin-textured leaves. They fall into two main floral groups—those growing cool, even chilly, from high elevations, with their flat-open pansy-like faces and very popular; others warmer-growing and sometimes simulating Odontoglossums to which they are closely related. All are best grown in rather small pots in osmunda or fern fiber with good drainage; they are kept moist the year around, at high relative humidity, and even the warm types do best at cooler night temperatures of 50 to 60° F., if possible under air-conditioning. Generally Miltonias are shade-loving, avoiding direct sun in summertime, the cool types 1000 to 1500 fc., the warmer Brazilian kinds 2000 to 3000 fc. of light intensity. There are many lovely hybrids. Division of plants grown large into smaller clumps is better than dividing too small and often unless propagation is being pushed. Miltonias last many weeks in beauty if kept cool and the flowers free from dampness.

clowesii (Brazil), showy epiphyte, with two-edged pseudobulbs carrying 2 to 3 sword-shaped, light green leaves 12 to 18 in. long, the inflorescence an arching wiry stalk with 7 to 10 glossy 2 to 3 in. flowers chocolate-brown barred with yellow, the lip with violet base and white in front; blooming August to November, the flowers lasting a month or more. **C**: 74   p. 474

laevis, sometimes known as Odontoglossum laeve (Mexico, Guatemala); a pretty Odontoglossum-like epiphyte of free-growing habit, with compressed 5 in. pseudobulbs topped by 2 to 3 folded leaves to 1½ ft. long; the inflorescence robust and erect to 2 or 3 ft. long, branched with clusters of spidery, long-lasting 2–2½ in. flowers greenish-yellow, transversely banded with reddish-brown, the lip white with purple; an abundant bloomer, mostly summer and fall. **C**: 24   p. 474

phalaenopsis (also known as Odontoglossum) (Colombia), a lovely "Pansy face" of smallish elegant habit, epiphyte with clustered, flat pseudobulbs topped by solitary grass-like leaves 8 to 12 in. long; a short stalk bears several beautiful flattened flowers 2 to 2½ in. across, with white sepals and petals, and a wide lobed white lip streaked with crimson and yellow base, blooming in spring and summer or later, lasting for 4 to 5 weeks. Reported to be growing on rocks at elevations of 16,000 ft. **C**: 24   p. 474

roezlii alba (Panama, Colombia); one of the loveliest "Pansy orchids", with solitary light green leaves on tightly clustered, flattened pseudobulbs; stalks loosely 2 to 6-flowered, the open faces 3 to 4 in. across, and an exquisite pure white, except for some yellow shading at the base of the lip; sweetly fragrant and blooming mainly from September to November, but sometimes again in spring. The well-known M. roezlii has white flowers with a large purple blotch on each petal; it is used in many hybrids. Requires free ventilation and fairly good light, but away from sun. **C**: 24   p. 474

spectabilis (Brazil); handsome epiphyte with flattened pseudobulbs bearing 2 thin, strapshaped leaves and showy large solitary flowers 3 to 4 in. across, the narrow petals and sepals pure white, large rosy lip with dark purple center and vein; blooming from spring to summer and into fall. This variable species has many color forms from white to deep purple. A spectacle when larger plants may bloom with dozens of flowers at the same time, lasting 6 weeks in beauty if kept cool and free from damp. **C**: 74   p. 474

vexillaria 'Volunteer', very floriferous form of the species vexillaria from Ecuador and Colombia, with the largest flowers in the genus, lavender pink veined purple, and a distinct pattern with white eye, stained in the center with deep yellow, blooming in May to July. Epiphytic with 1 or 2-leaved pseudobulbs, surrounded at base by 6 to 8 bluish-green two-ranked leaves; the inflorescence carrying 3 to 9 of the flat, pansy-like blooms, in many color variations, and long used in hybridization. Sturdy and easy to grow, this species is very popular because of its beautiful large flowers 3 or 4 in. or more across, and the generous number of blooms produced. **C**: 24   p. 474

MIMOSA (Leguminosae)

pudica (Brazil; naturalized in tropics of the world), the popular "Sensitive plant", also known as "Humble plant", "Live-and-die", or "Touch-me-not"; a most extraordinary low perennial 6 to 20 in. high; more or less hairy and slightly spiny, with base becoming woody; slender wiry branches bearing long-stalked feathery bipinnate, grass-green leaves; and small flowers in ball-like heads like purple puffs, blooming mainly in summer. Most remarkable because of the ability of its leaves to "go to sleep" at the slightest touch, causing the leaflets to fold upward and the petiole to drop. An ubiquitous weed throughout the tropics, and it is thought that its mechanism is actuated by heat. I have seen it not only in Brazil, but in the West Indies where it is called 'Mori-Vivi', across Africa, and in Viet Nam where it is known as the "Shame plant". Easily grown as a window sill plant in good light; usually from easily harvested seed, as an annual. A constant conversation piece also for children. **C**: 2 p. 529

speggazzinii (Argentina), "Sensitive-leaf tree"; spiny shrub or small tree about 6 to 8 ft. high or more, with woody branches bearing stalked leaves each consisting of a pair of pinnate feathery, dull green leaflets 2½ in. long, covered by sensitive hairs which when touched, trigger the mechanism at the base of the petiole that causes them to collapse and fold together, as if "going to sleep". Pretty clusters of small

lavender flowers in globular heads 1½ in. across; the stamens with rose-purple filaments and yellow anthers. A patio plant for the summer. **C**: 65   p. 529

MIMULUS (Scrophulariaceae)

aurantiacus, better known in horticulture as Diplacus glutinosus (Western U.S.: Oregon, California), "Monkey flower"; branching shrub 1 to 4 ft. high, with flexuous branches and narrow, leathery rich green leaves toothed and turned down at the toothed margins and sticky to the touch, pubescent beneath; the showy 1½ to 3 in. orange-salmon 2-lipped flowers having notched, spreading lobes impersonating a monkey-face; blooming freely over a long period extending from May to August. A sun-loving house plant; don't hesitate to prune to shape in spring. From cuttings. **C**: 13   p. 492

MOLUCCELLA (Labiatae)

laevis (Western Asia: Syria), "Bells of Ireland" or "Shell flower"; curious herbaceous aromatic summer-flowering annual, found in old-fashioned gardens, interesting because its 2 to 3 ft. stalks with roundish leaves to 1½ in. long are closely set with whorls of the showy oversized 1¼ in. shell-shaped, light-green calyx, netted with white, in which nestle the little pinkish, 2-lipped fragrant flowers. As cut flowers, the spikes of little apple-green bells are long lasting fresh or dried. From seed. **C**: 64   p. 416

MONADENIUM (Euphorbiaceae)

lugardae (So. Africa: Transvaal, Natal, Rhodesia), "Green desert rose"; spineless branching succulent with tuberous root and bright green, plumb stems 8 to 24 in. long, distinctly marked by raised rhombic fields; and with deciduous, fleshy, deep green, obovate leaves to 3 in. long; small bracted flowers toward the apex, pale green edged with orange. **C**: 13   p. 344

MONODORA (Annonaceae)

grandiflora (West Trop. Africa), "African orchid nut-meg"; tropical tree to 20 ft. high with oblong, glossy green leaves purplish beneath, and pendulous orchid-like yellow flowers spotted red, the crisped outer petals 3 to 4 in. long, strikingly beautiful, blooming early in summer. Mature trees bloom in the large warm conservatory; in tropical gardens outdoors. Propagated from ripened cuttings under covered moist-warm conditions. **C**: 52   p. 487

MONSTERA (Araceae); frequently called Philodendron; tropical American foliage plants more or less tree climbing by aerial roots, and exotic-looking foliage entire in the juvenile stage, later interestingly perforated or slashed in bizarre patterns. Inflorescence with fleshy, boat-shaped spathe surrounding the spadix with tiny petal-less flowers perfect and distributed over the whole column, distinguishing these from true Philodendron. The fleshy fruit of some Monsteras is edible and provides seed. The sap in the plants contains some calcium oxalate crystals that are irritating to tender tissues. Most Monsteras are not good house plants except M. deliciosa which is an excellent decorator. Propagated by cuttings or mossings of tips, also stem sections laid in damp moss; or seed.

deliciosa (Philodendron pertusum) (So. Mexico, Guatemala), "Ceriman", also known as "Mexican bread fruit"; stout, woody-stemmed, close-jointed treeclimber forming long hanging, cord-like aerial roots with large, thick, leathery leaves, glossy green, to 3 ft., pinnately cut and perforated with oblong holes; bisexual spadix and boat-shaped, white spathe; cone-like, edible fruit with pineapple aroma. The long-jointed, rapid-climbing juvenile stage with smaller, less perforated leaves is known as "Philodendron pertusum". Very durable, but needs strong light to maintain growth of large split leaves. Tolerates 20 fc. **C**: 4
                                                      p. 65*, 72, 509

deliciosa variegata (Philodendron pertusum variegatum), "Variegated monstera"; a mutant with irregular variegation where parts of the leaf may be entirely green, and other sections marbled cream to greenish-yellow, or entirely cream; new growth, of too happy, shows a tendency to revert back to green; such branches should be cut back to favor variegated growth. **C**: 53   p. 179

nechodomii: see Epipremnum pinnatum

obliqua expilata (leichtlinii), "Window-leaf", a form of obliqua from Alto Amazonas; the perforated leaves of this interesting "lace-leaf" are not much more than an ovate skeleton of veins, with very little green left of the blade. The leaves become 2 ft. large, the holes in the blade are usually arranged in double rows. **C**: 54 p. 72

pertusa (Panama, Guyana), also known as "Marcgravia paradoxa"; lush climber with soft-textured, unequal-sided leaves perforated and pinnatisect; a poor keeper, from tropical lowland forests. **C**: 54 p. 67, 72

standleyana (guttiferyum hort.) (Malaysia); close-jointed climber; thick leathery, dark green, small oval leaves, with depressed veins, on broadly winged petioles when young; broader and more smooth in maturity. The venation curves upwards, suggesting Pothos. **C**: 60 p. 83

MONSTERA: see also Philodendron

MONTBRETIA: see Tritonia

MORAEA (*Iridaceae*); the graceful "Peacock irises", also known as "Fortnight lilies" or "African iris" are cormous or rhizomatous plants forming clumps of stiff, evergreen iris-like leaves, and lovely flowers like miniature iris. Though the blooms are short-lived, opening one morning to die the next day, they are quickly followed by others, and few South African plants can rival Moraea for its peacock-like coloring. Beautiful in pots in cool greenhouse, sunroom, or patio, blooming from spring through summer and fall. After flowering, when the foliage wilts, keep dry for the winter to rest until January, then replant. By offsets of division.

glaucopsis (So. Africa: Cape Prov.), "Peacock iris"; cormous herb 1–2 ft. high, sometimes branching, with solitary, narrow-linear leaves, and short-lived white flowers to 2 in. long, with flaring outer segments each having a bluish-black blotch at base; blooming May-July. **C**: 75 p. 393

iridioides (Dietes grandiflora) (So. Africa: E. Cape, Natal), "African iris"; rhizomatous plant with linear dark green basal leaves arranged fan-like from creeping rootstock; large, somewhat fleeting 3 in. iris-like flowers white, with yellow or brown markings, crest of style marked with blue; summer-blooming. **C**: 75 p. 393

MORMODES (*Orchidaceae*)

igneum (Cent. America), a fantastic "Goblin orchid"; robust handsome epiphyte 18 to 24 in. high, with long spindle-shaped pseudobulbs carrying several plaited leaves which are deciduous; stout floral stalks from the body of the pseudobulb bearing elongate clusters of fleshy flowers, with sepals to more than 2 in. across; sepals and petals reddish-brown, the contorted lip orange and clawed at base; blooming winter to spring. Mormodes are displayed to advantage in baskets; they need abundant water while growing, but when the foliage drops, are kept quite dry. Propagated by division. **C**: 59 p. 472

lineata (Guatemala); showy epiphyte with plaited leaves on strong pseudobulbs; lightly arching raceme of very fragrant flowers, sepals and petals tongue-shaped, 1¼ in. long, outside greenish, inside yellow with purple stripes, lip yellowish with red dots, blooming January to March. Requires a considerable rest, at cooler temperature, after completion of growth, and the deciduous foliage drops in late fall. Propagated by division. **C**: 71 p. 472

MUEHLENBECKIA (*Polygonaceae*)

complexa (New Zealand), "Wire vine" or "Maidenhair vine"; with creeping, twining and wire-like purplish-brown stems furnished with scattered, tiny rounded, fresh-green leaves ½ to ¾ in. across; flowers greenish white, in small spikes; a graceful basket plant; in time becoming a regular thicket, especially if planted out, to cover walls, columns, or fences. **C**: 66 p. 266

MUEHLENBECKIA: see also Homalocladium

MURRAYA (*Rutaceae*)

exotica (India, Ceylon, Philippines to Australia), the "Orange jessamine"; handsome tropical ornamental evergreen shrub 10 to 12 ft. high, dense with whitish branches and glossy-green, odd-pinnate leaves, and small, ¾ in. bell-shaped, white flowers sweetly fragrant of jasmine, in clusters, succeeded by small ½ in., vivid-red berries; blooms several times a year. Propagated from seed sown immediately after harvesting; also by cuttings at 85° F. **C**: 16 p. 512

MUSA (*Musaceae*); the "Bananas" or "Plantains" are giant tropical herbs with huge fleshy, tree-like stems formed of the tightly packed sheaths of the leaf bases. They have usually large rhizomes, from which springs a trunk-like stem which flowers only once, and then is replaced by suckers from the base; the leaves are very large, with a stout midrib. Clusters of flowers with colored bracts appear from the crown, followed by the long green, yellow or reddish fruit with aromatic, edible flesh. Adaptable to growing in containers, for tropical effect, requiring plenty of water, although strong winds will shred the foliage into ribbons if exposed. Propagated from suckers. The non-suckering strictly ornamental bananas of the genus Ensete are grown from seed or cuttings. (See Ensete.)

ensete: see Ensete ventricosum

nana (cavendishii) (So. China), "Chinese dwarf banana" or "Dwarf banana", also known as "Dwarf Jamaica"; shapely stoloniferous plant of compact habit and much grown in tubs, then 3 to 6 ft. high, but to 10 ft. when planted out, with short stem formed of leafbases, topped by dense rosette of oblong, glaucous green, leathery leaves with satiny sheen, blotched with red when young, on short stout petioles; produces edible, deliciously fragrant yellow, 5 in. fruit; stands more cold and abuse than most bananas. The recommended livingroom banana. Propagate from suckers. **C**: 14 p. 203, 205, 507

x paradisiaca (syn. x sapientum) (acuminata x balbisiana) (India, Ceylon), "Bluefield banana", "Plantain", or "Common banana"; tree-like stoloniferous herb to 25 ft. high, with spirally arranged green foliage to 9 ft. long, having reddish midrib; becoming frayed or broken by the wind; the leaves forming a tall slender trunk by their sheathing bases; flowers yellow, bracts violet, followed by the well known yellow, edible fruit. After fruiting, the stem dies but is replaced by new suckers, which are used for propagation. The botanical status is somewhat confused; this may be the same as M. acuminata. **C**: 54 p. 205, 507

x paradisiaca 'Koae', "Variegated banana"; probably an Hawaiian bud mutation triploid of some Maoli banana of the Pacific; this most striking variegated-leaf banana is named after the Koae bird with "hair prematurely graying"; the leaves are striped or banded white and very light green on dark green, in feather fashion, the midrib, petiole and trunk alternating white and green, even the immature fruit is variegated, but ripens to yellow, is short and roundish, with yellow flesh. Beautiful but difficult, and subject to disease. **C**: 60 p. 205

zebrina (Java), "Blood banana"; usually known in horticulture as sumatrana; slender plant to 12 ft. high, with tall trunk bearing rather delicate long-stalked leaves, satiny bluish-green richly variegated with blackish blood-red, the channeled midrib brown-red, underside reddish-wine; maintaining its coloring with age, but sensitive to cold. **C**: 60 p. 205

MUSCARI (*Liliaceae*)

armeniacum 'Heavenly Blue' (Turkey), "Grape hyacinth"; charming May-blooming hardy bulbous plant 6 to 8 in. high suitable for forcing in pots, with narrow, channeled basal leaves, and delicate spires of nodding, urn-shaped flowers azure-blue, tipped white. Plant several bulbs in a pot in fall and set into a cool cellar or bury outdoors until well rooted; when brought indoors in a 50 to 60° F. cool room they will flower in 2 to 3 weeks. Wonderful long-lived subjects for naturalizing in the garden where it forms clumps. By offsets or division. **C**: 26 p. 373, 398

botryoides album (So. Europe), "Pearls of Spain" or "Starch hyacinths"; hardy bulbous flowering herb 8 to 10 in. high with soft-fleshy linear, channeled leaves slightly glaucous, and erect stalks with spike-like racemes of pretty, pure white and fragrant, tiny flowers blooming in spring; the white form of the "Bluebells" of our gardens. Very charming when flowered for the Easter season in pots, with several bulbs planted together. **C**: 26 p. 373

racemosum (wide-spread in Europe to Asia Minor), the "Starched grape-hyacinth"; low bulbous, hardy plant of the meadows and vineyards blooming in early spring; from small bulbs the slender channeled leaves, 6–9 in. long, stiff and fleshy; the 4 to 8 in. floral stalks with terminal racemes of tiny urn-shaped, dark blue, scented flowers, white at the mouth. Spreading by bulblets. **C**: 26 p. 398

MUSSAENDA (*Rubiaceae*)
luteola (flava) (Trop. Africa), "Yellow Buddha's lamp"; erect tropical shrub to 5 or 6 ft. high, with lanceolate leaves hairy beneath; small star-like tubular flowers 1 in. long, bright yellow with orange center but conspicuous for the several showy broad ovate, enlarged creamy-white, leaf-like sepals 3 in. long subtended from each flower cluster. For the warm conservatory; blooming from March through summer. From cuttings in spring. **C**: 54 p. 500

MYOPORUM (*Myoporaceae*)
laetum (New Zealand), "Mouse-hole tree" or "Ngaio"; voracious evergreen shrub of exceptionally fast growth to 30 ft. high and 20 ft. spread; branches first willowy and flexible, later stiff-woody, dense with soft-leathery, long elliptic, happy green, shining foliage 4 in. long, full of translucent oil glands. Small white bell-shaped flowers, with sticky buds, in clusters amongst leaves. Very decorative bush; wind and salt-resistant. In containers requires plenty of water when growing; likes fresh air and not too warm; for patio in summer. Much used for hedges in mild climate. Easy from cuttings or seed. **C**: 14 p. 298

MYOSOTIS (*Boraginaceae*)
sylvatica 'Compacta' (alpestris hort.), (Temperate Europe to Siberia), a dwarf "Forget-me-not"; exquisite small perennial herb grown as annual or biennial, of dense and bushy habit, 6 to 8 in. high, with soft, hairy leaves close along the numerous branches; the upper stalks loosely covered by the tiny ⅓ in. clear sky-blue flowers, delightfully charming with their yellow eyes. A favorite old bedding or pot plant blooming outside from spring through summer; where temperature permits, or in cool greenhouse also blooming in winter, from seed sown in summer. May be grown from cuttings but easiest from seed. **C**: 78 p. 409

MYRIOCARPA (*Urticaceae*)
stipitata (known in horticulture as Boehmeria argentea) (Mexico); ornamental tropical tree to 30 ft. high, with large and showy, bluish-green, quilted leaves to 1 ft. long, shading to silver toward the crenate-toothed margins, quilted and bristly on the surface, veined red-brown beneath; flowers without petals in long catkins. **C**: 54 p. 188

MYRIOPHYLLUM (*Haloragidaceae*)
prosperinacoides (braziliense) (Chile and Uruguay, north to Brazil and So. U.S.); "Parrot's feather", "Water milfoil" or "Water feather"; graceful fresh-water aquatic herb, normally rising about 6 in. above the surface, the rest submerged, with whorled, feathery leaves composed of many hair-like divisions, on weak shoots from a creeping rhizome-like stem in mud on the bottom of pond or aquarium. The commonest, easiest and best oxygenator for aquaria, their normally yellow-green or bluish leaves turning reddish in strong light; at night they go to sleep. Optimum temperature in summer 75° F., in winter 60° F. Propagated from long cuttings which may be planted directly in sand or soil. **C**: 22 p. 519

MYRSINE (*Myrsinaceae*)
africana (Azores, No. and So. Africa, Arabia, Afghanistan to the Himalaya in China), "African boxwood"; shrubby bush resembling boxwood but more graceful, to 4 ft. high, with angled, downy, slender red twigs, dense with small rounded, shiny dark green ½ in. leaves finely serrate; tiny pale brown axillary flowers and purplish-blue berries on female plants if pollinated. With its red stems and box-like glossy foliage attractive as a house plant or as cut foliage. From cuttings. **C**: 14 p. 300

MYRTILLOCACTUS (*Cactaceae*)
geometrizans (S. Mexico), "Blue myrtle"; branching tree type to 12 ft. high; smooth, slender columns six-ribbed, 3 to 4 in. dia., glaucous powder-blue; with practically no spines; small diurnal flowers greenish-white; edible blue fruit. **C**: 1 p. 152

MYRTUS (*Myrtaceae*); the "Myrtles" are subtropical aromatic evergreen shrubs, of which the classical or true myrtle, Myrtus communis, is at home in Greece and adjacent Mediterranean region. Its glossy leaves when rubbed are spicily scented; the small white flowers, with fuzzy stamens, are sweetly fragrant. Longtime house plants and used for decoration, they prefer a cool, sunny location. Propagated by cuttings of half-ripened wood.
communis boetica, "Citrus myrtle"; very attractive variety of the classic myrtle, to 6 ft. high, with stiff gnarled branches and smooth, dark green, rather large 1¼ in. boat-shaped thickish, ovate leaves with sharp tip; very fragrant, dense and symmetrical on erect branches, forming a compact, decorative bush. Sometimes, confused with Citrus aurantium 'Myrtifolia' which is known in southern Italy as Myrtus boetica; however, typical Myrtus have opposite leaf pairs whereas Citrus have alternate leaves. Wonderful for arid, sunny areas. **C**: 13 p. 292
communis compacta, "Compact myrtle" or "Dwarf myrtle"; slow-growing, small, compact form with densely set small black green leaves, ½ to 1 in. long, and which may be trimmed and sheared into various topiary or Bonsai forms; also for low edgings indoors or the patio. **C**:13 p. 312
communis microphylla, "German myrtle" or "Dwarf myrtle"; the compact form grown by European plantsmen in pots and sheared into little globes and much used for weddings and as house plant; densely leafy shrub with brown twigs covered by tiny lanceolate, almost needle-like, shining black-green overlapping leaves ½ to ¾ in. long, and white flowers of aromatic fragrance. **C**: 13 p. 300
communis variegata, "Variegated myrtle"; a variety of the Greek myrtle of the Mediterranean region; evergreen shrub eventually 15 ft. high, loosely leafy with leathery, rather broad-ovate, 2 in. leaves, dark lustrous green, attractively variegated or margined with creamy-white; the foliage smells spicy when bruised; fragrant white flowers with numerous stamens, followed by purple-black berries. **C**: 15 p. 187, 299

MYSTACIDIUM (*Orchidaceae*)
distichum, sometimes found as Angraecum distichum (West Africa: Sierra Leone, to Uganda); peculiar but neat little epiphyte, with ribbon-like, flexible and pendant stems to 8 in. long, clothed by two ranks of ½ in. fleshy, bright green, sickle-shaped leaves folded flat; small white, spurred ½ in. flowers upside-down, emerging from the axils of the foliage, blooming July to October. A favorite small botanical orchid of easy care in the house, faintly perfumed; best grown in a basket. By division. **C**: 11 p. 472

NAEGELIA: see Smithiantha

NANANTHUS (*Aizoaceae*)
malherbei (Aloinopsis) (So. Africa: Cape Prov.) "Miniature blue fan"; small branching 1½ in. rosette with tuberous rootstock, flat, bluish-gray fan-like, spatulate leaves to 1 in. long attractively decorated at the apex with a row of ivory warts; stalked, flesh or biscuit-colored 1 in. flowers. Slow-growing, striking species. **C**: 1 p. 355

NANDINA (*Berberidaceae*)
domestica (China, Japan), "Heavenly bamboo"; attractive shrub usually low, but may grow to 8 ft.; slender cane-like red-brown stems with 2 to 3-pinnate light green leaves, the ultimate leaflets narrow, to 2 in. long, turning fiery crimson in fall; small white flowers in large panicles, followed by bright red berries; fairly hardy, tolerating to about 10° F. Reminiscent of bamboo in its stems and lacy foliage, pinkish on expanding. From seeds or side branches. **C**: 76 p. 306

NARCISSUS (*Amaryllidaceae*); the "Lent lilies" include a large variety of mostly hardy bulbous plants belonging to the Amaryllis family, and have been known in cultivation since Theophrastus in Greece in 300 B.C.; with strap-shaped leaves and appealing pretty flowers mostly white or yellow, or both, the flower with a central trumpet or cup surrounded by 6 flaring perianth segments which comprise the sepals and petals. Narcissus have been subdivided into 11 divisions according to their characteristics. Best known are the large Trumpet narcissi derived primarily from the European, hardy N. pseudonarcissus, usually known as "Daffodils"; then the sweet-scented "Jonquils" with clusters of short-cupped flowers; the "Hoop-petticoats" (bulbocodium), the fragrant bunch-flowered "Polyanthus" or "Paper-whites" (tazetta), and also the fragrant "Poet's narcissus" (poeticus), so common in the Alps; and many miniature species. For flowering indoors, plant several bulbs into a pan, best in October, and bury outdoors, or put them in a dark, cool place such as a cellar at 35 to 45° F. for about 6 weeks or until wanted; when well rooted the pots may be moved to a cool window for gentle forcing, watering freely; it will then take 1 to 3 weeks, depending on season, to flower. Most narcissus may be planted in the garden after blooming where they easily naturalize.

bulbocodium (So. France, Spain, Portugal, Morocco), (Div. 10: Hoop-petticoats), "Petticoat-daffodil"; variable, weak little miniature only 5 in. high, in the Dutch variety 'Conspicuus', with slender, grass-like leaves; solitary up-facing flowers deep yellow, the inflated, funnel-shaped crown to 1 in. long, with starry perianth of narrow lobes, on stalk 4–8 in. high. **C**: 26   p. 398

cyclamineus 'February Gold'. a hybrid of N. cyclami-neus from Portugal with a "Trumpet" type; compact "Pot narcissus" about 10 in. high, with narrow linear leaves, and smallish but numerous flowers with lemon yellow perianth segments spreading 2 in. across, and orange yellow trumpet crenate at mouth. Normally early blooming, this pretty hybrid is ideal for pots from Valentine to Easter. **C**:76 p. 373

odorus 'Orange Queen' (Div. 7: Jonquils), a miniature daffodil to 12 in. tall, derived from odorus (Mediterranean, east to Yugoslavia); clustering, slender, narrow channeled leaves, and numerous stalks bearing 2–3 pretty, small 1¾ in. flowers with 6 segments and small cup, rich orange-yellow and very fragrant. Free-flowering and very striking. **C**: 26   p. 398

x poetaz 'Geranium', (1930) (poeticus x tazetta); a hybrid "Cluster narcissus", with the larger blooms of the Poet's narcissus and the several flowers on each stalk characteristic of tazetta; N. 'Geranium' is one of the most beautiful of the Poetaz with large white, rounded perianth segments and a shallow cup deep orange-red in its pleasantly scented, long-lasting blossoms on strong stems. Fine for late forcing. **C**: 26   p. 387

pseudo-narcissus 'Gold Medal' (1938), a true, short "Pot daffodil" and outstanding large-flowered trumpet variety almost exactly like 'King Alfred' except that it is very stocky and shapely, and ideally suited for growing in pans for the spring season, the large 3½ in. flowers clear yellow with richer colored trumpet, carried on solid stalks 8-12 in. high, 2 to 3 to a bulb; late flowering. **C**: 26   p. 375

pseudo-narcissus 'King Alfred' (Div. 1: Trumpet type), an old (1899) and well-known large-flowered "Lent lily" or "Yellow trumpet daffodil"; early spring-flowering robust bulbous plant 1½ to 2 ft. high with strap-like basal leaves, and large solitary, golden-yellow trumpet flowers 4 in. across; very substantial, with strong stalks, an out-standing florists forcing cut flower and pot daffodil, especially for the Easter season; probably the best-known cultivar derived from the W. European species N. pseudo-narcissus. Very hardy, and wonderful for naturalizing in gardens. **C**: 26   p. 399

'Silver Chimes' (Div. 5: triandrus hybrid), a charming miniature 8 to 10 in. high, from a cross of N. triandrus calathinus with N. tazetta; producing heads of 4 to 6 nodding flowers with flat white perianth 2 in. across, and small pale primrose-yellow cup, on sturdy stalks; strongly fragrant. Late-blooming. **C**: 26   p. 398

tazetta canaliculatus (Canary Islands), a tiny, sweet little miniature "Polyanthus narcissus" 6 to 9 in. high, with bluish foliage, and bunches of flowers 1 in. across, the perianth segments translucent white and with a small hollowed, orange-yellow cup; daintily fragrant. March-April blooming; not quite hardy. **C**: 26   p. 398

tazetta papyraceus (Mediterranean Reg.), the popular "Paper-white" or "Polyanthus narcissus"; small, white, fragrant starry flowers with short tube, several bunched together on each slender stem. Not winter-hardy. Planted in September, they are widely forced for early winter, blooming from November on. These sweet-scented Paper-whites are easily grown indoors in bowls of pebbles and water, or several in a pot; keep them dark and cool until growth is well advanced, then bring gradually into light. **C**: 26   p. 398

'W. P. Milner' (Trumpet miniature); an old but useful dwarf cultivar 8 to 10 in. high, similar to moschatus, with creamy-white 1½ in. flowers, having large expanded sulphur-white trumpet, serrated at apex, and rather soft segments; free-blooming and early. **C**: 26   p. 398

NAUTILOCALYX (Gesneriaceae); handsome tropical South American fibrous-rooted herbaceous plants with decorative, opposite leaves on erect, succulent stems; the yellow or cream-colored blooms appear in leaf axils, nearly hidden by foliage. Growing best in warmth with humidity, they soon become lanky and unattractive unless pinched back; root these tips for better looking new plants.

bullatus, known in horticulture as Episcia tessellata (Amazonian Peru); erect tropical plant to 2 ft. high, with large much-wrinkled, 5 to 8 in. bronzy leaves stiffly-hairy above, wine-red beneath; and small 1¼ in. pale yellow or straw-colored, hairy tubular flowers usually in dense clusters close in leaf axils. **C**: 10   p. 439

forgetii (Peru), nicknamed "Peruvian foliage plant"; ornamental foliage plant with erect, stout stems to 2 ft. high, and opposite, large glossy, wavy-margined, fleshy leaves about 6 in. long bright green, with a distinctive red pattern about the veins above and underneath; the hairy flowers pale yellow with reddish calyx lobes spotted green. **C**: 10 p. 439

lynchii, also known in horticulture as Alloplectus lynchii (Colombia), the "Black alloplectus"; erect, fibrous-rooted tropical plant with stout, succulent stems to 2 ft. high, at first purple and hairy, later turning brown and smooth, with handsome bronze to blackish-red, shiny elliptic leaves 6 in. long, toothed at margins; the not very showy flowers in axillary clusters, creamy with purplish hair, blooming in summer. **C**: 10   p. 439

NAVIA (Bromeliaceae)

nubicola (Venezuela: Amazonas Ter.), "Cloud plant"; symmetrical medium-sized rosette growing saxicolous in rocky regions; long, narrow, pointed leaves reddish at base; flowers white. **C**: 21   p. 137

NEANTHE: see Chamaedorea

NELUMBIUM (Nymphaeaceae)

nelumbo, now listed as Nelumbo nucifera, the "Lotus" or sacred "East Indian lotus"; (Trop. East Asia to N.E. Australia), large aquatic, stemless plant, a religious symbol of perpetual life in Buddhism and the seat of Buddha; long milky, prickly petioles, growing from tuberous, wide-spreading rhizomes, bearing shield-like round, peltate leaves 1 to 2 ft. across, high above the water, and bold stalks with large, delicate pink flowers 4 to 10 in. in diameter and with a haunting fragrance; flowers open for 2 days, closing each night; the petals then fall, leaving the prominent and peculiar flat-topped receptacle dotted with holes and bearing edible seed. The white or reddish tubers are starchy and are eaten in the Orient. Propagated from seed or rhizomes. The lotus is one of the few plants cultivated since ancient times for its flowers, and even now looks well when grown in containers for summer bloom. The "Egyptian lotus" is Nymphaea lotus. **C**: 64   p. 517

NEMATANTHUS (Gesneriaceae)

fluminensis (Brazil), "Thread-flower"; epiphytic climb-ing stems, with opposite, satiny green 2 to 4 in. fleshy elliptic leaves blotched red underneath; from the axils long thread-like stalks with 2 in. flowers suspended like small orange and red pouches; winter-blooming. Propagated by cuttings. **C**: 60   p. 446

NEMESIA (Scrophulariaceae)

strumosa (So. Africa), "Cape jewels" or "Pouch nemesia"; charming floriferous, densely branching tender annual herb 1 ft. or more high, with opposite sessile, linear dentate leaves, the slender leafy stems with masses of attractive 2-lipped flowers pouch-like at base and bearded throat 1 in. or more wide, borne in clusters in great profusion from February to December, depending on time of sowing the seed. With their cheerful, vivid colors from yellow and orange to scarlet and carmine-red, also white, often marked purple, Nemesias are attractive pot plants in the spring, or in the summer garden, and for cut flowers; dazzling in patio con-tainers. We have often admired them at their best planted along Emerald Lake and other resorts of the Canadian Northwest. **C**: 64   p. 414

NEOBENTHAMIA (Orchidaceae)

gracilis (Tropical E. Africa: Isle of Zanzibar); handsome branching terrestrial with swollen base, to 4 ft. high; a tall leafy stalk with numerous 2-ranked linear 8 in. leaves, rather limp, the lower ones falling in time; topped by a dense terminal cluster of fragrant, long-lived white flowers, to 1 in.

across, usually not opening fully, the lip spotted rose alongside a yellow center stripe, blooming winter-spring, again in fall. Culture similar to Phaius in a mixture of loam, peatmoss or osmunda, and fertilizer, and watered liberally while growing. In summer on the patio or under trees. Needs resting period. Propagated from lateral bulbs at the base; or offshoots from floral stalks after blooming. **C**: 60   p. 474

NEOBUXBAUMIA (Cactaceae)
polylopha (Cephalocereus) (Mexico), "Aztec column"; growing into an imposing column to 40 ft. high; close-ribbed, deep green at first, later dull green, yellow-spined; funnel-shaped dark red flowers. **C**: 1   p. 154

NEOFINETIA (Orchidaceae)
falcata (Angraecum falcatum) (Japan, Korea); lovely miniature epiphyte 3 to 6 in. high, with leathery linear light green, keeled leaves arranged in 2 ranks; freely blooming with clusters of pure white, waxy flowers $1\frac{1}{4}$ in. across, having a slender spur 2 in. long, and blooming in summer; intensely fragrant, mostly at night. Does best in small, well-drained pots; and succeeds in cooler temperature. **C**: 34   p. 457

NEOMARICA (Iridaceae)
gracilis (Mexico to Brazil), "Apostle plant"; curious tropical perennial with swordlike 10 to 15 in. fresh green leaves arranged flat as a fan in 2 ranks, from a short rhizome; and leafy stalks with showy iris-like but short-lived successive 2 in. flowers, their outer petals white, the small inner petals recurved and blue, base marked brown. An excellent and easy house plant. Propagate by division of the rhizomatous roots. **C**: 16   p. 393

longifolia: see Trimeza

northiana, known in horticulture as "Marica" (Brazil), the "Walking iris" or "Twelve apostles"; rhizomatous perennial herb 18 to 24 in. or more high, with flat, glossy green leaves arranged like fans, and with a leafy spike bearing very fragrant flowers 3 to 4 in. across, followed by young plants from the same point, bending down and rooting; the blooms with white outer petals marked brown at base, recurved inner segments tipped violet with base striped brown and yellow. Known as the "Apostle flower" because 12 leaves are said to form before one turns brown. Long popular as an easy-to-care-for house plant and for the strange beauty of its blooms. By division and from plantlets. **C**: 16   p. 393

x NEOMEA (Bromeliaceae)
'Popcorn' (Neoregelia spectablis and carolinae x Aechmea miniata); distinguished Hummel (California) hybrid; very spready, open rosette to $2\frac{1}{2}$ ft. across, the leathery leaves lacquered coppery olive-green and faintly tipped red at apex, deep wine-red underneath; small spines at margins; the inflorescence a dense compound head slightly raised, tubular white flowers tipped lilac-blue. **C**: 10   p. 141

NEOPORTERIA (Cactaceae)
heteracantha (North Chile), "Unkempt boy"; flattened globe to 5 in. dia., dark in color, with about 19 ribs, close areoles set with bristly radials and long gray central spines to 2 in. long; small carmine-rose flowers. **C**: 13   p. 159

NEOREGELIA (Bromeliaceae); the sturdy "Fingernail plants" or "Blushing bromels" are epiphytic or rock-dwelling, flaring nest-like rosettes mostly of low habit but with broad, leathery glossy leaves with spiny margins, often beautifully marbled, or banded and tipped with red; however, their most beautiful feature found in the best-known species is the striking central area turning a brilliant red around the floral cup and formed by the basal portion of the leaves and floral bracts, beginning in spring, which change to various shades of red at flowering time. The inflorescence itself is usually sunk deep in the base of the nest, with blue or white flowers peeking out of the water-holding cup. Neoregelias are good house plants; they are durable and fairly well tolerate the dry atmosphere of the living room but need good light. Propagated by suckers or seed.

ampullacea (Brazil: Espirito Santo and Guanabara); hard, channeled and recurving leaves to 1 ft. long and $1\frac{1}{2}$ in. wide, glossy olive-green with irregular brown-purple cross bands; purple teeth at margins; flowers clustered at base of "vase", lavender with center and base of petals white; each lasts one day only. **C**: 10   p. 141

carolinae (Brazil), spreading rosette of strap-shaped leaves metallic copper over green, toothed margins; inflorescence formed by brilliant lacquered orange-red bract leaves, surrounding the flowers with violet-purple petals edged white, deep in center. **C**: 10   p. 131*

carolinae 'Marechalii', "Blushing bromeliad"; spreading rosette of flattened metallic green leaves with toothed margins; at flowering time the inner leaves turn brilliant lacquered rosy-crimson, surrounding the nest of lilac flowers deep in center; the bract leaves remaining intensely colored for six months or more. **C**: 10   p. 140

carolinae 'Tricolor' (Brazil), "Striped blushing bromeliad"; very attractive mutant with glossy green leaves having ivory-white lengthwise bands which become rose-tinted in good light; at flowering time the inner leaves become shorter and turn carmine-red; flowers violet-purple edged white. **C**: 10   p. 140, 192

farinosa (Brazil: Espirito Santo, Estado do Rio); called "Crimson cup", because at flowering time the short inner leaves and bases of others turn a vivid crimson in advance of the appearance of the small purple flowers deep in their nest; outer leaves deep olive to purplish. **C**: 10   p. 140

x'Mar-Con' (marmorata x concentrica), "Marbled fingernail", beautiful bold rosette 18 in. across, the broad leaves leathery, and 4 in. wide, glossy light fresh green overlaid with red-purple design leaving a pattern of apple green blotches; tips typically wine-red; margins with scattered red-brown teeth; inflorescence very slightly raised in center, flowers lavender. **C**: 10   p. 140, 193

x'Marmorata hybrid' (spectabilis x marmorata), "Marbled fingernail plant"; bold rosette with light olive-green, stiff leaves blotched maroon and with red tips; lavender tinted flowers deep in center cup. Quite tough and much grown as a patio plant in Florida. **C**:10   p. 140

mooreana (Amazonian Peru), "Ossifragi vase"; introduced as "Ossifragi" by Lee Moore from near Iquitos; a very distinctive tubular epiphytic rosette 8—10 in. high, of numerous leathery, glossy, grass-green to deep green leaves with pale lines lengthwise, tapering to a slender point, and the margins armed with long black spines; in collected plants the tips markedly recoiling, in cultivated plants less so, when the rosette is more open and leaves arching; deep in center the almost hidden head of white 3-petaled flowers. **C**:10   p. 140

sarmentosa chlorosticta (Brazil: Estado do Rio); small rosette formed of bright-green leaves, painted maroon in such a way that the green shows as circular blotches; silver spotted or with a touch of silver beneath; sharp tips red; pale lavender flowers in center. **C**:10   p. 140

spectabilis (Brazil), an old favorite called "Fingernail plant", because of the distinctive red tip at the apex of the metallic olive green leaves; gray crossbands on the underside; blue flowers in low cushion. A leathery epiphytic rosette that has proved to be a good and interesting house plant. **C**: 10   p. 140

tristis (Brazil: Espirito Santo, Estado do Rio), "Miniature marble plant"; dwarf rosette of few leaves, deep olive to grayish green and mottled purplish maroon; gray-banded beneath and red-tipped; pale lavender flowers. **C**: 10   p. 140

zonata (Brazil: Espirito Santo); shapely rosette of hard-fleshy broad leaves olive green and heavily marbled and banded wine-red on both sides; flowers deep in center cushion, pale blue. **C**: 10   p. 141

NEPENTHES (Nepenthaceae); these tropical "Pitcher plants" are curious and interesting dwellers of Monsoon regions of the Eastern hemisphere, usually clamberers in bush country as well as climbers into jungle trees; a few others are growing as low herbs in the ground, in temperatures around 75° F. combined with high humidity, fog or mist. The prolonged midrib of the leathery leaf or broadened petiole acts as a clinging tendril from which develops a usually pendulous hollow pitcher with thickened rim, and a lid to keep out the rain. Insects are attracted inside by honey glands where they drown in pepsin liquid which digests them. Nepenthes usually produce two kinds of pitchers; near the base of the stems the larger, inflated urn-shaped juvenile form, beautifully colored; from the upper growth the more slender, trumpet-shaped huntsmen's horns, as a rule with less color. To encourage young shoots, branches should be sharply cut back in spring. The greenish flowers are not conspicuous, and males and females are on separate plants. Nepenthes require warmth and high air moisture and develop

lush young growth with perfect pitchers best in the tropical conservatory or small greenhouse, preferably under intermittent mist, with constant temperature of at least 70 to 80° F., in a house where such as Marantaceae, Bertolonia, or Selaginellas feel well. Usually grown in baskets or hanging pots, they like a mixture of osmunda or other fern fiber with sphagnum moss, and should be dunked or watered with lime-free or soft rain water. Propagation may be from fresh seed on chopped sphagnum. Cuttings are best taken from semi-hardened young growing stems, the base inserted through the enlarged hole of an inverted pot and stuffed around with fresh spahagnum moss, and set into a sweaty glass terrarium or under plastic hood at no less than 75° F., better 85° or 90° F., kept humid until rooted. (See p. 56.)

x atrosanguinea (distillatoria x sedenii), a willing grower, with slender pitchers rich maroon over greenish yellow, wings fringed. Rec. for the beginner. **C**: 60 p. cover*

x coccinea (distillatoria x mirabilis); a very satisfactory old hybrid (1882), with gracefully pendulous flask-shaped pitchers pale yellowish green, richly splashed with red-brown, the ring around the top lined with maroon, and with ciliate wings in back; inside bluish with red spots. We have found this plant very willing to produce its colorful pitchers, more so on young branches. **C**: 60 p. 523

x dicksoniana (rafflesiana x veitchii); robust hybrid (Edinburgh 1888), with big leathery leaves to 20 in. long, and large wing-shaped to heavy cylindrical pitchers, to 10 in. in extent, light green densely speckled with bright crimson, the broad rosy rim marked crimson. An excellent plant, with one of the largest pitchers in cultivation, in proper humid-warm surroundings. **C**: 60 p. 523

maxima, better known in horticulture as N. curtisii (Celebes, Borneo, N. Guinea); a sturdy "Pitcher plant"; slender, high-climbing species with firm, colorful pitchers; the lower ones flask-shaped, the upper ones from extended growth funnel-shaped and less colorful, pale green and heavily marbled blood-red. A free-growing, often-blooming plant recommended for trial as a house plant, as it tolerates temperatures lower than the sweaty-warm conditions required by most Nepenthes, and holding its pitchers remarkably long without shriveling. **C**: 10 p. 523

mirabilis (New Guinea, Queensland); remarkable terrestrial, sturdy species which we found growing in heavy red, pebbly volcanic soil in light, sunny forest on the highlands of Papua, at temperatures ranging from 50 to 88° F., in company of Cycads and Dischidias; short plants with beet-like roots, the leafy stems bearing at their tips on prolonged wiry threads the pretty yellowish green constricted pitchers 6 to 7 in. long, covered by rusty-red lids, and storing water; the small flowers brown-red and white, in erect clusters. These plants survive here, after 7 months without rain, in soil so dry that I was unable to cut them out with my bush knife. Inside their pitchers I found ants, flies and beetles. **C**: 60 p. 534

rafflesiana (Malaya, Sumatra to Borneo); remarkable straggling climber at home in the misty monsoon regions on low trees of the open savannah; the lower pitchers are urn-shaped, light green with red spots and markings, the upper ones large funnel-form on young, vigorous growth, greenish-yellow marked purplish-brown. In Malaya I have collected pitchers as long as 10 inches, a most magnificent sight, although it was difficult to get a good photograph because of the dense mist and warm drizzle of the monsoon hanging over the low jungle. **C**: 60 p. 521

'Superba'; vigorous tropical rambling epiphyte with long soft-leathery, deep green leaves, carrying pendulous variable pitchers 6 in. or more long, some urn-shaped, later ones funnel-shaped, yellow-green blotched with wine-red, the glossy ribbed rim with wine-red and crimson, inside spotted red, the lid striped red, the fringe in back with red hairs. This may be a hybrid; we received our plant from Lecoufle-France, who used it as a father plant for his beautiful N. 'Boissiense'. **C**: 60 p. 523

NEPETA (Labiatae)

mussinii (Caucasus, No. Iran), "Cat-mint" or "Catnip"; pungent hardy, much branched perennial to 2 ft. high, with tough whitish hairs; small wrinkled, oblong 1–2 in. leaves notched at margins, green above, whitish beneath; blue flowers having dark spots in clusters of loose spikes. Cat-mint tid-bits are attractive it seems to male cats who enjoy rolling in its soft, gray-green mounds. Leaves are used in preparing herb teas; the oil has value in sedatives. From seed or division. **C**: 13 p. 321

NEPETA: see also Glecoma

NEPHROLEPIS (Filices); the "Ladder" or "Sword-ferns" are an important genus in the family Polypodiaceae, widespread in the tropics and subtropics of the Eastern and Western hemispheres. The handsome pinnate fronds rise from crowns at varying intervals on long, thin rapid-growing wire-like rhizomes or "runners". Small plantlets develop from these buds which will cause the mother plant to become bushier, or they may be separated and used for propagation. This is important because the cultivars of the species exaltata, the father of our "Boston fern" do not develop viable spores. As most ferns, Nephrolepis prefer moist conditions and semi-shade, and the Boston ferns will endure house-plant conditions better than most ferns but dislike draft. Air too dry favors white and brown scale, white fly and mealy bug.

biserrata, known in horticulture as ensifolia, (Cuba to Peru, Polynesia, Queensland, So. China, So. India, Trop. Africa), "Bold sword fern"; tufted fern with spreading rhizome; large arching or pendant, fresh-green pinnate fronds 2–4 ft. long, the sickle-shaped leaflets somewhat leathery and lightly notched; sori near margin. Showy in the tropical fern house. **C**: 54 p. 240

biserrata furcans (Cuba to Brazil, Africa, Hong Kong to Queensland), "Boston fishtail fern"; massive fern with long arching pinnate fronds, the segments widely spaced, broad, leathery yellow-green, and distinguished by conspicuous forks toward their tips. Outstanding in the humid fern house. **C**: 60 p. 240

cordifolia 'Duffii' (New Zealand, Polynesia), "Pigmy swordfern"; a Nephrolepis that is different; densely crowded, compact fern with brown downy scales at base; the erect wiry fronds sometimes forked, and 1 to 2 feet long but less than 1 in. wide, closely set with tiny rounded, toothed, leathery leaflets. **C**: 22 p. 240

cordifolia 'Plumosa' (tesselata), (Japan to New Zealand), dubbed "Dwarf whitmanii" in horticulture, and though not directly related, is of similar feathery, bushy habit; tufted plant with stiff dark green rather narrow fronds, and bearing viable spores, on wiry blackish stalks, the pinnae more or less crenate or pinnately lobed. and lacy toward their tips, the runners forming tubers. A very attractive feather fern that may be propagated from spores as well as runners. **C**: 16 p. 239

ensifolia hort.: see biserrata

exaltata (wide-spread from Florida to Brazil, Africa, So. Asia, Australia); "Swordfern", tufted plant with simply pinnate, rather stiff erect, fresh green fronds usually 1 ft. high to 2 or even 5 ft. long but which can continue to grow in length nearly indefinitely, and bearing sori beneath; the root-stock sending out thread-like runners which produce buds, giving rise to new plants. From this variable species, much cultivated in Florida ferneries, have risen nearly 100 named cultivars, beginning with bostoniensis in 1894, and others in 1903, and for some 30 years these were the favorites with palms for decoration in home and business. **C**: 16 p. 239

exaltata bostoniensis, the "Boston fern", a variety found in Boston in 1894, and an old time house plant; rich green fronds simply pinnate larger and wider than the basic species, to 3 ft. long; the leaflets not lobed and nearly flat, more graceful and pendant; entirely without fertile spores; therefore propagation by division or runners. Still a favorite for decoration, appreciating good light, 25 fc. min., better a 100 fc., but not burning sun; not too warm, and all the atmospheric moisture possible. Air too dry favors white scale, brown scale, white fly and mealy bugs. Active growth is fairly fast from May to October. Tolerates air conditioning but dislikes drafts of air. **C**: 16 p. 233, 286

exaltata bostoniensis compacta, "Dwarf Boston fern"; long used as a house plant, of more compact habit, the wide simply pinnate fronds fresh green and spreading; freely clustering. **C**: 16 p. 239

exaltata 'Fluffy Ruffles', "Dwarf feather fern"; a dwarf variety with rather stiff upright fronds under 12 in. long, closely drawn together, rich dark green and almost leathery, yet finely and densely bipinnate with congested pinnae; close to the older ex. 'Muscosa'. Very attractive as a miniature feather fern in 3 or 4 in. pots. **C**: 10 p. 239

exaltata 'Hillii', the "Crisped feather-fern"; an excellent strong growing variety with sturdy, fresh-green pinnate fronds, the broad pinnae closely set and overlapping and wavy, or deeply lobed and crisped. **C**: 10 p. 239

exaltata 'Norwoodii' (1915), "Norwood lace fern"; a sport of 'Metropoli' cultivar similar to 'Whitmanii'; elegant commercial plant of short habit, with broad, tri-pinnate leaves to 18 in. long, the fresh-green, dense pinnae set in even ranks behind each other, on wiry brown axis. **C**:10
p. 239

exaltata 'Rooseveltii plumosa', "Tall feather-fern"; the tallest form of the feathered type, with rather leathery fronds, the longest arching or pendant, becoming 3 ft. or more in extent, the long wavy fresh-green pinnae lobed or monstrose at their tips; a noble fern ideal for large baskets. **C**: 10
p. 239

exaltata 'Verona', "Verona lace fern"; a dwarf tri-pinnate variety of Boston fern with delicate, small, very finely lacy fronds of drooping habit, the tips unfolding from pearly buds; one of the best of the lace type for house conditions. **C**: 10
p. 239

exaltata 'Whitmanii', a "feather fern", sport of 'Barrowsii'; old fashioned Lace fern of open habit, the broad, light green fronds are relatively short and arching, or pendant when older, the segments deeply and evenly cut and not bunched; tri-pinnate, with small segments, leaves up to 18 in. long; rarely reverts. **C**: 10
p. 239

NEPHTHYTIS (Araceae)

gravenreuthii (Cameroon); slow-growing plant, with broadly halberd-shaped, yellow-green leaves with dark veining, and wide, open sinus on slender, wiry petioles; spathe dotted, spadix white with green dots; fruit orange. **C**: 56
p. 97

NEPHTHYTIS: see also Rhektophyllum, Syngonium

NERINE (Amaryllidaceae)

; magnificent autumn-blooming bulbous plants, from South Africa, where they flower from March to May, with strap-shaped leaves, appearing usually after the long-stalked clusters of flowers with 6 spreading recurved segments. Important for prolific flowering is to keep the foliage growing all winter, to strengthen the bulb before going into dormancy in spring and summer when watering is stopped until new growth appears. In pots Nerine does best in the cool winter garden or greenhouse. Propagated by offsets.

curvifolia fothergillii (So. Africa), the "Curveleaf Guernsey lily"; a bulbous plant 18 in. high, with long-linear blue-green, glaucous leaves curved and thick, and usually appearing after the bloom; the showy inflorescence on a solid stalk in close terminal truss of 2 to 2½ in. flowers with long straight anthers and spreading, recurved petals, dark salmon or crimson red, and overlaid with shimmering gold-dust in fothergillii, glistening scarlet in the species, blooming in autumn. **C**: 25
p. 390

masonorum (So. Africa), "Dwarf pink Guernsey lily"; an exquisite small evergreen species, with grass-like leaves, and floral stalk hardly more than 6 in. high and bearing a small cluster of pale pink flowers with crisped, thread-like segments. Somewhat hardy in sheltered locations in warm-temperate zones, but very pretty in pots in the cool winter garden. **C**: 25
p. 390

NERIUM (Apocynaceae)

; the "Oleanders" are evergreen shrubs with slender willowy, flexuous branches, forming clumps, ideal for sunny, dry-hot locations, where with sun, warmth, fresh air, sufficient water and fertilizer they will thrive and bloom luxuriantly; in dark weather or fog they are shy. Oleander blooms on the growth of the year, so cut out old flowered-out wood early in spring to encourage new growth which must be well ripened in June to set buds. One of our oldest tub plants; ideal for the sunporch, foyers, or the sunny patio in summer, as it thrives in heat and strong light; keep cool in winter. All parts of the plant are poisonous if eaten. Cuttings root readily in a bottle of water.

oleander (So. Europe and No. Africa), "Oleander" or "Rose-bay"; evergreen shrub from 7 to 20 ft. high, with clustering, willowy branches set with pairs or whorls of linear-lanceolate, leathery leaves 4 to 8 in. long, dark green above, paler beneath and with a prominent midrib; the flexuous branches tipped with clusters of 1½ in. funnel-form flowers normally rosy-red, but ranging from white to purple; summer to autumn. Prune in early spring to control size, and cut down old wood. **C**: 13
p. 296, 489

oleander 'Album', "White oleander" or "Sister Agnes"; large flowered cultivar with single white 2 in. flowers and of

vigorous growing habit to 20 ft. tall, favored by So. California nurseries because the single flowered varieties have a way of "cleaning" themselves, or "shed" their faded blooms, rather than dry up and become brown and unsightly as in the doubles. **C**: 14
p. 296

oleander 'Carneum florepleno', oleander known in the trade as "Mrs. Roeding oleander"; somewhat shorter in growth, to 6 ft., and weaker in habit, and of a slightly weeping nature, the long flexuous branches loaded with double salmon-pink blossoms; double flowers have a tendency to "hang on". A favorite decorator tub plant, very resistant and freely flowering, even as a small pot plant in the sunny window. **C**: 14
p. 296

NERTERA (Rubiaceae)

granadensis, better known in horticulture as N. depressa (Andes of Peru to Cape Horn, New Zealand, Tasmania), the "Coral-bead plant" or "Hardy baby tears"; enchanting mat-forming, low creeping ground cover with angled branches and tiny vivid green, broad-oval, leathery, opposite leaves; inconspicuous, greenish flowers between April and June, followed from August on by the attractive, ¼ in. pea-size, translucent, orange-red berries. I have collected this species along cold Milford Sound, in the Fjordland of New Zealand where it is found growing on dripping rocks, frozen stiff in winter, and have also seen it in the mountains of New Guinea at 7,000 ft., under rippling water, in company with sphagnum moss. This is the key to getting the plants to set berries—except at blooming time, they must always be moist, and as cool and airy as possible during summer in soil not too rich. Charming in shallow pans. Propagated by division or seed. **C**: 72
p. 512

NICODEMIA (Loganiaceae)

diversifolia (Madagascar), "Indoor oak"; free growing bush with woody stems and thin-leathery quilted foliage having lobed and undulate margins, and looking remarkably like oak leaves; the surface has an iridescent metallic-blue sheen, and the petioles are bronzy. **C**: 4
p. 188

NICOLAIA (Zingiberaceae)

elatior, long known in horticulture as Phaeomeria magnifica (Amomum) (Indonesia), the magnificent "Torch ginger"; gigantic herb forming clumps of robust, leafy, arching canes to 20 ft. high, with alternate, pointed leaves 1 to 2 ft. long in 2 ranks; the striking inflorescence of large, torch-like heads of brilliant red, formed of innumerable waxen bracts, on separate leafless stems 5 ft. high or more, subtended by red basal bracts, margined white, forming a nest for the red cone, brightened by yellow-margined lips of the small red flowers. **C**: 52
p. 203

NICOTIANA (Solanaceae)

alata grandiflora (affinis) (Brazil, Uruguay, Paraguay), "Jasmine tobacco"; tender herbaceous perennial, grown as an annual 2 to 3 ft. high, usually planted for the fragrance of its blossoms; tall, sticky-hairy stalks set with large ovate, pubescent, soft leaves, terminated by loose clusters of long, trumpet-shaped flowers with long tube and flaring limb 2 in. across, white within, yellowish outside; closing in cloudy weather, and with a sugar-sweet perfume at night. From seed. **C**: 64
p. 416

x sanderae (alata x forgetiana), "Flowering tobacco"; herbaceous hybrid annual 2 to 3 ft. high with clammy-hairy, spoon-shaped pointed lower leaves to 1 ft. long and un-dulate; at the end of branches the fragrant, showy flowers to 3 in. long, salver form with greenish-yellow tube tinted with rose, the face carmine rose or shades of red, staying open all day. Very ornamental in the garden, and blooming continuously from June into late fall. From seed. **C**: 64   p. 416

NIDULARIUM (Bromeliaceae)

; these "Birdsnest" bromels are magnificent epiphytic rosettes from tropical Brazilian rainforests, forming flaring nests of broad, soft or thin-leathery leaves in metallic or purple colors, sometimes prettily mottled or variegated, and with prickly margins; this genus is distinguished from the similar Neoregelias by their low nested inflorescence rising slightly above the cup, or is carried on a short stalk, and its floral bracts are large and colorful in maturity, during winter the flowers appearing from between them. The shade-loving Nidulariums are durable decorators, prefer warmth but tolerate a more moderate temperature also; they like the light of a north window. They will grow easily in ordinary peatmoss or humusy soil. Propagated by suckers; also from seed.

amazonicum: see innocentii var. innocentii

fulgens (S.E. Brazil), "Blushing cup"; striking, bold rosette with numerous flattened shiny leaves pea green with dark mottling and conspicuous spines; inflorescence cup in center bright crimson tipped nile-green, fl. blue. **C**: 10   p. 141

innocentii var. innocentii (Brazil: Espirito Santo to Santa Catarina), "Black Amazonian birdsnest"; in the trade as Nid. "amazonicum"; large and striking rosette almost sinister with leaves metallic purple to almost black, the margins finely toothed; glossy beneath; inflorescence a short cup of rusty-red bract leaves with white flowers. **C**: 11   p. 141

innocentii lineatum (Brazil), "Striped birdsnest"; beautiful variety with the broad, pale to deep green foliage more striped and banded white along the leaf than it is green; the center cup, housing the white flowers, is edged in burnt-red. **C**: 60   p. 192

innocentii nana, "Miniature birdsnest"; small rosette of broad, thin-leathery leaves matte olive-green, finely toothed; underside glossy purple; at flowering time an inner nest of short leaves turns orange-red; flowers white. **C**: 10   p. 141

innocentii 'Striatum' (Brazil), "Striped birdsnest"; stocky rosette of broad light green recurved leaves striped lengthwise with yellow-ivory lines and bands; finely toothed margins; white flowers deep in carmine-tipped center cup. Very showy. **C**: 10   p. 141

NIEREMBERGIA (*Solanaceae*)

caerulea (Argentina), "Cup flower" or "Purple robe"; lovely small herbaceous perennial, but better treated as an annual; 6–12 in. high, freely branching with thin erect, hairy stems and linear leaves, and forming mounds; covered all summer with numerous, widely flaring saucer-shaped, lavender flowers 1 in. wide with violet lines, and bright yellow eye. For the garden or patio; somewhat hardy, but best kept in the cool greenhouse during winter; propagated by cuttings or seed. **C**: 78   p. 414

NOLINA: see Beaucarnea

NOPALEA (*Cactaceae*)

cochenillifera (Opuntia) (Puerto Rico, So. America), "Cochineal plant"; a tropical branching cactus, tree-like to 16 ft. high, forming cylindrical trunk and glossy dark green, long fleshy, flattened joints; usually spineless; flowers rosy-red. Host to the cochineal scale insect from which a valuable red dye is derived. **C**: 1   p. 167

NOPALXOCHIA (*Cactaceae*)

ackermanii (Epiphyllum) (Mexico: Chiapas, Oaxaca 6000–8000 ft.); a species "Orchid cactus"; in habitat mostly epiphytic, with flattened green branches, sometimes 3-angled, the angles notched (crenate); large and showy funnelform flowers glowing red; good flowering plant. **C**: 13   p. 165*

ackermannii 'Fire Glory'; a beautiful hybrid cultivar with large 4 in. flowers soft rosy-scarlet, not as deeply scarlet as the species. N. ackermannii is an epiphyte from Southern Mexico, with 2 to 3-angled pendant stems with step-like notches on the edges. **C**: 21   p. 169

phyllanthoides (Mexico); an old, free-flowering house plant widely grown under the name "Deutsche Kaiserin" (German Empress); an epiphyte from Puebla state at 5000 ft.; densely bushy with flattened, pendant, crenate branches bearing a profusion of day-flowering, carmine-rose flowers of medium size, and with pale tips and short tube; a lovely basket plant. Responds well to feeding. **C**: 21   p. 169

NOTOCACTUS (*Cactaceae*); the various "Ball cacti" are the finest, easiest to grow and most reliable blooming of cacti; they are mostly small-sized plants with highly colored bristle-like spines, a willing bloomer when quite small, from May to July, with relatively large, showy flowers predominantly yellow, sometimes red; favorite window-sill subjects attractive with or without blossoms. Notocactus grow best in nourishing, but gritty soil, kept moderately moist in summer, and with protection from burning sun; during winter not lower than 50° F. Grown from seeds.

haselbergii (So. Brazil), "White-web ball"; lovely small globe to 3 in., occasionally sprouting from the base; with about 30 low ribs, covered with soft glossy, silvery-white spines, pale yellow at the top; flowers orange-red to crimson. Of easy cultivation, but slow-growing. **C**: 13 p. 158

leninghausii (So. Brazil), "Golden ball"; attractive, small clustering, cylindrical column, to 3 ft., close-ribbed, covered with soft golden hair; flowers yellow at top. Beautiful, and of easy growth. **C**: 13   p. 158

mammulosus (Uruguay, Argentina), "Lemon ball"; simple plant nearly globose, to 3 in., shining green, with 18–25 high ribs, and yellowish to reddish spines from recesses on the knobs; beautiful, fragrant flowers yellow with red, appearing at an early stage. **C**: 13   p. 158

rutilans (Argentina) "Pink ball"; dark green globe to 2 in. high, with 25 rows of long knobs, the recessed areoles with straw-white radials and straight brownish central spines; large 2½ in. flowers bright pink with yellow throat, shimmering like silk. **C**: 13   p. 158

scopa (Brazil), "Silver ball"; globular to cylindrical, 1½ ft. high, closely ribbed and almost entirely covered with short, soft-hairy white radials and long brown needle-spines; flowers silky canary-yellow deeper in center. **C**: 13   p. 158

scopa cristata, "Spiralled silver-ball", a strangely beautiful, fasciated curiosity plant covered with glistening, pure-white hair. This is a cristate form of the "Silver ball" from Brazil and Paraguay, a normally globular plant with yellow flowers. Such crests are best grafted on such as Trichocereus. **C**: 13   p. 168, 532

NOTONIA: see Senecio jacobsenii

NOTYLIA (*Orchidaceae*)

sagittifera (Brazil); small epiphyte somewhat inconspicuous, but an interesting "botanical", having little pseudobulbs with solitary 3 in. leaves; from the base a pendant spray-like inflorescence 3 in. long with numerous tiny ¼ in. flowers greenish white; quite attractive because of their number, and faintly fragrant. Best grown on treefern cubes or small pots, receiving abundant moisture and moderate warmth. By division. **C**: 22   p. 474

NYCTOCEREUS (*Cactaceae*)

serpentinus (Mexico), a "Queen of the Night" or "Snake cactus"; slender erect or clambering nightbloomer; cylindric many-ribbed stems, deep green; woolly areoles and white to brownish spines; large white, sweet-scented funnel-form nocturnal flowers; red fruit. **C**: 1   p. 155

NYMPHAEA (*Nymphaeaceae*); the "Water lilies" are showy aquatic herbs with horizontal or erect rootstocks, sometimes tuberous; the leaves usually floating on the surface of the water; majestic flowers of great beauty are open in bloom either during the daytime, or at night; some species are hardy; others and their hybrids are tropical or tender, with showy, fragrant blooms, 8 to 14 in. across, including blue and purple shades, and with stiff stalks thrust well above the water. If outdoors, tropical Nymphaeas won't start to grow unless the water is warm enough and at least 70° F. Established in a greenhouse pool or a large wine-cask cut in half, such tender plants not exposed to frost may be left undisturbed for years, as long as the soil has not become impoverished. Old tubers may be taken up and carried over stored in damp sand but young plants perform better. In propagation the hardy varieties have rhizomes which may be taken up and divided into as many pieces as have roots with tops. The tropical kinds are more tricky; all may be grown from seed if available, but the more usual method is to use parts of rootstocks. The tropical varieties that are tuberous are increased by the small potato-like knobs forming at their base—the night-bloomers both at bottom or top—during summer, removed by early September, and stored in a jar with damp sand. Tropicals with short rhizome in the gigantea and stellata group are propagated from suckering sprouts appearing in spring, and which are taken off and potted like seedlings. The viviparous species are easily increased by the perfect little plantlets that sprout from the center of the floating leaves.

x daubeniana (hybrid of micrantha x caerulea); a tropical day-blooming "Pygmy water lily" for confined spaces and shallow water; very prolific with small fragrant, light lavender-blue flowers from 2 to possibly 6 in. across;

with sweet, abundant fragrance, and carried well above the water and the bright green, floating leaves. This charming plant is highly viviparous with perfect young plants developing at the junction of petiole and leaf, even miniature flowers appearing on these plantlets while still attached to the parent stem. Excellent for small containers or large aquariums, blooming continuously. **C**: 52   p. 517

x'Golden West', California 1936, a charming seedling of the Pring cultivar 'St. Louis'; floriferous tropical, day-blooming water lily with lovely red-mottled foliage; and with large flowers in autumn shades; on the opening day peach-pink, later changing light apricot-yellow, and deliciously fragrant. **C**: 52   p. 516

x helvola (tetragona x mexicana), known as "Pigmy yellow"; a hardy, rhizomatous "Pygmy water lily", French hybrid by Marliac 1890; one of the smallest in cultivation and best in its color; floating leaves only 2 to 4 in. wide, blotched all over with reddish brown; very free-blooming with cute canary-yellow flowers 1 to 2 in. across, afternoon blooming. The rootstock bears detachable tuber-like branches. Wonderful for culture in urns or oak tubs. **C**: 64   p. 516

x marliacea rosea (from hybrid group N. alba x odorata rosea); excellent robust, free-flowering hardy rhizomatous water lily with shining green or purple leaves, and large cup-shaped blooms often carried above the water, deep pink shading to rose, but growing paler as the flower ages. Pleasing as cut flower because of its considerable substance and sweet fragrance. This is one of the first crosses made by Marliac in France between N. alba, the English white water lily, and N. odorata rosea, the "Cape Cod lily" from Massachusetts. **C**: 64   p. 516

x'Missouri' (St. Louis 1933); "Queen of the Night" of the water lilies; also known as a "Husband's lily"; a spectacular tropical night-bloomer, with enormous and most gorgeous creamy-white flowers with broad petals, reaching a diameter of 14 in. across, displaying their column of erect stamens and carried nobly high above the leaf line, handsome mottled foliage tinted red. Such a fancy plant requires rich soil, warm surroundings, and ample room, at least 8 to 10 ft. Called "Husband's lily", or "Businessman's water lily" because men may enjoy them after they return home at night, opening about 6 in the evening and closing at 9 in the morning, their beauty a dream of white loveliness in the moonlight. **C**: 52   p. 516

x'Pink Pearl'; long-stemmed tropical day-bloomer, profuse with delicate pearl-pink, fragrant flowers of medium size, with showy yellow stamens surrounded by a ring of pink anthers, and carried on stems longer than ordinary, holding them well above the water. Excellent for a continuous display; the blooms opening on sunny days about 9 in the morning and closing at 4 in the afternoon. **C**: 52   p. 516

x'Somptuosa', from the French hybrid group N. alba x odorata rosea; Marliac 1909; a popular hardy water lily with large and very double fragrant flowers rosy pink deepening toward the center, and lightly spotted carmine, contrasting charmingly with the clear yellow stamens in the heart of the blooms. A meritorious hybrid these many years. **C**: 64   p. 516

OBREGONIA (Cactaceae)

denegrii (Mexico: Tamaulipas), "Artichoke-cactus"; low, interesting cactus with thick tap root, occasionally sprouting into groups; globular, flat on top, grayish to dark green, 3–5 in. across, with leaf-like keeled tubercles spirally arranged, stiff and angled, without spines except at tips; white 1½ in. flowers. **C**: 13   p. 159

OCHAGAVIA (Bromeliaceae)

carnea (lindleyana hort.) (Rhodostachys, Fascicularia) (Chile, Peru); small free-suckering terrestrial rosette with numerous linear, recurved, succulent channeled leaves dark green and spiny-toothed; gray scurfy lines underneath; inflorescence a short-stemmed tight head of pink bracts nesting in the center and covered with mealy-white dust; flowers pinkish-lavender. **C**: 19   p. 138

OCHNA (Ochnaceae)

serrulata (So. Africa: Natal), "Mickeymouse plant" or "Birdseye bush"; woody shrub to 5 ft. high and of spreading habit, with irregular branches; hard leathery, glossy green narrow-elliptic leaves 2 to 5 in. long, finely toothed at the margins, bronzy in spring; the flowers are the size of butter-

cups, but with yellow petals quickly falling, the greenish sepals turn vivid red, forming a persistent receptacle on which develop several green berry-like fruits which later turn a glossy jet black; Walt Disney fans see in this the bright eyes and big ears of his enchanting Mickey Mouse. Good pot or tub plant, relishing fresh air. From seed or cuttings. **C**: 13   p. 498

OCHROSIA (Apocynaceae)

elliptica, in horticulture better known as Kopsia arborea, (Queensland, New Caledonia); evergreen tree up to 20 ft. high, with milky juice; smooth green, obovate leathery leaves to 6 in. long; clusters of small ivory-white or yellow, fragrant flowers in summer-autumn; followed by scarlet-red, angled fruit 2 in. long, having violet-like odor when crushed. Salt-resistant plant for seaside homes. From seeds or cuttings. **C**:2   p. 485

OCIMUM (Labiatae)

basilicum (Trop. Asia, Africa, Pacific Isl.), "Sweet basil"; pretty, aromatic, branched annual tender herb 12–18 in. high, with shiny light green to purplish, broad ovate leaves 1–2 in. long, and small white flowers. The spicily scented leaves are one of the most popular of all sweetherbs used in cooking, giving mild flavor of anise and spice in seasoning of fish, shellfish, meats, game, stews, eggs, poultry stuffing, tomatoes and salads, and are used fresh or dried. Sow seeds on warm soil. **C**:14   p. 322

x ODONTIODA (Orchidaceae)

'Wey' (O. 'Laurette' x 'Cheer'); a compact plant with arching spray of a series of rounded, waxy flowers 2½ to 3 in. across, vivid rusty red with yellow on lip; blooming in early summer. This cultivar is typical of the bigeneric hybrids between Odontoglossum and Cochlioda, resulting in plants retaining the habit of Odontoglossum and its inflorescence, but introducing scarlet color, and making their flowers more colorful and longer lasting. Odontiodas prefer to be grown cool like Odontoglossum but tolerate somewhat higher temperatures. **C**: 72   p. 476

ODONTOGLOSSUM (Orchidaceae), tropical American epiphytic orchids, their rather short, usually flattened pseudobulbs with 1 or 2 soft-leathery leaves; and from the base with graceful sprays of exquisitely beautiful, long-lasting flowers. Most species come from high altitudes between 4000 and 12000 ft. with cool summer temperatures and bathed in fog or rain; consequently they do best in a cool, moist house with air-circulation, or under air-conditioning at 45 to 55° F.; in well-drained compost of ½ osmunda, or shredded tree fern, and ½ sphagnum moss; fairly shady; 1200 to 1500 fc. for Andean species, 2000 to 3000 fc. for Mexican kinds; most don't need a rest period; exceptions are such as O. citrosmum and grande. Propagated by separating the pseudobulbs just as they begin to grow.

crispum (Colombia); the "Lace orchid"; widely considered the most exquisitely beautiful of all orchids; lovely epiphyte at home in the Andes at 8 to 9000 ft., with 2-leaved stout flattened pseudobulbs, and arching sprays of daintily crisped and waxy, star-shaped 2½ to 3½ in. flowers of glistening pure white, except for the lip which is yellow at base and blotched reddish toward front; flowering variably at any time of year, mostly from winter to spring. This is one of the "cold" species, and if kept cool and moist enough will bloom freely and last in beauty a long time. **C**: 86   p. 455*, 476

grande (Guatemala, Mexico), called the "Tiger orchid"; very beautiful epiphyte and one of the finest of the genus; compact habit, with thick, compressed pseudobulbs topped by 1 to 3 heavy-textured leaves, and stiff-erect, 12 in. stalks bearing large waxy flowers 4 to 6 in. or more across, their sepals yellow barred with brown, the petals halfways from the base reddish-brown, yellow toward the tips, broad lip cream and spotted with brown, blooming anytime from fall to spring, and lasting in glorious perfection from 3 to 4 weeks. From mountains at 8500ft., this spectacular species needs more light than others, also a resting season after flowering; an easy grower. **C**: 22   p. 476

pulchellum (Mexico, Guatemala to Costa Rica), the "Lily-of-the-Valley orchid"; charming epiphytic orchid with dark green, slim pseudobulbs topped by 2 to 3 grass-like leaves and clustering; erect wiry stalks bearing sprays of small waxy, sweetly fragrant ½ to 1 in. flowers of crystalline

white, with a yellow crest and spotted crimson in center, blooming in winter and to spring. This charming species comes from altitudes to 8000 ft., and is easy to grow if kept cool enough, its exquisite flowers lasting for 5 weeks. **C**: 22
p. 476

OLEA (*Oleaceae*)
europaea (E. Mediterranean region), the classical "Olive tree"; small, sparry, evergreen tree to 30 ft. high, with age becoming gnarled and picturesque; stiff-leathery, narrow lanceolate willow-like leaves to 3 in. long, gray-green above, silvery scurfy beneath; axillary flowers yellowish-white and fragrant; oblong plum-like fruit green turning shining black when ripe. Olive trees are as old as history and economically very important in dry-hot regions, for the valuable oil of its fruit. Also adaptable as tubbed specimen for decoration, in sunny, dry locations. Somewhat frost-hardy. **C**: 13
p. 294, 505, 508, 526

OLEA: see also Osmanthus

ONCIDIUM (*Orchidaceae*); the "Dancing dolls" and "Butterfly" orchids are charming tropical American epiphytes, easy-to-grow and free-blooming mostly in yellow and brown, the individual flowers generally not large but very showy in large bouquet-like arching sprays, laterally from small pseudobulbs with 1 or 2 leaves. Oncidiums grow best in smallish containers or on osmunda or on treefern blocks, in good light between 2000 and 4000 fc., and while actively growing delight in abundant water; when new growth is completed a rest period of several weeks should be given to produce best flower production. Increased by division allowing several pseudobulbs to each section.

flexuosum (Brazil, Uruguay, Paraguay), a popular "Dancing doll orchid"; charming little epiphyte with ascending rhizome, producing pseudobulbs often at distant intervals and rooting profusely, usually bearing two glossy, linear 6 to 8 in. leaves; thin-wiry, arching stalks to 3 ft. long, carry airy clusters of 1 in. flowers in dainty showers, the bright golden yellow sepals and petals barred with brown, and the broad lip yellow with a few red dots, the whole resembling a miniature group of ballet dancers floating across the stage. Flowering with great freedom at different times of the year, mostly from July to winter. **C**: 21   p. 475

kramerianum (Costa Rica, Panama, Colombia, Ecuador); handsomely beautiful epiphyte with clustered, small angled pseudobulbs bearing a solitary rigidly leathery leaf 6 to 8 in. long and 3 in. wide, sometimes mottled with purple; erect flower stalks to 30 in. or more tall and round, producing a succession of very curious, highly colored 3 to 4 in. flowers, the long narrow dorsal sepal and linear petals chocolate-brown, the showy lateral sepals broad, orange-red mottled with yellow and exquisitely frilled, broad lip lemon-yellow bordered red-brown; blooming mostly in November-December, but again in spring or throughout the year. This "Butterfly orchid" is an outstanding plant in any collection and even if not prolific with masses of bloom, a succession of single spectacular flowers keeps blooming from the same persistent spike for years. **C**: 21
p. 455*, 475

lanceanum (Trinidad, Guyana, Venezuela, Colombia), the "Leopard orchid"; strikingly beautiful epiphyte with knotty rootstock, and but minute pseudobulbs; broad and thick, solitary green leaves mottled with brown; the stiff-erect floral spikes 1 to 1½ ft. tall with a cluster of large, vanilla-scented flowers about 2 in. across, the fleshy sepals and petals yellow shaded green, blotched with chocolate-brown, the large lip violet at base, rose in front, blooming May to August. A magnificent sight with the rose and violet lip in contrast against the yellow and brown of the rest of the flower; lasting 4 to 5 weeks. **C**: 9   p. 475

ornithorhynchum (Mexico to El Salvador and Costa Rica); a very pretty, small epiphyte with flexible, thin-leathery twin leaves on small pseudobulbs, prolific with arching sprays to 2 ft. long, dense with small and dainty, very fragrant ¾ in. flowers of soft rosy-lilac, with darker shading on lip and a yellow crest, freely blooming. An Oncidium that is different, with its masses of long-lasting rosy flowers like a flock of birds in the sky. Easy to grow, preferring cooler temperatures especially in winter. **C**: 21   p. 475

papilio (Trinidad, Venezuela, Brazil, Peru), the noble "Butterfly orchid"; a striking epiphyte with small pseudobulbs bearing a single rigid leaf 6 to 8 in. long mottled purplish-brown; the large, unusual flowers spreading 3 to 5 in. from tip to tip, and developing successively on erect stalks to 4 ft. long, flattened in the upper part; dorsal sepal and petals long linear antenna-like, reddish-brown marked with occasional yellow cross bands, the lateral sepals, broader and curving downward bright chestnut red with transverse yellow bars, lip yellow with brown border, blooming almost all year from old and new stalks, which should not be cut off. One of the finest, nobly curious orchid blooms, always commanding attention wherever seen. **C**: 21   p. 475

sarcodes (Brazil); a remarkably fine, vigorous epiphyte with dark green, tapering pseudobulbs topped by 2 to 3 shining green leaves; the handsome basal inflorescence 1½ to 3 or even 5 ft. long, erect or arching large 2 in. glossy, scalloped flowers on short branches, their sepals and petals yellow with blotches of brown-red, lip bright yellow with fewer red-brown spots, blooming from spring to summer, the waxy flowers long-lasting in beauty. Easily grown either in pot or basket. **C**: 21   p. 475

sphacelatum (Mexico to Honduras), the "Golden shower" orchid; a prolific epiphyte with elongate, flattened pseudobulbs carrying 2 to 3 rigid, linear leaves to 2 ft. long; the loosely branched inflorescence to 5 ft. long, with a veritable shower of many small, yellow flowers marked with brown, about 1 in. across, starting to bloom in November, but mainly spring and summer. The pretty flowers look like little dolls, when in bud the tips of the petals and sepals are bent back like recurved horns; the glorious sprays lasting 3 or 4 weeks. An easy plant for the amateur in the window, or under trees in summer. **C**: 21   p. 475

splendidum (Guatemala, Honduras, Panama); a noble epiphyte of robust habit, with short stubby pseudobulbs, bearing a single heavy stiff-fleshy mahogany-red, keeled leaf to 12 in. long; the stout erect, 3 ft. branched stem with large, substantial flowers 3 in. long, the small sepals and petals bright yellow barred with brown, the large lip golden yellow, blooming mostly from late winter to early summer; a handsome sight when crowded with long-lasting blossoms.
C: 21   p. 475

stipitatum (Honduras, Nicaragua, Panama); a distinct epiphyte with almost non-existent pseudobulbs and pendant, fleshy cylindrical leaves, often spotted with brown, 1 to 2 ft. long; the branched, densely-flowered inflorescence drooping and may carry 200 small 1 in. sulphur yellow blooms, marked red-brown on petals and sepals, the broad lip bright yellow; flowering early winter into summer and autumn. **C**: 9

ONOCLEA (*Filices; Polypodiaceae*)
sensibilis obtusilobata (Pennsylvania), "Sensitive fern"; an abnormal form in which the leaflets of barren fronds become again pinnatifid and contracted. The species O. sensibilis is wide-spread in Temperate regions from Newfoundland to Louisiana, Siberia to Japan, called "sensitive" because the herbaceous barren fronds are sensitive to cold or if cut, and fold their glaucous pale green leaflets face-to-face. Hardy, from underground creeping rhizome. **C**: 78   p. 249

OPHIOPOGON (*Liliaceae*)
jaburan (Japan), "White lily-turf"; evergreen clump forming perennial, with cord-like roots; long, grass-like but thick-leathery, dark green recurving leaves 1 to 2 ft. long and ½ in. wide; the pure white, drooping flowers on flattened stalks, in nodding clusters. Planted as casual ground cover or along walks; also durable house plant. Good in shady locations. By division. **C**: 16   p. 304

jaburan 'Variegatus', "Variegated Mondo"; attractive form which I collected in Java; the long linear, symmetrically arranged leaves gracefully recurving and friendly milky-green, striped and edged in white; the drooping little waxy-white flower bells on erect raceme. A pretty house plant.
C: 16   p. 304

japonicus (Japan, Korea), "Snake's-beard", "Mondo grass" or "Dwarf lily-turf"; evergreen, grass-like perennial with underground stolons and tuber-bearing roots, forming low tufts of very narrow-linear but leathery, blackish-green leaves 6 to 10 in. long and gracefully arching; small, pale lilac flowers. An excellent ground cover or edging for shady locations, forming turf-like mats. Good as durable house or terrarium plant. Easy by division. **C**: 16   p. 304

OPLISMENUS (*Gramineae*)
hirtellus variegatus, in horticulture as Panicum variegatum (West Indies), "Panicum" or "Basket grass"; creeping

tropical perennial grass, with weak thread-like stems, rooting at nodes, with their flowering culms generally erect, the rather broad, lanceolate, thin leaves 2–4 in. long. daintily striped white and pink; the inflorescence in purple bristly spikelets. A pretty grass for basket or conservatory ground cover. From stem sections. **C**: 2   p. 304

OPUNTIA (*Cactaceae*); the "Prickly pears" also known as "Tunas" or "Chollas", are widespread cacti native from the Canadian North to frigid Patagonia in South America. They are mostly grotesque and awkward plants characteristic of arid regions and may be small tree-like plants or low and spreading. Their stems or branches which are fleshy-succulent and loosely jointed, either flat and pad-like as in the "Bunny-ears" and "Tuna"; or cylindrical as in the "Chollas"; usually heavily armed with spines but without ribs. Their main differentiation from other tribes of cacti is the presence of glochids, or sharp, easily detached and irritating bundles of barbed bristles in the areoles. The flowers are wheelshaped and showy but short-lived; the reddish fruit of some species is edible and cultivated in warm-climate countries for food. Several genera resembling Opuntia are listed under Tephrocactus, Consolea and Nopalea. As container plants Opuntias are ideal for hot and sunny dry locations and for that "Mexican" effect; too much water will cause rot. Easy from joint cuttings, allowing cut to heal.

basilaris (S.W. U.S. from Utah south to No. Mexico), the ubiquitous "Beaver tail"; growing in clumps in arid regions; broadly obovate fleshy pads to 8 in. long a bluish coppery color; almost spineless; showy, variable flowers usually purple. **C**: 13   p. 167

bigelovii (Southwest U.S.), "Silver cholla" or "Teddy bear"; very spiny; a typical "Cholla cactus" characteristic of the stony deserts from New Mexico to California; erect cylindric trunks to 3 ft. high with short light green branches 2 in. thick, densely set with stiff, glistening cream spines; purple flowers and yellow fruit. The dead skeleton trunks can be used as ornamental support for other plants. **C**: 13   p. 164

brasiliensis (Brazil, Argentina, Bolivia), "Tropical tree-opuntia"; tree-like, to 12 ft. high; trunk and branches cylindrical, the terminal joints flat and leaf-like, glossy fresh green, with few spines; pale yellow flowers. A good and attractive house plant because of its tropical origin, resembling a miniature tree even as a young plant. **C**: 1   p. 164

clavarioides (Austrocylindropuntia), (Argentina, bordering Chile); a very interesting plant and a curiosity because of its weird shape and sinister coloring as "Black fingers", "Sea-coral", or "Fairy castles"; low, straggling plant with cylindrical grayish-brown joints but usually growing fan or staghorn-shaped or in other fasciated forms, covered with short white, hair-like spines; rarely blooming, the flowers pale greenish-brown. **C**: 13   p. 168, 532

cylindrica (Ecuador, Peru), "Emerald idol"; robust branching plant to 6 ft. high, the succulent cylindrical, dark green joints to 2 in. dia., regularly notched, and with deciduous little leaves; short white spines often wanting; flowers scarlet, only on old plants. **C**: 13   p. 167

elata elongata (Paraguay), "Green wax cactus"; robust, bushy plant to 6 or more ft. high, very attractive with elongate, smooth, fat but flattened joints 8 in. or more long, nearly spineless, rich nile green to dark olive, shaded purplish brown below each areole, and with satiny, waxy surface; pretty flowers with 2 layers of broad petals rich orange. **C**: 1   p. 165*

erectoclada (No. Argentina), "Dominoes"; creeping clusters of trapeziform, small flat pads to 2 in. long, covered with sharp brown spines, the belly side (ventral edge) of young pads always facing parent; flowers glossy carmine. **C**: 13   p. 167

erinacea ursina (California), the famous "Grizzly bear" of hidden reaches of the Mohave Desert. Forms low clumps with oblong, thick, grayish green flattened joints 4 to 6 in. long, densely covered with glistening white, wire-like spines usually 4 to 6 in. long, and I have seen them in the Mohave as long as 10 in. Flowers may be red or yellow. A beautiful sight in sunlight, but must be kept quite dry. **C**: 13   p. 164

ficus-indica (Trop. America: prob. Mexico), the "Indian fig"; a flat-jointed cactus that may grow bushy, or with woody stems to 10 ft. or more high; widely spread into warm-climate countries and cultivated for its pear-shaped juicy, orange-red fruit 2½–3 in. long, which is

peeled and the pulp eaten raw or cooked for its flavor and food value, from Mexico to Spain and Southern Italy and Eastward. The oblongish flat joints are green or glaucous bluish from 8 in. to almost 2 ft. long, in some forms spineless but with irritating yellow bristles; the flowers are yellow to 4 in. across. These are monstrous plants, but durable with sculptured and exotic decorative effect. **C**: 13   p. 509

ficus-indica 'Burbank', the "Spineless Indian fig"; a lightly glaucous grayish-green, spineless form and therefore easy to handle; tree-like to 15 ft. high, with long flattened, elegant joints to almost 2 ft. long; flowers canary yellow, and edible orange or red fruit. Such spineless cacti were being cultivated in North Africa and Sicily, also Argentina and Ceylon, at the time Luther Burbank developed this cultivar, by reselection carrying this species back in its history of evolution to a remote age before cacti were forced to develop spines. **C**: 13   p. 164

fulgida mamillata monstrosa (Arizona, Mexico), known in horticulture as Op. mamillata, the "Boxing-glove"; growing tree-like to 10 ft. tall; succulent cylindrical knobby joints to 2 in. thick growing somewhat irregular and contorted, sometimes splitting into crests at the tips; sharp yellowish spines; flowers pink. **C**: 13   p. 168

leptocaulis (frutescens) (S.W. U.S., Mexico), "Tooth-pick-brush"; brush-like, to 6 ft. high, the branches with pencil-thin joints dull green and woody, spines slender and with small deciduous leaves; flowers yellowish. **C**: 13   p. 167

leucotricha (Central Mexico), the beautiful "White-hair tree-cactus"; tree-like and branching, to 15 ft. high, the flat oblong, pubescent joints to 10 in. long, and covered with long bristle-like, flexible white spines; flowers yellow; aromatic yellowish, edible fruit. **C**: 13   p. 167

linguiformis 'Maverick'; "Maverick cactus"; very attractive California mutation, where a flat joint develops monstrose branches; the fleshy pad rich green, with multitudes of little adventitious cylinders and bulb-like growths sticking out from the areoles, tipped by beige bristles and glochids; the plant resembling a miniature pyramidal tree. **C**: 13   p. 168, 532

mamillata: see O. fulgida mamillata

microdasys (No. Mexico), called "Bunny ears", because of the young pads appearing ear-like near the apex of the older ones; plant low with rounded or oval flat, fleshy joints 6 in. long, satiny green and set with neat rows of yellow to light brown tufts of barbed bristles (glochids) which rub off easily and are painful to the skin; flowers pale yellow. Sensitive to rot and should be kept quite dry. **C**: 13   p. 167

microdasys albispina (albescens), "Polkadots"; vigorous cultivar with normal large type pad as in the species but its areoles bearing soft glochids (tufted barbed hairs) a pure white, prominently arranged in neat rows. **C**: 13   p. 167

monacantha: see vulgaris

rufida (Texas, No. Mexico), "Cinnamon-cactus"; bushy plant to 5 ft., eventually forming trunk; fleshy pads velvety grayish-green covered with tufts of short brown bristles (glochids), which rub off easily and cause itching under the skin; 2 in. flowers yellow. Will rot if too wet. **C**: 13   p. 167

schickendantzii (No. Argentina), "Lion's tongue"; shrub-like to 3 ft. high, and much branched, the elongate flattened joints warted, rather thin and narrow, dull green with reddish spines; flowers yellow. **C**: 13   p. 164

strobiliformis (Tephrocactus) (Argentina), "Spruce cones"; curious erect, branching succulent with blue-gray oval or conical joints having prominent tubercles appearing on spiraling folds and resembling a pine cone; spines usually absent, flowers yellow. **C**: 13   p. 164

subulata (Chile, Argentina), "Eve's-pin cactus"; tree-like, robust branching shrub to 12 ft. high; smooth cylindrical stems, bright green, with persistent, fleshy, awl-shaped leaves to 4 in. long, but few spines; flowers orange or greenish white. **C**: 13   p. 167

tomentosa (velutina) (Mexico), "Velvet opuntia"; tree-like, to 12 ft., very thick-fleshy long joints velvet green, with few pale yellow; flowers yellow; velvety carmine-red fruit along apex. Subject to rot if kept too wet. **C**: 13 p. 164

turpinii (glomerata) (Argentina), "Paperspine cactus"; low spreading clumps hardly more than 4 in. high, joints globular or ovoid, 1½ in. thick, smooth, dull grayish brown; the curious long spines flat, membranous and papery; flowers cream-colored. **C**: 13   p. 167

verschaffeltii (Northern Bolivia); clustering plant with cylindrical stems to 6 in. long, branching with little fingers of tubercled joints, light dull-green, the knobs supporting small fleshy, deciduous leaves, the areoles slightly woolly and with occasional straw-colored needle spines; deep red or orange flowers; very beautiful. **C**: 13   p. 167

vestita (Bolivia: La Paz), "Cotton-pole"; slightly branching erect or sprawling cylindrical stem to 1 in. thick, pale green and covered with long white hairs; small deciduous leaves persist for some time; deep red flowers. Fragile. **C**: 13   p. 167

vilis (Mexico: Zacatecas), "Little tree-cactus"; prostrate and spreading little plant with short cylindrical, club-shaped, pale green joints to 2 in. long and covered with whitish sharp spines; brilliant blood-red flowers. **C**: 1   p. 167

vulgaris (monacantha in horticulture) (So. Brazil to Argentina), "Irish mittens"; almost tree-like to 7 ft. high; flattened fleshy, glossy green joints nearly spineless; flowers deep yellow, and pear-shaped red fruit. The un-ripened fruit will root if planted, and grow, forming little ears, and offered in the trade as "Eared buds". One of the most attractive, best "pad" opuntias for the home. **C**: 1 p. 164

vulgaris variegata, better known as monacantha variegata, "Joseph's coat"; a pretty, variegated form having its long smooth joints beautifully marbled green, yellow and creamy-white, and sometimes also with pink. **C**:13 p. 164

OPUNTIA: see also Consolea, Nopalea

ORCHIDANTHA (*Musaceae*)
maxillarioides (Malaya), "Orchid flower"; small tropical perennial herb with tufted foliage of the feel of Aspidistra, and with flowers resembling the orchid Maxillaria; lanceolate leathery leaves 8 in. long on wiry petiole, light moss-green with some mottling; depressed midrib.; flowers with violet calyx and 1 in. green lip variegated purple. **C**: 10   p. 208

ORCHIS (*Orchidaceae*)
spectabilis (New Brunswick to Alabama, and west to Iowa), the "Showy woodland orchid"; a hardy terrestrial with tuberous roots, sprouting two obovate, shining leaves 4 to 8 in. long; a 4-angled spike to 1 ft. high, bearing variously from 1 to 10 flowers 1 in. long, sepals and petals purple and united in a hood, the ovate lip white; blooming according to climate between April and July. Growing best in compost with humus, peat, loam and sand, Orchis need plentiful moisture except for a period of rest after bloom. The tubers should be wintered in a cold frame or otherwise very cool. Propagated by division. **C**: 84   p. 476

OREOCEREUS (*Cactaceae*)
celsianus (Andes of Bolivia, Peru, Chile), the attractive "Old man of the Andes"; growing in clumps, creeping with young growth upright to 3 ft., areoles with long white hairs and long thin red spines; flowers dark red, day-flowering. **C**: 13   p. 154

OREODOXA: see Roystonea

OREOPANAX (*Araliaceae*)
peltatus (salvinii) (Mexico), "Mountain aralia"; evergreen tree with thin-leathery, palmately lobed leaves, the lobes fresh green with pale veins and toothed or lobed again; stalks with loose brown hairs. **C**: 2   p. 290

xalapensis (Mexico, C. America); evergreen shrub with palmately compound, thin-leathery leaves, the segments obovate, light green and corrugated, faintly serrate, on brown petioles. **C**: 2   p. 290

ORIGANUM (*Labiatae*)
dictamnus (Amaracus) (Crete), "Dittany of Crete" or "Hop marjoram"; procumbent semi-shrub to 1 ft. high, with ascending branches, small, thick, gray-woolly round-ovate leaves ¾ in. long; rosy blooms pendant with large pink bracts in hop-like heads. By division. Use: aromatic and tonic; leaves for seasoning, flowers for tea; medicinal for toothache. Interesting white-woolly house plant as miniature hops. **C**: 13   p. 320

ORNITHOCEPHALUS (*Orchidaceae*)
dolabratus (Ecuador), aptly named the "Mealybug orchid"; a small epiphyte, with pseudobulbs usually growing more or less downward; the fleshy, sickle-shaped leaves

arranged in an open fan; comparatively large, ¼ to ½ in. fragrant, clear white flowers, with pointed lip, along an ascending raceme rising from the axils of the leaves; blooming in autumn. A charming botanical curiosity with its diminutive flowers looking like an invasion of mealybugs. **C**: 24   p. 476

ORNITHOCHILUS (*Orchidaceae*)
fuscus (Nepal Himalayas, Burma to Vietnam, Hongkong); an interesting small epiphyte resembling Phalaenopsis in habit, with short stem and broad, fleshy leaves 3 to 9 in. long; arching or pendulous sprays of tiny ½ in. flowers, with sepals and linear petals greenish yellow streaked with red, the lobed lip fringed, yellowish with red midlobe, and yellow spur at the base; blooming spring and summer. **C**: 60 p. 476

ORNITHOGALUM (*Liliaceae*)
caudatum (So. Africa), the "False sea-onion" or "Healing onion"; an old-fashioned non-hardy window-sill plant with a big green bulb over 4 in. thick, usually showing above the soil; 5 to 6 basal strap leaves and a stalked raceme 1½–3 ft. long with 50–100 small white flowers 1 in. across, with petals having a green median stripe. The filament is wide at base (narrow in Urginea); the narrow channeled leaves 1½ in. wide, and form a tube at base. Keep dry after blooming for a rest. Crushed leaves of the "Healing onion" or "Meerzwiebel" are used by country people in Germany tied over cuts and bruises as a cauterizer, and are used cooked into syrup with rock-candy, against colds. Propagated by offsets. **C**: 13   p. 400, 529

thyrsoides (So. Africa: Cape), the well-known "Chincherinchee" or "African wonder flower"; from a tender bulb with fleshy, lanceolate basal leaves 1 ft. long, rise 2 or 3 strong floral stems 16 to 30 in. high carrying dense spikes of lovely white, cup-shaped flowers to 3 cm dia. with spreading segments and long yellow stamens; lasting up to 6 weeks. In South Africa they bloom from September to nearly Christmas, and are shipped to northern markets in our winter. Imported bulbs planted in our northern latitudes bloom in June or July, but they may be planted in pots between September and February to bloom, with gentle forcing from Christmas on. If properly rested after blooming, bulbs flower year after year. **C**: 13   p. 400

umbellatum (Mediterranean reg.), "Star of Bethlehem", also known as "Summer snowflake" or "Sleepy Dick"; small spring-blooming bulbous plant 6 to 10 in. high, with grass-like leaves, and 6 in. stems bearing large clusters of numerous star-like flowers 1 in. across, satiny-white inside, green and margined with white outside; opening only when the sun is shining just before mid-day and closing in the afternoon. Winter-hardy and forming clumps, naturalized in No. Europe and No. U.S., blooming in April-May. This lovely little plant may be grown in pots in a cool, light place; needs rest after foliage turns yellow. By division or from offsets. **C**: 76   p. 400

ORTHOPHYTUM (*Bromeliaceae*)
navioides (Brazil: Bahia); dense rosette with narrow glossy-green, flexible 1 ft. leaves tinged copper, and edged with small sharp spines; glossy beneath; leaves flattened back at flowering time to lay bare the inflorescence sunk in its center when foliage turns brilliant red; flowers white in tight cluster of petals. **C**: 9   p. 138

ORYZA (*Gramineae*)
clandestina, "Wild rice". I photographed this plant in the aquatic greenhouse of Nymphenburg Botanic Gardens in Munich where it is listed as from the North Temperate zone, growing in shallow water. However, Oryza are primarily tropical from Asia or Africa, as is the true rice. Oryza sativa; the temperate climate wild rice is Zizania aquatica which provided grain for American Indians. The handsome grass shown is perennial and clustering with slender canes about 3 ft. high, with narrow, matte rich green leaves, very ornamental at the edge of ponds. Its valid name may be Leersia oryzoides, the "Cut grass". **C**: 76   p. 306

OSCULARIA (*Aizoaceae*)
deltoides (in horticulture as Mesembryanthemum deltoides) (So. Africa: Cape Prov.), "Pink fig marigold"; pretty, somewhat shrubby succulent with erect or spreading branches, the short triangular-keeled and toothed ½ in. leaves in opposite pairs are a pretty blue-glaucous; the

fragrant ½ to 1 in. rosy purple flowers are small but plentiful, very charming on the window-sill in pots; also used in dish-gardens; easy to grow. From seed or cuttings. **C**: 13   p. 356

OSMANTHUS (Oleaceae)

fragrans (Olea) (Himalayas, China, So. Japan), "Sweet olive" or "Fragrant olive"; slow-growing evergreen shrub to 10 ft. or more high, with wiry, rather woody twigs and holly-shaped, stiff-leathery, olive-green glossy leaves 2 to 4 in. long, finely toothed or sometimes smooth at margins; the small white flowers in clusters, strongly and deliciously fragrant. An old house plant, preferably cool and moist, loved for its sweet jasmine perfume. From cuttings, treated with rooting hormones. **C**: 14   p. 296, 486

heterophyllus (ilicifolius) (Japan), "Holly osmanthus"; evergreen shrub of roundish habit, 15–20 ft. high, dense with leathery, holly-like leaves 1½ to 2½ in. long, glossy green deeply indented at the margins and with spiny teeth; fragrant white axillary flowers from fall to spring. A tough decorator plant favoring cool location. Cuttings root with the aid of hormones. **C**: 14   p. 297

heterophyllus 'Variegatus', in horticulture as ilicifolius variegatus (Japan), "Variegated false holly" or "Holly-leaf osmanthus"; extremely attractive, slow-growing, dense evergreen shrub resembling variegated holly but a better keeper indoors, the wiry stems with spiny, glossy-leathery leaves somewhat smaller, 1½–2½ in. long, fresh green to bluish gray-green, edged and variegated creamy-white and with pronounced teeth; tinted pink when young. A striking plant preferably for a cooler location. Grafted on privet for best growth, but cuttings can be used with the aid of rooting hormones. **C**: 15   p. 43*, 295

OSMUNDA (Filices; Osmundaceae)

cinnamomea (E. No. America, Mexico to Brazil, E. Asia), the "Cinnamon fern"; coarse but attractive, deep-rooted, rhizomatous fern with large crowns of 5 ft. pinnate fronds, rusty tomentose when young, the fertile fronds with reduced, narrower pinnae and becoming brown or cinnamon-colored as spores mature. Winter-hardy. The creeping rhizomes are densely covered with a felt of blackish-brown, wire-like and tough fibrous roots which spread widely around the plant. This root system is chopped out and used in cubes as a popular growing medium for orchids and other epiphytes. **C**: 84   p. 252, 539

OTHONNA (Compositae)

capensis (crassifolia) (So. Africa: Cape Prov.), "Ragwort vine"; succulent stem-rooting creeper with slender trailing branches to 3 ft. long, fleshy spindle-shaped, glossy green 1 in. leaves scattered or in clusters, hanging as if on strings; small daisy-like yellow flowers opening in sun. Excellent for hanging pots or baskets. From stem sections. **C**: 13 p. 351

OTOCHILUS (Orchidaceae)

fusca (Himalayas of Nepal, Burma); distinctive epiphyte with 4 to 6 in. pseudobulbs produced one above the other, not forming a rhizome, each bearing 2 narrow leaves; small ½ in. sweetly scented whitish flowers with brown column, the lip golden tinged with rose; in dusky, pendulous chains from the young growth. Best grown attached to treefern slabs or in hanging pots to allow its fascinating string of flowers to swing freely. Requires abundant moisture while growing. **C**: 72   p. 476

OUVIRANDRA: see Aponogeton

OXALIS (Oxalidaceae); low perennial herbs of the wood-sorrel family, some growing from bulbs, tubers or scaly bulbotubers, others have rhizomes, creeping or fibrous roots, a few are shrubby; the leaves palmately divided in the manner of clover into leaflets which close at night or in the dark. Practically all are good house plants in the sunny and airy window, or for baskets, having pretty foliage and often very showy, lovely flowers, which are opened only in sun-shine. Species with thickened roots tend to go into dormancy each year and should be somewhat rested after their blooming season.

acetosella (No. Europe; No. America: Québec to Ten-nessee), the "European wood-sorrel", often taken for the "Irish shamrock"; stemless perennial 3 in. high, from a slender scaly rhizome; the creeping roots producing long-stalked leaves of ¾ in. dia. with 3 obcordate, slightly hairy leaflets; ¾ in. flowers, their oval petals white with rosy veins; half the size of O. oregana. **C**: 26   p. 324

adenophylla (Chile, W. Argentina); hardy stemless perennial from the Cordilleras to 7000 ft., 4–6 in. high; from a roundish, bulb-like tuberous base rises a rosette of long-stalked leaves with 12–22 obcordate leaflets ½ in. long, glaucous grayish-green; 1 in. flowers lilac pink with deeper veins and orange throat, and with blackish-red spots at base of each of 5 petals. A pretty winter to springblooming house plant. When through flowering, tubers should be rested in a cool place. **C**: 13   p. 325

bowiei (purpurea) (So. Africa), "Giant pink clover"; stemless plant with thickened roots and scaly bulbs; the fleshy, long-stalked leaves with 3 large, obcordate, waxy, light green segments; strong flower stalks to 1 ft. long, topped by a cluster of giant, 1½ to 2 in., beautiful rosy-carmine flowers, during summer and fall. Propagated by bulblets. **C**: 13   p. 324

braziliensis (Brazil), one of the several "Shamrocks" sold by florists; bulbous plant 3–6 in. high, with substantial, rich shimmering green leaves of 3 bluntly obcordate, ciliate leaflets ½ in. long and wide, on leaf-bearing stem; beautiful large flowers 1 in. dia., bright magenta with darker crimson lines, in winter and spring. **C**: 13   p. 324

carnosa (Chile, Peru); persistent plant with horizontal rhizome and erect, thick stem up to 4 in., topped by small succulent, shiny grayish-green leaves with 3 obcordate ½ in. leaflets on long petioles, becoming pendant; the under-side glistening as if with dew; ¾ in. yellow flowers in spring. **C**: 13   p. 324

crassipes: see rubra
crassipes alba: see rubra alba

deppei (esculenta; Ionoxalis) (So. Mexico), "Lucky clover" or "Good-luck plant"; bulbous plant with big juicy, beet-like edible root; large leaves having 4 truncate (cut off straight across the apex) segments 1 in. long and crossed by purplish zone; summer flowers rosy-red with yellow base. Small bulblets form around the motherbulb, and which are used for propagation. The "Good-luck plant" is an attractive small pot plant for winter bloom. **C**: 13 p. 323, 536

flava (So. Africa); slow-growing bulbous plant, with long fleshy petioles and succulent leaves, digitate with 5–12 long narrow to oblong fingers; large trumpet-shaped flowers with a full, deep-yellow corolla, blooming in winter. Interesting with curled five-fingered foliage. **C**: 13   p. 325

hedysaroides rubra (Colombia, Venezuela, Ecuador), the "Firefern"; beautiful collection plant with erect, shrubby, wiry stem, and thin, fern-like foliage of glowing, satiny wine-red; each petiole with 3 stalked ovate leaflets which are sensitive to the touch; many little bright yellow flowers in attractive contrast to the showy leaves. Herbaceous shrub becoming 3 ft. tall, but usually only about 6 in. in pots. Propagated from cuttings. **C**: 13   p. 325

herrerae (Peru), "Succulent oxalis"; succulent, fibrous rooted species known in the trade as O. "Henrei", more diminutive than O. peduncularis; densely branched woody stem with swollen keeled, glaucous petioles topped by 3 tiny lightly notched fleshy light green leaflets; clusters of small yellow flowers on long stalks. Suitable for hanging pots. **C**: 13   p. 325

incarnata (Southwest Africa); fast-growing tufting perennial with masses of long ascending or procumbent thread-like stems forming at intervals rosettes of long thin-stalked leaves; the 3 heart-shaped leaflets light-green and deeply notched, forming small bulbs at the base of the rosettes; free-blooming with whitish ½ in. flowers tinged lavender-pink, yellow in tube, in winter; dormant in our summer. **C**: 13   p. 324

martiana 'Aureo-reticulata' (corymbosa var.) (Trop. America); "Gold-net sour clover"; an attractive ornamental "Sour-clover", with scaly bulb producing numerous pros-trate and ascending stalks, bearing large herbaceous leaves to 3 in. across, the 3 leaflets 1–1½ in. long, obcordate, fresh to deep green, and beautifully veined and reticulated with yellow; long-stemmed flowers carmine-rose with red lines from white throat; like others, goes "to sleep" at night. One of the most attractive for their beautiful foliage design, in pots. **C**: 15   p. 323

melanosticta (So. Africa), attractive low, bushy, bulbous plant, with dense foliage on green petioles, the 3 hard, grayish green leaf-segments covered with soft gray hair glistening like silver in the sun; fall-blooming with yellow flowers. Going dormant for a short 2 months rest period in summer, but the foliage should be kept growing as long as possible to encourage buds to form for best bloom. **C**: 13   p. 324

ortgiesii (Andes of Peru), "Tree oxalis"; robust perennial becoming 1½ ft. high; from a horizontal rhizome rises the leafy succulent stem, dense with olive-green to maroon-red leaves of 3 segments to 2½ in. long, and fish-tailed at apex, maroon underneath; small yellow, axillary flowers with deeper veins. Blooming at all seasons, but open only in the sun. Handsome curiosity plant with its unusual foliage. **C**: 13   p. 325

peduncularis (Ecuador); erect plant, with pretty, fresh green, waxy leaves of 3 small segments, borne on long, swollen petioles more or less arranged in rosettes along the thick fleshy stem; dense clusters of showy, orange-yellow flowers on long stalks, in continuous bloom. From cuttings. **C**: 13   p. 325

pes-caprae (cernua) (So. Africa), "Bermuda buttercup"; perennial with thickened roots and deep scaly bulbs; many basal, long-stalked leaves of 3 obcordate leaflets hairy beneath; nodding bell-shaped bright yellow flowers 1 to 1½ in. across, in winter-spring; a weed in Florida, Bermuda and other milder districts. Handsome flowering plant in pots or baskets. Spreads rapidly by rhizomes and bulbs. **C**: 13
p. 324

purpurea (variabilis) (So. Africa: Cape Prov.), "Grand duchess oxalis"; known in the cultivated large-flowered form 'Grand Duchess'; a low, spreading perennial from a blackish bulb with succulent, fresh green foliage 2½ to 3 in. across, of three ciliate leaflets not notched, on rosy-red stalks pubescent with soft white hair; barely topped by large and showy, pretty flowers to 2 in. across, bright rose with yellow base; winter-blooming in the sun. Spreads by bulbs and rhizome-like roots. **C**: 13   p. 324

regnellii (So. America), in horticulture sometimes as "rubra alba"; shapely, free-blooming perennial up to 10 in. high, with tuberculate rhizome, and leaves in a basal rosette, grass green with 3 deltoid, thin, finely ciliate segments up to 1 in. long, the apex as if cut off almost straight across; slender, white flowers. **C**: 14   p. 325

rubra (crassipes or "rosea" in horticulture) (So. Brazil); old-fashioned, free-blooming perennial, forming cylindric, erect tuber with brownish scales, the leaves with long hairy petioles, and 3 thin, coppery green, obcordate, somewhat hairy leaflets notched at apex; clusters of rosy-red flowers with red veins, opening to the sun; summer-blooming. **C**: 13   p. 325

rubra alba, also known as O. crassipes alba; free-blooming tuberous, wide-spreading, white flowered variety, the clusters of long-stalked small white blossoms borne well above the green foliage. A very pretty bloomer for pots. **C**: 13   p. 325

vulcanicola (siliquosa) (C. America); spreading, succulent bush, mat-forming in time; with fleshy red, leafy stems, the very attractive foliage conspicuous with 3 reddish obovate or obcordate leaflets dull, deep brown-red and purple beneath; golden yellow ½ in. flowers with maroon-brown lines ray-like from the throat. A gorgeous spring-flowering species in an impressive contrast of colors. From cuttings. **C**: 13   p. 325

PACHYCEREUS (Cactaceae)

pringlei (Mexico: Sonora, Baja California), "Mexican giant"; one of the most massive cacti, stout branching tree to 35 ft. high and 3 ft. thick at base; the olive green columns with 10–16 prominent but rounded ribs, closely studded with large oval areoles with short white or grayish wool, especially toward apex, numerous ash-gray radials and 1–3 long central spines at first reddish, afterwards gray; bell-shaped flowers greenish-red outside, white inside. **C**: 1   p. 154

PACHYCEREUS: see also Lemaireocereus

PACHYPHYTUM (Crassulaceae)

brevifolium (Mexico). "Sticky moonstones"; branched succulent stems to 10 in. high, which are sticky between the younger foliage; scattered blunt, thick-fleshy obovate leaves 1–1½ in. long, blue and diffused with red, covered with waxy bloom; bell-shaped flowers dark red in stalked clusters. **C**: 13   p. 348

compactum (Mexico: Hidalgo), "Thick plant"; small, slow-growing compact plant with thick cylindrical ¾ to 1 in. leaves, somewhat angled and keeled at pointed apex, green suffused with brown, and silvery white glaucous; flowers reddish on 15 in. stalk. A lovely little "hard" succulent. **C**: 13
p. 329

'Cornelius', "Moonstones"; robust succulent rosette of large fleshy, heavy obovate leaves glaucous olive green often tinged with pink and with red tips; the flowers with red petals. Slow-growing and durable, one of the "hard" succulents. From leaves or cuttings. **C**: 13   p. 329

oviferum (Mexico: San Luis Potosi) "Pearly moonstones"; short-stemmed, beautiful low succulent with swollen, long egg-shaped 1½ in. leaves exquisitely glaucous in delicate silver-white in winter, almost translucent reminiscent of moonstones; during summer intensifying to pink and amethyst; remarkable bell-shaped flowers intense red but nearly hidden by the whitish sepals. Slow-growing but precious **C**:13   p. 348

PACHYPODIUM (Apocynaceae)

lamieri (Madagascar), "Club foot"; weird succulent a thick, spiny gray column 3 to 8 ft. high and scarcely branched, at the base spindle-shaped; the pinkish spines 1 in. long; toward the top the spirally arranged strap-like leathery 8 in. leaves, dark shining green with white midrib; small funnel-shaped white flowers. A member of the queer vegetation prevailing in southern Madagascar. Keep dry in summer as it grows in our winter. From seed. **C**: 1   p. 353

PACHYSANDRA (Buxaceae)

terminalis variegata, the variegated leaf form of the "Japanese spurge"; a low growing evergreen perennial herb from Japan, spreading by means of creeping rootstocks or runners (stolons); soft-fleshy stems 6 to 12 in. high with fleshy obovate leaves 1–3 in. long, grouped whorl-like, coarsely toothed toward apex, glossy dark green edged and variegated ivory-white; small fluffy spikes of fragrant greenish-white flowers. Admirable winter-hardy ground cover in shade or sun; also for dishgarden plantings. Propagate by division or summer cuttings. **C**: 76   p. 299

PACHYSTACHYS (Acanthaceae)

coccinea, also known in horticulture as Jacobinia or Justicia coccinea (Brazil to Trinidad), "Cardinal's guard"; herbaceous shrub that may become 3 to 6 ft. high, with elliptic, rugose leaves to 8 in. long; scarlet flowers in dense terminal heads, the calyx with 5 linear segments, 2 in. corolla two-lipped and reflexed; blooming in February. Easy to grow and willing bloomer but likes fresh air with moderate temperature after having made good growth. From cuttings. **C**: 54   p. 493

x PACHYVERIA (Crassulaceae)

clavata cristata (Pachyphytum bracteosum x Echeveria x rosea); many-clustered miniature rosettes on a monstrose stem, the short fleshy leaves are glaucous silvery blue and very crowded. Slow-growing, charming plant best for magic coloring on the sunny window-sill. **C**:13   p.532

'Curtis', (Echeveria curtisii in horticulture), "White cloud"; nice, compact rosette of bluish-green, obovate, boat-shaped leaves with light green markings, and covered with silvery bloom, the tips red. A "hard" very pretty succulent. **C**: 13   p. 338

'E. O. Orpet' (Pachyphytum bracteosum x Echeveria flammea), "Large jewel plant"; large-growing, 8 in. spreading rosette of long-obovate, spoon-shaped, thick-fleshy leaves 3 in. or more long, exquisitely blue-glaucous with amethyst tinge, especially toward apex; flowers flesh-colored on coiling stalks. Propagates from leaves. **C**: 13
p. 338

scheideckeri (Pachyphytum bracteosum x Echeveria secunda), "Powdered jewel plant"; short-stemmed, clustering rosette of fleshy, narrow spoon-shaped, 2½ to 3 in. leaves softly colored glaucous bluish-gray, flushed with pink, and with reddish tips; bright red flowers tipped with yellow. Strikingly beautiful; freely forming offsets. **C**: 13   p. 338

PAEONIA (Ranunculaceae)

albiflora festiva (Siberia, China, Japan), "Double-flowered peony" or "Chinese peony"; outstandingly beautiful hardy perennial herb to 3 ft. high, with spindle-shaped dahlia-like tuberous roots, large ornamental, herbaceous, compound leaves, and magnificent rose-type fragrant flowers to 6 in. across, with double incurved white petals marked crimson in center; blooming normally in our gardens in late May, or early June. Peonies may be grown in large containers, transplanting them in September carefully and not too deep with a ball of roots. They bloom well only if exposed to a winter chilling period, and if so, may even be gently forced at 50–60° F. beginning late December to bloom in February or March. Propagation by division. **C**: 76   p. 402

PALISOTA (*Commelinaceae*)
barteri (Trop. Africa: Isle of Fernando Po); dark green nearly stemless rosette with spreading, stalked oblanceolate glossy leaves 1 to 2 ft. long, having parallel veins and edged with hairs; flowers purplish, in a dense oblong cluster near the base of the plant, followed by masses of striking lustrous, fiery orange-scarlet, pointed, fleshy berries, in summer or fall. Most interesting as a house plant. Propagated by division or seed. **C**: 4 or 54   p. 262, 514
elizabethae (pynaertii var.) (W. Equatorial Africa); ornamental leafy rosette becoming large, the lance-shaped, stalked foliage 1–3 ft. long and spreading; rich green surface with feathered pale yellow center band, and reddish hairs along margins; extended dense panicle of white flowers. Space-hungry but pretty foliage plant. **C**: 54   p. 262

PAMIANTHE (*Amaryllidaceae*)
peruviana (No. Peru), the "Peruvian daffodil"; a beautiful and distinctive tropical plant resembling Hymenocallis; with stoloniferous bulb having a false stem made up of the sheathing bases of its evergreen, 1 ft. linear keeled leaves; the angled floral stalk bearing several sweet-scented, showy flowers having a green tube, spreading into broad white outer segments with a yellow-green stripe and 4 in. long, the central staminal cap or corona 3 in. long, white with green median lines and bell-shaped, blooming in February-March. Warm house plant requiring humid atmosphere. Propagated by division of bulbs or from seed. **C**: 54   p. 391

PANAX: see Polyscias

PANDANUS (*Pandanaceae*); the "Screw pines" are tropical plants growing into branching trees related to palms and resembling Dracaenas, very decorative with their magnificent crowns of leathery, sword-shaped leaves arranged along the woody stems in spirals; rooting freely near the base and with age developing stilt-roots. The cultivated species require ample space but are very good house plants, especially the white-banded P. veitchii; this however, requires copious watering of the soil, if in containers, to maintain its variegation. Propagated by suckers sprouting from the base.
baptistii (New Britain Isl. near New Guinea), "Blue screw-pine"; symmetrical tropical plant with stiff channeled leaves spirally arranged, gracefully arching and tapering to a long point, blue-green with several yellow center stripes, smooth margins without thorns. From suckers. **C**: 3   p. 202
utilis (Madagascar), a "Screw-pine" of more economic than decorative value; spiral tropical rosette of long curving, strap-like, thick-leathery leaves to 3 in. wide, keeled beneath, deep olive-green but thorny with showy red spines at the margins; with age becoming a branching tree to 60 ft. high, supported by stilt-like brace roots; the leaves are used for making hats and baskets in their habitat. To get suckers for propagation, the center of the old plant may have to be removed. **C**: 53   p. 202
veitchii (Polynesia), "Variegated screw pine"; shapely and very satisfactory house plant; a bold rosette of thin-leathery, recurving leaves to 3 in. wide and narrowing to a long point, shining light to deep green, variously lined and broadly margined with creamy-white bands; the edges and keel beneath with small spines; with age developing stilt-like, thick aerial roots. When growing in a warm living room, frequent thorough soaking of the soil is necessary to maintain its variegation as otherwise the young foliage will come all-green. Propagated from suckers. **C**: 3   p. 191, 202

PANICUM: see Oplismenus

PAPHIOPEDILUM (*Orchidaceae*); the "Lady-slippers", commonly and erroneously, called "Cypripedium" in horticulture, are very popular tropical Asiatic terrestrial orchids without pseudobulbs; strap-shaped leathery basal leaves often attractively mottled and forming tufts; the curiously shaped pouch flowers with their flaring petals and showy dorsal sepal are very handsome and long lasting.

Their culture is easy, grown mostly in chopped fern-fiber and sphagnum, even with loam added, in semi-shade, somewhere between 600 and 2000 fc., and kept constantly damp and at high humidity. Definitely best for growing in the warmer atmosphere of the home are the mottled-leaved types such as P. maudiae with blooms that last 3 to 4 months. Propagated by division.
callosum splendens (Thailand, Cambodia, So. Vietnam); vigorous terrestrial orchid with bluish-green leaves marbled light and dark; solitary or pairs of large and elegant, 4 in. flowers carried on a 12 in. stalk, the beautiful white dorsal striped with green and crimson, the eyelashed petals greenish, and pouch brown-purple mahogany; blooming from February into summer, the attractive flowers lasting a long time. **C**: 12   p. 477
fairrieanum (Himalayas, Bhutan, Assam); one of the prettiest of "Cypripedium", in form as well as color; dwarf terrestrial with pale green leaves; the uniquely graceful flowers of medium size, 2½ to 3 in. across, the erect dorsal white, greenish at base and with purple lines; eyelashed, wavy petals sickle-shaped and coyly curved upward, whitish and streaked with purple, pouch green flushed red and veined with purple; the bloom begins in summer and may continue to January, the flowers lasting for six weeks. This distinctive species responds best if not too warm, and kept at 55 to 62° F., with steady high humidity and fresh air. **C**: 24   p. 455*, 477
godefroyae (Burma, Thailand, Vietnam); a charming small terrestrial, in habitat growing on rocks; of low habit, with dark green leaves checkered with gray above, and reddish beneath; the rounded, waxy and substantial 2½ to 3 in. flowers satiny white spotted with crimson, carried one or two on a short and hairy, purplish stalk, blooming during spring and summer time. This "Cypripedium" requires warmth and the benevolent surroundings of a humid greenhouse or terrarium; the fleshy, chaste flowers are long-lasting. **C**: 62   p. 477
x harrisianum (villosum x barbatum), a deep red "Lady slipper"; excellent dark hybrid, with leaves mottled light green on dark, and large 2½–3 in. flowers lacquered deep red marked with green, dorsal with purplish red veining, into the white apex; stem and petals hairy, blooming spring or summer. Many good forms have been selected and named from this cross, such as the well-known P. 'C.S. Ball' pictured, with waxy blooms 3 to 4 in. across, a very deep vibrant wine red almost sinister in appearance. A good house plant. **C**: 24   p. 477
insigne (Nepal Himalayas, Assam), terrestrial "Lady slipper", growing on rocks in the Himalayas at 6000 ft.; introduced and cultivated since 1819, best known as Cypripedium insigne; small plant with plain green leaves, soon forming tufts, the solitary flowers glossy-waxy, 3 to 4 in. across, the showy dorsal yellow-green with purple spots at base, white at apex, petals yellowish-green veined with brown, pouch reddish-brown; blooming dependably from October through winter to spring. This is the easiest to grow, with flowers lasting 6 weeks in perfection; with a minimum of care; no wonder that this has become the most popular for house growing, but it needs a spot on the cool side, steady moisture, and fresh air. This variable species has many named varieties. **C**: 24   p. 477
lowii (North Borneo); a most beautiful epiphyte, in habitat in tropical forest on tall trees or on limestone rocks; with strap-like, light green leaves and arching stalks to 3 ft. long carrying 2 to 6 hairy, exquisite flowers 4 to 6 in. across, the dorsal flag greenish with purple stripes; the long and slender, obovate petals greenish yellow blotched brown and tipped with purple, to 3 in. long; pouch coppery green with purple lines. This graceful and distinctive orchid blooms from February to July, its dainty flowers keeping in perfection for 2 or 3 months if kept cool after it appears. **C**: 62   p. 477
x maudiae (callosum x lawrenceanum); one of the most beautiful and certainly very popular, highly desirable hybrid "Cypripedium", its leaves marbled yellow-green and bluish-gray; the exquisite, usually solitary flowers about 4 in. long, and borne on slender stalks, its white dorsal striped with green; slender petals greenish white and closely lined with green, the waxy pouch yellowish-green, and blooming from March into August. An excellent house plant, even under air-conditioning, and its charming flowers remain in beauty for 3 to 4 months. **C**: 12   p. 477

rothschildianum (Papua, Borneo, Sumatra); remarkable terrestrial, strong growing, with leathery glossy green leaves, and carrying on a reddish hairy stalk to 2 ft. tall from 2 to 5 striking flowers to 5 in. across, having long almost tail-like pointed petals yellowish green and spotted red-purple, the flag and lateral sepal yellowish boldly striped with dark purple, a small pouch red-brown with yellow at the opening. This beauty normally blooms in summer, but may vary or not bloom at all one year. **C**: 62   p. 477

venustum (Nepal Himalayas); a handsome, compact, rather dwarf terrestrial with bluish-green leaves mottled light green; solitary flowers 3 in. across, with pointed dorsal white and striped with green; the whiskered petals greenish with darker lines and occasional purple tips and prominent warts; pouch yellow-green tinged rose or bronze and veined with green, blooming from November to March. **C**: 24 p. 477

PARODIA (*Cactaceae*); the charming "Tom Thumb cactus" are amongst the most beautiful of cacti; they are small flattened globes with colorful arching spines, and are everyone's favorite because of their willingness to bloom on the windowsill when quite young; the pretty flowers appearing on seedlings only 2 to 3 years old, on display from March to August and lasting a week or more. They grow best in porous, gritty soil with humus added, during summer moderately moist, in good sunlight; in winter cooler at 45 to 55° F. and dry. From seed.

aureispina (N. Argentina), "Golden Tom Thumb"; tiny bluish-green globe of $2\frac{1}{2}$ in. dia. with spiraled ribs, and covered with bright yellow spines. After being 2–3 years old this little beauty will easily bloom, with pretty golden flowers, but like with most globular cacti a period of cool nights will help to initiate buds. **C**: 13   p. 160

maassii (Bolivia), "Tom Thumb cactus"; small globe to 6 in. dia., yellow green, with 13 prominent ribs in spirals, long white radials and hooked central spines curving outward and set on knobs; flowers orange-red. **C**: 13   p. 160

sanguiniflora violacea (Argentina: Salta), "Crimson Tom Thumb"; solitary little soft green globes to 2 in. dia., woolly on top, the spiraled tubercles set with bristly white radials and brownish central spines, one of them hooked; numerous flowers a silky reddish-violet (blood-red in species). **C**: 13   p. 160

PARTHENOCISSUS (*Vitaceae*)

henryana (Ampelopsis) (China); sometimes erroneously known as "Cissus gongylodes". Ornamental, vigorous, tall climber with young branchlets 4-angled, the divided tendril clinging by adhesive tips; grape-like leaves of 5 slender, oblanceolate, thin-fleshy leaflets, toothed above middle, dull olive-green with broad silver band following midrib and feathering into lateral veins, reddish beneath. Atrractive in the home as a vine. **C**: 4   p. 258

quinquefolia vitacea (P. inserta), (E. Canada to Texas); a "Virginia creeper" known to California greenhouses as "Cissus sicyoides", and grown there on treefern poles; an attractive decorative plant, with lustrous deep green, palmately compound leaves, the 5 leaflets lanceolate and deeply toothed, vines climbing by tendrils without disks; fruit bluish-black. **C**: 66   p. 258

x PASCHIA (*Gesneriaceae*)

hybrida; alleged to be a bigeneric hybrid of Sinningia speciosa x Kohleria eriantha; known in the trade also as "Gloxiloma" or "Sinningiloma", originated by Federico Pasch in Guatemala in 1953; both supposed parents have 13 chromosomes, but this parentage is doubted by P. Arnold and others who suggest that this may possibly be a hybrid between S. speciosa and regina, and re-crossed. Resembling gloxinia but with oblique flowers facing sideways (zygomorphic), the large frilled corolla a glowing purplish red, in showy clusters; foliage silvery-hued typical of Kohleria eriantha. Grown like gloxinias, and propagated from seed or leaf cuttings. **C**: 10   p. 448

PASSIFLORA (*Passifloraceae*); the "Passion flowers" are tendril-climbing mostly tropical and subtropical vines with alternating leaves grown for their intriguing and usually showy flowers, a few also for their edible fruit. Their often strikingly colored flowers are regular and star-like, in the center of which is a crown composed of many free filaments, and a column bearing stamens and styles. P. caerulea is an example of the religious symbolism attached to its flowers.

According to size, for the sunny window on wire support, or the winter garden with fresh air, or on high wires under glass, with plenty of water. Plants in pots should not be moved while forming buds as otherwise they may not bloom. Propagated from cuttings. The short-lived, soon-closing flowers can be kept open by floating them in water.

x alato-caerulea (pfordtii) (alata x caerulea), "Showy passion flower"; free-blooming hybrid well-known because of its large and showy, fragrant, axillary 4 in. flowers larger than caerulea, with sepals white, petals pink, and a fringed crown purple, white and blue; with triloded leaves. A good house plant for the window, blooming when still young, all summer. **C**: 2   p. 281

caerulea (Brazil), a "Passion flower"; with showy $3\frac{1}{2}$ in. flowers; a religious symbol to early missionaries who saw in its ten greenish-white petal-like parts the ten apostles at the crucifixion of the Lord, in the blue, white and purple rays of the corona the crown of thorns, in the 5 anthers the wounds, the 3 stigmas the nails; cords and whips in the coiling tendrils of the vine, and the 5-lobed leaves the cruel hands of the persecutors. A willing bloomer even as a younger plant in pots. **C**: 52   p. 281

coccinea (Trop. So. America), a "Scarlet passion flower"; climbing with grooved purplish downy stems; leaves 3–6 in. long, ovate and coarsely crenate; free-blooming with flowers of medium size, the petals glowing scarlet, red sepals yellowish outside; the crown filaments deep purple and with pink to white base; pulpy edible fruit. **C**: 52   p. 281

coriacea (So. Mexicŏ to Peru), "Bat-leaf"; vigorous climber with red stems, interesting mainly because of its hard to being brittle, transversely oblong-peltate leaves, more broad than long like a butterfly, blue-green blotched with silver-gray; small 1 in. flowers in clusters, with pale green petals, a yellow ray-crown, and purplish-chocolate base. **C**: 54   p. 281

edulis (Brazil), the "Purple granadilla"; sturdy climber with angular stems, large 4–6 in. leaves deeply lobed, with wavy edges; $2\frac{1}{2}$ in. white flowers with corona white banded with purple, mostly summer-blooming; the 3 in. aromatic, edible fruit thickly purple-dotted and quite ornamental; widely grown in warm climates for its delicious flavor in beverages, fruit salads, sherbets, also jam and marmalade; fruit is produced in spring and fall. **C**: 52   p. 507

incarnata (Virginia to Florida and Texas), "Wild passion flower"; somewhat hardy, tall climber with 3-lobed leaves and axillary 2 in. flowers, the sepals pale lavender with a horn-like point, petals white, corona with purplish filaments; yellow edible, egg-shaped 2 in. fruit. Fairly winter-resistant. **C**: 64   p. 281

maculifolia (maculosa, organensis), (Venezuela), "Blotched-leaf passion vine"; tropical climber with attractive foliage, on lightly hairy, wiry stem, the curious, nearly triangular leaves with three brief lobes at the cut-off apex; quilted surface green, irregularly spotted and marbled with yellow; the underside purple and glandular-hairy; small creamy-white flowers. **C**: 54   p. 281

quadrangularis (Trop. America), "Granadilla"; robust tropical climber with winged stems and oval leaves, much grown in the tropics for its edible fruit, which is oblong, 5–10 in. long, yellowish-green and pulpy; 3–5 in. fragrant flowers with oval, white sepals, reddish petals, and crown with 5 rows of white and purple rays. Free-blooming, and setting fruit easily with artificial pollination. **C**: 52   p. 281

racemosa (princeps) (Brazil), "Red passion flower"; a red passion-vine, climbing by tendrils, with deeply 3-lobed or occasionally ovate leaves; 4 in. flowers in pendulous racemes, the narrow petals rosy-crimson and spreading, the fringed crown with outer rays purple tipped white, the short inner rays red. Good for pots, blooming in young stage. **C**: 52   p. 281

trifasciata (Venezuela, Brazil, Peru), "Three-banded passion vine"; ornamental climber grown for its colored foliage, with wiry stems and broad, beautifully colored, lightly 3-lobed leaves, satiny olive to deep bronze-green, with 3 broad, pink to silvery-green zones along the purple veins, and purple beneath; small, fragrant yellowish flowers. **C**: 54   p. 281

PAVONIA (*Malvaceae*)

intermedia rosea, hort.; probably only a form of P. multiflora from Brazil; tropical evergreen shrub inclined to grow single stems; lanceolate leaves 6 to 8 in. long, barely toothed and rough beneath; terminal clusters of axillary rosy flowers with prominent long stamens and bluish anthers; the species multiflora has leaves with pronounced teeth. An attractive winter-blooming plant best to flower without pinching and not too warm. Propagated by tip cuttings. **C**: 54   p. 495

PEDILANTHUS (*Euphorbiaceae*)
tithymaloides variegatus (W. Indies), "Ribbon cactus" or "Red bird", a variegated-leaved "Devil's backbone"; branching succulent bush to several feet high with poisonous milky juice, the gray-green, fleshy stems bent zigzag with each alternate waxy, pale green, ovate 2 to 3 in. leaf, beautifully variegated and edged white, tinged with carmine-red; the bracted, vivid red inflorescence is spurred, resembling a bird's head. Partially deciduous when too dry. Easy from cuttings. **C**: 54   p. 350

PELARGONIUM (*Geraniaceae*); subtropical herbs or shrubs of the Geranium family from semi-arid areas of South Africa, some strong growing erect, others trailing, even succulent; many different leaf forms with some very beautiful, some mimicking other plants and often aromatic. Irregular flowers usually in showy clusters, the calyx with a distinctive nectar spear; the seed is borne in pods resembling a storks-bill. Pelargoniums are roughly divided as follows: the "Lady Washington" or "Show geraniums" (P. x domesticum hybrids) with large, fancy flowers to over 3 in. across, spring blooming, for cool surroundings only; the "Zonal" or "Bedding geraniums" (P. x hortorum hybrids), the most popular group and beloved old window plants, where in good sunlight they will flower all year; the "Ivy geraniums" (peltatum), for baskets and window boxes; old-fashioned "Scented-leaved geraniums" with aromatic foliage in diverse patterns; and various "Odd" and "Succulent geraniums" intriguing to the collector, all of easy care as endearing house plants. Propagated from cuttings and sometimes seed.
acetosum (So. Africa), "Sorrel cranesbill"; remarkable shrubby plant of climbing habit, with slender stems, and waxy, cupped fan-shaped, silver green leaves shallowly lobed at apex; the curious flowers, very striking with rosy-salmon to blush-pink petals narrow-linear and spider-like. **C**: 13   p. 430
x ardens (fulgidum x lobatum), "Glowing storksbill"; an odd geranium with tuberous roots, short knobby stems and few stubby branches; large hairy, lobed leaves that may become 8 in. long; flowers glowing garnet-red with lighter edge, and dark red spots on upper petals; blooming in winter and spring. Goes dormant in summer, shedding its foliage for several months. **C**: 13   p. 430
capitatum 'Attar of Roses', strongly rose-scented; sprawling, shrubby plant, shorter then the species from So. Africa, and with leaves deeper lobed, light green with slightly hairy surface; flowers orchid-pink with upper petals conspicuously veined purple. **C**: 13   p.429
crithmifolium (from arid regions of So. Africa and S.W. Africa), "Succulent geranium"; stout succulent stems 6 to 8 in. high, with a tendency to spread along the ground; lacy bipinnate, grayish green, fleshy leaves scurfy and with nothing but ribs skeleton-like, clustered at end of branches and deciduous; umbels of small white flowers with a red spot on each narrow petal. **C**: 13   p. 353
dasycaulon (So. Africa), "Succulent geranium"; curious, bushy, low succulent shrub, forming a globose plant with numerous swollen branches; the fleshy leaves deeply cut and divided; flowers with narrow petals, creamy with deep red spot in throat, in small clusters. Easy as a house plant, though plant welcomes a dormancy period when it would shed its foliage. **C**: 13   p. 430
denticulatum (So. Africa), "Pine geranium"; pine-scented; branching, shrubby plant with clammy stems and dark green 2 in. leaves cut down almost to the bare pinnate ribs, sticky on the surface, the lobes with teeth; small pinkish flowers marked and veined with crimson on the notched upper petals. **C**: 13   p. 429
denticulatum filicifolium, "Fernleaf geranium"; pungent-scented, reminiscent of pine; green leaf blades very finely cut and reduced to bare ribs, very lacy in appearance and delicate fern-like. **C**: 13   p. 429
x domesticum 'Carlton's Pansy'; a small-flowered vigorous pansy type, densely bushy and to about 1 ft. high, with blooms 1½ in. dia., slightly larger than 'Mad. Layal'; the

upper petals dark violet shading through rose to the white border, more white on lower petals and veined with rose; the leaves are small and only about 1 in. across. Very pert and attractive covered with its pansy-like fl. **C**: 14   p. 426
x domesticum 'Chicago Market', lavender "Martha Washington geranium"; bushy, low-growing plant, producing an abundance of large 3 to 4 in. rosy lavender or orchid colored blooms, with dark red blotch in upper petals, early in spring. **C**: 76   p. 426
x domesticum 'Earliana'; a dwarf "Pansy geranium"; bushy plant with 1 in. leaves, and small flowers mainly orchid-colored, the upper petals maroon shading through rose to the white margins. Similar to 'Mad. Layal' but maroon takes place of violet. **C**: 14   p. 426
x domesticum 'Easter Greeting', introduced 1906 from Germany as 'Ostergruss'; popular early and long-blooming variety; medium-large 2½ in. rosy-carmine flowers, each petal with an elongate black blotch. One of the earliest flowering varieties, and excellent as spring pot plant. **C**: 76   p. 427
x domesticum 'Edith North'; a good, free-blooming, vigorous variety, with trusses of 2 in. flowers well above the foliage; salmon pink, shaded deeper on upper petals and with blotches dark brown. An old favorite pot plant for spring-blooming. **C**: 76   p. 427
x domesticum 'Firedancer'; magnificent hybrid cultivar, with plain green leaves, setting off the brilliant crimson of the well-rounded flowers, a dark maroon blotch in center, and a narrow white edge all around. **C**: 64   p. 426
x domesticum 'Grand Slam'; excellent, free-blooming, premium variety of bushy habit, and with handsome clusters of large 2½ in., good-holding flowers flame-red, with wavy margins, the top petals with velvety red-brown eye. **C**: 64
                                                                       p. 427
x domesticum 'Jessie Jarrett'; flowers of an exceptional, almost sinister shade of dark violet rose to purple, with black maroon blotch on all petals. **C**: 76   p. 427
x domesticum 'Josephine'; compact, low-growing plant of constant, free-blooming habit, the white flowers with rosy-pink blotches. **C**: 64   p. 427
x domesticum 'MacKay No. 12'; strong growing variety with pansy-like flowers lavender rose with white center, and purplish crimson upper petals. **C**: 76   p. 427
x domesticum 'Mad. Layal', "Pansy geranium"; charming compact, bushy French, small-flowered hybrid 1 ft. high with little waxy, toothed 1 in. leaves, and cute pansy-like flowers only 1 in. across, the upper petals violet with white margin, lower petals white, tinted and veined rose. A lovely house plant for the cool window or sunporch. **C**: 14   p. 426
x domesticum 'Marie Vogel'; a very popular and distinctive "Lady Washington pelargonium"; free-blooming with large 2½ in., showy flowers of salmon-rose, crimson blotches and feathering on upper petals, the margins ruffled; toothed, glabrous leaves on hairy, brown-woody stems. **C**: 76   p. 427
x domesticum 'Mrs. Mary Bard'; compact grower and early bloomer, with medium-size exquisite flowers pure white, striped purple at base of petals. **C**: 76   p. 427
x domesticum 'South American Bronze'; low-growing, conspicuous plant with flowers an unusual color combination of bronzy-red, or brown and maroon, with dark center and pale orchid border. **C**: 64   p. 426
x domesticum 'Springtime'; a lovely "Lady Washington geranium", long-blooming and gay with white 2½ in. flowers prettily zoned bright rose, and a daintily ruffled white margin. **C**: 76   p. 379, 427
echinatum (So. Africa: Cape Prov.), "Cactus geranium" or "Sweetheart geranium"; interesting low semi-shrub 6 to 18 in. high, having a succulent stem armed with hooked, soft, thorn-shaped stubbles cactus-like; gray-green, lobed deciduous leaves 1 to 3 in. long, white-hairy beneath; single white flowers, marked maroon in the shape of a heart. Willing bloomer as a pot plant, but this species welcomes a resting period during which it sheds its leaves. **C**: 13 p. 430
'Formosa', known in horticulture as "Fingered flowers", introduced from Japan; of compact habit with thick, succulent-woody stems, and curious small 2 in. leaves shaped like those of the ginkgo, deeply 5-lobed, and crenate at apex; clusters of small 1 in. flowers with 5 to 10 or more linear petals almost daisy-like, in flaming salmon except for the pale-colored or white, notched and pointed tips. Slow-growing oddity but a willing bloomer; of apparent hybrid origin. **C**: 13   p. 428

x fragrans; said to be a hybrid of exstipulatum x odoratissimum; recorded in Berlin about 1793, but found wild in the Karroo of So. Africa; the "Nutmeg geranium"; of branching, bushy and spreading habit with small rounded, lobed and ruffled but variable, grayish leaves, soft hairy, on wiry petioles; small white flowers marked purple. **C:** 13
p. 429

x fragrans 'Variegatum'; "Variegated nutmeg geranium"; low, compact plant with spreading branches and small rounded green 1 in. leaves irregularly variegated with white; strongly scented of nutmeg. **C:** 15   p. 430

gibbosum (Southwest Africa), "Knotted geranium" or "Gouty pelargonium"; curious plant with shrubby, scandent stem much swollen at the distant joints, and looking as if it suffered from arthritis; glaucous gray-green, pinnately lobed, ruffled leaves to 3 in. long; small, evening-scented, greenish-yellow clusters of flowers, blooming in June. **C:** 13 p. 430

graveolens (So. Africa: Cape Prov., and introduced into England in 1774), the old fashioned "Rose geranium"; an old favorite house plant with sweetly rose-scented leaves which are used in cookery, sachet, and for making perfume; bushy plant becoming 3 ft. or more high, with deeply lobed leaves 1½ to 4 in. across, appearing gray-green because of their covering of soft white hair; short-spurred flowers lavender pink marked with purple lines and spotted on upper petals, in short-stalked·axillary clusters, blooming in summer. **C:** 13   p. 429

graveolens 'Lady Plymouth', "Variegated rose geranium"; rose-scented, vigorous plant with spreading branches, and rather small, regularly lobed and toothed, soft white-hairy leaves gray-green, with a narrow white line around the edges. **C:** 15   p. 430

graveolens 'Minor'; "Little-leaf rose"; similar to the species but smaller in all its parts, with rough-hairy, deeply lobed, small leaves; flowers deeper in color, the narrow petals lavender rose with crimson markings. **C:** 13 p. 429

graveolens 'Variegatum', "Mint-scented rose-geranium"; an attractive plant having the largest leaf of all variegated rose-geraniums; pale grayish-green and rough to the touch, deeply lobed and prettily edged in white; fragrant of roses with a strong overtone of spicy mint. Excellent house plant. Rose geraniums are grown commercially for the fragrant oil distilled from them and used in soap and perfumes. Leaves are used to give rose flavor to desserts and jellies; also for sachets. Fresh flower petals add delicate exotic flavor to various foods and drinks. **C:** 13
p. 320, 429

grossularioides hort. (So. Africa), "Gooseberry geranium"; coconut-scented; small rounded, glossy leaves to 1½ in. long with closely toothed, scalloped margin, resembling those of a gooseberry; straggling or trailing on slender, wiry stems; small rosy-red flowers less than ¼ in. long; fairly hardy. **C:** 13   p. 429

x hortorum 'Alpha', a French hybrid of peltatum and a bronze-leaf hortorum; colored-leaf geranium of small, bushy habit, with thin, hard stems; smallish, shiny yellow-green leaves with narrow rust-red zone; attractive single scarlet flowers in great profusion. **C:** 13   p. 422

x hortorum 'Antares'; a spectacular miniature geranium with dark foliage, and large, single flowers in dark burning scarlet. **C:** 13   p. 425

x hortorum 'Appleblossom', (Mme. Jaulin), French type of the 1890's; habit small and bushy, with large heads of cupped, semi-double appleblossom flowers, delicate soft pink flushed with peach, almost white at margins. Very popular; free bloomer. **C:** 13   p. 428

x hortorum 'Appleblossom Rosebud', a "Rosebud geranium", charming bushy plant with dense clusters of beautiful double flowers, white edged with pink, and resembling miniature roses. Other varieties in rose-pink, scarlet, wine-purple. **C:** 13   p. 420

x hortorum 'Beauté Poitevine', an old favorite French geranium (1869), still in cultivation because of the pleasing pastel clear salmon pink of its large semi-double flowers, paler toward center; the growth is low and compact, the leaves with a pretty, dark zone. Early blooming in pots, but floriferous stock plants must be reselected to keep them from going blind. **C:** 13   p. 420

x hortorum 'Black Vesuvius'; a charming old English variety of 1889; very dwarf plant with dark olive-green small leaves 1 in. wide, zoned blackish brown; and showy single flowers conspicuously large, bright orange-scarlet. One of the oldest and darkest-leaved miniatures, it grows very slowly and keeps its size for years and years, about 4 in. high in bloom in the Northeast, taller to 10 in. in the sunny Southwest; the stems gnarling with age. **C:** 13   p. 424

x hortorum 'Bronze Beauty'; a favorite, slow-growing plant with large round, golden-green leaves, with a narrow red-brown zone and scalloped margins; single light salmon pink flowers. The 'Bronze Beauty' as grown in the West, also known as 'Bronze Beauty No. 1'. **C:** 13   p. 428

x hortorum 'Brooks Barnes' (N.J. 1952); slow-growing dwarf geranium to 6 in. high, with small 1½ in. very dark purple-blackish, wrinkled leaves, and medium large single flowers a lovely soft salmon pink, deepening toward the center. **C:** 13   p.425

x hortorum 'Brownie'; bushy miniature variety to 4 in. high, with leaves to 1 in. wide, green with purple-black ring, the new foliage golden brown, also zoned with chocolate; large clusters of single scarlet bl. **C:** 13 p. 424

x hortorum 'Carefree Deep Salmon'; All-American selection as an outstanding color-form of a new F_1 hybrid or heterosis seed strain of single and double garden geraniums by Pan-American Seed Co.; grown directly from seed, and available also in shades of pink and red, also bicolors and white. Geraniums from seed are not as early as those propagated from cuttings, but begin to bloom in June-July and are at their best from August to October. Seed has to be started warm at a constant 70–75° F.; for early bloom in Mid-May sow beginning of January; grow with all the light or sun possible, keep warm to initiate buds, then intermediate or cool, although these geraniums tolerate high summer temperatures. **C:** 13   p. 420

x hortorum 'Distinction'; old English variety of small, spreading habit, with smooth, roundish leaves light green, with a clean-cut brown-black circle just inside the toothed and crinkled margin; small single, cherry-red fl. **C:** 13 p. 422

x hortorum 'Dr. Margaret Sturgis'; beautiful large-flowered single, mottled and striped rosy-pink in outer areas with large white center. Strikingly attractive. **C:** 13 p. 428

x hortorum 'Fairy Tales'; free-blooming miniature; the large slightly ruffled, single flowers white tinted lavender pink, deepest in center; the small foliage with dark zoning. **C:** 13   p. 424

x hortorum 'Happy Thought' (1877), "Butterfly geranium"; old English mutant of vigorous, medium-large habit, with attractive leaves bright green painted with a large cream-yellow butterfly in center, faintly zoned with splashes of brown; single 1½ in. flowers soft carmine-red, scarlet on upper petals, the 3 in. clusters carried tall on wiry stalks. A modest yet gaily colored old-fashioned house plant steadily in bloom on the sunny window sill. **C:** 13   p. 423

x hortorum 'Honeymoon'; bushy, slow growing, free-blooming California variety, with large single flowers apricot-salmon, fading to lighter at margins. **C:** 13 p. 428

x hortorum 'Irene'; popular commercial var. almost everblooming with heads of large semi-double flowers cherry-red. **C:** 13   p. 420

x hortorum 'Improved Ricard'; excellent commercial cultivar, freely branching, of shapely habit, foliage only lightly zoned; with good heads of large semi-double flowers light brick-red or orange-scarlet. Fine pot plant blooming from Mother's Day through summer. **C:** 13 p. 420

x hortorum 'Jaunty'; a striking, willingly branching miniature geranium, blooming when only 2 to 3 in. high; the double flowers strawberry or carmine-red in large rounded clusters neatly above the prettily zoned foliage. **C:** 13   p. 424

x hortorum 'Mme. Salleron'; originated in France about 1845, and a unique, old foliage plant which I remember my father using in fancy carpet bedding; a small plant sprouting multiple short branches from the base, with little papery, crenate leaves glistening gray-green, with white border, on distinctive, long thin petioles; I have never seen it flowering; charming for edging. **C:** 13   p. 422

x hortorum 'Maxime Kovalevski', also known in horticulture as 'Santa Barbara' or 'Diablo'; eye-catching, medium-size, vigorous plant free-blooming with big heads of large single flowers in an unusual hue of glowing, clear bright orange. **C:** 13   p. 428

x hortorum 'Meteor miniature'; small bushy plant to 6 in. tall, with dark, black-zoned, scalloped leaves; and fully double, blood red flowers. **C:** 13   p. 424

x hortorum 'Miss Burdett Coutts', "Tricolor geranium"; a most beautiful, compact old English tricolor variety (1860), with gaily colored, firm, rounded leaves grayish-green in center, the margins ivory-white, and the circular zoning deep blood-red splashed with rose-pink; single scarlet flowers. An excellent fancy-leaved, slow-growing miniature for small pots in the sunny window, differing from all other tricolors by the silvery hue covering the foliage. **C**: 1   p. 423

x hortorum **'Mountain of Snow'**; an old English variety of 1855; low, bushy, spreading plant with thin leaves shimmering silver-green and with broad white margins; small single flowers scarlet. **C**: 13   p. 423

x hortorum **'Mrs. Parker'**; an excellent old English, free-blooming variety of compact habit, with shimmering green, rather thin, quilted leaves beautifully bordered in white; small semi-double, soft pink flowers. **C**: 13   p. 423

x hortorum **'Mrs. Pollock double'**, "Double tricolor"; an attractive, compact geranium with tricolor leaf, medium green to milky in center, with dark red and orange zone splashed irregularly, the scalloped margin creamy-yellow; semi-double and double flowers vermilion-red. **C**: 13
p. 190, 423

x hortorum **'North Star'**; very pretty semi-dwarf geranium of strong, bushy habit, the single flowers white, with pink veins radiating from center. **C**: 13   p. 425

x hortorum **'Olivia Kuser'** (N.J. 1950); excellent, vigorous, free-blooming seedling with large single flowers pure white, the petals rounded and overlapping. **C**: 13 p. 428

x hortorum **'Olympic Red'** (Alphonse Ricard x Radio Red 1939), popular commercial "Zonal geranium" of the French type, early-blooming, free-flowering and of self-branching habit; beautifully zoned leaves and large heads of semi-double, fiery red flowers. With sunshine, buds and blooms will form all year when pot grown indoors, most abundantly from Easter through summer. **C**: 13   p. 379

x hortorum **'Pastel'** (1954); medium-small, bushy, fancy-leaved tricolor geranium with silvery green foliage irregularly bordered ivory, the leaf surface widely zoned with coral pink splashed purplish-brown; single flowers salmon-pink. **C**: 13   p. 423

x hortorum **'Pride'**; semi-dwarf; robust but slow-growing plant, with relatively quite large single salmon blooms, free-blooming; stronger and larger growing than the 8 in. high 'Pixie'. **C**: 13   p. 424

x hortorum **'Prince Valiant'**; showy semi-dwarf, with dark foliage; single, vivid purple-crimson flowers flushed orange; good grower and free-blooming. **C**: 13   p. 425

x hortorum **'Pygmy'**; distinctive miniature about 6 in. high, self-branching, with ¾ in. rather light green, zoned and scalloped leaves; the small, vivid red, flat, double flowers blooming steadily through all sunny weather. Interesting because tiny flowers are in scale with plant. **C**: 13   p. 425

x hortorum **'Rober's Lavender'**; semi-dwarf type, strong plant with dark green foliage, and large single fully-rounded blooms lavender-pink. **C**: 13   p. 424

x hortorum **'Robin Hood'**; semi-dwarf plant, with prettily zoned foliage, and willing bloomer with long-stemmed fiery-red, double flowers; good for outdoor use also. **C**: 13   p. 425

x hortorum **'Royal Fiat'**; a lovely improvement over 'Fiat' (Calif. 1919); large 1¾ in. semi-double flowers light clear pink with darker, rosy lines; remarkable for their sharply notched petals; small green foliage with darker zoning. This fringed beauty from California with its bushy habit and carnation-like flowers is a charming window plant, but has a tendency to revert back to the plain rounded petal edges of 'Fiat'. **C**: 13   p. 428

x hortorum **'Salmon Comet'**; free blooming miniature with tiny foliage, zoned with a pretty, dark ring; the narrow-petaled, single salmon flowers in a large cluster larger than the little plant. **C**: 13   p. 424

x hortorum **'Salmon Supreme'**; distinctive commercial variety of the French type, more vigorous than 'Poitevine' and a free bloomer but with the same fine, large semi-double flowers of clear light salmon pink. **C**: 13 p. 420

x hortorum **'S.A. Nutt'**; old-fashioned Ohio commercial cultivar (1886), of compact habit, the dark foliage with bronzy zones, and free-blooming with numerous medium-sized clusters of smallish double flowers in deep velvety crimson-red. **C**: 13   p. 420

x hortorum **'Sirius'**; miniature free-blooming variety, with starry single salmon-pink flowers, changing to white in center. **C**: 13   p. 424

x hortorum **'Skies of Italy'**, "Maple-leaf tricolor geranium"; lovely small bushy plant with slender branches

dense with autumn-colored maple-like leaves having pointed lobes, green with creamy marginal variegation, and a brown zone tinted with orange-red and crimson, carried on wire-thin petioles; single scarlet flowers. **C**: 13   p. 423

x hortorum **'Souvenir de Mirande'** (Harriet Ann); the first blended flower with a large white center (France 1886); charming bushy plant, in constant bloom with single flowers of the apple-blossom type, rosy-red changing to a conspicuously large white center, the upper petals white and merely edged with carmine. **C**: 13   p. 428

x hortorum **'Sparkle'**; semi-dwarf type; bushy plant with dark-zoned foliage; prolific with semi-double, cheerful bright red flowers. **C**: 13   p. 425

x hortorum **'Tangerine'**; a free-blooming miniature geranium, with small foliage, and large single flowers deep salmon pink with orange glow, marked maroon on upper petals. **C**: 13   p. 424

x hortorum **'Tempter'**; bushy, free-blooming semi-dwarf with dark foliage, and large single purplish-crimson flowers. **C**: 13   p. 425

x hortorum **'Tiny Tim Pink'** (Pink form); a very miniature geranium, blooming when only 1 in. high, but spreading and forming mounds with tiny foliage as small as ¼ in. wide, purplish-green prettily zoned with brown; the single flowers light pink, red in another color form. Can be grown for years in the same pot, and lends itself to training as bonsai. **C**: 13   p. 424

x hortorum **'Tu-tone'**; shapely dwarf variety, dense with leaves, and semi-double pale pink flowers with white streaks; free-blooming. **C**: 13   p. 425

x hortorum **'Variegated Kleiner Liebling'**; small attractive, slow-growing plant to 8 in. high, with smallish, crinkled, grayish-green leaves edged in white; single rosy flowers. **C**: 13   p. 422

x hortorum **'Velma'**, "Tricolor geranium"; gaily-colored, very compact geranium with small tricolored foliage, the rounded, near waxy leaves grayish green in center, surrounded by a rosy-red zone and bordered in creamy-yellow; single salmon 1 in. flowers; of stronger growth, larger leaves than Mrs. H. Cox. **C**: 13   p. 190, 423

x hortorum **'Wilhelm Langguth'**; an old German variety of the 1890's, and still popular; spreading plant of low habit, with attractive, glistening grayish-green leaves bordered in white; showy double, cherry-red flowers distinguished by their size. **C**: 13   p. 423

x **limoneum**, "Lemon geranium"; a pretty crispum hybrid, and considered the true lemon-scented geranium used in old English finger bowls; small, three-lobed leaves crenate at apex, on longer wiry petioles than crispum; flowers small crimson-rose. **C**: 13   p. 429

'M. Ninon' (scabrum), "Apricot geranium"; bushy, branched plant with strong stems and apricot-scented coarse leaves deeply lobed, the lobes divided again and overlapping, with margins toothed and curled; bright pink flowers with carmine spot and purple veins on upper petals. **C**: 13   p. 429

x **nervosum**, "Lime geranium"; an old English hybrid (1820), of bushy growth; in habit similar to crispum, with small rounded, lime-scented green 1½ in. leaves sharply toothed at margins and lightly ruffled; showy clusters of lavender flowers with bright feathering in upper petals. **C**: 13   p. 429

odoratissimum (So. Africa: Cape Prov.), "Apple geranium"; low-growing, sprawling plant with trailing stems; and sweetly apple-scented, rounded, ruffled moss-green leaves, 1½ in. or more wide, silky-velvet to the touch; small white flowers with short spur, upper petals spotted and veined with red. **C**: 13   p. 429

peltatum (Eastern South Africa), called the "Ivy geranium", because of the ivy-like, pointed 5-lobed leaves which are fresh green, waxy and rather succulent, 2 to 3 in. across, on trailing, zigzag branches; flowers normally single and rose-carmine; but now highly hybridized into many types and colors with single or double flowers, in rounded clusters, in white, pink, rose, red and lavender; the 2 upper petals usually blotched or striped. Ivy geraniums are ideally suited for hanging baskets, window boxes, patio containers or as ground cover. **C**: 64   p. 422

peltatum 'L 'Elegante', "Sunset ivy-geranium"; an old French variety with beautiful waxy gray-green leaves edged in white, later adding pink tinting or a red border; small single white flowers with lavender tints, and darker markings in throat. Attractive in hanging basket. **C**: 64
p. 274, 422

peltatum 'New Dawn'; fine bushy variety, free trailing with lively green waxy leaves, and full clusters of double flowers vivid rose-cerise. **C**: 64   p. 422

'Prince of Orange' (syn. citriodorum), the "Orange geranium"; a bushy, branching plant with the scent of orange-peel; the 3-lobed light green 2 in. leaves broader than crispum and not as crisped, dentate at the margins; white flowers tinged pink, the upper petals feathered purple. **C**: 13   p. 429

'Prince Rupert'; lemon-scented; branched and densely erect plant with small, crinkly, parsley-like, lobed leaves larger than crispum, with shorter petioles; flowers orchid-lavender, the upper petals veined purple. **C**: 13   p. 429

'Prince Rupert variegated'; also known as 'French Lace'; lemon-scented; habit of P. crispum but bushier and more prolific, the small lobed and crisped leaves are light green and prettily edged in white. One of the prettiest miniature scented-leaf geraniums for the window sill, with age growing into a nice dense bush. **C**: 13   p. 429

quercifolium 'Fair Ellen', "Oakleaf geranium"; upright plant becoming shrubby, with medium-size leaves lobed in shape of an oakleaf, and with dark zoning; flowers bright magenta-pink with purple spot in upper petals; free flowering and the best true oakleaf variety. **C**: 13   p. 429

tomentosum (So. Africa: Cape of Good Hope); "Peppermint geranium"; strongly mint-scented species of sprawling habit, with large, soft velvety leaves, triangular heart-shaped and shallowly lobed, emerald green with a felt-like covering of white hairs; small, fluffy white or tinted flowers. **C**: 13   p. 429

'Toronto', "Ginger geranium"; sweetly ginger-scented; similar to nervosum, of bushy habit, with larger, light green, rounded leaves evenly dentate at margins; fair-sized rosy-lavender flowers with darker markings on upper petals. **C**: 13   p. 429

violarium (So. Africa), "Viola geranium"; odd species with low spreading branches; long-stalked powdery-gray, ovate, toothed leaves; and pansy-faced flowers, with lower petals white, the two upper petals ruby red with dark blotch. Nice in baskets and hanging pots, blooming in spring and summer. **C**: 13   p. 430

PELECYPHORA (Cactaceae)
aselliformis (Mexico: San Luis Potosi), "Hatchet-cactus"; curious, small globular to cylindric plant to 2 in. thick and 4 in. high, with grayish green tubercles strongly flattened sidewise hatchet-like, in spirals, the minute spines arranged comblike and not prickly; flowers purple. Clustering with age. **C**: 13   p. 163

PELEXIA (Orchidaceae)
maculata (W. Indies), attractive, bushy terrestrial orchid, looking like a foliage plant; with fleshy, metallic bronzy green leaves spotted white, running into a channeled petiole, reverse purplish; erect purple flower spikes with small greenish flowers tipped with pink, lip whitish; blooming April-May. **C**: 72   p. 454

PELLAEA (Filices); the "Cliffbrakes" are small rock-loving ferns of the family Polypodiaceae with thin-leathery fronds; the sori in a marginal band and protected by reflexed margins. They are attractive little pot plants, or for fern dishes, and add variety to the fernery, but like to be only moderately warm for best keeping quality. Propagated from spores; sometimes division.
rotundifolia (New Zealand), "Button fern", small rock-loving fern with creeping scaly rhizome and hairy stalks; the long, narrow fronds nearly uniform, to 12 in. long, and staying near ground; simply pinnate with evenly spaced leaflets round when young. later oblong, dark green and waxy leathery. A miniature fern that is different in its low habit, very attractive in small ferneries as ground cover, or in terrariums. **C**: 4   p. 238

viridis (So. Africa, Madagascar, Mascarenes), formerly in the trade as P. hastata; the "Green cliff-brake"; tufted fern with thin-leathery, dark-green bipinnate fronds 1½ to 2 ft. long, on shining black stalks, the pinnae at base with numerous segments, decreasing upwards. Needs to be kept moist in the pot. **C**: 4   p. 238

viridis macrophylla, better known in horticulture as Pteris adiantoides; variety with bipinnate fronds to 18 in. high, having thin-leathery leaflets much larger and broader than in the species, although less in number on each pinna,

and resembling Holly-fern, but leaves are not as leathery. Used in small fern-dishes as a "Table fern". **C**: 4   p. 238

PELLIONIA (Urticaceae)
daveauana (So. Vietnam, Malaya, Burma), "Trailing watermelon begonia"; tropical herbaceous creeper hugging the ground, or any support; with succulent, pinkish stems and 2-ranked alternate, flattened, thin-fleshy leaves oval when small, with oblique base, lanceolate-pointed to 2½ in. long in older plants, brown-purple to blackish with pale green to gray center area; small green flowers. Grown in baskets or on bark best in the greenhouse; also a fine ground cover indoors. **C**: 54   p. 263

pulchra (So. Vietnam: Cochin), "Satin pellionia"; attractive fleshy creeper hugging its support, or pendant from a basket, with pinkish stems; the obliquely oval leaves shingled like a roof, light green to grayish, entirely covered with a network of blackish or brownish veins, pale purple beneath on gray. **C**: 54   p. 263

PENSTEMON (Scrophulariaceae)
x gloxinioides, "Beard-tongue"; garden hybrid of the Mexican P. hartwegii; robust herbaceous perennial often grown as annual, 2 to 3 ft. high, with glossy green lance-shaped leaves and loose spikes of large, gloxinia-like bell flowers to 2 in. across, in many brilliant colors except blue and yellow, blooming from July to autumn. A colorful garden plant that may be grouped in containers for the patio. Not winter-hardy. Propagated from seed, also cuttings best in early fall. **C**:76   p. 416

PENTARHAPHIA: see Gesneria

PENTAS (Rubiaceae)
lanceolata (Trop. Africa, Arabia), "Egyptian star-cluster"; herbaceous flowering shrubby plant of scraggly habit with downy, slender branches and opposite soft or limp, ovate, bright green hairy leaves with sunken veins, 2 to 6 in. long, the tubular flowers 1¼ in. long, and hairy in the throat, in showy globose clusters, basically purplish rose but variably in other colors from pure white through pink and salmon to deep rose. Pretty, though somewhat soft flowering plants blooming from fall to late winter, also for the sunny window sill; keep pinched to shape. Best propagated in spring from soft-wooded cuttings; also seed, but seedlings often revert to poor colors. **C**: 14   p. 498

PEPEROMIA (Piperaceae); small tropical, usually succulent plants of the Pepper family, clustering, scandent or trailing, having minute bisexual flowers borne in slender, catkin-like spikes. Many species have interesting painted or variegated leaves and are good little house plants or used in dish-gardens; they generally need warmth and must not be kept too wet or they develop stem-rot, a disease caused by the fungus Phytophthora. Disease is controlled by steam sterilizing propagating materials. Propagation according to type by cuttings, division, from cut leaves, or joint cuttings.
acuminata (Mexico); fleshy, scandent stems with narrow ovate waxy grass-green leaves larger than glabella and having slender curved tip; double red rings at inter-nodes from which develop the young branchlets. **C**: 3 p. 269

bicolor (Ecuador), "Silver velvet peperomia"; exquisite small species with erect red-brown stem covered by white hair; and broadly oval, velvety 2¼ in. leaves olive to gray metal-green, with broad silver center band, as well as parallel silver stripes and edging, rosy-red underneath. This dainty beauty would grace every collection; best in warm greenhouse. **C**: 61   p. 269

blanda (Venezuela to Bolivia); shapely branching plant covered with short pubescence, the pinkish-brown, erect stems with opposite, broad-oval or obovate leaves, nice vivid-green with contrasting parallel light green veins. **C**: 3   p. 269

caperata (Brazil), "Emerald ripple"; sturdy, very useful little rosette with short, branching stem developing dense clusters of roundish, heart-shaped or peltate leaves about 1½ in. across, deeply corrugated and quilted like a washboard; waxy forest-green, the valleys tinted chocolate, the ridges often grayish; reverse pale green; the pink petioles striped red; slender flowering catkins greenish-white. Ideal for dishgarden and novelty planting. **C**: 3   p. 267

caperata 'Tricolor', "Tricolor ripple" (Roehrs 1957); a beautiful, variegated sport of sturdy habit, with corrugated leaves milky-green, broadly margined with creamy-white, red around the base and spreading out into the veins, on red petioles. The high variegation can only be maintained by careful nursing of similarly variegated cuttings. Keep on dry side. **C**: 55   p. 269

clusiaefolia (West Indies), "Red-edged peperomia"; stocky, slow-growing plant with thick-fleshy, rather narrow-obovate, concave leaves, 4 in. long and more, metallic olive-green with broad, red-purple margin, light green beneath except the purple midrib; the tiny flowers borne on long slender, catkin-like spikes. **C**: 3   p. 267

crassifolia (Trop. Africa), "Leather peperomia"; stiffly erect, fleshy green stem with alternate, rhombic or obovate succulent leaves 3½ in. long, waxy grass-green with pale midrib, light green beneath. Sturdy plant. **C**: 3   p. 268

dolabriformis (Peru), "Prayer peperomia"; curious small species with fleshy spatulate leaves folded together, showing the pale green underside, glued tight with a translucent layer and appearing like a swollen sickle. **C**: 3   p. 269

fosteri (Brazil), "Vining peperomia"; very attractive creeper with thick foliage in whorls of 2 or 3 along the slender red stems, rooting at nodes; the small, short 2½ cm leaves forest-green with light green parallel veins. Discovered by Mulford Foster growing as an epiphyte. **C**: 3   p. 268

glabella (Central America), "Wax privet peperomia"; branching plant smaller than acuminata, with slender, flexible, red stems and little, 1½ in., broadly elliptic leaves, waxy fresh-green, and tapered at the base. **C**: 3   p. 269

glabella 'Variegata', "Variegated wax privet"; dainty, freely branching plant, with slender rosy-red stems, and small elliptic leaves light green or milky green, broadly bordered or variegated creamy-white. **C**: 3   p. 267

griseo-argentea (Brazil), known in Brazil and introduced into Europe as hederaefolia, the "Ivy peperomia"; very attractive, bushy rosette with long petioles bearing round cordate, shield-like, thin, quilted leaves painted with glossy silver, while the sunken veins are purplish olive; long, erect, greenish-white catkins. Somewhat particular. **C**: 3   p. 268

incana (Brazil), "Felted pepperface"; amazing little plant that could be rated as a succulent; I have found this species growing on granite rocks between cacti and bromeliads on the exposed Restinga in Southeast Brazil; stiff green stem with broadly heart-shaped, hard fleshy, gray leaves entirely covered with white felt. Very tough, for the sunny window. **C**: 1   p. 269

maculosa (Santa Domingo), "Radiator plant"; ornamental, fleshy species with long spreading later pendant, narrow-lanceolate leaves to 7 in, long, waxy bluish gray-green, with silvery-green to ivory ribs, petioles prettily spotted red-purple; spikes to 1 ft. long. One of the largest species. **C**: 3   p. 267

magnoliaefolia (West Indies), "Desert privet"; robust species with large, fleshy, obovate-elliptic leaves 4-5 in. long, glossy fresh-green with depressed veins, on brownish stem; stalk of flower spike not hairy. **C**: 3   p. 268

marmorata (So. Brazil), "Silver heart"; attractive rosette with thin, heart-shaped leaves about 3 in. long, rich green to bluish, tapering to a long point, with basal lobes overlapping, the ridges prettily painted with silver-gray between the sunken, grass-green veins; gray-green beneath; pink stalks striped red; long, white catkins. **C**: 53   p. 267

obtusifolia (Venezuela), "Baby rubber plant"; branching and clustering plant with succulent stems on reclining base and short petioles, both striped maroon-brown, bearing the waxy rich green, fleshy, obovate or spatulate, 2-3 in. concave leaves, obtuse or notched at apex, pale green beneath; growing to 12 in. high; stalk of flower spike minutely hairy. Long cultivated as a good little plant for dishgarden and novelty planting, but often develops stem-rot, and Cissus rotundifolia is used as a more dependable substitute, being of somewhat similar appearance. **C**: 3   p. 268

obtusifolia 'Variegata', "Variegated peperomia"; beautiful, small and compact succulent plant with pale stems blotched bright red, and alternate, rounded or obovate-elliptic, waxy leaves, light green variegated with milky-green, and from the margin inward, a broad area of creamy-white. Popular for dishgarden planting, but sensitive to overwatering and basal rot; for this reason variegated Hoya is recommended as a similar looking but more dependable substitute. **C**: 3   p. cover*, 267

orba, "Princess Astrid peperomia"; lovely, bushy little plant introduced from Sweden, with stem dotted red and white pubescent; the numerous, waxy, light green, ovate leaves spoon-like on short stiff petioles, showing their true shape when few-branched; but if crowded, leaves stay small and densely clustered pixie-like. **C**: 3   p. 269

ornata (So. Venezuela); lovely succulent rosette-like species with short stout stem supporting a cluster of symmetrical, elliptic, fleshy leaves on stiff red stalks, silky-green above, with light veins lengthwise, paler beneath and with conspicuous, parallel, purplish-red ribs. **C**: 53   p. 268

pereskiaefolia (Brazil, Venezuela), "Leaf-cactus peperomia"; branching plant resembling Pereskia, with red stems first ascending then creeping, the small elliptic or obovate leaves succulent and waxy, dull green with pale veins, to 8 cm (3 in.) long. **C**: 3   p. 269

polybotrya (in horticulture as pericatii) (Colombia), "Coin leaf peperomia"; succulent species with stiff erect, fleshy green stem with wiry reddish petioles, supporting the thick, shining, waxy, smooth, peltate-pointed, shield-like leaves of vivid green 3½ in. long with fine purple edge, gray-green beneath; branching white catkins. A stately plant. **C**: 3   p. 268

pulchella (syn. verticillata) (Jamaica, Cuba), "Whorled peperomia"; white-hairy plant with erect, red stems, and small obovate, green leaves, arranged by tiers in whorls of 4-6, the young leaves with pale veins, older leaves thick and boat-shaped with round bottom which is pale green tinted red; tiny green catkins. **C**: 3   p. 269

puteolata (Peru), "Parallel peperomia"; gorgeous tropical plant with forking, pendant angled stems and slender, lanceolate, leathery leaves 4 in. long, waxy dark green with 5 contrasting, yellowish, sub-translucent, parallel veins depressed on the surface, and raised on the light green reverse. **C**: 53   p. 268

resedaeflora (fraseri) (Ecuador, Colombia), "Mignonette peperomia"; a "flowering" plant with small, quilted, round-cordate, begonia-like leaves, dull-green and finely white-pubescent as if frosted, red-ribbed beneath; the inflorescence on tall red stems with whorls of 3-4 leaves topped by fluffy white flower spikes resembling mignonette. **C**: 3 p. 268

rotundifolia (also known as nummularifolia) (Puerto Rico to Jamaica); low rambling creeper covering the ground, the thread-like green vines rooting at nodes, with alternate, tiny ⅓ in., very round and fat, waxy leaves evenly pale green; much seen in the Luquillo rainforest, Puerto Rico. **C**: 3 p. 268

rubella (Orizaba, Mexico), "Yerba Linda"; branched little bush with thin, upright, hairy, crimson stems, densely set with whorls of tiny obovate leaves, olive-green marked with a silver network; vivid crimson beneath. **C**: 3   p. 268

sandersii (argyreia) (Brazil), "Watermelon peperomia"; attractive rosette 6-8 in. high, almost stemless, with deep-red petioles bearing fleshy, broad peltate-pointed, concave leaves 3 to 4 in. across, glossy fresh-green to bluish, painted with showy bands of silver radiating from upper center, pale beneath; minute flowers in long whitish catkins. A favorite, beautiful house plant. Propagates from cut leaves. **C**: 3   p. 267

scandens (serpens) (Peru), "Philodendron peperomia"; with fleshy, reddish stem and petioles, the waxy fresh-green, small heart-shaped leaves resembling a philodendron, and well spaced, with long internodes. Too sparry to be attractive. **C**: 3   p. 269

scandens 'Variegata', "Variegated philodendron leaf"; scandent, semi-erect, fleshy creeper with small cordate leaves gracefully slender-pointed, 2 in. long, light green to milky-green, irregularly bordered with creamy-white, on red petioles and reddish stems. Somewhat rank in growth, but useful for baskets. **C**: 3   p. 267

tithymaloides (Santo Domingo), a "Pepper-face"; robust species with branching stems, with alternate, fleshy, 5 in. leaves broadly rhombic or elliptic, with tapering base and pointed apex, glossy dark green with paler veins, light green beneath; the petioles keeled beneath; greenish catkins. **C**: 3   p. 268

velutina (Ecuador), "Velvet peperomia"; dainty, branching, red-stemmed plant with small ovate, velvet-pubescent, deep bronzy green leaves marked with narrow, pale green midrib and parallel veins, red underneath. Very delicate; for the shady-warm greenhouse or terrarium. **C**: 61   p. 269

verschaffeltii (Brazil: Alto Amazonas), the "Sweetheart peperomia"; a beautiful and shapely plant which I rediscovered in 1954 on the upper Amazon; small, short-stemmed rosette of fleshy, oval-heart-shaped 3-4 in. leaves similar to marmorata but basal lobes not overlapping, and alternate on short, barely noticeable branching stem, the waxy surface is bluish-green with broad silver bands between the recessed yellowish veins, on petioles red with dots. **C**: 53   p. 43*, 189, 267

PERESKIA (*Cactaceae*); this genus belongs to the earliest and most primitive of cacti when this family still had leaves like other plants. The slender stems lend themselves as understock for grafting of such cacti as Schlumbergera, Zygocactus, and Rhipsalis.

grandifolia (Brazil), a "Rose cactus"; shrub or small tree to 15 ft. with very spiny trunk; characteristic of the most primitive of cacti having leaves; these are elliptic, fleshy waxy, rich green, to 6 in. long; flowers like wild roses rose-pink, in terminal cluster. **C**: 3   p. 149, 165*, 535

PERISTERIA (*Orchidaceae*)

guttata (Guyana: Demerara); curious species with conical pseudobulbs bearing 2 or 3 plaited leaves 12–15 in. long; fleshy flowers appearing from the base, the sepals and petals forming a cup, yellow with red, and spotted maroon, the lip tipped purple; blooming in autumn. This plant may be potted in compost of leafmold, peat or fern-fiber with loam, and perlite or broken pots for drainage; warmth and abundant water is needed during growth; thereafter a season of rest. Prop. by division. **C**: 60.   p. 478

PERISTROPHE (*Acanthaceae*)

angustifolia aureo-variegata (Java), "Marble-leaf"; soft tropical herb with small, narrow-lanceolate, thin, leathery fresh-green leaves to 3 in. long, variegated rich yellow in the middle and along primary veins; small rosy flowers in terminal clusters. **C**: 4   p. 189

PERNETTYA (*Ericaceae*)

mucronata (So. America: Magellan region to Chile), "Chilean pernettya"; bushy little evergreen shrub becoming 2–3 ft. high, with woody branches densely set with small stiff glossy green ½ in. ovate leaves lightly toothed, and tipped by a sharp translucent spine; numerous tiny urn-shaped nodding white or pink flowers in late spring, followed on female plants by little ½ in. stalked, depressed globose berries, persistent through the winter; depending on variety these may be white, pink, or brilliant red. Very pretty for tubs or window boxes, but note that berries are poisonous. More fruit is set if more plants are grouped together. Propagated by division of the underground runners; layering, cuttings, or seed. **C**: 82   p. 512

PERSEA (*Lauraceae*)

americana (W. Indies, Guatemala and Mexico), the "Avocado" tree; also called "Alligator pear"; a round-headed tropical and subtropical tree 20 to 30 ft. high or more, spreading wide with large leathery, elliptic or oval 4 to 8 in. leaves, glaucous beneath; and small ¼ in. greenish flowers, forming in winter. The Avocado tree grows fast in well-drained soil containing humus, beginning to bear when 4 to 8 years old, the large fleshy apple or pear-shaped edible fruit, for which it is usually cultivated, is about 4 in. across with green or purplish skin, its flesh is buttery and of high nutritional value rich in vitamins and containing 7 to 23% fat; it is served in salads. The large central seed of the avocado if suspended in a glass of water will easily germinate if kept at water level; also propagated from firm cuttings in May. Seedlings thrive as house plants in the sunny window. **C**: 63   p. 506

PETREA (*Verbenaceae*)

volubilis (Mexico to Panama, W. Indies), "Purple wreath"; one of the most beautiful of tropical twiners, climbing perhaps 30 ft. high, with woody or wiry stems, long brittle-hard, rough leaves to 8 in. long, and showy racemes of lovely, star-like flowers, long lilac-blue sepals and small violet corolla; in March-April. It takes a number of years before Petrea starts blooming regularly, and only with good light and air; best high up in a greenhouse. Young plants will not bloom. Drainage must be good; temperature during winter not over 60° F. nights. From cuttings. **C**: 54   p. 277

PETROCOSMEA (*Gesneriaceae*)

kerrii, from the forest floor in Thailand and Vietnam, "Hidden violets"; low, flattened rosette from a thick underground stem, with long-pointed cordate, quilted leaves to 4 in. long rather thick and fleshy, velvety green, attractively covered with white hairs, and with crenate margins, on brown petioles; small ½ in. violet-like flowers with upper lobes white blotched with yellow, the 3 lower lobes white, several to a short stalk usually hidden between or under the arching foliage, blooming in summer. This pretty species grows very well at warm temperature with high humidity but welcomes a cool-intermediate period to promote initiation of flowering. Propagated from rooted leaves, or division. **C**: 72   p. 446

PETROSELINUM (*Umbelliferae*)

crispum (C. and So. Europe to Sardinia, W. Asia), "Parsley"; small compact hardy biennial herb 6–12 in. high with fleshy root, and rosette of 3-pinnate, much divided and curly, rich green, somewhat fleshy leaves. Minute yellow flowers the second year. One of the most familiar of all herbs, used fresh or dried as flavoring or garnishing of eggs, fish, shellfish, meats, poultry, salads, sauces, soups and vegetables. For winter use, roots may be transplanted to boxes or pots and kept in a sunny window. Usually grown as an annual, seed is soaked in warm water 24 hours; germination takes several weeks. **C**: 14   p. 319

PETUNIA (*Solanaceae*); tender, clammy herbaceous, sun-loving perennials grown as annuals whose ancestry goes back to the South American species violacea and nyctagini-flora; of bushy or straggling growth with soft, often flabby leaves, and showy funnel-shaped, slightly fragrant flowers, in sunny position blooming continuously throughout the summer. Long popular as blooming plants for the window, in baskets, or for bedding; potted plants beginning indoors in May in time for Mother's Day. Much objective breeding has developed and perfected types for every purpose, especially since the introduction of F_1 or Heterosis hybrids which have proved particularly floriferous, more vigorous, and resistant in unfavorable climate. These are now available, in many gay or pastel colors, as well as bicolors, large singles and doubles suitable for pots (grandiflora), balcony types for window boxes and baskets (pendulas), and compact singles, and doubles, for bedding (multifloras). The large all-doubles are propagated from cuttings or seed, all others from seed.

x hybrida 'California Giant', a "Giant pot petunia"; herbaceous plant with small oval leaves, dwarfed by giant 4 to 6 in. single flowers ruffled toward the margin, in shades from white, rose, orchid, purple, combined with showy centers in contrasting colors or design; desirable potplant for Mother's Day. **C**: 63   p. 378, 419

x hybrida fl. pl. 'Caprice'; compact plant with clear carmine-rose flowers about 3 in. across, densely double with frilled petals pale beneath, early flowering; a named variety of the 'Panamerican' strain, which also comes in white, lavender, purple, or variegated forms, all of which make superb potplants from Mother's Day into summer; but not as free-blooming in garden plantings as the grandifloras and multifloras. **C**: 63   p. 379, 419

x hybrida multiflora 'Celestial Rose', a "Dwarf bedding petunia"; small floriferous plant of bushy habit, with shapely ovate light green, downy leaves and single, wide funnel-shaped round flowers carmine-rose with pale eye, 2 to 2½ in. across, produced continuously throughout the summer; excellent for bedding, window box or as pot-plant. **C**: 63   p. 419

x hybrida grandiflora 'Crusader', "Bicolored petunia'; large single flowers 3 to 3½ in. across, gaily striped or banded starlike with bright carmine-rose on white base with nicely frilled white edge. **C**: 63   p. 419

x hybrida grandiflora 'Elk's Pride', "Blue balcony petunia"; so reminiscent of the single deep blue balcony petunias in Swiss and Bavarian Alpine villages, but somewhat less scandent and of shorter habit; an old favorite for its large flowers 3 in. or more across, velvety deep violet-purple with ruffled margins. Very showy in window boxes and patio containers in combination with red geraniums. **C**: 63   p. 419

x hybrida 'Pink Magic' (Grandiflora), F_1 or first generation hybrid "Single petunia"; a favorite bedding and pot variety of uniform, dwarf and bushy habit, loaded with numerous medium size single, smooth 2½–3 in. flowers of bright carmine-rose with darker veining and small white eye; larger than 'Celestial Rose'. **C**: 63   p. 378

x hybrida grandiflora 'Popcorn', a giant ruffled single white petunia, good pot variety because of its short, bushy habit, yet with huge gleaming white blooms 2 to 4 in. across, its surface quilted and with wavy margins. Fine as a window plant or for bedding. **C**: 63   p. 419

x hybrida multiflora 'Satellite', a small-flowered bicolor petunia; lovely free-blooming F_1 hybrid literally covered in summer with brilliant bright rose flowers about 2 in. dia., strikingly painted with a beautiful, contrasting white star; excellent bedding or pot plant. **C**: 63   p. 419

x hybrida fl. pl. 'Sonata', "Double white petunia"; favorite commercial pot variety for large fully-double, pure white flowers about 3 in. across, beautifully fringed and crisped at margins; should be pinched to encourage bushy growth; Panamerican strain. Charming Mother's Day pot plant; not as floriferous in garden plantings as the double multifloras or the singles. **C**: 63   p. 419

x hybrida grandiflora 'White Cascade'; a robust Panamerican F_1 hybrid especially developed for a cascading growth habit suitable for window boxes, hanging baskets, and patio planters; early-blooming giant pure white single flowers 4 to 5 in. and more across; also in colors red, pink, and coral-rose. **C**: 63   p. 419

PHAEDRANASSA (Amaryllidaceae)

carmiolii (mountains of Costa Rica to Peru); remarkable "Queen-lily"; bulbous plant with 1 to 3 stalked oblanceolate, fleshy leaves to 15 in. long, appearing at or after blooming time, and brightly colored nodding, curved tubular 2 in. flowers crimson at base, pale green toward segmented tips, and with protruding stamens, clustered on 2 ft. stalk. Normally blooming in late spring. The foliage is kept growing until October when the pots are gradually dried off and kept cool to March; change soil yearly. Plants may be made to rest in summer. Propagate by offsets or bulb cuttings. **C**: 54
                                                              p. 392

PHAEOMERIA: see Nicolaia

PHAIUS (Orchidaceae)

wallichii (tankervilliae) (Sikkim Himalayas, S.E. Asia), the "Nun orchid"; majestic terrestrial plant of great beauty, with plaited leaves on stout pseudobulbs, and an erect spike 3 to 4 ft. high with spreading, fleshy flowers 4 to 5 in. across, having buff-brown sepals and petals powdery white in back, the large lip orange-yellow at base and beneath, wine-red in middle and white at outer fringes. A noble plant for winter and spring blooming, the fleshy, heavy-textured fragrant blooms long-lived, continuing to flower for 6 weeks. Grown in large pots in compost of rich loam and peatmoss with perlite, or fernfiber, with abundant moisture, adding fertilizer during growth. Need good light 2400–3600 fc; outdoors under trees in summer. Rest and divide after blooming. Propagated by division; or from dormant buds on the floral stalks—cut and lay in moist sand under cover. **C**: 22   p. 478

PHALAENOPSIS (Orchidaceae); the "Moth orchids" are

highly tropical Eastern hemisphere epiphytes without pseudobulbs, large handsome, fleshy leaves, and beautiful waxy flowers elegantly cascading on arching stalks. Phalaenopsis require 70 to 75° F. warmth with 70% humidity during the main growing season in summer and need less light than most orchids, from 600 to 1000fc., during summertime while in growth, in winter 1500 to 2000 fc. to toughen the plants to encourage flowering. Since they are without pseudobulbs, Phalaenopsis must never be permitted to dry out; the best growing medium is either osmunda, peatmoss, firbark, wood blocks or chunks of tree fern. The blossoms are mostly white, sometimes with rose or purple and largely spring-blooming; are highly prized as cut flowers for weddings. Blooms open in succession but the main flower stem should never be cut as new lateral stalks branch out developing new buds and consequently a plant may remain in bloom for the better part of a year. Phalaenopsis grow relatively fast from seed, blooming in about 3 years. Division is difficult but young plantlets tend to form on the nodes of the flower stalks following blooming. The flowering stems may also be cut into sections with eyes and laid into flasks on agar culture (Knudson solution C).

amabilis (Indonesia, New Guinea), "Moth orchid"; exquisite epiphyte without pseudobulbs; fleshy, light green, deflexed leaves to 12 in. long, and a pendant or arching spray of large 3 to 4 in. flowers, glistening snowy-white,

except for its yellow crest spotted with red; blooming mostly from October to March, but flowers may variously appear into summer, often blooming twice a year. This popular species has been a parent of excellent white-flowered hybrids; their exquisite flowers are long-lasting on the plant for several weeks, looking like clouds of butterflies floating in the air. **C**: 62   p. 455*

amabilis 'Summit Snow'; ('Doris' x 'Confirmation'); an outstanding "Moth orchid" with glistening, pure white flowers of exceptional size and good substance; an arching, branching spray of as many as 30 to 50 large flowers measuring 4 to 5-in. across, blooming in continuity from November to March and longer. The species P. amabilis is found epiphytic in lovely Indonesia, New Guinea and No. Australia; this magnificent plant has fleshy, light green deflexed leaves, and flowers 3 to 4 in. across snowy white except for their yellow crest spotted with red; the fleshy roots attach themselves to pots, baskets, slabs of wood or fern. **C**: 12   p. 479

lueddemanniana (Philippines); charming, rather compact epiphyte with shining yellow-green, fleshy leaves 6 to 8 in. long, and short clusters of thick, waxy flowers $1\frac{1}{2}$ in. across, whitish and beautifully marked with cinnamon-brown and cross bars of amethyst; blooming at different times of year, mainly in spring or summer. The highly fragrant, long lasting flowers keep for 2 months. Generally after flowering young plantlets develop at the tip of the spike which are used for propagation. **C**: 12   p. 479

schilleriana (Philippines), the "Rosy moth orchid"; beautiful epiphyte with flattened roots, and gorgeous fleshy tongue-like leaves 8 to 12 in. or more long, dark green and transversely blotched with silvery gray, purplish-red beneath; the arching, branched inflorescence with delicate, 2 to 3 in. scented flowers in a shade of dainty pinkish-rose in varying tints, paler toward the edges; blooming from February to May and into summer, lasting about one month. **C**: 12   p. 479

PHALERIA (Thymelaceae)

capitata (Java); a small tree with beautiful waxy-white flowers, sweet-scented like daphne, blooming in clusters directly from trunk or heavy branches. This is an example of tropical cauliflory, where flower buds form deep within the tissues of the tree and then burst out through the bark. From seed. **C**: 52   p. 528

PHANERA (Leguminosae)

williamsii (New Guinea), a fantastic "Chain liane"; leguminous tree climber that could be used as an arbor vine, with hairy, tendril-bearing branches and small cleft leaves; when with age woody stems develop up and down again from the trees they become curiously chain-like; and the thought occurred to me when I photographed these plants in the highland jungles of New Guinea that Tarzan might well find these dangling lianes useful as rope-ladders. p.533

PHASEOLUS (Leguminosae)

caracalla (Trop. So. America), "Snail flower"; perennial twiner with leaves usually of 3 leaflets; and fragrant, very curious flowers light purple, tinted yellow, and with contorted standard, the keel and wings spirally coiled like a small shell. Propagated from seed. **C**: 64   p. 275

PHILODENDRON (Araceae); handsome tropical Ameri-

can mostly scandent foliage plants with attractive leathery, usually glossy, sometimes velvety leaves that fall into two general types: the relatively fast climbers with aerial roots along the stems, attaching themselves to bark or other support, or swinging free with decorative but variable foliage; and the "self-heading", arborescent kinds that may be tree-perching when young but form tree-like trunks with age, topped by a great crown of large usually incised very ornamental leaves. Their inflorescence is typical with spadix inside a fleshy spathe, but the tiny petal-less flowers are unisexual, the males found toward the apex of the column, the female ones are grouped along the lower, basal part. Philodendrons are outstanding decorators for warm rooms, dark tolerant to 15 fc.; but better, more characteristic foliage is maintained at nearer 200 footcandles. Aerial roots should be tied to support, or inserted into soil, unless objectionable. Does well under airconditioning. Constant soil moisture and moss support must be maintained. Propagated by terminal cuttings, or mossings, stem joints with several leaves, or fresh seed.

andreanum (Colombia); "Velour philodendron"; beautiful climber from the moist coastal forest, with iridescent, velvety oblong sagittate leaves dark olive, suffused with copper, and ivory-white veins; translucent edge. **C**: 62   p. 76

'Angra dos Reis' (Brazil); slowly creeping stocky rosette with leathery, broad sagittate, green leaves so glossy that they appear to be lacquered; the fleshy leafstalks blotched red. **C**: 4   p. 80

asperatum (Brazil: Bahia); slowly climbing on thick stem, leaves broadly cordate, dull green, to 18 in. long, deeply corrugated by numerous sunken veins; petioles covered with dense, warty scales. **C**: 54   p. 75

auriculatum (biauriculatum) (Costa Rica, Honduras); compact rosette with oblong leaves dark green, with pale, broad midrib, carried stiffly upright on long, swollen, mottled petioles sheathed at base. **C**: 54   p. 80

auritum: see Syngonium auritum

x barryi (selloum x bipinnatifidum), decorative self-header, with the full head and attractive glossy leaves of bipinnatifidum, but not as cut and with broader segments and of more vigorous growth. **C**: 4   p. 73

bipinnatifidum (Brazil: Rio to Mato Grosso), "Tree philodendron"; showy rosette growing into stout tree with a formal head of upright, waxy green, stiff leaves to 3 ft. long; bipinnate with 10–12 lobes each side of prominent midrib, the lobes are narrow, and lobed again, with long lobe at apex. **C**: 4   p. 71

x'Burgundy'; fine commercial hybrid involving P. hastatum, erubescens, species No. 2, wendlandii, and imbe; compact grower slowly climbing, with arrow-shaped leathery leaves about 1 ft. long, deep green with reddish cast, on wine-red stem with winged petioles; base cordate or hastate; ribs beneath, and the young growth, burgundy red; thicker, tougher leaf than mandaianum. **C**: 4   p. 78

'Burle-Marx's Fantasy' (Brazil), beautiful species introduced by Bob Wilson-Florida; slowly climbing, with elegant foliage symmetrically shingled, leaves on winged petioles ovate-oblong with cordate base and slender twisted tip, waxy light olive-green, overlaid with blackish forest-green mosaic network of veins; the margins translucent; underside pale gray-green tinted purple. **C**: 62 p. 76

cannifolium (Guyana), "Flask philodendron"; epiphyte with slowly creeping stem bearing leathery, lanceolate leaves with tapering leaf base, swollen leafstalks are channeled. Very curious. **C**: 3   p. 77

cordatum: see oxycardium

x corsinianum (lucidum x coriaceum, Florence 1888), "Bronze shield", Italian hybrid; slow creeper with large, broadly cordate, coppery green, quilted leaves having sinuate edge, the veins light green, and purplish-red beneath. **C**: 56   p. 77

cruentum (Ecuador, Peru), "Red-leaf"; upright growing creeper with waxy green, oblong-pointed leaves with cordate base and depressed veins; back of leaf a beautiful wine-red, hence the name "Red-leaf"; petioles winged. **C**: 4   p. 80

domesticum (hastatum) (Brazil), known in horticulture as P. hastatum, the "Elephant's ear"; lush climber with fresh green, fleshy arrow-shaped leaves 1 ft. long, later hastate and undulate; pale veins raised and ascending; gorgeous inflorescence with tubular, pale green spathe, red inside. Subject to leafspot if kept too humid when chilly. **C**: 4   p. 65*, 75

dubia: see radiatum

eichleri (Brazil: Minas Geraes, Alto da Serra), "King of Tree-philodendrons"; with magnificent, pendant leaves to 7 ft. long, sagittate, metallic glossy and scalloped edges; spathe rosy-red, hooding a white spadix. **C**: 4   p. 73

elegans (Trop. So. America); high climber sending out aerial roots freely from internodes; large leaves thin-leathery, deep green, deeply pinnatifid with finger-like segments barely more than ribs. **C**: 54   p. 79

x'Emerald Queen'; an F₁ hybrid of two unidentified species; of good keeping quality, and disease resistant to "shot gun" fungus and bacterial rot; deep green, vigorous plant with short petioles and close internodes; hastate, medium sized, shiny leaves; an excellent totem pole subject because foliage stays about the same size; cold resistant. **C**: 4   p. 78

erubescens (Colombia), "Blushing philodendron"; clamberer rooting at every joint, with arrow-shaped, 10 in., waxy leaves bronzy-green edged red, wine-red beneath; petioles green with red, occasionally winged. **C**: 4   p. 75

x evansii (selloum x speciosum), semi-self-heading; showy plant with 3½ ft., glossy leaves like elephant's ears, lobed and wavy; young growth pinkish beneath. **C**: 4 p. 73

sp. "Fernleaf": see pinnatilobum

fibrillosum: see grazielae

x'Florida' (laciniatum x squamiferum); attractive hybrid with slender climbing stem having very little hairy fuzz; the petioles slender, round and rough-warty; soft leathery, deep green leaves usually cut into 5 pointed main-lobes; pale midrib, ribs depressed and brown-red on reverse. **C**: 4   p. 78

x'Florida compacta' (quercifolium x squamiferum), an excellent hybrid needing no support as it is non-vining, or only very slowly creeping; the tough petioles are round, rough and marked purplish-red, the interestingly lobed leaves are thick leathery, deep waxy-green with veins barely recessed. **C**: 4   p. 78

giganteum (Puerto Rico to Trinidad), "Giant philodendron"; a giant with climbing trunk and beautifully lacquered leaves to 4 ft., cordate ovate, pale veins, on closely bunched, fleshy petioles. **C**: 54   p. 74

gloriosum (Colombia), "Satin-leaf"; slow surface creeper with heart-shaped, stiff, satiny leaves a beautiful silver-green, and contrasting pinkish to white veins, margin reddish; slender petioles with pale stripes. **C**: 62 p. 76, 179

grazielae (fibrillosum) (N.E. Peru, Ecuador, Brazil); dwarf climber with small, shallow, cordate, broad leaves with cuspidate apex; thick waxy, dark green; petioles winged half way or at base; similar to microstictum but smaller. A little gem. **C**: 4   p. 75

guttiferum (Pará, Peru, Costa Rica), "Leather-leaf philodendron"; slow climber with stiff-leathery, elliptic oblong leaves on short, vaginate petioles. **C**: 54   p. 80

hastatum: see domesticum

(hastatum x imbe) x wendlandii; slow climber with stout petioles, and lush, leathery glossy dark green leaves oblique oblanceolate and irregularly wavy. **C**: 4   p. 78

ilsemannii (Brazil); may be a variegated form of sagittifolium; spectacular climber with leathery, sagittate leaves almost entirely white or cream, with gray and dark green marbling; petioles vaginate at base. **C**: 53
p. 43*, 76, 179

imbe (Brazil: Pernambuco to Sao Paulo); Imbe is the Brazilian Indio's name for most climbing philodendron. This species has leaves oblong sagittate, parchment-like and with veins nearly at right angles, some red beneath; petioles marked red. **C**: 4   p. 80

karstenianum (Mexico: Oaxaca), "Mexican philodendron"; climbing species known as "Mex", with brittle-glazed stem; oblong cordate, thin leathery leaves, fresh green, on reddish petioles, some round, some vaginate; the 8 in. leaves always stay about the same size. **C**: 16   p. 80

lacerum (Cuba, Haiti, Jamaica); stem climber with juvenile leaves ovate, entire and undulate margin; later leaves crenately lobed and mature leaves deeply incised, glossy-green with light green veins. **C**: 54   p. 79

laciniatum (Peru); climber with oddly-shaped, leathery leaves having five distinct lobes, and with elevated pale green center markings. **C**: 54   p. 79

lingulatum (Puerto Rico to Martinique); high tree climber with ovate, rich green, corrugated leaves; petiole with wings, and flattened near base of leaf. **C**: 54   p. 80

longistilum (Brazil); resembling a self-heading wendlandii but slowly creeping with close joints; glossy-green, leathery, oblanceolate leaves 12 in. long with bold midrib and distinctive wine-red back, on abbreviated stalks. **C**: 54   p. 77

lundii var. 'Sao Paulo' (Brazil: Sao Paulo); better than the type; cupped, and the richer green is less likely to yellow in the sun. Similar to selloum but leaves are broader, elaborately frilled. This may be a natural hybrid of lundii with selloum **C**: 4   p. 73

x'Lynette' (wendlandii x elaphoglossoides) "Quilted birdsnest"; unusually attractive rosette in form of a birdsnest, the leathery, fresh green leaves eye-catching with dense corrugations between lateral veins. **C**: 54   p. 78

mamei (Ecuador), "Quilted silver leaf"; slow creeper with large arrow-shaped, waxy, cordate-ovate, quilted leaves grass green to grayish green, marbled with silvery areas; flattened petioles green suffused pink, with horny edges, smooth and with whitish length-stripes, rounded at bottom of petiole. **C**: 60   p. 76, 179

x mandaianum (hastatum x erubescens), "Red-leaf philodendron"; the best clone of this hybrid cross, selected by Manda for darkest red coloring, both the glossy, arrow-shaped leaves and the stems being deep wine to metallic, purplish-red; the first philodendron hybrid in the United States 1936. **C**: 4   p. 75

martianum (Brazil), epiphytic clusters of waxy green, ovate leaves with pale, inflated stalks growing on procumbent stem; differing from cannifolium, according to Burle-Marx and Blossfeld, by the boat-shaped, merely flattened stalks, leaf with obtuse or cordate base, red edge. **C**: 3 p. 77

melanochrysum (Colombia, Costa Rica); "Black Gold"; so called because of the beautiful, almost black-olive, velvety leaves, shimmering with pink, veins pinkish; smaller than andreanum. Delicate. **C**: 62   p. 76

melanonii (Guyana, Surinam), "Red birdsnest"; shapely rosette of ovate, fresh green leaves turning red in sun, with pale veins, and marked by channeled, swollen, red leafstalks. I have seen these spots of red a far distance, high on 150 ft. forest trees, from a dugout canoe deep in the interior of Surinam. **C**: 54   p. 77

mello-barretoanum (Acre, N.W. Brazil); stout tree with large leaves broadly hastate, deeply cut, and the pointed segments lobed again and overlapping; wide basal sinus; has robust thorns on the trunk between leaf scars. **C**: 4   p. 73

micans (Dominica, Tobago), "Velvet-leaf vine"; leggy vine with small, heart-shaped leaves, glittering silky bronze above, reddish beneath; very susceptible to cold. This may be merely a juvenile form of P. oxycardium. **C**: 62   p. 76

microstictum, better known in horticulture as pittieri (Costa Rica); slow climber with broad heart-shaped, thick, glossy, apple-green leaves, attached to round petioles at edge of shallow sinus, giving a pleasing appearance. **C**: 4   p. 75

x'New Yorker'; slow climber with aerial roots along stem; similar to P. youngii, but more compact and its leathery leaves somewhat cupped with more natural gloss, and more red; joints close, petioles with blood-red elevations, leaves sagittate, rusty-red in younger foliage, dark green with red cast, and pale veins when older. **C**: 4   p. 78

oxycardium (Puerto Rico to Jamaica and Central America), known in horticulture as "cordatum"; in Europe as P. scandens; pre-Linnaean as hederaceum. The most popular and widely sold vining Philodendron, known as "Heartleaf philodendron" or "Parlor ivy", or simply "Cordatum vine". A tall tropical, rapid climber by aerial roots, with glossy deep green, broadly heart-shaped, soft-leathery leaves in juvenile stage 4 to 6 in. long; in maturity or flowering stage to 12 in. long. In habitat when very young the foliage is apparently velvety, from observations I made in Costa Rica. This species may be used in many ways, as a cascading vine in pots, baskets, window boxes, or room dividers; or it may be trained against support, preferably on mossed poles, bark slabs, or milled treefern pillars; also a wonderful indoor ground cover over soil beds. Philodendron vines love warmth and good light, but will subsist in poorly lighted locations down to 15 footcandles, one of the best keepers in dark areas, although the foliage will become smaller, and the growth rank. **C**: 4   p. 75, 80, 259

panduraeforme (bipennifolium) (So. Brazil); "Fiddle-leaf", "Horsehead", or "Panda"; handsome climber with unusual foliage, the basal lobes extended, central lobes narrowed toward middle; of leathery texture, dull olive green. Tolerates 15 fc. **C**: 4   p. 75

"pertusum" (bot. Monstera deliciosa, juv. form) (So. Mexico, Guatemala), popularly known as "Splitleaf", "Hurricane plant", "Cutleaf philodendron", or "Fruitsalad" plant. Widely used decorator; dark tolerant to 15 fc. but foliage becomes smaller with insufficient light. Preference 200 fc. **C**: 4   p. 72

pinnatilobum (Brazil: Alto Amazonas), "Fernleaf philodendron"; graceful tropical climber with leathery leaves, cut much finer into narrow-linear segments than distantolobium. Originally offered in Florida nurseries as 'Fernleaf'. **C**: 4   p. 81

pittieri: see microstictum

polytomum (Mexico: Vera Cruz); almost stemless, with large glossy leaves rounded in outline, and pinnately cut, lobes pointed, and occasionally lobed again. **C**: 54   p. 79

radiatum (dubium) (So. Mexico, Guatemala); lush climber with broad, rich green leaves deeply lobed; in the smaller and less incised juvenile stage known commercially as P. "dubia". **C**: 54   p. 79

selloum (S.W. Brazil); "Lacy tree philodendron"; a "self-header" tree-like or scandent on trees in habitat and forming aerial roots. Long remaining a large rosette of beautifully elegant foliage; a stout stem slowly develops after years of growth, first erect, later leaning and becoming scandent. The lush, glossy dark green leaves to 2 ft. or 3 ft. across, are bipinnate, deeply lobed and cut, with a short, stubby lobe at apex, carried on long slender stalks; juvenile leaves are smaller and simply lobed. The spathe greenish-white. Recommended for use as solitary specimen, where a large container plant can be properly placed for its magnificent decorative effect, in large lobbies of buildings, or interior courts in warmer climate. This tree philodendron is an all-around satisfactory ornamental plant, taking air-conditioning well, tolerates draft and some abuse, and while preferring good light, adapts itself to poorly lighted locations down to 15 fc.; however, with lack of light the leaf stalks will stretch. Keep moist at the roots always. **C**: 4 p. 68, 71

sellowianum (Brazil), fast growing self-header with deeply cut leaves which stand considerable cold. According to Blossfeld, similar to bipinnatifidum, but has green seed pod sheaths instead of black ones. **C**: 4   p. 73

sodiroi (laucheanum) (Brazil), "Silver-leaf philodendron"; in juvenile stage vining with small cordate, pointed, bluish-green glossy leaves, largely covered with silver; ribs are red underneath; petioles wine-red and winged; in later stage petiole becomes flat on top and rugose with green puckers, and without wings; leaves are larger and rounded, internodes close. **C**: 4   p. 76, 179

speciosum (Minas Geraes, S. Paulo, Mato Grosso), "Imperial philodendron"; majestic arborescent species becoming tree-like with age, with huge sagittate leaves 5 ft. or more long, rich green, thin-leathery, the veins sunken and margins wavy and almost frilled; flowers beautiful with fleshy spathe green with purple margins, carmine-red inside. **C**: 4   p. 74

squamiferum (Guyana); "Red-bristle philodendron"; from tropical Guyana; twisting vine with rich green, 5-lobed leaves, the center lobe broad ovate, lateral lobes pointed, basal lobes short; olive petioles covered with green to red bristles. **C**: 54   p. 75

trifoliatum (Venezuela); stem-rooting climber with fleshy, dark green, trifoliate leaves, and veins depressed; round petioles slender and with red mark at base. **C**: 54 p.79

tripartitum (Guatemala, Costa Rica, Panama), "Trileaf philodendron"; scraggly climber with trifoliate, leathery leaves glossy-green, the segments quite narrow and long pointed, and not quite cut to base. **C**: 54   p. 79

trisectum (Andes of Colombia); freely rooting climber, with leathery, trifoliate leaves, at maturity stage divided nearly to base, segments long and with sunken veins; petiole channeled. **C**: 54   p. 79

tweedianum (Argentina, Paraguay), probably the southern-most philodendron, which I found growing in the Paraná delta near Buenos Aires; large, hastate, deep green leaves irregularly lobed, and wavy margined; forming trunk. For cooler locations. **C**: 16   p. 74

verrucosum (lindenii) (Costa Rica), "Velvet-leaf"; long-vining, with delicate, undulate, heart-shaped leaves, shimmering velvety, dark, bronzy green; pale green vein areas and margins emerald green; salmon violet beneath; petioles red and covered with green hairs. Gorgeous. **C**: 62   p. 76, 179

warscewiczii (Mexico, Guatemala, Honduras, Panama); scandent species forming snaky trunks; with lush, very soft leaves fresh green, triangular sagittate and pinnately parted, the pointed segments well separated in mature leaves, and with overlapping lobes; goes deciduous in dry season. **C**: 54   p. 79

x'Wend-imbe' (wendlandii x imbe); semi-self-heading hybrid with many waxy-green, oblong-pointed leaves carried stiffly on flattened stalks; new growth pink underneath. **C**: 4   p. 78

wendlandii (Costa Rica, Panama), "Birdsnest philodendron"; self-heading rosette of thick, waxy-green, long obovate leaves with thick midrib, arranged like a birds-nest; the short petioles are spongy; spathe cream-white. **C**: 4   p. 77

williamsii (Brazil: Bahia, Espirito Santo), syn. 'Espiritu Santo'; a magnificent epiphyte, with arborescent stem, and fresh green, deeply hastate leaves almost 3 ft. long, and with undulate margins and reddish veins; spathe pale green outside, yellowish inside. **C**: 4   p. 74

PHLOX (Polemoniaceae)

drummondii (Texas), "Dwarf annual phlox"; pretty little bedding or pot plant for spring and summer bloom in a riot of gay colors; bushy herbaceous annual 6 to 8 in. high, with fresh green ovate leaves and flat terminal clusters of brightly colored, salverform 1 in. flowers in shades of rose-red, also white, often with contrasting eyes. This free-flowering plant is a joy to behold and the blooms are long-lasting and blooming until frost if faded flowers are removed. From seed, and Jan. sowings produce lovely gift plants for Mother's Day. **C**: 63   p. 409

paniculata (Eastern U.S.: New York to Arkansas), "Summer perennial phlox" or "Garden phlox"; showy herbaceous perennial to 4 ft. high, with lanceolate leaves on strong, stiff stems, the leafy stalks topped by large clusters of 1 in. normally purple flowers, but garden forms varying in color from white and pink to salmon, scarlet or lilac, some with contrasting eye, blooming in summer and into autumn. This winter-hardy Summer phlox is very popular in gardens and may be flowered in containers. Prop. by division, from spring cuttings or by root cuttings. **C**: 76 p. 417

PHOENIX (Palmae)

the "Date palms" are noble representatives of an ancient group of monoecious feather palms to 100 ft. high, having massive or slender trunks topped by luxurious spreading crowns of arching, pinnate fronds. Some are of vastly important economic value, Date palms having been cultivated for their fruit and other uses for at least 5500 years; other, especially P. roebelenii, are gracefully decorative, being container-grown in small pots as well as in large tubs, and today are the most important decorator palms in Japan; superseding Rhapis. Most Phoenix are warmth-loving and need to be regularly watered as, depending on their habit, they are always thirsty. However, when well taken care of, these palms are very durable, and with their rich green and glossy foliage may be kept in the same spot and tub for many years. Grown mostly from seed, a few from suckers.

canariensis (Canary Islands), "Canary Islands date"; stately, massive palm widely planted in subtropical regions as an ornamental; compact, robust and stiff when young, with age forming thick straight trunks, becoming 50 ft. high and covered with old leaf bases, crowned by masses of arching pinnate leaves to 20 ft. long, the short stalk armed with yellow spines which are actually modified pinnae, the pleated leaflets glossy-green, in several ranks and various directions; small yellow fruit in large clusters, edible but without flavor. Container grown younger specimens from suckers are stiffly formal; good for cool locations or the patio, and they require sun. **C**: 14   p. 222, 224

dactylifera (Arabia, No. Africa), the fruiting true "Date palm" of Egypt and North Africa, and its improved descendants planted in the Coachella Valley in the California desert; a massive tree becoming 100 ft. high, its trunk covered by leaf scars, and topped with stiff fronds spiny at the base, the narrow rigid, folded pinnae in double rows when older, bluish-glaucous, to 18 in. long and sharp-pointed; female trees will set delicious edible, sweet, oblong fleshy brown fruit in great, heavy clusters, if pollinated. As a decorator, not as attractive as P. canariensis, although its trunk is more slender, in growth more sparse, foliage rather grayish. **C**: 2   p. 224, 506

paludosa (India: Bengal; to Thailand and Vietnam), "Siamese date palm"; graceful tropical feather palm forming dense clumps or groups with several erect or reclining tree-like trunks 8 to 25 ft. long and 3 to 4 in. thick; the feathery, pinnate fronds 8 to 12 ft. long and arching, with soft leathery, light green leaflets, whitish or mealy beneath; the small ½ in. fruit red and becoming black-purple. A most curious palm forming impenetrable thickets with their stems turning complete loops, their tips subsequently growing erect. **C**: 54   p. 540

reclinata (Trop. Africa from Senegal south to Natal), "Senegal date palm"; a leaning date palm somewhat resembling Cocos in habit, and which will live in the subtropics; solitary trunks 40 ft. high, or shorter if allowed to cluster; the pinnate lustrous green leaves rather stiff and curving downward, the short leaflets to 12 in. long; small red fruit. Not very attractive nor as regular as P. canariensis as a container plant when young, nor as graceful and satisfactory as P. roebelenii. **C**: 4   p. 217, 224

roebelenii (humilis loureiri), (Assam, Burma, Vietnam), "Pigmy date palm" or "Dwarf date palm"; very graceful both as a miniature potplant or when with slender, rough trunk growing to 12 ft. tall, topped by a dense globular crown of arching feathery leaves, the soft pinnae narrow and folded, 5–8 in. long, dark green and glossy when rubbed; berry-like black fruit in large clusters; female trees often clustering. One of the most popular house plants for tropical effect, luxuriating in warmth and moisture, even airconditioning; for shady locations 20 to 100 footcandles. **C**: 4   p. 224, 226

PHOLIDOTA (Orchidaceae)

imbricata (Himalayas and So. China through Malaya and Philippines to No. Australia), the "Rattlesnake orchid"; free-blooming epiphyte allied to Coelogyne; pseudobulbs bearing solitary plaited leaves; with the new leaf a pendulous chain, snake-like to 12 in. long, with overlapping, small ½ in. shell-like flowers in two ranks, yellowish-white with 3 orange-yellow stripes in throat, musk-scented and not expanding well; mostly spring and summer-blooming. Thrives in suspended baskets or in pots with fir-bark or osmunda, and may grow near the window. Not requiring a rest period, plants need to be kept continuously moist. Prop. by division. **C**: 72   p. 478

PHORADENDRON (Loranthaceae)

flavescens (New Jersey to New Mexico), an American "Mistletoe"; a green parasite on many deciduous trees from New Jersey to Florida and westward, forming dense bunches 1 to 3 ft. across, with brittle-woody cylindrical, forking twigs, thick oval and opposite yellowish evergreen leaves to 2 in. long. Male and female flowers are on separate plants, borne in short spikes or catkins, the females developing amber-white, small round berries. Cut branches are used for Christmas decoration, a custom in Europe where the similar Old World mistletoe, Viscum album, is used and considered an invitation to a kiss, going back in history to the Druids.   p. 538

juniperum (U.S. Southwest, Arizona); another "Mistletoe", parasitic on coniferous trees such as Juniper and Cedar, mimicking the character of their host with their thick cylindric, jointed stems, having imbricated leaves that are reduced to yellowish or gray-green scales; thickly branched in clumps; tiny straw-colored berries.   p. 538

PHORMIUM (Liliaceae)

tenax 'Variegatum', (New Zealand), "Variegated New Zealand flax"; attractive variant with the usually brownish-green leaves striped and margined with creamy-yellow and white bordered red. The N.Z. flax, P. tenax, is a large tufting plant with 2-ranked, tough leathery leaves clasping at base and splitting at apex, normally 3 ft. but which may grow to 9 ft. long as in New Zealand where I have seen them grow right in the cold water of southern lakes; dull red flowers in tall clusters. Very dramatic with its stiff leaves in a fan pattern; good in containers, very tough and tolerant. By division. **C**: 64   p. 306

PHRAGMIPEDIUM (Orchidaceae)

the curious long-tailed "Mandarin orchids" are commonly known in horticulture as Selenipedium, and quite erroneously as Cypripedium. They are very handsome terrestrials, or epiphytes, of the American tropics, without pseudobulbs, and forming large clusters of lush foliage, and the fascinating, usually long-bearded flowers with often very long, ribbon-like petals, and which differ from true Paphiopedilum by being deciduous. Phragmipedium are kept warm and very moist at the roots during summer, cooler when going into winter. They are easy enough to grow in the wintergarden or in the home, but provision needs to be made for adequate humidity. Suitable potting materials are mixtures of such as osmunda, fernfiber or firbark, with loam or chopped sphagnum; they must be kept damp at all times. Propagated by division.

caudatum, better known as Selenipedium; (Mexico to Panama, Colombia to Peru), a "Mandarin orchid"; sensational species of robust habit, in habitat growing at 5000 to 7500 ft. in swamps or on rocks, or epiphytic in trees in decaying leaves, with long straplike, flexible, lush green leaves; the inflorescence may be 2 ft. tall, carrying from 1 to 4 remarkable flowers, the dorsal yellowish with green veins, the pendulous petals slender ribbon-like and twisted, to 2½ ft. long, brownish-crimson over yellow; pouch slipper-shaped bronzy green; normally blooming from February to August. **C**: 22   p. 479

x grande (Selenipedium) (longifolium x caudatum), "Spiralled lady slipper"; very striking plant of most robust growth, with glossy-green, flexible sword-shaped leaves 2 ft. long; with several flowers on stalk to 3 ft.; the ribbon-like, spiralled petals extending 12 in. or more, at the base yellowish-white, changing to carmine-red in the pendant portion; the narrow, incurved dorsal sepal pale yellow veined with green; a large waxy pouch greenish-yellow spotted red inside; blooming during spring and summer. An easy plant to grow in the moderate greenhouse or winter garden. **C**: 22   p. 479

PHRYGILANTHUS

aphyllus (Chile), a hostile parasite living by preference on the column cactus Trichocereus chilensis, there forming dense clusters of threadlike leafless branches 1½ in. long, with cochineal-red small flowers; later berrylike, reddish fruit turning white when ripe. These parasites attach themselves to the lifestream of the host plant stealing all its food from it. Parasites are most frequent among fungi.
C: 1 p. 538

PHYLLANTHUS (Euphorbiaceae)

speciosus (Xylophylla speciosa), "Woodyleaf flower"; curious evergreen shrub or tree to 15 ft. high, with broad elliptic or lanceolate leaflike branchlets (phyllodia) 2 to 3 in. long and 1 in. wide, arranged in two ranks on a branch, together looking as if they were a feathered, pinnate leaf; tiny flowers with whitish calyx arranged all along the margins of the "leaflets". Often and easily grown as specimen plants for its odd appearance; a conversation plant. Prop. under cover by greenwood cuttings at 85° F. **C**: 3   p. 528

PHYLLANTHUS: see also Breynia

PHYLLITIS (Filices: Polypodiaceae)

scolopendrium (Scolopendrium vulgare) (Europe, Madeira, No. Africa, Asia Minor, Japan, No. America), "Hart's-tongue fern"; rhizomatous, hardy fern with stout rhizome and long straight or curved strap-shaped leathery fronds 6–18 in. long, pale yellow green to bright green, quilted and undulate at margins; spore masses in thick strips at right angles to the bold midrib, running into a short black stalk. Very durable for cool locations. Propagated by spores. **C**: 28   p. 251
scolopendrium 'Crispum' (Scolopendrium vulgare), "Crisped hart's-tongue"; attractive fern with long narrow, strap-like light green fronds nicely and regularly undulated and wavy, 1 ft. long by 1–1½ in. wide, crested at the base, on stalks covered with hair-like brown scales. This is one of the many forms of the variable P. scolopendrium which generally do not produce spores and are sterile; they can be propagated by division or a piece of the rhizome with the basal end of a ripened leaf, placed in moist warm peat-sand. Very charming plant that prefers to be kept cool and moist. **C**: 30 p. 251

PHYLLOSTACHYS (Gramineae)

aurea (China) the "Golden bamboo" or "Fish pole bamboo"; a running bamboo; tall woody grass, with hollow canes flattened on one side, to 15 ft. or more high and 2 in. thick, brilliant yellow, the internodes at the base very short; straight and stiffly erect, very hard and bonelike when matured and used for fishing poles; leaves usually 2–4 in. and long-pointed, light green, and glaucous beneath. Usually hardy around New York City, down to zero ° F. Good in tubs, but requires frequent watering to keep foliage attractive. By division. **C**: 14   p. 302
bambusoides (China), "Japanese timber-bamboo", "Giant timber-bamboo" or "Madake"; a running bamboo; one of the largest and most valuable "winter-hardy" timber bamboos, with green culms 25 to 45 or even 80 ft. high, thick-walled and to 6 in. thick, striped or yellow in some forms; two unequal branches at each branch-bearing node, oblong pointed leaves 2½ to 6 in. long; culm-sheaths greenish to reddish. The very hard wood and straight growth have made this very versatile in its uses, especially for construction, and its strength is exceeded only by the Tonkin cane (Arundinaria amabilis). Hardy to about Norfolk, Virginia; it has withstood zero ° F. Edible shoots. Beautiful for tropical effect; commonest of the large timber bamboos. **C**: 64   p. 302
nigra (So. China), "Black bamboo"; graceful black-culmed running bamboo 4 to 8 and 15 ft. high, and 1½ in.

thick, green at first later speckled, then all black; slim branchlets with small leaves commonly 3 in. long, culm-sheaths greenish to buff; thin-walled; hardy to about Norfolk, Virginia, and tested to 5° F. Good in containers. **C**: 14   p. 305
sulphurea (viridis var.) (China, Japan), "Yellow running bamboo" or "Moso bamboo"; a very hardy running bamboo 15 to 30 ft. high, stems yellow or green and 2–3 in. thick; leaves to 5 in. long and ¾ in. wide, glaucous beneath, on purple petioles. Very handsome with arching yellow canes, and effective in containers in entryways or the patio. Hardy in the Northwest, or Washington D.C., tested to minus 10° F. **C**: 14   p. 305

PICEA (Pinaceae)

abies (excelsa) (Northern Europe), "Norway spruce" or "Living Christmas tree"; evergreen coniferous forest tree to 150 ft. high, of pyramidal habit, with reddish-brown scaly bark and whorled branches, the usually pendulous branchlets with linear 4-angled, shiny dark green needles ¾ in. long, spirally arranged; pendulous light brown cones 4 to 7 in. long. Young attractive pyramids are used growing in containers indoors, decorated as a Christmas tree, but are subsequently best planted outdoors in the spring. From seed. **C**: 76   p. 311

PIERIS (Ericaceae)

japonica, better known in horticulture as Andromeda japonica (Japan), "Lily-of-the-valley bush"; handsome evergreen, slow-growing bushy shrub, to 10 ft. high with age, having attractive, brown wood, gnarled with age; glossy green, leathery leaves 2 to 3 in. long in dense whorls; the waxy white ⅓ in. urn-shaped flowers in exquisite pendulous clusters, the calyx lobes tinted red; long-blooming and long-lasting from March to May. One of the most satisfactory and durable broadleaved evergreens, splendid for patio containers. From seed or cuttings.
C: 84   p. 504
japonica variegata, in horticulture as Andromeda, (Japan), the "Lily-of-the-valley bush"; compact evergreen shrub slowly growing to 5 ft. or more; the gnarled woody branches tiered and with leathery leaves 2 to 3 in. long, glossy green edged creamy white; loaded in spring with pendulous terminal clusters of waxy-white, urn-shaped flowers, the calyx lobes tinted red. A lovely and very sturdy plant used in containers in cool areas. Requires water and feeding, prefers acid soil; winter-hardy. From cuttings or division. **C**: 28   p. 295

PILEA (Urticaceae); small tropical herbs of the Nettle family, many with handsome, opposite leaves brocaded or painted with silver, others low creepers with tiny leaves perfect as ground-covers or hanging over the edge of pots. Their small flowers are greenish and insignificant, in axillary clusters. Pileas are of easy culture and thrive with some warmth and abundant moisture. Prop. by cuttings; the creepers by separation.
sp. 'Black Magic', (Colombia); dense, mat-forming creeper with small puckered, ½ in. roundish, herbaceous leaves deep metallic bronzy green with wavy margins lightly crenate. **C**: 10   p. 270
cadierei (Vietnam: Annam), the "Aluminum plant", also known as "Watermelon pilea"; a rapid growing, rather succulent plant, with thin fleshy, opposite, rather large 3 in., obovate, quilted foliage remotely toothed or crenate, attractively painted shining silvery aluminum over the vivid green to bluish-green blade; tiny flowers in stalked heads. **C**: 4   p. 173, 189
cadierei 'Minima', "Miniature aluminum plant"; a darling cultivar of dwarf, freely branching habit with pink stems and 1½–2 in. elliptic pointed, succulent, quilted leaves much smaller than the species, deep olive green, with raised areas covered with silver, and with crenate margins. Slow-growing. **C**: 4   p. 270
depressa (Puerto Rico), "Miniature peperomia"; freely branching, low, succulent creeper with tiny ¼ in. roundish obovate, fleshy leaves. light pea green and glossy, with apex crenate, opposite and dense on thin green stems, rooting at nodes where touching the ground. Excellent on the window sill, with a carpet of lively green, waxy leaves spilling over the side of the pot. **C**: 3   p. 270
involucrata (spruceana) (Peru), "Panamiga", the Pan-American "Friendship plant"; free-growing, small ornamental herb with ascending branches, dense with

oval, somewhat fleshy, deeply quilted leaves; deep green in the shade, coppery red-brown when exposed, of variable size from 1¼–3 in. long, hairy and with crenate margins, wine-red beneath; the tiny rosy-red flowers clustered closely in the axils of leaves. Considered a good house plant. **C**: 10   p. 270

   microphylla (known by some old horticulturists as muscosa), (West Indies), "Artillery plant"; small plant densely branched, with suberect, fleshy stems thick with tiny, watery-succulent, oblong, green leaves to ¼ in. long, having a tapering, cuneate base; flower clusters sessile with staminate flowers discharging a cloud of pollen when dry or shaken, hence the reference to artillery. **C**: 53 p. 270

   nummulariifolia (West Indies to Peru), "Creeping Charlie"; low creeping herb with thin reddish branches rooting at the nodes, with small, circular, ¾ in. quilted, crenate leaves, corrugated and hairy, light friendly green, paler beneath, flowers in tiny clusters. **C**: 10   p. 270

   pubescens 'Argentea' (E. Cuba), "Silver Panamiga"; a pretty, rather succulent small plant with opposite ovate or rhombic leaves, shining bluish-silver, with depressed veins, and deeply toothed at margins toward the apex, gray-green underneath and hairy at the nerves, brown stems covered with short hairs. **C**: 60   p. 270

   repens (Mexico), "Black leaf Panamiga"; low spreading herb with small 1¼ in. quilted leaves almost round, quite thin and glossy, coppery-brown with large crenations at the margin, purplish and hairy beneath on hairy brown branches; little greenish white flower-heads on long stalks. **C**: 60
                                                                           p. 270

   'Silver Tree' hort., (Caribbean), in the trade as "Silver and bronze", a copyrighted name 1957 by Mulford for this species from the Caribbean area; small herbaceous, densely branching plant with white-hairy stalks, quilted ovate but highly glossy leaves 1 to 3 in. long with depressed veins and crenate margins, bronzy-green, with broad silver band along center and silver dots on the sides, smooth above except for occasional white hairs; reddish beneath and tomentose. Pilea 'New Silver and Bronze' listed in the trade appears to be the same plant. Attractive miniature plant for dishgardens and terrariums. **C**: 4   p. 189, 270

PINGUICULA *(Lentibulariaceae)*

   caudata (bakeriana) (Mexico), "Tailed butterwort" or "Orchid fly catcher"; carnivorous perennial flattened rosette from moist bogs, with fleshy obovate pale green, clammy leaves covered with glistening glands exuding a sticky digestive fluid, which attract and capture insects, gradually absorbing them, providing the plant with nitrogenous matter difficult to obtain in sphagnum marshes. The pretty, autumn-blooming flowers deep carmine, with long spur. Grown in pots in lime-free sandy humus, or peat with sphagnum, stood in saucers of water on inverted pots; they will grow warm, but are better off cooler down to 50° F. Prop. by division, seed, or leaf cuttings but these must be removed without injuring them. I remember growing Pinguicula amongst pots of orchids to catch injurious Cattleya-flies, which were brought in with imported plants, at the Botanical gardens in Vienna. **C**: 72   p. 522

PINUS *(Pinaceae)*; the "Pines" are characteristic ever-green needle trees of the Northern hemisphere, with resinous trunks and spreading branches in whorls or tiers, seldom otherwise; the permanent leaves are needle-like, in bundles from 2 to 5; female flowers developing the familiar brown pine-cone. Very ornamental as decorators, especially the dwarf forms, and the primary dwarfed Bonsai trees in Japan. Pines need sun, well-drained, preferably sandy soil. Propagation from seed.

   densiflora (Japan), "Japanese red pine"; round-headed, irregular tree to 90 ft. high, often developing two or more trunks with reddish bark at ground level; the branches at first glaucous green, later brown; slender, flexible blue-green needles 2 to 5 in. long, in pairs. Handsome pine for informal effects. **C**: 26   p. 313

   mugo (mughus) (Eastern Alps and Balkans of S.E. Europe), "Dwarf Swiss mountain pine" or "Mugho pine"; low, shrubby, symmetrical little pine, spreading with age, the woody branches with stiff and twisted dark green crowded needles in bundles of two, 2–3 in. long. Very winter-hardy. Widely used because of low growth habit; an excellent container plant, and for Bonsai. **C**: 26   p. 313

   nigra (Corsica to Austria and W. Asia), the famous "Austrian pine"; pyramidal slow-growing pine tree 40 to 120 ft. high, rather dense and uniform in habit, the branches in regular whorls, becoming flat-topped with age; the very stiff, rigid needles dark green 3 to 6 in. long, in pairs. A tree of strong character in any decorative scheme, and smoke tolerant. **C**: 26   p. 313

   parviflora (pentaphylla) (Japan, Taiwan), "Japanese white pine"; normally growing 20 to 50 ft. high, but widely adopted in Japan for the culture of Bonsai, or dwarfed trees, becoming centuries old, yet, with care, rarely increasing in size to over 2 ft. The slender bluish-green needles 1½ to 2½ in. long are closely arranged in clusters of five at the ends of the twigs. Usually grafted on P. thunbergii, the black pine which tends to slow down the fast upright growth, developing into a multistem, low tree. Widely used for Bonsai or container planting. **C**: 26   p. 313, 541

   pinea (So. Europe and Turkey), "Italian stone pine"; the characteristic broad and flat-topped pine of the South Italian landscape, 40 to 80 ft. high; in youth a stout bushy globe; with stiff, bright to gray-green needles in two's, 4 to 8 in. long. In juvenile stage, needles are short, 1 in. long and silvery glaucous. Handsome decorator, when container grown; takes heat and drought. **C**: 64   p. 308

   thunbergii (Japan) "Japanese black pine"; handsome, spreading tree 20 to 100 ft. high, resembling, and perhaps a form, of the Austrian pine but quicker growing; with somewhat shorter, darker green leaves rarely over 4 in. long, the sharp-pointed stiff needles in pairs. May be pruned; excellent in planters and as Bonsai. **C**: 26   p. 313

PIPER *(Piperaceae)*; the "Peppers" are mostly tropical herbs or vines, erect bushes or root-climbers, with alternate, often aromatic leaves, and tiny flowers without corolla, borne in long catkin-like spikes; the fruit is a small berry. Some of the vining kinds of ornamental peppers have very handsome foliage of velvety texture, or gorgeously patterned and splashed with pink or silver; they require a combination of moderate to good warmth, humidity, and fair light to fully develop their exquisite beauty. Prop. from cuttings; the vines as stem sections with 1 or 2 eyes.

   betle (Bali, East Indies), "Betel-leaf pepper"; commonly used in Indonesia and India for chewing with betel nut; stems trailing; in Bali I have seen it climbing high on trees, with foliage in neat ranks; the leaves are fleshy, broadly heartshaped, 3–6 in. long, dark green with depressed veins; flowers in stalked catkins opposite the leaves. **C**: 54 p. 256

   crocatum (ornatum cv. crocatum) (Peru), "Peruvian ornamental pepper"; rich-looking beautiful, ornamental climber, sometimes confused with P. ornatum; thin-wiry stem, and peltate, highly glossy, slender-pointed leaves blackish olive-green, with silver-pink marbling and spotting along veins and veinlets, and corrugated; deep purple underneath. Delicate. **C**: 60   p. 256

   futokadsura (Japan), "Japanese pepper"; shrubby plant climbing by aerial roots, much like nigrum, but nearly hardy, and said to be deciduous, stands some frost; thick-leathery, ovate, slender-pointed leaves with 5 depressed veins, waxy, blackish green, and light green beneath; flowers greenish and berries red; slow grower. **C**: 16 p. 256

   magnificum (bicolor), (Peru), "Lacquered pepper tree"; magnificent, erect, branching foliage plant with corky stem and winged, clasping leaf-stalks bearing large fleshy, quilted, oval leaves to 8 in. long. lacquered forest-green of a metallic sheen, with ivory veins and edge; wine-red beneath. **C**: 60   p. 256

   nigrum (Malabar coast, Malaya, Java), "Black pepper"; tropical climber with flexuous stems dense with leathery, glossy blackish-green, ovate or elliptic leaves 3 to 6 in. long, and bearing long clusters of green berries turning first red, then black; furnishing, when dried, black pepper. More of a curiosity than ornamental. **C**: 4   p. 256

   ornatum (Celebes, New Britain), "Celebes pepper"; very attractive tall climber with slender reddish stems and petioles, and broad, peltate-pointed, shield-like, waxy leaves 3–4 in. long, deep green, beautifully etched with markings of silvery pink, becoming white in older leaves; pale green beneath. Exquisite vine for growing indoors. **C**: 10   p. 256

   porphyrophyllum (Cissus) (Indonesia), "Velvet cissus"; beautiful climber without tendrils, stems red with lines of white bristles, roundish-cordate, recurved, 3–4 in., quilted leaves, velvety moss-green with yellow veins and pink markings mainly along the veins, wine-red underneath. Striking vine, best in warm greenhouse. **C**: 60 p. 43*, 256

sylvaticum (Burma), "Silver cissus"; attractive ornamental vine with thin-wiry stems, and heavy leathery, ovate leaves more or less with cordate base, corrugated between the sunken dark veins, dark steel-green, the raised areas covered with stippled silver, and with metallic pink sheen. **C**: 60   p. 258

PIQUERIA (Compositae)

trinervia (Mexico, C. America, Haiti), the fragrant "Stevia serrata" of florists, grown for its profuse winter bloom; shrubby tropical herb 2 to 3 ft. high with slender flexible branches, opposite narrow, trinerved, glossy green leaves 1 to 3 in. long; the small fragrant, white disk flowers in clusters on twiglets from leaf axils, blooming from autumn on, and lovely in arrangements of cut flowers. Planted into 6 to 8 in. pots stevia require a nourishing soil for best production; best outdoors in summer; in winter good light with fresh air. Cut back after flowering. Propagated by division or from herbaceous cuttings in March. **C**: 14
                                                                  p. 504

PISONIA: see Heimerliodendron

PISTIA (Araceae)

stratiotes (Trop. America), "Water lettuce" or "Shell flower"; water-floating leaf rosettes bright green and fleshy, velvety hairy, with feathery roots hanging in the water; small, green flowers hidden between leaves. Apparently naturalized in other parts of the tropical world, I have seen them colonizing lagoons and rivers in West Africa, in Kenya and Uganda, and along the upper Nile. For ponds or aquariums. **C**: 54   p. 98, 519

PITCAIRNIA (Bromeliaceae)

andreana (Colombia: Choco); small terrestrial, with leafy stems to 8 in. long; recurving, narrow-lanceolate leaves, green and speckled with silver dots above, heavily frosted beneath with white scales; inflorescence with orange flowers tipped yellow. **C**: 10   p. 142

corallina (Colombia), "Palm bromeliad"; terrestrial plant to 3 ft high, with stalked leaves dark green and corrugated, petioles with brown spines; inflorescence in prostrate raceme with red stem, coral-red fl. **C**: 10   p. 142

tabulaeformis (Mexico); terrestrial rosette with oblong spatulate, papery 6 in. leaves light green showing darker, parallel veining, and lying flat on the ground; showy bright red flowers in sessile head. **C**: 21   p. 142

xanthocalyx (Mexico, W. Indies); I found this plant growing in Leningrad, Soviet Union, as a terrestrial rosette, with narrow flexible leaves 2–3 ft. long, tapering to a curly tip, light green, densely scurfy-white at back; inflorescence in many flowered raceme 18 in. long with bright yellow bracts and primrose-yellow petals. **C**: 21   p. 142

PITTOSPORUM (Pittosporaceae)

tobira (China, Japan), the "Japanese pittosporum" or "Mock Orange"; tough and handsome evergreen shrub densely branching into a rather sparry, flat-topped, shapely bush 6 to 15 ft. high, with thick-leathery obovate, dark lustrous green leaves 2 to 4 in. long, arranged in dense pseudo-whorls, and with clusters of small, creamy-white flowers at branch tips, with the fragrance of orange blossoms, in early spring. Very effective and dependable as a container plant where low size is desired, takes sun or shade; cool or warmer, air-conditioning or draft; full sun or darkness down to 20 to 30 fc. Propagated from seed, or half-ripened cuttings in late summer. **C**: 13   p. 293, 485

tobira 'Variegatum', "Variegated mock-orange"; attractive variegated form with leathery leaves slightly thinner, milky or grayish-green raggedly margined creamy-white; the little, fragrant flowers resembling orange-blossoms. Very tough and durable. **C**: 15   p. 295

viridiflorum (South Africa), "Cape pittosporum"; slow-growing evergreen shrub of erect habit to 20 ft. high, the woody branches with a tendency to reach straight up; thin-leathery long obovate leaves to 4 in. long, grayish green with pale midrib and semi-glossy; large clusters of greenish-white flowers with jasmine fragrance, followed by bright yellow fruit. **C**: 14   p. 294

PITYROGRAMMA (Filices); the "Gold" and "Silver ferns", often better known as Gymnogramma, are large and showy, mostly tropical or subtropical, members of the

Polypodiaceae family; from an ascending short rhizome rises a tufted rosette of large, feathery fronds to 3-pinnately cut, these strikingly covered with white or bright yellow powder on their underside. These powders are crystalline resinous secretions from glandular cells on the foliage, and indicate that such species originate from drier, xerophytic habitats. For this reason Goldferns may be used as indoor subjects with dry atmosphere, best at about 60° F.; they are also suitable for the patio, or in hanging baskets, when peatmoss or sphagnum should be added to fibrous soil. The fronds should not be watered overhead to keep them dry, but the roots need adequate watering. Old plants soon loose their beauty, but young ones grown from spores can rapidly take their place.

calomelanos (West Indies to Peru, Trop. West Africa), handsome "Silver fern"; large, robust, rhizomatous rosette of spreading dark green, bipinnate fronds to 3 ft. long, on long shining black stalks; the leathery segments toothed or cut, thickly powdered, white beneath. Being somewhat xerophytic, it is accustomed to dry air; avoid spraying the foliage. Old plants don't maintain their beauty; young plants from spores are fast-growing. **C**: 65   p. 250

chrysophylla (calomelanos aureo-flava), (So. America, West Indies), a "Gold fern"; showy fern of graceful habit, with wide-spreading tripinnate, somewhat fleshy fronds 2 to 3 ft. long, dull green above, thickly covered with golden yellow, waxy powder on the underside, carried on purplish-brown channeled stalks; the young shoots also waxy golden-yellow. Attractive fern for dryer locations, for ferneries or the patio in the mild south and southwest. **C**: 65   p. 250

pulchella (Venezuela), the "Silver-fern"; of upright habit; tufted powdery stalks carrying 3-pinnate, soft fronds 6–12 in. long, deep green and lightly powdered above; densely covered with pure white powder beneath. **C**: 66
                                                                  p. 194

triangularis (Gymnogramme), (Alaska, B.C., Oregon, California, south to Ecuador), the "California gold fern"; strikingly pretty species of dwarf habit, to 18 in. high, with elegant bipinnate, deltoid fronds to 7 in. long, soft herbaceous, dull fresh-green with crenate pinnules; the underside conspicuously coated with powder varying from orange to sulphur-yellow or almost white. **C**: 65   p. 250

PLATYCERIUM (Filices); unique epiphytic tropical ferns of the family Polypodiaceae, generally known as "Staghorn ferns"; always exciting the fancy of plant lovers. Most become large and they are characterized by two very different types of fronds: the sterile which are flat, generally rounded, shield-shaped, parchment-like or fleshy, enclosing the roots and clasping the tree or support on which they grow; the fertile fronds grow erect or pendant, more or less with antler-like segments 1 to 10 ft. long, which usually bear the brown spore masses. Because of their awkward size they are rarely grown in pots, but generally fastened on bark or nests and baskets of sphagnum and osmunda, and which should be dipped in water and occasional nutrient solution periodically; a greenhouse or winter garden is best for these striking exotics. Propagation by suckers which grow from adventitious buds at tips of roots if kept moist.

"alcicorne" hort. (Madagascar, Comores, Mauritius), "Elkhorn fern"; distinct from bifurcatum in having the fertile fronds shorter, more rigidly erect, a bright green, and only thinly hairy, widening to short forks, 2 to 3-lobed, and lightly recurving; the brown soral patches are located at or beside the last bifurcation, sometimes extending to the lower portions of the lobes instead of on the tip of the distal segments as in bifurcatum; the young basal fronds are rounded becoming somewhat crenulate with age, prolonged above into a few finger-like lobes. I have observed this very attractive variety in our Roehrs collection for some 40 years, since 1930, always distinctly different; a recommended, handsome and sturdy plant. **C**: 10   p. 246

andinum (Andes of E. Peru and E. Bolivia), "American staghorn"; this sole So. American, subandine species is a large epiphyte with mighty erect, nest-like and lobed barren fronds, and much forked long pendant fertile leaves to 10 ft. long, divided in pairs 3 or 5 times, ending in long ribbon-like lobes; the upper surface green and smooth, the underside densely covered with white hairs. Sporangia placed from the first to the third fork back. **C**: 60   p. 244

angolense (Trop. Africa), "Elephant's ear fern"; large epiphytic fern, close to stemaria, with ascending sterile fronds having purplish veining and wavy crest; broad wedge-shaped fertile fronds not divided into lobes, and with a felt-like covering of rust-colored wool underneath; the sporangia along the apex. I have encountered this species spread wide all across Africa from the Usambaras in Tanzania, on Lake Victoria, Uganda, the Ituri forest of the Congo to the Niger Delta in Nigeria. **C**: 60   p. 244

bifurcatum (E. Australia, New Guinea, New Caledonia, Sunda Isl.), the "Common staghorn fern"; easy growing epiphyte freely producing young plants on its roots; the basal, sterile fronds are kidney-shaped, in old specimen feathered and lobed; the usually laxly pendant, leathery, grayish dark green fertile fronds to 3 ft. long, are lightly covered with white, stellate hairs, and usually twice long-forked; soral patches only on distal segments, being the tips of the ultimate forks; reverse silvery or green. It is known as "Elkhorn" in Australia; on Mt. Boss, New South Wales, at 3000 ft., this species tolerates 15° F. cold, rainfall is 128 in. per year; near Sydney they are found growing between 800–1500 ft. Because this species is used to a variety of temperatures, it is the most common in cultivation, as it lends itself to home care better than most others, also tolerating some neglect. **C**: 22   p. 245

bifurcatum majus (Polynesia), "Greater staghorn"; magnificent form of more robust habit with larger foliage; the barren fronds are roundish, convex, and overlap each other; the broad, rich green fertile fronds tend to be erect, with the broad forking lobes elegantly pendulous. Widely found in cultivation. **C**: 10   p. 246

bifurcatum cv. 'Netherlands', "Regina Wilhelmina staghorn"; a cultivar originated in Holland, with soft-leathery fertile fronds bright green and moderately tomentose, broader but shorter than the type, and well divided into numerous lobes; its fronds arranged spreading in all directions star-like. One of the best all-around staghorns for house plant culture. **C**: 22   p. 244

coronarium (biforme), (Burma, Thailand, Malaya, Java, Philippines), "Crown staghorn"; a glorious epiphyte of which I have seen immense clusters growing high in trees of the rainforest in Malaya, the long, fresh green, pendulous fronds are to 15 ft. long, several times widely forked, and the lobes gracefully twisted; the thick barren fronds are tall and lobed; spore is curiously borne on a separate fertile reniform disk. An outstanding show species for winter garden or warm fernhouse. **C**: 60   p. 245

"diversifolium" hort. (Kivu, Congo), "Erect elkhorn"; dwarf epiphyte distinctively attractive because of the erect habit of its fronds which are broadly spreading into twice-divided, pendant lobes, covered with whitish stellate hairs; basal fronds kidney-shaped, neatly covering the fibrous roots. Because of its erect habit it lends itself to cultivation in pots. **C**: 10   p. 234, 246

ellisii (Madagascar); small epiphyte, with kidney-shaped basal fronds forming a circular cushion; the fertile fronds to 12 in. long, glossy green, with narrow base suddenly flaring into broad fans deeply lobed, the lobes broad and undulate; spores around sinus. **C**: 60   p. 246

grande (E. Australia, Singapore, Philippines), "Regal elkhorn"; magnificent epiphyte with a regal crown of upright spreading sterile fronds of glossy vivid green, the upper lobes doubly forked and staghorn-like with dark venation; pendulous, forked pairs of fertile fronds appear with age, holding between them the wedge-shaped disk bearing the sporangia. I have found P. grande as far south as New South Wales at 2000 ft., where the temperature ranges from 38° to 120°F., hosted on Stinging trees; also near Sydney at 600–800 ft. Mature specimen with fan-like sterile fronds spread 3 ft.; fertile fronds 5 ft. long. Spores are produced yearly after some 20 years of age. **C**: 60   p. 244

hillii (Queensland), "Stiff staghorn"; handsome, fresh green species with basal leaves always round and covering the rootstock; the several fertile fronds are rigidly erect, gradually broadening fanlike before dividing into numerous pointed lobes; green beneath, the sori carried at base of ultimate tips. An elegant fern for every collection. **C**: 10   p. 244

x lemoinei (said to be hybrid of willinckii x alcicorne or veitchii); prolific plant with basal fronds distinctly kidney-shaped, developing occasional erect lobes; gray fertile fronds very slender, erect spreading, later pendant, twice forked, very narrow in all parts, both sides densely white tomentose; sori V-shaped at tips. **C**: 10   p. 244

madagascariense (Madagascar); small, very attractive species mainly because of its deeply quilted, waffle-pattern, rounded basal frond, with the network of bluish veins raised high and forming tough elevated ridges, small leathery, bright green, wedge-shaped fertile fronds to 10 in. long and wide, lightly lobed to cleft and somewhat rudimentary, and set with sori toward the upper edge. This curious fern does well with warmth. **C**: 60   p. 246, 811

quadridichotomum (W. Madagascar); epiphytic plant with basal frond irregularly oblong and nest-like, appressed and round in the lower part, and erect and free, undulate and laciniate; the normal fertile fronds pendant, regularly in pairs 3 or 4 times, the divisions strap shaped, the underside thickly covered with yellowish stellate hairs. Sporangia area between the first and third forks. **C**: 62   p. 244

"ridleyii" hort. (Malaya, Borneo, Sumatra); epiphyte at home in high trees, allied to coronarium, but more compact; rounded basal fronds mostly appressed to the support; normal frond rather erect, fresh green 1 to 2 ft. long, about 5 times irregularly forked in pairs and sterile, but one branch sometimes carrying at its base a concave fertile lobe bearing the sporangia; the divisions are characteristically short and wide-spreading. The existence of this species is discussed in Baileya Sept. 64, but the photo shown is not yet proven correct. **C**: 60   p. 246

stemaria (aethiopicum), (W. Africa to Madagascar), "Triangle staghorn"; curious species with basal fronds convex and elongated into lobes; the triangular grayish green fertile fronds to 18 in. long, thick-leathery, with prominent ribs, and divided twice, the main fork spreading wide, with a sinus, around which follow the spore masses; the underside densely covered with silvery-white felt. **C**: 60   p. 246

sumbawense (Indonesia: Soembawa Isl.); smaller than alcicorne, and appears to be a variant of willinckii; basal fronds forked, the irregular fertile fronds tomentose; pendant, and deeply forked 2 to 3 times into narrow divisions bearing short patches of sporangia on the thickened last forks. Fronds with marked veins bluish-green on grass-green blade, gray-scaly beneath. **C**: 60   p. 246

vassei (Mozambique), "Antelope ears"; compact species, with large round, kidney-shaped basal fronds protecting its rootstock. The short antler-like fertile fronds are stiff upright, narrow at base and broadening upward, twice forked and thinly pubescent. Spores not quite at tips of fronds as in bifurcatum **C**: 60   p. 246

veitchii (Australia); very vigorous and prolific tree dweller, with rounded, cupping basal fronds and stiffly upright, leathery dark green fertile fronds, narrow at base, and forking into 6–8 broad lobes, white-hairy beneath; sporangia placed at tips. **C**: 60   p. 244

wallichii (Burma, Indochina); attractive epiphyte, with sterile basal fronds first circular and cupping, then with lobes upwards; the fertile fronds 1 to 2 ft. long, broad and several times slender-forked into concave segments, thick in all parts, very green, and slightly yellowish tomentose; looks like grande but more diminutive. **C**: 60   p. 246

willinckii (Java), "Silver staghorn fern"; a distinct epiphyte with uneven, forked basal leaves, and very conspicuous, densely silvery-pubescent fertile fronds, erect at first, later completely pendant, very narrow and several times forked into long slender lobes, sporangia-bearing at tips. Very attractive in its slender, silvery beauty. **C**: 60 p. 245

wilhelminae-reginae (New Guinea), "Queen elkhorn"; magnificent species with large crown of feathered sterile fronds spreading 5 ft.; not as deeply lobed as grande but fuller; the long, gracefully pendant normal fronds to 6 ft. long in pairs, each with 3–4 long lobes, flanked on both outsides by one or two sets of separate obliquely broad-triangular spore-blades; the fronds glossy dark green above, silvery beneath, with prominent dark veins; lettuce serration at base; young growth covered with silvery scales. I have found this species widespread from southern Papua near sea level to the forbidding mountain ranges of northern New Guinea, at 3500 ft. **C**: 60   p. 245

PLECTRANTHUS *(Labiatae)*

australis (parviflorus) (Australia, Pacific Islands), the "Swedish ivy"; vigorous creeping perennial herb with small leathery, thickish, metallic-green, waxy leaves almost round, $1\frac{1}{2}$–$2\frac{1}{2}$ in. across and deeply crenate, glaucous gray-green beneath and with purplish veins; small white 2-lipped flowers in spikes. A tough trailer tolerating abuse; good for hanging pots or wall containers. **C**: 15   p. 263

coleoides 'Marginatus' (tomentosus), "Candle plant"; very charming low, bushy plant, with 4-angled, erect stems, dense with opposite, ovate, herbaceous, hairy, 2 to 3 in. leaves dark green and grayish, the crenate and scalloped margins creamy-white; flowers in erect racemes white with purple. The green type is from the Nilghiris (So. India). **C**: 4   p. 263

nummularius: see australis

oertendahlii (Natal), "Prostrate coleus"; low, fleshy creeper with freely branching 4-angled reddish stem and small, broad ovate leaves 1½ in. long, friendly green to bronzy, patterned with an attractive network of silvery veins, the lightly crenate margins purple, the surface short-hairy; older leaves purple beneath, petioles purple; pale pink bilabiate flowers in erect racemes. An attractive and easy-to-please little foliage plant or trailer. **C**: 4   p. 263

purpuratus (So. Africa: Natal), "Moth king"; wiry creeper with small, lightly crenate, grayish to dark green, fleshy leaves covered with velvety pubescence, purplish underneath; small lavender flowers. Foliage when rubbed said to expel moths. **C**: 16   p. 263

tomentosus (South Africa), "Succulent coleus"; fleshy subshrub freely branching with a tendency to become scandent; square stems spotted with brown-red and resembling coleus but all covered with dense white hair as is the nettle-like foliage, the crenate leaves are succulent fleshy velvety olive green, silvery beneath; raceme of small purple flowers. Attractive house plant in its silvery habit. **C**: 15   p. 271, 350

PLECTROPHORA *(Orchidaceae)*

cultrifolia (Ecuador: Oriente); diminutive epiphyte with little flattened pseudobulbs bearing a single fleshy, knife-shaped leaf 2 in. long, arranged as in a fan; solitary trumpet-shaped, spurred flower ½ in. long on basal stalk; the narrow sepals whitish with greenish flush, petals cream, appressed to the cream-colored, trumpet-shaped lip which is striped orange inside. Grows best on a vertical treefern slab. Requires constant warmth and humidity, without becoming stagnant. Prop. from accessory leaf-fans when produced. **C**: 60   p. 478

PLEIONE *(Orchidaceae)*

formosana alba (pricei) (Mountains of Taiwan); called "Indian crocus", because they throw up their flowers before the foliage, just like crocus; delightful little terrestrial orchid forming clusters, with ovoid pseudobulbs bearing solitary, small 4 to 6 in. deciduous, folded leaves which immediately follow the bloom; the attractive, long-lasting 4 in. flowers rosy-lilac, with white fringed lip blotched pale brown. Spring-blooming, but various according to treatment. Being alpine plants, Pleione love moist, fresh air, and a location near the glass. Easy-growing and free-blooming, they do well in a compost of loam, humus or fernfiber and perlite or sand, several in a shallow pan, watering liberally while in active growth. When the foliage begins to die down it is time to rest them for several months. Divide and replant singly when new root tips appear. **C**: 71   p. 478

PLEIOSPILOS *(Aizoaceae)*

bolusii (So. Africa: Karroo), "Living rock cactus"; small, fairly fast growing stemless succulent with pairs of thick, stone-like keeled leaves 2 to 3 in. long, flattened inside, light gray-green with numerous dark-green dots; 2–3 in. flowers deep yellow. Growing during summer; when resting in winter withhold water. From seed. **C**: 13   p. 328

nelii (So. Africa: Cape Prov.), "Cleft stone", a succulent "Mimicry plant" in form of a split globe with modified thick leaves in pairs, 1¼–2 in. long, the top flat, lower side drawn forward like a chin, gray and with raised, dark dots; showy yellow 2 in. flowers, opening in the afternoon. Resting in winter. **C**: 13   p. 328

simulans (So. Africa: Cape Prov.), "African living rock"; small, highly succulent plant with pairs of thick triangular, sharp-angled leaves 2½–3 in. long, flat or trough-like and dotted on top, the apex pointed; dark green to bronzy in strong sun; fragrant yellow flowers. A picture of broken stones that miraculously bloom. From seed. **C**: 13 p. 355

PLEOMELE *(Liliaceae)*

reflexa (Dracaena) (Madagascar, Mauritius, India), "Malaysian dracaena"; ornamental rosette of densely clustering, short and narrow leathery leaves, deep glossy

green, without midrib, wavy and reflexed, persistently clasping the willowy, selfbranching stem, to 12 ft. high if given support; I have noticed them widely used in India and Thailand as a most satisfactory pot plant; flowers whitish. **C**: 4   p. 198

reflexa variegata, (South India, Ceylon), "Song of India", a beauty I fell in love with when I first saw it in Ceylon; a tropical evergreen; self-branching with slender, flexuous stems eventually becoming scandent to 10 ft. long, densely furnished with clasping, narrow lanceolate, leathery leaves beautifully margined by two wide bands of golden yellow or cream and framing the green center; very slow-growing. From cuttings. **C**: 54   p. 43*, 191, 199

thalioides (Dracaena), (Ceylon, Trop. Africa), "Lance dracaena"; shapely and rather unusual decorative plant of robust habit, with leathery, rich glossy green, abrupt lanceshaped leaves ribbed lengthwise; the base running into a long, clasping, erect, channel-like, gray-spotted stalk. Erect racemes with clusters of beige-white silky flowers, the tubular corolla spreading into linear petals **C**: 4   p. 198

PLEUROTHALLIS *(Orchidaceae)*; remarkable tropical

American epiphytes usually from cloud forests of high mountain ranges, without pseudobulbs, but instead with secondary slender stems from the rhizome, terminating in a single fleshy or leathery, often curious persistent leaf with flowers rising from the leaf base or resting on the blade. Though many of the flowers are small, they are interesting even if often diminutive, and very floriferous. Pleurothallis are of widely varied habit, and are of great interest to connoisseurs. The small species are best fastened to treefern slabs or boards; the larger kind in highly drained shallow pots in fern roots with sphagnum. All love a light location and constant dampness all year without pronounced rest, plus circulating air, and because of their chilly mountain homes generally like to be grown cool. Propagated by division, or the plantlets formed in the floral axis.

cardiothallis (Mexico, Nicaragua, Costa Rica); curious epiphyte with one-leaved rigid, slender stems, to 1½ ft. tall, each carrying a single cordate-lanceolate, deflexed leaf 6 to 10 in. long; tiny ½ in. fleshy flowers appear from the leaf axil one at a time, deep red-brown with green-yellow base; and blooming mostly in summer. As other Pleurothallis, this unusual and delicate species requires almost constant moisture. **C**: 72   p. 478

cogniauxiana (Costa Rica, Panama); curious epiphyte without pseudobulbs, but from the rhizome, secondary stems bear a single, 3 to 4 in. solitary fleshy leaf ovate in outline, most remarkable with the small round, malodorous but intriguing flowers emerging from the axis of the leaf and resting on it; sepals green densely spotted reddish brown; lip deep red in the rear, thickly covered with minute reddish-brown spots elsewhere, and margined with white; blooming in winter time. Pleurothallis like constant dampness or moisture throughout the year, with good drainage, and without pronounced rest, grow best in osmunda possibly with sphagnum or on fern slabs. Propagated by division of the rhizome. **C**: 72   p. 528

tuerckheimii (Guatemala to Panama); interesting epiphyte of the tropical cloud forest; without pseudobulbs, but carrying their solitary leaves on secondary stems 10 in. long; from the apex an arching raceme of odd, waxy 1 in. flowers carried 1-sided in pairs and not opening fully, the purplish-red lower sepals grown together to a long point, the short petals greenish-yellow with blood-red stripes; blooming in summer and fall. This curious species needs to be kept damp the year round, and not too warm. Propagate by division. **C**: 22   p. 478

PLUMBAGO *(Plumbaginaceae)*

capensis, as long known in horticulture; probably more correctly auriculata (So. Africa), the "Blue Cape plumbago"; widely cultivated shrubby perennial of straggling habit, with upright wiry stems from 2 to 8 ft. long and leafy with scattered small oblong foliage and terminal cluster of saucer-shaped azure or light blue flowers having a very slender tube and phlox-like spreading lobes 1 in. or more across, and blooming from spring to fall, and throughout the winter with strong sun. Long stems may be trained against walls; also used in window boxes; best planted out, but also a good house plant, or in larger containers for the patio where the lovely pale blue flowers bloom tirelessly; especially effective if trained into small standard trees. Cut back after flowering and set cool in winter for some rest. Propagated by tip or root cuttings, or division. **C**: 14   p. 492

indica coccinea (rosea) (So. Asia, India), "Scarlet leadwort"; showy tropical perennial, often sprawling or somewhat climbing, with wiry, zigzag stems, 2 to 3 ft. high, set with alternate elliptic 4 in. leaves, and extending into long terminal spike of scarlet-red salver-shaped flowers, with their slender tube 2 in. long; larger and more brightly colored than P. indica; in bloom from spring through summer into winter. Requires more warmth and is best planted out in a conservatory or large container. Propagated from cuttings of nearly ripe wood, by division or seed. **C**: 1   p. 492

PLUMERIA (Apocynaceae); the "Frangipanis" or "Temple trees" are tropical American more or less deciduous small trees widely planted in tropical regions because of the haunting sweet fragrance of their blossoms; in temple courts in Buddhist and Hindu Asia they are sacred. Stiff trees with milky sap, thick succulent branches, long leathery fleshy leaves and large waxy funnel-form flowers in terminal clusters Quite sensitive to cold, leaves tend to fall in winter; they are then kept on the dry side. Very handsome as a container plant on the warm patio, or planted out in the conservatory where they may spread; the flowers are used in Hawaii for leis. Spring cuttings of ripened wood should be allowed to heal their wounds before inserting in sand or peat-perlite and kept warm.

rubra (acuminata) (Mexico and C. America to Ecuador), the rosy "Frangipani trees"; with large waxy, single blossoms 2 in. or more across, carmine-rose with yellow eye, very fragrant, in clusters and blooming from June to September; stout, somewhat soft branches with latex-like, sticky juice, dark-green leaves 10 to 15 in. long, shedding during dry season. **C**: 53   p. 487

rubra acutifolia (Mexico), the sweetly fragrant "West Indian jasmine", and sacred as the "Temple tree" in India; the long leaves are wedge-shaped; exquisite 2 to 2½ in. flowers waxy-white, or tinted pink, with yellow throat, and deliciously scented; a gorgeous sight when older trees are in full bloom with their long-lasting blossoms in large clusters against the blue sky of the tropics. **C**: 51   p. 487

PODOCARPUS (Podocarpaceae); the "Buddhist pines" are coniferous evergreen resinous trees of the Southern hemisphere, native in the mountains of warm regions; very dense and formal in habit, with flat persistent, usually leathery and needle-like leaves, very handsome and noble in a somber way, and with catkin-like male flowers; berry-like fruit succeeding the greenish flowers on female trees. Generally best for cool locations, Podocarpus are very durable decorator plants provided thay have been well established in tubs; freshly dug trees are a risk and easily succumb. While preferring sun they will adjust down to 50 fc. of light. Soil should be kept moist if well drained. From seed or cuttings.

elatus (Australia), "Weeping podocarpus"; subtropical evergreen with willowy, furrowed branches arching and drooping with weight and age, loosely furnished with long-linear leathery leaves shiny dark green. Requires space for display **C**: 1   p. 308

elongatus (spinulosus) (Western and Southern Trop. Africa), "Weeping fern-podocarpus" or "African yellow-wood"; coniferous evergreen tree to 70 ft. high, with gracefully pendant branches, densely pinnate, with long soft-leathery, flat, narrow-linear, tapering leaves waxy dark green and 3 in. or more long; short male catkins, and globose crimson fruit. Very attractive decorator of weeping habit. Needles are longer than on P. gracilior. **C**: 1   p. 308

gracilior (Kenya, Uganda, Ethiopia), "African fern pine"; sub-tropical coniferous tree of erect habit to 60 ft. high, common on the slopes of Mt. Kenya at 7000 to 9000 ft., and a valuable timber tree; graceful willowy branches with long, narrow-lanceolate, needle-like leathery leaves glossy deep bluish green, to 4 in. long on young trees, and loosely arranged; 2½ in. long and dense on older specimen; glaucous, purple berries. This may be the same as the cultivated P. "elongatus" of horticulture; P. gracilior is apparently variable in habit and foliage with age, and plants from cuttings taken from older wood have a tendency to be more arching and pendulous; from vigorous seedlings more erect. **C**: 1   p. 308

macrophyllus (chinensis), (China, Japan), "Buddhist pine" or "Japanese yew"; dioecious coniferous tree to about 40 ft. high, with horizontal branches, and numerous crowded leafy twigs; the leathery, deep green, narrow linear-lanceolate leaves needle-like, 2–2½ in. long as seen in China, in cultivation 3 to 4¾ in. long; with a single midrib prominent on both sides; male, axillary flowers resembling catkins; berries bluish purple. **C**: 14   p. 309, 541

macrophyllus 'Maki' (Japan, China), "Southern yew" or "Bamboo-juniper" in Japan; widely cultivated evergreen shrub; for hedges in the South, and grown as a superb decorative container plant in the North; lends itself well to shearing and shaping; as a tree attaining 50 ft.; this cultivar is of more dense, compact habit, with rather erect branches, and needles smaller than macrophyllus, the waxy, blackish-green, linear-lanceolate leaves dense and spirally arranged, 1½ to 3 in. long and about ¼ in. wide, with distinct midrib; pale green beneath. Male flowers in 1½ in. catkins; the fleshy oval fruit on female trees glaucous-purple. **C**: 14 p. 309

Nagi (China, Japan, Taiwan), "Broad-leaf podocarpus"; tall conifer to 90 ft. high, with smooth, purplish bark, elegant spreading branches and slender, semi-pendant branchlets having shiny green, rigid-leathery, elliptic leaves 3 in. long by 1–1½ in. wide. The Japanese word Nagi stands for broad and short. Very durable display plant, more tree-like in youth than other Podocarpus and distinctly handsome in its decorative foliage pattern, for shade or sun, indoors or out. **C**: 13   p. 309

POINCIANA (Leguminosae)

pulcherrima, better known in horticulture as Caesalpinia pulcherrima, (W. Indies and other tropics),"Dwarf poinciana" or "Barbados pride"; sunloving, handsome shrub to 10 ft. high, with thorny, woody stems and twigs; delicate feathery bipinnate, mimosa-like fresh-green leaves, and flowers very showy, orchid-like, a gaudy orange-red with crisped, golden-edged 1 in. petals and protruding red stamens 2½ in. long; blooming throughout the hot sunny season, from March on. Foliage deciduous in winter if chilled. An excellent flowering decorator plant for sunny rooms or patio; not to be confused with the Royal Poinciana, Delonix regia. Propagated by seeds soaked in warm water for several hours; by layering, or soft cuttings moist-warm under glass **C**: 52   p. 487

POINSETTIA: see Euphorbia pulcherrima

POLIANTHES (Amaryllidaceae)

tuberosa (Mexico), the famed "Tuberose"; widely cultivated in the tropics where it is esteemed for the purity and powerful fragrance of its blooms as a cut flower and in gardens; a beautiful summer or fall blooming herb having a bulb-like tuberous root stock covered with the broadened bases of the grass-like, channeled leaves; the leafy floral spikes are wiry and 2 to 3 ft. high, bearing numerous funnel-shaped waxy-white flowers 1½ to 2½ in. long, in pairs. The tuberose may be had in flower throughout most of the year by potting bulbs in succession, taking 4 to 5 months to bloom; grow warm but keep dry until the leaves appear, then water freely; dry off after blooming when foliage turns yellow. By division of clumps or from offsets. **C**: 52   p. 390

POLYPODIUM (Filices); the "Polypodies" are a great genus of low-growing ferns mainly tropical from all parts of the world, many of them epiphytic and dwelling on rocks or trees, slowly creeping on slender or stout rhizomes covered by chaffy scales, and hugging their host or the ground and over the edge of pots, and from which rise wiry stalks carrying the usually thin-leathery fronds; these may be simple and sword-like, pinnate like a feather, broad and lobed, cut or compound; erect for pots or pendant and used for hanging baskets. Polypodium generally are very satisfactory as house plants if provision can be made to provide some humidity to the foliage such as misting them frequently or standing in a tray with water and pebbles or peatmoss; otherwise they are not particularly demanding. Some kinds are nice in baskets or fastened to treefern slabs, these can be dunked in water when becoming dry. Propagated by separating the rhizomes, or from spores.

angustifolium (Campyloneuron) (Cuba, Jamaica, Mexico, Peru, Brazil), "Ribbon fern"; distinctive epiphytic fern, with creeping rhizome covered by brown scales, bearing narrow-linear, ribbon-like leathery dark green fronds 1–1½ ft. long, with margins often rolled under, Becoming very bushy, and suitable for baskets or on blocks. **C**: 60 p. 251

aureum glaucum; a blue form of the "Hare's-foot fern" with stout creeping rhizomes clothed with bright rusty brown hair-like scales and resembling little feet; the wiry stalks bearing bold, thin-leathery fronds an attractive glaucous silvery blue, especially underneath; and lobed, with segments separated by a rounded sinus and not cut to center. The species aureum is wide-spread from the West Indies to Brazil, also in Australia where it grows epiphytic in trees, with leaves normally metallic light green. A very durable and pretty fern for the house, wanting good drainage. **C**: 10 p. 240

aureum 'Mandaianum' (Manda 1912), the "Crisped blue fern"; a beautiful crested form having graceful bluish glaucous fronds 12 to 18 in. long of broad, pendulous wavy pinnae, with margins irregularly lobed, crisped and lacerated. This beautiful form may be maintained by careful reselection; propagated by sectioning of the rhizome. A good, warmth-loving house and basket plant, retaining its shimmering blue color best, away from sun. **C**: 12 p. 240, 286

aureum 'Undulatum', the "Blue fern"; a sturdy cultivar of compact habit; the large fronds on stiff-wiry stalks from the furry rhizomes, gracefully arching, bluish glaucous and deeply lobed, the broad segments distinctly undulate and wavy. **C**: 10 p. 194

bifrons, (Ecuador); a very unusual creeping fern with wiry slender rhizomes carrying small oblanceolate, deeply lobed, both sterile and fertile fronds but besides, forming curious chestnut-like vessels hollow inside pocket-like, similar to Dischidia, separate from its leaves. Rare and difficult in cultivation, best trained to a piece of tree fern trunk and hung from a greenhouse rafter; needs warmth and humidity and should be dunked in water daily. **C**: 62 p.539

irioides: see polycarpon

musifolium (Phymatodes) (Malaya, Philippines, New Guinea), "Bananaleaf fern"; handsome epiphyte with woody rhizome bearing stalkless, oblanceolate, thin-leathery leaves to 3 ft. high, pea green, prettily marked with a network of dark veining. Grow in wide, shallow pot to contain rhizome. **C**: 62 p. 251

phyllitidis (Campyloneuron), (So. Florida, So. Brazil), "Strap fern"; found in the swampy Florida Everglades; strap-shaped, entire, thin-leathery, glossy fresh green, brittle-stiff fronds 1–3 ft. long, and to 4 in. wide, more or less wavy, borne stalkless on short creeping rhizome, clothed with brownish scales. **C**: 72 p. 251

polycarpon (irioides in horticulture), (New South Wales, Natal, Angola, Guinea); singular-looking, succulent fern with stout rhizome and stalkless, thick-fleshy yellow-green, simple fronds to 3 ft. long, gradually narrowed on both ends, and irregularly indented or undulate at margins. Propagated by division. **C**: 4 p. 251

polycarpon 'Grandiceps' (irioides grandiceps), "Fish-tail"; clustering fern with odd-shaped, thick leathery, almost succulent, waxy yellow-green fronds 1 to 2 ft. high with prominent midrib, and tips forking to points or broad crests. Very curious, yet attractive and durable house plant. **C**: 4 p. 251

polypodioides (Delaware to Florida and Texas, Trop. America, So. Africa), "Resurrection fern"; a drought-resisting fern epiphytic on trees in Southern states; small, interesting creeping fern, hardy in temperate climate, with thin, wiry rhizomes, bearing small leathery, pinnate fronds 2 to 6 in. long, matte grass green, the alternate pinnae with small brown, scaly dots evenly dispersed on the grayish reverse; the fronds fold and curl up during dry periods, reviving when again moist. **C**: 72 p. 240

subauriculatum (Goniophlebium), (Malaysia, Philippines), "Jointed pine fern"; elegantly decorative basket fern with long pendant, leathery, pinnate fronds to 10 ft. long, produced from creeping rhizomes; the well spaced linear, wavy segments are fresh glossy green. Should have moist-warm greenhouse; sensitive against frequent transplanting when in wire baskets: do not cover rhizomes. Propagation by division. **C**: 74 p. 248

subauriculatum 'Knightiae', (Australia), "Lacy pine fern"; an excellent and beautiful, slow growing and durable basket fern, with glossy yellow-green, pinnate fronds at first upright, later pendulous and long, the linear pinnae deeply serrate and sliced into narrow lacy, pointed lobes. **C**: 62 p. 248, 286

vulgare (Newfoundland to Alaska and Alabama; Eurasia), "Common polypody" or "Adder's fern"; ornamental, hardy evergreen fern, often epiphytic, growing on walls,

roofs or trees; mat-forming on stout, rusty-scaly rhizomes; straw-colored stalks bearing 6–12 in. pinnate fronds, the papery leaflets toothed and wavy. **C**: 72 p. 240

vulgare virginianum (E. No. America to Labrador), the American "Wall fern", is a hardy fern practically evergreen, growing on rocks, by creeping rhizomes, the pinnate leathery, vivid-green fronds with wavy segments on wiry stems. P. vulgare, the "Common polypody" of temperate regions of Europe, Asia, and Western No. America, often growing epiphytic, has larger fronds 6 to 15 in. long. The fibrous roots are used as planting material for young orchids. **C**: 72 p. 240

POLYPODIUM: see also Aglaomorpha, Drynaria

POLYSCIAS *(Araliaceae)*; Tropical evergreen free-branching shrubs with flexuous, willowy stems and exceedingly variable leaves, also between younger plants, and forms different on older specimen; often known in horticulture as Aralia or Panax, their foliage is not only interesting but also very attractive; widely used for hedges in the tropics, they have also been adapted as home and decorative plants where they have proven better keepers than expected; once acclimated, they prefer some shade, and need to be kept moist, watering when the surface of the soil shows dry. Propagated from tip cuttings, joint cuttings, or root sections.

balfouriana (Aralia) (New Caledonia), "Dinner plate aralia"; leafy, bushy tropical shrub in habitat to 25 ft. high, branching with willowy stems; the large leathery leaves variable but at first entire, later usually of 3 rounded, coarsely toothed glossy green leaflets to 4 in. across, on bronzy stems speckled gray. Needs moisture with nourishing soil and good drainage. **C**: 2 p. 291

balfouriana marginata, (New Caledonia), "Variegated Balfour aralia"; variegated form with the grayish-green, leathery leaflets having an irregular, white border. Much planted for hedges in tropical regions; a good house plant if properly kept moist. **C**: 4 p. 195, 291

balfouriana 'Pennockii', "White aralia"; attractive and sturdy Puerto Rican cultivar with large leathery, cordate oval 4 in. leaves having crenate margins, variegated and tinted creamy-white to pale green, irregularly dark green toward edge. **C**: 4 p. 195, 291

filicifolia (Aralia) (South Sea Islands) "Fernleaf aralia" or "Angelica"; evergreen shrub with flexuous stems and leathery but variable leaves, bright green and with purplish midrib, pinnate with leaflets cut into narrow lobes; fern-like in younger plants, broader and often oblong and entire when older. **C**: 2 p. 291

fruticosa (Aralia), (India, Malaysia, Polynesia), the "Ming aralia"; sometimes known in horticulture as P. filicifolia; tropical evergreen shrub freely branching with many willowy and twisting slender brownish stems to 8 ft. high; densely clothed, especially toward the tops with beautiful fern-like, lacy light green foliage 3 times pinnately cut into narrow serrate, stalked leaflets, on arching purplish-brown petioles. A decorator's delight with its artistic feel, and once acclimated a very good room plant tolerant to all locations not too cold; from 20 to 200 fc. of light. Water when surface of soil shows dry. Propagated from cuttings. **C**: 2 p. 287, 289, 291

fruticosa 'Elegans' (Aralia, Panax) (Polynesia); compact, branching evergreen bush dense with leathery leaves dark green and bipinnate, with the stalked pinnae deeply cut into broad, toothed lobes. Very elegant as a small indoor plant. **C**: 2 p. 291

guilfoylei 'Quinquefolia', "Celery-leaved panax"; a selected leaf form grown in Florida, with its hard-leathery leaves more or less irregularly cut with five divisions or lobes, in deep coppery olive green. **C**: 2 p. 291

guilfoylei victoriae (Polynesia), "Lace aralia"; charming tropical evergreen dense with slender, willowy branches and grayish-green, thin-leathery, lacy, bipinnate leaves, the small, pendant, feathery segments toothed and bordered white. Lovely small foliage plant enjoying warmth and moisture, with its tasseled variegated foliage the most exquisite of the variegated leaf forms. **C**: 54 p. 195, 291

paniculata 'Variegata', "Variegated roseleaf panax"; attractive variegated form of the species from Mauritius; willowy shrub with pinnate leaves, the leaflets leathery and deeply serrate, deep green and richly splashed with cream and greenish-white, glossy on both sides. **C**: 2 p. 195, 291

POLYSTACHYA (Orchidaceae)

cucullata (West Africa, Nigeria); charming epiphyte with slender cylindric pseudo-bulbs bearing solitary leaves to 6 in. long; from the apex an erect stalk carrying small clusters of 1 to 3 heavy-textured rather complex flowers 1½ in. across, their greenish-yellow sepals spotted purple, small greenish petals horn-like, the lip greenish and marked purple; blooming fall to winter. Best grown in small pots in osmunda or treefern fiber; also treefern slabs; need warmth with high humidity and liberal watering while growing; thereafter a brief rest. Propagated by division. **C**: 60   p. 478

POLYSTICHUM (Filices); evergreen ferns of temperate regions, including the group known in horticulture as Aspidium; belonging to the family Polypodiaceae; they are of small to medium habit, and with a short rhizome; their fronds are variously pinnate to 4 times divided, and usually of a thin-leathery texture, making them good keepers when grown in pots indoors. They are easy to grow but do however prefer a cooler location, constant moisture and shade. Propagated from spores, or division of root stocks. Some species are winter-hardy.

acrostichoides (No. America: Nova Scotia to Texas), "Christmas fern" or "Dagger fern"; hardy evergreen leather fern, with tufted pinnate fronds similar in appearance to Nephrolepis; 1 to 2 ft. long and 2 to 4 in. wide, thin-leathery and bright glossy green; green, stiff axis, brown near base, and covered with brown scales, on short creeping underground rootstock; the 24-30 pairs of pinnae oblique-halberd-shaped and finely serrate; spores set toward apex of frond. **C**: 77   p. 250

aristatum variegatum (Japan to Ceylon and Australia), the widespread "East Indian hollyfern"; creeping rhizomes with elegant leathery fronds 3 to 4-pinnate, 1 to 2 ft. long, the lower pinnae largest, on thin wiry stalks, the segments banded pale yellow along axis. **C**: 18   p. 194, 238

coriaceum: see Rumohra

setiferum proliferum (viviparum), (W. Indies), "Viviparous filigree fern"; cool-loving tufted fern, scaly at base; the fleshy, brown-woolly stalk bearing the pinnate, light green fronds to 1½ ft. long, the pinnae deeply and finely cut or lobed; bud-bearing in the axils of the pinnae, giving rise to young bulbils and plantlets. The species setiferum is wide-spread in Tropical and Temperate zones of both hemispheres, better known as the variable, winter-hardy "Hedge fern". Propagate by fastening the fronds bearing bulbils to the ground until rooted, or cut off with piece of stem and root in peat-sand in moist warm enclosure. **C**: 83   p. 249

setosum (discretum) (Japan), "Bristle fern"; low, spreading plant with stiff brown-scaly stalked, and glossy dark green, leathery, bipinnate fronds to 1½ ft. long, the pinnae dense with pinnules (secondary segments) overlapping, and ending in a bristle. **C**: 68   p. 250

tsus-simense (Aspidium), "Tsus-sima holly fern"; from the island of Tsus-sima in the Straits of Korea; dwarf and shapely tufted fern with small leathery, lanceolate, dark green fronds to 8 in. long, bipinnate in the lower part, the segments becoming gradually smaller toward the slender point and sharply toothed. As a miniature fern in 2 or 3 in. pots very suitable for terrariums, growing very slowly and with good keeping quality. Older specimen may be divided. **C**: 22   p. 238

PONCIRUS (Rutaceae)

trifoliata (North China), "Hardy orange" or "Trifoliate orange"; small, stiff-growing spiny, deciduous tree of vicious, sparry habit, to .15 ft. high; dark green flattened branches armed with long stout spines to 2½ in. long; blooms in spring on bare branches in axils of large spines; fragrant white flowers opening flat, to 2 in. across; trifoliate leaves with thin-leathery shining green leaflets, on winged petiole; small orange-like aromatic fruit to 2 in. dia., containing scant acid pulp, not edible. The hardiest of all citrus fruits; fairly so north to Philadelphia and New Jersey. Good as an impenetrable hedge; also used as grafting stock for more tender citrus. **C**: 75   p. 510

PONTHIEVA (Orchidaceae)

maculata (Venezuela, Colombia); terrestrial or epiphytic orchid of tufted habit, with usually two elliptic basal, thin leaves to 1 ft. long, pale green and soft-hairy; the inflorescence a hairy, spike-like cluster of 1 in. spreading flowers, the dorsal sepal pale brown with dark streaks, lateral sepals white with brown spots, petals yellow with brown streaks, the small fleshy lip yellow with brown stripes; blooming in March-April. Usually grown in rich, porous compost of loam and peatmoss, leafmold or fern fiber. Prop. by division. **C**: 60   p. 478

PORTEA (Bromeliaceae)

petropolitana extensa (Brazil: Espirito Santo, Rio); tall rosette of shiny yellow-green leaves to 2 ft. long, broadened at base and with blackish spines; showy panicles on a striking coral-red arching stalk, the brilliant coloring extending to the slender green ovaries tipped purple; flowers lilac. **C**: 9   p. 142

PORTULACA (Portulacaceae)

grandiflora (Brazil), "Rose moss" or "Sun plant"; annual succulent herb to 6 in. high with low spreading, reddish branches; scattered, cylindrical 1 in. leaves and lustrous, rose-like flowers 1 to 1½ in. across, normally carmine red with white center, but also in rose, cerise, orange, yellow, or white forms, and in double-flowered strains. Free-blooming in their brilliant colors all summer and fall, the flowers open fully only in sun, close in late afternoon. Wonderful in pots at the sunny window or patio. Grown as an annual from seed; the doubles safer from cuttings. **C**: 63   p. 353

PORTULACARIA (Portulacaceae)

afra variegata, "Rainbow bush" or "Elephant bush" in variegated form; lovely little succulent with fleshy red-brown stems and sparry branches dense with opposite, pretty ¾ in. leaves milky-green broadly margined creamy-white, and with a thin carmine-red edge. The green species is from South Africa, a dense succulent shrub 6 to 12 ft. high, with dainty pink flowers. Very durable plant; from cuttings. **C**: 15   p. 329

POTHOS (Araceae)

argyraeus: see Scindapsus pictus

aureus: see Scindapsus aureus

hermaphroditus (Java), a "True pothos"; characterized by the leaves appearing as if constricted through the middle; shrubby tree-climber with hard, narrow leaves arranged on both sides of branches. The Greek word "Pothos" personifies desire, identified with the winged god Eros. **C**: 60   p. 83

jambea (Java); tree-climber of the rain-forest, with slender branches and the alternate, ovate leaves relatively large, and set on short, winged stalks; a true Pothos. **C**: 54   p. 83, 259

POTOMATO (Solanaceae); this scientific curiosity, also known as Topato or Pomato, is the fanciful result of grafting a tomato scion (Lycopersicon esculentum) on a sprouted potato tuber (Solanum tuberosum), producing a true chimera, the meeting of the tissues of scion and understock. Theoretically this union is supposed to produce potatoes in the ground and tomatoes above, but this does not necessarily happen; however it is fun to try. My own experiments were confined to plants in large pots grown in the greenhouse. The process of grafting is not difficult: a potato is started in a big pot of soil and set into a sunny location until it sprouts; when strong enough it is cut down to about 6 in. above the soil, split into the middle, and a wedge-graft, made of the top of a tomato plant, is inserted into the understock and tied together with string or raffia, until both cuts heal and unite, growing together. **C**: 52 p. 537

PRIMULA (Primulaceae)

malacoides (China: Yunnan), "Fairy primrose" or "Baby primrose"; small bushy herbaceous plant grown as an annual, with numerous light green, smallish, papery leaves, white-hairy beneath and toothed at margins; several straight stalks bear pretty, long lasting flowers in successive umbels above each other, in pastel colors from lavender and rose-pink to crimson-red or white, from late winter into spring. Much progress has been made in breeding to increase size of flowers from the original ½ in. to now 1 in. or more across and into shades of salmon-rose with deeper eye and orange-yellow center. Grow best cool 45° F. Very pretty as small plants in 4 or 5 in. pots. Non-irritating to skin. From seed. **C**: 16   p. 370

obconica (China: Hupeh), "German primrose"; a winter-blooming pot-primrose grown as an annual; with fresh-green, brittle, broad-cordate strong-scented leaves sparsely covered with irritating hairs, and showy umbels of large flowers 1 to 2 in. or more across, in pastel shades or rose-pink to carmine-pink, lilac to almost blue, or even white, with greenish eye. Primroses like cool semi-shade, and bloom best from early winter to May, but in cool enough climates the year round. The glandular hairs of this species may cause irritation of the skin, or dermatitis, when touched, but not everyone is allergic to it. From very fine seed. **C**: 16   p. 370

x polyantha (a hybrid group of P. vulgaris probably with veris and elatior), the hardy "Polyanthus primrose"; "Lady's fingers" or "English primrose"; popular in American gardens and blooming in early spring; clump forming perennial 8-12 in. high with long obovate leaves narrowed into winged petioles; the 1 to 2 in. flowers sweetly fragrant like roses, in many colors, yellow, red and yellow, orange, bronze, maroon, or with white, borne in clusters well above the foliage; (those of vulgaris are solitary). Easily grown, this beautiful primrose in a riot of colors presents a brilliant display at flower shows or as a small pot plant if taken from frost-free cold frame into a cool location to gently force but never over 50° F. From seed or division. **C**: 78   p. 370

sinensis (China, Himalayas), "Chinese primrose"; winter-blooming pot plant with a rosette of long-stemmed, hairy, strong-scented leaves attractively lobed, the flower clusters first low and small, later pushing up on heavy stalks and becoming larger; the blooms mostly mauve with yellow eye, but also in scarlet to pink and white. Somewhat brittle and subject to bud-drop, and "poor keepers"; very sensitive to over-watering, requiring good drainage. Perennial in habit, but in cultivation as annual grown from seed. **C**: 71   p. 370

PROTEA (*Proteaceae*); singularly distinctive African members of the Australian oak family; slow-growing evergreen shrubs characteristic to South Africa, with woody branches bearing a spectacular and wondrous terminal inflorescence, consisting of long slender tubular flowers in large tight clusters, often disclosing numerous colored stamens, and surrounded by showy, leathery or rigid brightly-hued or silky petal-like bracts overlapping like scales and resembling a large colored artichoke or thistle, lasting for several months. When the heads first open they are full of honey and are known to the Africans as "honeypots". Proteas are superb cut flowers but away from their home are not easy to grow; in the cool greenhouse growing in a large tub, they delight in full sunlight and abundant ventilation. Out-of-doors in summer most like an acid soil and moisture with perfect drainage, but dislike being disturbed. Propagated from seed which germinates best during low temperatures; cuttings from firm wood, with the aid of hormone powder, may take 3 months to root, best under constant mist.

barbigera (So. Africa: S.W. Cape Prov.), "Giant woolly-beard" or "Frilled panties"; most handsome, sparry, slow-growing evergreen shrub to 5 ft. high; white-hairy, light gray-green, oval, leathery leaves 6 in. long, undulate and hairy at margins; the beautiful inflorescence to 8 in. across, balls resembling pine cones; the outer bracts soft pink or rose, tipped with fine silvery-white hairs, and surrounding a soft mass of white-woolly flowers which become black-violet in the raised center of the flower heads. Blooming in South Africa from June to November, their mid-winter to early summer. **C**: 63   p. 489

cynaroides (So. Africa: Southwest Cape Prov.), the famous "King protea" or "Honeypot sugarbush"; characteristic, most spectacularly showy shrub of spreading habit, 2 to 6 ft. high, with long-stalked, varying roundish to ovate thick-leathery stalked leaves 2 to 5 in. long, edged red, on woody red stem; the snowy peak of hairy flowers packed in the center of a large inflorescence 8 to 12 in. across, surrounded by numerous shingled series of stiff leathery pointed bracts as in a cup, white to delicate pink and silvery silky-downy, blooming in May-June in our Northern hemisphere, from December to May south of the equator. **C**: 63   p. 489

PRUNUS (*Rosaceae*)
caroliniana cv. 'Bright'n Tight', (Laurocerasus) a compact and very dense form of the "American cherry-

laurel" also called "Mock-orange"; the native species is at home from No. Carolina to Texas, an evergreen tree to 40 ft. or more; dense with oblong-lanceolate, leathery, glossy green leaves 2 to 4 in. long; small, cream-white flowers in racemes; little shining black fruits but inedible. Very handsome evergreen, and shaped specimen with their lustrous foliage are wonderful on the patio or cool entrance halls or the winter garden. From cuttings. **C**: 14   p. 288

laurocerasus (S.E. Europe to Iran), "English laurel" or "Cherry-laurel"; handsome decorative, quick-growing evergreen bush up to 20 ft. high with smooth, pale green shoots, and broad, leathery, dense foliage, dark glossy green, oblong-pointed, to 6 in. long; small, fragrant white flowers between leaves followed by dark purple fruit. With its grand foliage an impressive decorator for the solarium and other cool locations, and widely used as a tub plant in Europe; tolerant to sun or shade, stands shearing well, but requires lots of water and fertilizer. Somewhat frost-hardy. Easily from cuttings. **C**: 78   p. 292

triloba plena (China); known as the double "Flowering almond" more correctly a "Double flowering plum"; a graceful ornamental shrub 6 ft. or more high with woody, erect branches and serrate, deciduous leaves; sometimes trilobed; the dainty, double pink flowers flushed with rose, 1½ in. across, unfolding previous to the foliage from tight round buds, set tightly along woody branches of last year's growth, blooming in spring about April. The species normally has 1 in. single flowers. Fairly winter-hardy. Grown as a small standard it is charming for Easter forcing. One of several "Flowering almonds", the true Almond is P. amygdalus, and which produces the edible kernel within its stone; the dwarf Almond is P. tenella. Prunus triloba is a good subject for blooming in pots; it flowers on previous summer's wood and the little trees are cut hard after blooming, the leaders to 3 eyes, branches to 2 eyes; pots are plunged outdoors in summer, kept in cold frames between salt hay until needed; they force willingly from January on (see page 30). Propagated by grafting on plum understock, also from own-rooted plants by layering or softwood cuttings in summer. **C**: 76   p. 504

PSEUDERANTHEMUM (*Acanthaceae*)
alatum (Mexico), "Chocolate plant"; low growing, somewhat sensitive herb with copper-brown papery leaves, on flat, winged petioles, silver blotching near midrib; gray beneath; small salverform, purple flowers in racemes. **C**: 60   p. 184

atropurpureum tricolor (Polynesia); colorful tropical shrubby plant with small, elliptic, leathery leaves metallic purple, variously splashed with green, white and pink. **C**: 4   p.186

reticulatum (New Hebrides); tropical shrub with attractive lanceolate smooth foliage, slightly fleshy, green with reticulation of golden-yellow veins; 1¼ inch flowers with wine-purple in throat, and dots of same color on lower lip. **C**: 54   p. 186

PSEUDOPANAX (*Araliaceae*)
lessonii (New Zealand), "False panax"; small multiple-stemmed evergreen tree 8 to 20 ft. high, with slender branches and palmately compound shining green leaves, the obovate segments are about 4 in. long, thick and leathery, and toothed toward apex. An excellent, tough decorator for cooler locations, the solarium or the patio, withstanding draft, but requiring moisture. **C**: 16   p. 291

PSEUDORHIPSALIS (*Cactaceae*)
macrantha (Mexico: Oaxaca), "Fragrant moondrops"; epiphytic cactus with thin flattened joints waxy green, scalloped at the margins and branching toward tip; the 1 in. flowers from the areoles with linear outer petals yellow, the inner petals white, of intense lemon fragrance; the fruit a red berry. **C**: 10   p. 172

PSEUDOSASA (*Gramineae*)
japonica (Bambusa Metake), (Japan), the "Female arrow bamboo" or "Hardy Metake bamboo"; a running bamboo of moderate size; with hollow round stems 6 to 15 ft. high and to ¾ in. dia., from creeping rootstocks, with broad deep green foliage 4 to 12 in. long, glaucous beneath. Fairly hardy in New Jersey, has withstood freezing to zero° F. One of the best for decoration in tubs, holding leaves better than other species; for cool bright rooms; keep moist. Propagated by division. **C**: 26   p. 302

PSIDIUM (Myrtaceae)

cattleianum (Brazil), "Strawberry guava"; dense shrub 8 to 10 ft. or more high, the smooth branches greenish-gray to golden brown, with obovate leathery, glossy leaves 2–3 in. long; white 1 in. flowers composed principally of a brush of many stamens, and 1½ in. berry-like dark purplish-red fruit with white flesh of sweet-tart strawberry flavor, in fall and winter; used in making jellies. Excellent in containers. From seed or layers, also root sprouts. **C**: 14   p. 508

guajava (W. Indies, Mexico to Peru), "Common guava" or "Guava"; small branching tree with brownish, scaly bark, reaching 30 ft.; 4-angled branchlets with light green deeply nerved, relatively thin elliptic leaves 4 to 6 in. long, hairy beneath, an attractive salmon when young; large 1 in. white flowers; and producing edible globose or pear-shaped yellow, sweet-flavored fruit with purplish flesh 1 to 3 in. across; used for making jam. From seed, layers, root sprouts. **C**: 14   p. 508

PSILOTUM (Psilotaceae)

triquetrum (New Zealand and subtrop. and trop. So. and No. Hemisphere), the "Whisk fern"; curious club-moss botanically interesting as a very primitive vascular plant, supposedly the first step of evolution from liverworts toward higher plants; they resemble more closely than any other living forms the extinct Psylophytales, the most primitive vascular plants yet discovered, and regarded as the parent stock from which all other vascular plants may have evolved; the plant is a herbaceous perennial devoid of roots and true leaves, growing both as epiphyte and terrestrial, with the underground and ribbed aerial stems with numerous scale leaves; erect or pendant branched green shoots to 20 in. long, with yellow, globose sporangia looking like flowers, in leaf axils. We see them in Florida and Hawaii, on tree ferns, and other host trees. **C**: 62   p. 538

PTERIS (Filices); known as "Table ferns"; these are generally small, mostly tropical ferns of the family Polypodiaceae, with short scaly rhizome and a wide variety of foliage, some to 4 ft. tall, others short, but mainly popular because of their wide use as "Table ferns", in fern dishes or terrariums, in combination with flowering plants; also as pot plants or in ferneries. These rather robust ferns are not too particular as to soil mixtures or watering, while shade-loving also tolerate strong light and some neglect. But after a while they will outgrow their usefulness and become unsightly unless given fertilizer or transplanted. Many attractive or variegated cultivars have been introduced. Propagated from spores. Keep cooler in winter.

adiantoides: see Pellaea viridis macrophylla

cretica 'Albo-lineata', "Variegated table fern"; a very pretty, useful, variegated form of low habit, with small, clean-cut, leathery fronds, differing from the species only in the broad band of creamy white, down the center of each linear lanceolate leaflet, which are toothed and wavy-margined; the fertile fronds are taller to 2 ft. high and more slender. The very variable species cretica has given rise to most of the attractive horticultural cultivars, being of strong, robust habit with plain green, parchment-textured fronds 6 to 12 in. long, on wiry stalks. **C**: 6   p. 243

cretica 'Rivertoniana' (Dreer 1915), "Lacy table fern"; a bushy, symmetrical form of wimsettii, and improved gauthieri; with stiff-erect fresh green fronds 12 to 16 in. high; of firm texture, the brown stems set with 4–5 pairs of lateral pinnae and terminal, the lower ones compound, all deeply cut almost to center, into pointed toothed lobes. **C**: 6   p. 243

cretica 'Wilsonii', (ouvrardii, cristata), "Fan table fern"; an excellent commercial low "Table fern" basically 6–8 in. high, the fresh-green thin-leathery fronds of a young plant spreading; the fertile fronds 12 to 16 in. high, with segments tending to form a fan-shape, and forking toward tips into broad, dense crests. **C**: 6   p. 243

cretica 'Wimsettii', "Skeleton table fern"; a robust, variable form of medium height to 18 in. tall, making the first break into cresting away from the plain species, and desirable because of its almost leathery toughness; the slender fresh-green leaf segments irregularly toothed or pointedly lobed, and some of the tips terminate in small forks or crests. Tolerates abuse, chills, dryness, and varying light; in a greenhouse nearly becoming a persistent weed. **C**: 4   p. 243

dentata (Trop. and So. Africa), "Sleepy fern"; bushy, large fern with broad bipinnate fronds 2–3 ft. long, on straw-colored stalks; the pinnae fresh green and somewhat soft, with linear lobes finely serrate. Grown in California mainly for use in ferneries and on the patio, not as a "Table fern". **C**: 54   p. 243

ensiformis 'Victoriae', "Silver table fern" or "Victoria table fern"; an elegant, graceful little fern 6 to 14 in. high with both the short, broad sterile fronds and the abundant, erect, slender fertile fronds having leathery leaflets beautifully banded white, bordered by a wavy margin of rich green. Quite sturdy, and beautiful in fern dish or terrarium, remaining fairly small. The parent species ensiformis is widespread from the Himalayas to Ceylon, to Queensland and Samoa, growing to 12 in. high, plain deep lacquer-green. **C**: 6   p. 194, 243

multifida (serrulata) (Japan, China), the "Chinese brake"; in commercial use since 1825; it has graceful, light-green fronds on short stems which have wings, gradually narrowing downward; the segments are slender-narrow, and distantly spaced. **C**: 6   p. 243

quadriaurita 'Argyraea', (Central India), "Silver bracken"; robust fern with beautiful large, herbaceous fronds to 3 ft. high, the pinnae deeply lobed, in pairs along stalks, the lower ones forked, and large terminal leaflet, all attractively light to bluish-green and with a center band of silvery white along the midrib. **C**: 6   p. 194, 243

quadriaurita 'Flabellata', (So. Africa to Ethiopia, Fernando Po), "Leather table fern"; an elegant strong fern 12–15 in. high with blackish wiry stalks bearing numerous pairs of evenly pinnate, thin-leathery leaflets a bright glossy green. **C**: 4   p. 243

semipinnata (Japan, Hong Kong, Himalayas, Philippines), "Angel-wing fern"; singular-looking species, bearing on erect, black wiry stalks fronds to 1½ ft. long, with pairs of distantly spaced, pale-green pinnae, of which the larger ones are deeply lobed only on one side, reminiscent of angels' wings. **C**: 56   p. 243

tremula (New Zealand, Tasmania, New South Wales), "Trembling brake fern", or "Australian bracken"; robust grower with large, attractive, bright green, herbaceous, broad, 3 to 4-pinnate fronds to 3 or 4 ft. high and spreading; lower pinnae often compound, upper segments linear and finely crenate, on stiff brown stems. I have found this species in northern New South Wales, growing 4 ft. high, in dry forest gullies. Used in larger Poinsettia combinations for underplanting of decorative foliage, but also grown in 6 or 7 in. pots as "Poor man's cibotium". **C**: 6   p. 243

vittata (Old World Tropics or Subtropics of Africa, Asia, Europe, Australia, Polynesia), the plant usually in the trade as Pt. "longifolia"; rapid-growing, long-leafed, graceful fern with arching pinnate fronds to 2½ ft. long, the simple dark green leaflets are fewer, more distantly spaced and broader than in longifolia, long and tapering to a slender point, their bases running down the stem, and these pinnae are usually turned upward at an obvious short angle from the hairy stalks; the marginal teeth of the sterile pinnae are curved. **C**: 54   p. 243

PTYCHOSPERMA (Palmae)

elegans (syn. Seaforthia elegans), (Queensland), "Alexander palm", "Solitair" or "Princess palm"; handsome medium-sized, solitary feather palm to 22 ft. tall, with gracefully slender trunk to 3–4 in. dia., topped by 6 to 8 rather short pinnate fronds 3 to 6 ft. long; about 20 pairs of bright-green pinnae, jagged as if cut off at apex; bushy inflorescence of white, fragrant flowers, and small bright-red fruits. **C**: 4   p. 220

macarthurii (formerly Actinophloeus, or Kentia) (New Guinea), "Hurricane palm" or "MacArthur palm"; a clustering feather palm constantly sprouting additional slender grayish trunks, usually only 10 ft. high but occasionally growing to 25 ft.; pinnate leaves in a sparse crown, the leaflets glossy-green and rather soft, 6–15 in. long, with the apex jagged and toothed as if bitten off; clusters of fruit first green, then yellow, finally red. Very effective and satisfactory as compact clusters in tubs. **C**: 4   p. 221

PUNICA (Punicaceae); the "Pomegranates" are shrubs or small trees from So. Europe and Asia with evergreen or deciduous showy flowers, some bearing delicious edible fruit with juicy flesh; the double-flowered kinds are grown for ornament in pots or tubs and do not produce edible

fruit. Both types are characteristic decorator plants in containers, sometimes grown into old, contorted specimen; they prefer cooler locations indoors but tolerate heat with sun, and are ideal for the patio or terrace, having been cultivated since ancient times. Ripening of young shoots is important for next season's flowering. Propagated from ground suckers; also cuttings or layers.

granatum (S.E. Europe to Himalayas), "Pomegranate"; shrub or small tree becoming 20 ft. high, with shining oblong leaves $\frac{3}{4}$ to 3 in. long; $1\frac{1}{4}$ in. flowers orange-red, with wrinkled petals and numerous stamens, the leathery calyx purple, and its lobes persistent on the edible fruit, which is fleshy, several-chambered, deep yellow to red, and as large as an orange, the juicy flesh is crimson, of delicious, somewhat acid flavor. An excellent ornamental container plant tolerating heat, for the summer patio, later the cool winter garden; can be pruned to shape. Intriguing as a Bonsai subject. From hardwood cuttings or suckers.
C: 14   p. 314

granatum legrellei, "Double-flowering pomegranate"; popular, free-blooming, ornamental form of the pomegranate tree which is found at home from S.E. Europe to the Himalayas, but distinguished by its fully-double, showy $1\frac{1}{2}$ in. flowers of coral-red, striped yellowish-white outside and pendant at the end of wiry, somewhat spiny branches, blooming from June to September; the lanceolate, wavy-margined, glossy leaves to 3 in. long. **C**: 14   p. 488

granatum nana (Iran to Himalayas), "Dwarf pomegranate"; a miniature version of the pomegranate tree, in form of a slow-growing shrub to 6 ft. high but usually smaller, with woody stem and wiry branches, shining vivid-green, narrow leaves, and fiery scarlet flowers with salmon calyx at end of thin branchlets, followed by orange-red fruit with hard rind and juicy, edible pulp. This dwarf form is charming as a small potplant for the window sill or winter garden, very free-blooming especially from July to September. Propagated from pieces of leafless branches in early spring, with bottom heat; also from softer cuttings.
C: 14   p. 508

PUYA (Bromeliaceae)
alpestris 'Marginata' (Chile), "Banded puya"; dense rosette of narrow spiny, recurving leaves shiny light gray-green banded light yellow at the margins; silvery-gray beneath; branched inflorescence, with metallic blue-greenish flowers and orange anthers, in up to 20 branches to 4 ft. high **C**: 19   p. 138

PYCNOSTACHYS (Labiatae)
dawei (Tropical Africa: Uganda), "Blue porcupine"; stout herbaceous tropical perennial 3 to 5 ft. high, of loosely pyramidal habit, with opposite lanceolate, pubescent, toothed leaves to 1 ft. long; cobalt-blue, $\frac{3}{4}$ in. flowers in whorls in dense terminal spikes to 5 in. long, the calyx teeth needle-like; a desirable winter-bloomer for the moderately warm greenhouse or winter garden; responds to short-day treatment for fuller or advanced flowering. From soft cuttings in late winter. **C**: 70   p. 493

PYRACANTHA (Rosaceae); the "Firethorns" are Asiatic
evergreen, usually thorny shrubs with alternate leathery leaves, spring blooming with small white fragrant flowers, in pretty clusters on spurs of the previous year's wood, followed in autumn by masses of brilliant scarlet or orange fleshy berries which hang on; where protected so the birds can't get them, late through winter into spring. Firethorns are of sprawling habit and the branches may be tied to stakes, as espaliers or into pyramids, or clipped into globes or topiary shapes. Of easy care, sun-loving, they are excellent as large container plants, or grown as small pot plants for their vivid fruit during winter. Propagated best from cuttings; seedlings often are poor bloomers.

coccinea 'Lalandii', a fairly winter-hardy "Firethorn"; robust early-blooming French cultivar of P. coccinea, the red-orange berried "Everlasting thorn", from So. Europe and Asia Minor; evergreen, woody, thorny shrub 10 to 15 ft high with oval oblong, shiny leathery, dark green 1 to 2 in. leaves, finely toothed; the numerous small, May-blooming white flowers followed in fall by dense clusters of waxy $\frac{1}{3}$ in., yellow-orange colored berries which stand out from the stem. The best known species in cultivation in gardens.
C: 26   p. 513

crenulata flava (Himalayas), "Yellow Nepal firethorn". evergreen shrub or small tree 10 to 15 ft. high, with young twigs rusty-hairy, oblanceolate, wavy-toothed, shining leaves 1 to 3 in. long; $\frac{3}{4}$ in. white flowers in May-June followed by round $\frac{1}{2}$ in. yellow fruit; orange-red in the species. Not hardy North. **C**: 14   p. 513

x duvalii; seedling of P. koidzumii (formosana) from Taiwan; very vigorous "Red firethorn", of spreading habit and growing to 10 ft. tall, evergreen with grayish pubescent branchlets, handsome narrow-oblong leathery leaves, and dense with white flowers followed by heavy clusters of showy orange-red, depressed-globular berries holding well into late winter. **C**: 14   p. 513

koidzumii 'Victory' (crenato-serrata cv.), an excellent "Red firethorn"; robust evergreen shrub of Chinese (Taiwan) origin, becoming large and spreading to 8 ft. with thorny, rambling branches and deep green leathery foliage, very ornamental as a smaller container plant bearing large clusters of glistening brilliant, scarlet, long-lasting $\frac{1}{2}$ in. berries into winter and for Christmas. Small white, fragrant flowers in May, on spurs along wood of last year's growth. Timely decorator for the Christmas season, the branches often trained into pyramids or against trellis. Not as hardy as P. coccinea. Our selection in the New York area as the best decorator variety in containers for interior display through the winter; the vivid, late-ripening berries are held for months, sometimes until new blooms appear in spring.
C: 14 or 64   p. 386, 513

PYRROSIA (Filices: Polypodiaceae)
lingua (Cyclophorus) (Japan, China, Vietnam, Taiwan), "Tongue fern", known in horticulture as Diplazium lanceum; creeping fern close to Asplenium, with thin red-scaly rhizome; stalked lanceolate, wavy, thick-leathery leaves 9 in. long, deep green above, only lightly hairy; fuzzy brown and densely covered, with stellate hairs and brown scales underneath; becoming pendant. Propagation by division. Neat little fern needing good drainage. **C**: 74   p. 251

QUERCUS (Fagaceae)
suber (So. Spain, Portugal, No. Africa), the "Cork oak"; an evergreen tree related to our California "Live oak", to 50 ft. or more high, with broad round-topped head and thick, deeply furrowed bark which is spongy and possessing elastic properties; shining dark green, ovate 3 in. leathery leaves with toothed margins, grayish-tomentose beneath. Their light-weight bark is removed in sections around the Western Mediterranean for use in insulation and other economic purposes, and trees are selected in rotation or when ready for the stripping of their grayish bark down to the cambium layers about every 8 to 10 years, following which it grows back again without seriously harming the tree. A curiosity plant which may be grown in containers. From seed. **C**: 13   p. 526

QUESNELIA (Bromeliaceae)
liboniana (Billbergia quintitissima), (Brazil: Bahia, Estado do Rio, Guanabara); tubular plant with stiff-leathery concave leaves dark green, edged with small brown spines, reverse banded with gray scales; inflorescence few-flowered in simple spike, the sepals coral-red, petals dark blue. **C**: 10   p. 142

marmorata (Aechmea marmorata) (Brazil: Espirito Santo to Sao Paulo); called "Grecian vase", because of its tall formal shape, the ends of leaves recurved; bluish and mottled green and maroon; pendant spike, rose-pink bract leaves and blue flowers. **C**: 10   p. 129

testudo (So. Brazil); spreading rosette with leathery, glossy light green, channeled leaves to 2 ft. long by $1\frac{1}{2}$ in. wide, grayish scaly beneath; the margins dense with fine brown spines; small erect brush-like inflorescence with surrounding bracts rosy-pink, flowers electric blue.
C: 10   p. 142

QUISQUALIS (Combretaceae)
indica (Burma to New Guinea and Philippines), "Rangoon creeper"; woody tropical clamberer with liane-like vining stems without tendrils; soft, light green pubescent 5 in. leaves, and drooping clusters of beautiful very fragrant flowers having slender green calyx tube to 3 in. long with petals red when in bud, opening white, but later changing to pink and crimson-red; blooming mostly in summer from May to August. With its changing colors one of the most

beloved of tropical flowering shrubs that may be kept in bush form by pruning. Best planted out in large conservatory; responds to some rest in winter, after which it may be cut back. Propagated warm from soft cuttings with heel. **C**: 54   p. 490

RAFFLESIA (*Rafflesiaceae*)

arnoldii (Sumatra), the world's largest flower; enormous fleshy parasite, with a solitary giant flesh-colored flower 3 ft. across, rising from a superficial rhizome, and without leaves; the foul-smelling inflorescence weighs 15 pounds, the central cup is intense purple and holds 6 quarts of water. Its seeds lodge in the surface roots of Cissus angustifolia where they penetrate the bark and germinate. **C**: 62 p. 538

RAMONDA (*Gesneriaceae*)

myconii (pyrenaica) (in rock crevices of the High Pyrenees between Spain and France), "Alpine gesneria"; small alpine perennial rosette to 8 in. across, with toothed, deep green, softly hairy, wrinkled $2\frac{1}{2}$ to 3 in. leaves; showy potato-like violet flowers with broad, overlapping lobes and yellow eye; blooming May and June. Winter-hardy in the rockgarden. Propagated from seed, branch divisions or leaf cuttings. This species is also found in the isolated mountains of Montserrat, to the north of Barcelona. **C**: 83   p. 447

RANUNCULUS (*Ranunculaceae*)

asiaticus (S.E. Europe, Syria, Iran), "Turban" or "Persian buttercup"; the Garden ranunculus is a slender spring-blooming perennial with fleshy, claw-like tuberous roots; the alternate leaves scalloped and deeply lobed on erect stalks, each bearing 1 to 4 gay flowers that look like large buttercups, $1\frac{1}{2}$ in. across, very variable in many colors; the wild type only 6 to 12 in. high; the cultivated hybrids are taller, stalks 18 in. or more tall bearing showy camellia-like flowers usually double or semi-double $2\frac{1}{2}$ to 4 in. in dia., in many shades of yellow, orange, red, pink, cream and white; large tubers producing many stalks, normally blooming in spring. Not winter-hardy. Superb cut flowers. Ranunculus may be grown in large pots, planted in October and grown at cool 40° F. and sunny, to start flowering in January indoors. Strong well-rested tubers may be planted in August, and may be forced into bloom from October on. After blooming, and the leaves turn yellow, keep dry and store in cool, airy place. Propagate by separating the offsets, or seed. **C**: 64   p. 402

RAPHIOLEPIS (*Rosaceae*)

indica 'Enchantress'; "India-hawthorne"; charming evergreen shrub to 5 ft. high; dark green, leathery elliptic leaves to 3 in. long, bluntly toothed at margins; large pretty flowers appleblossom-like, rosy-pink with white eye, 1 in. across, in loose clusters, carried in profusion from late winter to late spring and into summer; dark blue berries follow. A compact-growing form of the species indica which has smaller white flowers tinged with pink and comes from South China. Excellent and durable decorator plant in containers, growing in sun or partial shade. Propagated by ripened cuttings in late summer, or by layers. **C**: 14 p. 485

RATHBUNIA (*Cactaceae*)

alamosensis (Mexico: Sonora), "Rambling Ranchero" or "Mexican cina"; columnar species of vigorous habit with straggling stems, to 10 ft. long and 3 in. dia., erect at first but becoming bent and rooting at tips; 5–8 irregular ribs, the recessed areoles armed with awl-like whitish spines; diurnal flowers scarlet. **C**: 1   p. 155

RAVENALA (*Musaceae*)

madagascariensis (Madagascar), the remarkable "Travelers tree"; striking tree with palm-like trunk to 90 ft. high, topped by leathery, banana-like leaves with pale midrib, arranged in two ranks like a fan, and sheltering the great flower bracts, with white blooms and sky-blue seed; the cup-shaped leaf bases hold healthy drinking water for thirsty travelers; a landmark in Madagascar, its fan of leaves pointing east and west. **C**: 53   p. 207, 534

REBUTIA (*Cactaceae*); the "Crown cacti" or "Pigmy cacti" are charming High Andine miniature globes with tubercled, not ribbed bodies, loved by cactus fanciers because of their dwarf habit, grown in tiny pots, with a profusion of relatively large, brilliant red to yellow blossoms

from April to July, on the smallest window sill. Airy and with good light, short of burning in summer; in gritty, porous soil, with even moisture; cool in winter (45° F.) and dry. Propagated by offsets, or seed.

kupperiana (Aylostera) (Bolivia), "Red crown"; small globe with depressed top, glaucous gray, acute tubercles in about 20 rows covered with thin spines; grows to 4 in. high but blooms freely even as a tiny 1 inch plant with showy scarlet flowers. **C**: 13   p. 157, 165*

RECHSTEINERIA (*Gesneriaceae*); the "Helmet flowers" formerly known botanically as Corytholoma, are tropical, American herbs with tuberous roots, sending up one or more stems with leaves opposite, or whorled in threes, the showy usually nodding flowers tubular and often two-lipped or helmet-shaped, in brilliant reds to salmon, pink or yellow, essentially summer-blooming. Normally the tubers welcome a dormant period after flowering but they may be kept growing and blooming nearly all year by removing old stems. It is better to rest them by keeping them dry and warm for several months. Propagation may be by tip cuttings, leaf petiole cuttings, or leaf sections laid flat which will form bulblets with roots and eyes. Large tubers also may be divided; also sown from seed.

cardinalis (Gesneria macrantha) (C. America), "Cardinal flower"; brilliantly flowered, tuberous plant with stout white-hairy stems, under 1 ft. high; round-cordate, emerald green, velvety leaves 3 to 6 in. long, topped by a cluster of large upward-curved tubular 2 in. bilabiate, hooded flowers, white downy over brightest scarlet; throat marked purple, and with protruding red filaments; blooming from tubers in late winter and spring; splendid as a Christmas plant or for Valentine's Day; may be flowered from seed in July-August. **C**: 10   p. 447

cyclophylla (So. Brazil), "Roundleaf helmet flower"; tuberous plant sprouting one or more succulent stems to 1 or 2 ft. high, with fleshy nearly round 5 to 6 in. leaves, lightly crenate, rich green and somewhat hairy; tubular nodding flowers 1 to $1\frac{1}{4}$ in. long, scarlet-red with deep crimson dots inside flaring lobes. Very similar to R. macropoda, but considered of more robust habit with greener, more velvety leaves. Dr. Moore distinguishes cyclophylla as having flower disk composed of 2 glands, against macropoda with 5 glands. An excellent house plant producing flowers over a long period, mostly in spring. **C**: 9   p. 447

leucotricha (Brazil: W. Paraná), a breathtaking new species which I saw when first on exhibition in Sao Paulo in 1954; huge 1 ft. tubers sprouting happily without soil in glazed bowls, the glistening silvery foliage suggesting to me the name "Brazilian edelweiss". Found on cliffs near the waterfall 'Salto Apucarazinho' at 3300 feet, it is called locally "Rainha do Abismo". The stout, densely matted, white, later brown hairy stems to 10 in. high, carry one or two whorls of 3–4 large obovate leaves to 6 in. long, densely covered with shimmering, long silvery-white hair, with margins entire or obscurely crenate; slender tubular inflated $1\frac{1}{4}$ in. flowers soft rosy coral, entirely covered outside with silky white hair, the lobes sometimes marked with crimson; blooming spring and summer. Easy from seed. **C**: 9   p. 447

macropoda (So. Brazil), "Vermilion helmet flower"; charming tuberous herb with unbranched hairy stem 6 to 9 in. high, bearing opposite, rather thin, rugose, velvety bright green leaves almost round, 3 to 5 in. broad; small nodding flowers in clusters, the slender 1 to $1\frac{1}{4}$ in. tubes vermilion-red with the lower lobes marked brown-red; blooming March-April. **C**: 9   p. 447

verticillata (Dircaeo-Gesneria) (Brazil), "Double-decker plant"; tuberous plant with wiry-erect, wine-red stem bearing whorls of three dark pubescent, crenate leaves, and dense clusters of slender tubular, dusky-rose flowers marked throughout with little stripes of fuchsia-red, inside yellow marked red; blooming in winter. **C**: 9   p. 447

REHMANNIA (*Gesneriaceae*)

angulata (C. China), "Foxglove gloxinia"; perennial herb for the cool greenhouse, with a rosette of soft, irregularly lobed, obovate leaves; a sticky-hairy, leafy stalk 1 to 3 ft. high with showy, large, bell-shaped swollen, bilabiate flowers $2\frac{1}{2}$ to 3 in. long, rosy-red with 2 yellow lines in throat and spotted purple. Formerly included with the foxgloves (Digitalis). Somewhat frost-hardy under cover; in the

East and North often grown as a biennial; seed sown in March blooms in the greenhouse beginning the following March, and lasting for 6 weeks. **C**: 76   p. 447

REINECKIA (*Liliaceae*)
   carnea (China), "Fan grass"; creeping rhizome bearing narrow, matte-green, thin-leathery, keeled leaves ½ in. wide and 8 to 18 in. long, standing in 2 ranks and clasping at base, gracefully deflexed. Durable house plant for cooler locations. By division. **C**: 16   p. 304

REINHARDTIA (*Palmae*)
   gracilis gracilior (Mexico: Chiapas; B. Honduras), the exotic "Little window palm"; very attractive diminutive palm, 3–5 ft. high; slender, suckering multiple canes ¼ in. thick, with waxy-green bilobed foliage having elevated ribs and window-like apertures in the leaf plane adjoining the midrib, each fenestrate leaf divided into 4 pinnae, the apex coarsely toothed as if chewed off. A curious collection plant with its little windows. **C**: 60   p. 230
   simplex (Costa Rica), "Fanleaf reinhardtia"; a dwarf palm of unusual appearance with its simple oblong or sometimes two-lobed, corrugated leaves; dark green, and jagged at the upper margins, borne on slender stem. Very attractive in a 6 in. pot. Much like R. gracilis but without the window-like holes. **C**: 60   p. 230

REINWARDTIA (*Linaceae*)
   indica (trigyna), (Mountains of No. India), "Yellow flax"; bushy shrub-like perennial to 4 ft. high, spreading by underground roots, wiry branches and obovate, membranous, dark green leaves; large brilliant yellow 2 in. cupped flowers form in great profusion in late fall and early winter; blooms do not last long but for weeks new ones open every day. A good flowering house plant if kept light and airy; pinch several times in summer to keep bushy. Propagated from rooted suckers or seeds. **C**: 14   p. 498

RENANTHERA (*Orchidaceae*)
   monachica (Philippines), the "Fire orchid"; a striking, colorful epiphyte without pseudobulbs, stiff leafy stems to 1½ ft. tall simulating Vandas, alternate short and thick bluish-green leaves mottled with brown; from the axils an arching elongate cluster of bright orange long-lasting spreading 1½ in. flowers boldly marked with blood-red blotches and bars; blooming mostly late winter to spring. Renantheras require good light without burning sun, and constantly warm and moist, less so in winter; they have rampant roots and do well on treefern blocks or sufficiently large pots in osmunda or firbark mixed with sphagnum. Propagated by cutting the stems into sections, preferably with roots that form along old leaf axils. **C**: 10   p. 479

RESEDA (*Resedaceae*)
   odorata (No. Africa, Red Sea, Egypt), "Mignonette"; a charming branching annual or perennial herb, at first upright but becoming spreading and sprawling, with oblanceolate leaves, and small yellowish-white, inconspicuous flowers showing contrasting saffron-red anthers, in terminal, spike-like clusters; much loved for its sweet fragrance, and a vase or pot filled with honeyed Mignonette spreads delightful perfume everywhere. Sow directly into pots, depending on time wanted. Winter-flowering in the cool greenhouse from August sowing; July to September bloom in the garden or on the patio, from April-July sowing. **C**: 26   p. 415

RESTREPIA (*Orchidaceae*)
   xanthophthalma (Pleurothallis) (Mexico, Guatemala, Panama, Venezuela); quaint dwarf epiphyte 3 to 6 in. high with tufted stems entirely hidden by papery sheaths, each stem bearing a single leaf, and also a short-stalked small solitary 1 in. flower near the base of the leaf, yellow, blotched red-brown, with dorsal sepal and petals running into pointed, fleshy tips, the broader, lower sepals grown together; blooming mostly in summer. Grown similar to Odontoglossum, wanting a moderate supply of water the year round. Increased from divisions with leading eye. **C**: 72   p. 460

RHAPHIDOPHORA (*Araceae*)
   aurea: see Scindapsus aureus
   celatocaulis (Monstera latevaginata) (Borneo), "Shingle plant"; also known as Marcgravia paradoxa in juvenile stage, when the rounded to heartshaped, fleshy leaves, in climbing, cling close to tree, the blades overlapping; maturity-stage leaves 2½–3 ft. long, irregularly pinnately cut and perforated. **C**: 60   p. 81, 259, 533
   "laciniosa" (Monstera subpinnata) (East Indies); slender but wiry creeper with daintily divided, emerald-green, leathery leaves on wiry petioles, the segments attached alternately to both sides of midrib. **C**: 60   p. 81
   decursiva (Ceylon to Vietnam); stem stiffly scandent, with large, glossy dark green leaves pinnately divided to midrib; decorative, but slow, sparry and stubborn, and not easy to train. **C**: 54   p. 81

RHAPIS (*Palmae*)
   excelsa (flabelliformis) (So. China), the "Lady palm"; a very small fan-palm with bamboo-like canes 1–2 in. thick, growing upward very slowly and eventually reaching 12 ft. or more; the thin stems densely matted with coarse fiber, forming clumps from underground suckers, the leathery leaves glossy dark-green, divided into 3–10 broad segments 9–12 in. long. Widely used in China and Japan as a durable potted palm; ideal for cool locations but actually enjoys warmer temperature also; considered temperamental; best to keep on dry side if not too well rooted; heavy feeder. Light preference 50 fc. up. Easiest propagation by division. **C**: 22   p. 227
   excelsa 'Variegata', "Variegated lady palm"; very attractive Japanese cultivar having palmate leathery leaves with its segments banded and striped lengthwise with ivory-white, alternating with green. Extremely slow-growing and highly prized in Japan, and known there as "Fuiri Kannon Chiku", or the "Lined Kannon bamboo". **C**: 60 p. 230
   humilis (So. China, Thailand), "Slender lady palm"; more graceful than excelsa, the reed-like stems thinner, only ½ to ¾ in. dia., normally less vigorous and even slower growing; likewise and densely covered with dark brown fibers; the deep green palmate leaves more slender, and divided into 9 to 20, narrower segments. Surprisingly, this lady palm responds beautifully to warm living-room temperatures, freely sprouting young cane after cane. **C**: 22   p. 227

RHEKTOPHYLLUM (*Araceae*)
   mirabile (Nephthytis picturata), (Nigeria, Cameroon, Congo); creeping plant sending out long runners rooting at nodes; the large arrowshaped leaves are thin-leathery, dark green between the veins, variegated silvery-cream in form of a fernleaf; older leaves plain glossy green. **C**: 4 p. 98, 180

RHIPSALIDOPSIS (*Cactaceae*); floriferous epiphytic and rock-dwelling cacti of compact shrubby habit, from moist-warm Brazil; admirably suited for pot culture on the window sill; the dense little bushes with 2 to 5-angled joints, later flattened, and at their tips the beautiful starry pink or red flowers covering the plant in spring. A cool period during winter at 50° F. or less is important for the best initiation of buds. Propagated by stem joints and small branches.
   gaertneri (Schlumbergera, Epiphyllopsis) (So. Brazil), the "Easter cactus"; bushy epiphyte with stiffish spreading branches of long flattened joints, dull green with purplish crenate margins, a few bristles at apex; starlike regular flowers, deep scarlet in March-April; the ovaries angled. **C**: 10   p. 170
   x graeseri (Ephiphyllopsis) (gaertneri x rosea), a free-blooming hybrid which I first saw in Brazil where it flowered in their spring (Sept.), while in the Northern hemisphere it blooms in March, indicating its being influenced by day-length; wide open, starshaped regular flowers rosy-red with double row of broad petals; freely branching compact plant. **C**: 10   p. 170
   x graeseri 'Rosea' (Epiphyllopsis), a lovely cultivar with flowers of a lighter shade: clear pink with the center flushed deep rose. **C**: 10   p. 165*, 170
   rosea (Brazil; Paraná) "Dwarf Easter cactus"; densely bushy, shrub-like dwarf cactus rather erect, with small ¾ in. flattened joints 3 to 5-angled or keeled, almost pencil-like, waxy green tinted purple, with minute hair-like bristles; small wide open regular flowers of delicate rosy-pink with orchid edges, 1 in. dia. **C**: 10   p. 170

RHIPSALIS (*Cactaceae*); the "Mistletoe cactus", also known as "Willow cactus", belongs to a strange group of leafless epiphytes dwelling in moist-warm tropical forests perching on trees from Florida to Argentina, with one species escaped to Africa. Their pendulous stems are slender, pencil or rope-like, densely branching, provided with bristles but without spines. The small flowers are followed by depressed globular, berry-like fruit resembling that of mistletoe; the berries are sticky and probably, adhering to birds or driftwood, brought to Africa in this manner thousands of years ago. They may be grown in pots but when older are most effective in hanging baskets in porous compost of peat and humus, or osmunda, with broken brick, sand and sphagnum, the soil must be kept moderately moist, best by dipping. Easy from cuttings, or division.

capilliformis (E. Brazil), "Old man's head"; a hanging, epiphytic cactus with long pale-green, branching, cylindrical, string-like, pendant stems like a wig; many small, greenish-white flowers almost at the end of stems; white berry-like fruit. **C**: 10   p. 171

cassutha (wide-spread from Florida to Peru to Brazil, and even Ceylon and trop. Africa), "Mistletoe-cactus"; growing on trees or rocks hanging in many strands to 30 ft. long; branches thin-cylindrical, light green, somewhat bristly when young; flowers cream, borne along the joints, followed by mistletoe-like white fruit, as have many Rhipsalis. In Africa, this is found on Lake Kivu, and I have seen them as well in eastern Kenya, and in the forests of the Usambara Mts. of Tanzania. **C**: 10   p. 171

cereuscula (Brazil to Argentina), "Coral cactus"; a much branched bush to 2 ft. high, with long lively green, thin cylindrical stems and branches often erect, each crowned by a cluster of another type of short joints, and set with tiny bristles; flowers pinkish to white, borne at the tips; fruit white. **C**: 10   p. 172

houlletiana (S.E. Brazil), "Snowdrop cactus"; high altitude epiphyte with both cylindrical, stiff and erect branches, and hanging, flat thin-leathery ones leaf-like and notched; flowers cream with reddish eye; the fruit a carmine-red berry. **C**: 10   p. 172

mesembryanthemoides (Brazil: Estado do Rio) "Clumpy mistletoe cactus"; pretty epiphyte erect with two kinds of branches, a long slender woody axis 4 to 8 in. long, densely set with numerous short fleshy, light green, needle-like fruiting joints covered sparsely with white bristles near tips; small white flowers; fruit white. Attractive as a miniature tree when grafted on the slender stem of Pereskia. **C**: 10   p. 171

paradoxa (Brazil: Sao Paulo), the "Chain cactus"; curious, moderately branching epiphyte, with hanging, jointed stems to 3 ft. long, glossy green with 3 acute angles twisted into zigzag links, and producing numerous aerial roots; flowers white; fruit reddish. **C**: 10   p. 172

quellebambensis (Peru), "Red mistletoe"; attractive epiphytic cactus with pendant thin cylindric branches, dull green with occasional purple markings, and lightly grooved; creamy flowers, and glossy, rich carmine-red, berry-like fruit at the tips. **C**: 10   p. 165*, 171, 509

rhombea (Brazil: Coastal Sao Paulo), "Copper-branch"; bushy plant to 2 ft. high first erect but later hanging, with the joints leaf-like and flat, to 2 in. broad, sometimes 3 angled, and strongly notched at margins, dark green or reddish; flowers creamy-yellow with red spots inside; round red fruit. **C**: 10   p. 172

RHODODENDRON (Ericaceae); the "Rose-bays" are a large genus primarily woody evergreen shrubs of the Heath family; with big leathery leaves and stunning, rounded clusters of beautiful funnel-form flowers of good substance, highly hybridized with many named cultivars and in colors of white, pink, red, purple and even yellow or orange; there are also others of smaller habit. As understood by botanists today, *Rhododendron* includes as subgroup many small-leaved evergreen and deciduous kinds known by long horticultural usage as Azaleas, a name now difficult to segregate on technical characters except perhaps based on hairtype on the foliage (Dr. Encke), and in this Manual listed separate from Rhododendron. "True" Rhododendrons are primarily used in garden plantings, in moist, cool situations, and some of the tender hybrids are forced for flower shows or Easter. After blooming, the spent flower trusses should be broken out to allow new buds to form for the succeeding year's bloom. Propagated under mist from cuttings of young growth or from leafbud cuttings with the aid of rooting hormones, also by grafting on vigorous understock of Rhododendron ponticum for the evergreens; R. luteum for deciduous types.

x'Bow Bells' ('Corona' x williamsianum), a "Miniature rhododendron"; dainty miniature evergreen about 12 in. high, with hard-leathery, 1½ in. ovate-oval, smooth leaves, pale beneath; the new growth bronzy; loose clusters of 1½ in. open bell-shaped single flowers of good substance, dainty salmon-pink with red spots in throat. Photographed in bloom in April; a darling plant for pots. With age this plant may grow to 4 ft. high. **C**: 70   p. 502

charitopes (Burma), charming miniature evergreen shrub to 18 in. high, with obovate leaves 1 to 2½ in. long, glossy green with yellowish shining scales beneath; clusters of flat bell-shaped flowers 1 in. wide, purplish-pink spotted with crimson on upper lobes, petals wavy; blooming April-May. **C**: 70   p. 502

indicum (lateritium) (Japan), "Satsuki azalea"; low evergreen or semi-evergreen densely branched shrub, seldom to 6 ft. high, and lending itself to dwarfing as bonsai; small 1—1½ in. lanceolate, slightly hairy leaves with finely toothed margins; large funnel-form flowers 3 in. broad, bright red or rosy, June blooming unless forced; hardy. This is not the Azalea (Rhod.) "indica" of florists, which is a group name for many large-flowered tender hybrids derived from Azalea (Rhod.) simsii, the so-called "Belgian indica", forced in greenhouses during winter and spring in many named cultivars in a wide assortment of lovely colors. **C**: 82   p. 312

'Jean Marie de Montague'; excellent griffithianum cultivar suitable for Easter forcing indoors; compact, shapely plant budding well with firm, solid heads of bell-shaped flowers to more than 3 in. across, glowing bright crimson and very charming with wavy petals; smallish, dull-green elliptic foliage. Beautiful color for the outdoor garden on the Pacific Coast, but not hardy enough in the northern Atlantic States. **C**: 82   p. 369

'Pink Pearl' (Geo. Hardy x Broughtonii 1897); an old and widely popular English hybrid derived from the large-flowering griffithianum of Sikkim and the hardy catawbiense; a shapely plant with very large trusses of good-textured, soft rose-pink 2½—3 in. flowers with darker shading and spotted maroon in throat; midseason. Good for early forcing for Easter or Spring flower shows, and quite at home in the Pacific Northwest but not bud-hardy in the Northeast. **C**: 82   p. 369

racemosum (China), evergreen shrub with elliptic leaves 1 to 2 in. long, densely scaly on the glaucous underside; small 1 in. funnel-like flowers soft pink with the five lobes tipped rose, and with white throat; blooming in March-April. There are 3 forms of this species; 6 in. dwarf, 2½ ft. compact, and tall to 7 ft. **C**: 70   p. 502

'Roseum elegans', a rosy-lilac "Rose-bay"; well-known old hybrid considered the best winter-hardy for gardens as far north as Boston; bred with the dominant R. catawbiense, a good evergreen species from mountainous regions of Virginia and North Carolina to Georgia; compact globular bush 6 ft. or more high with leathery decurved, olive-green leaves; a heavy budder with numerous 5 to 6 in. trusses of rosy lavender flowers of good substance, shaded deeper purple at margins, marked with purple spots on upper petals, individual blooms measuring 2¼ in. or more across; mid-season. This hardy catawba hybrid is occasionally forced for March or Easter, although its flowers are not as exciting in comparison to the more tender hybrids that succeed on the Pacific Coast. Excellent even under poor conditions, always reliable and willing to bloom and struggle along. **C**: 82   p. 502

RHODOSPATHA (Araceae)
picta (Brazil); thin-leathery plant with rooting branches; elliptic, deep green leaves shaded yellow, irregularly marked orange-yellow; petioles sharply bent at base of blade; spathe flesh-colored. **C**: 60   p. 98

RHOEO (Commelinaceae)
discolor: see spathacea

spathacea, known in horticulture as R. discolor, (Mexico and West Indies), "Moses-in-the-cradle"; attractive rhizomatous clustering foliage plant with stiff, fleshy lance-shaped, waxy leaves 8 to 12 in. long, tightly arranged in a rosette, the clasping foliage metallic-dark-green, and vivid glossy purple beneath; in the leafbases, little white flowers are peeking from shell-shaped bracts. Very pretty and popular plant for window or the patio, and very tolerant. Prop. by seed or suckers. **C**: 14 or 16   p. 191, 262

spathacea 'Vittata', "Variegated boat lily"; multicolored form, with leaves striped lengthwise on the surface pale-yellow, tinted red; red-purple beneath. Not quite as robust and more particular than the species, inclined to revert back to the green type, if too dark or fertilized too much. For propagation, select highly colored suckers only, which form when center is cut out. **C**: 4   p. 262

RHOICISSUS (*Vitaceae*)
capensis (Vitis) (So. Africa), "Cape grape" or "Evergreen grape"; strong clambering vine with globular ground tubers; brown-hairy, somewhat woody stems, and long-stalked, thickish leathery, metallic green, glossy leaves nearly round or kidney-shaped, to 8 in. across, deeply lobed and wavy-toothed; rusty-tomentose beneath and at the margin; red-black glossy fruit. Sturdy indoor vine with interesting foliage, requiring fairly good light 50 fc. up, temperate to warm; water when surface becomes dry. **C**: 13   p. 257

RHOMBOPHYLLUM (*Aizoaceae*)
dolabriforme (So. Africa: Cape Prov.), "Hatchet plant"; tufting succulent freely branching, with opposite hatchet-shaped leaves 1 to 2 in. long, the upper side curving upward, the lower surface with a wedge-shaped widened keel and a tooth-like projecting tip, grass green with translucent dots; golden-yellow flowers $1\frac{1}{2}$ in. across. A curious yet easy little house plant. From seed, division or also cuttings. **C**: 13 p. 354
nelii, in horticulture as Hereroa nelii (So. Africa: Cape Prov.), "Elkhorns"; clustering small succulent with opposite spreading, pale bluish, gray-green leaves $\frac{1}{2}$ to $\frac{3}{4}$ in. long, two-lobed at apex and not wedge-shaped; $1\frac{1}{2}$ in. yellow flowers. From division or seed. Used in dishgardens where it proves both odd and durable. **C**: 13   p. 354

RHOPALOSTYLIS (*Palmae*)
baueri (South Pacific: Norfolk Is.), "Norfolk Island shaving brush"; medium tall, elegant feather palm to 60 ft. high, taller and stronger in all parts than R. sapida; the pinnate fronds 6–9 ft. long, robust-leathery dark green, and resembling Howeia as decorative tub plant, best for medium-warm locations. Much more graceful than R. sapida, with longer fronds nicely down-curved. **C**: 16   p. 230
sapida (New Zealand, Norfolk Is.), the "Nikau palm", also known as the "Shaving-brush palm" or "Feather-duster"; in habitat representing the southern limit of palms; attractive feather palm to 30 ft. or less high, usually with straight trunk strongly ringed, 4–8 in. thick, and topped by a prominent bulbous crownshaft, from which rises a crown of pinnate fronds 4 to 8 ft. long, standing stiffly erect like a brush-like tuft; the erect, channeled leaflets glossy green, with split apex; purplish flowers at base of crownshaft, and small vivid red fruit. **C**: 64   p. 223

RHYNCHOSPERMUM: see Trachelospermum

RHYNCHOSTYLIS (*Orchidaceae*); the "Foxtail orchids" are tropical epiphytes of rather small vandaceous habit, without pseudobulbs, clothed by fleshy leaves in ranks; and cord-like aerial roots from their base, their beautiful pendant inflorescence dense with small fragrant, waxy flowers. These fine orchids dislike being disturbed, and require warmth and constant atmospheric moisture, less so in winter. They respond best planted in baskets or low pans, in porous osmunda, fernfiber or firbark chips, with sphagnum and potsherds added. Propagated by rooted sections of the stem.
coelestis (Saccolabium) (Thailand); a "Foxtail orchid"; delightful dwarf epiphyte of vandaceous habit, stout stem to about 8 in. tall, clothed with rigidly fleshy leaves closely spaced in ranks; large cord-like roots developing from the lower part; the inflorescence dense with $\frac{3}{4}$ in. fragrant, waxy flowers white and marked with purplish blue; blooming spring, summer or fall. **C**: 60   p. 479
gigantea (syn. Saccolabium) (Burma, Thailand, Laos, Vietnam), a "Foxtail orchid"; handsome epiphyte with stout stem densely clothed with leathery leaves, the gorgeous inflorescence pendulous, dense with fragrant, waxy 1 in. flowers, white and spotted violet, the mauve lip spurred, blooming late autumn and winter. **C**: 59   p. 481

RICINUS (*Euphorbiaceae*)
communis coccineus (Trop. Africa), "Red castor-bean" or "Palma Christi"; bold and striking shrub-like plant,

becoming woody and tree-like to 20 or even 40 ft. high. The giant leaves are peltate, palmately divided into 5–11 lobes, 1 to 3 ft. across, beautifully bronzed to dark red, on stout brownish branches. Small flowers in terminal clusters, followed by brown-prickly husks holding poisonous seeds from which castor-oil is extracted. Usually grown as an annual foliage plant, it may reach to 6–15 ft. in one season; requiring full sun, plenty of heat and moisture. **C**: 52 p. 301

RIVINA (*Phytolaccaceae*)
humilis (Trop. and Subtrop. America: West Indies, Mexico, etc.), "Rouge plant" or "Baby pepper"; soft-leaved tropical herb with thin stem and branches; membranous green, lightly pubescent, ovate foliage, and pendant slender sprays of tiny pinkish-white flowers, forming clusters of lustrous little $\frac{1}{4}$ in. berries of bright crimson, soon dropping. Propagated from seed or cuttings. **C**: 66 p. 512

ROCHEA (*Crassulaceae*)
coccinea (well known in horticulture as Crassula rubicunda) (So. Africa); branched shrubby succulent 8 to 12 in. high, grown primarily for its showy scarlet red, fragrant flowers numerous in beautiful clusters from spring to summer; borne on slender stems, with opposite green leaves $\frac{1}{2}$–1 in. long, and reddish beneath. To succeed as a pot plant, it must be kept near 40° F. in winter and as long as possible, with good light and fresh air. An old-time, long-lasting blooming plant. Propagate from spring cuttings. **C**: 75   p. 335
falcata: see Crassula falcata

RODRIGUEZIA (*Orchidaceae*)
batemanii (Burlingtonia rubescens) (Brazil, Peru); small-growing but showy epiphyte with clustered little pseudobulbs from a creeping rhizome carrying a solitary, rigid leaf 4 in. long; the arching inflorescence with 4 to 10 flowers $2\frac{1}{2}$ in. across, sepals and petals white, flushed with rose, the lip veined rose-red, blooming spring and summer. They need good light 2000 to 3500 fc.; grow best in pots with osmunda or bark chips, in baskets or on treefern slabs, kept warm and with liberal watering at all times. Propagated by division. **C**: 10   p. 479

ROHDEA (*Liliaceae*)
japonica (Japan), "Sacred lily of China"; extremely durable, modest foliage plant; clustering basal rosette of oblanceolate, arching, channeled or plaited, thick-leathery, matte green leaves 2 to 3 in. wide, densely arranged somewhat in two ranks, and growing from thick rhizome; white flowers aroid-like; fruit a red berry. An old favorite of the Japanese, where it is cultivated in hundreds of named varieties. Propagated by division. **C**: 27   p. 202
japonica marginata (Japan), variegated "Sacred lily of China" or "Sacred Manchu-lily"; rhizomatous fleshy rosette of sheathing, channeled, thick leathery recurving foliage black-green, prettily bordered with white, 6 to 18 in. long. A favorite in Japan where many named variegated forms are being cultivated by fanciers. In China, it was adapted by the Manchu emperors as their national "flower". **C**: 3   p. 304

RONNBERGIA (*Bromeliaceae*)
morreniana (S.W. Colombia), "Giant spoon"; handsome terrestrial plant with long stalked upright lance-shaped leaves, thin leathery, bright green, zoned with bars and spots of very deep green; flowers blue. Difficult. **C**: 60
                                                              p. 142
ROSA (*Rosaceae*); Roses are hardy, bushy or climbing
woody plants mostly thorny and deciduous, the best loved and probably most widely planted shrubs in gardens of Temperate zones. They have evolved from primitive mostly 5-petaled native roses, through mutation and hybridization into a number of distinct classes with flowers large or small, single or double, usually sweetly fragrant in almost all colors except blue. Leading groups are the 'Hybrid Teas' or "Monthly roses", most popular of all, large double flowers blooming continuously throughout the season, and with the fragrance of the Tea rose (R. odorata hyb.); the Flori-bunda class (Hybrid Tea with Polyantha), a free-flowering bush rose with small tearose-like flowers larger than polyantha, in clusters, blooming most all summer; Poly-anthas (rehderiana), the "Baby roses", (hybrids of multi-flora x chinensis), vigorous bushes with clusters of small

flowers under 2 in., not fragrant, for long-continued summer bloom and long used for growing in containers; the 'Grandifloras' (Hybrid Tea x Floribunda), very vigorous with Hybrid-Tea-like flowers in clusters; 'Climbers' with small to medium-sized flowers in clusters; "Miniature roses", lilliput-type derivatives of R. chinensis. For Easter forcing, bushes with well-ripened wood should be potted after they have had a good frost; then pruned lightly and started slowly in a cool 50° F., sunny location in January, misting the canes until the sleeping eyes break, when the temperature may be increased, with liberal watering. Roses may be propagated from well-ripened shoots in fall with the aid of rooting hormones, but this is a slow process; most roses are available from nursery shops as budded or grafted dormant plants, ready to bloom the same season.

'Bonfire' (1928) (multiflora x wichuriana), an old-fashioned "Rambler rose"; a good, vigorous climbing rose for Easter forcing in pots as a "trained rambler", with long, scandent woody canes more upright and blooming slightly later than 'Eugene Jacquet', but its densely double small flowers, under 1 in. dia., are more crimson-red and not as bluish carmine; the heavy flower clusters appearing as short axillary branches toward the upper end of canes. To prepare for forcing, plants with strong wood only must be selected, properly frost-ripened, then potted if possible before December; in January the tips of canes only should be pruned slightly, trained into globes, baskets or trellis, and laid down at 45° F. covered with moist burlap until the eyes break. **C**: 64   p. 377

borboniana 'Magna Charta' (1876), an old "Hybrid Perpetual" rose; long favored for forcing in pots because of its prolific, bushy habit, producing masses of erect canes with light green foliage, topped by clusters of great, very double, fragrant, globular flowers, with numerous, carmine-rose petals; usually opening more or less at the same time and making a grand show. May be pruned hard to keep low and bushy, or its long canes may be trained to trellis. **C**: 52   p. 376

chinensis minima (R. roulettii) (China), "Pygmy rose"; well-loved miniature form of the Bengal rose, averaging only about 8 to 10 in. high, vigorous, hardy and long-lived, best grown from cuttings; with appealing, 1½ in. double flowers of lively rose-pink with pale eye, in continuous bloom, especially from June to fall; once thought lost to cultivation it turned up 1918 on the window-sill of a Swiss cottage. This plant may be the same as the cultivar 'Pompon de Paris' (1839) which is usually found in horticulture as R. roulettii, the "Fairy Rose". **C**: 14   p. 488

x floribunda 'Fashionette'; a spreading, rather sparry bush with bright green foliage and long-stemmed replicas of fine tea roses in smaller size about 3 in. dia., in soft shell pink veined with red and flushed with yellow, quickly opening; delightfully fragrant, but we have found it a shy bloomer in pots. **C**: 64   p. 376

x floribunda 'Spice'; vigorous bush of sparry habit; stiff-erect branches with large shiny foliage, and bearing heavy clusters of fully double, rather flat 2¾ in. flowers, the buds crimson, opening to salmon scarlet, on sturdy wood **C**: 64   p. 376

x grandiflora 'Queen Elizabeth'; vigorous hybrid of Hybrid Tea and Floribunda, to 5 ft. or more high, with Hybrid Tea-type 3½ in. flowers borne singly or in long-stemmed clusters, soft rose-pink and very fragrant; very close to Hybrid Tea roses, this type is valuable for its large number of flowers appearing at one time, and that lend themselves to cutting. **C**: 52   p. 376

odorata 'Golden Rapture' ('Geheimrat Duisberg') (1933), an exquisite yellow Hybrid tea rose; seedling of 'Rapture' (sport of 'Butterfly' which in turn was a sport of 'Ophelia') x 'Julien Potin'; the first easy growing greenhouse cut-flower yellow; with fresh-green glossy foliage, and gracefully slender, clear golden yellow flowers of 40 petals, medium size with deeper shading in the center, not fading, and with old-rose fragrance. **C**: 52   p. 488

odorata 'Happiness' ('Rouge Meilland') (1951) (Hybrid tea); magnificent seedling involving 'Rome Glory', 'Tassin', etc.; an exhibition bloom and a long-stemmed 'fancy' red rose typical of a superior greenhouse cut flower; large double flowers having 35–40 heavy, substantial petals glowing deep red, unfolding to velvety crimson with blackish sheen, the outer petals curl to points in layer after layer about the solid center, 5 to 6 in. across; very long lasting and slightly fragrant; not too free-growing but very rewarding. **C**: 52   p. 488

odorata 'Mrs. W. C. Miller' (1909); a dependable old "Hybrid Tea" rose of robust and bushy habit with leathery leaves on strong canes stiffly bearing their elegant, tightly double, many-petaled blooms, a dainty, soft pink flushed with rose; the petals are of good substance and unfolding gracefully like a true Tea rose, with its typical intense fragrance, and blooming "monthly" into autumn; the flowers when open spreading 5 in. across. A top performer for blooming in pots for the spring season even under forcing, with strong necks able to produce flowers even with insufficient sunlight, instead of going blind like many other roses. **C**: 52   p. 376

x polyantha 'Margo Koster' (Sunbeam) (1935), excellent commercial "Baby-rose"; sport of 'Dick Koster' of short compact habit about 1 ft. high, and large rather globe-shaped double flowers with incurved petals like 'Dick Koster' but a delicate soft light orange, shading to salmon-red in the heart; good pot-forcer for Easter. **C**: 64   p. 376

x polyantha 'Mothers-day' (1949), excellent "Baby rose" for pots, sport of 'Dick Koster'; well-shaped and of small compact habit, never rambling, with glossy, pinnate leaves and short wiry branches profusely bearing clusters of relatively large globular flowers of firm texture, deep crimson-red. An ideal potplant for Easter and Mother's Day. **C**: 64   p. 376

ROSCOEA (Zingiberaceae)

alpina (Himalayas to Burma); small perennial herb with fleshy roots about 4–6 in. high; parallel-veined leaves 3–4 in. long, not fully developed at flowering-time, sessile, oblong-lanceolate; spikes with flowers having white corolla-tube and dark purple limb. **C**: 26   p. 205

ROSMARINUS (Labiatae)

officinalis (So. Europe, Asia Minor), "Rosemary"; rugged, evergreen aromatic shrub to 2 ft. or more high; downy shoots with leathery, linear leaves ¾ to 1½ in. long, needle-like and slightly bent, shiny grayish green above, white-downy beneath; axillary lavender-blue flowers. Used since ancient times; the spicy resinous leaves fresh or dry give exciting flavor to appetizers, meats, dressings, sauces, fish, game, poultry, fried potatoes and vegetables. The fragrant leaves when dried give a piny scent to sachets and moth preventatives. Distilled oil is used for medicine and perfumes. From seed, cuttings or layers. May be taken indoors as a pot plant. **C**: 13   p. 320

ROYSTONEA (Palmae)

regia (Oreodoxa) (Cuba), "Cuban Royal palm"; a majestic palm with column-like, smooth gray trunks, somewhat swollen above the middle, to 70 ft. high or more, topped by a terminal crown of gracefully arching feathery fronds regularly pinnate, 8 to 10 ft. long, the pinnae to 2½ ft. long and 1 in. wide, dark glossy green and prominently ribbed and arranged in double rows, standing out in 2 planes on either side of the axis. Indoors, primarily a show palm for the tropical conservatory. **C**: 54   p. 221

RUBUS (Rosaceae)

reflexus (So. China, Hong Kong), "Trailing velvet plant"; attractive robust, somewhat sparry woody clamberer with rambling stems having occasional thorns, the young growth, petioles and underside of the foliage covered with cinnamon-colored wood; sturdy, pubescent, toothed and lobed leaves 4 to 8 in. long, vivid emerald-green painted with chocolate-brown, along primary ribs, followed by a zone of splashed silver. Handsome foliage vine especially striking in younger shoots and when kept warm; best planted out but if in pots then trained on trellis or wire frame. **C**: 54   p. 266

RUELLIA (Acanthaceae)

amoena (graecizans) (So. America), "Red-spray ruellia" or "Red Christmas pride"; herbaceous sub-shrub 1 to 2 ft. high, with lanceolate, glossy green leaves 3 to 5 in. long and becoming smaller toward the top; brilliant red, swollen tubular 1¼ in. flowers, streaked yellow inside, blooming in summer or winter. Sun-loving flowering house plant best grown from fresh cuttings annually, or from seed. **C**: 8   p. 493

blumei (Java); small herb with opposite sessile, narrow obovate leaves, thin yet stiff, bluish-green, with silver-gray between the raised ribs; the little bell-shaped, purplish-pink flowers on axillary stalks; winter-blooming. **C**: 60   p. 186

macrantha (Brazil), "Christmas pride"; bushy herbaceous shrub to 3 to 5 ft. high with erect angled branches; handsome opposite, lanceolate leaves 4 to 6 in. long, matte dark green, with veins depressed; pretty trumpet-shaped, axillary flowers 3 in. across, carmine-rose with pale throat lined purplish red, and blooming from winter into spring. Large-flowered species for the conservatory, but blossoms easily drop. From herbaceous cuttings. **C**: 60   p. 493

makoyana (Brazil), "Monkey plant"; low-spreading herbaceous plant with small, ovate leaves satiny olive-green and shaded violet, with silvery veins; rosy-carmine flowers. **C**: 60

strepens, grown in horticulture as ciliosa (U.S.: New Jersey to Florida and Texas); hairy herbaceous perennial 1 to 3 ft. high, with ovate, rugose 3 in. leaves, the margins with eyelash hairs; large 2 in. trumpet flowers, variably blue or white with purple midveins on lobes. Summer-blooming August-September, in habitat, but I have photographed plants grown from cuttings, in flower, in January. Fairly winter-hardy. From seeds, cuttings or division. **C**: 22 p. 493

RUMOHRA (Filices: Polypodiaceae)
adiantiformis, known as Polystichum coriaceum or Aspidium capense in horticulture, (widely distributed from So. America to So. Africa, New Zealand, Polynesia), the "Leather fern" or "Leatherleaf"; a spreading fern with creeping brown rhizome, in dense clusters, similar to Davallia; fresh green fronds to 3 ft. high, thick-leathery, one to three-pinnate, with the oblong segments coarsely toothed. A durable fern for both warm or cold locations; when warm will grow faster. Cut fronds are much used by florists in flower arrangements because of their decorative beauty and long-lasting qualities. **C**: 4   p. 238

RUSCUS (Liliaceae)
aculeatus (England to Mediterranean region and Iran), "Butchers broom"; evergreen shrub with creeping rootstock and erect grooved stems 1½ to 3 ft. high, tiny greenish flowers curiously borne along midrib of the broad-ovate, stiff, spine-tipped, leaf-like branches (cladodes), ¾-1½ in. long and dull dark green; separate male and female flowers present on same plant, occasionally producing bright red berries. Dried, bleached or brightly dyed branches are used for flower arrangements of "everlastings", or for Christmas. Propagated by division or suckers. **C**: 26   p. 300

hypoglossum (So. Europe: Spain to the Balkans), "Mouse thorn"; evergreen tufted shrub with creeping rootstock and rigid stems 18 in. high, the flexible leathery, leaf-like branches (cladodes) narrow and tapering to both ends; to 4 in. long, not spine-tipped, the small yellowish flowers curiously sit on the centers of their upper surface. An unusual conversation plant also for indoors; or as a ground cover. Propagated by division. **C**: 15   p. 300, 528

RUSCUS: see also Semele

RUSSELIA (Scrophulariaceae)
equisetiformis (Mexico), "Coral plant", or "Fountain bush"; shrubby plant with whip and rush-like, 4-angled stems to 4 ft. long, arching or pendulous; the normally lanceolate dentate leaves mostly reduced to small scaly bracts on the branches; tubular two-lipped flowers with fiery scarlet-red corolla 1 in. long, in nearly continuous bloom. Also good basket plant. Prop. by cuttings. **C**: 13
       p. 276

RUTA (Rutaceae)
graveolens (So. Europe), the "Herb-of-grace" or "Rue"; handsome, strongly aromatic sub-shrub almost evergreen to 3 ft. high, with fern-like, much divided, glaucous blue-green 3 to 5 in. leaves, and small yellowish flowers in terminal clusters; fairly hardy. Cultivated for centuries in herb gardens, its name a symbol of repentance and regret. The bitter, musty-flavored leaves may be used, but sparingly, in many foods such as vegetables, stews, salads, vegetable juices, or sandwiches. From seed or root division. **C**: 13
       p. 321

SABAL (Palmae)
minor (No. Carolina, Florida to Georgia and Texas), the "Dwarf palmetto" or "Scrub palmetto"; wide-ranging dwarf fan palm usually without trunk, the rigid palmate leaves 3–5 ft. wide, stiff and flat, glaucous or grayish green, the segments cut halfway or more; fronds generally upright but older blades kink at the junction of petioles, hanging downward and folding; long erect inflorescence and black-glossy fruit. **C**: 14   p. 228

palmetto (Carolina coast to Florida), "Cabbage palm" or "American palmetto"; variable fan palm 20 to 90 ft. high, with stout trunk either almost smooth and brown, or covered with a criss-cross pattern of leaf-bases, slightly curving; the palmate leaves 3–6 ft. long, in a relatively small head, divided into many slender, hanging segments, green or bluish above and gray on the underside, with numerous thread-like fibers; the midrib extending through into the blade; white flowers, and blackish fruit. Widely planted on the southern fringe of the temperate zone, surviving some freezing weather; seldom used as a tub plant, although it is a decorative fan palm for cool locations. **C**: 14   p. 228

SACCHARUM (Gramineae)
officinarum (China, East Indies), the widely cultivated and ornamental "Sugar cane"; stout perennial grass to 15 ft. tall, with solid yellowish-green canes ¾ to 2 in. thick; rich green, arching, clasping leaves to 3 ft. long, with broad midrib and rough edge; inflorescence in spikelets in large terminal fluffy, silky plumes. Its saccharine sap is a major source of sugar, extracted by crushing the canes; the fermented and distilled juice becomes a well-known intoxicating drink, otherwise known as rum. Requires warmth, moisture and nourishment. Propagated by stem pieces with nodes. **C**: 52   p. 306, 537

SACCOLABIUM—see Rhynchostylis.

SADLERIA (Filices: Polypodiaceae)
cyatheoides (Hawaii), "Pigmy cyathea"; attractive, vigorous small tree fern forming a trunk 4 to 5 ft. high, with a crown of fleshy, soft-leathery light green, bipinnate fronds to 3 ft. long; neatly regular, linear segments crenate and turned under at edges; leaf stalks stout and fleshy. **C**: 60   p. 236

SAGITTARIA (Alismaceae)
montevidensis (Argentina to Brazil, Chile and Peru), "Giant arrowhead"; very large tender perennial aquatic herb sometimes growing to 4 ft. long, the thick, soft-fleshy leaves arrow-shaped, mostly broad, sometimes narrow to 1 or 2 ft. long and with long, spreading pointed basal lobes; flowers very large, 2 to 2½ in. across, the rounded petals white with a purple blotch at the base. Only for large heatable greenhouse ponds on aqua-terraria the year round, also outdoor ponds in summer, for their magnificent flowers. This tropical species is propagated from seed; the hardy kinds form tubers or runners. **C**: 64   p. 518

SAINTPAULIA (Gesneriaceae); well beloved small tropical East African herbaceous plants known as "Usambara" or "African violets"; usually symmetrical rosettes of pubescent, spoon-shaped, brittle leaves and basal stalked flowers from the leaf axils, the corolla is flattened and two-lipped, with 5 large rounded lobes; the upper lip is 2-lobed and usually smaller than the 3-lobed lower lip. The primary color is in shades of violet-blue. The first two species S. ionantha and confusa were discovered in Tanzania, then German East Africa in 1892. The first seedlings from one or hybrids from both species began to appear the following year. Since Saintpaulias are also easily propagated vegetatively from leaf cuttings, the miracle of a single somatic or body cell has given rise to thousands of variations and cultivars including pink and white sports, mainly from one species, S. ionantha. These mutants, and further intense hybridizing, have resulted in many different horticultural types and forms including double-flowered and bicolors, most of them named and recorded. In addition, numerous new species have been discovered, some miniatures, others creeping. The relative ease of cultivation in the home, and their simple propagation by leaf cuttings, have made Saintpaulias the most popular house plant in America, well suited to our warm homes. Adequate moisture and humidity, and a minimum of 600 fc. of light to not over 1500 fc. for 16 hours daily are basic rules for success. Until fully grown, or wanted, buds should be removed continuously, followed by adequate feeding when pot bound. Experienced growers recommend to remove a ring of leaves near the top to allow the buds to push through; and with fertilizer added when they are hungry, the little plants will be quite happy. Saintpaulias, like most gesneriads, are unaffected by daylength but candle power is important and with light intensity high enough will flower the year around. Easily propagated by leaf-petiole cuttings, also by seed; older plants may be divided and planted in new soil.

AFRICAN SPECIES:

Saintpaulia species are all at home in tropical East Africa though found at varying elevations, from near sea-level to chilly mountains at 7000 ft. Although 20 or more species are now known, practically only S. ionantha and confusa have given rise to the thousands of named hybrids and cultivars recorded in horticulture.

amaniensis (Tanzania), from the E. Usambara Mts., near Amani at 3000 ft.; low-growing miniature creeper with very small convex, heart-shaped pointed, green 1½ to 2 in. leaves covered with gray hairs, and small ⅝ in. lively intermediate violet-blue flowers. **C**: 10   p. 434

confusa, originally known as kewensis in horticulture, later on as diplotricha; (Tanzania); an original "Usambara violet"; miniature species from the moist E. Usambara Mts. near Amani at 3000 ft.; erroneously called diplotricha but now correctly identified as confusa, having more firm, light green, and thinner leaves 1½ to 2 in. long, toothed at margins, covered with flattened, not erect hairs, silvery green beneath; small medium-blue flowers with darker violet center, 1 in. across. **C**: 22   p. 435

difficilis (E. Africa: Tanzania), pretty species from the East Usambara Mts. bordering Kenya; robust plant with long-petioled, pleasing light yellow-green, thin quilted leaves 3 to 4 in. long and with sawtooth margins; prolific with light violet-blue, dark-eyed flowers conspicuously showing their bright yellow anthers, the limb 1 in. across. **C**: 10   p. 435

diplotricha (Tanzania), rock-dweller from Maweni near low, coastal Tanga; very similar to S. confusa, with 1 in. fully round flowers a lively sky-blue; the limb is not flat but the lower lip forms an angle with the erect upper lip; the leaves are quilted and incurved, thicker than confusa and bronzy above, reddish purple beneath, with long hairs spreading and short hairs erect in a rather dense but inconspicuous covering; the margins with pronounced rounded teeth. **C**: 10   p. 435

grandifolia (Tanzania); in its West Usambara Mts. habitat found growing on clay; giant species of very robust growth with large 4 to 5 in. fresh green thin leaves, cordate at base, both the erect or curved hairs all long and well spaced; long 3 in. stalks bearing clusters of 10 to 12 single flowers 1 to 1¼ in. across, blue-violet with darker center and with crenate edges. **C**: 10   p. 434

grotei (Tanzania); from deep shade in the E. Usambara Mts., near Amani at 3000 ft., on rocks close to a waterfall; robust trailing or straggling species growing into a large plant full of fresh-green, short-hairy, rounded, fleshy leaves with notched edges, on long, flexible, brown petioles; flowers small, 1 in. across, pale violet-blue with darker center and edges, in axillary clusters. Excellent for hanging baskets. **C**: 24   p. 434

intermedia (Tanzania: East Usambara Mts.); in habitat found growing on rock; a compact plant slowly stalk-forming (caulescent), later with multiple crowns and creeping with short internodes; the 2 in. leaves purplish-green above but grayish with silver felt, reddish underneath, the margins sawtoothed; the hairs are all long and bent over in the upper half; pretty flowers medium-blue with violet eye, ⅞ to 1 in. across. **C**: 10   p. 435

ionantha (Tanzania), the original "African violet"; found near warm-humid, sweltering Tanga at about 100 ft. above sea-level in 1892; most of our named cultivars are mutant strains of this species; robust plant with large flowers a pretty light violet-blue, 1 to 1¼ in. across, very prolific above dark, coppery-green, pubescent 2 to 3 in. leaves, red beneath. On the steep and breezy rocks towering up from the Amboni Caves, north of Tanga. I have seen plants hanging from steep cliffs or narrow crevices, barely providing foothold, surviving the seasonal dry periods by means of rhizomes almost bare of foliage. **C**: 10   p. 435

magungensis (Tanzania), from shady and humid foothills of the West Usambara mountains; creeping species with procumbent brown stems and 1½ to 2½ in. rounded, convex, medium green leaves with depressed veins, the underside pale green; small-size ⅞ in. flowers medium blue with darker eye. May be grown in hanging baskets. **C**: 62   p. 435

nitida (Tanzania: Nguru Mts. at 3000 ft.); handsome small rosette with shining, dark green almost hairless, waxy leaves nearly round and lightly cupped, the few scattered hairs all short, gray-green beneath; 1 in. flowers violet blue. **C**: 10   p. 434

orbicularis (Tanzania), from cool elevations of the W. Usambara Mts. between 4000 and 7000 ft.; upright, somewhat weak species with thin, light green, rounded 1¾ in. leaves having depressed veins, on long, thin brown petioles; small ¾ in., dainty flowers very pale lavender blue with dark purple center, on wiry brown petioles. Blooms from multiple crowns, but is otherwise shy. **C**: 22   p. 434

pendula (Tanzania: Eastern Usambara Mts.): rock-dweller in habitat; robust, spreading plant with exuberantly creeping, strong stems; light green, roundish and cupped plate-like, pliable leaves 1½–2 in. long, covered on the upper surface with almost erect, long hairs, with notched margins, and pale beneath; 1⅛ in. flowers medium blue, one or two on each stalk. **C**: 10   p. 434

rupicola (East Africa: S.E. Kenya), the "Kenya violet"; a robust species which I found growing in 1960 high on bare rock or in clefts of isolated cliffs at Kaloleni 25 miles northwest of Mombasa, near sea-level; robust rosette of light green, heart-shaped glossy leaves to 2½ in. long, covered with erect hairs both short and long; whitish beneath; lightly crenate at margins; prolific with 1 to 1⅛ in. single flowers wisteria or purplish-blue. The persistent stem forms a thick rhizome to 8 in. long through generations of succeeding leaves, later dropped, and this sustains the plant in habitat through periods of drought. A promising house plant with a tendency to hold its blossoms. **C**: 10
p. 435

shumensis (Tanzania), from dry forest on cliffs in W. Usambara Mts. at 6000 ft.; attractive miniature species with shiny fresh green, rather thick, round 1½ in. leaves sparsely covered with long white hairs, neatly notched at margins; the small ¾ in. flowers very pale gray-blue nearly white, with a violet eye, and showy blue pistil, carried on thin stalks. The smallest of the miniatures, and not very floriferous. **C**: 22   p. 434

tongwensis (Tanzania), from Mt. Tongwe in the East Usambaras at 2300 ft.; robust species of pleasing habit, with long-stalked stiff, pubescent, bronzy elongate leaves 3 in. or more long, with light green central vein; the margins scalloped; prolific bloomer with 1¼ in. flowers pale amethyst-blue, on strong stalks. **C**: 3   p. 435

velutina (Tanzania), from rain forest in the W. Usambara Mts. at 4000 ft.; shapely, dwarf plant, with small thinnish, crenate 2 in. leaves blackish velvety green above, reddish-purple underneath, and covered on the leaf surface with numerous short-erect hairs together with some long ones; tiny ⅝ to ¾ in. flowers deep violet with lobes sometimes tipped with white, in abundance. **C**: 10   p. 434

HORTICULTURAL CULTIVARS:

Several thousand named hybrids, mutations and variants of African violets have been created by intent or by accident, and new and often better ones are being introduced continuously. It is difficult to know which cultivars selected today will not be outdated tomorrow, and for this reason this Manual shows a collection of present favorites not for their names but representing the various types and forms that have been developed and introduced in horticulture, types that will remain with us though their names may change. Single-flowered varieties more resemble true violets and are favored by many plant enthusiasts but they tend to drop their flowers more quickly than the doubles which hold their blossoms, and also bloom better during hot weather in the summertime.

'Azure Beauty', typical "double white with violet center"; free-blooming and large-growing hybrid, prolific with beautiful double 1 in. flowers, white with violet in base, and carried erect on firm stems; smooth, shiny foliage. **C**: 10.   p. 436

'Blue Boy-in-the-Snow', typical variety with leaves mottled cream; meritorious leaf sport of 'Blue Boy' having light green foliage splashed with creamy-white throughout the leaf and maintaining this variegation in new growth; 1 to 1½ in. flowers violet-blue on pink petioles. S. 'Blue Boy' was the first choice selected from primary hybrids between ionantha and confusa, made by Armacost in Los Angeles about 1930 from Benary (Germany) seeds received in 1927; an excellent, profuse bloomer with violet-blue, single flowers over a fresh green rosette of leaves. **C**: 10   p. 436

'Blue Caprice', typical "double light blue"; charming stocky, compact plant and willing early bloomer, with "girl" type scalloped leaves dark green, except for the characteristic pale center near the base, and with $1\frac{1}{4}$ to $1\frac{1}{2}$ in. light lavender, double flowers in profusion. One of the best commercial cultivars, with flowers that stand up and last. **C**: 10 p. 437

'Blushing Bride', a superb, typical "Double pink African violet"; charming Roehrs (1966) hybrid; an open rosette with fleshy, crinkled leaves, and beautiful cupped, double and semidouble $1\frac{1}{4}$–$1\frac{1}{2}$ in. flowers, good clear pink with deep rose center and prettily frilled petals; a welcome addition with its dainty charm. **C**: 10 p. 380

'Calypso', typical "single purple with white edge"; sturdy plant with fleshy, dark green, quilted leaves; substantial, large $1\frac{1}{2}$ in. blooms with lower petals lilac, the smaller upper petals in purple, adding distinction and richness, and the whole flower prettily edged with a narrow white, wavy border. **C**: 10 p. 437

'Diana', typical "European single violet" of the Englert 'Harmonie' strain; bushy, evenly formed, and compact cultivar much grown in Europe, with rich green leaves, and a full bouquet of large single flowers, velvety purple with darker violet center, carried very prolifically on stiff erect stalks. This 'Harmonie' strain is distinguished by its vigorous growth and floriferous blooming habit, combining with it the non-dropping feature of its long-lasting flowers, even the single ones; the visible pollen creates a charming contrast. Roots are comparatively not salt-sensitive and help them to grow under adverse conditions. **C**: 10 p. 437

'Fair Elaine', typical "double pink with white edge"; robust grower with large, rather stiff, glossy and quilted leaves, and large $1\frac{1}{2}$ in. double flowers a pretty deep pink, with pale or white edging. **C**: 10 p. 436

'Icefloe', typical "large double white"; an excellent Granger introduction of flat habit, with large dark green leaves, and a prolific bloomer with double $1\frac{1}{2}$ to $1\frac{3}{4}$ in. flowers pure white but changing to bluish pink with aging. **C**: 10 p. 436

'Kenya violet-blue' (Roehrs 1964), typical long-lasting double violet-blue, the "Kenya double violet"; a new breakthrough in Saintpaulia hybrids, incorporating parentage of the blue S. rupicola which I found in Kenya; with the wine-red cv. 'Flash', resulting in a vigorous, shapely plant with large, $3\frac{1}{2}$ in., shiny green, spoon-type leaves, and $1\frac{1}{2}$ to $1\frac{3}{4}$ in. flowers double like a rose and a clear violet-blue; the non-dropping blossoms drying up on the plant instead of rotting when past blooming. A fast grower flowering in 4 in. pots from leaf propagation, and splitting, in 6 months. **C**: 10 p. 380, 437

'My Flame', typical "double pinkish flowers with violet margin"; robust plant with small, $1\frac{1}{2}$ in. foliage, and 1 in. lavender pink blooms dressed very effectively by the contrasting purplish violet along the wavy border. Flower colors variable from one plant to another, or even on the same one. **C**: 10 p. 437

'Pink Amour', typical "single frilled pink with dark edge"; showy, compact grower with waxy bronzy-green, scalloped "holly-type" leaves, and sturdy, medium-large rose-pink crested single flowers, prettily frilled at the cerise margins. **C**: 10 p. 436

'Pooty-cat', typical "star-type"; a remarkable plant because of its regular flowers with 5 even-sized, jointed petal-lobes forming a perfect star; the large single 2 in. blooms violet-purple outlined by a border of pale purple, in clusters from between the dark green, pointed foliage. **C**: 10 p. 437

'Red Spark', typical "semi-double red"; a good Roehrs hybrid 1964, with heavy, dark, somewhat brittle foliage, and large $1\frac{3}{4}$ in. semi-double flowers outstanding for its glowing, bright red-purple blooms. **C**: 10 p. 436

'Rhapsodie'; a German strain (Holtkamp) of triploid hybrids commercially grown under license in Europe, America and overseas, primarily for their prolific bloom together with good keeping qualities. By purposeful implosive breeding, and doubling of chromosomes following colchicine treatment, a series of excellent, named cultivars have evolved that produce an abundance of long-lasting, thick-textured non-dropping blooms, and that must be snipped off with scissors; originally in single-flowered, now also semi-doubles and doubles in several colors and bicolors. For the most bloom, the distributors recommend the removal of the third row of leaves around the crown to allow the heavy buds to push through to form a veritable bouquet. The pictured cultivar 'Elfriede' is an intense dark blue in fully

round $1\frac{1}{2}$ in. blooms which form a dense formal mound above the shapely rosette of shiny dark green, quilted leaves. **C**: 10 p. 436

'Savannah Sweetheart', typical "holly-leaf, double bright rose"; low rosette with fleshy rich green leaves holly-like curly and ruffled, with margins turned under forming irregular points; free-blooming with heavy, large double flowers $1\frac{7}{8}$ in. across, with frilled petals rosy pink; hanging because of their weight. **C**: 10. p. 437

'Sea Foam' (Roehrs 1964), typical "Semi-double rose with white edge"; sturdy Roehrs hybrid, with dull dark green, quilted foliage, and stiff-stalked carmine pink semi-double $1\frac{1}{2}$ to $1\frac{1}{2}$ in. flowers, very charming with pale pink or white border. **C**: 10 p. 437

'Spring Sky', typical "single pale blue"; (Blue Eyes x Blue Boy), compact, free-blooming Roehrs hybrid (1959) with $1\frac{1}{4}$ in. light blue flowers and dark eye; light green to coppery satiny leaves; more prolific than 'Blue Eyes'. **C**: 10 p. 436

'Star Girl', typical "Girl-leaf single"; attractive variety, with white flowers streaked and edged with medium-blue as if framing a white star, carried on stiff stems well above the "girl-type" leaves, distinguished by their scalloped margin and area of pale yellow in the leaf base. **C**: 10 p. 436

SALVIA (Labiatae)

leucantha (Mexico), "Mexican bush-sage"; floriferous shrub of graceful habit, about 2–4 ft. tall, with woolly branches and narrow wrinkled leaves 6 in. long, white-downy beneath; long slender, velvety purple racemes of small flowers white and woolly, the calyx covered with dense purple wool; young stems bloom continuously from summer to fall. Not quite hardy. Very attractive in their woolly dress and aromatic foliage. Division of old clumps. **C**: 64 p. 322

officinalis (So. Europe, Mediterranean region), "Garden sage"; hardy shrubby white-woolly perennial to 2 ft. high, with slender oblong gray-green more or less hairy, pebbly textured 2 in. leaves; light purple or white two-lipped, beautiful flowers in terminal raceme. A familiar seasoning herb for poultry, dressings, cheese, veal, roast pork, sausage, tomatoes; leaves also medicinal as a condiment or tea for sore throats. Oil distilled from whole plant perfumes soaps. From seed, cuttings or layers. **C**: 25 p. 322

sclarea (So. Europe), "Clary sage"; handsome hardy biennial with clammy-hairy herbaceous stem 2–3 ft. high; pubescent broad-ovate, scalloped, pebbly gray-green leaves often to 9 in. long, smaller higher up the 3 ft. flowering stalk, the flowers whitish-blue in clusters, and with floral bracts colored rose and white. Use: foliage for flavoring wine, beer and ale; leaves may be eaten in omelettes or as fritters or to flavor soups, also employed in sachets. Flowers make tea. Oil from seeds used in perfumes. From seed. **C**: 76 p. 322

splendens (Brazil), the "Scarlet sage"; striking herbaceous perennial usually cultivated as an annual 1 to 3 ft. tall, freely branching, with squarish stems, and ovate, rich green, glabrous leaves; the showy inflorescence in erect spikes bearing scarlet-red flowers with bell-like red calyx, and tubular corolla $1\frac{1}{2}$ to $2\frac{1}{2}$ in. long, blooming from June to frost. A favorite for the sunny garden, or for window boxes. From seed, or cuttings when pinching. **C**: 64 p. 411

SALVINIA (Filices: Salviniaceae)

auriculata (Trop. America), "Floating fern"; small, flowerless aquatic fern, floating on water with $\frac{3}{8}$ in. oval leaves pale yellowish-green and warty-haired on the surface, set along thread-like floating rhizomes, soon forming colonies. Used in aquaria or indoor pools, rapidly multiplying by division if kept warm and sunny; during winter warm and light in shallow dishes on loamy soil in $\frac{1}{2}$ in. water. **C**: 2 p. 253, 519

SANCHEZIA (Acanthaceae)

nobilis glaucophylla (Ecuador); handsome tropical shrub cultivated for its large lanceolate, soft-leathery leaves to 9 in. long, glossy-green with bold, contrasting yellow veins; large yellow flowers with bright red bracts in showy terminal panicle. **C**: 4 p. 187

SANSEVIERIA (Liliaceae); succulent members of the lily family known variously as "Bow-string hemp", "Snake plant" or "Zebra lily", with short rhizomatous rootstocks sprouting thick fleshy, mostly erect sword-shaped leaves,

often variegated or otherwise patterned; inflorescence a long cylindrical raceme with small whitish, fragrant flowers. They are amongst our most resistent house plants in difficult locations; all are tough and tolerate poor light, drafts and dryness. Propagation from division or from cross-sections of leaves.

cylindrica (sulcata) (So. Trop. Africa, Natal), "Spear sansevieria"; arching, but rigid cylindric leaves to 5 ft. long and 1¼ in. thick, grouped several to a growth, usually furrowed or grooved, and tapering to a point, dark green with gray-green crossbanding which disappears with age; flowers pinkish. **C**: 1   p. 360

ehrenbergii (Ethiopia, Kenya, Tanzania), "Blue sansevieria" or "Seleb"; elegant plant with angular leaves arranged as in a large fan, 2 to 5 ft. high, concealing the stem; the blue-green foliage above with triangular channel bordered by white papery edge; flat on sides and rounded below. **C**: 3   p. 360

grandis (Somaliland), "Grand Somali hemp"; epiphytic succulent sending out runner-like rhizomes producing 2–4 broad-obovate leaves to 6 in. wide and 10 in. long, spreading to near the ground; dull green with broad bands of deeper green, margins red; white flowers in dense raceme. **C**: 4   p. 360

guineensis (West Africa: Guinea coast of Nigeria, Cameroun), a very bold "Bowstring hemp"; robust, attractive rosette with 8 to 10 extremely broad, sword-shaped leaves 3 to 4 in. wide, and 1½ to 3 ft. high, arranged in birdsnest fashion, glossy deep green irregularly cross-banded light grayish green. I photographed this striking species in Southern Nigeria. **C**: 3   p. 360

intermedia (E. Trop. Africa), "Pigmy bowstring"; stiff, shapely, dense rosette of very thick, recurved leaves about 10 to 12 in. high, gray-green, with deep triangular channel inside, and keeled beneath, marked with numerous pale crossbands; the inner edges horny. **C**: 3   p. 360

"Kirkii" hort., "Star sansevieria"; probably S. grandicuspis from the Congo; shapely, starry rosette with about 14 narrow, thick, angled or concave leaves ¾ to 1¼ in. wide, slightly furrowed on back, and gently arching outward, 16–20 in. long, ending in a long soft, awl-shaped tip, dull grass to deep green, alternately cross-banded light grayish green, the margins with fine, brown-horny edge. **C**: 3 p. 360

parva (E. Africa), "Flowering parva sansevieria"; dense rosette to 1½ ft. high, of narrowly lanceolate, recurved and spreading, very concave leaves fresh green with dark cross-banding, and with long green circular tip; flowers pinkish white. Interesting when as a hanging plant long pendulous stolons carry offspring like monkeys on a rope. **C**: 1 p. 361

senegambica (S. cornui in horticulture), (Senegal, Gambia), sub-erect, broadly oblanceolate leaves, with a concave channel at the base, flattening out higher up and ending in a slender point; matte medium green, only very little striped on the outside. **C**: 3   p. 360

suffruticosa (Kenya), "Spiral snake plant"; stemforming succulent rosette sending out thick stolons bearing spirally arranged, lightly arching, thick leaves to 2 ft. long, channeled and clasping at the lower end, becoming cylindrical toward the tapering apex, rich green banded with pale green and grooved. **C**: 1   p. 361

trifasciata, erroneously called "zeylanica" in horticulture (Transvaal, Natal, E. Cape), "Snake plant", "Mother-in-law's-tongue" or "Lucky plant"; robust succulent rosette with stiffly erect, linear-lanceolate, thick-leathery, concave leaves usually 15–18 in., but under favorable conditions to 4 ft. long, deep grass to almost blackish-green with irregular light green to gray-white cross bands, on fleshy rhizomes bearing from 2 to 8 leaves; small greenish white flowers in a loose raceme, fragrant at night. Next to its variegated form laurentii, the most widely grown Sansevieria in the world; as small 6 to 12 in. plants in dishgardens; larger plants or combinations in jardinieres in every shop window or difficult corner of the house. **C**: 3   p. 361

trifasciata 'Bantel's Sensation' (sport of laurentii 1927), "White sansevieria"; unusual cultivar with slender, oblanceolate leaves about 15 in. long, some of which have one side with the typical crossbars of trifasciata, the other half a golden yellow band like laurentii, and adjoining this are stripes of white, edged with green, in other leaves either one or the other character may predominate. **C**: 3   p. 360

trifasciata 'Hahnii' (U.S. Pat. 1941), "Birdsnest sansevieria", sport of trifasciata laurentii found at New Orleans in 1939; entirely different in habit, forming a low, vase-like rosette 5 to 12 in. across of broad, elliptic leaves 4 to 6 in. long, spirally arranged, spreading and reflexed, dark green with pale green crossbanding; robust and freely suckering. Ideal for novelty dishes. **C**: 3   p. 328, 360

trifasciata 'Golden hahnii' (pat. 1953), "Golden birdsnest"; very decorative succulent plant when fully variegated; a sport of hahnii with firm, broad-elliptic leathery leaves in a low, flattened rosette, grayish-green with broad cream to golden-yellow bands alongside the margin, and more or less cross-banded in gray. From suckers. **C**: 3
                                                                p. 196, 360

trifasciata laurentii (N.E. Congo), "Variegated snake plant" or "Goldband sansevieria"; the most cultivated of decorative sansevierias, with elegant, stiff-erect, sword-shaped leaves 2 to 5 ft. high, rising from the fleshy rhizome; deep green with light crossbanded center and broad yellow bands bordering both sides of each concave leaf; its good keeping qualities and nicely turned rosettes, soon clustering, have made it an easy to maintain, most satisfactory house plant. **C**: 3   p. 196, 361

trifasciata 'Silver hahnii marginata', "Gilt-edge silver birdsnest"; charming Pennsylvania (Hahn) cultivar; rosette of fat habit with spreading leaves silvery gray, irregularly banded cross-wise with dark green; the margins creamy yellow. Most attractive. **C**: 3   p. 360

zeylanica (Ceylon), "Devil's tongue"; the true species, and easily distinguished from trifasciata which is known as "zeylanica" in the trade; as compared in habitat in Ceylon, the true species forms a much more shapely, spreading rosette only about 1 ft. high, with numerous, gracefully recurving, thick-fleshy leaves, grayish green with dark green crossbands, channeled in the middle, keeled beneath and grooved, ending in a long circular tip. **C**: 3   p. 360

SANTOLINA (Compositae)

chamaecyparissus (incana) (So. Europe), "Lavender cotton"; low shrubby plant with brittle, woody stems 1 to 2 ft. high, densely clothed with rough, finely divided whitish-gray tomentose, aromatic leaves; small button-like ¾ in. yellow flower heads without ray-florets in profusion on unclipped plants. Attractive with its silvery lacy foliage in the sunny window or for edging in carpet bedding, sheared to keep in shape and looking like coral. From cuttings. **C**: 13 or 63   p. 322, 410

SARCANTHUS (Orchidaceae)

lorifolius (Burma: Tenasserim); exquisite epiphytic orchid with leafy, rooting stem; the 2-ranked fleshy, channeled leaves 10 in. long but less than 1 in. wide, ending in a long fleshy point; lovely ⅓ in. waxy flowers in dense pendant, grape-like clusters of 40 or more, the sepals and petals yellowish with purple length-stripes; lip with yellow side lobes and violet-flushed front-lobe; and a milky-white spur with 2 short points; blooming November to February. These tropical orchids continuously need warm and humid conditions with free-moving air; as growing material treefern fiber, firbark, or osmunda with sphagnum. Propagated from offshoots or rooted stem sections. **C**: 10   p. 481

SARCOCOCCA (Buxaceae)

ruscifolia (W. China), "Sweet box"; spreading slow-growing Asian evergreen shrub 4 or 5 ft. high, related to boxwood, but with longer, ovate 2–3 in. leaves lustrous dark green and leathery, with wavy margins; small milk-white flowers nearly hidden in the foliage, fragrant enough to be noticed many feet away. Useful as a dishgarden plant when small; a decorator plant for shade or sun. Winter-hardy. From cuttings. **C**: 16   p. 298

SARCOGLOTTIS (Orchidaceae)

metallica 'Variegata' (Costa Rica); terrestrial orchid with pretty, elliptic to oblanceolate leaves semi-matte black-green with greenish-silver bands lengthwise; reverse gray-green; whitish flowers. Prop. by division. **C**: 60 p. 454

SARCOPODIUM: see Dendrobium

SARRACENIA (Sarraceniaceae); the North American Pitcher plants, also known as "Hunter's horn", "Trumpet leaf", or "Sidesaddle flower", are curious insectivorous herbaceous perennials dwelling in bogs from Labrador to Florida, with basal hollow fluted leaves which are winged or keeled on one side and with a lid on top, and holding

liquid as in a water-pitcher, serving as fly-traps, the plants making use of the insects caught as food. Handsome nodding flowers with red or yellow, fleeting petals and spreading, leathery persistent sepals. The striking and beautiful pitchered leaves are effective carnivorous traps, a sugary secretion near the mouth attracting flies, beetles, and spiders and other small insects which enter the pitcher but can't get out because of downward-pointed hairs inside; the captives fall into the liquid at the bottom of the pitcher where they are digested and absorbed. In cultivation, Sarracenias are grown best in a cool greenhouse, a glass terrarium or bowl, or a plastic tent; the rhizomes may be planted directly into several inches of sphagnum moss, with washed gravel for a base; or also planted into pots with acid peat, sand and sphagnum mixed, and kept cool and moist; watering should be with lime-free distilled or rain water. Propagate by division of rootstocks in March-April.

drummondii (abundant in swamps in Georgia, Florida, Alabama), the showy "Lace trumpets"; attractive colorful, tall fluted tubes 1 to 2½ ft. long, fresh green and veined purple except toward apex, with rolled margins and the wavy lid, both of which are beautifully marbled with white between the netted green veins; flowers greenish purple. **C**: 72   p. 520

flava (Virginia to Florida), the "Yellow pitcher plant" or "Umbrella trumpets"; tall and slender tubes to 3 ft. long, light green or sometimes crimson, with prominent veins, mouth and lid edged yellow-green and with crimson throat; large nodding yellow flowers 2 to 4 in. across, April-May blooming. **C**: 84   p. 520

minor (No. Carolina to Florida), "Hooded pitcher plant" or "Rainhat trumpet"; pitchers a short 5 in. or also growing to 2 ft. high, fresh-green to purple, beautiful with white translucent spots near the yellowish top, the lid arching spoon-like over mouth, purple netted inside; 2 in. flowers pale yellow, blooming in May. **C**: 72   p. 520

psittacina (in sandy swamps of Georgia, Florida, Alabama and Louisiana), the "Parrot pitcher plant"; low rosette with leaves 2 to 6 in. long, the tube club-shaped, with an erect, broad obovate, flat wing, green with red and white veins, the inflated top with incurved beak like a parrot; flowers green and red. **C**: 72   p. 520

purpurea (Labrador to Maryland and the Rocky Mts.), the "Northern pitcher plant" or "Sidesaddle flower"; low hardy perennial rosettes usually found in sphagnum bogs; green to dark brownish-red, prostrate pitchers 3 to 12 in. long, broadly winged, the throat and lid hairy inside, and beautifully veined crimson; nodding purple flowers, blooming in April-May. The most common kind found in cultivation. **C**: 84   p. 520

SASA (Gramineae)

fortunei (Arundinaria variegata) (Japan), "Miniature variegated arundinaria" or "Dwarf white-stripe bamboo"; having first seen this plant in Brazil, I consider it the most attractive small bamboo in cultivation; its growth habit is low, with gracefully arching branches 1–3 ft. high, and finely pubescent, showy leaves 3–5 in. long and 1 in. wide, much broader than pygmaea, and beautifully banded and striped white, alternating with green. Its running, wandering habit makes this a beautiful ground cover, and with sufficient supply of water, possibly even a good house plant. Somewhat hardy. **C**: 66   p. 305

pygmaea (Arundinaria; Pleioblastus viridistriatus vagans) (Japan), "Pigmy bamboo" or "Dwarf bamboo"; one of the smallest known bamboos, with canes rarely taller than 10 in. high and ⅛ in. thick, from creeping rootstocks, forming dense clumps; small leathery, deep green leaves to 2 in. long, smooth beneath. Running bamboo and aggressive spreader; hardy to zero °F. **C**: 64   p. 305

SAXIFRAGA (Saxifragaceae)

sarmentosa (China, Japan), "Strawberry geranium" or "Mother-of-thousands"; loosely tufted perennial spreading near the ground, strawberry-like, by threadlike runners bearing young plantlets; the soft, fleshy, rounded, coarsely toothed, bristly-hairy leaves 1 to 2 in. across and more, deep olive green with silver-gray areas following the veins, densely spotted and shaded purple beneath; numerous flowers on erect panicle, white, with 2 petals longer than others. Useful basket or pot plant for the house, easy to care for but preferring good light and not too warm. **C**: 13
                                                          p. 263

sarmentosa 'Tricolor', the "Magic carpet"; beautiful variety smaller and more tender than the type, with leaves dark green and milky green, conspicuously variegated inward from the margin with ivory-white, and tinted pink or even rosy-crimson in younger leaves, with red edging; purplish-rose beneath. Likes fresh air-movement and good light. **C**: 75   p. 189

SCHEFFLERA (Araliaceae)

actinophylla: see Brassaia actinophylla

digitata (New Zealand), "Seven fingers"; bush or small tree 10–20 ft. high, densely branching, sometimes growing epiphytic, with thin-leathery leaves palmately compound, usually seven obovate foliolate leaflets to 7 in. long, dull satiny green above, shiny light green beneath, densely ciliate and undulate at margins and with yellowish depressed veins; greenish-yellow flowers in panicled clusters, the purplish-black fruit berry-like. S. digitata prefers cool locations, and in New Zealand I saw large colonies of this evergreen bush in the southern Fiordland, especially near beautiful Milford Sound, revelling in foggy atmosphere and rain within sight of icy glaciers. Not for warm indoor decoration. **C**: 76   p. 290

venulosa, in horticulture, especially in Australia as Heptapleurum (Queensland, China, Vietnam, India); branching tree with palmately compound leaves; the 7–8 stalked leaflets lanceolate when young, obovate or elliptic in maturity, soft-leathery, semi-glossy on both sides, to 6 in. long; mature leaves entire, dark green, but lightly toothed in juvenile stage; inflorescence in panicles with whitish flowers, followed by small red fruits. Resembles Brassaia but of more slender habit. **C**: 3   p. 290

SCHINUS (Anacardiaceae)

terebinthifolius (Brazil, Paraguay), "Brazilian pepper-tree" or "Christmas-berry tree"; ornamental evergreen tree 20 to 30 ft. high of more rigid habit and less pendulous than S. molle, the California pepper tree; broadly spreading with willowy to woody branches densely clothed with pinnate leaves 4–7 in. long, of 5 to 9 broad leathery leaflets, dark glossy green and long-persistant. Small white flowers, followed by bright red berries on female trees, very showy in winter. Very decorative as a patio tree and also as a small house plant, requiring large container and plentiful watering. From cuttings or seed. **C**: 14   p. 301

SCHISMATOGLOTTIS (Araceae)

neo-guineensis variegata (Papua); herbaceous plant with underground stem; papery, heart-shaped leaves medium green with yellow-green blotching, on slender stalks. **C**: 54   p. 98

SCHISMATOGLOTTIS: see also Aglaonema

SCHIZANTHUS (Solanaceae)

x wisetonensis, "Butterfly-flower" or "Poor man's orchid"; beautiful hybrid of pinnatus x grahamii from Chile; a bushy herbaceous annual plant with brittle, slender, sticky branches; pale green, divided fern-like leaves, and a profusion of showy, irregular, orchid or pansy-like flowers 1½–2 in. across in shades of lilac, purple, pink, carmine, reddish-brown, or white, the upper lip often marked with purple and yellow, blooming spring and summer. A source of pleasure for the enthusiastic horticulturist, these charming flowers are especially suited for sunny but cool-moist locations, and are much favored in Canada and Britain. From seed sown in August for spring bloom. **C**: 64   p. 371, 415

SCHIZOCENTRON (Melastomaceae)

elegans (Mexico), "Spanish shawl"; creeping herb forming a dense mat, the reddish stems rooting at nodes; small ½ in. ovate leaves, deep green and lightly hairy; covered with purple flowers 1 in. dia. during summer; good basket plant, or by a window if humid. **C**: 72   p. 264

SCHLUMBERGERA (Cactaceae); epiphytes of the tropical rainforest, densely branching with glossy green, flattened leaf-like joints having scalloped margins and blunt rounded tips; blooming in winter with usually regular (actinomorphic) flowers. Culture see p. 151. Propagated by joint cuttings rooted in moist peatmoss; also grafting (see p. 62). As mentioned in the introduction to Cacti, the Christmas cactus is a short-day plant. It needs two rest

periods with reduced temperature and moisture; one for two months after blooming to gather new strength, and the other from August to early October to prepare for bud initiation if wanted in bloom for Christmas. 55 to 60° F. temperature from middle of September on should begin to show buds after 4 to 5 weeks; if the weather at this time is unusually sunny and warm, short-day treatment will be necessary by limiting the day length to 9 hours until buds show. Schlumbergera is often cleft-grafted on Pereskia, resulting in attractive little tree-forms.

bridgesii (Epiphyllum truncatum) (Bolivia?) Grandmother's "Christmas cactus"; branching epiphyte with small glossy green leaf-like joints, crenate and with blunt apex; pendant flowers starting in December with flaring petals carmine-red tinged purple in center; angled ovaries. The flowers are usually but not always zygomorphic (irregular), which indicates probable hybrid origin with Zygocactus, while other Schlumbergeras have normally regular flowers. **C**: 9   p. 165*, 170

russelliana (Brazil: Organ Mts.), "Shrimp cactus"; dwarf growing epiphyte and one of the oldest known species; will respond well to grafting on such understock as Selenicereus or Pereskia; small joints with crenate edge; regular, star-shaped small flowers with several rows of orange-red petals; spring (March) blooming. **C**: 9   p. 170

SCHLUMBERGERA: see also Rhipsalidopsis

SCHOMBURGKIA (*Orchidaceae*)

undulata (Laelia) (Trinidad, Venezuela, Colombia); handsome, bold epiphyte with tall, spindle-shaped, 2–3 leaved pseudobulbs, bearing the 3 ft. reedy stalk topped by a large clustered inflorescence, in the 1½ in. waxy flowers with narrow sepals and petals wine-purple, beautifully twisted and crisped, the rosy lip with raised white ridges; blooming between December and July. Schomburgkia grow in firbark chips, or a compost of coarse osmunda, fernfiber with pebbles, perlite or charcoal, and loam added; they bloom best in good light near the glass almost without shade, in about 3000 fc.; on the patio in summertime. Water liberally while growing, less after bloom. Propagated by division. **C**: 21   p. 481

SCILLA (*Liliaceae*);

bulbous mostly small herbs of the lily family, some with very attractive basal leaves and others with cheery or showy urn or star-shaped flowers mostly in spring. Delightful plants for a cool room. Since Scilla bulbs have no protective coat, they should be left in the soil during their rest period, or stored in peat or sand. Propagated by offsets; seedlings flower 3 to 4 years old.

peruviana (Mediterranean reg.: Algeria), erroneously but popularly called the "Cuban lily"; handsome bulbous plant 1 ft. high, very showy with broadly dome-shaped cluster to 6 in. across of small star-like lilac-blue ½–1 in. flowers having petals edged in rose, most attractive with blue stamens and yellow pollen, carried on a thick-fleshy stalk above long broad, soft-succulent, fresh-green or slightly glaucous, strap-shaped leaves in basal rosette; blooming in May-June, after which the bulbs go dormant for a short time. A charming house plant looking different from other scillas; not winter-hardy. **C**: 16   p. 401

sibirica 'Spring Beauty' (Yugoslavia, C. and So. Russia), the "Siberian hyacinth" is not from Siberia; low-growing bulbous herb 4 to 6 in. high, with basal leaves ¾ in. wide, and dainty flaring bells porcelain blue ½ in. across, in several flowered clusters in the species; the popular cultivar 'Spring Beauty' is more robust and taller 6 to 8 in. high, the flowers larger 1¼ in. dia. and brilliant blue with deep-blue stripe down center of each petal, and suitable for gentle forcing in pots; plant in September and keep dark and cold until January, then bring to light. Winter-hardy and normally blooming from February to April, this Scilla is ideal for naturalizing in the garden. **C**: 28   p. 401

tubergeniana (Mountains of N.W. Iran), "Persian squill"; a choice miniature bulbous plant barely 6 in. high, each bulb producing 3 or more stalks with beautiful starry flowers 1 in. across, pale blue with deeper mauve backs and a deep blue median line on each segment; the blooms appear just before the broad, bright green basal leaves, in February-March, earlier than S. siberica which it resembles. Perfectly winter-hardy, this Persian squill can be grown and flowered in pots in a cool room. **C**: 15   p. 401

violacea (So. Africa), "Silver squill"; small suckering bulbous plant with swollen base, very attractive with its variegated foliage; strap-like fleshy 3 to 5 in. leaves olive-green, with silver blotching and banding, glossy wine-red beneath; small green and blue flowers on slender racemes; in winter. **C**: 16   p. 186, 401

SCINDAPSUS (*Araceae*);

the "Ivy-arums" are tropical climbers from the Malaysian monsoon area, high climbing or creeping by rootlets, and usually attractive variegated leaves, small in the juvenile state, but of huge size in maturity. The best known Scindapsus aureus, commonly known as "Pothos", has recently been transferred to the genus Rhaphidophora. Scindapsus thrive in warm, even superheated temperatures and kept fairly dry at the roots, especially the highly variegated varieties, making them ideal house plants or basket vines, or for hydro-culture. Propagated easily by tips or joint cuttings; seed is rarely available.

aureus (bot. Rhaphidophora aurea, Birdsey 62) (Solomon Islands), "Devils ivy", "Hunter's robe", "Ivy arum", "Golden Ceylon creeper", "Golden pothos"; known commercially as "Pothos"; fleshy vine climbing tall by rootlets; juvenile leaves broad ovate, waxy, dark green with yellow variegation, 3 to 5 in. long; mature leaves to 2 ft., the blades becoming lobed or slashed; bisexual flowers on short spadix within the boat-shaped spathe. Very popular; loves warmth. **C**: 9   p. 83

aureus 'Marble Queen', "Taro-vine", "Variegated philodendron"; mutant with the green leaves richly variegated and streaked nearly pure white when grown in good light; resents chills and wetness. **C**: 9   p. 43*, 66*, 83

aureus 'Tricolor', "Devil's-ivy pothos"; a colorful mutant having its succulent, ovate leaves highly variegated, marbled and spotted pale green, deep yellow and cream, on medium green. **C**: 9   p. 179

aureus 'Wilcoxii', "Golden pothos"; a California form found and selected by Wilcox, with sturdy green leaf variegated with golden-yellow, the variegations not blending into the green portion of the leaf but terminating abruptly; petioles and portions of stems often ivory-white in color. More durable than the more highly variegated cultivars. **C**: 9   p. 259

pictus (Indonesia, Philippines); tree-climber, clinging close to bark; thick-leathery, waxy leaves obliquely ovate, dark green overlaid with greenish silver blotching. **C**: 60   p. 83

pictus argyraeus (Pothos argyraeus) (Borneo), "Satin pothos"; beautiful creeper with the smaller, cordate leaves satiny, bluish-green with markings and edge of silver; probably a juvenile stage of pictus. **C**: 60   p. 83, 179

SCIRPUS (*Cyperaceae*)

cernuus, long known in horticulture as Isolepis gracilis (E. Indies, naturalized in So. Europe). "Miniature bulrush"; grass-like, graceful tufted plant to 8 in. high with numerous round, threadlike, glossy fresh green stems becoming pendant, tipped by little white flower heads as in bulrush. Excellent for hanging baskets, for terrariums, or the window sill standing in saucer of water. By division.   **C**: 4   p. 304

SCOLOPENDRIUM: see Phyllitis

SCUTELLARIA (*Labiatae*)

mociniana (Mexico), the "Scarlet skullcap"; magnificent herbaceous shrub of robust habit, to 20 in. high, with purplish-brown square stems; opposite, ovate-cordate, quilted leaves 4 in. long, dark metallic green, thin-leathery, and with crenate margins, grayish green beneath; striking erect, terminal spikes with a dense burst of brilliant scarlet-red, tubular flowers 1½ to 2 in. long, with orange-yellow lip and showing white stamens; the individual blossoms will last only 6–10 days but succeeding clusters of flowers appear from upper leaf axils, extending blooming period from January to July. Easy from cuttings. **C**: 53   p. 493

SCUTICARIA (*Orchidaceae*)

steelei (hadwenii) (Brazil to Ven.); peculiar but charming epiphyte with branching rhizome bearing on each branch a solitary channeled, whiplike, pendant cylindric leaf to 4 ft. long, and short stalked clusters of handsome fragrant, waxy flowers 2 to 3 in. or more across, pale yellow spotted with brown-purple, the lip with an orange crest; the long-lasting flowers appearing in summer or fall. Because of their pendulous habit, plants must be grown on rafts, treefern slabs or in baskets hung sideways; best in osmunda or treefern fiber. Kept moist and warm while growing; they are given a short rest after blooming. Propagated by division. C: 60   p. 482

SEAFORTHIA: see Archontophoenix and Ptychosperma

x SEDEVERIA (*Crassulaceae*)
derenbergii, "Baby echeveria"; trim and good little bigeneric succulent combining blood of Echev. derenbergii and Sedum allantoides; stem-forming rosette 1½–2 in. across, of keeled bluish-gray leaves flattened on surface. Propagated from cuttings only. C: 13   p. 338
'Green Rose' (Sedum aureum x Echeveria derenbergii), attractive hybrid in the California nursery trade; firm little 2 in. rosette forming brown stem, dense with small obovate light green leaves covered by bluish glaucescence, waxy smooth and lightly keeled beneath, the tips tinted salmon in the sun; flowers yellow, tipped with orange. Ideal for dish-gardens. C: 13   p. 348

SEDUM (*Crassulaceae*); the "Stone crops" or "Live-forever" are a large genus of attractive, tender or hardy low succulents with decumbent stems or trailing branches rooting at the joints; their plump or fleshy leaves frequently arranged as a rosette; inflorescence is in showy terminal clusters of small starry flowers usually white or yellow. Sedums are widely cultivated; winter-hardy kinds in rock gardens; the tender, subtropical species in dishgardens or hanging pots. They all like and need sun to bring out the best in the coloring of their foliage, and a collection of little pots in the sunny window, or in outdoor containers during summer can be a breathtaking sight when with intense light, their waxy often glaucous foliage actually glows in lovely pastel shades and shimmering iridescence. Established plants need practically no care, but they enjoy cool nights; too much dampness then may spot their foliage, which on occasion is also very brittle. Propagated by cuttings or from individual leaves, and suitable for mass propagation of cheap succulents.
adolphii (Mexico), "Golden sedum" or "Butter plant"; small branching later sprawling rosettes of plumb, fleshy, boat-shaped and keeled leaves 1 to 1½ in. long, waxy yellowish green with reddish margins; white flowers star-like in dense clusters sideways from between the foliage. C: 13   p. 329
'Adolphii hybrid' (S. adolphii x Graptopetalum amethystinum), known in the trade as "Golden Glow" or "Honey sedum"; Hummel hybrid of robust habit with broader waxy golden coppery or red-tinted leaves pointed and somewhat boat-shaped, and which do not shatter as easily as the species. A good dishgarden plant exhibiting its glowing golden color best if kept exposed to sunlight. C: 13   p. 348
adolphii 'Peach blossom', a selected clone of the California hybrid complex of Sedum adolphii with Grapto-petalum (Pachyphytum) and possibly others; sturdy little plant with obovate almost boat-shaped leaves narrower and smaller than Graptopetalum paraguayense and holding foliage better; waxy glaucous greenish gray with peachy (pink) tinting toward apex in good light. A favorite dish-garden plant. C: 13   p. 348
guatemalense: see rubrotinctum
lineare 'Variegatum' (Japan, China), "Carpet sedum"; turf-forming low succulent to 6 in. high, with watery rooting branches and small linear, soft fleshy leaves to 1 in. long, gray-green with white margins; star-like yellow flowers in summer. May be used in carpet-bedding clipped to shape, or as pretty ground cover. C: 15   p. 349
lucidum, in the horticultural trade as S. "tortuosum" (Eastern Mexico: Orizaba); small branching succulent with leafy twisting, rooting stems, the dense fleshy, obovate almost boat-shaped leaves 1 to 2 in. long, yellowish pea-green with bronzy cast, glossy with some waxy powder; flowers white. C: 13   p. 348
mexicanum (Mexico), "Mexican stone crop"; freely branching succulent creeper with flexible, rooting stems dense with needle-like, fresh-green flattened leaves ½ in. long; clusters of golden-yellow flowers. Pretty as a hanging pot plant. C: 13   p. 349
morganianum (Mexico), "Burrow tail", a lovely hanging succulent plant, with tassels of short spindle-

shaped leaves yellowish-green covered with silvery-blue bloom; terminal flowers pale pink. The heavy branches are quite pendulous and a beautiful sight when grown quite long. C: 13   p. 271, 349
multiceps (Algeria), "Pigmy Joshua tree"; small shrubby succulent to 8 in. high, resembling a Joshua tree (Yucca) in miniature; thin woody stems branching tree-like to 4 in. high, topped by tiny terminal rosettes of gray-green needle-like, ciliate ¼ in. leaves; starry yellow flowers. The plant rests in summer, the leaves whither but persist. Excellent for miniature landscapes. C: 13   p. 348
x orpetii (treleasei x morganianum), "Giant burrow tail" or "Lamb's tail"; an excellent succulent for hanging pots and baskets, similar to morganianum but with the fatter, glaucous ¾ in. leaves more spreading open on their tasseled branches, and holding on more firmly, whereas morganianum leaflets tend to drop on touch, and scatter. C: 13 p. 271, 349
pachyphyllum (Mexico: Oaxaca), "Jelly beans"; small shrubby succulent 10 to 12 in. high, with weak stems sometimes rooting, close-set with cylindric, club-shaped fleshy leaves 1½ in. long and curved upward, glaucous blue over light green, with red tips; flowers bright yellow, in large clusters. C: 13   p. 348
platyphyllum (So. Mexico: Oaxaca), "Powdered jade plant"; a robust highly succulent shrub to 6 in. high, resembling a Jade plant, with fleshy stalk and obovate leaves up to 3 in. long; unusual because of its very pale yellow-green color throughout, covered with bluish-silvery glaucescence; flowers greenish-white, dotted red, in a loose cluster. C: 13   p. 348
praealtum cristatum (Mexico), "Green coxcomb"; a weird looking fasciated succulent with fleshy, flat as well as round stems, bearing at their fan-shaped apex a crest of thick, shining light green oblanceolate leaves 2 to 3 in. long; the normal species grows 5 ft. tall and has bright yellow flowers. C: 15   p. 349
rubrotinctum (long known in horticulture as S. guatemalense) (Mexico), "Christmas cheers"; branching small succulent, rooting at the joints; with thickly clustered fleshy club-shaped, ½ in. leaves looking like jelly-beans, glossy green with red-brown tips; turning all coppery red when exposed to sun; pretty clusters of starry yellow flowers. Dropping leaves root easily where they fall. C: 13   p. 329
sieboldii (Japan), "October plant" or "October daphne"; graceful perennial creeper with unbranched, wiry red stems 6 to 10 in. long set with whorls of 3 roundish, notched fleshy leaves ¾ in. wide, glaucous-bluish-gray changing to copper, and edged in red; starry pink flowers in clusters at the end of each vine, blooming in October. Handsome in hanging baskets; keep on dry side. Propagated by division or cuttings. Limited winter-hardy. C: 13   p. 349
spectabile (Japan, Central China), hardy "Live forever" or "Showy sedum"; strong growing, tough winter-hardy perennial 1 to 2 ft. high, freely suckering at base with erect thick fleshy stems and soft leathery, glaucous light green or grayish obovate leaves 2½ to 3 in. long, toothed toward apex, and generally set in twos and threes along the stems, which terminate in large flat clusters of rosy-lavender or red flowers in late summer. Propagation by division of the root-stock or cuttings. Probably the showiest of Sedums, good for tubs which may be left outdoors during the cold season with plants going dormant in winter; or if left indoors, nearly so, until spring. C: 25   p. 349
spectabile 'Variegatum' (Japan), "Variegated live-forever"; robust succulent perennial 12 to 20 in. high, suckering from the rootstock, with upright leafy stems well set with fleshy leaves yellow, with the margins bluish-green and toothed; large pink flowers. Very handsome even as a foliage plant. C: 25   p. 349
telephium (W. Europe), "European ice plant" or "European live-forever"; robust perennial with thickened carrot-like roots, strong stems erect to 18 in. high, set with oblong-ovate bluish-green soft-leathery, toothed leaves 2 to 4 in. long and rarely opposite, not in twos and threes as spectabile; in early autumn each leafy stem is topped by a large cluster of bronzy, rosy or red-purple flowers. Winter-hardy and going dormant. By division. C: 25   p. 349
treleasei (Mexico), called "Silver sedum", because of the mealy-white bloom on the bluish-green leaves, which remain quite blue with age; small, first erect, later sprawling succulent with fleshy jelly-bean-like leaves about 1 in. long, somewhat flattened and curving upward; clusters of bright yellow flowers. Good for dishgardens. C: 13   p. 348

"Weinbergii", in the California horticultural trade, the "Mother-of-pearl plant"; probably a clone from a hybrid of Graptopetalum paraguayense with Sedum, one of many created by California specialists with a history now difficult to determine. A sturdy, stem-forming little rosette with thick leaves ½ in. wide and lightly keeled beneath; shimmering bronzy green with rosy tips and overlaid with bluish glaucescence. **C**: 13   p. 348

weinbergii: see also Graptopetalum paraguayense

SEEMANNIA (*Gesneriaceae*)
latifolia (Bolivia at 2000 to 3600 ft.); a pretty, very floriferous branching plant from scaly rhizomes, closely related to Kohleria; by appearance smooth, but covered with a coating of very fine hairs; the stems first erect, then reclining with tips erect; elliptic to lanceolate bright green leaves 5 to 6 in. long, on clasping petiole; from the axils in whorls around the main stem the bell-shaped swollen ¾ in. flowers brick-colored or cinnabar-red outside, the short lobes deep red inside, with the throat yellow speckled red-brown; in constant bloom for months, beginning in early spring. Propagated from tip cuttings or the scaly rhizomes. **C**: 60   p. 447

SELAGINELLA (*Selaginellaceae*); Cryptogamous fern-allies; much branching flowerless herbs with numerous scale-like leaves; the spores are produced on the tips of branches in the axils of fruiting spikes; some species are moss-like or creeping, or forming rosettes, or climbing; almost all requiring high humidity and warmth, favoring shade. Propagation of the creeping species is simply by division, or sections of their stems which root at every joint. The rosette types are layered by pegging down their fronds on constantly damp peatmoss, and young plants will form on branches; occasionally spores will drop from them and germinate if left undisturbed. Leaf cuttings may be made of those with harder fronds, inserted in moist peat. Rhizomatous kinds are increased by division.
caulescens (Sunda Islands, China, Japan, India), "Stalked selaginella"; stiff, upright growing mountain dweller with erect wiry, greenish stalks to 2 ft. high, bearing on their upper half close, deltoid branchlets with narrow, bright-green leaf-segments of firm texture and good keeping quality. Propagated from branches or leaves, also division in older plants. **C**: 60   p. 254
denticulata: see kraussiana
emmeliana (cuspidata emiliana) (So. America), "Moss fern" or "Sweat plant"; tufted, small rosette of fern-like, erect, lacy bright green fronds, usually 3 to 6 in. high, reveling in high humidity; if allowed to dry the tips will curl and turn brown and won't recover. They require shade and warmth with constant close, moist atmosphere. Propagation by pegging down mature fronds on a damp, mossy surface, or layering older leaves or sections puddled into the surface of moist peatmoss, even cinders, and young plantlets will rise. **C**: 62   p. 254
kraussiana, known in horticulture as denticulata (Cameroons to So. Africa), "Trailing Irish moss" or "Club moss"; a charming, moss-like herb matforming, creeping stems rooting as they grow, with tiny, crowded, bright green, pinnate, scale-like leaves. Very useful as a quickly spreading, moist-tropical ground cover in terrarium or conservatory, even at moderate temperatures. **C**: 62   p. 254
kraussiana brownii (Azores), "Cushion moss", "Irish moss" or "Dwarf clubmoss"; shapely, moss-like cushions of densely clustering, short branches of vivid emerald green, supported by translucent aerial roots. Very attractive and pretty in terrariums, forming low, green mounds. **C**: 74   Cover*, p. 254
lepidophylla (Texas, Mexico, Peru), the legendary "Resurrection plant", "Rose of Jericho", or "Birdsnest moss"; densely tufted perennial, about 6 in. across, their branched stems spirally arranged, with the flattened, fairly hard, scale-like leaves looking like fern fronds, first green later red-brown with age. When the plants dry out, the branches curl up into a tight ball; when placed into lukewarm water, the fronds loosen and will unfold again to fresh emerald-green. This character remains even when the plants are already dried up and dead. Collected in Texas and sold in variety stores as a curiosity. In cultivation a drier atmosphere is preferable, as under high humidity young rosettes will form on the fronds. **C**: 71   p. 254, 539

martensii divaricata, "Zigzag selaginella"; slender form with fewer branches and shorter lateral leaves, on stiff zigzag stems. **C**: 74   p. 254
martensii 'Watsoniana', "Variegated selaginella"; attractive form of firm texture with the grass-green growth tending upright, but ends of branchlets pendant, and with pale or silvery tips. **C**: 62   p. 254
uncinata (caesia) (So. China), "Rainbow fern"; exquisite low creeper with straw-colored, slender, rambling stems rooting along the ground, alternately set with branched lateral leaves which are metallic blue and iridescent in the shade. Requiring warmth with shade and high humidity to bring out its shimmering beauty. **C**: 62   p. 194, 254
willdenovii (caesia arborea) (Vietnam, Malaysia, Himalayas), "Peacock fern"; robust growing and most beautiful, shade-loving rambler at first erect, but soon climbing between shrubs to 20 ft. high, the light brown stems supported by stiff stilt-roots and bearing spreading fronds of magnificently shimmering peacock-blue. Requires more than others a warm greenhouse with high humidity and shade. **C**: 62   p. 254, 286

SELENICEREUS (*Cactaceae*); beautiful tropical "Night-blooming cereus", with slender ribbed or angled clambering or trailing stems forming aerial roots which help them to climb; spines are few or small. The stunning, large and fragrant, exquisite flowers to 14 in. across are mainly white, opening with beginning of darkness and closing before dawn. Night-blooming cereus should have rich soil; and they like more humidity than other cacti, with a place in the south window, or sunporch. In need of liberal waterings while growing, especially when forming buds, before blooming time in summer; drier in winter. In home or winter garden they may be trained on a trellis or wire support. Propagated by joint or tip cuttings.
grandiflorus (Jamaica, Cuba), the best-known "Queen of the Night"; climbing epiphyte with large trumpet flowers salmon outside, white inside, and blooming by moonlight, with a powerful vanilla perfume; the flowers expand at sunset and fade off in the morning; stems 1 in. thick, light grayish green to purple. **C**: 10   p. 156
hamatus (So. and E. Mexico, Lesser Antilles) "Big-hook cactus climber"; epiphytic climber with vigorous stems glossy grass-green, ½ in. thick and ribbed, with few aerial roots; nocturnal flowers white, pale green outside, to 10 in. long and fragrant; a "Moon-cactus". **C**: 10   p. 156
urbanianus (Cuba, Haiti), "Moon-cereus"; clambering nightbloomer, epiphytic with aerial roots, and 4 to 5-angled joints of 2 in. dia.; giant 12 in., fragrant flowers white inside, tan outside. **C**: 10   p. 156
wercklei (C. America), "Werckle's moon-goddess"; night-blooming epiphyte with much branched cylindrical, green stem ½ in. thick, with about 12 low ribs, spineless; large white, fragrant flowers, and yellow fruit. **C**: 10   p. 156

SELENIPEDIUM: see Phragmipedium

SEMELE (*Liliaceae*)
androgyna (Ruscus) (Canary Islands), "Climbing butcher's broom"; shrubby evergreen vine with wire-like stems twining 60 ft. high, the true leaves represented by scales; those looking like glossy dark green, leathery leaves being leaf-like branches (cladodes), to 4 in. long, alternately 2-ranked along the lateral stems; tiny greenish-yellow, purple-centered flowers borne in clusters on the margins of the cladodes, followed by small red berries. A curious collection plant trained to trellis or wire support, with "leaves" lasting for years; water moderately. Propagated by division or seed. **C**: 13   p. 298

SEMPERVIVUM (*Crassulaceae*)
arachnoideum (Mts. of So. Europe), "Cobweb house leek" or "Cobweb hen-and-chicks"; a small plant with about 50 tiny leaves in a dense globular rosette to 1 in. across; the pale green tips of nearly all the leaves connected with white hairs; outer leaves tinted red-brown; clustering by stolons and forming mounds. Bright red flowers in a tall cluster. Quite winter-hardy. By offsets. Clusters of cobwebby rosettes are nice for pans at the cold window. **C**: 25   p. 350, 530
arachnoideum glabrescens; a variety of the Cobweb plant with small ¾ to 2 in. rosettes of more loose habit, with leaves somewhat spreading and covered by shorter white hairs not spun across like a web; beautiful with starry red flowers on erect, leafy inflorescence. **C**: 25   p. 350

tectorum calcareum (Europe: Alps), the brown-tipped "House leek", "Hen-and-chickens" or "Old man and woman" because of its many offsets; attractive, small leathery 3 to 4 in. rosette forming clusters; glaucous light gray-green obovate leaves broadly painted red-brown at bristle-pointed apex; flowers pale red on spikes. This hardy succulent is known as the "Roof houseleek" because it grows on thatched roofs of cottages in Europe. From offsets. **C**: 13   p. 329

SENECIO (*Compositae*); the "Groundsels" are a varied genus of herbaceous or fleshy erect or creeping herbs, shrubs, and even small trees, including also those succulent species known as Kleinias. What they have in common are their composite daisy-like ray and disk flowers, which are white or red in the succulent Kleinia group, various but basically yellow in the true Senecios. The members of this genus are of such vastly different appearance that it is difficult to recognize their relationship; they include the florists' Cineraria and the Dusty Millers, the German ivy, and succulents resembling Echeveria, Sedum, Crassula or Stapelia. Senecios are easily grown and practically all of them like cool surroundings with fresh air. They are propagated according to their habit, from seed, cuttings, or division.

articulatus, better known in horticulture as Kleinia articulata (So. Africa: Cape Prov.), the "Candle plant" or "Candle cactus"; succulent with erect swollen, jointed stems to 2 ft. high, the joints vary from 2 to 6 in. long, glaucous blue with darker lines; fleshy 2 in. light gray leaves deeply lobed; thistle-like flower heads yellowish-white. During period of dryness the stems will lose their foliage, which makes the plant look like a wax candle. Easily propagated from the falling joints. One of the best known succulents, often grown in baskets. **C**: 13   p. 351

cineraria, in seed catalogs as Cineraria maritima 'Diamond' (Mediterranean region), "Silver groundsel"; beautiful snow-white woolly perennial to 2½ ft. or more tall, with elegant, thick leaves at first oak-leaved, later pinnately cut, the pinnae well separated and broad, crenate at their broadening apex; inflorescence on white-felted stalks, in clusters of small ¾ in. thistle-like flower heads with short bright yellow rays, blooming with age in July to September, or later in subtropical climate. Much used as a border or bedding plant; equally striking in the sun or in cool rooms when under lights at night. Best by seed, also cuttings. **C**: 13   p. 190, 410

confusus (Mexico), "Orange-glow vine"; glabrous vine or scandent shrub with arrow-shaped soft and smooth, thickish leaves having toothed margins, 2 in. or more long, fresh-green turning purple; pretty flowers in small heads, with flaring orange-red florets in terminal clusters. **C**: 66   p. 264

cruentus 'Grandiflora nana', large flowered florists' "Cineraria", a compact herbaceous plant with watery stem and large hairy, soft-fleshy, turgid leaves which at blooming time are literally covered by a large globular head of daisy-like, medium large 2 in. flowers in many vivid red, salmon, or blue shades, usually with white eye. Grown as an annual pot plant from seed, blooming from late winter to late spring; they are cool-loving plants and must never be allowed to dry out as the foliage would quickly collapse. Very showy. **C**: 78   p. 371

cruentus 'Multiflora nana', the small-flowered "Cineraria" of florists; bushy herbaceous pot plant of low, compact habit and with smaller foliage; the many smallish, often white-eyed flowers about 1½ in. dia. or more, predominantly in bright reds, purples and true blues; this class makes up in big showy clusters what the flowers lack in individual size; a favorite commercial and widely cultivated group for early spring bloom and Mother's Day in May; later if climate is cool enough. Fast, easy grower; for largest specimen, seedlings *must* be transplanted to larger pots as soon as the new roots show. From seed. **C**: 78   p. 371

fulgens (So. Africa: Natal), "Scarlet kleinia"; succulent plant with tuberous root and thick fleshy stem to 1½ ft. long, wholly covered with glaucous bloom; the fleshy obovate leaves 3 to 5 in. long and toothed toward apex, light green and waxy-powdered violet-gray; flower heads with orange red florets. Promising house plant. **C**: 13   p. 351

haworthii, in horticulture as Kleinia tomentosa (So. Africa), "Coccoon plant"; striking small succulent semi-shrub about 1 ft. high, entirely clothed with appressed soft

pure white wool; the fleshy cylindrical-pointed leaves 1 to 2 in. long; flower heads orange-yellow. Beautiful in their snowy dress, the foliage unfortunately drops off easily when handled. Keep dry in winter. **C**: 13   p. 351

herreianus (Kleinia gomphophylla) (S.W. Africa), "Green marble vine"; clustering succulent with creeping, rooting branches, the distant, fleshy green leaves berry-like, to ¾ in. long and pointed, with translucent stripes and lines serving as windows. **C**: 13   p. 350

jacobsenii (petraeus, introduced as Notonia) (Kenya, Tanzania), "Weeping notonia"; mat-forming succulent plant erect when small, later creeping upwards, or pendulous with weight if in baskets; with alternate obovate fleshy, sessile nerveless leaves 3 in. long, glossy green; the stems rooting at the joints; flower heads orange. Very satisfactory for the sunny window or in hanging pots. **C**: 13   p. 271, 351

kleiniaeformis (So. Africa), "Halberd kleinia" or "Spearhead"; branching succulent to 18 in. high, with glaucous stems that have a pungent odor when they are bruised; the long obovate, fleshy 3–4 in. leaves have a pointed lobe on both sides which are rolled inward and folded together resembling a medieval pike; the apex is sometimes split. Flowerheads white or yellow. Attractive for its blue-glaucous and peculiar foliage. **C**: 13   p. 351

leucostachys, in horticulture as Centaurea or S. cineraria candidissima (Argentina: Patagonia), "Tall dusty miller"; striking semi-shrubby perennial 2 to 3 ft. high, with stems and foliage a beautiful silvery-white; the slender herbaceous branches with leaves white-tomentose, pinnatifid and deeply cut fern-like into lobes; inflorescence in clusters with small heads of yellow flowers. An attractive plant for patio containers, porch boxes and bedding, kept in shape for borders by shearing. From soft cuttings. **C**: 64   p. 410

macroglossus variegatus (Kenya), "Variegated jade vine" or "Variegated wax-ivy"; very attractive succulent creeper, of which the green type comes from Eastern Cape Province, but I found this lovely variegated form on the tableland of Kenya; densely branching and mat-forming, with small 1 to 1½ in. ivy-like lobed, waxy, thin-succulent leaves, with cordate base, green to milky-green and bordered or variegated with cream; pretty daisy-like ray-flowers with 12–14 white florets with yellow center. A colorful house plant for baskets or on trellis, with the homey feel of ivy; very prolific grower, and should be pinched to keep in shape. **C**: 15   p. 271, 351

mikanioides (So. Africa), "Parlor" or "German ivy"; bushy herbaceous plant, in time climbing, with glossy fresh-green, ivy-shaped, lobed leaves rather soft-fleshy; fragrant, yellow disk-flowers. A favorite old house plant for the window, best on the cooler side, in good light. **C**: 16   p. 263

scaposus (So. Africa: Cape Prov.), "Silver coral kleinia"; short-stemmed branching low succulent 12 in. high, with long 2 to 3 in. cylindrical leaves crowded at the apex of stems arranged in rosettes, silver cobwebby when young, the older leaves gray, marbled olive green and smooth; flower heads yellow. A living coral in a flower pot. **C**: 13   p. 351

serpens (better known as Kleinia repens) (So. Africa: Cape Prov.), "Blue chalk sticks"; low succulent branching shrub 8–12 in. high, with fleshy, nearly cylindrical leaves to 1½ in. long, grooved above, bluish-gray with blue waxy coating; flower heads pale yellow. Attractive but has a tendency to drop leaves. From cuttings. **C**: 13   p. 328

stapeliaeformis, known in horticulture as Kleinia stapeliaeformis (So. Africa: Cape Prov.), "Candy stick"; branching succulent to 10 in. high, with erect 5 to 7-angled fleshy, dark green to purple stems ¾ in. thick, painted silver between ribs, the angles toothed and tipped with awl-shaped, withering leaves; solitary scarlet flower heads. Keep dry during dormancy period in summer. A curious, easy house plant; from cuttings. **C**: 13   p. 351

tropaeolifolius (E. Africa: Malawi), "Succulent nasturtium"; tuberous rooted vining herb to 1 ft high, with irregularly rounded peltate, fleshy leaves 3 in. across, pale green above, violet beneath, and nostalgically resembling a Tropaeolum on the window sill; cylindrical flower heads with yellow ray florets. **C**: 3   p. 271

SEQUOIA (*Taxodiaceae*)
sempervirens (U.S.: Oregon and California coast), "Redwood tree", "California redwood" or "Coast redwood"; the tallest tree in the Western hemisphere, to 340 ft. high. A majestic and gigantic coniferous, evergreen tree with red-

brown, fibrous-barked trunk 10 to 20 ft. (to 28 ft.) in dia.; horizontal branches, spreading in flat sprays, with needles in two ranks deep green, and bluish beneath, nearly 1 in. long. Their knotty burls, cut from trunks, will sprout young growth when placed in a dish of water. Young conical trees are attractive container plants. From seed. **C**: 64   p. 310, 536

SEQUOIADENDRON (*Taxodiaceae*)

giganteum (California), the famous "Giant sequoia", "Giant redwood" or "Big tree"; at home on the high western slopes of the Sierra Nevadas at an elevation of 4600 to 6000 ft., where venerable old trees have lived for 5000 years; diameter of trunk to 38 ft., and reaching a height of 290 ft.; the pendulous green shoots are smooth and cord-like; closely covered by spine-like leaf bases, leaves widely awl-shaped, spirally densely arranged closely overlapping, deep green, glaucous when young; grayish hanging cones 2 to $3\frac{1}{2}$ in. long. The wood of the Giant sequoia is a beautiful red brown, and extremely durable, never invaded by destructive insects; the fissured bark is divided into cinnamon-brown ridges and to 20 in. thick. Young trees may be grown in pots from small seedlings; it is fairly winter-hardy in temperate regions. **C**: 14   p. 525

SERENOA (*Palmae*)

repens (So. Carolina to Florida Keys), "Saw palmetto"; scrubby, variable palm with creeping stems often underground, and forming wide-spread colonies; heads of palmate leaves 3 ft. across, deeply cut almost to base into 18–24 widely separated rigid, pendant segments, powdery blue-green in coastal types or bright yellow-green, on thorny petioles; fragrant white flowers, and edible blackish fruit. **C**: 13   p. 228

SERISSA (*Rubiaceae*)

foetida variegata (Japan, China), "Yellow-rim serissa"; small shrub to 2 ft. high with opposite, elliptic, small $\frac{3}{4}$ in. thin-leathery leaves dark green with ivory-white margin and mid-rib; quite floriferous with little white $\frac{3}{4}$ in. flowers. Grown mainly for its attractive foliage, sometimes in dish-gardens. From cuttings. **C**: 16   p. 488

SERJANIA (*Sapindaceae*)

glabrata (communis) (Peru); decorative twining shrub with herbaceous, bipinnate, almost fern-like leaves with ovate, stalked leaflets fresh-green, overlaid with silver in the center, and crenate at margins; yellowish flowers in axillary clusters. Attractive foliage plant that can be kept relatively bushy. **C**: 10   p. 266

SETARIA (*Gramineae*)

palmifolia (Panicum plicatum hort.) (Malaysia), the "Palm-grass"; slender perennial grass 4 to 6 ft. tall, the stalk knotted and branching, with plaited leaves 2 ft. long and 3 in. wide, nearly emerald green, and slightly hairy; the spikelets subtended by persistent long bristles, in terminal panicles. There is also a form with leaves striped white. Very showy in containers, and may be kept indoors for some time; outdoors in summer. Propagated from sideshoots. **C**: 52   p. 306

SETCREASEA (*Commelinaceae*)

pallida (Mexico); creeping plant with young growth erect, having fleshy, lanceolate clasping leaves waxy bright green, with gray bloom beneath; flowers lavender. **C**: 65   p. 260

purpurea (Mexico), "Purple heart"; so named because of the striking purple color of this plant in strong sun, aided by a pubescence of pale hair covering the lance-shaped leaves, 5 to 7 in. long, carried on erect fleshy stems; large 3-petaled orchid-colored lilac flowers. Outstanding in its coloring as an indoor plant, it becomes even more effective if set outdoors during summer in good sun. **C**: 13   p. 260

SETCREASEA: see also Callisia

SIBTHORPIA (*Scrophulariaceae*)

europaea 'Variegata', "Cornish moneywort"; variegated form of the green species native in England, France, Spain and Portugal; slender herbaceous, hairy trailer resembling Glecoma hederacea but leaves are alternate;

thread-like vines with rounded or kidney-shaped leaves to $\frac{3}{4}$ in. across, shallowly lobed, variegated or edged creamy-white; fairly hardy. Ground-cover or basket plant. **C**: 28   p. 265

SIDERASIS (*Commelinaceae*)

fuscata (Tradescantia, Pyrrheima) (Brazil), "Brown spiderwort"; very pretty clustering rosette of broad and oblong olive green leaves to 8 in. long, with silvery center band down the middle, and covered with brown hair as is the purple reverse; large lavender blue flowers at base. **C**: 53   p. 191, 262

SINNINGIA (*Gesneriaceae*); tropical herbaceous, shade-loving pubescent plants, commonly known as "Gloxinias"; with tuberous corm-like roots, opposite leaves and large tubular trumpet or bell flowers usually with 5 lobes; after blooming they need a resting period in the dark; with care the tubers may be grown 10 years or more. The large "Bell gloxinias" of florists with erect showy blooms in various colors are botanically grouped together as S. speciosa fifyana and may be enjoyed in flower under a light intensity preferably of 2400 fc. from January into summer from tubers started in succession from October to March; Gloxinias grow nicely in home temperatures of 65 to 70° F. at night, higher in daytime. Seeds may be sown the year round and will bloom in 6 to 7 months; December sowings will bloom in June; a June sowing should make Christmas. Aside from the fine seed, propagation may be by leaf cuttings, sprouts, and also slices of the tuber with eye.

barbata (Brazil); pretty subshrub with ascending red-brown stems and shining bluish-green, opposite oblong-lanceolate leaves 4–6 in. long, with sunken veins and toothed margins, the hairy underside and petioles red-purple; axillary flowers with leafy calyx and hairy white corolla $1\frac{1}{2}$ in. long, streaked with red inside, the limb cream-colored, the tube pouched below; in bloom continuously almost the year round, and not going dormant in winter. **C**: 60   p. 448

'Doll Baby' (S. pusilla x eumorpha); a pretty miniature rosette to 3 in. across, with thin, dark olive-green crenate leaves with bronze veins; from the heart rise charming small $1\frac{1}{4}$ in. slipper flowers lavender-blue with purple eye and pale lemon throat; blooming for a long time, best in summer-time. **C**: 10   p. 448

eumorpha (erroneously grown as S. maximiliana) (S.E. Brazil), a comely, free-blooming tuberous herb, with short reddish, hairy stem and large, glossy, thin, lightly downy 4 in. leaves of bronzy green; from the leaf axils rise the milky-white, inflated tubular $1\frac{1}{2}$ in. flowers like nodding bells, lined with lilac and yellow in the throat; blooming mainly from spring to summer. **C**: 60   p. 448

pusilla (Brazil), "Miniature slipper plant"; a darling miniature rosette only 1 to 2 in. high, of little, $\frac{1}{4}$ to $\frac{1}{2}$ inch oval, puckered leaves olive-green with brown veins, hugging the ground; slender thin-wiry stalks each bearing an exquisite oblique-tubular flower with 5 lobes spreading $\frac{1}{2}$ in. across, orchid-colored pale purple with violet marking, and lemon-yellow throat, blooming constantly if grown with high humidity. We grow this little charmer successfully in small glass brandy snifters covered with clear plastic. If grown more open, the little plants will go into a short dormancy period in autumn until January. Seed sets easily, or propagated from the knobby tubers. **C**: 60   p. 431, 448

regina (Brazil), "Cinderella slippers"; beautiful compact species, related to speciosa, but distinguished by having its ovate pointed, bronzy green, red-backed, velvety 4 to 6 in. leaves beautifully patterned with ivory or silvery veins; free-blooming with a profusion of nodding 2 in., oblique "slipper" type, violet flowers; shorter and more slender than speciosa macrophylla; normally blooming in summer. **C**: 10   p. 43*, 448

schiffneri (Brazil); erect-growing softly-hairy plant with fleshy stems and thinnish, pointed leaves to 9 in. long, velvety green with crenate margins and unequal base, the mid-vein pale green; the tubular trumpet flowers in clusters of 3 to 4 from the axils, $1\frac{1}{4}$ in. long, white with purple dots in throat. **C**: 60   p. 448

speciosa (from moist-warm rocky slopes in southern Brazil), the "Violet slipper gloxinia"; more or less compact tropical tuberous herb, nearly stemless, with stalked oblong-ovate, fleshy, white-velvety, fresh-green leaves crenate at margins; showy, foxglove-like oblique bell flowers $1\frac{1}{2}$ to 2 in. long, usually velvety violet-blue with lilac lobes, but quite variable, and wild forms have been found with white or red flowers; normally summer blooming. From these, hybridization has developed the present-day fancy slippers and florists gloxinias. **C**: 10   p. 448

speciosa 'Emperor Frederick', a bicolor "Gloxinia" of florists; leading commercial hybrid bred with large upright bell-shaped flowers 3½ to 4 in. across, velvety dark crimson and bordered white; the fleshy, velvety leaves are large and somewhat oversized, typical of the horizontal foliage of the "crassifolia" type which is quite brittle and requires much space; however, its large tubers may be held over, year after year. Grown in quantity in Belgium for export. **C**: 10
p. 365*, 380, 449

speciosa florepleno 'Chicago'; a "Double gloxinia" of the habit of the 7- to 8-petaled frilled 'Switzerland'; produced following meticulous breeding in U.S.; very exciting in having large 3½ in. flowers featuring two or more rows of frilled petals glowing crimson with white borders, blooming best from spring to summer. **C**: 60   p. 380, 449

speciosa florepleno 'Monte Cristo' (1957), a "Double gloxinia" produced as a result of long hybridizing in Germany and subsequently in Louisiana; this named cultivar has large double flowers with 2, 3, 4 or 5 diminishing rows of petals or crowns in glowing velvety crimson, lighter toward the wavy margins and white outside; shapely plant of compact habit. As in other doubles, the corolla holds persistently to the calyx for the duration of the bloom and usually dries without dropping. Reliable double fl. tubers are obtained only by leaf, stem or tip cuttings. While seed can be obtained by hand-pollination, the resulting seedlings vary in degree of doubleness, only about 60% being double, the balance more or less single. **C**: 60   p. 402, 449

speciosa 'Princess Elizabeth', "Blue bell gloxinia"; large-leaf 'crassifolia' type gloxinia, with attractive flowers having a white throat and a broad light blue border over the spreading lobes; midseason to late blooming. **C**: 10   p. 449

speciosa 'Switzerland', "Ruffled grandiflora gloxinia"; a superb 'grandiflora' gloxinia, inheriting its prolific blooming habit from the soft-leaved regina hybrid 'Gierth's Red', and its intensive scarlet with white border from 'Emperor Frederick'; in addition, the large erect flowers have 7-8 lobes, daintily ruffled, and spreading 4 in. across. The tubers of this fancy type are relatively smaller than in the 'crassifolia' group, and are usually grown from seed. **C**: 10   p. 449

speciosa 'Tigrina', "Tigered bell-gloxinia"; gorgeous large-flowered 'grandiflora' hybrid with regina blood; the attractive leaves are flexible and are patterned with light veins; erect, huge flowers measure 4½ to 5 in. across, having white throat densely speckled in the same color as the spreading lobes, which are usually red or blue. **C**: 10   p. 449

'Tom Thumb'; the most charming little Baby gloxinia that ever appeared for the pleasure of the gesneriad fancier; in every detail resembling its big brother the (fyfiana) gloxinia, except for its diminutive size no larger than an African violet. First of its kind to flower in 2½ and 3 in. pots, and a perfect window sill plant with little bell-shaped, velvety red blossom edged in white, 1¾ in. across. The plant is only 4 in. high, and its small ovate, crenate leaves 2 in. long, deep green above, silvery-haired beneath; the bell flowers are erect and not oblique and slipper-like as in speciosa. This hybrid was developed by Fischer's in New Jersey. Originally a red slipper gloxinia (S. speciosa) with 2 in. tube and 3 in. leaves was selfed at U. of Minnesota and a seedling 'Corsage' sent to Fischer; there it was inbred, then crossed with a Standard size gloxinia resulting in various sized offspring; one of the smallest was then crossed w.S. pusilla as the male parent; the few plants from this cross were all small and the growers believe this to be partly due to the dwarfing influence of the little pusilla. **C**: 10   p. 449

SISYRINCHIUM (*Iridaceae*)

striatum (Chile, Argentina), "Satin-flower", "Yellow-eyed grass" or "Rush-lily"; hardy perennial 1 to 2 ft. high, with short rootstock and grass-like, 2-ranked glaucous leaves, and leafy stalks bearing a spike of flowers, yellow and veined with purplish brown, ¾ in. long, June-July blooming. An easy grower for a moist-cool location, attractive in clumps in large containers, the pretty flowers opening in sunshine. From seed or division. **C**: 76   p. 394

SKIMMIA (*Rutaceae*)

japonica 'Nana', "Dwarf Japanese skimmia"; handsome evergreen, low growing, compact shrub, dioecious with separate female and male plants, dense with thick-leathery,

elliptic, glossy green 2½ in. leaves clustered at end of branchlets, and tipped by erect clusters of small creamy-white, fragrant flowers, followed by coral-red ⅓ in. berries on female plants, and which last for months during winter and spring. Originally from Japan, this dwarf type is a selected seedling resembling reevesiana but not as tall and leggy; developed by Teufel-Oregon. A thrifty long-lasting decorative pot plant; winter-hardy in our garden in New Jersey. Best from cuttings to reproduce the female plants; from seed also. **C**: 16   p. 512

SMILAX (*Liliaceae*)

ornata (Jamaica), "Sarsaparilla"; tall-growing vine climbing by tendrils, with angled, green thorny stem, and thin-leathery, lanceolate to cordate leaves to 10 in. long, shiny fresh-green with pale blotching. A big ornamental vine primarily for the conservatory. **C**: 54   p. 266

SMITHIANTHA (*Gesneriaceae*); the "Temple bells" are tropical American herbs growing from scaly rhizomes, long known in horticulture as Naegelia. They are prized as potted plants both for their handsome velvety foliage, and their spires of beautiful nodding, obliquely inflated bell-flowers in bright reds to vivid orange, yellowish or rose, blooming from late summer through early winter. Smithianthas like it warm and humid; rhizomes are planted in spring when new growth shows, growing best in good light or under fluorescent tubes. After blooming the plant should be dried off and kept dry and warm for 3 months. For propagation, the rhizomes are broken into pieces; also from seed; or from leaves which root best if cut without petiole.

cinnabarina (Naegelia) (Mexico, Guatemala), "Red velvet-leaf gesneriad"; typical "Temple bells" of compact habit about 1 ft. or more high, presenting a stunning display of 1 to 1½ in. nodding bells, bright scarlet red outside, the inflated belly in cream, and red-spotted throat, borne on a red central stalk, above the gorgeously beautiful red plush leaves; blooming most profusely in spring. **C**: 10   p. 451

'Exoniensis'; a prolific growing and free-blooming hybrid with robust red-hairy stems, the leaves patterned plush red on green, and topped by masses of orange-yellow nodding "slipper" flowers in bold terminal clusters. **C**: 10
p. 451

x hybrida 'Compacta' (zebrina x multiflora hybrid); low-growing bushy "Temple bells"; having large, deep green leaves covered with purplish velvety hair; free blooming with a profusion of nodding, foxglove-like bells lemon-yellow, shading into red on top, and veined with red. **C**: 10
p. 451

multiflora (So. Mexico: Oaxaca); charming soft-hairy plant 12 in. or more high with velvety green, cordate, crenate leaves topped by a graceful crown of nodding digitalis-like flowers 1¼ to 1½ in. long, distinctively creamy-white, and blooming mainly in summer. Easy to hybridize, this is an early parent introducing creamy shades into hybrids. **C**: 60   p. 451

'Orange King'; excellent hybrid of multiflora and zebrina parentage, sturdy with stiff stems and beautiful large velvety, emerald-green leaves crenate at edge, and overlaid with a pattern of red along veins; the flower bells in elongate clusters orange-red, yellow beneath; inside yellow with red spots. I photographed this robust beauty coming into bloom in September at Berkeley, California. **C**: 60   p. 451

'Rose Queen' (1949); lovely commercial zebrina hybrid of open habit, free growing with sturdy stem and large velvety green leaves mottled with purple, and slender, obliquely inflated bells an exquisite rose-pink, and spotted purple inside. **C**: 60   p. 451

zebrina (Naegelia) (Eastern Mexico: Vera Cruz); a beautiful species of tall habit to 2 ft. high, introduced about 1840 and a parent of most hybrids; pale-hairy stems with dark green, 4 to 6 in. flexible leaves prettily traced with red-brown and covered with silky hairs; a willing bloomer, with spires of numerous nodding flowers on slender stalks, the bells 1½ in. long, brilliant scarlet above, yellow below, the throat pale yellow spotted with red; blooming fall to winter. **C**: 60   p. 451

'Zebrina discolor', known in horticulture as S. x hybrida; a cultivar with more contrastingly colored purplish-red and green foliage; flowers scarlet with yellow, brown-spotted inside. These commercial hybrids combine the beauty and colors of S. zebrina and multiflora, and are easy to grow and willing to bloom. **C**: 60   p. 451

SOBRALIA (*Orchidaceae*)

decora (galleottiana) (Mexico, Guatemala to Costa Rica); vigorous terrestrial orchid, forming clumps, the slender, reedlike stems to 2 ft. high, clothed with scattered, handsome plaited leaves, and at their top bearing 1 or 2 large and showy, cattleya-like fragrant flowers, 3 to 4 in. across, but lasting only a short time; sepals and petals creamy-white with light rose blush, lip purplish rose, and streaked with yellow and brown; blooming from April to July, or various, on stems of the previous year. Sobralias are not difficult to grow but require a fair amount of space. They like fresh air; ideal for the patio in summer. Grown in large pots in compost of loam, humus, or shredded fir-bark, with good drainage, they enjoy liberal watering and feeding while active. Propagated by division, but don't disturb unless necessary. **C**: 60			p. 480

SOLANDRA (*Solanaceae*)

longiflora (Jamaica), "Chalice vine" or "Trumpet plant"; clambering evergreen shrub to 6 ft. high; the woody branches with small hard, oval or obovate leaves on purple petioles; and large and showy, stiff upright trumpet-like flowers 9–12 in. long and contracted at the throat, greenish-white and showing purplish-brown venation, the limb turned back and frilled. Requiring much space, light and water to complete growth, then keep dry until leaves begin to drop resulting in proliferous blooms. From cuttings. **C**: 54			p. 277

SOLANUM (*Solanaceae*); the "Nightshades" are a varied group of herbs, shrubs, vines and even trees of over 2000 species, mostly tropical, and which include the economically important potato and tomato, but also many ornamentals. They are generally of easy care. Flowers are usually small and wheel-shaped but often showy; the fruit is a small or larger berry, some edible, others poisonous. Propagated according to character from seed, tubers, or tuber sections, or soft wood cuttings.

jasminoides (Brazil), "Potato vine"; handsome shrubby evergreen or deciduous twiner with twiggy stems 8 to 10 ft. long, purplish-green ovate leaves 2 to 3 in. long, sometimes 2 to 5-parted; showy star-shaped flowers 1 in. wide, white tinged with blue, in branching clusters, blooming nearly perpetually, heaviest in spring. Useful and decorative against walls; if too rampant may be cut to prevent tangling. From seed or cuttings. **C**: 63			p. 488

pseudo-capsicum (Madeira), "Jerusalem cherry", "Christmas cherry" or "Cleveland cherry"; a popular old house plant with flexible branches dense with lanceolate, turgid-firm, deep green 2 to 2½ in. leaves with velvety feel above, and smooth beneath, wavy at the margins; small white, star-like flowers in late summer; in late autumn loaded with large globular, lustrous orange-scarlet, cherry-like fruit to 1 in. dia.; there is also a yellow-fruited strain. Much grown as a pot plant for Christmas, with fruit hanging on all winter, creating visions of miniature cherry trees. Easy from seed, but also cuttings. **C**: 13			p. 386, 514

rantonnetii (Paraguay, Argentina), "Blue potato tree"; rambling evergreen or deciduous shrub to 6 ft. high, unarmed and nearly smooth; ovate or oval, bright green undulate leaves to 4 in. long; charming 1½ in. flowers dark blue or violet with yellow eye, in clusters blooming throughout the warm summer and fall, sometimes nearly all year; the red fruit-like small apples. Blooms as small plant but may be grown into treeform by staking, or trained on support as a vine. From cuttings or seed. **C**: 64			p. 488

SOLISIA (*Cactaceae*)

pectinata (Mexico: Tehuacan) "Lace-bugs"; very small globe only one inch thick, with dense warts almost entirely covered with white spines arranged comb-like; flowers around the sides yellow; juice, milky. Thrives better when grafted. **C**: 13			p. 163

SONERILA (*Melastomaceae*); small tropical plants from India, Malaysia and Indonesia, with exquisitely beautiful, herbaceous foliage often bristly and puckered with pearly spots, and clusters of showy, rosy flowers in summertime. Little jewels for collections of exotic tropicalia, these delicate charmers are not quite as fussy as Bertolonias, but still should be grown, in open, porous, humusy soil in the moist-warm greenhouse, a terrarium or under a plastic hood. Propagated by seeds or cuttings in spring.

margaritacea (Java), "Pearly sonerila"; small, tropical herbaceous plant with pubescent red stem and ovate, deep copper-green leaves with bristly surface, puckered with pearly spots and bristles that glisten in the sunlight like frosted silver; glowing red-purple beneath; flowers rosy lavender. **C**: 62			p. 185

margaritacea 'Mme. Baextele', "Frosted sonerila"; an attractive miniature herb with ovate leaves smooth silvery gray, densely covered with small pearly spots often confluent; red-purple underneath; small pink flowers. This beautiful Belgian cultivar has leaves pronouncedly more silvery than the species. **C**: 62			p. 185

SOPHRONITIS (*Orchidaceae*)

coccinea (grandiflora) (So. Brazil), colorful dwarf epiphyte with small pseudobulb and a stiff, 3 in. leaf; strikingly colored solitary flowers 1½ to 3 in. across, vivid scarlet with salmon sheen, the throat yellow with red stripes, blooming from September to February. I will never forget these brilliant spots of red in the chilly rain-forest of the Serra do Mar near Sao Paulo. Sophronitis are somewhat frail, growing throughout the year during which they need high humidity, without rest. Propagated by division. **C**: 22			p. 455*, 481

SPARAXIS (*Iradiceae*)

tricolor alba (Steppes of South Africa), the "Harlequin flower" or "Wand flower"; tender, small perennial plant 18 in. high with bulbous corm; narrow sword-shaped leaves mostly basal, with parallel veins, and flowers close to Ixia but generally larger, to 2 in. across, opening star-like wide and flat, cream-white with sulphur-yellow throat and with darker center stripe on each petal. The species also comes in other colors, from orange to pink, red or purple, always with a black blotch at base of each segment. Sparaxis begin growing in winter and flower over a long period in late spring, attractive when grown in containers; pot in November and keep dark and cool until growth begins. After blooming reduce watering; and when dormant keep dry until January at 40–50° F. By offsets. **C**: 63			p. 395

SPARMANNIA (*Tiliaceae*)

africana (So. Africa), "African linden" or "Indoor-linden"; tender African shrub growing fast to 10–20 ft. high, usually as a thicket; many first soft, later woody branches from base, with dense coarse foliage; the large soft leaves angled and lobed, 6 to 10 in. across, light green and white-hairy on both sides and on the sparry stems; flowers with white petals and yellow filaments in terminal clusters. A decorative and willing house plant widely used in Europe, where it prefers the often cooler homes; also for winter garden, solarium, as well as the sunny window. Requires ample water and feeding; prune to control height. From cuttings or seed. **C**: 16			p. 301

SPARTIUM (*Leguminosae*)

junceum (Genista) (Mediterranean region, Canary Islands), "Spanish" or "Weaver's broom"; ornamental shrub 6 to 8 ft. high, with grooved almost leafless, reed-like branches, the small leaves if any, linear; fragrant yellow, butterfly-like 1 in. flowers in loose, showy terminal clusters to 15 in. long; very handsome when in bloom from May to September; in California even later. May be pruned for bushiness after flowering. Somewhat frost-hardy. From seed. **C**: 75			p. 503

SPATHICARPA (*Araceae*)

sagittifolia (Brazil: Bahia), "Fruit-sheath plant"; interesting herb, with tuberous rhizome, having small, arrow-shaped, membranous leaves waxy green; the inflorescence on stiff stalks with recurved green spathe and spadix, with male and female flowers attached in rows along its center. A conversation piece. **C**: 4			p. 97, 534

SPATHIPHYLLUM (*Araceae*); the "Peace lilies" are easy-growing, generally vigorous tropical American herbs, forming clumps, with happy-green foliage, and a succession of "flowers" resembling Anthurium but with showy spathes white to greenish; the actual minute flowers arranged along a club-shaped spadix, willingly setting seed when pollinated. Spathiphyllum belong to our best indoor decorative plants as they adapt readily to low intensity light; as foliage plants in interior boxes or containers they tolerate 20 fc. but to flower will need at least 50 fc.; excellent under air-conditioning. Propagated by division or seed.

blandum (gardneri) (W. Indies, Jamaica, Surinam); robust plant with large, lanceolate, leathery, deep green leaves and sunken veins; spathe spoon-shaped and pointed, pale green; spadix white with elevated knobs. **C**: 4 p. 88

cannaefolium (dechardii) (Venezuela, Guyana); leathery plant with thick, dull, black-green, corrugated leaves tapering toward base, ribbed petioles; thick, fleshy spathe green outside, white inside; long, free, cream spadix. **C**: 4 p. 88

'Clevelandii' (kochii), "White flag" or "White anthurium"; freely branching and free flowering commercial plant close to wallisii but larger in all parts; thin-leathery, glossy-green, lanceolate leaves with undulate margin; the inflorescence on reed-like stems with ovate white, papery 4 to 6 in. spathe turning apple green with age, and having a green line on back; maze-like spadix white. Satisfactory under various conditions; plant tolerates low light to 20 fc. **C**: 4 p. 88

cochlearispathum (So. Mexico, Guatemala); large plant to 6 ft. high, with long, heavy, corrugated leaves, light dull green and with undulate margin; inflorescence on stiff stalk, with spathe large, almost leathery, fresh yellow-green; maze-like, white spadix. **C**: 4 p. 88

floribundum (multiflorum) (Colombia), "Spathe flower"; dwarf, compact plant with matte, satiny green, leathery leaves obovate or elliptic, with pale center band, on broadly winged petioles; small 2 to 3 in. spathe white; short spadix green and white. Very floriferous. **C**: 4 p. 88

'Mauna Loa', "White anthurium"; a diploid hybrid developed by Griffith-Los Angeles, and so far as I could determine in discussing its origin with him, a seedling of S. floribunda x a Hawaiian hybrid, probably S. 'McCoy'; robust plant of compact habit, tending to divide from the base; leaves dark glossy green; very floriferous over an extended period, with pure white spathes 4 to 5 in., and even 8 in. long in older plants in large pots, somewhat cupping and of soft-leathery texture, slightly scented. We have found the strain variable, leaning toward 'Clevelandii'. **C**: 4 p. 88

x'McCoy'; Takahashi Hawaiian hybrid of probably cochlearispathum x 'Clevelandii'; vigorous, large-growing, very showy plant to 5 ft. high, with long glossy green leaves; the spathes white or creamy changing to light green with age, 8–18 in. long, of good texture; the slender club-like white spadix attached 1 in. up from base of spathe, having pointed knobs. Recommended as a tub plant. **C**: 4 p. 88

patinii (Colombia), "White sails"; graceful plant, with papery, narrow oblanceolate, waxy leaves with depressed veins on thin vaginate petioles; inflorescence on long thin stems with a pendant, slender, white spathe tipped green; thin spadix green and white. **C**: 4 p. 88

phrynifolium (Costa Rica, Panama), "Bold Peace lily"; found in the trade as S. friedrichsthalii; a robust plant with fresh-green, corrugated, fleshy leaves having long-cuneate base; the inflorescence with broad, pale yellow-green, papery spathe, and enclosed, white, knobby spadix. **C**: 4 p. 66*, 88

SPATHOGLOTTIS (Orchidaceae)

plicata (Malaysia, Pacific Islands), handsome terrestrial orchid of prolific growth, with small corm-like pseudobulbs and grass-green plaited leaves, 2 to 3 ft. tall; erect stalks topped by a showy cluster of wide-open 1 to 1½ in. flowers of a pleasing rosy-purple, blooming almost throughout the year, mostly from April to June. Spathoglottis are accustomed to good light, and are nice on the patio or under a tree during summer time; growing in large pots in porous compost of rich loam, peatmoss and osmunda or treefern fiber; freely supplied with water and fertilizer while in active growth; when resting after bloom the bulbs may be divided. **C**: 8 p. 480

SPIRAEA (Rosaceae)

x vanhouttei (cantoniensis x trilobata), the exquisite "Bridal wreath"; beautiful deciduous, winter-hardy shrub to 6 ft. high, dense with numerous, gracefully arching slender stems, coarsely toothed leaves bluish beneath, and tiny white flowers in masses of tight, lacy clusters forming a veritable blanket of snow in May and June in outdoor gardens. Much used for forcing into early bloom from February on, taking 3 weeks. Flowering from new branchlets along 1 year or older wood, tips only are cut out previous to forcing. Propagated from hardwood cuttings. **C**: 76 p. 504

SPIRANTHES (Orchidaceae); the "Ladies-tresses" are usually terrestrial orchids broadly dispersed from snow-bound regions to tropical habitats; growing from rhizomes or tubers, with slender stems bearing leaves mostly toward base, and small, hooded flowers in twisted terminal spikes. The hardy species thrive in rich, porous loam; the tender kinds in well-drained pots with a compost of loam and fibrous peat or fernfiber, and sphagnum added. When actively growing, they need liberal amounts of water and fresh air; outside in summer. After the inflorescence has died down a strict resting period is beneficial, to prepare for future flowering. Propagation by dividing the rootstalks before new growth starts.

elata (Florida, West Indies, Mexico, Costa Rica, Brazil), "Ladies tresses"; charming terrestrial orchid with fleshy roots, forming a rosette of lanceolate leaves to 6 in. long, prettily banded lengthwise with silver; an erect pubescent spike bearing numerous greenish-white tubular, nodding flowers ¼ in. long, blooming April to July. **C**: 60 p. 480

longibracteata in. hort. (bracteosa; Cyclopogon longibracteosa) (Brazil); terrestrial orchid 1 ft. high, with erect leafy spike, the oblong leaves in rosettes; and topped by flowers hooded on top, lateral sepals free, brownish-yellow and white, and hairy; blooming late spring. Grown in humus compost containing peat, sphagnum or fern fiber with loam added. **C**: 60 p. 462

SPIRONEMA: see Callisia, Hadrodemas

SPREKELIA (Amaryllidaceae)

formosissima (Mexico and Guatemala), the "Jacobean lily", also known as "Orchid amaryllis", "St. James' lily" or "Aztec lily"; beautiful bulbous herb with daffodil-like linear leaves; one or two hollow pink floral stalks 12 in. long bear a solitary, gorgeous orchid-like crimson-red flower to 5 in. across and with bright red filaments, normally appearing in May or June before or with the foliage. If grown in pots, keep dry through the winter as soon as leaves begin to yellow; with alternating moisture and drying out, Sprekelia may flower several times a year. Repot only every 3 to 4 years. Propagated by offsets or seed. **C**: 51 p. 392

STACHYS (Labiatae)

officinalis (Europe, Asia Minor), "Wood betony" or "Wound-wort"; hardy, tufting coarse perennial herb 1–2 ft. high, commonly known as "Betony", with ovate-oblong, herbaceous, pubescent, ornamental leaves 2 to 4 in. long, fresh-green and quilted, the margins irregularly wavy and crenate; small purple two-lipped flowers in dense whorls, forming an oblong terminal spike. Now primarily ornamental, but once cultivated for use in domestic medicine. By division. **C**: 25 p. 321

STANGERIA (Cycadaceae)

eriopus (S. paradoxa in horticulture) (So. Africa: Natal), "Hottentot's-head"; unique small cycad with trunk to 1 ft. long above ground, partially subterranean; the leaves strangely fern-like, pinnate with broad leaflets to 18 in. long, leathery, grayish green and spiny-toothed. Male cones cylindrical, the female ones shorter. Differing from all other cycads in having venation on leaflets running from midrib to margin as in ferns, rather than parallel. Very exotic decorator plant. **C**: 54 p. 317

STANHOPEA (Orchidaceae); fascinating epiphytic orchids of tropical America, having pseudobulbs with 1 single plaited leaf, and large and showy extraordinary, waxy flowers intensely fragrant; these appear from the base of the bulbs and push out through the growing medium to hang downward through a hanging basket or perforated pot. The massive flowers are incredibly beautiful with reflexed sepals and petals, some with two horns projecting from the lip, but rarely last more than 2 to 3 days. Best grown in baskets, the plants are kept moist at all times while active, but after completion of the new pseudobulbs, water is withheld for 1 month. Propagated by division.

eburnea (Trinidad, Guyana), a curious epiphytic "Horned orchid"; small furrowed pseudobulbs bearing a single, strongly ribbed leaf; the peculiar but beautiful waxy-white, strongly perfumed, pendulous flowers 6 in. across, spotted on the lip, and from its base two little horns, in small clusters of 2 to 3 blooms which push out through the bottom of a hanging basket. **C**: 73 p. 533

tigrina, probably more correctly S. hernandezii (Mexico), "Horned orchid"; from apprentice days in my father's greenhouses have I remembered this fantastic epiphyte as ,my most impressive orchid; the memory of its pendant, large 6 in., waxy flowers with sepals and petals deep blood-red marked with yellow, the orange-yellow lip blotched with maroon and its ivory horns, together with an over-powering fragrance of vanilla, haunts me still. Blooming May to July. **C**: 73   p. 455*, 481

wardii (Mexico, Guatemala, Panama, Venezuela), known in habitat as "El Toro"; magnificent epiphyte with conical pseudobulbs and shiny ribbed leaves; the pendulous, robust inflorescence with spicily fragrant flowers to 5 in. across, yellow sepals and petals, beautifully painted with maroon rings and spots; in the center two maroon-black eyes, or one large, confluent blotch; in front are two charac-teristic, sickle-shaped, light yellow horns; blooming between July and September. Although the flowers are waxy, they are relatively short-lived; however while they last will present a spectacle worthy of inviting all the neighbors to see. **C**: 73   p. 481

STAPELIA (Asclepiadaceae); African dry-region cactus-like, leafless stem succulents of the milkweed family; low growing, with mostly 4-sided, grooved thick-fleshy stems, and forming clumps, and large fleshy-grotesque flowers shaped like 5-pointed stars or wheels, usually with offensive odor which attracts flies that in the hope of finding ripe meat, succeed only in affecting pollination with perhaps a drop of nectar as reward. Keep cool, dry and sunny in winter, more moist with some shade in summer. Easily propagated by division or cuttings.

clavicorona (So. Africa: Transvaal), "African crown"; leafless, succulent plant more or less cactus-like, to 1 ft. high, with thick-fleshy, deeply furrowed, 4-angled velvety pubescent green stems, in branching clusters, the ridges with prominent teeth; flowers with 5 short-pointed petal-lobes around the wheel-like corona, 2½ in. across, yellowish with transverse purple lines. **C**: 1   p. 352

gigantea (So. Africa: Zululand to Zambia), "Zulu giants" or "Giant toad plant"; clustering fat, deeply 4-ribbed stems 6 to 10 in. high, pale green and velvety; from the angles appear the gigantic star-shaped flowers measuring from 10 to 15 in. across pale yellow with transverse crimson lines, and long white hairs at the margins; unfortunately with unpleasant odor. **C**: 1   p. 352

hirsuta (So. Africa: Cape Prov.), "Hairy star fish flower"; stubby 4-angled, clustering fingers from 3 to 8 in. high only, soft-hairy and sooty-green, with crenate ridges; the flowers 4 to 5 in. wide, wrinkled, cream-colored with transverse bands of purple, margins of the slender lobes hairy and the center shaggy with long white and purple hairs. A favorite dishgarden plant of compact size. **C**: 1
                                                          p. 328, 352

nobilis (So. Africa: Transvaal; Mozambique), "Noble star flower"; branching tufted, velvety light green stems 4 to 8 in. high, with angles much compressed, and with minute teeth; flowers star-like, near the base of young stems, to 8 but never over 12 in. across, with a big depression in center, the ciliate petals recurved, outside purple, inside ochre-yellow with thin transverse crimson lines, and thinly covered with purple hairs. **C**: 1   p. 352

variegata (So. Africa: Cape Prov.), the "Carrion flower" or "Spotted toad cactus"; branching succulent with finger-like, angled soft-fleshy stems to 4 in. high, green or grayish mottled with purple, and armed with teeth; from near the base very showy, greenish yellow star flowers 2 to 3 in. across with purple-brown spots on the broad petals and the wheel-shaped corona. Most common in cultivation these little clusters are, because of their small size, very suitable for dishgardens. **C**: 13   p. 352

STELIS (Orchidaceae)
ciliaris (Mexico to Costa Rica); charming small epiphyte lacking pseudobulbs, with broadly oblong, leathery leaves to 6 in. long, narrowed downward to a stalk; arching or pendant chains of small maroon or purple ½ in. flowers, the sepals ciliate, petals and lip fleshy, the stalk naked for half length, then densely flowered; blooming in February and spring. At home perching on trees in humid forests at 3000 ft., this species is used to moderate temperatures at night and constant moisture. Propagated by division. **C**: 34   p. 480

STENANDRIUM (Acanthaceae)
lindenii (Peru); low tropical herb spreading sideways with broad elliptic leaves papery-smooth, metallic coppery green with beautifully contrasting yellow-green vein area, purplish beneath; flowers yellow, in 3 in. spikes. **C**: 60 p. 184

STENOCACTUS: see Echinofossulocactus

STENOCARPUS (Proteaceae)
sinuatus (Queensland, New South Wales), "Fire-wheel tree"; an Australian wonder tree slowly to 30 ft., in habitat to 100 ft. high, with evergreen foliage, in young trees like oak-leaves 1–1½ ft. long, lobed or pinnately cut into 1 to 4 pairs of oblong lobes, hard-leathery, glossy light green with pale midrib, and lighter beneath; in older trees leaves are smaller and usually unlobed; with age and when well established blooming with 2–3 in. scarlet and yellow flowers curiously arranged in clusters like the spokes of a wheel; orange-red when young, the inflorescence explodes like fireworks into fiery-red when mature. Decorative in pots or tubs with beautiful juvenile foliage. From cuttings or seed. **C**: 13
                                                          p. 298, 501

STENOCHLAENA (Filices: Polypodiaceae)
palustris (India, So. China, Australia), "Liane fern"; tropical epiphytic fern climbing by its slender woody rhizome, which is covered with occasional brown scales, and bearing leathery, pinnate fronds shining green, finely serrate at margins, coppery when young; as a pot plant likes to climb on tree-fern slabs. **C**: 74   p. 248

STENOGLOTTIS (Orchidaceae)
longifolia (So. Africa: Natal); African terrestrial orchid from the forest floor, with fleshy tuberous roots producing a tuft of brown-purple, spotted soft leaves, from the center of which springs an erect spike to 2 ft. tall, with small, pretty, star-shaped flowers, lavender-pink and spotted purple, fall-blooming from August to November. Grown in porous compost of osmunda or fern-fiber, fibrous loam, humus, perlite and sphagnum. Stenoglottis need liberal watering while active; after blooming, with foliage withering, they should be rested for several months. Propagated by division of the fleshy roots. **C**: 16   p. 480

STENOMESSON (Amaryllidaceae)
incarnatum (Andes of Peru and Ecuador to 10,000 ft.), "Crimson stenomesson"; pretty bulbous plant 1½ to 2 ft. high, with succulent, strap-shaped leaves narrowed above, appearing with or after bloom; the floral stalk bearing a cluster of large, bright red flowers with tube 3 to 4 in. long and with spotted, flaring segments, blooming from early summer to August; after flowering, the bulb requires a complete rest in winter. Easily increased from offset-bulbs in spring. **C**: 63   p. 392

STENORRHYNCHUS (Orchidaceae)
navarrensis (Spiranthes) (Costa Rica); shapely terres-trial rosette of obovate soft-fleshy leaves, light grass-green with pale blotching; inflorescence a stout erect, bracted stalk topped by an elongate cluster of rosy pink flowers, sheltered by colorful salmon-rose bracts. Grown in well-drained compost of treefern fiber, sandy loam and sphagnum or peatmoss; after inflorescence dies down, a resting period is needed to ensure flowering for next season. Nice for the patio in summer, with liberal watering. Propagated by division while dormant. **C**: 60   p. 480

STENOSPERMATION (Araceae)
popayanense (Andine Ecuador, Colombia); slow-climbing by roots from the nodes; small, leathery, green, lanceolate leaves on sheathed petioles; boat-shaped, white spathe and white spadix. **C**: 4   p. 98

STENOTAPHRUM (Gramineae)
secundatum variegatum (So. Carolina to Texas, Trop. America), the variegated "St. Augustine grass"; creeping stoloniferous grass with flattened stems and firm linear leaves to 6 in. long, prettily banded creamy white, the tips round. The green species is much used for lawns in the Southern states, and as cattle feed in the tropics. The varie-gated form is very attractive as house plant or for baskets, where it will make long pendant stolons with shorter leaves. From cuttings. **C**: 2   p. 304

STEPHANOTIS (*Asclepiadaceae*)
   floribunda (Madagascar), "Madagascar jasmine" or "Wax flower", evergreen wiry climber twining to 15 ft. high, with opposite elliptic, thick-leathery glossy dark green leaves to 4 in. long, producing axillary clusters of very beautiful, exquisite waxy, white tubular flowers 2 in. wide, and intensely fragrant. A favorite for pots on wire or trellis, for the winter garden, or light, warm window; keep cooler and drier in winter. Propagate by cuttings. **C**: 2   p. 279

STEVIA serrata of florists: see Piqueria trinervia

STIGMAPHYLLON (*Malpighiaceae*)
   ciliatum (W. Indies, Brazil), "Brazilian gold vine" or "Golden vine"; slender tropical twiner with thread-like stems, and glabrous, heart-shaped, thin-leathery leaves, oblique at base and with ciliate margins; lovely 1½ in. golden yellow flowers in clusters of 3–7. Propagation by cuttings of new ripened wood. **C**: 52   p. 276

STRELITZIA (*Musaceae*)
   alba (augusta) (S.E. Africa), "Great white strelitzia"; palm-like tree of which I have seen extensive groves in inland Natal, west of Durban, to 30 ft. high, with woody trunk bearing shining green, leathery leaves 4–6 ft. long in two ranks, frequently cut into ribbons by the wind; the curious, large inflorescence on short stalks between the foliage; from a rigid, boat-shaped, pointed purplish bract or spathe, rises a row of white sepals and petals. Must be of fairly large size and age before blooming. **C**: 63   p. 207
   parvifolia juncea (So. Africa: East Cape), the "Rush-like strelitzia"; from the Port Elizabeth area; a most curious, slow-growing "Bird of Paradise" amazing to find because it has no leaves at all, just a dense cluster of spiky tufts 4 to 5 ft. high, of cylindrical, fleshy but rigid, reed-like grayish stems tapering to a needle-point; inflorescence as in the typical S. reginae with bright orange-yellow sepals and blue-tongue; blooming mainly in spring. **C**: 13   p. 534
   reginae (So. Africa: Transkei), the spectacular "Bird-of-Paradise"; trunkless, compact, clustering but slow-growing plant with fleshy roots; usually to 3 ft. high, sometimes 5 ft.; stiff-leathery, concave, oblong, bluish-gray leaves with pale or red midrib; strikingly exotic, long-stemmed flowers emerge from the green boat-shaped bracts which are bordered in red, the numerous pointed petals brilliant orange, contrasting with an arrow-shaped tongue of vivid blue. Begins to bloom 3 years from seed, best if well rooted in the container, normally in winter and spring; by keeping the plant dry and cool in spring, flowering can be delayed until summer. **C**: 13   p. 207

STREPTANTHERA (*Iridaceae*)
   cuprea (So. Africa), "Kaleidoscope flower"; low cormous herb about 9 in. high, with leaves in fan-shaped rosettes; 2 or 3 slender flexuous stalks with several flat, wheel-shaped flowers 1½ in. across glowing orange, the center deep violet with a ring of pale yellow spots; blooming in spring into June. For pots, plant corms in November and keep cool until growth begins. Water till flowers fade then rest until January. By offsets. **C**: 64   p. 394

STREPTOCALYX (*Bromeliaceae*)
   holmesii (Amazonian Peru); wonderful discovery, and most spectacular with its inflorescence measuring to 2 ft., lush with tiers of cherry-red 6 in. bracts, and crowned by panicles of flowers with fleshy orange calyx and white petals, over rich green leaves. **C**: 10   p. 142
   poeppigii (Brazil, Peru, Colombia); multileaved large rosette with spreading linear foliage to 4 ft. long and 2 in. wide, egg-shaped toward base, gray-green with reddish mottling, the margins thorny. Long-lasting inflorescence an erect raceme white-mealy, bracts brilliant red, and flowers violet. **C**: 10   p. 148
   poitaei (Amazonian Peru); dense rosette of linear leaves coming to a short point at apex, the underside somewhat scaly; the bold inflorescence with showy crimson bracts. **C**: 10   p. 148

STREPTOCARPUS (*Gesneriaceae*); the "Cape primroses" are handsome, long-lived principally African perennial herbs of low habit, with fibrous roots, unusual foliage, and lovely flowers in many colors and markings; their seed pods are characteristically twisted. The genus falls into two groups:

those forming erect stems bearing pairs of opposite leaves and flowers axillary; and others nearly stemless and forming basal rosettes of several leaves, or sometimes an apparent large single leaf. The inflorescence in the stemless types arises from the base of the midrib. Single leaves may be from 3 in. to 5 ft. long according to species. The better known Streptocarpus make good house plants, almost ever-blooming, especially if seed is sown at intervals. Culture is similar to Saintpaulias but they can take stronger light and cooler night temperatures. Propagation either from the dust-fine seed, sown on moist milled sphagnum; from leaf cuttings or leaf sections by cutting into the midrib in one or more places and placing the leaf, or part, on the rooting medium; by division, or from stem cuttings.
   caulescens (Trop. East Africa), "Violet nodding bells"; succulent, soft-hairy, branching plant to 1½ or 2ft. high, with satiny light green, fleshy purplish stem covered with white hair; opposite, small blunt-cordate leaves, slender axillary stalks bearing forked clusters of pretty, nodding ½ in. two-lipped slipper flowers with spreading petals, beautiful violet or lilac with white throat; summer-blooming. An easy grower as a house plant. **C**: 10   p. 450
   holstii (Trop. East Africa); erect branching herbaceous, somewhat weedy plant to 1½ ft. high, with slender fleshy stems swollen at the nodes; the opposite ovate, slightly hairy, almost glossy, wrinkled leaves 1½–2 in. long; slender reddish stalks carry clusters of the 1 in. lively purple flowers, with open, white throat extending into a large lower lip; summer-blooming, and easy to grow and propagate. **C**: 10   p. 450
   x hybridus, the showy "Hybrid Cape primrose"; complex group of hybrids with a long line of parents including S. dunnii for color, rexii for bushy habit, and wendlandii for length of stalk; light green fleshy, quilted leaves in partial rosette; free-blooming with large trumpet-like flowers, in a wide range of color, from white with purple veining, through rose, orchid, mauve, blue, to purple. Exceptionally handsome are the 'Wiesmoor hybrids' with shorter leaves, and fringed and crested flowers 3 to 4 in. across. These will bloom in pots 6 to 8 months from seed; from January sowing in summer, from June in spring; continuing in the living room for months. **C**: 21   p. 450
   polyanthus (So. Africa: Natal), "Sky-blue fountain"; curious stemless plant remarkable with large apparent single quilted oblong leaf, reaching a length of 12 in. and half as broad; hairy above, and wine-red underneath; from the base of the midrib of the leaf, actually the foreshortened stem confluent with the leaf, arise one or several wiry stalks about 1 ft. high or more, bearing a branching inflorescence of numerous smallish flowers with 1½ in. yellow tube and lavender-blue, flat oblique, spreading lobes, 1½ in. across, with yellow throat; summer-blooming. A collector's plant. **C**: 65   p. 450
   rexii (So. Africa: Cape Prov.), the first species "Cape primrose" brought to Europe in 1824; small fibrous-rooted, stemless plant with long narrow, quilted and pubescent leaves in rosette hugging the ground, with several flower stalks bearing charming trumpets 2 in. long, pale lavender lined with purple in the throat; summer-blooming. **C**: 21   p. 450
   saxorum (Trop. East Africa), "False African violet"; a very pretty, small plant with branches spreading flat on the ground, the thick-fleshy, vivid yellow green, pubescent, elliptic, 1 in. leaves in crowded whorls; exquisite, relatively large flowers with white tube and oblique limb of large spreading pale lilac lobes, 1½ in. across, on long thin stalks, from leaf axils, blooming over many months, especially in spring and summer. Have seen them clinging to perpendicular, exposed cliffs, at 3500 ft. in the humid Usambara Mountains of Tanzania, from the distance resembling Saintpaulia in color of foliage and flowers. Because of its spreading or trailing habit suitable for baskets. **C**: 71   p. 450
   wendlandii (So. Africa: Natal), "Royal nodding bells"; fantastic stemless plant characterized by an apparent single huge leaf, with age and space becoming 3 ft. long and 2 ft. wide, the corrugated surface olive green and densely hairy, a beautiful purple beneath; tall, forking flowering stalks to 2½ ft. high sprouting from the base of the leafblade (actually from a foreshortened stem confluent with the leaf), may bear as many as 30 small, bright blue flowers with violet markings, white in throat, the oblique limb 1½ in. across. A curiosity plant for botanical collections with ample space; long-blooming throughout the summer. Propagated from leaf sections or seed. **C**: 65   p. 450

x STREPTOGLOXINIA (*Gesneriaceae*)
'Lorna' (Streptocarpus x Sinningia); a lovely hybrid originating in California in 1947 as 'Stroxinia', supposedly bigeneric between Streptocarpus x Sinningia; resembling "gloxinia" in type of leaves and flowers, which may be nodding and foxglove-like, or erect true bell-shaped, except that upper lobes overlap and are curled back; flower texture is rich and velvety, color deep rose in 'Lorna', blue in 'Nancy', with speckled throat. **C**: 10   p. 448

STREPTOSOLEN (*Solanaceae*)
jamesonii (Colombia, Ecuador at 6000 ft.), the "Fire-bush" or "Marmalade bush", also known as the "Orange browallia"; rough-pubescent, floriferous, sprawling, shrubby herb with small 1½ in. oval, wrinkled leaves on rambling branches, and terminal clusters of tubular, bell-shaped, orange flowers 1¼ in. long, in spring. Wonderful flowering plant requiring sun and warmth in daytime, cool at night. Propagated from cuttings. **C**: 16   p. 278

STROBILANTHES (*Acanthaceae*)
dyerianus (Perilepta) (Burma), "Persian shield"; beautiful tropical herbaceous shrub with magnificent iridescent 6 in. leaves, long-ovate and toothed, purple with silver above and curiously shimmering; glowing purple beneath; pale blue flowers. Best in a moist-warm green-house. **C**: 62   p. 187
isophyllus (Goldfussia) (N.E. India), "Bedding cone-head"; floriferous, densely branched small sub-shrub 1 to 2 ft. high with shiny, willow-like opposite toothed leaves 2 to 3 in. long; swollen tubular flaring, 1 in. flowers lavender or blue with white. Blooms profusely either in winter or summer according to treatment. Charming for its mass of flowers in summer, but of short duration; used for bedding in the summer garden or on the patio; a house plant in winter. From soft cuttings. **C**: 16   p. 492

STROPHANTHUS (*Apocynaceae*)
gratus (West Trop. Africa), "Climbing oleander"; a robust clambering evergreen shrub, with opposite leathery, ovate to obovate olive-green leaves somewhat puckered, 4 in. or more long, on brown-purple woody stems; the waxy, trumpet-shaped flowers 1¾ in. dia., with crinkled lobes, flushed purplish-red outside and pinkish-white inside, and with a prominent pale rose-purple inner crown, blooming late spring to summer. From seed or cuttings. **C**: 52   p. 489

SWAINSONA (*Leguminosae*)
galegifolia (Queensland, New South Wales), "Winter sweet pea", "Swan flower" or "Darling pea"; ornamental subshrub with flexuous, scandent branches nearly climbing, bearing unequally pinnate leaves, and small pea-like ¾ in. flowers in long-stalked showy racemes, deep red, or in other shades from blue to white. Free-blooming as a container plant in a cool winter garden or outdoors. From half-hard cuttings. **C**: 64   p. 278

SYAGRUS (*Palmae*)
weddelliana (Microcoelum martianum) (Brazil: Guana-bara), long known in horticulture as Cocos weddelliana, a "Terrarium palm"; attractive little feather palm which grows in the humid Organ Mts. very slowly, becoming to 7 ft. high, forming slender, solitary trunks elegantly bearing arching pinnate fronds to 5 ft. long, the narrow, stiff segments glossy yellow-green and neatly spaced; small orange fruit. Grown in greenhouses as Cocos weddelliana, usually in 2 in. pots only 6–8 in. high, this graceful "Dwarf coconut palm" has long been a favorite for glass terrariums but it requires high warmth and humidity, and has been overtaken in popularity by the much more satisfactory Chamaedorea elegans. **C**: 54   p. 230

SYNADENIUM (*Euphorbiaceae*)
grantii rubra (Euphorbia) (Tanzania), "Red milk bush"; ornamental shrub with ascending, thick, succulent branches with milky juice; leathery-fleshy obovate leaves 3 to 6 in. long, beautiful wine red, the margins finely toothed, reverse vivid red purple; small red flowers. A beautiful, easy-growing decorative house plant. **C**: 1   p. 344

SYNGONIUM (*Araceae*); tropical American creeping or climbing vines having milky juice, the string-like stems root-ing at the joints; the long-stalked, decorative foliage usually arrow-shaped and prettily painted in the juvenile stage, in later stages becoming palmately divided and usually plain green. Inflorescence not showy, with spathe pale green to yellowish. Small plants are long lasting in dishgardens; or they may be trained against slabs of redwood or cedar bark. Excellent for hydro culture. Propagated easily by cuttings.
auritum (Philodendron auritum) (Jamaica), "Five fingers"; climber with fleshy, rich, dull or glossy green, 3 to 5-sected leaves to 10 in. long; petioles vaginate. **C**: 54 p. 82
erythrophyllum (Panama), "Copper syngonium"; dainty creeping plant with small, arrow-shaped, waxy leaves having 2 ear-like basal lobes; blade metallic coppery-green and covered with tiny silvery pink dots, reddish beneath; mature leaf trifoliate. **C**: 60   p. 82, 259
hoffmanii (C. America), "Goose foot"; attractive creep-ing plant with the glabrous young leaves arrow-shaped, matte grayish-green with silver-white center area and veins. **C**: 4   p. 82
macrophyllum (Mexico to Panama), "Big leaf syn-gonium"; climber with large, heart-shaped, emerald-green, fleshy leaves with velvet sheen, becoming divided in the maturity stage, much larger than podophyllum. **C**: 54   p. 82
podophyllum (Nephthytis liberica) (Mexico to Costa Rica), "African evergreen"; in juvenile stage a small plant with arrow-shaped, thin green leaves, on slender petioles, later starting to creep; in succeeding stages leaves become lobed and then palmately divided into 5 to 9 segments, to 25 cm (10 in.) across. **C**: 4   p. 82, 259
podophyllum albolineatum (Mexico, Nicaragua) (earliest name: angustatum), "Arrow-head vine"; known commercially as Nephthytis triphylla; the juvenile leaves heart-shaped or 3-lobed, very ornamental with silver white center and veining; the mature palmate leaves are green. **C**: 4   p. 81, 82
podophyllum albolineatum 'Ruth Fraser'; a horti-cultural selection showing a more distinct improvement in variegation over the type, also variegation lasts longer. **C**: 4   p. 82
podophyllum 'Albo-virens'; a mutant with slender juvenile hastate leaves shaded ivory to greenish-white; blade edged green. **C**: 4   p. 82
podophyllum 'Atrovirens'; variety with the juvenile leaves hastate or lobed, and the vein areas ash green to cream, on a dark green background. **C**: 4   p. 82
podophyllum 'Emerald Gem', "Arrowhead plant"; with 4–6 in. juvenile leaves arrow-shaped, more fleshy, quilted, dark green and glossy than the species; the petioles are shorter making this form more compact, and soon inclined to creep. **C**: 4   p. 82
podophyllum 'Trileaf Wonder', cultigen with a varying amount of ashgreen on the leaf, principally on the midrib and lateral veins; also the mature, segmented leaves are produced more quickly. **C**: 4   p. 81
podophyllum xanthophilum (Mexico), cultivated under the name of "Green Gold"; the juvenile, arrow-shaped leaves are suffused and marbled with yellow-green. Very durable and attractive. **C**: 4   p. 81, 178
triphyllum 'Lancetilla' (Honduras, Costa Rica); in juvenile state the papery, dull green upright leaves are unequal-sided, ovate with sunken veins, later becoming increasingly 3-lobed. **C**: 54   p. 82
wendlandii (Costa Rica); dainty creeper with tri-lobed, deep green, velvety leaves, and sharply contrasting white veins in the juvenile foliage; the divided maturity-stage leaves plain green. **C**: 4   p. 43*, 82

SYRINGA (*Oleaceae*)
vulgaris 'Marie Legraye', "White lilac", an old French hybrid (1879); with large pure white flowers in open, narrow, medium-large clusters, and much used for early winter forcing. This and several other such forcing varieties are descended from the common lilac at home in S.E. Europe, a bushy shrub 10 ft. or more high, with ovate leaves to 5 in. long, blooming outdoors in May with lateral clusters of fragrant flowers, normally pale purple tinged with violet. In the absence of freezing winter climate, plants may be compelled into dormancy by drying them off gradually but completely, starting in August (for forcing see p. 30). Propagated from summer cuttings or by grafting on Ligus-trum ovalifolium. **C**: 76   p. 504

SYZYGIUM: see Eugenia

TABERNAEMONTANA: see Ervatamia

TACCA (*Taccaceae*)
chantrieri (Malaya), "Bat flower" or "Cat's whiskers"; tropical stemless perennial herb with swollen rhizome; corrugated olive-green leaves with oblique base; from the base appears the long-stalked curious inflorescence bat-like in both shape and color, with wide-spreading, wing-like bracts of rich maroon-black, accompanied by long trailing filaments or "whiskers" 1 ft. long; the somber, 2 in. long black flowers are succeeded by heavy berries; blooming late winter-spring. A conversation piece of weird and sinister appearance. Propagated by division or separation of suckers. **C:** 54   p. 494

TAENIOPHYLLUM (*Orchidaceae*)
fasciola (syn. Epidendrum, Limodorum, Vanilla) (Fiji Islands); one of the tiniest of orchids, a leafless epiphyte which I found on the trunk of a large tree in the rain forest of Naduruloulou, on Viti Levu; the highly developed, brittle strands of roots only 1/10 in. wide but 3 to 6 in. long, containing chlorophyll and delegated to carry out the functions of the absent leaves; these roots are flattened against the bark, and radiate from a common center from which rise stems reduced to minimum, bearing little pale cream flowers under ¼ in. long; capsule bright yellow. **C:** 60   p. 533

TAGETES (*Compositae*); the marigolds are robust, free branching sun-loving herbaceous plants with finely divided ferny foliage, and handsome, long lasting flowers blooming from early summer to frost, if old flowers are picked off. The popular French marigolds are forms of the perennial T. patula from Mexico, grown as annuals for summer bedding or as gift plants in small pots in spring. Propagated by seed, or rooting prostrate branches.
patula 'Naughty Marietta', "Single French marigold"; very pretty single marigold of semi-dwarf habit, to 16 in. high, bushy, with finely netted leaves; 2 in. flowers golden-yellow, painted in the center with a deep red velvety eye. **C:** 64   p. 409
patula 'Spry', "Dwarf French marigold"; popular free-blooming, dwarf bedding marigold to 10 in. high, the 1½ in. double flowers with a brightly contrasting golden yellow anemone crest over deep rich mahogany guard petals. A pretty, easy-grown little pot plant for Mother's Day in May. **C:** 64   p. 409

TANACETUM (*Compositae*)
vulgare (W. Europe to Siberia; naturalized along the roadsides of Eastern U.S.), "Common tansy", "Tansy" or "Buttons"; a strong-scented hardy perennial herb to 3 ft. high, the robust angular stems with alternate pinnate bright green aromatic leaves 4 to 5 in. long, segments again cut or toothed; numerous yellow flower heads the size of small buttons. An old medicinal plant; the leafy tips of Tansy are used in cosmetics and ointments; also in the liqueur Chartreuse. In the kitchen the fresh young leaves may be sparingly used in an omelet or baked fish. The ancients used tansy to preserve meat and keep flies off it. By division of roots. **C:** 26   p. 321
vulgare crispum (Europe), "Crisped tansy", "Jesus-wort" or "Fernleaf tansy"; a very pretty variety of lacy appearance, to 2½ ft. high, with its beautiful dark green foliage finely divided fern-like, alternate from robust reddish stems; the golden-yellow hard, button-like flowers bloom in large clusters during summer. Foliage and flowers keep well in bouquets; the flowers even when dried. To dry, hang in bunches in an airy place in the shade. **C:** 26   p. 320

TAVARESIA (*Asclepiadaceae*)
grandiflora (Decabelone) (Tropical S.W. Africa), "Thimble flower"; clustering leafless succulent to 8 in. tall, branching with thick, 10 to 14-angled, green stems, the ridges with tubercles each bearing 3 white bristles; large flowers funnel-shaped with spreading lobes, about 4 in. across, lemon-yellow densely spotted purplish-red, and looking like giant thimbles. From cuttings. **C:** 1   p. 352

TAXODIUM (*Taxodiaceae*)
distichum (New Jersey, Delaware to Florida, west to Texas), the "Bald cypress" or "Swamp cypress", also known as the "Tidewater red cypress"; tall deciduous coniferous tree, becoming 150 feet high, with buttressed trunk usually 4–5 ft., but sometimes 12 ft. or more in diameter, usually hollow in old age; the spreading branches with delicate and feather-like light yellow green ½ in. needles in graceful sprays. At home in cypress swamps and tidewater bayous along the Gulf of Mexico and the Florida everglades; this is not only a very decorative tree but valuable as the "Red" or "Yellow cypress" for its durable red or yellow lumber used in greenhouse construction and benches, for boats, piling, shingles, and wherever wood must withstand warm and wet conditions. The base of the tree is flared to help absorb oxygen from the water; it is also known for its curious "Cypress knees" which are modified roots of a very light, soft spongy wood, growing out of the water, and sometimes attaining a height of 10 ft., breather roots supplying the root system with air when the tree stands completely in water. When with age they become hollow they are used as bee-hives. Propagated from seed. **C:** 64   p. 526

TAXUS (*Taxaceae*)
cuspidata 'Capitata' hort. (Japan, Korea, Manchuria), "Upright Japanese yew"; somber evergreen coniferous tree to 50 ft. high, more or less branching, with needle-like flat leaves 1 in. long, blackish green and leathery arranged on one plane along either side of the flexuous branchlets; berry-like scarlet fruit. The term 'Capitata' is used by the nursery trade to distinguish the type grown in upright, sheared pyramidal form from the purposely low-trimmed hedge or globular forms. Widely used for formal plantings and in containers as pyramids, needs moisture but with good drainage. Perpetuated from seed or from leadertip cuttings for the 'Capitata' type, other cuttings for the wide-branching cuspidata. **C:** 75   p. 311

TECOMA: see Campsis

TECOMARIA (*Bignoniaceae*)
capensis (So. Africa to Transvaal and north), "Cape honey-suckle"; rambling, evergreen shrub 6–8 ft. high, with leaves of 7–9 ovate, shining green, toothed leaflets to 2 in. long, and bearing bunched masses of curved funnel-form flowers fiery orange scarlet, 2 in. long, and with prominent, protruding stamens. Propagation by layering of branches. **C:** 13   p. 276

TECTARIA (*Filices: Polypodiaceae*)
cicutaria (West Indies), "Button fern"; rhizomatous fern, with brown, scaly stalks 15 in. long, bearing tripinnate fresh-green herbaceous fronds triangular in outline, to 2½ ft. long; toward the apex simply lobed, softly downy in the fertile, spore-bearing leaflets, conspicuous by their circular elevations on the upper surface; in addition, the fronds bear occasional bulbils on their surface, looking like tiny buttons. **C:** 60   p. 249

TELOPEA (*Proteaceae*)
speciosissima (Australia: New South Wales, Queensland), "Crimson waratah"; magnificent flowering shrub related to Protea, of erect habit, 10–12 ft. high, with narrow obovate, leathery leaves, toothed in upper part; striking terminal inflorescence of coral-red flowers with protruding styles, packed in a dense cone 3 to 4 in. across, and ringed with numerous narrow bracts of brilliant crimson; blooming "down under the equator" in their late spring. **C:** 65   p. 489

TEPHROCACTUS: see Opuntia

TETRAPANAX (*Araliaceae*)
papyriferus (Aralia) (China, Taiwan), the "Rice paper plant"; a small multi-stemmed tree 10–25 ft. high with big ornamental foliage to about 1 ft. across; the lobed leaves are gray-green above, and covered with white felt beneath while young. Creamy-white flowers on tan furry stems in big clusters. The slender stems are filled with white pith from which the Chinese rice-paper is made. Handsome plant for cooler locations but requiring considerable space. From seed or cuttings. **C:** 64   p. 291

TETRAPLASANDRA (*Araliaceae*)
meiandra (Hawaii), "Hawaiian Ohe tree"; evergreen tree with stout willowy branches and thick, soft-leathery, glossy green leaves with pale veining, pinnately divided in 7 oblique-ovate leaflets, pale beneath, on coppery petioles. Known as 'Ohe' by the Hawaiians. On Hilo, trees grow to 80 ft. tall. Promising decorator plant. **C:** 3   p. 290

TETRASTIGMA (*Vitaceae*)

voinierianum (formerly Vitis voinieriana) (Vietnam), "Lizard plant" or "Chestnut vine"; robust climber with woody stems, and clambering, fleshy, brown-hairy branches having coiled wiry tendrils and gigantic, digitate, thick-fleshy leaves with 3–5 shining green, stalked, broad-obovate or oblique leaflets to 10 in. long, wavy-toothed at margins and pale green, pubescent underneath. Magnificent vine for the winter garden or even the home if space permits and light is good, taking care to keep moist enough at the roots. **C**: 54   p. 257

TEUCRIUM (*Labiatae*)

flavum (Eastern Mediterranean: Syria), "Yellow germander"; much branched shrubby perennial 1 to 2 ft. high woody at base, with thick-ovate, crisply gray-hairy leaves, crenate above middle; axillary yellow flowers somewhat two-lipped, along leafy stalk. Ornamental for its yellow flowers and silvery foliage. Once used as a gout remedy, to relieve painful inflammation of the joints. **C**: 13   p. 320

fruticans (W. Mediterranean to So. Italy), "Tree germander"; evergreen shrub 2–4 ft. high, yellowish-woolly with small ovate, boxwood-like revolute leaves to 1¼ in. long, white or reddish-brown hairy beneath; bluish 2-lipped flowers. Use: medicinal against fevers. Not hardy. From cuttings. **C**: 13   p. 320

THELOCACTUS (*Cactaceae*)

bicolor (So. Texas to Central Mexico) "Glory of Texas"; variable globular to cylindrical plant to 4 in. thick, bluish-green, with 8 oblique notched ribs, furnished with long straight as well as arched spines variously colored white, yellow and red; flowers purplish. **C**: 13   p. 163

THEOBROMA (*Sterculiaceae*)

cacao (C. America, Trinidad, Guayana), the prized "Cacao" or "Chocolate tree"; wide branching evergreen tree to 25 ft. high, with attractive oblong, slender-pointed satiny, hard-papery, pendant leaves to 1 ft. long; the tiny, yellowish flowers sometimes in axillary clusters, but curiously, mostly directly from the trunk and heavy branches, where they develop into the large, ribbed pods 6 to 10 in. long, containing imbedded in its pulp 50 to 100 bean-like seeds which, when roasted, yield the chocolate and cocoa of commerce. **C**: 54   p. 528

THEVETIA (*Apocynaceae*)

peruviana (W. Indies, Mexico), "Yellow oleander" or "Be-still tree"; tropical evergreen shrub 6 to 8 ft. high with linear, shining green, 4 to 6 in. leaves with edges rolled under; large, funnel-shaped lemon-yellow flowers 2 to 3 in. long, shading to pinkish or orange-apricot, and sweetly fragrant like a tea-rose; blooming anytime, mostly June to November. Takes heat and sun with ample water; may be trained into small tree. Poisonous like their relatives, the oleanders. Propagated by cuttings or seed. **C**: 2   p. 489

THRINAX (*Palmae*)

microcarpa (So. Florida and nearby Keys, Greater Antilles, Yucatan, B. Honduras, Panama), "Key palm"; a thatch palm more robust than parviflora, with trunk 9–12 ft. but reaching 30 ft., and 8–12 in. dia.; the petiole with fibrous webbing, supporting palmate leaves about 3 ft. across, divided for about ½ their length, the straight segments forming a semi-circle, shiny grayish-green above, light gray beneath; abundant, small white sessile fruit. **C**: 54   p. 228

parviflora (Florida Keys, Bahamas, Cuba, Haiti, Jamaica), "Florida thatch palm" or "Jamaica thatch palm"; slender solitary fan-palm to about 25 ft. tall, and 4–6 in. thick, enlarged at base by root-like growths; 2-edged petioles reddish at base and with hairy fiber; palmate leaves 3–4 ft. across, cut halfway to base into about 50 segments and forming almost a complete circle, green on both sides, joined by radiating prominent yellow ribs; small berry-like, white fruit in large clusters. Shapely, attractive tub plant. **C**: 4   p. 228

THRIXSPERMUM (*Orchidaceae*)

arachnites (Indonesia, Malaya); curious and fascinating small epiphyte with leafy stems to 8 in. long, clinging close to their support, and flowers 2 to 3 in. across in clusters, usually opening one at a time, having slender pointed pale yellow segments, and a white lip sac-like at base, and spotted

with red; blooming spring and summer. The flowers are short-lived, but new ones continue to open at intervals for several months. This small vandaceous collector's plant needs constant warmth and moisture; best grown on osmunda or treefern blocks. **C**: 60   p. 480

THRYALLIS (*Malpighiaceae*)

glauca (Galphimia) (Mexico to Panama), "Gold shower thryallis"; handsome bush of open habit, 3 to 5 ft. high, with scandent branches, opposite thin-leathery oblong bluish glaucous leaves to 2 in. long, and small ¾ in. yellow flowers in showy, elongate clusters. Very floriferous evergreen for the warm greenhouse or conservatory. From cuttings under glass. **C**: 52   p. 485

THUJA (*Pinaceae*)

occidentalis (E. No. America: Quebec to Hudson Bay, New Jersey to No. Carolina, west to Minnesota), the "American arbor-vitae", or "White cedar"; ornamental coniferous tree of pyramidal habit to eventually 60 ft. high, with reddish buttressed trunk divided near ground into several secondary trunks, the branches densely arranged, flat fan-like, branchlets like fern fronds but with hard scale-like, shingled leaves bright green to yellowish green, with a strong resinous odor; small oblong cones. A popular, quick-growing symmetrical evergreen much planted around homes, requiring moisture; also used in tubs for temporary decoration. From cuttings. **C**: 76   p. 311

THUNBERGIA (*Acanthaceae*)

alata (So. E. Africa, but naturalized in Tropics), the "Black-eyed Susan"; twining perennial herb, with herbaceous, toothed, triangular ovate leaves to 3 in. long, on winged petioles; funnel-shaped showy flowers 1½ in. long, creamy-yellow or orange with black-purple throat, blooming late summer to autumn. Attractive vine for the cool, light window. Can be grown as an annual from seed; also cuttings. **C**: 14   p. 280

THYMUS (*Labiatae*)

hirsutus (Balkans; Soviet Union: Crimea to Caucasus), "Woolly thyme"; low, cushion-forming shrubby plant, with woody prostrate shoots and ascending branches bearing dense tufts of tiny linear gray-hairy leaves; floriferous with ¼ in. rosy flowers having reddish calyx, in oblong clusters. With strong mint-like odor. **C**: 13   p. 322

vulgaris (Mediterranean reg. from Portugal to So. France, So. Italy to Greece), the "Common thyme" or "French thyme"; aromatic low-growing, wiry-stemmed shrubby perennial 8 to 12 in. high, the white-pubescent stems with tiny lemon-scented, gray-green, hairy oval, recurved leaves ½ in. long; small clusters of ¼ in. rosy-lilac flowers in May-June. Hardy. The leaves, dried or fresh are widely used as subtle seasoning for food: soups, sausages, chowders, sauces, vegetables. Dried flowers in sachets, and add flavor to herb teas. Oil of thyme is used in cough medicines and perfumes. Famous as a source of honey. Propagated by seed, cuttings or division. **C**: 13   p. 321

vulgaris aureus (So. Europe), "Golden lemon thyme"; hardy perennial subshrub 6 to 12 in. high, forming mounds of silvery green aromatic oval ¼ in. foliage variegated or edged with gold; purplish flowers. In addition to being very ornamental, its aromatic branches find culinary use, fresh or dried for seasoning vegetable juices, fish, shellfish, poultry stuffing, soups. By division. **C**: 13   p. 320

TIBOUCHINA (*Melastomaceae*)

semidecandra (So. Brazil), "Glory-bush" or "Princess flower"; beautiful free-branching herbaceous shrub of open habit, with age becoming woody and tree-like, growing to 10 ft. or more high; with 4-angled stems and fresh-green velvety 3 to 5 in. ovate leaves densely covered with soft white hairs, and often edged red; new growth bronzy red; large brilliant violet-purple flowers 3 in. or more across, blooming over a long period of time; from May to January and intermittently into spring. With its brilliant royal purple blossoms and handsome foliage, one of the showiest flowering plants of the warm-temperate greenhouse; or the patio in summer. From half-hardened cuttings in spring. **C**: 66   p. 493

TIGRIDIA (*Iridaceae*)

pavonia (mountains of Mexico and Guatemala), "Tiger flower" or "Mexican shell flower", also known as "One-day lily"; gay, sun-loving, summer-blooming cormous plant with ribbed, sword-like leaves at base and along the forked floral stalks, 18 to 30 in. high, carrying showy bright flowers to 6 in. across, with 3 large segments forming a triangle, joined with 3 smaller segments to form a center cup. The typical form is orange-scarlet spotted with crimson at the base, but horticultural varieties come in colors from white to pink, red, yellow, orange to purple, with variously painted center cup. The blooms last only a day but are succeeded by a series of others along the stem. After flowering, and when foliage turns yellow, keep dry to rest and remove corms from pots. By offsets. **C**: 63   p. 393

TILLANDSIA (*Bromeliaceae*); the "Silver birds" or "Wild pines" are epiphytes of the American subtropics and tropics, the largest and most widely distributed genus in the bromeliad family. It had to adjust to many conditions and by evolution many types have evolved, from the tiny, ghost-like chains of the Spanish moss to elegant kinds with spectacular flowers beautiful as orchids; and dwelling on trees and house tops as epiphytes or on rocks, and even perched on telephone wires. Their bodies are often rigid and usually covered with fuzzy, silvery scales which can capture and absorb atmospheric humidity, allowing it to enter the leaf cells. The flowers, appearing from fall to spring, are nearly always in two ranks, in one or more spikes. Some Tillandsias do not take kindly to indoor growing, and prefer a cooler sunporch or the winter garden, where they often thrive best if left growing as collected on a tree branch or planted on a treefern block or basket; dunked every few days in a bucket of water. Unless in a greenhouse they always appreciate periodic misting of the foliage. Propagated by offsets or division.

araujei (Brazil: Rio, Sao Paulo); an exciting small air plant growing on rocks along the ocean, with scandent stems dense with spirals of short tapering leaves, concave, short and hard, light green with purplish base and silvery scurf; growing into crawling or pendant strands 4–5 ft. long; flowers in short spikes, delicately colored pink and white. **C**: 7   p. 143

balbisiana (So. Florida, Yucatan, Costa Rica); xerophytic, ball-shaped rosette with bulbous base, densely furnished with numerous narrow linear, twisted or recurved succulent leaves about 6 in. long, and to $\frac{1}{2}$ in. wide, grayish-powdery over gray-green; long stalked, small inflorescence with waxy-powdery bracts green, rosy, and yellow; the flowers purple marked with white. **C**: 7   p. 144

bulbosa (West Indies, So. Mexico, to Colombia, Brazil), "Dancing bulb"; bulbous type of epiphyte, with onion-like swollen base, leaves rolled up tight and turning away from silvery base and stem at sharp angles, then twist; glossy olive green with coppery tinting; short inflorescence with violet flowers tipped white. **C**: 19   p. 143

cyanea (morreniana) (Ecuador: Manabi to Loja Prov.), "Pink quill"; excellent suckering rosette of linear, channeled leaves grooved with red-brown pencil lines lengthwise; beautiful short spike with broad clear pink bracts in flattened ranks, lasting for months, and large, pansy-like violet-blue flowers darker than lindenii, in succession all year. **C**: 60   p. 131*, 144

fasciculata (Florida, W. Indies, C. America), "Wild pine"; large epiphyte, dense rosette with hard linear-lanceolate 18 in. leaves gray and recurved; the branched inflorescence showy with flattened greenish bracts tinged red; flowers blue. **C**: 7   p. 144

flabellata (Guatemala), "Red fan"; beautiful rosette of narrow, recurved leaves fresh green turning red in good light; giving rise to a 2 ft. inflorescence branching into flattened spikes arranged fan-like, the bracts vivid red, flowers blue. **C**: 19   p. 144

flexuosa (aloifolia) (So. Florida to So. America), "Spiralled airplant"; hard rosette with leaves starting off at the base with a twist, broad but tapering, thick leathery, concave, silvery gray over green, with indistinct silver bands outside; 2-ranked inflorescence with rose bracts and white flowers. **C**: 19   p. 143, 538

imperialis (Mexico: Oaxaca, Puebla, Veracruz), "Christmas candle"; showy epiphyte at home at 5000 to 8000 ft. altitude; dense, formal rosette of broad, smooth leathery leaves 1$\frac{1}{2}$ ft. long, pleasing light green, from which rises a spectacular flaming red central cone with purple flowers, an inflorescence looking like a candle radiating a festive spirit, and the Mexicans use them at Christmas time to decorate for their 'Natividad'. **C**: 10   p. 144

ionantha (erubescens) (So. Mecixo to Nicaragua), "Sky plant"; as though growing in blue sky, this charmer clings to tree branches; one of the tiniest bromels, seldom growing over 2 in. dia., miniature rosette of slender silvery leaves; closely overlapping, recurving, thick-fleshy, channeled, fresh green but covered on outside with silvery bristles, the whole center turning fiery red, followed by small violet flowers. **C**: 19   p. 144

lindenii (N.W. Peru: Piura), "Blue-flowered torch"; attractive, formal rosette of recurved linear channeled leaves green with red-brown pencil lines becoming more prominent toward base; inflorescence a long spike of flattened carmine-rose bracts with magnificent royal-blue flowers and white eye. **C**: 60   p. 144

"mooreana" (Amazonian Peru); open olive-green rosette with giant branched inflorescence similar to T. wagneriana but larger, the inflated bracts a deeper shade of pink, and with large lavender-blue flowers. **C**: 60   p. 148

recurvata (Subtropical and Tropical America: Florida to Arizona and Texas, to No. Argentina and Chile), "Ball moss"; epiphytic rosettes arranging themselves in spiral fashion around a twig, forming dense masses; on occasion adapting also to a strange life perched high on old, fabric-covered telephone wires, a familiar sight in Florida; thread-like, stiff-leathery curving leaves 1 to 7 in. long, covered with silver-white hairs or hair-like scales, on short, tufting stems, with some roots present; few-flowered inflorescence with scurfy bracts, and narrow pale violet or white petals. **C**: 19   p. 143

streptophylla (Jamaica), "Twist plant"; curious tight basal rosette formed of sharply recurved leathery leaves gradually tapering to a coiling tip; they are gray-green but covered with silvery scurf, and turning red-purple in strong light; the branched inflorescence with rosy bracts and lilac flowers. **C**: 19   p. 143

stricta (Brazil, Paraguay, Argentina, Guayana, Venezuela), "Hanging torch"; small rosette becoming caulescent (stalked), the narrow, thin-leathery, tapering leaves recurving, green with silvery scurf; attractive short-stalked inflorescence pendant, with red-tinged bracts, and flowers deep violet turning to red. **C**: 7   p. 143

usneoides (S.E. United States from New Jersey to Texas, on to Argentina and Chile), "Spanish moss"; a true epiphyte growing from trees in silvery gray thread-like masses to 25 ft. long, densely covered by the gray scales which are a means of receiving and holding atmospheric moisture, and which help to enable the plant to dispense with roots; small axillary flowers with petals $\frac{3}{8}$ in. long, yellowish green or blue. Can be hung in the living room window but must be fogged frequently to supply needed humidity. **C**: 72   p. 125, 143, 533

utriculata (Florida, W. Indies), "Big wild pine"; largest bromeliad in U.S.; rosette of spreading linear leaves 2 ft. long, gradually tapering from an ovate base forming pockets to hold water, and recurved; long compound spike with two-ranked green bracts edged red, bearing erect flowers with greenish white petals; plant dies after fruiting without offsetting. **C**: 7   p. 144

wagneriana (Amazonian Peru), "Flying bird"; possibly the most outstanding and beautiful of all Tillandsias with its branched inflorescence of long-lasting flattened bracts in silver pink, and blue flowers, rising 20 in. high from a rosette formed by the soft, thin-leathery, shining green leaves. **C**: 10   p. 143

TITANOPSIS (*Aizoaceae*)

calcarea (So. Africa: Cape Prov.), "Limestone mimicry"; cluster-forming stemless little succulent with fleshy roots; rosette of grayish spatulate triangular leaves to 1$\frac{1}{4}$ in. long, resembling stones; grayish-green and covered with gray-white warts or tubercles; $\frac{3}{4}$ in. flowers deep yellow. Looking so much like the surrounding limestone, the plants are indistinguishable in habitat where they grow. Rest in winter. From seed. **C**: 1   p. 328, 355

TITANOTRICHUM (*Gesneriaceae*)

oldhamii (Taiwan, So. China); showy, erect, tall-growing perennial herb 1 to 3 ft. high, from an underground scaly, fleshy rhizome; rough green, white-hairy, toothed, ovate leaves to 7 in. long, decreasing in size up the elongate bristle-haired stem to the nodding, swollen tubular, hairy golden flowers 1$\frac{1}{2}$ to 2 in. long, bold brown-red inside on spreading lobes; blooming in spring and summer; scale-like reproductive bodies often replace flowers toward the arching apex of the inflorescence, or in autumn after flowering. Grows best in large but shallow pan with adequate moisture, warm as well as to the cooler side. Keep dry in a light place during winter to rest. Propagated by seed, leaf cuttings, the scaly rhizomes, or bulbils from the flower stalks. **C**: 60   p. 450

TOLMIEA (*Saxifragaceae*)
   menziesii (Alaska, B.C. down to California Coast), "Piggy-back plant"; pubescent perennial herb with soft, fresh-green, lobed and toothed leaves to 3½ in. across, and covered with scattered white bristles, carried in a basal rosette, grown in pots as a curiosity, as it produces young plantlets from adventitious buds out of the base of the mature leaves, which can be cut off and rooted even in a glass of water. Small greenish nodding flowers, lined with maroon, in long erect, slender raceme. Interesting on the window sill; also winter-hardy in sheltered nooks outdoors. **C**: 16   p. 275, 529

TORENIA (*Scrophulariaceae*)
   fournieri (Vietnam), "Wish-bone plant"; dainty small, herbaceous perennial usually grown as an annual 10 to 12 in. high, rapid-growing with 4-angled branches bearing ovate, fresh-green, serrate leaves and scattered, attractive 2-lipped pale violet flowers, looking like miniature gloxinias, with the 3 lobes of the lower lip deep velvety violet, a yellow blotch in the middle of the lower lobe. Blooming almost continuously, best in gardens or window boxes from June until fall, but very pretty as a small pot plant flowering in mid-winter when sown in early autumn. Propagated also from cuttings. **C**: 22   p. 415

TOUMEYA (*Cactaceae*)
   papyracantha (New Mexico), "Paper-spine pee-wee"; small ovoid or short cylindric cactus 2 in. high, occasionally clustering, with 8–13 spiralled ribs cut into prominent knobs; spines thin, flat and papery, glossy white; spreading radials, and 3–4 curved central spines; flowers with silky-white petals. **C**: 13   p. 158

TRACHELOSPERMUM (*Apocynaceae*)
   jasminoides (Rhynchospermum) (Himalayas), the "Star jasmine" or "Confederate jasmine" of the South; small woody evergreen slowly climbing and twining with 2–3 in. leathery leaves; small white, star-like ¾ in. flowers with wavy lobes. A pretty, free-blooming jasmine for pots and intensely fragrant; can be kept bushy by trimming. **C**: 65   p. 279

TRACHYCARPUS (*Palmae*)
   fortunei, often grown as Chamaerops excelsa in horticulture (China, Japan), the "Windmill palm"; a somewhat hardy fan palm planted as far north as Scotland; Oregon; and North Carolina; withstanding 10° F. The solitary, shaggy trunk 10 to 40 ft. high, is covered with a mat of long, dark brown fibers; the tough palmate dark green leaves 3 ft. across, are divided into stiffish, folded segments nearly to base and hanging at the ends, shiny green or glaucous beneath; the petiole toothed; small fruit lustrous blue. Better known as a "hardy" palm outdoors than as indoor plant. **C**: 28, better 16   p. 223, 229

TRADESCANTIA (*Commelinaceae*); the "Spider-worts" belong to the most liked small house plants called "Inch plants" with their creeping or hanging branches, because they "inch" along very quickly. Easily grown on every window shelf, or in baskets, they offer a wide variety of attractive foliage colors. When the branches become too long and unsightly, they are pinched off and the tips are stuck unrooted into new pots in sandy soil where they root quickly, producing shapely young plants. Several other genera are similar in appearance and habit to Tradescantia, such as Zebrina, Setcreasea, Callisia, Commelina, Cyanotis, Gibasis, Dichorisandra.
   albiflora 'Albo-vittata' (Central America), "Giant white inch plant"; vigorous branching plant with fairly large and fleshy lance-shaped leaves 3 to 4 in. long, delicate bluish-green, with white bands and margins; 3-petaled white flowers. **C**: 15   p. 261
   albiflora 'Laekenensis', "Rainbow inch plant"; somewhat watery creeping stems with small delicate ovate, pale green leaves striped and banded lengthwise with white, and tinted purplish; 3-petaled white flowers. **C**: 3   p. 261

   blossfeldiana (Brazil), "Flowering inch plant"; robust branching, hairy plant with elliptic, waxy, almost leathery olive green leaves 4 in. long, the underside purplish and densely silver-hairy, and purple stem; free blooming with 3-petaled flowers, white tipped pale purple. As much a flowering as well as a foliage plant for the light window. **C**: 1   p. 261
   blossfeldiana 'Variegata'; an attractive sport with the fleshy leaves striped cream to yellow and pale green, midrib purple; underneath deep purplish and white-hairy. **C**: 3   p. 261
   dracaenoides: see Callisia fragrans
   fluminensis (Trop. So. America), the "Rio tradescantia" or "Wandering Jew"; a sturdy little trailer with somewhat wiry branches purplish-brown; small ovate, clasping leaves 1½ to 2 in. long, at base with some hairs; shining, almost bluish green, purplish beneath; ⅜ in. white flowers with 3 equal petals. **C**: 13   p. 261
   fluminensis 'Variegata' (Argentina, Brazil), "Variegated wandering Jew" or "Speedy Henry"; fast-growing, lively little creeper rooting at nodes, generally smaller and weaker than albiflora, with shining ovate leaves 1½ in. long, fresh green and striped and banded yellow and creamy white; flowers with 3 white petals. One of the most loved, pretty little trailers for the window, very tolerant whether light or shady, cool or warm. Pinch back to keep in shape, and the growing shoots may be stuck, several together, in light soil right in a pot to root. **C**: 15   p. 261
   multiflora: see Gibasis geniculata
   navicularis (Peru), "Chain plant" or "Boatleaf tradescantia"; half-creeping succulent with closely 2-ranked, short boat-shaped pointed, clasping, very fleshy leaves coppery green, mottled purple beneath; stalked clusters of rosy-purple flowers. **C**: 13   p. 260
   reginae: see Dichorisandra
   sillamontana (N.E. Mexico), introduced by the trade as Tradescantia "White velvet" and "White gossamer"; fleshy trailer growing erect at first, clasping ovate 1½–2½ in. leaves in ranks, deep green with parallel veins but entirely covered with fluffy white wool, underside and stems purplish; tripetaled flowers rich orchid; erroneously listed as Cyanotis veldthoutiana. In strong light the leaves turn reddish, contrasting with the snowy white hair. **C**: 1   p. 261
   velutina (Guatemala), "Velvet tradescantia". Softly white-hairy herb with branching stems up to 18 in. long; leaves oblong-lanceolate, 5 in. long, dense softly hairy above, short silky beneath; terminal and axillary umbels of flowers with downy sepals and purplish-rose petals. **C**: 15   p. 260

TRADESCANTIA: see also Callisia, Dichorisandra, Gibasis, Siderasis

TREVESIA (*Araliaceae*)
   palmata sanderi (Vietnam), "Vietnam snowflake plant"; attractive tree with large rich green leaves to 2 ft. across, digitately compound and lobed similar to micholitzii, the segments and basal sinus very much corrugated and quilted. Very unusual for its fantastic foliage; slow-growing. May be used for decoration for limited periods, unless in solarium or winter-garden. Soil must have good drainage. From cuttings. **C**: 54   p. 289

TRICHANTHA minor: see Columnea teuscheri

TRICHOCAULON (*Asclepiadaceae*)
   cactiforme (S.W. Africa, So. Africa: Cape Prov.), quaint club-shaped cylindrical, cactus-like succulent to 4 in. high and 2 in. thick, the little gray-green body grooved into irregular flat tubercles, more regular and transversely toward base; not freely branching; the globose head covers itself with tiny ½ in. yellow starry flowers spotted red-brown. Rest from January to June and keep dry. Best from seed; cuttings are difficult. **C**: 13   p. 352

TRICHOCENTRUM (*Orchidaceae*)
   orthoplectron (Ecuador); curious and beautiful little epiphyte with small, one-leaved pseudobulbs, the fleshy, partly folded, red-spotted leaves to 3½ in. long; flowers usually single, 2 in. across, and strongly fragrant; sepals and petals yellow with large maroon spots, lip white with a narrow purple spot on each side; blooming in autumn. These small plants may be grown in small, well-drained

pots, in baskets in osmunda and sphagnum; or on treefern sections, with constant moisture and high humidity with circulating air. Do not disturb more than necessary. **C**: 22
p. 480

TRICHOCEREUS (*Cactaceae*)
pachanoi (Ecuador), "Night-blooming San Pedro"; slender column forming straight branches, to 18 ft., light to dark green, glaucous while young, 6—8 rounded ribs, wanting spines; large white nocturnal flowers very fragrant. **C**: 13
p. 152, 165*
pasacana (Bolivia, Argentina), "Torch cactus"; giant, stout tree to 30 ft. high, sparingly branched; columns dull green, closely ribbed, freely yellow-spined; nocturnal flowers white. Edible greenish fruits called "Pasacana". Barrel-shaped in seedling stage. **C**: 1
p. 154
spachianus (W. Argentina), the beautiful "Torch cactus"; short slender columns to 4 ft. high, close-ribbed, clustering, short brown spines; flowers white. **C**: 1
p. 153

TRICHODIADEMA (*Aizoaceae*)
densum (So. Africa: Karroo), "Miniature desert rose"; shrub-like small succulent with fleshy roots and short stems forming tufts, the crowded green, thick leaves ¾ in. long covered with glands and glistening in the sun, the apex with radiating white hairs; free-blooming with daisy-like flowers carmine-red 2 in. across. Best from seed. **C**: 13
p. 355
stellatum (So. Africa: Cape Prov.); turf-forming shrubby succulent to 4 in. high, with long fleshy roots, tiny leaves ⅓ in. long and nearly cylindric, gray-green and glistening with glands on the surface; stiff white bristles at apex; 1 in. daisy-like flowers violet-red; free-blooming. From cuttings or seed. With the curious roots elevated, a bonsai-like effect may be created. **C**: 13
p. 353

TRICHOPILIA (*Orchidaceae*)
suavis (Costa Rica, Panama, Colombia); beautiful small epiphyte with thin pseudobulb and solitary broad leaves; from the base showy clusters of relatively large flowers to 4 in. across, delicately hawthorn-scented, creamy-white, with large frilled lip, yellow in throat and spotted purplish rose, usually blooming between December and May, occasionally in October, lasting weeks in perfection. Responding willingly to home culture; best grown in pots with osmunda or treefern fiber, with broken crockery for drainage. Trichopilia seem to flower more profusely if divided into small clumps. **C**: 22
p. 481

TRICHOSPORUM: see Aeschynanthus

TRIFOLIUM (*Leguminosae*)
dubium (minus) (native in Ireland and elsewhere in Europe; naturalized in No. America), the "Yellow clover", and widely grown as "Irish shamrock"; an annual trefoil with branching, creeping stems 6—18 in. long and hued brown; the 3 small leaflets matte satiny green, obovate and obcordate ⅜ in. long, the terminal one attached by individual stalklets to the petiole; the petioles and stipules being hairy, unlike in repens which are not hairy and has leaflets rounded at summit. T. dubium has canary-yellow or greenish-yellow small flowers ½ in. long, in loose heads, while repens has corolla white or tinged with pink. In Ireland this is believed by most the "genuine Irish shamrock". From seed. **C**: 26
p. 323
repens minus (Ireland and elsewhere in Europe), "Irish shamrock" or "Little white clover"; a dwarf form of the perennial white clover grown in little pots and bought for St. Patrick's Day by the Irish of North America; starting as a miniature rosette with 3 obovate, fresh-green ¼ in. leaflets on each petiole, it later creeps perennially and has small round heads of white flowers. The Legume family to which Trifolium belongs is known for its ability to accumulate and fix available Nitrogen by the presence of minute soil-bacteria in nodules on their roots. Winter-hardy; propagated from runners or seed. This species is considered by many as the "genuine Irish shamrock". Leaves with 4 leaflets occasionally found are considered "lucky". **C**: 26
p. 323

TRIMEZA (*Iridaceae*)
caribaea (Cipura martinicensis) (West Indies), "Long House iris"; in horticulture as Neomarica longifolia; tropical iris-like perennial over 2 ft. high, with sword-like, flattened bluish green leaves 1½ in. wide from bulb-like base; the inflorescence on long stiff-erect, wiry stalk bearing several

small, short-lived flowers 2 to 2¼ in. across, resembling Neomarica or Moraea, the 3 flags lemon-yellow, the short inner segments recurved and spoon-like spotted brown in center, and opening in succession. By division. **C**: 16   p. 393

TRIPOGANDRA: see Hadrodemas

TRITHRINAX (*Palmae*)
acanthocoma (So. Brazil). "Webbed trithrinax"; fan-palm with solitary trunk to 12 ft. high, and 3—4 in. thick, distinctive because of its intricate fibrous web-like covering formed by leaf bases and long 6 in. spines; palmate leaves to 3 ft. across, deeply cut into rigid segments which are split at apex, dark grayish-green above, lighter green beneath. **C**: 54
p. 228

TRITOMA: see Kniphofia

TRITONIA (*Iridaceae*)
deusta, known in horticulture also as Montbretia (So. Africa: S.W. Cape), "Blazing stars"; perennial cormous summer-flowering plant 12 in. high, with 4 to 10 linear leaves; the flowers are funnel-shaped and 4½ in. across, all facing upward, on a thin wiry stem bending over like a Freesia; similar to crocata but differing in having a dark brown-red blotch at the base of each of the outer segments of the cinnabar-red corolla. These brightly colored flowers are long-lasting as cut flowers; corms may be grown in pots by planting in November and keeping them quite cool until growth begins; after blooming let them rest until January at 40—50° F. Propagate by offsets. **C**: 64
p. 394

TRITONIA: see also Crocosmia

TROPAEOLUM (*Tropaeolaceae*)
majus florepleno, "Double English nasturtium", a pretty variety with showy double or semi-double flowers of rich yellow, of T. majus, the Garden nasturtium or "Indian cress" (the flower buds are used for seasoning). A quick growing, somewhat succulent glabrous annual herb, climbing by means of coiling petioles, with waxy peltate leaves and long stemmed, fragrant 2—2½ in. irregular flowers usually orange, sometimes red. For the sunny window or window box. Propagated from cuttings; the single variety from seed. **C**: 75
p. 280
tricolor (Bolivia, Chile), "Tricolored Indian cress"; climbing perennial with small tubers, the leaves divided usually into 6 leaflets, and beautiful 1 in. flowers with fiery scarlet spur, purplish calyx lobes and yellow petals. Very pretty summer-flowering twiner grown on wireframes, for the sunny cool window, or winter garden. Rest in winter. Propagate from cuttings, or bulbs that form when branches are layered. **C**: 75
p. 280

TSUGA (*Pinaceae*)
canadensis (E. No. America: Nova Scotia to Alabama), "Hemlock-spruce" or "Canada hemlock"; a very prolific coniferous often branching evergreen tree with slender, horizontal branches, gracefully drooping in age; the flat needles lustrous dark green above, bluish beneath, about ⅝ in. long mostly arranged in opposite rows on branchlets; small brown pendulous cones. Prefers moisture, sun and wind protection but very tolerant to situations wet or dry, sunny or shade, and may be clipped into dense columns, or hedges. Best from seed. **C**: 76
p. 310

TULBAGHIA (*Liliaceae*)
fragrans (So. Africa: Transvaal), popularly known as the "Pink agapanthus"; herbaceous perennial with rhizomatous rootstock, glaucous basal leaves 1 in. wide and about 1 ft. long, the slender 1½ ft. stalks bearing clusters of numerous small star-shaped ½ in. lavender pink, very fragrant flowers, nearly everblooming. The foliage does not have the garlic odor of T. violacea. An exceptional plant for cool pot culture in the house because of its fragrant, long-lasting flowers and winter-blooming habit; during summer on the patio. By division. **C**: 14
p. 401

TULIPA (*Liliaceae*); the "Tulips" are horticulturally most important and have been cultivated in Europe since the 16th century, where they became a craze and worth their weight in gold. They are hardy spring-blooming bulbous plants of the Eastern hemisphere, having basal and stem leaves, and bell-shaped erect flowers, widely planted in gardens. The garden tulips, descendants of T. gesneriana have been,bred in thousands of named varieties. Tulips are divided into two main groups, the species tulips or "Botanicals" in 4 different divisions, and the "Garden tulips" in 11 divisions of different characters (1969 Revision), more or less derived from T. gesneriana. Commercial growers largely prefer the "Triumph" tulips (Div. 4) for flowering in pots for Easter, they are handsome, stocky clones with large substantial, long-lasting, noble flowers in many colors, a group, with good weather resistance, resulting from crosses of the beautiful but taller "Darwin" and the short "Early" tulips; normally blooming late April. "Darwin tulips" (Div. 6) are also old favorites for planting into pans, but usually on the tall side, 20–24 in. high, with medium-large flowers for Easter and May. For early forcing the "Single early" (Div. 1) and "Double Early" (Div. 2) may be flowered from January on. Bulbs are potted several together in late fall, stored in cool, dark cellar or buried outside, until wanted for forcing; this is done best at 55–65° F. For good results, obtain aged blooming size bulbs in autumn, though tulips can be propagated by offsets which have to be 3 to 4 years old to be of flowering size.

'Blizzard' (Div. 4: Triumph); heavy-bodied Triumph tulip 15–18 in. high, with stout stem and heavy foliage, the substantial large, full flowers creamy-white 3 in. long; a good white for Easter pots. **C**: 26   p. 399

fosteriana 'Princeps' (Div. 13: Fosteriana hybrid) (Central Asia: Samarkand, Turkestan), "Scarlet rockery tulip"; an early flowering (March-April), short and sturdy bulbous plant about 12 in. high with solitary flower 4 to 6 in. across when open, the flaring petals intense scarlet, in the base black outlined in yellow; with glossy sheen as if lacquered, and very beautiful. **C**: 76   p. 418

'Kees Nelis' (Div. 4: Triumph); mid-season Triumph tulip about 14 to 18 in. tall, of somewhat flexuous habit but responding fast to forcing; flowers very striking in dark flame red, edged with orange-yellow. **C**: 26   p. 399

'Livingstone' (Div. 11: Double Late), a late "Peony-flowered tulip" normally blooming late May; double flowers bright crimson-red with yellow in base; becoming enormous and spreading like huge peonies; long-lasting. A gorgeous, strong pot plant and suitable for forcing. **C**: 26   p. 399

'Makassar' (Div. 4: Triumph); excellent late-blooming Triumph tulip for Easter pots, about 20 in. tall, with stiff stem and foliage, the cupped flowers 3 in. high, clear yellow, often producing more than one bloom per stalk. **C**: 26   p. 399

'Red Giant' (Div. 4: Triumph); a fine, stocky, late blooming 'Triumph' tulip about 12–15 in. high, for Easter; large substantial, long-lasting, deep scarlet flowers, often 2 or 3 on each stiff robust stem. Unfortunately bulbs are subject to fungus disease (Fusarium) under unfavorable conditions, but where the strain is clean and healthy, one of best for growing in pots during spring season. **C**: 26   p. 399

'Robinea' (Div. 4: Triumph), a superior "Triumph tulip" for Easter flowering; a healthy companion to 'Red Giant', even stockier, 12 to 14 in. above pot, with firm-fleshy, large flowers deep crimson, often producing more than one bloom per stalk; resistant to "rotting off" or "fire". **C**: 26   p. 375

TUPIDANTHUS (Araliaceae)
calyptratus (Assam: Khasia Hills; Burma), "Mallet flower"; small evergreen tree which later becomes a tall climber to 20 ft. high; leaves palmately divided into stalked, obovate segments to 7 in. long, glossy green and fleshy; reddish petioles and stem. A trim patio plant or for the solarium, resembling Brassaia but branches from base and grows into broad, denser shrub, and more tough-leathery. **C**: 16   p. 290

TURNERA (Turneraceae)
ulmifolia angustifolia (West Indies, Mexico to Argentina), "West Indian holly" or "Sage rose"; herbaceous shrub 2 to 4 ft. high, with scandent stems and alternate narrow elliptic leaves 3 to 4 in. long and nettle-like, deep glossy green, white-hairy beneath; axillary golden-yellow 2 in. flowers wtih 5 petals, blooming from March to September. Rather weedy but pretty in flower, and best grown light and airy, repeatedly pinched, and not too warm. From cuttings. **C**: 14   p. 498

ULMUS (Ulmaceae)
parvifolia (China, Japan), "Evergreen elm" or "Chinese elm"; spreading open-headed small tree to 40 ft. high, ever-

green in mild climate, but deciduous in colder regions; the branches gracefully arching and weeping, with elliptic, firm leaves 1 to 3 in. long, shining green and smooth above, hairy beneath at first, on thinly pubescent branchlets. Very pretty tree for patio shading, and very charming and satisfactory for sunny areas provided sufficiently watered. **C**: 26   p. 294

URBINIA agavoides: see Echeveria agavoides.

URBINIA purpusii: see Echeveria purpusorum

URGINEA (Liliaceae)
maritima (Canary Islands to Syria; colonized in Brittany and Normandy), the "Sea-onion" or the apothecary's "Squills"; an old house plant, forming a very large ovoid red-brown-skinned bulb 4 to 6 in. in diameter partially above ground; in the spring with 10 to 20 fleshy, glaucous green, strap-shaped basal leaves 12 to 18 in. long and 2 to 4 in. wide; as they wilt the bulb is kept nearly dry and cool, the old leaf-bases remaining for a time. When after the leafless time in summer, the bud of the inflorescence shows from the center, the dry soil is again moderately watered, and a slender reddish stalk to 3 ft. or more long will form in late summer to bear a pyramidal raceme of starry, whitish flowers $\frac{1}{2}$ in. wide, each segment with an indistinct green median stripe, the filament is thread-like, anthers green. The bulb may weigh 15 pounds, containing 22% sugar, also emetic and cathartic properties and is made into syrup of squills as cough medicine. Whether in pots or just laying bare on the ground, or on cupboards as in Italy, the bulbs produce their long spikes of flowers lasting to perfection for weeks. Propagated by offsets, seeds, or bulb scales. **C**: 13   p. 400

UTRICULARIA (Lentibulariaceae)
alpina (Mountains of the West Indies and So. America), the "Mountain bladderwort"; a most interesting insectivorous plant without roots living epiphytic on trees; with stalked, thin-leathery elliptic leaves 4 to 6 in. long; the plant sending out glassy tuberous branches 2 in. long which enter into the bark of the host tree and act as water-storage organs; these bear at their tips tiny $\frac{1}{2}$ in. bladder-like organs which can only be opened from the outside and into which insects and other small creatures are sucked, then close and the plant absorbs the results of their decay. Pretty, orchid-like $1\frac{1}{2}$ in. white flowers blooming in summer. Best grown in suspended baskets, in fern fibers, peat and live sphagnum moss; using soft rain water. Propagated by division. **C**: 72   p. 522

VACCINIUM (Ericaceae)
ovatum (Brit. Columbia to California), florists' "Huckleberry"; evergreen branching shrub 6 to 12 ft. high, with stiff wiry branches, attractively set with glossy green leathery, ovate elliptic foliage all facing flat to one side, the small leaves $1\frac{1}{4}$ to $1\frac{1}{2}$ in. long, with prominent midrib, and margins lightly dentate; tiny white or pink flowers, followed by black berries, good in jellies or syrup. A favorite long-lasting ornamental florists' "greens" used with cut flower arrangements, especially during winter. From seed or cuttings. **C**: 64   p. 300

VALLOTA (Amaryllidaceae)
speciosa (purpurea) (So. Africa: Cape Prov.), the "Scarborough lily"; a charming evergreen plant with large brown bulb; strap-shaped bright green leaves 18–24 in. long; the fleshy, hollow 2–3 ft. stalk carrying a cluster of funnel-shaped, bright scarlet, long-lasting flowers 3 in. across, blooming from June on. Strong undisturbed bulbs produce several flower stalks in succession. An old, good house plant, but which must be kept moderately moist even during its cool rest period in winter. From bulblets or seed. **C**: 14   p. 390

VANDA (Orchidaceae); (Asia, Australia); "Moon Orchids";
erect, leafy, frequently clambering or climbing stems and fleshy roots, the evergreen foliage in two ranks, and beautiful fleshy or waxy, long-lasting and usually fragrant flowers from lateral stalks. There are two important categories of Vandas—the terete (cylindric)-leaved and the strap (flat)-leaved species. Both types are easily grown and flowered provided they receive plenty of water with humidity and circulating air continuously. They grow in chunks of tree-fern fiber, or osmunda, but should not be repotted more than necessary. Vandas are active throughout the year and are not given any rest period, although some species such as coerulea like somewhat cooler conditions for best flower

production. Most Vandas freely produce aerial roots from the stem and these should be allowed to hang free since they assist in absorbing moisture from the atmosphere. Exposure to very bright light, not less than 4000 fc. for at least 6 hours, short of burning, is necessary to the production of flowers, and they may bloom 2 to 3 times a year. When Vandas get too tall they may be cut back. Propagated by cutting the stem, or taking off young basal growths.

coerulea (Himalayas at 4000 ft., Assam, Burma, Thailand), the famous beautiful "Blue orchid"; with stems 1 to 3 ft. high, two-ranked with 8 in., strap-type, leathery channeled leaves; the axillary elongate floral racemes with exquisite, large round, membranous flowers 3 to 4 in. across, sepals and petals in shades of blue from pale to cobalt, with a network of deep purple, the small lip dark purple-blue; blooming from late summer to February and lasting 6 weeks in perfection. Coming from mountainous regions, this species is grown cooler than other Vandas; no collection should be without this singular beauty. **C**: 72   p. 482

x'Miss Agnes Joaquim' (hookeriana x teres) (Singapore 1893), the Hawaiian "Corsage orchid"; prolific, epiphytic hybrid, with terete (cylindric) leaves, pencil-like stems to 7 ft. long, climbing trees or fern trunks in hot regions by aerial roots; large 3 in. blooms from the axils of the leaves, their sepals white tinged with rose, the larger petals mauve purple, the broad lip purple with yellow throat spotted red. The famous orchid grown in Hawaii for leis and corsages; this hybrid flowering in succession throughout the year, the exquisite blooms lasting for 6 weeks. **C**: 60
p. 482

tricolor (Java, Bali), "Tricolor strap vanda", a "Strap vanda"; spectacular, free-growing epiphyte with stems to 3 ft. long, dense with two ranks of recurving strap-shaped leaves; the inflorescence in lateral racemes of very fragrant waxy, long-lasting flowers 2–3 in. across. sepals and petals lemon-yellow spotted with reddish brown, lip white with mid-lobe rosy-purple; blooming at different times of the year, mostly in winter to spring. **C**: 10   p. 455*, 482

VANILLA (Orchidaceae); unusual leafy, or leafless tropical vining orchids of vigorous growth, with succulent zigzag stems and alternate fleshy leaves in ranks, fast climbing and clinging to bark or treeferns by means of aerial roots from the leaf nodes; the axillary flowers, produced on older plants are cattleya-like and showy but of short duration, succeeded, however, by others over a long period. From the long seed pods the vanilla used for flavoring is extracted. The flowers must be hand pollinated to produce the fruit, but before the pollen can be placed on the pistil, a covering shield protecting the stigma must be lifted. Vanilla are grown more as a curiosity; moderately warm and moist, in compost of leaf-mold, loam and osmunda or treefern fiber, and probably firbark chips, on trellis or other support. Propagated from stem sections in sphagnum or on osmunda.

fragrans, better known as V. planifolia (or aromatica) (Eastern Mexico), the common "Vanilla" of commerce; tall climbing orchid said to attain a length of 300 ft., the light green, cylindrical stem bears 2 ranks of succulent, green fleshy, elliptic leaves 6 to 8 in. long as well as aerial roots, the 2 in. flowers in axillary clusters, sepals and petals greenish-yellow, and wavy-edged lip almost white, deep yellow in throat; blooming almost throughout the year. This is the commonly cultivated vanilla, its cured seedpod yielding the economically important extract used for flavoring. **C**: 10
p. 482

fragrans 'Marginata', "Variegated vanilla"; a striking epiphytic climber, very attractive for ornamental purposes, especially when grown on bark, the nicely draped fleshy leaves of milky-green, arranged alternately along the clinging stem, bordered on both margins by a broad band of creamy-white. An epiphyte difficult to get going, and best grown undisturbed on treefern slab, humid-warm but never wet. **C**: 60   p. 192, 274, 482

pompona (lutescens or grandiflora) (Mexico to Panama, Peru and Brazil); a large-leaved climber with stout cylindrical stem to many ft. long; broad ovate succulent, thick-fleshy, dark green leaves 6 to 12 in. long distinguished by marked veining on the surface; large 6 in. flowers greenish yellow with bright yellow lip, more willingly blooming than fragrans; mostly in summer, but in well-established plants almost everblooming. **C**: 60   p. 482

VEITCHIA (Palmae)
merrillii (formerly Adonidia) (Philippines), the "Christmas palm" or "Manila palm"; a medium-sized, rather formal, well-groomed erect palm only 20 ft. high, with slender, prominently ringed single trunk; the 6 ft. fronds above a glossy green crown-shaft in handsome rigidly arching crown; broad leaflets many and closely placed, feathered almost to base of petiole, the leathery pinnae bright green and sword-shaped, their tips oblique and jagged-toothed; enhanced by lustrous red fruits in pendulous clusters below the crown almost all year, but especially so during wintertime. Attractive as a compact tub plant which tolerates abuse and deep shade down to 25 fc. **C**: 4   p. 220, 221

VELTHEIMIA (Liliaceae); beautiful and unique tender South African bulbous plants winter-blooming in Northern latitudes from December to March; the fleshy-wavy leaves to 3 in. wide form a rosette at the base of the stout floral stalk which grows 1 to 2 ft. high and carries a dense conical truss aloe-like, of pendant tubular flowers in bluish shades of pink and very long-lasting. They are easy house plants but should not be kept too warm; when the leaves turn yellow, watering should be stopped and the bulbs kept dry until September. Propagated by offsets; also by leaves pulled off close to bulbs, in spring.

glauca 'Rosalba' (So. Africa), "Pinkspot veltheimia"; charming bulbous herb with about 7 lanceolate-oblong glaucous leaves 1½ in. wide having depressed nerves, in a basal rosette, from which rises the red-spotted stalk bearing a conical cluster of pendant tubular whitish flowers dotted with rose; blooming from November into late winter. **C**: 15
p. 396

viridifolia (capensis) (So. Africa: Coastal Cape Prov.), "Forest-lily"; handsome bulbous rosette of 3 in. broad lance-shaped arching leaves intensely green as though varnished, with undulate margins; numerous long tubular, nodding 1½ in. flowers yellowish-green shading to dusty-red, dotted and tipped green, carried in a conical truss on long red-spotted fleshy stalk; late winter blooming to March or April in Northern latitudes; in South Africa July to September; the inflorescence lasts for a month. **C**: 16 p. 396

VERBENA (Verbenaceae)
x hortensis, the "Garden verbena"; hybrids of V. teucrioides and others from Brazil to Chile; herbaceous perennial bedding plants with procumbent spreading stems more or less rooting near the base; hairy, toothed leaves, soft to the touch; broad, showy 2 to 3 in. clusters of salver-form flowers often fragrant, in gay colors from pink and red to yellow, white, blue, salmon, purple and lilac; some with white eye; blooming profusely from spring to October. Watch out for White fly. From cuttings or seed. **C**: 63 p. 411

peruviana 'Chiquita', "Peppermint-stick verbena"; pretty, carpet-forming shrubby perennial originally at home in Peru, Uruguay and So. Brazil; the first creeping and rooting, then ascending branches with crenate, rough leaves 1 to 2 in. long; showy clusters of ½ in. salverform lavender flowers gaily striped white like a peppermint-stick. With its trailing branches ideal as a basket plant nearly everblooming, especially in summer. From cuttings. **C**: 63   p. 410

VERONICA: see Hebe

VIBURNUM (Caprifoliaceae)
odoratissimum (India, China, Japan), "Sweet viburnum"; large evergreen shrub to 10 ft., with willowy, reddish, rugose branches, and opposite, flexible leathery, elliptic leaves 4 to 6 in. long, glossy dark green with pale midrib and lightly crenate margins; small pure white fragrant flowers in conical panicles. A frugal, old-fashioned and durable decorator tub plant, as house plant or the summer patio; may be shaped by trimming. From cuttings. **C**: 16
p. 294

suspensum (Ryukyu Islands), "Sandankwa viburnum"; handsome evergreen shrub of broad-spreading habit to 6 ft. high, with rough branches and dark green, leathery oval, netted leaves 2 to 4 in. long and crenate toward apex, paler beneath; loose clusters of tiny white flowers tinted rose, and fragrant; globose red fruit. A dense, low bush for decoration in a tub, in cool location. **C**: 16   p. 294, 484

tinus (S.E. Europe, Mediterranean reg.), "Laurestinus"; evergreen thickly branched and luxuriantly leafy shrub, with ovate deep green, stiff-leathery 3 in. foliage with rough underside, on reddish petioles; dense 2 to 3 in. clusters of tiny pinkish-white, very fragrant blooming; blooming May to August. Known in Europe as "Laurustinus", it is one of the most popular durable decorator tub plants, and with its handsome foliage and compact shape ideal for cooler areas, the patio, terrace or roof garden. They may be forced into flower; and shaped like laurel. From cuttings. **C**: 16   p. 484

VICTORÍA (*Nymphaeaceae*)
cruziana, in horticulture sometimes as trickeri; (Brazil, Paraguay, No. Argentina), the "Santa Cruz water lily" or "Water platter"; spectacular perennial aquatic with a thick rhizome, thorny petioles and round, floating leaves 2 to 5 ft. across, not as large as the 6 ft. foliage of V. regia but with broader upturned margins to 8 in. high and green; night-blooming, fragrant flowers white turning deep pink on second day. Getting along on somewhat less heat, 75° F., this is the kind usually grown in sufficiently warm outdoor ponds, always imposing wherever seen, in the North mostly in larger conservatory pools in 75 to 85° F. water, it is not as sensitive to chills as V. regia which is at home in the steamy igarapes of Amazonas. Projecting air-filled ribs on the underside of the leaf give great buoyancy, sufficient to support great weight—up to 150 pounds. Propagated from hand-pollinized seed sown during winter in soil with shallow water at 80 to 95°F. **C**: 52   p. 517, 536

VINCA (*Apocynaceae*)
major variegata (So. Europe, No. Africa), "Band plant"; trailing evergreen basket plant, or for window boxes, with shrubby base; long, thin, wiry vines creeping or pendulous; bearing opposite, oval, green waxy-glossy leaves 2 in. long, beautifully edged in cream; large blue flowers. Trim back for wintering; propagation by division or cuttings. **C**: 14   p. 265
minor (C. Europe to Asia Minor), "Periwinkle", "Virginia flower" or "Running myrtle"; the German "Immergruen"; small shrubby, evergreen trailer with thin-wiry stems rooting at internodes; glossy leathery, dark green oblong leaves 2 in. long, and pale, azure blue flowers with white throat. Winter-hardy, and primarily a matforming ground cover for shady locations. **C**: 78   p. 265
rosea: see Catharanthus

VIOLA (*Violaceae*)
cornuta 'Bluette'; large-flowered type of the "Horned violet", much planted for spring blooming, and whose ancestry is in the Pyrenees of Spain; these little, short-lived herbaceous perennials, 6 in. high, have oval, wavy-toothed smooth leaves, and long-stemmed, medium-small, bright purplish-violet $1\frac{1}{2}$ in. flowers, the yellow throat extending back and forming a spur. These miniature pansies are hardy, and very floriferous over a long season into late summer, also come in other colors yellow, red, apricot and white. From late summer sowing. **C**: 78   p. 411
hederacea (Erpetion reniforme) (New South Wales, Victoria, Tasmania), "Australian violet", "Trailing violet", or "Ivy-leaved violet"; attractive small trailing species; the vertical rhizome putting out long, thread-like stolons with well separated tufts of leaves; these kidney-shaped or rounded, $\frac{3}{4}$ to $1\frac{1}{2}$ in. across, fresh green and herbaceous; small $\frac{5}{8}$ in. flowers with petal-tips white, center area violet except for white eye; scarcely spurred. For the cool window, in summer outdoors. Propagated from runners. **C**: 78 p. 280
tricolor hortensis (x wittrockiana), the "Garden pansy"; charming short-lived herbaceous perennial usually grown as a biennial or annual, 4–8 in. high; highly developed hybrids in many forms and sizes from the small-flowered V. tricolor or "Johnny-jump-up" from Europe and C. Asia which reseeds itself in our gardens; long branching, prostrate stems, soft, crenate foliage, and with nodding, flat flowers usually in 3 colors violet, blue or crimson, yellow or white, and with a blackish "face" as well as other colors including brown; as now developed with giant, striking blooms to 4 in. across. Mainly blooming during the cool season from winter to spring, or even later, pansies are favorite garden plants, normally planted in fall, to precede other summer annuals in flower. Fairly winter-hardy with protection. Best from seed. **C**: 78   p. 411

VITIS (*Vitaceae*); the "Grapevines" are tendril-climbing woody deciduous vines of the northern hemisphere with ornamental foliage but insignificant greenish flowers in

clusters, followed by typical juicy berries as in the "Persian wine grape", V. vinifera. True Vitis are usually of coarse habit, often winter-hardy, and grown as screens for walls, over pergolas, up on columns, or on trellis support. To keep under control vines are cut back hard in late winter rather than spring. to avoid bleeding; pinching once or twice in summer is also recommended. Propagated by hardwood cuttings in winter, or layering. Some "Vitis" are properly Ampelopsis or Cissus, etc.
vinifera, the "Wine grape"; originally believed from the Caucasus region; already known in ancient Egypt, and cultivated for centuries in Europe but not reliably hardy in north temperate climate. Woody deciduous vine moderately climbing by tendrils, with rather thin, coarsely toothed, 3 to 5-lobed leaves and with intermittent tendrils; small greenish unisexual flowers in long clusters; free-fruiting with large bunches of delicious, fleshy, glaucous berries, tender and sweet, black, red or green; for growing under glass such modified cultivars as 'Black Alicante.' 'Black Hamburg', or the white 'Muscat' have long been favorites. Glasshouse grapes are usually planted out but may be grown in big pots plunged in beds; good drainage, light, and ventilation, and evenly warm day temperature of 75° F. while growing, to avoid mildew, are conditions for success. Careful pruning is practiced; it is important to know that fruit is borne only on new season young laterals. When first planting, a leader vine is allowed to grow up a wire or rafter to a designated height, and two to six branches sidewise on wires if so desired. The following year the leader growth is stopped and young shoots encouraged to break from eyes along last year's woody vine, and these may fruit if strong enough. If grown indoors the plants are rested and kept cool and dry going into winter; before the eyes rise in February all the new laterals are cut back to two eyes, and from these the vines will grow that produce the grapes, usually after the 4th leaf. During summer this new growth may be pinched 4 leaves beyond the fruit. The hardy American vineyard grapes of the north such as 'Concord' are forms of the native Fox grape, Vitis labrusca. Propagated by hardwood cuttings from 1 year wood made when vines are cut back in February-March. **C**: 63   p. 507

VITIS: see also Ampelopsis, Cissus, Parthenocissus, Rhoicissus, Tetrastigma

VITTARIA (*Filices: Polypodiaceae*)
lineata (Pan tropic, in Florida, West Indies, to Peru, and east to Asia), the odd "Shoestring fern" or "Old-man's beard"; in Florida, where it grows on palmetto palms, from short rhizomes, locally known as the "Florida ribbon-fern"; rather hard-leathery, rush-like blackish green fronds like flattened ribbons, less than $\frac{1}{8}$ in. wide and 6 to 18 in. long and pendant; narrowed downward to a stout stem; the spores along edges turned under. Not very attractive but remarkable and unique in collections. **C**: 60   p. 251

VRIESEA (*Bromeliaceae*); the "Painted feathers" or "Lobster claws" are a spectacular genus of mostly epiphytic rosettes remarkable in having either the soft-leathery, smooth foliage banded or mottled in fantastic designs, strikingly beautiful; or featuring a spike carrying during spring and summer a two-ranked inflorescence of brightly colored boat-shaped bracts lasting from 3 to 6 months in beauty; the peeking flowers are short-lived and barely noticed. Vrieseas are excellent house plants and decorators, but need warmth and humidity; preferring some shade, they are ideal for brightening the north window. Propagated from offshoots; also parachute-like seed. (See p. 61.)
altodaserrae (Brazil: Sao Paulo, Paraná, Santa Catarina); robust, large rosette of stiff green, smooth leaves with faint dark green marbling; erect 3 ft. inflorescence with slender branches, the green bracts along the stalk tipped with salmon. **C**: 10   p. 148
barilletii (Ecuador); small rosette of soft green leaves tinted copper and with the smooth edge of all Vrieseas; inflorescence a flattened head of spreading bracts solid purple at the base to purple-spotted and a yellow apex; flowers yellow. **C**: 9   p. 146
carinata (S.E. Brazil), "Lobster claws"; dainty plant with pale green foliage; flattened spike with spreading bracts deep yellow and with crimson base, the yellow dotted green; flowers yellow. **C**: 9   p. 146

x erecta (poelmannii x rex), "Red feather"; shapely rosette of glossy light green leaves; the inflorescence a flattened spike with dense keel-shaped, deep lacquer-red bracts, the tips separated; the peeking flowers yellow. **C**: 9   p. 146

x'Favorite' (ensiformis hybrid); vigorous hybrid rosette of shiny rich-green leaves; inflorescence a tall, slender stem, usually branched, with maroon keel-shaped bracts darker than poelmannii, and arranged separated along stalk; flowers yellow. **C**: 9   p. 147

fenestralis (Brazil: Espirito Santo, Guanabara, Paraná), "Netted vriesea"; magnificent rosette of broad recurved foliage arranged spirally, the yellow-green leaves to 20 in. long, ornamented by numerous dark green lines and net-work of cross lines, purplish circles underside; simple inflorescence with night-blooming sulphur yellow, fragrant flowers scattered on pale spike. **C**: 62   p. 146, 193

glutinosa (stenostachya) (Trinidad); dainty rosette with arching thin-leathery linear leaves 12 in. long, smooth, fresh-green, to bluish green at base, with red-brown, more or less pronounced crossbands; inflorescence with slender, flattened branches, the bracts flaming red, flowers spreading. **C**: 10   p. 148

guttata (So. Brazil), "Dusted feather"; small compact rosette of glaucous bluish-green leaves liberally marked with maroon spots; inflorescence a pendant, lightly flattened spike of greenish beige bracts covered with silver pink bloom, flowers lemon-yellow. **C**: 9   p. 147

heliconioides (Guatemala, Costa Rica, Guyana, Colombia, Brazil, Bolivia), "Heliconia bromeliad"; striking flowering plant, with rosette of plain, glossy green leaves 8 in. long, suffused with red underneath; the erect flattened inflorescence heliconia-like, with lateral triangular boat-shaped floral bracts bright red above the middle, greenish-yellow at the apex; the flowers peeking out with creamy white petals. **C**: 9   p. 147

hieroglyphica (Brazil: Espirito Santo to Paraná), "King of bromeliads"; large epiphytic regal rosette with broad yellow-green leaves, to 20 in. long, fantastically cross-banded with hieroglyphic marks and lines running length-wise which are dark green above and purplish-brown beneath; inflorescence a tall branched spike with sulphur-yellow flowers. **C**: 9   p. 145

hieroglyphica zebrina (Brazil: Santa Teresa), "Zebra king"; a form still more showy with darker green leaves and the fantastic irregular crossbanding very pronounced and purplish black. **C**: 60   p. 192

imperialis (Brazil: Estado do Rio), "Giant vriesea"; gigantic terrestrial rosette which I found growing on the dry west slopes of the Organ Mountains; leathery green leaves in good light are deep wine-red, and even young plants produce seedling-like suckers at the base; the inflorescence a tall branched spike 6 ft. or more, the large bract leaves glossy maroon-red, and from which extend the arching secondary bracted spikes with yellow flowers. **C**: 9   p. 145

incurvata (S.E. Brazil), "Sidewinder vriesea"; light green rosette of soft leaves; the inflorescence on a leaning stem with the bract head recurving outward, fleshy bracts red and edged yellow; flowers yellow. **C**: 9   p. 146

x mariae (magnifica), "Painted feather"; carinata hybrid larger than carinata, its light green rather flattened foliage tinted pink and showy; the striking featherlike, long-lasting spike with distichous bracts salmon-rose at base, and yellow dotted brown toward apex; yellow flowers. Very pretty, and colorful for months. **C**: 9   p. 131*, 145

philippo-coburgii vagans (S.E. Brazil), "Vagabond plant"; small rosette which sends out its offshoots from traveling stems; stiff leaves light green, curious with black bases; inflorescence a flattened spike with yellow and red bracts, and yellow flowers. **C**: 9   p. 146

platynema variegata (So. Brazil); stiff rosette of broad recurved leaves, attractive for its foliage which is bluish green with yellow-green lines and reticulated by darker green cross-netting; the tips of leaves are wine-red beneath; erect inflorescence with purplish bracts and greenish flowers. **C**: 60   p. 148

x poelmannii (gloriosa x vangeertii); vigorous, shapely rosette of light green leaves, with flattened spike, bracts crimson-red with greenish-yellow apex, and yellow flowers. **C**: 9   p. 147

x'Polonia'; small shapely rosette of little leaves; the showy inflorescence an erect stem branching into several miniature flattened heads of glowing rosy-red bracts and deep yellow flowers. **C**: 9   p. 147

psittacina (Brazil: Bahia to Guanabara), "Dwarf painted feather"; small epiphytic rosette of yellowish-green, thin-leathery, recurving leaves 8 in. long; the simple, pinnate inflorescence loosely set feather-like, with 2 ranks of fleshy, inflated red bracts edged with yellow; the flowers yellow and spotted with green. **C**: 9   p. 146

x retroflexa (Brazilian hybrid of psittacina x simplex); small rosette with thin-leathery, waxy, pale green leaves; inflorescence spike pendant then reflexing upward at an angle, with open or loose, yellow floral bracts on sides depressed; flowers yellow. **C**: 9   p. 147

rodigasiana (So. Brazil), "Wax-shells"; dwarf rosette of soft, dull green leaves with base tinged purple; inflorescence on leaning branched stem with waxy bracts yellow tinged red; flowers lemon-yellow. **C**: 9   p. 146

saundersii (botafogensis; Encholirion) (Brazil: Estado do Rio), "Silver vriesea"; small rosette with broad, stiffly recurved leaves, silvery gray and well spotted purplish-red, especially beneath; inflorescence on arching branched stem with waxy lemon-yellow bracts, and yellow flowers. **C**: 7
p. 146

scalaris (Brazil: Espirito Santo to Santa Catarina); small rosette of thin-leathery, obovate light green leaves 6 in. long; the charming inflorescence pendulous with scattered flowers attached as if to a snaky wire; floral bracts red-brown, flowers yellow. **C**: 9   p. 147

splendens longibracteata (Venezuela, Trinidad, Guyana); small rosette with broad, recurving leaves all olive green without crossbands, spike yellow-green at base merging into orange-red. **C**: 9   p. 148

splendens 'Major' (Guyana); spectacular rosette of slender bluish-green leaves marked with broad, deep purplish crossbands, underneath grayish with the purple bands very bold; flower spike long and sword-shaped with flattened fiery-red bracts suffused with copper; the flowers yellow. This clone 'Major' is more robust than the species, and similar to the cultivar "Flaming Sword" which is a cross with V. splendens longibracteata. **C**: 9   p. 43*, 145

WASHINGTONIA (*Palmae*)

filifera (So. California desert, S.W. Arizona, Baja California), "Desert fan palm" or "Petticoat palm"; in habitat in Agua Caliente canyon, Palm Springs, California; bold solitary, erect fan palm with massive grayish trunk to 3 ft. thick and 50 to 75 ft. tall, usually clothed by the densely shingled older leaves and looking like a skirt, unless burnt off; the top with a crown of palmate, gray-green, leathery fronds 6 ft. or more across, divided more than half-way to base, and with many long threads attached to seg-ments, on thorny, long, green petioles. For dry-hot climate, best planted out. **C**: 2   p. 227

robusta (N.W. Mexico: Sonora, Baja California) the "Mexican fan palm"; more slender than filifera and faster growing, to nearly 100 ft. tall, the upper part dense with brown dead, and living glossy bright green foliage, the plaited fan-leaves are stiff and lightly cut, to 4 ft. long and with some fibrous threads in juvenile stage; fruit black-brown. Rarely used as container plant, but much planted in warm-arid climate along avenues and homes for tropical effect. **C**: 14   p. 227

WATSONIA (*Iridaceae*)

humilis maculata (So. Africa: Cape Prov.), "Southern bugle-lily"; handsome small spring or summer-blooming cormous plant closely allied to Gladiolus, with smaller, regular flowers and more tubular; the sword-shaped leaves basal and along the 12–18 in. stalk; flowers 2 in. long, deep rose with a purple blotch between the lobes at base. Plant bulbs into pots from September on, keep cold till growth is well advanced during winter, then place into greenhouse or window, and feed. After flowering reduce watering till foliage yellows then keep dry and dormant until repotted. By bulblets. **C**: 64.   p. 395

WEDELIA (*Compositae*)

trilobata (West Indies, N. So. America); branching herb with slender, flexible trailing stems; elliptic, fresh green, notched and lightly lobed, somewhat fleshy leaves, 2 to 4 in. long; and attractive marigold-like flowers with golden yellow florets. An attractive basket plant or ground cover. **C**: 16   p. 264

WEIGELA (*Caprifoliaceae*)
   praecox (Korea, Manchuria), "Early rose weigela"; deciduous shrub to 6 ft. with leafy branches, elliptic leaves soft pubescent; 3 to 5 funnel-shaped flowers 1 in. or more long, on lateral shoots, rose with yellow throat, blooming outdoors in May. May be used for forcing; flowering on wood of the previous summer. Propagated by cuttings of half-ripe shoots in summer, by autumn cuttings, or suckers.
C: 76   p. 504

WELWITSCHIA (*Gnetaceae*)
   mirabilis (bainesii) (S.W. Africa, So. Angola); curious, ancient succulent, the "Tree tumbo" or "Mr. Big" of the foggy Namib desert; from a low base to 4 ft. thick, with big taproot, spread 2 thick-fleshy glaucous or waxy green, monstrous, strap-like leaves very slowly stretching to 20 ft. long and 8 in. wide, corrugated lengthwise, with parallel veining, and continuing to grow from the base, not from the apex, over the surface of the ground or curling; the ends usually become split and shredded into leathery thongs if exposed to desert storms. Inflor. cone-like, on female plants green, 2 in. long, on branched stalk, the male cones red-brown, only 1 in. long. Said to become 1000–2000 years old, these desert dwellers survive on the surface dew which daily settles along the south-western coast of Africa, with cold nights and hot days. **C**: 63   p. 530

WISTERIA (*Leguminosae*)
   floribunda (Glycine) (Japan), "Japanese wisteria"; deciduous woody high-climbing twiner, which may be dwarfed into bush form by pruning; with long, pinnate leaves; beautiful pendulous racemes to 1½ ft. long, of fragrant, peashaped flowers in shades of violet-blue. Winter-hardy. Older, well established plants bloom best, on support, in full sun. To develop dwarf, weeping forms, prune back hard in late winter, and again in summer pinch out most young straggling shoots and head back some half to induce bud spurs to develop. Propagate by layering, or root cuttings. **C**: 76   p. 276

WITTROCKIA (*Bromeliaceae*)
   superba (Nidularium splendens) (S.E. Brazil); large robust, well-formed rosette with leathery concave, yellow-green waxy leaves tinged bronze and occasionally blotched dark green, the pointed apex tipped red, margins with prominent reddish teeth; the green inflorescence a cup as in Nidularium but slightly raised, with petals blue but completely enclosed by the sepals. **C**: 9   p. 147

WOODWARDIA (*Filices: Polypodiaceae*)
   radicans (Canary Isl., So. Europe, China, No. India, Java), the giant "Chain fern"; a large fern with bipinnate fronds 3 to 6 ft. long, thin-leathery, fresh green, heavy and pendant; the pinnae alternate and not cut to base, with wavy margins; large buds or tubers form toward the apex of the leaf which root to form new plants. Decorative in outdoor ferneries; requiring much moisture. **C**: 66   p. 249

XANTHORRHOEA (*Liliaceae*)
   arborea (Australia: No. New South Wales, Queensland), the ancient and curious "Grass tree"; long-lived perennial plant with a strong, palm-like woody trunk 6–9 in. thick, rough with old leaf-bases, topped by a dense multitude of narrow whip-like leathery, dark green leaves, flat or tri-angular, 3 to 4 ft. long, and less than ¼ in. wide. Have noticed colonies of grass-trees 6 to 10 feet. high on grassy slopes in eucalypt forests from south of Brisbane to the table lands above Cairns and it takes centuries for them to reach such dimensions.   p. 525

XANTHOSOMA (*Araceae*); the "Yautias" are tropical American herbs, large and gigantic, tree-like; or small, having beautiful foliage; with thick rootstocks, occasionally forming trunks, or tuberous rhizomes; succulent stalks carry the showy, rather fleshy leaves, usually arrow-shaped; the unisexual inflorescence with spadix having male flowers toward the apex, females along the basal part, enclosed by the rolled-in spathe. Requiring warmth and high humidity, Xanthosomas are best grown in the greenhouse, in fibrous compost of humus and loam. Propagated from suckers, or by cutting up the stem or rootstock.
   lindenii 'Magnificum' (Phyllotaenium), "Indian kale"; horticultural form of X. lindenii (Colombia) differing by

having its yellowish to deep-green leaves beautifully and broadly veined cream to white, and with pale line just inside of margin. This ornamental, evergreen herb, growing from rhizome, with its showy, arrow-shaped, thin-leathery leaves, is one of the most beautiful of warm greenhouse exotics.
C: 56   p. cover*, 66*, 97, 180
   violaceum (Puerto Rico to Jamaica), "Blue taro", "Violet-stemmed taro", "Yautia"; producing edible rhizome; fleshy leafstalks brownish-purple with blue wax-covering; sagittate leaves dark green with purple margin, veins beneath purple. The main root (tuber or corm) is too acid for food, but the lateral tubers growing from it are eaten after cooking. **C**: 54   p. 97

XYLOBIUM (*Orchidaceae*)
   powellii (Panama, Nicaragua, Costa Rica); handsome small epiphyte with ovoid pseudobulbs bearing plaited leaves to 2 ft. long; short-stalked clusters of small ½ in. faintly fragrant, charming flowers not opening widely with deep buff-yellow sepals and petals, and cream-colored lip; blooming during summertime. Grown moderately warm and moist during active growth; cooler when resting. Planting materials may be treefern fiber or osmunda, also firbark mixes. Propagated by division. **C**: 22   p. 482

XYLOPHYLLA: see Phyllanthus

YUCCA (*Liliaceae*); the "Palm lilies" or "False agaves" are bold succulent American members of the lily family, forming persistent rosettes of more or less stiff-leathery, sword-shaped leaves, many slowly developing into spectacular trees with woody trunks to 50 ft. high; their inflorescence is showy, with striking clusters of cup-shaped nodding, waxy white flowers fragrant at night. Yuccas are excellent decorator plants in containers for that stiff, desert-landscape effect; they prefer sun but don't mind poor light down to 20 fc.; they tolerate abuse and may be placed in warm or cool locations, and take air-conditioning well. Propagation of the stemless kinds by division or root pieces; the stem-forming from seed or the occasional branch shoots; Y. elephantipes also from suckers.
   aloifolia 'Marginata' (S.E. United States, W. Indies, Mexico), the "Variegated Spanish bayonet"; attractive, tree-forming, stiff rosette of thick-fleshy, sharp-pointed, dagger-like leaves 2½ in. wide and 2½ ft. long, glaucous green with creamy-yellow margins, on usually single trunks to 20 ft. high; cup-shaped flowers creamy-white tinged with purple, in 2 ft. panicles. The type species is plain glaucous green. **C**: 13   p. 196, 200, 362
   brevifolia, native to high deserts in Southern California, Nevada, Utah and Arizona, the extraordinary "Joshua tree"; hard to believe but this is a member of the lily family, growing into a succulent tree to 65 ft. high, with palm-like trunk to 4 ft. thick branching into tortuous arms dense with rosettes of short and rigid dagger-shaped leaves; in bloom from February to April with greenish white 2 in. cup-shaped flowers in dense foot-long clusters. Its grotesque silhouette is very effective in dry-sunny desert landscapes. **C**: 63 p. 527
   elephantipes (guatemalensis), in California nurseries and in the Azores as Y. gigantea; (S.E. Mexico, Guatemala), "Spineless yucca" or "Bulb stem palm-lily"; round-headed "palm-lily" with leafy trunks having a distinct swollen base, with age branching and reaching a height of 30 ft. or more, topped by rosettes of sword-shaped, soft-leathery leaves, normally 18 to 36 in., but becoming to 4 ft. long and 3 in. wide, glossy grass-green with rough margins and soft tip; cylindric 2 ft. inflorescence with ivory-white flowers directly from the crown. A striking silhouette plant with that stiff yet tropical look, taking much abuse, preferring sun but tolerating dark areas to 20 fc.; warm or cool, also air-conditioning. Propagated mostly from seed. Y. gigantea may validly be a larger and lax-leaved variety of elephantipes.
C: 13   p. 200, 362
   elephantipes 'Variegata', "Giant variegated palm-lily"; an attractive variegated cultivar of the spineless yucca; the wide, glossy-green soft-leathery 2½ to 3 ft. leaves prettily banded with creamy-white. In California this type with larger, more arching and hanging leaves is grown under the name Y. gigantea variegata, and according to Dr. Encke this lax-leaved form should be recognized as Y. elephantipes var. gigantea 'Variegata'. **C**: 13   p. 362
   gigantea: see elephantipes

gloriosa (U.S.: shores of Carolina to Florida), "Palm-lily"; to 8 ft. high, with short, thick trunk topped by dense rosette of sword-shaped, flat and leathery, glaucous gray-green, rough leaves 2 in. wide, with reddish margins and spiny point; white bell-like flowers striped purple outside, opening and fragrant at night. **C**: 13　　p. 200

gloriosa 'Variegata', "Variegated mound-lily", "Spanish dagger" or "Soft-tip yucca"; decorative variegated form of Y. gloriosa native along the shores from South Carolina to Florida, forming multiple trunks and reaching 8 ft. tall; white and purple bell-like 4 in. flowers on elongate inflorescence. A formal shapely rosette of stiffish-fleshy sword-shaped, spreading leaves 2 ft. long, deep pea-green, lightly glaucous and bordered milky-green and creamy-white, and tinged with red in sun, obscurely toothed and with a soft harmless spine tip. Blends well with lush tropical plants without being fussy. **C**: 13　　p. 362

ZAMIA (Cycadaceae)

floridana (So. Florida), "Coontie" or "Seminole bread"; a dwarf plant with underground tuber-like trunk, sending up many slender, pinnate, leathery dark green fronds to 2 ft. long and somewhat twisted, the segments with margins rolled under, and without midrib. Compact grower and a good house plant. **C**: 20　　p. 316

furfuracea (Mexico and Colombia), "Jamaica sago-tree"; stem more or less tuberous, sometimes branched, bearing a tangled profusion of pinnate leaves 3–4 ft. long, on prickly stalks, the thick-leathery leaflets oblanceolate 2–8 in. long, more or less toothed and overlapping, densely brown scurfy beneath, or on both sides when young; male cones 4 in. long, the female ones shorter. An excellent, "different" decorator plant, hard and durable as iron, and of relatively small size. **C**: 22　　p. 316

ZAMIOCULCAS (Araceae)

zamiifolia (Zanzibar, E. Africa); unusual, curious tropical herb with thick horizontal rhizome, sending up swollen stalks with pinnately arranged, small, dark, waxy leaves with yellow-green veins; short inflorescence near base with boat-shaped, green spathe. **C**: 60　　p. 97

ZANTEDESCHIA (Araceae); the "Calla lilies" are well-beloved tender winter-blooming African herbs with thick rhizomes, and attractive arrow- or lance-shaped soft-fleshy leaves, cultivated for their long-stalked, striking funnel-shaped inflorescence, consisting of a showy flower-like, coiled and flaring spathe like a trumpet, surrounding the orange-yellow unisexual spadix which carries, as typical in all aroids, the tiny true flowers; the males along most of the upper part of the column, the female organs only near the base. Dormant "Calla" tubers are planted in large enough pots in early fall and grown at 55 to 60° F. more; or when in full growth they need ample water and feeding. During summer it is best to give them a rest period of 2 to 3 months allowing the foliage to dry off. The white Callas are good house plants provided they receive full sunlight in winter. Propagation by division of rhizomes, baby tubers, suckers, or seeds.

aethiopica (Calla, Richardia), (So. Africa and north to Egypt); "White calla", "White arum-lily", "Pig lily", "Trumpet lily", "Common calla", "Calla lily"; robust marsh-loving herb with thick rhizome; forming a tuft of fleshy-stalked, glossy-green succulent leaves 2–3 ft. high; a stout basal stalk 3 to 5 ft. high bearing the large, funnel-shaped, rolled-flaring waxy-white spathe 8–10 in. long, surrounding a bright yellow spadix. Also used as cut flowers. For winter-spring flowering, rest in summer to August. **C**: 66　　p. 96, 403

aethiopica 'Compacta', "Dwarf white calla"; more compact growing form, in the trade as 'Godefreyana'; a rhizomatous plant with shining green leaves and fragrant flowers, usually 18–24 in. high as against aethiopica which is 2 to 3 ft. tall; more floriferous but the waxy-white, funnel-shaped spathes only 5 to 6 in. long; a better plant for pot culture, and if kept growing, watered and nourished adequately, blooming best from September through winter to July; then allow the plant a dormancy period of 2 to 3 months. **C**: 64　　p. 403

albo-maculata (So. Africa: Cape, Natal, Transvaal), "Spotted calla"; rather slender plant 1½–2 ft. high, with long arrow-shaped green leaves ornamented by white-translucent oblong spots; the spathe trumpet-shaped 4 to 5 in. long with the flaring limb pointed, creamy-white, crimson in the base. Somewhat more cold-tolerant than other species. **C**: 54　　p. 403

elliottiana (So. Africa: Transkei), the showy "Yellow calla" or "Golden calla"; growing from a flattened rhizome; the succulent, cordate, bright green leaves have translucent white spots; the obliquely flaring, tubular spathe to 6 in. long is rich yellow. To flower from Easter to early summer tubers are rested to January, then potted. **C**: 54　　p. 66*, 96, 374, 403, 535

rehmannii (So. Africa: Natal), the shapely "Pink calla"; a small tuberous species about 18 in. high having narrow tapering, rather firm leaves bright green with translucent linear white spots; very distinctive with lovely pale rosy-purple spathe, shaped like an obliquely flaring 3 to 4 in. tube, cream inside. Blooming in South Africa from November to January; in northern latitudes in May and June; grown in pots and forced to bloom for Easter. **C**: 54　　p. 403

ZEBRINA (Commelinaceae)

pendula (Mexico), "Silvery wandering Jew"; fleshy, trailing plant rooting at joints, the small ovate leaves fairly succulent, about 2 in. long, deep green to purple with two broad, glistening silver bands, vivid purple beneath; clustered flowers rosy-purple. Wonderful little trailer for the window sill, a red counterpart to a white Tradescantia, which it much resembles. Tolerant to warm or cool, light or shade; pinch to keep in shape, or one can take the tips and easily root several together in pot of sandy soil. **C**: 13　　p. 261

pendula 'Discolor', "Tricolor wandering Jew"; fleshy trailer, with large glossy coppery and nile-green foliage 2–3 in. long, overlaid and edged with metallic purple and splashed rusty red, two narrow silver bands down the long and thinnish leaf, purple beneath; purple flowers. **C**: 13　　p. 261

pendula 'Discolor multicolor'; a strong growing variegated form which tends to keep its variegation; large, long ovate 2½ in. leaves with green background are striped and banded pinkish cream, and splashed with rusty red and a bit of silver; purple beneath. **C**: 3　　p. 261

pendula 'Quadricolor', "Gay wandering Jew"; an exquisitely colorful form with the small leaves purplish green with silver stripes, broadly banded glistening white, and shaded pink and carmine red; margins and underside are purple. Pretty but delicate, requiring warmth and moderate moisture; only the best colored shoots must be selected for propagation to maintain the balance of beautiful colors. **C**: 13　　p. 261

purpusii (Mexico), "Bronze wandering Jew"; vigorous creeper with fleshy, oblique, broad ovate leaves about 2–2½ in. long, overlapping shingle-like, olive to purplish brown with faint green stripes; shiny vivid purple beneath; flowers lavender. Foliage almost black if exposed to good light; a good addition to the various Tradescantia-type window trailers. **C**: 13　　p. 261

ZEPHYRANTHES (Amaryllidaceae)

grandiflora (carinata) (Jamaica, Mexico, Guatemala); the common "Zephyr-lily", also called "Rain-lily" because in habitat the flowers appear after a rain; a pretty bulbous herb with grass-like, linear basal leaves to 1 ft. long and large solitary, deep rose-pink funnel-form flower 3 to 4 in. across, at the end of each hollow stalk; blooming through spring and summer. After 10 weeks of slight rest after the flowering is over, and the leaves have turned yellow, if grown in pots, the bulbs may be flowered again within the same year. Durable house plant for the cool window or winter garden. If the pots are plunged outdoors in the spring, the lovely flowers may be enjoyed outdoors in autumn. Propagated by bulblet offshoots. **C**: 63　　p. 392

ZINGIBER (Zingiberaceae)

zerumbet (India, Malaya to Polynesia), the "Wild ginger" or "Bitter ginger"; well-known and widely cultivated tropical ginger with knobbed rootstock, at first taste aromatic, then becoming bitter; leafy shoots 1½–2 ft. high, the lanceolate, thin leaves 4–8 in. long, more or less hairy beneath; in the late summer an oblong flowering head, 2–3 in. long, appears on a stalk about a foot long, separate from the leaves, consisting of large green to red overlapping bracts, and small inconspicuous yellowish flowers. **C**: 54　　p. 205

ZINNIA (*Compositae*)
 elegans 'Pumila', "Youth-and-Old-Age", "Lilliput zinnia" or "Cut-and-come-again"; sun-loving herbaceous annual of bushy habit with bright green, opposite leaves and with smallish, double flowers about 2 in. across, the broad rays in bright colors from yellow to pink and scarlet, also white; particularly adapted for dry hot climate as in Mexico where the original species came from. With its vivid green foliage and showy gayly colored flower heads an attractive and willing pot plant, or for the garden where it blooms from July on. Several colors may be planted together in larger containers for the patio, or they are ideal cut flowers and will keep coming with new blooms all the more. From seed with warmth. **C**: 63   p. 411

ZYGOCACTUS (*Cactaceae*); the "Crab" or "Claw cactus" has long been cultivated as an old-fashioned house plant, and has lived through all kinds of abuse; often seen in neglected shop windows, and yet this is where it often blooms best, most likely in fall and winter. Zygocactus are small epiphytes densely branched, with bright green, flattened joints, each of which is distinguished by having prominent teeth toward the tip. Flowers are zygomorphic (irregular) with round ovaries. Culture on p. 151. Propagated by single joints rooted in peatmoss; also by grafting (see p. 62).

 truncatus (Brazil: Organ Mts., Estado do Rio), "Thanksgiving cactus", also called "Crab cactus"; branching epiphyte with flattened joints dark glossy green distinguished by two prominent teeth or claws at apex; October-November blooming with irregular (zygomorphic) scarlet flowers having round ovaries. **C**: 10   p. 170
 truncatus delicatus, "White crab-cactus"; a natural variety, of upright habit, but reluctant to branch; long dark green joints sharply toothed; the irregular white flowers delicately tinged pink, ovaries round. November-December bloom. **C**: 10   p. 170

ZYGOPETALUM (*Orchidaceae*)
 mackayi (So. Brazil); strikingly beautiful, robust orchid which I have seen growing as a terrestrial in quantity in red clay on high savannah of the Serra do Mar above Santos; clustered pseudobulbs with 18 to 24 in. leaves arranged fan-like; with the new growth a tall stalk bearing 6 to 8 very colorful, waxy flowers 2 to 3 in. across and very fragrant, petals and sepals yellow-green blotched with purplish brown, the broad lip white with a network of veins in blue; blooming from fall into spring. The stunning flowers are long-lasting, about 4 weeks. Grown in a compost of chopped fiber, shredded bark, humus and loam; they require good light, fresh air and copious watering during growth, drier for a rest from July to September. Propagated by division. **C**: 21   p. 482

Victorian Window Gardening, 1863.

Plant Decor and Landscaping Indoors with Exotics

A little Philodendron panduraeforme in glazed dish is fully content under frequently used desk light, adding a touch of life without being possessive.

Kentia palm (Howeia forsteriana), proven for durability under unfavorable conditions, in a Los Angeles living room. Kentias can take much abuse, and low light intensity from 200 foot candles down to 20 foot candles.

725

Foliage plants accented and blessed by concealed light in a recessed alcove. To many foliage plants, 15 to 20 foot candles of light intensity, equivalent to fair reading light, is sufficient if combined with reduced moisture.

Exotic Plants to Live With

Living room planting in Pretoria, South Africa under fluorescent lights. The planters are covered with grass mats made by Africans as sleeping mats, and adorned with their tribal carvings.

Solitary Philodendron pertusum (juv. Monstera deliciosa) traces exotic patterns in an otherwise rather bare room. Such Monsteras tolerate as low as 15 foot candles of light but prefer as high as 200 foot candles which enables them to maintain their large character leaves.

There is no better place for plants indoors, excepting a greenhouse, than this comfortable Winter garden, filled with light, in our New Jersey home. On a broad window sill we can probe first-hand the merits of numerous house plant candidates; if the sun becomes too burning hot we can control the light by letting down Venetian blinds. The temperature drops to a refreshing 50°F on winter nights; in summer we have airconditioning. A cozy room, very much enjoyed by us and by our plants, especially when deep snow covers outdoor patio and gardens.

An ideal location for all kinds of flowering plants on a broad shelf next to the good daylight of the East windows, at the Ross home, in Camden, Maine. Such "Picture Windows" and even more so the modern sliding glass doors wondrously combine the outdoors with the living room and bring Nature indoors and closer to daily life.

Exotic Plants to Live With

This area of bay windows permits the grouping of indoor plants in decorative stands but each will need daily attention which, however, may give a plant lover much pleasure.

A fancier's collection of house plants ingeniously distributed next to three windows where they get the best light. Daily inspection is needed here also, but such care will have its own rewards.

A snug window planting in the library of a home in Cologne, Germany. The exotics framing the single pane of glass invite the view out into the wintry landscape.

Exotic Plants to Live With

Window planting of tropicals demonstrate the modern trend in Germany. A variety of attractive plants grows in the recessed space built at low level right into the wall of the house.

Attractive indoor window box with Monstera deliciosa, Dieffenbachia amoena, and Dracaena warneckei, at Rutgers, New Brunswick, New Jersey.

Exotic Plants to Live With

Home-made window greenhouse along the lagoon at Venice, Italy, where the nights sometimes become quite chilly.

Solar window greenhouse of aluminum which can be ventilated, and heated by thermostatic control. An aluminum tray filled with wet vermiculite or peatmoss provides the necessary moisture.

The typical Dutch living room window with house plants, at Aalsmeer, Holland. Such windows are large, and curtains never obscure the view.

730

Rooftop Solarium

This penthouse in New York provides just about ideal conditions for plants that flower and bear fruit. The glass-roofed greenhouse adjoins a glass-walled living room offering day to day enjoyment of this secluded garden room. A splashing fountain framed by luxuriant vines on marble columns add a touch of Southern charm.

The Conservatory for Atmosphere and Study

A restful Winter garden conservatory invites guests to the gracious way of life of the Empress Hotel in Victoria, British Columbia.

Naturalistic planting of an Exotic garden at Roehrs Exotic Nurseries in New Jersey not only presents a composite and idealistic picture of a tropical landscape, but also houses various mother plants needed for propagation.

Delicate tropicalia luxuriate so much more easily in their own microclimate when grouped and planted together in ground beds. In this manner they will tolerate adverse conditions and temperature changes without harm.

The main showhouses of the New York Botanic Garden are enhanced by large ponds of tropical water lilies in many glorious colors when in bloom during summer time.

A botanical garden is dedicated to research and education and plants are brought together from all over the world to be carefully labeled and cultivated as here at the great palm house of the New York Botanic Garden in the Bronx. (Cycas circinalis, right).

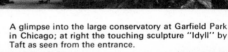

A landscape of succulent plants endemic to the Western hemisphere, at Missouri Botanic Gardens, Saint Louis. There is also a companion house devoted to the succulents of the Eastern hemisphere.

A glimpse into the large conservatory at Garfield Park in Chicago; at right the touching sculpture "Idyll" by Taft as seen from the entrance.

Conservatories – Old and New

The beginning of greenhouses, the early orangerie at Florence, Italy. Simply an orthodox building with very tall windows and primitive heating, dating from the 18th century (U.S.D.A. photo).

The ultra-modern Climatron, a geodesic dome structure of aluminum tubing, covered with $\frac{1}{4}$ inch thick Plexiglass, and 175 feet in diameter, at Missouri Botanic Gardens, Saint Louis, and built based on the principles conceived by Buckminster Fuller.

In mechanically controlled, air-conditioned temperature and climate areas, epiphytic bromeliads luxuriate with climbers and ferns under optimum light conditions inside the futuristic Climatron at Saint Louis.

Commercial Greenhouses

When winter's ice and snow sweep down from the North Pole, delicate exotics are kept cozily comfortable in steam-heated glasshouses. With reduced light, growth comes to a near standstill and plants hibernate, awaiting the waxing sun of spring to awaken them.

Exotic plants including bromeliads, orchids, Nepenthes and Anthuriums are commercially grown as here at a "stove house" of Roehrs Exotic Nurseries in New Jersey.

Giant glasshouse for large container plants suitable for interior decor as maintained by Roehrs in New Jersey. Some of these specimen plants come from Florida, California or Puerto Rico, to be grown on and acclimated before being sent out for use in interior landscaping.

The sunny days and cool nights of southern California are naturally suited to the commercial production of Cacti and other succulents as here at one of the author's nurseries, Los Angeles Plant Co., Manhattan Beach.

Plastic Greenhouses

Various plastic materials are used as substitutes for glass in greenhouses. At Paul Ecke's Poinsettia ranch, in Encinitas, California, numerous houses were built utilizing rigid, translucent corrugated panels of Poly-vinyl Chloride, clearer than fiberglass and comparatively long lasting, at the same time diffusing strong sunlight.

Structures of lightweight, as these in Pittsburgh, Pennsylvania, may be easily covered by sheets of polythene film which is inexpensive but deteriorates after each single season when exposed to the ultra-violet rays of the sun. Polythene allows beneficial carbon dioxide molecules to exchange freely but water does not pass through and is held as condensation.

An easily constructed house covered with plastic material reinforced by webbing grows good fuchsias and geraniums for Merry Gardens, Camden, Maine.

An unobtrusive little lean-to greenhouse snuggled away by the side of a vine-covered home in Massachusetts. Slats of wood may be rolled down to provide partial shade on bright days.

The Small Home Greenhouse

Inside a trim Lord and Burnham glasshouse featuring automatic ventilation and heat, a plant lover has a wide choice of plants, especially the flowering kind, that may be grown under ideal conditions of light and humidity. Seasonable blooming plants may be brought into the living room when needed for temporary decoration.

A small greenhouse 12 x 15 ft. is big enough to have given a lifetime of happy avocation to a fancier of cacti in New Jersey. With so many varieties there is bloom the year round, requiring a minimum of care.

The Small Home Greenhouse

A greenhouse knows no seasons. Cozily protected from winter's snow in Connecticut, the fragrance and beauty of out-of-season flowering bulbs or plants may be enjoyed even more because of the contrasts in climate indoors and outside.

In creating a comfortable climate, in their little greenhouse, plant hobbyists may have the pleasure of spending much leisure time surrounded by a collection of their favorite indoor plants.

Sheltered Interior Courts

Lovely Exotic planting featuring West Indian treeferns and many species of tropicalia, in the glass-covered peristyle of a distinguished suburban residence in Westchester, New York. Through windows and glass doors all the rooms of the house look out into the square interior court.

A corner of the peristyle planting with a quiet refreshing indoor pool. Left Cyperus and Monstera, right the leathery Dracaena hookeriana.

Sheltered Interior Courts

Spectacular display of palms, ferns, and large Ficus trees in the vast, glass-covered garden court of the research laboratories of Bell Telephone Company at Holmdel, New Jersey. From galleries 700 feet long, a stunning view is had over the lavish plantings, some of which are periodically changed according to the season. **Planned by Eero Saarinen Associates.**

Open-air patio court, accented by palms and cycads of David Barry's office building on San Vicente Boulevard, West Los Angeles. The various offices face into the planted area, offering both quiet and stimulation off the busy boulevard.

The sober lines at the street entrance to the foyer of a large insurance company building in downtown Newark, New Jersey, are enlivened by solitary container plants that will tolerate occasional chills. Left Dracaena marginata; right Fatsia japonica.

Plantscaping in Office Buildings

The spacious streetfloor of Union Carbide Corp., on Park Avenue, New York is attractively decorated by the tracery of the handsome foliage of giant Philodendron selloum in tubs. Although they tolerate low-light locations down to 15 foot candles, overhead ceiling lights are responsible for their luxuriant condition.

741

Plantscaping in Office Buildings

Modern skyscrapers, architectural giants of great esthetic beauty, need the warmth and softening decor of live plants properly placed. Below left: Philodendron erubescens; above right: Monstera deliciosa and Ficus lyrata.

The center area on the 38th floor of a New York Insurance Company building is simply but attractively planted with creeping Rhoicissus capensis and weeping Ficus exotica trees. Ficus are light-hungry and should have a minimum of 50 foot candles to subsist.

Plantscaping in Office Buildings

An ideally planned built-in planter space in the corner of an executive office in Los Angeles is designed for an elaborate grouping of interesting exotics. Each pot or tub is imbedded in peat-perlite mix over charcoal drainage, and can be removed and changed at any time.

Plantscaping in Public Buildings

The General Assembly building of the United Nations is carefully airconditioned. Plunged under overhead spotlights into sunken beds alongside the stairways facing the tall glass windows south toward the Plaza are large groupings of decorative plants. Despite low light readings between 16 and 50 foot candles, their behavior has demonstrated that many ornamentals adapt themselves with the same pleasure to airconditioned surroundings as humans do. Amongst the most satisfactory plants long observed are Dracaena warneckei, Spathiphyllum, Aglaonema and Dieffenbachia.

THE EQUITABLE LIFE

Window planting of a wide variety of tropicalia on 7th Avenue in New York. Four overhead banks of incandescent spot lights not only keep all plants in happy growth but also attract the astonished attention of the people on the street.

A naturalistic indoor planting behind borders of cut stone, aptly named a "winter garden" in a modern office in Germany. It seems that a greater variety of rather sensitive plants are used in this manner, but it may be remembered that German rooms are generally kept cooler than in America.

745

Accent Plants in Modern Offices

Ficus benjamina in cypress tub and saucer at Chase Manhattan Bank in downtown New York. Ficus are light-hungry and need at least 50 foot candles of light to subsist.

Reception area on the 16th floor, with sole Brassaia actinophylla, the Queensland umbrella tree, in clay pot lacquered black. Brassaias, known in the trade as Schefflera, tolerate extremely dry-warm conditions, but need fair light 50 foot candles and up, preferably sun.

Combination touffs of durable Kentia palms, Howeia forsteriana, in the air-conditioned conference room of a manufacturing firm in Los Angeles. These palms can take much abuse and get along on little light, as low as 20 foot candles.

Podocarpus macrophylla, the coniferous "Buddhist pine", with dark green leathery needles, in a Park Avenue office, New York. Podocarpus take aircon-ditioning provided they are well established, they prefer cool conditions and need light, 50 foot candles up.

The artistic "Seagrape" from Carib-bean shores, Coccoloba uvifera, with flexuous branches and stiff-leathery foliage with red to ivory veins. To maintain their large foliage, they re-quire at least 50 foot candles of light.

748

Ficus elastica 'Decora', the "Red leaf rubber plant" (left) does best in good light preferring sunlight, but has been found to tolerate poor light, down to 20 foot candles, more than other Ficus. Rhoicissus capensis, the "Evergreen grapevine" from South Africa (right) as a low centerpiece on tables, is satisfied with medium light 50 foot candles up.

Accent Plants in Offices

Yucca elephantipes, the "Spineless palm lily" for that desert look, takes much abuse. Light tolerance 20 to 500 foot candles and up.

Ficus lyrata, better known as F. pandurata, the "Fiddle-leaf fig" with its large leathery leaves is one of the most favored decorators but requires more exacting attention to even moisture and light, preferably 200 foot candles, though it can occasionally get along on 40 foot candles.

Imposing barrier of tubbed Ficus lyrata (left) standing on shallow trays with pebbles, on the light-flooded 60th floor of a Wall Street Bank, and viewing the harbor, downtown Manhattan.

Room Dividers

Attractive room-divider in the gracious manner, with various foliage plants and ferns, in a country club lobby near Los Angeles, California.

749

Plants Enhance Stairways

An otherwise bleak floor level of glass and marble is admirably softened by a solid border of Brassaia actinophylla, plunged in a built-in planting area atop the black marble railing.

This happy planting in a Fifth Avenue bank in the airconditioned main banking room situated next to the escalators, enjoys full natural light from the avenue. The deep plant container conceals a well-drained copper liner with false bottom and containing fluffy peatmoss into which the plants are plunged with their pots.

Brassaia and Pandanus dress up an otherwise useless and unsightly corner alongside the escalator leading up to the Midway Restaurant on the Will Rogers Turnpike in Oklahoma.

750

Plants in Dining Rooms

A corner of the elegant Four Seasons Restaurant on midtown New York's Park Avenue, dominated by several bold Ficus 'Decora' trees in the midst of the dining area and nearly reaching to the ceiling, as placed by Everett Conklin.

The dining area in a New York bank is tastefully divided into inviting nooks by small plantings of durable exotics, primarily Dieffenbachia amoena.

Plants in Dining Rooms

The relaxed, cozy atmosphere of this California restaurant is created in large part by the presence of live plants at the junctures of the contoured banquettes, flooded by warm light. The leathery, long-suffering Dracaena hookeriana are principally used.

Large Monstera deliciosa, individually lighted from above, beautifully enliven an otherwise stark, empty wall in the lobby of Hollywood Park Race track in California.

Plant Decor in Clubs

The circular walls of the Clubhouse lounge are planted throughout with vining Philodendron and Monsteras, forming an effective screen of greenery.

Architects designing this restaurant waiting room in California thoughtfully made provision to include a large planting area for foliage plants, with pleasing results, and a topic of conservation amongst waiting guests.

Restaurant Areas

754

Even small corners may be attractively planted, at the same time concealing undesirable views.

Home-like charm pervades throughout this country club lounge where room dividers with Monsteras, Philodendron and Dieffenbachias create secluded hide-aways. Additional light is a necessity to plants in this cozily dark room, well supplied by various exquisite lamps.

Exotics for Decor

The gracious elegance of Victorian times can be instinctively felt on entering the main lobby of the Biltmore Hotel in Los Angeles. In true tradition the palms used for decoration here are long-time favorite Kentias, botanically Howeia forsteriana.

755

Elaborate step planting with tropical plants wonderfully and effectively draping an otherwise routine stairway, in California. Plants used are Fatsia, Sansevieria, Syngonium, Philodendron, Dieffenbachia, Phoenix roebelenii and Asplenium ferns.

Public Use of Plants in Europe

Tropical plants and vines brought to the public in an unusual manner by having them grow inside a series of semicircular glass enclosures, as seen in the dining room of the famous Palmengarten, with its rich botanical collections, in Frankfurt, Germany.

The well-known affection for indoor plants and flowers in northern Europe is very much in evidence here in this "Rasthaus Steigerwald" along the Autobahn Frankfurt-Nuremberg, in Bavaria.

Indoor Decor Overseas

Contrary to our recommendations in America, one sees "Philodendron pertusum" (Monstera deliciosa) satisfactorily growing all over Europe on mere bamboo sticks, as here at this parador in 12th century Ciudad Rodrigo, in Western Spain.

When I stayed at the Golden Dragon Inn in China, the focal center of the lobby was a life-size gilded dragon imaginatively planted all around with a wide selection of tropical plants.

In Japan, the Matsuya Department store in Asakusa, Tokyo, displays both Western dress as well as the glamorous kimono. But at strategic spots one sees palms in attractive pots glazed deep blue, as here this traditional Rhapis excelsa.

The lately so ubiquitous decorator palm of Japan, Phoenix roebelenii, grown in classical glazed pots, can be seen everywhere in public buildings, offices and restaurants, supplied and changed monthly by plant rental service, as here at the Imperial Hotel in Tokyo.

Tropical
Plantings
in Business

A steamship company office in Los Angeles is immensely enhanced by the central sculpture of a Polynesian god enclosed in a planting of climbing Monstera deliciosa, Dieffenbachia picta, and Ficus elastica.

"Electric typewriters take to the Tropics" was the theme of this spectacular window display of the IBM Corporation on Madison Avenue in New York one winter. The vari-hued machines were beautifully highlighted in an exotic setting of hundreds of lush, tropical plants, colorful orchids and tall treeferns.

Plant Decor in Business

A modern glass and concrete reception area of a California factory where floor plantings of Ficus lyrata and various Philodendron magically relieve the coldness of this futuristic building.

An awkward corner of an automobile showroom in Los Angeles has become a center of attraction by enclosing the area with a low retaining wall, filling it with soil and planting it with decorative ornamentals.

Shopping Centers Beautified

A group of the decorative, semi-dwarf Christmas palms, Veitchia merrillii, in a commendable scheme to create an attractive parking area, in Miami, Florida.

Tropical planting indoors at this large shopping mall near Chicago was created by Lange Florist who understood to give prospective shoppers the exhilarating atmosphere of swaying palms and exotic greenery.

Tree philodendron, Philodendron selloum, attractively frame the entrance of Sears Department store at Torrance, in Southern California. The temperature here is very similar to their native Paraná, in Southern Brazil.

Flowers Promote Sales

New Jersey's largest department store, Bamberger's of Newark, turned various floors into veritable gardens on the occasion of Mother's Day week during May. Vast numbers of decorative and flowering plants were installed by Roehrs Exotic Nurseries creating tropical scenes with pools and waterfalls, treeferns and palms; elsewhere were massed beds of azaleas, gloxinias, chrysanthemums and hydrangeas, even cacti. Shoppers were thrilled to walk the garden paths or under pergolas decorated with cut flowers, a wonderful tribute to mothers!

*Shopping Places
for Plants*

House plants as well as accessory supplies, soils, pesticides, tools and seeds are merchandised at moderate prices in the garden department of most shopping centers, some with elaborate display houses as here at Paramus, New Jersey.

The retail florist has the advantage of life-long experience enabling him to give horticultural advice and personal service. Where he maintains his own greenhouses, plants are likely home-grown and acclimated. Specialist growers of tropical plants such as Roehrs greenhouses shown here may be able to supply species not easily found.

Plant Vacations Outdoors

All house plants and some greenhouse dainties enjoy being moved outdoors for the warm season onto the patio, or garden by the side of the house. This will renew their vigor and strengthen tissues, and they enhance the outdoor living space. A simple shade roof, as above, in New England, or under a tree is ideal. Left, a step-stand for the summer vacation of begonias and ivies.

Coniferous trees generally are not indoor plants. The container-grown dwarfed bonsai-like Japanese larch (Larix leptolepis), being hardy, is happiest on the sun-porch or the patio, as here in suburban New York.

Plant Vacations Outdoors

Above, the back of a garden wall is utilized for shade plants, and serves to decorate the backyard. Right, Begonia rex, which as all begonias relish fresh air, plunged in the shade of a peach tree. Below, container-grown, cool-loving Camellias, are in their natural element under the diffused light of tall trees.

Shade Houses in Warm Climate

Wherever possible, commercial growers also move many of their foliage plants out under lathhouses or the open sun. After a few weeks the foliage takes on new lustre and firmness, as have these Brassaias and Japanese boxwood at Roehrs in New Jersey.

The famous "Estufa Fria" in Parque Verde Eduardo VII in Lisbon is a veritable Botanical garden and probably the largest slatted open air conservatory of its kind in the world. More than 500 ft. in extent and 40 ft. high, it is open to the public, and thousands of happy Portuguese daily stroll to the sound of music in the midst of vast naturalistic plantings of tropical and subtropical flora under outdoor conditions, hardly surpassed in Europe; flowering climbers and tropical shrubs bloom profusely in the diffused light and fresh air, while noble treeferns and stately palms are reflected in quiet ponds of water-lilies and sacred lotus.

Commercial lathhouses shelter acres of Kentia palms (Howeia forsteriana) grown by Wilcox in Southern California. The pots are plunged into the ground; irrigation is by rotary sprinklers, and oil-burning stoves stand ready in case of an occasional frost.

Garden pool in Florida by Holmes Nurseries, Tampa. Boulders
and driftwood form a naturalistic setting for colorful epiphytic
bromeliads.

Exotic Landscaping Outdoors

In balmy, tropical Hawaii, large specimen of Spathiphyllum hybrid 'McCoy'
luxuriate shaded by trees on a secluded patio near Diamond Head, Honolulu.

767

Exotic Landscaping Outdoors

During New York's tropical summer, when the temperature on a typical humid-warm day may be 85° F and more, the Channel Gardens at Rocketeller Center are transformed into a flourishing tropical landscape, extending from Fifth Avenue to the splashing Prometheus Fountain (above). Pools of rippling water are bordered by thousands of colorful exotic plants and trees, including Caladium, Crotons, Dracaenas and bromeliads, even pineapples in fruit (below).

Tropical Plantings in New York

Tree-dwelling bromeliads dominate the view of Radio City from Fifth Avenue; in the background the 70-story R.C.A. Building. Left, the tropical Brassaia actinophylla thrives particularly well under warm conditions.

Tropical Plantings in New York

A grouping of big and little Cacti and other succulents form part of the mid-summer show at the Channel Gardens in Rockefeller Center. Carefully labelled, they are studied by many of the millions of sightseers and passing strollers along the Promenade. And when the summer show is over, it is no small job to change the settings (right) to make room for the glories of autumn.

The most famous roofgardens in London, are unbelievably located on top of the sixth floor of one of England's leading department stores, 120 ft. up, at Derry and Toms, in Kensington High Street. Exquisitely landscaped, various vistas surprise the visitor, the Castilian garden with formal pools and Moorish architecture, the quaint Tudor garden, and a shady English woodland scene. Tired shoppers and children find here a wonderful opportunity to relax, and take tea, as in a big garden, yet in the heart of busy London.

Gardens in the Sky

An excellent example of roofgardening in New York were the International gardens on the 11th floor of Rockefeller Center. Above, the Spanish Garden; left, Japanese Garden and teahouse.

771

Container Planters Outdoors

The curious, frugal Beaucarnea recurvata, a succulent water-storing tree from arid Mexico —here in a terra-cotta pot on a patio in Del Amo, California.

Variegated New Zealand flax, Phormium tenax "Variegatum", in terra-cotta urn, for sunny locations.

Outdoor planter box with bright, easy-going Coleus and scented-leaf Peppermint geranium, Pelargonium tomentosum.

Terra-cotta strawberry jar attractively studded with rosettes of bluish-white Echeveria elegans, in the center an Aloe succotrina; requiring little care the year round.

Dominating a group of smaller containers, a large "Hollywood juniper" (Juniperus chinensis "Torulosa") with characteristically twisted branches at the Los Angeles Art Museum on Wilshire Boulevard (right).

Pyramidal Taxus cuspidata capitata, bordered by Emerald Ripple ivy, planted in 4 ft. fiberglass container, as seen in New York City.

Oversize planter bowl photographed on the center island of a thoroughfare in Durban, South Africa. Everyone of these colorful containers is planted differently, some with annuals, such as Tagetes, others with Azaleas, Dracaena fragrans, Kalanchoe, even Haemanthus.

Along the bank of the broad Nile River in Cairo, I found this sun-toughened Schefflera growing in an exquisite container of Islamic tile.

Plants Add Charm

That plants add charm is evident to every visitor that has ever seen the hanging baskets on the lamp posts of queenly Victoria, British Columbia. The northerly latitude and moist-cool climate brings out the best in ivy geranium, petunias and other vari-colored annuals, although they still must be periodically watered.

Bromeliads planted into corkballs from a fisherman's net dangle from nylon threads in the corner of a sheltered patio in New Orleans, Louisiana. In this humid climate these epiphytes require practically no other care.

Even dainty Caladiums allow themselves to be paraded down the front steps during summer time in southern New England.

The semi-tropical climate of Florida of course favors such charming plant corners outdoors, combining succulent agaves with bromeliads and cycads.

Travelling on the Soviet ship "Maria Uljanova" to Leningrad, I was pleased to observe the use of living plants along the glassed-in gangways of the ship. Though they were mostly simple myrtles, primulas, begonias, fuchsias, agave and opuntia, they were carefully tended.

The living beauty and decor of a variety of large ornamental plants, in the gracious tradition and romantic mood of the Victorian era, charmingly blends with modern architectural styling of the promenade decks, foyers and public halls of the Italian luxury liner "Raffaello". Here this "Kentia palm", Howeia forsteriana, nursery-grown in galvanized metal pots, and embellished around the base with Pteris ferns and variegated Chlorophytum, is placed inside a glazed ceramic jardiniere, feeling fine as if at home on a floating island back and forth across the Atlantic Ocean.

A plant lover would notice with pleasure that fondness for plants can be found in any surroundings, even Mobile estates. Attractive succulent gardens and container-grown ornamental plants help to beautify every home on wheels here in Paramount, California.

The popular affection in Europe for plants especially indoors is strikingly demonstrated by this display of potted plants in the picture windows of this "Woon Ark", a flat-bottomed house boat at anchor in a quiet canal of the sprawling harbor of Aalsmeer, Holland. During summertime this barge will be circled by window boxes overflowing with trailing geraniums, petunias, fuchsias and nasturtiums.

In the Moorish-Spanish Tradition

It is difficult to believe the number of potted plants maintained in Spanish homes until one has seen the Moorish-style patio courts in old Cordoba, dating back to the 8th century. Pots seem to be climbing up the walls, and occupy every free nook and corner.

The Moorish-Spanish tradition has been carried over into America, and this humble dwelling near Vera Cruz in Mexico gives proof of the deep-rooted love for plants of the Mexican people.

When the Moors ruled Spain, between 712 and 1492, they interpreted their visions of the paradise of Islam by building gardens with cool waters, fruiting trees, and flowering and fragrant plants, a tradition preserved here at the Alhambra, in Granada, under the azure skies of Andalusia.

Container gardening around the Mediterranean is simplified by the mild climate which lets most plants live and exist without much tedious care. Preserved since the 16th century, the Boboli Gardens in Florence, Italy, are a classic example in the use of miniature fruit trees in containers, as these citrus in terracotta jars.

A flowering bromeliad, Aechmea fasciata, is growing contentedly under the Adriatic sun in an interesting vividly colored ceramic vessel, by the side of an obscure canaletto in nostalgic Venice, Italy.

Under the Southern Sun

Widely used as a house or porch plant in the hot central Indian plains is this self-branching Pleomele reflexa, grown in hand-made clay pots, here at Benares, India.

A window in a modest section of Istanbul with its lavish use of potted plants and vines, testifies to the fondness for flowers also in Turkey.

Plant Arrangements in Containers

A redwood box of plants that can take some abuse. Such long-lived foliage plants as these Dracaena warneckei, Ficus 'Decora' and F. doescheri, Brassaia, Dieffenbachia amoena, and Philodendron pertusum, are suitable for buildings with much traffic. Plunged with their pots into moist peatmoss, over a layer of charcoal to keep from souring, watering may be postponed for days.

These tropicalia are somewhat more fussy, requiring warmth and an even moisture. But groupings such as these help to sustain each other, plunged together in moist peat. Here are Dracaena, Calathea, Rhektophyllum, Stromanthe, Dieffenbachia, Begonia masoniana, Maranta and Syngonium

Brass planter serving as a table with plate glass top, which also shields the plants, an attractive and practical combination seen near Sydney, Australia.

778

Naturalistic ceramic root with opening for a few plants effectively arranged: Dracaena godseffiana, variegated Peperomia and the dwarf palm, Chamaedorea elegans.

Plastic bowl with brass legs planted with Aechmea and Nidularium, Phoenix roebelenii palm, Dieffenbachia, Maranta, Scindapsus, Dracaenas, Philodendron. An unobtrusive white umbrella with fluorescent coil shines light just where it is wanted.

"Dixieland" dishgarden of a type made in the New York area, planted rather full for show, and combining succulents which tolerate some moisture such as Crassula and Sansevieria, with proven foliage plants including Brassaia, Cryptanthus, Dracaena, Peperomia.

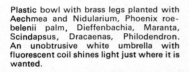

Centerpiece in shallow dish with the showy Dieffenbachia 'Rud. Roehrs' ringed by Peperomias sandersii and obtusifolia, all plants which prefer limited moisture.

Bromeliad Arrangements

Skeleton stem of a Cholla cactus (Opuntia) from our Western deserts provides just the right foothold for these dwarf bromeliads. Tillandsia ionantha left and right center; left center Cryptanthus bivittatus minor; right Cryptanthus lacerdae. Submerge periodically in a bucket of water.

Hand decorated pottery bowl with crooked manzanita branch, sandblasted and cemented in to hold firm. Epiphytic bromeliads are fastened in osmunda-sphagnum, preferably with nylon thread (or soft galvanized wire). Bottom: Vriesea splendens; top left to right: Guzmania, Neoregelia, Vriesea, Tillandsia lindenii.

Simulated ceramic root looking surprisingly like real driftwood, planted with exotic bromeliads; left to right, Cryptanthus bivittatus minor, Billbergia nutans, Vriesea rodigasiana, Cryptanthus zonatus. Such bromeliads require only a minimum of space to anchor their roots, but their centers should be kept filled with fresh water.

The fawn Bambi in a highly glazed novelty planter, planted with candelabra-like Euphorbia lactea, colorful Sedum and Crassula, and several cacti.

Terra-cotta birdbath, very showy if displayed on its stand, and ideal to move outdoors in summer. White-haired cacti and succulent rosettes surround the dominant old Aloe arborescens. Gravel surface in various colors provide interesting contrast.

Old antique pieces such as this copper pan planted in California as a miniature desert scape with many contrasting types of cacti and other succulents—all excellent keepers under dry conditions and requiring little attention.

Naturalistic planting of "Living stones" (Lithops), peculiar to the deserts of Southern Africa, succulents requiring the best of light, and extremely sensitive to wetness. These 3 year old plants are growing in an 8 inch clay pan in porous soil mixed of coarse sand, clay, leafmold and perlite; pebbles of quartz or weathered stone cover the surface not only to imitate their natural habitat but also to shield the delicate rootlets from heat-burn. In their native home these xerophytes mimic the surrounding gravel and hide between it, filtering sunlight through their windowed tops. Lithops are actually only a pair of leaves filled with water, and we lightly sprinkle, rather than soak them, only when they are beginning to shrivel, about every week or two. They respond well to additional fluorescent light, even throughout the night.

781

Various Plant Arrangements

A "Fern dish" very popular during the early 1900's, then made as centerpieces for decorating the dining room tables all winter long. This glazed bowl is planted with various dwarf "table ferns" (Pteris) and Adiantum, and must be kept constantly moist.

"Valentine heart", a thoughtful gift for Sweetheart Day, February 14. Bright red-glazed novelty planter with variegated Hoya, Billbergia nutans, and Aglaonema 'White Rajah', all plants that need very little water.

Seasonable planting can be made for Christmas in many ways with miniature plants resembling wintery trees. Araucaria, Arizona cypress, Euonymus, Wax privet, Ardisia for holly are used and lightly sprayed with wax or synthetic snow, creating a fascinating winter scene.

One of the many containers that may be used for unusual effect: A real birdcage on its stand lends itself to planting with small table ferns and vines, but all have to be kept moist.

The Arranging of Flowers as an Art

With the possible exception of a painting, nothing quite permits one's expression of artistic feeling so well as to take a few branches from a tree and put them together with some suitable plants, cut flowers or foliage at hand to fashion a tasteful arrangement of natural beauty to enjoy in the home. There are countless delightful and decorative effects that can be created by the exercise of a little skill, sense of beauty, and imagination.

Harmony is the sum of any successful simple flower array, taking into account color relation, proportion, symmetry and natural growth, as well as the suitability of the intended container. But beyond the fashioning of mere natural bouquets one may reach out further into the classic concept of Oriental styling in the art of Ikebana. In its cult, the Japanese are inspired by the study of the moods of nature, and their simple creations follow rules laid down by age-old tradition, based on three main lines symbolically called Heaven, Man, and Earth. Since Heaven is tallest, Man of lesser height, and Earth lowest, formal Japanese arrangements are necessarily asymmetrical or unequal. The three different classes of plants or branches are placed so that their tips would theoretically form an unequal-sided triangle, and the stems, leaves and flowers are placed accordingly. The primary line of Heaven, the tallest, is placed first. forming the axis against which the other lines are arranged to result in a pleasing unit of balance and harmony. Wire or pin-holders with weighted bases, or modeling clay is used to anchor the stems if the dish is shallow. The arrangement shown above consists of Oleander for the Heaven line, Aphelandra and Euphorbia for the Man line and Primula signifying the Earth. At left, holly-branches are placed to represent Heaven.

A rectangular aquarium with frame of stainless steel can be made into a "Mini-garden", a small green world with miniature plants, twigs, rocks, and moss. Delicate tropicals thrive in the humid atmosphere generated through moisture transpired by the leaves and also evaporated from the soil, reducing the need for watering to a minimum.

Terrariums to hold Moisture

Miniature greenhouse consisting of a brightly colored plastic tray and a glass-like cover of transparent lucite. While normally this is an excellent container to start seeds, in this case it is used for tiny pots of cacti, helping them to grow faster.

Called "Table hothouse" and made in Russia, this glass container framed in wood is hermetically sealed. The inventor, Dr. Vadilo of Leningrad claims that the plants, such as Maranta, Saintpaulia and Eucharis inside, never need watering.

784

Glass Containers as Terrariums

A bottle garden with a variety of tropical plants, using a demijohn or carboy 2 feet high of white or tinted glass always raises astonished wonderment because of the seeming impossibility to place and arrange the plants through a 1¾ inch neck. However, the tools shown explain: a brass tube to lift out a core of soil, the tongs for placing of plants, and a long spatula for setting soil and moss around them.

Because of its simple elegance, the glass globe is most popular as a glass garden. Many otherwise delicate miniature plants and ferns may be used, such as Selaginella, Adiantum, Saintpaulias, Episcias, Marantas and Pileas, but hardier species may be included. Dracaena sanderiana, Acorus, Syngonium, boxwood, Peperomias and Chamaedorea palms, even succulents such as Haworthias, are old favorites.

Containers of sparkling crystal-like glass come in many shapes, as the water pitcher above.

PLANTING and CARE

A large brandy inhaler is shown being planted, here with little ferns. Plants are arranged on an incline, placed in holes scooped out with a spoon, and planted with the aid of a pair of grippers made of bamboo canes.

Once the planting is completed, it is watered well with a rubber clothes sprinkler, then not again until the surface moss shows signs of drying. Excessive humidity favors diseases and fogs the glass; if covered, the glass jar may not need watering for 6 months.

785

"Life in the Tropics" was the topic of this small but life-like planting of many of the rarest of exotic plants, always a special point of interest for indoor plant fanciers. This exhibit intended to depict the terrific competition for living space in the tropical rain-forest, where plant life flourishes even on the thatch of this stilted jungle hut.

"Mission Bells" shows a landscape of succulent plants of arid regions of both hemispheres bringing together many types of cacti, Euphorbias and other succulents, collector's items for those that want plants easy to care for.

A Los Angeles visitor to the spring Flower Show in New York, caught here with one of the most beautiful of all tropical foliage plants, Dracaena goldieana from Equatorial Africa.

786

Exotics at Flower Shows

The International Flower Show in New York City Coliseum has always been anticipated by hundreds of thousands of professional and amateur horticulturists every year. Never awed by the competition of the larger exhibitors with their skilled gardeners and thousands of plants, hundreds of just plain plant lovers bring to the Show their prized plants, and men and women alike in work clothes busy themselves all night tidying up their exhibits for the judges next day. A wonderful spirit prevails amongst all those that work together in one common interest, to enjoy and appreciate the beauty of nature's creations. Above, a Roehrs 1959 exhibit of exotic plants. In the center, tikis carved of treefern, or hapu, as would be seen in Hawaii. At left, Dracaena goldieana; back, from left to right, Licuala grandis, Cycas media, Ptychosperma macarthurii (tall), Phoenix roebelinii and Coccoloba grandifolia.

Plant Growing in the Tropics

by Sir Henry Pierre, Port of Spain, Trinidad

There is no attempt in this short article, to cover the field of horticulture in the tropics as such. It is only meant to serve as a guide for those interested, and as a stimulus for further investigation and research, to gardeners who have the wish, the will and the time, to devote to the further development of horticulture in the various parts of the tropics.

Whereas in the colder climates, one has to contend with Spring, Summer, Autumn and Winter as definite seasons—in the tropics generally, the variations are far from showing any definite line of demarcation. One may describe them as the Wet Season and the Dry Season; the Wet Season occurring for about six months of the year, with a high rainfall, roughly from June to November, and the Dry Season for the other six months of the year, with a low rainfall, one merging gradually into the other. Not infrequently one experiences a wet Dry Season or a dry Wet Season. In any case the night to day temperature variation in the Dry Season—the cooler part of the year—is 66 degrees to 88 degrees Fah. with a comparatively low humidity, and the temperature variation in the rainy season is from 68 degrees to 92 degrees Fah. with a high relative humidity; these are from our location 10 deg. North of the Equator. South of the Equator the changes are a month or two later.

These of course are the averages for the low lying districts. There are however, many factors which contribute to much wider variations in rainfall, temperature and humidity, in the same or different countries, governed by the presence of mountains, valleys or proximity to the sea.

Even at the Equator, the height above sea level makes a great difference to the type of plants one can grow satisfactorily. From 3,000 feet upwards it begins to become sub-tropical. One can make rough calculation that for every 300 feet rise in altitude, there is a fall of about 1 degree Fah. in temperature.

The position of the sun in the Temperate Zones shows a definite Northern or Southern variation, whereas in the tropics the sun varies little from overhead, except in East-West direction. Light intensity in the tropics is not much different from that in the Temperate Zone during summer, even at high altitudes; ranging from 12,000 foot-candles in bright sunlight to 500 foot-candles on a day with an overcast sky with no shadows. It does change a little from one season to the next, as sunshine is less intense in the rainy season due to water vapor in the sky.

In the cold climates there are marked differences in the length of daylight according to the seasons, whereas in the tropics, daylight variations are limited to an hour one way or the other throughout the year—sunrise to sunset.

Protection against Tropical Elements

Besides the Wet and Dry Seasons in the Tropics—that is the extremely heavy rains, and a merciless sun—one has to contend with and should be protected against winds, an element which can be most destructive.

Pot plants thrive in the tropics and are blessed with natural temperatures which make hot houses unnecessary. Some form of shelter should however be provided for protection against the elements.

This may be provided by means of a frame eight feet high, constructed of hardwood, bamboo or metal. It is then covered over by wooden slats or bamboo strips, or palm leaves running north and south, with the sides and ends similarly protected. The development of plastics—open meshed—(like saran) or otherwise, has made such protection much less laborious, although perhaps not as decorative or attractive as rustic designs, or a hedge well trimmed at the sides, or a trailing flowering vine as an overhead cover. In this environment, house plants are of course in their element—but they must be house plants and not sun lovers which should only be brought into the house for short periods.

A moist atmosphere could be maintained within the shade house by having the ground below the stands covered by a creeper which is kept moist. The plants themselves also are kept close to each other.

In the open garden there is little that one can do about heavy rainfall, except the planting out of seedlings in time, so that they will have attained sufficient size and strength to withstand the rain. Seeds do germinate more quickly in the tropics, but the seedlings are less sturdy. One should also select plants which will have a better chance of survival during the rainy season.

During the dry season, the watering of pot plants is still the bugbear, as it has to be more frequent in the absence of regular rains. There are however systems which are being developed for an intermittent or continuous administration of water, such as overhead mist nozzles or revolving sprinklers, or trickle-watering through low caliper plastic tubing.

Drying winds may be subdued by natural hedges of hibiscus or sweet lime, or artificial hedges of wood, bamboo, plaited straw, teak laths or a brick wall. These "windbreaks" need only to be placed on the side of the prevailing winds.

Plant Growing in the Tropics

Essential to success with plants under tropical conditions — SHADE for tender plants from burning sun, as provided here by individual umbrellas, fashioned of Pandanus leaves in South India — and perfect DRAINAGE, as illustrated by these orchids growing in broken brick and perforated pots, exposed to long periods of Monsoon rain in Southeast Asia.

Growing Materials for Warm Climates

Soil in its preparation should have incorporated with it large quantities of unrotted organic matter—a most important factor. With this, heavy mulching should be practised twice a year, before the rains and before the dry season. Experience has shown that mulching prevents the washing away of the top soil during rains, and the unrotted organic matter checks cracking of the soil, and allows free drainage. In the dry season the mulch retains moisture, keeps the top soil cool, and slows up leaching. It is surprising to see how quickly organic material disappears in the tropics—as snow does under the sun.

The addition of finely ground limestone does help. Stable manure if obtainable, should be well matured, or if fresh, buried deeply under the surface. Small amounts of ammonium sulphate added to the soil will be found useful. Normal fertilizing practices will usually suffice, but should be applied in small amounts at more frequent intervals. It has been my experience that applications of dilute solutions of organic fertilizer appear to produce better results.

Soil for plants in pots or other containers, and also hanging baskets must be porous and have incorporated with it more organic material in the form of rough scrapings from forest soil, tree-fern fiber, coconut bass—(the left overs after the extraction of the fibre from the coconut husk); chopped bagasse—(the waste fibre from the sugar cane), or any other form of organic material which may be obtained locally from coffee hull, cocoa, rice or corn husks. These could also replace peat moss, a very expensive commodity in the tropics.

Pots may be devised from bamboo cut into sections, or plaited straw or even well decorated containers from the vast number of tinned foods that are now widely consumed.

Underpotting is far less harmful than overpotting, and it has been my experience that certain plants that are difficult to flower in pots, do so readily when pot bound. Of course at that stage it will be necessary to fertilize regularly to keep them going, depending on the type of plant.

To accommodate the plants, step-like stands of three tiers, each eighteen inches high, twelve inches wide and as long as convenient, appear to provide more economy in space, and are more convenient than flat benches.

Pest Control and Propagation

Soil sterilization is almost a must, particularly to lessen nematode infestation. Formaldehyde in this respect is of considerable value—1 part of 40% Formalin to 39 parts of water.

Seed boxes may be made comparatively free from infection by pouring boiling water into the soil previous to sowing and covering it with newspaper. When cool, and the seeds

Plant Growing in the Tropics

are sown, again a newspaper covering kept moist by watering the surface, encourages a very satisfactory germination. At the first sign of germination the cover is removed.

Air-layering (bagging or marcotting) is one of the commonest forms of propagation in the tropics. Removing a ring of bark from a semi mature part of the stem for about 1 inch, below a joint or leaf—the bared area is surrounded with wet moss, enclosing it in a translucent plastic sheet; tie firmly above and below the unbarked strip. Watch for roots appearing through the moss (in about 4-6 weeks), cut off and transplant immediately; shade for about a week and keep the area damp. It is surprising how many plants can be propagated in this manner.

Insect pests are a natural hazard and need even more constant attention in the tropics. The usual insecticides should be applied, diluted but more frequently. Scale insects are easily scrubbed off with a tooth brush and soap suds—sprays containing soapy water also help. There is a particular ant called the Parasol ant, which is a leaf cutting insect attacking plants chiefly at night. It is easier then to track them on their trails with a light and follow, as they carry pieces of leaf to their nests, as their path may be obliterated during the day when the ants are all within the colony. Ant bait can be used successfully by placing it in their tracks or around the base of the plant which is being attacked, and this is carried by them back to the nests, destroying the colony.

Plants for Tropical Conditions

It is said that certain types of plants will not thrive or flower satisfactorily above or below certain levels of altitude, but when one finds the English Forget-me-nots thriving with as much vigour at the top of Mount Pilatus in Switzerland, as on the shores of Trinidad and Tobago in the West Indies, it gives one food for a lot of thought.

Tropical blooms are unfortunately short-lived. There are compensations however, in that there are flowers all the year round, and shrubs such as Bougainvillea, Caesalpinia, Hibiscus, Ixora, Oleander, are but a few examples. Poinsettias (a short-day flowering shrub) continue to hold their colored bracts for three to four months. Annuals like Red Salvias, Snapdragons (Antirrhinum), Zinnias give much flowering pleasure for fairly long periods. Brightly coloured foliage plants including Acalyphas, Coleus, Cordylines, Crotons, Heliconias, Pandanus, are some that give much colour to a garden and need little attention.

Plants requiring pruning should be cut following their flowering and fruiting stage, or else during their resting period if any.

Choose plants suitable to your local conditions. Rock gardens or ferns for damp shaded areas. Cacti and succulents for dry open parts. As a general rule, plants considered hardy in cold climates do not do well in the tropics.

Reap the benefit from the experience of local inhabitants or neighbours, but keep in mind the primary requirements for plant growing in the tropics—protection from heavy rainfall, the midday sun, and the drying winds.

HOUSE PLANTS (little or no sun)—
Ferns, Palms; Aglaonema, Asparagus, Aspidistra, Rex Begonias, Calatheas, Dieffenbachias, Dracaenas, Ficus elastica, Pandanus, Philodendrons, Sansevierias.

HOUSE or VERANDA PLANTS (adaptable for partial sun 2-3 hrs.)—
all the above; also Anthurium, Begonias incl. tuberous kinds, Coleus, Cyclamen, Cordylines, Crotons, Dracaenas, Fuchsia, Marantas, Saintpaulias, Vincas.

POT AND GARDEN PLANTS (full sun at least 5 hrs. daily)—
Amaryllis, Azalea 'Indica', Beloperone, Bougainvilleas; Cape bulbs and Dutch bulbs; Cactuses; Calceolarias, Calla lilies; Cannas, Cinerarias, Euphorbias, Gardenias, Geraniums, Gerberas, Helianthus, Impatiens, Poinsettias, Zinnias.

PLANTS tolerating the RAINY SEASON—
Balsam, Browallia, Calliopsis, Celosia, Coleus, Cosmos, Dahlia, Dwarf Marigold, Dianthus, Gaillardia, Salvia, Verbena, Vinca, Zinnias.

FLOWERS for PARTIAL SHADE—
Alyssum, Antirrhinum, Centaurea, Cynoglossum, Dianthus, Godetia, Myosotis, Nicotiana, Pansies, Petunias, Phlox, Saintpaulias, Tagetes.

FLOWERS around TREES—
Ground orchids; Impatiens, Salvias, Zinnias.

LITTLE WATER—
Alyssum, Coreopsis, Cosmos, Gaillardias, Petunias, Verbenas, Zinnias.

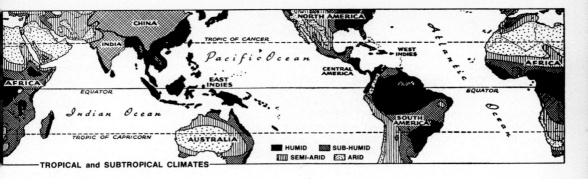

—TROPICAL and SUBTROPICAL CLIMATES—

PLANT GEOGRAPHY

To give a potential indoor plant the best possible chance of success, we should know something about the climatic backgrounds that prevail in their native habitats. Environment has shaped characteristics in plant types that make them either tolerant or difficult to acclimate when taken into cultivation. If too sensitive it may be best not to try them as a house plant, but many tropicals are far too beautiful not to tempt us to experiment with them just the same.

This is where knowledge of their origin helps us to provide for them those conditions under which such plants feel at home. Each climatic zone has favored and subsequently evolved plant populations that are peculiar to it, and the pages following try to sketch just what is characteristic in the different floristic regions of the world.

As if by magic carpet, the fascinating world of the Collector of Exotics extending East from the tropics of the Western hemisphere to subtropical Europe, the vast expanse of Africa and Monsoon Asia and on to Australasia and the islands of the Southern Seas—these are the areas where our house plants were originally found.

Temperature and Rainfall at typical locations in the Tropic and Subtropic Zones.

NORTH AMERICA	LAT. deg.	ELEV. feet	TEMP. °F. min.	max.	RAIN in.
California, San Diego	32.7 N	131	35	88	11
Florida, Miami	25.8 N	10	27	95	56
Mexico, Mexico City	19.2 N	7575	24	92	29
Mexico, Vera Cruz	19.1 N	52	49	96	63
WEST INDIES					
Cuba, Habana	23.8 N	161	50	95	48
Puerto Rico, San Juan	18.2 N	100	62	94	61
Jamaica, Kingston	18.1 N	24	57	98	33
CENTRAL AMERICA					
Guatemala, Guatemala City	14.3 N	4855	41	90	51
Costa Rica, San José	9.5 N	3760	47	94	71
Panama, Colón	9.2 N	25	66	95	127
SOUTH AMERICA					
Venezuela, Caracas	10.3 N	3420	45	91	32
Venezuela, Ciudad Bolívar	8.9 N	125	66	97	35
Guyana, Georgetown	6.5 N	70	68	92	90
Colombia, Buenaventura	3.5 N	39	73	92	390
Ecuador, Quito (Sierra)	0.1 S	9350	36	78	49
Ecuador, Mendez (Oriente)	2.4 S	2290	61	89	102
Brazil, Manaos (Amazonas)	3.0 S	147	66	101	72
Brazil, Rio de Janeiro	22.5 S	210	52	102	43
Brazil, Sao Paulo	23.3 S	2690	28	101	56
Peru, Iquitos (Amazon)	3.7 S	295	64	88	103
Peru, Lima	12.3 S	512	40	90	2
Peru, Cuzco	13.3 S	11.319	28	80	32
Bolivia, La Paz	16.3 S	12.001	27	75	22
Chile, Santiago	33.2 S	1706	24	99	14
Argentina, Buenos Aires	34.3 S	82	28	103	38
EUROPE					
France, Marseilles	43.1 N	246	12	100	23
Italy, Palermo (Sicily)	38.1 N	229	37	97	30
Spain, Seville (Andalusia)	37.2 N	98	22	124	19

AFRICA	LAT. deg.	ELEV. feet	TEMP. °F. min.	max.	RAIN in.
Egypt, Cairo	30.3 N	98	31	113	1
Cameroon, Douala	4.0 N	33	66	90	156
Equat. Africa, Brazzaville	4.2 S	951	53	101	49
East Africa, Nairobi	1.1 S	5450	36	89	38
Tanzania, Amani (Usamb.)	4.5 S	3100	45	86	50
Tanzania, Tanga	5.1 S	98	64	93	61
Madagascar, Tamatave	18.9 S	13	55	100	125
So. Africa, Johannesburg	26.1 S	5750	23	90	33
South Africa, Cape Town	33.5 S	40	31	104	25
ASIA					
Israel, Haifa	32.6 N	33	35	99	27
Japan, Nagasaki	32.4 N	436	22	98	79
China, Yunnan-Fu	25.2 N	6371	24	91	42
Sikkim, Manjitar, Rangit R.	27.1 N	818	51	95	175
India, Cherrapunji (Assam)	25.2 N	4226	49	90	426
India, Madras	13.4 N	22	57	113	48
Taiwan, Keelung (Teipei)	20.1 N	33	37	92	135
Burma, Mandalay	21.6 N	248	48	107	33
Philippines, Baguio	16.5 N	4790	46	77	183
Philippines, Manila	14.3 N	47	58	101	80
Thailand, Bangkok	13.4 N	14	52	106	52
Vietnam, Saigon	10.4 N	37	59	104	70
Ceylon, Colombo	6.5 N	24	62	97	80
Borneo, Sandakan	5.5 N	10	69	97	120
Sumatra, Toba	2.5 N	3773	57	80	90
Malaya, Singapore	1.2 N	8	66	97	95
Java, Jakarta	6.1 S	26	66	96	72
Java, Bogor	6.6 S	920	64	90	172
New Guinea, Port Moresby	9.3 S	128	68	98	41
AUSTRALASIA					
Hawaii, Honolulu	21.2 N	13	52	90	28
Hawaii, Hilo	19.4 N	40	51	91	137
Fiji Is., Suva	18.8 S	44	57	98	112
Australia, Brisbane	27.3 S	137	36	109	45
New Zealand, Auckland	36.5 S	152	32	90	44

791

North along the Tropic of Cancer stretches the subtropical rain forest, an extension of the primeval forest of the tropical zone, found in the lowlands of northern Mexico, the southern tip of Texas, and southern Florida to approximately 30 deg. north latitude. Here evergreen oaks predominate, with epiphytic Tillandsia and orchids, mainly Epidendrums, climbing aroids and figs, Magnolias, Sabal-palms, and a few woody lianas. In swampy ground the oak gives place to the Bald cypress (Taxodium), and on sandy ground to the Long-leaf pine, Pinus palustris, and caribaea. Many of the trees are hung with the gray festoons of the epiphytic Spanish moss, Tillandsia usneoides, which is found as far south as northern Chile and central Argentina. Western Mexico and southwest United States is a region of high plateaus, deep canyons and arid deserts, transversed by wooded mountain ranges. The dry Pacific slopes, especially, are inhabited by a xerophytic flora including many cacti which coincide in bloom with numerous showy annuals at the end of the winter rains. Coryphantha vivipara, and Opuntia fragilis and polyacantha have even adapted themselves to the cold-temperature climate as far north as western Canada, where temperatures may drop to — 40 deg. below zero F.

The vegetation of the Everglades of South Florida is typical of the subtropical forest of swamps and bayous along the Gulf of Mexico. Predominant are the Bald cypress, Taxodium, draped with the gray beards of Spanish moss, mingling with the Sabal palmetto, and many native Ficus and other hardwood trees, their branches host to a multitude of silvery Tillandsias and other epiphytes. These islands of jungle are edged by landbuilding mangroves.

The arid desert stretching away on the slopes of 11,485 ft. Mt. Gorgonio, and facing sheer Mt. San Jacinto (10,805 ft.) in inland Southern California, receiving between 3 and 12 inches of rain a year, is known as the "Devil's garden" because its gravelly soil is planted with a multitude of cacti, primarily wickedly armed Chollas (Opuntias) and Ferocactus acanthodes, a barrel cactus when young, cylindrical and as tall as man in maturity. This is a region of severe climatic contrasts, and having lived here for some time, I have experienced burning days of 122° F oven heat, and clear chilly nights well below freezing, even snow. But following sparse rain in late winter, the desert mysteriously bursts into glorious though fleeting bloom.

The Americas: Mexico

The Rio Suchiate forms the border between Chiapas in Mexico, and Guatemala, a green area humid-warm and lush with palms and other tropical vegetation, and a good grade of coffee, sugar cane and bananas are cultivated. On cooler mountain slopes in Chiapas at 2200 ft. and further west in Oaxaca, at 3,000-4,000 feet, where there is a dry season during December-January, and the temperature drops to 50° F, is the habitat of the sturdy tree-climbing Monstera deliciosa.

Mexico is the country richest in cacti, and of the 125 genera comprising this family, 61 are represented there. Four principal types each favour particular regions: the zone of Opuntias or 'Tunas' centers on the semi-arid plateaux of northern and central Mexico; the globular cacti including Mammillaria are most abundant in the northern desert regions from San Luis Potosi to Chihuahua; the zone of epiphytic cacti are the humid forests of Vera Cruz, Tabasco, and Chiapas; and the tall columnar types such as various Cereus range from the southeast of Puebla south to Oaxaca, the Pacific slopes, and western plains of Sonora and Baja California. The slender columns of Lemaireocereus marginatus, or pipe organ cactus, are widely planted in villages as a natural fence growing to 20 ft. high; large groups inhabit the mountain-ringed, 8,000 ft. tableland of the Valley of Mexico (left), in sight of the gigantic, snow-clad volcanic cones of the Popocatepetl, towering 17,883 ft., and the 'Sleeping Woman', Ixtaccihuatl, 17,388 ft.

Aside from a small area in Vera Cruz, the moist tropical zone in Mexico is most pronounced in Chiapas and Tabasco and then, following the Pacific coast, into Central America. On the dry plateau of Mexico, cacti are the dominant feature of the landscape. Opuntias alternate with columnar cacti, and Agave, Beaucarnea, Dasylirion, Dioon, Furcraea, and Yucca. Cacti represented are Lemaireocereus, the Organ pipe; Cephalocereus, the Old Man; Echinocereus, Lemaireocereus, Echinocactus, Ferocactus, Myrtillocactus, Opuntia; the smaller Astrophytum, and more than 300 Mammillaria; well-known succulents are Dudleya, Echeveria, Euphorbia, including the native "Poinsettia"; Graptopetalum, Pachyphytum, Sedum, and Urbinia. Some of our garden plants are Cobaea, Cuphea, Dahlia, Fuchsia, Penstemon, Tagetes, Oxalis; bulbous Polianthes, Sprekelia and Tigridia. As moisture increases, we find various epiphytic cacti; in commelinads, Commelina, Rhoeo, Callisia, Spironema, and Zebrina; a profusion of colorful or decorative trees and shrubs: Aphelandra, Boehmeria, Bouvardia, Datura, Eupatorium, Ficus petiolaris, Malvaviscus, Miconia, Oreopanax, Plumeria, Pseuderanthemum, Thevetia; of palms the dwarf Chamaedorea. Well-known climbers are Antigonon, Asarina, Passiflora, Petrea, Smilax, Solandra; little Allophyton, many Begonias; Calathea and Maranta, Hoffmannia, Peperomia, Pilea, Pinguicula, Schizocentron; in gesneriads Achimenes, Columnea, Episcia, Smithiantha. In ascending forests we run into aroids, as Anthurium, Philodendron, Syngonium, and large Monstera deliciosa; the moist forest grows various ferns, like Adiantum, Polypodium as well as Selaginella, and in tree ferns Alsophila, Cyathea, and the well-known decorator Cibotium schiedei; bromeliads are represented by Aechmea and Hechtia; orchids with such good genera as Brassavola, Epidendrum, Gongora, Laelia, Lycaste, Maxillaria, Odontoglossum, Sobralia, Stanhopea and Vanilla.

In Central America, especially Costa Rica and Guatemala, the condensed flora rivals that of South America. The high plateau of Mexico narrows southward and competes with the chain of the Cordilleras forming numerous terraces, valleys and narrow coastal areas, all with different climatic conditions. On the eastern slopes of the Sierra with various volcanoes which tower above 12,000 feet, exists a rich, moisture-loving vegetation topped in the "Tierra Fria" with oaks, pines, and firs (Abies religiosa), followed downward toward the sea by tree fern, and mahogany; cycads such as Dioon and Ceratozamia; Heliconia, Macaranga, Plumeria, and Theobroma, the cacao; climbing Passiflora and Petrea; the dwarf palms Chamaedorea and Neanthe. In bromeliads: Guzmania and Tillandsia; many other well-known house plants are found here: Abutilon, Begonia, Calathea, Costus, Fuchsia, Peperomia, Piqueria, Salvia, Tripogandra; in ferns Tectaria; and numerous aroids, notably Anthurium, including the beautiful A. scherzerianum, Dieffenbachia, Monstera, Philodendron, Spathiphyllum, Syngonium. The flora is especially rich in orchids. Costa Rica alone has the greatest concentration of orchid species for its size in the world and these grow all the way to the cloud forests. On drier slopes we find many xerophytes: cacti, such as Lemaireocereus; and Yucca, Dasylirion, Agave, and Pedilanthus.

After crossing the Continental Divide east from San José, Costa Rica, the road winds downward to Turrialba. The countryside is green savannah alternating with banana plantations, and patches of prolific jungle. Here at 2,000 ft., temperature can range from 50 to 90° F. We meet many old acquaintances: Philodendron, Syngonium, gesneriads, Begonias, Selaginellas, Adiantum, Costus; various epiphytic orchids and bromeliads, and the interesting Monstera friedrichsthalii (right). Colorful 2-wheeled oxcarts leisurely creak their way and easygoing peones have as ready answer: "Quién sabe que sera!"

In the Pacific lowlands of Guatemala, the rich black humus seems bottomless. Bananas and sugar cane are much in cultivation. Near Chicacao, Guatemala, on a volcanic slope at 4,000 ft., it is moist and cool with daily rains starting at 1 p.m. Deep barrancos are the best hunting grounds for interesting plants, and here we find various orchids, also Anthuriums, Achimenes, bromeliads, Heliconias, Tripogandra, Begonias, Philodendrons, and magnificent tree-ferns. The trees are densely covered with Philodendron lacerum and many epiphytes.

The Americas – West Indies

The West Indies are but an insular extension of the American tropical flora, differing from the continent mainly by the greater profusion of ferns and orchids, aided by the more uniform climate peculiar to tropical islands and modified only by their high mountains.

Those islands that are traversed by mountain chains or sierra, such as Cuba, Santo Domingo, Puerto Rico, and Trinidad, are favored by ample rains which create, mainly on the side exposed to the northeast tradewinds, an abundant vegetation, with true rain forests in the mountains. There, at the higher altitudes, the tropical vegetation of both Central and South America meet. The steep mountains are covered with mixed forests of tree ferns, mainly the slender Cyathea and Alsophila; and the mountain palms Euterpe globosa and Coccothrinax; Cecropia, Tabebuya, Persea, Coccoloba, Clusia, Psidium, Heliconia. The trees are loaded with epiphytes: small orchids, bromeliads, and aroids; Marcgravias clothe the trunks. Orchids are many but not particularly showy. Aroids are strong with Anthurium, Dieffenbachia, Philodendron, and Xanthosoma. In moist locations we find gesneriads such a Columnea, Gesneria, and Kohleria; delicate Begonias crawl on the forest floor; Aphelandra, Calathea, Oplismenos, Peperomia, Pilea, Wedelia; small ferns: Adiantum, Doryopteris, Hemionitis, Pellaea, and Polystichum. There are many flowering shrubs such as Cestrum and Malpighia; Heliconia; climbing Cissus, Rhoicissus,, Ipomoea, and Solandra. The deep red loam, flattening out toward the northern coast, produces tall sugar cane and is also the home of the stately Royal palm, Oreodoxa, and the spreading Saman tree. Along the shores are thickets of Coccoloba and landbuilding mangroves on stilts.

In drier regions, especially on the south side, is rolling savannah where we find the gorgeous Poinciana trees. In the southwest, and in Cuba northeast, are even desert-like areas, as in Mexico, with various cacti such as the cochineal, the Turk's cap; and in succulents Agave.

The flora of the lower islands of the Lesser Antilles, including the Virgin Islands, suffers from lack of beneficial rains, and savannah mingles with low thorn forest, Juniperus, Ericaceae, Agave and some ferns.

Such bromeliads as Vriesea (Thecophyllum) sintenisii and Guzmania berteroniana grow in great profusion on the trunks of Mountain palms, Euterpe globosa, in the Sierra of Puerto Rico, together with other epiphytes such as small orchids, Anthurium, Begonias, Peperomias, gesneriads, Pileas, many ferns, and climbing Marcgravias, the "Shingle plant".

The steep faced northern slopes of the Sierra Luquillo in tropical Puerto Rico, ascending to 3,500 feet, receive a rainfall of 200 inches yearly often pouring down like hammer blows, producing true rainforest with great tree ferns, mainly the slender Cyathea arborea, growing to 40 ft. high and mingling with mountain palms (Euterpe) and Cecropia, and inhabited by many epiphytes, climbers and ferns.

The Americas –
Northern South America

The northern parts of South America, including Colombia, Venezuela, and the forests of Guiana, serve to link the Central American flora with the Amazon region, or Hylaea. This northern area, its shores often lined with Cereus, is partly rainy, partly dry. Irregularly forested mountains hold many epiphytes such as orchids, especially the showy Cattleya; bromeliads and aroids, and we also find attractive Melastomaceae, Piperaceae, Dioscorea, Marantaceae, Passiflora, Begonia; Gesneriaceae include Alloplectus, Diastema, Episcia, Koellikeria, Kohleria, Nautilocalyx. Where the winds have been deprived of their moisture by coastal mountain ranges, there stretch the savannahs, known as "Llanos" in Venezuela and Colombia, and "Campos" in Brazil, and few trees grow, with occasional acacia and cacti. But gallery forests with mahogany, palms, Carludovica and aroids follow the broad rivers which are lined with Montrichardia growing right into the water. Common aroids are Anthurium, Dieffenbachia, Monstera, Philodendron, Rhodospatha, Spathiphyllum, Stenospermation, Xanthosoma. We also find Allamanda, Aphelandra, Browallia, Cyclanthus, Dioscorea, Eucharis, Heliconia, and Peperomias.

As if on stilts, a forest giant at the edge of the Surinam river in Guiana, bears his share of epiphytes, but most prominent, as if they were colorful birds, Philodendron melinonii in shapely rosettes, with their brilliant red, swollen leafstalks which can be spotted at great distance. The primeval jungle is immense. Great trees, draped with falling lianas and bearing fungi, ferns, orchids, bromeliads, philodendron, and many other plants that can find a footing, struggle for life with one another. Graceful palms fringe the river banks and break the mass of small foliage, and above it the giants, perhaps 200 ft. high, whose crowns have burst through the canopy into the sun. At 1 p.m. the temperature measures 84° F., with an oppressive humidity of 81%; the daily range is only 10 deg. Even the leaves hang away from the brilliant sun. A temperature above the 80's from January to December, heavy with evaporation from a land that is mostly leaves, swamps and rivers, drains the body and all vivacity, and when leaving the river, it needs an effort of will and a compelling purpose, to keep walking along all day in this steamy heat.

Venezuela is the home of the Easter orchid Cattleya mossiae, growing epiphytic in the high, cool forests of the Cordillera de la Costa at 3,500 ft. or higher, where the temperature drops to 49° F., and also further west in the eastern extensions of the Andes. From December to April it is very dry and warm, and rainy from May to November. Growing on live trees (left) makes a big difference in the plants, and I had counted one specimen producing 22 flowers at the same time.

The Americas — Tropical Amazonas

The greatest vegetation is found in the Hylaea, the interior of Amazonas, the most pronounced territory of moist forest in the world, where regular rains to 120 inches fall during a year and extending to where the upper Amazon bends to the south in Peru. In the lofty rain forests we find some 2,000 varieties of both hardwood and softwood trees, Brazil-nuts, and lactiferous Euphorbiaceae of which Hevea, the Para-rubber, is economically most important. Large areas are covered with ground water; this region exhibits the greatest variety of palms in the world. The giant water lily, Victoria regia, forming colonies in quiet warm backwaters of rivers or "igarapes", is strikingly characteristic.

Most of the smaller plants, bromeliads, orchids and aroids have taken to the giant trees, while long lianas form a jungle often hard to penetrate. Scattered, on higher ground, grow showy Gesneriaceae, Marantaceae, and Melastomaceae. Some important aroids come from here: the fancy-leaved Caladium; Dieffenbachia, and Philodendron. We also find Bignonia, Calathea, Cissus, Dioscorea, Echites, Passiflora, Peperomia and other Piperaceae, but generally this is a land of trees, climbers, and epiphytes.

The trees of the Amazonian forest are tall and often scattered, many with latex, and the sunlight plays in lacy patterns along the level floor. The heat makes me weak in the knees and dizzy sweat pours out, as it does from the naked body of the madero, our guide. Long lianas are reaching for the ground where Marantas and Calatheas grow, with Heliconias, Selaginellas, Miconia, Dieffenbachia and Passifloras. On the trees live various Philodendron and bromeliads, while tiny hummingbirds, which the Brazilians call "Beija-flor" or "Flower-kissers", are whirring from one to another.

Dieffenbachia picta grows in well drained red clay in company of climbing ferns, selaginellas, and slender Asai palms, by the side of the wide Amazon. Here the mighty river is 250 feet deep and dolphins can be seen frolicking. Although the thermometer showed only 85°F., I was gasping for breath and the sweat was running down my ears, bleeding from insect bites and scratches, so that in spite of piranhas I took a chance and jumped into the refreshing waters to swim. During the rainy season the river rises 30 feet and thousands of miles are inundated, forcing vicious ants and many snakes into the trees with the epiphytic plants. Since there are no roads, the only way to travel any distance is by river. A little Amazon steamer takes one to some opening in the forest, to transfer to a small canoe. Paddling, to explore closer to the shore, and landing for short stabs inland, one is received by swarms of monkeys, gorgeous parrots and macaws, toucans, cicadas, and brilliant butterflies, but of course one must not overlook an occasional snake.

The Americas - Brazil South

In southeastern Brazil the wet coastal ranges of the Serra do Mar extend near the coast from Bahia south through Petropolis and the wild Organ Mountains, to Sao Paulo and down to Paraná. These escarpments catch the fogs and rains carried by the prevailing southeast winds and are covered with low, dripping, often cool forest; Alto da Serra has the greatest rainfall in Brazil, 140-160 inches per year; a veritable paradise for epiphytic orchids, bromeliads, cacti and climbing aroids. Philodendron and Anthurium abound on the trees and, along with them, moisture-loving cacti: Hatiora, Rhipsalis, Schlumbergera, and Zygocactus. Climbers are Allamanda, Aristolochia, Bignonia, Bougainvillea, Cissus, Dipladenia, Ipomoea, Manettia, Mikania, and Passiflora. On the ground one of the richest collections of attractive plants are at home here: in gesneriads: Columnea, Rechsteineria, and Sinningia; of Marantaceae: Calathea, Ctenanthe, Maranta and Stromanthe; showy Melastomaceae with Bertolonia, Miconia, and Tibouchina. Foremost among terrestrial aroids are various beautiful Dieffenbachias. Many palms include Butia, Cocos, Syagrus. Ferns are represented by Adiantum, Blechnum, Dryopteris, Hemionitis, Nephrolepis; also the fern-ally Selaginella.

Vriesea hieroglyphica is one of the most spectacular bromeliads ever collected. This beauty grew in the dripping, low jungle of the rainforest on the Serra do Mar, in southern Brazil, at about 3,000 feet elevation. Chilly, drizzly clouds were sweeping in from the ocean, and the humidity measured from 95 to 100%, the temperature 59° F. at midday. Rain falls 300 days a year, more than anywhere else in Brazil, and epiphytes flourish on every limb. Together with bromeliads such as Neoregelias, Billbergias, Quesnelias, Tillandsias, Aechmeas, Catopsis and many Vrieseas revel Philodendrons — especially imbe and hastatum; Epiphyllums, Begonias, Anthuriums, Vanilla and other orchids, particularly the brilliant red Sophronitis. It is a surprise to see, in the densest thicket, the heavy Vriesea hieroglyphica clinging like birds to thin twigs, only a few feet from the ground. In grassy clearings of the forest, in the red clay soil, grow Zygopetalum, Drosera, Isolepis, Mimosa, and amongst Vriesea bituminosa and many Nidulariums, the sky-blue Neomarica coerulea.

The Banderantes coast south of Rio, along the Atlantic, washed by breaking salty waves, is a strip of white sand and matted, thorny brambles, with occasional pepper trees, Eugenia uniflora and Ipomoea, the morning glory; and a strange assortment of Philodendron glaziovii, imbe and longilaminatum; rosettes of Anthurium, the silver-felted Peperomia incana, Neomarica, Furcraea gigantea and whole stretches of the dwarf palms Diplothemium maritima. On smooth slippery granite rocks grow procumbent snaky cacti, Cephalocereus fluminensis, right alongside such orchids as Cyrtomium and Epidendrum; saxicolous bromeliads Billbergia amoena, Aechmea, Quesnelia and Tillandsia, all out in glaring sun but filled with refreshing water in their center cups...

Southward into Paraná occur pure groups of noble Araucaria angustifolia, and forests of Mimosas extend into Paraguay. South, in Argentina, grow beautiful flowering Cassia, and the well-known Petunia, Acanthostachys, Brodiaea, Dyckia, Lippia, Nierembergia, Tillandsia including the "Spanish moss", Tradescantia, and Zephyranthes. In northwest Argentina, where grasses do not flourish, the thorn-savannah or "Espinal" region is rich in cacti: Echinopsis, Gymnocalycium, Lobivia, Malacocarpus, Notocactus, Parodia, Pseudolobivia, Rebutia, many of which are very free-blooming even as small house plants. Other genera represented are Opuntia, various Cereus such as Oreocereus trollii, Harrisia, Cleistocactus, and Rhipsalis; some tree-like Philodendron, with Acacia, dot the landscape. But gradually, with moderate rainfall, the land stretches out into the grassy pampas.

Northwest Argentina is cradled in the High Cordilleras, with icy 20,000 ft. peaks and plunging valleys, and more cacti are concentrated there than anywhere else south of the equator. Coming from Bolivia into Jujuy, the giant Trichocereus pasacana begin to appear at 16,000 feet, and as we descend to the "Espinal" at 3,000 ft., between rock and gravel and thorn, cacti become most numerous and varied, extending on through Salta down to Mendoza. Familiar types are Oreocereus trollii, Cleistocactus straussii, various Opuntias and Echinopsis, many Parodias and Lobivias, especially in friendly Salta province, where Gymnocalycium begin to appear, becoming more common further south.

On the western frontier of Brazil where the Rio Paraná forms the border of Paraguay, spreads the vast rainforest of the Mato Grosso. Here the Rio Iguassu, 2 miles wide, drops in mighty cascades, higher than Niagara, down to the Parana. The foaming water tosses between gigantic rocks, and clouds of mist drift widely over the area. The thermometer reads 67° F., the humidity 90-100%. Because of this abundant moisture in the atmosphere, the giant trees, perhaps 150 feet high, are covered with epiphytes and climbers — bromeliads, orchids, Rhipsalis, Peperomias, Bignonia, Passiflora, Begonias and Philodendrons. The handsome Philodendron selloum with its lobed leaves, and forming trunks six feet high, is growing both on trees and on the ground. This specimen, growing in slithery clay, looks over into Argentina from the brink of the cataracts of Iguassu; and which are fringed by stately Arecastrum romanzoffianum palms, better known as "Cocos plumosa".

South America – Andine Region

The well watered eastern slope of the Cordilleras probably holds the richest still unexplored treasure of tropical plants in this hemisphere. Beni Province of eastern Bolivia, the Montana of Peru, the selva of Oriente Province in Ecuador, and Amazonian Colombia, as well as the Choco, promise to yield many a new beautiful plant. Already, the territory along the Rio Dagua and similar locations on the slopes of the Andes in Colombia, Ecuador, and Peru have given us many well-known plants. In aroids we know some of our most attractive Anthuriums from that region, as well as Dieffenbachia and Philodendron in great variety, also many showy bromeliads and orchids. At 4,000 feet and higher we find tree-ferns, Adiantum, Bertolonia, Brunfelsia, Calathea, Carludovica, Cinchona, Coca, Datura, Fittonia, Fuchsia, Heliconia, Hippeastrum, Heliotropium, Hymenocallis, Passiflora, Peperomia and Sanchezia. Going higher there grow cool-loving Calceolarias and Philodendron verrucosum, at an altitude of 10,000 feet in Ecuador and Peru. At this elevation, however, the forest gradually gives way to the grassy altiplano, the home of our potato. Pepper trees mingle with Agave. In drier areas grow many cacti. On the chilly Puna is the region of the Gentian, and south, in Bolivia, grows the colorful Cantua, the sacred flower of the Incas, which I saw blooming in an altitude of 13,000 feet in the shadow of snowy 21,000-foot Pico Illimani.

Leaving behind the bleak Paramo in Ecuador, flanked by sheer crags of the Andes and the row of glistening volcanoes highest of which is El Chimborazo (20,557 ft.), and traveling down one day along a one-way road toward the low jungles of the Pacific slope, I met a native girl, and telling her that I was looking for "orquideas y plantas ornamentales". she quickly scrambled up the steep side and brought out this beautiful "Dancing doll" orchid with crisped flowers. probably Oncidium grandiflorum or macranthum, here at 4,000 feet.

Cereus hexagonus, or "Tuna" as known to the natives, growing along the roads and as windbreaks around coffee haciendas, in the Department Valle del Cauca, Colombia, particularly in the vicinity of Palmira, a region of everlasting spring where the temperature never goes below 70° nor above 85° F. The attractive columns are a beautiful bluish color and almost spineless, reaching a height of 40 feet or more. Further southwest is the habitat of Anthurium andraeanum, in the region between Pasto and Tumaco, in southwest Colombia, at an altitude of 3,000-5,500 ft., where the temperature ranges between 58° and 86° F. They grow epiphytical by preference, on high branches, receiving morning and evening sun, in very wet forests, on trees completely clothed in tree mosses, over ground eternally soggy. The rainfall is very high, probably six months almost continuous, while during the dry months daily fine drizzles drift in from the Pacific Ocean, bringing moisture close to saturation.

Of the cool Andine flora, no less than a quarter of the species belong to the Compositae; others characteristic are Fuchsia, Gunnera, Libertia, Oxalis, Salpiglossis, Schizanthus, Tropaeolum, Lapageria. In northern and central Chile much of the area is from 9,000 to 18,000 feet elevation. Here, with Calceolaria, which I also saw abundant on the eastern Cordillera of Peru, is the home of Escallonia, Greigia, Ochagavia, and Puya. In lower regions, where the climate resembles that of the Mediterranean, but with longer periods of drought, some tropical forms remain, such as Cissus, Oncidium, Passiflora, Peperomia, Solenomelus, Tillandsia, and Verbena, but spinous bushes and cacti predominate. Along the coast in northern Chile around Arica there is absolute desert, rich in nitrates, and no rain has been recorded there in the history of man. To the south, beech trees (Nothofagus) extend over into New Zealand, presumably over an earlier bridge through Antarctica, where fossil remains of early plantlife have been discovered.

The ancient Inca City of Machu Picchu, remote in the eastern Cordillera of Peru, and the last refuge of the Virgins of the Sun, was built on a mountain top with sheer granite walls dropping thousands of feet to the Rio Urubamba below, and surrounded by some of the most wildly beautiful scenery of rugged peaks in the world. There on these rock walls perch literally millions of red bromeliads making the rocky side look red with the flame color of their leaves, or the gray of Puyas. Further below we meet Fuchsias, Begonias, Calceolarias, Nicotiana, Passiflora, Solanum, Masdevallia, Sobralia, and many ferns.

Chile, narrow but 2,600 miles long; and climbing above 20,000 ft. in the High Andes, with rainfall from nothing to 200 in. a year, it is understandable that its flora is remarkably diverse. On stony mountain slopes and sandy lomas grow numerous species of cacti, plants that are hard and flower well. Columns of Trichocereus chilensis occur amidst snow, while in the absolute desert of North Chile only such cacti as Copiapoa and Neoporteria can persist. By contrast, in the south, from Valdivia to Magellan Strait, a chilly climate and rain the year 'round developed rich, temperate rain forest with Araucaria araucana, the "Monkey-puzzle" (right). Nothopanax, Podocarpus, and Libocedrus trees.

Mediterranean Subtropics

The most unique creation of the flora of the Canary Islands is the Dragon tree, Dracaena draco, which grows on rocky precipices along the coast of Tenerife. Reddish resin appears in cracks of the bark which created the legend that the sap turns to dragon's blood, with curious properties. The climate is dry like the Sahara but rich in humidity from the sea.

The MEDITERRANEAN region belongs to the northern zone of hot summers and is remarkable for the dense thickets of fragrant, often prickly, shrubs with rigid leaves such as Myrtus communis, Ruscus aculeatus, Viburnum, Laurus nobilis, Smilax, Santolina, Limonium, the olive Olea europaea, and the edible fig Ficus carica; evergreen oaks and needle-leaved conifers like Juniperus, Cupressus, and the Italian pine Pinus pinea. In stony river beds grows the oleander, Nerium; Acanthus and Dianthus are typical. Erica and Pelargonium are common with Africa. There is more variety, with Vinca major, Arum, Convolvulus, Centaurea, Mathiola, Cheiranthus, Cyclamen, Antirrhinum, Mentha, Helxine, Scirpus, Selaginella, Drosophyllum, Pteris, and the Algerian ivy Hedera canariensis; Lathyrus, and numerous sweet-scented Labiatae like Rosmarinus and Lavandula, and the fan-palm Chamaerops humilis which is wild on both sides of the Mediterranean. Much of the vegetation is xerophytic, including many succulents: Caralluma, Sempervivum; bulbs: Scilla, Ornithogalum, Hyacinthus. It is interesting that the cacti which are prominent in this region, are not native but were introduced after the discovery of America.

In a sunny land such as Italy, shade is much appreciated, and so, most Italian gardens are planned with shady walks under great trees- and between tall hedges of tediously sheared laurel and occasional marble statuary; juniper, ilex, boxwood, pine and slender cypress — but few flowers; cool grottoes, splashing fountains and reflecting pools, surrounded by dwarf citrus trees in terracotta jars.

Earlier in history, during the apex of Roman civilization, the Peristyle type of garden was an open air room and an integral part of every Roman villa, providing a sheltered refuge to the family, the forerunner of today's patio. As may be seen in Pompeii (below), buried intact by 20 ft. of volcanic ash in 79 A.D. and now unearthed, these gardens would feature sculptured statues, marble tables, water fountains and pools supplied by a carefully constructed system of lead pipes, allowing the cultivation of native trees, annuals and creepers, including cypress, roses, daffodils and lilies, cyclamen and others, some grown in colored pots. The walls were painted in soft colors with murals of flowers and beautiful women, to soften the feeling of being enclosed and confined.

Africa – Northern Subtropics

NORTHERN AFRICA, with exception of occasional oases and along river courses, of which the most important is the Nile, and few mountains, as the Atlas in Morocco, home of Cedrus atlantica, is largely occupied by the greatest desert in the world, the Sahara, a region which extends to Somaliland and Arabia. Although largely inside the tropic zone,, the character of its vegetation is determined by a minimum of water supply. Where scanty spring rain occurs the water-storing ice plant, Mesembryanthemum crystallinum, a member of the desert family Aizoaceae, survives only because of its roots spreading enormous distances; so does the annual gourd. At the springs of the oases congregate the date palms, Phoenix dactylifera. Nearby are spinous Acacias, broom-like Genista, cactus-like Euphorbia; bulbous Ornithogalum and Allium; fleshy Stapelia and Sedum. At higher elevations we find Cedrus groves. There are occasional Iris and, eastward, Ficus sycomorus, oleander (Nerium), Jasminum, Reseda; the Cyperus and Nymphaea are characteristic of the Nile.

The Tree of Life, Phoenix dactylifera, is much prized as the staple food and chief source of wealth from remotest antiquity, in the desert regions of North Africa where it is indigenous. "Their feet in the water and their heads in the fire" — as the Arabs say — date palms flourish whenever subterranean water is found, bearing nourishing fruit containing more than half its weight of sugar, on female trees. This palm is a beautiful tree growing to a height of 80 feet, and all its parts yield products of economic value to the dwellers of the desert. Each palm can produce 200 lbs. or more of fruit. Its trunk furnishes timber for house building; the midribs of the leaves supply materials for crates and furniture; the leaflets for basketry; the leaf bases for fuel; the fruit stalks for rope and fuel; the fibre for cordage; the ground seeds for stock feed; vinegar and a strong liquor are made from the fermented fruit.

A feature of the melancholy landscape on both sides of the western Mediterranean, an area of burning hot summers and often chilly though sunny winters, is the relatively hardy, stiff-leaved European fan palm, Chamaerops humilis. Growing on stony ledges and pebbly semi-desert, I have seen it in habitat in Andalusia and Mallorca, as well as on sandy tracts in Morocco, usually forming clusters and remaining low and scrublike, with an occasional arborescent variation forming trunks to 12 ft. high.

Africa - Western Equatorial Belt

The countries bordering the Gulf of Guinea, the Ivory Coast, Ghana, Nigeria, Cameroon, Gabon, and large parts of the Congo, are a region of wet rain forest and intermittent grassland, with up to 80 inches of rain annually. It is the home of the true Dracaena and many beautiful Leguminosae with showy flowers. Aroids are represented by the African equivalent of the American philodendron, Rhaphidophora; other aroids are the true Nephthytis, and Rhektophyllum. Some beautiful Dracaena are at home here, especially D. goldieana, godseffiana, and sanderiana; other interesting plants are Acanthus, Monodora, Mussaenda, Palisota, Thunbergia; there are Begonia, Clerodendrum, Cyperus, some Polypodium, Sansevieria, and the orchids Angraecum and Mystacidium. Of trees we know Ficus, Phoenix palms, and Podocarpus. In general, however, there is less variety than in the similar South American forests. The trees of the primeval forest are 200 and 250 ft. high, including mahogany, ebony, teak and other valuable timber, their trunks covered with epiphytes; there is also the Kola tree, the Kapok tree, the bamboo palm and oil palm; and the coffee shrub is found wild. Opposite Fernando Po, washed by the Atlantic, rises the magnificent volcanic cone of Mt. Cameroon, 13,370 ft. high, receiving 430 in. of precipitation, one of the rainiest regions in the world, causing tree-ferns of great beauty to flourish at 4,000-7,000 ft. The only bromeliad in Africa, Pitcairnia filiciana, is found in Guinea.

The famous and very ornamental oil palm of tropical Western and Equatorial Central Africa, Elaeis guineensis, furnishes the palm-oil of commerce, extracted from the pulp of the clustered fruit by crushing and boiling its kernels. An important export product, it is used in making soap, for margarine and candles, and as a cooking oil. Sugar-rich palm wine is also tapped by Africans from the rising sap of the young flower stalks.

The area known as the Guinea Coast is a mysterious land of rivers; an entrancing landscape with a thousand streams coming down to lose themselves along the coast. Here is the western extreme of the vast evergreen rainforest which stretches without a break to Equatorial Africa, and rising in tiers to Sierra Leone and Liberia to the very Nimba Mountains above 6,000 feet. The motionless air seems to hang in space, overcharged with humidity and the smell of rotting plants; limpid waters reflect a sparkling sky. The valuable oil palm, Elaeis guineensis is abundant. Many of the peculiar genera found in this primeval forest are also found in the Guianas and Brazil on the opposite side of the Atlantic. I photographed the scene at left in the heart of the rainforest near Ibadan, in southern Nigeria, tall native oil palms, mingling with Raphia palms, Dracaena fragrans, and with the large foliage of Anthocleista.

Africa - The Eastern Highlands

EAST AFRICA is a land of contrasts. The equator passes over or near icy Mt. Kenya and Kilimanjaro to heights of 19,000 feet, while below, where yearly rain only averages 40 inches, large areas are occupied by savannah forest, with grasses that slash and leaves that sting, Euphorbia, Aloe, and Sansevieria steppes, and thorn forests with Acacias.

On moist slopes and humid valleys of the mountains, especially the Usambaras and west to the Ruwenzori, an unbelievably varied vegetation grows in brick-red soil, with forests of Podocarpus, the fir of East Africa; palm-like Dracaena, the little Callopsis, and many ferns like Nepholepis, Platycerium, and Asplenium, the tree fern Cyathea; Selaginella, Schefflera, Impatiens, Streptocarpus, and several species of the African violet — Saintpaulia, which often grows in crevices of rocks to elevations of 6,500 feet. In drier locations we find Acanthus, Cissus, Coffea, Cyanotis, Encephalartus, Ficus sycomorus, Ricinus. Along the river courses are marginal forests with spreading Raphia palms, the source of raffia fiber; various tropical bananas; Papyrus, epiphytic Peperomia, and a few orchids such as Angraecum, Bulbophyllum, and Vanilla. Strangely, the only cactus native in Africa, Rhipsalis cassutha, is found on Zanzibar, southwest to Malawi and on to Mt. Kilimanjaro.

Although just below the equator, Africa's highest volcanic cone, Mt. Kilimanjaro, towers 19,565 ft., its snowy cap glistening above perpetual mists from the rainforests below. The flora of this magnificent Afro-alpine mountain is most remarkable because of its isolation, and it is possible to traverse almost as varied a succession of types of vegetation as might be encountered on a journey from the equator to the vicinity of the poles. The country of the plateau from 3,000 to 6,000 ft. is typical grass land, with acacia forest along the watercourses, the rising slopes planted with coffee. Then begin the wet, dense forests which clothe the mountain to 10,000 ft., with great camphor trees and hardwoods, Podocarpus, Schefflera, Dracaenas; Cyatheas and bamboo but no palms. The Alpine zone to 14,500 ft. is characterized by weird forests of 10 ft. treelobelias, giant palm-like Senecio, and tree Erica. Right, "Bwana" Graf with true Schefflera, found growing as a liane at 9,000 ft.

After so much ochreous bush and steppe, to climb the tortuous road into the romantically beautiful, humid Usambara Mountains, rising to 8,424 ft., is an experience I shall never forget. The moist valleys such as that of the quietly flowing Sigi river from the Shambala Mountains produce an unbelievably luxuriant vegetation in the brick-red gneiss soil. Serious Raphia palms with giant leaves over 30 ft. long, furnishing raffia fiber, hang heavily over an undergrowth of dracaenas, selaginellas, costus, gloriosas and terrestrial orchids, with lianas climbing up to the highest tops of dense Ficus sycomorus trees. Several species of saintpaulias are found here and on up to Amani at 3,000 ft., in company of rhipsalis. We finally reach the sheer rock outcroppings of Bomele Peak, where from one open clearing one can see an assortment of schefflera, cussonias, dracaenas, aloe, begonias, treeferns and polypodiums; and clinging to the cliffs by rhizomes, the light blue Streptocarpus saxorum (left).

805

Where the mighty Zambesi river divides Zambia and Rhodesia and its foaming waters are tossing down 350 ft. into the mile-long chasm of Victoria Falls, its constantly rising clouds of mist surging upward gave rise to a narrow strip of rainforest, dominated by clusters of Phoenix reclinata, the "False date palm", their lush fronds diamond-dusted by the pearly droplets which ethereally form a gorgeous rainbow against the sky of brilliant blue.

The Bottle tree, or Baobab, Adansonia digitata, is a landmark on grassy plains of tropical and subtropical Africa, from Senegal to the Transvaal, and east to Tanzania. It is one of the longest-lived—by some reckoned to be 5,000 years old—and also largest trees, and though not more than 60 ft. high, its swollen trunk will attain a diameter of 30 feet. By casting its leaves and storing water in its huge trunk the Baobab maintains life through the driest season.

From the Ruwenzori, legendary Mountains of the Moon, and known since Ptolemy nearly 2,000 years ago, the Congo forest of Central Africa stretches away westward as far as the eye can see, to Equatorial West Africa and the Guinea coast. Inhabited by the pygmy people, herds of elephants, and chattering monkeys, the vast high forest is only occasionally broken by open savannah. Typical of all Central Africa are the euphorbias, like the tree-like Euphorbia candelabrum, a gathering place of elephants, here on the Semliki river (left).

Africa - The Plant-rich South

The desert of the Karroo, 150 miles northeast of Capetown, has an intensely xerophytic flora mainly of succulents of weird form, thorny acacia bushes and numerous bulbous plants. Most remarkable are the stone-mimicry plants, imitating the gravel amidst which they exist, and one may walk on them without seeing them. Notable in the landscape are the many species of handsome Aloe, growing as low rosettes to imposing 60 ft. trees, a gorgeous sight when in bloom. On left, the "Dragon tree" aloe, Aloe dichotoma, growing into monstrous trees to 30 ft. high.

Some of the remaining colonies of the prehistoric Cycad family, largely extinct since they flourished during the age of the dinosaurs, are still found in southern Africa. At the Royal Kraal of Mujaji in the northern Transvaal I was privileged to witness the veneration of native Africans for the majestic Encephalartos transvenosus, shown here, the living talisman of the Rain Queen.

The South African region occupies a series of great plateaus and steep slopes. An extra-ordinary variety in vegetation resulted from the irregular distribution of the rainfall. There are no large or dense forests, no climbers nor luxuriant trees. The flora is mostly xerophytic, spinous acacias are everywhere. In a southwesterly direction, where the winters are wet, the silver-tree, Leucadendron, is typical. The east, with dry winters, produces savannah, with forest growth restricted to the river valleys. Characteristic here are certain tree ferns (Hemitelia, Todea), Cycads, Phoenix reclinata, and Musa.

Natal has Encephalartus, Podocarpus, Cussonia trees, Dracaena hookeriana, Carissa, the bulbous Clivia, Watsonia; the succulent Ceropegia, Kalanchoe, Crassula, Senecio, Cotyledon; a variety of Pelargonium, Streptocarpus, Plectranthus, Peristrophe, Asparagus plumosus, and Strelitzia.

Below the stony Kalahari desert, where palms cease, the southernmost terrace or "bush" is the richest area for species of its size in the world; about 16,000 species of flowering plants have been recorded. Native trees or shrubs are Erica, Pelargonium, Sparmannia. This is the home of many good succulents: Ceropegia, Cotyledon, Crassula, Faucaria, Gasteria, Haworthia, Kalanchoe, Huernia, Lampranthus, Rochea, Senecio, Stapelia, no less than 400 species of Mesembryanthemum, and numerous species of Aloe are most characteristic. Many well-known bulbous plants are South African: Agapanthus, Amaryllis, Crinum, Freesia, Gladiolus, Haemanthus, Ixia, Lachenalia, Nerine, Ornithogalum, Scilla, Veltheimia. Our "Calla" lilies, Zantedeschia, are found here, also Gerbera, Gazania, Chlorophytum, Asparagus, Plectranthus, Moraea, Oxalis, Pteris, Leonotis, Aristea, Dimorphoteca, Helichrysum, the curious cycad Stangeria, the gesneriad Streptocarpus; Begonia and the orchids Disa, Polystachia, and Stenoglottis.

The southwest Cape region, especially is the region of the heaths, with 600 species of Erica, as well as many glorious Proteas, which have on predominantly acid soils to 30 inches of rain in winter and, while the summers are dry, the humidity remains relatively high. This is also the home of numerous Pelargonium, Oxalis, Liliaceae, and Iridaceae; however, this region is notably poor in trees.

Asia - the Near East

From Central Asia west through Iran into Asia Minor are a succession of mountains and plateaux with deserts receiving very few rains because the surrounding ranges shut out the rain clouds. Where irrigated, some of the land is very fertile. Tamarisk and spinous Acacias are some of the few shrubs withstanding gales so characteristic of deserts. Cupressus, Olea, the figtree, Ficus carica; Punica, Laurus and Populus also are typical trees, and we find shrubby Jasminum, Ruscus, Molucella, Prunus laurocerasus, Pyracantha; the attractive and well-loved Anemone and Ranunculus, Delphinium, Cyclamen, Bellis, Matricaria, Eremurus, Helleborus, Hedera, and various Arum. Here is the historic habitat of many good bulbous species, foremost the tulips, Tulipa; also Hyacinthus, Muscari, Crocus, Fritillaria, Chionodoxa, Colchicum, and Lilium candidum and other lilies. This is also the home of more species of all the types of Iris than anywhere else in the world, many growing up in the rugged Taurus range. On mountains of Syria and Lebanon scattered groves of the historic Cedrus libani survive.

The Lebanon rises steeply from the almost treeless Bika plain to a tableland more than 8,000 ft. high, with peaks rising to 10,131 ft. The upper slopes are bleak and covered by snow in winter. The stately Cedar of Lebanon, Cedrus libani, grows in scattered groves at about 6,000 feet, chiefly on the western slopes, about 85 miles northeast of Beirut, (above); also found on the Taurus. They are beautiful evergreen trees, 80 ft. high, with fragrant wood, the timber which King Solomon took to build the Temple in Jerusalem, by some reported 6,000 years old.

The ancient city of Jerusalem, where history goes back 4,000 years, stands on a rocky 2,500 ft. plateau projecting southward from the Judean hills. On the east the valley of Kidron separates this plateau from the ridge of the Mount of Olives 200 ft. higher. While the temperature may go to 112° F., the seabreeze tempers the heat during summer, and there is usually a sharp drop at night, to as low as 25° F. in winter, with occasional snow. The rainfall is about 26 in., mainly occurring from November to April, while mid-June to mid-September is almost completely rainless. As a consequence the flora has a xerophytic character, but in the rainy season the bare land becomes a garden of brilliant color. About 2500 species of flowering plants have been identified in the region of Palestine. The flower that first strikes the eye, growing in bright profusion by the wayside, is Anemone coronaria. Also widespread are Ranunculus asiaticus, poppy, Salvia, Cyclamen, Muscari and Ornithogalum umbellatum, the "Star-of-Bethlehem", which may be found even in the Temple area. On the Mount of Olives the olive is the most characteristic tree, and the slopes are brightened by snowy Amygdalus, the flowering almond. In the far south at Elath the temperature may rise to 114° F. but the nights are cool with dew.

Asia - Himalayas and North India

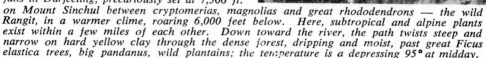

Dawn is breaking over the Himalayas. The deep valleys are filled with white layers of clouds, tossed by a chilly wind. Soon the East turns pink with the coming of the sun, and the light spreads north and west gradually revealing a breath-taking landscape. As far as the eye can see unfolds the colossal chain of the "Home of the Snows", as they are known in Sanskrit, now glowing in rose. Serenely, Mount Kanchenjunga, 28,146 ft., the highest massif in the Sikkim Himalayas, stands in sheer majesty, and fleeting flashes of sun reflect from the icy slopes as he stands guard over one of the richest concentrations of ferns, orchids and alpine plants in the world. Snow sometimes falls in Darjeeling, precariously set at 7,500 ft. on Mount Sinchul between cryptomerias, magnolias and great rhododendrons — the wild Rangit, in a warmer clime, roaring 6,000 feet below. Here, subtropical and alpine plants exist within a few miles of each other. Down toward the river, the path twists steep and narrow on hard yellow clay through the dense forest, dripping and moist, past great Ficus elastica trees, big pandanus, wild plantains; the temperature is a depressing 95° at midday.

North in India is the broad Terai, a low malarial belt which skirts the base of the Himalayas. This giant jungle holds many Ficus, Bamboo, Dendrobium orchids, and vining Hoya. But at higher elevation, the monsoon region is succeeded by an area of the world's highest mountains, the Himalayas; melting snows, and rains totalling 180 inches annually, are the source of rushing rivers. The deep valleys are immensely rich in plant life and, in Sikkim alone, 4,000 species of flowering plants are found. Perpetual moisture nourishes the dripping forest, favoring ferns, of which 250 species are known, including small Adiantum, Pteris, Pellaea, Lygodium, as well as tree ferns such as Alsophila gigantea. The steep sides of the mountains are clothed with lofty trees, and masses of jungle with 12 species of bamboo, Calamus and 20 other palms; Musa, Holmskioldia, Sauromatum, Colocasia, Homalomena, Gonatanthus, Kaempferia, and Begonia are plentiful. There are 600 species of orchids, largely in the subtropic zone up to 6,500 feet, which is the limit of palms in Sikkim, and the best known are Cymbidium, Coelogyne, Paphiopedilum, Dendrobium, Vanda, Phaius, Eria, Aerides, and the velvet leaf Anoectochilus. Most of these genera are also found eastward in Assam. In the temperate zone, from 7,000-11,500 feet, we find first Magnolia, oak, and chestnut, with Osmanthus, Trachelospermum, Aralia, Aucuba, Hydrangea, Bergenia, and Camellia. At the higher elevations is the region of Rhododendron represented by 134 species, and which grow here in forests of giant trees, up to 17,000 feet. At still greater heights they give place to conifers such as Cedrus deodara. Beyond the tree line stretches a great alpine flora with Leontopodium, the edelweiss, Primula and Saxifraga, up to the bare granite walls and the everlasting snows of the world's mightiest mountains. Westward, in Kashmir is the home of many of our well-known perennials, including numerous Primula, Iris and Delphinium.

The Jains are dissenters from Hinduism, dating back to about 700 B.C. Living a disciplined life, they vow to preserve all living things, be chaste, abstain from alcohol, flesh and honey, and to pursue purity of thought, contemplation and repentance. Jains excel in architecture and the art of carving in stone. Their magnificent temple at Calcutta has gardens, with beds laid out in cement and mosaic stone, leaving only small openings for such carpet plants as Coleus, Impatiens and Pileas. In taller plants there are Cannas with variegated foliage, sheared bamboo, sunflowers, and in glazed urns, Pandanus. A singularly beautiful bed is designed in the shape of an open lotus blossom, the symbol of truth and purity (left).

809

Asia - India South

Fabled India, contrary to general belief, is mostly hot and dry; I remember, in Benares, a stifling heat of 115° F in the shade; each summer the inland plains heat up to 120° F, the relative humidity drops to 1%, and plants don't thrive at such temperatures. Most of its natural vegetation is in regions with abundant moisture where the southwest monsoon touches mountain ranges or the escarpments of the 'Ghats' ringing the central plateau of the Deccan, as on the western coast from Bombay down the Malabar to Cochin, the Neilgherries, and the Ganges delta and Assam fronting the Himalayas and Khasia Hills, as well as the west coast of Ceylon. The more fertile parts of the peninsula have been so long and so densely inhabited by man that there is little of the original vegetation left. Characteristic are the Palmyra palms Borassus, Phoenix and Corypha palms; Cycas, Artocarpus, Tupidanthus, and Musa. India loves the fragrance of its Dianthus, Ervatamia, Hedychium, Jasminum, and Murraya as well as its native Tea roses and ramblers; there are showy Hibiscus, Acalypha, Datura, Kaempferia, Costus, Eranthemum, Gloriosa, Thunbergia, Elettaria, Strobilanthes, Barleria; the little Hoya, Serissa, Cyanotis, Selaginella; Zingiber, Arisaema, Colocasia, and Alocasia; in ferns Cyrtomium, Adiantum, and many others. In drier regions, especially South India, are thickets of Euphorbia. A feature of the landscape everywhere are the spreading, pillar-rooted Banyan trees, Ficus benghalensis; the drip-tipped peepul or Bo-tree, Ficus religiosa, and the India rubber tree, Ficus elastica. In drier West Pakistan, is the home of Ficus krishnae with its curious, cup-shaped leaves.

Southern India is the seat of ancient Dravidian civilization and its legacy of art and architecture is best expressed in its temples. The towering Gopuram of the Kapaleswarar Koil temple in Madras, chiseled in every detail with infinite patience, and painted in gay colors, is a marvel of industry, elaboration and exuberant Hindu imagination. Mingled with the wonderfully life-like sculptures of the Vedic deities of the Hindu Pantheon are such symbolic plants as the bo-tree and the lotus, as depicted above.

Along the Coromandel coast of South India, often pitifully dry, infested with Euphorbias and other thorny vegetation but with few trees, firewood is extremely scarce. For this reason dried manure is used traditionally for fuel, depriving the worn-out soils of much-needed organic fertilizer. One sees graceful Tamil women actually follow the sacred cows to scoop up fresh dung, and paste it against the mud-walls of their homes to dry, to serve for slow cooking fires later. In the background, left, is a row of massive Palmyra palms, Borassus flabellifer, common in this hot region, host to epiphytal plants like orchids, ferns, and even fig trees, as well as monkeys, mongoose and bats. They are extremely valuable and have, according to the Tamils, 801 uses, for food, timber, and toddy tasting like mild champagne.

Tropical Madagascar is considered as an ancient continental island. With seven months rain, much is covered by rainforest, rich in tree ferns, lianas, bamboos, Cyperus, and epiphytic orchids. The lofty granitic interior bears a savannah vegetation and, towards the south, it is drier and more thorny. With a flora of over 4,000 species, of which three-quarters are endemic, the Compositae and heaths as well as the succulent Euphorbia and Kalanchoe, suggest the generally African character of the flora, whilst the bamboo and the pitcher plants (Nepenthes) connect it with the Indian monsoon region. Typical Madagascar plants known in cultivation are: Angraecum orchids, Aponogeton, Chrysalidocarpus palms, Dracaena marginata, Euphorbia, Hypoestes, Kalanchoe, Nicodemia, Pandanus utilis, Platycerium ferns, Stephanotis, and the Traveller's tree, Ravenala. The peculiar Platycerium madagascariense (above) is found growing as an epiphyte in rainforest at 2,500 to 3,000 ft. altitude in the eastern region.

Ceylon, with mountains above 8,000 feet high, is a paradise of changing vegetation on its southwest side, from coconut-fringed shores, forests of teak, and Caryota palms; tea plantations up to cool regions where Rhododendrons grow. In general, the flora is both Indian and Malaysian, as evidenced by the presence of Ficus, Aglaonema, Alocasia, Rhaphidophora, Pleomele, true Pothos, Coleus, Impatiens, and showy Chirita, many crotons (Codiaeum), Sansevieria, and even Rhipsalis. However, the almost complete absence of epiphytic ferns and orchids is remarkable.

The botanical wealth of Ceylon, with some 4,000 species of plants is, in proportion to its size, the equal of that of any other country in the world. The southwestern side of the island which receives the heavy rains of the monsoon, has an exuberant vegetation. Nowhere flourish palms so luxuriantly as here, and tall, graceful Coconut palms stretch from the sandy shores inland to elevations of 2,000 feet; in the drier north, Palmyra palms predominate. Climbing higher into the mountains, and where not supplanted by tea plantations, the equatorial rainforest becomes more varied. Banyan, teak and other giants of the forest, some of them 200 feet high are hung with lianas and rattans — as well as flying dogs; buttressed roots are host to the true Pothos scandens. Stately Fishtail palms and slender Arecas stand like sentinels on promontories and in clearings. The fecundity of tropical growth, its savage strength and remorseless sense of purpose, its wild beauty and teeming life is impossible to forget. As we near the heights, the character of the forest changes. Flowering trees add color, such as Cassias, the Golden Shower trees. There are 150 orchids and 300 ferns, with groves of tree fern, and bamboo, until at 6,000 feet begin the rhododendrons, a brilliant red in July, and reaching the proportions of a tree, together with magnolias, myrtles and native tea.

The famed Botanical Gardens of Peradeniya, near Kandy, are reached by a scenic railway winding into picturesque mountains 75 miles from Colombo. Blessed by a moderate moist-warm climate, the extensive grounds contain a great collection of luxuriant plants. I look with envy at the perfection of such delicate exotics as Alocasia growing in bamboo sections serving as flower pots; Gloriosas are rambling happily, rare Aglaonemas, endemic crotons in bright color, durable Pleomele, and the "King of the Forest" orchids, Anoectochilus, with maroon-black velvet leaves and veins of gold, are lovingly cared for.

Asia: Indo-Malaysia and the East Indies

South and Eastward from the continental Southeast Asia of Assam, Burma, Thailand and Vietnam stretch the fabulous "East Indies" including Malaya, the Philippines, and Java, Sumatra, Borneo, the Moluccas and New Guinea to the South. The southwest monsoon brings beneficial rains, producing a moist-warm climate and some of the wettest areas on earth, as in the eastern Himalayas, mountainous Assam, Malakka, and the Sunda Islands, producing typical rain forest. Cherrapunji in Assam, the wettest place on earth, has received as much as 82 feet of rain a year. The less continuously wet monsoon forest takes in most of the rest of the area. This region is characterized by its mangrove swamps, and striking forms of

Along the many brown rivers and network of canals in central Thailand most people live and trade. Here they catch fish, wash laundry, bathe babies, go swimming, and fetch water, all in the same spot. Here a little palm-thatched hut on stilts is displaying quite a collection of orchids in clay pots, some perforated for drainage, carefully and lovingly tended and enjoyed.

Ficus, including the weeping fig, Ficus benjamina. There is a profusion of warm epiphytic orchids, Nepenthes, Platycerium, Zingiber and various Musaceae, giant bamboo, Alocasia, Croton, Hoya, the many Nymphaea and Nelumbo. Leading palms are coconut, Areca, Caryota, Licuala, Phoenix, and the climbing Calamus. Instead of the American philodendron we find Epipremnum; Pleomele take the place of African Dracaena, and for the cacti of America there are cactus-like Euphorbia.

BURMA is primarily known for its orchids such as Aerides, Bulbophyllum, Calanthe, Cymbidium, Dendrobium, Paphiopedilum, Renanthera, Thunia, and Vanda. Typical also are Curcuma, Kaempferia, Leea, Pellionia, Steudnera, Strobilanthes. From THAILAND, we know Aglaonema, Chlorophytum, Jasminum, and Petrocosmea.

VIETNAM has given us some interesting plants in Hydrosme, Musa coccinea, Pellionia, Pilea, Serissa, Tetrastigma, Torenia, and in orchids Aerides, Coelogyne, Cymbidium, Paphiopedilum, Phalaenopsis, Saccolabium. It is probable that most Citrus originated in the region of Indochina and South China.

The ancient Khmers, 1,000 years ago, conceived the most wonderful stone temples in existence. Built without the use of mortar and lacking keystones, the unbelievable tropical vegetation of the forest has been struggling with these massive buildings, sending slender, expansive roots between the masonry. The destructive force of the jungle trees now hold together what they once pried apart. Left, the roots of a gigantic Ficus tree embrace the granite carvings of the central sanctuary of the Neak Pean temple in Cambodia.

Asia – the East Indies

The fabled "East Indies" are immensely wealthy in fascinating exotic plants. Gesneriads are represented by Aeschynanthus, Agalmyla, Monophyllaea. Typical aroids are Alocasia, Colocasia, Cryptocoryne, Epipremnum, Homalomena, Lasia, Rhaphidophora, Scindapsus, Spathiphyllum. Brightly colored Crotons, Acalypha and Hibiscus, Celosia, Amaranthus, Ixora, Plumbago; Phaeomeria and Amomum, Costus, Phrynium, Pandanus, Impatiens, Quisqualis, Sonerila, Vinca, Hoya and Piper, creeping Ficus and Freycinetia; numerous ferns with Adiantum, Asplenium, Davallia, Nephrolepis, Polypodium, and Selaginella.

MALAYSIA has a number of aroids, including Aglaonema, Alocasia, Schismatoglottis, Typhonium; typical are Begonia, Crotons, the giant Dendrocalamus, Dischidia, Nepenthes, Orchidantha, Piper, Platycerium and Tacca. In gesneriads we find Chirita and Didymocarpus; in orchids Arachnis, Bulbophyllum, Haemaria, Paphiopedilum, Phalaenopsis, Spathoglottis.

The SUNDA ISLANDS are rich in orchids, especially the Spider flowers. Java has Arachnis; other orchids are Anoectochilus, Bulbophyllum, Coelogyne, Macodes, and Vanda. Well-known trees are Schefflera and Trevesia. There are several hundred palms in the Sundas, with coconut along the coasts.

Java, the most important of the Sunda Islands, is an emerald jewel of great tropical beauty and luxuriant vegetation, as here along the Tjiliwong river. 606 miles long, it has a teeming population of over 50 million, and every possible spot is densely cultivated. A high mountain chain with surmounting volcanoes (Mt. Salak above), rise to 12,060 ft. The tall trees of the primeval rainforest, ascending the slopes of the western range at 5,000 ft. form a canopy of reflected light, home to an abundance of orchids, Medinillas, Nepenthes, and tree-dwelling Rhododendrons.

The PHILIPPINES are an archipelago of approximately 7,000 tropical islands, of which the largest ones are mountainous reaching, both in Luzon and Mindanao, over 9,600 feet. At these elevations, even in the tropics, it is cool and the mountain tops are covered with forests of pine. Approaching 5,000 feet some very beautiful orchids occur, especially Phalaenopsis, while Vanda and Dendrobium are found at lower, moist-warm elevations; above 5,000 feet grows a multitude of more inconspicuous or 'botanical' orchids.

The island of Bali, in the Lesser Sundas, has rare beauty and charm. Its culture was derived direct from India, before the 8th century. Blessed by extremely rich rainfall, abundant rainfall, and very equable climate, a luxuriant vegetation takes increasingly possession of every foothold as we go higher into the mountains, and in the saturated moisture, ferns and mosses grow everywhere. Remarkable are the showy orchids, used by the people even around their little huts, amongst them Dendrobium, Arundinaria, Coelogyne, and Vandas. By the wayside near Kitamani, a naked boy offered me a collected flowering plant of the exquisite Vanda tricolor suavis for 2 rupiahs (then 13¢).

813

Asia - New Guinea

New Guinea is one of the most mountainous regions in the world. It was given the name by the Spanish in the 16th century because the Melanesian natives resembled the negroes of West Africa. This immense island, just below the equator, is covered by high, often inaccessible, mountains in the central range to 16,404 ft. high, and covered with perpetual snow and glaciers above 14,500 feet. Separated by one series after another of steep razorback walls the natives speak 750 different languages, and deadly hostility prevails between one village and the next. In areas of copious rainfall, almost vertical ridges, screened in mist and cloud, are covered with dense jungle, noted for the presence of many fantastic orchids. Thousands of Araucarias, 150 ft, .high, inhabit southern slopes. Millions of Pandanus grow from 5,000 ft. up nearly to the line of frost at about 10,000 feet. In the lower lands, with brooding humidity, there is luxuriant tropical growth, with creepers and lianes, and palms are very abundant; along the tidal courses of rivers Nipa palms are common; the sago palms, Metroxylon, are numerous in seasonal swamps, and coconut palms fringe the coast.

For Papuans living in the high mountains, the pandanus is their coconut; the pineapple-like fruit provides food, the trunks and leaves building material and household goods. Mainly on the northern slopes, between 5,000 and 9,000 ft., in invigorating climate, are forests of millions of these pandanus "palms" (above), giants to 100 feet high, leaves 17 ft. long, and with stilt roots springing from the trunk forty feet above the ground, and difficult for travel.

There are about 9,000 species of plants in New Guinea; the island has one of the richest Orchid floras in the world, with some 130 genera including about 2600 species. On expedition into the forbidding 13,000 ft. Finisterre Range, we recruited 60 native bearers, and for days on end had to scramble up and slither down the perpendicular mountain sides, usually in exhausting steamy heat alternating with chilly rain, worn out by day and freezing at night. For a handful of salt, Papuan "Grasshopper" boys helped us in gathering an undreamed-of variety of plants, including many small and large orchids, Nepenthes, various unknown Hoyas and other showy climbers, gesneriads and begonias, ferns, Cordylines and even epiphytic Rhododendron (photo left: Hoya on left, Medinilla right).

East Asia - Subtropical China

A Chinese garden is a place of contemplation, and part of China's ancient civilization and culture. From the time of Confucius and Laotse in the 5th century B.C., with increasing sophistication, her people have been inspired to greater appreciation of the beauties of nature. Their gardens are intended to be a painting of the landscape of nature with its inner meanings and rhythm, representing in its hills and ponds, in stones and woods, the active and the passive principles, the male and female forces of Yin and Yang, to remind us of the harmony and order of the universe. The five elements have a deep religious meaning, and are symbolized in the Oriental garden so that the element of earth is represented in hills and islands, that of water in ponds and waterfalls, that of fire in glowing flowers, that of wood in trees, that of wind in the power which swings tree branches and scatters petals, and that of metal in stones. The Chinese love of rockery, in grotesque shapes, amounts to worship, possessing as it does the quality of unchangeable solidity, which human character often lacks. A Chinese garden is a place for relaxation, romance, and meditation, planned at every step with new surprises. Plants are not selected for their looks but for their symbolism.

Horticulture in China is part of the struggle for existence and resultant skill born of 40 centuries of handed-down experience. The maintenance of soil fertility under constant in-

tensive cultivation for such an immense period has been secured in the main by the use of "night soil" and every other kind of manure, and regular crop rotation, alternating with legumes for fertilizing the soil. The growing of economic as well as ornamental plants long ago had reached the level of an art. Air-layering for propagation has been practiced for 1,000 years, and potted plants were grown in China during the Yuan dynasty in the 13th century. With the favorable climate of southern China, the many little nurseries there, though primitive, employ skilled fundamental practices. The frequent monsoon rains in Kwangtung would soon stagnate the roots of many plants in ordinary soil, but for centuries Chinese gardeners have prevented water-logging of plants in pots, by the use of containers of porous clay with perforated sides and by planting into burnt earth as potting material. Large pieces of stiff clay are piled on a log fire and burned slowly for several days, resulting in crumbly, porous pebbles with the capacity of absorbing water yet leaving space for quick drainage and aeration. Dry dung is added near the top, and such as Crotons, Celosias, Araucarias, and orchids (left) respond by luxuriant growth.

East Asia - Japan

The eastward extensions of the Himalayan region, marked by the occurrence of Cycas, also found in Japan; of nutmeg, sandalwood, and palms, are interesting as linking the flora of the southeast of Asia with tropical Australia and the western Pacific.

Southern China, west in Yunnan, and southern Japan is a region containing the remainder of warm-temperate rainforests largely destroyed due to intense cultivation except in the temple groves or on mountains. Here originate not only most bamboo, but many showy, broadleaved, tender evergreens such as Ardisia, Aucuba, Azalea, Buxus, Camellia, Cotoneaster, Daphne, Elaeagnus, Euonymus, Eurya, Fatsia, Gardenia, Gilibertia, Ligustrum, Osmanthus, Pittosporum, Raphiolepis, Rhododendron, Sarcococca, Skimmia, Thea, Trachelospermum, Trevesia, Viburnum, and most Citrus including Fortunella. In palms, Trachycarpus, Livistona, and the reed-like Rhapis which is also the oldest cultivated palm in the world. Other well-known trees and shrubs are Lagerstroemia, Rosa, Hibiscus, Buddleja, Prunus, Hydrangea, Kerria, Clerodendrum; the vining Lonicera, Rubus, Parthenocissus, Ampelopsis, Hoya,

Chinese culture and learning have influenced and shaped Japan since early times. Confucius' ideals of life, the pine for long life, and the bamboo for purity of mind, has been transplanted to Japan, where was added the plum, the symbol of delicacy and feminine elegance, to form a trinity of moral purpose. That is why bamboo, not necessarily beautiful, is widely used as a decorative plant. In addition, it symbolizes virtues such as fidelity, humility, wisdom, and gentleness, and because it becomes stronger as it grows—also promises long life. Pictured above is a "male" Otake, or "strong" bamboo, in Kyoto. A "Komoso," or hooded Buddhist priest and wandering musician, often erroneously called a "beggar monk," who made a vow to carry a basket over his head, plays on his flute soliciting food.

Piper, and Wisteria. In perennials we know Liriope, Chrysanthemum, Dianthus, many Primula, Astilbe, Dicentra, Iris, Sedum sieboldii, Hosta; Hemerocallis and many species of Lilium. Dicentra is endemic in Japan.

In the south we find Ficus, both the graceful Ficus retusa tree as well as the creeping Ficus pumila. There grows also the dwarf banana Musa nana, Cycas and Tetrapanax; in smaller plants Aspidistra, Alpinia, Ligularia, Ophiopogon, Saxifraga, Carex, Acorus, Arisaema, Rohdea, Reineckia, and even Peperomia. Many cool ferns include Cyrtomium, Polystichum, Davallia, Adiantum, Woodwardia, and the fern-like Selaginella. As far south as Taiwan and west, especially in Yunnan, many orchids occur, mainly Cymbidium and Dendrobium.

Rhapis (left) are known to be the oldest cultivated palms in the world, and are immensely popular in Japan and China. They are well adapted to the sparsely heated homes of the Orient, often bitterly cold in winter, where a tropical palm could not survive. The specimen above graces the entrance to the Japanese inn "Kikukawa", or "River of Chrysanthemum" in the historic city of Nara. A sign above the entrance, where you must leave your shoes, invites: "Lodging, stay and rest!" The window is shaped like a wooden mallet, symbolizing prosperity, carried by the God of Prosperity, one of the seven lucky gods of Japan — it brings money and good fortune every time he shakes it. This inn is one of those delightful places, where a charming hostess on her knees, serves warm rice wine, and appetizing sukiyaki cooked right on your low table, or delicious "tempura", those large prawn or shrimp, deep-fried in camellia oil — and she also arranges for a geisha, for pleasant entertainment.

East Asia – Japan

Simplicity has become a cult to the Japanese, best expressed in their dwellings, particularly since the Tokugawa period. A great and abiding love of natural wood takes full advantage of its ingrained beauty, achieving pleasing contrasts of wood against wood or white plaster. To withstand frequent earthquakes, each beam is fitted without relying on nails, and supports a heavy roof of gray tiles or thatch. The size of rooms is determined by the number of rice-straw mats, or tatami, it accommodates. 2 in. thick and measuring about 3 x 6 ft., they are placed on the polished floor and used for sleeping. The house is divided into rooms by sliding screens; light is admitted by a paper-covered lattice mounted as a sliding window. These screens are removable so as to throw all rooms into one. At night the house is shuttered by wooden panels running in grooves along the outside. Bare of major furniture, at one end of the room is the "tokonoma", a raised alcove, framed by a pillar of some exceptional wood, and featuring a hanging scroll and perhaps a single azalea bonsai, or a styled arrangement of flowers, intended to direct the attention of all eyes to the beauty of nature, and its inner meanings. An all absorbing love of nature makes a garden an integral part of the Japanese home. When opening wide the "shoji" or sliding screens, there is revealed, as on a stage, and changing with the seasons, a make-believe landscape in miniature. Deceivingly, great mountains are expressed by hills or rocks, distant forests by a few bamboo, the sea with its rolling waves by neatly combed sand, a lake by a pool, grassy meadows by varying types of moss, individual trees by dwarfed bonsai, and a distant road by stepping stones. The half-moon bridge reflects a full moon, and each treasured tree has its symbolic meaning. In meditation, and by capturing its feeling, the restless mind finds harmony with nature, and with it a deep serenity.

Shaped Bonsai, or dwarfed trees, grown in containers, are objects of loving care and admiration in every home in Japan. Many little nurseries devote long years to their culture, an art which was brought over from China before the 13th century. Numerous trees or shrubs are used for this purpose, and while the White pine is most favored, I have seen Ginkgo, Cedrus, Pomegranate, Prunus, Maple, Azalea, Camellia, Oleander, Cypress, Juniper, Beech and Birch, and even Parthenocissus, no two of which are alike, and some are known to be more than 300 years old. The picture (below) shows Bonsai trees in pottery, Black pine on the left, 5-leaf pine center and right, outside of an oil shop in Kyoto.

Australia - Tropics and Subtropics

Located in the Southern hemisphere, between the Indian Ocean and the South Pacific, this continent has, as a whole, a level surface with only 5% of its area above 2,000 feet. Mountains along the coast, though not very high, are sufficient to keep out rains, and the interior is largely a vast, dry plain. The highest peak is Mt. Kosciusko, 7,330 ft.

Northern Australia and, with its summer rains, Queensland especially are tropical and practically an extension of the Indian monsoon region, with much of the flora of the Sunda Islands and New Guinea. Well-known Queensland trees are Araucaria, Agathis, Brachychiton, and Brassaia; also represented are Cycas and Macrozamia, showy Hibbertia and Swainsona, and climbing Hoya, plus a wide variety of ferns including Platycerium, Nephrolepis, Polypodium, Davallia and Asplenium; and Dendrobium orchids. In the northwest we find the amusing bottle tree. At latitude 19 deg. south begins a sub-tropical desert.

Most of the wooded area of east, southeast, and southwest Australia is of an open savananh type, with exceptionally lofty Eucalyptus over 300 feet high, constituting three-fourths of the forest. These and the silk oaks, Grevillea, graduate into scrub, with thickets of shrubby but showy Acacia, Melaleuca, Eugenia, Pittosporum, Hakea, Callistemon, Casuarina, Epacris, Boronia, Leptospermum, Chamaelaucium, Chorizema. This group is often referred to as 'New Holland' plants, adapted to cultivation in southwest United States, Spain and Portugal. Curious grass-trees, with cycads, grow on the borders of the grassy savannah of the interior, adorned during the rainy season with numerous Liliaceae, iridaceous bulbs, and terrestrial orchids.

The flora of south-western Australia is the largest and most interesting in Australia, with more than 5,000 species. Most remarkable in the drier interior are the "Blackboys", or grass-trees, Xanthorrhoea preissii, which are conspicuous in the landscape. They are long-lived perennials with thick woody, palm-like trunks and terminate in a rosette of narrow linear leaves and an upright spike of whitish flowers. Tradition has it that these "Blackboys" grow only one inch in 100 years.

This fairyland of leaf and stem is a jungle of treeferns, banyan, fan palms and wild bananas in great profusion, at Kuranda, in the far north of Queensland, 21 miles inland from Cairns. With moist heat ranging between 58 degrees to 88 degrees F. along the coast, and the highest rainfall in Australia, 144 to 200 in. annually, true rainforest flourishes with feathery palms including lofty Archontophoenix alexandrae, Ptychosperma elegans, and Calamus; Brassaia actinophylla, the Kauri pine Agathis, Pandanus monticola, Alocasia macrorhiza, creeping Ficus and Hoya, and climbing Rhaphidophora, epiphytic Platycerium and Asplenium, and orchids.

New Zealand - Beauty Primeval

Few floras are as remarkable in their geographical affinities as is that of New Zealand. Forming part of the 'Australasian festoon' and built up of rocks of every geological period including pumice and other volcanic rocks, with mountain ranges rising over 12,000 feet far above a snow-line of about 3,000 feet with glaciers and boiling springs, there is almost every variety of soil condition. The islands are, however, so narrow as to have a distinctly insular climate, although the rainfall is not high. Hard frosts are, in general, absent and the resultant flora is largely arborescent and evergreen. Annual species and bulbs are very few in number. In dry situations "scrub" occurs, such as Leptospermum, Metrosideros, Coprosma, and Pittosporum, also subalpine xerophytes like Hebe, Celmisia, and Dracophyllum; more general, however, is the forest. Tree-like araliads are typical: Meryta, Nothopanax, Pseudopanax, and Schefflera; so are the flat-leaved conifers, Agathis, Libocedrus, and Podocarpus. Epiphytes are limited in the number of genera, Astelia being the most conspicuous; well-known lianas are Muehlenbeckia and Rubus. Ferns of all sizes abound although there are less than 200 species; they have no equal in the world in wealth of form, luxuriance of growth and manifold interest. Growing on salt-sprayed coastal cliffs to humid forest dell, they range from large tree ferns, Dicksonia and Cyathea to 60-foot high, to Gleichenia, Nephrolepis, Pellaea, Pteris, Blechnum, Polystichum, Adiantum, Microsorium, Asplenium, Davallia, the plume-like Leptopteris, and the delicate Hymenophyllum, or filmy-fern.

The forest flora has Melanesian affinities, including the southernmost of palms, Rho-

Rhopalostylis sapida, the exotic-looking Nikau palm, marks the southern limit of palms in the Eastern Hemisphere, growing in colonies down the subtropical northwest tip of South Island at latitude 42. Their association with luxuriant King treeferns, large branched cabbage trees, Cordyline australis, Phormium and Pseudopanax, together with many perchers complete a lush, tropical similitude.

palostylis, reaching into South Island, but cycads are absent. While the arborescent Cordyline is also Melanesian, the New Zealand flax, Phormium tenax, is more limited and endemic. Curiously interesting is the bridge with the South Andine flora, by the presence of many Compositae, and Oxalis, Gunnera, Fuchsia, Calceolaria, Nertera, and numerous Nothofagus beeches. On the other hand, the N. Z. flora has few relations with Australia; there are no eucalypts.

Primeval New Zealand is spectacularly beautiful. In the southwest region of South Island there are many deep fjords, carved out by glacial action, and at Milford Sound the waters of the Tasman Sea cut deeply into the crags and peaks of the Southern Alps. Yet, on the sheer mountain sides, and along glacial rivers, in the valleys between, dwells an amazing flora, supported by a yearly rainfall of 280 inches. I have spent many crisp, cool days of hiking along steep, saturated trails, thrilling to old and new acquaintances, from Fuchsia excorticata trees, Nothofagus trees, Dicksonias and single-trunked Dracaena indivisa, to thickets of Schefflera digitata, Phormium and Griselinia; in the dripping undergrowth, Nertera depressa, Dracophyllum, Filmy ferns and a multitude of other ferns; also the little orchid, Earina autumnalis.

Islands of the South Pacific

The Pacific Islands include thousands of islands from Australia to beyond the mid-Pacific, mostly in the tropic zone and warm throughout the year. But rainfall conditions vary greatly, especially on the low coral reefs. The islands of volcanic origin are mountainous, receiving heavy rains. Ringed by sheltered lagoons, and shores fringed with graceful coconut palms, the steep mountains rise like vast green gardens, with waterfalls and secret pools, from the green-blue waters. At night the air is fragrant with the perfume of flowers. Yet the flora of Polynesia, with the exception of Hawaii, New Caledonia, and perhaps Fiji, is comparatively unremarkable.

Aside from coconut and other palms, Araliaceae, Compositae, Euphorbiaceae, Myrtaceae, Rutaceae, and Rubiaceae are well represented. Plants we know from cultivation include Cycas, Dizygotheca, Polyscias; native Araceae are Cyrtosperma and Scindapsus; many Cordyline, especially the green variety 'Ti', Dianella, Alpinia, Gunnera, Heliconia, Homalocladium, Muehlenbeckia, Zingiber; in Euphorbiaceae: Croton, Breynia, Acalypha; Pseuderanthemum, Pandanus, Freycinetia, Plectranthus, Ficus parcelli, and Artocarpus, the Breadfruit tree. Ferns are further represented by Davallia, the climbing Lygodium, Nephrolepis, Blechnum, Adiantum, and Platycerium, the staghorn fern. Norfolk Island is known for its Araucaria heterophylla; Lord Howe Island for the Howeia palms (Kentia).

Although equally distant between America and Asia, HAWAII's endemic genera are mostly represented also in the Old World, while very few are found in America. Of Australian and Asiatic affinity are Coprosma, Freycinetia, Metrosideros, Pandanus, and Pittosporum. Native Cibotium glaucum and C. chamissoi are prolific at elevations of 4,000 feet, and vast forests of these robust tree ferns cover the slopes of Hawaii's snow-covered volcano, Mauna Loa, which towers 13,680 feet. Further endemics are only 3 genera of Orchids; Tetraplasandra, Hillebrandia, true Pleomele (Dracaena), and 30 species of Pritchardia palms; but no Ficus. Natural vegetation is distinctly influenced by the greatly differing rainfall throughout the archipelago, from 12 in. on the Oahu coast to 150 in. on its mountain slopes, while on the northside of Mt. Waialeale on Kauai an average of 480 in. a year have been recorded.

When we reach Nuuanu Pali Pass on the island of Oahu in Hawaii, a vast panoramic beauty spreads out windwards. To the left is the rugged backbone of the island, a characteristic, forbidding chain of lava-carved mountains clothed in green, plunging steeply into a rich cultivated plain, blessed by abundant rains of the northeast tradewinds. "Ti" plants (Cordyline), and Pandanus odoratissimus are everywhere.

The larger part of the volcanic, mountainous, beautiful island of Samoa is covered with dense jungle which reach down close to the shores of the South Pacific Seas. The temperature averages from 74 to 84° F., and with a heavy yearly rainfall of 193 in., falling almost daily, the vegetation is unbelievably lush. Large banyans form a spreading roof, near which spreads a chaos of leaves, with treeferns, Hibiscus, Cordyline, Casuarina, Artocarpus, festooned with lianes, and porous lava rocks wildly overgrown with orchids, ferns and mosses, along frequent streams. A boyhood dream fulfilled, I could never fail to remember the sea so blue, lapping against the coral shores of the green, steep mountains of this romantic South Seas island.

Secluded Lord Howe is a small island of the Tasman Sea, 436 miles northeast of Sydney, Australia; discovered on February 17, 1788. Of volcanic formation, it is only 5½ miles long, and fronted by coral reefs. With an even, subtropical climate and abundant moisture it is covered from the shore to its highest elevation at 2,840 ft., with endemic forests of graceful Howeia (Kentia) forsteriana, known as the Paradise palm, to 60 ft. high. Seed of this later so important decorative palm was first exported to England in 1871 during the opulent era of Queen Victoria (1830-1900). Each annual flush of flowering strands takes 6 years to develop into viable seed that is ripe and ready for harvest. This seed is then gathered, as a community enterprise, and exported, particularly to temperate zone countries where this palm has long been grown. Its grace and dark green fronds of leathery texture, with excellent keeping qualities, have made it a favorite of florists and for indoor decorations in containers.

The Fijian archipelago forms the eastern end of Melanesia, with the same negroid people that extend west through New Guinea. The moist region of the islands supports a luxuriant tropical vegetation which reaches 4,000 ft. on the central range of the interior of Viti Levu. Along the coral reefs, Pandanus, Alstonias and coconuts ring the lagoons, but inland and up higher thrives a typical flora with numerous orchids, wild citrus, Mussaenda, Acalypha, Miconia, Costus, Alpinia, Crinum, coppery-red Cordylines; the Fijian kauri, Agathis vitiensis, and tall Veitchia joannis palms. The trees are laden with climbers and lianas of Hoya, Stenochlaena, Freycinetia, Passiflora, frayed Scindapsus aureus (right), Rhaphidophora and Epipremnum. Aside from tall, gracefully slender treeferns, Cyathea lunulata, both ground and trees abound with multitudes of selaginellas, tassel ferns (Lycopodium), and creeping ferns, Lygodium, Angiopteris, Polypodium, Davallia, and Drynaria rigidula.

Common Names of Exotic Plants

Some descriptive

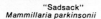

"Sadsack"
Mammillaria parkinsonii

Some humorous

"Smiling Jack"
Mammillaria nejapensis

Finding List to the Manual's Botanical Names

(more names may be found in the cyclopedia EXOTICA)

Page numbers of photos of individual species, varieties and cultivars will be found alphabetically at the end of each listing in the text, serving as a quick-reference to plants illustrated.
Asterisk (*) indicates photo in color.

*by the
same
author*

More Information

for the serious Plant Fancier,
the Study of Botany, and for Research
the World's most comprehensive and universally accepted
taxonomic reference work on Decorative Plants,
constantly revised and updated through several editions

EXOTICA

Pictorial Cyclopedia of Exotic Plants

12,000 illustrations, 200 in color;
more than 1800 pages, in large 8½ x 11 in. format,
picturing and describing nearly 10,000 different
Tropical and Subtropical plants, including important
variations and cultivars, classified by families.
Common and scientific names; Cultural recommendations,
Propagation, Pest controls;
Use of Ornamentals in Decor or as a Hobby.
Key to Care in five languages.
Family characteristics; Bibliography;
Botanical Glossary and Charts.
Historic, Climatic, and Geographic backgrounds, with maps.
Horticultural Color Guide.

ROEHRS COMPANY - *Publishers*

East Rutherford, New Jersey 07073, U. S. A.

QUICK REFERENCE KEY TO CARE (see chart on page 19)

ENVIRONMENT		TEMPERATURE			LIGHT			SOIL		WATERING		Explanations of the 5 elements combined in the **PICTOGRAPH SYMBOLS** used in this Manual
HOUSE PLANT	GREENH. PLANT	WARM	TEMPERATE	COOL	SUN	FILTERED	NO SUN	LOAM	HUMUS	DRY	MOIST	
1	51	W			S			L		D		
2	52	W			S			L			M	
3	53	W				F		L		D		
4	54	W				F		L			M	
5	55	W					N	L		D		
6	56	W					N	L			M	
7	57	W			S				H	D		
8	58	W			S				H		M	
9	59	W				F			H	D		
10	60	W				F			H		M	
11	61	W					N		H	D		
12	62	W					N		H			
13	63		T		S			L		D		
14	64		T		S			L			M	
15	65		T			F		L		D		
16	66		T			F		L			M	
17	67		T				N	L		D		
18	68		T				N	L			M	
19	69		T		S				H	D		
20	70		T		S				H		M	
21	71		T			F			H	D		
22	72		T			F			H		M	
23	73		T				N		H	D		
24	74		T				N		H		M	
25	75			C	S			L		D		
26	76			C	S			L			M	
27	77			C		F		L				
28	78			C		F		L			M	
29	79			C			N	L		D		
30	80			C			N	L			M	
31	81			C	S				H	D		
32	82			C	S				H		M	
33	83			C		F			H	D		
34	84			C		F			H		M	
35	85			C			N		H	D		
36	86			C			N		H		M	

Environment (pg. 16)

⌐⌐ = can be used as House Plant

⌐∧ = will do best in humid greenhouse

Temperature (pg. 27-30)

W = **WARM:** 62-65°F (16-18°C) at night; can rise to 80 or 85°F (27-30°C) in daytime before vents must be opened

T = **TEMPERATE:** 50-55°F (10-13°C) at night, rising to 65° or 70°F (18-21°C) on sunny day, or higher with air

C = **COOL:** 40-45° (5-7°C) at night, 55-60°F (13-15°C) when sunny, with air; 50°F (10°C) in cloudy weather

Light (pg. 22-26)

= **MAXIMUM LIGHT, SUNNY:** Preference 4000-8000 ft.-candles; tolerance 500-2000 ft.-candles.

= **PARTIAL SHADE:** Preference 1000-3000 ft.-candles; tolerance 100-1000 ft.-candles.

= **SHADY OR AWAY FROM SUN:** Preference 50-500 ft.-candles; tolerance, as low as 10 ft.-candles.

Soil (pg. 37-42)

= **LOAM,** clay or good garden soil; humus added, peatmoss, or fibers

= **HUMUSY SOIL,** peatmoss, leafmold, or fibers

Watering (pg. 31-36)

= **DRENCH** thoroughly then allow to become dry

= **KEEP MOIST** but not constantly wet